CAMBRIDGE EARTH SCIENCE SERIES

Editors
A. H. Cook, W. B. Harland, N. F. Hughes,
A. Putnis and M. R. A. Thomson

Plankton stratigraphy

Plankton stratigraphy

VOLUME 1
*Planktic foraminifera,
calcareous nannofossils
and calpionellids*

Edited by

HANS M. BOLLI

JOHN B. SAUNDERS

KATHARINA PERCH-NIELSEN

assisted by Karin E. Fancett

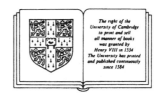

The right of the
University of Cambridge
to print and sell
all manner of books
was granted by
Henry VIII in 1534
The University has printed
and published continuously
since 1584

CAMBRIDGE UNIVERSITY PRESS
Cambridge
New York   New Rochelle
Melbourne   Sydney

CAMBRIDGE UNIVERSITY PRESS
Cambridge, New York, Melbourne, Madrid, Cape Town, Singapore, São Paulo

Cambridge University Press
The Edinburgh Building, Cambridge CB2 8RU, UK

Published in the United States of America by Cambridge University Press, New York

www.cambridge.org
Information on this title: www.cambridge.org/9780521367196

© Cambridge University Press 1985

First published 1985
Reprinted 1987
First paperback edition (in two volumes) 1989
Re-issued in this digitally printed version 2007

A catalogue record for this publication is available from the British Library

Library of Congress Catalogue Card Number: 83-25170

ISBN 978-0-521-36719-6 paperback

Contents

Preface

We have long had the desire to produce a handbook dealing exclusively with planktic microfossils in their role as stratigraphic markers. Though several very useful reference works have appeared in the last decade, we feel that there is still a need for such a book for the use of stratigraphers in industry and academia and for advanced students. The amount of information now available on the various fossil groups is enormous, particularly as a result of the wealth of material that has come from the Deep Sea Drilling Project over the last 15 years. It is also widely spread and not always easy to find, hence our endeavour to get as much as possible together in a single volume. Such a project could only be undertaken by a team of specialists and so 18 paleontologists have worked together to produce this book.

We have restricted ourselves to those groups that are used in the dating of deeper water sediments. However, in the case of many taxa (particularly the calcareous ones) their presence in neritic environments allows correlation of these facies also. In fact, because many biostratigraphic studies were completed before the availability of such a wealth of deep sea cores, zonations were largely erected on surface and subsurface sections where the sediments were mostly deposited under bathyal conditions.

To keep the length of the book within acceptable limits, authors had to be selective in the number of taxa that could be included and also in the degree of illustration possible. In as many cases as possible we have included not only the marker species but have also drawn attention to morphologically similar forms with which they might be confused.

Of necessity, the interpretation of taxa and choice of zonal schemes largely reflects the views of the authors, with which some other authorities may disagree to a certain extent. For reasons of space we have been forced to concentrate on certain key regions where the planktic succession has been worked out best.

The editors would like to acknowledge the very good cooperation that they have received from all the contributors.

It has been a pleasure to work with Cambridge University Press and we have been exceedingly fortunate in our Press subeditor, Karin Fancett, whose dedication and expertise have been vital to the project.

Dr J. P. Beckmann was good enough to read large portions of the text for us.

We would like to thank both the Federal School of Technology in Zurich and the Museum of Natural History in Basel for their help. The illustrative material for the majority of the chapters was prepared in these two institutions – the drawings by Mr A. Uhr and photographs by Mr U. Gerber in Zurich and the wording for figures and charts by Mr R. Panchaud in Basel. We are most grateful to all these people for their skill and good cooperation.

Note added in 1989 reprinting
We have taken the opportunity to divide this book into two volumes to facilitate its use.

1989   H. M. B.
                          J. B. S.
                        K. P.-N.

Contributors

John A. Barron, US Geological Survey, Menlo Park, California 94025, USA

Hans M. Bolli, Geological Institute, ETH, 8092 Zürich, Switzerland

Jonathan P. Bujak, 1–2835 19th Street NE, Calgary, Alberta T2E 7AZ, Canada

Michèle Caron, Geological Institute, University of Fribourg, 1700 Fribourg, Switzerland

Patricia S. Doyle, Scripps Institution of Oceanography, La Jolla, California 92093, USA

Juliane Fenner, Holländerey 31, D2300, Kiel-Kronshagen, German Federal Republic

Silvia Iaccarino, Geological Institute, University of Parma, 43100 Parma, Italy

D. Graham Jenkins, Department of Earth Sciences, The Open University, Milton Keynes, MK7 6AA, England

Hanspeter Luterbacher, Geological Institute, University of Tübingen, 7400 Tübingen, German Federal Republic

Katharina Perch-Nielsen, Geological Institute, ETH, 8092 Zürich, Switzerland

Jürgen Remane, Geological Institute, University of Neuchâtel, 2000 Neuchâtel 7, Switzerland

William R. Riedel, Scripps Institution of Oceanography, La Jolla, California 92093, USA

Fred Rögl, Natural History Museum, 1014 Vienna, Austria

Annika Sanfilippo, Scripps Institution of Oceanography, La Jolla, California 92093, USA

John B. Saunders, Natural History Museum, 4001 Basel, Switzerland

Monique Toumarkine, Geological Institute, ETH, 8092 Zürich, Switzerland

M. Jean Westberg-Smith, Scripps Institution of Oceanography, La Jolla, California, 92093, USA

Graham L. Williams, Geological Survey of Canada, Dartmouth, Nova Scotia B2Y 4A2, Canada

1

Introduction

The last 15 years have been of vital importance in the history of the use of planktic microfossils for the dating of Cretaceous and Cenozoic deep sea sediments. During this time, the *Glomar Challenger* of the Deep Sea Drilling Project (DSDP) has drilled at a total of 624 sites in almost every ocean basin in the world and brought up Cenozoic and Mesozoic sediment cores from the great majority of them. As the oceans occupy 72% of the surface of the Globe, the results were bound to have a profound effect on geologic thinking. A review of some of the important findings of the first ten years of the project is given in Warme *et al.* (1981).

Zonal schemes based on planktic foraminifera and calcareous nannofossils were well established and already proven in many land sections, both surface and subsurface, by the time the DSDP began in 1968. They provided the tool for the dating not only of the sedimentary sequences but also of the underlying oceanic crust that was reached at many of the sites. The latter dating has relied heavily on a knowledge of the age of the sediments immediately above the crust because direct dating of oceanic basalts has hardly ever proved possible. Magnetostratigraphy has lately developed into a most important tool but as a magnetic reversal has no unique signature, it has to be identified by other means – for example micropaleontologic – or be recognized in an established sequence that has previously been dated, perhaps by reference to a section on land that has radiometric tie-points.

The stratigraphic significance of the siliceous microfossils – radiolaria, silicoflagellates and diatoms – was less well known at the beginning of the Deep Sea Drilling Project. For these groups the zonal schemes that have since been established are based to a large degree on DSDP cores. For the calcareous groups the wealth of new sections has enabled a considerable refinement in resolution. The overall result is that there is now a much closer correlation between the fossil groups, both calcareous and siliceous, and this subject is covered in Chapter 2.

A recent book edited by B. U. Haq & A. Boersma (1978), *Introduction to Marine Micropaleontology*, is useful for its discussion of the biology of the various fossil groups.

However, species are not treated from a stratigraphic viewpoint in any depth. Mention should also be made of *Oceanic Micropalaeontology* edited by A. T. S. Ramsay (1977). In this collection of papers the chapters are much more individually designed with rather unequal coverage in a stratigraphic sense but with additional topics reviewed (living forms, biogeography, Quaternary climates, etc.). Neogene foraminifera are not covered.

The arrangement of the present volume

As the purpose of the book is strictly biostratigraphic, the accent is on species rather than genera and comparative notes largely replace taxonomic descriptions as the latter can be found elsewhere. To aid quick identification, the taxa are named on the illustrations in most cases. Also, it has been possible to standardize magnifications for some fossil groups to facilitate comparison between species.

Stratigraphic ranges are given against standard zonal schemes. This causes some oversimplification with too many first and last occurrences happening at exactly the same time. This is known to be the case, particularly since refinement has become possible as better continuous, expanded sections have become available, but it is inherent in charts designed regionally rather than for single sections. For the majority of foraminiferal species we have also included charts arranged on last occurrences or 'tops' as it is with this kind of information that oil company biostratigraphers usually have to operate.

Standardization can only be carried so far and con-

tributors have found it practical to treat their groups in different ways. Thus, for example, evolutionary lineages are given prominent consideration in the radiolaria, while in the dinoflagellate species, descriptions were considered superfluous and the accent was placed on good illustrations (wherever possible of type material), ranges and comparison of zonal schemes.

The chapter on ichthyoliths is included because these small skeletal fragments of fish made principally of apatite are often the only fossils remaining in deep sea clays and thus the only means to date them biostratigraphically. Their stratigraphic significance has been established by comparison with other microfossils when they occur together or otherwise by reference to dated sediments above and below those that contain ichthyoliths exclusively.

Reference lists are placed at the end of each chapter rather than being combined, as the convenience outweighs the advantage of avoiding a small amount of overlap. On the other hand, the indices for all chapters have been placed together at the end of the book but divided into fossil groups.

References

Haq, B. U. & Boersma, A. 1978. *Introduction to Marine Micropaleontology*. Elsevier, 376 pp.

Ramsay, A. T. S. (ed.) 1977. *Oceanic Micropalaeontology*. Academic Press, vol. 1, pp. 1–808, vol. 2, pp. 809–1453.

Warme, J. E., Douglas, R. G. & Winterer, E. L. (eds.) 1981. The Deep Sea Drilling Project: A decade of progress. *Spec. Publ. Soc. Econ. Paleontol. Mineral.*, **32**, 1–564.

2

Comparison of zonal schemes for different fossil groups

The purpose of this chapter is to present comparative charts of the various zonal schemes for different fossil groups as used in the present work. The results for the Late Mesozoic are given in Fig. 1 and for the Cenozoic in Fig. 2.

The subdivision of Late Mesozoic and Cenozoic rock sequences began in Europe with the erection of the Cretaceous Period by A. d'Halloy in 1822, of the Eocene, Miocene and Pliocene Epochs by Lyell in 1832, and of the Pleistocene by Lyell in 1836. The Oligocene was added by Beyrich in 1854 and the Paleocene by Schimper in 1874. The subdivision of these major units into a series of stages was also undertaken in Europe in the latter half of the last century. A few additional subdivisions have been proposed more recently, particularly for the younger part of the Neogene. Stages have also been proposed in other parts of the world to fit local stratigraphic conditions; for example in California and New Zealand.

To tie later work into the classical stratigraphy requires comparison with European stratotypes, many of which are poor in microfossils, particularly planktic species, as they were chosen for their macrofossil content and were usually rich molluscan horizons. Recent work using all fossil groups, particularly including calcareous nannofossils, and now adding other tools such as magnetostratigraphy, is yielding more reliable placement of the stratotypes. Recent state-of-the-art reviews are given in Harland *et al.* (1982), Haq (1983) and Berggren *et al.* (1983a, b, c).

Stage designations are generally not considered in the present book, though exceptions are made for the Mediterranean (Chapter 8) and Paratethys (Chapter 9) where they are particularly frequently used by stratigraphers.

Foraminifera

The Cretaceous stratotypes were defined in sediments of epicontinental sea origin rich in megafossils including pelagic forms such as ammonites and belemnites but poor in planktic foraminifera. It has been necessary to go to other, more favourable areas to correlate from ammonite assemblages to planktic foraminiferal assemblages. Such regions include: the Vocontian trough in SE France, the North Tunisian platform and the platform of the western interior of the United States of America.

In the Cenozoic, first attempts were made to compare the European stratotypes with other facies in different parts of the world using molluscs. Though this produced a gross correlation, in detail it has caused considerable confusion over the years as, for example, in the placing of the Miocene/Pliocene boundary in tropical areas. The presence of larger foraminifera such as the nummulites and the orbitoids in many type areas has enabled these to be used extensively to bridge the gap and larger foraminiferal zones are still in wide use.

Planktic foraminiferal species were almost entirely ignored as markers until the 1940s because morphologic differences between species were not appreciated, leading to the description of relatively few, mostly long-ranging taxa. A change in attitude was forecast by Grimsdale (1951) who compared the ranges of 41 Tertiary planktic species from the Gulf of Mexico and the Caribbean with their equivalents in the Middle East. This was the beginning of a strong commitment by micropaleontologists first in oil companies but soon also in universities and geological surveys, to the zonation of Cretaceous and Tertiary rock sequences using planktic foraminifera. Sediments laid down under tropical to warm temperate paleoconditions were the first to be tackled due to their high species diversity but later work has spread to higher northern and southern latitudes.

The low latitude zonal schemes used in Chapters 5 and 6 were erected largely in Trinidad with additional zones added from Venezuela and from Caribbean DSDP sites 29, 31, 147, 148 and 152. Zones used in Chapter 8 for Mediterranean Neogene faunas were erected in that area while those for southern mid latitudes (Chapter 7) were almost all erected in New Zealand.

In the Caribbean, the first dating in both Late Mesozoic and Cenozoic was done using megafossils, predominantly molluscs but also other groups such as corals, echinoids and sponges, with larger foraminifera also playing a prominent role. Against this framework, the rock sequences were dated and then, wherever possible, zoned using foraminifera – first benthic and then planktic forms. Though the correlation with megafossils worked reasonably well at older levels, some confusion was caused in the later Neogene where greater provinciality of megafossils made long-range correlation less reliable.

Calcareous nannofossils

For the Jurassic, the correlation scheme used is the one of Barnard & Hay (1974), who correlated calcareous nannofossil events with the ammonite zones in England, France and Germany.

For the Cretaceous, the calcareous nannofossil zones of Sissingh (1977) were correlated by him with stages based on the coccolith content of the stratotypes in France and northwest Europe and nearby sections as well as material from western Asia and New Jersey.

For the Cenozoic, correlation between the CN/CP zones of Okada & Bukry (1980, based on Bukry, 1973, 1975) and the NN/NP zones of Martini (1971) is taken from the authors themselves. Bukry (1973, 1975) based his zonation mainly on deep sea sections from low to mid latitudes, whereas Martini (1971) mainly used land sections from Europe, California and the Caribbean. Correlation with the N/P foraminiferal zones is taken from Berggren *et al.* (1983a).

Calpionellids

There is as yet no detailed ammonite zonation of the Late Tithonian with which the calpionellid zones can be correlated. First attempts in this direction have been made by Barthel *et al.* (1966) and by Enay & Geyssant (1975) in the Subbetic Zone of southern Spain. In the Berriasian, on the other hand, calpionellid and ammonite zones have been calibrated in great detail using material collected bed by bed from the same sections, in SE France by Le Hégarat & Remane (1968) and in the Subbetic Zone by Allemann, Grün & Wiedmann (1975). Valanginian zonations were correlated in the same manner by Busnardo, Thieuloy, Moullade, *et al.* (1979).

Radiolaria

The Cretaceous radiolarian zones are rather tenuously aligned with stages via co-occurring calcareous microfossils as determined from the results of Deep Sea Drilling Project Legs 3, 10, 14, 32, 41 and 62 (Maxwell, von Herzen, *et al.*, 1970; Hayes, Pimm, *et al.*, 1972; Worzel, Bryant, *et al.*, 1973; Larson, Moberly, *et al.*, 1975; Lancelot, Seibold, *et al.*, 1978; Thiede, Vallier, *et al.*, 1981), and from a new investigation of the Cismon section (appendix to Chapter 13 of this volume).

Since practically none of the stratotypes of the Cenozoic stages contain siliceous microfossils, radiolarian zonal boundaries must be related to epochs indirectly by correlations with calcareous microfossil zonations. The most consistent body of information for this purpose is provided by Bukry's routine contributions to the Initial Reports of the Deep Sea Drilling Project, and this source has been used in relating radiolarian events to nannofossil zones (Sanfilippo, Westberg & Riedel, 1981). That compilation, based on the published results from DSDP Legs 10 through 50, forms the basis for positioning most of the radiolarian zonal boundaries in Fig. 2. At the upper end of the column, however, Quaternary zonal boundaries (two of them with braces indicating intervals of uncertainty) are based on results from DSDP Leg 68 (Prell, Gardner, *et al.*, 1982). At the lower end of the column, the bottom of the *Bekoma bidartensis* Zone is positioned according to information from DSDP Legs 10 and 15 (Worzel, Bryant, *et al.*, 1973; Edgar, Saunders, *et al.*, 1973).

Also in Fig. 2, the Early/Middle Eocene boundary is

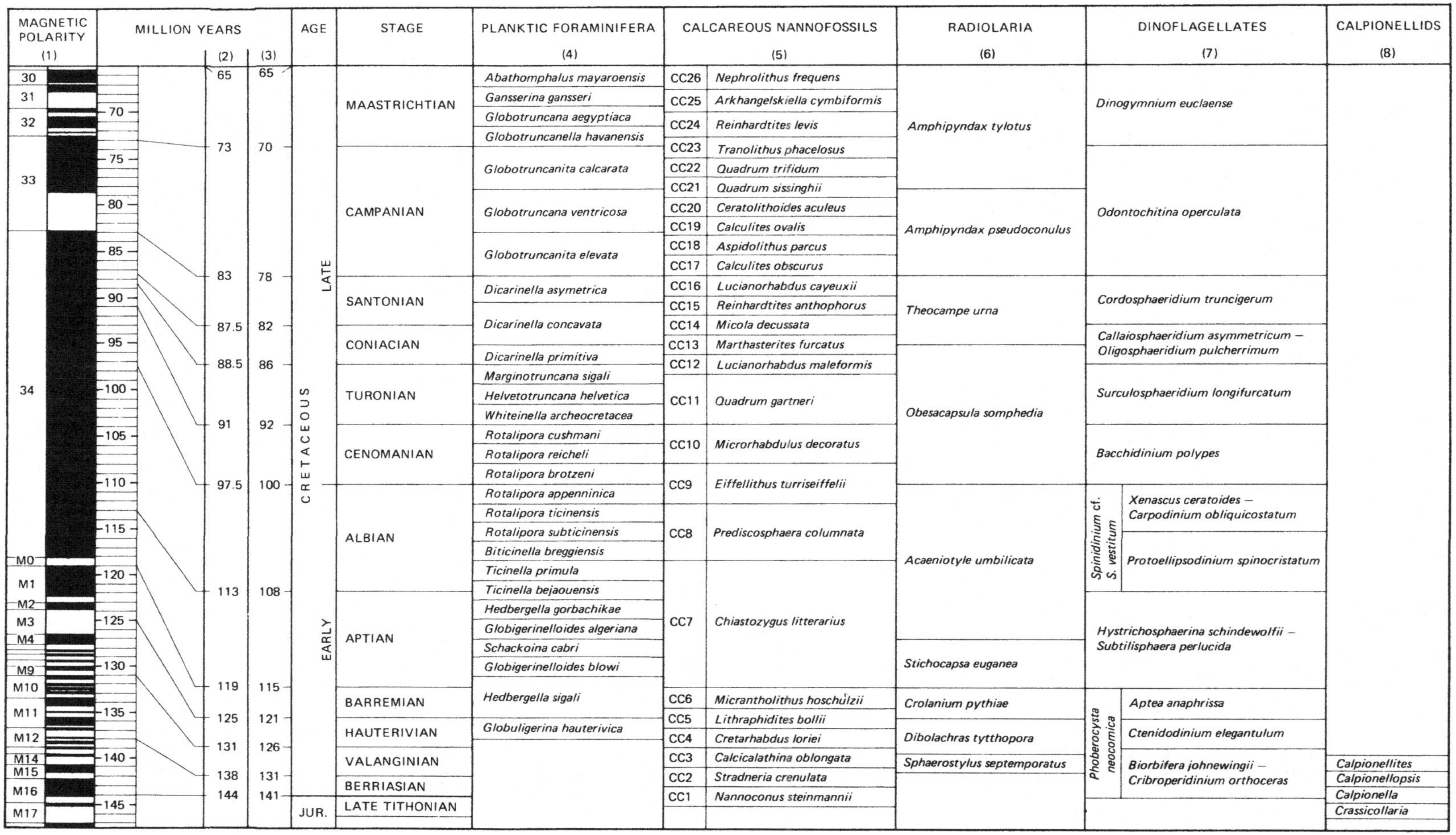

Fig. 1. Correlation of the biostratigraphic zonal schemes for the Late Jurassic and the Cretaceous as used in this volume. For the sake of clarity, the zones are not shown on a linear scale but the proportional time span of the stages and their correlation with the magnetic polarity scale is shown by oblique connecting lines. The figures used for the connection (column 2) are from Harland et al. (1982). Column (3) gives the earlier figures of van Hinte (1976) as these were used in individual chapters.

The zonal schemes are those of the following authors: (4) Caron, this volume; (5) Sissingh, 1977; (6) Sanfilippo & Riedel, this volume; (7) Williams, 1977; (8) Allemann et al., 1971.

The magnetic polarity scale (column 1) is from Harland et al. (1982); black intervals represent normal and white intervals reversed polarity.

Correlation chart (columns left to right):

(1) MAGNETIC POLARITY: 30, 31, 32, 33, 34, M0, M1, M2, M3, M4, M9, M10, M11, M12, M14, M15, M16, M17

(2) MILLION YEARS: 65, 70, 73, 83, 87.5, 88.5, 91, 97.5, 100, 113, 119, 125, 131, 138, 144

(3) MILLION YEARS: 65, 70, 78, 82, 86, 92, 100, 108, 115, 121, 126, 131, 141

AGE: CRETACEOUS (LATE / EARLY); JUR.

STAGE: MAASTRICHTIAN, CAMPANIAN, SANTONIAN, CONIACIAN, TURONIAN, CENOMANIAN, ALBIAN, APTIAN, BARREMIAN, HAUTERIVIAN, VALANGINIAN, BERRIASIAN, LATE TITHONIAN

(4) PLANKTIC FORAMINIFERA: Abathomphalus mayaroensis; Gansserina gansseri; Globotruncana aegyptiaca; Globotruncanella havanensis; Globotruncanita calcarata; Globotruncana ventricosa; Globotruncanita elevata; Dicarinella asymetrica; Dicarinella concavata; Dicarinella primitiva; Marginotruncana sigali; Helvetotruncana helvetica; Whiteinella archeocretacea; Rotalipora cushmani; Rotalipora reicheli; Rotalipora brotzeni; Rotalipora appenninica; Rotalipora ticinensis; Rotalipora subticinensis; Biticinella breggiensis; Ticinella primula; Ticinella bejaouensis; Hedbergella gorbachikae; Globigerinelloides algeriana; Schackoina cabri; Globigerinelloides blowi; Hedbergella sigali; Globuligerina hauterivica

(5) CALCAREOUS NANNOFOSSILS: CC26 Nephrolithus frequens; CC25 Arkhangelskiella cymbiformis; CC24 Reinhardtites levis; CC23 Tranolithus phacelosus; CC22 Quadrum trifidum; CC21 Quadrum sissinghii; CC20 Ceratolithoides aculeus; CC19 Calculites ovalis; CC18 Aspidolithus parcus; CC17 Calculites obscurus; CC16 Lucianorhabdus cayeuxii; CC15 Reinhardtites anthophorus; CC14 Micola decussata; CC13 Marthasterites furcatus; CC12 Lucianorhabdus maleformis; CC11 Quadrum gartneri; CC10 Microrhabdulus decoratus; CC9 Eiffellithus turriseiffelii; CC8 Prediscosphaera columnata; CC7 Chiastozygus litterarius; CC6 Micrantholithus hoschulzii; CC5 Lithraphidites bollii; CC4 Cretarhabdus loriei; CC3 Calcicalathina oblongata; CC2 Stradneria crenulata; CC1 Nannoconus steinmannii

(6) RADIOLARIA: Amphipyndax tylotus; Amphipyndax pseudoconulus; Theocampe urna; Obesacapsula somphedia; Acaeniotyle umbilicata; Stichocapsa euganea; Crolanium pythiae; Dibolachras tytthopora; Sphaerostylus septemporatus

(7) DINOFLAGELLATES: Dinogymnium euclaense; Odontochitina operculata; Cordosphaeridium truncigerum; Callaiosphaeridium asymmetricum – Oligosphaeridium pulcherrimum; Surculosphaeridium longifurcatum; Bacchidinium polypes; Xenascus ceratoides – Carpodinium obliquicostatum; Protoellipsodinium spinocristatum; Spinidinium cf. S. vestitum; Hystrichosphaerina schindewolfii – Subtilisphaera perlucida; Phoberocysta neocomica; Aptea anaphrissa; Ctenidodinium elegantulum; Biorbifera johnewingii – Cribroperidinium orthoceras

(8) CALPIONELLIDS: Calpionellites; Calpionellopsis; Calpionella; Crassicollaria

Fig. 2. Correlation of the biostratigraphic zonal schemes for the Paleocene to the Holocene as used in this volume.

The connection with a linear magnetic scale is arranged as in Fig. 1 with values for columns (1) and (2) taken from Berggren *et al.* (1983a). The earlier values (column 3) used in the individual chapters are from Ness *et al.* (1980). Wide spacing of the oblique lines indicates high resolution and close spacing indicates low resolution in zonal subdivision.

The zonal schemes are those of the following authors: (4) Bolli, 1957a, b, 1970; Bolli & Bermudez, 1965; Bolli & Premoli Silva, 1973; Bolli & Saunders, this volume; (5) Banner & Blow, 1965; Blow, 1969; Berggren & Van Couvering, 1974; (6) Bukry, 1973; (7) Martini, 1971; (8) Nigrini, 1971; Sanfilippo *et al.*, 1981; (9) Barron, this volume; Gombos, 1976, 1982; Fenner, this volume; (10) Bukry, 1981; (11) Williams, 1977.

On this chart the base of *Globorotalia margaritae* Zone is taken as the base of the Pliocene. This differs from Berggren *et al.* (1983a, b) who placed the Miocene/Pliocene boundary within the *Neogloboquadrina dutertrei* s.l. Zone (= *Globorotalia humerosa* Zone of Fig. 2).

The Late Eocene *Theocyrtis bromia* radiolarian Zone has been replaced by three new zones since Chapter 14 was written. These are from oldest to youngest: '*Carpocanistrum*' *azyx*, *Calocyclas bandyca*, *Cryptoprora ornata* (Riedel & Sanfilippo *in* Saunders *et al.*, 1984).

The Early/Middle Eocene boundary is placed at the base of the *Hantkenina nuttalli* foraminiferal zone which is equated with a position within the NP 14 nannofossil zone. On radiolaria, the Early/Middle Eocene boundary is placed somewhat lower, within the *Phormocyrtis striata* Zone, and would correlate with the NP 13/NP 14 zonal boundary. This convention has been used in Chapter 14.

Note. In earlier printings the nannofossil names for the zones were placed incorrectly against the CP/NP zonal numbers in the Paleocene. This is corrected here.

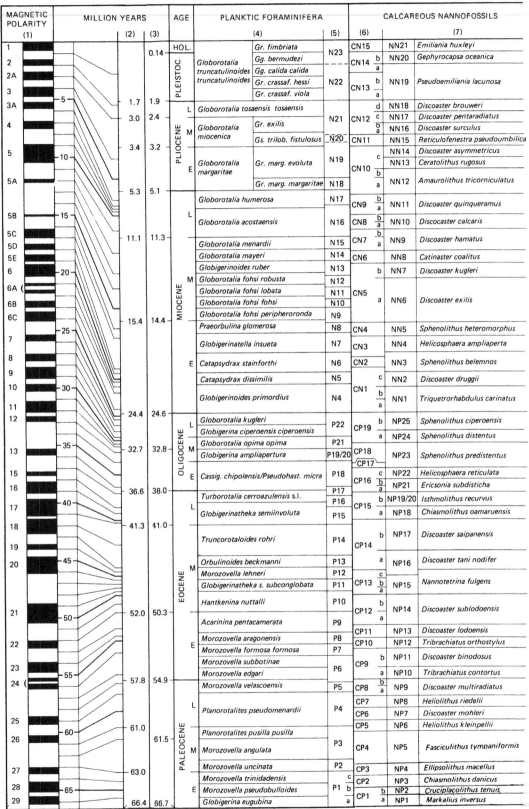

RADIOLARIA (8)	DIATOMS (9)	SILICOFLAGELLATES (10)		DINOFLAGELLATES (11)	AGE	
Buccinosphaera invaginata	Pseudoeunotia doliolus	Dictyocha aculeata			HOL.	
Collosphaera tuberosa						
Amphirhopalum ypsilon	Nitzschia reinholdii	Mesocena quadrangula				PLEISTOC.
Anthocyrtidium angulare		Dictyocha stapedia stapedia	Dictyocha delicata			
Pterocanium prismatium	Rhizosolenia praebergonii		Dictyocha ornata	Achomosphaera ramulifera	L	PLIOCENE
					M	
Spongaster pentas	Nitzschia jouseae	Dictyocha fibula	Dictyocha aspera aspera		E	
	Thalassiosira convexa					
Stichocorys peregrina	Nitzschia miocenica		Dictyocha neonautica			
	Nitzschia porteri					
Didymocyrtis penultima					L	
Didymocyrtis antepenultima	Coscinodiscus yabei	Dictyocha brevispina		Hystrichosphaeropsis obscura – Dapsilidinium pastielsii		MIOCENE
Diartus pettersoni	Actinocyclus moronensis					
	Craspedodiscus coscinodiscus					
	Coscinodiscus gigas var. diorama		Distephanus stauracanthus		M	
Dorcadospyris alata	Coscinodiscus lewisianus	Corbisema triacantha		Lejeunecysta fallax – Sumatradinium hispidum		
	Cestodiscus peplum		Cannopilus schulzii			
Calocycletta costata	Denticulopsis nicobarica	Naviculopsis ponticula				
Stichocorys wolffii	Triceratium pileus	Naviculopsis quadrata			E	
Stichocorys delmontensis	Craspedodiscus elegans	Naviculopsis lata		Cordosphaeridium cantharellum		
Cyrtocapsella tetrapera	Rossiella paleacea		Distephanus speculum haliomma			
Lychnocanoma elongata	Rocella gelida	Naviculopsis biapiculata				
Dorcadospyris ateuchus	Bogorovia veniamini		Corbisema tricantha mediana	Cordosphaeridium funiculatum – Thalassiphora pelagica	L	OLIGOCENE
	Rocella vigilans				M	
Theocyrtis tuberosa	Cestodiscus reticulatus	Corbisema apiculata		Deflandrea heterophlycta	E	
Cryptoprora ornata	Coscinodiscus excavatus					
	Baxteriopsis brunii					
Calocyclas bandyca	Asterolampra marylandica			Areosphaeridium diktyoplokus – Pentadinium laticinctum	L	
"Carpocanistrum" azyx						
Podocyrtis goetheana		Dictyocha hexacantha				
Podocyrtis chalara	Brightwellia imperfecta					
Podocyrtis mitra	Hemiaulus gondolaformis			Eatonicysta ursulae – Distatodinium ellipticum	M	EOCENE
Podocyrtis ampla	Hemiaulus alatus					
Thyrsocyrtis triacantha	Pyxilla caput avis					
Dictyoprora mongolfieri	Triceratium kanayae	Naviculopsis foliacea	Dictyocha spinosa			
Theocotyle cryptocephala						
Phormocyrtis striata	Craspedodiscus oblongus					
Buryella clinata			Naviculopsis robusta	Homotryblium tenuispinosum – Hafniasphaera septata	E	
Bekoma bidartensis	Craspedodiscus undulatus ?	Naviculopsis constrictus				
	Hemiaulus inaequilateralis			Ceratiopsis speciosa – Apectodinium parvum	L	PALEOCENE
	Sceptroneis sp. A					
UNZONED		Corbisema hastata		Ceratiposis diebelii – Palaeoperidium pyrophorum	M	
	Odontotropis klavensii				E	

Chiropteridium dispersum

placed within the NP 14 nannofossil zone whereas on radiolaria it would be placed within the *Phormocyrtis striata* Zone and thus at the NP 13/NP 14 zonal boundary.

Diatoms

Diatom stratigraphy in the Paleocene (Gombos, 1976) is still at a very early stage. Further study of the paleogeographic distribution of the proposed marker species and their stratigraphic ranges might result in changes.

Early Early Eocene diatomaceous sediments have not yet been found in deep sea cores. Because of that, the top of the *Hemiaulus inaequilateralis* Zone and the base of the *Craspedodiscus undulatus* Zone (Gombos, 1976, 1982) have not been defined.

Low and middle latitude diatom zones for the late Early Eocene and Middle Eocene (Fenner, 1984) are correlated with radiolarian and calcareous nannofossil zones in Atlantic and Caribbean DSDP sites. Correlation with planktic foraminiferal zones was only possible at a few of these sites (94, 356, 390A).

For the Late Eocene, diatom ranges have come entirely from DSDP Site 366 on the Sierra Leone Rise (Fenner, 1984).

Diatom zones for the low latitude Oligocene (Fenner, 1984) are based on many DSDP sites with long sediment sections in the Atlantic and Pacific Oceans and in the Caribbean. The stratigraphic ranges of the diatom marker species seem reliable. They were placed against radiolaria, calcareous nannofossils and planktic foraminifera.

Correlation of latest Oligocene to Quaternary low latitude diatom zones with planktic foraminiferal, calcareous nannofossil and radiolarian zones is after Barron (1984) and Barron *et al.* (1984) and is synthesized from correlations at DSDP Sites 71, 77, 158, 495, 503, 504, 572, 573, 574 and 575 in the eastern and central equatorial Pacific Ocean.

Silicoflagellates

The silicoflagellate zones were correlated with the calcareous nannofossil CN/CP zones by Bukry (1981) and his suggestions are followed here. He erected his zonation for tropical and subtropical oceanic areas based on cosmopolitan and low latitude silicoflagellate distribution in DSDP cores.

Dinoflagellates

The dinoflagellate zonation shown in Figs. 1 and 2 was erected by Williams (1977) as a worldwide synthesis of several zonations previously applied by authors to various areas. Age justification for these zones is as follows.

Late Triassic dinoflagellate zones have been recognized in the Arctic, New Zealand and Australian Karnian–Rhaetian, and in the northwest European Rhaetian. The ages of these zones are based primarily on spores and pollen, with some additional data from *Monotis* faunas in the Arctic.

Age control for Jurassic dinoflagellate zones is based

extensively on the occurrence of ammonites and hence the Jurassic dinoflagellate zones erected for northwest Europe, Portugal and the Arctic correlate directly both with the standard ammonite zonations of Arkell (1956) and Casey (1973) and also with local ammonite zonations such as that of Mouterde *et al.* (1965) in areas such as Portugal. Additional data on the ages of particular sections, especially in the Arctic, are furnished by other fossil groups including pelecypods, ostracodes, foraminifera and spores.

Age control for the Cretaceous dinoflagellate zones in surface sections is based extensively on ammonites in England, France, Romania and the Arctic. However, in some surface sections such as Speeton in England, ammonites only occur at certain horizons so that other fossil groups including belemnites, brachiopods and pelecypods are used locally to determine the age of strata. Cretaceous subsurface sections rely exclusively on microfossils for age control. These mostly comprise planktic and benthic foraminifera, tintinnids, calcareous nannofossils, ostracodes, spores, pollen and ammonite aptychi. In addition to the fossils listed above, echinoids have been used extensively to date Late Cretaceous surface sections upon which dinoflagellate zonations are based.

Age control for Paleogene and Neogene dinoflagellate zones is based largely on the occurrence in these sections of planktic foraminifera and calcareous nannofossils. Thus there is direct correlation with the standard Cenozoic planktic foraminiferal and calcareous nannofossil zonations. In high latitude areas such as the Norwegian–Greenland Sea, age control is primarily provided by siliceous microfossils including silicoflagellates, diatoms and radiolarians.

Radiometric and magnetic polarity data

To correlate zonal schemes with radiometric ages and magnetic polarity reversals, authors of the various chapters used the latest data available at the time when they prepared their charts. Van Hinte (1976) was used for the Cretaceous and Ness, Levi & Couch (1980) for the Cenozoic. Just before completion of the book, newer data became available from Harland *et al.* (1982) and Berggren *et al.* (1983a, b, c). This new information, along with the earlier figures, has been incorporated in Figs. 1 and 2 of the present chapter where we have included, together with the paleomagnetic data, the million year values of Harland *et al.* for the Mesozoic (Fig. 1) and of Berggren *et al.* for the Cenozoic (Fig. 2).

References
Allemann, F., Catalano, R., Farès, F. & Remane, J. 1971.
    Standard calpionellid zonation (Upper Tithonian–
    Valanginian) of the western Mediterranean Province.
    Proceedings II Planktonic Conference, Roma, 1970, **2**, 1337–40.
Allemann, F., Grün, W. & Wiedmann, J. 1975. The Berriasian of
    Caravaca (Prov. of Murcia) in the subbetic zone of Spain
    and its importance for defining this stage and the

Jurassic–Cretaceous boundary. Colloque sur la limite Jurassique–Crétacé, Lyon, Neuchâtel, sept. 1973. *Mem. Bur. Rech. geol. minieres*, **86**, 14–22.

Arkell, W. J. 1956. *Jurassic Geology of the World.* Hafner Publishing Co., New York, 806 pp.

Banner, F. T. & Blow, W. H. 1965. Progress in the planktonic foraminiferal biostratigraphy of the Neogene. *Nature*, **208**, 1164–6.

Barnard, T. & Hay, W. W. 1974. On Jurassic Coccoliths: A tentative zonation of the Jurassic of Southern England and North France. *Eclog. geol. Helv.*, **67**, 563–85.

Barron, J. A. 1984. Late Eocene to Holocene diatom biostratigraphy of the equatorial Pacific Ocean, DSDP Leg 85. *Initial Rep. Deep Sea drill. Proj.*, **85** (in press).

Barron, J. A., Keller, G., Dunn, D. A., Kennett, J. P., Lombari, G., Burkle, L. H. & Vincent, E. 1984. A multiple microfossil biochronology for the Miocene. In: S. M. Savin (ed.), Cenozoic Paleoceanography Synthesis (CENOP). *Mem. geol. Soc. Am.* (in press).

Barthel, K. W., Cediel, F., Geyer, O. F. & Remane, J. 1966. Der subbetische Jura von Cehegin (Provinz Murcia, Spanien). *Mitt. Bayer. Staatssamml. Palaeontol. hist. Geol.*, **6**, 167–211.

Berggren, W. A., Kent, V., Flynn, J. J. & Van Couvering, J. A. 1983a. Cenozoic Geochronology, (preprint).

Berggren, W. A., Kent, V. & Van Couvering, J. A. 1983b. Neogene geochronology and chronostratigraphy. In: N. J. Snelling (ed.), Geochronology and the Geological Record. *Geological Society of London, Special Paper* (in press).

Berggren, W. A., Kent, V. & Flynn, J. J. 1983c. Paleogene geochronology and chronostratigraphy. In: N. J. Snelling (ed.), Geochronology and the Geological Record. *Geological Society of London, Special Paper* (in press).

Berggren, W. A. & Van Couvering, J. A. 1974. The Late Neogene. Biostratigraphy, geochronology and paleoclimatology of the last 15 million years in marine and continental sequences. *Palaeogeogr. Palaeoclimatol. Palaeoecol.*, **16**, 1–215.

Blow, W. H. 1969. Late Middle Eocene to Recent planktonic foraminiferal biostratigraphy. *Proceedings First International Conference on Planktonic Microfossils, Geneva 1967*, **1**, 199–422.

Bolli, H. M. 1957a. Planktonic foraminifera from the Oligocene–Miocene Cipero and Lengua formations of Trinidad, B.W.I. *Bull. U.S. natl. Mus.*, **215**, 97–123.

Bolli, H. M. 1957b. Planktonic foraminifera from the Eocene Navet and San Fernando formations of Trinidad, B.W.I. *Bull. U.S. natl. Mus.*, **215**, 155–71.

Bolli, H. M. 1970. The foraminifera of Sites 23–31, Leg 4. *Initial Rep. Deep Sea drill. Proj.*, **4**, 577–643.

Bolli, H. M. & Bermudez, P. J. 1965. Zonation based on planktonic foraminifera of Middle Miocene to Pliocene warm-water sediments. *Boletino Informativo, Asoc. Ven. Geol., Min. y Petr.*, **8**, 119–49.

Bolli, H. M. & Premoli Silva, I. 1973. Oligocene to Recent planktonic foraminifera and stratigraphy of the Leg 15 Sites in the Caribbean Sea. *Initial Rep. Deep Sea drill. Proj.*, **15**, 475–97.

Bukry, D. 1973. Low-latitude coccolith biostratigraphic zonation. *Initial Rep. Deep Sea drill. Proj.*, **15**, 685–703.

Bukry, D. 1975. Coccolith and silicoflagellate stratigraphy, north-western Pacific Ocean, Deep Sea Drilling Project Leg 32. *Initial Rep. Deep Sea drill. Proj.*, **32**, 677–701.

Bukry, D. 1981. Synthesis of silicoflagellate stratigraphy for Maestrichtian to Quaternary marine sediment. *Spec. Publ. Soc. Econ. Paleontol. Mineral.*, **32**, 433–44.

Busnardo, R., Thieuloy, J.-P., Moullade, M., *et al.* 1979. Hypostratotype mésogéen de l'étage valanginien (Sud-Est de la France). *Ed. Cent. natl. Rech. sci.*, **6**, 1–143.

Casey, R. 1973. The ammonite succession at the Jurassic–Cretaceous boundary in eastern England. In: R. Casey & P. F. Rawson (eds.), The Boreal Lower Cretaceous. *Geological Journal, Special Issue*, **5**, 193–266.

Edgar, N. T., Saunders, J. B., *et al.* 1973. *Initial Rep. Deep Sea drill. Proj.*, **15**, 1–1137.

Enay, R. & Geyssant, J. R. 1975. Faunes tithoniques des chaînes bétiques (Espagne méridionale). Colloque sur la limite Jurassique–Crétacé, Lyon, Neuchâtel, sept. 1973. *Mem. Bur. Rech. geol. minieres*, **86**, 39–55.

Fenner, J. 1984. Eocene–Oligocene planktic diatom stratigraphy in high and low latitudes. *Micropaleontology* (in press).

Gombos, A. M. Jr 1976. Paleogene and Neogene diatoms from the Falkland Plateau and Malvinas Outer Basin, Leg 36, Deep Sea Drilling Project. *Initial Rep. Deep Sea drill. Proj.*, **36**, 575–687.

Gombos, A. M. Jr 1982. Early and middle Eocene diatom evolutionary events. *Bacillaria*, **5**, 225–42.

Grimsdale, T. F. 1951. Correlation, age determination and the Tertiary pelagic foraminifera. *Proceedings Third World Petroleum Congress, The Hague*, sec. 1, 463–75.

Haq, B. U. 1983. Jurassic to Recent nannofossil biochronology: An update. In: B. U. Haq (ed.), Nannofossil Biostratigraphy. *Benchmark Papers in Geology*, **78**, 358–78.

Harland, W. B., Cox, A. V., Llewellyn, P. G., Pikton, C. A. G., Smith, A. G. & Walters, R. 1982. *A Geologic Time Scale.* Cambridge Earth Science Series, Cambridge University Press, 131 pp.

Hayes, D. E., Pimm, A. C., *et al.* 1972. *Initial Rep. Deep Sea drill. Proj.*, **14**, 1–975.

Lancelot, Y., Seibold, E., *et al.* 1978. *Initial Rep. Deep Sea drill. Proj.*, **32**, 1–980.

Larson, R. L., Moberly, R., *et al.* 1975. *Initial Rep. Deep Sea drill. Proj.*, **32**, 1–980.

Le Hégarat, G. & Remane, J. 1968. Tithonique supérieur et Berriasien de bordure cévenole. Corrélation des ammonites et des calpionelles. *Geobios*, **1**, 7–70.

Martini, E. 1971. Standard Tertiary and Quaternary calcareous nannoplankton zonation. In: A. Farinacci (ed.), *Proceedings II Planktonic Conference Roma, 1970*, **2**, 739–85.

Maxwell, A. E., von Herzen, R., *et al.* 1970. *Initial Rep. Deep Sea drill. Proj.*, **3**, 1–806.

Mouterde, R., Ruget, C. & Moitinho de Almeida, F. 1965. Coupe du Lias au sud de Condeixa. *Com. Serv. geol. Portugal*, **48**, 16–91.

Ness, G., Levi, S. & Couch, R. 1980. Marine magnetic anomaly time-scales for the Cenozoic and Late Cretaceous: A précis, critique, and synthesis. *Rev. Geophys. Space Phys.*, **18**, 753–70.

Nigrini, C. 1971. Radiolarian zones in the Quaternary of the Equatorial Pacific Ocean. In: B. M. Funnell & W. R. Riedel (eds.), *The Micropalaeontology of Oceans*, pp. 443–61, Cambridge University Press.

Okada, H. & Bukry, D. 1980. Supplementary modification and introduction of code numbers to the low-latitude coccolith biostratigraphic zonation. *Marine Micropaleontol.*, **5**, 321–5.

Prell, W. L., Gardner, J. V., *et al.* 1982. *Initial Rep. Deep Sea drill. Proj.*, **68**, 1–495.

Sanfilippo, A., Westberg, M. J. & Riedel, W. R. 1981. Cenozoic Radiolarians at Site 462, Deep Sea Drilling Project Leg 61, Western Tropical Pacific. *Initial Rep. Deep Sea drill. Proj.*, **61**, 495–505.

Saunders, J. B., Bernoulli, D., Müller-Merz, E., Oberhänsli, H., Perch Nielsen, K., Riedel, W. R., Sanfilippo, A. & Torrini, R. Jr 1984. Stratigraphy of the late Middle Eocene to Early Oligocene in the Bath Cliff section, Barbados, West Indies. *Micropaleontology*, **30**, 390–425.

Sissingh, W. 1977. Biostratigraphy of Cretaceous calcareous nannoplankton. *Geol. Mijnbouw*, **56**, 37–65.

Thiede, J., Vallier, T. L., *et al.* 1981. *Initial Rep. Deep Sea drill. Proj.*, **62**, 1–1120.

van Hinte, J. E. 1976. A Cretaceous time scale. *Bull. Am. Assoc. Petrol. Geol.*, **60**, 498–516.

Williams, G. L. 1977. Dinocysts. Their paleontology, biostratigraphy and paleoecology. In: A. T. S. Ramsay (ed.), *Oceanic Micropalaeontology*, pp. 1231–1325. Academic Press, London.

Worzel, J. H., Bryant, W., *et al.* 1973. *Initial Rep. Deep Sea drill. Proj.*, **10**, 1–748.

3
Introduction to the foraminiferal chapters

The planktic foraminifera are treated in Chapters 4 to 9. Chapter 4 deals with the Cretaceous on a worldwide basis. Chapter 5 covers the Paleocene and Eocene in low latitudes and northern mid latitudes, including the Mediterranean and Alpine areas, with the southern mid latitudes being dealt with in a part of Chapter 7.

Chapter 6 covers the Oligocene to Holocene of low latitudes only, as progressive latitudinal diversification becomes increasingly apparent through the Neogene. Thus, distinct regions have been treated separately, the southern mid latitudes in a part of Chapter 7, the Mediterranean in Chapter 8 and Central Paratethys in Chapter 9.

Particular features of some chapters

Though all the foraminiferal chapters give the same type of information as regards discussion and illustration of taxa, definition of zones and presentation of distribution charts, it has been found necessary to treat different regions and different parts of the stratigraphic column each in its own way.

In Chapters 4, 5 and 6, in particular, we have used more space for discussion, illustration and comparison of taxa as these are the forms with the greatest worldwide application in stratigraphy. Taxa have not been treated in strict alphabetic order in Chapters 5 and 6 but have been discussed under groups showing similar morphologic characters.

In Chapter 7 on the southern mid latitudes, additional information has been given on paleoceanography, paleo-temperature and species diversity to explain some of the particular conditions that pertain to that region.

In Chapter 8 on the Mediterranean region, the history and stratigraphic position of Neogene stages is given briefly with a calibration of the biostratigraphic zonation of their stratotypes.

Chapter 9, dealing with the Oligo-Miocene of Central Paratethys, shows that planktic foraminifera have their stratigraphic value even where conditions for their existence are marginal. Here too, space is given to Paratethyan stages as these are poorly known outside Europe but are the most widely used stratigraphic tools for many workers.

Discussion and illustration of taxa

The basic style of the chapters as regards information on taxa is similar throughout. There is a reference to the type and, where applicable, to any neotype or lectotype; important synonyms are listed. The taxonomic notes are not primarily descriptive as such information can be found in the original publication or reproduced in the *Catalogue of Foraminifera* (Ellis & Messina, 1940 with continuing supplements). Instead, greater weight is given to a discussion of those features of a taxon that distinguish it from morphologically similar forms.

The optimal stratigraphic use of any taxon depends on its accurate identification. The inclusion of specimens that do not conform to the established limits of any taxon may adversely affect its use. The great variability found in many planktic foraminiferal species resulting from morphologic variation and also differences between growth stages may make the delineation of a species difficult. This is especially so in lineages where change is gradational through time, in which case assignment to a particular taxon may be somewhat arbitrary.

The central form on which the identification of a taxon depends has to be the primary type and therefore, as a general rule, we have figured holotypes, neotypes and lectotypes. Additional specimens are illustrated if the original figure is particularly poor or if a range of variability needs to be shown.

The degree of variability around the central type allowed in a taxon varies from one worker to another. We try to discuss this aspect though, for reasons of space, the details may have been published elsewhere. Examples of lineages or groups of species dealt with in this way include *Globorotalia mayeri*, *Globorotalia opima*, *Globorotalia fohsi* and *Turborotalia cerroazulensis*. The last two examples we consider to be lineages where the developmental changes through time have been well documented and can be considered to be proven. Some other cases of phylogenetic relationship between taxa we consider to be still speculative. Some of those proposed by Jenkins (1971) and Blow (1979) are discussed in Chapter 6, but others proposed by Kennett & Srinivasan (1983) appeared after we had completed our work.

With few exceptions a taxon is discussed and illustrated only once, in most cases in the chapter that covers the area from which it was originally described. If it is widely used it will also appear on the range charts or on the zonal schemes in other chapters.

We consider that a uniform magnification for illustrations is a basic requirement for reliable interspecific comparison. To make the resulting picture easily comparable with what is seen down the microscope yet big enough to show essential detail, we have chosen a magnification of × 60. For a few forms we have added an enlarged photograph while for others, details are shown at a considerably higher magnification.

To aid quick identification and comparison of taxa,

names are placed beneath the figures where the primary type is also indicated. Full details regarding the provenance of the specimen is to be found in the appropriate legend.

Generic assignments

The grouping of species within genera has shown many changes over the years, but this has not affected the stratigraphic value of the taxa. In the present volume we do not generally discuss generic matters for Cenozoic taxa as other works are readily available (Stainforth *et al.*, 1975; Blow, 1979). Cretaceous genera are less well served particularly as numerous new ones have been erected in recent years. It has therefore been considered helpful to include in Chapter 4 a description and brief discussion of the Cretaceous genera recognized therein.

The use of generic names in Chapter 6 on low latitude Oligocene to Holocene forms is more conservative than that adopted in Chapter 5 on the equivalent Paleocene and Eocene forms. This is because the recent generic changes and additions in the lower part of the Cenozoic are more widely accepted than those in younger horizons.

Direction of coiling

The preferred direction of coiling in trochospiral planktic foraminifera can be used as a stratigraphic indicator in certain instances.

Through the Cretaceous up to about the end of the Albian all species show random coiling. A rapid change takes place around the base of the Cenomanian and from there to the top of the Maastrichtian virtually all trochospiral species show a very strong preference of well over 90% for dextral coiling.

The more complex patterns for the Cenozoic were compiled by Bolli (1971) and are here updated in Fig. 1. In the Paleocene to Eocene interval coiling is random at the beginning, after which many species display a strong preference for either sinistral or dextral coiling. There is a repetition of this pattern in the Oligocene to Holocene interval where, again, random coiling is replaced by a preferred direction in many species. During the latter phase some taxa, such as *Pulleniatina* and the menardiform group of Globorotalias, may show abrupt reversals in direction of coiling. For certain groups, such as the menardiform Globorotalias, the detailed pattern of coiling reversals is known to be different from one ocean to another.

Where coiling trends are stratigraphically significant they are noted in Chapter 6 but will not be found in other chapters.

Stratigraphic ranges

The distribution of all taxa discussed in the text is shown on one or more of the range charts. In Chapters 7, 8 and 9, where the number of taxa is limited, ranges are given on a

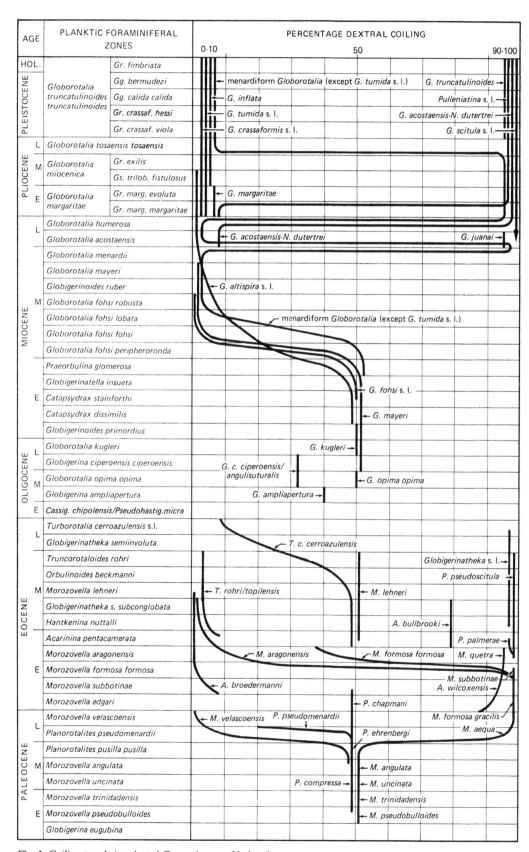

Fig. 1. Coiling trends in selected Cenozoic taxa. Updated
from Bolli (1971)

AGE	MY	LOW LATITUDES ZONES	LOW LATITUDES SUBZONES	N/P ZONES	MEDITERRANEAN ZONES	MEDITERRANEAN SUBZONES	SOUTHERN MID-LATITUDES ZONES
HOL.		Globorotalia truncatulinoides truncatulinoides	Gr. fimbriata	N 23			G. truncatulinoides
PLEISTOCENE	0.14		Gg. bermudezi				
			Gg. calida calida	N 22	G. truncatulinoides excelsa		
			Gr. crassaf. hessi		G. cariacoensis		
			Gr. crassaf. viola				
PLIOCENE L	1.9	Globorotalia tosaensis tosaensis		N 21	G. inflata		G. inflata
PLIOCENE M	2.4	Globorotalia miocenica	Gr. exilis		G. aemiliana		
			Gs. trilob. fistulosus	N 20			
PLIOCENE E	3.2	Globorotalia margaritae	Gr. marg. evoluta	N 19	G. puncticulata / G. punct.-G. margaritae		G. puncticulata
			Gr. marg. margaritae	N 18	G. margaritae / S. seminulina s. l.		
MIOCENE L	5.1	Globorotalia humerosa		N 17	Non-distinctive Zone / G. conomiozea	G. suterae	G. conomiozea
		Globorotalia acostaensis		N 16	G. obliquus extremus / G. acostaensis	G. obliquus extremus-G. bulloideus	G. miotumida
MIOCENE M	11.3	Globorotalia menardii		N 15	G. menardii s. l.		
		Globorotalia mayeri		N 14	G. siakensis	G. siakensis-G. obliquus obliquus	G. mayeri
		Globigerinoides ruber		N 13		G. subquadratus	
		Globorotalia fohsi robusta		N 12		G. altispira altispira	
		Globorotalia fohsi lobata		N 11			
		Globorotalia fohsi fohsi		N 10	O. suturalis-G. peripheroronda	G. praemenardii-G. peripheroronda	O. suturalis
		Globorotalia fohsi peripheroronda		N 9		O. universa / O. suturalis	
MIOCENE E	14.4	Praeorbulina glomerosa		N 8	P. glomerosa		P. glomerosa curva
		Globigerinatella insueta		N 7	G. trilobus s. l.		G. trilobus
		Catapsydrax stainforthi		N 6	G. dehiscens dehiscens-C. dissimilis	G. altiaperturus	
		Catapsydrax dissimilis		N 5		G. dehiscens dehiscens	
		Globigerinoides primordius		N 4			
OLIGOCENE L	24.6	Globorotalia kugleri		P 22/N3	G. kugleri		G. woodi connecta
		Globigerina ciperoensis ciperoensis			G. ciperoensis ciperoensis		G. woodi woodi
OLIGOCENE M		Globorotalia opima opima		P 21/N2	G. opima opima		G. dehiscens / G. euapertura
	32.8	Globigerina ampliapertura		P 20/N1	G. sellii		G. angiporoides
OLIGOCENE E		Cassig.chipolensis/Pseudohastig.micra		P 18/19	G. ampliapertura-G. euapertura		G. brevis
EOCENE L	38	Turborotalia cerroazulensis s.l.		P 17 / P 16	T. cerroazulensis s. l.	T. c. cunialensis / T. c. cerroazulensis/cocoaensis	
		Globigerinatheka semiinvoluta		P 15	G. semiinvoluta		G. linaperta
EOCENE M	41	Truncorotaloides rohri		P 14	T. rohri	T. c. pomeroli/cerroazulensis	T. inconspicua
		Orbulinoides beckmanni		P 13	O. beckmanni		
		Morozovella lehneri		P 12	M. lehneri	T. c. possagnoensis	G. index
		Globigerinatheka s. subconglobata		P 11	G. s. subconglobata		
		Hantkenina nuttalli		P 10	H. nuttalli		A. primitiva
EOCENE E	50.3	Acarinina pentacamerata		P 9	A. pentacamerata	T. c. frontosa	M. crater
		Morozovella aragonensis		P 8	M. aragonensis		
		Morozovella formosa formosa		P 7	M. formosa formosa		P. wilcoxensis
		Morozovella subbotinae		P 6	M. subbotinae		
		Morozovella edgari			M.edgari		
PALEOCENE L	54.9	Morozovella velascoensis		P 5	M. velascoensis		G. triloculinoides
		Planorotalites pseudomenardii		P 4	P. pseudomenardii		
PALEOCENE M		Planorotalites pusilla pusilla		P 3	P. pusilla pusilla		
	61.5	Morozovella angulata			M. angulata		
		Morozovella uncinata		P 2	M. uncinata		
PALEOCENE E		Morozovella trinidadensis		P 1 d	M. trinidadensis		G. pauciloculata
		Morozovella pseudobulloides		P 1 c / b	M. pseudobulloides		
	66.7	Globigerina eugubina		P 1 a	G. eugubina		

Fig. 2. Correlation between Cenozoic low latitude, Mediterranean and southern mid latitude zonal schemes

single chart according to first occurrences. The stratigraphic distribution of the more numerous taxa treated in Chapters 4, 5 and 6 is presented in three different ways: (a) species arranged according to their generic grouping, (b) species in order of first occurrence, (c) species in order of last occurrence. Each method has its advantages. Grouping by genera allows one to see the range of a particular genus and also the development of the whole planktic foraminiferal fauna through time. Presentation by first or last occurrence facilitates identification of which species may be expected to be found at any particular horizon or zone. In addition, presentation in order of last occurrence aids oil company stratigraphers who have to work with tops, particularly when using ditch cuttings.

Zonal schemes

Varying zonal schemes are used in different regions of the world and those identified by the same species names or the same numbering system may not be used in a similar way by all workers. This is particularly the case in the younger Cenozoic where zones are more restricted in their geographic extent. To avoid stratigraphic confusion, it is necessary to know exactly how a particular zone is being used and so each chapters carries a list of zones used therein with information on category, age, author and definition of boundaries. Major zonal schemes are figured alongside each other for easy comparison.

In the present chapter, comparison is made across latitudes by a correlation of low latitude, Mediterranean and southern mid latitude schemes (Fig. 2).

Recent reference works on planktic foraminifera

A number of publications dealing with planktic foraminifera and their stratigraphic significance have appeared in recent years. Some of the information contained in them is additional to what will be found in the present book making them complementary. Some important works are listed below:

The *Manual of Planktonic Foraminifera* by Postuma (1971) covers both Cretaceous and Tertiary index forms and includes cross-sections as well as photographs and drawings.

In 'Cenozoic planktonic foraminiferal zonation and characteristics of index forms' Stainforth *et al.* (1975) give a concise review of genera and species with a good treatment of stratigraphic significance and zonal applications that is clearly directed towards the practical stratigrapher.

Oceanic Micropalaeontology (ed. Ramsay, 1977) contains two major chapters, one by W. A. Berggren on a number of important Paleogene genera and one by B. A. Masters on Mesozoic taxa.

In 1979 Blow's *The Cainozoic Globigerinida* appeared. This major work, which also includes a reprint of his 1969 paper, contains exhaustive descriptions and discussions both taxonomic and stratigraphic that very much express the author's own viewpoint.

Two important works on Cretaceous planktic foraminifera have recently appeared. They are the 'Atlas de Foraminifères planctoniques du Crétacé moyen' edited by Robaszynski & Caron (1979) and the 'Atlas of Late Cretaceous planktonic foraminifera' by Robaszynski *et al.* (1984). These give the conclusions of a recent working group on Cretaceous taxa.

A *Systematic Index of Recent and Pleistocene Planktonic Foraminifera* was published by Saito *et al.* in 1981 and, after our chapters had already been completed, we received a copy of *Neogene Planktonic Foraminifera, A Phylogenetic Atlas* by Kennett & Srinivasan (1983).

Two important publications documenting the planktic genera are those by Loeblich & Tappan (1964) and Banner (1982). The latter shows the great proliferation in genera in recent years and has a key for their separation.

References

Banner, F. T. 1982. A classification and introduction to the Globigerinacea. In: F. T. Banner & A. R. Lord (eds.), *Aspects of Micropaleontology (papers presented to Professor Tom Barnard)*, pp. 142–239. Allen & Unwin.

Berggren, W. A. 1977. Atlas of Palaeogene planktonic foraminifera. Some species of the genera *Subbotina, Planorotalites, Morozovella, Acarinina* and *Truncorotaloides*. In: A. T. S. Ramsay (ed.), *Oceanic Micropalaeontology*, vol. 1, pp. 205–300. Academic Press.

Blow, W. H. 1969. Late Middle Eocene to Recent planktonic foraminiferal biostratigraphy. *Proceedings First International Conference on Planktonic Microfossils, Geneva 1967*, **1**, 199–422.

Blow, W. H. 1979. *The Cainozoic Globigerinida*. Brill, 3 vols., 1412 pp.

Bolli, H. M. 1971. The direction of coiling in planktonic foraminifera. In: B. M. Funnell & W. R. Riedel (eds.), *The Micropaleontology of Oceans*, pp. 639–48. Cambridge University Press.

Ellis, B. R. & Messina, A. 1940. *Catalogue of Foraminifera*. American Museum of Natural History New York. Supplements post 1940.

Jenkins, D. G. 1971. New Zealand Cenozoic planktonic foraminifera. *Palaeontol. Bull. geol. Surv. N.Z.*, **42**, 1–278.

Kennett, J. P. & Srinivasan, S. 1983. *Neogene Planktonic Foraminifera, A Phylogenetic Atlas*. Hutchison Ross Publishing Company, 265 pp.

Loeblich, A. R. Jr & Tappan, H. 1964. Foraminiferida. In: R. C. Moore (ed.), *Treatise on Invertebrate Paleontology, C. Protista*, vols. 1 and 2, pp. 1–868. The Geological Society of America and the University of Kansas Press.

Masters, B. A. 1977. Mesozoic planktonic foraminifera. A world-wide review and analysis. In: A. T. S. Ramsay (ed.), *Oceanic Micropalaeontology*, vol. 1, pp. 301–731. Academic Press.

Postuma, J. 1971. *Manual of Planktonic Foraminifera*. Elsevier Publishing Company, 420 pp.

Robaszynski, F. & Caron, M. (coord.) 1979. Atlas de Foraminifères planctoniques du Crétacé moyen. *Cah. Micropaleontol.*, **1**, 1–185, **2**, 1–181.

Robaszynski, F., Caron, M., Gonzales, J. M. & Wonders, T. 1984.

Atlas of Late Cretaceous planktonic foraminifera. *Rev. Micropaleontol.*, **26** (3–4), 145–305.

Saito, T., Thompson, P. R. & Breger, D. 1981. *Systematic Index of Recent and Pleistocene Planktonic Foraminifera.* University of Tokyo Press, 190 pp.

Stainforth, R. M., *et al.* 1975. Cenozoic planktonic foraminiferal zonation and characteristics of index forms. *Univ. Kansas Paleontol. Contrib.*, article **62**, 1–425, 2 parts.

4

Cretaceous planktic foraminifera

MICHÈLE CARON

CONTENTS

Introduction

In 1927 Cushman introduced the genus *Globotruncana* into which until the early 1940s all trochospiral, single and double keeled Cretaceous planktic species were placed. A large number of taxa, distinguished by number and position of peripheral keels, shape of test and number and shape of chambers, were included in this genus. Other distinguishing criteria like position and category of apertures, additional structural elements like tegilla and portici were at that time not yet taken into consideration for taxonomic subdivision.

A number of *Globotruncana* species were also described in thin sections from limestones (de Lapparent, 1918; Renz, 1936; Vogler, 1941; Bolli, 1945). General test shape and number and position of peripheral keels are readily identifiable from section but not the position of apertures and the presence of supplementary and accessory structures which later became used for generic distinction. The comparison of specimens in sections with isolated forms is therefore often problematic. Useful illustrations for identifying sections of Globotruncanidae can be found in Postuma (1971) where all species are portrayed not only as isolated specimens but also as oriented axial thin sections. In addition, Fleury (1980) has a large number of drawings of oriented thin sections of Turonian to Maastrichtian Globotruncanidae.

Though numerous species, some of them excellent index forms like *calcarata* or *contusa*, had already been established for some time, it was not until the 1930s that the stratigraphic significance of the Cretaceous planktic foraminifera became more widely realized (Viennot, 1930; Thalmann, 1934). Authors like Renz (1936), Gandolfi (1942), Bolli (1945) and Cita (1948) already made full use of them in their biostratigraphic studies.

The inclusion at this stage of all trochospiral and keeled taxa into the genus *Globotruncana* had no detrimental effect on their biostratigraphic application. This is a fine example to show that it is the species and not the genus concept that is of basic importance in biostratigraphic studies. The generic separation of certain species previously included in *Globotruncana* began with the introduction by Brotzen (1942) of the genus *Rotalipora*, characterized by umbilical sutural supplementary apertures. The years 1950 to 1960 became the most prolific for the introduction of new taxa, particularly genera. Several of those proposed during this period such as *Abathomphalus*, *Hedbergella* and *Ticinella*, contain significant index forms.

In 1964 Loeblich & Tappan integrated in their treatise all generic names valid at that time. Pessagno (1967) demonstrated the importance of the position of the primary aperture in the Globotruncanidae, its gradual migration through time towards the umbilicus, the delicate structures that became progressively more important in the umbilical area and the complexity of the imperforate peripheral margin. As a result, he erected the genera *Archaeoglobigerina* and *Whiteinella* and substantially emended *Marginotruncana*. From 1970 to 1982 the wealth of data produced during these years by the scanning electron microscope, particularly from well preserved specimens obtained from Deep Sea Drilling Project (DSDP) sites, confirmed the earlier observations which were often based on less well preserved specimens.

Most workers now became aware of the complexity caused by the multitude of previously established taxa and began to exercise restraint in creating new ones or in further splitting existing species. Only few new genera and species were therefore proposed during this last period. The trend in recent years has moved towards a stabilization of established taxa. This includes more accurate definitions on both the generic and specific level including documentation of the variability in individual species. In turn this has led to the identification and elimination of numerous synonyms within the great number of taxa proposed during the last decades and thus a much needed reduction of the valid taxa.

Such tasks have been carried out since 1976 largely by the European Working Group on Planktic Foraminifera. Their first results included the zonation of the 'Middle Cretaceous' (Albian to Turonian) accompanied by careful diagnoses and SEM micrographs of the index species. This first part appeared in 1979 as an Atlas in two volumes (Robaszynski & Caron, 1979). A second part, covering the Late Cretaceous (Santonian to Maastrichtian), was published in spring 1984.

The chapter on Cretaceous planktic foraminifera departs from those on the Tertiary in this volume in that the generic classification used here, the characteristics and phylogeny of the trochospiral genera and a brief discussion of the genera themselves are all included here. The reason is that these criteria are not yet accessible in a condensed form within one publication as is the case for the Tertiary, for instance in Stainforth *et al.*, 1975.

Classification used in this chapter

The Cretaceous planktic foraminifera are grouped into the three families Heterohelicidae, Planomalinidae and Globotruncanidae, based on their chamber arrangement. In addition, the trochospiral forms with a typically cancellate chamber surface are separated from the Globotruncanidae and are placed in the family Favusellidae. The chamber arrangement in the Heterohelicidae is very variable and may change during growth stages. Initially it may be planispiral, bi-, tri or trochospiral, becoming biserial later. Final stages may show reduction to a uniserial state or there may be proliferation of chambers. The planispiral forms are included in the Planomalinidae with generic distinctions based largely on apertural features and chamber shape. Despite its brief juvenile trochospiral stage, the genus *Schakoina* is also included in this family.

By far the most dominant group of Cretaceous planktic foraminifera are those with a trochospiral chamber arrangement. Most of them are placed in the family Globotruncanidae. All genera included here are thought to have been derived from *Hedbergella*, the oldest and most primitive of the genera. Generic distinctions within the Globotruncanidae are based on chamber shape, apertural and accessory structural features and number of peripheral keels. Virtually all index forms used for the biostratigraphic subdivision of the Cretaceous belong to the Globotruncanidae.

The recent findings by Grigelis & Gorbachik (1980) of a connection between the Protoglobigerinas of the Middle Jurassic and the miniscule *Globuligerina hoterivica* of the Early Cretaceous has shown the existence of an ancestral stock from which the first *Hedbergella* evolved. A new family was therefore needed to include these primitive trochospiral forms. Longoria (1974) proposed the family Caucasellidae for them but Grigelis & Gorbachik (1980) have instead chosen the family Favusellidae (Longoria 1974) whose diagnosis, after emendation, combines the characters common to the two families: *Globigerina*-like chambers, surface covered by polygonal sculptures, umbilical primary aperture.

Table 1. Classification of the Cretaceous Globigerinacea

1. Test bi-, tri-, multiserial:	HETEROHELICIDAE	*Heterohelix* *Pseudotextularia* *Pseudoguembelina* *Guembelitria* *Racemiguembelina*
2. Test planispiral	PLANOMALINIDAE	*Planomalina* *Globigerinelloides* *Hastigerinoides* *Schackoina*

3. Test trochospiral
  '*Globigerina*-like', primitive
  forms: FAVUSELLIDAE
      primary aperture entirely umbilical and chamber surface roughly reticulated   *Globuligerina*
      primary aperture varies from umbilical to extraumbilical and chamber surface
      regularly reticulated   *Favusella*
  '*Hedbergella*-lineage',
  evolved forms: GLOBOTRUNCANIDAE
      primary aperture umbilical–extraumbilical–nearly peripheral, protected
      by lip or flap
          without supplementary apertures on umbilical side
              periphery without keel   *Hedbergella*
              pustulose angular periphery   *Praeglobotruncana* p.p.
              periphery with 2 keels   *Falsotruncana*
          with supplementary apertures on umbilical side   *Ticinella*
              periphery without keel   *Biticinella*
              periphery with 1 keel   *Rotalipora*
      primary aperture umbilical–extraumbilical
          protected by portici
              periphery without keel   *Whiteinella*
              periphery with 1 keel   *Helvetoglobotruncana*
              with pustulose, truncate periphery   *Praeglobotruncana* p.p.
              periphery with 2 keels
                  umbilical sutures depressed, radial or curved   *Dicarinella*
              periphery with 2 keels, sometimes fused in 1 keel on the last whorl
                  umbilical sutures raised, curved   *Marginotruncana*
              only with imperforate, angular periphery or passing gradually to 1 keel   *Globotruncanella*
          protected by tegilla
              periphery with 2 keels
                  umbilical sutures depressed, radial   *Abathomphalus*
      primary aperture umbilical
          protected by portici
              periphery with 1 keel
                  umbilical sutures raised, radial or curved   *Globotruncanita*
              periphery with 2 keels
                  umbilical sutures raised, curved   *Rosita*
          protected by tegilla
              periphery truncated by 2 keels
                  umbilical sutures raised, curved   *Globotruncana*
              periphery with inflated rugose chambers
                  umbilical sutures depressed, radial
                      with rugose surface
                          imperforate peripheral band and weakly developed double
                          keel   *Archaeoglobigerina*
                          1 keel on the last whorl, 2 keels weakly visible in
                          section on the previous whorl   *Gansserina*
                    with costellae arranged in meridional pattern
                        globulose periphery   *Rugoglobigerina*
                        imperforate peripheral band and well developed double
                        keel   *Rugotruncana*

Characteristics and Phylogeny of Trochospiral Genera

Within the families Globotruncanidae and Favusellidae the different genera are defined by criteria shown on the key (Fig. 1). It allows for a rapid characterization of a given genus, but reference to its phylogenetic development (Fig. 2) is needed to show its relation to other genera and its place within the evolution of the family.

The general idea of the new classification of the planktic foraminifera (Loeblich & Tappan, 1974b; Steineck & Fleisher, 1978) is that morphology and phylogeny are not dissociated. It is shown (Tappan, 1971; Hart & Bailey, 1979; Wonders 1980) that the interaction of form and function forces the tests of Globigerinacea to develop morphological characters which re-occur several times during the Cretaceous in members of different lineages. The classification here applied (Table 1) is based on a hierarchy of morphological criteria based on a better knowledge of the finest details of the tests and their evolution as well as the phylogenetic relations recognized between certain species (Caron, 1983a). Fig. 2 combines the origin, the evolution and the diversification of all the genera which are here included in the Globotruncanidae and Favu-

sellidae. This evolutionary scheme not only shows the principle of the systematic treatment, but can also be applied to biostratigraphy. In the absence of zonal markers it allows the determination of the stratigraphic position of a sample through the evolutionary stage of the assemblage. This corresponds to the subdivision of the Cretaceous into the four episodes proposed by Longoria & Gamper (1975): (1) hedbergellid, (2) ticinellid, (3) marginotruncanid, (4) globotruncanid.

The episodes 2 to 4 terminate with the extinction of the taxa they represent. The last event is complete in that all representatives of the respective genera, including the less evolved, disappear. However, it cannot be fully excluded that the earliest Tertiary forms may have as their ancestors certain primitive and therefore more resistant terminal Maastrichtian forms of the genera *Heterohelix* and *Guembelitria* respectively (Bang, 1979).

The evolutionary lineages shown in Fig. 2 are based on detailed investigations of the sequences and on large numbers of specimens. By these means the following lineages could be recognized that also proved to be of considerable biostratigraphic value:

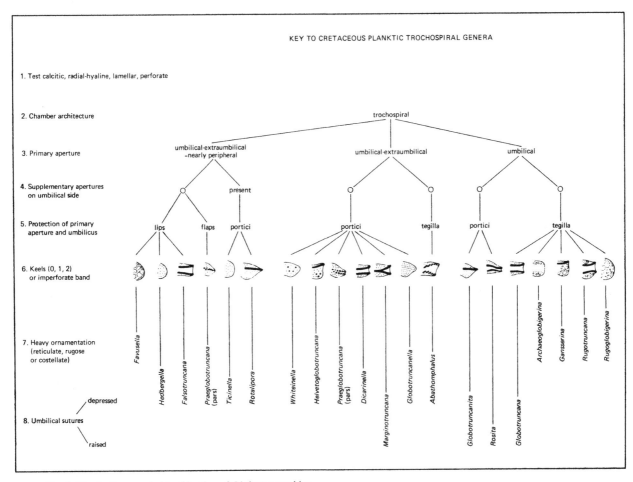

Fig. 1. Key to the generic identification of Globotruncanidae and Favusellidae

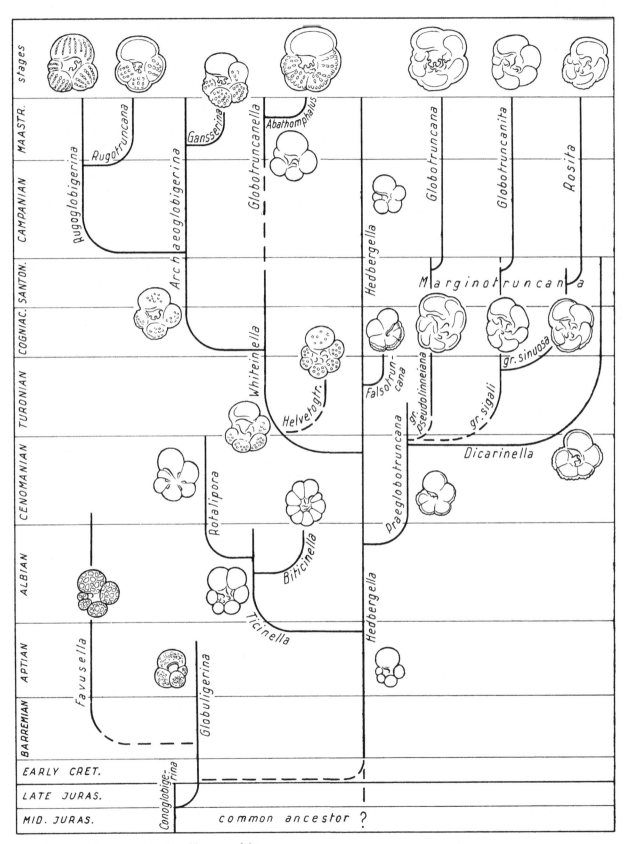

Fig. 2. Phylogeny and stratigraphic range of the
Globotruncanidae and Favusellidae genera

Hedbergella – Ticinella   Rotalipora

Hedbergella – Praeglobotruncana – Dicarinella

Marginotruncana – Globotruncana

Whiteinella – Archaeoglobigerina – Rugoglogigerina

Marginotruncana sinuosa – group *fornicata– contusa = Rosita*

Marginotruncana sigali – group *stuartiformis = Globotruncanita*

Globotruncanella – Abathomphalus

Rugoglobigerina – Rugotruncana

In addition to numerous earlier publications dealing with evolutionary lineages, reference is here made to the more recent works by Linares-Rodriguez (1977) and Wonders (1975, 1978, 1980).

Discussion of genera

ABATHOMPHALUS Bolli, Loeblich & Tappan, 1957

Type species: *Globotruncana mayaroensis* Bolli, 1951.

The genus differs from *Globotruncana* in the umbilical– extraumbilical position of the primary aperture and in the radial sutures on the umbilical side. *Globotruncanella* is regarded as the ancestor of *Abathomphalus*, based on the evolutionary lineage *G. citae – A. intermedius – A. mayaroensis.*

ARCHAEOGLOBIGERINA Pessagno, 1967

Type species: *Archaeoglobigerina blowi* Pessagno, 1967.

A wide imperforate peripheral band bordered by two faint keels are present on the globular early chambers of the last whorl. This is typical for the genus which evolved from *Whiteinella* (*W. archaeocretacea – A. cretacea*). *Archaeoglobigerina* itself is probably the ancestor of the inflated and ornamented (rugose, costellate) Late Cretaceous genera *Rugoglobigerina, Rugotruncana* and *Gansserina.*

BITICINELLA Sigal, 1956

Type species: *Anomalina breggiensis* Gandolfi, 1942.

The genus differs from *Globigerinelloides* in the presence of true supplementary apertures like those of *Ticinella* in one umbilicus and relict apertures in the other. In comparison, *Globigerinelloides* has relict apertures in both umbilici. *Biticinella* differs from *Planomalina* in the absence of a keel.

DICARINELLA Porthault, 1970

Type species: *Globotruncana indica* Jacob & Sastry, 1950; synonym of *Dicarinella hagni* Scheibnerova, 1962 (see Robaszynski & Caron, 1979).

The genus differs from *Praeglobotruncana* in its well developed keels and portici; from *Marginotruncana* in the umbilical sutures which are depressed and usually radial. In the *D. primitiva – D. concavata – D. asymetrica* lineage, the genus

shows a trend towards curved umbilical sutures, a tendency already present in certain specimens of the ancestral *D. algeriana.*

FALSOTRUNCANA Caron, 1981

Type species: *Falsotruncana maslakovae* Caron, 1981.

The species belonging to this genus have long been confused with *Praeglobotruncana* from which they differ in the presence of a long primary aperture with a long, slim lip and the absence of a flap or porticus. They evolved from a *Hedbergella* ancestor, retaining its rudimentary primary aperture and developing a stout double keel.

FAVUSELLA Michael, 1973

Type species: *Globigerina washitensis* Carsey, 1926.

The test surface is reticulated into coarse regular polygonal areas by ridges forming a honeycomb-like pattern; each polygonal area has more than twenty minute pores. *Favusella* differs from *Globigerina* in its regular reticulate sculpture and in the position of the primary aperture which may become umbilical–extraumbilical. It differs from *Hedbergella* in the reticulate sculpture.

GANSSERINA Robaszynski, Caron, Gonzalez & Wonders, 1984

Type species: *Globotruncana gansseri* Bolli, 1951.

The test is plano-convex, flat on the spiral side, inflated umbilically. Early whorls are globigeriniform with weakly developed double keel, visible in thin sections (Pessagno, 1967). Rugosities and sometimes faint, discontinuous costellae are present mainly on the umbilical side. Like *Rugotruncana*, *Gansserina* is thought to have originated from an *Archaeoglobigerina* ancestor through the development of a single keel and coarse rugosities.

GLOBIGERINELLOIDES Cushman & Ten Dam, 1948

Type species: *Globigerinelloides algeriana* Cushman & Ten Dam, 1948.

The successive apertures remain as relict apertures on both sides of the typically planispiral test. *Globigerinelloides* differs from *Biticenella* in the lack of supplementary apertures and from *Planomalina* in the absence of a peripheral keel.

GLOBOTRUNCANA Cushman, 1927

Type species: *Pulvinulina arca* Cushman, 1926.
Synonym: *Rosalinella* Marie, 1941.
Globotruncana differs from *Marginotruncana* in having its umbilical primary aperture covered by a series of tegilla; from *Rugotruncana* in the absence of meridionally arranged

rugosities; from *Abathomphalus* in the umbilical position of the primary aperture and in having sigmoidal and raised umbilical sutures.

GLOBOTRUNCANELLA Reiss, 1957

Type species: *Globotruncana citae* Bolli, 1951.
Emendation: Pessagno, 1967.

Globotruncanella differs from *Hedbergella* in the presence of portici, and from *Archaeoglobigerina* in the absence of tegilla and of a large imperforate peripheral band. *Globotruncanella* is a homeomorph of *Whiteinella* but is segregated because of the wide chronological gap that separates them.

GLOBOTRUNCANITA Reiss, 1957

Type species: *Rosalina stuarti* de Lapparent, 1918.

The genus was proposed by Reiss to separate the Late Cretaceous species with an umbilical system of portici and a single peripheral keel from the genus *Globotruncana*. Some authors still contest this separation and maintain the name *Globotruncana* for all keeled species of the Late Cretaceous.

GLOBULIGERINA Bignot & Guyader, 1971

Type species: *Globigerina oxfordiana* Grigelis, 1958.
Synonyms: *Polskanella* Fuchs, 1973.
  Caucasella Longoria, 1974.
  Globuligerina Grigelis, 1974.

The trochoid test is very small with 4–6 inflated and compacted chambers forming the last whorl; the surface is sculptured as tubercles or sometimes an irregular reticulation with 3 to 4 minute pores in each polygon. *Globuligerina* is regarded as a descendant from the Jurassic genus *Conoglobigerina* Morozova & Moskalenko, 1961, from which it differs in its trochoid test and its reticulate sculptured chamber surface. *Globuligerina* seems itself to be a form intermediate to the genus *Favusella* which differs from it essentially in the great variability of the number of chambers, height of spire, position of primary aperture and more regular reticulate sculpture. According to Grigelis & Gorbachik (1980), it is possible to establish a phylogenetic sequence extending from Middle Jurassic to Cenomanian. This would include in the family Favusellidae all members of the three genera *Conoglobigerina* Morozova, *Globuligerina* Bignot & Guyader and *Favusella* Michael.

GUEMBELITRIA Cushman, 1933

Type species: *Guembelitria cretacea* Cushman, 1933.

The very small (150–200 μm), triserial test consists of globular chambers. The walls are finely perforate, each pore is surrounded by a blunt cone (sometimes there are two pores per cone).

HASTIGERINOIDES Brönnimann, 1952

Type species: *Hastigerinella alexanderi* Cushman, 1931.

The test is planispiral, the periphery lobate to stellate. Relict apertures are visible in both umbilici. *Hastigerinoides* differs from *Schackoina* in having digitiform last chambers instead of chambers extended as tubulospines and in being planispiral from its initial stage.

HEDBERGELLA Brönnimann & Brown, 1958

Type species: *Anomalina lorneiana* d'Orbigny, var. *trocoidea* Gandolfi, 1942.
Emendation: Longoria, 1974.

Hedbergella is regarded as the common ancestor of all genera belonging to the family Globotruncanidae. Its characters changed so slowly during the Cretaceous that only minor morphological differences developed between *H. delrioensis* of Albian age and *H. holmdelensis* of Maastrichtian age. *Hedbergella* differs from *Globuligerina* and *Favusella* by the lack of a reticulate ornamentation; from *Ticinella* by the absence of umbilical supplementary apertures and from *Whiteinella* by the absence of portici.

HELVETOGLOBOTRUNCANA Reiss, 1957

Type species: *Globotruncana helvetica* Bolli, 1945.

The test is plano-convex, flat on the spiral side, strongly inflated umbilically. A peripheral keel may be present. The umbilicus is covered by a system of portici with infralaminal accessory apertures. Numerous pustules cover particularly the umbilical side of the test.

Helvetoglobotruncana is a short-lived genus of Early to Middle Turonian age. It evolved from a *Hedbergella* stock in the Late Cenomanian, via a *Whiteinella* stage. *Helvetoglobotruncana* is a close homeomorph of the Middle Maastrichtian *Gansserina gansseri*. However, the absence of a phyletic relationship and the wide chronological gap clearly separate them.

HETEROHELIX Ehrenberg, 1843

Type species: *Spiroplecta americana* Ehrenberg, 1844. (Now considered a junior synonym of *Heterohelix navarroensis* Loeblich, 1951, on the basis that '*H.* [*Spiroplecta*] *americana* Ehrenberg, 1844, is a junior secondary homonym of *H.* [*Textilaria*] *americana* Ehrenberg, 1843' (Brown, 1969).)
Synonym: *Guembelina* Egger, 1899.

Test either with a minute initial planispiral coil followed by a biserial stage, or biserial throughout. *Heterohelix* differs from *Pseudotextularia* in the width of chambers exceeding thickness.

MARGINOTRUNCANA Hofker, 1956

Type species: *Rosalina marginata* Reuss, 1845.
Emendations: Pessagno, 1967. Porthault, 1970.

Marginotruncana differs from *Dicarinella* in its umbilical, sigmoidal and raised sutures; from *Globotruncana* in the extraumbilical position of its primary aperture and lack of tegilla; from *Rosita*, whose primary aperture opens in the umbilical area, in its umbilical–extraumbilical position. *Marginotruncana* may possess two keels, widely spaced as in *marginata* or closely spaced as in *sinuosa*, or a single one as in *sigali* and *marianosi*.

PLANOMALINA Loeblich & Tappan, 1946

Type species: *Planomalina aspidostroba* Loeblich & Tappan, 1946. (Junior synonym of *Planomalina buxtorfi* (Gandolfi), 1942.)
Emendations: Reiss, 1957. Wonders, 1975.

Planomalina differs from *Globigerinelloides* in a truncated test and the presence of a peripheral keel resulting in an acute edge view.

PRAEGLOBOTRUNCANA Bermudez, 1952

Type species: *Globorotalia delrioensis* Plummer, 1931.
Synonym: *Rotundina* Subbotina, 1953.
Emendations: Reiss, 1957. Banner & Blow, 1959. Porthault, 1970.

The umbilical–extraumbilical primary aperture carries a flap (*P. delrioensis*) or a porticus (*P. stephani*). Relict flaps or portici may be visible in the umbilical area. *Praeglobotruncana* evolved from *Hedbergella* by acquisition of an acute, imperforate periphery and pustules. Through development of a true double keel it gave rise to *Dicarinella* with *D. algeriana* as transitional form. *Praeglobotruncana* differs from *Marginotruncana* in its umbilical, radial and depressed sutures, and in the absence of a true double keel; from *Falsotruncana* in its relict flaps or portici in the umbilical area and in the absence of a true double keel.

PSEUDOGUEMBELINA Brönnimann & Brown, 1953

Type species: *Guembelina excolata* Cushman, 1926.

A sutural supplementary aperture, present only in the last chambers, opens backwards at the point where the base of the chamber meets the median suture. It is sometimes covered by a tubular flap. Longitudinal costae are fine or coarse. *Pseudoguembelina* differs from *Heterohelix* in the presence of sutural supplementary apertures along the median suture of the last chambers.

PSEUDOTEXTULARIA Rzehak, 1891

Type species: *Cuneolina elegans* Rzehak, 1891.

Pseudotextularia differs from all other Heterohelicidae in having a greater number of chambers (typically 8–10 pairs). The primary aperture is a low wide arch at the base of the septal face.

RACEMIGUEMBELINA Montanaro-Galitelli, 1957

Type species: *Guembelina fructicosa* Egger, 1899.

With its proliferation of chambers and conically-shaped test this monospecific genus is readily distinguished from all other heterohelicid genera.

ROSITA Robaszynski, Caron, Gonzalez & Wonders, 1984

Type species: *Globotruncana fornicata* Plummer, 1931.

The spiral side of the test can be moderately high as in *R. fornicata* or extremely high as in *R. contusa*. The umbilical area is covered by portici in helicoidal arrangement, situated deep in the umbilical cavity in high spired forms. The last chambers have a characteristic undulating surface. The two keels are closely spaced.

ROTALIPORA Brotzen, 1942

Type species: *Rotalipora turonica* Brotzen, 1942. (Now junior synonym of *Rotalipora cushmani* (Morrow), 1934.)
Synonyms: *Thalmanninella* Sigal, 1948.
        Anaticinella Eicher, 1972.
        Pseudoticinella Longoria, 1973.
Emendations: Reichel, 1950. Sigal, 1958.

Rotalipora differs from *Ticinella* in the presence of a keel and from *Globotruncanita* in the presence of umbilical sutural, supplementary apertures.

RUGOGLOBIGERINA Brönnimann, 1952

Type species: *Globigerina rugosa* Plummer, 1927.
Synonyms: *Trinitella* Brönnimann, 1952.
        Plummerita Brönnimann, 1952.
        Kuglerina Brönnimann & Brown, 1956.
Emendation: Pessagno, 1967.

Rugoglobigerina differs from *Rugotruncana* in the lack of a double keel. According to Pessagno (1967), *Archaeoglobigerina* could have given rise to *Rugoglobigerina* by acquisition of meridionally arranged costellae. Later, *Rugotruncana* evolved from *Rugoglobigerina* by flattening of its chambers and by development of a true double keel.

RUGOTRUNCANA Brönnimann & Brown, 1956

Type species: *Rugotruncana tilevi* Brönnimann & Brown,
1956. (Junior synonym of *Rugotruncana subcircumnodifer* Gandolfi, 1955.)
Emendation: Pessagno, 1967.

Rugotruncana differs from *Rugoglobigerina* in the presence of
a double keel; from *Globotruncana* in the presence of globigeriniform early chambers and costellae.

SCHACKOINA Thalmann, 1932

Type species: *Siderolina cenomana* Schacko, 1897.
Synonym: *Leupoldina* Bolli, 1957.

Schackoina differs from *Hastigerinoides* in its tubulospines or
bulbous extensions of chambers instead of digitiform chambers,
in the minute initial trochospiral stage and absence of relict
apertures.

TICINELLA Reichel, 1950

Type species: *Anomalina roberti* Gandolfi, 1942.

Ticinella differs from *Rotalipora* in the lack of a peripheral keel.

WHITEINELLA Pessagno, 1967

Type species: *Whiteinella archaeocretacea* Pessagno, 1967.

Whiteinella differs from *Hedbergella* in the presence of portici;
from *Archaeoglobigerina* in the absence of tegilla and absence
of the wide imperforate peripheral band; from *Praeglobotruncana* in the absence of imbricated pustules along the peripheral
margin. *Whiteinella* is a homeomorph of *Globotruncanella* but
is segregated because of the wide chronological gap that
separates them.

Zonal schemes

Historical development

Viennot (1930) and Thalmann (1934) were the first to
point out the value of Globotruncanidae as stratigraphic
markers. Renz (1936), Marie (1938), Gandolfi (1942), Bolli
(1945) and Cita (1948) used Globotruncanidae species to date
and subdivide Alpine–Mediterranean sections without pro-
posing actual zonal schemes. Local zonations based on Globotruncanidae appeared first in the 1950s, mostly from North
Africa (Sigal, 1952, 1955; Dalbiez, 1955) and the Caribbean
(Bolli, 1951, 1957; Brönnimann, 1952). More generalized
zonal schemes were subsequently proposed by Bolli (1959),
Van Hinte (1965) and Sigal (1967). Bolli (1966) made the first
attempt to fit the available data into a global scale from Aptian
to Maastrichtian, based on 19 zones. Neither the principles of
zonation nor the degree of resolution have since fundamentally
been modified though a somewhat larger number of zones (28)
are proposed from Hauterivian to Maastrichtian in the present
chapter.

Several stages, like the terminal Albian, the Cenomanian,
the Turonian and the Late Maastrichtian, have preserved
their marker species and in most cases these species have also
survived the taxonomic reshuffling of the last decades. The
results of Moullade (1966) have allowed the refinement of the
Barremian–Aptian zonation, as did the schemes of Longoria
(1974) for the Aptian. Sigal's work (1966) on *Ticinella* improved
the subdivision of the Albian.

The recognition of certain phyletic lines (*Rotalipora
subticinensis – R. ticinensis – R. appenninica*, or *Dicarinella
primitiva – D. concavata – D. asymetrica*) formed an acceptable
base for a further subdivision of the Late Albian and the
Coniacian–Santonian.

In spite of the wealth and quality of the data at hand,
no further attempts were made to improve the zonal subdivision
of the Campanian and Maastrichtian. The problem here is
that although many species or even genera appeared over this
interval, their gradual morphological evolution through
numerous intermediate morphotypes (*Globotruncanita stuartiformis*) or their long range (*Globotruncana linneiana*) made
them poor index species. For the Campanian, a rather widely
spaced zonation exists therefore which is characterized by
morphologically very distinct markers like *Globotruncanita
elevata*, *Globotruncana ventricosa* and *G. calcarata*. If needed,
a further subdivision of the Campanian–Maastrichtian could
readily be realized by using additional taxa and evolutionary
trends, in particular, some of the characteristic Heterohelicidae
which so far have not been applied to zonal schemes.

Planktic foraminiferal zonation used in this Chapter
(Fig. 3)

To subdivide the time interval of some 55 m.y. from the
Hauterivian to the end of the Maastrichtian, 28 zones are used
in this chapter. They are based on first and last occurrences of
marker species, some of which represent well established
lineages. Their correlation with the classical stages is obtained
through ammonite zones, using the schemes proposed by
Moullade (1966), Porthault (1974), Sigal (1977), Colloque sur
le Cénomanien, Paris (1978), Aspekte der Kreide Europas
Symposium, Münster (1978) and Colloque sur le Turonien,
Paris (1981). All these schemes are essentially based on simul-
taneous investigations of the macro- and microfaunas from
particularly well suited sections rich in faunas.

The dating of the macrofossil zones by radiometric
methods is still far from exact with large variations from one
author to another. Even so, this allows some evaluation of the
absolute duration of each macrofossil zone and, as a consequence, also of the microfossil zones. Based on such evidence
the 'Middle Cretaceous' (Albian to Turonian) represents a
privileged interval during which the pace of species replacement

has allowed the establishment of datum planes which sometimes succeed each other at intervals of less than half a million years.

In the last column of Fig. 3 are plotted the successive evolutionary events of first and last occurrences of genera. They provide the following more widely spaced datum planes:

(a) first occurrence of small globular trochospiral forms: *Hedbergella*;

(b) first occurrence of supplementary apertures on umbilical side: *Ticinella*;

(c) first occurrence of keels in the more evolved forms: *Rotalipora*;

(d) first occurrence of forms with a pustulose periphery: *Praeglobotruncana*;

(e) first occurrence of double keeled forms with depressed umbilical sutures: *Dicarinella*;

(f) last occurrence of *Rotalipora*;

(g) first occurrence of double keeled forms with raised umbilical sutures and with portici: *Marginotruncana*;

(h) first occurrence of forms with double keel and tegillae: *Globotruncana*;

(i) first occurrence of forms with a single keel and portici: *Globotruncanita*;

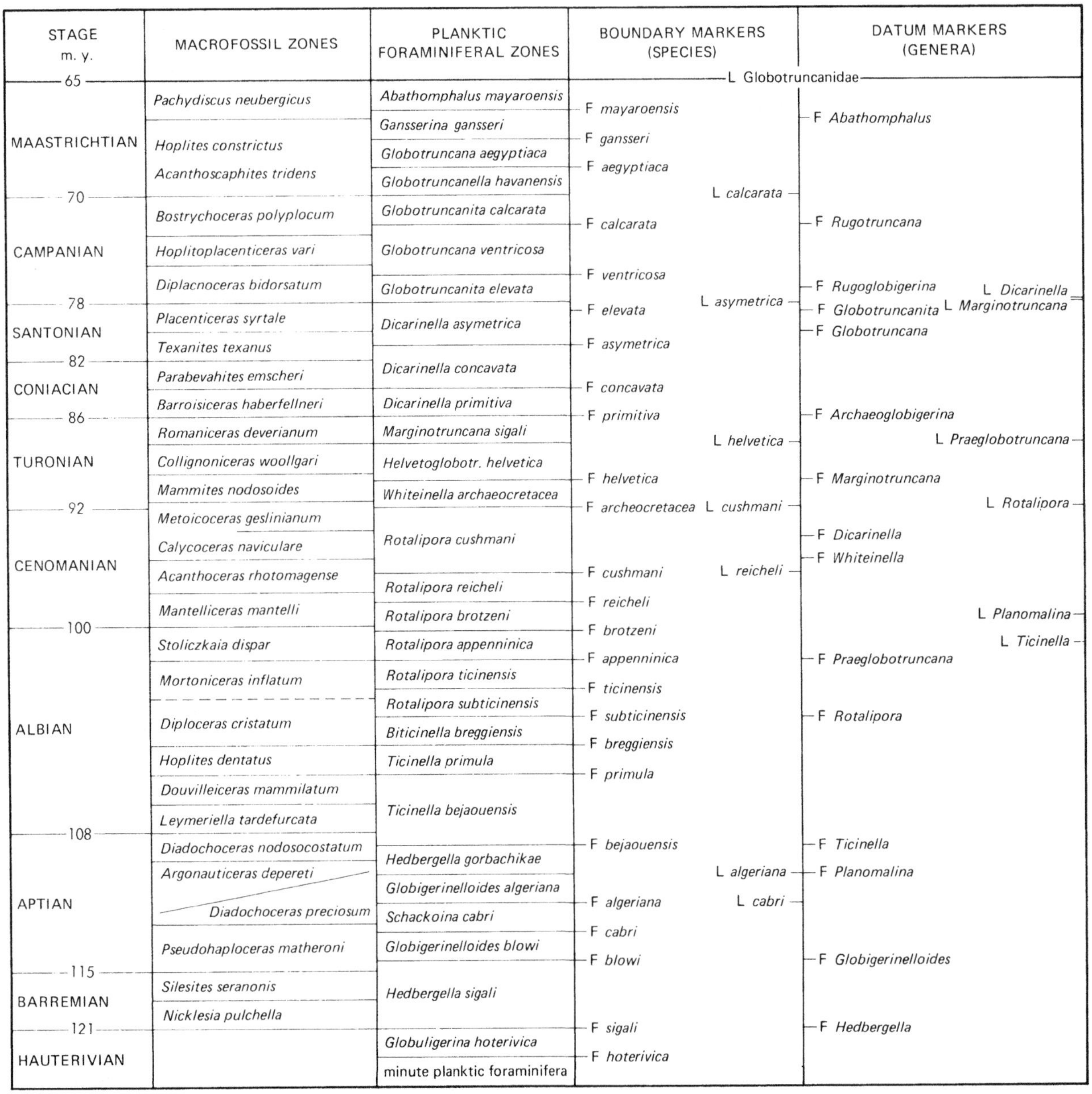

Fig. 3. Correlation of the zonal scheme used in this chapter with ammonite zones and first and last occurrences of zonal and other index species

(j) last occurrence of forms with double keel and porcel: *Dicarinella* and *Marginotruncana*;

(k) first occurrence of globular forms with costellae: *Rugoglobigerina*;

(l) first occurrence of keels in the forms with globular chambers: *Rugotruncana*, *Gansserina* and *Abathomphalus*;

(m) last occurrence of Cretaceous planktic foraminiferal taxa.

Comparison of zonal schemes
(Fig. 4)

The comparison of zonal schemes proposed during the past 15 years for use on a global scale shows a general

STAGE m. y.	CARON this chapter	WONDERS 1980	VAN HINTE 1976	SIGAL 1977	POSTUMA 1971	PESSAGNO 1967	BOLLI 1966
—65—	mayaroensis	mayaroensis	mayaroensis	mayaroensis	mayaroensis	mayaroensis	mayaroensis
MAASTRICHTIAN	gansseri	contusa	contusa	gansseri	gansseri	gansseri	gansseri
			stuarti				
		gansseri	gansseri				
	aegyptiaca	"tricarinata"	scutilla	stuarti + falsostuarti	stuartiformis	subcircumnodifer	lapp. tricar.
—70—	havanensis						
	calcarata	calcarata	calcarata	calcarata	calcarata	calcarata	calcarata
CAMPANIAN	ventricosa	ventricosa	subspinosa	elevata + stuartiformis	elevata	elevata	stuarti s. l.
			stuartiformis				
—78—	elevata	elevata	elevata			A. blowi	
SANTONIAN	asymetrica	carinata	elevata + concavata	concavata + carinata	carinata	fornicata	fornicata
—82—	concavata	concavata	concavata	concavata	concavata	concavata	concavata
CONIACIAN		primitiva	concavata + sigali		schneegansi		schneegansi
—86—	primitiva		sigali + renzi	sigali + schneegansi			
TURONIAN	M. sigali	sigali				archaeocretacea	helvetica
	helvetica	helvetica	helvetica	helvetica	helvetica	sigali	
—92—	archaeocretacea	archaeocretacea	lehmanni	cushmani			gigantea
	cushmani	cushmani	cushmani		cushmani	cushmani + greenhornensis	cushmani
CENOMANIAN	reicheli	globotruncanoides	reicheli + gandolfii	globotruncanoides + brotzeni	greenhornensis	evoluta	reicheli
			gandolfii + greenhornensis		appenninica		
—100—	brotzeni	appenninica s. l.			buxtorfi		brotzeni
	appenninica	appenninica + buxtorfi	appenninica + buxtorfi	appenninica + buxtorfi			appenninica
		ticinensis + buxtorfi	ticinensis + buxtorfi		breggiensis		
		praebuxtorfi					
ALBIAN	ticinensis	ticinensis	breggiensis	breggiensis			ticinensis
	subticinensis	subticinensis					
	breggiensis		primula + bejaouensis	reicheli + primula	subticinensis		
	primula			planispira	roberti		roberti
			bejaouensis	bejaouensis			
—108—	bejaouensis		ferreolensis + bejaouensis				
	gorbachikae		ferreolensis	trochoidea			
	algeriana		algeriana	algeriana			
APTIAN	cabri		cabri	ferreolensis			
				cabri			
	blowi		G. blowi	blowi/ maridalensis			
				gottisi/ duboisi			
—115—	H. sigali		sigali	similis			
BARREMIAN —121—				sigali			
HAUTERIVIAN	hoterivica		hoterivica	gr. hauterivica			
	minute planktic foraminifera						

(Vertical annotation between CARON and WONDERS columns: "Zonation proposed in Atlas Mid Cretaceous events, 1979". Vertical annotations in PESSAGNO 1967 column: contusa, forn.-stuartiformis, bull., renzi, helv., Rotalipora s. s.)

Fig. 4. Correlation of Cretaceous planktic foraminiferal zonal schemes

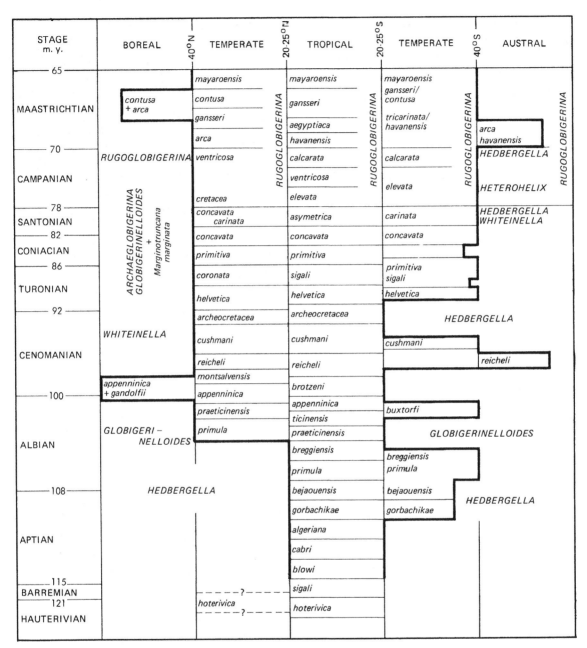

Fig. 5. Zonal correlation across latitudes in the Atlantic realm

For each area, data are summarized according to the following authors:

Boreal province
Bang (1979), Berggren (1962), Douglas & Rankin (1969), Hart (1979).
Northern temperate province
Ascoli (1976), Bailey & Hart (1979), Caron (1972), Carter & Hart (1977), Hart & Bailey (1979), Luterbacher (1972), McNulty (1979), Pozaryska & Peryt (1979), Price (1977), Robaszynski (1979), Robaszynski & Amédro (1980).
Tropical province
See references on Atlantic realm in legend of Fig. 6.
Southern temperate province
Caron (1978), Premoli-Silva & Boersma (1977).
Austral province
Sliter (1977).

STAGE m.y.	PACIFIC REALM — Japan area	W. Pacific	E. Pacific	ATLANTIC REALM — Gulf	Caribbean	E. Atlantic	TETHYAN (s.s.) REALM — W. Tethys Alpine area	Central Tethys	E. Tethys
MAASTRICHTIAN — 65		mayaroensis		mayaroensis	mayaroensis	mayaroensis	mayaroensis	mayaroensis	mayaroensis
		gansseri		gansseri	contusa / gansseri	caliciformis + citae	gansseri	gansseri	stuarti
70		subcircumnodifer	rugosa / havanensis	subcircumnodifer	lap. tricarinata		subcircumnodifer	stuarti + falsostuarti	
CAMPANIAN	japonica robusta	calcarata	marisi	calcarata	calcarata		calcarata	calcarata	morozovae
		stuartiformis + elevata	rosetta / churchi	elevata			ventricosa	elevata + stuartiformis	
78			stuartiformis + elevata	blowi	elevata	stuartiformis	elevata		elevata
SANTONIAN	hanzavae	carinata	coronata + linneiana	fornicata	conc. carinata	conc. carinata	carinata	concavata carinata	fornicata
82		concavata		concavata	concavata concavata				concavata
CONIACIAN	japonica	primitiva	cachensis			primitiva	concavata	concavata	primitiva
86		renzi		archaeocretacea	schneegansi	schneegansi	schneegansi	sigali	angusticarinata / lapparenti
TURONIAN	rugosa	helvetica	helvetica	sigali		helvetica	helvetica	helvetica	helvetica
92			cushmani			"grandes globigerines"	archaeocretacea	aumalensis / paradubia	cushmani
CENOMANIAN	greenhornensis gandolfii brotzeni		stephani	cushmani + greenhornensis		cushmani + greenhornensis	cushmani / reicheli	cushmani	deekei
	brotzeni	greenhornensis	appenninica	evoluta	appenninica	globotruncanoides + brotzeni	montsalvensis / brotzeni	globotruncanoides + brotzeni	appenninica
100		evoluta + appenninica / appenninica + buxtorfi	buxtorfi	ticinensis	ticinensis	washitensis / appenninica + buxtorfi	appenninica buxtorfi	appenninica + buxtorfi	ticinensis
ALBIAN	breggiensis	breggiensis	ticinensis	breggiensis		ticinensis / subticinensis	ticinensis / praeticinensis	breggiensis	infracretacea
			breggiensis	primula		breggiensis			
		roberti / primula	roberti	bejaouensis	rohri	primula	rischi	rischi + primula	planispira
108								planispira	
APTIAN		gorbachikae cheniourensis	pacifica	trocoidea	barri	bejaouensis	bejaouensis	bejaouensis	roberti + cheniour.
				gorbachikae		cheniourensis	trocoidea	trocoidea	cheniourensis trocoidea
		algerianus	ferreolensis	algerianus		ferreolensis	algeriana ferreolensis	algeriana ferreolensis	algeriana protuberans
		aptica	blowi	ferreolensis / cabri	protuberans	cabri	cabri	cabri maridal. + blowi	blowi + subcretacea
		gottisi	sigali	maridalensis gottisi		blowi + gottisi	blowi	gottisi + duboisi	aptica
115		sigali				aptica + similis		similis	tuschepsensis
BARREMIAN						tuschepsensis	sigali	sigali	sigali
121						sigali			
HAUTERIVIAN								hauterivica	

Fig. 6. Zonal correlations within north-tropical latitudes
For each area, data are summarized according to the following authors:

PACIFIC REALM

Japan area
Maiya & Takayanagi (1977), Takayanagi (1960), Takayanagi & Iwamoto (1962), Takayanagi & Okamura (1977).

Western Pacific
Bukry *et al.* (1971), Douglas (1971, 1973), Caron (1975), Hofker (1978), Krasheninnokov & Hoskins (1973), Luterbacher (1975), McNulty (1976), Premoli-Silva & Sliter (1981).

Eastern Pacific
Dailey (1973), Douglas (1969, 1972), Hamilton (1953), Marianos & Zingula (1966), Pessagno & Longoria (1973a, b), Sliter (1968), Takayanagi (1965).

ATLANTIC REALM

Gulf area
Eicher & Worstell (1970), Longoria (1970), Longoria & Gamper (1974), Michael (1973), Pessagno (1967), Smith & Pessagno (1973).

Caribbean and Florida
Bartenstein, Bettenstaedt & Bolli (1966), Bolli (1951, 1957, 1959, 1966), Brönnimann (1952), Gradstein (1978), Grandstein *et al.* (1978), Guillaume, Bolli & Beckmann (1972), Premoli-Silva & Bolli (1973).

Eastern Atlantic
Butt (1979), Dupeuble (1979), Lehmann (1963), Pflaumann & Cepek (1982), Pflaumann & Krasheninnikov (1977), Sigal (1978), Sliter (1980), Wiedmann, Butt & Einsele (1978).

TETHYAN REALM

Western Tethys and Alpine area
Caron (1966), Corminboeuf (1961), Herm (1962), Hermes (1969), Klaus (1959), Kuhry (1972), Linares-Rodriguez (1977), Luterbacher & Premoli-Silva (1964), Marks (1972), Moullade (1966, 1974), Porthault (1974, 1978a, b), Premoli-Silva & Luterbacher (1966), Risch (1969), Salaj & Gasparikova (1979), Salaj & Samuel (1966), Sturm (1969), Van Hinte (1963), Wonders (1979, 1980).

Central Tethys
Barr (1972), El-Naggar (1966, 1970), Fleury (1980), Saint-Marc (1974), Sigal (1977).

Eastern Tethys
Drushtchitz & Gorbachik (1979), Gorbachik (1971a, b), Maslakova (1971).

agreement in the use of markers including those that Bolli had already proposed in 1966. The greater resolution of the zones subdividing the Aptian and Albian as represented by Sigal (1977) is based on his numerous regional investigations, as well as on Moullade's (1966) and Longoria's (1974) work. The assemblage zones of Van Hinte (1976) and Sigal (1977) have served as a base for most of the biostratigraphic works during the last years. We can expect a progressive stabilization of zonal schemes, in particular since the publication of the Mid-Cretaceous Atlas.

Zonal correlations across latitudes and within the North Tropical Belt
(Figs. 5, 6)

During the Cretaceous, the temperature gradient from pole to equator was less than today, and in the North Atlantic only two provinces, Boreal and Tethyan, were separated from each other by a transitional belt created by the action of warm surface water currents (Douglas & Sliter, 1966; Bailey & Hart, 1979). This pattern was mirrored in the Southern Hemisphere, where the cold austral province was separated from the tropical province by a temperate transitional zone widely open to warm surface waters flowing from the Tethys (Fig. 5).

The cold boreal and austral provinces have associations characterized by forms with globular chambers and thin walls. The following genera, in stratigraphical order, compose almost the whole population of these regions: *Hedbergella, Globigerinelloides, Heterohelix, Whiteinella, Archaeoglobigerina, Rugoglobigerina.*

The warm water provinces are characterized by diversified associations rich in thick-walled species, ornamented by keels. The keeled taxa, *Rotalipora, Planomalina, Marginotruncana, Globotruncana* and *Globotruncanita*, occupy the Tethyan province s.l.

The northern and southern intermediate zones contain a blended microfauna due to mixing of warm and cold waters by oceanic currents.

Warm surface water currents may well not have been the only cause of temporary Tethyan faunal incursions into the cold boreal and austral provinces which were predominantly epicontinental seas. A global sea-level rise of some 50 metres would have sufficed for the outer-shelf to be populated by keeled taxa, as Eicher (1969) and Douglas (1972) have described for the Cenomanian Texas platform. Such sea-level changes may explain the Early Cenomanian, and mid-Maastrichtian progression of warm and deep-water associations over the boreal and austral continental shelves. In general, the disparity between the warm and cold provinces is indicated by the diminution in numbers and complexity of taxa towards the poles. This explains the choice of tropical zonal schemes for the definition of generalized zonal schemes (see details in Robaszynski & Caron, 1979).

No major differences appear in planktic foraminiferal associations when following the northern tropical belt across the oceans. During the Late Cretaceous, the warm belt gave rise to a constant set of ecological parameters affecting these populations over a considerable area. Nevertheless, it is possible to distinguish (Fig. 6) a 'Tethyan province' s.l., comprising the Tethys s.s. and the early N–S elongated Atlantic Ocean, and a 'Pacific province', whose microfauna was generally similar to the Tethyan one although it appears to be somewhat endemic in the area of Japan.

Before the DSDP sections became available it was customary to contrast the 'Tethyan province' to a 'Pacific province' regarding, for instance, its richness in Early Cretaceous associations or the exclusive presence of *Globotruncana stuarti* in the Maastrichtian. The multitude of newer data now at hand has removed some of these differences but confirms, for example, the dominance of globular forms with costellae (*Rugoglobigerina* and *Rugotruncana*) in the Maastrichtian associations of the Caribbean and Pacific.

Zones used in this chapter
(Figs. 3, 7–9)

GLOBULIGERINA HOTERIVICA ZONE

Category: Interval zone
Age: Late Hauterivian
Author: Van Hinte (1972)
Definition: Interval from first occurrence of *Globuligerina hoterivica* to first occurrence of *Hedbergella sigali.*

HEDBERGELLA SIGALI ZONE

Category: Interval zone
Age: Barremian to early Aptian
Author: Moullade (1966)
Definition: Interval from first occurrence of *Hedbergella sigali* to first occurrence of *Globigerinelloides blowi.*
Remarks: Joint occurrence of *Hedbergella sigali* and *Globuligerina hoterivica.*
First appearance of the genus *Hedbergella* as very primitive small sized and rare forms.

GLOBIGERINELLOIDES BLOWI ZONE

Category: Interval zone
Age: Early Aptian
Author: Moullade (1974)
Definition: Interval from first occurrence of *Globigerinelloides blowi* to first occurrence of *Schackoina cabri.*
Remarks: Joint occurrence of *Globigerinelloides blowi, Hedbergella sigali, Globuligerina hoterivica* and *Hedbergella bizonae.*
First appearance of the genus Globigerinelloides.

SCHACKOINA CABRI ZONE

Category: Total range zone
Age: Early Aptian to Late Aptian
Author: Bolli (1959). By synonymy = *Leupoldina protuberans* Zone.
Definition: Interval of total range of *Schackoina cabri*.

GLOBIGERINELLOIDES ALGERIANA ZONE

Category: Total range zone
Age: Late Aptian
Author: Moullade (1966)
Definition: Interval of total range of *Globigerinelloides algeriana*.
Remarks: Joint occurrence of *Globigerinelloides algeriana*, *Hedbergella bizonae* and *Globigerinelloides ferreolensis*. In the middle part of the zone: extinction of *Globigerinelloides blowi*. Top of zone: first occurrence of *Planomalina cheniourensis*.

HEDBERGELLA GORBACHIKAE ZONE

Category: Partial range zone
Age: Late Aptian
Author: Longoria (1974)
Definition: Interval, with *Hedbergella gorbachikae*, from last occurrence of *Globigerinelloides algeriana* to first occurrence of *Ticinella bejaouensis*.
Remarks: Joint occurrence of *Hedbergella gorbachikae*, *Globigerinelloides ferreolensis* and *Planomalina cheniourensis*.

TICINELLA BEJAOUENSIS ZONE

Category: Interval zone
Age: Late Aptian to Early Albian
Author: Moullade (1966)
Definition: Interval from first occurrence of *Ticinella bejaouensis* to first occurrence of *Ticinella primula*.
Remarks: In the lower part of the zone, extinction of *Planomalina cheniourensis* and *Globigerinelloides ferreolensis* and first occurrence of *Ticinella roberti*.

TICINELLA PRIMULA ZONE

Category: Interval zone
Age: Middle Albian
Authors: Longoria & Gamper (1974)
Definition: Interval from first occurrence of *Ticinella primula* to first occurrence of *Biticinella breggiensis*.
Remarks: Joint occurrence of *Hedbergella gorbachikae*, *Ticinella roberti* and *Ticinella primula*.

BITICINELLA BREGGIENSIS ZONE

Category: Interval zone
Age: Late Albian
Author: Postuma (1971)
Definition: Interval from first occurrence of *Biticinella breggiensis* to first occurrence of *Rotalipora subticinensis*.

ROTALIPORA SUBTICINENSIS ZONE

Category: Interval zone
Age: Late Albian
Author: Postuma (1971)
Definition: Interval from first occurrence of *Rotalipora subticinensis* to first occurrence of *Rotalipora ticinensis*.

ROTALIPORA TICINENSIS ZONE

Category: Interval zone
Age: Late Albian
Author: Dalbiez (1955)
Definition: Interval from first occurrence of *Rotalipora ticinensis* to first occurrence of *Rotalipora appenninica*.

ROTALIPORA APPENNINICA ZONE

Category: Interval zone
Age: Late Albian
Author: Brönnimann (1952)
Definition: Interval from first occurrence of *Rotalipora appenninica* to first occurrence of *Rotalipora brotzeni*.

ROTALIPORA BROTZENI ZONE

Category: Interval zone
Age: Early Cenomanian
Author: Lehmann (1966)
Definition: Interval from first occurrence of *Rotalipora brotzeni* to first occurrence of *Rotalipora reicheli*.

ROTALIPORA REICHELI ZONE

Category: Total range zone
Age: Early to Middle Cenomanian
Author: Bolli (1966)
Definition: Interval of total range of *Rotalipora reicheli*.

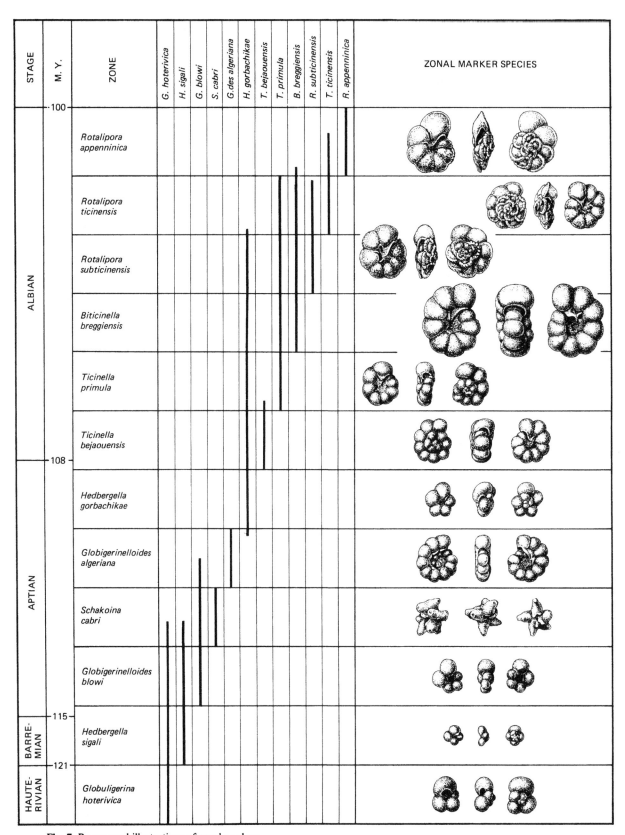

STAGE	M.Y.	ZONE	G. hoterivica	H. sigali	G. blowi	S. cabri	G. des algeriana	H. gorbachikae	T. bejaouensis	T. primula	B. breggiensis	R. subticinensis	T. ticinensis	R. appenninica	ZONAL MARKER SPECIES
ALBIAN	100	*Rotalipora appenninica*													
ALBIAN		*Rotalipora ticinensis*													
ALBIAN		*Rotalipora subticinensis*													
ALBIAN		*Biticinella breggiensis*													
ALBIAN		*Ticinella primula*													
ALBIAN	108	*Ticinella bejaouensis*													
APTIAN		*Hedbergella gorbachikae*													
APTIAN		*Globigerinelloides algeriana*													
APTIAN		*Schakoina cabri*													
APTIAN		*Globigerinelloides blowi*													
BARRE-MIAN	115	*Hedbergella sigali*													
BARRE-MIAN	121														
HAUTE-RIVIAN		*Globuligerina hoterivica*													

Fig. 7. Ranges and illustrations of zonal markers
(Hauterivian–Albian)

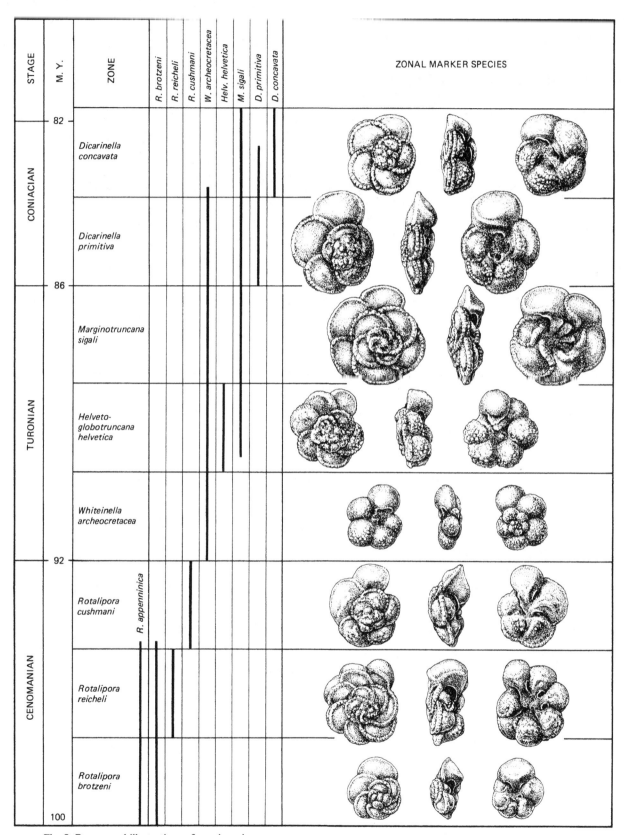

Fig. 8. Ranges and illustrations of zonal markers
(Cenomanian–Coniacian)

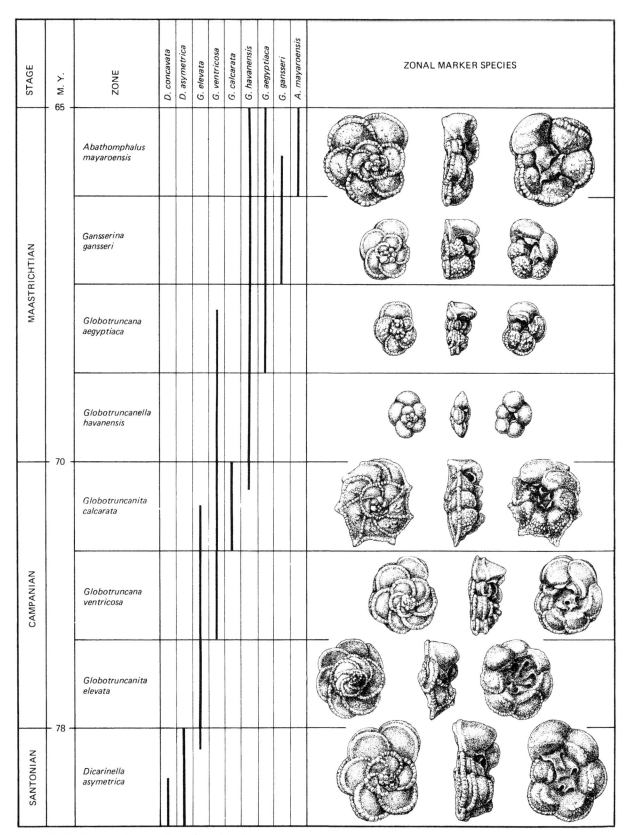

Fig. 9. Ranges and illustrations of zonal markers
(Santonian–Mastrichtian)

ROTALIPORA CUSHMANI ZONE

Category: Total range zone
Age: Middle to Late Cenomanian
Author: Borsetti (1962)
Definition: Interval of total range of *Rotalipora cushmani*.
Remarks: First occurrence, in the lower part of the zone, of *Whiteinella baltica* and, in the upper part of the zone, of *Whiteinella paradubia, Helvetoglobotruncana praehelvetica* and *Dicarinella algeriana*.
Extinction of the genus *Rotalipora* marks the upper limit of the zone.

WHITEINELLA ARCHAEOCRETACEA ZONE

Category: Partial range zone
Age: Early Turonian
Author: Bolli (1966). By synonymy = *Praeglobotruncana gigantea* Zone.
Definition: Interval with *Whiteinella archaeocretacea*, from extinction of *Rotalipora cushmani* to first occurrence of *Helvetoglobotruncana helvetica*.

HELVETOGLOBOTRUNCANA HELVETICA ZONE

Category: Total range zone
Age: Middle Turonian
Author: Sigal (1955)
Definition: Interval of total range of *Helvetoglobotruncana helvetica*.

MARGINOTRUNCANA SIGALI ZONE

Category: Partial range zone
Age: Late Turonian
Author: Barr (1972)
Definition: Interval, with *Marginotruncana sigali*, from extinction of *Helvetoglobotruncana helvetica* to first occurrence of *Dicarinella primitiva*.

DICARINELLA PRIMITIVA ZONE

Category: Interval zone
Age: Early Coniacian
Author: Caron (1978)
Definition: Interval from first occurrence of *Dicarinella primitiva* to first occurrence of *Dicarinella concavata*.

DICARINELLA CONCAVATA ZONE

Category: Interval zone
Age: Late Coniacian to Early Santonian
Author: Sigal (1955)
Definition: Interval from first occurrence of *Dicarinella concavata* to first occurrence of *Dicarinella asymetrica*.

DICARINELLA ASYMETRICA ZONE

Category: Total range zone
Age: Upper part of Early Santonian to Late Santonian
Author: Postuma (1971). By synonymy = *Globotruncana concavata carinata* Zone.
Definition: Interval of total range of *Dicarinella asymetrica*.

GLOBOTRUNCANITA ELEVATA ZONE

Category: Partial range zone
Age: Early Campanian
Author: Postuma (1971)
Definition: Interval, with *Globotruncanita elevata*, from last occurrence of *Dicarinella asymetrica* to first occurrence of *Globotruncana ventricosa*.

GLOBOTRUNCANA VENTRICOSA ZONE

Category: Interval zone
Age: Upper part of Early Campanian to Late Campanian
Author: Dalbiez (1955)
Definition: Interval from first occurrence of *Globotruncana ventricosa* to first occurrence of *Globotruncanita calcarata*.

GLOBOTRUNCANITA CALCARATA ZONE

Category: Total range zone
Age: Upper part of Late Campanian
Author: Herm (1962)
Definition: Interval of total range of *Globotruncanita calcarata*.

GLOBOTRUNCANELLA HAVANENSIS ZONE

Category: Partial range zone
Age: Lower part of Early Maastrichtian
Author: Caron (1978)
Definition: Interval, with *Globotruncanella havanensis*, from last occurrence of *Globotruncanita calcarata* to first occurrence of *Globotruncana aegyptiaca*.

GLOBOTRUNCANA AEGYPTIACA ZONE

Category: Interval zone
Age: Early Maastrichtian
Author: Caron (this paper)
Definition: Interval from first occurrence of *Globotruncana aegyptiaca* to first occurrence of *Gansserina gansseri*.

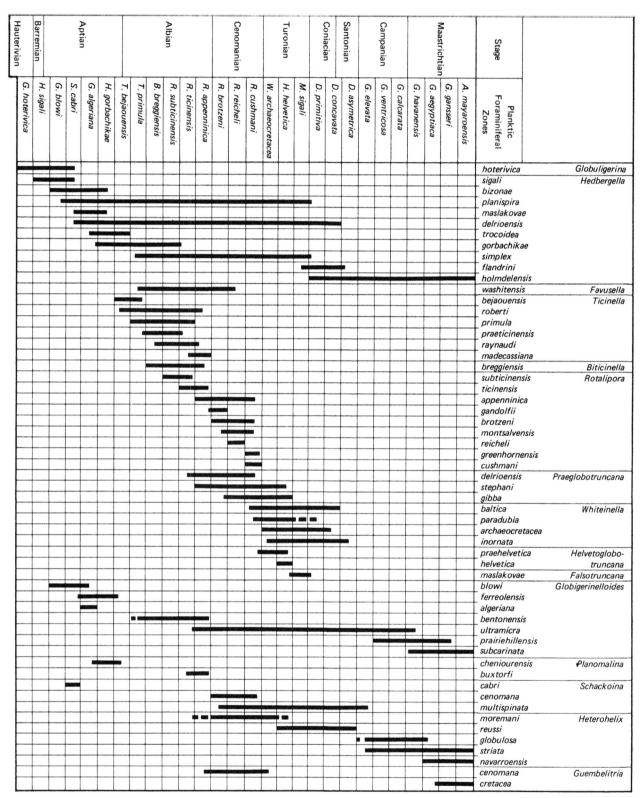

Figs. 10–11. Distribution of selected Cretaceous planktic foraminifera arranged by genera

Stage (left to right): Hauterivian | Barremian | Aptian | Albian | Cenomanian | Turonian | Coniacian | Santonian | Campanian | Maastrichtian | Stage

Planktic Foraminiferal Zones (columns, left to right): G. hoterivica | H. sigali | G. blowi | S. cabri | G. algeriana | H. gorbachikae | T. bejaouensis | T. primula | B. breggiensis | R. subticinensis | R. ticinensis | R. appenninica | R. brotzeni | R. reicheli | R. cushmani | W. archaeocretacea | H. helvetica | M. sigali | D. primitiva | D. concavata | D. asymetrica | G. elevata | G. ventricosa | G. calcarata | G. havanensis | G. aegyptiaca | G. gansseri | A. mayaroensis

Species ranges (right column, genus group):

algeriana	Dicarinella
canaliculata	
imbricata	
hagni	
primitiva	
concavata	
asymetrica	
sigali	Marginotruncana
renzi	
schneegansi	
marianosi	
pseudolinneiana	
coronata	
marginata	
sinuosa	
fornicata	Rosita
contusa	
arca	Globotruncana
lapparenti	
bulloides	
linneiana	
ventricosa	
falsostuarti	
aegyptiaca	
stuartiformis	Globotruncanita
elevata	
subspinosa	
calcarata	
stuarti	
conica	
havanensis	Globotruncanella
petaloidea	
citae	
intermedius	Abathomphalus
mayaroensis	
cretacea	Archaeoglobigerina
blowi	
bosquensis	
rugosa	Rugoglobigerina
scotti	
hexacamerata	
macrocephala	
reicheli	
subcircumnodifer	Rugotruncana
subpennyi	
gansseri	Gansserina
alexanderi	Hastigerinoides
watersi	
subdigitata	
costulata	Pseudoguembelina
excolata	
palpebra	
elegans	Pseudotextularia
fructicosa	Racemiguembelina

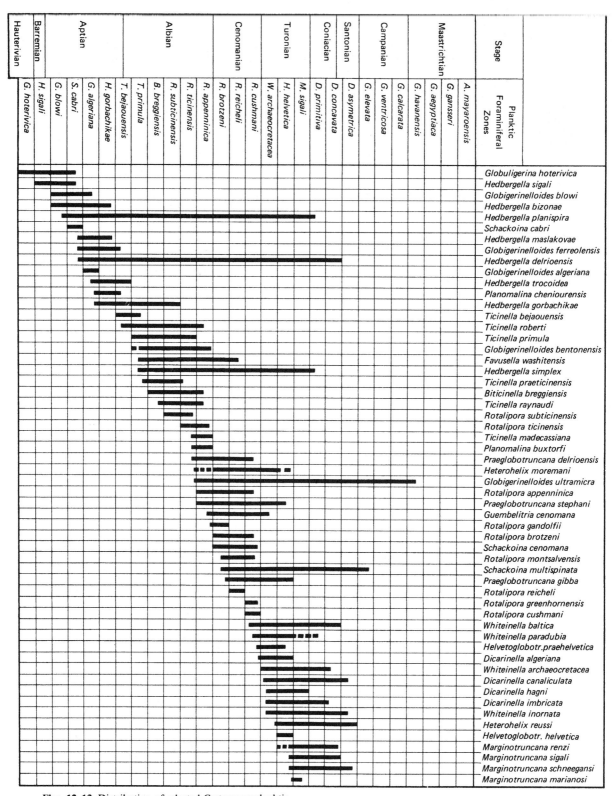

Figs. 12–13. Distribution of selected Cretaceous planktic foraminifera in order of first occurrence

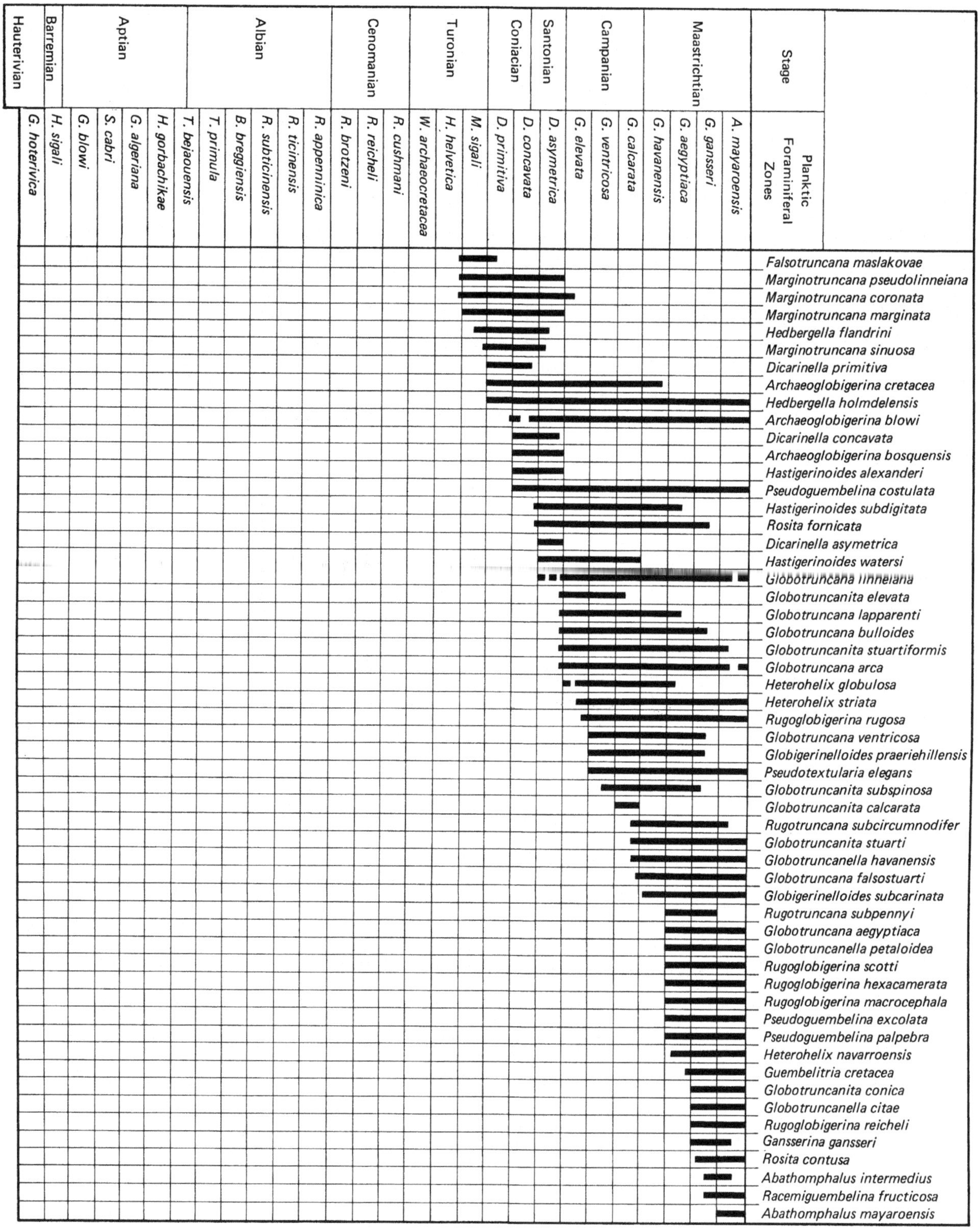
Stage
Planktic Foraminiferal Zones
Hauterivian
Barremian
Aptian
Albian
Cenomanian
Turonian
Coniacian
Santonian
Campanian
Maastrichtian
G. hoterivica
H. sigali
G. blowi
S. cabri
G. algeriana
H. gorbachikae
T. bejaouensis
T. primula
B. breggiensis
R. subticinensis
R. ticinensis
R. appenninica
R. brotzeni
R. reicheli
R. cushmani
W. archaeocretacea
H. helvetica
M. sigali
D. primitiva
D. concavata
D. asymetrica
G. elevata
G. ventricosa
G. calcarata
G. havanensis
G. aegyptiaca
G. gansseri
A. mayaroensis
Falsotruncana maslakovae
Marginotruncana pseudolinneiana
Marginotruncana coronata
Marginotruncana marginata
Hedbergella flandrini
Marginotruncana sinuosa
Dicarinella primitiva
Archaeoglobigerina cretacea
Hedbergella holmdelensis
Archaeoglobigerina blowi
Dicarinella concavata
Archaeoglobigerina bosquensis
Hastigerinoides alexanderi
Pseudoguembelina costulata
Hastigerinoides subdigitata
Rosita fornicata
Dicarinella asymetrica
Hastigerinoides watersi
Globotruncana linneiana
Globotruncanita elevata
Globotruncana lapparenti
Globotruncana bulloides
Globotruncanita stuartiformis
Globotruncana arca
Heterohelix globulosa
Heterohelix striata
Rugoglobigerina rugosa
Globotruncana ventricosa
Globigerinelloides praeriehillensis
Pseudotextularia elegans
Globotruncanita subspinosa
Globotruncanita calcarata
Rugotruncana subcircumnodifer
Globotruncanita stuarti
Globotruncanella havanensis
Globotruncana falsostuarti
Globigerinelloides subcarinata
Rugotruncana subpennyi
Globotruncana aegyptiaca
Globotruncanella petaloidea
Rugoglobigerina scotti
Rugoglobigerina hexacamerata
Rugoglobigerina macrocephala
Pseudoguembelina excolata
Pseudoguembelina palpebra
Heterohelix navarroensis
Guembelitria cretacea
Globotruncanita conica
Globotruncanella citae
Rugoglobigerina reicheli
Gansserina gansseri
Rosita contusa
Abathomphalus intermedius
Racemiguembelina fructicosa
Abathomphalus mayaroensis

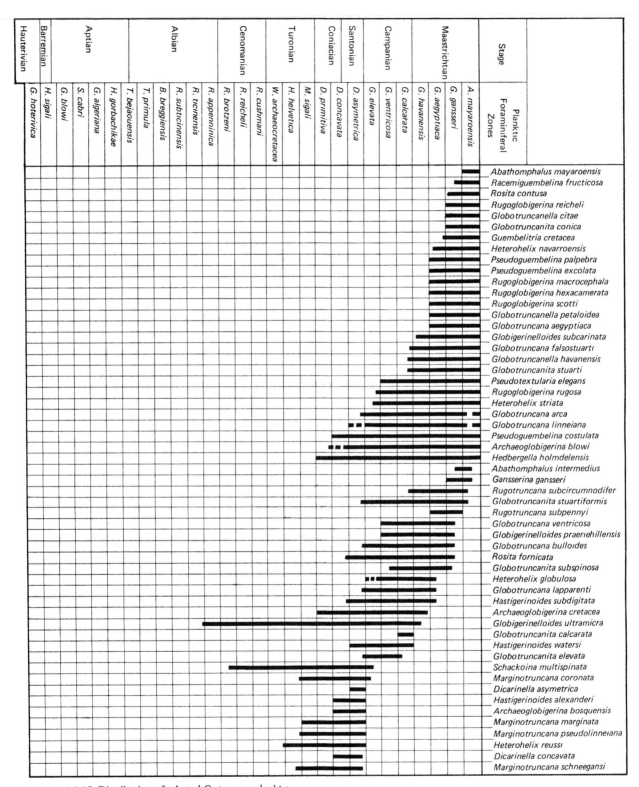

Figs. 14–15. Distribution of selected Cretaceous planktic foraminifera in order of last occurrence

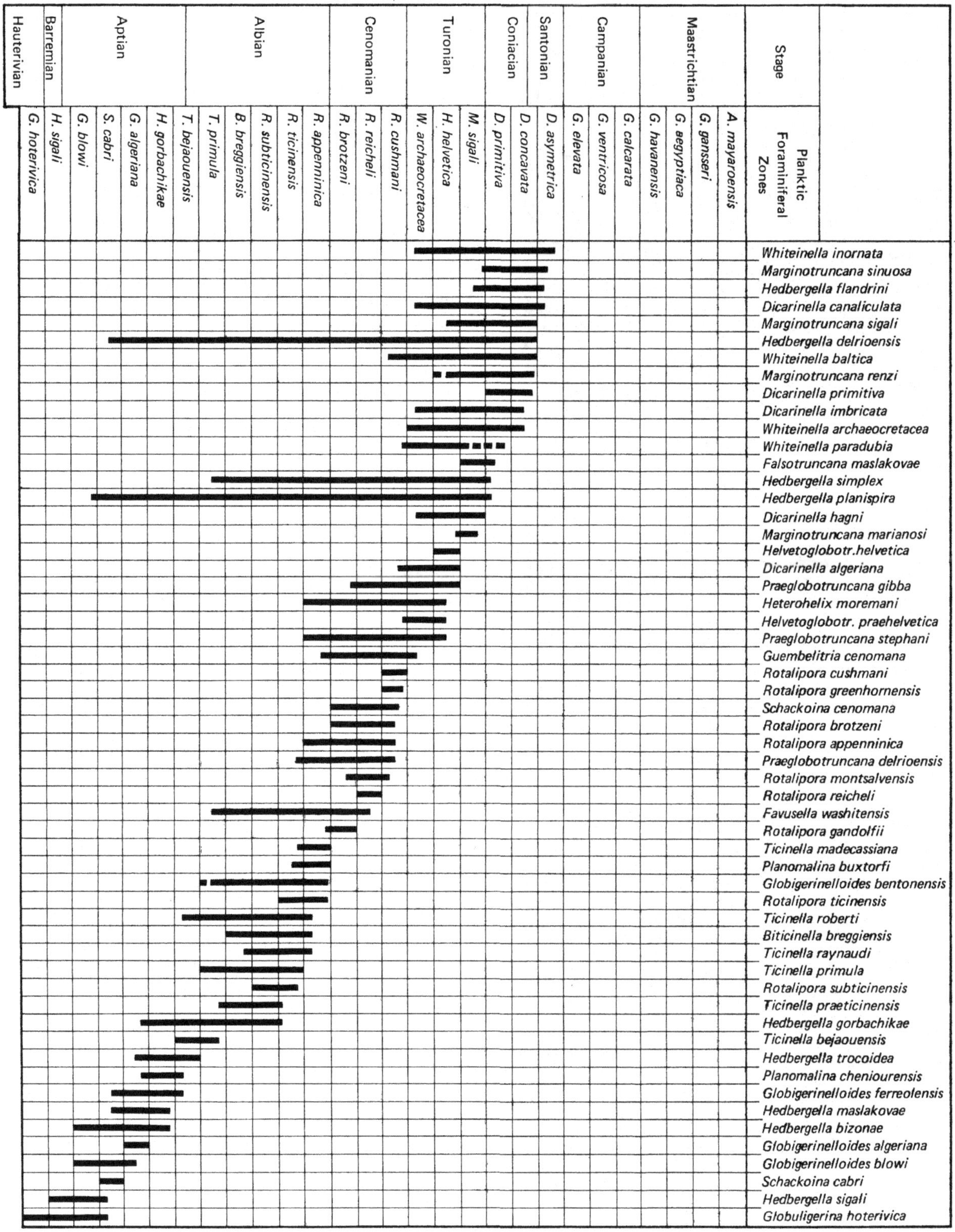
Stage
Hauterivian
Barremian
Aptian
Albian
Cenomanian
Turonian
Coniacian
Santonian
Campanian
Maastrichtian
Planktic Foraminiferal Zones
G. hoterivica
H. sigali
G. blowi
S. cabri
G. algeriana
H. gorbachikae
T. bejaouensis
T. primula
B. breggiensis
R. subticinensis
R. ticinensis
R. appenninica
R. brotzeni
R. reicheli
R. cushmani
W. archaeocretacea
H. helvetica
M. sigali
D. primitiva
D. concavata
D. asymetrica
G. elevata
G. ventricosa
G. calcarata
G. havanensis
G. aegyptiaca
G. gansseri
A. mayaroensis
Whiteinella inornata
Marginotruncana sinuosa
Hedbergella flandrini
Dicarinella canaliculata
Marginotruncana sigali
Hedbergella delrioensis
Whiteinella baltica
Marginotruncana renzi
Dicarinella primitiva
Dicarinella imbricata
Whiteinella archaeocretacea
Whiteinella paradubia
Falsotruncana maslakovae
Hedbergella simplex
Hedbergella planispira
Dicarinella hagni
Marginotruncana marianosi
Helvetoglobotr. helvetica
Dicarinella algeriana
Praeglobotruncana gibba
Heterohelix moremani
Helvetoglobotr. praehelvetica
Praeglobotruncana stephani
Guembelitria cenomana
Rotalipora cushmani
Rotalipora greenhornensis
Schackoina cenomana
Rotalipora brotzeni
Rotalipora appenninica
Praeglobotruncana delrioensis
Rotalipora montsalvensis
Rotalipora reicheli
Favusella washitensis
Rotalipora gandolfii
Ticinella madecassiana
Planomalina buxtorfi
Globigerinelloides bentonensis
Rotalipora ticinensis
Ticinella roberti
Biticinella breggiensis
Ticinella raynaudi
Ticinella primula
Rotalipora subticinensis
Ticinella praeticinensis
Hedbergella gorbachikae
Ticinella bejaouensis
Hedbergella trocoidea
Planomalina cheniourensis
Globigerinelloides ferreolensis
Hedbergella maslakovae
Hedbergella bizonae
Globigerinelloides algeriana
Globigerinelloides blowi
Schackoina cabri
Hedbergella sigali
Globuligerina hoterivica

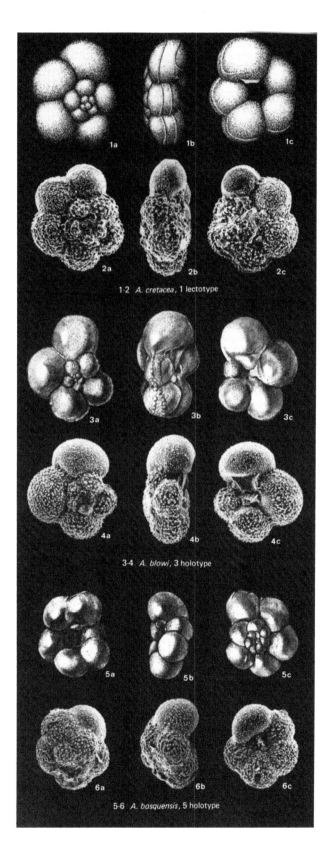

1·2 *A. cretacea*, 1 lectotype

3-4 *A. blowi*, 3 holotype

5-6 *A. bosquensis*, 5 holotype

GANSSERINA GANSSERI ZONE

Category: Interval zone
Age: Late Maastrichtian
Author: Brönnimann (1952)
Definition: Interval from first occurrence of *Gansserina gansseri* to first occurrence of *Abathomphalus mayaroensis*.

ABATHOMPHALUS MAYAROENSIS ZONE

Category: Total range zone
Age: Uppermost part of Late Maastrichtian
Author: Brönnimann (1952)
Definition: Interval of total range of *Abathomphalus mayaroensis*.

Discussion of taxa

Genus ABATHOMPHALUS Bolli, Loeblich & Tappan, 1957

ABATHOMPHALUS INTERMEDIUS (Bolli)
Figures **21.7–9**; 11, 13, 14

Type reference: *Globotruncana intermedia* Bolli, 1951; pl. 35, figs. 7–9.

This species has been described as a transitional form from *Globotruncanella havanensis* to *A. mayaroensis*. It has a wide umbilical area and inflated chambers on the spiral side; the ventral keel may be absent on the last chambers of the final whorl.

ABATHOMPHALUS MAYAROENSIS (Bolli)
Figures **21.10–11**; 11, 13, 14

Type reference: *Globotruncana mayaroensis* Bolli, 1951, pp. 190, 198, pl. 35, figs. 10–12.

This species is one of the more distinctive among the *Globotruncanidae*. The spiral side is slightly convex, the umbilical side slightly concave. The two peripheral keels are variably spaced, greatest separation occurring at the midpoints of the chambers. The keels are composed of short, radially oriented costellae. *A. mayaroensis* differs from *A. intermedius* in the presence of a double keel on all chambers of the last whorl.

Fig. 16. *Archeoglobigerina* (all figures × 60)

1–2 *A. cretacea* (d'Orbigny)
   1a–c Lectotype. Specimen obtained by d'Orbigny from white chalk of St-Germain with *Belemnitella mucronata, G. ventricosa* Zone, Late Campanian. **2a–c** specimen from sample DW 15, *G. elevata* Zone, Early Campanian, Lower Taylor clay, Mc Lennan County, Texas, USA (from Robaszynski & Caron, 1979).

3–4 *A. blowi* Pessagno
   3a–c Holotype. Sample TX 252, Lower Taylor Marl, *G. elevata* Zone, Early Campanian, Mc Lennan County, Texas, USA. **4a–c** Topotype. From sample DW 15, same locality as holotype (from Robaszynski & Caron, 1979).

5–6 *A. bosquensis* Pessagno
   5a–c Holotype. Sample TX 226, Austin chalk, Atco member, *D. concavata* Zone, Early Santonian, South Bosque river, Waco, Mc Lennan County, Texas, USA. **6a–c** Specimen from R. Herb collection, DSDP Leg 26-258-7-3, Coniacian, Naturaliste Plateau, Indian Ocean.

Genus ARCHAEOGLOBIGERINA Pessagno, 1967

ARCHAEOGLOBIGERINA BLOWI Pessagno
Figures **16.3–4**; 11, 13, 14

Type reference: *Archaeoglobigerina blowi* Pessagno, 1967, p. 316, pl. 59, figs. 5–7.

The species differs from *A. cretacea* in having only 4 instead of 5–6 chambers in the last whorl. The periphery is rounded, marked by an imperforate band which occurs mainly on the first one or two chambers of the last whorl.

ARCHAEOGLOBIGERINA BOSQUENSIS Pessagno
Figures **16.5–6**; 11, 13, 14

Type reference: *Archaeoglobigerina bosquensis* Pessagno, 1967, pp. 316–17, pl. 60, figs. 10–12.

The species differs from *A. cretacea* by its high trochospiral coil and absence of any keels, even on the first chambers of the last whorl.

ARCHAEOGLOBIGERINA CRETACEA (d'Orbigny)
Figures **16.1–2**; 11, 13, 14

Type reference: *Globigerina cretacea* (d'Orbigny), 1840, p. 34, pl. 3, figs. 12–14.
Lectotype: *Globotruncana cretacea* (d'Orbigny); Banner & Blow, 1960, pp. 8–10, pl. 7, figs. 1a–c.
Homonym: Senior homonym of *Globotruncana cretacea* Cushman, 1938, which was re-named *Globotruncana mariei* by Banner & Blow, 1960.

The species differs from *A. blowi* in having 6 instead of 4 chambers in the last whorl. The rounded periphery is marked not only by an imperforate band, but also by two faint keels, which may be absent on the last chamber.

Genus BITICINELLA Sigal, 1956
BITICINELLA BREGGIENSIS (Gandolfi)
Figures **36.16–17**; 10, 12, 15

Type reference: *Anomalina breggiensis* Gandolfi, 1942, p. 102, pl. 3, figs. 6a–c.
Emendation: *Biticinella breggiensis* (Gandolfi): Luterbacher & Premoli-Silva, 1962, p. 272, pl. 23, figs. 2–4.

Test planispiral in last whorl; chambers inflated, not keeled, generally broader than high; surface coarsely perforate. Relict apertures on spiral side like those of *Globigerinelloides*, supplementary apertures on umbilical side like those of *Ticinella*.

Genus DICARINELLA Porthault, 1970

DICARINELLA ALGERIANA (Caron)
Figures **17.1–2**; 11, 12, 15

Type reference: *Praeglobotruncana algeriana* Caron, 1966, pp. 74–5, holotype, figured by Reichel, 1950, pp. 612–13, pl. 16, fig. 8, pl. 17, fig. 8, as *Globotruncana (Globotruncana)* aff. *renzi* Thalmann & Gandolfi.

This species is the most primitive of the genus *Dicarinella*. It is transitional from *Praeglobotruncana stephani* to *D. imbricata* and differs from the former in the presence of a narrow imperforate peripheral band between the two lines of pustules and from the latter in the absence of a true double keel with its imbricated pattern.

DICARINELLA ASYMETRICA (Sigal)
Figures **17.3–4**; 11, 13, 14

Type reference: *Globotruncana asymetrica* Sigal, 1952, p. 35, fig. 35.
Synonyms: *Globotruncana lobata* de Klasz, 1955, p. 43, pl. 7, figs. 2a–c.
Globotruncana fundiconulosa Subbotina, 1953, p. 200, pl. 14, figs. 4a–c.
Globotruncana (Globotruncana) ventricosa White subsp. *carinata* Dalbiez, 1955, p. 171, figs. 8a–c, is a junior synonym; see Robaszynski & Caron, 1979, vol. 2, pp. 61 and 71.

This species is the last representative of the genus *Dicarinella*. Two closely spaced keels give a typical profile with prolongation of the second keel surrounding the umbilicus. *D. asymetrica* differs from *D. concavata* in the presence of this periumbilical ridge and from *G. ventricosa* in the presence of portici covering the umbilical–extraumbilical primary aperture.

DICARINELLA CANALICULATA (Reuss)
Figures **17.5–6**; 11, 12, 15

Type reference: *Rosalina canaliculata* Reuss, 1854, p. 70, pl. 26, figs. 4a–b.
Neotype: *Marginotruncana canaliculata* (Reuss): Pessagno, 1967, p. 303, pl. 74, figs. 5–8.

This species differs from *D. imbricata* in having two parallel, widely spaced keels; from *M. pseudolinneiana* in having radial umbilical sutures and portici; from *M. marginata* for the same reasons and also due to the absence of inflated chambers.

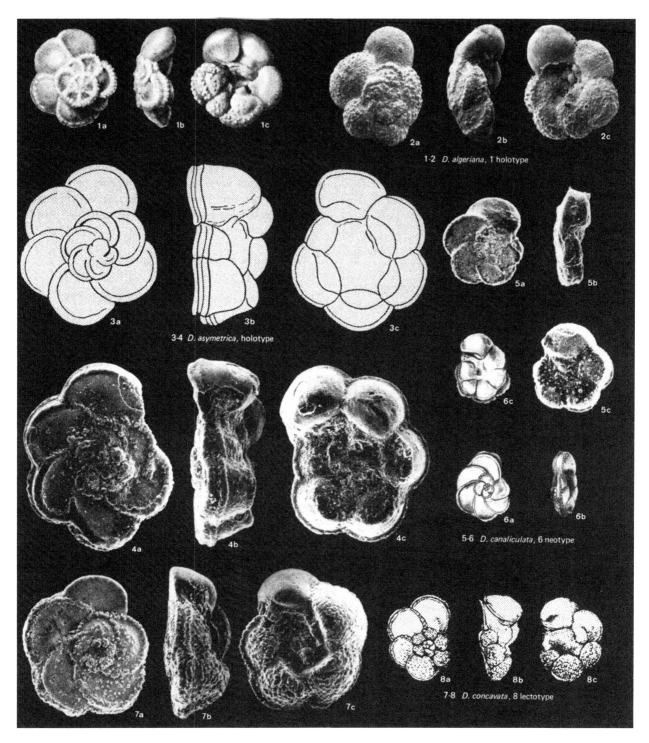

Fig. 17. *Dicarinella* (all figures × 60)

1–2 *D. algeriana* (Caron)
  1a–c Holotype, **2a–c** Topotype. Specimens from sample Sigal no. 10574, *H. helvetica* Zone, Middle Turonian, Sidi Aissa, south of Aumale, Algeria.

3–4 *D. asymetrica* (Sigal)
  3a–c Holotype, **4a–c** Holotype, SEM photos; from sample Sigal no. 13.614, *D. asymetrica* Zone, Santonian, Aïn-ed-Deffa, E Algeria (4 from Robaszynski & Caron, 1979).

5–6 *D. canaliculata* (Reuss)
  5a–c Specimen from Bellone well, at 112 m, *H. helvetica* Zone, Middle Turonian, N France (from Robaszynski & Caron, 1979).
  6a–c Neotype. From Gosau beds, *D. concavata* Zone, Upper Coniacian–Lower Santonian, Edelbachgraben, Austria.

7–8 *D. concavata* (Brotzen)
  7a–c Specimen from sample Lehman T 406, *D. asymetrica* Zone, Santonian, Tarfaya, Marocco (from Robaszynski & Caron, 1979).
  8a–c Lectotype. From Brotzen collection, slide A 265 NR, specimen 16, *D. asymetrica* Zone, Late Santonian, Wadi Madi section, Mount Carmel, Israel.

DICARINELLA CONCAVATA (Brotzen)
Figures **17.7–8**; 11, 13, 14

Type reference: *Rotalia concavata* Brotzen, 1934, p. 66, pl. 3, fig. b.

Lectotype: *Globotruncana concavata* (Brotzen): Kuhry, 1970, pp. 299–300, pl. 2, figs. 16–18.

Lateral view clearly asymmetrical, concavo-convex, with early chambers globular shaped and late chambers of the outer whorl with two close keels. Compared with *D. asymetrica* no keel surrounds the umbilicus. It differs from *D. primitiva* in having more strongly inflated chambers on the umbilical side.

DICARINELLA HAGNI (Scheibnerova)
Figures **18.1–3**; 11, 12, 15

Type reference: *Praeglobotruncana hagni* Scheibnerova, 1962, p. 219, figs. 6a–c.

Synonyms: *Globotruncana roddai* Marianos & Zingula, 1966, p. 340, pl. 39, figs. 5a–c.

Globotruncana indica Jacob & Sastry, 1950, p. 267, figs. 2a–c, not retained as senior synonym (see Robaszynski & Caron, 1979, vol. 2, p. 85).

The species differs from *D. concavata* in having a slightly convex spiral side and no globular early chambers; from *D. imbricata* it differs in its lower trochospire.

DICARINELLA IMBRICATA (Mornod)
Figures **18.4–5**; 11, 12, 15

Type reference: *Globotruncana (Globotruncana) imbricata* Mornod, 1949–50, pp. 589–90, fig. 5 (III a–d).

Neotype: *Dicarinella imbricata* (Mornod): Caron, 1976, pp. 332–3, figs. 3a–c.

Test convexo-concave, accentuated by a shift towards the umbilicus of the last formed two chambers, which may be without a keel. The keel-band diverges obliquely from one chamber to the next, forming an imbricated sequence. *D. imbricata* differs from other *Dicarinella* species by this characteristic imbricated aspect of the sequence of its chambers.

DICARINELLA PRIMITIVA (Dalbiez)
Figures **18.6–8**; 11, 13, 15

Type reference: *Globotruncana (Globotruncana) ventricosa* White subsp. *primitiva* Dalbiez, 1955, p. 171, fig. 6.

Lectotype: *Dicarinella primitiva* (Dalbiez): Robaszynski & Caron, 1979, vol. 2, p. 96, pl. 60, figs. 1a–c (Dalbiez's specimen 1955, slide FG 468).

The species has typical globular early chambers on the spiral side and only slightly inflated ones on the umbilical side. It differs from *D. concavata* in its less inflated chambers on the umbilical side; from *Marginotruncana schneegansi* in having two well developed keels.

Genus FALSOTRUNCANA Caron, 1981
FALSOTRUNCANA MASLAKOVAE Caron
Figures **30.14**; 10, 13, 15

Type reference: *Falsotruncana maslakovae* Caron, 1981, pp. 67–8, pl. 2, figs. 1a–d.

The species is characterized by an imperforate wide peripheral band separating two keels which may be absent on the last chamber. The profile is similar to that of *D. caniculata*, but differs from that species in its long, umbilical – extraumbilical – nearly peripheral primary aperture, bordered by a narrow lip. The umbilicus is shallow, without protections like portici or tegilla.

Genus FAVUSELLA Michael, 1973
FAVUSELLA WASHITENSIS (Carsey)
Figures **25.25–26**; 10, 12, 15

Type reference: *Globigerina washitensis* Carsey, 1926, p. 44, pl. 7, fig. 10.

Neotype: *Globigerina washitensis* Carsey: Plummer, 1931, p. 193, pl. 13, figs. 12a–b.

A second neotype was selected by Longoria (1974), p. 75, pl. 26, figs. 4–6, on the grounds that Plummer chose a 3- instead of 4-chambered specimen as was originally represented by the holotype. This action does not conform with the International Code of Zoological Nomenclature.

All species of *Favusella* erected by Michael (1973) are here considered as synonymous. Only *F. washitensis* is retained as representative for this genus. 3–5 globular chambers form the last whorl. The primary aperture is arched, umbilical-extraumbilical, nearly peripheral, with a narrow lip. The test has a heavily reticulated ornamentation.

Genus GANSSERINA Robaszynski, Caron, Gonzalez & Wonders, 1984
GANSSERINA GANSSERI (Bolli)
Figures **30.11–13**; 11, 13, 14

Type reference: *Globotruncana gansseri* Bolli, 1951, p. 196, pl. 35, figs. 1–3.

Synonyms: *Globotruncana lugeoni* Tilev, 1951, pp. 41–6, pl. 1, fig. 5.

Globotruncana monmouthensis Olsson, 1960, pp. 50–1, pl. 10, figs. 22–24.

Globotruncana arabica El-Naggar, 1966, pp. 81–3, pl. 6, figs. 3a–d.

Test plano-convex; 4–7 chambers in the last whorl, single-keeled, strongly convex on the umbilical side, where faint discontinuous costellae appear on the early chambers. According to Brönnimann & Brown's observations (1956, text-fig. 23, p. 550) the presence of a faint double keel on a chamber of the penultimate whorl proves the origin of this species from an

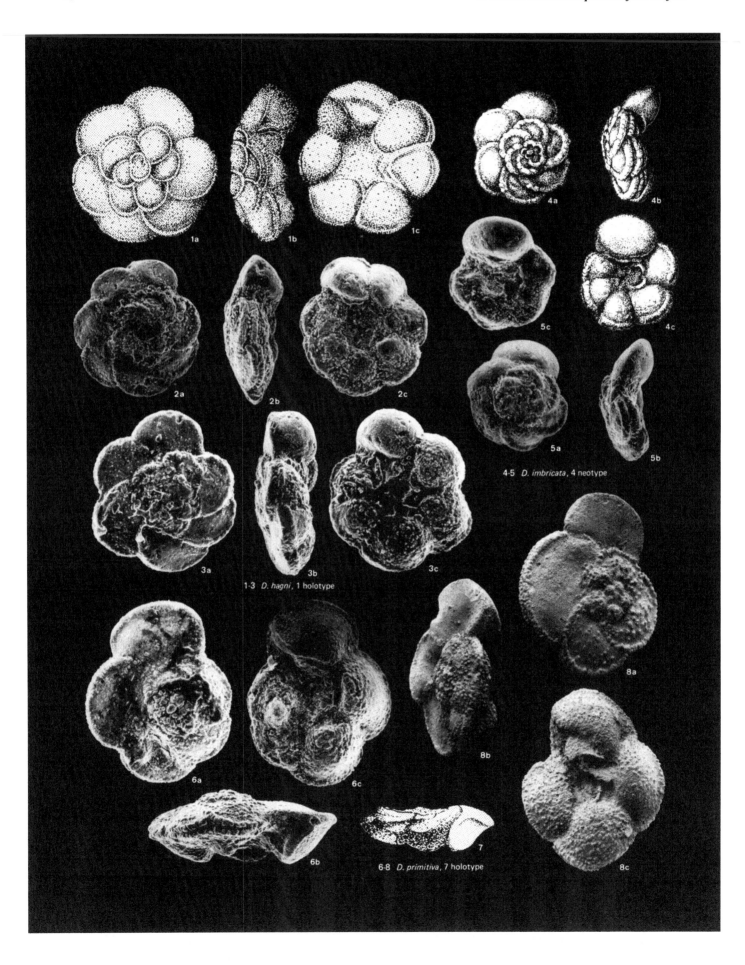

4-5 *D. imbricata*, 4 neotype

1-3 *D. hagni*, 1 holotype

6-8 *D. primitiva*, 7 holotype

Archaeoglobigerina ancestor (see also Pessagno, 1967, p. 342) rather than *Rugoglobigerina. G. gansseri* is readily distinguished from other species of this age, but is nearly homeomorphic with *H. helvetica* of mid-Turonian age. However, no phylogenetic relationship exists between these two species.

Genus GLOBIGERINELLOIDES Cushman & Ten Dam, 1948

GLOBIGERINELLOIDES ALGERIANA Cushman & Ten Dam
Figures **29.5–7**; 10, 12, 15

Type reference: *Globigerinelloides algeriana* Cushman & Ten Dam, 1948, p. 43, pl. 8, figs. 4–6.

Synonym: *Planomalina pustulosa* Umiker, 1952, p. 9, fig. 1.

G. algeriana differs from *G. ferreolensis* in the greater number of chambers, from *Planomalina cheniourensis* in the absence of a keel and from *Biticinella breggiensis* in its completely planispiral test with relict apertures on both sides.

GLOBIGERINELLOIDES BENTONENSIS (Morrow)
Figures **29.8–9**; 10, 12, 15

Type reference: *Anomalina bentonensis* Morrow, 1934, p. 201, pl. 30, figs. 4a–b.

Synonym: *Planomalina caseyi* Bolli, Loeblich & Tappan, 1957, pl. 1, figs. 4–5.

The chambers are closely arranged, the periphery only slightly lobate. The last chamber is usually longer than high; residual apertures are clearly visible in both umbilici.

GLOBIGERINELLOIDES BLOWI (Bolli)
Figures **29.10–11**; 10, 12, 15

Type reference: *Planomalina blowi* Bolli, 1959, p. 260, pl. 20, figs. 2a–b.

Synonyms: *Planomalina maridalensis* Bolli, 1959, p. 261, pl. 20, figs. 6a–b.
Globigerinella duboisi Chevalier, 1961, p. 205, pl. 1, figs. 15a–b.
Globigerinella gottisi Chevalier, 1961, p. 205, pl. 1, figs. 9a–b.

Bolli (1959) distinguished two species on the basis of spherical as against laterally compressed chambers, as did Chevalier (1961) with his two species. *G. blowi* differs from *G. subcarinata* in the absence of an acute axial margin, from *G. ultramicra* in having fewer chambers.

GLOBIGERINELLOIDES FERREOLENSIS (Moullade)
Figures **29.12–13**; 10, 12, 15

Type reference: *Biticinella ferreolensis* Moullade, 1961, p. 14, pl. 1, figs. 1–5.

Synonym: *Biglobigerinella sigali* Chevalier, 1961, p. 33, pl. 1, figs. 19a–b.

The species differs from *G. algeriana* in having fewer chambers and from *G. bentonensis* in having spherical chambers in edge view.

GLOBIGERINELLOIDES PRAIRIEHILLENSIS Pessagno
Figures **29.14–15**; 10, 13, 14

Type reference: *Globigerinelloides prairiehillensis* Pessagno, 1967, pl. 90, figs. 1–2.

The species differs from *G. ferreolensis* in possessing fewer chambers, which increase more rapidly in size in the last whorl.

GLOBIGERINELLOIDES SUBCARINATA (Brönnimann)
Figures **29.16–17**; 10, 13, 14

Type reference: *Globigerinella messinae subcarinata* Brönnimann, 1952, pp. 44–5, pl. 1, figs. 10–11.

This species possesses an imperforate peripheral band in all chambers of the last whorl, a feature which is absent in the other *Globigerinelloides* species.

GLOBIGERINELLOIDES ULTRAMICRA (Subbotina)
Figures **29.18–19**; 10, 12, 14

Type reference: *Globigerinella ultramicra* Subbotina, 1949, p. 33, pl. 2, figs. 17–18.

Fig. 18. *Dicarinella* (all figures × 60)

1–3 *D. hagni* (Scheibnerova)
　1a–c Holotype, from *M. sigali* Zone, Late Turonian, quarry at Horné Srnie, Klippen belt of the West Carpathians, Slovakia, Czechoslovakia. **2a–c** Specimen from Bellone well, at 130 m, Early Turonian, N France (from Robaszynski & Caron, 1979). **3a–c** Specimen from Middle Turonian, Angola.

4–5 *D. imbricata* (Mornod)
　4a–C Neotype, from Mornod's type-locality along the Ruisseau des Covayes, sample MC 1495, *M. sigali* Zone, Late Turonian, Montsalvens, Préalpes fribourgeoises, Switzerland. **5a–c** Specimen Z 1393, *M. sigali* Zone, Late Turonian, Pont du Fahs, N Tunisia.

6–8 *D. primitiva* (Dalbiez)
　6a–c Topotype, from El Kef section, Coniacian, Tunisia (from Robaszynski & Caron, 1979). **7** Holotype, *D. primitiva* Zone, Coniacian, El Kef section, N Tunisia. **8a–c** Specimen from sample Siga. 13.311, *D. concavata* Zone, Late Coniacian, Le Keb, Tunisia.

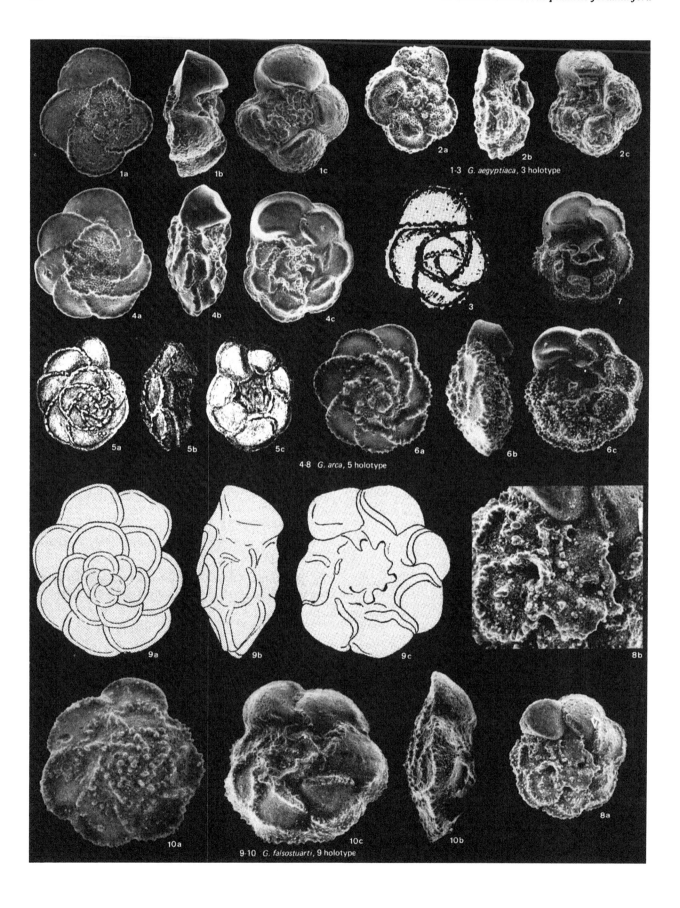

1-3 *G. aegyptiaca*, 3 holotype

4-8 *G. arca*, 5 holotype

9-10 *G. falsostuarti*, 9 holotype

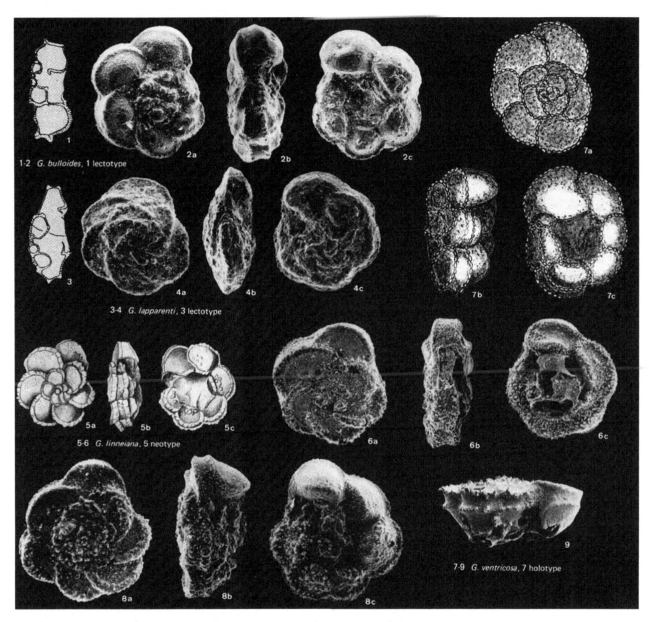

Fig. 19. *Globotruncana* (all figures × 60)

1–3 *G. aegyptiaca* Nakkady
    1a–c Sample TX 20 XB, *G. gansseri* Zone, Late Maastrichtian, Corsicana Formation, Falls County, Texas, USA. **2a–c** DSDP Leg 32-305-19-1, *G. aegyptiaca* Zone, Early Maastrichtian, Shatsky Rise, Northwestern Pacific. **3** Holotype. Anglo-Egyptian Oilfields Ltd., sample Dar. 290, Maastrichtian, Abu Durba, Western Sinai, 174 km SE Suez, Egypt.

4–8 *G. arca* (Cushman)
    4a–c Sample TX 4 AD, *G. gansseri* Zone, Late Maastrichtian, Corsicana Formation, Limestone County, Texas, USA. **5a–c** Holotype. From upper portion of the Papagallos shales (Mendez shale), Late Maastrichtian, Hacienda El Limon, San Luis Potosi, Mexico. **6a–c** DSDP Leg 3-21-3-6, 130–150 cm, *A. mayaroensis* Zone, Late Maastrichtian, NE Rio Grande Rise, South Atlantic. **7** DSDP Leg 11-98-13-c.c., *G. calcarata* Zone, Late Campanian, NE Providence Channel, Bahama Islands. **8a–b** El Burrueco section, *G. calcarata* Zone, Late Santonian, S Spain. 8b magnification × 300.

9–10 *G. falsostuarti* Sigal
    9a–c Holotype. Sample Sigal 12596, Route Munier-Lamy, Maastrichtian, Constantine, Algeria. **10a–c** Late Maastrichtian, El Kef, N Tunisia.

Fig. 20. *Globotruncana* (all figures × 60)

1–2 *G. bulloides* Vogler
    1 Lectotype. Maastrichtian?, Island of Misol, East Indonesia. **2a–c** Sample DW 15-5, Lower Taylor clay, *G. elevata* Zone, Early Campanian, Mc Lennan County, Texas, USA.

3–4 *G. lapparenti* Brotzen
    3 Lectotype. Maastrichtian, Hendaye Region, Basses-Pyrénées, France. **4a–c** Topotype from La Loya, Bay, Hendaye Region, SW France.

5–6 *G. linneiana* (d'Orbigny)
    5a–c Neotype. From recent beach sand containing re-deposited Late Cretaceous material. Right bank of Rio Martin Perez, Bahia de la Habana, Cuba. **6a–c** DSDP Leg 15-152-15-2, 138–140 cm, *G. gansseri* Zone, Late Maastrichtian, S Haiti, Caribbean Sea.

7–9 *G. ventricosa* (White)
    7a–c Holotype, **8a–c** Topotype. From Loc. 4 of White. Papagallos Formation, Late Campanian, El Barranco, road to Aldama, Tampico embayment, Mexico. **9** DSDP Leg 11-98-13-2, 99–101 cm, *G. calcarata* Zone, Late Campanian, NE Providence Channel, Bahama Islands.

The species differs from other *Globigerinelloides* species essentially by its small size and great number of chambers.

Genus GLOBOTRUNCANA Cushman, 1927

GLOBOTRUNCANA AEGYPTIACA Nakkady
Figures **19.1–3**; 11, 13, 14

Type reference: *Globotruncana aegyptiaca* Nakkady, 1950, p. 690, pl. 90, fig. 20, spiral side.

Synonyms: *Globotruncana aegyptiaca duwi* Nakkady, 1950.
  Globotruncana gagnebini Tilev, 1951, p. 50, pl. 3, figs. 2a–c.
  Rugotruncana skewesae Brönnimann & Brown, 1956, p. 550, pl. 23, figs. 4–6.

The test is typically 4-chambered in the last whorl, crescent shaped and truncated by two closely spaced keels. The lower keel merges into a generally well raised periumbilical rim. On the umbilical side the chamber surfaces are decorated with pustules.

GLOBOTRUNCANA ARCA (Cushman)
Figures **19.4–8**; 11, 13, 14

Type reference: *Pulvinulina arca* Cushman, 1926, p. 23, pl. 3, figs. 1a–c.

Synonym: *Globotruncana churchi* Martin, 1964, p. 79, pl. 9, fig. 5.

The two keels are separated by a large imperforate peripheral band. The periumbilical rim is well marked. *G. arca* differs from *G. falsostuarti* in the presence of two widely separated keels, present throughout the last whorl.

GLOBOTRUNCANA BULLOIDES Vogler
Figures **20.1–2**; 11, 13, 14

Type reference: *Globotruncana linnei* (d'Orbigny) subsp. *bulloides* Vogler, 1941, p. 287, pl. 23, figs. 32–39.

Lectotype: *Globotruncana bulloides* Vogler: Pessagno, 1967, p. 325, fig. 33 of Vogler 1941.

Synonyms: *Globotruncana culverensis* Barr, 1962, p. 569, pl. 71, figs. 1a–c.
  Globotruncana fresnoensis Martin, 1964, p. 80, pl. 9, figs. 8a–c.

The chamber walls of *G. bulloides* are distinctly inflated on both the spiral and umbilical sides. The two well developed keels are widely spaced. The species differs from *G. lapparenti* and *G. linneiana* in the pronounced inflation of its chambers.

GLOBOTRUNCANA FALSOSTUARTI Sigal
Figures **19.9–10**; 11, 13, 14

Type reference: *Globotruncana falsostuarti* Sigal, 1952, p. 43, pl. 46.

Synonyms: *Globotruncana leupoldi* Bolli, 1945, p. 235, pl. 9,

fig. 17 (possible synonymy, but holotype known only in thin section).
  Globotruncana stephensoni Pessagno, 1967, pp. 354–6, pl. 69, figs. 4–6.

This species is almost a homeomorph of *Globotruncanita stuarti* but differs from it in the presence of two keels on the earliest chambers, which merge in later ones into a single keel. In contrast to *G. stuarti* it also differs in the presence of tegilla covering the umbilical area.

GLOBOTRUNCANA LAPPARENTI Brotzen
Figures **20.3–4**; 11, 13, 14

Type reference: *Globotruncana lapparenti* Brotzen, 1936, pp. 175–6, numerous figures of thin sections in de Lapparent's thesis, 1918.

Lectotype: *Globotruncana lapparenti* Brotzen: Pessagno, 1967, p. 314, pl. 2, of de Lapparent, 1981.

Homonym: *Rosalinella lapparenti* Marie, 1941, invalid.

This species is very close to *G. linneiana* from which it only differs in having a narrower imperforate peripheral band and less raised keels.

GLOBOTRUNCANA LINNEIANA (d'Orbigny)
Figures **20.5–6**; 11, 13, 14

Type reference: *Rosalina linneiana* d'Orbigny, 1839, p. 101, pl. 15, figs. 10–12.

Neotype: *Globotruncana linneiana* (d'Orbigny): Brönnimann & Brown, 1956, p. 542, pl. 20, figs. 13–15, selected from beach sand of Habana Bay, Cuba (d'Orbigny's type-locality).

Synonym: *Pulvinulina tricarinata* Quereau, 1893 (probable synonymy, this species was described from a thin section).

The species is characterized in side view by its box-like shape with two well raised and widely spaced keels. It differs from *G. lapparenti* in its stout keels and from *Marginotruncana pseudolinneiana* in the umbilical position of the primary aperture, always protected by a well developed tegillum.

GLOBOTRUNCANA VENTRICOSA White
Figures **20.7–9**; 11, 13, 14

Type reference: *Globotruncana canaliculata* var. *ventricosa* White, 1928, p. 284, pl. 38, figs. 3a–c.

The species differs from *G. linneiana* in the strongly inflated umbilical side, from *Dicarinella asymetrica* in the umbilical position of the primary aperture, protected by a tegillum and not by a porticus.

Genus GLOBOTRUNCANELLA Reiss, 1957

GLOBOTRUNCANELLA CITAE (Bolli)
Figures **21.1–2**; 11, 13, 14

Type reference: *Globotruncana citae* Bolli, 1951, p. 197, pl. 35, figs. 4–6.

The test appears distinctly compressed in edge view. It differs from *G. havanensis* in the presence of a rudimentary peripheral keel compared with only an imperforate margin and aligned pustules in that species. Phylogenetically *G. citae* is regarded as the descendant of *G. havanensis* and the ancestor of *Abathomphalus intermedius* which in turn gave rise to *A. mayaroensis*.

GLOBOTRUNCANELLA HAVANENSIS
(Voorwijk)
Figures **21.3–4**; 11, 13, 14

Type reference: *Globotruncana havanensis* Voorwijk, 1937, p. 195, pl. 1, figs. 25, 26, 29.
Synonym: *Globotruncana petaloidea subpetaloidea* Gandolfi, 1955, p. 52, pl. 3, figs. 12a–c.

The species differs from *G. citae* only in having an imperforate margin or aligned pustules instead of a true keel. Typical is the long umbilical–extraumbilical primary aperture protected by a triangular porticus.

GLOBOTRUNCANELLA PETALOIDEA
(Gandolfi)
Figures **21.5–6**; 11, 13, 14

Type reference: *Globotruncana (Rugoglobigerina) petaloidea* subsp. *petaloidea* Gandolfi, 1955, p. 52, pl. 3, figs. 13a–c.

The species differs from *G. havanensis* and *G. citae* in having only 4 instead of 5 chambers in the last whorl, which gives it a petaloid aspect in equatorial view. In edge view, chambers appear compressed, lined by an imperforate peripheral band.

Genus GLOBOTRUNCANITA Reiss, 1957

GLOBOTRUNCANITA CALCARATA (Cushman)
Figures **23.6–7**; 11, 13, 14

Type reference: *Globotruncana calcarata* Cushman, 1927, p. 115, pl. 23, fig. 10.

This plano-convex species differs from all other *Globotruncanidae* species in possessing a tubulospine visible on each chamber of the last whorl. This characteristic feature combined with its short range makes the species an outstanding index form for the Late Campanian.

GLOBOTRUNCANITA CONICA (White)
Figures **22.1–2**; 11, 13, 14

Type reference: *Globotruncana conica* White, 1928, p. 285, pl. 38, figs. 7a–c.

The species differs from *Globotruncanita stuarti* in the asymmetric conical test with a high spiral side. From the high trochospiral but double keeled *Rosita contusa* it is distinguished in possessing only a single keel and in the more numerous chambers not being indented on the high conical spiral side as is the case in *R. contusa*.

GLOBOTRUNCANITA ELEVATA (Brotzen)
Figures **22.3–4**; 11, 13, 14

Type reference: *Globorotalia elevata* Brotzen, 1934, p. 66, pl. 3, fig. c.
Lectotype: *Globotruncana elevata* (Brotzen): Kuhry, 1970, pp. 292–4, 296–9, pl. 1, figs. 1–3.

The species differs from *Globotruncanita stuartiformis* in its plano-convex profile and crescent-shaped instead of triangular chambers. *G. subspinosa* has a more irregular outline and crescent-shaped chambers, which tend to extend backwards.

GLOBOTRUNCANITA STUARTI (de Lapparent)
Figures **23.1–3**; 11, 13, 14

Type reference: *Rosalina stuarti* de Lapparent, 1918, p. 11, pl. 1, figs. 5, 6, 7; pl. 4 (p. 12); pl. 5 (p. 13).
Lectotype: *Globotruncana stuarti* (de Lapparent): Pessagno, 1967, p. 356; as pl. 4 (p. 12), lower three figures in de Lapparent, 1918.

The species differs from *G. elevata* in its biconvex test, from *G. stuartiformis* in the regular trapezoidal instead of triangular-shaped chambers. *G. conica* differs from *G. stuarti* in a higher and smoother trochospire and a less elevated umbilical side.

GLOBOTRUNCANITA STUARTIFORMIS
(Dalbiez)
Figures **23.4–5**; 11, 13, 14

Type reference: *Globotruncana (Globotruncana) elevata stuartiformis* Dalbiez, 1955, p. 169, text-figs. 10a–c.

The species differs from *G. elevata* in its biconvex test and from *G. stuarti* in having triangular instead of trapezoidal chambers.

GLOBOTRUNCANITA SUBSPINOSA (Pessagno)
Figures **22.5–8**; 11, 13, 14

Type reference: *Globotruncana (Globotruncana) subspinosa* Pessagno, 1960, p. 101, pl. 1, figs. 4–6.

The species differs from *G. stuartiformis* in having crescent-shaped chambers with undulating surfaces and a tendency to extend backwards. These features result in an extremely irregular outline of the test.

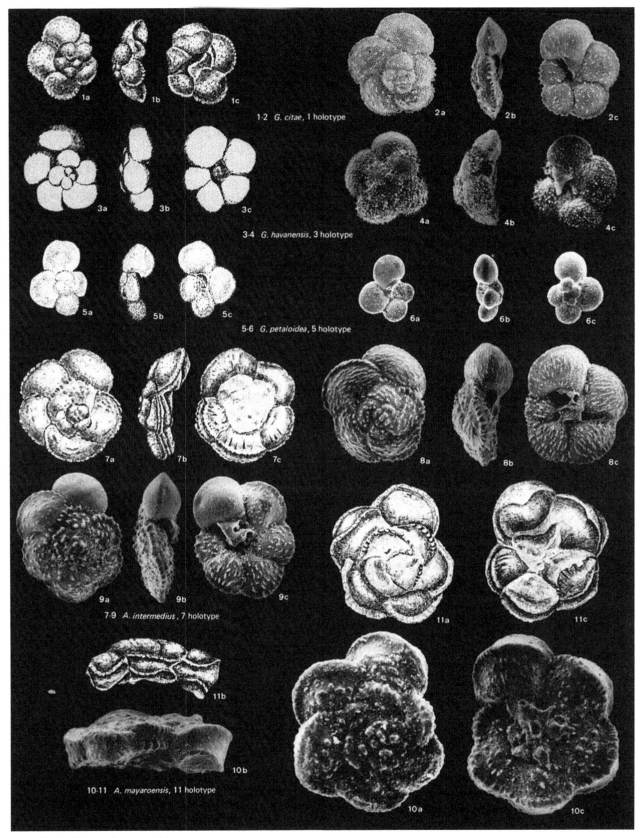

1-2 *G. citae*, 1 holotype

3-4 *G. havanensis*, 3 holotype

5-6 *G. petaloidea*, 5 holotype

7-9 *A. intermedius*, 7 holotype

10-11 *A. mayaroensis*, 11 holotype

Fig. 21. For caption see p. 56.

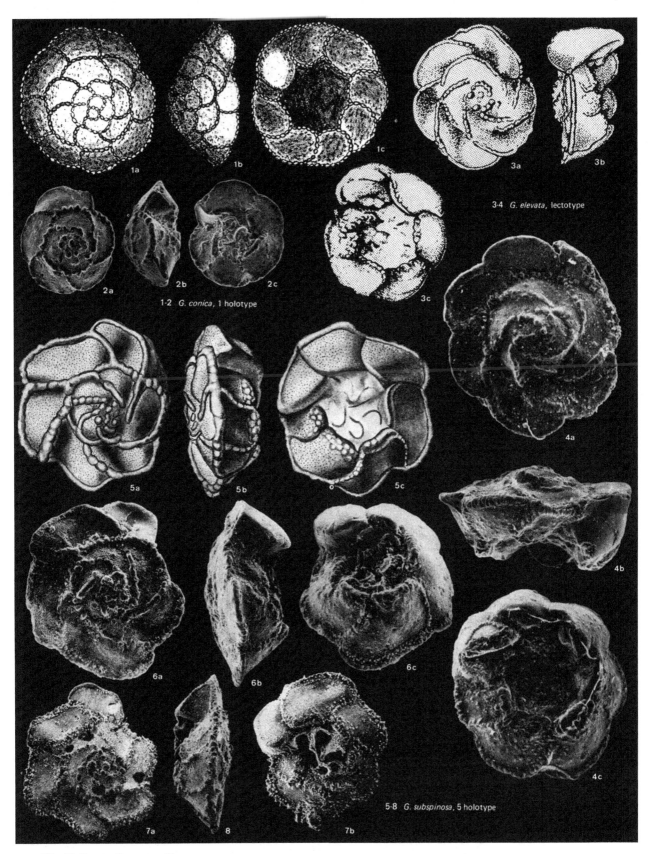

Fig. 22. For caption see p. 56.

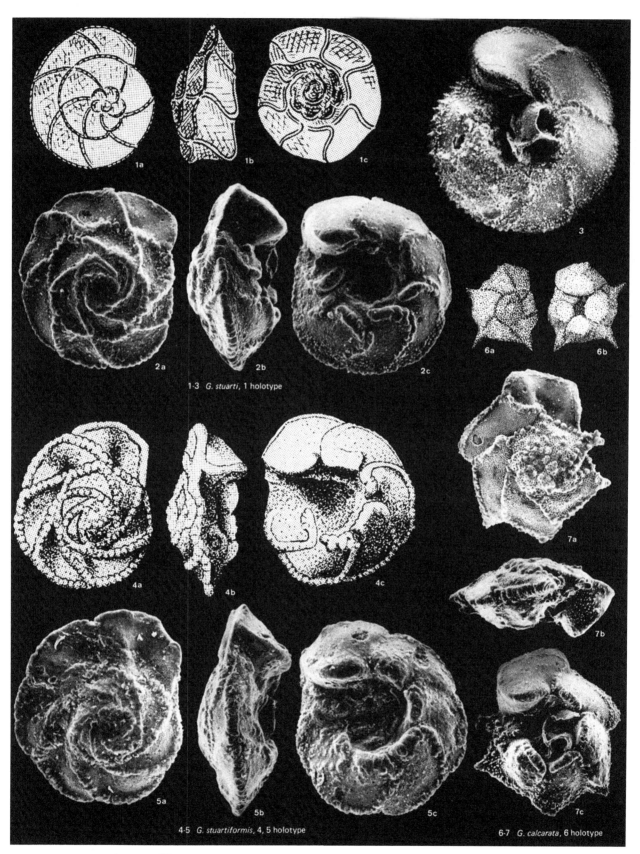

1-3 *G. stuarti*, 1 holotype

4-5 *G. stuartiformis*, 4, 5 holotype

6-7 *G. calcarata*, 6 holotype

Fig. 23. For caption see p. 56.

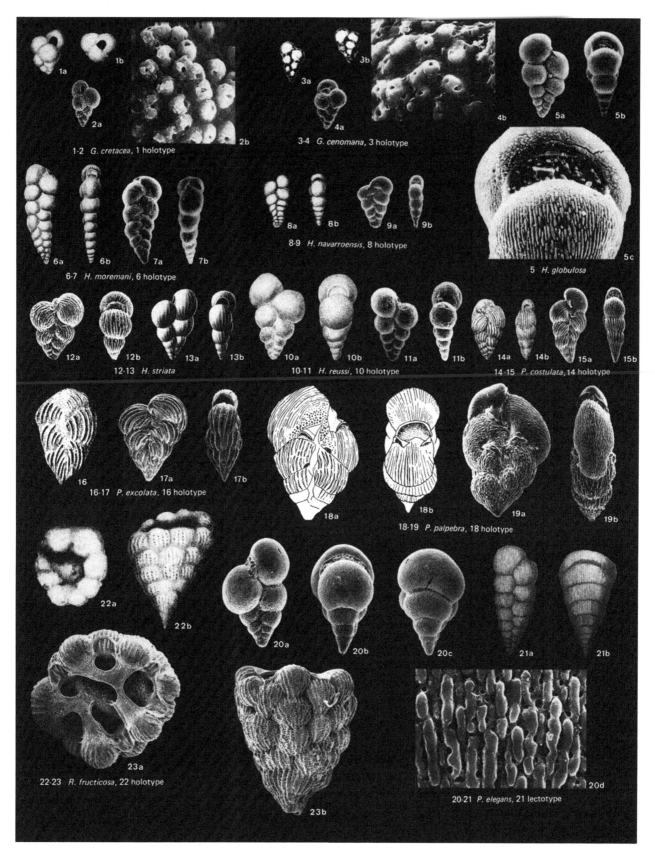

Fig. 24. For caption see p. 56.

Fig. 21. *Globotruncanella, Abathomphalus* (all figures × 60)

1–2 *G. citae* (Bolli)
1a–c Holotype. Maastrichtian, Lantern Estate, Central Range, Trinidad. **2a–c** Sample DSDP Leg 32-313-15-3, *G. gansseri* Zone, Late Maastrichtian, Northeastern Mid-Pacific Mountains.

3–4 *G. havanensis* (Voorwijk)
3a–c Holotype. Maastrichtian, Locality 64, Havana, Cuba. **4a–c** Sample from *G. havanensis* Zone, Early Maastrichtian, Semsales, Canton of Fribourg, Switzerland.

5–6 *G. petaloidea* (Gandolfi)
5a–c Holotype. Sample from lower part of the Colon formation, Maastrichtian, Papayel well no. 1, 30 km north of Fonseca, State of Magdalena, Northern Colombia. **6a–c** Sample TX 4 AC, *G. gansseri* Zone, Late Maastrichtian, Corsicana Formation, Limestone County, Texas, USA (from Smith & Pessagno, 1973).

7–9 *A. intermedius* (Bolli)
7a–c Holotype. Maastrichtian, Lantern Estate, Central Range, Trinidad. **8a–c, 9a–c** Sample from DSDP Leg 32-313-15, c.c., *G. gansseri* Zone, Late Maastrichtian, Northern Mid-Pacific Mountains.

10–11 *A. mayaroensis* (Bolli)
10a–c Sample from *A. mayaroensis* Zone, Late Maastrichtian, Marnes de Nay, Pyrénées occidentales, France. **11a–c** Holotype. Late Maastrichtian, Guayaguayare area, County of Mayaro, Southeastern Trinidad.

Fig. 22. *Globotruncanita* (all figures × 60)

1–2 *G. conica* (White)
1a–c Holotype. Specimen from Mendez Formation, Late Maastrichtian, E of Gueirero, Tampico embayment, Mexico. **2a–c** Sample TX 9 AG, *G. gansseri* Zone, Late Mastrichtian, Corsicana Formation, along Walkers Creek, Milam County, Texas, USA (from Smith & Pessagno, 1973).

3–4 *G. elevata* (Brotzen)
3a–c, 4a–c Lectotype, drawings and SEM photos. Specimen A 264 NR, collected by Brotzen, from Wadi Madi chalk, *D. asymetrica* Zone, Late Santonian, SE slope Mount Carmel, Israel.

5–8 *G. subspinosa* (Pessagno)
5a–c Holotype. Sample from Rio Yauco mudstone, Mayaguez group, Maastrichtian, Mayaguez-Yauco district, Southern Puerto Rico. **6a–c** Sample from *G. calcarata* Zone, Late Campanian, El Kef section, Tunisia. **7a–b, 8** Specimens from DSDP Leg 11-98-13-2, 99–101 cm and 131–133 cm, *G. calcarata* Zone, Late Campanian, NE Providence Channel, Bahama Islands, North Atlantic.

Fig. 23. *Globotruncanita* (all figures × 60)

1–3 *G. stuarti* (de Lapparent)
1a–c Holotype. Sample from lower part of Sainte-Anne levels, Maastrichtian, La Pointe Sante-Anne, Hendaye, W. Pyrénées, France. **2a–c** Dalbiez collection, Late Maastrichtian, Zebbeus Formation, N Tunisia. **3** Specimen from DSDP Leg 32-305-18-5, *G. aegyptiaca* Zone, Early Maastrichtian, Shatsky Rise, Northwestern Pacific.

4–5 *G. stuartiformis* (Dalbiez)
4a–c, 5a–c Holotype, drawings and SEM photos. Early Maastrichtian, El Kef–Melleque section, NW Tunisia.

6–7 *G. calcarata* (Cushman)
6a–b Holotype. Sample from Pecan Gap chalk, *G. calcarata* Zone, Late Campanian, Pecan Gap Formation, Famerville, Collin County, Texas, USA. **7a–c** Specimen from *G. calcarata* Zone, Late Campanian, Tanzania.

Fig. 24. *Guembelitria, Heterohelix, Pseudoguembelina, Pseudotextularia, Racemiguembelina* (all figures × 60, except 2b, 4b × 1200, 5c × 240, 20d × 500)

1–2 *G. cretacea* Cushman
1a–b Holotype. Navarro group, Late Maastrichtian, sample from pit of Seguin Brick and Tile Company, Guadelupe County, Texas, USA. **2a–b** Sample TX 3 BA, *G. gansseri* Zone, Late Maastrichtian, Corsicana Formation, Falls County, Texas, USA (from Smith & Pessagno, 1973).

3–4 *G. cenomana* (Keller)
3a–b Holotype. Cenomanian, northern edge of the Dniepr-Donetz Basin, USSR. **4a–b** Sample PTX 329, Britton formation, Late Centomanian, Dallas County, Texas, USA (from Smith & Pessagno, 1973).

5 *H. globulosa* (Ehrenberg)
5a–c Specimen from sample DW 15/4, *G. elevata* Zone, Early Campanian, Lower Taylor clay, Tradinghouse Creek, NNW of Harrison, Mc Lennan County, Texas, USA.

6–7 *H. moremani* (Cushman)
6a–b Holotype. Specimen from Eagle Ford Formation, Late Cenomanian, south bank of small stream, N of Itasca, Hill County, Texas, USA. **7a–b** Specimen from sample D 2/7, *R. cushmani* Zone, Late Cenomanian, Eagle Ford Formation, Irving, Dallas County, Texas, USA.

8–9 *H. navarroensis* Loeblich
8a–b Holotype. Specimen from Navarro group, Kemp clay, Late Maastrichtian, from the base of a pit of the Seguin Tile and Brick Company, Mc Queney, Guadelupe County, Texas, USA. **9a–b** Sample TX 4 AC, *G. gansseri* Zone, Late Maastrichtian. Corsicana formation, Limestone County, Texas, USA (from Smith & Pessagno, 1973).

10–11 *H. reussi* (Cushman)
10a–b Holotype. Sample from lower part of Austin chalk, Coniacian, ditch, west side of Sherman–Dennison highway, Texas, USA. **11a–b** Specimen from Sample D 6b, Middle member of Austin chalk, east bank of Ten Mile creek, *D. concavata* Zone, Early Santonian, Dallas County, Texas, USA.

12–13 *H. striata* (Ehrenberg)
12a–b Sample TX 3 BA, Falls County, *G. gansseri* Zone, Corsicana Formation, Late Maastrichtian, Texas, USA. **13a–b** Specimen from Kjolby Gaard Marl at Kjolby Gaard Jutland, Denmark, designated as type-locality by Pessagno, 1967 (13 from Berggren, 1962).

14–15 *Pseudoguembelina costulata* (Cushman)
14a–b Holotype. Specimen from lower beds of Upper Taylor marls, Campanian, Kickapoa Creek, NW Annona, Red River County, Texas, USA. **15a–b** Sample TX 4 AC, *G. gansseri* Zone, Late Maastrichtian, Corsicana Formation, Limestone County, Texas, USA (from Smith & Pessagno, 1973).

16–17 *Pseudoguembelina excolata* (Cushman)
16 Holotype. Specimen from upper portion of the Papagallos shales (= Mendez shale), Late Maastrichtian, SE of Guerrero, San Luis Potosi, Mexico. **17a–b** Sample TX 3 BA, *G. gansseri* Zone, Late Maastrichtian, Corsicana Formation, Falls County, Texas, USA (from Smith & Pessagno, 1973).

18–19 *Pseudoguembelina palpebra* Brönnimann & Brown
18a–b Holotype. Specimen from construction pit of the Gran Templo Nacional Masonico, Havana, Cuba. **19a–b** Sample TX 4 AC, *G. gansseri* Zone, Late Maastrichtian, Corsicana Formation, Limestone County, Texas, USA (from Smith & Pessagno, 1973).

20–21 *Pseudotextularia elegans* (Rzehak)
20a–d Sample TX 4 AC, *G. gansseri* Zone, Late Mastrichtian, Corsicana Formation, Limestone County, Texas, USA (from Smith & Pessagno, 1973). **21a–b** Lectotype, designated by White, 1929, as the specimen figured by Rzehak, 1895, figs. 1a–b under the name *Pseudotextularia varians*. Specimen from Late Cretaceous, Bruderndorf, Niederösterrehch.

22–23 *R. fructicosa* (Egger)
21a–b Holotype. Nierenthal-Schichten, Maastrichtian, Bavarian Alps, Germany. **22a–b** Sample DSDP Leg 3-21-2-1, 0–2 cm, *A. mayaroensis* Zone, Late Maastrichtian, Rio Grande Rise, South Atlantic (from Smith & Pessagno, 1973).

Genus GLOBULIGERINA Bignot & Guyader, 1971

GLOBULIGERINA HOTERIVICA (Subbotina)
Figures **25.1–3**; 10, 12, 15

Type reference: *Globigerina hoterivica* Subbotina, 1953, p. 50, pl. 1, figs. 1a–c.
Synonym: *Globigerina kugleri* Bolli, 1959, p. 270, pl. 23, figs. 3a–c.

This is the only *Globuligerina* species retained here. The test is a small trochospire consisting of only few subglobular chambers, four of them forming the last whorl. The surface is covered by a tubercular or reticulate sculpture. The species represents a link between the Middle to Late Jurassic *G. oxfordiana* and *G. helvetojurassica* and early forms belonging to the genus *Favusella*, which first appeared during the Barremian.

Genus GUEMBELITRIA Cushman, 1933

GUEMBELITRIA CENOMANA (Keller)
Figures **24.3–4**; 10, 12, 15

Type reference: *Guembelina cenomana* Keller, 1935, p. 547, pl. 3, figs. 13–14.
Synonym: *Guembelitria harrisi* Tappan, 1940, p. 115, pl. 19, figs. 2a–b.

The species differs from *G. cretacea* only in its low apertural arch.

GUEMBELITRIA CRETACEA Cushman
Figures **24.1–2**; 10, 13, 14

Type reference: *Guembelitria cretacea* Cushman, 1933, p. 37, pl. 4, fig. 12.

The test is triserial, elongate and its surface decorated with pore mounds. The species differs from *G. cenomana* in having a higher apertural arch.

Genus HASTIGERINOIDES Brönnimann, 1952

HASTIGERINOIDES ALEXANDERI (Cushman)
Figures **35.14–15**; 11, 13, 14

Type reference: *Hastigerinella alexanderi* Cushman, 1931, p. 87, pl. 11, figs. 6–9.

The holotype differs from *H. subdigitata* and *H. watersi* in having all but the last chamber of the last whorl strongly elongated, up to five times as long as wide, with pointed ends instead of clavate tips as is typical for the last chambers in *H. subdigitata* and *H. watersi*.

HASTIGERINOIDES SUBDIGITATA (Carman)
Figures **35.18–20**; 11, 13, 14

Type reference: *Globigerina subdigitata* Carman, 1929, p. 315, pl. 34, figs. 4–5.

The last whorl is formed by 7 chambers in the holotype compared with 5–6 in *H. alexanderi* and *H. watersi*. It further differs from *H. alexanderi* in the chambers being much less elongate and from *H. watersi* in having rounded instead of clavate tips.

HASTIGERINOIDES WATERSI (Cushman)
Figures **35.16–17**; 11, 13, 14

Type reference: *Hastigerinella watersi* Cushman, 1931, p. 86, pl. 11, figs. 4–5.

This species exhibits great diversity in length and presence or absence of a clavate tip in the chambers of the last whorl.

Genus HEDBERGELLA Brönnimann & Brown, 1958

HEDBERGELLA BIZONAE (Chevalier)
Figures **25.4–5**; 10, 12, 15

Type reference: *Hastigerinella bizonae* Chevalier, 1961, p. 34, pl. 1, figs. 24a–c.
Synonyms: *Hedbergella bollii* Longoria, 1974, p. 53, pl. 13, figs. 12–14.
Hedbergella kuhryi Longoria, 1974, p. 60, pl. 14, figs. 4–6.

The species is characterized by its distinctly stellate equatorial periphery with the chambers of the last whorl increasing only very gradually in size, becoming more elongated as added.

HEDBERGELLA DELRIOENSIS (Carsey)
Figures **25.6–7**; 10, 12, 15

Type reference: *Globigerina cretacea* d'Orbigny var. *delrioensis* Carsey, 1926, p. 43, type-figure not given.
Neotype: 1. *Hedbergella delrioensis* (Carsey, 1926): Longoria, 1974, pp. 54–5, pl. 10, figs. 1–3.
2. Second neotype selected by Masters, 1976, is not valid.
Synonyms: *Globigerina infracretacea* Glaessner, 1937, p. 28, text-fig. 1.
Globigerina portsdownensis Williams-Mitchell, 1948, pl. 8, figs 4a–c.

The species with usually 5–6 globular chambers forming the last whorl shows little variability. It is often abundant from Aptian to Cenomanian.

HEDBERGELLA FLANDRINI Porthault
Figures **25.12–14**; 10, 13, 15

Type reference: *Hedbergella flandrini* Porthault, 1970, pp. 64–5, pl. 10, figs. 1–3.

The species has a similar stellate outline as *H. bizonae* but differs in possessing only 4–5 instead of 6 chambers in the last whorl. The chambers also increase more rapidly in size.

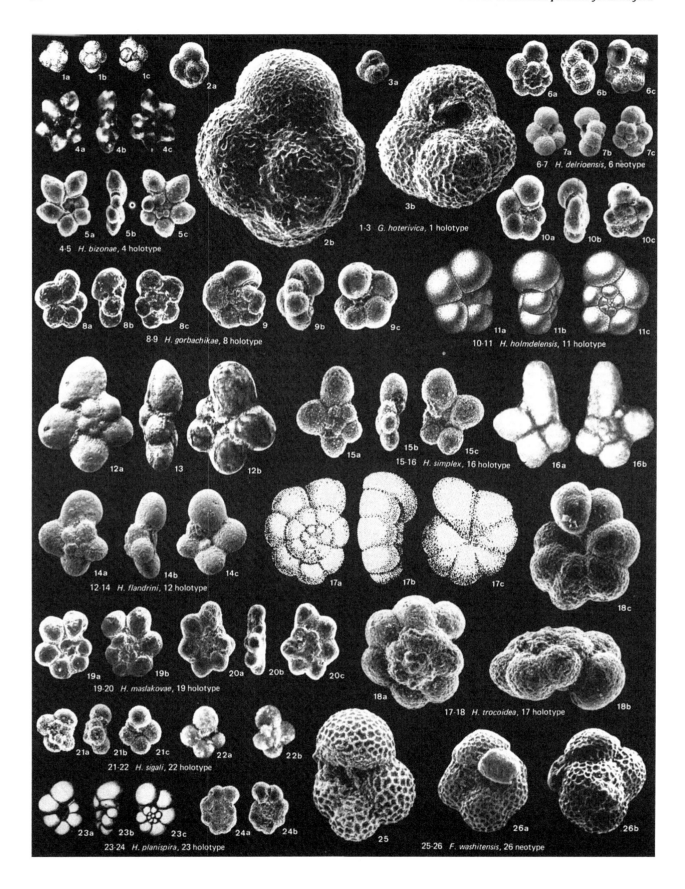

1a 1b 1c
2a
3a
6a 6b 6c
4a 4b 4c
7a 7b 7c
6-7 *H. delrioensis*, 6 neotype
5a 5b 5c
4-5 *H. bizonae*, 4 holotype
1-3 *G. hoterivica*, 1 holotype
2b
10a 10b 10c
8a 8b 8c
9 9b 9c
11a 11b 11c
8-9 *H. gorbachikae*, 8 holotype
10-11 *H. holmdelensis*, 11 holotype
12a
13
12b
15a 15b 15c
16a 16b
15-16 *H. simplex*, 16 holotype
12-14 *H. flandrini*, 12 holotype
14a 14b 14c
17a 17b 17c
18c
19a 19b 20a 20b 20c
19-20 *H. maslakovae*, 19 holotype
18a
18b
17-18 *H. trocoidea*, 17 holotype
21a 21b 21c 22a 22b
21-22 *H. sigali*, 22 holotype
23a 23b 23c 24a 24b
23-24 *H. planispira*, 23 holotype
25
26a 26b
25-26 *F. washitensis*, 26 neotype

Fig 25. *Globuligerina, Hedbergella, Favusella* (all figures × 60, except 2b, 3b × 250)

1–3 *G. hoterivica* (Subbotina)
   1a–c Holotype. Sample from *G. hoterivica* Zone, Hauterivian, along the Pshish River, Krasnodar Kray, N Caucasus, USSR. **2a–b, 3a–b** Specimens from A. Butt collection, DSDP Leg 47-397-50 and -47, *G. hoterivica* Zone, Early Barremian, SE Canary Islands, N Atlantic.

4–5 *H. bizonae* (Chevalier)
   4a–c Holotype. Sample from *S. cabri* Zone, Late Aptian, Le Bédoule, near Cassis, SE Marseille, SE France. **5a–c** Sample from Allemann collection, *G. algeriana* Zone, Late Aptian, Iran.

6–7 *H. delrioensis* (Carsey)
   6a–c Neotype. Specimen IFL-906 from an exposure of the Grayson Formation, Early Cenomanian, on Shoal Creek, just south of the 34th Street Bridge, Austin, Travis County, Texas, USA. **7a–c** Specimen from *R. appenninica* Zone, Late Albian, Folkstone, Kent, England (from Robaszynski & Caron, 1979).

8–9 *H. gorbachikae* Longoria
   8a–c Holotype. Specimen JFL-1018, sample L-43 from the La Peña Formation at La Boca Canyon, *H. gorbachikae* Zone, Late Aptian, S–SE of Monterrey, Mexico. **9a–c** DSDP Leg 40-864-31, c.c., *T. primula* Zone, Early Albian, Angola Basin, South Atlantic.

10–11 *H. holmdelensis* Olsson
   10a–c DSDP Leg 40-364-16, c.c., Campanian, Angola Basin, South Atlantic. **11a–c** Holotype. Specimen from the Navesink Formation, Early Maastrichtian, in a small tributary to Hop Brook, near Holmdel, New Jersey.

12–14 *H. flandrini* Porthault
   12a–b Holotype. **13** Paratype. Specimens from sample 94.5 m above the base of a section in the gully of the Lara, on the left bank of the Roudoule, Sample no. 3880, *D. asymetrica* Zone, Santonian, N Puget-Théniers, Alpes-Maritimes Department, SE France. **14a–c** Specimen from sample TC 16-10, *D. asymetrica* Zone, Santonian, San Miguel-Ubeda, Jaen, S Spain (from Robaszynski & Caron, 1979).

15–16 *H. simplex* (Morrow)
   15a–c DSDP Leg 40-364-21, c.c., *D. primitiva* Zone, Coniacian, Angola Basin, South Atlantic. **16a–b** Holotype. Sec. 31, T. 216, R. 22 W, *R. cushmani* Zone, Late Cenomanian, Huntland Shale, Greenhorn Formation, Hodgeman County, Kansas, USA.

17–18 *H. trocoidea* (Gandolfi)
   17a–c Holotype. **18a–c** Topotype. From bed 14, Scaglia variegata, *T. bejaouensis* Zone, Early Albian, Breggia river, Canton of Tessin, Switzerland.

19–20 *H. maslakovae* Longoria
   19a–b Holotype. Specimen JFL-720, sample L-29 from the La Peña Formation at La Boca Canyon, *H. gorbachikae* Zone, Late Aptian, S–SE of Monterrey, Mexico. **20a–c** DSDP Leg 40-364-35-3, 65–66 cm, *H. gorbachikae* Zone, Late Aptian, Angola Basin, South Atlantic.

21–22 *H. sigali* Moullade
   21a–c DSDP Leg 40-363-26, c.c., from lower Aptian levels reworked in Albian sample, Walvis Ridge, South Atlantic. **22a–b** Holotype. Sample from *H. sigali* Zone, Barremian, section at Saint Cyrice, near Orpierre, Hautes-Alpes Department, SE France.

23–24 *H. planispira* (Tappan)
   23a–c Holotype. Sample from Washita group, Grayson Formation, Early Cenomanian, Denton Creek, in south-facing bluff near the NW end of Grapevine Lake, Roanoke, Denton County, Texas, USA. **24a–b** DSDP Leg 40-363-36-1, 74–76 cm, Early Albian, Walvis Ridge, South Atlantic.

25–26 *F. washitensis* (Carsey)
   25 Specimen from an exposure of the Grayson Formation, type-locality of *H. delrioensis*, Early Cenomanian, on Shoal Creek, just south of the 34th Street Bridge, Austin, Texas, USA. **26a–b** Neotype. Specimen JFL-1092, same locality as above.

HEDBERGELLA GORBACHIKAE Longoria
Figures 25.8–9; 10, 12, 15

Type reference: *Hedbergella gorbachikae* Longoria, 1974, pp. 56–8, pl. 15, figs. 11–13.

The last 2–3 chambers are less regularly rounded compared with *H. delrioensis*, from which it also differs in the last chamber protruding into and covering most of the umbilical area.

HEDBERGELLA HOLMDELENSIS Olsson
Figures 25.10–11; 10, 13, 14

Type reference: *Hedbergella holmdelensis* Olsson, 1964, p. 160, pl. 1, figs. 2a–c.

The species differs from *H. delrioensis* in the more robust, larger test size and the more compact arrangement of the subglobular chambers. Perforations on the chamber surface are exceptionally sparse and small.

HEDBERGELLA MASLAKOVAE Longoria
Figures 25.19–20; 10, 12, 15

Type reference: *Hedbergella maslakovae* Longoria, 1974, pp. 61, 63, pl. 24, figs. 11–12.

Size, number and arrangement of chambers are similar to *H. bizonae*, from which it differs in the more rounded and less radially elongated chambers. Relict apertures may be preserved on the spiral side.

HEDBERGELLA PLANISPIRA (Tappan)
Figures 25.23–24; 10, 12, 15

Type reference: *Globigerina planispira* Tappan, 1940, p. 122, pl. 9, figs. 12a–c.

The species differs from *H. delrioensis* in its very low spiral side and greater number of chambers, 7–8 compared with 5–6, in the last whorl. It shows very little variation from Aptian to Turonian.

HEDBERGELLA SIGALI Moullade
Figures 25.21–22; 10, 12, 15

Type reference: *Hedbergella* (*Hedbergella*) *sigali* Moullade, 1966, p. 87, pl. 7, figs. 24–25.

The species differs from the slightly larger *H. delrioensis* in having only 4 instead of 5–6 globular chambers forming the last whorl.

HEDBERGELLA SIMPLEX (Morrow)
Figures 25.15–16; 10, 12, 15

Type reference: *Hastigerinella simplex* Morrow, 1934, p. 198, pl. 30, fig. 6.

Synonyms: *Hastigerinella simplicissima* Magné & Sigal, 1954, pl. 14, figs. 11a–c.

Hedbergella amabilis Loeblich & Tappan, 1961, pl. 3, figs. 1a–c.

This species shows great morphological variability particularly in the degree of radial chamber elongation in the last whorl. It is similar to *H. flandrini*, from which it differs in the still stronger and more distinct elongation of the last chambers.

HEDBERGELLA TROCOIDEA (Gandolfi)
Figures **25.17–18**; 10, 12, 15

Type reference: *Anomalina lorneiana* var. *trocoidea* Gandolfi, 1942, p. 99, pl. 2, figs. 1a–c.

Synonym: *Praeglobotruncana rohri* Bolli, 1959, p. 267, pl. 22, figs. 5–7.

Lectotype: *Hedbergella trocoidea* (Gandolfi): Caron & Luterbacher, 1969, in Gandolfi, 1942, pl. 4, fig. 2.

The species with usually 7 chambers forming the last whorl is distinctly larger than morphologically similar *Hedbergella* species like *H. planispira* and also possesses a slightly higher trochospire. *H. trocoidea* possibly gave rise to *Ticinella bejaouensis* by acquisition of umbilical supplementary apertures.

Genus HELVETOGLOBOTRUNCANA Reiss, 1957
HELVETOGLOBOTRUNCANA HELVETICA (Bolli)
Figures **30.7–8**; 10, 12, 15

Type reference: *Globotruncana helvetica* Bolli, 1945, p. 226, pl. 9, fig. 6.

Synonyms; *Globotruncana carpathica* Scheibnerova, 1963, p. 140, text-figs. 2a–c.
Globotruncana helvetica posthelvetica Hanzlikova, 1963, pp. 325–7, pl. 1, figs. 1–4.

The species differs from *H. praehelvetica* in having a true keel throughout all chambers of the last whorl. It resembles *G. gansseri* in its plano-convex test with a typical staircase-like imbricate chamber arrangement on the spiral side. This feature provides the spiral side with its distinct morphology, which is easily visible in the thin section of the holotype.

HELVETOGLOBOTRUNCANA PRAEHELVETICA (Trujillo)
Figures **30.9–10**; 10, 12, 15

Type reference: *Rugoglobigerina praehelvetica* Trujillo, 1960, p. 340, pl. 49, fig. 6.

The species shows the same plano-convex profile but differs from *H. helvetica* in the absence of a true keel in the last whorl. It is regarded as an evolutionary link from some Late Cenomanian *Whiteinella* to *H. helvetica*.

Genus HETEROHELIX Ehrenberg, 1843
HETEROHELIX GLOBULOSA (Ehrenberg)
Figure **24.5**; 10, 13, 14

Type reference: *Textularia globulosa* Ehrenberg, 1840, p. 135, pl. 4, figs. 2, 4, 5, 7, 8.

Lectotype: *Heterohelix globulosa* (Ehrenberg, 1840): Pessagno, 1967, p. 260, as fig. 5.

The globular chambers increase gradually in size, except for the late chambers, which increase more rapidly. Faint costae are present on all chambers.

HETEROHELIX MOREMANI (Cushman)
Figures **24.6–7**; 10, 12, 15

Type reference: *Guembelina moremani* Cushman, 1938, p. 10, pl. 2, figs. 1–3.

The 6–9 pairs of chambers increase slowly and regularly in size, giving the test a slender aspect. The test surface is smooth.

HETEROHELIX NAVARROENSIS Loeblich
Figures **24.8–9**; 10, 13, 14

Type reference: *Heterohelix navarroensis* Loeblich, 1951, p. 107, pl. 12, figs. 3a–b.

This species commonly displays a planispiral initial stage. It differs from *H. striata* in the laterally more compressed and distinctly finer costate character of the test, which is also smaller in size and in which the chambers increase less rapidly.

HETEROHELIX REUSSI (Cushman)
Figures **24.10–11**; 10, 12, 14

Type reference: *Gümbelina reussi* Cushman, 1938, p. 11, pl. 2, figs. 6a–b.

The species differs from *H. globulosa* in the slightly more compressed chambers and from *H. striata* in the much finer costae.

HETEROHELIX STRIATA (Ehrenberg)
Figures **24.12–13**; 10, 13, 14

Type reference: *Textularia striata* Ehrenberg, 1840, p. 135, pl. 4, figs. 1a, 2a, 3a.

Lectotype: *Heterohelix striata* (Ehrenberg): Pessagno, 1967, p. 264, as fig. 2a of Ehrenberg, 1840.

The species differs from its probable ancestor *H. globulosa* in possessing fewer and much more strongly developed costae.

Genus MARGINOTRUNCANA Hofker, 1956
MARGINOTRUNCANA CORONATA (Bolli)
Figures **26.1–2**; 11, 13, 14

Type reference: *Globotruncana lapparenti* Brotzen, subsp.

coronata Bolli, 1945, text-fig. 1, figs. 21–22, pl. 9, figs. 14–15.

Lectotype: *Marginotruncana coronata* (Bolli): Pessagno, 1967, p. 305, pl. 9, fig. 15 of Bolli, 1945.

The species differs from *M. pseudolinneiana* in its generally larger test size and more strongly compressed profile, apparent most clearly in the last chamber.

MARGINOTRUNCANA MARGINATA (Reuss)
Figures 26.3–4; 11, 13, 14

Type reference: *Rosalina marginata* Reuss, 1845, p. 36, pl. 13, figs. 18a–b.

Neotype: *Globotruncana linneiana marginata* (Reuss): Jirova, 1956, pp. 241–2, pl. 1, figs. 1a–c.

Lectotype: *Globotruncana marginata* (Reuss): Bolli, Loeblich & Tappan, 1957 (Jirova's neotype has priority).

The species differs from other *Marginotruncana* species in having inflated chambers both spirally and umbilically; from *Dicarinella canaliculata* in umbilical sigmoidal and raised sutures; from *Globotruncana bulloides*, which has similarly inflated chambers, in the presence of portici.

MARGINOTRUNCANA MARIANOSI (Douglas)
Figures 26.5–6; 11, 12, 15

Type reference: *Globotruncana marianosi* Douglas, 1969, p. 183, text-figs. 5a–c.

The species differs from other *Marginotruncana* taxa in having only one peripheral keel in the last whorl and from *M. sigali* in its plano-convex profile with chambers inflated on the umbilical side.

MARGINOTRUNCANA PSEUDOLINNEIANA
Pessagno
Figures 26.7–8; 11, 13, 14

Type reference: *Marginotruncana pseudolinneiana* Pessagno, 1967, pl. 65, figs. 24–27.

The species has the same general shape as *Globotruncana linneiana*, but differs in having an umbilical–extraumbilical primary aperture. Its portici tend to develop into tegilla with infra- and intra-laminal accessory apertures. From *M. coronata* it differs in its two more strongly developed keels, divided by a wider peripheral band and in its typical rectangular axial profile.

MARGINOTRUNCANA RENZI (Gandolfi)
Figures 27.1–2; 11, 12, 15

Type reference: *Globotruncana renzi* Gandolfi, 1942, pp. 124–5, pl. 3, fig. 1; pl. 4, fig. 16. No holotype was designated by Gandolfi, 'holotype' written on his original slide refers to the specimen illustrated in his pl. 3, fig. 1 (Caron, 1966, p. 79).

It may be difficult to separate *M. renzi* from *M. sigali*, *M. schneegansi* or *M. sinuosa*. Typically, *M. renzi* has a biconvex, fairly low trochospiral test, with 5–6 chambers forming the last whorl and increasing slowly in size. The two closely spaced keels merge into a single one on the last or the last few chambers. Sutures are curved and raised in both spiral and umbilical sides. *M. renzi* differs from *M. sigali* in its double keel, from *M. schneegansi* in its raised umbilical sutures and from *M. sinuosa* in its petaloid-shaped chambers on the spiral side. For a more complete diagnosis and remarks on the status of *M. renzi* reference is made to Robaszynski & Caron, 1979, vol. 2, p. 133.

MARGINOTRUNCANA SCHNEEGANSI (Sigal)
Figures 27.3–6; 11, 12, 14

Type reference: *Globotruncana schneegansi* Sigal, 1952, p. 33, pl. 34.

Neotype: *Marginotruncana schneegansi* (Sigal): Caron, 1974, pp. 330–3, figs. 1a–c.

M. schneegansi differs from *M. sigali* in having more inflated and petaloid-shaped chambers, increasing more rapidly in size, and umbilical sutures which are not raised. It differs from *Dicarinella primitiva* in its typical keel formed by two closely spaced parallel rows of pustules (Caron, 1974, pl. 1, figs. 4–5). Some specimens show intermediate characters between the genera *Dicarinella* and *Marginotruncana*.

MARGINOTRUNCANA SIGALI (Reichel)
Figures 27.7–8; 11, 12, 15

Type reference: *Globotruncana* (*Globotruncana*) *sigali* Reichel, 1950, p. 610, figs. 5a–c.

This species is widely recognized for its typical features and consequently as a Turonian zonal marker. The original description by Reichel (1950) is excellent and remains a model for observation, drawing and diagnosis. Like *M. renzi*, *M. schneegansi* and *M. sinuosa*, *M. sigali* has a biconvex test but differs from these in its roughly pustulose keels merging into a single one in the last whorl. As is the case in all *Marginotruncana* species, the double keel on the early whorls remains visible in thin section.

MARGINOTRUNCANA SINUOSA Porthault
Figures 27.9–11; 11, 13, 15

Type reference: *Marginotruncana sinuosa* Porthault, 1970, p. 81, pl. 11, figs. 12a–b.

This species is generally considered as ancestral and transitional to *Rosita fornicata* from which it differs in a lower trochospire and the absence of prominent globular chambers in the early whorls.

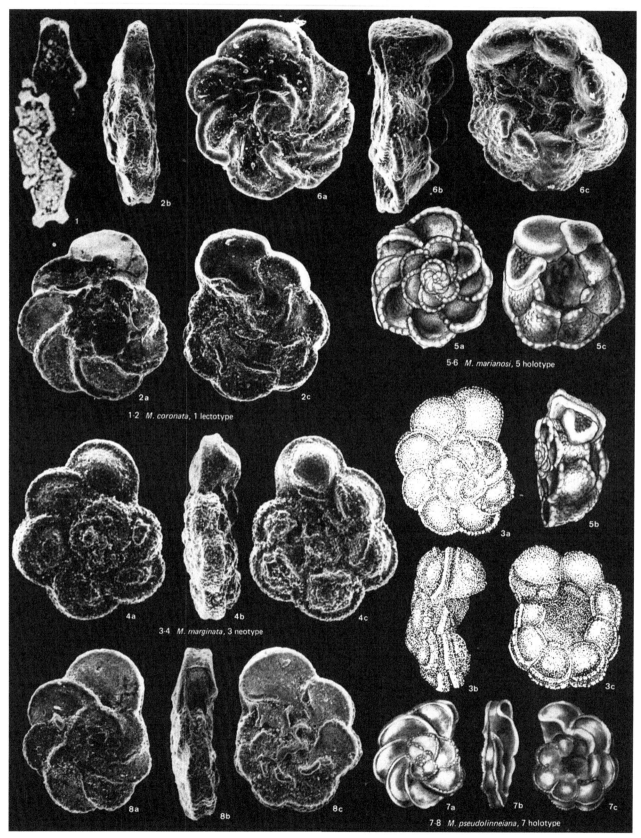

Fig. 26. For caption see p. 64.

1-2 *M. renzi*, 1 holotype

3-6 *M. schneegansi*, 3 neotype

7-8 *M. sigali*, 7 holotype

9-11 *M. sinuosa*, 9 holotype

Fig. 27. For caption see p. 64.

Fig. 28.

Fig. 26. *Marginotruncana* (all figures × 60)

1–2 *M. coronata* (Bolli)
  1 Lectotype. Sample no. 889, Hornmatt section, NW branch of riverlet between Hornmatt and Wissflueli, SW of Dallenwil, Canton of Nidwalden, Switzerland. **2a–c** Sample from *D. primitiva* Zone, Coniacian, El Kef section, Tunisia (from Robaszynski & Caron, 1979).

3–4 *M. marginata* (Reuss)
  3a–c Neotype. Sample from Plänersmergel, Luznice, Bohemia, Czechoslovakia. **4a–c** Topotype (from Robaszynski & Caron, 1979).

5–6 *M. marianosi* (Douglas)
  5a–c Holotype. Sample from *M. helvetica* Zone, Middle Turonian, San Miguel Island, Santa Barbara County, California, USA. **6a–c** Specimen from Lamolda collection, sample MCP-355, *H. helvetica* Zone, Middle Turonian, Spain.

7–8 *M. pseudolinneiana* Pessagno
  7a–c Holotype. Sample G 19, *D. coronata* Zone, Coniacian, Lauer Gosau Formation, Gosauthal, Edelbachgraben, Austria. **8a–c** Topotype (from Robaszynski & Caron, 1979).

Fig. 27. *Marginotruncana* (all figures × 60)

1–2 *M. renzi* (Gandolfi)
  1a–c Holotype. From an isolated exposure of flysch in the Breggia river section, *M. sigali* Zone, Late Turonian, Canton of Tessin,

Switzerland. **2a–c** Sample from Allemann collection, *D. primitiva* Zone, Coniacian, Lurestan, Iran.

3–6 *M. schneegansi* (Sigal)
  3 Neotype. **4–6** Neoparatypes. All from sample Si.13311. *D. concavata* Zone, Late Coniacian, Le Keb, Tunisia.

7–8 *M. sigali* (Reichel)
  7a–c Holotype. Sample Si.10574, *H. helvetica* Zone, Middle Turonian, Sidi Aïssa, S Aumale, Algeria. **8a–c** Postuma collection, sample from *M. sigali* Zone, Late Turonian, El Kef section, Tunisia.

9–11 *M. sinuosa* Porthault
  9a–b Holotype. **10.** Paratype. Sample from *D. concavata* Zone, Early Santonian, La Roudoule river, near Pujet-Théniers, Alpes-Maritimes, SE France. **11a–c** Calandra collection, *D. primitiva* Zone. Late Coniacian, Djebel Negilet, Tunisia.

Fig. 28. *Rosita* (all figures × 60)

1–2 *R. contusa* (Cushman) **1a–b** Holotype. Mendez Shale, Late Maastrichtian, San Luis Potos, Mexico. **2a–c** Sample from *A. mayaroensis* Zone, Late Campanian, Marnes de Nay, Pyrénées occidentales, France.

3–4 *R. fornicata* (Plummer)
  3a–c Holotype. Sample 226-T-8, *G. havanensis* Zone, Early Maastrichtian, Onion Creek, near Austin, Texas County. Texas, USA. **4a–c** DSDP Leg 32-305-24, *G. ventricosa* Zone, Middle Maastrichtian, Northeastern Mid-Pacific.

Genus PLANOMALINA Loeblich & Tappan, 1946

PLANOMALINA BUXTORFI (Gandolfi)
Figures **29.1–2**; 10, 12, 15

Type reference: *Planulina buxtorfi* Gandolfi, 1942, p. 103, pl. 3, figs. 7a–c.

Emendation: *Planomalina buxtorfi* (Gandolfi): Wonders, 1975, pl. 1, fig. 4; text-fig. 4, figs. 3a–b, 4a–b.

Synonym: *Planomalina apsidostroba* Loeblich & Tappan, 1946, p. 258, pl. 37, figs. 22–23.

Transitional forms between the assumed ancestor *Globigerinelloides bentonensis* and *P. buxtorfi* were described by Wonders (1975) as *P. praebuxtorfi* (not illustrated in this chapter), which differs in lacking a complete keel along the equatorial periphery.

PLANOMALINA CHENIOURENSIS (Sigal)
Figures **29.3–4**; 10, 12, 15

Type reference: *Planulina cheniourensis* Sigal, 1952, p. 21, text-fig. 17.

The species evolved from *G. algeriana* by progressive acquisition of a peripheral keel. It differs from *P. buxtorfi* in its more numerous chambers and absence of a peripheral keel in the last 3–4 chambers. No apparent phylogenetic relationship exists between *P. cheniourensis* restricted to the Aptian and *P. buxtorfi* confined to the Albian.

Genus PRAEGLOBOTRUNCANA Bermudez, 1952

PRAEGLOBOTRUNCANA DELRIOENSIS (Plummer)
Figures **30.1–2**; 10, 12, 15

Type reference: *Globorotalia delrioensis* Plummer, 1931, pp. 199–200, pl. 13, fig. 2.

This species is the oldest representative of the genus *Praeglobotruncana*, evolving from *Hedbergella delrioensis* by acquisition of numerous pustules along the periphery, at least on early chambers of the last whorl. It differs from *P. stephani* in its lower trochospire and acute pustulose equatorial ridge.

PRAEGLOBOTRUNCANA GIBBA Klaus
Figures **30.5–6**; 10, 12, 15

Type reference: *Praeglobotruncana stephani* Gandolfi var. *gibba* Klaus, 1960, pp. 304–5, holotype designated in Reichel, 1950, pl. 16, fig. 6, pl. 17, fig. 6.

This species differs from *P. stephani* in its higher trochospire and in the presence of portici in a deeper umbilicus.

PRAEGLOBOTRUNCANA STEPHANI (Gandolfi)
Figures **30.3–4**; 10, 12, 15

Type reference: *Globotruncana stephani* Gondolfi, 1942, p. 130, pl. 3, figs. 4a–c.

P. stephani is intermediate between *P. delrioensis* and *P. gibba* with regard to the height of the trochospire and the thickness of the pustulose peripheral band. It shows imbricate flaps around the umbilicus, sometimes well preserved in the last chamber. *P. stephani* gave rise to the genus *Dicarinella* (*D. algeriana*, which is the most primitive) by a splitting of the pustulose peripheral band into two closely spaced keels and later, independently, to the genus *Marginotruncana* by an increase in both the overlap of chambers and the curved, raised umbilical sutures.

Genus PSEUDOGUEMBELINA Brönnimann & Brown, 1953

PSEUDOGUEMBELINA COSTULATA (Cushman)
Figures **24.14–15**; 11, 13, 14

Type reference: *Guembelina costulata* Cushman, 1938, p. 16, pl. 3, figs. 7a–b.

Each chamber has two supplementary apertures near the median line on either side of the test, covered by arched and tubuliform lobes. *P. costulata* differs from *P. excolata* in having a smaller test, narrower in lateral view, finer costae and narrower flaps above the supplementary apertures.

PSEUDOGUEMBELINA EXCOLATA (Cushman)
Figures **24.16–17**; 11, 13, 14

Type reference: *Guembelina excolata* Cushman, 1926, p. 20, pl. 2, fig. 9.

Compared with *P. costulata*, *P. excolata* is characterized by fewer but coarser longitudinal costae and larger flaps covering the supplementary apertures.

PSEUDOGUEMBELINA PALPEBRA Brönnimann & Brown
Figures **24.18–19**; 11, 13, 14

Type reference: *Pseudoguembelina palpebra* Brönnimann & Brown, 1953, p. 155, text-figs. 9–10.

Specimens of *P. palpebra* are generally larger than those of *P. excolata* and costae are finer and more numerous. The supplementary sutural apertures near the median line are protected by low and broad lips.

Genus PSEUDOTEXTULARIA Rzehak, 1891

PSEUDOTEXTULARIA ELEGANS (Rzehak)
Figures **24.20–21**; 11, 13, 14

Type reference: *Cuneolina elegans* Rzehak, 1891, p. 4.

Lectotype: *Guembelina elegans* (Rzehak): White, 1929, in Rzehak, 1895, pl. 7, figs. 1a–b.

Synonym: *Pseudotextularia varians* Rzehak, 1895, pl. 7, figs. 1a–b.

Chambers increase only slowly in size as added and are broader than high and finely costate.

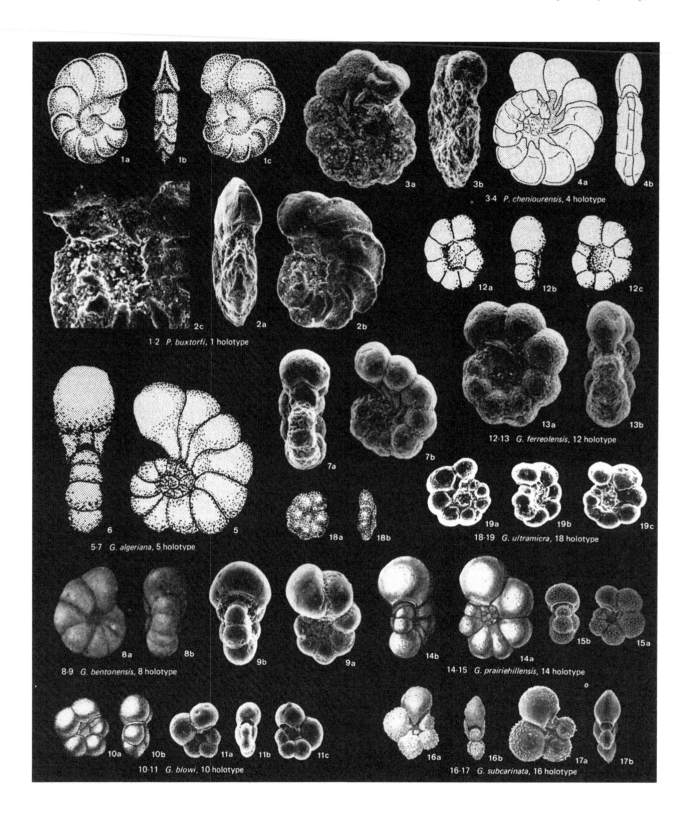

3-4 *P. cheniourensis*, 4 holotype

1-2 *P. buxtorfi*, 1 holotype

12-13 *G. ferreolensis*, 12 holotype

5-7 *G. algeriana*, 5 holotype

18-19 *G. ultramicra*, 18 holotype

8-9 *G. bentonensis*, 8 holotype

14-15 *G. prairiehillensis*, 14 holotype

10-11 *G. blowi*, 10 holotype

16-17 *G. subcarinata*, 16 holotype

Genus RACEMIGUEMBELINA Montanaro-Galitelli, 1957

RACEMIGUEMBELINA FRUCTICOSA (Egger)
Figures **24.22–23**; 11, 13, 14

Type reference: *Guembelina fructicosa* Egger, 1899, p. 36, pl. 14, figs. 8–9, not figs. 24–26.

Synonym: *Pseudotextularia varians* Rzehak, 1895, p. 217, pl. 7, figs. 2–3; not figs. 1a–b.

This readily recognizable species has a biserial early stage, which later becomes multiserial by development of supplementary, inflated and coarsely costate chambers. The test form is conical. Each terminal chamber possesses a basal arched aperture opening into the central cavity. Apertures are protected by ponticuli, successive ponticuli being arranged in a planar mode.

Fig. 29. *Planomalina, Globigerinelloides* (all figures × 60, except 2c × 120)

1–2 *P. buxtorfi* (Gandolfi)
1a–c Holotype. **2a–c** Topotype. Bed 31, Scaglia bianca, *R. ticinensis* Zone, Late Albian, Breggia river, Canton of Tessin, Switzerland.

3–4 *P. cheniourensis* (Sigal)
3a–b Sample M.C. 1684, *H. gorbachikae* Zone, Late Aptian, Djebel Oust section, N Tunisia. **4a–b** Holotype. Oued Cheniour section, Aptian, N Algeria.

5–7 *G. algeriana* Cushman & Ten Dam
5 Holotype. **6** Paratype. *G. algeriana* Zone, Late Aptian, Djebel Menaouer, Relizane, W Algeria. **7a–b** Caron Collection, *G. algeriana* Zone, Late Aptian, Iran.

8–9 *G. bentonensis* (Morrow)
8a–b Holotype. Sec. 31, T.21, S., R.22 W., *R. cushmani* Zone, Late Cenomanian, Hartland Shale, Greenhorn Formation, Hodgeman County, Kansas, USA. **9a–b** DSDP Leg 40-364-34-3, 11–12 cm, *T. bejaouensis* Zone, Early Albian, Angola Basin, Southeastern Atlantic (from Caron, 1978).

10–11 *G. blowi* (Bolli)**10a–b** Holotype. *S. cabri* Zone, Middle Aptian, Cuche Formation, Piparo river, Central Range, Trinidad. **11a–c** DSDP Leg 40-364-28-1, 49–50 cm, reworked in Late Albian, Angola Basin, Southeastern Atlantic (from Caron, 1978).

12–13 *G. ferreolensis* (Moullade)
12a–c Holotype. *G. algeriana* Zone, Late Aptian, Saint-Ferréol section, Drôme, SE France. **13a–b** Sample M.C. 1684, *G. gorbachikae* Zone, Late Aptian, Djebel Oust section, N Tunisia.

14–15 *G prairiehillensis* Pessagno
14a–b Holotype. Sample TX 251-c, *G. calcarata* Zone Late Campanian, Taylor Formation, Prairie Hill, Limestone County, Texas, USA. **15a–b** Sample TX-3-BA, *G. gansseri* Zone, Late Maastrichtian, Corsicana Formation, Texas, USA (from Smith & Pessagno, 1973).

16–17 *G. subcarinata* (Brönnimann)
17a–b Holotype. *A. mayaroensis* Zone, Late Maastrichtian, Guayaguayare Formation, Trinidad. **17a–b** Sample TX 20-xc, *G. gansseri* Zone, Late Maastrichtian, Corsicana Formation, Texas, USA (from Smith & Pessagno, 1973).

18–19 *G. ultramicra* (Subbotina)
18a–b Holotype. Cenomanian, from Kapustnaya Gorge, southern slope of Caucasus, USSR. **19a–c** Specimen from Grayson Formation, Early Cenomanian, Denton County, Texas, USA (from Michael, 1973, figured as *Globigerinelloides caseyi* Bolli).

Genus ROSITA Robaszynski, Caron, Gonzalez & Wonders, 1984

ROSITA CONTUSA (Cushman)
Figures **28.1–2**; 11, 13, 14

Type reference: *Pulvinulina arca* Cushman var. *contusa* Cushman, 1926, p. 23, no type-figure – *nomen nudum*.

Holotype: later illustrated in Cushman, 1946, p. 150, pl. 62, figs. 6a–c.

Synonyms: *Globotruncana linnei* d'Orbigny *caliciformis* Vogler, 1941, p. 288, pl. 24, fig. 23.
Globotruncana contusa galeoidis Herm, 1962, pp. 74–5, pl. 1, figs. 3–4.

The species is characterized by its highly trochospiral test, concave umbilical side and undulating chamber surface outlined by heavy beading. The narrow double keel is most visible in umbilical view. The umbilicus is wide and deep, covered by helicoidally arranged portici.

ROSITA FORNICATA (Plummer)
Figures **28.3–4**; 11, 13, 14

Type reference: *Globotruncana fornicata* Plummer, 1931, p. 130, pl. 13, figs. 4a–c.

This moderately trochospiral species with slightly undulating surfaces of the last chambers is thought to have developed from *Marginotruncana sinuosa*. Through gradual increase in the spire height, size of chambers and plication of chamber surfaces, it gave rise through intermediate forms to *R. contusa*.

Genus ROTALIPORA Brotzen, 1942

ROTALIPORA APPENNINICA (Renz)
Figures **31.1–4**; 10, 12, 15

Type reference: *Globotruncana appenninica* Renz, 1936, p. 20; p. 14, fig. 2; p. 71, fig. 7a; pl. 6, figs. 1–11; pl. 8, fig. 4.

Lectotype: *Globotruncana appenninica* Renz: Marie 1948, in Renz, 1936, p. 14, fig. 2 (left section).

Synonyms: *Globotruncana appenninica* Renz var. α Gandolfi, 1942, p. 118, text-fig. 40.
Globotruncana (Rotalipora) appenninica Renz sub. *balernaensis* Gandolfi, 1957, p. 60, pl. 8, figs. 3a–c.

This species differs from *R. gandolfii* in lacking periumbilical flanges on all chambers; from *R. ticinensis* in its more lobulate periphery and in fewer chambers forming the last whorl.

ROTALIPORA BROTZENI (Sigal)
Figures **31.5–7**; 10, 12, 15

Type reference: *Thalmanninella brotzeni* Sigal, 1948, p. 102, pl. 1, figs. 5a–c.

This species differs from *R. appenninica* in a less lobulate periphery and *R. gandolfii* in its raised umbilical sutures.

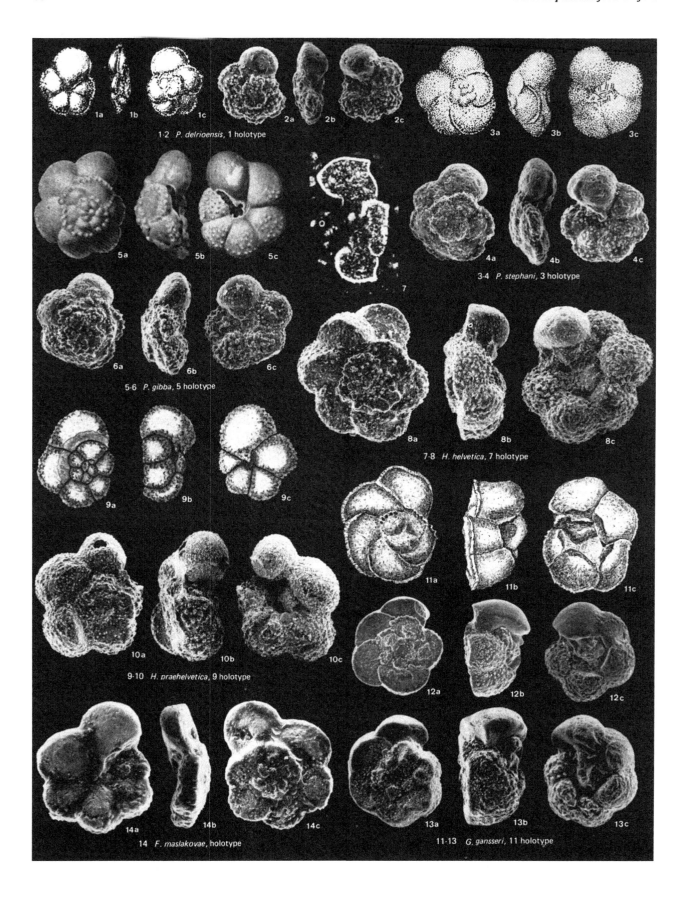

1-2   *P. delrioensis*, 1 holotype

3-4   *P. stephani*, 3 holotype

5-6   *P. gibba*, 5 holotype

7-8   *H. helvetica*, 7 holotype

9-10   *H. praehelvetica*, 9 holotype

14   *F. maslakovae*, holotype

11-13   *G. gansseri*, 11 holotype

ROTALIPORA CUSHMANI (Morrow)
Figures **31.8–11**; 10, 12, 15

Type reference: *Globorotalia cushmani* Morrow, 1934, p. 199, pl. 31, figs. 4a–b (no holotype designated, original specimen refigured by Brönnimann & Brown, 1956, pl. XX, figs. 10–12).

Synonyms: *Rotalipora turonica* Brotzen, 1942, p. 32, text-figs. 10–11.

Rotalipora alpina Bolli, 1945, pp. 224–5, pl. 9, figs. 3–4.

The acute periphery is accentuated by a thick, beaded and protruding keel. A concentration of pustules, forming a typical Y is present on the umbilical side of each chamber. The species differs from its ancestor *R. montsalvensis* in having a more protruding keel, raised sutures on the spiral side and typical beads on the umbilical side.

ROTALIPORA GANDOLFII Luterbacher & Premoli-Silva, 1962
Figures **33.5–7**; 10, 12, 15

Type reference: *Rotalipora appenninica gandolfii* Luterbacher & Premoli-Silva, 1962, p. 267. Holotype designated by its authors in Gandolfi, 1942, pl. II, fig. 5c.

Fig. 30. *Praeglobotruncana, Helvetoglobotruncana, Gansserina, Falsotruncana* (all figures × 60)

1–2 *P. delrioensis* (Plummer)
1a–c Holotype. Station 226-T-10, Shoal Creek, *R. appenninica* Zone, Late Albian, Del Rio Formation, Austin, Texas County, Texas, USA. **2a–c** Topotypic specimen (from Robaszynski & Caron, 1979).

3–4 *P. stephani* (Gandolfi) **3a–c** Holotype. Bed 56, Scaglia rossa, *R. reicheli* Zone, Middle Cenomanian, Breggia river, Canton of Tessin, Switzerland. **4a–c** Topotypic specimen (from Robaszynski & Caron, 1979).

5–6 *P. gibba* Klaus
5a–c Holotype, designated by Klaus 1960, after Reichel 1950 (pl. 16, fig. 6; pl. 17, fig. 6, for *Globotruncana stephani turbinata*), bed 43, top of Scaglia bianca, *R. brotzeni* Zone, Early Cenomanian, Breggia river, Canton of Tessin, Switzerland. **6a–c** Bed 56, Scaglia rossa, *R. reicheli* Zone, Middle Cenomanian, Breggia river, Canton of Tessin, Switzerland.

7–8 *H. helvetica* (Bolli)
7 Holotype. Sample no. 952, 15–20 m above the top of the 'Knollen-Schichten', *H. helvetica* Zone, Middle Turonian, Säntis range, Canton of St-Gallen, Switzerland. **8a–c** Sigal collection, sample 10574, *H. helvetica* Zone, Middle Turonian, Sidi-Aïssa, near Aumale, Algeria.

9–10 *H. praehelvetica* (Trujillo)
9a–c Holotype. *W. archaeocretacea* Zone?, Early Turonian?, Salt Creek, W of Redding, Shasta County, California. **10a–c** Robaszynki collection, Saint-Bénin bore-hole, *W. archaeocretacea* Zone, Early Turonian, N France.

11–13 *G. gansseri* (Bolli)
11a–c Holotype. Maastrichtian marls, Brighton area, Pitch lake, Southwestern Trinidad. **12a–c** Sample TX-3-BA, *G. gansseri* Zone, Late Maastrichtian, Corsicana Formation, Texas, USA (from Smith & Pessagno, 1973). **13a–c** Calandra collection, *G. gansseri* Zone, Late Maastrichtian.

14 *F. maslakovae* Caron
14a–c Holotype. Sample 2.1392, *M. sigali* Zone, Late Turonian, Pont du Fahs, Tunisia (from Caron, 1981).

Synonym: *Rotalipora appenninica* (Renz), subsp. *marchigiana* Borsetti, 1962, pp. 36–7, pl. III, figs. 4, 6.

This species evolved from *R. appenninica* by acquisition of a more inflated umbilical side and periumbilical flanges.

ROTALIPORA GREENHORNENSIS (Morrow)
Figures **32.1–2**; 10, 12, 15

Type reference: *Globorotalia greenhornensis* Morrow, 1934, p. 199, pl. 31, figs. 1a–c.

Synonym: *Rotalipora globotruncanoides* Sigal, 1948, p. 100, pl. 1, figs. 4a–c.

Typical for the species is the high number (8–10) of elongated and crescent-shaped chambers forming the last whorl, and strongly curved and raised sutures on both the spiral and umbilical side. *R. greenhornensis* is a descendant of *R. brotzeni* from which it differs mainly in the higher number of chambers in the last whorl.

ROTALIPORA MONTSALVENSIS Mornod
Figures **32.3–4**; 10, 12, 15

Type reference: *Globotruncana (Rotalipora) montsalvensis* Mornod, 1949–50, pp. 584–5, fig. 4 (I a–c).

Neotype: *Rotalipora montsalvensis* Mornod: Caron, 1976, pp. 329–30, figs. 1a–c.

Synonym: *Globotruncana (Rotalipora) montsalvensis minor* Mornod, 1949–50, p. 586, text-fig. 4 (II a–c).

The species is characterized by the inflated chambers. On the umbilical side their surfaces dip gradually towards the umbilicus and are without ornamentation. The intercameral sutures are depressed and not beaded on each side. *R. montsalvensis* differs from *R. cushmani* in the more inflated chambers, the absence of beaded sutures and a concentration of pustules on the central part of each chamber.

ROTALIPORA REICHELI Mornod
Figures **32.5–6**; 10, 12, 15

Type reference: *Globotruncana (Rotalipora) reicheli* Mornod, 1949–50, pp. 583–4, text-fig. 5 (IV a–c).

Neotype: *Rotalipora reicheli* Mornod: Caron, 1976, pp. 330–1, text-figs. 2a–c.

This species is readily recognizable by its distinct plano-convex test shape. The spiral side is plane to slightly concave, the umbilical one high with the last chambers often slightly inflated. The chambers of the last whorl display a peripheral ridge surrounding the umbilical area. *R. reicheli* differs from *R. gandolfii* in the more definitely plano-convex shape and more inflated chambers on the umbilical side.

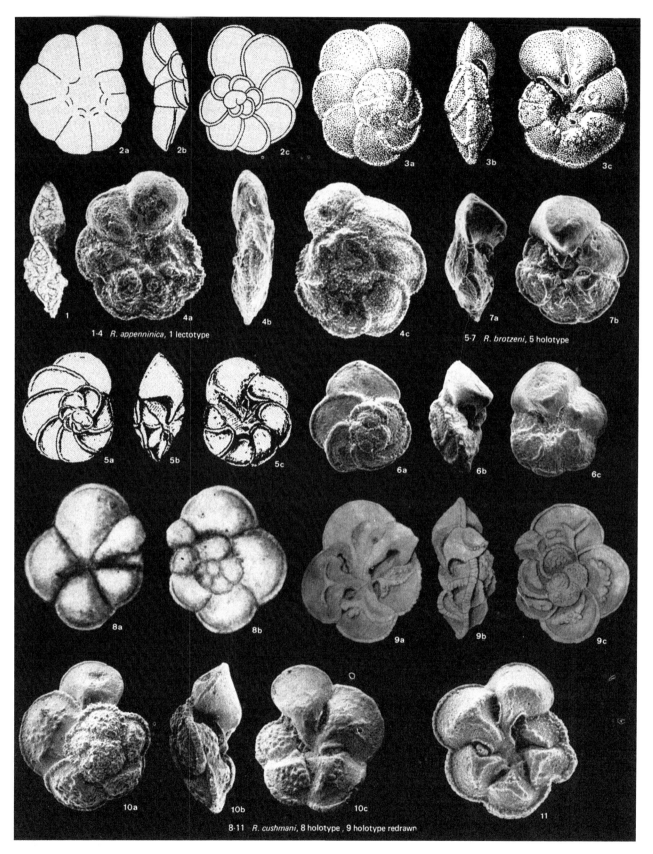

Fig. 31. For caption see p. 72.

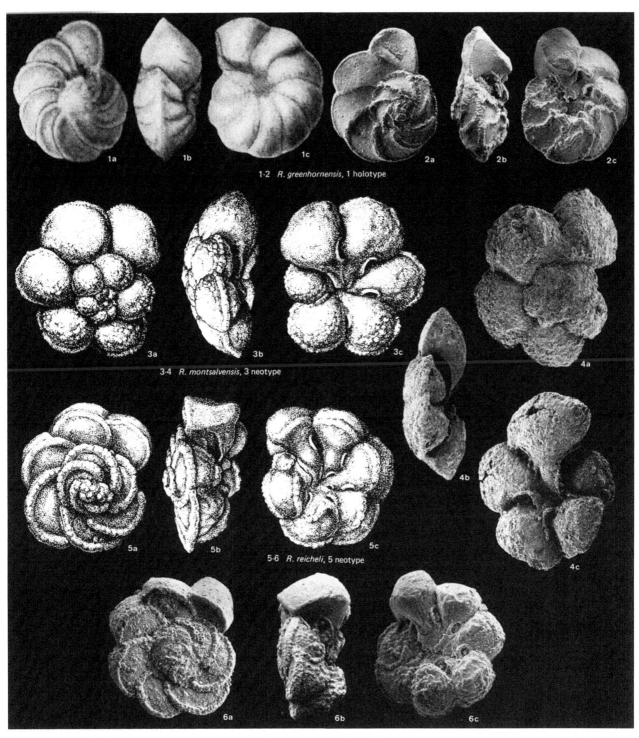

1-2 *R. greenhornensis*, 1 holotype

3-4 *R. montsalvensis*, 3 neotype

5-6 *R. reicheli*, 5 neotype

Fig. 32. For caption see p. 72.

ROTALIPORA SUBTICINENSIS (Gandolfi)
Figures **33.1–2**; 10, 12, 15

Type reference: *Globotruncana* (*Thalmanninella*) *ticinensis*
subsp. *subticinensis* Gandolfi, 1957, p. 59, pl. 8, figs.
1a–c.

This species differs from *R. ticinensis* only in the absence of a
keel on the last few chambers.

ROTALIPORA TICINENSIS (Gandolfi)
Figures **33.3–4**; 10, 12, 15

Type reference: *Globotruncana ticinensis* Gandolfi, 1942, pl. 2,
figs. 3a–c.

R. ticinensis is considered to be a link in the lineage *Ticinella
praeticinensis* – *Rotalipora subticinensis* – *R. ticinensis* – *R.
appenninica*. The morphological changes taking place during
evolution are a gradual reduction in the number of chambers
and a progressive acquisition of a peripheral keel on all
chambers of the last whorl.

Fig. 31. *Rotalipora* (all figures × 60)

1–4 *R. appenninica* (Renz)
1 Lectotype. Level n. 6, profil I, *R. appenninica* Zone. Late Albian,
Bottacione, NE Gubbio, Italy. **2a–c** Specimen selected by Gandolfi
(1942) for his *G. appenninica* Renz var. α, from sample Gandolfi no.
34, *R. appenninica* Zone, Late Albian, Scaglia bianca, Breggia river,
Canton of Tessin, Switzerland. **3a–c** Specimen designatd by Gandolfi
(1957) for his (*Rotalipora*) *appenninica* subsp. *balernaensis*, selected
and drawn by Reichel (1950) from sample Gandolfi no. 36, Late
Albian, Breggia river. **4a–c** Topotypic specimen, from level G. 112,
collected by Luterbacher & Premoli-Silva, 1962, *R. appenninica* Zone,
Late Albian, Bottacione profil, NE Gubbio, Italy.

5–7 *R. brotzeni* (Sigal)
5a–c Holotype. *R. brotzeni* Zone, Early Cenomanian, Sidi Aïssa,
Algeria. **6a–c** Section Q1, Arnayon, SE France (from Robaszynski &
Caron, 1979). 7a–b Sample M.C. 1581, *R. brotzeni* Zone, Early
Cenomanian, Pont du Fahs section, Tunisia.

8–11 *R. cushmani* (Morrow)
8a–b Holotype. Sec. 31, T. 21S., R. 22 W., *R. cushmani* Zone, Late
Cenomanian, Hartland Shale, Greenhorn Formation, Hodgeman
County, Kansas, USA. **9a–c** The refigured holotype by Brönnimann
& Brown, 1956, pl. XX, figs. 10–12. **10a–c** Topotypic specimen (from
Robaszynski & Caron, 1979). 11 Sample M.C. 1586, *R. cushmani*
Zone, Late Cenomanian, Pont du Fahs section, Tunisia.

Fig. 32. *Rotalipora* (all figures × 60)

1–2 *R. greenhornensis* (Morrow)
1a–c Holotype. Sec. 31, T. 21S, R. 22W., *R. cushmani* Zone, Late
Cenomanian, Hartland Shale, Greenhorn Formation, Hodgeman
County, Kansas, USA. **2a–c** Topotypic specimen (from Robaszynski
& Caron, 1979).

3–4 *R. montsalvensis* Mornod
3a–c Neotype. **4a–c** Topotype. From type-locality, Bed 13, profile II,
R. brotzeni Zone, Early Cenomanian, Montsalvens, Canton of
Fribourg, Switzerland (4 from Caron, 1976).

5–6 *R. reicheli* Mornod
5a–c Neotype. From type-locality, Bed 18, profile III, *R. reicheli*
Zone, Middle Cenomanian, Ruisseau des Covayes, Montsalvens,
Canton of Fribourg, Switzerland. **6a–c** Coupe R1, *R. reicheli* Zone,
Middle Cenomanian, Montmorin section, SE France (from
Robaszynski & Caron, 1979).

Genus RUGOGLOBIGERINA Brönnimann, 1952

RUGOGLOBIGERINA HEXACAMERATA
Brönnimann
Figures **34.1–2**;11, 13, 14

Type reference: *Rugoglobigerina reicheli hexacamerata* Brönnimann, 1952, pp. 23–5, pl. 2, figs. 10–12.
Synonyms: *Rugoglobigerina rugosa pennyi* Brönnimann, 1952,
p. 34, pl. 4, figs. 1–3.
Globotruncana (*Rugoglobigerina*) *hexacamerata* subsp.
subhexacamerata Gandolfi, 1955, p. 34, pl. 1, fig. 11 (the
holotype does not possess a double keel).

R. hexacamerata differs from other *Rugoglobigerina* species in
generally having 6 globular chambers forming the last whorl,
increasing only slowly in size. The species evolved from *R.
rugosa* which differs in having only 4–5 chambers, increasing
more rapidly in size.

RUGOGLOBIGERINA MACROCEPHALA
Brönnimann
Figures **34.3–4**; 11, 13, 14

Type reference; *Rugoglobigerina* (*Rugoglobigerina*) *macrocephala* Brönnimann, 1952, pl. 2, figs. 1–3.

This species is very characteristic with 4 inflated chambers
forming the last whorl. They increase very rapidly in size as
added, the final one comprising about half of the total test.

RUGOGLOBIGERINA REICHELI Brönnimann
Figures **34.5–6**; 11, 13, 14

Type reference: *Rugoglobigerina* (*Rugoglobigerina*) *reicheli*
Brönnimann subsp. *reicheli* Brönnimann, 1952, pp.
18–19, pl. 3, figs. 10–12.

The species differs from *R. rugosa* and *R. hexacamerata* in the
markedly spinose character and often slightly clavate shape of
the early chambers of the final whorl.

RUGOGLOBIGERINA RUGOSA (Plummer)
Figures **34.9–10**; 11, 13, 14

Type reference: *Globigerina rugosa* Plummer, 1926, p. 38, pl.
2, figs. 10a–d.
Lectotype: *Rugoglobigerina rugosa rugosa* (Plummer): Brön-
nimann, 1952, p. 29, in Plummer, 1926, pl. 2, fig. 10a.
Synonym: *Globotruncana* (*Rugoglobigerina*) *rugosa subrugosa*
Gandolfi, 1955, p. 72, pl. 7, figs. 5a–c.

It is likely that *R. rugosa* gave rise to *R. macrocephala* in the
Gansserina gansseri Zone. It differs from *R. macrocephala* in
having 5 instead of 4 chambers in the last whorl surrounding a
deep umbilicus.

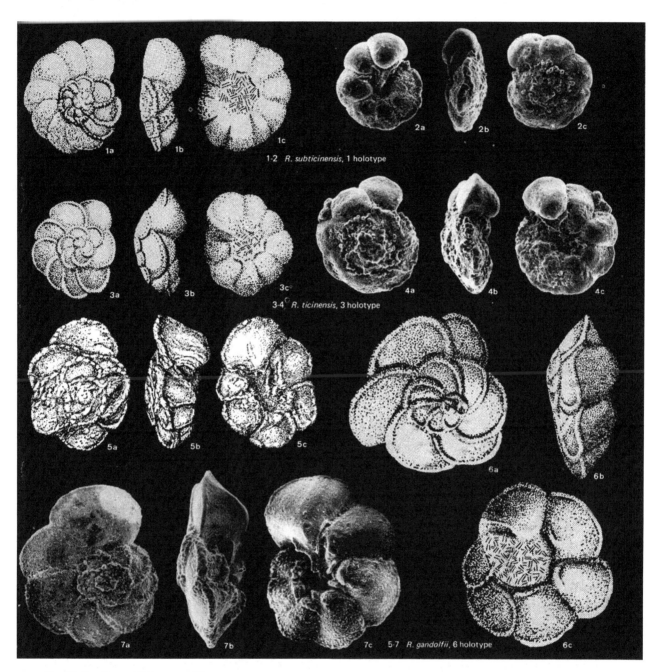

Fig. 33. *Rotalipora* (all figures × 60)

1–2 *R. subticinensis* (Gandolfi)
  1a–c Holotype, designated by Gandolfi, 1957 (after Gandolfi, 1942, pl. 2, fig. 4, for *G. ticinensis* var. α), from Bed 28, Scaglia bianca, *R. subticinensis* Zone, Late Albian, Breggia river, Canton of Tessin, Switzerland. **2a–c** Topotypic specimen (from Robaszynski & Caron, 1979).

3–4 *R. ticinensis* (Gandolfi)
  3a–c Holotype. **4a–c** Topotype. Bed 29, Scaglia bianca, *R. ticinensis* Zone, Late Albian, Breggia river, Canton of Tessin, Switzerland (from Robaszynki & Caron, 1979).

5–7 *R. gandolfii* Luterbacher & Premoli-Silva
  5a–c Topotype, after Luterbacher & Premoli-Silva, 1962. **6a–c** Holotype, designated by its authors after Gandolfi, 1942 (pl. 2, figs. 5a, c, d, for *G. appenninica* s.s.), Bed 55, Scaglia bianca, *R. brotzeni* Zone, Early Cenomanian, Breggia river, Canton of Tessin, Switzerland. **7a–c** Specimen from Bed 33, Scaglia bianca, Breggia river, Canton of Tessin, Switzerland, *R. appenninica* Zone, Late Albian.

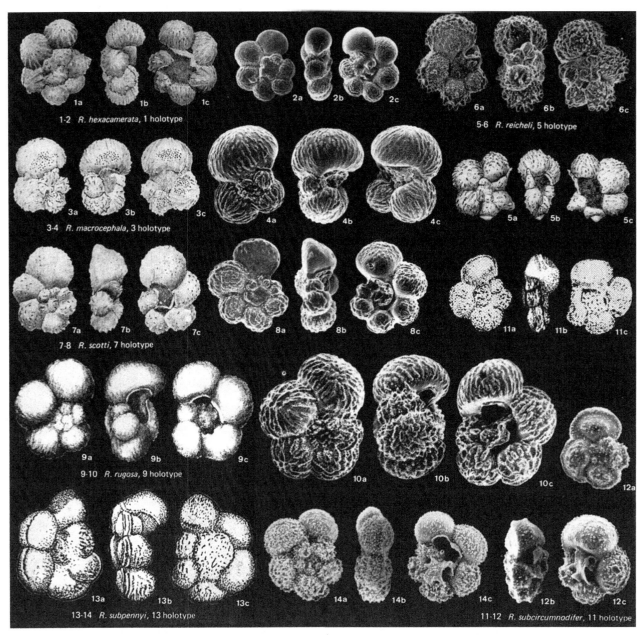

Fig. 34. *Rugoglobigerina, Rugotruncana* (all figures × 60)

1–2 *Rugoglobigerina hexacamerata* Brönnimann
1a–c Holotype. *A. mayaroensis* Zone, Late Maastrichtian, Guayaguayare Formation, Trinidad. **2a–c** Sample TX-4-AC, *G. gansseri* Zone, Late Maastrichtian, Corsicana Formation, Texas, USA (from Smith & Pessagno, 1973).

3–4 *Rugoglobigerina macrocephala* Brönnimann
3a–c Holotype. *A. mayaroensis* Zone, Late Maastrichtian, Guayaguayare Formation, Trinidad. **4a–c** Sample TX-20-XA, *G. gansseri* Zone, Late Maastrichtian, Corsicana Formation, Texas, USA (from Smith & Pessagno, 1973).

5–6 *Rugoglobigerina reicheli* Brönnimann
5a–c Holotype. *A. mayaroensis* Zone, Late Maastrichtian, Guayaguayare Formation, Trinidad. **6a–c** Sample TX-20-A, *G. gansseri* Zone, Late Maastrichtian, Corsicana Formation, Texas, USA (from Smith & Pessagno, 1973).

7–8 *Rugoglobigerina scotti* (Brönnimann)
7a–c Holotype. *A. mayaroensis* Zone, Late Maastrichtian,

Guayaguayare Formation, Trinidad. **8a–c** Sample TX-9-XE, *G. gansseri* Zone, Late Maastrichtian, Corsicana Formation, Texas, USA (from Smith & Pessagno, 1973).

9–10 *Rugoglobigerina rugosa* (Plummer)
9a–c Holotype. Sample about 5 feet below Midway Greensand, Upper Navarro Formation, Maastrichtian, Milam County, Texas, USA. **10a–c** Sample TX-20-XA, *G. gansseri* Zone, Late Maastrichtian, Corsicana Formation, Texas, USA (from Smith & Pessagno, 1973).

11–12 *Rugotruncana subcircumnodifer* (Gandolfi)
11a–c Holotype. Sample from Rancheria Valley, Early Maastrichtian, Colon Formation, Magdalena State, Northern Colombia. **12a–c** DSDP Leg 32-313-19-3, *R. subcircumnodifer* Zone, Early Maastrichtian, Northeastern Mid-Pacific.

13–14 *Rugotruncana subpennyi* (Gandolfi)
13a–c Holotype. Sample from Rancheria Valley, Early Maastrichtian, Colon Formation, Magdalena State, Northern Colombia. **14a–c** DSDP Leg 32-313-16cc, *G. gansseri* Zone, Middle Maastrichtian, Northeastern Mid-Pacific.

RUGOGLOBIGERINA SCOTTI (Brönnimann)
Figures **34.7–8**; 11, 13, 14

Type reference: *Trinitella scotti* Brönnimann, 1952, p. 57, pl. 4, figs. 4–6.

R. scotti resembles *R. hexacamerata* but differs in that its final chamber is compressed and has an imperforate peripheral band. The compression of only the last chamber was taken by Brönnimann as the reason to create a new genus *Trinitella*. Following Pessagno (1967) we consider that this feature alone does not justify a new genus.

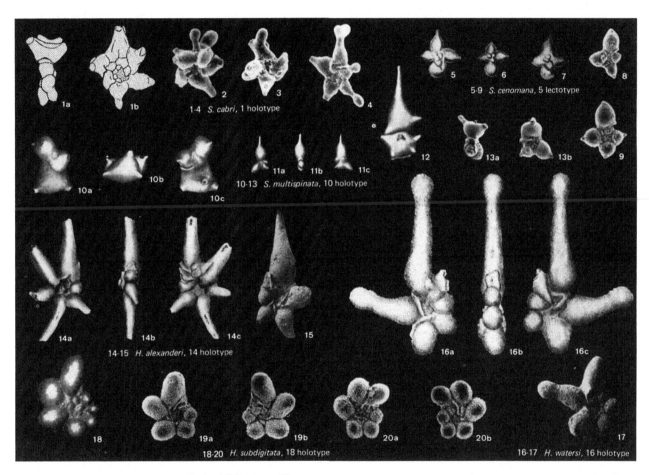

Fig. 35. *Schackoina, Hastigerinelloides* (all figures × 60)

1–4 *S. cabri* Sigal
1a–b Holotype. Aptian, Djebel Rhazouane, Tunisia. **2–4** Specimens from sample MC 1268, *S. cabri* Zone, Early Aptian, Kacha River, SE of Bakhchisarai, Crimea, USSR.

5–9 *S. cenomana* (Schacko)
5 Lectotype. **6–7** Paratypes. Specimens from Helle-Mühle, near Moltzow, Mecklenburg, Germany, Cenomanian. **8–9** From DSDP Leg 40-364-26 cc and 364-28-1, 49–50 cm, Cenomanian, reworked by drilling in Albian samples. Angola Basin, Southeastern Atlantic (from Caron, 1978).

10–13 *S. multispinata* (Cushman & Wickenden)
10a–c Holotype. **11a–c, 12** Paratypes. From NE 1/4, sector 11, T. 6N, R.8.W. of principal Meridian on north bank of Boyne River, Southern Manitoba, Canada, chalk equivalent to Taylor group of Texas, Campanian. **13a–b** Specimen from calcareous shales, Greenhorn Formation, Bull Creek section in Black Hills, Wyoming, USA (from Eicher & Worstell, 1970).

14–15 *H. alexanderi* (Cushman)
14a–c Holotype. **15** Topotype. From yellow calcareous clay, Austin Chalk, Coniacian–Santonian, Grayson County, Texas, USA (15 from Masters, 1977b).

16–17 *H. watersi* (Cushman)
16a–c Holotype. From Eastern slope of a deep road cut between two railroad underpasses near the northern edge of the town of Howe, Grayson County, Texas, USA. **17** Topotype, ephebic stage, ultimate chamber broken at constriction. Austin Chalk, Grayson County, Texas, USA (from Masters, 1977b).

18–20 *H. subdigitata* (Carman)
18 Holotype. From east flank of the Centennial syncline on the west slope of Sheep Mountain, T. 15.N, R.77.W., Niobrara Chalk, Late Cretaceous, Wyoming, USA. **19a–b, 20a–b** Specimens from sample D 6b, middle member of Austin Chalk, east bank of Ten Mile Creek, *D. concavata* Zone, Early Santonian, Dallas County, Texas, USA.

Genus RUGOTRUNCANA Brönnimann & Brown, 1956

RUGOTRUNCANA SUBCIRCUMNODIFER (Gandolfi)

Figures **34.11–12**; 11, 13, 14

Type reference: *Globotruncana* (*Rugoglobigerina*) *circumnodifer subcircumnodifer* Gandolfi, 1955, p. 44, pl. 2, figs. 8a–c.

Synonyms: *Rugotruncana tilevi* Brönnimann & Brown, 1956, p. 547, pl. 22, figs. 1–3.
      Rugotruncana ellisi Brönnimann & Brown, 1956, p. 547, pl. 2, figs. 7–9.

For comparison with *R. subpennyi* see below.

RUGOTRUNCANA SUBPENNYI (Gandolfi)

Figures **34.13–14**; 11, 13, 14

Type reference: *Globotruncana* (*Rugoglobigerina*) *pennyi subpennyi* Gandolfi, 1955, p. 73, pl. 7, figs. 7a–c.

R. subpennyi differs from *R. subcircumnodifer* in having 6 instead of 4–5 chambers forming the last whorl, a flat spiral side and two more closely spaced keels, often absent in the two final chambers.

Genus SCHACKOINA Thalmann, 1932

SCHACKOINA CABRI Sigal

Figures **35.1–4**; 10, 12, 15

Type reference: *Schackoina cabri* Sigal, 1952, pp. 20–1, text-fig. 18.

Synonym: *Leupoldina protuberans* Bolli, 1959, p. 264, pl. 20, figs. 20a–b.

S. cabri has bulbous protuberances on the last 1 to 3 chambers; these are paired or, very occasionally, in threes. In *S. cenomana* the protuberances are single tubulospines. *S. cabri* has 4–5 chambers in the last whorl, whereas *S. multispinata* has only 3, each of which has several tubulospines.

SCHACKOINA CENOMANA (Schacko)

Figures **35.5–9**; 10, 12, 15

Type reference: *Siderolina cenomana* Schacko, 1897, p. 166, pl. 4, figs. 3–5.

S. cenomana differs from the other *Schackoina* specifies in having 4 chambers in the last whorl, each with a single tubulospine.

SCHACKOINA MULTISPINATA (Cushman & Wickenden)

Figures **35.10–13**; 10, 12, 14

Type reference: *Hantkenina multispinata* Cushman & Wickenden, 1930, p. 40, pl. 6, figs. 4–6.

Synonym: *Schackoina cushmani* Barr, 1962, p. 565, pl. 69, fig. 3, text-figs. 5a–f.

S. multispinata differs from the other *Schackoina* species in having only 3 chambers in the last whorl, the first bearing a single tubulospine, the later two chambers with two or more (up to 7) tubulospines each.

Genus TICINELLA Reichel, 1950

TICINELLA BEJAOUENSIS Sigal

Figures **36.1–3**; 10, 12, 15

Type reference: *Ticinella roberti* var. *bejaouensis* Sigal, 1966, pl. 5, figs. 5–7.

Emendation: *Ticinella bejaouensis* Sigal, emend. Moullade, 1966, p. 103.

The species differs from *Hedbergella trocoidea*, its direct ancestor, in the presence of supplementary apertures in the umbilical area and from *T. roberti*, its descendant, in having a flat spiral side.

TICINELLA MADECASSIANA Sigal

Figures **36.4–5**; 10, 12, 15

Type reference: *Ticinella madecassiana* Sigal, 1966, p. 197, pl. 3, figs. 7a–b.

T. madecassiana differs from other *Ticinella* species in the lower trochospire and small number of chambers.

Fig. 36. *Ticinella, Biticinella* (all figures × 60)

1–3 *T. bejaouensis* Sigal
   1a–c Holotype. Diego-Suarez well, at 215.2 m, *T. bejaouensis* Zone, Early Albian, Mount Raynaud, Madagascar. **2a–c** Sample Allemann 69.127, *T. primula* Zone, Early Albian, Caravaca, Spain. **3a–b** DSDP Leg 40-364-34-3, *T. bejaouensis* Zone, Early Albian, Angola Basin, Southeastern Atlantic (from Caron, 1978).

4–5 *T. madecassiana* Sigal
   4a–c Holotype. Diego-Suarez well at 121.2 m, *R. appenninica* Zone, Late Albian, Mount Raynaud, Madagascar. **5a–c** Sample Allemann 69.132, *R. ticinensis* Zone, Late Albian, Caravaca, Spain

6–7 *T. primula* Luterbacher
   6a–c Holotype. Le Maley well at 61 m, Middle Albian, west of Cornaux, Canton of Neuchâtel, Switzerland. **7a–c** Sample Allemann 68.63, *R. ticinensis* Zone, Late Albian, Caravaca, Spain.

8–9 *T. praeticinensis* Sigal
   8a–c Sample Allemann 69.127, *T. primula* Zone, Early Albian, Caravaca, Spain. **9a–c** Holotype. Diego-Suarez well at 184.9 m, *B. breggiensis* Zone, Middle Albian, Mount Raynaud, Madagascar.

10–12 *T. raynaudi* Sigal
   10a–b Sample Allemann 69.132, *R. ticinensis* Zone, Late Albian, Caravaca, Spain. **11a–c** Holotype. Diego-Suarez well at 166.1 m, *R. ticinensis* Zone, Late Albian, Mount Raynaud, Madagascar. **12a–c** DSDP Leg 40-364-26-cc, *R. ticinensis* Zone, Late Albian, Angola Basin, Southeastern Atlantic (from Caron, 1978).

13–15 *T. roberti* (Gandolfi)
   13a–c Holotype, **14, 15a–b** Topotypes. Bed 27, top of Scaglia variegata, *T. primula* Zone, Early Albian, Breggia river, Canton of Tessin, Switzerland.

16–17 *B. breggiensis* (Gandolfi)
   16a–c Holotype, **17a–b** Topotype. Bed 31, base of Scaglia bianca, *R. appenninica* Zone, Late Albian, Breggia river, Canton of Tessin, Switzerland.

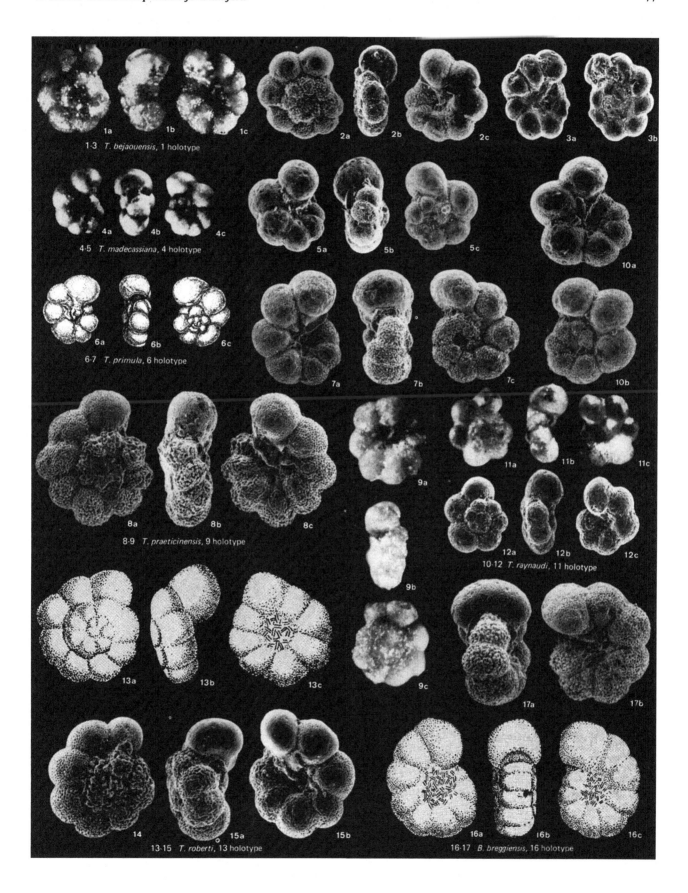

1-3 *T. bejaouensis*, 1 holotype

4-5 *T. madecassiana*, 4 holotype

6-7 *T. primula*, 6 holotype

8-9 *T. praeticinensis*, 9 holotype

10-12 *T. raynaudi*, 11 holotype

13-15 *T. roberti*, 13 holotype

16-17 *B. breggiensis*, 16 holotype

TICINELLA PRAETICINENSIS Sigal
Figures **36.8–9**; 10, 12, 15

Type reference: *Ticinella praeticinensis* Sigal, 1966, pp. 195–6, pl. 2, figs. 3–5.

T. praeticinensis differs from *T. roberti* in the subrounded periphery. Where pustules occur on the early chambers of the last whorl, they gradually decrease in size and number to finally disappear altogether on the ultimate chambers. This species gave rise to *R. subticinensis* by the progressive development of a pseudo-keel.

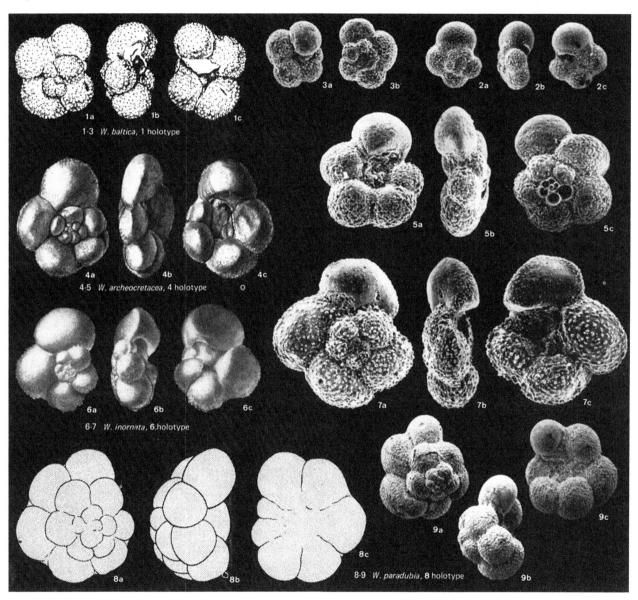

Fig. 37. *Whiteinella* (all figures × 60)

1–3 *W. baltica* Douglas & Rankin
  1a–c 'Holotype. Sample DBS-8, Early Santonian, Bawnodde greensand, southwestern Bornholm Island, Baltic Sea. **2a–c** DSDP Leg 26-258, Coniacian–Santonian, Naturaliste Plateau, Indian Ocean (from Herb, 1974). **3a–b** DSDP Leg 40–364, *D. primitiva* Zone, Angola Basin, Southeastern Atlantic (from Caron, 1978).

4–5 *W. archaeocretacea* Pessagno
  4a–c Holotype. Sample TX-105, *W. archaeocretacea* Zone, Early Turonian, South Bosque Formation, Austin, Travis County, Texas, USA. **5a–c** Bellone well, Northwestern France, *H. helvetica* Zone, Early Turonian (from Robaszynski & Caron, 1979).

6–7 *W. inornata* (Bolli)
  6a–c Holotype. Trinidad Petroleum Development well Moruga 15, core 6802-6827, *D. primitiva* Zone, Coniacian, Naparina Hill Formation, Trinidad. **7a–c** Sample Caron WA-16/9, *H. helvetica* Zone, Early Turonian, South Bosque Formation, Mc Lennan County, Texas, USA.

8–9 *W. paradubia* (Sigal)
  8a–c Holotype. *R. cushmani* Zone, sample probably from northern Algeria. **9a–c** Sample from *R. cushmani* Zone, Late Cenomanian, Colorado group, Greenhorn Formation, Hodgeman County, Kansas, USA.

TICINELLA PRIMULA Luterbacher
Figures **36.6–7**; 10, 12, 15

Type reference: *Ticinella primula* Renz, Luterbacher &
Schneider, 1963, p. 1085, text-fig. 4.

T. primula differs from *T. bejaouensis* in its pseudo-planispiral
aspect and in having fewer chambers.

TICINELLA RAYNAUDI Sigal
Figures **36.10–12**; 10, 12, 15

Type reference: *Ticinella raynaudi* Sigal, 1966, pp. 200–2, pl.
VI, figs. 1 (a–b) – 3(a–b).

T. raynaudi is similar to *T. primula* from which it differs in the
slightly higher trochospire, more lobate periphery and a trend
for the later chambers to become radially elongate, even
digitiform (var. *digitalis* of Sigal, 1966).

TICINELLA ROBERTI (Gandolfi)
Figures **36.13–15**; 10, 12, 15

Type reference: *Anomalina roberti* Gandolfi, 1942, pp. 100–1,
pl. 2, figs. 2a–c.

T. roberti differs from *T. primula* in a higher trochospire and a
roughened surface on the chambers; from *Hedbergella trocoidea* in the presence of umbilical supplementary apertures
and from *T. bejaouensis* in a generally larger test and a higher
trochospire.

Genus WHITEINELLA Pessagno, 1967

WHITEINELLA ARCHAEOCRETACEA Pessagno
Figures **37.4–5**; 10, 12, 15

Type reference: *Whiteinella archaeocretacea* Pessagno, 1967,
pp. 298–9, pl. 51, figs. 2–4.
Synonym: *Praeglobotruncana* ? *gigantea* Lehmann, 1963, p.
140, pl. 2, figs. 4a–c, renamed *Praeglobotruncana
lehmanni* by Porthault, 1969, pp. 538–9, pl. 2, figs. 6a–c
(for reasons of homonymy).

W. archaeocretacea differs from *W. inornata* in lacking an
acute and true imperforate peripheral margin.

WHITEINELLA BALTICA Douglas & Rankin
Figures **37.1–3**; 10, 12, 15

Type reference: *Whiteinella baltica* Douglas & Rankin, 1969,
p. 198, text-figs. 9A–C.
Synonyms: *Rugoglobigerina* ? *alpina* Porthault, 1969, pl. 2,
figs. 2a–c.
Hedbergella bornholmensis Douglas & Rankin, 1969, p.
194, text-figs. 6A–C.

W. baltica differs from *W. archaeocretacea* in having 4 instead
of 5 chambers forming the last whorl, they are more globular
in *W. baltica* than in *W. archaeocretacea*.

WHITEINELLA INORNATA (Bolli)
Figures **37.6–6**; 10, 12, 15

Type reference: *Globotruncana inornata* Bolli, 1957, p. 57, pl.
13, figs. 5a–c.

W. inornata differs from *W. archaeocretacea* in the acute and
imperforate peripheral margin.

WHITEINELLA PARADUBIA (Sigal)
Figures **37.8–9**; 10, 12, 15

Type reference: *Globigerina paradubia* Sigal, 1952, p. 28,
text-Fig. 28.

W. paradubia differs from all other *Whiteinella* species in its
high trochospire and in 6–7 instead of 4–5 chambers forming
the last whorl.

References

Ascoli, P. 1976. Foraminiferal and ostracod biostratigraphy of the
Mesozoic–Cenozoic Scotian Shelf, Atlantic Canada. In:
C. T. Schafer & B. R. Pelletier (eds.), *First International
Symposium on Benthonic Foraminifera of Continental
Margins*, Part B, Paleoecology and Biostratigraphy, pp.
653–743, Halifax, Nova Scotia, Canada.

Aspekte der Kreide Europas. 1978 (published 1979). International
Union of Geological Sciences, ser. A, no. 6, ed. by
J. Wiedmann. E. Schweizerbart'sche Verlagsbuchhandlung,
Stuttgart, 680 pp.

Bailey, H. W. & Hart, M. B. 1979. The correlation of the Early
Senonian in Western Europe using foraminiferida. *Aspekte
der Kreide Europas*, IUGS ser. 1, no. 6, pp. 159–69.

Bang, I. 1971. Planktonic foraminifera of the lowermost Danian.
Proceedings II Planktonic Conference, Roma, 1970, **1**, 17–26.

Bang, I. 1979. Foraminifera from the type section of the *eugubina*
Zone and Cretaceous/Tertiary boundary localities in
Jylland, Denmark. In: P W. K. Christensen & T. Birkelund
(eds.), *Proceedings Cretaceous-Tertiary Boundary Events
Symposium*, **2**, 127–30.

Banner, F. T. 1982. A classification and introduction to the
Globigerinacea. In: F. T. Banner & A. R. Lord (eds.),
Aspects of Micropaleontology (papers presented to Professor
Tom Barnard), pp. 142–239, Allen & Unwin, London.

Banner, F. T. & Blow, W. H. 1959. The classification and
stratigraphical distribution of the Globigerinaceae.
Palaeontology, **2**, 1–27.

Banner, F. T. & Blow, W. H. 1960. Some primary types of species
belonging to the superfamily Globigerinaceae. *Contrib.
Cushman Found. foramin. Res.*, **11**, 1–41.

Barr, F. T. 1962. Upper Cretaceous planktonic foraminifers from
the Isle of Wight, England. *Paleontology*, **4**, 552–80.

Barr, F. T. 1972. Cretaceous biostratigraphy and planktonic
foraminifera of Libya. *Micropaleontology*, **18**, 1–46.

Bartenstein, H., Bettenstaedt, F. & Bolli, H. M. 1966. Die
Foraminiferen der Unterkreide von Trinidad, W.I. Part 2:
Maridale-Formation (Typlokalität). *Eclog. geol. Helv.*, **59**,
129–77.

Bé, A. W. & Tolderlund, D. S. 1971. Distribution and ecology of
living planktonic foraminifera in surface water of the
Atlantic and Indian Oceans. In: B. M. Funnell &
W. R. Riedel (eds.), *The Micropalaeontology of Oceans*, pp.
105–49, Cambridge University Press, Cambridge.

Berggren, W. A. 1962. Some planktonic foraminifera from the Maestrichtian and type Danian stages of Southern Scandinavia. *Stockholm Contrib. Geol.*, **9**, 1–106.

Bermudez, P. J. 1952. Estudio sistematico de los Foraminiferos Rotaliformes. *Bol. Geol. Minist. Minas Venezuela*, **2**, 3–230.

Bignot, G. & Guyader, J. 1971. Observations nouvelles sur *Globigerina oxfordiana* Grigelis. *Proceedings II Planktonic Conference, Roma*, 1970, **1**, 79–83.

Bolli, H. 1945. Zur Stratigraphie der Oberen Kreide in den höheren helvetischen Decken. *Eclog. geol. Helv.*, **37**, 217–328.

Bolli, H. 1951. The genus *Globotruncana* in Trinidad, B.W.I. *J. Paleontol.*, **25**, 187–99.

Bolli, H. M. 1957. The genera *Praeglobotruncana, Rotalipora, Globotruncana*, and *Abathomphalus* in the Upper Cretaceous of Trinidad, B.W.I. In: A. R. Loeblich, Jr, Studies in Foraminifera. *Bull. U.S. natl. Mus.*, **215**, 51–60.

Bolli, H. 1959. Planktonic foraminifera from the Cretaceous of Trinidad, B.W.I. *Bull. Am. Paleontol.*, **39**, 257–77.

Bolli, H. M. 1966. Zonation of Cretaceous to Pliocene marine sediments based on planktonic foraminifera. *Boletin Informativo Asociacion Venezolana de Geologia, Mineria y Petroleo*, **9**, 3–32.

Bolli, H. M., Loeblich, A. R. Jr & Tappan, H. 1957. Planktonic foraminiferal families Hantkeninidae, Orbulinidae, Globorotaliidae and Globotruncanidae. *Bull. U.S. Natl. Mus.*, **215**, 3–50.

Borsetti, A. M. 1962. Foraminiferi planctonici di una serie cretacea dei Dintorni di Piobbico (Prov. di Pesaro). *G. Geol.*, ser. 2, **29**, 19–75.

Brönnimann, P. 1952. Globigerinidae from the upper Cretaceous (Cenomanian–Maestrichtian) of Trinidad, B.W.I. *Bull. Am. Paleontol.*, **34**, 5–71.

Brönnimann, P. & Brown, N. 1953. Observations on some planktonic Heterohelicidae from the Upper Cretaceous of Cuba. *Contrib. Cushman Found. foramin. Res.*, **4**, 150–6.

Brönnimann, P. & Brown, N. K., Jr 1956. Taxonomy of the Globotruncanidae. *Eclog. geol. Helv.*, **48**, 503–61.

Brönnimann, P. & Brown, N. K., Jr 1958. *Hedbergella*, a new name for a Cretaceous planktonic foraminiferal genus. *J. Washington Acad. Sci.*, **48**, 15–17.

Brotzen, F. 1934. Foraminiferen aus dem Senon Palästinas. *Zeitschrift des deutschen Vereins zur Erforschung Palästinas*, **57**, 28–72.

Brotzen, F. 1936. Foraminiferen aus dem schwedischen untersten Senon von Eriksdal in Schonen. *Sver. geol. Unders.*, **30**, 1–206.

Brotzen, F. 1942. Die Foraminiferengattung *Gavelinella* nov. gen. und die systematik der Rotaliiformes. *Sver. geol. Unders.*, **36**, 5–61.

Brown, N. K. 1969. Heterohelicidae Cushman, 1937, amended, a Cretaceous planktonic foraminiferal family. *Proceedings First International Conference on Planktonic Microfossils, Geneva, 1967*, **2**, 21–67.

Bukry, D., Douglas, R. G., Kling, S. A. & Krasheninnikov, V. 1971. Planktonic microfossil biostratigraphy of the Northwestern Pacific Ocean. *Initial Rep. Deep Sea drill. Proj.*, **6**, 1253–1300.

Butt, A. 1979. Lower Cretaceous foraminiferal biostratigraphy, paleoecology, and depositional environment at DSDP Site 397, Leg 47A. *Initial Rep. Deep Sea drill. Proj.*, **47**, 257–71.

Carman, K. W. 1929. Some foraminifera from the Niobrara and Benton formations of Wyoming. *J. Paleontol.*, **3**, 309–15.

Caron, M. 1966. Globotruncanidae du Crétacé supérieur du synclinal de la Gruyère (Préalpes médianes, Suisse). *Rev. Micropaleontol.*, **2**, 68–93.

Caron, M. 1971. Quelques cas d'instabilité des caractères génériques chez les foraminifères de l'Albien. *Proceedings II Planktonic Conference, Roma*, 1970, **7**, 145–57.

Caron, M. 1972. Planktonic foraminifera from the Upper Cretaceous of Site 98, Leg 11, Deep Sea Drilling Project. *Initial Rep. Deep Sea drill. Proj.*, **11**, 551–9.

Caron, M. 1974. Sur la validité de quelques espèces de *Globotruncana* du Turonien et du Coniacien. *Actes du VIe Colloque Africain de Micropaleontologie, Tunis*, **28**, 329–45.

Caron, M. 1975. Late Cretaceous planktonic foraminifera from the Northwestern Pacific: Leg 32 of the Deep Sea Drilling Project. *Initial Rep. Deep Sea drill. Proj.*, **32**, 719–24.

Caron, M. 1976. Révision des types de Foraminifères planctoniques décrits dans la région du Montsalvens (Préalpes fribourgeoises). *Eclog. geol. Helv.*, **69**, 327–33.

Caron, M. 1978. Cretaceous planktonic foraminifers from DSDP Leg 40, Southeastern Atlantic Ocean. *Initial Rep. Deep Sea drill. Proj.*, **40**, 651–78.

Caron, M. 1981. Un nouveau genre de foraminifère planctonique du Crétacé: *Falsotruncana* nov. gen. *Eclog. geol. Helv.*, **74**, 65–73.

Caron, M. 1983a. Taxonomie et phylogénie de la famille des Globotruncanidae. *2nd Kreide Symposium, Munchen, 1982, Zitteliana, München*, **10**, 677–81.

Caron, M. 1983b. La spéciation chez les foraminifères planctiques: une réponse adaptée aux contraintes de l'environnement. In: *2nd Kreide Symposium, Munchen, 1982, Zitteliana, München*, **10**, 671–6.

Caron, M. & Luterbacher, H. 1969. On some type specimens of Cretaceous planktonic foraminifera. *Contrib. Cushman Found. foramin. Res.*, **20**, 23–9.

Carpenter, W., Parker, W. & Jones, T. 1862. *Introduction to the Study of the Foraminifera*. Ray Society Publications. 319 pp., 22 pl.

Carsey, D. O. 1926. Foraminifers of the Cretaceous of Central Texas. *Texas University Bulletin*, **2612**, 1–56.

Carter, D. J. & Hart, M. B. 1977. Aspects of Mid-Cretaceous stratigraphical micropaleontology. *Bull. Br. Mus. nat. Hist.*, ser. Geol., **29**, 1–135.

Chevalier, J. 1961. Quelques nouvelles espèces de Foraminifères dans le Crétacé inférieur méditerranéen. *Rev. Micropaleontol.*, **4**, 33–6.

Cita, M. B. 1948. Ricerche stratigrafiche e micropaleontologiche sul Cretacico e sull'Eocene di Tignale (Lago di Garda). *Riv. Ital. Paleontol. Stratigr.*, **54**, no. 2, 49–64; no. 3, 117–33; no. 4, 143–68.

Cita, M. B. 1963. Tendance évolutive des foraminifères planctiques (Globotruncanae) du Crétacé supérieur. In: G. H. R. von Koenigswald *et al.* (eds.), *Evolutionary Trends in Foraminifera*, pp. 112–38, Elsevier Publishing Co., Amsterdam.

Colloque sur le Cénomanien, Paris, 6–7 septembre 1976 (published 1978). *Geologie Mediterraneenne*, **5** (1), 3–224.

Colloque sur le Turonien, Paris, 26–27 octobre 1981. *Mém. Mus. nat Hist. nat.*, new ser., **49**, 15–241.

Corminboeuf, P. 1961. Tests isolés de *Globotruncana mayaroensis* Bolli, *Rugoglobigerina, Trinitella* et Heterohelicidae dans le Maestrichtien des Alpettes. *Eclog. geol. Helv.*, **54**, 107–22.

Cushman, J. A. 1926. Some foraminifera from the Mendez

Shale of Eastern Mexico. *Contrib. Cushman Lab. foramin. Res.*, **2**, 16–28.

Cushman, J. A. 1927. An outline of a reclassification of the foraminifera. Contrib. Cushman Lab. foramin. Res., 3, 1–105.

Cushman, J. A. 1931. *Hastigerinella* and other interesting foraminifera from the Upper Cretaceous of Texas. *Contrib. Cushman Lab. foramin. Res.*, **7**, 83–9.

Cushman, J. A. 1933. Some new foraminiferal genera. *Contrib. Cushman Lab. foramin. Res.*, **9**, 32–7.

Cushman, J. A. 1938. Some new species of rotaliform foraminifera from the American Cretaceous. *Contrib. Cushman Lab. foramin. Res.*, **14**, 66–71.

Cushman, J. A. 1946. Upper Cretaceous foraminifera of the Gulf Coastal Region of the United States and adjacent areas. *Prof. Pap. U.S. geol. Surv.*, **206**.

Cushman, J. A. & Ten Dam, A. 1948. *Globigerinelloides*, a new genus of the Globigerinidae. *Contrib. Cushman Lab. foramin. Res.*, **24**, 42–3.

Cushman, J. A. & Wickenden, R. T. 1930. The development of *Hantkenina* in the Cretaceous with a description of a new species. *Contrib. Cushman Lab. foramin. Res.*, **6**, 39–43.

Dailey, D. H. 1973. Early Cretaceous foraminifera from the Budden Canyon Formation, Northwestern Sacramento Valley, California. *Univ. California Publ. Geol. Sci.*, **106**, 1–111.

Dalbiez, F. 1955. The genus *Globotruncana* in Tunisia. *Micropaleontology*, **1**, 161–71.

Douglas, R. G. 1969. Upper Cretaceous planktonic foraminifera in northern California. Part 1 – Systematics. *Micropaleontology*, **15**, 151–209.

Douglas, R. G. 1971. Cretaceous foraminifera from the Northwestern Pacific Ocean: Leg 6, Deep Sea Drilling Project. *Initial Rep. Deep Sea drill. Proj.*, **6**, 1027–53.

Douglas, R. G. 1972. Paleozoogeography of Late Cretaceous planktonic foraminifera in North America. *J. foramin. Res.* **2**, 14–34.

Douglas, R. G. 1973. Planktonic foraminiferal biostratigraphy in the central North Pacific Ocean. *Initial Rep. Deep Sea drill. Proj.*, **17**, 673–94.

Douglas, R. G. & Rankin, C. 1969. Cretaceous planktonic foraminifera from Bornholm and their zoogeographic significance. *Lethaia*, **2**, 185–217.

Douglas, R. & Sliter, W. V. 1966. Regional distribution of some Cretaceous Rotaliporidae and Globotruncanidae (Foraminiferida) within North America. *Tulane Stud. Geol.*, **4**, 89–130.

Drushtchitz, V. & Gorbachik, T. 1979. Zonengliederung der Unteren Kreide der südlichen UdSSR. The zonal concept of the Lower Cretaceous of Southern USSR, based on ammonites and foraminifera. *Aspekte der Kreide Europas*, IUGS ser. A, no. 6, pp. 107–16.

Dupeuble, P. A. 1979. Mesozoic foraminifers and microfacies from holes 400A, 401 and 402A of the DSDP Leg 48. *Initial Rep. Deep Sea drill. Proj.*, **48**, 451–73.

Egger, J. G. 1899. Foraminiferen und Ostrakoden aus den Kreidemergeln der oberbayerischen Alpen. *Abh. bayer. Akad. Wiss. Munchen, Math.-Naturw. Kl.*, **21**, 1–230.

Ehrenberg, C. G. 1840. Über die Bildung der Kreidefelsen und des Kreidemergels durch unsichtbare Organismen. *Abh. K. Akad. Wiss. Berlin, Physik.*, Berlin, Deutschland (1838), p. 135.

Ehrenberg, C. G. 1843. Verbreitung und Einfluss des mikroskopischen Lebens in Sid- und Nord-Amerika. *Abhandlungen der K. preuss. Akademie der Wissenschaften Phys.-math. Kl.* (1841), **1**, 291–446.

Ehrenberg, C. G. 1844. Eine Mitteilung über zwei neue Lager von Gebirgsmassen aus Infusorien als Meeres-Absatz in Nord-Amerika und eine Vergleichung desselben mit den organischen Kreide-Gebilden in Europa und Afrika. *Bericht der K. preuss. Akademie der Wissenschaften*, Berlin, 57–8.

Eicher, D. L. 1969. Cenomanian and Turonian planktonic foraminifera from the Western Interior of the United States. *Proceedings First International Conference on Planktonic Microfossils, Geneva, 1967*, **2**, 163–74.

Eicher, D. 1972. Phylogeny of the Late Cenomanian foraminifer *Anaticinella multiloculata* (Morrow). *J. foramin. Res.*, **2**, 184–90.

Eicher, D. L. & Worstell, P. 1970. Cenomanian and Turonian foraminifera from the Great Plains, United States. *Micropaleontology*, **16**, 269–324.

El-Naggar, Z. R. 1966. Stratigraphy and planktonic foraminifera of the Upper Cretaceous – Lower Tertiary succession in the Esna-Idfu region, Nile Valley, Egypt, U.A.R. *Bull. Br. Mus. nat. Hist.*, ser. Geol., **2**, 1–291.

El-Naggar, Z. R. 1970. The genus *Rugoglobigerina* in the Maastrichtian Sharawna shale of Egypt. *Proceedings II Planktonic Conference, Roma, 1970*, **1**, 477–537.

Fleury, J. J. 1980. Les zones de Gavrovo-Tripolitza et du Pinde-Olonos (Grèce continentale et Péloponnèse du Nord). Evolution d'une plate-forme et d'un bassin dans leur cadre alpin. *Societe Geologique du Nord*, **4**, 1–648.

Frerichs, N. E. 1974. A reexamination of the holotype of *Globigerina subdigitata* Carman. *J. foramin. Res.*, **4**, 109–11.

Fuchs, W. 1973. Ein Beitrag zur Kenntnis der Jura-Globigerinen und verwandten Formen an Hand polnischen Materials der Callovien und Oxfordien. *Verh. geol. Bundesanst.*, **3**, 445–87.

Gandolfi, R. 1942. Ricerche micropaleontologiche e stratigrafiche sulla Scaglia e sul flysch cretacici dei Dintorni di Balerna (Canton Ticino). *Riv. Ital. Paleontol.*, **48**, 1–160.

Gandolfi, R. 1955. The genus *Globotruncana* in Northeastern Colombia. *Bull. Am. Paleontol.*, **36**, 7–118.

Gandolfi, R. 1956. Notes on some species of *Globotruncana*. *Contrib. Cushman Found. foramin. Res.*, **8**, 59–65.

Glaessner, M. F. 1937. Planktonforaminiferen aus der Kreide und dem Eozän und ihre stratigraphische Bedeutung. *Moscow Univ. Paleontol. Lab., Etyusy Mikropaleontol.*, **1**, 27–46.

Gorbachik, T. N. 1971a. On Early Cretaceous foraminifera of the Crimea. *Voprosy Mikropaleontol.*, **14**, 125–216.

Gorbachik, T. N. 1971b. A brief characteristic of Cretaceous and Paleogene deposits of the Mountain Crimea. *XII European Micropaleontological Colloquium*, 13–28.

Gradstein, F. M. 1978. Biostratigraphy of Lower Cretaceous Blake Nose and Blacke Bahama basin foraminifera DSDP Leg 44, Western North Atlantic Ocean. *Initial Rep. Deep Sea drill. Proj.*, **44**, 663–701.

Gradstein, F. M., Bukry, D., Habib, D., Renz, O., Roth, P. H., Schmidt, R. R., Weaver, F. M. & Wind, F. H. 1978. Biostratigraphic summary of DSDP Leg 44: Western North Atlantic Ocean. *Initial Rep. Deep Sea drill. Proj.*, **44**, 657–62.

Grigelis, A. 1958. *Globigerina oxfordiana*, sp. n. – occurrence of Globigerines in the Upper Jurassic deposits of Lithuania. *Nauk Dokl. Vyssh. Shk., Geol.-Geogr. Nauki*, Moscow, **3**, 109–11.

Grigelis, A. 1974. On the Jurassic stage of the development of plankton foraminifera. *Transactions of the Academy of Science of the USSR*, **219**, 1203–5.

Grigelis, A. & Gorbachik, T. 1980. Morphology and taxonomy of Jurassic and early Cretaceous representatives of the superfamily Globigerinacea (Favusellidae). *J. foramin. Res.*, **10**, 180–90.

Guillaume, H. A., Bolli, H. M. & Beckmann, J. P. 1972. Estratigrafia del Cretaceo inferior en la Serrannia del Interior, oriente de Venezuela. Memoria IV Ccongreso Geologico Venezolano, vol. III, *Bol. Geol. Minist. Minas Venezuela*, **5**, 1619–55.

Hamilton, E. L. 1953. Upper Cretaceous, Tertiary, and recent planktonic foraminifera from Mid-Pacific flat-topped seamounts. *J. Paleontol.*, **27**, 204–37.

Hanzlikova, E. 1963. *Globotruncana helvetica posthelvetica* n. subsp. from the Carpathian Cretaceous. *Vestnik Cestnedniko Ustava Geologickecho Rocnik*, **38**, 325–7.

Hart, M. B. 1979. Biostratigraphy and palaeozoogeography of planktonic Foraminiferida from the Cenomanian of Bornholm, Denmark. *Newsletters on Stratigraphy*, **8**, 83–96.

Hart, M. B. 1980. A water depth model for the evolution of the planktonic Foraminiferida. *Nature*, **286**, 252–4.

Hart, M. B. & Bailey, H. W. 1979. The distribution of planktonic Foraminiferida in the Mid-Cretaceous of NW Europe. *Aspekte der Kreide Europas*, IUGS ser. A, no. 6, pp. 527–42.

Herb, R. 1974. Cretaceous planktonic foraminifera from the Eastern Indian Ocean. *Initial Rep. Deep Sea drill. Proj.*, **26**, 745–55.

Herm, D. 1962. Stratigraphische und miktopaläontologische Untersuchungen der Oberkreide im Lattengebirge und im Nierental. *Abh. bayer. Akad. Wiss., Munchen*, new ser., **104**, 1–119.

Hermes, J. J. 1969. Late Albian foraminifera from the Subbetic of Southern Spain. *Geol. Mijnbouw*, **48**, 35–66.

Hofker, J. 1956. Die Globotruncanen von Nordwest-Deutschland und Holland. *Neues Jahrb. Geol. Paleontol. Abhandlungen*, **103**, 312–40.

Hofker, J. 1978. Analysis of a large succession of samples through the Upper Maastrichtian and the Lower Tertiary of Drill Hole 47.2, Shatsky Rise, Pacific, Deep Sea Drilling Project. *J. foramin. Res.*, **8**, 46–75.

Jacob, K. & Sastry, M. V. A. 1950. On the occurrence of *Globotruncana* in Uttatur stage of the Trichinopoly Cretaceous, South India. *Science and Culture, Calcutta*, **16**, 267.

Jirova, D. 1956. The genus *Globotruncana* in Upper Turonian and Emscherian of Bohemia. *Universitas Carolina Geologica*, **2**, 239–55.

Keller, B. M. 1935. Microfauna of the Upper Cretaceous of the Dnieper-Donetz basin and adjoining areas. *Moskov Obshch. Ispytaley Prirody Byull. Otdel Geol.*, **13**, 522–58.

Klasz, I. de. 1955. A new *Globotruncana* from the Bavarian Alps and North Africa. *Contrib. Cushman Found. foramin. Res.*, **6** (1), 43–4.

Klaus, J. 1959. Le 'Complexe schisteux intermédiaire' dans le synclinal de la Gruyère (Préalpes médianes). Stratigraphie et micropaléontologie, avec l'étude spéciale des Globotruncanidés de l'Albien, du Cénomanien et du Turonien. *Eclog. geol. Helv.*, **52**, 753–851.

Klaus, J. 1960. Etude biométrique et statistique de quelques espèces de Globotruncanidés 1. Les espèces du genre *Praeglobotruncana* dans le Cénomanien de la Breggia. *Eclog. geol. Helv.*, **53**, 285–308.

Krasheninnikov, V. A. & Hoskins, R. H. 1973. Late Cretaceous, Paleogene and Neogene planktonic Foraminifera. *Initial Rep. Deep Sea drill. Proj.*, **20**, 105–203.

Kuhry, B. 1970. Some observations on the type material of *Globotruncana elevata* (Brotzen) and *Globotruncana concavata* 'Brotzen'. *Rev. Esp. Micropaleontol.*, **2**, 291–304.

Kuhry, B. 1972. Stratigraphy and micropaleontology of the Lower Cretaceous in the subbetic south of Caravaca (Province de Murcia, SE Spain). *Proc. Koninklijke Nederlandse Akademie van Wetenschappen*, ser. B, **75**, 193–222.

Lambert, G. 1974. Zululand Cretaceous foraminifera and the Tethys. *Actes du VIe Colloque Africain de Micropaleontologie, Tunis*, **28**, 13–27.

Lamolda, M. A. 1975. Helvetoglobotruncaninae subfam. nov. y consideraciones sobre los globigeriniformes del Cretacico. *Rev. Esp. Micropaleontol.*, **8**, 395–400.

Lapparent, J. de. 1918. *Etude lithologique des terrains crétacés de la région d'Hendaye. Mémoire pour servir à l'explication de la carte géologique de la France.* 115 pp.

Lehmann, R. 1963. Etude des Globotruncanidés du Crétacé supérieur de la province de Tarfaya (Maroc occidental). *Notes Mem. Serv. geol. Maroc*, **21**, 133–79.

Lehmann, R. 1966. Les foraminifères pélagiques du Crétacé du Bassin côtier de Tarfaya. 1. Planomalinidae et Globotruncanidae du Sondage de Puer Cansado (Albien supérieur, Cénomanien inférieur). *Notes Mem. Serv. geol. Maroc*, no. 175, *Paleontologie*, 153–67.

Linares-Rodriguez, D. 1977. Foraminiferos planctonicos del Cretacico superior de las Cordilleras Beticas (Sector Central). Thesis, Université de Malaga, 410 pp. (unpublished).

Loeblich, A. R. 1951. Coiling in the Heterohelicidae. *Contrib. Cushman Found. foramin. Res.*, **2**, 106–10.

Loeblich, A. R. & Tappan, H. 1946. New Washita foraminifera. *J. Paleontol.*, **20**, 238–58.

Loeblich, A. R., Jr - Tappan, H. 1961. Cretaceous planktonic foraminifera: Part I – Cenomanian. *Micropaleontology*, **7**, 257–304.

Loeblich, A. R., Jr & Tappan, H. 1964a. Foraminiferida. In: R. C. Moore, *Treatise on Invertebrate Paleontology*, C. Protista, vols. 1 & 2, pp. 1–868. The Geological Society of America and the University of Kansas Press.

Loeblich, A. R., Jr & Tappan, H. 1964b. Foraminiferal classification and evolution. *J. geol. Soc. India*, **5**, 6–40.

Longoria, J. F. 1970. Estudio en seccion delgada de algunas especies del genero *Globotruncana* Cushman del este de Mexico. *Instituto Mexicano del Petroleo*, **70**, 1–135.

Longoria, J. F. 1973. *Pseudoticinella*, a new genus of planktonic foraminifera from the Early Turonian of Texas. *Rev. Esp. Micropaleontol.*, **5**, 417–23.

Longoria, J. F. 1974. Stratigraphic, morphologic and taxonomic studies of Aptian planktonic foraminifera. *Rev. Esp. Micropaleontol.*, numero. extraord., 1–107.

Longoria, J. F. & Gamper, M. A. 1974. Albian planktonic foraminifera from the Sabinas Basin of Northern Mexico. *Actes du VIe Colloque Africain de Micropaleontologie, Tunis*, **28**, 39–71.

Longoria, J. F. & Gamper, M. A. 1975. Classification and

evolution of Cretaceous planktonic foraminifera. Part I: The super-family Hedbergelloidea. *Rev. Esp. Micropaleontol.*, numero especial, 61–96.

Luterbacher, H. P. 1972. Foraminifera from the Lower Cretaceous and Upper Jurassic of the Northwestern Atlantic. *Initial Rep. Deep Sea drill. Proj.*, **11**, 561–93.

Luterbacher, H. P. 1975. Early Cretaceous foraminifera from the Northwestern Pacific: Leg 32 of the Deep Sea Drilling Project. *Initial Rep. Deep Sea drill. Proj.*, **32**, 703–18.

Luterbacher, H. P. & Premoli-Silva, I. 1962. Note préliminaire sur une révision du profil de Gubbio. *Riv. Ital. Paleontol.*, **68**, 253–88.

Luterbacher, H. & Premoli-Silva, I. 1964. Biostratigrafia del limite Cretaceo–Terziario nell'Appennino Centrale. *Riv. Ital. Paleontol.*, **70**, 67–128.

McNulty, C. L. 1976. Cretaceous foraminiferal stratigraphy, DSDP Leg 33, Holes 315A, 316, 317A. *Initial Rep. Deep Sea drill. Proj.*, **33**, 369–81.

McNulty, C. L. 1979. Smaller Cretaceous foraminifers of Leg 43, Deep Sea Drilling Project. *Initial Rep. Deep Sea drill. Proj.*, **43**, 487–505.

Magné, J. & Sigal, J. In: Cheylan, G., Magné, J., Sigal, J. & Grekoff, N. 1954. Résultats géologiques et micropaléontologiques du sondage d'El Krachem (Hauts Plateaux algérois), Description de quelques espèces nouvelles. *Bull. Soc. geol. Fr.*, ser. 6, **3**, 471–88.

Maiya, S. & Takayanagi, Y. 1977. Cretaceous foraminiferal biostratigraphy of Hokkaido. *Spec. Pap. palaeontol. Soc. Japan*, **21**, 41–51.

Marianos, A. W. & Zingula, R. P. 1966. Cretaceous planktonic foraminifers from Dry Creek, Tehama County, California. *J. Paleontol.*, **40**, 328–42.

Marie, P. 1938. Zones à Foraminifères de l'Aturien dans la Mésogée. *C.r. Somm. Soc. geol. Fr.*, séance du 5.12.1938, 341–2.

Marie, P. 1941. Les Foraminifères de la Craie à *Belemnitella mucronata* du Bassin de Paris. *Mem. Mus. natl. Hist. nat. Paris*, new series, **12**, pp. 237, 256, 258.

Marie, P. 1948. A propos de *Rosalinella cushmani* (Morrow). *Bull. Soc. geol. Fr.*, **5** (18), 39–42.

Marks, P. 1972. Late Cretaceous planktonic foraminifera from Prebetic tectonic elements near Jaen (Southern Spain). *Rev. Esp. Micropaleontol.*, numero extraord., 99–123.

Martin, L. 1964 Upper Cretaceous and Lower Tertiary foraminifera from Fresno County, California. *Jahrbuch Austria Geol. Bundesanstalt*, special volume, **9**, 1–128, pls. 1–16.

Maslakova, N. I. 1964. Contribution to the systematics and phylogeny of the Globotruncanids. *Voprosy Mikropaleontol.*, **8**, 102–17.

Maslakova, N. I. 1978. Globotruncanidae du S de la partie européene de l'URSS. *Academie des Sciences URSS, Editions Sciences, Moscou*, 1–164.

Maslakova, N. I. 1978. Globotruncanidae du S de la partie européene de l'URSS. Academie des Sciences URSS, Editions Sciences, Moscou, 1–164.

Masters, B. A. 1976. Planktic foraminifera from the Upper Cretaceous Selma Group, Alabama. *J. Paleontol.*, **50** (2), 318–30.

Masters, B. A. 1977a. The neotype of *Globigerina cretacea* var. *delrioensis* Carsey. *J. Paleontol.*, **51**, 643.

Masters, B. A. 1977b. Mesozoic planktonic foraminifera. A world-wide review and analysis. In: A. T. S. Ramsay, *Oceanic Micropaleontology*, pp. 301–731, Academic Press, London.

Michael, F. Y. 1973. Planktonic foraminifera from the Comanchean series (Cretaceous) of Texas. *J. foramin. Res.*, **2**, 200–20.

Montarro-Galitelli, E. 1957. A revision of the foraminiferal family Heterohelicidae. *Bull. U.S. Natl Mus.*, **215**, 133–54.

Mornod, L. 1949–50. Les Globorotalidés du Crétacé supérieur du Montsalvens (Préalpes fribourgeoises). *Eclog. geol. Helv.*, **42**, 573–96.

Morozova, V. G. & Moskalenko, T. A. 1961. Planktonic foraminifers of the Bajocian and Bathonian boundary layers of Central Dagestan (Northeastern Caucasus). *Voprosy Mikropoleontol.*, no. 5, 3–30 (in Russian).

Morrow, A. L. 1934. Foraminifera and Ostracoda from the Upper Cretaceous of Kansas. *J. Paleontol.*, **8**, 186–205.

Moullade, M. 1961. Quelques foraminifères et ostracodes nouveaux du Crétacé inférieur vocontien. *Rev. Micropaleontol.*, **3**, 213–16.

Moullade, M. 1966. Etude stratigraphique et micropaléontologique du Crétacé inférieur de la 'Fosse vocontienne'. *Doc. Lab. Geol. Fac. Sci. Lyon*, **15**, 1–369.

Moullade, M. 1974. Zones de Foraminifères do Crétacé inférieur mésogéen. *C.r. Seances Acad. Sci., Paris*, ser. D, **278**, 1813–16.

Nakkady, S. E. 1950. A new foraminiferal fauna from the Esna shales and Upper Cretaceous chalk of Egypt. *J. Paleontol.*, **24**, 675–92.

Olsson, R. 1960. Foraminifera of latest Cretaceous and earliest Tertiary age of the New Jersey Coastal Plain. *J. Paleontol.*, **34**, 1–58.

Olsson, R. 1964. Late Cretaceous planktonic foraminifera from New Jersey and Delaware. *Micropaleontology*, **10**, 157–88.

d'Orbigny, A. 1839. Foraminifères. In: R. de la Sagra (ed.), *Histoire physique, politique et naturelles de l'île de Cuba*, Bertrand, Paris, 224 pp.

d'Orbigny, A. 1840. Mémoire sur les foraminifères de la craie blanche du bassin de Paris. *Mem. Soc. geol. Fr.*, **4**, 1–51.

Pessagno, E. A. 1960. Stratigraphy and micropaleontology of the Cretaceous and lower Tertiary of Puerto Rico. *Micropaleontology*, **6**, 87–110.

Pessagno, E. A., Jr 1967. Upper Cretaceous planktonic foraminifera from the Western Gulf Coastal Plain. *Palaeontogr. Am.*, **5**, 259–441.

Pessagno, E. A., Jr & Longoria, J. F. 1973a. Shore laboratory report on Mesozoic planktonic foraminifera, DSDP Leg 16. *Initial Rep. Deep Sea drill. Proj.*, **16**, 893.

Pessagno, E. A., Jr & Longoria, J. F. 1973b. Shore laboratory report on Mesozoic Foraminiferida, Leg 17. *Initial Rep. Deep Sea drill. Proj.*, **17**, 891–4.

Pflaumann, U. & Cepek, P. 1982. Cretaceous foraminiferal and nannoplankton biostratigraphy and paleoecology along the West African continental margin. In: U. von Rad, K. Hinz, M. Sarnthein & E. Seibold (eds.), *Geology of the Northwest African Continental Margin*, pp. 309–53, Springer Verlag, Berlin, Heidelberg.

Pflaumann, U. & Krasheninnikov, V. A. 1977. Early Cretaceous planktonic foraminifers from Eastern North Atlantic, DSDP Leg 41. *Initial Rep. Deep Sea drill. Proj.*, **41**, 539–64.

Plummer, H. J. 1926. Foraminifera of the Midway formation in Texas. *University of Texas Bulletin*, **2644**, 1–206.

Plummer, H. J. 1931. Some Cretaceous Foraminifera in Texas. *University of Texas Bulletin*, **3101**, 109–203.

Porthault, B. 1969. Foraminifères planctoniques et biostratigraphie du Cénomanien dans le Sud-Est de la France. *Proceedings First International Conference on Planktonic Microfossils, Geneva, 1967*, **2**, 526–46.

Porthault, B. 1970. In: P. Donze, B. Porthault & O. de Villoutreys, Le Sénonien inférieur de Puget-Theniers (Alpes-Maritimes) et sa microfaune. *Geobios*, **3**, 41–106.

Porthault, B. 1974. Le Crétacé supérieur de la 'Fosse vocontienne' et des régions limitrophes (France, Sud-Est). Thesis, Université Claude Bernard, Lyon, 342 pp. (unpublished).

Porthault, B. 1978a. Foraminifères caractéristiques du Cénomanien à faciès pélagique dans le Sud-Est de la France. *Geologie Mediterraneenne*, **1**, 183–94.

Porthault, B. 1978b. Conclusions générales: III. Rapport sur les corrélations. *Geologie Mediterraneene*, **5**, 206 and 213.

Postuma, J. 1971. *Manual of Planktonic Foraminifera*. Elsevier Publishing Co., Amsterdam, 420 pp.

Pozaryska, K. & Peryt, D. 1979. The Late Cretaceous and Early Paleocene foraminiferal 'Transitional Province' in Poland. *Aspekte der Kreide Europas*, IUGS ser. A, **6**, 293–303.

Premoli-Silva, I. & Boersma, A. 1977. Cretaceous planktonic foraminifers – DSDP Leg 39 (South Atlantic). *Initial Rep. Deep Sea drill. Proj.*, **39**, 615–31.

Premoli-Silva, I. & Bolli, H. 1973. Late Cretaceous to Eocene planktonic foraminifera and stratigraphy of Leg 15 sites in the Caribbean Sea. *Initial Rep. Deep Sea drill. Proj.*, **15**, 499–547.

Premoli-Silva, I. & Luterbacher, H. P. 1966. The Cretaceous–Tertiary boundary in the Southern Alps (Italy). *Riv. Ital. Paleontol.*, **72**, 1183–1266.

Premoli-Silva, I. & Sliter, W. V. 1981. Cretaceous planktonic foraminifers from the Nauru Basin, Leg 61, Site 462, Western equatorial Pacific. *Initial Rep. Deep Sea drill. Proj.*, **61**, 423–37.

Price, R. J. 1977. The stratigraphical zonation of the Albian sediments of northwest Europe, as based on foraminifera. *Proc. Geol. Assoc. London*, **88**, 65–91.

Quereau, E. C. 1893. Die Klippenregion von Iberg (Sihltal). *Beitr. geol. Karte Schweiz*, **3**, 1–158.

Reichel, M. 1950. Observations sur les *Globotruncana* du gisement de la Breggia (Tessin). *Eclog. geol. Helv.*, **42**, 596–617.

Reiss, Z. 1957. The *Bilamellidea*, nov. superfam. and remarks on Cretaceous globorotaliids. *Contrib. Cushman Found. foramin. Res.*, **8**, 127–52.

Renz, O. 1936. Stratigraphie und mikropalaeontologische Untersuchung der Scaglia (Obere Kreide-Tertiär) im zentralen Apennin. *Eclog. geol. Helv.*, **29**, 1–149.

Renz, O., Luterbacher, H. & Schneider, A. 1963. Stratigraphisch-paläontologische Untersuchungen im Albien und Cénomanien der Neuenburger Jura. *Eclog. geol. Helv.*, **56**, 1073–1116.

Reuss, A. E. 1845. *Die Versteinerungen der böhmischen Kreideformation*, vol. 1, 58 pp., Verlagsbuchhandlung E. Schweizerbart'sche, Stuttgart.

Reuss, A. E. 1854. Beiträge zur Charakteritik der Kreide-Schichten in den Ostalpen, besonders in Gosauthale und aus Wolfgangsee. *Sitzaber. Kon. Akad. Wissens. Wien, Math.-Naturw. Kl., Denkschrift*, **7**, 1–156.

Risch, H. 1969. Stratigraphie der Höheren Unterkreide der Bayerische Kalkalpen mit Hilfe von Mikrofossilien. Thesis, Université Munich, 180 pp. (unpublished).

Robaszynski, F. 1979. Comparison between the Middle Cretaceous of Belgium and some French regions. *Aspekte der Kreide Europas*, IUGS ser. A, no. 6, pp. 543–61.

Robaszynski, F. & Amédro, F. 1980. Synthèse biostratigraphique de l'Aptien au Santonien du Boulonnais à partir de sept groupes paléontologiques: foraminifères, nannoplancton, dinoflagellés et macrofaunes. *Rev. Micropaleontol.*, **22**, 195–321.

Robaszynski, F. & Caron, M. (coordinators). 1979. Atlas de Foraminifères planctoniques du Crétacé moyen. Parts 1–2. *Cah. Micropaleontol.*, **1** and **2**, 1–185 and 1–181.

Robaszynski, F., Caron, M., Gonzalez, J. M. & Wonders, A. 1984. Atlas of Late Cretaceous planktonic foraminifera. *Rev. Micropaleontol.*, **26**, fasc. 3–4, 145–305.

Rzehak, A. 1891. Die Foraminiferenfauna der alttertiären Ablagerungen von Brudendorf in Niederösterreich mit Berücksichtigung des angeglicken Kreidevorkommens von Leitzerdorf. *Annalen des K.K. naturhistorischen Hofmuseums*, **6**, 1–12.

Rzehak, A. 1895. Ueber einige merkwürdige Foraminiferen aus dem österreichischen Tertiär. *Annalen des K.K. naturhistorischen Hofmuseums*, **10**, 213–30.

Saint-Marc, P. 1974. Biostratigraphie de l'Albien, du Cénomanien et du Turonien du Liban. *Actes du VIe Colloque Africain de Micropaléontologie, Tunis*, **28**, 111–18.

Salaj, J. & Gasparikova, V. 1979. Microbiostratigraphy of the Upper Cretaceous of the West Carpathians based on foraminifers and nannofossils and the question of relations and migrations of Boreal and Tethyan elements. *Aspekte der Kreide Europas*, IUGS ser. A, no. 6, pp. 279–92.

Salaj, J. & Samuel, O. 1966. Foraminifera der Westkarpaten – Kreide. *Geologicky ustav dionyza stura*, Bratislava, 1–291.

Schacko, G. 1897. Beitrag über Foraminiferen aus der Cenoman-Kreide von Moltzow in Mecklenburg. *Ver. Freunde Naturg. Meckelenburg Archiv*, **50** (1896), 161–8.

Scheibnerova, V. 1962. Stratigraphy of the Middle and Upper Cretaceous of the Mediterranean province on the basis of the Globotruncanids. *Geologicky Sbornik, Bratislava*, **13**, 197–226.

Scheibnerova, V. 1963. Some new Foraminifera from the Middle Turonian of the Klippen Belt of West Carpathians in Slovakia. *Geologicky, Sbornik, Bratislava*, **14**, 139–43.

Scheibnerova, V. 1971. Palaeoecology and palaeogeography of Cretaceous deposits of the Great Artesian Basin (Australia). *Rec. geol. Surv. New South Wales*, **13**, 5–48.

Scheibnerova, V. 1973. Possible Cretaceous dispersal routes for austral foraminifera. *Quarterly Notes geol. Surv. New South Wales*, **12**, 7–18.

Sigal, J. 1948. Notes sur les genres de foraminifères *Rotalipora* Brotzen, 1942 et *Thalmanninella*. Famille des Globorotaliidae. *Rev. Inst. Fr. Pet. Paris*, **3**, 95–103.

Sigal, J. 1952. Aperçu stratigraphique sur la micropaléontologie du Crétacé. *XIXe congres geologique international, Monographies regionales, 1 ere serie: Algerie*, **26**, 3–43.

Sigal, J. 1955. Notes micropaléontologiques nord-africaines. 1. Du Cénomanien au Santonien: zones et limites en faciès pélagiques. *C.r. Somm.. Soc. geol. Fr.*, no. 8, 157–160.

Sigal, J. 1956. Notes micropaléontologiques nord-africaines: 4.

Biticinella breggiensis (Gandolfi), nouveau genre. *C.r. Somm. Soc. geol. Fr.*, no. 3, 35–7.

Sigal, J. 1958. La classification actuelle des familles de Foraminifères planctoniques du Crétacé. *C.r. seances Acad. Sci. Paris*, 16 juin, 262–5.

Sigal, J. 1966. Contribution à une monographie des Rosalines. 1. Le genre *Ticinella* Reichel, souches des Rotalipores. *Eclog. geol. Helv.*, **59**, 187–217.

Sigal, J. 1967. Essai sur l'état actuel d'une zonation stratigraphique à l'aide des principales espèces de Rosalines (Foraminifères). *C.r. Somm. Soc. geol. Fr.*, fasc. 2, p. 48.

Sigal, J. 1977. Essai de zonation du Crétacé méditerranéen à l'aide des foraminifères planctoniques. *Geologie Mediterraneenne*, **4**, 99–108.

Sigal, J. 1978. Chronostratigraphy and ecostratigraphy of Cretaceous formations recovered on DSDP Leg 47B, Site 398. *Initial Rep. Deep Sea drill. Prof.*, **40**, 287–326.

Sliter, W. V. 1968. Upper Cretaceous foraminifera from Southern California and northwestern Baja California, Mexico. *Univ. Kansas Paleontol. Contrib.* **49**, Protozoa, 1–141.

Sliter, W. V. 1977. Cretaceous foraminifers from the Southwestern Atlantic Ocean, Leg 36, Deep Sea Drilling Project. *Initial Rep. Deep Sea drill. Proj.*, **36**, 519–73.

Sliter, W. V. 1980. 9. Mesozoic foraminifers and deep-sea benthic environments from Deep Sea Drilling Project Sites 415 and 416, Eastern North Atlantic. *Initial Rep. Deep Sea drill. Proj.*, **50**, 353–427.

Smith, C. C. & Pessagno, E. A. 1973. Planktonic foraminifera and stratigraphy of the Corsicana Formation (Maestrichtian) North-Central Texas. *Spec. Publ. Cushman Found. foramin. Res.*, **12**, 5–68.

Stainforth, R. M. *et al.* 1975. Cenozoic planktonic foraminiferal zonation and characteristics of index forms. *Univ. Kansas Paleontol. Contrib.*, article **62**, 1–425, in 2 parts.

Steineck, P. L. & Fleisher, R. L. 1978. Towards the classical evolutionary reclassification of Cenozoic Globigerinacea (Foraminiferida). *J. Paleontol.*, **52**, 618–35.

Sturm, M. 1969. Zonation of Upper Cretaceous by means of planktonic foraminifera, Attersee (Upper Austria). *Ann. Soc. geol. Pol.*, **39**, 103–32.

Subbotina, N. N. 1949. Microfauna of the Cretaceous of the southern slope of the Caucasus (Russian). *Leningrad, Vses Neft. Nauchno-Issled. Geol.-Razved Inst. Vnigri. (All-Union Petroleum Scient. Research Geol. Prospecting Institute). Microfauna of the oil fields of the USSR*, sbornik 2, Trudy, new series, vypusk 34, p. 33.

Subbotina, N. N. 1953. Fossil foraminifers of the USSR: Globigerinidae, Globorotaliidae, Hantkeninidae. *Trudy VNIGRI*, no. **76**, 1–291 (in Russian). Translated into English by E. Lees, published by Collet's Ltd., London and Wellingborough.

Takayanagi, Y. 1960. Cretaceous foraminifera from Hokkaido, Japan. *Sci. rep. Tohoku Univ., Sendai*, ser. 2 (geol.), **32**, 1–154.

Takayanagi, Y. 1965. Upper Cretaceous planktonic foraminifera from the Putah Creek subsurface section along the Yolo-Solano County Line, California. *Sci. rep. Tohoku Univ., Sendai*, ser. 2 (geol.), **36**, 161–237.

Takayanagi, Y. & Iwamoto, H. 1962. Cretaceous planktonic foraminifera from the Middle Yezo group of the Ikushumbetsu, Miruto, and Hatonosu areas, Hokkaido. *Trans. Proc. Palaeontol. Soc. Japan*, new series, **45**, 183–96.

Takayanagi, Y. & Okamura, M. 1977. Mid-Cretaceous planktonic microfossils from the Obira area, Rmoi, Hokkaido. *Spec. Pap. palaeontol. Soc. Japan*, **21**, 31–9.

Tappan, H. 1940. Foraminifera from the Grayson formation of northern Texas. *J. Paleontol.*, **14**, 122.

Tappan, H. 1943. Foraminifera from the Duck Creek formation of Oklahoma and Texas. *J. Paleontol.*, **17**, 476–517.

Tappan, H. 1971. Microplankton, ecological succession and evolution. *N. Am. Paleontol. Conv. Chicago*, 1969, Proc. H, pp. 1058–1103.

Tappan, H. & Loeblich, A. R., Jr 1973. Evolution of the oceanic plankton. *Earth Sci. Rev.*, **9**, 207–40.

Thalmann, H. E. 1932. Die Foraminiferen-Gattung *Hantkenina* Cushman, 1924, und ihre regional-stratigraphische Verbreitung. *Eclog. geol. Helv.*, **25**, 287–92.

Thalmann, H. E. 1934. Die regional-stratigraphische Verbreitung der oberkretazischen Foraminiferen-Gattung *Globotruncana* Cushman 1927. *Eclog. geol. Helv.*, **27**, 413–28.

Tilev, N. 1951. Etude des Rosalines maestrichtiennes (genre *Globotruncana*) du Sud-Est de la Turquie (Sondage de Ramandag), (in French, with Turkish summary). *Turkey, Maden Tetkik ve Arama Enstitüsü Yayinharindam (Mining Research and Exploration, Institute of Turkey Publ.), Ankara*, ser. B, no. 16.

Trujillo, E. F. 1960. Upper Cretaceous foraminifera from near Redding, Shasta County, California. *J. Paleontol.*, **34**, 290–346.

Umiker, R. 1952. Geologie der westlichen Stockhornkette mit besonderer Berücksichtigung der Kreidestratigraphie. Dissertation, Universität Bern, Switzerland, 77 pp. (unpublished).

Van Hinte, J. E. 1963. Zur Stratigraphie und Mikropaläontologie der Oberkreide und des Eozäns des Krappfeldes (Kärnten). *Jahrb. geol. Bundesanst.*, special volume, **8**, 1–147.

Van Hinte, J. E. 1965. The type Campanian and its planktonic foraminifera I. *Proc. Koninkl. Nederl. Akad. van Wetenschappen*, ser. B, no. 1, 8–28.

Van Hinte, J. E. 1972. The Cretaceous time scale and planktonic foraminiferal zones. *Proc. Koninkl. Nederl. Akad. van Wetenschappen*, ser. B, 75, no. 1, 1–8.

Van Hinte, J. E. 1976. A Cretaceous time scale. Bull. Am. Assoc. Petrol. Geol., **60**, 498–516.

Viennot, P. 1930. Sur la valeur stratigraphique des Rosalines. *C.r. Somm. Soc. geol. Fr.*, **7**, 60–2.

Vogler, J. 1941. Ober-Jura und Kreide von Misol (Niederländisch-Ostindien). *Palaeontographica*, **4** (suppl.), 243–93.

Voorwijk, G. H. 1937. Foraminifera from the Upper Cretaceous of Habana, Cuba. *Proc. Koninkl. Nederl. Akad, Wetenschappen*, **40**, 190–8.

White, M. P. 1928. Some index foraminifera of the Tampico embayment area, part 2. *J. Paleontol.*, **2**, 280–313.

White, M. P. 1929. Some index foraminifera of the Tampico embayment area of Mexico, part 3. *J. Paleontol.*, **3**, 30–58.

Wiedmann, J., Butt, A. & Einsele, G. 1978. Vergleich von marokkanischen Kreide-Küstenaufschlüssen und Tiefseebohrungen (DSDP): Stratigraphie, Paläoenvironment und subsidenz an einem passiven Kontinentalrand. *Geol. Rdsch.*, **67**, 454–508.

Williams-Mitchell, E. 1948. The zonal value of foraminifera in the chalk of England. *Proc. Geol. Assoc. London*, **59**, 91–112.

Wonders, A. A. H. 1975. Cretaceous planktonic foraminifera of the

Planomalina buxtorfi group from El Burrueco, Southern Spain. *Proc. Koninkl. Nederl. Akad. Wetenschappen*, ser. B, **78**, 83–93.

Wonders, A. A. H. 1978. Phylogeny, classification and biostratigraphic distribution of keeled Rotaliporinae. *Proc. Koninkl. Nederl. Akad. Wetenschappen*, ser. B, **81**, 113–44.

Wonders, A. A. 1979. Middle and late Cretaceous pelagic sediments of the Umbrian sequence in the Central Appennines. *Proc. Koninkl. Nederl. Akad. Wetenschappen*, ser. B, **82**, 171–205.

Wonders, A. A. 1980. Middle and Late Cretaceous planktonic Foraminifera of the Western Mediterranean area. *Utrecht Micropaleontology Bulletin*, **24**, 1–158.

5
Paleocene and Eocene planktic foraminifera

MONIQUE TOUMARKINE &
HANSPETER LUTERBACHER

CONTENTS

Introduction

The planktic foraminifera assemblages of the earliest Paleocene consist of very small and primitive forms from which all Paleogene species evolved. Early Paleocene assemblages are characterized by small forms with globular to ovate chambers; these soon increase in size and in the complexity of their tests. Most of the lineages flourishing during the Middle Paleocene to Late Eocene can be traced back to Early Paleocene ancestors. At the end of the Early Paleocene, such genera as *Planorotalites*, *Acarinina* and *Morozovella* were already established. During the Middle and Late Paleocene they dominated the rich and well-diversified tropical and subtropical assemblages. In higher latitudes, generally only a few representatives of the genus *Acarinina* are present and a subdivision based on planktic foraminifera becomes difficult. The faunal change from the Paleocene to the Eocene is gradual. The boundary is placed here at the top of the *Morozovella velascoensis* Zone and corresponds for all practical purposes to the extinction of the large and heavily ornamented forms of the Middle and Late Paleocene *M. velascoensis* group and also to the first occurrence of forms belonging to the genus *Pseudohastigerina*.

The basal part of the Early Eocene contains only relatively small and lightly ornamented *Morozovella* species; large and well-ornamented forms such as *Morozovella formosa formosa*, *M. aragonensis* and *M. caucasica* do not appear until within the Early Eocene. During the same period, large *Acarinina* are widespread and may dominate the assemblages. Some representatives of the latter genus reach into higher latitudes where they may form almost monospecific assemblages.

The Middle and Late Eocene contain very characteristic species such as *Hantkenina*, *Clavigerinella*, *Globigerinatheka* and the *Turborotalia cerroazulensis* group. In tropical and subtropical regions, the base of the Middle Eocene is clearly indicated by the appearance of the first representatives of the genus *Hantkenina*. The transition from the Early to the Middle Eocene is a time of gradual change within the planktic foraminifera associations. It sets in during the topmost Early Eocene *Acarinina pentacamerata* Zone with the first occurrence of *Turborotalia cerroazulensis frontosa*, *Acarinina bullbrooki*, *Morozovella spinulosa*, 'Globigerinoides' higginsi and representatives of the genus *Clavigerinella*. It continues within the earliest Middle Eocene *Hantkenina nuttalli* Zone with the appearance of the genera *Hantkenina* and *Globigerinatheka*.

The end of the Middle Eocene is marked in lower latitudes by the disappearance of virtually all spinose forms (*Morozovella*, *Acarinina*, *Truncorotaloides*). In the zonal scheme used here, the Eocene–Oligocene boundary is placed at the extinction level of the last representatives of the *Turborotalia cerroazulensis* group, which practically coincides with the extinction of the genus *Hantkenina*.

In higher latitudes, some of the Middle and Late Eocene zonal markers are often missing. For this reason Toumarkine & Bolli (1970) established an alternative zonal scheme for mid-latitudes, based on the evolutionary changes in the *Turborotalia cerroazulensis* lineage, which is usually well represented in more temperate regions (Fig. 2).

Virtually all conspicuous index forms of low latitudes are not found in higher latitudes. There, the assemblages consist frequently of long-ranging species with poorly characterized morphology. Therefore, Paleogene planktic foraminiferal zonations proposed for use in higher latitudes (e.g. Chapter 7, this volume; Berggren, 1972) provide considerably reduced stratigraphic resolutions. In higher latitudes, other groups of planktic microfossils like dinoflagellates give in general much better stratigraphic results than planktic foraminifera.

In Europe, the low latitude planktic foraminiferal zonation discussed in this chapter has been successfully applied in the Mediterranean area with a northern limit corresponding roughly to the Aquitaine Basin and the alpine Helvetic Nappes. Further to the north, in such areas as the North Sea Basin, it is no longer applicable.

In the preparation of this chapter, H. P. Luterbacher has been largely responsible for the Paleocene and Early Eocene and M. Toumarkine for the Middle and Late Eocene.

Definition and discussion of zones

Subdivisions of the Paleogene were developed initially in two widely separated geographic areas. In the eastern hemisphere, it was mainly in the northern foothills of the Caucasus Mountains that planktic foraminifera were already used to solve stratigraphic problems in oil and gas exploration during the 1930s. A first comprehensive résumé of the use of Paleogene planktic foraminifera as a stratigraphic tool was given by Subbotina (1953). Subsequent papers by Soviet specialists refined and partially revised Subbotina's fundamental paper (see Krasheninnikov, 1969, for references). Due to lack of communication and the language barrier, the work by Soviet micropaleontologists was ignored in the western hemisphere for a long time. Berggren (1960b) was mainly responsible for bringing it within the reach of workers outside the Socialist countries.

In the western hemisphere, the first detailed subdivision of the Paleogene based on planktic foraminifera was established in Trinidad. Tribute must be paid to H. G. Kugler who, in a managerial position, greatly contributed to the development and publication of what may be called the 'Trinidad zonation' (Bolli, 1957a, b). This zonation was rapidly applied outside the Caribbean and represents the base for the low latitude planktic foraminiferal zonation discussed in this chapter (see also Bolli, 1966; Stainforth et al., 1975; Blow,

1979). The virtually independent development of two pioneer Paleogene planktic foraminiferal subdivisions based on differently defined sets of taxa has led to confusion, some of which still persists.

A discussion of all subsequent modifications of the original zonal scheme proposed by Bolli in 1957 is not intended here.

The zonation adopted by us will be discussed briefly together with some of the modifications; an exhaustive review of the existing literature will not be attempted (for correlations with other zonal schemes, reference is made to Fig. 4). Some modifications are due to local peculiarities in the Paleogene stratigraphy of Trinidad (e.g. non-carbonate facies in the basal Paleocene, hiatus at the Paleocene–Eocene boundary); others are caused by subsequent taxonomic revisions which force us to abandon well-established names, or else by the emendation of the definition of some taxa with subsequent changes in their stratigraphic range. Our approach was to maintain established zonal names and definitions whenever possible and to make only those changes which are useful and unavoidable.

RAD. AGE	AGE		PLANKTIC FORAMINIFERAL ZONES	DATUM MARKERS
38	EOCENE	L		L *Turborotalia cerroazulensis* s. l.
			Turborotalia cerroazulensis s. l.	
				L *Globigerinatheka semiinvoluta*
			Globigerinatheka semiinvoluta	
41				L *Truncorotaloides rohri*
			Truncorotaloides rohri	
				L *Orbulinoides beckmanni*
			Orbulinoides beckmanni	
				F *Orbulinoides beckmanni*
		M	*Morozovella lehneri*	
				L *Morozovella aragonensis*
			Globigerinatheka s. subconglobata	
				F *Globigerinatheka mexicana mexicana*
			Hantkenina nuttalli	
50.3				F *Hantkenina*
			Acarinina pentacamerata	F *Turborotalia cerroazulensis frontosa*
			Morozovella aragonensis	F *Acarinina pentacamerata*
		E	*Morozovella formosa formosa*	F *Morozovella aragonensis*
			Morozovella subbotinae	
			Morozovella edgari	L *Morozovella edgari*
54.9	PALEOCENE			L *Morozovella velascoensis*
			Morozovella velascoensis	
		L		L *Planorotalites pseudomenardii*
			Planorotalites pseudomenardii	
				F *Planorotalites pseudomenardii*
			Planorotalites pusilla pusilla	
61.5				F *Planorotalites pusilla pusilla*
			Morozovella angulata	F *Morozovella angulata*
			Morozovella uncinata	
		E	*Morozovella trinidadensis*	F *Morozovella uncinata*
			Morozovella pseudobulloides	F *Morozovella trinidadensis*
				F *Morozovella pseudobulloides*
			Globigerina eugubina	
66.7				F *Globigerina eugubina* / L *Globotruncana*

Fig. 1. Paleocene to Eocene planktic foraminiferal zonal scheme, datum markers and radiometric ages used in this chapter

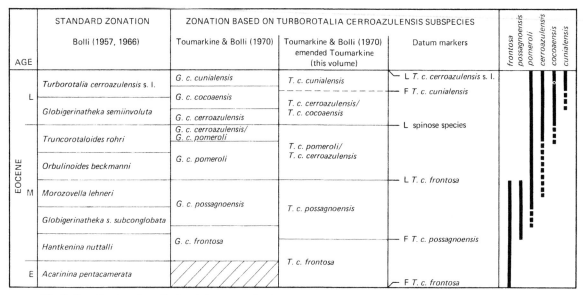

Fig. 2. Zonation based on *Turborotalia cerroazulensis*

	Blow (1979) (Manuscript completed 1972)		Bolli (1957a,b, 1966) Premoli Silva & Bolli (1973)		Blow, 1969 Berggren & Van Couvering, 1974		
AGE	Datum Markers				Datum Markers		AGE
L	L *C. inflata* / F *G. gortanii gortanii*	P 17	*Turborotalia cerroazulensis* s.l.	P 17	L *Hantkenina*		L
	F *C. inflata*	P 16	*Globigerinatheka semiinvoluta*	P 16	L *G. mexicana*		
	F *P. semiinvoluta*	P 15		P 15	L *G. lehneri – T. rohri*		
	F *P. semiinvoluta*	P 15	*Truncorotaloides rohri*	P 14			
M	L *G. beckmanni*	P 14	*Orbulinoides beckmanni*	P 13			M
	F *G. beckmanni*	P 13	*Morozovella lehneri*	P 12			
	L *S. boweri*	P 12	*Globigerinatheka s. subconglobata*	P 11	F *G. lehneri – T. topilensis*		
	F *G. kugleri kugleri*	P 11	*Hantkenina nuttalli*	P 10	F *Hantkenina*		
	F *S. frontosa frontosa*	P 10	*Acarinina pentacamerata*	P 9			
E	F *G. (A.) aspensis*	P 9	*Morozovella aragonensis*	P 8			E
	F *G. (M.) aragonensis*	b / a P 8	*Morozovella formosa formosa*	P 7	F *G. aragonensis*		
	F *G. (M.) formosa*		*Morozovella subbotinae*	P 6	b		
		P 7	*Morozovella edgari*		a	F *Pseudohastigerina*	
L	F *G. (A.) wilcoxensis berggreni*		*Morozovella velascoensis*	P 5	L *G. pseudomenardii*		L
	F *G. (M.) subbotinae subbotinae* F *G. (M.) soldadoensis soldadoensis*	P 6 / P 5	*Planorotalites pseudomenardii*	P 4	F *G. pseudomenardii*		
	F *G. (G.) pseudomenardii*	P 4	*Planorotalites pusilla pusilla*	P 3			
M	F *G. (M.) angulata angulata*	P 3	*Morozovella angulata*		F *G. angulata*		M
	F *G. (A.) praecursoria praecursoria*	P 2	*Morozovella uncinata*	P 2	L *G. daubjergensis*		
E	F *G. (T.) compressa compressa* F *G. (T.) pseudobulloides*	b / a P 1	*Morozovella trinidadensis*	d			E
			Morozovella pseudobulloides	P 1 c / b			
	L *Rugoglobigerina* sp.	M 18 Pα	*Globigerina eugubina*	a	L *Globotruncana – Rugoglobigerina*		

Fig. 3. Comparison of the Paleocene to Eocene planktic foraminiferal zonal scheme used in this chapter with the P zonal schemes of Blow, 1969/Berggren & Van Couvering, 1974, and Blow, 1979

We believe that we have now reached a general consensus concerning the main steps involved in the evolution of Paleogene planktic foraminifera and also, thereby, concerning the major stratigraphic divisions based on them. Unfortunately, this basic agreement is hidden by a thicket of divergent definitions of taxa and zones leading to some confusion which is often semantic rather than real.

Blow (1969), in his work on the late Middle Eocene to Recent planktic foraminiferal stratigraphy, introduced the zones P 13 to P 17 for the subdivision of the late Middle and Late Eocene. In the same year Berggren published a complete zonation for the Paleocene and Eocene numbered from P 1 to P 17, making reference in his paper to unpublished work by Blow & Berggren. However, such a joint paper never appeared in print.

In 1974 Berggren & Van Couvering published the same zonal scheme again but with slight emendations. Blow (1979)

published yet another scheme P 1–P 17, which differed considerably from that of Berggren & Van Couvering (1974). Changes concern the subdivision of the Early Paleocene (P 1) and the youngest Late Paleocene to the top of the Early Eocene (P 6–P 9). Furthermore, a minor difference also concerns the boundary P 14/P 15.

The differences between the Berggren & Van Couvering (1974) and Blow (1979) zonal schemes are illustrated on Figs. 3 and 4, and both are correlated with the zonation used in this volume.

Until the publication of Blow's 1979 scheme, the Berggren & Van Couvering (1974) scheme was applied by workers using P zones. With the additional scheme attributable to Blow (1979) it is now necessary for authors to quote which scheme is followed, to avoid confusion and erroneous zonal interpretations and correlations. When reference is made in this volume to a P zonal scheme, such as on the range charts

		Bolli (1957a,b,1966), Premoli Silva & Bolli (1973), this volume	Krasheninnikov (1965, 1969)		Hillebrandt (1974)	Blow, 1969 Berggren & van Couvering (1974)		Blow (1979)	
EOCENE	Late	T. cerroazulensis s. l.	G. corpulenta		G. cerroazulensis	P 17	G. gortanii/G. centralis	P 17	G. g. gortanii/G. (T.) centralis
						P 16	C. inflata	P 16	C. inflata
		G. semiinvoluta			G. semiinvoluta	P 15	G. mexicana	P 15	P. semiinvolutus
	Middle	T. rohri	T. rohri		T. rohri	P 14	T. rohri - G. howei	P 14	G. (M.) sp. spinulosa
		O. beckmanni	H. alabamensis		O. beckmanni	P 13	O. beckmanni	P 13	G. beckmanni
		M. lehneri	A. rotundimarginata		G. lehneri	P 12	G. lehneri	P 12	G. (M.) lehneri
		G. s. subconglobata	A. bullbrooki	G. kugleri	G. s. subconglobata	P 11	G. kugleri	P 11	G. kugleri/S. frontosa boweri
		H. nuttalli		H. aragonensis	H. aragonensis	P 10	H. aragonensis	P 10	S. f. frontosa/G. (T.) pseudomayeri
	Early	A. pentacamerata	G. aragonensis	A. pentacamerata	G. caucasica / G. palmerae	P 9	A. densa	P 9	G. (A.) aspensis/G. lozanoi prolata
		M. aragonensis		G. aragonensis	G. aragonensis	P 8	G. aragonensis	P 8 b	G. (M.) aragonensis/G. (M.) formosa
		M. formosa formosa	G. subbotinae	G. marginodentata	G. formosa/A. angulosa	P 7	G. formosa	P 8 a	G. (M.) formosa/G. (M.) lensiformis
		M. subbotinae			G. lensiformis	P 6 b	G. subbotinae/P. wilcoxensis	P 7	G. (A.) wilcoxensis berggreni
		M. edgari		G. subbotinae	G. marginodentata/G. subbotinae				
PALEOCENE	Late	M. velascoensis	G. velascoensis	G. velascoensis	G. velascoensis	P 6 a	G. velascoensis/G. subbotinae	P 6	G. (M.) s. subbotinae/G. (M.) velascoensis acuta
						P 5	G. velascoensis	P 5	M. s. soldadoensis/G. (M.) velascoensis pasionensis
		P. pseudomenardii		G. pseudomenardii	G. pseudomenardii	P 4	G. pseudomenardii	P 4	G. (G.) pseudomenardii
	Middle	P. pusilla pusilla	G. angulata	G. conicotruncata	G. pusilla	P 3	G. pusilla/G. angulata	P 3	G. (M.) a. angulata
		M. angulata		G. angulata	G. angulata				
		M. uncinata	A. uncinata		G. inconstans/G. uncinata	P 2	G. uncinata/G. spiralis	P 2	G. (A.) p. praecursoria
	Early	M. trinidadensis	G. triloculinoides/ G. pseudobulloides		G. trinidadensis	P 1 d	G. compressa/G. inconstans/ G. trinidadensis	P 1 b	G. (T.) c. compressa/E. eobulloides simplissima
		M. pseudobulloides				P 1 c	G. pseudobulloides		
			E. eobulloides		G. edita	P 1 b	G. triloculinoides	P 1 a	G. (T.) pseudobulloides/G. (T.) archaeocompressa
		G. eugubina				P 1 a	G. eobulloides	P α	G. (T.) longiapertura
						M 18			R. hexacamerata

Fig. 4. Correlation of major Paleocene and Eocene zonal schemes

Figs. 5–10, the scheme by Blow (1969)/Berggren & Van Couvering (1974) is followed.

Zones used in this chapter
Early Paleocene
GLOBIGERINA EUGUBINA ZONE

Category: Total range zone
Age: Early Paleocene
Author: Luterbacher & Premoli Silva (1964)
Definition: Total range of *Globigerina eugubina*.
Remarks: The assemblage of the *Globigerina eugubina* Zone consists exclusively of very small globigerinids with average diameters below 0.1 mm. Besides the zonal marker, typical species include *Globigerina fringa* and early representatives of *Globoconusa daubjergensis*. Small and undifferentiated heterohelicids may be frequent.

This zone was first described from central Italy where it overlies the assemblages with rugoglobigerinids and globotruncanids of the latest Maastrichtian. Subsequently, assemblages of the *Globigerina eugubina* Zone have been found to have a worldwide distribution (e.g. Premoli Silva, 1977).

A few authors (e.g. Berggren, 1965; Hofker, 1978; Blow, 1979) consider the faunal assemblages of the *Globigerina eugubina* Zone to represent impoverished and relict faunules composed of dwarfed representatives of such Late Cretaceous genera as *Rugoglobigerina*, *Planomalina*, *Globigerinelloides* and *Hedbergella* (Zone M 18, *Rugoglobigerina hexacamerata* partial range zone in Blow, 1979). Nevertheless, well-preserved assemblages from the *Globigerina eugubina* Zone recovered by the Deep Sea Drilling Project (DSDP) are rather in favour of the original interpretation by Luterbacher & Premoli Silva (1964) who regarded them as representing the rootstock of globigerinids, from which the Paleocene planktic foraminifera evolved. The assemblages of the zone P α (*Globorotalia* (*Turborotalia*) *longiapertura* partial range zone) described and discussed by Blow (1979) seem to be close to those of the *Globigerina eugubina* Zone. The assemblages of the type-locality (basal Paleocene of the Shatsky Rise, northwestern Pacific) have been placed by Krasheninnikov & Hoskins (1973) within the *Globigerina eugubina* Zone.

Smit (1977, 1982) describes from SE Spain a planktic foraminiferal association (*Guembelitria cretacea* Zone) between the *Abathomphalus mayaroensis* Zone (latest Cretaceous) and the *Globigerina eugubina* Zone. This association has also been reported from several other localities (e.g. Gamper, 1977). For a more detailed discussion, reference is made to Smit's papers.

Likewise, Herm, Hillebrandt & Perch-Nielsen (1981) recognized in SE Bavaria a *Globigerina fringa* Zone which is still older than the *Globigerina eugubina* Zone. Bang (1980) described assemblages equivalent to those of the *Globigerina eugubina* Zone in the Lower Danian of the type area.

MOROZOVELLA PSEUDOBULLOIDES ZONE

Category: Interval zone
Age: Early Paleocene
Author: Leonov & Alimarina (1961) as *Globigerina pseudobulloides – Globigerina daubjergensis* Zone. Name shortened by Bolli (1966).
Definition: Interval from first occurrence of *Morozovella pseudobulloides* to first occurrence of *Morozovella trinidadensis*.
Remarks: Within this zone a differentiation of the stock of the relatively simple forms of the previous zone takes place. Most of the lineages of Paleocene planktic foraminifera have become established by the end of this zone.

Morozovella pseudobulloides and *Globigerina triloculinoides*, are typical species of the *Morozovella pseudobulloides* Zone. *Globoconusa daubjergensis* is a conspicuous form of assemblages of this age, especially in higher latitudes. *Planorotalites compressa* is represented by typical specimens in the younger part of this zone.

This zone was not recognized in the original Trinidad zonation (Bolli, 1957a), probably because of the unfavourable lithology of this part of the section.

The rapid evolution of the planktic foraminiferal assemblages within the *Morozovella pseudobulloides* Zone allows further subdivisions (Figs. 3, 4) which provide detailed correlations within areas with well developed Early Paleocene deposits (e.g. North Sea).

MOROZOVELLA TRINIDADENSIS ZONE

Category: Interval zone
Age: Early Paleocene
Author: Bolli (1957a)
Definition: Interval between first occurrences of *Morozovella trinidadensis* and *Morozovella uncinata*.
Remarks: This zone, which is the youngest zone of the Early Paleocene (Stainforth *et al.*, 1975), is characterized by the presence of *Morozovella trinidadensis*, *Morozovella pseudobulloides*, *Planorotalites compressa* and *Globoconusa daubjergensis*.

Blow (1979) placed this zone, with its type-locality in Trinidad, as Subzone 1b of his more comprehensive *Globorotalia pseudobulloides* Zone (P 1).

Middle Paleocene

MOROZOVELLA UNCINATA ZONE

Category: Interval zone

Age: Middle Paleocene

Author: Bolli (1957a) emended Bolli (1966)

Definition: Interval from first occurrence of *Morozovella uncinata* to first occurrence of *Morozovella angulata*.

Remarks: This zone is marked by forms with angular-conical chambers in the initial portion of the last whorl (*Morozovella uncinata, Morozovella praecursoria*).

Blow's Zone P 2 (*Globorotalia (Acarinina) praecursoria praecursoria* Zone) has the same type-locality as this zone. *Morozovella praecursoria* and *Morozovella uncinata* are considered as separate species (Luterbacher, 1964) and not as synonyms (Blow, 1979).

MOROZOVELLA ANGULATA ZONE

Category: Interval zone

Age: Middle Paleocene

Author: Alimarina (1963); see also Hillebrandt (1965)

Definition: Interval from first occurrence of *Morozovella angulata* to first occurrence of *Planorotalites pusilla pusilla*.

Remarks: In this zone, species of *Morozovella* with angular-conical chambers throughout their youngest whorl become predominant (*Morozovella angulata, Morozovella conicotruncata* and others). Typical representatives of the genus *Acarinina* form a conspicuous part of the planktic foraminiferal assemblages. This zone is recognized by most authors, but its delimitation is often treated in different ways. The present definition adheres to the one given by Bolli (1966), whereas the *Globorotalia (Morozovella) angulata angulata* partial range zone as used by Blow (1979) includes also the overlying *Globorotalia (Planorotalites) pusilla pusilla* Zone.

PLANOROTALITES PUSILLA PUSILLA ZONE

Category: Interval zone

Age: Middle Paleocene

Author: Bolli (1957a)

Definition: Interval from first occurrence of *Planorotalites pusilla pusilla* to first occurrence of *Planorotalites pseudomenardii*.

Remarks: The zonal marker may be poorly represented or absent, as frequently observed in assemblages of higher latitude. In this case, it may become difficult to separate this zone although its assemblages are characterized by an abundance of *Morozovella conicotruncata* and the appearance of *Planorotalites chapmani* within the zone.

Blow (1979) has cast some doubt on the stratigraphic range of *Planorotalites pusilla pusilla* as recognized by other authors indicating that it would have more or less the same range as *Morozovella angulata*. However, this postulated extended range of the marker species is not confirmed by our own experience.

Late Paleocene

PLANOROTALITES PSEUDOMENARDII ZONE

Category: Total range zone

Age: Late Paleocene

Author: Bolli (1957a)

Definition: Total range of *Planorotalites pseudomenardii*.

Remarks: The planktic foraminiferal assemblages of this zone are rich in heavily ornamented representatives of the genus *Morozovella*. Outside the tropical and subtropical realm, these heavily ornamented *Morozovella* species are absent in general and globular representatives of the genus *Acarinina* (e.g. *Acarinina mckannai*) are dominant. In this case, it may be difficult or even impossible to separate the Late Paleocene into two zones.

Planorotalites pseudomenardii is considered by most authors (e.g. Berggren, 1977) as an excellent marker and easily recognizable species. However, Blow (1979) finds representatives of *Planorotalites pseudomenardii* ranging into Early Eocene assemblages of his Zone P 7 which is equivalent to the *Morozovella subbotinae* Zone as used in this report. It is probable that Blow uses a somewhat wider definition of this species including forms classified by others within the longer-ranging *Planorotalites chapmani*. Correlations of coccolith zones with planktic foraminiferal zones (Hay & Mohler, 1969; Caro *et al.*, 1975) consistently suggest a short range for *Planorotalites pseudomenardii* as advocated by the majority of authors.

MOROZOVELLA VELASCOENSIS ZONE

Category: Interval zone

Age: Late Paleocene

Author: Bolli (1957a)

Definition: Interval from last occurrence of *Planorotalites pseudomenardii* to last occurrence of *Morozovella velascoensis*.

Remarks: Typical representatives of *Morozovella velascoensis* may be absent even in rich assemblages from the tropical and subtropical realms. In this case, the zone may be recognized based on the co-occurrence of *Morozovella acuta* and *Acarinina soldadoensis*.

Several species make their first appearance within this zone, but become dominant only within the basal Early Eocene assemblages (namely *Morozovella edgari*, *M. subbotinae*, *M. formosa gracilis*). The overlap of the ranges of *Morozovella subbotinae* and *M. velascoensis* is used by some authors for the definition of an additional zone or subzone (e.g. Berggren, 1971).

It is established usage to place the boundary between the Paleocene and the Eocene at the top of the *Morozovella velascoensis* Zone (e.g. Bolli, 1957a, 1966; Krasheninnikov, 1965b, 1969; Postuma, 1971; Stainforth *et al.*, 1975). Berggren (1960a, 1971) places the same boundary at the first occurrence of *Pseudohastigerina wilcoxensis* ('*Pseudohastigerina* datum') which for all practical purposes is about age-equivalent to the top of the *Morozovella velascoensis* Zone as used here.

Early Eocene
MOROZOVELLA EDGARI ZONE

Category: Interval zone
Age: Early Eocene
Author: Premoli Silva & Bolli (1973)
Definition: Interval between last occurrence of *Morozovella velascoensis* and last occurrence of *Morozovella edgari*.
Remarks: The planktic foraminiferal assemblages of this zone are dominated by lightly built representatives of the genus *Morozovella* as, for example, *Morozovella subbotinae* and *M. formosa gracilis*. *Acarinina soldadoensis soldadoensis* and its relatives are abundant within assemblages from this zone.

The subdivision of the interval corresponding more or less to the time between the disappearance of *Morozovella velascoensis* and the first occurrence of *Morozovella aragonensis* is handled in different ways by authors because most faunal changes in this interval are very gradual.

The *Morozovella edgari* Zone is missing in the original Trinidad zonation. This is due to a gap in the succession of planktic foraminiferal assemblages in Trinidad as shown by Bolli (1957a); Luterbacher (1964); Premoli Silva & Bolli (1973); Stainforth *et al.* (1975) and also by Blow (1979).

The *Morozovella edgari* Zone corresponds to the *Globorotalia aequa* Zone in Luterbacher (1964), to the older part of the *Globorotalia subbotinae* Zone in Stainforth *et al.* (1975), to the lower part of the Subzone P 6b (*Globorotalia subbotinae/Pseudohastigerina wilcoxensis*) in Berggren & Van Couvering (1974) and probably to the lower part of the Zone P 7 (*Globorotalia (Acarinina) wilcoxensis berggreni* partial range zone) in Blow (1979).

MOROZOVELLA SUBBOTINAE ZONE

Category: Interval zone
Age: Early Eocene
Author: The name *Globorotalia subbotinae* Zone has been used by Soviet authors (e.g. Anonymous, 1963); the

definition given here follows Luterbacher & Premoli Silva (*in* Caro *et al.*, 1975).
Definition: Interval between last occurrence of *Morozovella edgari* and first occurrence of *Morozovella aragonensis*.
Remarks: Typical assemblages of this zone are characterized by the co-occurrence of such species as *Morozovella subbotinae*, *M. marginodentata*, *M. formosa gracilis* and *Acarinina soldadoensis soldadoensis*. *Morozovella formosa formosa* has its first appearance within this zone.

The *Morozovella subbotinae* Zone as defined here corresponds approximately to the *Globorotalia rex* Zone of Bolli, 1957a. *Morozovella subbotinae* is a senior synonym of *M. rex* (see, for example, Berggren, 1977).

MOROZOVELLA FORMOSA FORMOSA ZONE

Category: Interval zone
Age: Early Eocene
Author: Bolli (1957a) named this zone, but the definition adopted here is the one given by Beckmann *et al.* (1969)
Definition: Interval between first occurrence of *Morozovella aragonensis* and first appearance of *Acarinina pentacamerata*.
Remarks: A typical planktic foraminiferal assemblage from the *Morozovella formosa formosa* Zone comprises *M. quetra*, *M. lensiformis*, *M. formosa formosa*, *Acarinina soldadoensis soldadoensis*, *A. soldadoensis angulosa* and *A. primitiva*.

Bolli (1966) and Premoli Silva & Bolli (1973) used the first appearance of *Globigerina taroubaensis* and *G. turgida*, two rather indistinct species, for the definition of the base of the zone. Beckmann *et al.* (1969) recognized the top of the *Morozovella formosa formosa* Zone by the first appearance of typical *Acarinina pentacamerata* (= *Globigerina aspensis* auct.) and by the presence of numerous and typically developed *Morozovella aragonensis*. The Zone P 7 (*Globorotalia formosa* Zone) of Berggren (1971) corresponds to the same interval, whereas only the older subzone of the Zone P 8 (*Globorotalia (Morozovella) formosa* total range zone) of Blow (1979) is equivalent to the present zone.

It is within the *Morozovella formosa formosa* Zone that the surface temperature of the Paleogene oceans reached its maximum (e.g. Douglas & Savin, 1973) and that the heavily ornamented *Morozovella* species made their greatest penetration into higher latitudes. In the higher latitudes, temperatures deteriorated at the end of the *Morozovella formosa formosa* Zone (Kennett, 1982; I. Premoli Silva, personal communication) and the well-ornamented representatives of the genus *Morozovella*, with the exception of *M. aragonensis*, withdrew again from middle latitudes where the planktic foraminiferal assemblages became dominated by spiny and globular representatives of the genus *Acarinina*.

MOROZOVELLA ARAGONENSIS ZONE

Category: Interval zone

Age: Early Eocene

Author: Bolli (1957a)

Definition: Interval between first occurrence of *Acarinina pentacamerata* and first occurrence of *Turborotalia cerroazulensis frontosa* or *Planorotalites palmerae*.

Remarks: Typical planktic foraminifera of this zone are *Morozovella aragonensis*, *Acarinina pentacamerata* and *A. soldadoensis angulosa*. *Morozovella formosa formosa* and *M. subbotinae* persist, usually in small numbers, into the older part of this zone whereas *M. caucasica* appears within its younger part.

In one sense or another, a *Globorotalia aragonensis* Zone is recognized by most authors, but its definition is treated in different ways. The *Morozovella aragonensis* Zone as used here correlates with Zone P 8 of Berggren & Van Couvering (1974) and with the Subzone P 8b of Blow (1979).

ACARININA PENTACAMERATA ZONE

Category: Interval zone

Age: Early Eocene

Author: Introduced by Krasheninnikov (1965a, b) as a subzone

Definition: Interval from first occurrence of *Turborotalia cerroazulensis frontosa* to first occurrence of representatives of the genus *Hantkenina*.

Remarks: The nominate species generally dominates the assemblages of this zone. Other common constituents are *Morozovella aragonensis* and *Turborotalia cerroazulensis frontosa*. *Planorotalites palmerae* is restricted to this zone, but has a very patchy geographic distribution. *Acarinina soldadoensis soldadoensis* disappears close to the top of the zone.

Because of the rarity of *Planorotalites palmerae*, the name *Acarinina pentacamerata* Zone is used as a substitute for the *Globorotalia palmerae* Zone of the original Trinidad zonation. Krasheninnikov (1965a, b) subdivided an extended *Globorotalia aragonensis* Zone, which corresponds to the entire younger part of the Early Eocene, into a *Globorotalia aragonensis* Subzone and an *Acarinina pentacamerata* Subzone.

Benjamini (1980) stated that the extinction of *Planorotalites palmerae* and the first occurrence of *Hantkenina aragonensis* are separated by a considerable gap for which he introduces a 'Sphaeroidinellopsis' senni Zone.

The top of the *Acarinina pentacamerata* Zone corresponds to the boundary between the Early and the Middle Eocene as adopted by the majority of authors.

It corresponds to the first occurrence of representatives of the genus *Hantkenina* and of typical Middle Eocene representatives of the genera *Acarinina* and *Morozovella* (e.g. *A. bullbrooki*, *M. spinulosa*).

Middle Eocene

HANTKENINA NUTTALLI ZONE

Category: Interval zone

Age: Middle Eocene

Author: Bolli (1957b) emended Stainforth *et al.* (1975), renamed Toumarkine (1981)

Definition: Interval with zonal marker from first occurrence of representatives of the genus *Hantkenina* to first occurrence of *Globigerinatheka mexicana mexicana*.

Remarks: The base of this oldest zone of the Middle Eocene is marked by the first occurrence of representatives of the genus *Hantkenina*. The first forms belonging to the genus *Globigerinatheka* appear close to the same level. Some genera and species occur already before this zone, but become frequent only at the base of the Middle Eocene (*Truncorotaloides* spp., *Turborotalia cerroazulensis*, *Morozovella spinulosa*, *Clavigerinella* spp., *Acarinina bullbrooki*, *A. spinuloinflata*, *Globigerina senni*). Together with these typically Middle Eocene species and genera, some Early Eocene species persist (e.g. *Acarinina pentacamerata*, *Morozovella aragonensis*, *Globigerina inaequispira*). Forms of the *Globigerina lozanoi*–'*Globigerinoides' higginsi* group and of the *Turborotalia griffinae*–'*Hastigerina' bolivariana* group are widespread during the late Early and the early Middle Eocene. The genus *Hantkenina* does not reach into higher latitudes and may also show a rather patchy distribution in some subtropical regions. According to Stainforth *et al.* (1975), alternative criteria for the recognition of the base of the Middle Eocene are the extinction of *Acarinina soldadoensis soldadoensis* and the appearance of common and typical *Acarinina bullbrooki*.

In the alternative zonation based on the evolution of the *Turborotalia cerroazulensis* lineage, parts of the *Hantkenina nuttalli* Zone and the *Globigerinatheka subconglobata subconglobata* Zone are characterized by the co-occurrence of *Turborotalia cerroazulensis frontosa* and *T. cerroazulensis possagnoensis*.

Blow (1979) places the base of the Middle Eocene (base of Zone P 10, *Subbotina frontosa frontosa*/ *Globorotalia* (*Turborotalia*) *pseudomayeri* Zone) at the first occurrence of *Subbotina frontosa frontosa* which we placed at the base of the *Acarinina pentacamerata* Zone. However, Blow (1979) correlates the base of his Zone P 10 approximately with the middle part of the *Globorotalia palmerae* Zone of Bolli (1957b) which is equivalent to the *Acarinina pentacamerata* Zone of this report.

GLOBIGERINATHEKA SUBCONGLOBATA SUBCONGLOBATA ZONE

Category: Concurrent range zone

Age: Middle Eocene

Author: Bolli (1957b) emended Proto Decima & Bolli (1970), renamed Bolli (1972), redefined Stainforth *et al.* (1975)

Definition: Interval with zonal marker from first occurrence of *Globigerinatheka mexicana mexicana* to last occurrence of *Morozovella aragonensis*.

Remarks: The zone was originally named *Globigerapsis kugleri* Zone by Bolli (1957b), but Proto Decima & Bolli (1970) stated that the marker species is not identical with '*Globigerapsis*' *kugleri* and named it *G. subconglobata curryi*. Bolli (1972) renamed the zone as *Globigerinatheka subconglobata subconglobata* Zone in order to avoid further confusion. Typical assemblages of this zone are characterized by the rapid evolution of the genus *Globigerinatheka* into several species and subspecies. *G. subconglobata curryi* is the first representative of the *G. subconglobata curryi–G. subconglobata euganea–Orbulinoides beckmanni* lineage. *Morozovella lehneri*, a conspicuous species of the Middle Eocene, appears near the base of the zone.

The *Globigerinatheka subconglobata subconglobata* Zone is correlated with the middle part of the *Turborotalia cerroazulensis possagnoensis* Zone of Toumarkine & Bolli (1970) emended Tourmarkine, this volume, and corresponds to the P 11 Zone of Blow (1979).

MOROZOVELLA LEHNERI ZONE

Category: Interval zone

Age: Middle Eocene

Author: Bolli (1957b)

Definition: Interval with the zonal marker from last occurrence of *Morozovella aragonensis* to first occurrence of *Orbulinoides beckmanni*.

Remarks: The zonal markers *Morozovella lehneri* and *Orbulinoides beckmanni* are generally well represented in tropical and subtropical regions, but they are often missing in temperate assemblages. In this case, the coexistence of *Turborotalia cerroazulensis frontosa* and *T. cerroazulensis possagnoensis* together with the first representatives of *T. cerroazulensis pomeroli* could be an alternative criterion for the recognition of the zone.

Globigerinatheka subconglobata euganea and the *Hantkenina alabamensis* group appear in tropical and subtropical areas during this zone.

Blow's (1979) Zone P 12 and the younger part of the *Turborotalia cerroazulensis possagnoensis* Zone (Toumarkine & Bolli, 1970) correspond to the *Morozovella lehneri* Zone.

ORBULINOIDES BECKMANNI ZONE

Category: Total range zone

Age: Middle Eocene

Author: Bolli (1957b), renamed by Cordey (1968) and Blow & Saito (1968)

Definition: Total range of *Orbulinoides beckmanni*.

Remarks: The original name of this zone was *Porticulasphaera mexicana* Zone (Bolli, 1957b), but it was changed into *Orbulinoides beckmanni* Zone by Cordey (1968) and Blow & Saito (1968) for taxonomic reasons.

Orbulinoides beckmanni is the youngest member of the *Globigerinatheka subconglobata curryi–G. subconglobata euganea–Orbulinoides beckmanni* evolutionary lineage. Typical assemblages of the zone contain, besides the zonal marker, *Globigerinatheka subconglobata euganea*, *G. mexicana barri*, *G. mexicana kugleri*, *Morozovella lehneri*, *M. spinulosa*, *Hantkenina alabamensis*, *Truncorotaloides topilensis*, *T. rohri*, *Planorotalites pseudoscitula* and '*Hastigerina*' cf. *bolivariana*.

Though well represented in lower latitudes (including the Mediterranean area), *Orbulinoides beckmanni* is generally missing in temperate regions (e.g. Alps). An alternative possibility for recognition of this time interval is given by the *Turborotalia cerroazulensis* lineage: *T. cerroazulensis pomeroli* is generally the only representative of the lineage present, but in some areas (e.g. southern Atlantic, Toumarkine, 1978), *T. cerroazulensis cerroazulensis* may appear already within the *Orbulinoides beckmanni* Zone.

The *Orbulinoides beckmanni* Zone corresponds to the P 13 Zone (*Globigerapsis beckmanni* Zone) of Blow (1979) and also to the older part of the *Turborotalia cerroazulensis pomeroli/Turborotalia cerroazulensis cerroazulensis* Zone (Toumarkine & Bolli, 1970, emended Toumarkine, this volume).

TRUNCOROTALOIDES ROHRI ZONE

Category: Interval zone

Age: Middle Eocene

Author: Bolli (1957b)

Definition: Interval with the zonal marker from last occurrence of *Orbulinoides beckmanni* to last occurrence of *Truncorotaloides rohri*.

Remarks: The extinction of nearly all spinose forms (*Morozovella, Truncorotaloides, Acarinina*) of the Middle Eocene marks the boundary between the Middle and the Late Eocene. Most authors follow this definition. In open-marine conditions of the tropical and subtropical areas, this extinction occurs abruptly, but it is less distinct in cooler regions where the same boundary is often somewhat difficult to trace.

Planorotalites pseudoscitula and '*Hastigerina*' cf. *bolivariana* disappear at the top of the zone. The genus *Globigerinatheka* is represented by the subspecies of *G. mexicana* and *G. index*; in temperate regions (Alpine – Mediterranean area, Caucasus, southern Atlantic), the first representatives of *G. subconglobata luterbacheri* are observed. *Hantkenina alabamensis* and *Pseudohastigerina micra* persist. In the zonation based on the *Turborotalia cerroazulensis* lineage, the *Truncorotaloides rohri* Zone corresponds to the younger part of the *T. cerroazulensis pomeroli/T. cerroazulensis cerroazulensis* Zone which is characterized by the appearance of common *T. cerroazulensis cerroazulensis* close to its base and of *T. cerroazulensis cocoaensis* near its top, whereas *T. cerroazulensis pomeroli* ranges throughout the interval.

Blow (1979) placed the top of his Zone P 14 at the first occurrence of *Globigerinatheka semiinvoluta* which he assumed to be slightly prior to the last occurrence of *Truncorotaloides rohri*.

Late Eocene

GLOBIGERINATHEKA SEMIINVOLUTA ZONE

Category: Interval zone

Age: Late Eocene

Author: Bolli (1957b) modified by Proto Decima & Bolli (1970)

Definition: Interval with zonal marker from last occurrence of *Truncorotaloides rohri* to last occurrence of *Globigerinatheka semiinvoluta*.

Remarks: Tropical and subtropical assemblages of this zone are characterized by *Globigerinatheka semiinvoluta* and advanced forms of the *Turborotalia cerroazulensis* lineage (*T. cerroazulensis cerroazulensis*, *T. cerroazulensis cocoaensis*). They occur together with *T. cerroazulensis pomeroli*, which becomes less frequent, *Hantkenina alabamensis*, *H. primitiva*, *Globigerinatheka index tropicalis* and, near the top of the zone, *Cribrohantkenina inflata*. Other characteristic elements of the assemblages are numerous large species of *Globigerina* (*G. eocaena, G. venezuelana* and others); *Globigerina angiporoides* and representatives of the *G. ampliapertura* group make their first appearances.

In temperate regions, *Globigerinatheka subconglobata luterbacheri* and *G. index index* are abundant whereas the zonal marker may be absent. In some areas (e.g. Alps, southern Atlantic), *Globigerinatheka semiinvoluta* is represented by forms which are considerably smaller than those of fully developed tropical assemblages. In assemblages from higher latitudes, it may become nearly impossible to define the top of the *Globigerinatheka*

semiinvoluta Zone because of the absence of the zonal marker, of the evolved members of the *Turborotalia cerroazulensis* lineage and of *Cribrohantkenina inflata*. In these assemblages, representatives of the genus *Globigerinatheka* may persist into the *Turborotalia cerroazulensis* s.l. Zone or even to the top of the Late Eocene whereas in tropical assemblages they disappear at the top of the *Globigerinatheka semiinvoluta* Zone.

Most of the spinose forms among the planktic foraminifera disappear at the limit between the Middle and the Late Eocene. However, a few generally very small spinose globigerinids and acarininids such as *Globigerina danvillensis*, *G. medizzai*, *Acarinina rugosoaculeata*, '*Globorotalia*' *aculeata* persist into the Late Eocene (e.g. Howe & Wallace, 1932; Jenkins, 1971; Berggren, 1972; Toumarkine & Bolli, 1975).

The *Globigerinatheka semiinvoluta* Zone corresponds to part of the *Turborotalia cerroazulensis cerroazulensis/T. cerroazulensis cocoaensis* Zone (Toumarkine & Bolli, 1970, emended Toumarkine, this volume) and to part of Zone P 15 in Blow (1979).

TURBOROTALIA CERROAZULENSIS S.L. ZONE

Category: Interval zone

Age: Late Eocene

Author: Bolli (1957b), renamed by Bolli (1966, 1972)

Definition: Interval with zonal marker from last occurrence of *Globigerinatheka semiinvoluta* to last occurrence of *Turborotalia cerroazulensis* s.l.

Remarks: The slightly keeled and very flattened *T. cerroazulensis cunialensis* which is the most advanced member of the *Turborotalia cerroazulensis* lineage appears within this zone. In addition to the most highly evolved forms of the *Turborotalia cerroazulensis* lineage, the assemblages of this zone are characterized by the abundance of large species of *Globigerina* (*G. yeguaensis, G. eocaena, G. corpulenta, G. gortanii praeturritilina, G. gortanii gortanii, G. tripartita, G. venezuelana, G. cryptomphala*), *G. angiporoides, G. ampliapertura, Turborotalia increbescens* and *T. opima nana* which continue into the Early Oligocene together with some very small forms (*Pseudohastigerina naguewichiensis, P. barbadoensis, Turborotalia postcretacea*).

The *Turborotalia cerroazulensis cunialensis* Zone of Toumarkine & Bolli (1970) corresponds to the upper part of the *Turborotalia cerroazulensis* s.l. Zone.

A correlation of the *Turborotalia cerroazulensis* s.l. Zone with the zones defined by Blow (1979) is somewhat difficult. It probably corresponds in part to his Zone P 16 which is defined by the total range of *Cribrohant-*

kenina inflata. The last occurrence of this species is somewhat older than that of *Turborotalia cerroazulensis* s.l. and the youngest part of our nominate zone corresponds therefore entirely or partly to the Zone P 17 of Blow.

The Eocene/Oligocene boundary is drawn here at the top of the *Turborotalia cerroazulensis* s.l. Zone, i.e. at the extinction level of the last representatives of the *Turborotalia cerroazulensis* lineage. In tropical areas, the extinction of the genus *Hantkenina* takes place at approximately the same level, but it may occur earlier in subtropical and temperate regions.

Blow (1979) placed the Eocene/Oligocene boundary at the top of his Zone P 17 which he considered to range higher than the *Turborotalia cerroazulensis* s.l. Zone.

Zonation of the Eocene based on the evolutionary lineage of *Turborotalia cerroazulensis*

Because of the absence or patchy distribution of most of the marker species of the standard planktic foraminiferal zonation in assemblages from higher latitudes, Toumarkine & Bolli (1970) introduced an alternative zonation of the Eocene which is based on the succession of subspecies in the evolutionary lineage of *Turborotalia cerroazulensis*. These subspecies have a wider paleogeographic distribution and are also more resistant to dissolution than most of the tropical markers.

The zonation discussed below is a modified version of the original one, because subsequent investigations have shown that some of the subspecies have longer ranges than originally thought. Only five zones are now recognized instead of the original seven (Fig. 2).

The boundaries between the zones are somewhat variable, since it is often difficult to place the transitional forms into one or another subspecies. The zonation is best used in conjunction with parallel evolutionary lineages, e.g. the succession of species and subspecies of the genus *Globigerinatheka*.

TURBOROTALIA CERROAZULENSIS FRONTOSA ZONE

Category: Interval zone
Age: Early to Middle Eocene
Author: Toumarkine & Bolli (1970), emended Toumarkine, this volume
Definition: Interval with the zonal marker from its first occurrence to first occurrence of *Turborotalia cerroazulensis possagnoensis*.
Remarks: The zonal marker appears always in the late Early Eocene at the base of the *Acarinina pentacamerata* Zone. *Turborotalia cerroazulensis possagnoensis* occurs first within the *Hantkenina nuttalli* Zone.

TURBOROTALIA CERROAZULENSIS POSSAGNOENSIS ZONE

Category: Concurrent range zone
Age: Middle Eocene
Author: Toumarkine & Bolli (1970)
Definition: Interval with the zonal marker from its first occurrence to the last occurrence of *Turborotalia cerroazulensis frontosa*.
Remarks: The top of the zone corresponds rather well to the top of the *Morozovella lehneri* Zone.

T. cerroazulensis pomeroli appears near the base of the *Morozovella lehneri* Zone. Its coexistence with the subspecies *frontosa* and *possagnoensis* characterizes the *Morozovella lehneri* Zone and the youngest part of the *Globigerinatheka subconglobata subconglobata* Zone.

TURBOROTALIA CERROAZULENSIS POMEROLI/T. CERROAZULENSIS CERROAZULENSIS ZONE

Category: Interval zone
Age: Middle Eocene
Author: Toumarkine & Bolli (1970) emended Toumarkine, this volume
Definition: Interval with the zonal markers from the last occurrence of *Turborotalia cerroazulensis frontosa* to the extinction of the genus *Truncorotaloides* and other spinose forms.
Remarks: In 1970, Toumarkine & Bolli established a *T. cerroazulensis pomeroli* Zone ranging from the last occurrence of the subspecies *frontosa* and a *T. cerroazulensis pomeroli/T. cerroazulensis cerroazulensis* Zone ranging from the first occurrence of the subspecies *cerroazulensis* to the extinction level of the spinose forms. However, additional investigations have shown that the subspecies *cerroazulensis* often occurs already at the top of the *Morozovella lehneri* Zone (e.g. Ben Ismail-Lattrache, 1981) and the two zones have been merged into a more comprehensive *T. cerroazulensis pomeroli/ T. cerroazulensis cerroazulensis* Zone. The presence alone of the subspecies *pomeroli* is a good indicator of the *Orbulinoides beckmanni* Zone (e.g. Toumarkine & Bolli, 1970).

TURBOROTALIA CERROAZULENSIS CERROAZULENSIS/T. CERROAZULENSIS COCOAENSIS ZONE

Category: Interval zone
Age: Late Eocene
Author: Toumarkine & Bolli (1970) emended Toumarkine, this volume.
Definition: Interval with the zonal markers from the extinction

of *Truncorotaloides* and other spinose forms to the first occurrence of *Turborotalia cerroazulensis cunialensis*.

Remarks: The first occurrence of the subspecies *cocoaensis* can be older than thought by Toumarkine & Bolli in 1970 and the zonation has to be modified accordingly.

TURBOROTALIA CERROAZULENSIS CUNIALENSIS ZONE

Category: Total range zone

Age: Late Eocene

Author: Toumarkine & Bolli (1970)

Definition: Interval with the zonal marker. Its last occurrence corresponds to the extinction of all subspecies of *T. cerroazulensis*.

Remarks: This zone can only be recognized if the ecological conditions favour the presence of the zonal marker. The first representatives of the subspecies *cunialensis* may occur already in the youngest part of the *Globigerinatheka semiinvoluta* Zone, but the presence of numerous and well developed specimens is always indicative of the youngest part of the Late Eocene.

Discussion of taxa

Remarks on the generic classification of Paleogene planktic foraminifera

The classification of planktic foraminifera adopted by Bolli, Loeblich & Tappan (1957), which has been used with minor modifications by a majority of authors during the decade following its publication, was based on exclusively morphological criteria, e.g. all forms with a trochospiral test, an umbilical–extraumbilical aperture and without additional openings are placed within the genus *Globorotalia* which thus includes a wide variety of forms which are not interrelated phylogenetically. From a practical point of view, this purely morphological classification has its advantages, since the stratigraphically useful taxonomic unit is at the species level whereas the generic level can be neglected and needs not to be encumbered with an ever increasing number of generic names based on a more or less subjective interpretation of the phylogenetic relations among the *Globigerinacea* (Luterbacher, 1964).

However, this purely typological approach is rather unsatisfactory from a more academic and theoretical point of view and an evolutionary reclassification of the planktic foraminifera is now advocated by many authors (see Steineck & Fleisher, 1978). Somewhat reluctantly, we have tried to adapt to this trend and abandoned the purely typological generic classification used by us until now (e.g. Stainforth *et al.*, 1975). We have placed within the same genus all the species which are believed to belong to the same phylogenetic lineage. This approach is similar to the one used by Premoli Silva &

Lohmann (personal communication). It is somewhat startling to place '*Globigerina*' *pseudobulloides* and *Globigerina inconstans* as representatives of the genus *Morozovella* but both species are at the base of lineages which lead to typical representatives of this genus.

The same remark holds also for the end-members of the *Turborotalia cerroazulensis* evolutionary lineage. The early subspecies (*T. cerroazulensis frontosa* and *T. cerroazulensis possagnoensis*) have not yet acquired the characteristics of the genus *Turborotalia* whereas the end-form *T. cerroazulensis cunialensis* differs from the strictly morphological definition of the genus by the acquisition of a keel.

For a recent listing of all available genus groups and family groups of planktic foraminifera, reference is made to Banner (1982).

Paleocene to Early Eocene

Early Paleocene small species of *Globigerina* and *Globoconusa*

Planktic foraminiferal faunas from the base of the Paleocene are characterized by a flood of small globigerinids (average size 100 μm) which form the rootstock of the Cenozoic planktic foraminifera. These forms have been treated differently by various authors (for a more detailed discussion see Premoli Silva (1977)). Some of them (Berggren, 1965; Hofker, 1978; Blow, 1979) considered at least the older part of this assemblage (*Globigerina eugubina* Zone) as juvenile or relict specimens of Late Cretaceous rugoglobigerinids, whereas others (e.g. Bolli, 1966) placed them at the base of the Cenozoic planktic foraminiferal assemblages. These differing views are, at least in part, caused by the bad preservation and poor illustration of the assemblages of the *Globigerina eugubina* Zone from the Central Apennines (Luterbacher & Premoli Silva, 1964) and by the widespread gaps in the stratigraphic record at the Cretaceous–Tertiary boundary. In many classical areas, the earliest preserved Paleocene planktic foraminiferal assemblages already belong to a higher evolutionary level.

It seems that generic and specific characteristics are not yet well stabilized and fluctuate somewhat in these early Paleocene assemblages. This has led to the introduction of a large number of species belonging to these early Paleocene planktic foraminifera (Morozova, 1961; Luterbacher & Premoli Silva, 1964; Blow, 1979 and others).

Conclusions on synonymies as well as on stratigraphic positions and evolutionary trends have to be based on careful examination of type-specimens and their succession in complete and undisturbed sections (Bang, 1980). Such an endeavour is beyond the scope of this somewhat pragmatic compilation and we, therefore, stick to a few taxonomic notes on a small number of readily recognizable and stratigraphically useful species.

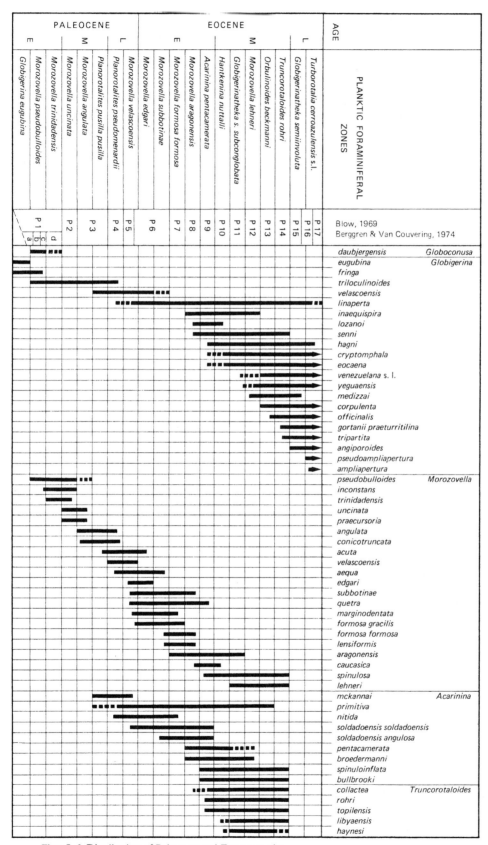

Figs. 5–6. Distribution of Paleocene and Eocene species arranged by genera

	PALEOCENE			EOCENE			AGE
	E	M	L	E	M	L	PLANKTIC FORAMINIFERAL ZONES

Column species (left to right):
Globigerina eugubina, *Morozovella pseudobulloides*, *Morozovella trinidadensis*, *Morozovella uncinata*, *Morozovella angulata*, *Planorotalites pusilla pusilla*, *Planorotalites pseudomenardii*, *Morozovella velascoensis*, *Morozovella edgari*, *Morozovella subbotinae*, *Morozovella formosa formosa*, *Morozovella aragonensis*, *Acarinina pentacamerata*, *Hantkenina nuttalli*, *Globigerinatheka s. subconglobata*, *Morozovella lehneri*, *Orbulinoides beckmanni*, *Truncorotaloides rohri*, *Globigerinatheka semiinvoluta*, *Turborotalia cerroazulensis s.l.*

Zones (Blow, 1969; Berggren & Van Couvering, 1974):
P 1 (a, b, c, d), P 2, P 3, P 4, P 5, P 6, P 7, P 8, P 9, P 10, P 11, P 12, P 13, P 14, P 15, P 16, P 17

Species ranges (right column labels):

compressa	Planorotalites
pusilla pusilla	
chapmani	
pseudomenardii	
pseudoscitula	
palmerae	
wilcoxensis	Pseudohastigerina
micra s. l.	
naguewichiensis s. l.	
eocanica eocanica	Clavigerinella
akersi	
colombiana	
eocanica jarvisi	
nuttalli	Hantkenina
mexicana s. l.	
dumblei	
alabamensis s. l.	
primitiva	
inflata	Cribrohantkenina
bolivariana	"Hastigerina"
cf. bolivariana	
griffinae	Turborotalia
cerroazulensis frontosa	
cerroazulensis possagnoensis	
cerroazulensis pomeroli	
cerroazulensis cerroazulensis	
cerroazulensis cocoaensis	
cerroazulensis cunialensis	
opima nana	
increbescens	
postcretacea	
subconglobata micra	Globigerina-
s. subconglobata	theka
subconglobata curryi	
subconglobata euganea	
subconglobata luterbacheri	
mexicana mexicana	
mexicana barri	
mexicana kugleri	
index rubriformis	
index index	
index tropicalis	
semiinvoluta	
beckmanni	Orbulinoides
higginsi	"Globigerinoides"
carcoselleensis	Globorotaloides
suteri	
dissimilis s. l.	Catapsydrax
cubensis s. l.	Chiloguembelina

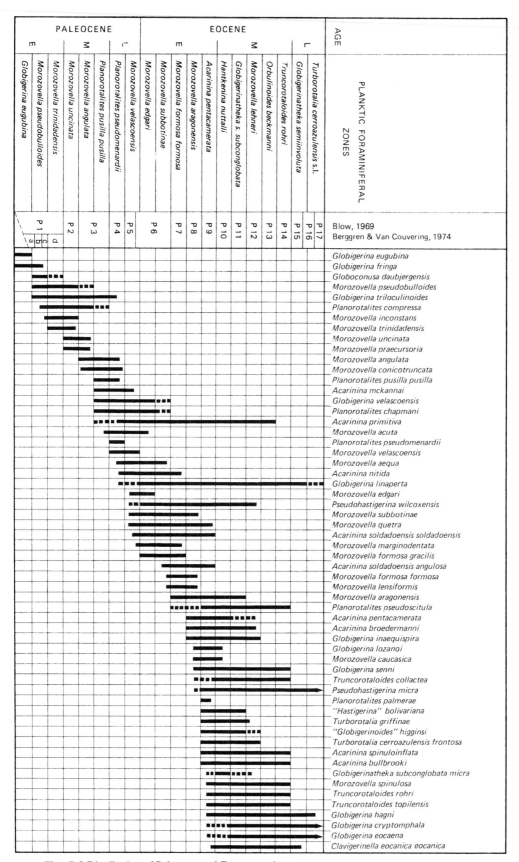

Figs. 7–8 Distribution of Paleocene and Eocene species in order of first occurrence

PALEOCENE			EOCENE			AGE

Zone headers (left to right):

PALEOCENE — E | M | L

EOCENE — E | M | L

Species columns (top, PALEOCENE–EOCENE):

- *Globigerina eugubina*
- *Morozovella pseudobulloides*
- *Morozovella trinidadensis*
- *Morozovella uncinata*
- *Morozovella angulata*
- *Planorotalites pusilla pusilla*
- *Planorotalites pseudomenardii*
- *Morozovella velascoensis*
- *Morozovella edgari*
- *Morozovella subbotinae*
- *Morozovella formosa formosa*
- *Morozovella aragonensis*
- *Acarinina pentacamerata*
- *Hantkenina nuttalli*
- *Globigerinatheka s. subconglobata*
- *Morozovella lehneri*
- *Orbulinoides beckmanni*
- *Truncorotaloides rohri*
- *Globigerinatheka semiinvoluta*
- *Turborotalia cerroazulensis s.l.*

PLANKTIC FORAMINIFERAL ZONES

Blow, 1969
Berggren & Van Couvering, 1974

Zones: P 1 (a b c d) | P 2 | P 3 | P 4 | P 5 | P 6 | P 7 | P 8 | P 9 | P 10 | P 11 | P 12 | P 13 | P 14 | P 15 | P 16 | P 17

Species ranges (right-hand list):

- *Hantkenina nuttalli*
- *Clavigerinella akersi*
- *Clavigerinella colombiana*
- *Clavigerinella eocanica jarvisi*
- *Hantkenina mexicana s. l.*
- *Turborotalia cerroazulensis possagnoensis*
- *Hantkenina dumblei*
- *Globigerinatheka subconglobata subconglobata*
- "*Hastigerina*" cf. *bolivariana*
- *Truncorotaloides libyaensis*
- *Truncorotaloides haynesi*
- *Morozovella lehneri*
- *Globigerinatheka mexicana mexicana*
- *Globorotaloides carcoselleensis*
- *Globigerinatheka index rubriformis*
- *Chiloguembelina cubensis s. l.*
- *Globigerinatheka subconglobata curryi*
- *Globigerinatheka mexicana barri*
- *Globigerinatheka mexicana kugleri*
- *Globigerinatheka index index*
- *Turborotalia cerroazulensis pomeroli*
- *Globigerina venezuelana s. l.*
- *Globigerina yeguaensis*
- *Catapsydrax dissimilis*
- *Globigerina medizzai*
- *Globigerinatheka subconglobata euganea*
- *Hantkenina alabamensis s. l.*
- *Turborotalia cerroazulensis cerroazulensis*
- *Orbulinoides beckmanni*
- *Globigerina corpulenta*
- *Globorotaloides suteri*
- *Globigerinatheka subconglobata luterbacheri*
- *Hantkenina primitiva*
- *Globigerina officinalis*
- *Turborotalia cerroazulensis cocoaensis*
- *Globigerina gortanii praeturritilina*
- *Globigerina tripartita*
- *Globigerinatheka index tropicalis*
- *Turborotalia opima nana*
- *Turborotalia increbescens*
- *Globigerinatheka semiinvoluta*
- *Globigerina angiporoides*
- *Cribrohantkenina inflata*
- *Pseudohastigerina naguewichiensis*
- *Globigerina pseudoampliapertura*
- *Turborotalia postcretacea*
- *Turborotalia cerroazulensis cunialensis*
- *Globigerina ampliapertura*

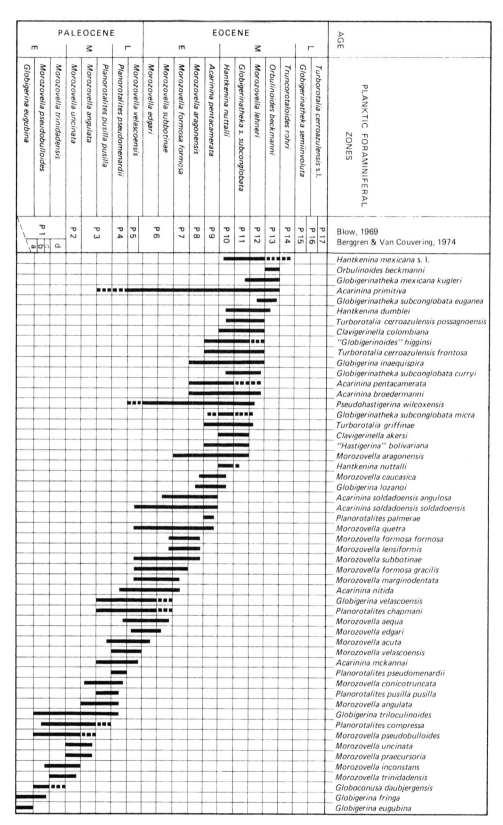

Figs. 9–10. Distribution of Paleocene and Eocene species in order of last occurrence

Range chart. Column headers — age subdivisions:

PALEOCENE			EOCENE			AGE
e	m	l	e	m	l	

Zone-marker species (vertical columns, left to right):

- Globigerina eugubina
- Morozovella pseudobulloides
- Morozovella trinidadensis
- Morozovella uncinata
- Morozovella angulata
- Planorotalites pusilla pusilla
- Planorotalites pseudomenardii
- Morozovella velascoensis
- Morozovella edgari
- Morozovella subbotinae
- Morozovella formosa formosa
- Morozovella aragonensis
- Acarinina pentacamerata
- Hantkenina nuttalli
- Globigerinatheka s. subconglobata
- Morozovella lehneri
- Orbulinoides beckmanni
- Truncorotaloides rohri
- Globigerinatheka semiinvoluta
- Turborotalia cerroazulensis s.l.

PLANKTIC FORAMINIFERAL ZONES

Zones: P1 (a, b, c, d), P2, P3, P4, P5, P6, P7, P8, P9, P10, P11, P12, P13, P14, P15, P16, P17

Blow, 1969
Berggren & Van Couvering, 1974

Species ranges (right-hand list):

- Globigerina ampliapertura
- Globigerina pseudoampliapertura
- Turborotalia postcretacea
- Pseudohastigerina naguewichiensis
- Globigerina angiporoides
- Turborotalia increbescens
- Turborotalia opima nana
- Globigerina tripartita
- Globigerina gortanii praeturritilina
- Globigerina officinalis
- Globigerina corpulenta
- Globorotaloides suteri
- Catapsydrax dissimilis s. l.
- Globigerina yeguaensis
- Globigerina venezuelana
- Chiloguembelina cubensis s. l.
- Globigerina cryptomphala
- Globigerina eocaena
- Pseudohastigerina micra s. l.
- Turborotalia cerroazulensis cunialensis
- Cribrohantkenina inflata
- Turborotalia cerroazulensis cocoaensis
- Hantkenina primitiva
- Turborotalia cerroazulensis cerroazulensis
- Hantkenina alabamensis s. l.
- Turborotalia cerroazulensis pomeroli
- Globigerinatheka index index
- Globigerina linaperta
- Globigerina hagni
- Globigerinatheka index tropicalis
- Globorotaloides carcoselleensis
- Globigerinatheka semiinvoluta
- Globigerinatheka subconglobata luterbacheri
- Globigerinatheka index rubriformis
- Globigerina medizzai
- Globigerinatheka mexicana barri
- Globigerinatheka mexicana mexicana
- Clavigerinella eocanica jarvisi
- Clavigerinella eocanica eocanica
- Morozovella lehneri
- Truncorotaloides haynesi
- Globigerinatheka subconglobata subconglobata
- Truncorotaloides libyaensis
- "Hastigerina" cf. bolivariana
- Truncorotaloides rohri
- Truncorotaloides topilensis
- Morozovella spinulosa
- Acarinina bullbrooki
- Acarinina spinuloinflata
- Truncorotaloides collactea
- Globigerina senni
- Planorotalites pseudoscitula

GLOBIGERINA EUGUBINA Luterbacher & Premoli Silva
Figures **11.1–3**; 5, 7, 10

Type reference: *Globigerina eugubina* Luterbacher & Premoli Silva, 1964, pp. 105–6, pl. 2, fig. 8.

A small form (diameter generally 0.1 mm) with a low trochospire and 5 to 6 subglobular chambers in the last whorl increasing gradually in size. The umbilicus is open, but shallow, and the aperture a low arch in an umbilical to extraumbilical position.

Globigerina taurica is considerably larger and more tightly and regularly coiled. Detailed descriptions and illustrations are given by Smit (1977, 1982). *Globigerina eugubina* probably gave rise to *Planorotalites compressa* (see Premoli Silva, 1977).

Hofker (1978) selected *Globigerina eugubina* as type-species of the genus *Parvularugoglobigerina*, but the types of *Globigerina eugubina* do not show the characteristics attributed by Hofker to this genus (see also Premoli Silva, 1977; Bang, 1980). The generic position of *Globigerina eugubina* is still uncertain.

GLOBIGERINA FRINGA Subbotina
Figures **11.4–5**; 5, 7, 10

Type reference: *Globigerina fringa* Subbotina, 1950, p. 104, pl. 5, figs. 19–21.
Synonym: *Globigerina (Eoglobigerina) eobulloides* Morozova, 1959, p. 1113, fig. 1.

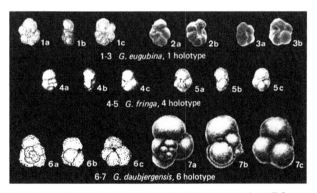

Fig. 11. Early Paleocene small *Globigerina* species (all figures × 60)

1–3 *G. eugubina* Luterbacher & Premoli Silva
1a–c Holotype. Sample 3, earliest Paleocene Scaglia Rossa, Ceselli section, Italy. **2a–b** Topotypes (from Stainforth *et al.*, 1975). **3a–b** DSDP Leg 15-152-10, cc., basal Paleocene, Caribbean Sea (from Stainforth *et al.*, 1975).

4–5 *G. fringa* Subbotina
4a–c Holotype. **5a–c** Paratype. Lower Paleocene, N Caucasus, USSR (from Subbotina, 1953).

6–7 *G. daubjergensis* Brönnimann
6a–c Holotype. Danian of Daubjerg, Denmark. **7a–c** Specimen from the Lower Paleocene of Pamietowo, Poland (=holotype of *Globigerina koslowskii*, from Brotzen & Pozaryska, 1961).

A small and tightly coiled trochospire with 4 chambers in the last whorl increasing rapidly in size. *Globigerina fringa* is probably the ancestor of *Morozovella pseudobulloides*.

GLOBOCONUSA DAUBJERGENSIS (Brönnimann)
Figures **11.6–7**; 5, 7, 10

Type reference: *Globigerina daubjergensis* Brönnimann, 1953, p. 340, fig. 1.
Synonyms: *Globoconusa conusa* Khalilov, 1956, p. 249, pl. 5, fig. 2.
Globigerina kozlowskii Brotzen & Pozaryska, 1961, p. 161, pl. 1–3.

Globoconusa daubjergensis stands here for an entire plexus of high-trochospiral forms with a relatively small and tapering inner whorl and 3 to 4 chambers in the last whorl. In typical specimens, the wall-structure is hispid with scattered small pores.

Early representatives of the *Globoconusa daubjergensis* plexus are often somewhat transitional to *Guembelitria* and display a smooth surface. The evolution of *G. daubjergensis* has been studied by Hofker (1962) and Hansen (1970). It consists of an increase in height of the spire, the appearance of sutural openings on the spiral side and acquisition of an umbilical bulla.

'*Globigerina kozlowskii*' is representative for the final evolutionary stages of *Globoconusa daubjergensis*. According to Premoli Silva (1977), the *Globoconusa daubjergensis* plexus evolves from a long-axis '*Guembelitria*' stage to a short-axis '*Guembelitria*' stage, to a '*Globigerina*' stage and finally to a '*Globigerinoides*' and to a '*Catapsydrax*' stage.

The range of *Globoconusa daubjergensis* does not extend beyond the base of the Middle Paleocene and its lineage seems to end there.

Globorotalia with compressed and smooth tests (*Planorotalites pseudomenardii* group)

The forms of the *Planorotalites pseudomenardii* group are a good example of the problems of a phylogenetically meaningful generic classification as advocated by Steineck & Fleisher (1978) and others and its consequent proliferation of generic names based on a more or less subjective interpretation of the iterative evolution of the Globigerinacea.

Several authors (e.g. McGowran, 1968; Berggren, 1977) placed the forms attributed here to the *Planorotalites pseudomenardii* group within the genus *Planorotalites* Morozova, 1957. However, the type-species of this genus – *Globorotalia pseudoscitula* Glaessner, 1937 – is phylogenetically unrelated to the smooth lenticular forms derived from *Planorotalites compressa*. The generic name *Planorotalia* Morozova, 1957 which was erected to include the forms of the *pseudomenardii* group is not available at present (misidentified type-species,

ICZN Article 70b), until the problem is resolved by the International Commission of Zoological Nomenclature. Blow (1979) attributed forms of the *Planorotalites pseudomenardii* group to the subgenus *Turborotalia* which has the phylogenetically unrelated *Globorotalia centralis* Cushman & Bermudez, 1937 as type-species.

Among the numerous species and subspecies belonging to the *Planorotalites pseudomenardii* group (e.g. Blow, 1979), we have retained for this handbook *P. compressa, P. chapmani, P. pseudomenardii, P. pusilla pusilla* and *Pseudohastigerina wilcoxensis* since they are thought to be stratigraphically useful and readily recognizable.

PLANOROTALITES COMPRESSA (Plummer)
Figures 12.1–2; 6, 7, 10

Type reference: *Globigerina compressa* Plummer, 1926, p. 125, pl. 8, fig. 11.

A fragile, smooth-shelled test with a low trochospire and a rounded to slightly subacute periphery in lateral view. The chambers increase rather rapidly in size. *P. compressa* is thought to be the predecessor of *P. chapmani* with which it is linked by a series of intermediate forms in the older part of the Middle Paleocene.

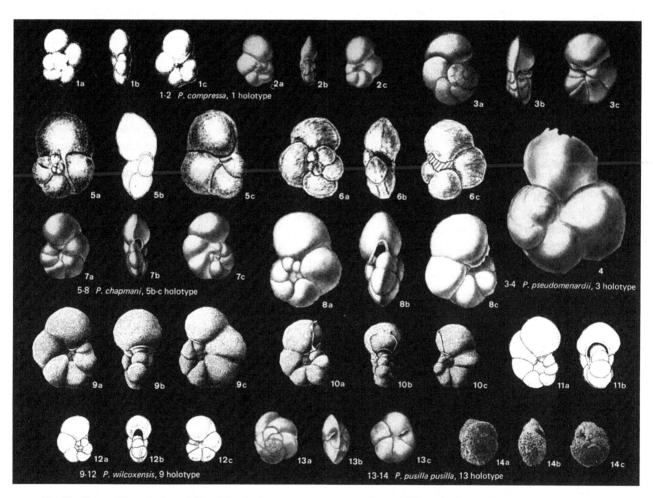

Fig. 12. *Planorotalites* species and *Pseudohastigerina wilcoxensis* (all figures × 60)

1–2 *P. compressa* (Plummer)
   1a–c Holotype. Paleocene Midway Fm., Navarro County, Texas, USA. **2a–c** Specimen from the *P. pusilla pusilla* Zone, Lizard Springs Fm., Trinidad (from Bolli, 1957a).

3–4 *P. pseudomenardii* (Bolli)
   3a–c Holotype. **4** Spiral view of large paratype, *P. pseudomenardii* Zone, Paleocene, Lizard Springs Fm., Trinidad (from Bolli, 1957a).

5–8 *P. chapmani* (Parr)
   5a Paratype, **5b–c** Holotype. 'Upper Eocene' of W Australia. **6a–c** Specimen from the Paleocene King's Park Shale, Perth, W Australia (from McGowran, 1964). **7a–c** Specimen from the *P. pseudomenardii* Zone, Paleocene, Lizard springs Fm., Trinidad (from Bolli, 1957a,

given as '*Globorotalia elongata* Glaessner'). **8a–c** Specimen from the Paleocene Boongeroda Greensand, Perth, W Australia (from Berggren, Olsson & Reyment, 1967).

9–12 *P. wilcoxensis* (Cushman & Ponton)
   9a–c Holotype of '*Hastigerina eocaenica*' Berggren, Early Eocene 'Tarras Ton', Fehmarn, W Germany. **10a–c** Specimen from the Early Eocene Roesnaes Clay, Rögle Klint, Denmark (9–10 from Berggren, 1960a). **11a–b** Specimens from the Early Eocene Bashi Fm., Alabama, USA. **12a–c** Specimens from the Early Eocene Manasquan Fm., New Jersey, USA (11–12 from Berggren *et al.*, 1967).

13–14 *P. pusilla pusilla* (Bolli)
   13a–c Holotype. *P. pusilla pusilla* Zone, Paleocene, Lizard Springs Fm., Trinidad. **14a–c** DSDP Leg 14-144A-3-4, 120–122 cm, Paleocene, northwestern South Atlantic (from Stainforth *et al.*, 1975).

PLANOROTALITES CHAPMANI (Parr)
Figures **12.5–8**; 6, 7, 10

Type reference: *Globorotalia chapmani* Parr, 1938, p. 87, pl. 9, figs. 8–9.

Synonyms: *Globorotalia ehrenbergi* Bolli, 1957a, p. 77, pl. 20, figs. 8–20.

Globorotalia troelseni Loeblich & Tappan, 1957, p. 196, pl. 60, fig. 11, pl. 63, fig. 5.

The very low biconvex test is almost equilateral with the umbilical side only slightly more convex than the flat spiral side. The periphery is subacute in axial view. *P. chapmani* is intermediate between *P. compressa* and *P. pseudomenardii* with its imperforate keel. Within the Late Paleocene, *Pseudohastigerina wilcoxensis*, a completely equilateral form, develops from *P. chapmani* (see Berggren, Olsson & Reyment, 1967).

Blow (1979, p. 1059) asserts that the holotype of *P. chapmani* has to be attributed to the *Valvulineridae* rather than to the planktic foraminifera. If this view is substantiated, the Paleocene globorotaliids hitherto named *P. chapmani* could be named *P. ehrenbergi* Bolli, 1957a.

PLANOROTALITES PSEUDOMENARDII (Bolli)
Figures **12.3–4**; 6, 7, 10

Type reference: *Globorotalia pseudomenardii* Bolli, 1957a, p. 77, pl. 20, figs. 14–17.

The biconvex, lenticular test has acquired an imperforate keel, but otherwise resembles closely *P. chapmani*. Blow (1979) extended drastically the stratigraphic range of this species, but this view does not coincide with our own experience. *P. pseudomenardii* is a reliable and easily recognizable marker for its nominate zone in the Late Paleocene.

PSEUDOHASTIGERINA WILCOXENSIS (Cushman & Ponton)
Figures **12.9–12**; 6, 7, 10

Type reference: *Nonion wilcoxensis* Cushman & Ponton, 1932, p. 64, pl. 8, fig. 11.

Synonym: *Hastigerina eocaenica* Berggren, 1960a, pp. 85–91, pl. 5, figs. 1–2.

The small tests are biumbilicate, planispirally coiled and evolute. In lateral view, the periphery is rounded, although in some specimens it has a tendency to become subacute. The species intergrades with *Planorotalites chapmani* in the Late Paleocene and with *Pseudohastigerina micra* in the early Middle Eocene.

PLANOROTALITES PUSILLA PUSILLA (Bolli)
Figures **12.13–14**; 6, 7, 10

Type reference: *Globorotalia pusilla pusilla* Bolli, 1957a, p. 78, pl. 20, figs. 8–10.

The small, biconvex tests are tightly coiled with 5 to 6 crescentic chambers in the last whorl increasing gradually in size. On the spiral side, the sutures are strongly curved backwards. The umbilicus is narrow. The surface of the test in well-preserved specimens is coarsely perforate and pitted, but it may be smooth in abraded specimens.

The generic attribution and the position of *Planorotalites pusilla pusilla* within the lineages of Paleogene planktic foraminifera are still uncertain. It is placed here in the genus *Planorotalites*, but it probably should be placed in a genus of its own.

Keeled *Morozovella* species of the Paleocene to Early Eocene
The *Morozovella conicotruncata* lineage

The *Morozovella* species and their predecessors of the Paleocene and the Early Eocene can be subdivided into several lineages (Berggren, 1966; Luterbacher, 1966), which consist of a succession of stratigraphically useful species. Whereas there exists a fairly good agreement on the general morphologic trends within these lineages, the subdivision of the forms into species and the definition of these species is handled differently by various authors.

One of the most conspicuous 'lineages' is the one which has *Morozovella velascoensis* as its end-member. It starts with forms with globular chambers and ends with angular-conical forms with a heavy peripheral 'keel' (carina or muricocarina).

Fig. 13. *Morozovella conicotruncata* and related species (all figures × 60)

1–2 *M. inconstans* (Subbotina)
1a–c Holotype. **2a–c** Paratype. Paleocene, N Caucasus, USSR (from Subbotina, 1953).

3–4 *M. trinidadensis* (Bolli)
3a–c Holotype. **4a–c** Transitional form between *Morozovella trinidadensis* and *Morozovella praecursoria*, *M. trinidadensis* Zone, Paleocene, Lizard Springs Fm., Trinidad (from Bolli, 1957a).

5 *M. praecursoria* (Morozova)
5a–c Holotype. Early Paleocene, N Caucasus, USSR.

6–10 *M. conicotruncata* (Subbotina)
6a–c Holotype. **7a–c, 8a–c** Paratypes. Paleocene, N Caucasus, USSR (from Subbotina, 1953). **9a–c, 10a–c** Specimens from the *P. pusilla pusilla* Zone, Paleocene, Lizard Springs Fm., Trinidad. (9 holotype of *Globorotalia angulata abundocamerata* Bolli, 10 given as *Globorotalia angulata* (White), from Bolli, 1957a.)

11–12 *M. velascoensis* (Cushman)
11a–c Specimen from the *P. pseudomenardii* Zone, Paleocene, Lizard Springs Fm., Trinidad (from Bolli, 1957a). **12a–c** Holotype. Paleocene, Velasco Fm., Tampico Embayment, Mexico.

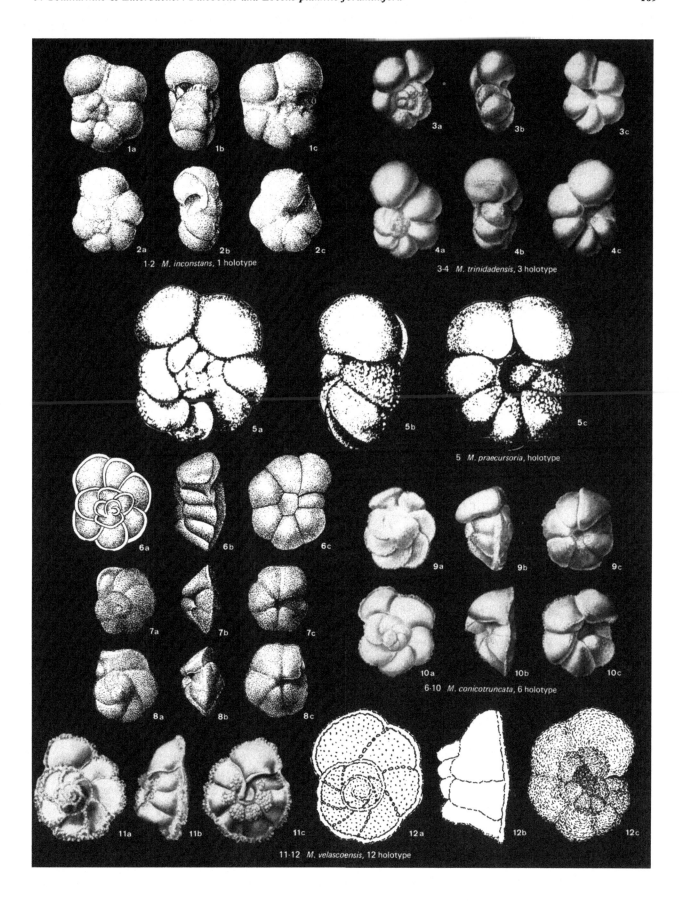

1-2 *M. inconstans*, 1 holotype

3-4 *M. trinidadensis*, 3 holotype

5 *M. praecursoria*, holotype

6-10 *M. conicotruncata*, 6 holotype

11-12 *M. velascoensis*, 12 holotype

MOROZOVELLA INCONSTANS (Subbotina)
Figures **13.1–2**; 5, 7, 10

Type reference: *Globigerina inconstans* Subbotina, 1953, p. 58, pl. 3, figs. 1–2.

The low trochospiral test has 5 to 6 globular chambers in the last whorl increasing fairly gradually in size, and a wide and shallow umbilicus. The spiral side is flattened with the small early portion of the test slightly depressed. The sutures are depressed.

Morozovella inconstans differs from *M. trinidadensis* by its globular chambers without any differentiation in the ornamentation and by the generally smaller number of chambers in the last whorl. Contrary to Blow (1979), we prefer to keep the two species separated since they represent successive steps in the evolution of the lineage (see also Berggren, 1965; Luterbacher, 1964). *M. pseudobulloides* has fewer chambers in the last whorl and a narrower umbilicus.

MOROZOVELLA TRINIDADENSIS (Bolli)
Figures **13.3–4**; 5, 7, 10

Type reference: *Globorotalia trinidadensis* Bolli, 1957a, p. 73, pl. 16, figs. 19–23.

M. praecursoria is similar to *M. trinidadensis*, but the initial chambers of its final whorl are more strongly angular. *M. trinidadensis* and *M. praecursoria* represent two successive stages in the acquisition of the conical shape leading to the typical *Morozovella* species.

MOROZOVELLA PRAECURSORIA (Morozova)
Figures **13.5**; 5, 7, 10

Type reference: *Acarinina praecursoria* Morozova, 1957, p. 1111, fig. 1.

Morozovella praecursoria differs from *M. uncinata* by the larger number of chambers in the last whorl, which results in a wider umbilicus. The two species represent the same evolutionary step in two different evolutionary branches leading to the typical *Morozovella* species. The trend is for the acquisition of subangular chambers and the concentration of the spines at the periphery which starts in the early chambers of the last whorl.

MOROZOVELLA CONICOTRUNCATA (Subbotina)
Figures **13.6–10**; 5, 7, 10

Type reference: *Globorotalia conicotruncata* Subbotina, 1947, pp. 115–17, pl. 4, figs. 11–13, pl. 9, figs. 9–11.
Synonym: *Globorotalia angulata abundocamerata* Bolli, 1957a, p. 74, pl. 17, figs. 4–6.

The test is conicotruncate in lateral view with 5 to 8 tightly joined angular-conical chambers which increase only slowly in size. The spines on the surface of the test tend to be more concentrated on the periphery, but do not fuse to form a heavy 'keel' as in *Morozovella velascoensis*.

MOROZOVELLA VELASCOENSIS (Cushman)
Figures **13.11–12**; 5, 7, 10

Type reference: *Pulvinulina velascoensis* Cushman, 1925c, p. 19, pl. 3, fig. 5.

Typical representatives of *Morozovella velascoensis* are close to their predecessor *Morozovella conicotruncata* to which they are linked by series of transitional forms. Characteristic for the species are the raised and beaded intercameral sutures on the spiral side and the heavy spines on the periphery which fuse to form a heavy 'keel' (muricocarina). Such spines are also present on the umbilical chamber tips.

The *Morozovella angulata* lineage

The *Morozovella angulata* lineage is parallel to the *Morozovella conicotruncata* lineage and demonstrates a virtually identical and contemporaneous evolution from globigeriniid (*M. pseudobulloides*) to morozovellid forms (*M. acuta*). At all levels, typical representatives of the two lineages are linked by intermediate forms. This parallel and interlinked development probably indicates that we are not dealing with two separated phylogenetic lineages, but rather with a uniform plexus of which the two lineages described here represent the easily distinguishable forms.

MOROZOVELLA PSEUDOBULLOIDES (Plummer)
Figures **14.1–2**; 5, 7, 10

Type reference: *Globigerina pseudobulloides* Plummer, 1926, p. 33, pl. 8, fig. 9.

The low trochospiral test generally has 5 chambers in the last whorl. The chambers are spherical to ovate and increase rapidly in size as added. The surface of the test does not display any differentiation towards the periphery.

Globigerina fringa is smaller and more tightly coiled, whereas *Morozovella uncinata* has subangular chambers in the early part of the last whorl.

MOROZOVELLA UNCINATA (Bolli)
Figures **14.3–4**; 5, 7, 10

Type reference: *Globorotalia uncinata* Bolli, 1957a, p. 74, pl. 17, figs. 13–15.

The main characteristics of *Morozovella uncinata* are the angular-conical shape of the early chambers in the last whorl and the strongly backwardly curved sutures on the spiral side (see also taxonomic notes on *M. praecursoria*).

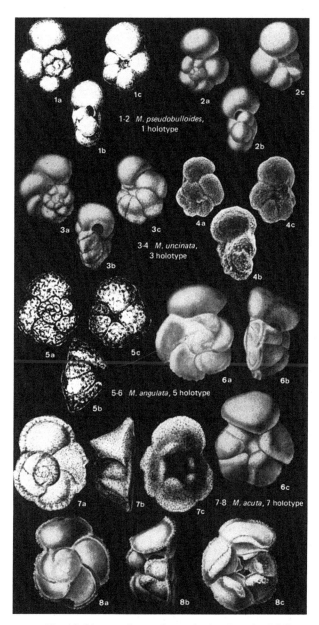

Fig. 14. *Morozovella angulata* and related species (all figures × 60)

1–2 *M. pseudobulloides* (Plummer)
> **1a–c** Holotype. Paleocene, Midway Fm., Navarro County, Texas, USA. **2a–c** Specimen from the *M. trinidadensis* Zone, Paleocene, Lizard Springs Fm., Trinidad (from Bolli, 1957a).

3–4 *M. uncinata* (Bolli)
> **3** Holotype. *M. uncinata* Zone, Paleocene, Lizard Springs Fm., Trinidad. **4** Specimen from the Paleocene of Crimea, USSR (from Stainforth *et al.*, 1975).

5–6 *M. angulata* (White)
> **5a–c** Holotype. Paleocene, Velasco Fm., Tampico Embayment, Mexico. **6a–c** Specimen from the *P. pusilla pusilla* Zone, Paleocene, Lizard Springs Fm., Trinidad (from Bolli, 1957a).

7–8 *M. acuta* (Toulmin)
> **7a–c** Holotype. Late Paleocene (to Early Eocene?), Salt Mountain Limestone, Alabama, USA. **8a–c** Specimen from Vincentown Formation, New Jersey, USA (from Loeblich & Tappan, 1957).

MOROZOVELLA ANGULATA (White)
Figures **14.5–6**; 5, 7, 10

Type reference: *Globigerina angulata* White, 1928, p. 191, pl. 27, fig. 13.

In this species – the successor of *Morozovella uncinata* – all chambers of the last whorl are angular-conical, but it does not yet have a peripheral 'keel', although the delicate spines are concentrated around the periphery and the umbilical shoulders. The 'parallel' species, *M. conicotruncata*, has more, and therefore less rapidly enlarging, chambers in the last whorl.

MOROZOVELLA ACUTA (Toulmin)
Figures **14.7–8**; 5, 7, 10

Type reference: *Globorotalia wilcoxensis* Cushman & Ponton var. *acuta* Toulmin, 1941, p. 608, pl. 82, figs. 6–8.

The conicotruncate test of this species has a well developed 'keel' and ornamented umbilical shoulders at least in the early chambers of the last whorl. The last chamber occupies up to one-third of the whorl. *M. velascoensis* and *M. acuta* are linked by intermediate forms, but the forms attributed to *M. acuta* have a wider geographic and statigraphic distribution.

The *Morozovella subbotinae* lineage

The *Morozovella subbotinae* lineage comprises several branches which can be traced back into the Middle to Late Paleocene to ancestral forms which are probably close to *Morozovella aequa*. End-forms of the different branches are *Morozovella formosa formosa*, *M. caucasica* and, probably, *M. marginodentata*. The original authors of *M. edgari* (Premoli Silva & Bolli, 1973) regard this taxon as closely related to *M. velascoensis* and not as belonging to the *M. aequa–M. marginodentata* branch.

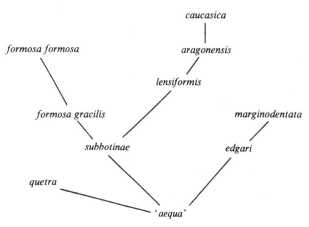

Hypothetical relations among Early Eocene morozovellas

MOROZOVELLA AEQUA (Cushman & Renz)
Figures **15.1–3**; 5, 7, 10

Type reference: *Globorotalia crassata* (Cushman) var. *aequa* Cushman & Renz, 1942, p. 12, pl. 3, fig. 3.

The last whorl has $3\frac{1}{2}$ to $4\frac{1}{2}$ angular-conical chambers which increase rapidly in size as added. The lobate periphery has a faint 'keel' at least in the early part of the whorl. The sutures on the spiral side are depressed or flush, the chambers somewhat imbricated.

Morozovella aequa differs from *M. angulata* in being less lobate and more tightly coiled.

MOROZOVELLA SUBBOTINAE (Morozova)
Figures **15.9–11**; 5, 7, 10

Type reference: *Globorotalia subbotinae* Morozova, 1929, pp. 80–1, pl. 2, fig. 16.
Synonym: *Globorotalia rex* Martin, 1943, p. 117, pl. 8, fig. 2.

The last whorl has generally 4 to 5 (rarely 6) chambers which increase fairly rapidly in size with the last chamber occupying one-third to one-fourth of the whorl. The periphery has a spinose 'keel', but the spines forming the 'keel' are not fused. The umbilicus is narrow and deep. *Morozovella subbotinae* differs from *M. aequa* by the more fragile, generally tightly coiled aspect and by the distinct spinose 'keel'.

MOROZOVELLA FORMOSA GRACILIS (Bolli)
Figures **15.12**; 5, 7, 10

Type reference: *Globorotalia formosa gracilis* Bolli, 1957a, p. 75, pl. 18, figs. 4–6.

The last whorl consists of 5–6 chambers which increase rapidly, but regularly in size. The periphery has a distinct 'keel' throughout the entire last whorl. The sutures on the spiral side are strongly curved and frequently beaded. The umbilicus is open and deep, the umbilical shoulders of the chambers are rounded and without bundles of spines.

M. subbotinae has less chambers in the last whorl and these increase more rapidly in size. There is only a light 'keel'. *M. lensiformis* is more tightly coiled.

MOROZOVELLA FORMOSA FORMOSA (Bolli)
Figures **15.13**; 5, 7, 10

Type reference: *Globorotalia formosa formosa* Bolli, 1957a, p. 76, pl. 18, figs. 1–3.

The last whorl has 6 to 7 (rarely 8) chambers which increase regularly and relatively slowly in size. The periphery has a spinose 'keel' throughout. The umbilicus is open and deep, surrounded by smooth umbilical shoulders. The sutures on the spiral side are raised, generally beaded and pass smoothly without any break into the peripheral 'keel'.

MOROZOVELLA LENSIFORMIS (Subbotina)
Figures **16.1**; 5, 7, 10

Type reference: *Globorotalia lensiformis* Subbotina, 1953, p. 214, pl. 18, figs. 4–5.

The test is tightly coiled, with a narrow umbilicus and 4 to 5 chambers in the last whorl increasing rapidly in size. Sutures on spiral side are curved, depressed or flush. The development of the peripheral 'keel' is weak.

Morozovella lensiformis differs from *M. subbotinae* by its more compact shape and tighter coiling.

MOROZOVELLA ARAGONENSIS (Nuttall)
Figures **16.4–6**; 5, 7, 10

Type reference: *Globorotalia aragonensis* Nuttall, 1930, p. 288, pl. 24, figs. 6–11.

The test is rather tightly coiled and has 5 to 7 chambers in the last whorl. The periphery is almost circular and displays a distinct 'keel'. The sutures on the spiral side are flush or slightly raised and form a distinctive, abrupt angle with the peripheral 'keel'.

Morozovella aragonensis differs from *M. lensiformis* mainly by the larger number of chambers in the last whorl, the almost circular periphery and the heavier 'keel'.

Fig. 15. *Morozovella subbotinae* and related species (all figures × 60)

1–3 *M. aequa* (Cushman & Renz)
　　1a–c Holotype. Paleocene, Soldado Fm., Soldado Rock, Trinidad. **2a–c, 3a–c** Specimens from the *M. velascoensis* Zone, Paleocene, Lizard Springs Fm., Trinidad (from Bolli, 1957a).

4–5 *M. quetra* (Bolli)
　　4a–c Holotype, **5a–c** Paratype. *M. formosa formosa* Zone, Eocene, Lizard Springs Fm., Trinidad (from Bolli, 1957a).

6 *M. edgari* (Premoli Silva & Bolli)
　　6a Holotype, **6b–c** Paratypes. DSDP Leg 15-152-3-2, 136–138 cm, *M. edgari* Zone, Eocene, Caribbean Sea (from Premoli Silva & Bolli, 1973).

7–8 *M. marginodentata* (Subbotina)
　　7a–c Holotype. Late Paleocene to Early Eocene, N Caucasus. **8a–c** Specimen from the Early Eocene, Caucasus (from Stainforth *et al.*, 1975).

9–11 *M. subbotinae* (Morozova)
　　9a–c Specimen from the Late Paleocene to Early Eocene Lodo Fm., California, USA (holotype of *Globorotalia rex* Martin, from Martin, 1943). **10a–b** Holotype. Lower Eocene, Emba region, USSR. **11a–c** Specimen from the Early Eocene, NW Crimea, USSR (from Subbotina, 1953).

12 *M. formosa gracilis* (Bolli)
　　12a–c Holotype. *M. subbotinae* Zone, Eocene, Lizard Springs Fm., Trinidad.

13 *M. formosa formosa* (Bolli)
　　13a–c Holotype. *M. formosa formosa* Zone, Eocene, Lizard Springs Fm., Trinidad.

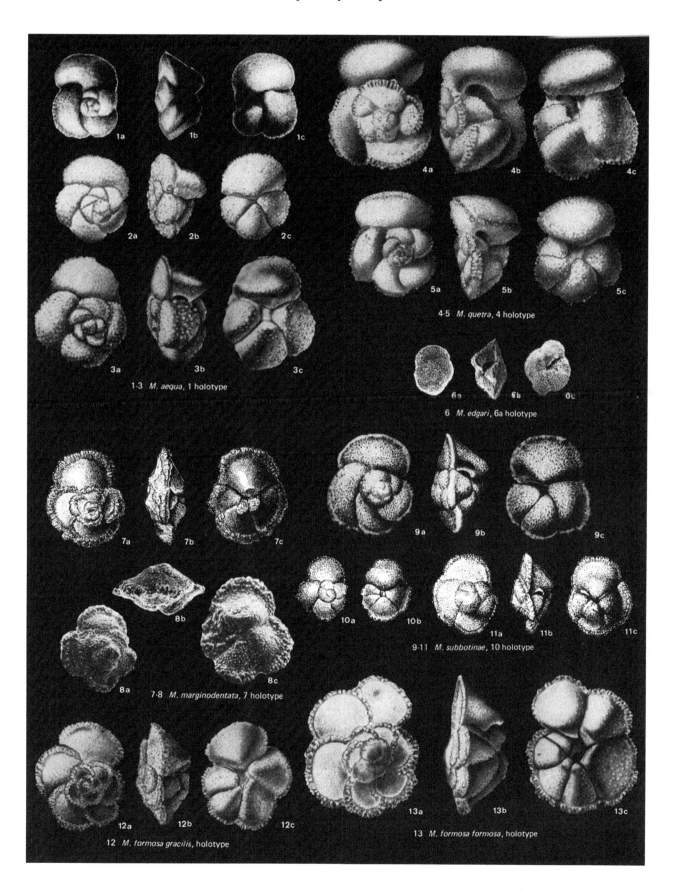

4-5 *M. quetra*, 4 holotype

6 *M. edgari*, 6a holotype

1-3 *M. aequa*, 1 holotype

7-8 *M. marginodentata*, 7 holotype

9-11 *M. subbotinae*, 10 holotype

12 *M. formosa gracilis*, holotype

13 *M. formosa formosa*, holotype

MOROZOVELLA CAUCASICA (Glaessner)
Figures **16.**2–3; 5, 7, 10

Type reference: *Globorotalia aragonensis* Nuttall var. *caucasica* Glaessner, 1937, p. 31, pl. 1, fig. 6.

The conicotruncate test is strongly ornamented with a heavy 'keel' and bundles of spines on the umbilical shoulders and is an Early Eocene homeomorph of the Late Paleocene *Morozovella velascoensis*. The last whorl is composed of 5 to 8 chambers which increase regularly in size. The umbilicus is wide and deep.

Single specimens of *Morozovella velascoensis* and *M. caucasica* are often virtually identical but can be distinguished by the co-occurrence of transitional forms to *M. acuta* and *M. aragonensis* respectively.

MOROZOVELLA QUETRA (Bolli)
Figures **15.**4–5; 5, 7, 10

Type reference: *Globorotalia quetra* Bolli, 1957a, pp. 79–80, pl. 19, figs. 1–6.

Last whorl with 4 to 5 angular-conical to subangular chambers which are longer than broad and attached to each other at almost right angles. The periphery is strongly lobate, acute to subacute in early chambers, becoming subacute to rounded in younger chambers. The test is spinose, with spines more densely concentrated at the periphery in early chambers.

MOROZOVELLA EDGARI (Premoli Silva & Bolli)
Figures **15.**6; 5, 7, 10

Type reference: *Globorotalia edgari* Premoli Silva & Bolli, 1973, p. 526, pl. 7, figs. 10–12, pl. 8, figs. 1–12.

This species is characterized by its small size, fragile aspect with only a weakly developed 'keel', and the relatively large number of chambers ($4\frac{1}{2}$ to 6) in the last whorl.

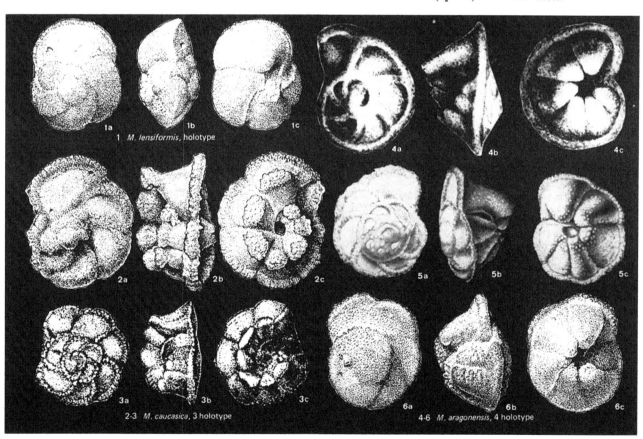

Fig. 16. *Morozovella aragonensis* and related species (all figures × 60)

1 *M. lensiformis* (Subbotina)
  1a–c Holotype. Early Eocene, N Caucasus, USSR.
2–3 *M. caucasica* (Glaessner)
  2a–c Specimen from the Early Eocene, N Caucasus, USSR (from Subbotina, 1953). **3a–c** Holotype. Early Eocene, N Caucasus, USSR.

4–6 *M. aragonensis* (Nuttall)
  4a–c Holotype. Early Eocene, Aragon Fm., Tampico Embayment, Mexico. **5** Specimen from the *M. aragonensis* Zone, Eocene, Lizard Springs Fm., Trinidad (from Bolli, 1957a). **6a–c** Specimen from the Early Eocene, Caucasus (from Subbotina, 1953).

It differs from *Morozovella subbotinae* by the larger number of chambers in the last whorl and the less lobate peripheral outline. *M. formosa gracilis* is more robust and has a better-developed 'keel'.

MOROZOVELLA MARGINODENTATA
(Subbotina)
Figures 15.7–8; 5, 7, 10

Type reference: *Globorotalia marginodentata* Subbotina, 1953, pp. 212–13, pl. 17, figs. 14–16, pl. 18, figs. 1–3.

The test is umbilico-convex to lenticular, with acute, lobate periphery with broad and frilled 'keel'. The 4 to 6 chambers of the last whorl increase rapidly in size. The umbilicus is very narrow and surrounded by rounded to sharp and ornamented umbilical chamber tips.

Morozovella marginodentata differs from *M. subbotinae* and other coeval *Morozovella* species by the broad 'keel' which gives the periphery a very characteristic pinched look.

The *Acarinina soldadoensis* group
ACARININA SOLDADOENSIS
SOLDADOENSIS (Brönnimann)
Figures 17.1–2; 5, 7, 10

Type reference: *Globigerina soldadoensis* Brönnimann, 1952a, p. 7, pl. 1, figs. 1–9.

The spiral side of the compact test is flattened, the umbilical side inflated. The 4 to 5 chambers of the final whorl increase gradually in size and are somewhat imbricated on the spiral side. The entire surface of the test is covered with spines which are somewhat more prominent on the umbilical side.

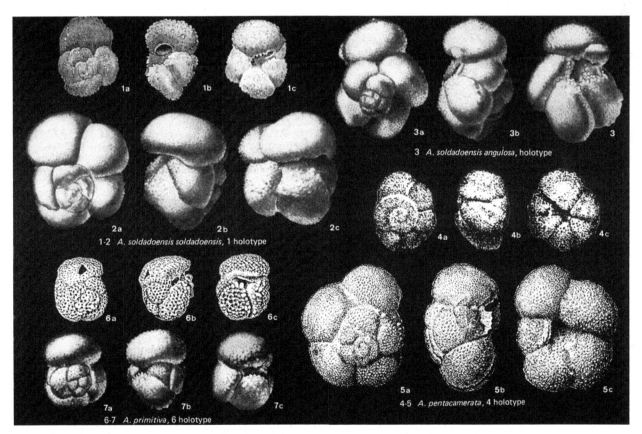

Fig. 17. *Acarinina soldadoensis* and related species (all figures ×60)

1–2 *A. soldadoensis soldadoensis* (Brönnimann)
  1a–c Holotype. Early Eocene part of Lizard Springs Fm., Trinidad.
  2a–c Specimen from the *M. formosa formosa* Zone, Eocene, Lizard Springs Fm., Trinidad (from Bolli, 1957a).
  3 *A. soldadoensis angulosa* (Bolli)
  3a–c Holotype. *M. formosa formosa* Zone, Eocene, Lizard Springs Fm., Trinidad.

4–5 *A. pentacamerata* (Subbotina)
  4a–c Holotype. Early Eocene, N Caucasus (from Subbotina, 1947).
  5a–c Specimen from the Early Eocene, N Caucasus (from Subbotina, 1953).
6–7 *A. primitiva* (Finlay)
  6a–c Holotype. Early (or Middle?) Eocene, New Zealand (from Jenkins, 1971). **7a–c** Specimen from the *M. subbotinae* Zone, Eocene, Lizard Springs Fm., Trinidad (from Bolli, 1957a).

ACARININA SOLDADOENSIS ANGULOSA
(Bolli)
Figures 17.3; 5, 7, 10

Type reference: *Globigerina soldadoensis angulosa* Bolli, 1957a, p. 71, pl. 16, figs. 4–6.

This subspecies is characterized by the axially elongate chambers in the last whorl which are strongly imbricated on the spiral side. *Acarinina soldadoensis angulosa* differs from *Morozovella quetra* by the rounded periphery.

ACARININA PENTACAMERATA (Subbotina)
Figures 17.4–5; 5, 7, 10

Type reference: *Globorotalia pentacamerata* Subbotina, 1947, pp. 128–9, pl. 7, figs. 12–17, pl. 9, figs. 24–26.

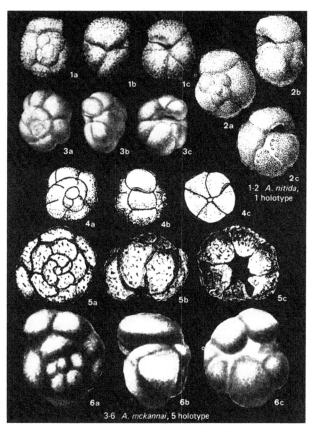

Fig. 18. *Acarinina mckannai* and related species (all figures × 60)

1–2 *A. nitida* (Martin)
  1 Holotype. Early Eocene part of the Lodo Fm., California, USA. **2a–c** Specimen from the Early Eocene, N Caucasus, USSR (holotype of *Acarinina acarinata* Subbotina, from Subbotina, 1953).
3–6 *A. mckannai* (White)
  3a–c Specimen from the *P. pseudomenardii* Zone, Paleocene, Lizard Springs Fm., Trinidad (from Bolli, 1957a). **4a–c** Specimen from the Paleocene, N Caucasus, USSR (holotype of *Acarinina subsphaerica* (Subbotina), from Subbotina, 1953). **5a–c** Holotype. Paleocene, Velasco Fm., Tampico Embayment, Mexico. **6a–c** Specimen from the Paleocene, N Caucasus, USSR (from Shutskaya, 1958).

The relatively large test has a flattened spiral and an inflated umbilical side. The 5 to 8 globular chambers of the last whorl increase gradually in size. The umbilicus is wide and deep.

Acarinina aspensis (Colom, 1954) and *A. pentacamerata* (Subbotina, 1947) are closely interrelated by transitional forms and are considered by several authors (e.g. Stainforth *et al.*, 1975) as synonyms, whereas others (e.g. Hillebrandt, 1976; Blow, 1979) keep them separated.

ACARININA PRIMITIVA (Finlay)
Figures 17.6–7; 5, 7, 10

Type reference: *Globoquadrina primitiva* Finlay, 1947, p. 291, pl. 8, figs. 129–134.

The last whorl of the compact test has 3 to 4 somewhat compressed chambers which are longer than wide on the spiral side. The umbilicus is small, but deep. The aperture, which is umbilical to extraumbilical–umbilical in position has a distinct lip. The surface is covered with coarse blunt spines or pustules.

The *Acarinina mckannai* group

The representatives of this group have a very spinose and tightly coiled test with a high trochospire. They may become very frequent in middle Paleocene to basal Eocene assemblages outside the tropical realm in which typical *Morozovella* species are missing.

ACARININA MCKANNAI (White)
Figures 18.3–6; 5, 7, 10

Type reference: *Globigerina mckannai* White, 1928, p. 194, pl. 27, fig. 16.
Synonym: *Globigerina subsphaerica* Subbotina, 1947, pp. 108–9, pl. 5, figs. 23–28.

The tightly coiled test has 5 to 7 chambers in the last whorl. These chambers are considerably larger than those of the initial whorls and encroach on the narrow and deep umbilicus. The outline of the hispid to spinose test is almost subglobular.

Acarinina nitida has only 4 to 5 chambers in the last whorl.

ACARININA NITIDA (Martin)
Figures 18.1–2; 5, 7, 10

Type reference: *Globigerina nitida* Martin, 1943, p. 115, pl. 7, fig. 1.
Synonym: *Acarinina acarinata* Subbotina, 1953, pp. 229–30, pl. 22, figs. 4–10.

The compact test is almost subglobular in lateral view and has an equatorial periphery which is only very slightly lobate. The last whorl has 4 to 5 chambers, which enclose the narrow umbilicus. The surface of the test is very spinose, especially on the umbilical side.

The stratigraphically useful representatives of the genus *Globigerina*

GLOBIGERINA TRILOCULINOIDES Plummer
Figures **19.1–2**; 5, 7, 10

Type reference: *Globigerina triloculinoides* Plummer, 1926, p. 134, pl. 8, fig. 10.

The test is tightly coiled and low trochospiral with 3 to $3\frac{1}{2}$ globular chambers in the last whorl which increase rapidly in size, the last one occupying up to one half of the entire whorl. The equatorial periphery is trilobate. The surface of the test has a distinct honeycomb pattern.

Globigerina triloculinoides Plummer has been selected by Brotzen & Pozaryska (1961) as type-species of the genus (respectively subgenus) *Subbotina*.

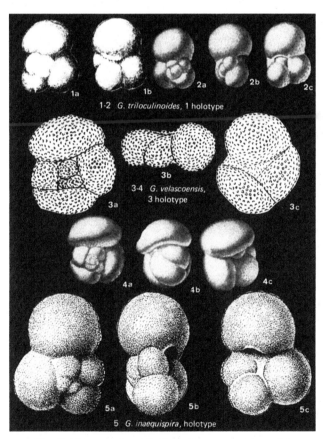

Fig. 19. Three stratigraphically useful representatives of the genus *Globigerina* (all figures × 60)

1–2 *G. triloculinoides* Plummer
    1a–b Holotype. Paleocene, Midway Fm., Navarro County, Texas, USA. **2a–c** Specimen from the *P. pusilla pusilla* Zone, Paleocene, Lizard Springs Fm., Trinidad (from Bolli, 1957a).

3–4 *G. velascoensis* Cushman
    3a–c Holotype. Paleocene, Velasco Fm., Tampico Embayment, Mexico. **4a–c** Specimen from the *P. pseudomenardii* Zone, Paleocene, Lizard Springs Fm., Trinidad (from Bolli, 1957a).

5 *G. inaequispira* Subbotina
    5a–c Holotype. Early Eocene, N Caucasus, USSR.

GLOBIGERINA VELASCOENSIS Cushman
Figures **19.3–4**; 5, 7, 10

Type reference: *Globigerina velascoensis* Cushman, 1925c, p. 19, pl. 3, fig. 6.

The umbilical side of the compact test is inflated whereas the spiral side is flat or even somewhat depressed. The 4 chambers of the last whorl increase rapidly in size with the last one occupying almost one half of the whorl. The umbilicus is deep and narrow and the aperture in general has a distinct lip. The surface of the test is coarsely perforated, but rather smooth.

The somewhat similar *Acarinina primitiva* has a strongly spinose to pustulose surface and a more compact aspect.

GLOBIGERINA INAEQUISPIRA Subbotina
Figures **19.5**; 5, 7, 10

Type reference: *Globigerina inaequispira* Subbotina, 1953, p. 69, pl. 6, figs. 1–4.

Typical for this species is the discrepancy between the rather small and tightly coiled inner whorl and the rather loosely coiled and rapidly expanding outer whorl. The surface of the test is rather smooth, finely cancellate or slightly hispid.

Globigerina linaperta Finlay is more tightly coiled and does not show the contrast between the aspect of the inner and the outer whorl.

Early to Late Eocene

Eocene representatives of the genus *Planorotalites*

PLANOROTALITES PALMERAE (Cushman & Bermudez)
Figures **20.14–29**; 6, 7, 10

Type reference: *Globorotalia palmerae* Cushman & Bermudez, 1937, p. 26, pl. 2, figs. 51–53.

This easily recognizable species is very much compressed and has a stellate axial periphery. The 5 to 7 chambers of the last whorl are radially elongated and prolonged by a spine formed by the extension of the faint keel bordering them.

This very distinctive Paleogene planktic foraminiferal species has been frequently confused with small specimens of the benthic genus *Pararotalia* (see, for example, Blow, 1979). Schmidt & Raju (1973) clearly demonstrated the difference between stellate representatives of the genus *Pararotalia* and *Planorotalites palmerae*. *Pararotalia* has a prominent umbilical knob and an areal aperture (see Figs. 20.30–32 of *Pararotalia spinigera*). The 'plug' visible sometimes in *Planorotalites palmerae* is composed of secondary incrustations and has nothing to do with the true umbilical plug in *Pararotalia*.

Another criterion in favour of the planktic nature of *Planorotalites palmerae* is the evident evolutionary link with *P. pseudoscitula* which has been illustrated by Schmidt & Raju (1973) and Hillebrandt (1976) (see also Brönnimann & Rigassi, 1963; Hillebrandt, 1965).

PLANOROTALITES PSEUDOSCITULA
(Glaessner)
Figures **20.2–10**; 5, 7, 9

Type reference: *Globorotalia pseudoscitula* Glaessner, 1937, p. 32, figs. 3a–c.

Synonym: *Globorotalia renzi* Bolli, 1957b, p. 168, pl. 38, figs. 3a–c.

This small species (0.2 to 0.3 mm) is lenticular, with 5 to 7 tightly coiled chambers in the last whorl. The periphery is circular, slightly lobate and with a faint keel. The surface of the test is rather smooth, but the surface of the chambers of the initial part of the last whorl may be somewhat rugose. The perforation is coarse in the older part of the test, fine in the youngest chambers.

Planorotalia renzi (Fig. 20.1) was placed in synonymy with *P. pseudoscitula* by Hillebrandt (1976) and by Blow (1979).

The genus *Pseudohastigerina*

PSEUDOHASTIGERINA MICRA (Cole)
Figures **21.1–8**; 5, 7, 9

Type reference: *Nonion micrus* Cole, 1927, p. 22, pl. 5, fig. 12.

Synonym: *Nonion danvillensis* Howe & Wallace, 1932, p. 51, pl. 9, figs. 3a–b.

This planispiral, biumbilicate, rather small species (0.2 to 0.4 mm) differs from its ancestral form *P. wilcoxensis* in being more laterally compressed. During the late Early Eocene and the early Middle Eocene both species occur together with transitional forms.

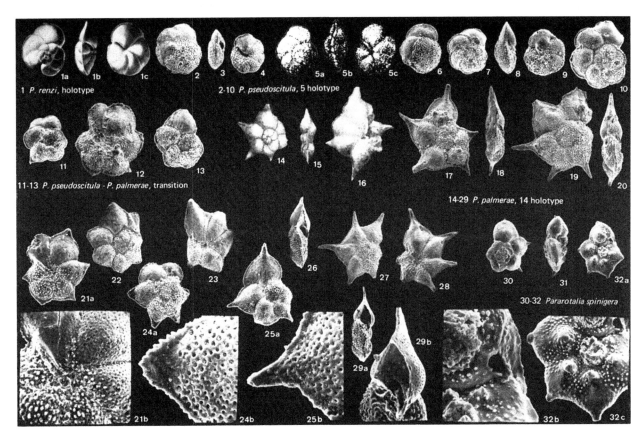

Fig. 20. Late Early and Middle Eocene *Planorotalites* species and *Pararotalia spinigera* (all figures × 60, except 21b, 29b, 32c × 150; 24b, 25b, 32b × 300)

1 *P. renzi* (Bolli)
   1a–c Holotype. Sample Hg 8581, *O. beckmanni* Zone, Middle Eocene, Navet Fm., Trinidad.
2–10 *P. pseudoscitula* (Glaessner)
   2–4 Sample Hk408, *O. beckmanni* Zone, Middle Eocene, Navet Fm., Trinidad. **5a–c** Holotype. Lower foraminiferal layers at Il'sk station, N Caucasus, USSR. **6** Sample ER 1001, *A. pentacamerata* Zone, Early Eocene, Richmond Fm., Jamaica. **7–10** Sample MNG-1, CC 18, *A. pentacamerata* Zone, Early Eocene, Cauvery Basin, S India.
11–13 *P. pseudoscitula–P. palmerae*, transition

11–12 Sample ER 1001, *A. pentacamerata* Zone, Early Eocene, Richmond Fm., Jamaica. **13** Sample MNG-1, CC 18, *A. pentacamerata* Zone, Early Eocene, Cauvery Basin, S India.
14–29 *P. palmerae* (Cushman & Bermudez)
   14 Holotype. **15–16** Paratypes. Eocene, Tejar 'Cuba', Arroyo Naranjo, Havana Province, Cuba. **17–20, 27–28** Sample ER 999. **24a–b, 25a–b** Sample ER 1001, *A. pentacamerata* Zone, Early Eocene, Richmond Fm., Jamaica. **21a–b, 29a–b** Sample MNG-1, CC 18, *A. pentacamerata* Zone, Early Eocene, Cauvery Basin, S India. **22–23** Sample Hillebrandt 700924/1, *A. pentacamerata* Zone, Early Eocene, Agost, Province Alicante, Spain. **26** Sample Br. 1005-77, *A. pentacamerata* Zone, Early Eocene, Cuba.
30–32 *Pararotalia spinigera* (Le Calvez)
   30–32c Sample Le Calvez, late Early Eocene, Pierrefonds, France.

The periphery of *P. micra* is generally rounded but becomes subacute in larger specimens.

P. micra is larger than *P. naguewichiensis* and has less chambers in the last whorl (5 to 7 instead of 6 to 8).

We include *P. danvillensis* (Fig. 21.9) in the range of variability of *P. micra*. *P. micra* has a worldwide distribution. This apparently fragile species is in fact very resistant to bad ecologic conditions and may be a dominant species in high latitude assemblages and in Early Oligocene assemblages of the Alps and Eastern Europe.

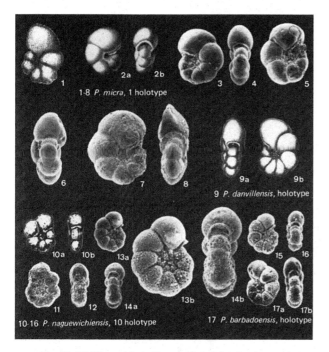

Fig. 21. Middle and Late Eocene *Pseudohastigerina* species (all figures × 60, except 13b and 14b × 120)

1–8 *P. micra* (Cole)
   1 Holotype. Middle Eocene, Guayabal Fm., Potrero, Vera Cruz, Mexico. **2a–b** Sample Hg 8581, *O. beckmanni* Zone, Middle Eocene, Navet Fm., Trinidad (from Bolli, 1957b). **3–4** Sample 655, *T. cerroazulensis* s.l. Zone, Late Eocene, Possagno, Italy (from Toumarkine & Bolli, 1975). **5–6** Sample K 146, Middle Eocene, El Midawarah Fm., Fayoum Area, Egypt (6 from Shamah *et al.*, 1982). **7–8** Samples HMB 67/19 and Tou 484, *T. cerroazulensis* s.l. Zone, Late Eocene, Chapapote Fm., Rio Tuxpan, Tampico Embayment, Mexico.

   9 *P. danvillensis* (Howe & Wallace)
   9a–b Holotype. Eocene, Jackson Fm., Louisiana, USA.

10–16 *P. naguewichiensis* (Myatliuk)
   10a–b Holotype. W Ukraine (from Subbotina, 1953). **11–12** Sample 4526, Early Oligocene, Marnes à Foraminifères, Vacherie, Haute-Savoie, France (from Charollais *et al.*, 1980). **13a–b, 14a–b, 15, 16** DSDP Leg 32-313-5, 0, 2–3 cm, *C. chipolensis/P. micra* Zone, Early Oligocene, Central Pacific (13, 16 from Toumarkine, 1975).

   17 *P. barbadoensis* Blow
   17a–b Holotype. Sample Cb. 1964, P 19 Zone, early Oligocene, Cipero Fm., Trinidad.

PSEUDOHASTIGERINA NAGUEWICHIENSIS (Myatliuk)
Figures **21.10–16**; 5, 8, 9

Type reference: *Globigerinella naguewichiensis* Myatliuk, 1950, p. 281, pl. 4, figs. 4a–b.

Synonym: *Pseudohastigerina barbadoensis* Blow, 1969, p. 409, pl. 53, figs. 7–9.

This very small species (diameter 0.2 mm) differs from *P. micra* by its smaller size, the numerous chambers in the last whorl (often 8), the straighter sutures and rounder periphery.

P. naguewichiensis and *P. barbadoensis* (Fig. 21.17) are very close. *P. naguewichiensis* has possibly a more lobate equatorial outline and is less tightly coiled than *P. barbadoensis*. As the two species have the same stratigraphic range, we include *P. barbadoensis* in the variability of *P. naguewichiensis*.

The genus *Clavigerinella*

The main features of the genus are the radially elongate to clavate shape of the chambers in the last whorl, the planispiral mode of coiling and the equatorially elongate aperture bordered by lateral flanges. Hillebrandt (1976) assumed that *Clavigerinella* originate from the *Globigerina inaequispira* plexus. This is also confirmed by Blow (1979) and our own observations. However, transitional forms are rare and in many sections the first representatives of the genus *Clavigerinella* appear rather abruptly.

Representatives of the genus *Clavigerinella* range from the youngest part of the Early Eocene to within the Late Eocene, but their main occurrence is in the older part of the Middle Eocene. The specimens of *Clavigerinella* are usually broken and rare. Exceptionally they may occur in large numbers as is the case in some localities of northern South America. Nevertheless, they are very useful stratigraphic markers because of their distinct shape, which is typical even in fragments, and because of their restricted stratigraphic range.

CLAVIGERINELLA AKERSI Bolli, Loeblich & Tappan
Figures **22.10–14**; 5, 8, 10

Type reference: *Clavigerinella akersi* Bolli, Loeblich & Tappan, 1957, p. 30, pl. 3, figs. 5a–b.

The most important feature of the species is the bulbous, clavate shape of the chambers of the last whorl. It has only 4 chambers in the last whorl instead of 4 to 5 as in the subspecies of *C. eocanica*.

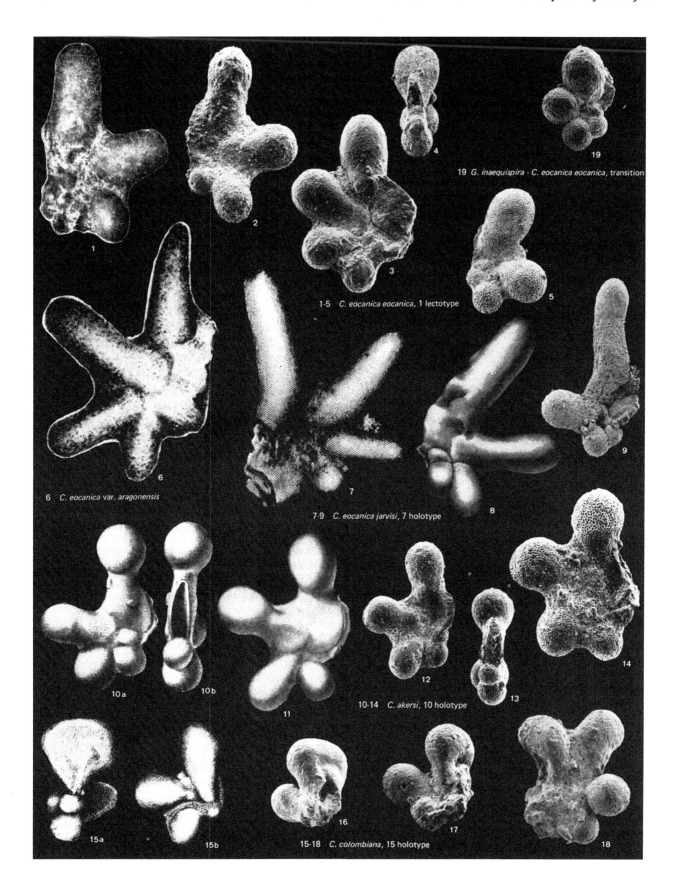

19 *G. inaequispira - C. eocanica eocanica*, transition

1-5 *C. eocanica eocanica*, 1 lectotype

6 *C. eocanica* var. *aragonensis*

7-9 *C. eocanica jarvisi*, 7 holotype

10-14 *C. akersi*, 10 holotype

15-18 *C. colombiana*, 15 holotype

CLAVIGERINELLA COLOMBIANA (Petters)
Figures **22.15–18**; 5, 8, 10

Type reference: *Hastigerinella colombiana* Petters, 1954, p. 40, pl. 8, figs. 10a–b.

The flattening of the chambers of the last whorl in the direction of coiling gives the test the aspect of a paddle-wheel and clearly differentiates it from the other species of the genus.

Complete tests are rare; loose broken-off chambers, however, may be common and are easily recognizable.

The taxon is found in only a few areas (Colombia, Ecuador, Peru, Venezuela and also the Indo-Pacific), but when present, it can be very abundant.

CLAVIGERINELLA EOCANICA EOCANICA (Nuttall)
Figures **22.1–5**; 5, 7, 9

Type reference: *Hastigerinella eocanica* Nuttall, 1928, p. 376, pl. 50, figs. 9–11.

Lectotype: no holotype was designated. We propose herewith as *lectotype* the specimen figured by Nuttall, 1928, p. 50, fig. 9.

For taxonomic remarks see *C. eocanica jarvisi.*

CLAVIGERINELLA EOCANICA JARVISI (Cushman)
Figures **22.7–9**; 5, 8, 9

Type reference: *Hastigerinella jarvisi* Cushman, March 1930, p. 18, pl. 3, fig. 8.

Synonym: *Hastigerinella eocanica* var. *aragonensis* Nuttall, September 1930, p. 290, pl. 24, figs. 16–17.

Fig. 22. *Clavigerinella* species (all figures × 60)

1–5 *C. eocanica eocanica* (Nuttall)
1 Lectotype (here designated). Eocene, Upper Uzpanapa River, Isthmus of Tehuantepec, Mexico (fig. 9, pl. 50, Nuttall, 1928). **2–3** Sample PJB 168, *M. lehneri* Zone, Middle Eocene, El Datil Fm., Punta Mosquito, Margarita, Venezuela. **4** Sample PJB 136, *H. nuttalli* Zone, Middle Eocene, El Datil Fm., Venezuela. **5** Sample JB 1412, *H. nuttalli* Zone, Middle Eocene, Navet Fm., Trinidad.

6 *C. eocanica* var. *aragonensis* (Nuttall)
Arroyo La Laja, La Antigua, Aragon Fm., Tampico Embayment, Mexico (from Nuttall, 1930).

7–9 *C. eocanica jarvisi* (Cushman)
7 Holotype. Eocene, Trinidad. **8** *M. lehneri* Zone, Middle Eocene, Navet Fm., Trinidad (from Bolli, 1957b). **9** Hk 408, *O. beckmanni* Zone, Middle Eocene, Navet Fm., Trinidad.

10–14 *C. akersi* Bolli, Loeblich & Tappan
10a–b Holotype. Sample K 8820, Middle Eocene, Navet Fm., Trinidad. **11** Sample HGK 8820, *H. nuttalli* Zone, Middle Eocene, Navet Fm., Trinidad (from Bolli, 1957b). **12–14** Sample JB 1406, *H. nuttalli* Zone, Middle Eocene, Navet Fm., Trinidad.

15–18 *C. colombiana* (Petters)
15a–b Holotype. Upper Carreto Fm., Middle Eocene, Bolivar Dept., Colombia. **16–18** Sample PJB 136, *H. nuttalli* Zone, Middle Eocene, El Datil Fm., Punta Mosquito, Margarita, Venezuela.

19 *G. inaequispira–C. eocanica eocanica* transition
Sample PJB 136, *H. nuttalli* Zone, Middle Eocene, El Datil Fm., Punta Mosquito, Margarita, Venezuela.

According to Cushman (1930, p. 19) the chambers of *Clavigerinella eocanica jarvisi* are more slender than those of *C. eocanica eocanica* and their outer ends are often pointed. The two subspecies have the same stratigraphic range.

Hastigerinella eocanica var. *aragonensis* Nuttall (Fig. 22.6), published in September 1930, is here considered as a junior synonym though Nuttall distinguished it from *H. jarvisi* by the more rounded distal ends of the chambers. *H. eocanica* var. *aragonensis* is possibly an intermediate form between *C. eocanica eocanica* and *C. eocanica jarvisi.*

The genera *Hantkenina* and *Cribrohantkenina*

The unmistakable, planispiral tests of *Hantkenina* and *Cribrohantkenina* with their distinctive spines are among the most typical constituents of Middle and Late Eocene planktic foraminiferal assemblages. The stratigraphic importance of these two genera is emphasized by the appearance of *Hantkenina* at the base of the Middle Eocene and the disappearance of both genera at the top of the Late Eocene.

Although several authors (Thalmann, 1942b; Brönnimann, 1950; Ramsay, 1962; Blow, 1979) have attempted a restudy of the entire group, the taxonomy of the various species of the genera *Hantkenina* and *Cribrohantkenina* is still somewhat confused. The number of species described amounts to about fifteen, but we have retained here only five of them which we judge to be the most representative and stratigraphically useful ones. The morphology of the representatives of the genus *Hantkenina* seems to be very much subject to ecologic conditions; it is possible that the remaining ten species represent merely variants caused by environmental changes. The genus *Cribrohantkenina* with its peculiar cribrate aperture is thought to be monospecific in accordance with Stainforth *et al.* (1975) and Blow (1979).

Banner & Blow (1959), Blow & Banner (1962) consider that the genus *Hantkenina* has *Pseudohastigerina*-like forms as ancestors, but its first representatives always appear abruptly and without transitional forms. In contrast Benjamini & Reiss (1979) see *Hantkenina* evolving from the *Clavigerinella* stock.

The evolutionary trends within the genus *Hantkenina* lead from strongly stellate forms to more and more compact tests and to a narrowing of the initially rather wide triradiate aperture and finally to transitional forms to *Cribrohantkenina.*

HANTKENINA NUTTALLI Toumarkine
Figures **23.1–5**; 5, 8, 10

Type reference: *Hantkenina nuttalli* Toumarkine, 1981, p. 112, pl. 1, fig. 4 (from Nuttall, 1930, pl. 24, fig. 3, paratype of *Hantkenina mexicana* var. *aragonensis*) and figs. 5 and 6 (fig. 3, pl. 24 from Nuttall, 1930, redrawn in Bolli *et al.*, 1957).

Hantkenina nuttalli, the earliest representative of the genus *Hantkenina*, differs from the other species of the genus by its large size (0.5 to 1.0 mm) and the stellate shape of the test. The chambers (4 to 5) of the last whorl increase rapidly in size and are well separated from each other and radially elongate.

Hantkenina mexicana Cushman is very close to *H. nuttalli*, but it is generally smaller and has longer spines. The general shape of the test is not as stellate as in *H. nuttalli* and the chambers become slightly imbricated. *H. lehneri* Cushman & Jarvis (Fig. 23.7) has more (5 to 7) chambers in the last whorl and these are more cylindrically finger-shaped.

Nuttall (1930) described a variety of *H. mexicana*, *H. mexicana* var. *aragonensis*, but he illustrated three specimens without designating a holotype. Thalmann (1942b) raised the variety to specific rank and designated the specimen figured by

Nuttall on pl. 24, fig. 1 as lectotype. Bolli *et al.* (1957) erroneously re-illustrated the specimen of fig. 3 of Nuttall (1930) instead of the one of his fig. 1 as the lectotype of *H. aragonensis*. Ramsay (1962) noticed this mistake and remarked that the real lectotype of *H. aragonensis* (pl. 24, fig. 1 of Nuttall, 1930) is very close to the holotype of *H. mexicana*. He, therefore, placed *H. aragonensis* into the synonymy of *H. mexicana*. His conclusions were also followed by Blow (1979). Toumarkine (1981) has come partly to the same conclusions: the holotype of *H. aragonensis* is a junior synonym of *H. mexicana*, but the paratype re-illustrated by Bolli *et al.* (1957, pl. 2, figs. 3a–b = Nuttall, 1930, pl. 24,

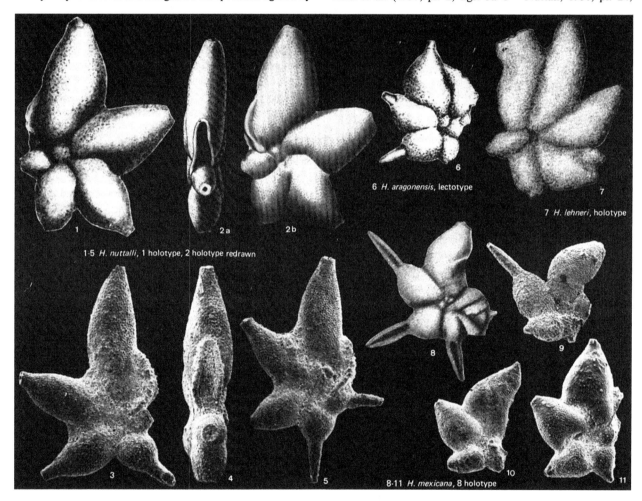

Fig. 23. *Hantkenina nuttalli* and related species (all figures × 60)

1–5 *H. nuttalli* Toumarkine
   1 Holotype. *H. nuttalli* Zone, Middle Eocene, Aragon Fm., 1200 m N 48° W of La Antigua, Mexico (from Nuttall, 1930, as *H. mexicana* var. *aragonensis*). **2a–b** Holotype redrawn (from Bolli *et al.*, 1957, figured erroneously as lectotype of *H. aragonensis*). **3–5** Sample ER 1545, *H. nuttalli* Zone, Rio Sambre section, Font Hill Fm., Jamaica (from Toumarkine, 1981).

   6 *H. aragonensis* Nuttall
   Lectotype (designated by Thalmann, 1942b), Middle Eocene, Aragon Fm., Arroyo Puentitla, 2600 m N 73° E of El Tule, Mexico.

7 *H. lehneri* Cushman & Jarvis
   Holotype, Eocene, near source of Moruga River, Trinidad.
8–11 *H. mexicana* Cushman
   8 Holotype. Eocene, yellowish brown clay, La Laja, Zardo Creek, 1 km SW of Tierra Colorado, Mexico. **9–10** Sample PJB 152, *G. subconglobata subconglobata* to *M. lehneri* Zone, Middle Eocene, El Datil Fm., Punta Mosquito, Margarita, Venezuela. **11** Sample JS 1406, *H. nuttalli* Zone, Middle Eocene, Trinidad.

fig. 3) is considered as a separate taxon which she named *H. nuttalli.*

When Bolli (1957b) used *Hantkenina aragonensis* as the zonal marker of the oldest zone of the Middle Eocene, he used this name for the large stellate forms which are now called *H. nuttalli.* Consequently, the *Hantkenina aragonensis* Zone is renamed *Hantkenina nuttalli* Zone.

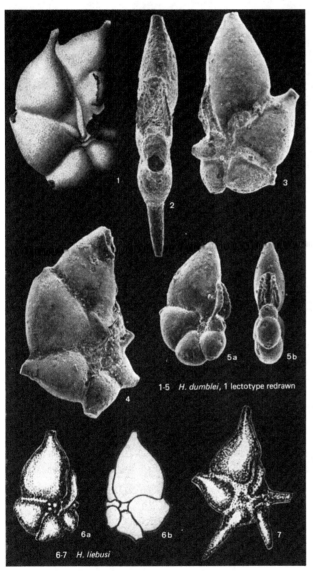

Fig. 24. *Hantkenina dumblei* and *H. liebusi* (all figures × 60)
1–5 *H. dumblei* Weinzierl & Applin
  1 Lectotype redrawn (from Bolli *et al.*, 1957, lectotype designated by Brönnimann, 1950), Middle Eocene, Claiborne Fm., Rio Bravo Oil Co., Deussen B 1 4, 010 ft, South Liberty Dome, Liberty Co., Texas, USA. **2** Sample 637, *G. subconglobata subconglobata* Zone, Middle Eocene, Possagno, N Italy. **3–4** Sample Ba 24, Middle Eocene, Bou Arada, Tunisia (from Mèchmèche & Toumarkine, 1983). **5a–b** Sample PJB 166, *M. lehneri* Zone, Middle Eocene, El Datil Fm., Punta Mosquito, Margarita, Venezuela.
6–7 *H. liebusi* Shokhina
  6a–b, 7 Middle Eocene, foraminiferal beds, Ilskaya, N Caucasus, USSR.

HANTKENINA MEXICANA Cushman
Figures **23.8–11**; 5, 8, 10

Type reference: *Hantkenina mexicana* Cushman, 1925a, p. 3, pl. 2, fig. 2.

In *Hantkenina mexicana*, the stellate aspect of the test is less pronounced than in *H. nuttalli. H. liebusi* Shokina (Figs. 24.6–7) has a larger final chamber and a more tightly coiled and laterally compressed test. Its chambers become imbricated.

Many authors use the name *H. mexicana* in a broad sense for all stellate representatives of *Hantkenina* from the earlier part of the Middle Eocene.

HANTKENINA DUMBLEI Weinzierl & Applin
Figures **24.1–5**; 5, 8, 10

Type reference: *Hantkenina dumblei* Weinzierl & Applin, 1929, p. 402, pl. 43, figs. 5a–b.
Lectotype: *Hantkenina dumblei* Weinzierl & Applin, 1929: Brönnimann, 1950, p. 409 (specimen figured by Weinzierl & Applin, 1929, pl. 43, fig. 5b).

Hantkenina dumblei has a more compact and more compressed test than *H. nuttalli* and *H. mexicana.* The chambers encroach on each other and tend to fuse. The periphery is less lobate than in *H. liebusi* but more so than in *H. alabamensis.*

HANTKENINA ALABAMENSIS Cushman
Figures **25.1–10**; 5, 8, 0

Type reference: *Hantkenina alabamensis* Cushman, 1925a, p. 3, pl. 1, fig. 1.

Hantkenina alabamensis is the most advanced representative of the genus. Its chambers increase slowly in size and are closely pressed against each other. The contour of the periphery is rounded.

H. longispina Cushman, 1925a (Fig. 25.11) has longer spines arranged in an apical position, whereas *H. brevispina* Cushman, 1925a (Fig. 25.12) differs only by the shorter and blunter spines. We include both species in the range of variability of *H. alabamensis.*

HANTKENINA PRIMITIVA Cushman & Jarvis
Figures **25.13–15**; 5, 8, 9

Type reference: *Hantkenina alabamensis* var. *primitiva* Cushman & Jarvis, 1929, p. 16, pl. 3, fig. 3.

Hantkenina primitiva differs from *H. alabamensis* mainly by the ontogenetically late development of the spines which are present only in the last 2 or 3 chambers. According to Blow (1979), this phenomena probably represents a regressive feature. In addition, *H. primitiva* is distinguishable from *H. alabamensis* by its more embracing and laterally more compressed chambers.

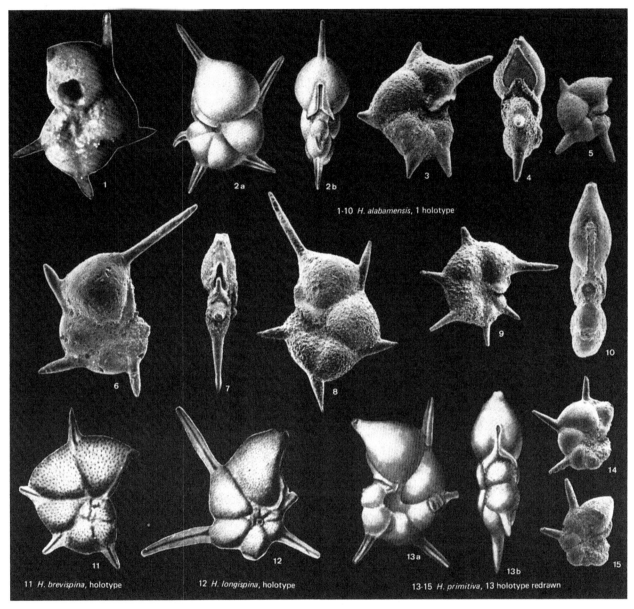

Fig. 25. *Hantkenina alabamensis* and related species (all figures × 60)

1–10 *H. alabamensis* Cushman
  1 Holotype, Eocene, Zeuglodon bed, Cocoa Post Office, Alabama, USA. **2a–b** Late Eocene, Pachuta Fm., Jacksonian, Alabama, USA (from Bolli *et al.*, 1957). **3–4, 9** DSDP Leg 40-363-9-4, 58–60 cm, *T. cerroazulensis* s.l. Zone, Late Eocene, S Atlantic (9 from Toumarkine, 1978). **5** DSDP Leg 32-305-9, cc., *G. semiinvoluta* Zone, Late Eocene, Central Pacific (from Toumarkine, 1975). **6** Sample Tou 484, *T. cerroazulensis* s.l. Zone, Late Eocene, Chapapote Fm., Rio Tuxpan, Tampico Embayment, Mexico. **7–8** Sample Hk 408, *O. beckmanni* Zone, Middle Eocene, Navet Fm., Trinidad. **10**

Sample HMB 68/68, *G. semiinvoluta* Zone, Late Eocene, Possagno, N Italy.

11 *H. brevispina* Cushman
  Holotype. Eocene, Rio Pantepec, Mexico.

12 *H. longispina*
  Holotype. Eocene, Rio Tuxpan, Mexico.

13–15 *H. primitiva* Cushman & Jarvis
  13a–b Holotype redrawn (in Bolli *et al.*, 1957), Late Eocene, Mt Moriah beds, Vistabella Quarry, Trinidad. **14–15** DSDP Leg 40-363-9-4, 58–60 cm, *T. cerroazulensis* s.l. Zone, Late Eocene, S Atlantic (from Toumarkine, 1978).

CRIBROHANTKENINA INFLATA (Howe)
Figures **26.1–7**; 5, 8, 9

Type reference: *Hantkenina inflata* Howe, 1928, p. 14, pl. 14, fig. 2.

Synonyms: *Hantkenina danvillensis* Howe & Wallace, 1934, p. 37, pl. 5, figs. 14, 17.

Hantkenina (Cribrohantkenina) bermudezi Thalmann, 1942b, p. 812, pl. 1, fig. 6.

Hantkenina lazzarii Pericoli, 1959, p. 82, pl. 1, figs. 1–3.

Cribrohantkenina inflata is readily recognized by its 4 to 6 inflated chambers each carrying a stout spine in the last whorl and by its peculiar accessory areal apertures. The primary aperture is trilobate as in *Hantkenina* whereas the accessory areal apertures are tuberculate holes.

The monospecific genus *Cribohantkenina* has been studied in detail by several authors (Thalmann, 1942b; Blow & Banner, 1962; Dieni & Proto Decima, 1964).

The specimens illustrated on Figs. 26.1–4, 6–7 from the Shubuta Member of the Yazoo Formation are exceptionally well preserved. The specimen on Fig. 26.5 shows a more common state of preservation.

The *Turborotalia griffinae–'Hastigerina' bolivariana* group

Some relatively large species generally with 4 globular chambers in the last whorl and an umbilical–extraumbilical aperture may be very prominent in late Early Eocene and Middle Eocene assemblages. The aperture can extend towards the spiral side and be very narrow or may be more open. These species show a tendency towards involute coiling. The extreme forms appear nearly planispiral. Their specific as well as generic assignment is doubtful. They were called *Globigerina wilsoni* Cole, 1927 (*Globorotalia wilsoni* by Beckmann, 1971, *Globorotaloides wilsoni* by Poore & Brabb, 1977), *Globigerina wilsoni bolivariana* Petters, 1954 (*Globorotalia bolivariana* by Bolli, 1957b, '*Hastigerina*' *bolivariana* by Blow, 1979), *Pseudohastigerina globulosa* Hillebrandt, 1976 (modified by Hillebrandt for taxonomic reasons to *P. sphaeroidalis*, 1978), *Globorotalia (Turborotalia) griffinae*, Blow, 1979.

The holotypes of *Globigerina wilsoni* Cole and *G. wilsoni bolivariana* Petters have been re-examined. As a consequence *Turborotalia griffinae* Blow and '*Hastigerina*' *bolivariana* (Petters), and also '*Hastigerina*' cf. *bolivariana* are retained. We agree with Hillebrandt (1976) and Blow (1979) in thinking that these species evolved from *Globigerina inaequispira*, which at the same time produced the *Clavigerinella* species. It is very interesting to note that in some samples from Venezuela both '*H.*' *bolivariana* and *Clavigerinella* represent the main part of the assemblages of the planktic foraminifera in the early Middle Eocene or perhaps the late Early Eocene. *T. griffinae* evolved from *G. inaequispira* during the *Acarinina pentacamerata* Zone and apparently it is the ancestral form of both '*H.*' *bolivariana* and '*H.*' cf. *bolivariana*. It is possible that '*H.*' cf. *bolivariana* needs to be renamed. *G. wilsoni* (Figs. 27.1–4) seems to belong to a different lineage, perhaps to the *T. opima nana* group. It is smaller than the other species studied here (*T. griffinae*, '*H.*' *bolivariana* and '*H.*' cf. *bolivariana*), but it certainly does not belong to the *G. increbescens* group where it was included by Blow (1979).

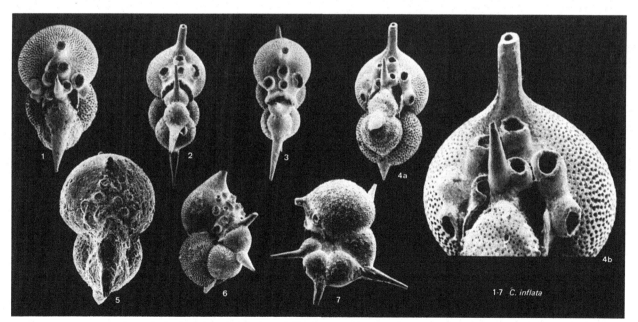

Fig. 26. *Cribrohantkenina inflata* (all figures × 60, except 4b × 150)

1–7 *C. inflata* Howe

1–4b, 6–7 Late Eocene, Yazoo Fm., Mississippi, USA (4a from Pomerol, 1973). 5 Sample Tou 489, *T. cerroazulensis* Zone, Chapapote Fm., Rio Tuxpan, Tampico Embayment, Mexico.

'HASTIGERINA' BOLIVARIANA Petters
Figures **27.24–29**; 5, 7, 10

Type reference: *Globigerina wilsoni bolivariana* Petters, 1954, p. 39, pl. 8, figs. 9a–c.

Petters, who described the species from Colombia, noticed that the test is globular, completely involute on the ventral side, nearly so on the dorsal side, and that the aperture is a narrow slit extending more on the ventral side than on the dorsal side. It can be a very large species (0.45–1 mm). Petters compared

his subspecies to Cole's *Globigerina wilsoni* (Fig. 27.1) but only with the illustrations and not with the actual specimens. Petters also compared his subspecies to the figures of *G. wilsoni* given by Stainforth (1948) but they probably illustrate *bolivariana* rather than *wilsoni*. In fact, the species and the subspecies are very different (except perhaps for the texture of the test). As explained by Weiss (1955) the difference between the two species is that *Globigerina wilsoni* is a small species (0.25 mm) which shows a larger portion of the penultimate whorl on the spiral side. Weiss observed the two forms

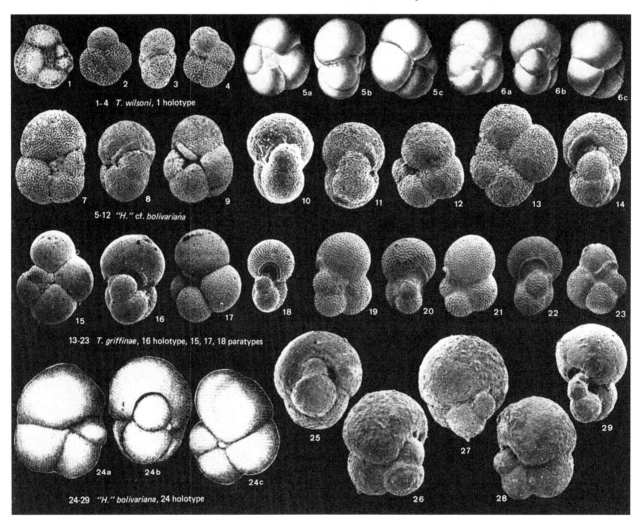

Fig. 27. *Turborotalia* species and '*Hastigerina*' *bolivariana* (all figures × 60)

1–4 *T. wilsoni* (Cole)
  1 Holotype. Middle Eocene, Guayabal Fm., near Potrero, State of Vera Cruz, Mexico. **2–4** Sample HMB 67/24, *M. lehneri* Zone, Middle Eocene, Guayabal Fm. (type-locality), near Potrero, State of Vera Cruz, Mexico.

5–12 '*H.*' cf. *bolivariana*
  5a–c, 6a–c Sample Hg 8581, *O. beckmanni* Zone, Middle Eocene, Navet Fm., Trinidad (from Bolli, 1957b). **7–9** Sample Hk 408, *O. beckmanni* Zone, Middle Eocene, Navet Fm., Trinidad. **10** Core Halimba 34, Middle Eocene, Bakony Mountain, Hungary (from Toumarkine, 1971). **11–12** Sample PJB 152, *G. subconglobata*

subconglobata to *M. lehneri* Zone, Middle Eocene, El Datil Fm., Punta Mosquito, Margarita, Venezuela.

13–23 *T. griffinae* Blow
  13–14 Sample PJB 152, *G. subconglobata subconglobata* to *M. lehneri* Zone, Middle Eocene, El Datil Fm., Punta Mosquito, Margarita, Venezuela. **15, 17, 18** Paratypes, **16** Holotype. Kane 9, core 42, Zone P9, N of Cape Verde Islands, Atlantic. **19–23** Sample Mèchmèche Ba 19, Middle Eocene, Bou Arada, Tunisia.

24–29 '*H.*' *bolivariana* (Petters)
  24a–c Holotype. Middle Eocene, Upper Carreto Fm., Dept. Bolivar, Colombia. **25–29** Sample PJB 136, early Middle Eocene, El Datil Fm., Punta Mosquito, Margarita, Venezuela.

G. wilsoni and *G. wilsoni bolivariana* in samples from Peru, but it is probable that he described as *G. wilsoni bolivariana* some species which belong to the *Turborotalia opima nana* group (described later by Bolli, 1957c). Weiss also noticed in *Globigerina wilsoni bolivariana* a *Globigerinella*-like tendency to uncoil (see Fig. 27.29). Bolli (1957b) changed the generic name to *Globorotalia* and raised the subspecies to specific rank. However, the specimens of *Globorotalia bolivariana* he illustrated from the *Orbulinoides beckmanni* Zone of Trinidad are not exactly identical to the forms described by Petters: they are not as large and not as planispiral and it is always possible to see some of the chambers of the penultimate whorl. Also the aperture extends only slightly onto the spiral side. They are considered here as '*Hastigerina*' cf. *bolivariana*. Hillebrandt (1976) described a new species *Pseudohastigerina globulosa* (later renamed *P. sphaeroidalis* by Hillebrandt (1978) due to the preoccupation of the earlier name). Hillebrandt described the form from the *Globorotalia palmerae* Zone of Southern Spain as a large and completely planispiral species with globular chambers. He claimed that this species arises from *Globigerina inaequispira*. *P. sphaeroidalis* seems to be very close to '*H*.' *bolivariana*. The relations between the two forms require more investigations.

The generic name of the species was changed by Blow (1979) to *Hastigerina*? We use here '*Hastigerina*' for these forms, although they are completely separated from the Neogene *Hastigerina*.

'HASTIGERINA' cf. BOLIVARIANA
Figures **27.5–12**; 5, 8, 9

The best illustrations of this species are those given by Bolli (1957b, pl. 37, figs. 14a–16) under the name *Globorotalia bolivariana*. This form differs from typical '*H*.' *bolivariana* by its smaller size (see Fig. 27), less planispiral coiling and a less globular last chamber. In assemblages from the early Middle Eocene of Venezuela it is possible to find both '*H*.' *bolivariana* and '*H*.' cf. *bolivariana* and they belong to the same group, but possibly it would be better to give a new specific or subspecific name to the forms described here as '*H*.' cf. *bolivariana*.

'*H*.' cf. *bolivariana* has a longer stratigraphic range: it is present during the whole Middle Eocene, but it is often very abundant during the *Orbulinoides beckmanni* Zone. Blow (1979, p. 1178) wrote that the forms illustrated by Bolli (1957b) as *Globorotalia bolivariana* belong probably to *G. increbescens* Bandy (1949), but *G. increbescens* has a rather wide aperture and belongs to another lineage.

TURBOROTALIA GRIFFINAE Blow
Figures **27.13–23**; 5, 7, 10

Type reference: *Globorotalia (Turborotalia) griffinae* Blow, 1979, p. 1072, pl. 96, fig. 8.

This species has 4, exceptionally $4\frac{1}{2}$, globular chambers in the last whorl. The surface of the test is cancellate. The aperture is a widely open arch bordered by a distinct lip, umbilical–extraumbilical in position and extending somewhat towards the spiral side.

It differs from '*H*.' *bolivariana* by its non-planispiral coiling and from '*H*.' cf. *bolivariana* by its widely open aperture instead of a narrow slit.

This species arises from *Globigerina inaequispira* according to Blow (1979). It occurs from the late Early Eocene *Acarinina pentacamerata* Zone to the Middle Eocene *Morozovella lehneri* Zone.

Early to Middle Eocene species of *Globigerina* and '*Globigerinoides*'
GLOBIGERINA SENNI (Beckmann)
Figures **28.1–5**; 6, 7, 9

Type reference: *Sphaeroidinella senni* Beckmann, 1953, p. 394, pl. 26, fig. 2.

This compact and thick-walled species is readily distinguished from all other Paleogene globigerinas by the slightly embracing chambers of the last whorl and by the nearly closed umbilicus, which is surrounded by distinct granulations.

The wall structure of *G. senni* shows strong affinities to the genus *Globigerinatheka* (see e.g. Bolli, 1972) and some authors (e.g. Blow, 1979) consider it as its possible ancestor.

The particular value of *G. senni* lies in its resistance to solution. In very deep water assemblages it may be about the only planktic species remaining, in which case its upper limit at the top of the Middle Eocene can be useful.

GLOBIGERINA LOZANOI Colom
Figures **28.6–11**; 6, 7, 10

Type reference: *Globigerina lozanoi* Colom, 1954, p. 149, pl. 2, figs. 1–48.

Lectotype: *Globigerina lozanoi* Colom, 1954: Blow, 1979, p. 854, specimen figured by Colom, pl. 2, fig. 45.

G. lozanoi is characterized by a relatively large number of chambers in the last whorl ($4\frac{1}{2}$ to 6), which increase slowly in size, a high trochospire and a loosely, somewhat irregular coiling pattern. The large number of chambers in the last whorl is similar to *Acarinina pentacamerata*, but *A. pentacamerata* has a spinose test instead of a smooth one as in *G. lozanoi*.

Hillebrandt (1976) and Blow (1979) claim that *G. lozanoi* is the ancestral form of '*Globigerinoides*' *higginsi*. Intermediate forms between the two species exist (Figs. 28.12–13).

'GLOBIGERINOIDES' HIGGINSI Bolli
Figures **28.14–16**; 5, 7, 10

Type reference: *'Globigerinoides' higginsi* Bolli, 1957b, p. 164, pl. 36, figs. 11a–b.

This very high, loosely coiled, trochospiral form is one of the most conspicuous species of the late Early and early Middle Eocene. The main feature is the secondary sutural aperture on the spiral side between the last two chambers, which is visible in well-preserved specimens. Sometimes secondary sutural apertures may occur also between the earlier chambers. '*G.*' *higginsi* has no phyletic relationship with the Neogene *Globigerinoides* and its generic position is uncertain. It is the end-form of an evolutionary lineage beginning with *Glo-*

bigerina lozanoi. The size of the species and the shape of its spire are variable (Figs. 28.15–16).

El Naggar (1971) erected the new genus *Guembelitrioides* with '*Globigerinoides*' *higginsi* as type species. This name is, however, not used here.

The *Acarinina bullbrooki–Acarinina spinuloinflata* group

Numerous spinose *Acarinina* species were described from the late Early and Middle Eocene. As common features they all possess 4 to 5 subangular-conical chambers in the last whorl, a rounded to subacute periphery in side view, and measure 0.25 to 0.5 mm. They include the species *spinuloinflata*,

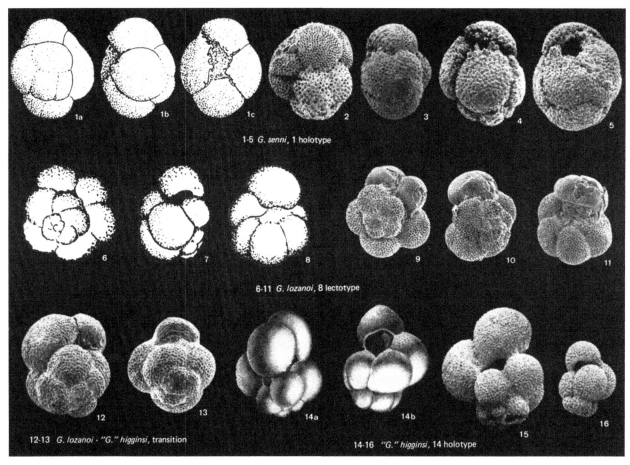

Fig. 28. *Globigerina senni, Globigerina lozanoi* and '*Globigerinoides*' *higginsi* (all figures × 60)

1–5 *G. senni* (Beckmann)
1a–c Holotype. Sample S 209, Middle Eocene, Mount Hillaby, Barbados. **2–3** Sample Hk 408, *O. beckmanni* Zone, Middle Eocene, Navet Fm., Trinidad. **4–5** DSDP Leg 32-313-5-2, 123–125 cm, *M. lehneri* to *O. beckmanni* Zone, Middle Eocene, Central Pacific (from Toumarkine, 1975).

6–11 *G. lozanoi* Colom
6–7 Paratypes, **8** Lectotype (designated by Blow, 1979). Late Ypresian–Early Lutetian, Carretera de Aspe-Crevillente, Alicante,

Spain. **9–11** Sample ER 1002, *A. pentacamerata* Zone, Early Eocene, Richmond Fm., Jamaica.

12–13 *G. lozanoi*–'*G.*' *higginsi*, transition
12 Sample PJB 152, *G. subconglobata subconglobata* to *M. lehneri* Zone, Middle Eocene, El Datil Fm., Punta Mosquito, Margarita, Venezuela. **13** Sample Ba 21, Middle Eocene, Bou Arada, Tunisia (from Mèchmèche & Toumarkine, 1983).

14–16 '*G.*' *higginsi* Bolli
14a–b Holotype. Early Middle Eocene, core at lat. 30°43′ N, long. 62°28′ W, Atlantic Ocean. **15–16** DSDP Leg 32-313-7, cc, *M. lehneri* Zone, Middle Eocene, Central Pacific (from Toumarkine, 1975).

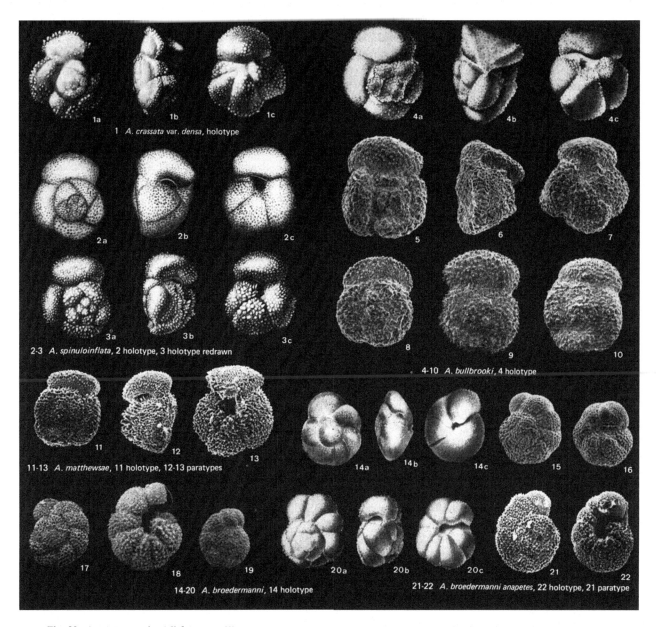

Fig. 29. *Acarinina* species (all figures × 60)

1 *A. crassata* var. *densa* (Cushman)
1a–c Holotype (figured in Cifelli, 1972). Eocene, East bank of Moctezuma River near mouth of Rio Tamuin, Vera Cruz, Mexico.

2–3 *A. spinuloinflata* (Bandy)
2a–c Holotype, **3a–c** Holotype redrawn (Cifelli, 1972). Middle Eocene, Tallahatta Fm., Little Stave Creek, Clarke County, Alabama, USA.

4–10 *A. bullbrooki* (Bolli)
4a–c Holotype. Sample K 9077, *H. nuttalli* Zone, Middle Eocene, Navet Fm., Trinidad. **5–7** Sample 626, **9**, Sample 629, *H. nuttalli* Zone, **10** Sample 634, *G. subconglobata subconglobata* Zone, all from the Middle Eocene, Possagno, N Italy. **8** Core Pénzesgyor 9, Middle Eocene, Bakony Mountain, Hungary (from Toumarkine, 1971).

11–13 *A. matthewsae* Blow
11 Holotype, **12–13** Paratypes. Sample RS 24, Zone P 11, Middle Eocene, Kilwa area, Tanzania, E Africa.

14–20 *A. broedermanni* (Cushman & Bermudez)
14a–c Holotype. Early Eocene, Capdevila Fm., Havana Province, Cuba. **15–16** Sample Ba 19, Middle Eocene, Bou Arada, Tunisia (from Mèchmèche & Toumarkine, 1983). **17–18** DSDP Leg 32-313-7, cc, **19** DSDP Leg 32-305-10, cc, Middle Eocene, Central Pacific. **20a–c** Sample KTO 145, *A. pentacamerata* Zone, Early Eocene, Navet Fm., Trinidad (from Bolli, 1957b).

21–22 *A. broedermanni anapetes* Blow
21 Paratype, **22** Holotype. Sample RS 24, Zone P 11, Middle Eocene, Kilwa area, Tanzania, E Africa.

bullbrooki, densa, rotundimarginata and *matthewsae*. Of these, only the first two are considered here. *A. rotundimarginata* Subbotina, 1953, is generally regarded as a junior synonym of *Truncorotaloides collactea* (Berggren, 1977; Blow, 1979). *Pulvinulina crassata* var. *densa* Cushman, 1925d, from which Cifelli (1972) figured the holotype (Fig. 29.1), has 4½ chambers in the last whorl, a more lobate general shape than *A. bullbrooki* and *A. spinuloinflata*, and the pustules on the spiral side are concentrated around the periphery forming a 'pseudo-keel'. This last character makes it questionable if it belongs to *Acarinina. Pulvinulina crassata* var. *densa* possibly belongs to *Morozovella* or *Truncorotaloides*. More investigations are needed to clarify its generic position. The relationships between *A. matthewsae, A. bullbrooki* and *A. spinuloinflata* are discussed below.

ACARININA BULLBROOKI (Bolli)
Figures **29.4–10**; 6, 7, 9

Type reference: *Globorotalia bullbrooki* Bolli, 1957b, p. 167, pl. 38, figs. 5a–b.

Acarinina bullbrooki differs from *A. spinuloinflata* by a flatter spiral side, a much more conical umbilical side and more subangular chambers which increase more rapidly in size. The general shape of the test is more quadrangular. *Globorotalia (Acarinina) matthewsae* Blow, 1979 (Figs. 29.11–13) is very close to *A. bullbrooki* but has a less flat spiral side, a less conical umbilical side and a more rounded periphery in side view, but nevertheless its periphery is not as rounded as in *A. spinuloinflata*.

    A. bullbrooki, which appears near the base of the *Acarinina pentacamerata* Zone, can be the dominant species in Middle Eocene assemblages with a reduced species diversity. The species is used as zonal marker by Krasheninnikov (1965b, 1969, Fig. 4) for the base of the Middle Eocene.

ACARININA SPINULOINFLATA (Bandy)
Figures **29.2–3**; 6, 7, 9

Type reference: *Globigerina spinuloinflata* Bandy, 1949, p. 122, pl. 23, figs. 1a–c.

This spinose species is very close to *A. bullbrooki* but has a less convex umbilical side, a rounder periphery and a less quadrangular general shape of the test. The two species are linked by intermediate forms.

ACARININA BROEDERMANNI (Cushman & Bermudez)
Figures **29.14–20**; 6, 7, 10

Type reference: *Globorotalia (Truncorotalia) broedermanni* Cushman & Bermudez, 1949, p. 40, pl. 7, figs. 22–24.

The large number of chambers in the last whorl (6 to 8), which increase very gradually in size, is the main feature of this rather small, hispid and biconvex *Acarinina*. The general aspect of the test is compact and the periphery nearly circular.

    Blow (1979) distinguished two subspecies, a more tightly coiled, *Globorotalia (Acarinina) broedermanni broedermanni*, and a *G. (A.) broedermanni anapetes* (Figs. 29.21–22) which is more loosely coiled, has a flatter spiral side and a more widely open umbilicus. According to Blow (1979) *G. (A.) broedermanni anapetes* is restricted to the early Middle Eocene and has a shorter stratigraphic range than *G. (A.) broedermanni broedermanni*, which appears during his P 8a Zone (equivalent to the *Morozovella formosa formosa* Zone). Because of the small difference between the two subspecies we use *A. broedermanni* in a broad sense.

Middle Eocene representatives of the genus *Morozovella*

Among the several species of *Morozovella* described from the Middle Eocene we retain only three: *M. aragonensis, M. spinulosa* and *M. lehneri*. It is probable that *M. aragonensis* is the ancestral form of *M. spinulosa* which in turn gives rise to *M. lehneri* (Berggren, 1977). *M. aragonensis* is essentially an Early Eocene species but ranges into the Middle Eocene.

MOROZOVELLA SPINULOSA (Cushman)
Figures **30.1–8**; 6, 7, 9

Type reference: *Globorotalia spinulosa* Cushman, 1927, p. 114, pl. 23, figs. 4a–c.

Several relatively small spinose *Morozovella* species with delicate to robust conicotruncate tests and a faint to distinct 'keel' occur during the Middle Eocene and have been described under a variety of names. Many of these forms have been assigned to *Globorotalia spinulosa*. The poorly figured holotype of *Morozovella spinulosa* is a small and very delicate form with a faint 'keel' (H. M. Bolli, personal communication). The more robust forms with a strongly convex umbilical side and sometimes slightly overlapping chambers which increase rapidly in size in the last whorl are here excluded from *M. spinulosa*. Probably, they are transitional forms to *M. aragonensis* or may even belong to *Truncorotaloides*. However, the taxonomic status of the Middle Eocene *Morozovella* species needs revision.

    Some authors (e.g. Berggren, 1977) consider *Pulvinulina crassata* Cushman, 1925d (Fig. 30.9) from which Bandy (1964) selected and illustrated a lectotype (Fig. 30.10), as a senior synonym of *M. spinulosa*. Blow (1979) noted that the holotype has been lost and proposed that the name *Pulvinulina crassata* should be removed from the list of valid zoological names and that the name '*spinulosa*' should be used. He distinguished two subspecies: *M. spinulosa spinulosa* and *M. spinulosa coronata* (Figs. 30.11–12) which has a more widely open umbilicus surrounded by a 'coronet of muricae'. We do not differentiate between these two varieties.

M. spinulosa differs from *M. aragonensis* in being more delicate, smaller, more loosely coiled, more spinose and in having a less convex umbilical side. *M. lehneri* has radially elongate chambers.

MOROZOVELLA LEHNERI (Cushman & Jarvis)
Figures 31.1–13; 6, 8, 9

Type reference: *Globorotalia lehneri* Cushman & Jarvis, 1929, p. 17, pl. 3, figs. 16a–c.

The 5 to 7, or even more, radially elongate chambers, the strongly lobate and frilled periphery with a spiny 'keel' and the very much compressed, almost lenticular test, are the main characters of the species and distinguish it from other *Morozovella* species and particularly from *M. spinulosa*. However, the two species are linked by intermediate forms.

M. lehneri is a conspicuous species and a useful marker restricted to the Middle Eocene of tropical to subtropical areas.

Representatives of the genus *Truncorotaloides*

The main characteristics of the genus are the small openings on the spiral side at the junctions of the intercameral and the spiral sutures. In specimens not well preserved it may be difficult to recognize these small secondary apertures. Such specimens may then be attributed to either *Morozovella* or *Acarinina*. All forms belonging to the genus have a hispid test. They are of stratigraphic significance from the late Early Eocene (*Acarinina pentacamerata* Zone) to the top of the Middle Eocene (*Truncorotaloides rohri* Zone).

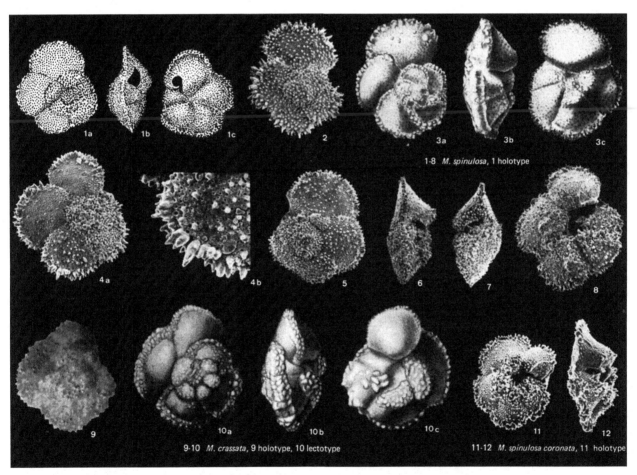

1–8 *M. spinulosa*, 1 holotype

4a 4b 5 6 7 8

9 10a 10b 10c 11 12

9–10 *M. crassata*, 9 holotype, 10 lectotype   11–12 *M. spinulosa coronata*, 11 holotype

Fig. 30. *Morozovella spinulosa* and related species (all figures × 60, except 4b × 150)

1–8 *M. spinulosa* (Cushman)
  1a–c Holotype. Middle Eocene, Alazan Clay, Rio Tuxpan, Vera Cruz, Mexico. 2 Sample HMB 67/24, *M. lehneri* Zone, Middle Eocene, Guayabal Fm., Tampico Embayment, Mexico. 3a–c Sample K 8820, *H. nuttalli* Zone, Middle Eocene, Navet Fm., Trinidad (from Bolli, 1957b). 4–8 Sample Hk 408, *O. beckmanni* Zone, Middle Eocene, Navet Fm., Trinidad.

9–10 *M. crassata* (Cushman)
  9 Holotype. Eocene, Moctezuma River, Vera Cruz, Mexico. 10a–c Lectotype (Bandy, 1964). Eocene shale on east bank of Moctezuma River, near Hacienda Romance, Vera Cruz, Mexico.

11–12 *M. spinulosa coronata* Blow
  11 Holotype, 12 Paratype. Sample RS 24, Zone P 11, Middle Eocene, Kilwa area, Tanzania, E Africa.

TRUNCOROTALOIDES COLLACTEA (Finlay)
Figures 6, 7, 9 (for illustrations see Chapter 7, this volume)

Type reference: *Globorotalia collactea* Finlay, 1939, p. 29, figs. 164–165.

T. collactea differs from *T. rohri* in being much smaller, having a less spinose wall and a more compact test. *T. collactea* is dextrally coiled as opposed to the sinistrally coiled *T. rohri*. The small sutural apertures on the spiral side at the base of the final chamber occur only occasionally and are visible only in well-preserved specimens.

TRUNCOROTALOIDES HAYNESI Samanta
Figures **32.2–3, 33.10–11**; 6, 8, 9

Type reference: *Truncorotaloides haynesi* Samanta, 1970, p. 205, pl. 3, fig. 24.

T. haynesi is the largest of the *Truncorotaloides* species and differs from *T. topilensis* and *T. libyaensis* by having 6 chambers in the last whorl instead of 4 and 5. The chambers are more angular than those of *T. rohri*. Very large specimens (Fig. 33.11) are frequent but are usually associated also with relatively small ones (Fig. 33.10).

TRUNCOROTALOIDES LIBYAENSIS
EL Khoudary
Figures **32.4–6, 33.8–9**; 6, 8, 9

Type reference: *Truncorotaloides libyaensis* El Khoudary, 1977, p. 330, pl. 2, fig. 1.

T. libyaensis differs from *T. topilensis* in having 5 chambers in the last whorl instead of the characteristic 4. The general shape of the chambers is similar in the two species but they are less angular in *T. libyaensis*. *T. haynesi* has 6 chambers in the last whorl; *T. rohri* has more globular chambers and a more rounded periphery.

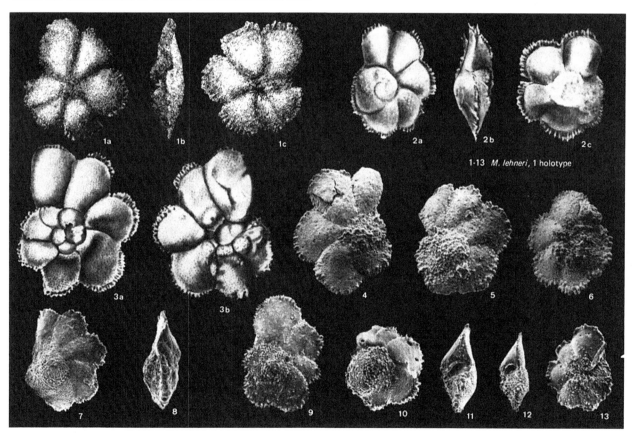

Fig. 31. *Morozovella lehneri* (all figures × 60)

1–13 *M. lehneri* (Cushman & Jarvis)
  1a–c Holotype. Middle Eocene, Lower Marl, near source of Moruga River, Trinidad. 2a–c Eocene, San Luis Fm., 2 km SW of Mir, Oriente Province, Cuba (from Cushman & Bermudez, 1949). 3a–b Sample BB 124, *O. beckmanni* Zone, Middle Eocene, Navet Fm., S Trinidad (from Bolli, 1957b). 4–6 Sample PJB 166, *M. lehneri* Zone, Middle Eocene, El Datil Fm., Punta Mosquito, Margarita, Venezuela. 7–8 Sample Allemann 77303, *M. lehneri* Zone, Middle Eocene, Beluchistan, Pakistan. 9–13 Sample HMB 67/24, *M. lehneri* Zone, Middle Eocene, Guayabal Fm., Tampico Embayment, Mexico.

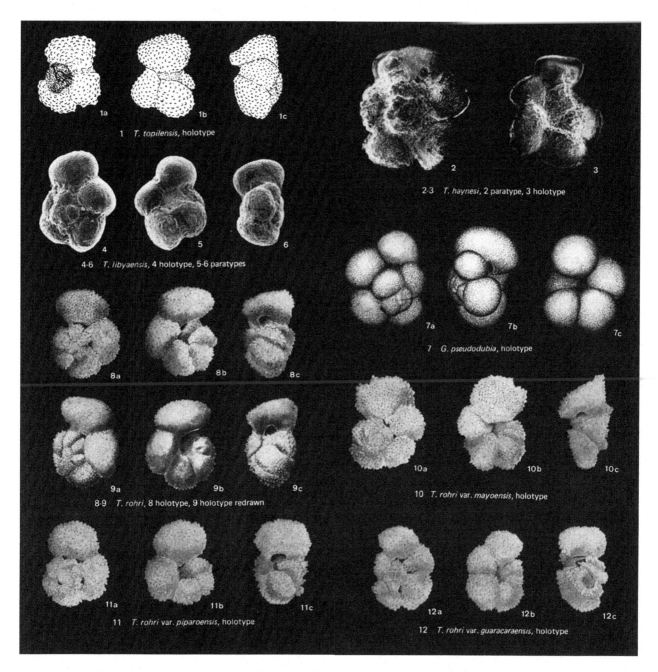

Fig. 32. *Truncorotaloides* species, holotypes (all figures × 60)

1 *T. topilensis* (Cushman)
  1a–c Holotype. Eocene, Palacho Hacienda, Tampico Embayment, Mexico.

2–3 *T. haynesi* Samanta
  2 Paratype, **3** Holotype. *O. beckmanni* Zone, Middle Eocene, Lakhpat, Cutch, India.

4–6 *T. libyaensis* El Khoudary
  4 Holotype, Sample G 49, **5–6** Paratypes, Sample G 53. All from the *G. subconglobata subconglobata* Zone, Middle Eocene, Jabal Al Akhdar, S of Al Hilal village, Libya.

7 *Globigerinoides pseudodubia* Bandy
  7a–c Holotype. Locality 7, Middle Eocene, Tallahatta Fm., Claiborne, Little Stave Creek, Clarke County, Alabama, USA.

8–9 *T. rohri* Brönnimann & Bermudez
  8a–c Holotype, **9a–c** Holotype redrawn (in Bolli *et al.*, 1957). Sample KR 23842, Middle Eocene, Duff road area, Trinidad.

10 *T. rohri* var. *mayoensis* Brönnimann & Bermudez
  10a–c Holotype. Information as for Figs. 32.8–9.

11 *T. rohri* var. *piparoensis* Brönnimann & Bermudez
  11a–c Holotype. Information as for Figs. 32.8–9.

12 *T. rohri* var. *guaracaraensis* Brönnimann & Bermudez
  12a–c Holotype. Information as for Figs. 32.8–9.

TRUNCOROTALOIDES ROHRI Brönnimann &
Bermudez
Figures **32.8–9, 33.12–18**; 6, 7, 9

Type reference: *Truncorotaloides rohri* Brönnimann & Bermudez, 1953, p. 818, pl. 87, figs. 7–9.

T. rohri is strongly spinose, especially on the umbilical shoulders and around the peripheral margin and has a flattened spiral side. Brönnimann & Bermudez (1953) described, together with the species *rohri*, three varieties: *T. rohri* var. *guaracaraensis* (Fig. 32.12), *T. rohri* var. *piparoensis* (Fig. 32.11), *T. rohri* var. *mayoensis* (Fig. 32.10). They differ from each other by the shape of the chambers of the last whorl (completely rounded for *guaracaraensis* to angular for *mayoensis*), but they are linked by transitional forms which are difficult to distinguish. We retain only the species *rohri*. *T. rohri* has a rounder axial periphery than *T.*

topilensis, T. libyaensis or *T. haynesi*. Its chambers are less angular and increase more regularly in size. It differs from *T. collactea* in its larger size, and the more spinose and less compact test.

The relationship between *T. rohri* and *Globigerinoides pseudodubia* Bandy (1949) (Fig. 32.7) needs to be investigated, as Bandy's species is possibly identical and would have priority (Stainforth *et al.*, 1975). Blow (1979) claimed that *G. pseudodubia* is synonymous with one of the varieties described by Brönnimann & Bermudez, *T. rohri* var. *guaracaraensis*, but belongs to the genus *Acarinina*.

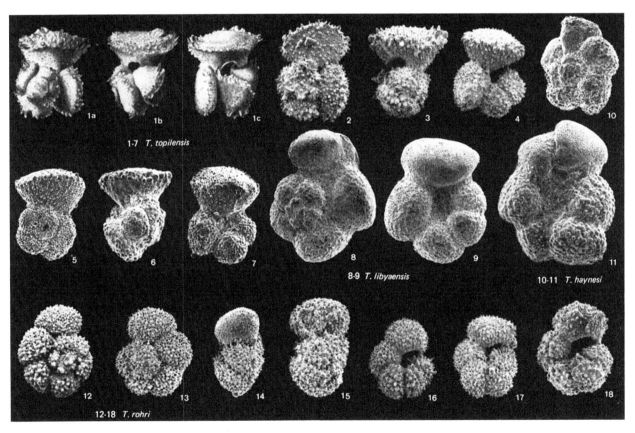

Fig. 33. *Truncorotaloides* species (all figures × 60)

1–7 *T. topilensis* (Cushman)
    1a–c Sample Hg 8581, *O. beckmanni* Zone, Middle Eocene, Navet Fm., Trinidad (from Bolli, 1957b). **2–4** Sample Hk 408, *O. beckmanni* Zone, Middle Eocene, Navet Fm., Trinidad. **5** Sample 146, Middle Eocene, El Midawarah Fm., Wadi El Rayan, Fayoum area, Egypt (from Shamah *et al.*, 1982). **6–7** Sample Boukhazy S 85, Middle Eocene, Nile Valley, Egypt.
8–9 *T. libyaensis* El Khoudary
    Sample Z 17, Middle Eocene, Beni Mazar, Nile Valley, Egypt (from Boukhary, Toumarkine & Khalifa, 1982).

10–11 *T. haynesi* Samanta
    10 Sample 146, Middle Eocene, El Midawarah Fm., Wadi El Rayan, Fayoum area, Egypt (from Shamah *et al.*, 1982). **11** Sample Ba 24, Middle Eocene, Bou Arada, Tunisia (from Mèchmèche & Toumarkine, 1983).
12–18 *T. rohri* Brönnimann & Bermudez
    12 DSDP Leg 32-313-7, cc, *M. lehneri* Zone, Middle Eocene, Central Pacific. **13–18** Sample Hk 408, *O. beckmanni* Zone, Middle Eocene, Navet Fm., Trinidad.

TRUNCOROTALOIDES TOPILENSIS (Cushman)
Figures **32.1, 33.1–7**; 6, 7, 9

Type reference: *Globigerina topilensis* Cushman, 1925b, p. 7, pl. 1, fig. 9.

This is one of the most distinctive species of the Middle Eocene. *T. topilensis* differs from *T. rohri* by its angular profile accentuated by spines, and in having only 4 chambers in the last whorl. These 4 chambers are at right angles to each other and increase rapidly in size. The younger chambers, at least the last two, have a very characteristic shape with flattened peripheral margins emphasized by a rim of spines and pustules. The shape of the chambers of *T. haynesi* and *T. libyaensis* is close to that of *T. topilensis* but they have 6 and 5 instead of 4 chambers in the last whorl.

The *Turborotalia cerroazulensis* lineage

As already noted by Bolli (1957b), the evolutionary lineage of '*Globorotalia centralis*' can be subdivided into several subspecies of stratigraphic value (see also Blow & Banner, 1962; Berggren, 1966; Samuel & Salaj, 1968). Toumarkine & Bolli (1970) described and figured in detail the Middle to Late Eocene *Turborotalia cerroazulensis* lineage in which they distinguished a sequence of six subspecies. The gradual change of the chambers of the last whorl, from a distinctly inflated, rounded periphery in lateral view in *Turborotalia cerroazulensis frontosa* to forms with strongly compressed chambers and imperforate peripheral keel in *T. cerroazulensis cunialensis*, was taken as the most significant character to distinguish the subspecies. In contrast to other characters like wall texture or surface features, it is particularly the overall morphological shape of the test that remains recognizable even in more poorly preserved specimens. In such instances this character may remain the only criterion by which a specimen can reliably be determined and be applied to stratigraphy.

In their taxonomic treatment of the lineage Toumarkine & Bolli followed the same principles as have previously been applied to similar sequences. A typical example for this is the Miocene *Globorotalia fohsi* lineage. The principle is to leave the whole plexus in the same genus and species and distinguish the morphological changes that take place in time merely on the subspecies level. As is the case in such evolutionary lineages with gradual morphological changes, a clear separation between two successive taxa remains to some degree subjective.

The lineage was originally based by Toumarkine & Bolli on observations in the Possagno and Cava Zillo sections of northern Italy, Barbados and Trinidad, and was later confirmed in a number of DSDP sites such as 305 and 313 of the central Pacific Leg 32, and 360, 362A and 363 in Leg 40 of the southeastern Atlantic. From this it is concluded that the lineage has a worldwide distribution wherever ecological conditions allowed its development.

The *Turborotalia cerroazulensis* lineage and its application to biostratigraphy as proposed by Toumarkine & Bolli has been widely followed by authors such as Stainforth *et al.* (1975). Nevertheless, Blow (1979) objected to this concept and also to the placing by Toumarkine & Bolli of *Turborotalia centralis* in synonymy with *T. cerroazulensis pomeroli*. The reasons for this synonymy are given in the taxonomic notes under *cerroazulensis pomeroli*.

Blow distinguished the three species *centralis*, *cerroazulensis* and *cunialensis*, and also admitted a relationship between them. He did not, however, accept *Globigerina frontosa* as the ancestral form of the *T. cerroazulensis* plexus. As such a form he described a *Globorotalia (Turborotalia) praecentralis* (Figs. 34.14–16). Compared with *T. centralis* this form is very small and distinctly more compressed equatorially. In shape and arrangement of its chambers it more closely resembles *Turborotalia pseudomayeri* (Figs. 34.13a–b) of which it is also regarded by Blow as a possible ancestor.

In his attempt to distinguish the taxa here discussed, Blow placed little emphasis on the changing degree of angularity of the periphery, the criterion applied by Toumarkine & Bolli, with the exception of the end-form *cunialensis*. His distinction, for example, of the holotypes of *Globorotalia centralis* and '*Globigerina*' *cerroazulensis* (the latter was not seen by Blow) is based especially on the wall texture. In contrast to *T. centralis*, Blow attributed to *T. cerroazulensis* a 'shiny and polished appearance', a feature that in our experience depends much on the state of preservation of a specimen, and cannot, therefore, be used as a distinguishing character.

In the present work the forms of the *cerroazulensis* lineage are attributed to the Paleogene genus *Turborotalia* Cushman & Bermudez, 1949 (type-species *Globorotalia centralis* Cushman & Bermudez, 1937 = *Globorotalia cerroazulensis pomeroli* Toumarkine & Bolli, 1970).

TURBOROTALIA CERROAZULENSIS
FRONTOSA (Subbotina)
Figures **34.11, 35.16–18**; 5, 7, 10

Type reference: *Globigerina frontosa*, Subbotina, 1953, p. 84, pl. 12, figs. 3a–c.
Synonym: *Globigerina boweri* Bolli, 1957b, p. 163, pl. 36, figs. 1a–c.

The rather high arch-like aperture with a faint lip separates this form from species of Eocene *Globigerina* with which it may share the general morphology of the test (see e.g. illustrations of *Globigerina inaequispira* (Fig. 19.5) and *G. eocaena* (Figs. 42.1–4)). The aperture has an umbilical–extraumbilical position but does not reach the periphery of the test. Within the *Turborotalia cerroazulensis* lineage, the aperture extends to the periphery in the subspecies *possagnoensis* and *pomeroli*, it then migrates back towards the umbilicus in the subspecies *cerroazulensis* and finally reoccupies the same position as in the

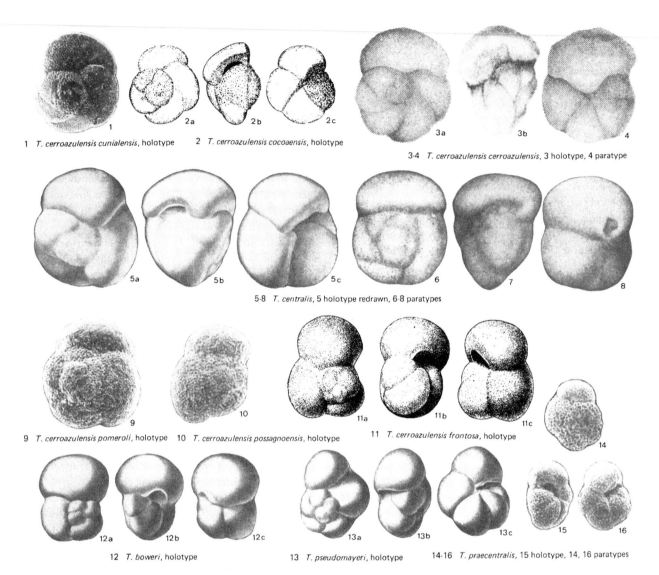

1 *T. cerroazulensis cunialensis*, holotype   2 *T. cerroazulensis cocoaensis*, holotype

3-4 *T. cerroazulensis cerroazulensis*, 3 holotype, 4 paratype

5-8 *T. centralis*, 5 holotype redrawn, 6-8 paratypes

9 *T. cerroazulensis pomeroli*, holotype   10 *T. cerroazulensis possagnoensis*, holotype   11 *T. cerroazulensis frontosa*, holotype

12 *T. boweri*, holotype   13 *T. pseudomayeri*, holotype   14-16 *T. praecentralis*, 15 holotype, 14, 16 paratypes

Fig. 34. Holotypes of subspecies of *Turborotalia cerroazulensis, T. boweri, T. pseudomayeri* and *T. praecentralis* (all figures × 60)

1 *T. cerroazulensis cunialensis* (Toumarkine & Bolli)
Holotype. Sample 655, *T. cerroazulensis* s.l. Zone, Late Eocene, Possagno, N Italy.

2 *T. cerroazulensis cocoaensis* (Cushman)
2a–c Holotype. Eocene, Cocoa Sand, Cocoa Post Office, Alabama, USA.

3–4 *T. cerroazulensis cerroazulensis* (Cole)
3a–b Holotype, **4** Paratype. *T. cerroazulensis* s.l. Zone, Late Eocene, Chapapote Fm., Rio Tuxpan, near village Chapapote, Mexico.

5–8 *T. centralis* (Cushman & Bermudez)
5a–c Holotype redrawn (in Bolli *et al.*, 1957). **6–8** Paratypes. Sample Bermudez Sta. 92, Eocene, Jicotea, Santa Clar Province, Cuba.

9 *T. cerroazulensis pomeroli* (Toumarkine & Bolli)
Holotype. Sample 643, Middle Eocene, Possagno, N Italy.

10 *T. cerroazulensis possagnoensis* (Toumarkine & Bolli)
Holotype. Sample 638, *M. lehneri* Zone, Middle Eocene, Possagno, N Italy.

11 *T. cerroazulensis frontosa* (Subbotina)
11a–c Holotype. Foraminiferal layer, Green Series, *Acarinina* Zone (lower part), Middle Eocene, River Kuban, N Caucasus, USSR.

12 *T. boweri* (Bolli)
12a–c Holotype. Sample HGK 8820, *H. nuttalli* Zone, Middle Eocene, Nariva River, Navet Fm., Trinidad.

13 *T. pseudomayeri* (Bolli)
13a–c Holotype. Sample K. 8817, *H. nuttalli* Zone, Middle Eocene, Navet Fm., Trinidad.

14–16 *T. praecentralis* Blow
14, 16 Paratypes, **15** Holotype. DSDP Leg 6-47/2-8-1, 77–79 cm, Zone P 8b, Early Eocene, Shatsky Plateau, Pacific.

subspecies *frontosa* in the Late Eocene subspecies *cocoaensis* and *cunialensis*.

Another important characteristic of the subspecies *frontosa* is the size and shape of the last chamber which is globular and occupies about one half of the entire test.

TURBOROTALIA CERROAZULENSIS POSSAGNOENSIS (Toumarkine & Bolli)
Figures **34.10, 35.13–15**; 5, 8, 10

Type reference: *Globorotalia cerroazulensis possagnoensis* Toumarkine & Bolli, 1970, p. 139, pl. 1, fig. 4.

This subspecies differs from its ancestor, the subspecies *frontosa*, by the flattening of the last chamber which gives the test a quadrangular aspect and leads to a stretching of the aperture which becomes slit-like rather than arch-like but does not reach the periphery.

Blow (1979) rejected the validity of this subspecies and indicated without further discussion that the specimens described as *Turborotalia cerroazulensis possagnoensis* are close to *Subbotina inaequispira* and *S. pseudoeocaena*. However, the latter two species have globular chambers throughout the entire test and arch-like apertures in a completely umbilical position. Blow also denied the existence of transitional forms between the subspecies *frontosa* and *possagnoensis* as well as between the subspecies *possagnoensis* and *pomeroli*. However, such transitional forms have been observed by us (Figs. 35.10–12).

TURBOROTALIA CERROAZULENSIS POMEROLI (Toumarkine & Bolli)
Figures **34.9, 35.4–9**; 5, 8, 9

Type reference: *Globorotalia cerroazulensis pomeroli* Toumarkine & Bolli, 1970, p. 140, pl. 1, fig. 13.
Synonym: partim: *Globorotalia centralis* Cushman & Bermudez, 1937, p. 26, pl. 2, fig. 65 (holotype, refigured by Bolli *et al.*, 1957, pl. 36, figs. 4a–c; see Figs. 34.5a–c).

Turborotalia cerroazulensis pomeroli differs from its ancestor *T. cerroazulensis possagnoensis* by the larger number of chambers in the last whorl (4 to 6 instead of $3\frac{1}{2}$), and by its larger size (maximum diameter 0.5 to 0.55 mm instead of 0.45 mm) and more rounded periphery.

T. cerroazulensis pomeroli is probably the most abundant and widely distributed subspecies of the *T. cerroazulensis* lineage, since it is generally very well represented in assemblages from tropical as well as temperate regions.

The holotype of *T. centralis* is lost. The remaining three paratypes were found to be virtually identical with *T. cerroazulensis* (Cole) in that their peripheral outline in lateral view is slightly more angular and less inflated compared with the figure of the holotype of *T. centralis* (see Toumarkine & Bolli, 1970). Subsequently, Cifelli & Belford (1977) erected from one

of the three paratypes of *centralis* a lectotype (should have been named neotype because it replaces a lost holotype). For such forms with a less angular periphery Toumarkine & Bolli (1970), however, had already proposed the subspecies *T. cerroazulensis pomeroli*, which name is retained here. Later, Blow (1979) maintained that the holotype of *T. cerroazulensis pomeroli* is a junior synonym of *T. centralis*. This would be correct if the subspecies *pomeroli* is compared with the figures of the lost holotype of *T. centralis* in that both display in lateral view a virtually identically inflated periphery. This is not the case, however, when *pomeroli* is compared with the lectotype resp. paratypes which, as pointed out above, do not closely compare with the holotype figures.

TURBOROTALIA CERROAZULENSIS CERROAZULENSIS (Cole)
Figures **34.3–4, 36.16–18**; 5, 8, 9

Type reference: *Globigerina cerro-azulensis* Cole, 1928, p. 217, pl. 1, figs. 11–13.
Synonym: partim: *Globorotalia centralis* Cushman & Bermudez, 1937, p. 26, pl. 2, figs. 62–64 (paratypes), but not fig. 65 (holotype).
Lectotype: Cifelli & Belford, 1977, p. 104, pl. 1, figs. 16–18.

Turborotalia cerroazulensis cerroazulensis differs from *T. cerroazulensis pomeroli* by its flattened spiral side which gives a more angular aspect in lateral view. The aperture is higher, it tends to migrate towards the umbilicus and does not reach the periphery. However, the two subspecies are linked by an uninterrupted series of intermediate forms.

Traditionally, *Globigerina cerroazulensis* and *Globorotalia cocoaensis* have been regarded as synonyms. Bolli (1957b) considered *G. cocoaensis* to be the valid one and used it to name the youngest zone of the Eocene, but changed it into *Globorotalia cerroazulensis* Zone in 1959 because the latter species has priority by a few months. In 1970, J. P. Beckmann compared the holotypes of the two species and stated that, contrary to the illustrations, the holotype of *Globorotalia cocoaensis* has a considerably more angular periphery in lateral view than the holotype of *Globigerina cerroazulensis*. Based on this observation, Toumarkine & Bolli (1970) retained both names as valid and considered the two forms as consecutive subspecies of the *Turborotalia cerroazulensis* lineage.

Sztrákos (1973) considered *T. cerroazulensis* as a junior synonym of *Globigerina applanata* Hantken 1883, but the latter name was considered later as a nomen oblitum (Stainforth, Sztrákos & Jeffords, 1982).

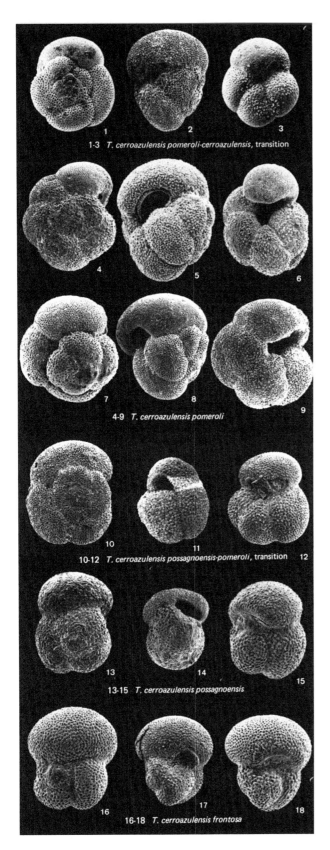

1-3 *T. cerroazulensis pomeroli-cerroazulensis*, transition

4-9 *T. cerroazulensis pomeroli*

10-12 *T. cerroazulensis possagnoensis-pomeroli*, transition

13-15 *T. cerroazulensis possagnoensis*

16-18 *T. cerroazulensis frontosa*

TURBOROTALIA CERROAZULENSIS COCOAENSIS (Cushman)
Figures **34.2, 36.10–12**; 5, 8, 9

Type reference: *Globorotalia cocoaensis* Cushman, 1928, p. 75, pl. 10, figs. 3a–c.

Turborotalia cerroazulensis cocoaensis shares with the subspecids *cerroazulensis* the almost flat spiral side, but in lateral view its periphery is considerably more acute. The aperture is a high arch between the umbilicus and the periphery but does not reach the latter.

TURBOROTALIA CERROAZULENSIS CUNIALENSIS (Toumarkine & Bolli)
Figures **34.1, 36.1–6**; 5, 8, 9

Type reference: *Globorotalia cerroazulensis cunialensis* Toumarkine & Bolli, 1970, p. 144, pl. 1, fig. 37.

Turborotalia cerroazulensis cunialensis is the final form of the *T. cerroazulensis* lineage. It is characterized by a very acute profile in lateral view caused by the highly compressed test and by a faint imperforate keel.

Blow & Banner (1962) discussed forms of *T. cerroazulensis* with an acute periphery but with a perforate 'pseudo-keel'. Blow (1979) recognized the existence of forms with a true imperforate keel in the *T. cerroazulensis* lineage, but he thought that their distribution was very patchy and that they occurred always in very small numbers. In fact, the subspecies *cunialensis* is found less frequently than the other subspecies of the *T. cerroazulensis* lineage because its test is more fragile and more sensitive to dissolution and other environmental factors. In addition, continuous sections across the Eocene–Oligocene boundary in a suitable facies are rare.

The original specimens described from Northern Italy

Fig. 35. *Turborotalia cerroazulensis* lineage (all figures × 60)

1–3 *T. cerroazulensis pomeroli–T. cerroazulensis cerroazulensis*, transition **1** Sample HMB 68/72, **3** Sample 651, *T. cerroazulensis* s.l. Zone, Late Eocene, Possagno, N Italy. **2** Sample Hk 408, *O. beckmanni* Zone, Middle Eocene, Navet Fm., Trinidad.

4–9 *T. cerroazulensis pomeroli* (Toumarkine & Bolli) **4–6, 8, 9** Sample HK 408, *O. beckmanni* Zone, Middle Eocene, Navet Fm., Trinidad (8, from Toumarkine & Bolli, 1970). **7** Sample Ba 21, Middle Eocene, Bou Arada, Tunisia (from Mèchmèche & Toumarkine, 1983).

10–12 *T. cerroazulensis possagnoensis–T. cerroazulensis pomeroli*, transition **10** Sample HK 408, *O. beckmanni* Zone, Middle Eocene, Navet Fm., Trinidad. **11** DSDP Leg 32-313-7, cc, *M. lehneri* Zone, Middle Eocene, Central Pacific. **12** Sample HMB 67/23, early Middle Eocene, Aragon Fm., Tampico Embayment, Mexico.

13–15 *T. cerroazulensis possagnoensis* (Toumarkine & Bolli) **13** Sample 638, *M. lehneri* Zone, Middle Eocene, Possagno, N Italy. **14** DSDP Leg 40-363-11-1, 58–60 cm, Middle Eocene, S Atlantic (from Toumarkine, 1978). **15** Sample Ba 19, early Middle Eocene, Bou Arada, Tunisia (from Mèchmèche & Toumarkine, 1983).

16–18 *T. cerroazulensis frontosa* (Subbotina) **16, 18** Sample Ba 19, early Middle Eocene, Bou Arada, Tunisia (from Mèchmèche & Toumarkine, 1983). **17** Sample HMB 67/23, early Middle Eocene, Aragon Fm., Tampico Embayment, Mexico.

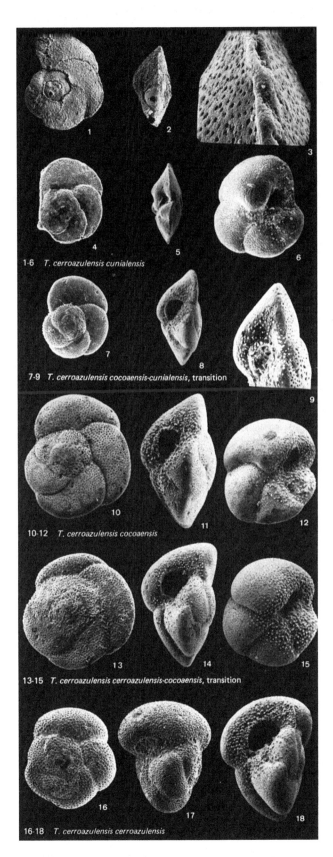

1-6 *T. cerroazulensis cunialensis*

7-9 *T. cerroazulensis cocoaensis-cunialensis*, transition

10-12 *T. cerroazulensis cocoaensis*

13-15 *T. cerroazulensis cerroazulensis-cocoaensis*, transition

16-18 *T. cerroazulensis cerroazulensis*

àre rather small probably due to prevailing near-shore conditions. In more open-marine sections such as encountered in the South Atlantic DSDP leg 40, Site 363 (Toumarkine, 1978) specimens are of larger size. The taxon was also observed in such widely separated areas as India (Fleisher, 1974) and Mexico (Butterlin *et al.*, 1977).

The genera *Globigerinatheka* and *Orbulinoides*

The genus *Globigerinatheka* was defined in 1952 by Brönnimann to include globular globigerinids with a large enveloping final chamber covering the umbilicus and with secondary sutural apertures with bullae. In its original definition, it did not include forms with an enveloping final chamber and sutural apertures but without bullae; these forms were considered as belonging to the genera *Globigerinoides* or *Globigerina*.

Bolli *et al.* (1957) introduced the genus *Globigerapsis* for all forms with a large enveloping final chamber but without bullae and the genus *Porticulasphaera* for very large spherical forms with numerous sutural and areal apertures. In 1968, Blow & Saito re-examined the type species of *Porticulasphaera* and came to the conclusion that its generic characteristics are those of the genus *Globigerapsis*. Consequently, they introduced the generic name *Orbulinoides* as a replacement for *Porticulasphaera*. Cordey (1968) was the first to publish this new name. Proto Decima & Bolli (1970) demonstrated that the presence or absence of bullae covering the secondary sutural apertures should not carry generic rank and, therefore, placed *Globigerapsis* in synonymy with *Globigerinatheka* which is the bullate form of *Globigerapsis*. Both forms coexist in all species without significant differences in their stratigraphic distribution. In addition, it is often impossible to distinguish a bulla from a small final chamber.

The classification of the species and subspecies of the

Fig. 36. *Turborotalia cerroazulensis* lineage (all figures × 60, except 3 × 600, 9 × 120)

1–6 *T. cerroazulensis cunialensis* (Toumarkine & Bolli)
  1 DSDP Leg 40-363-9-3, 105–107 cm, *T. cerroazulensis* s.l. Zone, Late Eocene, S Atlantic (from Toumarkine, 1978). 2 DSDP Leg 40-363-9, cc, *G. semiinvoluta* Zone, Late Eocene, S Atlantic (from Toumarkine, 1978). 3–6 Sample 655, *T. cerroazulensis* s.l. Zone, Late Eocene, Possagno, N Italy (5–6 from Toumarkine & Bolli, 1970).

7–9 *T. cerroazulensis cocoaensis–T. cerroazulensis cunialensis*, transition
  7–8 Sample 655, 9 Sample 656, *T. cerroazulensis* s.l. Zone, Late Eocene, Possagno, N Italy (8–9 from Toumarkine & Bolli, 1970).

10–12 *T. cerroazulensis cocoaensis* (Cushman)
  Information as for Figs. 36.3–6 (from Toumarkine & Bolli, 1970).

13–15 *T. cerroazulensis cerroazulensis–T. cerroazulensis cocoaensis*, transition
  13, 15 Sample 655, 14 Sample HMB 68/71. Information as for Figs. 36.3–6 (13, 15 from Toumarkine & Bolli, 1970).

16–18 *T. cerroazulensis cerroazulensis* (Cole)
  16–17 Sample Ba 26, Middle Eocene, Bou Arada, Tunisia (from Mèchmèche & Toumarkine, 1983). 18 Sample HMB 68/71. Information as for Figs. 36.3–6.

genera *Globigerinatheka* and *Orbulinoides* used hereafter is discussed in detail by Bolli (1972). The definition of the species and subspecies is based on easily recognizable features such as the general form and size of the test, the shape of the apertures and the thickness of the wall.

Blow published in 1979 (written in 1972 before the publication of Bolli's paper) a different approach to the taxonomy of the same forms. He based the definition of the species mainly on the structure of the test as seen in the scanning electron microscope.

Bolli (1972) recognized among the representatives of the genus *Globigerinatheka* four morphological groups which also represent four different evolutionary lineages. With the exception of the monospecific *Globigerinatheka semiinvoluta* group and of *Orbulinoides beckmanni*, the end-form of the *G. subconglobata* group, all members of the same lineage are treated as subspecies of a single species.

The four lineages discussed by Bolli (1972) have the following main characteristics:

Globigerinatheka index group (4 subspecies): general shape of the tests somewhat elongate, wall robust and with deeply incised sutures, apertures usually not covered by bullae.

Globigerinatheka mexicana group (3 subspecies): more or less globular delicate tests of small size, presence of bullae frequent.

Globigerinatheka semiinvoluta group (no subspecies): large irregularly dome-shaped final chamber making up half or more of the entire volume of the specimens, aperture a high arch to circular. The overall size of the tests is very variable.

Globigerinatheka subconglobata group (5 subspecies): globular tests of large size (except in the ancestral *G. subconglobata micra*), wall robust, occasional presence of bullae. The end-form of the *G. subconglobata* lineage is the monospecific genus *Orbulinoides* which has an embracing final chamber and numerous sutural and areal apertures (see Proto Decima & Bolli, 1970).

Bolli (1972) gave a tentative discussion of the evolution of *Globigerinatheka*. The evolutionary trends and the paleogeographic distribution of its taxa are of significance for the Middle and Late Eocene biostratigraphy and paleoecology. Their paleogeographic distribution shows considerable variation from low to higher latitudes. Bolli (1972) provided a pattern based on observations from the Caribbean, East Africa, the Caucasus, the Alpine–Mediterranean area and New Zealand. These observations became more complete by the addition of later data from the tropical Pacific (Toumarkine, 1975), the South Atlantic (Toumarkine, 1978) and the Alpine–Mediterranean area (Toumarkine & Bolli, 1975).

Of the more significant taxa the *G. subconglobata–G. euganea–O. beckmanni* lineage is well developed through-out the tropical Atlantic and Caribbean, the tropical Pacific and the Mediterranean regions. *G. semiinvoluta* is represented by small and large specimens in low latitudes, by smaller specimens only in more temperate areas such as the Alpine–Mediterranean, to become absent altogether in higher latitudes. *G. subconglobata luterbacheri* is not known from low latitudes but may become frequent and very large in size in mid-latitudes such as the Alps and the South Atlantic (DSDP Leg 40, Sites 360, 363). Similarly, *G. index* is also largely confined to mid-latitudes where it is used for instance as a zonal marker in the New Zealand–Austral area. *G. mexicana mexicana* on the other hand is more cosmopolitan in its occurrence and as such has become an index to define the base of the *G. subconglobata subconglobata* Zone.

Fig. 37. *Globigerinatheka* holotypes (all figures × 60)

1–2 *G. subconglobata luterbacheri* Bolli
  1 Holotype, **2** Paratype. Sample AA 3858 of Eckert (1963), Late Eocene, Looegg, Central Switzerland.

3 *Orbulinoides beckmanni* Saito
  3a–c Holotype. Eocene, Ogosawara Group, Sea cliff of Onion Beach, at the east side of the small embayment of Oki-mura, Haha-jima Island, Japan.

4–5 *G. subconglobata euganea* Proto Decima & Bolli
  4 Holotype, **5** Paratype. Sample FPD 69/17, *O. beckmanni* Zone, Middle Eocene, Cava Zillo section, Euganei Hills, N Italy.

6 *G. subconglobata curryi* Proto Decima & Bolli
  6a–b Holotype. Sample K 9071, *M. lehneri* Zone, Middle Eocene, Navet Fm., Nariva River, Trinidad (from Bolli, 1957b, described as *Globigerapsis kugleri*).

7 *G. subconglobata subconglobata* (Shutskaya)
  Lectotype (designated by Bolli, 1972). Eocene, Kerestin horizon, District of Nal'tchik, Khéu River, Caucasus, USSR.

8–10 *G. subconglobata micra* (Shutskaya)
  8 Holotype, **9–10** Paratypes, Nagout Core, Eocene (layers with *Morozovella aragonensis*), Tcherkess horizon, USSR.

11 *G. semiinvoluta* (Keijzer)
  11a–c Holotype. Sample F286, Late Eocene, section E of Mayari, Cuba.

12 *G. index tropicalis* (Blow & Banner)
  12a–b Holotype. Sample FCRM 1645, Late Eocene, Lindi area, Tanganyika, E Africa.

13 *G. index rubriformis* (Subbotina)
  13a–c Holotype. *G. conglobatus* Subzone, Late Eocene, White Clay horizon, right tributary of the River Sukhaya Tsetse, USSR.

14 *G. mexicana kugleri* (Bolli, Loeblich & Tappan)
  14a–c Holotype. Sample from a block of Eocene Navet Fm. (*O. beckmanni* Zone) in Oligocene Nariva Fm., Pointe-à-Pierre, Trinidad.

15 *G. mexicana barri* Brönnimann
  15a–c Holotype. Core from well Harmony Hall 2, Eocene, S of Pointe-à-Pierre, Trinidad.

16–17 *G. mexicana mexicana* (Cushman)
  16a–c Holotype redrawn (in Blow & Saito, 1968), **17a–b** Holotype. Palacho Hacienda, probably from a block of Guayabal Fm. of Middle Eocene *M. lehneri* Zone age in Late Eocene Tantoyuca Fm., State of Vera Cruz, Mexico.

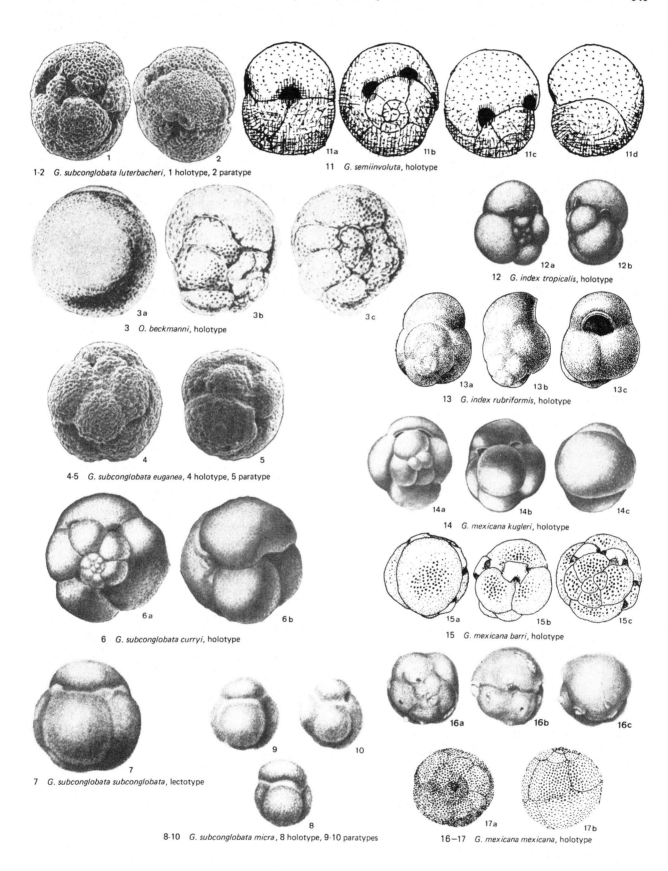

1-2 *G. subconglobata luterbacheri*, 1 holotype, 2 paratype

11 *G. semiinvoluta*, holotype

3 *O. beckmanni*, holotype

12 *G. index tropicalis*, holotype

13 *G. index rubriformis*, holotype

4-5 *G. subconglobata euganea*, 4 holotype, 5 paratype

14 *G. mexicana kugleri*, holotype

6 *G. subconglobata curryi*, holotype

15 *G. mexicana barri*, holotype

7 *G. subconglobata subconglobata*, lectotype

8-10 *G. subconglobata micra*, 8 holotype, 9-10 paratypes

16-17 *G. mexicana mexicana*, holotype

GLOBIGERINATHEKA INDEX INDEX (Finlay)
Figures **38.20–24**; 5, 8, 9; Chapter 7, Figure **6.6**

Type reference: *Globigerinoides index* Finlay, 1939, p. 125, pl. 14, fig. 85.

Globigerinatheka index index has a fairly round test with rather large apertures at the base of the final chambers opposite the sutures. *G. index index* has thicker walls, a more rugose surface and more incised, crevice-like sutures than *G. index tropicalis*. *G. index rubriformis* has a higher initial spire, but the three subspecies are all linked by intermediate forms. Some specimens may be difficult to assign to *G. index index* or to *G. subconglobata subconglobata*, but *G. index index* has a larger final chamber.

   G. index index is fairly common in Austral areas (where it was described) and in northern regions (Caucasus, Alpine areas, northern Europe). It rarely occurs in tropical regions, where it stops at the end of the Middle Eocene, whereas it persists during the *Globigerinatheka semiinvoluta* Zone in middle latitudes and perhaps till the Eocene/Oligocene boundary in higher latitudes (Berggren, 1971; Jenkins, 1971; Blow, 1979).

GLOBIGERINATHEKA INDEX RUBRIFORMIS (Subbotina)
Figures **37.13, 38.18–19**; 5, 8, 9

Type reference: *Globigerinoides rubriformis* Subbotina, 1953, p. 92, pl. 14, fig. 6.

This subspecies is characterized by a high initial spire. The walls are not as thick and not as granular as those of *G. index index*. The sutures are less incised. The subspecies is morphologically similar to *G. index tropicalis* except for the high spire.

   G. index rubriformis occurs in the Middle Eocene of temperate and tropical regions.

GLOBIGERINATHEKA INDEX TROPICALIS (Blow & Banner)
Figures **37.12, 38.14–17**; 5, 8, 9

Type reference: *Globigerapsis tropicalis* Blow & Banner, 1962, p. 124, pl. 15, figs. D, E.

The general shape of the test and the arrangement of the chambers are the same as in the other subspecies of *G. index*. *G. index tropicalis* differs from *G. index index* in having a more delicate wall and less deeply incised sutures, but specimens intermediate in character exist and according to Bolli (1972) it is likely that *G. index tropicalis* has evolved from *G. index index*.

   G. index tropicalis differs from *G. semiinvoluta* by its more individualized chambers, by its less globular final chamber and its more elongated general shape of the test.

   Blow (1979) claimed that the species is no longer valid

and is a junior synonym of *G. mexicana*. However, according to Bolli (1972) the representatives of the *G. mexicana* plexus are very different from *G. index tropicalis*: the forms related to *G. mexicana* are always rather small, delicate and globular, whereas *G. index tropicalis* is more elongated, more robust and larger.

   Sztrákos (1973) claimed that *Globigerina globosa* Hantken would be a senior synonym of *G. tropicalis*, but Stainforth *et al.* (1982) recommended that Hantken's species be considered as a nomen oblitum.

   G. index tropicalis is common during the Late Eocene in temperate as well as in tropical regions.

GLOBIGERINATHEKA MEXICANA MEXICANA (Cushman)
Figures **37.16–17, 39.33–39**; 5, 8, 9

Type reference: *Globigerina mexicana* Cushman, 1925b, p. 61, pl. 1, figs. 8a–b.

G. mexicana mexicana is characterized by its small globular test and its small sutural apertures with distinct rims. It differs from *G. mexicana barri* by the lack of bullae and from *G. mexicana kugleri* by its more tightly coiled initial spire. *G. semiinvoluta* has a more embracing and hemispherical last chamber, high-arched apertures and indistinct sutures. All the other subspecies of *G. mexicana* are generally larger (except for the small, primitive *G. subconglobata micra*) and have a more robust test.

   G. mexicana has often been misinterpreted. Cushman himself was at the origin of the confusion because he figured under the name '*mexicana*' different species of *Globigerinatheka*; some of them certainly belong to the *G. subconglobata curryi–euganea* lineage. Bolli, Loeblich & Tappan (1957)

Fig. 38. The *Globigerinatheka index* subspecies and *Globigerinatheka subconglobata luterbacheri* (all figures × 60)

1–13 *G. subconglobata luterbacheri* Bolli
   1–3 DSDP Leg 73-522-38; 1: section 2, 10–12 cm; 2: section 2, 70–72 cm; 3: section 1, 70–72 cm. *G. semiinvoluta* Zone, Late Eocene, S Atlantic. **4** DSDP Leg 40-363-10-4, 58–60 cm, *T. rohri* Zone, Middle Eocene, S Atlantic (from Toumarkine, 1978). **5–11** Sample AA 3858 of Eckert (1963), Late Eocene, Looegg, Central Switzerland (5–8, from Bolli, 1972). **12–13** DSDP Leg 32-305-9, cc, *G. semiinvoluta* Zone. Late Eocene, Central Pacific (from Toumarkine, 1975).

14–17 *G. index tropicalis* (Blow & Banner)
   14–15 Sample Bo 536, type locality of *G. semiinvoluta* Zone, Late Eocene, Hospital Hill Marl, Navet Fm., Trinidad (from Bolli, 1972). **16–17** DSDP Leg 40-363-10-1, 58–60 cm, *G. semiinvoluta* Zone, Late Eocene, S Atlantic (17 from Toumarkine, 1978).

18–19 *G. index rubriformis* (Subbotina)
   DSDP Leg 40-363-11-1, 58–60 cm, *M. lehneri* to *G. subconglobata subconglobata* Zone, Middle Eocene, S Atlantic (from Toumarkine, 1978).

20–24 *G. index index* (Finlay)
   20–21 DSDP Leg 32-305-1-2, 18–20 cm, *G. semiinvoluta* Zone, Late Eocene, Central Pacific (20 from Toumarkine, 1975). **22–24** DSDP Leg 40-363-10-4, 58–60 cm, *T. rohri* Zone, Middle Eocene, S Atlantic (from Toumarkine, 1978).

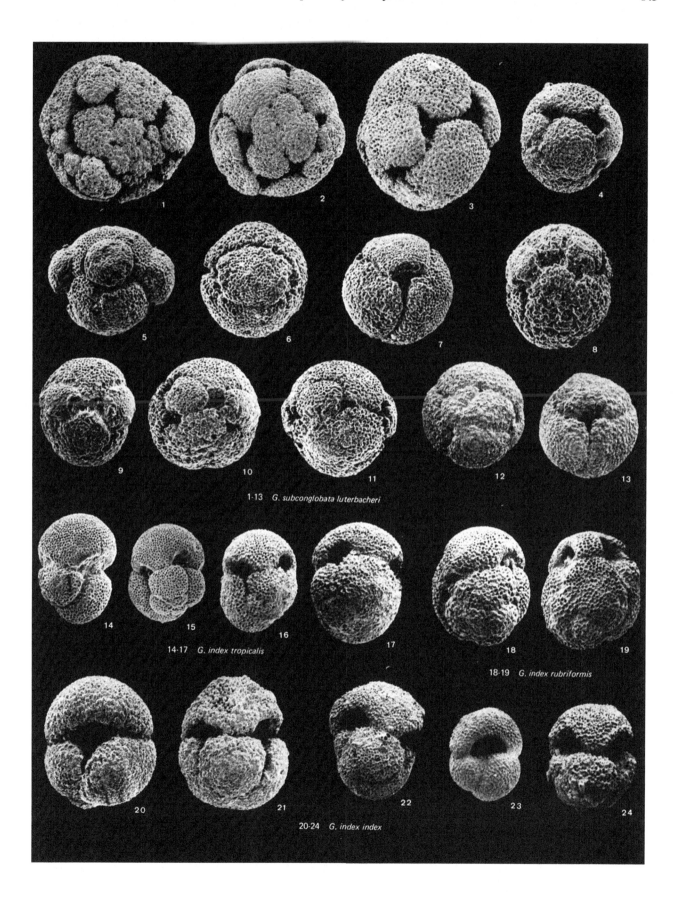

1-13 *G. subconglobata luterbacheri*

14-17 *G. index tropicalis*

18-19 *G. index rubriformis*

20-24 *G. index index*

erroneously took *G. mexicana* as type species for the genus *Proticulasphaera*. Blow & Saito (1968) claimed that *G. semiinvolutus* Keijzer, 1945 is a junior synonym of *G. mexicana*. Bolli (1972) discussed the status of *G. mexicana mexicana* and pointed out the differences between it and the other species and subspecies of *Globigerinatheka*.

G. mexicana mexicana occurs in tropical as well as in temperate regions.

GLOBIGERINATHEKA MEXICANA BARRI
Brönnimann
Figures **37.15, 39.23–32**; 5, 8, 9

Type reference: *Globigerinatheka barri* Brönnimann, 1952b, p. 27, text-figs. 3a–c.

G. mexicana barri is the bullate form of the *G. mexicana* plexus. It differs from *G. mexicana kugleri* in having bullae but also a smaller, tightly coiled initial spire.

G. mexicana barri is generally small and delicate. The subspecies is common in the mid and late part of the Middle Eocene in tropical and temperate regions.

GLOBIGERINATHEKA MEXICANA KUGLERI
(Bolli, Loeblich & Tappan)
Figures **37.14, 39.18–22**; 5, 8, 10

Type reference: *Globigerapsis kugleri* Bolli, Loeblich & Tappan, 1957, p. 34, pl. 6, figs. 6a–c.

The subspecies is characterized by a larger and looser coiled initial spire and more globular chambers resulting in a less compact test than the two other subspecies of *G. mexicana*.

G. mexicana kugleri is particularly frequent in the *Orbulinoides beckmanni* Zone of the Middle Eocene in tropical areas.

GLOBIGERINATHEKA SEMIINVOLUTA
(Keijzer)
Figures **37.11, 39.1–17**; 5, 8, 9

Type reference: *Globigerinoides semiinvolutus* Keijzer, 1945, p. 206, pl. 4, figs. 58a–e.

The main character of *G. semiinvoluta* is the final hemispherical chamber which embraces nearly half of the earlier test and the high-arched to circular sutural apertures with distinct rims. However, a great amount of intraspecific variability can be seen in the dimensions of the test as well as the size and the number and shape of the apertures. This variability is due to ecologic conditions: the typical specimens found in tropical areas (Figs. 39.8, 15, 17) are generally larger with better developed apertures than those of the temperate areas (Figs. 39.1–6, 12–14). The number of apertures (1 to 4, commonly 2 or 3) is not directly related to the size of the specimen.

G. semiinvoluta differs from *G. index tropicalis* by its larger and more embracing final chamber and its less incised sutures. It differs from *G. mexicana mexicana* by the same characters and also by its generally larger size, at least in tropical regions.

The stratigraphic range of *G. semiinvoluta* is restricted to the nominate zone of the Late Eocene.

GLOBIGERINATHEKA SUBCONGLOBATA SUBCONGLOBATA (Shutskaya)
Figures **37.7, 40.16–20**; 5, 8, 9

Type reference: *Globigerinoides subconglobatus* Chalilov var. *subconglobata* Chalilov (ms.) in Shutskaya, 1958, pp. 86–7, pl. 1, figs. 4–11 (no holotype designated).

Bolli (1972) designated the specimen figured by Shutskaya (1958) on pl. 1, fig. 8, as lectotype.

G. subconglobata subconglobata differs from *G. index index* by its more compact test and smaller apertures. It resembles the other subspecies of *G. subconglobata* but it is smaller.

G. subconglobata subconglobata can be very abundant in temperate regions during the Middle Eocene. According to Shutskaya (1958) *G. subconglobata subconglobata* evolved from *G. subconglobata micra* close to the base of the Middle Eocene.

Fig. 39. The *Globigerinatheka mexicana* subspecies and *Globigerinatheka semiinvoluta* (all figures × 60)

1–17 *G. semiinvoluta* (Keijzer)
1 Sample Campredon 57/68, *G. semiinvoluta* Zone, Late Eocene, Menton, Alpes Maritimes, France. 2 Sample Campredon 197/68, *G. semiinvoluta* Zone, Late Eocene, Contes Section, Alpes Maritimes, France. 3–4 Sample HMB 68/60, *G. semiinvoluta* Zone, Late Eocene, Possagno, N Italy (3 from Toumarkine & Bolli, 1975). 5–6 Sample Charollais RF 2/4, *G. semiinvoluta* Zone, Late Eocene, Les Combes section, Haute-Savoie, France. 7 DSDP Leg 73-522-38-2, 70–72 cm, *G. semiinvoluta* Zone, Late Eocene, S Atlantic. 8–11, 16, 17 Sample Bo 536, *G. semiinvoluta* Zone (type-locality), Late Eocene, Hospital Hill Marl, Navet Fm., Trinidad (from Bolli, 1972). 12–14 DSDP Leg 40-363-10-1, 58–60 cm, *G. semiinvoluta* Zone, Late Eocene, S Atlantic (from Toumarkine, 1978). 15a–c *G. semiinvoluta* Zone (type-locality), Late Eocene, Hospital Hill Marl, Navet Fm., San Fernando, Trinidad (from Bolli *et al.*, 1957).

18–22 *G. mexicana kugleri* (Bolli, Loeblich & Tappan)
Sample Hk 408, *O. beckmanni* Zone, Middle Eocene, Navet Fm., Pointe-à-Pierre, Trinidad (18–19 from Bolli, 1972).

23–32 *G. mexicana barri* Brönnimann
23–27, 29a–c, 30–32 Information as for Figs. 39.18–22 (23–27, 30–32 from Bolli, 1972; 29a–c from Bolli *et al.*, 1957). 28 DSDP Leg 40-363-10-1, 58–60 cm, *G. semiinvoluta* Zone, Late Eocene, S Atlantic (from Toumarkine, 1978).

33–39 *G. mexicana mexicana* (Cushman)
33, 34 Information as for Figs. 39.18–22 (from Bolli, 1972). 35 Sample Ba 22, Middle Eocene, Bou Arada, Tunisia (from Mèchmèche & Tourmarkine, 1983). 36–39 DSDP Leg 32-305-10, cc, *T. rohri* to *M. lehneri* Zone, Middle Eocene, Central Pacific (39 from Toumarkine, 1975).

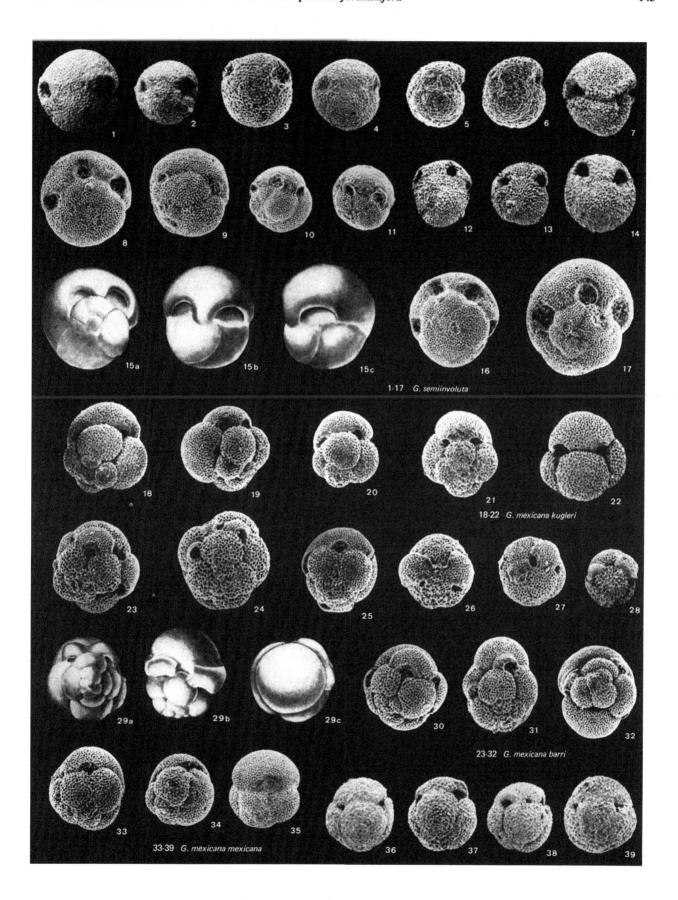

1-17 *G. semiinvoluta*

18-22 *G. mexicana kugleri*

23-32 *G. mexicana barri*

33-39 *G. mexicana mexicana*

GLOBIGERINATHEKA SUBCONGLOBATA
MICRA (Shutskaya)
Figures 37.8–10; 5, 7, 10

Type reference: *Globigerinoides subconglobatus* Chalilov var. *micra* Shutskaya 1958, p. 87, pl. 1, fig. 1.

As noticed by Bolli (1972) and Blow (1979), *G. subconglobata micra* has affinities to *Globigerina senni* by its relatively thick wall and the fine spinose projections in the umbilical area ('muricae', Blow, 1979). A possible synonymy of *G. subconglobata micra* with *G. orbiformis* Cole would still have to be investigated (see Bolli, 1972).

GLOBIGERINATHEKA SUBCONGLOBATA
CURRYI Proto Decima & Bolli
Figures 37.6, 40.11–15; 5, 8, 10

Type reference: *Globigerapsis kugleri* Bolli, Loeblich & Tappan in Bolli, 1957b, pl. 36, figs. 21a–b, designated as holotype by Proto Decima & Bolli, 1970, p. 889.

Proto Decima & Bolli described the evolutionary lineage *G. subconglobata curryi–G. subconglobata euganea–Orbulinoides beckmanni* (first observed by Beckmann, 1953). They regarded this lineage as having its origin in *G. subconglobata subconglobata*.

This lineage is very useful for the recognition of the Middle Eocene *Globigerinatheka subconglobata subconglobata*, *Morozovella lehneri* and *Orbulinoides beckmanni* zones and was found in many tropical to warm-temperate regions (e.g. Caribbean, North Africa, northern Italy, South Atlantic, Central Pacific), but it is absent in more temperate regions. *G. subconglobata curryi* differs from *G. mexicana kugleri* in the larger size of the test (average 0.5–0.6 mm as against 0.35–0.40 mm), and the more robust wall-structure. It has a more restricted stratigraphic range. It differs from *G. mexicana barri* in the larger test size and the more deeply incised sutures which produce a less spherical test. Bullate specimens are not nearly as common as in *G. mexicana barri*. A continuous transition exists between *G. subconglobata curryi* and its successor *G. subconglobata euganea*. The former differs from the latter in having distinctly more incised sutures, more inflated chambers and consequently a less spherical form of the whole test. The coiling of the spire is looser.

GLOBIGERINATHEKA SUBCONGLOBATA
EUGANEA Proto Decima & Bolli
Figures 37.4–5, 40.8–10; 5, 8, 10

Type reference: *Globigerinatheka subconglobata euganea* Proto Decima & Bolli, 1970, pp. 894–5, pl. 1, fig. 7.

G. subconglobata euganea is considered to have evolved from *G. subconglobata curryi* and to be the ancestral form of *Orbulinoides beckmanni*. It possesses a more spherical test,

more numerous but smaller apertures in the final chamber and is more tightly coiled than *G. subconglobata curryi*. Occasionally it can present spiral sutural apertures in the preceding 1 to 3 large chambers. In the overall shape of the test *G. subconglobata euganea* is similar to *Orbulinoides beckmanni* from which it differs in the lack of areal apertures and spiral sutural apertures between the early chambers. It differs from *G. subconglobata luterbacheri* by a more spherical and regular test, by less pronounced sutures and by a smoother aspect of the test. The sizes of the tests are more or less the same (0.5 to 0.6 mm).

GLOBIGERINATHEKA SUBCONGLOBATA
LUTERBACHERI Bolli
Figures 37.1–2, 38.1–13; 5, 8, 9

Type reference: *Globigerinatheka luterbacheri* Bolli, 1972, pp. 132–3, pl. 7, fig. 13.

G. subconglobata luterbacheri and *G. subconglobata euganea* are very large forms with diameters between 0.4 and 0.6 mm, occasionally up to 0.7 mm. The subspecies *luterbacheri* differs from *euganea* in a less regular test shape, which can be subglobular or slightly elongate. The wall is thicker and more coarsely perforate. Apertures, particularly in larger specimens, are often covered by bullae which can become as large as a rudimentary end-chamber. The test of *G. subconglobata euganea* appears more compact and more spherical. *G. subconglobata luterbacheri* is also close to *G. subconglobata curryi* but can be distinguished by its narrower initial spire.

Fig. 40. *Globigerinatheka subconglobata curryi–euganea–Orbulinoides beckmanni* lineage and *G. subconglobata subconglobata* (all figures × 60)

1–6 *O. beckmanni* (Saito)
1a–b, 2a–b Sample Hg 8581, *O. beckmanni* Zone, Middle Eocene, Navet Fm., Trinidad (1a–b from Bolli, Loeblich & Tappan, 1957, 2a–b from Bolli, 1957b). 3 Sample S 181. 4 Sample S 206. *O. beckmanni* Zone, Middle Eocene, Mount Hillaby River section, Barbados (from Proto Decima & Bolli, 1970). **5–6** DSDP Leg 32-313-5-1, 5–8 cm, *O. beckmanni* Zone, Middle Eocene, Central Pacific (from Toumarkine, 1975).

7 *G. subconglobata euganea–O. beckmanni*, transition Sample S 206. Information as for Figs. 40.3–4 (from Proto Decima & Bolli, 1970).

8–10 *G. subconglobata euganea* Proto Decima & Bolli
8 Information as for Figs. 40.5–6. **9** DSDP Leg 32-313-7, cc, *M. lehneri* Zone, Middle Eocene, Central Pacific. **10** DSDP Leg 40-363-10, cc, *O. beckmanni* Zone, Middle Eocene, S Atlantic (from Toumarkine, 1978).

11–15 *G. subconglobata curryi* Proto Decima & Bolli
11–12 Sample S 204, *M. lehneri* Zone, Middle Eocene, Mount Hillaby River section, Barbados (from Proto Decima & Bolli, 1970). **13–14** Information as for Fig. 40.9 (from Toumarkine, 1975). **15** Sample Allemann 77303, *M. lehneri* Zone, Middle Eocene, Beluchistan, Pakistan.

16–20 *G. subconglobata subconglobata* (Shutskaya)
16 Sample Boukhary Z 17, Middle Eocene, Beni Mazar, Nile Valley, Egypt. **17** Sample 633. **18–20** Sample 631. *G. subconglobata subconglobata* Zone, Middle Eocene, Possagno, N Italy.

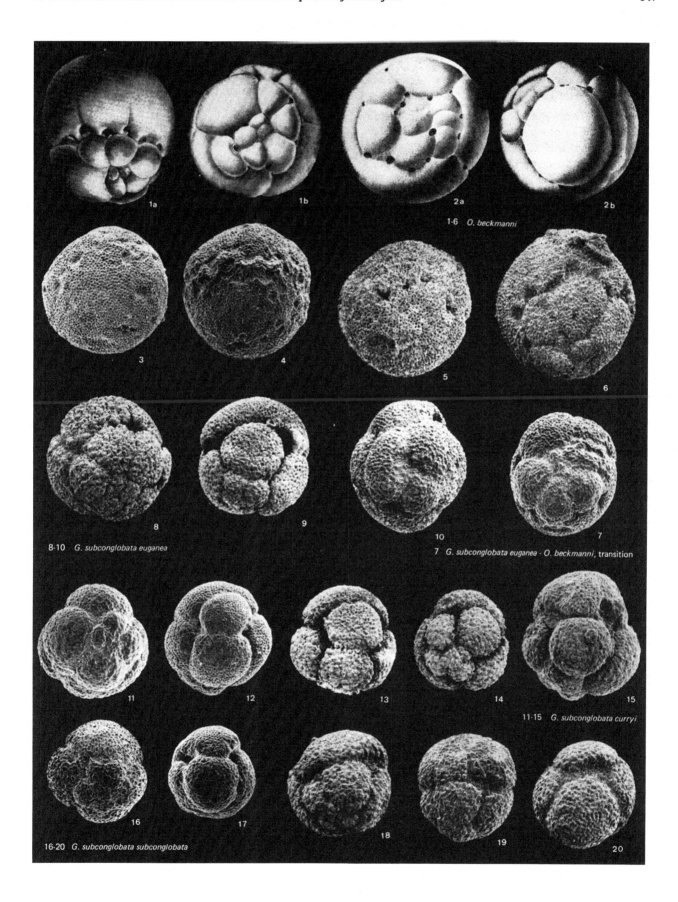

1-6   *O. beckmanni*

8-10   *G. subconglobata euganea*

7   *G. subconglobata euganea - O. beckmanni*, transition

11-15   *G. subconglobata curryi*

16-20   *G. subconglobata subconglobata*

ORBULINOIDES BECKMANNI (Saito)

Figures **37.3, 40.1–6**; 5, 8, 10

Type reference: *Porticulasphaera beckmanni* Saito, 1962, pp. 221–2, pl. 34, figs. 1a–c.

The genus *Orbulinoides* is monotypic. *Orbulinoides* differs from *Globigerinatheka* in possessing spiral areal apertures in the large globular end-chamber which connect via vestibules with apertures of the covering thick wall. *Orbulinoides* possesses a larger number of sutural apertures in the last two or three chambers and it may have areal apertures in the last chambers which are not known in *Globigerinatheka*.

O. beckmanni has a similarly large test (0.4 to 0.7 mm) as the large species of *Globigerinatheka* (*G. subconglobata euganea* and *G. subconglobata luterbacheri*) but it can be differentiated from them by the almost spherical shape of the test and the final chamber which strongly embraces the umbilical region and the multiple sutural openings.

As discussed under *Globigerinatheka mexicana mexicana*, the true nature of that taxon was misunderstood. This necessitated the renaming of Beckmann's *Globigerinoides mexicana* by Saito (1962) who gave it the specific name *beckmanni*. Cordey (1968) and Proto Decima & Bolli (1970) discussed details of the test morphology and the relationship with *Globigerinatheka*. Proto Decima & Bolli (1970) showed that *Orbulinoides beckmanni* is the last form of the evolutionary lineage *Globigerinatheka subconglobata curryi–G. subconglobata euganea–Orbulinoides beckmanni*. This lineage has been found in many tropical to warm-temperate regions (Caribbean, northern Italy, South Atlantic and central Pacific).

Middle to Late Eocene species of *Globigerina* and *Globorotaloides*

Large representatives of the genus *Globigerina* play an important part in Eocene planktonic foraminiferal assemblages. They become prominent in late Early and Middle Eocene assemblages and usually dominate them during the Late Eocene and the Early Oligocene. These large *Globigerina* species form a group in which it is difficult to delimit the various species. This holds especially for the Eocene forms in which the characters of the species or groups of species are not yet fully established. Consequently, a large number of species have been described which are in part synonymous.

Remarks on some species which occur already in the Eocene are to be found in the taxonomic notes on Oligocene planktic foraminifera. They are *Globigerina gortanii*, *G. corpulenta*, *G. tripartita*, *G. venezuelana* and *G. yeguaensis*.

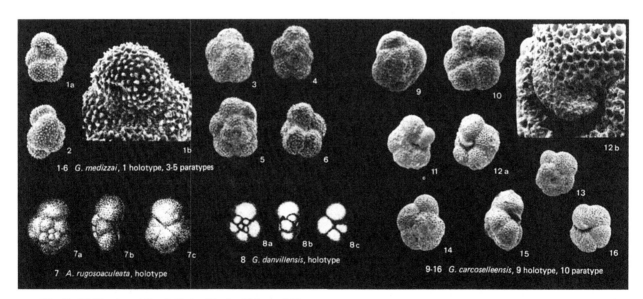

Fig. 41. *Globigerina medizzai*, *G. danvillensis*, *Globorotaloides carcoselleensis* and *Acarinina rugosoaculeata* (all figures × 60, except 1b × 180, 12b × 300)

1–6 *G. medizzai* Toumarkine & Bolli
  1a–b Holotype, 3–5 Paratypes. **2–6** Sample HMB 68/60, *G. semiinvoluta* Zone, Late Eocene, Possagno, N Italy.

7 *Acarinina rugosoaculeata* Subbotina
  7a–c Holotype. Zone of Buliminids, Late Eocene, Kiev stage, upper substage Don Basin, Voroshilovgrad region, River Krasnaya, USSR.

8 *G. danvillensis* Howe & Wallace

8a–c Holotype. Late Eocene, Jackson Fm., Upper Horizon (Bed 2), Danville Landing, Ouachita River, Catahoula Parish, Louisiana, USA.

9–16 *G. carcoselleensis* Toumarkine & Bolli
  9 Holotype, **10** Paratype. Sample 638, *G. subconglobata subconglobata* Zone, Middle Eocene, Possagno, N Italy. **11, 12a–b, 15** DSDP Leg 40-363-10, cc, *O. beckmanni* Zone, Middle Eocene, S Atlantic (11, 12a–b, 15 from Toumarkine, 1978). **13–14, 16** DSDP Leg 32-313-7, cc, *M. lehneri* Zone, Middle Eocene, Central Pacific (13, 16 from Toumarkine, 1975).

GLOBIGERINA CRYPTOMPHALA Glaessner
Figures **42.5–6**; 6, 7, 9

Type reference: *Globigerina bulloides* d'Orbigny var. *cryptomphala* Glaessner, 1937, p. 29, pl. 1, fig. 1.

The main feature of *G. cryptomphala* is the last chamber which covers the umbilicus like an irregularly-shaped bulla.

GLOBIGERINA EOCAENA Guembel
Figures **42.1–4**; 6, 7, 9

Type reference: *Globigerina eocaena* Guembel, 1868, p. 662, pl. 2, figs. 109a–c.

Neotype: *Globigerina eocaena* Guembel, 1868: Hagn & Lindenberg, 1969, p. 23, pl. 1, figs. 1a–b.

G. eocaena is the most primitive species of an evolutionary lineage which leads to *G. gortanii gortanii* (with *G. corpulenta* and *G. gortanii praeturritilina* as intermediate forms). *G. eocaena* differs from *G. corpulenta* and the following species of this lineage by a flat spiral side. *G. corpulenta* has a raised spire and the *gortanii* subspecies has a still higher one.

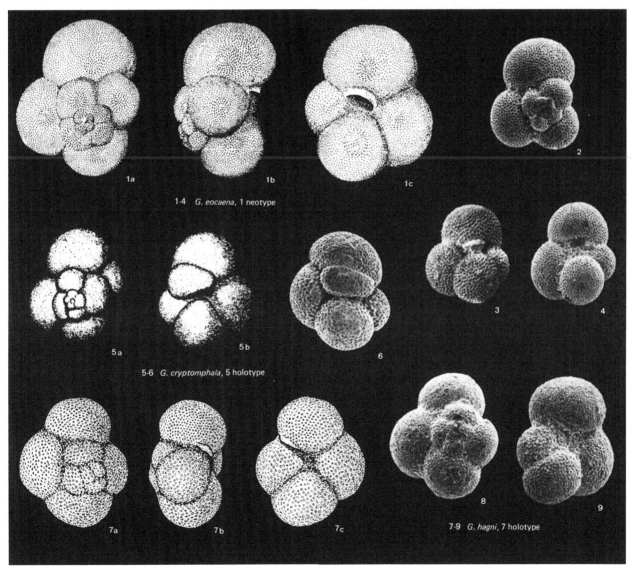

Fig. 42. Large Eocene *Globigerina* species (all figures × 60)

1–4 *G. eocaena* Guembel
1a–c Neotype (Hagn & Lindenberg, 1966). Late Eocene, Gerhartsreiter Graben near Siegsdorf. **2, 4** Sample 648, *T. cerroazulensis* s.l. Zone, Late Eocene, Possagno, N Italy. **3** Sample 65/322, *T. cerroazulensis* s.l. Zone, Late Eocene, Puget-Théniers-Entrevaux, France (from Campredon & Toumarkine, 1972).

5–6 *G. cryptomphala* Glaessner
5a–b Holotype. Late Eocene, N Caucasus, USSR. **6** Sample 4526, Early Oligocene, Marnes à Foraminifères, Vacherie, Haute-Savoie, France (from Charollais *et al.*, 1980, described as *Catapsydrax pera*).

7–9 *G. hagni* Gohrbandt
7a–c Holotype. Sample 64/1, 36/9 (Go), Middle Eocene, Salzburg area, Austria. **8–9** Sample 629, *H. nuttalli* Zone, Middle Eocene, Possagno, N Italy.

GLOBIGERINA HAGNI Gohrbandt
Figures **42.7–9**; 6, 7, 9

Type reference: *Globigerina hagni* Gohrbandt, 1967, p. 324, pl. 1, figs. 1–3.

Globigerina hagni is close to *G. eocaena*. However, the chambers of the last whorl increase only slowly in size. *G. hagni* has 4 distinct chambers in the last whorl instead of $3\frac{1}{2}$ as in *G. eocaena*. These chambers are more or less of the same size, whereas the last chamber of *G. eocaena* is much larger than the preceding ones. *G. hagni* differs also from *G. eocaena* by its umbilical–extraumbilical aperture which is quite umbilical in *G. eocaena*.

GLOBIGERINA MEDIZZAI Toumarkine & Bolli
Figures **41.1–6**; 6, 8, 9

Type reference: *Globigerina medizzai* Toumarkine & Bolli, 1975, p. 77, pl. 5, fig. 16, pl. 6, figs. 1, 5, 7, 8.

The small species (diameter 0.2 to 0.4 mm) is one of the few forms which still persist into the Late Eocene. Other similar but perhaps not closely related forms are *Acarinina rugosoaculeata* Subbotina (Figs. 41.7a–c) described from the Southern USSR, *Globorotalia aculeata* Jenkins (see Chapter 7) from austral areas and *Globigerina danvillensis* Howe & Wallace (Figs. 41.8a–c) found in the northern hemisphere from America, the Atlantic and Europe. *A. rugosoaculeata* differs from *G. medizzai* by its extraumbilical aperture. *G. aculeata* and *G. danvillensis* have more globular chambers and more lobate peripheries. *G. medizzai* is more regularly and more finely spinose than *G. aculeata* which has more-pronounced, but scarcer spines.

GLOBOROTALOIDES CARCOSELLEENSIS
Toumarkine & Bolli
Figures **41.9–16**; 5, 8, 9

Type reference: *Globorotaloides carcoselleensis* Toumarkine & Bolli, 1975, p. 81, pl. 5, fig. 24.

Globorotaloides carcoselleensis, which appears during the *Globigerinatheka subconglobata subconglobata* Zone, is the oldest representative of the genus. In spiral view, the last chamber often looks smaller than the preceding one, but on the umbilical side it extends over the umbilicus and partly covers it as is typical for the genus. *G. carcoselleensis* differs from *G. suteri* mostly by a lower rate of increase of chambers of the last whorl and by the shape of the last chamber which is clearly elongate towards the umbilicus but does not cover it completely.

References

Alimarina, V. P. 1963. Some peculiarities in the development of planktonic foraminifers in connection with the zonal subdivision of the Lower Paleogene in the northern Caucasus. *Academy Nauk SSSR Voprosy Mikropaleontologii*, 7, 158–95 (in Russian).

Anonymous. 1963. Decision of the permanent interdepartmental commission on the Paleogene of the USSR. *Sov. Geol.*, 6, 145–51 (in Russian).

Bandy, O. L. 1949. Eocene and Oligocene foraminifera from Little Stave Creek, Clarke County, Alabama. *Bull. Am. Paleontol.*, 32, 1–211.

Bandy, O. L. 1964. The type of *Globorotalia crassata* (Cushman). *Contrib. Cushman Found. foramin. Res.*, 15, 34–5.

Bang, I. 1980. Foraminifera from the type section of the *eugubina* Zone compared with those from Cretaceous/Tertiary boundary localities in Jylland, Denmark. In: W. K. Christensen & T. Birkelund, Cretaceous–Tertiary Boundary Events Symposium. *Arbok. Dan. geol. Unders.*, 1979, 139–65.

Banner, F. T. 1982. A classification and introduction to the Globigerinacea. In: F. T. Banner & A. R. Lord (eds.) *Aspects of Micropaleontology* (papers presented to Professor Tom Barnard), pp. 142–239, Allen & Unwin, London.

Banner, F. T. & Blow, W. H. 1959. The classification and stratigraphical distribution of the Globigerinaceae. *Palaeontology*, 2, 1–27.

Beckmann, J. P. 1953. Die Foraminiferen der Oceanic Formation (Eocaen–Oligocaen) von Barbados, Kl. Antillen. *Eclog. geol. Helv.*, 46, 301–412.

Beckmann, J. P. 1971. The foraminifera of sites 68 to 75. *Initial Rep. Deep Sea drill. Proj.*, 8, 713–25.

Beckmann, J. P., El-Heiny, I., Kerdany, M. T., Said, R. & Viotti, C. 1969. Standard planktonic zones in Egypt. *Proceedings First International Conference on Planktonic Microfossils, Geneva*, 1967, 1, 92–103.

Ben Ismail-Lattrache, K. 1981. Etude micropaléontologique et biostratigraphique des séries paléogènes de l'anticlinal du Jebel Abderrahman (Cap Bon. Tunisie Nord Orientale). Thèse 3ème cycle. Tunis (unpublished), 234 pp.

Benjamini, Ch. 1980. Planktonic foraminiferal biostratigraphy of the Avedat Group (Eocene) in the northern Negev, Israel. *J. Paleontol.*, 54, 325–58.

Benjamini, C. & Reiss, Z. 1979. Wall-hispidity and -perforation in Eocene planktonic foraminifera. *Micropaleontology*, 25, 141–50.

Berggren, W. A. 1960a. Some planktonic Foraminifera from the Lower Eocene (Ypresian) of Denmark and northwestern Germany. *Stockholm Contrib. Geol.*, 5, 41–108.

Berggren, W. A. 1960b. Paleogene biostratigraphy and planktonic Foraminifera of the southwestern Soviet Union – an analysis of recent Soviet investigations. *Stockholm Contrib. Geol.*, 6, 64–128.

Berggren, W. A. 1965. Paleocene – a micropaleontologist's point of view. *Bull. Am. Assoc. Petrol. Geol.*, 44, 1473–84.

Berggren, W. A. 1966. Phylogenetic and taxonomic problems of some Tertiary planktonic foraminiferal lineages. *Akademy Nauk SSSR Voprosy Mikropaleontologii*, 10, 309–32.

Berggren, W. A. 1969. Rates of evolution in some Cenozoic planktonic foraminifera. *Micropaleontology*, 15, 351–65.

Berggren, W. A. 1971. Multiple phylogenetic zonations of the Cenozoic based on planktonic Foraminifera. *Proceedings II Planktonic Conference, Roma*, 1970, 41–56.

Berggren, W. A. 1972. Cenozoic biostratigraphy and paleobiogeography of the North Atlantic. In: A. S. Laughton *et al.*, *Initial Rep. Deep Sea drill. Proj.*, 12, 965–1001.

Berggren, W. A. 1977. Atlas of Paleogene planktonic foraminifera. Some species of the genera *Subbotina, Planorotalites*,

Morozovella, Acarinina and *Truncorotaloides.* In:
A. T. S. Ramsay (ed.), *Oceanic Micropaleontology*, pp. 205–99, Academic Press, London.

Berggren, W. A., Olsson, R. K. & Reyment, R. A. 1967. Origin and development of the foraminiferal genus *Pseudohastigerina* Banner & Blow, 1959. *Micropaleontology*, **13**, 265–88.

Berggren, W. A. & Van Couvering, J. A. 1974. The Late Neogene. Biostratigraphy, geochronology and paleoclimatology of the last 15 million years in marine and continental sequences. *Palaeogeogr. Palaeoclimatol. Palaeoecol.*, **16** (1–2), 1–215.

Blow, W. H. 1969. Late Middle Eocene to Recent planktonic foraminiferal biostratigraphy. *Proceedings First International Conference on Planktonic Microfossils, Geneva, 1967*, **1**, 199–422.

Blow, W. H. 1979. *The Cainozoic Globigerinida*, 3 vols., E. J. Brill, Leiden, 1413 pp.

Blow, W. H. & Banner, F. T. 1962. The Mid-Tertiary (Upper Eocene to Aquitanian) Globigerinaceae. In: F. E. Eames *et al.*, *Fundamentals of Mid-Tertiary Stratigraphical Correlation*, pp. 61–151, Cambridge University Press, Cambridge.

Blow, W. H. & Saito, T. 1968. The morphology and taxonomy of *Globigerina mexicana* Cushman, 1925. *Micropaleontology*, **14**, 357–60.

Bolli, H. M. 1957a. The genera *Globigerina* and *Globorotalia* in the Paleocene–Lower Eocene Lizard Springs Formation of Trinidad, B.W.I. *Bull. U.S. natl Mus.*, **215**, 61–81.

Bolli, H. M. 1957b. Planktonic Foraminifera from the Eocene Navet and San Fernando formations of Trinidad, B.W.I. *Bull. U.S. natl Mus.*, **215**, 155–72.

Bolli, H. M. 1957c. Planktonic Foraminifera from the Oligocene–Miocene Cipero and Lengua Formations of Trinidad, B.W.I. *Bull. U.S. natl Mus.*, **215**, 97–123.

Bolli, H. M. 1966. Zonation of Cretaceous to Pliocene marine sediments based on planktonic foraminifera. *Boletino Informativo Asociacion Venezolana de Geologia, Mineria y Petroleo*, **9**, 3–32.

Bolli, H. M. 1972. The Genus *Globigerinatheka* Brönnimann. *J. foramin. Res.*, **2**, 109–36.

Bolli, H. M., Loeblich, A. R. & Tappan, H. 1957. Planktonic foraminifera families *Hantkeninidae, Orbulinidae, Globorotaliidae* and *Globotruncanidae*. *Bull. U.S. natl Mus.*, **215**, 3–50.

Boukhary, M. A., Toumarkine, M. & Khalifa, H. 1982. Etude biostratigraphique à l'aide des foraminifères planctiques de l'Eocène de Beni Mazar, Vallée du Nil, Egypte. 8ème Colloque Africain de Micropaléontologie, Paris, 1980. *Cah. Micropaleontol.*, **1**, 53–64.

Brönnimann, P. 1950. The genus *Hantkenina* Cushman in Trinidad and Barbados, B.W.I. *J. Paleontol.*, **24**, 397–420.

Brönnimann, P. 1952a. Trinidad Paleocene and lower Eocene Globigerinidae. *Bull. Am. Paleontol.*, **34** (143), 1–34.

Brönnimann, P. 1952b. *Globigerinoita* and *Globigerinatheka*, new genera from the Tertiary of Trinidad, B.W.I. *Contrib. Cushman Found. foramin. Res.*, **3**, 25–8.

Brönnimann, P. 1953. Note on planktonic Foraminifera from Danian localities of Jutland, Denmark. *Eclog. geol. Helv.*, **45**, 339–41.

Brönnimann, P. & Bermudez, P. J. 1953. *Truncorotaloides*, a new foraminiferal genus from the Eocene of Trinidad, B.W.I. *J. Paleontol.*, **27**, 817–20.

Brönnimann, P. & Rigassi, D. 1963. Contribution to the geology and paleontology of the area of the City of La Habana, Cuba, and its surroundings. *Eclog. geol. Helv.*, **56**, 193–480.

Brotzen, F. & Pozaryska, K. 1961. Foraminifères du Paléocène et de l'Eocène inférieur en Pologne septentrionale; remarques paléogéographiques. *Rev. Micropaleontol.*, **4**, 155–66.

Butterlin, J., Perch-Nielsen, K., Premoli Silva I. & Toumarkine, M. 1977. Livret-guide pour le 'Field Conference sur le Paléogène au Mexique'.

Campredon, R. & Toumarkine, M. 1972. Les formations paléogènes du synclinal de Puget-Théniers-Entrevaux (Basses Alpes, France). *Rev. Micropaleontol.*, **15**, 134–48.

Caro, Y., Luterbacher, H., Perch-Nielsen, K., Premoli Silva, I., Riedel, W. R. & Sanfilippo, A. 1975. Zonations à l'aide de microfossiles pélagiques du Paléocène supérieur et de l'Eocène inférieur. *Bull. Soc. geol. Fr.*, **17**, 125–47.

Charollais, J., Hochuli, P., Oertli, H., Perch-Nielsen, K., Toumarkine, M. Rögl. F. & Pairis, J.-L. 1980. Les Marnes à Foraminifères et les Schistes à Meletta des Chaînes Subalpines septentrionales (Haute-Savoie, France). *Ecol. geol. Helv.*, **73**, 9–69.

Cifelli, R. 1972. The holotypes of *Pulvinulina crassata* var. *densa* Cushman and *Globigerina spinuloinflata* Bandy. *J. foramin. Res.*, **2**, 157–9.

Cifelli, R. & Belford, D. J. 1977. The types of several species of Tertiary planktonic foraminifera in the collections of the U.S. National Museum of Natural History. *J. foramin. Res.*, **7**, 100–5.

Cole, W. S. 1927. A foraminiferal fauna from the Guayabal Formation in Mexico. *Bull. Am. Paleontol.*, **14**, 1–46.

Cole, W. S. 1928. A foraminiferal fauna from the Chapapote Formation in Mexico. *Bull. Am. Paleontol.*, **14**, 3–32.

Colom, G. 1954. Estudio de las biozonas con foraminiferos del Terciario de Alicante. *Bol. Esp. Inst. Geol. y Minero*, **66**, 1–279.

Cordey, W. G. 1968. Morphology and phylogeny of *Orbulinoides beckmanni* (Saito, 1962). *Palaeontology*, **11**, 371–5.

Cordey, W. G., Berggren, W. A. & Olsson, R. K. 1970. Phylogenetic trends in the planktonic foraminiferal genus *Pseudohastigerina* Banner & Blow, 1959. *Micropaleontology*, **16**, 235–42.

Cushman, J. A. 1925a. A new genus of Eocene foraminifera. *Proc. U.S. natl Mus.*, **66**, 1–4.

Cushman, J. A. 1925b. New foraminifera from the Upper Eocene of Mexico. *Contrib. Cushman Lab. foramin. Res.*, **1**, 4–9.

Cushman, J. A. 1925c. Some new foraminifera from the Velasco Shale of Mexico. *Contrib. Cushman Lab. foramin. Res.*, **1**, 18–23.

Cushman, J. A. 1925d. An Eocene fauna from the Moctezuma River, Mexico. *Bull. Am. Assoc. Petrol. Geol.*, **9**, 298–303.

Cushman, J. A. 1927. New and interesting foraminifera from Mexico and Texas. *Contrib. Cushman Lab. foramin. Res.*, **3**, 111–19.

Cushman, J. A. 1928. Additional foraminifera from the Upper Eocene of Alabama. *Contrib. Cushman Lab. foramin. Res.*, **4**, 73–9.

Cushman, J. A. 1930. Fossil species of *Hastigerinella*. *Contrib. Cushman Lab. foramin. Res.*, **6**, 17–19.

Cushman, J. A. & Bermudez, P. J. 1937. Further new species of foraminifera from the Eocene of Cuba. *Contrib. Cushman Lab. foramin. Res.*, **13**, 1–29.

Cushman, J. A. & Bermudez, P. J. 1949. Some Cuban species of *Globorotalia*. *Contrib. Cushman Lab. foramin. Res.*, **25**, 26–44.

Cushman, J. A. & Jarvis, P. W. 1929. New foraminifera from Trinidad. *Contrib. Cushman Lab. foramin. Res.*, **5**, 6–17.

Cushman, J. A. & Ponton, G. M. 1932. An Eocene foraminiferal fauna of Wilcox age from Alabama. *Contrib. Cushman Lab. foramin. Res.*, **8**, 51–72.

Cushman, J. A. & Renz, H. H. 1942. Eocene, Midway, Foraminifera from Soldado Rock, Trinidad. *Contrib. Cushman Lab. foramin. Res.*, **18**, 1–20.

Dieni, I. & Proto Decima, F. 1964. *Cribrohantkenina* et altri Hantkeninidae nell'Eocene superiore di Castelmoro (Colli Euganei). *Riv. Ital. Paleontol. Stratigr.*, **70**, 555–92.

Douglas, R. G. & Savin, S. M. 1973. Oxygen and carbon isotope analysis of Cretaceous and Tertiary foraminifera from the Central North Pacific. In: E. L. Winterer *et al.*, *Initial Rep. Deep Sea drill. Proj.*, **17**, 591–605.

Eckert, H. R. 1963. Die obereozänen Globigerinen-Schiefer (Stad- und Schimbergschiefer) zwischen Pilatus und Schrattenfluh. *Eclog. geol. Helv.*, **56**, 1001–72.

El Khoudary, R. H. 1977. *Truncorotaloides libyaensis*, a new planktonic foraminifer from Jabal Al Akhdar (Libya). *Rev. Esp. Micropaleontol.*, **9**, 327–36.

El-Naggar, Z. R. 1971. On the classification, evolution and stratigraphical distribution of the Globigerinacea. *Proceedings II Planktonic Conference, Roma, 1970*, **1**, 421–76.

Finlay, H. J. 1939. New Zealand foraminifera: key species in stratigraphy. No. 2. *Trans. Proc. R. Soc. N.Z.*, **69**, 89–128.

Finlay, H. J. 1947. New Zealand foraminifera: key species in stratigraphy. No. 5. *N.Z. J. Sci. Technol.*, section B, **28**, 259–92.

Fleisher, R. L. 1974. Cenozoic planktonic foraminifera and biostratigraphy, Arabian Dea Deep Sea Drilling Project, Leg 23A. In: R. B. Whitmarsh *et al.*, *Initial Rep. Deep Sea drill. Proj.*, **23**, 1001–72.

Gamper, M. A. 1977. Acerca del limite Cretacico–Terciario en Mexico. *Rev. Inst. Geol. Univ. nac. Autonoma. Mexico*, **1**, 23–7.

Glaessner, M. F. 1937. Planktonforaminiferen aus der Kreide und dem Eozaen und ihre stratigraphische Bedeutung. *Moscow Univ. Lab. Paleontol. Studies Micropaleontology*, **1**, 27–46.

Gohrbandt, K. H. A. 1967. Some new planktonic foraminiferal species from the Austrian Eocene. *Micropaleontology*, **13**, 319–26.

Guembel, C. W. 1868. Beiträge zur Foraminiferenfauna der nordalpinen älteren Eocängebilde oder der Kressenberger Nummulitenschichten. *Abhandlungen Bayerische Akademie der Wissenschaften, Math.-Physik Kl.*, **10**, 579–730.

Hagn, H. & Lindenberg, H. G. 1966. Revision of *Globigerina (Subbotina) eocaena* Guembel from the Eocene of the Bavarian Alps. *Akademy Nauk SSSR Voprosy Mikropaleontologii*, **10**, 342–50 (in Russian).

Hagn, H. & Lindenberg, H. G. 1969. Revision der von C. W. Guembel 1868 aus dem Eozän des bayerischen Alpenvorlandes beschriebenen planktonischen Foraminiferen. *Proceedings First International Conference on Planktonic Microfossils, Geneva, 1967*, **2**, 229–49.

Hansen, H. J. 1970. Danian foraminifera from Nugssuaq, West Greenland. *Medd. Gronland*, **193**, 1–132.

Hantken, M. v. 1883. Die Clavulina szaboi-Schichten im Gebiete der Euganeen und der Meeralpen und die cretacische Scaglia in den Euganeen. *Math.-natw. Ber. Ungarn*, **2**, 121–69.

Haq, B. U., Premoli Silva, I. & Lohmann, G. P. 1977. Calcareous plankton paleobiogeographic evidence for major climatic fluctuations in the early Cenozoic Atlantic Ocean. *J. geophys. Res.*, **82**, 3861–76.

Hay, W. W. & Mohler, H. P. 1969. Paleocene–Eocene calcareous nannoplankton and high-resolution biostratigraphy. *Proceedings First International Conference on Planktonic Microfossils, Geneva, 1967*, **2**, 250–53.

Herm, D., Hillebrandt, A. v. & Perch-Nielsen, K. 1981. Die Kreide/Tertiär-Grenze im Lattengebirge (Nördliche Kalkalpen) in mikropaläontologischer Sicht. *Geologica Bavarica*, **82**, 319–44.

Hillebrandt, A. v. 1962. Das Paleozän und seine Foraminiferenfauna im Becken von Reichenhall und Salzburg. *Abhandlungen Bayerische Akademie der Wissenschaften, Mathematisch-Naturwissenschaftliche Klasse*, **108**, 1–182.

Hillebrandt, A. v. 1965. Foraminiferen-Stratigraphie im Alttertiär von Zumaya (Provinz Guipuzcoa, N.W.-Spanien) und ein Vergleich mit anderen Tethys-Gebieten. *Abhandlungen Bayerische Akademie der Wissenschaften, Mathematisch-Naturwissenschaftliche Klasse*, **123**, 1–62.

Hillebrandt, A. v. 1974. Bioestratigrafía del Paleógeno en el Sureste de España (Provincias de Murcia y Alicante). *Cuadernos Geologicos*, **5**, 135–53.

Hillebrandt, A. v. 1976. Los foraminíferos planctónicos, nummulítidos y coccolitofóridos de la zona de *Globorotalia palmerae* del Cuisiense (Eoceno inferior) en el SE de España (Provincias de Murcia y Alicante). *Rev. Esp. Micropaleontol.*, **8**, 323–94.

Hillebrandt, A. v. 1978. *Pseudohastigerina sphaeroidalis* nom.nov. for *Pseudohastigerina globulosa* v. Hillebrandt, 1976, non *Pseudohastigerina wilcoxensis globulosa* (Gohrbandt, 1967). *Rev. Esp. Micropaleontol.*, **10**, 337.

Hofker, J. Sr. 1962. Correlation of the Tuff Chalk of Maestricht (type Maestrichtian) with the Danske Kalk of Denmark (type Danian), the stratigraphic position of the type Montian, and the planktonic foraminiferal break. *J. Paleontol.*, **36**, 1051–89.

Hofker, J. Sr. 1978. Analysis of a large succession of samples through the Upper Maastrichtian and the Lower Tertiary of drill hole 47.2, Shatsky Rise, Pacific, Deep Sea Drilling Project. *J. foramin. Res.*, **8**, 46–75.

Howe, H. V. 1928. An observation on the range of the genus *Hantkenina*. *J. Paleontol.*, **2**, 13–14.

Howe, H. V. & Wallace, W. E. 1932. Foraminifera of the Jackson Eocene at Danville Landing on the Ouachita, Catahoula Parish, Louisiana. *Bull. Louisiana Dept. Conservation Geol.*, **2**, 18–79.

Howe, H. V. & Wallace, W. E. 1934. Apertural characteristics of the genus *Hantkenina*, with description of a new species. *J. Paleontol.*, **8**, 35–7.

Jenkins, D. G. 1971. New Zealand Cenozoic planktonic foraminifera. *Palaeontol. Bull. geol. Surv. N.Z.*, **42**, 1–278.

Keijzer, F. G. 1945. Outline of the geology of the eastern part of the Province of Oriente, Cuba (E of 76° WL), with notes on the geology of other parts of the island. *Geographische en Geologische Mededeelingen*. Publicaties uit het Geographisch en uit het Mineralogisch-Geologisch Instituut der Rijksuniversiteit te Utrecht. Physiographisch-Geologische Reeks. Ser. II, no. 6, 1–239.

Kennett, J. P. 1982. *Marine Geology*. Prentice Hall Inc., Englewood Cliffs, N.J., 813 pp.

Khalilov, D. M. 1956. Pelagic foraminifers of the Paleogene deposits of Azerbaydzhan. *Akademy Nauk Azerbaydzhan, SSR, Inst. Geol. Trudy*, **17**, 234–61 (in Russian).

Krasheninnikov, V. A. 1965a. Zonal stratigraphy of Paleogene deposits. *International Geological Congress, 21st, Norden, 1960, Doklady Soviet Geologists*, Problem 16, Problems of Cenozoic stratigraphy, Akademy Nauk SSSR Izd., Moscow, 37–61 (in Russian).

Krasheninnikov, V. A. 1965b. Zonal stratigraphy of the Paleogene in the eastern Mediterranean. *Akademy Nauk SSSR Geol. Inst. Trudy*, **133**, 1–76 (in Russian).

Krasheninnikov, V. A. 1969. Geographical and stratigraphical distribution of planktonic foraminifers in Paleogene deposits of tropical and subtropical areas. *Akademy Nauk SSSR Geol. Inst. Trudy*, **202**, 1–190 (in Russian).

Krasheninnikov, V. A. & Hoskins, R. H. 1973. Late Cretaceous, Paleogene and Neogene planktonic foraminifera. In: B. C. Heezen *et al.*, *Initial Rep. Deep Sea drill. Proj.*, **20**, 105–203.

Leonov, G. P. & Alimarina, V. P. 1961. Stratigraphy and planktonic foraminifers of the 'transitional' Cretaceous to Paleogene beds of the central Precaucasus. *Moskov. Univ. Trudov Geol. Fak. Sbornik* (k 21 sessii mezhdunarod, geol kongr.), Moskov Univ. Izd., pp. 29–53 (in Russian).

Loeblich, A. R., Jr & Tappan, H. 1957. Planktonic foraminifera of Paleocene and early Eocene age from the Gulf and Atlantic coastal plains. *Bull. U.S. natl Mus.*, **215**, 173–98.

Luterbacher, H. P. 1964. Studies in some *Globorotalia* from the Paleocene and Lower Eocene of the central Apennines. *Eclog. geol. Helv.*, **57**, 631–730.

Luterbacher, H. P. 1966. Remarks on evolution of some globorotalias in the Paleocene of the central Apennines. *Akademy Nauk SSSR Voprosy Mikropaleontologii*, **10**, 334–41.

Luterbacher, H. P. & Premoli Silva, I. 1964. Biostratigrafia del limite Cretaceo–Terziario nell'Appennino centrale. *Riv. Ital. Paleontol. Stratigr.*, **70**, 67–128.

McGowran, B. 1964. Foraminiferal evidence for the Paleocene age of the King's Park Shale (Perth basin, Western Australia). *J. R. Soc. Western Australia*, **47**, 81–8.

McGowran, B. 1968. Reclassification of early Tertiary *Globorotalia*. *Micropaleontology*, **14**, 179–98.

Martin, L. T. 1943. Eocene foraminifera from the type Lodo Formation, Fresno County, California. *Stanford Univ. Publ. Geological Sciences*, **3**, 93–125.

Mèchmèche, R. & Toumarkine, M. 1983. La Formation Souar dans la région de Bou Arada, Tunisie. *Premier Congrès National des Sciences de la Terre, Tunis, 1981. Actes du Congrès* (in press).

Morozova, V. G. 1939. On the stratigraphy of the Upper Cretaceous and Paleogene of the Emba region according to the foraminiferal faunas. *Moskov. Obshch. Ispytateley Prirody Byull., Otdel. Geol.*, **17**, 59–86 (in Russian).

Morozova, V. G. 1957. Foraminiferal superfamily Globigerinidea, superfam. nov. and some of its representatives. *Akademy Nauk SSSR Doklady*, **114**, 1109–11 (in Russian).

Morozova, V. G. 1959. Stratigraphy of the Dano-Montian deposits of Crimea according to the foraminifers. *Akad. Nauk SSSR Doklady*, **124**, 1113–16 (in Russian).

Morozova, V. G. 1960. Zonal stratigraphy of the Dano-Montian deposits of the USSR and the Cretaceous–Paleogene boundary. *International Geological Congress, 21st, Norden, 1960, Doklady Soviet Geologists*, Problem 5, The Cretaceous–Tertiary boundary, Akademy Nauk SSSR Izd., Moscow, 83–100 (in Russian).

Morozova, V. G. 1961. Planktonic foraminifers from the Danian–Montian of the southern Soviet Union. *Akad. Nauk SSSR Paleont. Zhur.*, **2**, 8–19 (in Russian).

Myatliuk, E. V. 1950. The stratigraphy of the flysch deposits of the northern Carpathian Mountains according to the foraminiferal faunas. *Trudy VNIGRI*, new ser., **51** (Microfauna of the USSR 4), 225–87 (in Russian).

Nuttall, W. L. F. 1928. Notes on the Tertiary foraminifera of Southern Mexico. *J. Paleontol.*, **2**, 372–6.

Nuttall, W. L. F. 1930 (September). Eocene foraminifera from Mexico. *J. Paleontol.*, **4**, 271–93.

Parr, W. J. 1938. Upper Eocene foraminifera from the deep borings in King's Park, Perth, Western Australia. *J. R. Soc. Western Australia*, **24**, 69–101.

Pericoli, S. 1959. Sulla presenza del genere *Hantkenina* Cushman nell 'scaglia' dell'Urbinate. *Boll. Soc. Nat. Napoli*, **67** (1958), 68–89.

Petters, V. 1954. Tertiary and Upper Cretaceous foraminifera from Colombia, S.A. *Contrib. Cushman Found. foramin. Res.*, **5**, 37–41.

Plummer, H. J. 1926 (1927). Foraminifera of the Midway formation in Texas. *Bull. Texas Univ. Bur. Econ. Geol.*, **2644**, 1–206.

Pomerol, Ch. 1973. *Stratigraphie et paléogéographie, Ere Cénozoique (Tertiaire et Quaternaire)*. Doin eds., Paris, 269 pp.

Poore, R. Z. & Brabb, E. E. 1977. Eocene and Oligocene planktonic foraminifera from the Upper Butano Sandstone and type San Lorenzo Formation, Santa Cruz Mountains, California. *J. foramin. Res.*, **7**, 249–72.

Postuma, J. A. 1971. *Manual of planktonic foraminifera*. Elsevier Publishing Co., Amsterdam, 420 pp.

Premoli Silva, I. 1977. The earliest Tertiary *Globigerina eugubina* Zone: paleontological significance and geographical distribution. *Mem. 2nd Congress Latinoam. Geol.*, vol. 3, spec. publ. 7, 1541–55.

Premoli Silva, I. & Bolli, H. M. 1973. Late Cretaceous to Eocene planktonic foraminifera and stratigraphy of Leg 15 sites in the Caribbean Sea. In: N. T. Edgar, J. B. Saunders *et al.*, *Initial Rep. Deep Sea drill. Proj.*, **15**, 449–547.

Proto Decima, F. & Bolli, H. M. 1970. Evolution and variability of *Orbulinoides beckmanni* (Saito). *Eclog. geol. Helv.*, **63**, 883–905.

Ramsay, W. R. 1962. Hantkenininae in the Tertiary rocks of Tanganyika. *Contrib. Cushman Found. foramin. Res.*, **13**, 79–89.

Saito, T. 1962. Eocene planktonic foraminifera from Hahajima (Hillsborough Island). *Trans. Proc. Paleontol. Soc. Japan*, news series, **45**, 209–25.

Samanta, B. K. 1970. Middle Eocene planktonic foraminifera from the Lakhpat, Cutch, Western India. *Micropaleontology*, **16**, 185–215.

Samuel, O. & Salaj, J. 1968. Microbiostratigraphy and foraminifera of the Slovak Carpathian Paleogene. *Geologiska ustav. Dionyza Stura* (Bratislava), 232 pp.

Schmidt, R. & Raju, D. S. N. 1973. *Globorotalia palmerae* Cushman & Bermudez and closely related species from the

Lower Eocene, Cauvery Basin, South India. *Proc. Koninkl. Nederlandse Akad. Wetenschappen*, ser. B., **76**, 167–78.

Shamah, K., Blondeau, A., Le Calvez, Y., Perch-Nielsen, K. & Toumarkine, M. 1982. Biostratigraphy of the Middle Eocene El Midawarah Formation, Wadi El Rayan Region, Fayoum area, Egypt. 8ème Colloque Africain de Micropaléontologie, Paris, 1980, *Cah. Micropaleontol.*, 1982, fasc. **1**, 91–104.

Shokhina, V. A. 1937. The genus *Hantkenina* and its stratigraphical distribution in the north Caucasus. (Problems of Paleontology.) *Publications of the Laboratory of Paleontology, Moscow University, USSR*, **2–3**, 425–41.

Shutskaya, E. K. 1958. Variations of some lower Paleogene planktonic foraminifers of the northern Caucasus. *Akad. Nauk SSSR Voprosy Mikropaleontologii*, **2**, 84–90. Translated into French in *Questions de Micropaléontologie*, *B.R.G.M., Service d'Inf. Géol.*, 1960, 93–103.

Smit, J. 1977. Discovery of a planktonic foraminiferal association between the *Abathomphalus mayaroensis* Zone and the *Globigerina eugubina* Zone at the Cretaceous/Tertiary boundary in the Barranco del Gredero (Cavavaca, SE Spain). A preliminary report I & II. *Proc. Koninkl. Nederlandse Akad. Wetenschappen*, ser. B, **80**, 280–301.

Smit, J. 1982. Extinction and evolution of planktonic foraminifera after a major impact at the Cretaceous–Tertiary boundary. *Spec. Pap. Geol. Soc. Am.*, **190**, 329–52.

Stainforth, R. M. 1948. Applied micropaleontology in coastal Ecuador. *J. Paleontol.*, **22**, 113–51.

Stainforth, R. M., Lamb, J. L., Luterbacher, H., Beard, J. H. & Jeffords, R. M. 1975. Cenozoic planktonic foraminiferal zonation and characteristics of index forms. *Univ. Kansas Paleontol. Contrib.*, article **62**, 1–425.

Stainforth, R. M., Sztrákos, K. & Jeffords, R. M. 1982. *Globigerina cerroazulensis* Cole, 1928, and *Globigerapsis tropicalis* Blow & Banner, 1962 (Foraminiferida): proposed conservation. *Bull. zool. Nomencl.*, **39**, 45–9.

Steineck, P. L. & Fleisher, R. L. 1978. Towards the classical evolutionary reclassification of Cenozoic Globigerinacea (Foraminiferida). *J. Paleontol.*, **52**, 618–35.

Subbotina, N. N. 1947. Foraminifers of the Danian and Paleogene deposits of the northern Caucasus. Microfauna of the Caucasus Emba region, and central Asia, *Trudy VNIGRI*, 39–160 (in Russian).

Subbotina, N. N. 1950. Microfauna and stratigraphy of the Elburgan Horizon and the Horizon of Goriatchii Klintch. *Trudy VNIGRI*, new ser., **51** (Microfauna of the USSR 4), 5–112 (in Russian).

Subbotina, N. N. 1953. Fossil foraminifers of the USSR: Globigerinidae, Hantkeninidae, and Globorotaliidae. *Trudy VNIGRI*, new series, **76**, 296 pp. (in Russian). Translated into English by E. Lees, *Fossil foraminifera of the USSR; Globigerinidae, Hantkeninidae and Globorotaliidae.* Collet's Ltd., London and Wellingborough, 321 pp.

Sztrákos, K. 1973. Révision des espèces '*Globigerina*' *aplanata* et '*Globigerina*' *globosa* décrites par M. Hantken d'Euganea (Italie). *Rev. Micropaleontol.*, **16**, 224–8.

Thalmann, H. E. 1942a. *Hantkenina* in the Eocene of East Borneo. *Stanford Univ. Publ., Geological Sciences*, **3**, 1–24.

Thalmann, H. E. 1942b. Foraminiferal genus *Hantkenina* and its subgenera. *Am. J. Sci.*, **240**, 809–23.

Toulmin, L. D. 1941. Eocene smaller foraminifera from the Salt Mountain limestone of Alabama. *J. Paleontol.*, **15**, 567–611.

Toumarkine, M. 1971. Etude des foraminifères planctoniques de deux sondages (H-849 et PGYT-31) dans l'Eocène de la Montagne du Bakony (Transdanubie, Hongrie). Colloquium on Eocene Stratigraphy, Budapest, 1969. *Ann. Inst. Publ. Hung.*, **54**, 285–99.

Toumarkine, M. 1975. Middle and Late Eocene planktonic foraminifera from the North-western Pacific, Leg 32 of the Deep Sea Drilling Project. In: R. L. Larson, R. Moberly *et al.*, *Initial Rep. Deep Sea drill. Proj.*, **32**, 735–51.

Toumarkine, M. 1978. Planktonic foraminiferal biostratigraphy of the Paleogene of Sites 360 to 364 and the Neogene of Sites 362A, 363 and 364 Leg 40. In: H. M. Bolli, W. B. F. Ryan *et al.*, *Initial Rep. Deep Sea drill. Proj.*, **40**, 679–721.

Toumarkine, M. 1981. Discussion de la validité de l'espèce *Hantkenina aragonensis* Nuttall, 1930. Description de *Hantkenina nuttalli*, n. sp. *Cah. Micropaleontol.*, Livre Jubilaire en l'honneur de Madame Y. Le Calvez, fasc. **4**, 109–19.

Toumarkine, M. & Bolli, H. M. 1970. Evolution de *Globorotalia cerroazulensis* (Cole) dans l'Eocène moyen et supérieur de Possagno (Italie). *Rev. Micropaleontol.*, **13**, 131–45.

Toumarkine, M. & Bolli, H. M. 1975. Foraminifères planctoniques de l'Eocène moyen et supérieur de la coupe de Possagno. *Schweiz. Palaontol. Abh.*, **97**, 69–185.

Weinzierl, L. L. & Applin, E. R. 1929. The Claiborne Formation on the Coastal Domes. *J. Paleontol.*, **3**, 384–410.

Weiss, L. 1955. Planktonic index foraminifera of northwestern Peru. *Micropaleontology*, **1**, 301–19.

White, M. P. 1928. Some index foraminifera of the Tampico Embayment area, Mexico. *J. Paleontol.*, **2**, 177–215, 280–317.

6
Oligocene to Holocene low latitude planktic foraminifera

HANS M. BOLLI & JOHN B. SAUNDERS

CONTENTS

Introduction

We have made only sparing reference to generic designation in the present chapter except where grouping at that taxonomic level is, itself, useful in stratigraphy. For discussions on the various genera the reader is referred to such works as Loeblich & Tappan (1964) and, for more recent treatment, to Stainforth *et al.* (1975) and Banner (1982). If species designations can be argued, generic grouping comes under even greater attack and there is unlikely ever to be wholesale agreement. One hopes that species have some basis of biological support; the same can hardly be said for some genera.

As regards the inclusion of species, we have had to make a choice. Practical use over the years has led to this choice but, again, it is a personal one as is the degree of subdivision to subspecific level and the grouping of such subspecies; stratigraphic utility is the ultimate aim. If a species of no great stratigraphic value is included, it is to point out the danger of confusing it with a morphologically similar one. Species are not formally described. The taxonomic notes are largely restricted to information bearing on differentiation between morphologically confusing taxa.

The attempt to treat species in groups has revealed a considerable amount of confusion hidden among the taxonomy that has its effect on ranges and lineages. Due to space limitations, it is not possible to treat all the questions that have arisen. Two specific problems encountered during the preparation of the chapter have been dealt with in separate papers (Bolli & Saunders, 1981, 1982a).

The grouping of taxa has departed from the alphabetic approach. We have attempted a 'natural' progression beginning with the Oligocene Globigerinas and moving as far as possible through the most closely connected groups, and generally in an upward stratigraphic sequence, ending with Pliocene and Pleistocene markers. The order of discussion is given in the list of contents at the head of this chapter.

The approach we have taken is centred round a careful review of the original type of each species. If older taxonomic names are to be retained, the clearest understanding possible of the original concept is essential if confusion is to be kept to a minimum. Even more recently proposed species as, for example, *Globigerina ampliapertura* and *Globigerina ciperoensis* can easily be used incorrectly if the central concept is not clearly defined. The amount of variation allowed within a species is treated differently by different workers and has led to great disagreement between the 'lumpers' and the 'splitters'. We have tried to steer a middle course always keeping in mind the biostratigraphic utility of a taxon.

However, to do justice to each species is beyond the scope of this handbook as the range of variation needs to be shown at length, as we have tried to do for *Globorotalia mayeri*, for example (Bolli & Saunders, 1982a).

The placing of illustrations has been designed as far as possible to bring together taxa that can most easily be confused. The holotype may be figured alone if this seems sufficient, though if a lectotype or neotype has been erected, this has also been illustrated. Additional figures have been used as necessary, either from publications or from newly photographed specimens. Almost all figures have been brought to a common magnification of × 60, though this may only be approximate in the case of early taxa. We feel it extremely important that the worker be able to make direct comparison of taxa *at the same magnification* and to view neotypes at the same size as the original figure for the taxon. A statement of varying magnifications in the figure descriptions is, in our opinion, not enough. Some of the anomalies that have arisen due to a misunderstanding of size differences are referred to in the taxonomic notes.

The range charts (Figs. 6–12) are arranged as follows: Figs. 6–8 show the ranges of the species discussed in the chapter in a taxonomic grouping. Figs. 9 and 10 are arranged on first occurrences and Figs. 11 and 12 on last occurrences. The inclusion of a pair of charts arranged on 'tops' rather than 'bottoms' is designed particularly for use in drilling where the succession must, of necessity, be viewed from youngest to oldest and where 'bottoms' are of less reliability when ditch samples (cuttings) have to be used. An added advantage of the chart arranged on last occurrences is that it graphically indicates times of major extinctions while that on first occurrences shows periods of major evolutionary expansion. Either chart can be used to give a quick overview of the diagnostic taxa likely to be found in any zone and therefore the richness of that zone under ideal conditions.

The range charts have a stratigraphic scale that has, of necessity, to be compressed. This makes it impossible to show the first and last occurrences of one species in exact relationship to those of neighbouring taxa. The result is that too many taxa give the impression of appearing or becoming extinct at exactly the same moment; this is known to be an over-simplification, particularly since results have become available from Deep Sea Drilling Project (DSDP) core holes.

Low latitude zonal schemes

Stainforth *et al.* (1975) provided a concise review of zonations based on planktic foraminifera. They discussed their general features, limitations of their application, influence of climatic factors, possibilities of interrelation of planktic foraminiferal and nannofossil zones for higher biostratigraphic resolution, radiometric dating of zones and the impact of planktic foraminiferal zones on classical Tertiary stratigraphy. Little new information can be added to these statements by Stainforth *et al.* to which reference is made here.

The same authors also provided a comparison of the different schemes proposed by various workers, illustrating them with zonal correlation charts. They provided concise descriptions of the zonal system chosen by them for the Oligocene to Recent and they gave species distribution charts linked to this system. Furthermore, they discussed problems of the age assignment of individual zones and of the placement of major boundaries, particularly those between the Eocene/Oligocene, Oligocene/Miocene and Pliocene/Pleistocene. Reference to Stainforth *et al.* is, therefore, also made for this information.

In discussing Oligocene to Holocene low latitude planktic foraminiferal zones and their application we can therefore restrict ourselves to a review of the historical development of the zonal scheme followed in the present chapter. This scheme was proposed by Bolli (1957 and later papers) and is basically the same as the one introduced subsequently by Banner & Blow (1965b) and given a letter/number notation by them. It was also adopted by Stainforth *et al.* (1975) with few exceptions.

The review of the historical development of the zonal scheme used here is followed by discussions of Late Miocene, Pliocene and Pleistocene zonal schemes which in recent years have undergone modifications. This is followed by reviews of the influence and zonal resolution caused by the evolutionary trends in certain taxa and of the effects on zonation of the geographic restriction of some low latitude species.

The zones from base Oligocene to Holocene are listed with category, age and author. Definitions for each are followed by additional criteria for their identification.

Historical development

Most of the Tertiary zonation based on planktic foraminifera, both for the Paleogene and Neogene, was originally developed in Trinidad, West Indies. A first proposal came from Cushman & Stainforth (1945), who subdivided the Oligocene–Miocene Cipero Formation of that island into three zones. Cushman & Renz (1947) added another zone to this scheme, and Stainforth (1948) proposed a zone for the Middle Miocene Lengua Formation (Fig. 1).

During this early phase of using planktic foraminifera in biostratigraphy, Grimsdale (1951) offered the first intercontinental correlation by age/stage (though not by zones) of 41 Tertiary planktic foraminiferal species from the Gulf of Mexico and the Caribbean with equivalent species from the Middle East.

Encouraged by the promising and successful initial results in the use of planktic foraminifera in the younger Tertiary, a more concerted effort to investigate and apply them began in Trinidad in the late 1940s and early 1950s. It led to the zonal scheme for the Paleocene to Middle Miocene as was published by Bolli in 1957.

Because Late Miocene and younger formations are almost lacking planktic foraminifera in Trinidad, other, more suitable sections elsewhere had to be investigated to subdivide this interval. They were found partly in neighbouring Venezuela, from where Bolli & Bermudez (1965) introduced several new zones from the Late Miocene to the Pleistocene. A more detailed subdivision of the Pliocene to Holocene was established by Bolli & Premoli Silva (1973) based on core holes from the Caribbean DSDP Legs 4 and 15.

The planktic foraminiferal low latitude zonal scheme for the whole Tertiary, established with few exceptions mainly in Trinidad and also in Venezuela and the Caribbean and applied in this volume, is shown on Fig. 2. This development is discussed at greater length in Saunders & Bolli (1983).

Bolli (1959) in his paper on 'Planktonic foraminifera as index fossils in Trinidad, West Indies and their value for worldwide stratigraphic correlation' summarized the studies in Trinidad and pointed out their interregional application. The same author in 1966 discussed the Cretaceous to Pliocene zonal schemes based on planktic foraminifera, listing the authors of zones and giving definitions and additional faunal remarks useful in characterizing the zones. Furthermore, correlation of the zones established in the Caribbean area with other, later schemes proposed from higher latitudes (Israel, New Zealand) were included in that paper.

Banner & Blow (1965b) proposed a planktic foraminiferal zonal scheme that very closely matched the zonal intervals distinguished by Bolli (1957) for the Oligocene to Middle Miocene. Most of the zones were defined slightly

PRESUMED AGE					ZONE	FORMATION	AUTHOR, YEAR
today		when erected					
MIOCENE	M	MIOCENE	E		*Globorotalia menardii*	Lengua	Stainforth, 1948
		OLIGOCENE	L		*Globorotalia fohsi*	Cipero	Cushman & Stainforth, 1945
	E		M–L		*Globigerinatella insueta*		Cushman & Stainforth, 1945
					Globigerina dissimilis		Cushman & Renz, 1947
OLIGOCENE	L		E		*Globigerina concinna*		Cushman & Stainforth, 1945
EOCENE	L	EOCENE	L		*Hantkenina alabamensis*	Hospital Hill	Stainforth, 1948

Fig. 1. The first planktic foraminiferal zones erected in Trinidad

AGE			PLANKTIC FORAMINIFERAL ZONES and SUBZONES	
HOL.				● *Gr. fimbriata*
PLEISTOCENE			*Globorotalia truncatulinoides truncatulinoides*	● *Gg. bermudezi*
				● *Gg. calida calida*
				● *Gr. crassaf. hessi*
				● *Gr. crassaf. viola*
PLIOCENE	L		● *Globorotalia tosaensis tosaensis*	
	M		● *Globorotalia miocenica*	● *Gr. exilis*
				● *Gs. trilob. fistulosus*
	E		*Globorotalia margaritae*	*Gr. marg. evoluta*
				Gr. marg. margaritae
MIOCENE	L		*Globorotalia humerosa*	
			Globorotalia acostaensis	
	M		*Globorotalia menardii*	
			Globorotalia mayeri	
			Globigerinoides ruber	
			Globorotalia fohsi robusta	
			Globorotalia fohsi lobata	
			Globorotalia fohsi fohsi	
			Globorotalia fohsi peripheroronda	
	E		*Praeorbulina glomerosa*	
			Globigerinatella insueta	
			Catapsydrax stainforthi	
			Catapsydrax dissimilis	
			Globigerinoides primordius	
OLIGOCENE	L		*Globorotalia kugleri*	
			Globigerina ciperoensis ciperoensis	
	M		*Globorotalia opima opima*	
			Globigerina ampliapertura	
	E		*Cassig. chipolensis/Pseudohast. micra*	
EOCENE	L		*Turborotalia cerroazulensis* s.l.	
			Globigerinatheka semiinvoluta	
	M		*Truncorotaloides rohri*	
			Orbulinoides beckmanni	
			Morozovella lehneri	
			Globigerinatheka s. subconglobata	
			Hantkenina nuttalli	
			Acarinina pentacamerata	
	E		*Morozovella aragonensis*	
			Morozovella formosa formosa	
			Morozovella subbotinae	
			● *Morozovella edgari*	
PALEOCENE	L		*Morozovella velascoensis*	
			Planorotalites pseudomenardii	
	M		*Planorotalites pusilla pusilla*	
			Morozovella angulata	
			Morozovella uncinata	
			Morozovella trinidadensis	
	E		*Morozovella pseudobulloides*	
			Globigerina eugubina	

differently, however, and zonal names mostly given by two taxa; for example 'Globorotalia (Turborotalia) acostaensis (s.s.) – Globorotalia (Globorotalia) merotumida, partial range Zone'. To avoid such lengthy and cumbersome zonal names, the authors added a letter/number designation to each zone, using P for Paleogene and N for Neogene and numbering the zones sequentially within each subdivision.

The present authors agree with the statement of Stainforth *et al.* (1975) that such a 'scheme is convenient but open to the general criticism that a code designation contains no inherent clue to stratigraphic level. A more specific defect, arising from controversy over the Oligocene–Miocene boundary, is that the oldest Neogene zone is coded N 4 instead of N 1.' In addition to these drawbacks, its rigidity makes further subdivisions cumbersome. Despite these disadvantages, this abbreviated scheme has found wide acceptance merely for the convenience of avoiding writing out fossil names as zonal markers and because of the fact that the numbers, increasing from oldest to youngest, gives the non-paleontologist/stratigrapher a clue to the sequence of zones.

Blow (1969) emended several of the zonal definitions originally proposed by Banner & Blow (1965b). Other changes to the scheme have more recently been proposed by Srinivasan & Kennett (1981), who emended zones N 12 and N 13 and subdivided N 17 into an N 17A and N 17B.

Such changes to the definitions of existing zones, retaining the same names or letter/number, as well as the applications of zonal names already in use to zones differently defined, have repeatedly occurred in the literature. Because the application of zones of the same name but with different definitions may lead to erroneous biostratigraphic interpretations and correlations such practices should be avoided. The rigidity of the letter/number system is particularly felt in the Pleistocene and to a lesser degree also in the Pliocene, where, since the inception of the scheme in 1965, a considerably more detailed subdivision by planktic foraminifera has been found possible.

Fig. 3 gives the Oligocene to Holocene zonation as used in our work together with the names of the species whose first (F) and last (L) occurrence define the intervals. For those workers who wish to convert this scheme to the P/N numbered system of Banner & Blow, Fig. 4 gives the correlation as closely as possible with the zonal markers as used by Blow (1969).

Oligocene and Miocene zonation has remained fairly stable since the original schemes were set up. On the other hand, the Pliocene and Pleistocene have received considerably

Fig. 2. The Tertiary planktic foraminiferal zones developed in the Caribbean area
Heavier line: Zone established in Trinidad
Lighter line: Zone established in Eastern Venezuela
Black dot: Zone established in Caribbean deep sea cores

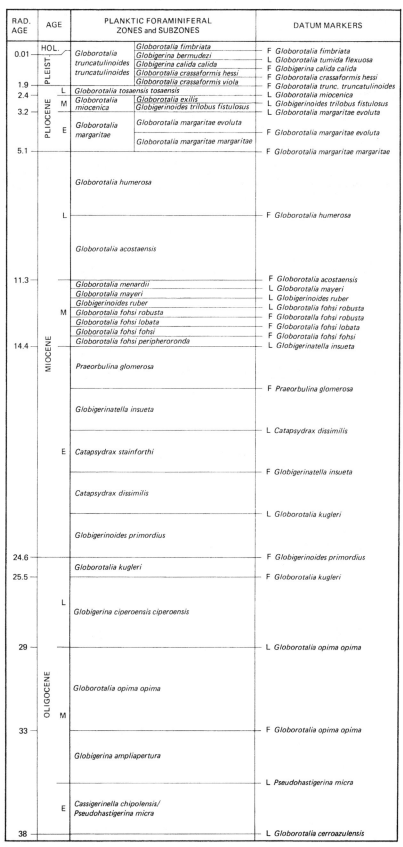

Fig. 3. Oligocene to Holocene planktic foraminiferal zonal scheme, datum markers and radiometric ages used in this chapter

more attention in recent years and these advances are discussed under their appropriate headings in the following sections.

Late Miocene and Pliocene zonal schemes

Fig. 5 contains among others the Late Miocene to Holocene zonal schemes proposed by Bolli & Premoli Silva (1973), Stainforth *et al.* (1975) and Lamb & Beard (1972). The Bolli & Premoli Silva scheme is based on Caribbean land sections and DSDP cores, that of Lamb & Beard on Caribbean land sections and on well sections from the continental slope of the northern Gulf of Mexico. The Lamb & Beard and Stainforth *et al.* schemes are basically identical, with the difference that in Lamb & Beard there is an additional sub-

division of the *Globorotalia acostaensis*, *Globorotalia margaritae* and *Globorotalia truncatulinoides* zones into subzones. These two zonal schemes are therefore here treated as a unit and compared with the Bolli & Premoli Silva scheme.

The differences between the two schemes lie in the use of some different zonal markers, in different ranges being given for some of the key species and in a different placing of the Pliocene/Pleistocene boundary. From the Middle Miocene *Globorotalia menardii* Zone to the Middle Pliocene *Globigerinoides trilobus fistulosus* or the approximately equivalent *Pulleniatina obliquiloculata* Subzone there exists good agreement between the two zonal schemes.

The definitions of the *Globorotalia menardii* Zone are

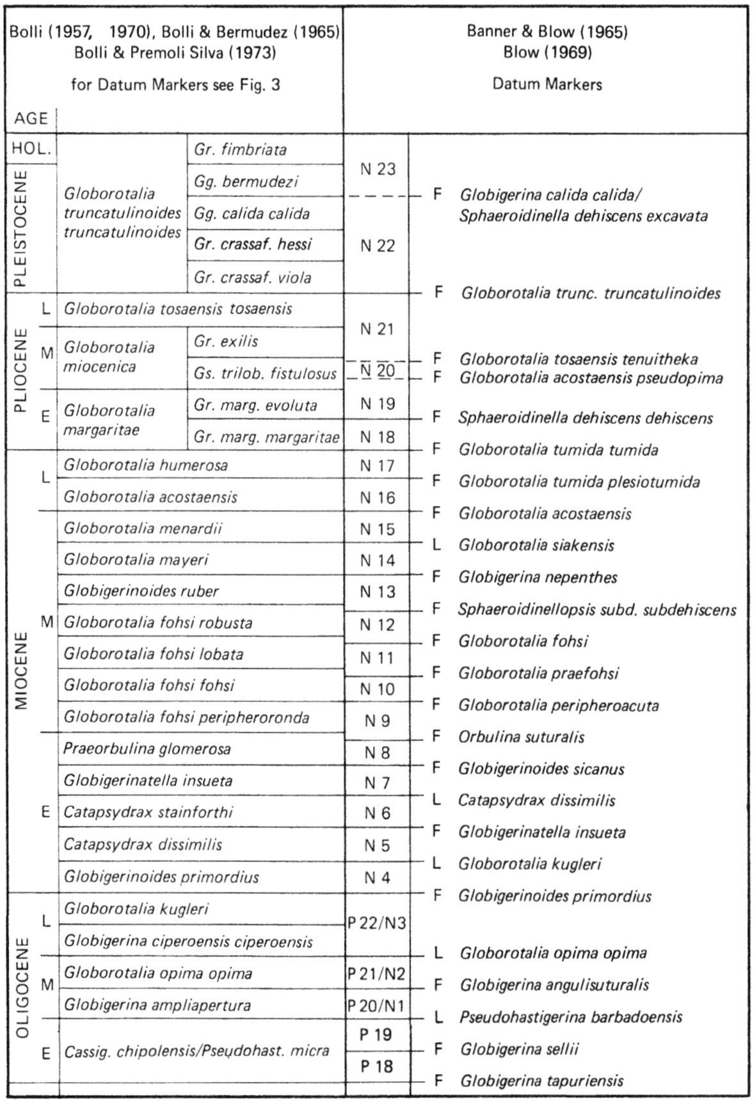

Bolli (1957, 1970), Bolli & Bermudez (1965) Bolli & Premoli Silva (1973) for Datum Markers see Fig. 3			Banner & Blow (1965) Blow (1969) Datum Markers		
AGE					
HOL.		*Gr. fimbriata*	N 23		
PLEISTOCENE	*Globorotalia truncatulinoides truncatulinoides*	*Gg. bermudezi*		F	*Globigerina calida calida/ Sphaeroidinella dehiscens excavata*
		Gg. calida calida			
		Gr. crassaf. hessi	N 22		
		Gr. crassaf. viola		F	*Globorotalia trunc. truncatulinoides*
PLIOCENE L	*Globorotalia tosaensis tosaensis*		N 21		
M	*Globorotalia miocenica*	*Gr. exilis*		F	*Globorotalia tosaensis tenuitheka*
		Gs. trilob. fistulosus	N 20	F	*Globorotalia acostaensis pseudopima*
E	*Globorotalia margaritae*	*Gr. marg. evoluta*	N 19	F	*Sphaeroidinella dehiscens dehiscens*
		Gr. marg. margaritae	N 18	F	*Globorotalia tumida tumida*
MIOCENE L	*Globorotalia humerosa*		N 17	F	*Globorotalia tumida plesiotumida*
	Globorotalia acostaensis		N 16	F	*Globorotalia acostaensis*
	Globorotalia menardii		N 15	L	*Globorotalia siakensis*
	Globorotalia mayeri		N 14	F	*Globigerina nepenthes*
	Globigerinoides ruber		N 13	F	*Sphaeroidinellopsis subd. subdehiscens*
M	*Globorotalia fohsi robusta*		N 12	F	*Globorotalia fohsi*
	Globorotalia fohsi lobata		N 11	F	*Globorotalia praefohsi*
	Globorotalia fohsi fohsi		N 10	F	*Globorotalia peripheroacuta*
	Globorotalia fohsi peripheroronda		N 9	F	*Orbulina suturalis*
	Praeorbulina glomerosa		N 8	F	*Globigerinoides sicanus*
	Globigerinatella insueta		N 7	L	*Catapsydrax dissimilis*
E	*Catapsydrax stainforthi*		N 6	F	*Globigerinatella insueta*
	Catapsydrax dissimilis		N 5	L	*Globorotalia kugleri*
	Globigerinoides primordius		N 4	F	*Globigerinoides primordius*
OLIGOCENE L	*Globorotalia kugleri*		P 22/N3		
	Globigerina ciperoensis ciperoensis			L	*Globorotalia opima opima*
M	*Globorotalia opima opima*		P 21/N2	F	*Globigerina angulisuturalis*
	Globigerina ampliapertura		P 20/N1	L	*Pseudohastigerina barbadoensis*
E	*Cassig. chipolensis/Pseudohast. micra*		P 19	F	*Globigerina sellii*
			P 18	F	*Globigerina tapuriensis*

Fig. 4. Correlation of the Oligocene to Holocene planktic foraminiferal zonal scheme with that proposed by Banner & Blow (1965b) and Blow (1969)

identical. The *Globorotalia acostaensis* Zone of Stainforth *et al* (1975) comprises the *Globorotalia acostaensis* and *Globorotalia dutertrei* (now *Globorotalia humerosa*) zones of Bolli & Premoli Silva. In Lamb & Beard the *Globorotalia acostaensis* Zone is however subdivided into a lower *Sphaeroidinellopsis seminulina* and an upper *Sphaeroidinellopsis sphaeroides* Subzone. The boundary between the two subzones is determined by the first occurrence of *S. sphaeroides*, which is slightly earlier than the first occurrence of *G. humerosa*.

Full agreement also exists in the extent of the *Globorotalia margaritae* Zone, subdivided in Bolli & Premoli Silva into a lower *Globorotalia margaritae margaritae* Subzone and an upper *Globorotalia margaritae evoluta* Subzone. Lamb & Beard subdivide the *Globorotalia margaritae* Zone into a lower *Globorotalia multicamerata* Subzone (first *G. margaritae* to first *P. primalis*) and an upper *Pulleniatina primalis* Subzone (first *P. primalis* to extinction of *G. margaritae*).

In practice, a clear distinction of *G. margaritae margaritae* and *G. margaritae evoluta* can be difficult. The distribution of these taxa may also be ecologically controlled, which is true also for the occurrence in the Atlantic province of *P. primalis* within the lower part of the *Globorotalia margaritae* Zone.

There exists some inconsistency on the first occurrence of *P. primalis* in that the species appears, following Lamb & Beard, within the *Globorotalia margaritae* Zone but in Stainforth *et al.* already within the upper part of the *Globorotalia acostaensis* Zone. Based on Caribbean DSDP evidence (Sites 148 and 154A of Leg 15), Bolli & Premoli Silva place the first occurrence at or slightly above the base of the *Globorotalia margaritae* Zone, i.e. in-between the two first occurrences as given by Lamb & Beard and Stainforth *et al.*

The extinction level of *Globoquadrina altispira* determines the *Pulleniatina obliquiloculata/Globorotalia truncatu-*

AGE		Bolli & Premoli Silva 1973 and present volume		Blow 1969	Parker 1973	Berggren 1973	Lamb & Beard 1972		Stainforth *et al* 1975	Poag & Valentine 1976
HOL.			*G. fimbriata*	N 23	VI		*G. tumida*			*G. ungulata*
PLEISTOCENE		*G. truncatulinoides truncatulinoides*	*G. bermudezi*		V			*P. finalis*	*G. truncatulinoides*	*G. crassaformis*
										T. inflata I
			G. calida calida	N 22						*G. flexuosa*
										T. inflata II
			G. crassaformis hessi				*G. truncatulinoides*	*G. dutertrei*		*G. cultrata* I
										T. inflata III
										G. cultrata II
										T. inflata IV
			G. crassaformis viola							*G. cultrata* III
										T. inflata V
										G. cultrata IV
										T. inflata VI
PLIOCENE	L	*G. tosaensis tosaensis*		N 21	IV	Pl 6		*G. tosaensis*		*G. incisa*
	M	*G. miocenica*	*G. exilis*		III	Pl 5	*P. obliquiloculata*		*P. obliquiloculata*	
			G. trilobus fistulosus	N 20		Pl 4				
						Pl 3				
	E	*G. margaritae*	*G. margaritae evoluta*	N 19	II	Pl 2	*G. margaritae*	*P. primalis*	*G. margaritae*	
			G. margaritae margaritae	N 18		Pl 1		*G. multicamerata*		
MIOCENE	L	*G. humerosa*		N 17	I		*G. acostaensis*	*S. sphaeroides*	*G. acostaensis*	
		G. acostaensis		N 16				*S. seminulina*		

Fig. 5. Correlation of Late Miocene to Holocene low latitude planktic foraminiferal zonal schemes

linoides boundary in Stainforth *et al.* and Lamb & Beard. This species disappears in the Caribbean at virtually the same level as does *Globigerinoides trilobus fistulosus*, which is taken by Bolli & Premoli Silva as the marker species for the subzone following the *Globorotalia margaritae* Zone. According to Stainforth *et al.*, *G. trilobus fistulosus* continues within the Pleistocene. In our experience the fistulose specimens, which continue after the extinction of *G. altispira* are forms determined by Bolli (1970, pl. 1, figs. 6, 7) as *G. trilobus* cf. *fistulosus*. They are specimens with some fistulose extensions on the ultimate chamber only. These are not nearly as strongly extended and finger-like as is the case in typical forms, where they are often present also on earlier chambers (Fig. 22.4 compared with Figs. 22.5–11). *G. trilobus* 'A' of Bolli (1970) is restricted to the Pleistocene (Figs. 22.1–3).

Though some different zonal/subzonal markers may be applied, the subdivision of the interval from the base of the *Globorotalia menardii* Zone to the top of the *Globigerinoides trilobus fistulosus* Subzone and its approximate equivalent the *Pulleniatina obliquiloculata* Subzone, is virtually identical in the three zonal schemes.

The approach to zonal subdivision immediately following the top of the *Globigerinoides trilobus fistulosus* Subzone and its approximate equivalent the *Pulleniatina obliquiloculata* Sub-zone differs considerably between the schemes of Bolli & Premoli Silva, Lamb & Beard and Stainforth *et al.* The first disagreement concerns the placing of the Pliocene/Pleistocene boundary. Both Stainforth *et al.* and Lamb & Beard place it at 2.8 m.y., at the extinction level of *Globoquadrina altispira*. Like most other authors including Berggren & Van Couvering (1974) we adhere to an approximate 1.9 m.y. date (Ness, Levi & Couch, 1980) for the boundary which coincides with the first occurrence of *Globorotalia truncatulinoides truncatulinoides* and approximately also with the extinction of *Discoaster brouweri*.

The approximately 1 m.y. interval between the extinction of *Globigerinoides trilobus fistulosus* and *Globoquadrina altispira* at 2.8 m.y. and the first appearance of *G. truncatulinoides truncatulinoides* at 1.9 m.y. was divided by Bolli & Premoli Silva into a *Globorotalia exilis* Subzone below and a *Globorotalia* cf. *tosaensis* Zone above, the boundary between the two zones being determined by the extinction of *Globorotalia miocenica*, which occurs only very slightly prior to that of *Globorotalia exilis*.

The extinction of *G. miocenica* took place according to Stainforth *et al.* at about 2.2 m.y. and according to Bolli & Premoli Silva at about 2.4 m.y., or approximately about half-way between the extinction of *G. altispira/G. trilobus fistulosus* and the appearance of *G. truncatulinoides truncatulinoides*. The presence of an interval above the extinction of *G. miocenica/G. exilis* and before the appearance of *G. truncatulinoides truncatulinoides* has been observed in several Caribbean DSDP sections

(29, 30, 31, 148, 154A) and is thus regarded as established for that area.

In contrast, Lamb & Beard and Stainforth *et al.*, indicate a short overlap of these species. This is based apparently on overlaps determined in the Gulf of Mexico Core Holes A and B and in a Sigsbee Knoll piston core. On the other hand their Gulf of Mexico Core Hole C and land sections on Jamaica do not show such overlaps. They explain the lack of overlap as due to the presence of a hiatus which would not be the case according to our interpretation.

Pleistocene zonation

Placing the Pliocene/Pleistocene boundary in tropical–subtropical areas on the first occurrence of *Globorotalia truncatulinoides truncatulinoides* has widely been practised and accepted in both the Atlantic and Indo-Pacific provinces. It is also followed here. Exceptions to this in more recent publications are found in Lamb & Beard (1972) and Stainforth *et al.* (1975). In these the Pliocene/Pleistocene boundary is placed at the extinction level of *Globoquadrina altispira*, that is well below the first occurrence of *Globorotalia truncatulinoides truncatulinoides* (Fig. 5).

Climatic changes within the Pleistocene caused by glacial and interglacial periods have affected the marine plankton to various degrees in all latitudinal belts including the tropics. A subdivision of the Pleistocene by planktic foraminiferal zonal schemes is, therefore, even more restricted to limited latitudinal intervals than is the case for the Late Miocene and Pliocene.

While in higher but still temperate latitudes, such as the Mediterranean area, only few faunal changes have taken place during the past 1.9 m.y, certain faunal developments that allow for a subdivision took place in lower latitudes. Particularly in tropical–subtropical areas, climatic changes are expressed by alternations of warmer and colder water faunal associations, which allow the erection of so-called eco-zones.

A combination of biostratigraphic and eco-zoning based on foraminifera, in particular planktic forms, and on calcareous nannoplankton, has been worked out in considerable detail in the Gulf Coast area. There, Poag & Valentine (1976) subdivide the Pleistocene of the Texas–Louisiana Basin into 12 biostratigraphic and ecostratigraphic zones based on planktic foraminifera. A subdivision of the Pleistocene into 4 biozones has been established in the Caribbean by Bolli & Premoli Silva (1973), based on DSDP sections, and an attempt has been made by Rögl & Bolli (1973) for the definition of eco-intervals in the upper Pleistocene of the Gulf of Cariaco, based on alternations of lower and higher latitude planktic foraminifera. Rögl (1974) has found it possible to apply the same four-fold biostratigraphic subdivision as established in the Caribbean also to the DSDP Site 262 in the Timor Trough.

Evolutionary trends and their influence on zonal resolution

The low latitude zonal scheme used in this volume distinguishes a sequence of 29 zones/subzones from the base of the Oligocene to the Holocene. Using the radiometric dating of Ness *et al.* (1980), the average duration of an individual zone during this time span of 38 m.y. is thus 1.3 m.y. The degree of resolution of a given time span however depends on the number of recognizable taxa, their ranges and rate of evolution during such a period, factors which may vary considerably. This is very much the case for the Oligocene to Holocene interval, where comparatively long periods with few faunal events alternate with intervals with more numerous taxa and more rapid, almost explosive, evolutionary developments (Fig. 3).

From this point of view the Oligocene to Holocene can be divided into the following intervals:

(1) The Oligocene with a duration of approximately 13.4 m.y. (38–24.6) is a comparatively long period with few events; it is divided into only 5 zones with an average duration of 2.7 m.y.

(2) The duration of the Miocene is evaluated at about 19 m.y., divided here into 14 zones of an average length of 1.4 m.y. However, the duration of individual zones is very variable. The 5 zones of the Early Miocene (10.2 m.y.) have average durations of 2 m.y., the 7 zones of the Middle Miocene (3.1 m.y.) only 0.4 m.y. In contrast, the 2 zones distinguished in the Late Miocene (6.2 m.y.) occupy a time span of as much as 3.1 m.y. each.

(3) The Pliocene, with a duration of about 3.2 m.y. is divided into 5 zones/subzones with an average duration of 0.6 m.y.

(4) Finally, the average duration of one of the 4 zones into which the 1.9 m.y. of Pleistocene is divided is a mere 0.5 m.y.

The considerable differences in duration of zones during the listed intervals reflect the pattern of the faunal composition, new occurrences and extinctions, and development and speed of morphological changes at the various times.

After the abundance of forms in the Middle and Late Eocene, the Oligocene represents a time of comparative monotony and slow development strongly dominated by representatives of the genus *Globigerina* with, in addition, a few non-carinate *Globorotalia* species. Their distribution has only allowed a rather crude subdivision of this comparatively long time span.

A sudden proliferation of taxa caused primarily by the appearance of the genus *Globigerinoides* and the rapid development of *Globoquadrina* species characterizes the beginning of the Early Miocene, within which interval also occurs the short-ranging index form *Globigerinatella insueta* and the lineage of *Praeorbulina* to *Orbulina*. The first appearance of *Orbulina* determines the Early–Middle Miocene boundary.

The still closer subdivision of the first half of the Middle Miocene is based on the very rapid evolution of the *Globorotalia fohsi* plexus from the subspecies *peripheroronda* to *robusta*. The sudden disappearance of this group, and also the disappearance, slightly later, of other index forms like *Globigerinoides ruber*, *Globorotalia mayeri* and *Cassigerinella chipolensis*, mark the onset of a distinct slowing down or almost a stagnation in the faunal development in the later part of the Middle Miocene. This trend continued for about 6 m.y. to nearly the top of the Late Miocene, where one of the most spectacular and prolific developments of new genera and species known in planktic foraminiferal history began.

The magnitude of the events of appearance and extinction of low latitude planktic foraminiferal taxa from the latest Miocene and throughout the Pliocene is second only to that at the Cretaceous/Tertiary boundary.

This sudden increase of events clearly shows on the distribution chart of Figure 10. Within this brief time interval several new genera appear, among them the significant *Candeina*, *Pulleniatina* and *Sphaeroidinella*. Of particular stratigraphic significance is the proliferation of *Globorotalia* species, some of them only short-lived. Representatives of several other genera also either appear or disappear within the Pliocene.

The effects on zonation of the geographic restriction of marker species

With such a multitude of taxa to choose from it is not surprising that several different Pliocene/Pleistocene zonal schemes have been proposed. These schemes, shown here on Fig. 5, were commented on by Bolli & Krasheninnikov (1977).

Especially when using some of the Pliocene index forms on a worldwide scale, it is essential to be aware of the following anomalies in their geographic distribution: a number of Pliocene index forms are restricted to the Atlantic province in their typical development. They include *G. pseudomiocenica*, *G. miocenica*, *G. multicamerata*, *G. pertenuis* and *G. exilis*. These forms developed in the Atlantic and, because of the closing of the Isthmus of Panama within the Early Pliocene, were unable to spread into the Indo-Pacific due to lack of a low latitude connection. On the other hand, *Pulleniatina praespectabilis* and *P. spectabilis* were restricted to the Indo-Pacific, where they also developed in the Early Pliocene but were never able to colonize the Atlantic.

Zones based on these taxa are consequently geographically restricted. Affected are the Middle Pliocene zones based on *G. miocenica* and *exilis* proposed by Bolli & Premoli Silva (1973) and by Berggren (1973) (Zone Pl 5). Further, Banner & Blow's zone N 20 and Berggren's Pl 4, both rely on *G. multicamerata*. All these zones are only applicable in the

Atlantic province, including the Gulf of Mexico and the Caribbean.

On observations in Caribbean DSDP sites and on sections in Jamaica, Bolli & Krasheninnikov (1977) pointed out that during the development of the *G. multicamerata–pertenuis–exilis–pseudomiocenica–miocenica* suite in the Atlantic Province, *G. menardii* s.l., *G. tumida* s.l. and species of *Pulleniatina* are virtually absent from the area. *Pulleniatina*, represented by *P. primalis*, made a first restricted appearance within the *Globorotalia margaritae margaritae* Subzone to reappear again only in Late Pliocene/Early Pleistocene as *P. praecursor* and *P. obliquiloculata*. Similarly, the *G. plesio-tumida–tumida* lineage made a short appearance in the latest Miocene to reappear again as *G. tumida tumida* only in the uppermost Pliocene. Zones relying on these taxa, as do Banner & Blow's zones N 18 and N 20, are therefore difficult to apply in the Atlantic province.

Zones used in this chapter

CASSIGERINELLA CHIPOLENSIS/ PSEUDOHASTIGERINA MICRA ZONE

Category: Concurrent range zone

Age: Early Oligocene

Author: Blow & Banner (1962), renamed by Bolli (1966a)

Definition: Joint occurrence of the two zonal markers.

Remarks: The zone originally named *Globigerina oligocaenica* Zone was defined by the range of the zonal marker and also by the overlap of *Pseudohastigerina micra* and *Cassigerinella chipolensis* which, according to Blow & Banner, coincides with the range of *G. oligocaenica*. This species was subsequently found to be a junior synonym of *G. sellii*; the name of the zone was consequently changed by Blow (1969) to Zone P 19, *Globigerina sellii/Pseudohastigerina barbadoensis* Zone. *G. sellii* is now known to range upwards in the Oligocene higher than had been originally thought. Furthermore, it can be difficult to distinguish the species from morphologically related and similar forms like *G. tapuriensis*, *G. tripartita* or *G. binaiensis*. The most reliable way to identify the Early Oligocene is therefore by using the overlap of the ranges of *Cassigerinella chipolensis* and *Pseudohastigerina micra*, based on which Bolli (1966a) renamed the zone. Our concept of the zone includes the interval covered by Banner & Blow's Zone P 19 together with the underlying Zone P 18, *Globigerina turritillina turritillina* Zone (renamed *Globigerina tapuriensis* Zone by Blow, 1969), originally considered by its authors as uppermost Late Eocene.

GLOBIGERINA AMPLIAPERTURA ZONE

Category: Interval zone

Age: Middle Oligocene

Author: Bolli (1957), redefined by Bolli (1966a)

Definition: Interval between last occurrence of *Pseudohastigerina micra* and first occurrence of *Globorotalia opima opima*.

Remarks: In Trinidad, where the zone was established, the zonal marker disappears within its upper part. The interval between its last occurrence and the first occurrence of *Globorotalia opima opima* is there dominated by forms close to *Globigerina venezuelana*. For internal purposes a distinction of this interval was found useful and was therefore separated as *Globigerina* cf. *venezuelana* Zone, but this has never been formalized. The base of *Globorotalia opima opima* marks the upper boundary of the zone, and at about the same time the characteristic *Globigerina ciperoensis ciperoensis* also appears.

GLOBOROTALIA OPIMA OPIMA ZONE

Category: Taxon range zone

Age: Middle Oligocene

Author: Bolli (1957)

Definition: Range of zonal marker.

Remarks: In addition to the restriction of the nominate taxon to the zone, the interval is also characterized by typically developed *Globigerina ciperoensis ciperoensis* and *G. ciperoensis angulisuturalis*, both of which however continue through the next younger zone.

GLOBIGERINA CIPEROENSIS CIPEROENSIS ZONE

Category: Interval zone

Age: Late Oligocene

Author: Cushman & Stainforth (1945), emended by Bolli (1957)

Definition: Interval with zonal marker, from last occurrence of *Globorotalia opima opima* to first occurrence of *G. kugleri*.

Remarks: The zonal marker was originally placed in *Globigerina concinna* Reuss and the zone, as it was erected in the Cipero Formation of Trinidad by Cushman & Stainforth (1945), was called 'Zone I, Lower (*Globigerina concinna*) Zone'. Its base was determined by the top of the Upper Eocene Hospital Hill Marl, the upper limit by the base of the *Globigerinatella insueta* Zone in the sense of Cushman & Stainforth. This includes a much wider stratigraphic interval than that defined later by Bolli (1957) and used subsequently. The top of the zone is characterized by the disappearance of the zonal marker and *Globigerina ciperoensis angulisuturalis*. In the upper part of the zone may be found under favourable conditions the characteristic *G. binaiensis*, which apparently developed from *G. sellii* within the zone.

GLOBOROTALIA KUGLERI ZONE

Category: Interval zone

Age: Late Oligocene

Author: Bolli (1957), emended in this volume

Definition: First occurrence of zonal marker to first occurrence of frequent *Globigerinoides primordius* and/or *G. trilobus* s.l.

Remarks: The zone was originally defined by the total range of the zonal marker. Subsequently it became apparent that older samples with *Globorotalia kugleri* were completely or almost lacking in *Globigerinoides*, whereas in later ones representatives of that genus could become fairly frequent. As discussed in the section on *Globigerinoides*, the first occurrence of the genus, originally thought to be an excellent datum for the identification of the Oligocene/Miocene boundary, is now regarded as dependent on local environmental conditions and therefore stratigraphically less reliable. Study of widely separated sections indicates that *G. kugleri* may occur without or with only rare early *Globigerinoides* specimens in some samples while in others representatives of that genus may become frequent. These observations have prompted us to divide the interval spanning the range of *G. kugleri* into two zones, an older *Globorotalia kugleri* Zone, where *Globigerinoides primordius* is either absent altogether or, when present, only very rare, and a younger *Globigerinoides primordius* Zone, where the zonal marker becomes frequent and in the upper part of the zone develops into *G. trilobus* with its subspecies. In some sections a clear separation of these two zones may be difficult or not practical, but in others the distinction proves useful.

GLOBIGERINOIDES PRIMORDIUS ZONE

Category: Concurrent range zone

Age: Early Miocene

Author: Blow (1969), emended in this volume

Definition: Interval from first occurrence of frequent *Globigerinoides primordius/trilobus* s.l. to last occurrence of *Globorotalia kugleri*.

Remarks: See *Globorotalia kugleri* Zone.

CATAPSYDRAX DISSIMILIS ZONE

Category: Interval zone

Age: Early Miocene

Author: Cushman & Renz (1947), emended by Bolli (1957)

Definition: Interval with zonal marker, from last occurrence

of *Globorotalia kugleri* to first occurrence of *Globigerinatella insueta*.

Remarks: The zone was originally described as *Globigerina dissimilis* Zone, falling between the *Globigerina concinna* and *Globigerinatella insueta* zones of Cushman & Stainforth (1945). *Globigerinoides ruber* and *G. altiapertura* appear first in this zone, while *Globigerina binaiensis* disappears within it.

The zone has been named *Globigerinita dissimilis* Zone on many recent charts, as *Catapsydrax* was taken to be a junior synonym of *Globigerinita*. However, studies undertaken for this work indicate to us that the genera are both valid and thus the zone reverts to its earlier designation. This question is discussed under the taxonomy of the species involved.

CATAPSYDRAX STAINFORTHI ZONE

Category: Concurrent range zone

Age: Early Miocene

Author: Bolli (1957)

Definition: Interval with zonal marker, from first occurrence of *Globigerinatella insueta* to last occurrence of *Catapsydrax dissimilis*.

Remarks: The zone is based on the joint occurrence of *Catapsydrax dissimilis* and *Globigerinatella insueta*. The latter is often rare, which may make it difficult to separate the zone from the *Catapsydrax dissimilis* Zone on this criterion alone. On the other hand, several taxa occur for the first time in the *Catapsydrax stainforthi* Zone including *Globorotalia archeomenardii*, earliest forms attributable to *G. fohsi peripheroronda* and *Globoquadrina altispira altispira*.

For the generic designation of the zonal marker species see remarks for the previous zone.

GLOBIGERINATELLA INSUETA ZONE

Category: Interval zone

Age: Early Miocene

Author: Cushman & Stainforth (1945), emended by Bolli (1957)

Definition: Interval with zonal marker, from the last occurrence of *Catapsydrax dissimilis* to the first occurrence of *Praeorbulina glomerosa*.

Remarks: Under the name 'Zone II. Middle (*Globigerinatella insueta*) Zone' it included originally a thicker, not precisely defined interval. In its redefined usage it comprises only the middle portion of the range of the zonal marker. In addition to the taxa which define the extent of the zone, the following are of significance: continuing presence of *Catapsydrax stainforthi*, appearance of the small sized *Globigerinoides diminutus*, appearance of *G.*

bisphericus, which in the upper part of the zone begins to evolve into *Praeorbulina glomerosa curva* through the intermediate *P. sicana*. Specimens of *Globorotalia fohsi peripheroronda* are particularly well developed at this level.

PRAEORBULINA GLOMEROSA ZONE

Category: Concurrent range zone
Age: Early Miocene
Author: Blow (1959), renamed by Bolli (1966a)
Definition: Interval with zonal marker, from its first occurrence to last occurrence of *Globigerinatella insueta*.
Remarks: The zone was originally named *Globigerinatella insueta/Globigerinoides bispherica* Subzone and its lower boundary taken slightly deeper stratigraphically than in the above definition. Jenkins (1966) introduced a *Praeorbulina glomerosa curva* Zone for an interval between the first occurrence of *P. glomerosa curva* and the first occurrence of *Orbulina suturalis*. In the same year Bolli called the zone the *Praeorbulina glomerosa* Zone, the name also used by us. Blow (1969) renamed his *Globigerinatella insueta – Globigerinoides bispherica* Subzone of 1959 as Zone N 8, *Globigerinoides sicanus–Globigerinatella insueta* Zone.

In addition to the joint occurrence of the taxa which define the zone it is also recognizable by the absence of the genera *Catapsydrax*, *Clavatorella* and *Sphaeroidinellopsis* and the continued presence of *Globigerinoides diminutus*, which becomes extinct in the upper part of the zone. The evolutionary lineage within *Praeorbulina glomerosa* with the subspecies *curva*, *glomerosa* and *circularis* takes place within the zone, the final stage of which is the development of *Orbulina suturalis* very close to the top of the zone.

GLOBOROTALIA FOHSI ZONE s.l.

Cushman & Stainforth (1945) introduced a 'Zone III Upper (*Globorotalia fohsi*) Zone' without formally defining it. Based on stratigraphically successive subspecies of *Globorotalia fohsi*, Bolli (1951) subdivided the zone into the subzones *G. fohsi barisanensis* (now *peripheroronda*), *G. fohsi fohsi*, *G. fohsi lobata* and *G. fohsi robusta*; these were elevated by him to zonal rank in 1957 and are also maintained here. Blow & Banner (1966) proposed a slightly modified zonation based on the evolutionary lineage of *G. fohsi*, comprising their zones N 9–N 12. Reference is made to Bolli (1967), where Blow & Banner's zonal scheme is discussed and compared with that previously established by Bolli (1957). Boundaries between the *fohsi* zones are based on the first appearance of successive subspecies. Because morphological changes from one subspecies to the next younger are gradual, the placing of the zonal boundaries

remains to some extent subjective. For further discussions on zonal subdivisions based on the subspecies of *G. fohsi* reference is made to Bolli (1967).

Except for the rapidly evolving *fohsi* lineage, the interval remains markedly poor in other events which might serve for subdivision in the absence of the marker subspecies.

GLOBOROTALIA FOHSI PERIPHERORONDA ZONE

Category: Interval zone
Age: Middle Miocene
Author: Bolli (1957), renamed by Blow & Banner (1966), Bolli (1967)
Definition: Interval with zonal marker, from last occurrence of *Globigerinatella insueta* to first occurrence of *Globorotalia fohsi fohsi*.
Remarks: The appearance of the genus *Orbulina* is important as a marker for the approximate base of the zone. Furthermore, the zone is marked by the first appearance of the genus *Sphaeroidinellopsis* represented by small specimens of *S. disjuncta*. Typically developed and, in the Caribbean area, mainly restricted to the zone is *Clavatorella bermudezi*.

GLOBOROTALIA FOHSI FOHSI ZONE

Category: Lineage zone
Age: Middle Miocene
Author: Cushman & Stainforth (1945), emended by Bolli (1957)
Definition: Interval with zonal marker, from its first occurrence to first occurrence of *Globorotalia fohsi lobata*.
Remarks: The first occurrence of *Globorotalia praemenardii* and *Globigerinoides mitra* are recorded in this zone in the Caribbean area.

GLOBOROTALIA FOHSI LOBATA ZONE

Category: Lineage zone
Age: Middle Miocene
Author: Bolli (1957)
Definition: Interval with zonal marker, from its first occurrence to first occurrence of *Globorotalia fohsi robusta*.
Remarks: Some authors (for example, Stainforth *et al.*, 1975) combine this zone with the next younger one. In some geographic areas this stage within the lineage can be used; however, its non-recognition does not necessarily indicate a hiatus.

GLOBOROTALIA FOHSI ROBUSTA ZONE

Category: Taxon range zone
Age: Middle Miocene
Author: Bolli (1957)

Definition: Range of zonal marker.

Remarks: The only characteristic event in addition to the presence throughout the zone of the marker is the appearance of larger sized *Sphaeroidinellopsis multiloba* with up to 5 chambers in the last whorl.

GLOBIGERINOIDES RUBER ZONE

Category: Interval zone

Age: Middle Miocene

Author: Bolli (1966a)

Definition: Interval with zonal marker, from last occurrence of *Globorotalia fohsi robusta* to last Miocene occurrence of zonal marker.

Remarks: The zone represents a stratigraphic interval that was previously not, or only very indistinctly, recognized in the Caribbean region. There, *Globigerinoides ruber* disappears simultaneously with *Globorotalia fohsi robusta*, or very shortly afterwards. In Java the distribution of the Bodjonegoro-1 planktic foraminifera (Bolli, 1966b) shows that *Globigerinoides ruber* may persist for a considerable time after the extinction of *Globorotalia fohsi robusta* (in Bodjonegoro-1 for over 200 metres). The *Globigerinoides ruber* Zone was therefore introduced to divide an otherwise very thick *Globorotalia mayeri* Zone in the Java section. Its recognition also stresses a hiatus which exists elsewhere, between the *Globorotalia fohsi robusta* and the *Globorotalia mayeri* Zone, for instance in the Caribbean area. Such a hiatus had already been indicated in Trinidad by an abrupt lithological change and also by the presence of pebble beds. It is known that the widespread disappearance of *Globigerinoides ruber* in the Middle Miocene is only a temporary one, probably caused by the onset of conditions not tolerated by the species. It is assumed that the species continued to live locally isolated or in an endemic way until its widespread re-appearance within the Early Pliocene *Globorotalia margaritae margaritae* Subzone.

No additional useful criteria based on other taxa than those given for the definition are known to identify the zone.

GLOBOROTALIA MAYERI ZONE

Category: Interval zone

Age: Middle Miocene

Author: Brönnimann (1951a), emended by Bolli (1966a)

Definition: Interval with zonal marker, from last Miocene occurrence of *Globigerinoides ruber* to last occurrence of zonal marker.

Remarks: The zone as originally defined also included the *Globigerinoides ruber* Zone. Brönnimann (1951a) mentioned the *Globorotalia mayeri* Zone, but only in his later paper (1951b) gave the basis for its separation from

the overlying *Globorotalia menardii* Zone and the underlying *Globorotalia fohsi* Zone. See also remarks under *Globigerinoides ruber* Zone. *Globorotalia continuosa* becomes extinct together with the zonal marker at the top of the zone, *Cassigerinella chipolensis* within it. *Globorotalia lenguaensis* and *Globigerina nepenthes* appear at the base of the zone.

GLOBOROTALIA MENARDII ZONE

Category: Interval zone

Age: Middle Miocene

Author: Stainforth (1948), emended by Brönnimann (1951a), redefined by Bolli (1966a)

Definition: Interval with zonal marker, from last occurrence of *Globorotalia mayeri* to first occurrence of *G. acostaensis*.

Remarks: The zone as defined originally included the whole Lengua Formation of Trinidad. Brönnimann (1951a) subdivided Stainforth's zone into a lower, *Globorotalia mayeri* Zone and an upper, *Globorotalia menardii* Zone, with the boundary between the two zones marked by the extinction of *G. mayeri*.

Environmental changes in Trinidad caused a rapid disappearance of planktic foraminifera at the top of the Lengua Formation, hence the top of the *Globorotalia menardii* Zone could not be established there. The above definition of the zone is based on more favourable sections in coastal Eastern Venezuela (Bolli & Bermudez, 1965) and in Java (Bolli, 1966b).

The zone is only recognizable by the negative criteria given in its definition, and in addition by the absence of *Cassigerinella chipolensis* and *Globorotalia continuosa* compared with the zones below and of *G. juanai*, *G. menardii* 'B' and *Polyperibola christiani*, which first appear in the zone above.

GLOBOROTALIA ACOSTAENSIS ZONE

Category: Interval zone

Age: Late Miocene

Author: Bolli & Bermudez (1965)

Definition: Interval with zonal marker, from its first occurrence to first occurrence of *Globorotalia humerosa*.

Remarks: The lower and upper boundaries of the zone are not always easy to determine. Early *Globorotalia acostaensis* as, for example, those present in the basal part of the zone in Bodjonegoro-1 of Java, are somewhat atypical in that they possess only 4 chambers in the last whorl compared with 5 in the holotype. The upper boundary of the zone is based on the first occurrence of *G. humerosa*, which gradually evolves from *G. acostaensis*; this makes the placing of the boundary somewhat subjective. Where conditions are favourable, *G. juanai* and *Polyperibola christiani* may be present in the zone, at least in the Atlantic province.

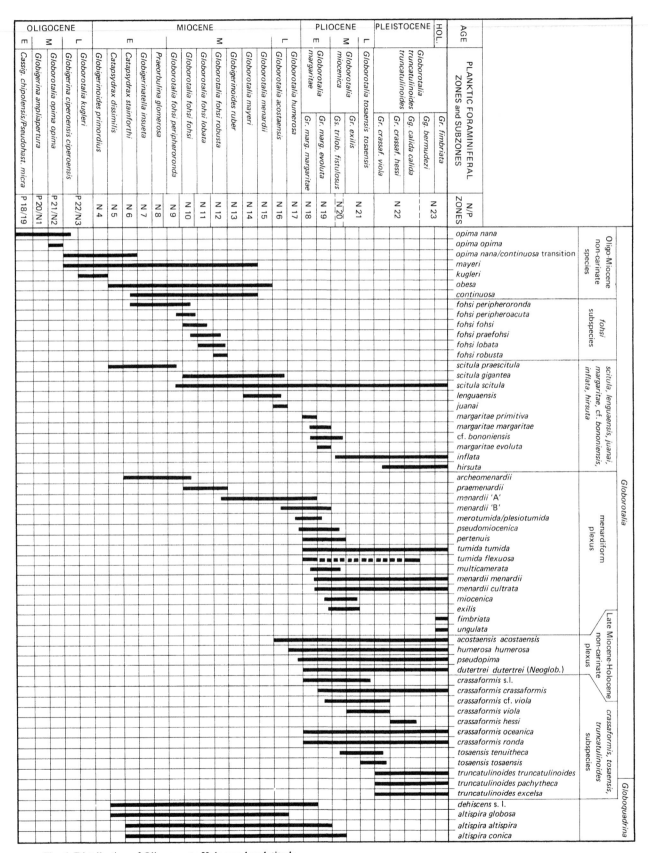

Fig. 6. Distribution of Oligocene to Holocene low latitude
planktic foraminifera (*Globorotalia* and *Globoquadrina*)

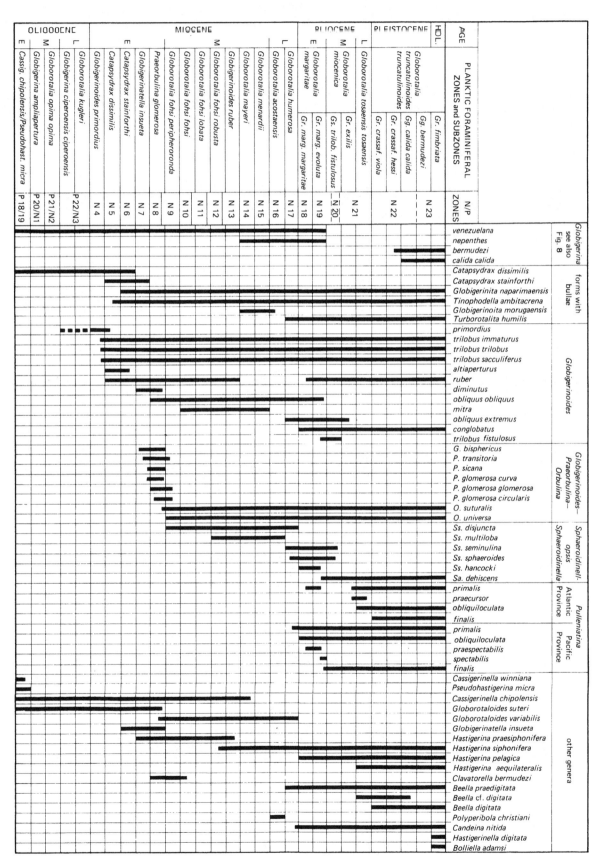

Fig. 7. Distribution of Oligocene to Holocene low latitude planktic foraminifera (genera other than *Globorotalia* and *Globoquadrina*)

GLOBOROTALIA HUMEROSA ZONE

Category: Interval zone

Age: Late Miocene

Author: Bolli & Bermudez (1965), renamed in this volume

Definition: Interval with zonal marker, from its first occurrence to first occurrence of *Globorotalia margaritae* s.l.

Remarks: The zone was originally named *Globorotalia dutertrei/Globigerinoides obliquus extremus* Zone. To simplify the terminology of the zonal scheme, only the first of the two markers was retained by Bolli (1966a). Bolli & Premoli Silva (1973) renamed it *Neogloboquadrina dutertrei* Zone, following the erection of the genus by Bandy, Frerichs & Vincent (1967). Takayanagi & Saito (1962) separated *Globorotalia humerosa* as an evolutionary earlier stage from *Neogloboquadrina dutertrei*. Developing from *G. acostaensis* it appears first at the base of the zone to continue throughout the zone, with typical *N. dutertrei* gradually developing from it only in the *Globorotalia margaritae* Zone. Renaming of the zone is thus necessary.

The Late Miocene *Globorotalia humerosa* Zone marks the beginning of an explosive appearance of new taxa that becomes even more spectacular in the Early Pliocene. In addition to the zonal marker, the following diagnostic taxa appear at about the base of the zone: *Globigerinoides obliquus extremus* and *Sphaeroidinellopsis seminulina*. Within the zone appear *Globorotalia merotumida/plesiotumida*, *G. pseudomiocenica*, *G. pseudopima*, *Sphaeroidinellopsis sphaeroides*, *Pulleniatina primalis* (Pacific realm) and *Candeina nitida*. *Sphaeroidinellopsis disjuncta* and *Globorotaloides variabilis* disappear at or near the top of the zone.

GLOBOROTALIA MARGARITAE ZONE

Category: Taxon range zone

Age: Early Pliocene

Author: Bolli & Bermudez (1965)

Definition: Range of zonal marker.

Remarks: The zone was subdivided by Cita (1973) into the *Globorotalia margaritae margaritae* Subzone and the *Globorotalia margaritae evoluta* Subzone. Because morphological changes are gradual between the two subspecies and their development appears to be influenced by local ecological conditions, the placing of the subzonal boundary remains to some degree subjective and to some extent variable from area to area. However, many

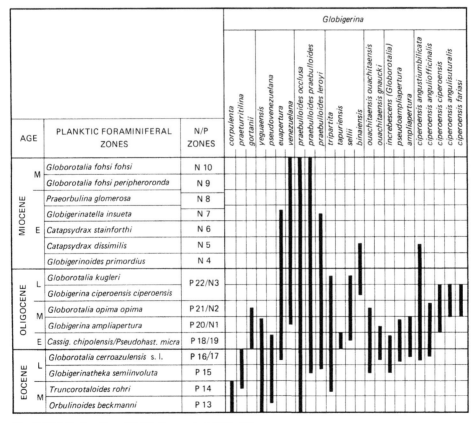

Fig. 8. Distribution from Middle Eocene to Middle Miocene of selected low latitude *Globigerina* species and subspecies

first and last occurrences take place around the bound
aries and within the subzones and these can aid considerably in the identification of the two subzones.

GLOBOROTALIA MARGARITAE
MARGARITAE SUBZONE

Category: Lineage zone
Age: Early Pliocene
Author: Cita (1973), redefined by Bolli & Premoli Silva (1973)
Definition: Interval with zonal marker, from first occurrence
of *Globorotalia margaritae* s.l. to the first occurrence of
G. margaritae evoluta.
Remarks: At the base of the subzone, approximately at the
level of the first occurrence of *Globorotalia margaritae*
s.l. appear the following characteristic taxa: *G. crassaformis* s.l., *G. pertenuis* (Atlantic realm), *G. tumida
tumida* (mainly in the Pacific realm), *N. dutertrei*,
Globigerinoides conglobatus and *Sphaeroidinellopsis hancocki*. Within the subzone appear *Globorotalia
multicamerata* (Atlantic realm), *G.* cf. *bononiensis*, *Pulleniatina primalis* (Atlantic realm) and *P. praespectabilis*
(Pacific realm). *Globoquadrina dehiscens* s.l. becomes
extinct at about the upper boundary of the subzone.

GLOBOROTALIA MARGARITAE EVOLUTA
SUBZONE

Category: Taxon range zone
Age: Early Pliocene
Author: Cita (1973)
Definition: Range of zonal marker, from its development from
Globorotalia margaritae margaritae to its extinction.
Remarks: Together with the presence of the nominate subspecies, the following events characterize the subzone:
at or near the base appears *Globorotalia crassaformis
crassaformis*, within the subzone appear *G. miocenica*
and *G. exilis* (both mainly in the Atlantic realm), *G.
crassaformis* cf. *viola*, *Globigerinoides trilobus fistulosus*,
Sphaeroidinellopsis dehiscens and *Pulleniatina spectabilis*
(restricted to upper part of the subzone in the Pacific
realm). *Globigerina venezuelana* and *G. nepenthes*
become extinct at or near the top of the subzone.

GLOBOROTALIA MIOCENICA ZONE

Category: Interval zone
Age: Middle Pliocene
Author: Bolli (1970), renamed by Bolli & Premoli Silva (1973),
redefined in this volume
Definition: Interval with zonal marker, between last occurrence
of *Globorotalia margaritae evoluta* and last occurrence
of *G. miocenica* or *G. exilis*.
Remarks: The zone was originally named *Globorotalia exilis/*

Globorotalia miocenica Zone and defined as the interval
with *Globorotalia exilis* and/or *G. miocenica* between the
extinction of *Globorotalia margaritae* and the extinction
of the zonal markers. The subdivision of the zone into
two subzones by Bolli & Premoli Silva (1973) and the
recognition that the extinction level of *G. exilis* is slightly
above that of *G. miocenica* necessitated a redefinition.

GLOBIGERINOIDES TRILOBUS FISTULOSUS
SUBZONE

Category: Interval zone
Age: Middle Pliocene
Author: Bolli & Premoli Silva (1973)
Definition: Interval with subzonal marker, between last occurrence of *Globorotalia margaritae evoluta* and last occurrence of the subzonal marker.
Remarks: *Globorotalia inflata* first appears within the subzone
while *G. multicamerata* last occurs within it. *G. pertenuis*,
Globoquadrina altispira altispira and *G. altispira conica*
last occur at or near the top of the zone, at about the
level of extinction of the zonal marker.

GLOBOROTALIA EXILIS SUBZONE

Category: Interval zone
Age: Middle Pliocene
Author: Bolli & Premoli Silva (1973)
Definition: Interval with zonal marker, from extinction of
Globigerinoides trilobus fistulosus to extinction of
Globorotalia miocenica or *G. exilis*.
Remarks: The subzonal marker may extend slightly above
Globorotalia miocenica, which is usually fairly common
up to the point of its extinction while *G. exilis* becomes
scarce, in particular when it is found above *G. miocenica*.
G. crassaformis viola appears at the base of the subzone
and *Globigerinoides obliquus extremus* last occurs within
it.

GLOBOROTALIA TOSAENSIS TOSAENSIS
ZONE

Category: Interval Zone
Age: Late Pliocene
Author: Bolli (1970), renamed in this volume
Definition: Interval with zonal marker, from extinction of
Globorotalia miocenica/G. exilis to first occurrence of *G.
truncatulinoides truncatulinoides*.
Remarks: The zone was originally named *Globorotalia truncatulinoides* cf. *tosaensis*. It is here changed to *Globorotalia
tosaensis tosaensis* in accordance with the taxonomic
treatment followed here. The zonal marker is often not
typically developed or may be very rare or even absent,
as is the case in many of the investigated sections in the

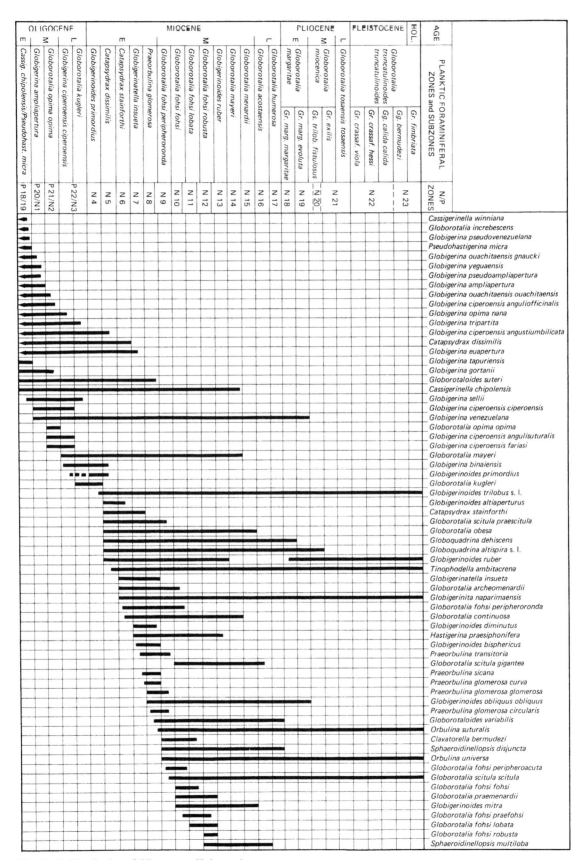

Figs. 9–10. Distribution of Oligocene to Holocene low latitude planktic foraminifera (in order of first occurrence)

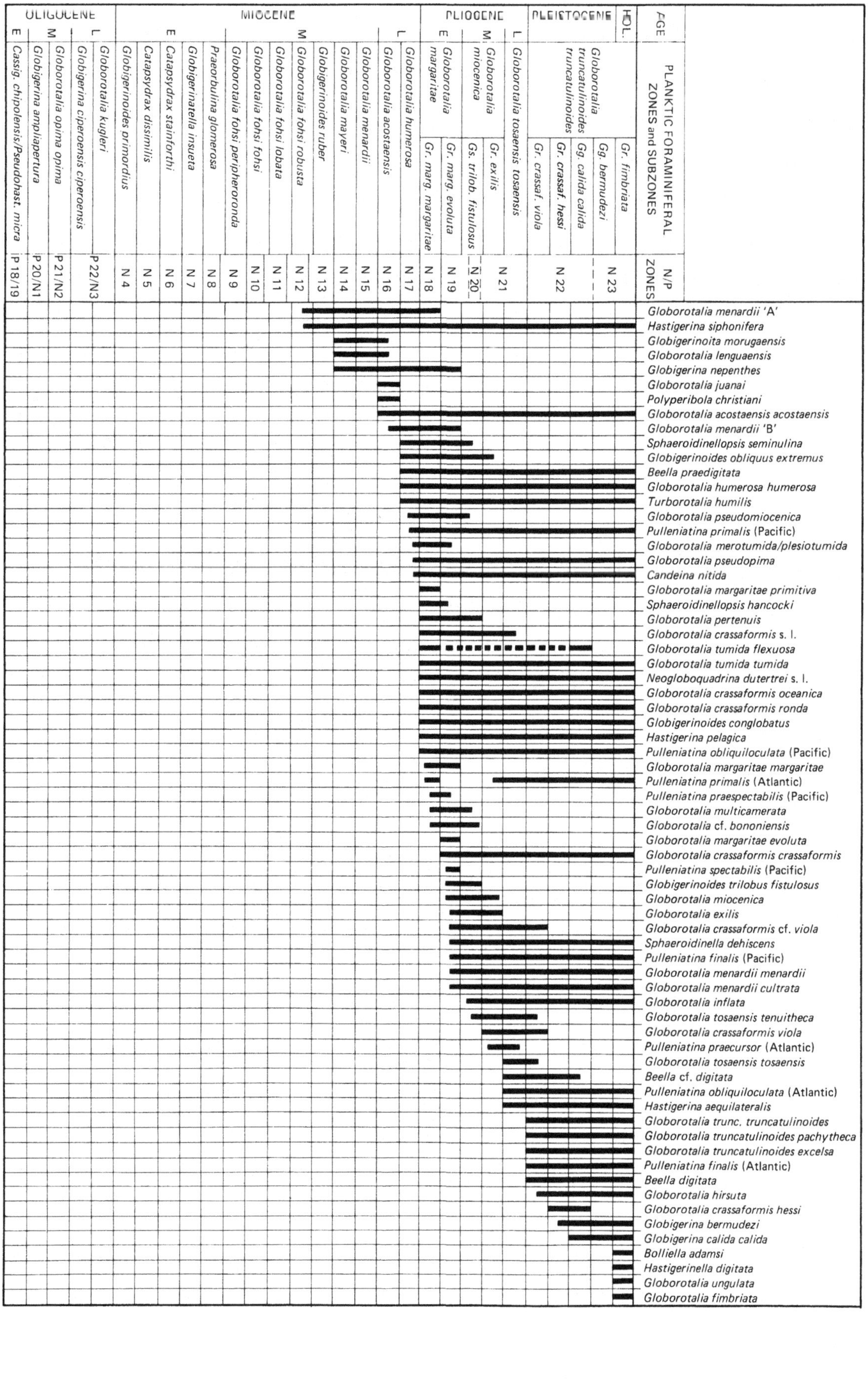
AGE
OLIGOCENE
MIOCENE
PLIOCENE
PLEISTOCENE
HOL.
E
M
L
PLANKTIC FORAMINIFERAL ZONES and SUBZONES
Cassig. chipolensis/Pseudohast. micra
Globigerina ampliapertura
Globorotalia opima opima
Globigerina ciperoensis ciperoensis
Globorotalia kugleri
Globigerinoides primordius
Catapsydrax dissimilis
Catapsydrax stainforthi
Globigerinatella insueta
Praeorbulina glomerosa
Globorotalia fohsi peripheroronda
Globorotalia fohsi fohsi
Globorotalia fohsi lobata
Globorotalia fohsi robusta
Globigerinoides ruber
Globorotalia mayeri
Globorotalia menardii
Globorotalia acostaensis
Globorotalia humerosa
Globorotalia margaritae
Gr. marg. margaritae
Gr. marg. evoluta
Globorotalia miocenica
Gs. trilob. fistulosus
Gr. exilis
Globorotalia tosaensis tosaensis
Globorotalia truncatulinoides truncatulinoides
Gr. crassaf. viola
Gr. crassaf. hessi
Gg. calida calida
Gg. bermudezi
Gr. fimbriata
N/P ZONES
P 18/19
P 20/N1
P 21/N2
P 22/N3
N 4
N 5
N 6
N 7
N 8
N 9
N 10
N 11
N 12
N 13
N 14
N 15
N 16
N 17
N 18
N 19
N 20
N 21
N 22
N 23
Globorotalia menardii 'A'
Hastigerina siphonifera
Globigerinoita morugaensis
Globorotalia lenguaensis
Globigerina nepenthes
Globorotalia juanai
Polyperibola christiani
Globorotalia acostaensis acostaensis
Globorotalia menardii 'B'
Sphaeroidinellopsis seminulina
Globigerinoides obliquus extremus
Beella praedigitata
Globorotalia humerosa humerosa
Turborotalia humilis
Globorotalia pseudomiocenica
Pulleniatina primalis (Pacific)
Globorotalia merotumida/plesiotumida
Globorotalia pseudopima
Candeina nitida
Globorotalia margaritae primitiva
Sphaeroidinellopsis hancocki
Globorotalia pertenuis
Globorotalia crassaformis s. l.
Globorotalia tumida flexuosa
Globorotalia tumida tumida
Neogloboquadrina dutertrei s. l.
Globorotalia crassaformis oceanica
Globorotalia crassaformis ronda
Globigerinoides conglobatus
Hastigerina pelagica
Pulleniatina obliquiloculata (Pacific)
Globorotalia margaritae margaritae
Pulleniatina primalis (Atlantic)
Pulleniatina praespectabilis (Pacific)
Globorotalia multicamerata
Globorotalia cf. bononiensis
Globorotalia margaritae evoluta
Globorotalia crassaformis crassaformis
Pulleniatina spectabilis (Pacific)
Globigerinoides trilobus fistulosus
Globorotalia miocenica
Globorotalia exilis
Globorotalia crassaformis cf. viola
Sphaeroidinella dehiscens
Pulleniatina finalis (Pacific)
Globorotalia menardii menardii
Globorotalia menardii cultrata
Globorotalia inflata
Globorotalia tosaensis tenuitheca
Globorotalia crassaformis viola
Pulleniatina praecursor (Atlantic)
Globorotalia tosaensis tosaensis
Beella cf. digitata
Pulleniatina obliquiloculata (Atlantic)
Hastigerina aequilateralis
Globorotalia trunc. truncatulinoides
Globorotalia truncatulinoides pachytheca
Globorotalia truncatulinoides excelsa
Pulleniatina finalis (Atlantic)
Beella digitata
Globorotalia hirsuta
Globorotalia crassaformis hessi
Globigerina bermudezi
Globigerina calida calida
Bolliella adamsi
Hastigerinella digitata
Globorotalia ungulata
Globorotalia fimbriata

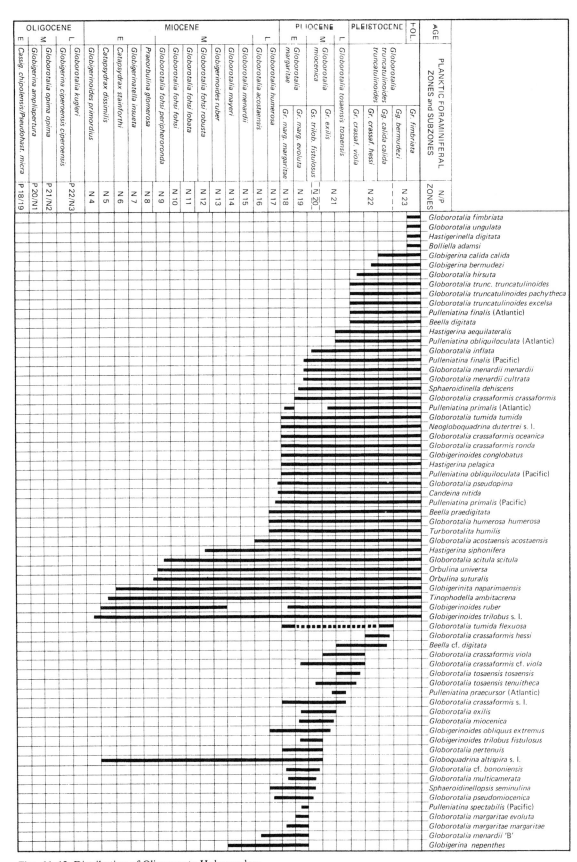

Figs. 11–12. Distribution of Oligocene to Holocene low
latitude planktic foraminifera (in order of last occurrence)

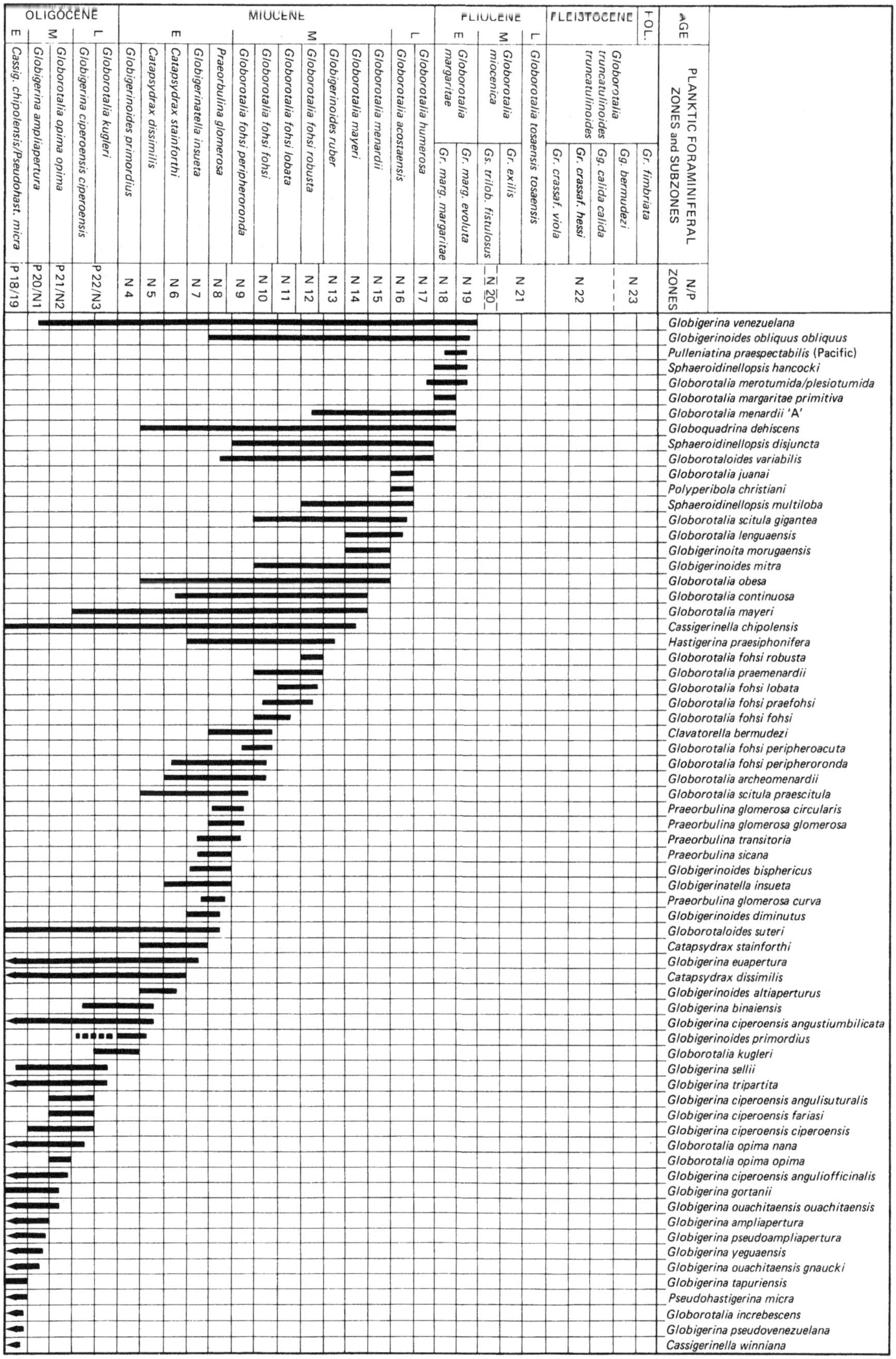
AGE
OLIGOCENE
MIOCENE
PLIOCENE
PLEISTOCENE
HOL.
E
M
L
PLANKTIC FORAMINIFERAL ZONES and SUBZONES
Cassig. chipolensis/Pseudohast. micra
Globigerina ampliapertura
Globorotalia opima opima
Globigerina ciperoensis ciperoensis
Globorotalia kugleri
Globigerinoides primordius
Catapsydrax dissimilis
Catapsydrax stainforthi
Globigerinatella insueta
Praeorbulina glomerosa
Globorotalia fohsi peripheroronda
Globorotalia fohsi fohsi
Globorotalia fohsi lobata
Globorotalia fohsi robusta
Globigerinoides ruber
Globorotalia mayeri
Globorotalia menardii
Globorotalia acostaensis
Globorotalia humerosa
Globorotalia margaritae
Gr. marg. margaritae
Gr. marg. evoluta
Globorotalia miocenica
Gs. trilob. fistulosus
Gr. exilis
Globorotalia tosaensis tosaensis
Globorotalia truncatulinoides truncatulinoides
Gr. crassaf. viola
Gr. crassaf. hessi
Gg. calida calida
Gg. bermudezi
Gr. fimbriata
N/P ZONES
P 18/19
P 20/N1
P 21/N2
P 22/N3
N 4
N 5
N 6
N 7
N 8
N 9
N 10
N 11
N 12
N 13
N 14
N 15
N 16
N 17
N 18
N 19
N 20
N 21
N 22
N 23
Globigerina venezuelana
Globigerinoides obliquus obliquus
Pulleniatina praespectabilis (Pacific)
Sphaeroidinellopsis hancocki
Globorotalia merotumida/plesiotumida
Globorotalia margaritae primitiva
Globorotalia menardii 'A'
Globoquadrina dehiscens
Sphaeroidinellopsis disjuncta
Globorotaloides variabilis
Globorotalia juanai
Polyperibola christiani
Sphaeroidinellopsis multiloba
Globorotalia scitula gigantea
Globorotalia lenguaensis
Globigerinoita morugaensis
Globigerinoides mitra
Globorotalia obesa
Globorotalia continuosa
Globorotalia mayeri
Cassigerinella chipolensis
Hastigerina praesiphonifera
Globorotalia fohsi robusta
Globorotalia praemenardii
Globorotalia fohsi lobata
Globorotalia fohsi praefohsi
Globorotalia fohsi fohsi
Clavatorella bermudezi
Globorotalia fohsi peripheroacuta
Globorotalia fohsi peripheroronda
Globorotalia archeomenardii
Globorotalia scitula praescitula
Praeorbulina glomerosa circularis
Praeorbulina glomerosa glomerosa
Praeorbulina transitoria
Praeorbulina sicana
Globigerinoides bisphericus
Globigerinatella insueta
Praeorbulina glomerosa curva
Globigerinoides diminutus
Globorotaloides suteri
Catapsydrax stainforthi
Globigerina euapertura
Catapsydrax dissimilis
Globigerinoides altiaperturus
Globigerina binaiensis
Globigerina ciperoensis angustiumbilicata
Globigerinoides primordius
Globorotalia kugleri
Globigerina sellii
Globigerina tripartita
Globigerina ciperoensis angulisuturalis
Globigerina ciperoensis fariasi
Globigerina ciperoensis ciperoensis
Globorotalia opima nana
Globorotalia opima opima
Globigerina ciperoensis anguliofficinalis
Globigerina gortanii
Globigerina ouachitaensis ouachitaensis
Globigerina ampliapertura
Globigerina pseudoampliapertura
Globigerina yeguaensis
Globigerina ouachitaensis gnaucki
Globigerina tapuriensis
Pseudohastigerina micra
Globorotalia increbescens
Globigerina pseudovenezuelana
Cassigerinella winniana

Caribbean region. The identification of the zone then rests largely on negative criteria, that is on the absence of *Globorotalia miocenica*, *G. exilis* and *G. truncatulinoides truncatulinoides*. In the Indo-Pacific realm, where *G. miocenica* and *G. exilis* are not developed or only atypically so, it may be difficult to separate the zone from the underlying *Globorotalia exilis* Subzone.

GLOBOROTALIA TRUNCATULINOIDES TRUNCATULINOIDES ZONE

Category: Taxon range zone
Age: Pleistocene–Holocene
Author: Bolli & Bermudez (1965), renamed by Bolli (1966a) and Bolli & Premoli Silva (1973)
Definition: Range of zonal marker.
Remarks: The zone was originally named *Globorotalia truncatulinoides/Globorotalia inflata* Zone. To simplify the terminology Bolli (1966a) changed it to *Globorotalia truncatulinoides* Zone. Bolli & Premoli Silva (1973) named it *Globorotalia truncatulinoides truncatulinoides* Zone. In line with the taxonomic treatment followed in the present paper, this is also adopted here. Bolli & Premoli Silva (1973) subdivided the zone into 4 subzones.

GLOBOROTALIA CRASSAFORMIS VIOLA SUBZONE

Category: Interval zone
Age: Pleistocene
Author: Bolli & Premoli Silva (1973)
Definition: Interval with subzonal marker, from first occurrence of *Globorotalia truncatulinoides truncatulinoides* to first occurrence of *G. crassaformis hessi*.
Remarks: *Globorotalia tosaensis tosaensis* and *G. tosaensis tenuitheca* disappear within the subzone and *G. crassaformis* cf. *viola* together with the zonal marker at the top.

GLOBOROTALIA CRASSAFORMIS HESSI SUBZONE

Category: Interval zone
Age: Pleistocene
Author: Bolli & Premoli Silva (1973)
Definition: Interval with subzonal marker, from its first occurrence to first occurrence of *Globigerina calida calida*.
Remarks: *Globorotalia crassaformis* cf. *viola* and *G. crassaformis viola* do not range into the subzone.

GLOBIGERINA CALIDA CALIDA SUBZONE

Category: Concurrent range zone
Age: Pleistocene
Author: Bolli & Premoli Silva (1973)
Definition: Interval with subzonal marker, from its first occurrence to extinction of *Globorotalia tumida flexuosa*.
Remarks: The zone is largely characterized by the presence of *Globorotalia tumida flexuosa* as has been observed in numerous sections from the low latitude Atlantic realm. However, this subspecies is apparently controlled by environment rather than time (in low latitude Pacific sections it is present for only a brief interval in the Early Pliocene, as is the case in well Bodjonegoro-1, Java). The *Globigerina calida calida* Subzone is therefore not applicable in the Pacific region.

GLOBIGERINA BERMUDEZI SUBZONE

Category: Interval zone
Age: Pleistocene
Author: Bolli & Premoli Silva (1973)
Definition: Interval with subzonal marker, from last occurrence of *Globorotalia tumida flexuosa* to first occurrence of *G. fimbriata*.
Remarks: Because of the restricted occurrence and lateral distribution of *Globorotalia tumida flexuosa*, which determines the base of the subzone, the recognition of this interval, like the earlier subzones, is restricted. It has been documented in the Caribbean, the Gulf of Mexico and in the adjoining low latitude Atlantic.

GLOBOROTALIA FIMBRIATA SUBZONE

Category: Taxon range zone
Age: Holocene
Author: Bolli & Premoli Silva (1973)
Definition: Range of subzonal marker.
Remarks: The subzone represents approximately the last 11 000 years (Holocene) and thus corresponds to the postglacial warming period. *Globorotalia ungulata* is another readily recognizable taxon that is restricted, or nearly so, to this subzone. Very rare *Hastigerinella digitata* and *Bolliella adamsi* may also be present in this subzone, both are almost unknown from earlier sediments.

Discussion of taxa

This chapter deals with the typical low latitude species with the following exceptions:

All species of the genera *Globigerinoides* and *Clavatorella*, including *G. bollii*, *G. bulloideus*, *G. tenellus* and *C. sturanii*, originally described from the Mediterranean area, are discussed and figured in the present chapter. Only *G. grilli,* characteristic for the Parathethys region, is treated in Chapter 9. *Globigerina calida calida,* a subzonal marker used in this chapter is also an index form in the Mediterranean area; taxonomic notes and illustrations are therefore found in Chapter 8. Though *Globigerina juvenilis* was originally described from low latitudes, it is considered to be of more value as an index species in southern mid-latitudes and therefore is illustrated and discussed in Chapter 7.

Oligocene *Globigerina* species

The planktic foraminiferal fauna of the Oligocene is comparatively monotonous compared with that below and above this interval. The characteristic Eocene forms like *Hantkenina, Globigerinatheka* and the carinate Globorotalias are no longer present and the diversified Neogene fauna with *Globigerinoides,* the younger carinate Globorotalias, *Globoquadrina, Orbulina* and other new genera has not yet developed. Instead, the interval is dominated by numerous taxa of the genus *Globigerina.* Some of them range from the underlying Eocene into or through the Oligocene, others are restricted to part of it or commence here to continue into the Miocene. Also present in the Oligocene are other less dominant taxa including a few non-carinate *Globorotalia* spp., typical for the Middle to Late Oligocene, and further representatives of the genera *Catapsydrax, Globorotaloides* and *Cassigerinella.* Typical representatives of the genus *Globoquadrina* appear towards the end of the Oligocene.

While planktic foraminiferal associations are largely identical in low to mid-latitudes throughout the Paleocene and Eocene, as between the Caribbean and Mediterranean, a differentiation commences in the Oligocene that continues to become gradually more accentuated through the Miocene and younger sediments. The Oligocene index species, with few exceptions originally described from low latitudes, are also still present and applicable in mid-latitudes. On the other hand, some taxa erected in mid-latitudes, like *G. gortanii,* are less typically developed in lower latitudes.

Within the dominant Oligocene *Globigerina* population occur a few short-ranging and morphologically distinct forms like *G. ampliapertura* and *G. ciperoensis ciperoensis* which proved to be valuable index fossils and were proposed as zonal markers by Bolli (1957). Later, Blow & Banner (1962) and Blow (1969) dealt in considerable detail with the Oligocene Globigerinas, erected numerous new taxa and used some of them to define their latest Eocene and Early Oligocene zones P 17–P 19 (*G. gortanii, G. tapuriensis* and *G. sellii*) with the first and last mentioned described from the Mediterranean.

The Oligocene Globigerinas form a morphologically closely related group of species many of which are of little stratigraphic significance. To provide an overview of the whole group, it is necessary to include here holotype figures and

ranges for the less significant taxa in order to draw the distinctions between these and the important ones.

The comparatively large number of Oligocene Globigerinas can be divided into the following groups of related taxa within which certain evolutionary trends are apparent, with their origin mostly going back into the Eocene:

(1) *Globigerina corpulenta–G. praeturritilina–G. gortanii*

(2) *Globigerina yeguaensis–G. pseudovenezuelana–G. euapertura–G. venezuelana*

(3) *Globigerina praebulloides occlusa–G. praebulloides praebulloides–G. praebulloides leroyi*

(4) *Globigerina tripartita–G. tapuriensis–G. sellii–G. binaiensis*

(5) *Globigerina ouachitaensis ouachitaensis–G. ouachitaensis gnaucki*

(6) *Globorotalia increbescens–Globigerina pseudoampliapertura–G. ampliapertura*

(7) *Globigerina ciperoensis angustiumbilicata–G. ciperoensis anguliofficinalis–G. ciperoensis ciperoensis–G. ciperoensis angulisuturalis–G. ciperoensis fariasi*

It is evident from the distribution chart that the Eocene/Oligocene boundary does not represent a break within these evolutionary lineages.

Taxonomic notes

(1) GLOBIGERINA CORPULENTA Subbotina–G. PRAETURRITILINA Blow & Banner–G. GORTANII (Borsetti)
Figures **13.17–19**; 8, 9, 12

Type references: *Globigerina corpulenta* Subbotina, 1953, p. 76, pl. 9, figs. 5a–c.
Globigerina praeturritilina Blow & Banner, 1962, p. 99, pl. 13, figs. A–C.
Catapsydrax gortanii Borsetti, 1959, p. 205, pl. 1, figs. 1a–d.

This group is characterized by its relatively high spired, compact arrangement of globular to ovoid chambers, increasing moderately in size. Typical for the group is also the frequent presence of a small, often flattened, delicate, bulla-like rudimentary chamberlet, partly or wholly covering the deep umbilical area. Such a small chamber is seen for example in the holotype figure of *G. gortanii* (Figs. 13.17b–c) and a broken off one in the holotype figure of *G. turritilina* (which according to its author Blow is a synonym of *G. gortanii*). The height of the spire increases from *G. corpulenta* through *G. praeturritilina* to *G. gortanii*.

G. corpulenta is here regarded as morphologically intermediate between *G. eocaena* and *G. praeturritilina.* It differs from *G. eocaena* in a higher spire and in the above-mentioned bulla-like small chambers which are almost always present. *G. praeturritilina* differs from *G. corpulenta* in a still higher spire.

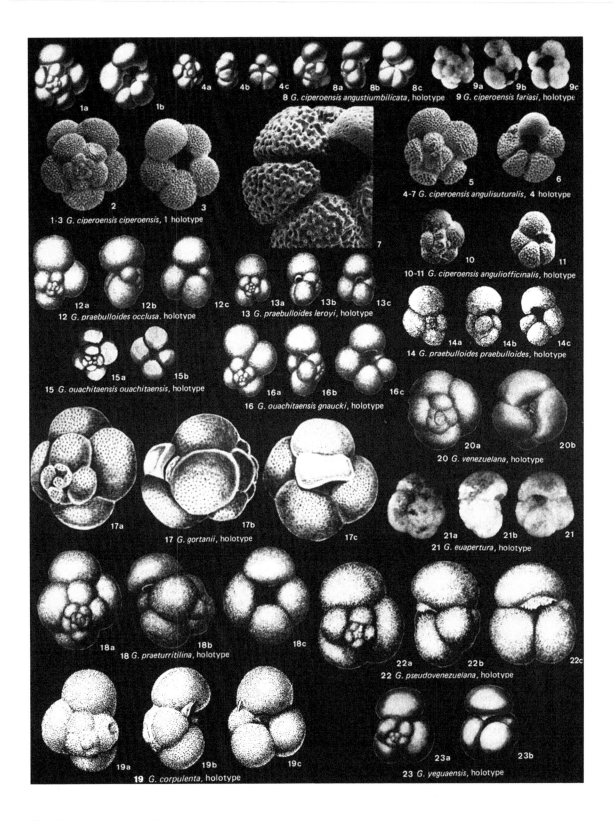

Fig. 13. For caption see p. 180.

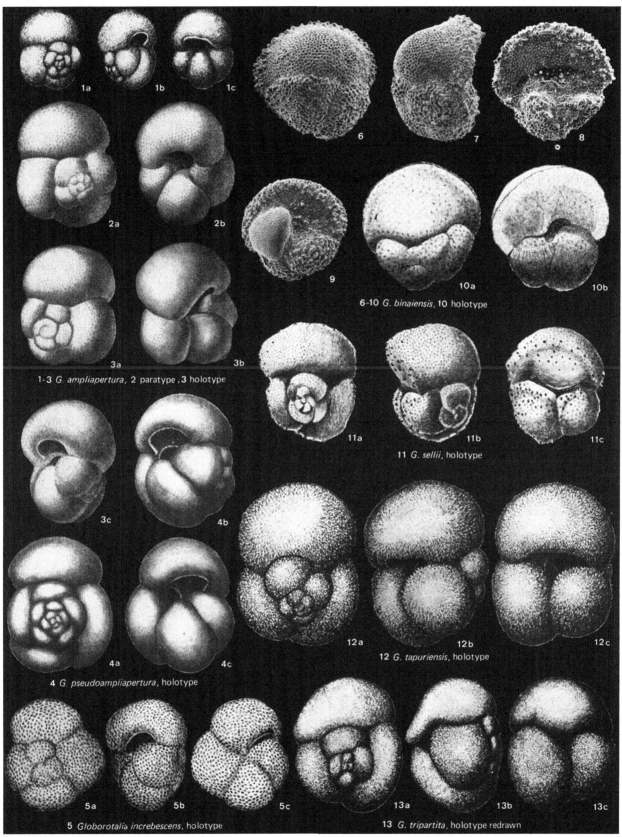

6-10 *G. binaiensis*, 10 holotype

1-3 *G. ampliapertura*, 2 paratype, 3 holotype

11 *G. sellii*, holotype

4 *G. pseudoampliapertura*, holotype

12 *G. tapuriensis*, holotype

5 *Globorotalia increbescens*, holotype

13 *G. tripartita*, holotype redrawn

Fig. 14. For caption see p. 180.

Here the aborted small end-chambers often occur in larger specimens. They are irregularly shaped but usually do not cover the umbilical area as in *G. gortanii*. *G. gortanii* is distinguished from *G. praeturritilina* in the spire being again higher, the test usually being larger and the bulla-like chamberlets when present covering most or all of the umbilicus. Because of this feature, *G. gortanii* was originally placed by Borsetti in the genus *Catapsydrax*.

Every transitional stage exists between *G. eocaena–G. corpulenta–G. praeturritilina* and *G. gortanii*.

(2) GLOBIGERINA YEGUAENSIS Weinzierl & Applin–G. PSEUDOVENEZUELANA Blow & Banner–G. EUAPERTURA Jenkins– G. VENEZUELANA Hedberg
Figures 13.20–23; 7, 8, 9, 12

Type references: *Globigerina yeguaensis* Weinzierl & Applin, 1929, p. 408, pl. 43, figs. 1a–b.

Globigerina pseudovenezuelana Blow & Banner, 1962, p. 100, pl. 11, figs. J–L.
Globigerina euapertura Jenkins, 1960, p. 351, pl. 1, figs. 8a–c.
Globigerina venezuelana Hedberg, 1937, p. 681, pl. 92, figs. 7a–b.

The group is characterized by $3\frac{1}{2}$ or more often 4 low trochospirally arranged and rather appressed chambers forming the last whorl. The apertures in this group are low arched and, in well preserved specimens, may show a triangular tooth-like lip which is more strongly developed in *G. pseudovenezuelana*. The specimens are usually quite large sized and are a typical and often dominant component within the monotonous Oligocene fauna.

G. yeguaensis and *G. galavisi* Bermudez, 1961, have nearly the same range and are morphologically very close. In our opinion differences fall within the species variability. *G. galavisi* is therefore here placed in synonymy with *G. yeguaen-*

Fig. 13. Oligocene *Globigerina* species (all figures × 60, except 2, 3, 5, 6 × 90; 7 × 220)

1–3 *G. ciperoensis ciperoensis* Bolli
1a–c Holotype. Sample Bo 273, *G. opima opima* Zone, Middle Oligocene, Cipero Fm., Cipero coast, Trinidad. **2–3** DSDP Leg 40-363-3-1, 58–60 cm, *G. ciperoensis ciperoensis* Zone, Late Oligocene, Walvis Ridge, Southeast Atlantic (from Toumarkine, 1978).

4–7 *G. ciperoensis angulisuturalis* Bolli
4a–c Holotype. Sample Bo 306A, *G. opima opima* Zone, Middle Oligocene, Cipero Fm., Cipero coast, Trinidad. **5–7** Same origin as 2–3, 7 detail of 6 (from Toumarkine, 1978).

8 *G. ciperoensis angustiumbilicata* Bolli
8a–c Holotype. Sample Bo 291A, *G. ciperoensis ciperoensis* Zone, Late Oligocene, Cipero Fm., Cipero coast, Trinidad.

9 *G. ciperoensis fariasi* Bermudez
9a–c Holotype. Sample JPB 209, Middle Oligocene, Tingnaro Fm., Matanzas Province, Cuba.

10–11 *G. ciperoensis anguliofficinalis* Blow
10–11 Holotype. Sample Cb. 1964, Zone P 19, Early Oligocene, Cipero Fm., San Fernando, Trinidad.

12 *G. praebulloides occlusa* Blow & Banner
12a–c Holotype. Sample FCRM 1922, Zone P 19, Early Oligocene, Lindi area, East Africa.

13 *G. praebulloides leroyi* Blow & Banner
13a–c Holotype. Sample FCRM 1965, Zone P 19, Early Oligocene, Lindi area, East Africa.

14 *G. praebulloides praebulloides* Blow
14a–c Holotype. Sample RM 19285, *G. insueta* Zone, Early Miocene, Pozon Fm., Eastern Falcon, Venezuela.

15 *G. ouachitaensis ouachitaensis* Howe & Wallace
15a–b Holotype. Late Eocene, Jackson Fm., Danville Landing, Ouachita River, Louisiana, USA.

16 *G. ouachitaensis gnaucki* Blow & Banner
16a–c Holotype. Sample FCRM 1965, Zone P 19, Early Oligocene, Lindi area, East Africa.

17 *G. gortanii* (Borsetti)
17a–c Holotype. Oligocene, Marne variegate di Vigoleno, Piacenza Province, Italy.

18 *G. praeturritilina* Blow & Banner
18a–c Holotype. Sample FCRM 1645, *G. semiinvoluta* Zone, Late Eocene, Lindi Area, East Africa.

19 *G. corpulenta* Subbotina
19a–c Holotype. *G. conglobatus* Zone, Late Eocene, Northern Caucasus, USSR.

20 *G. venezuelana* Hedberg
20a–b Holotype. Sample E-4032, Middle Tertiary, Carapita Fm., Anzoategui State, Venezuela.

21 *G. euapertura* Jenkins
21a–c Holotype. At 1188 feet in Lakes Entrance oil shaft, *G. dehiscens* Zone, Late Oligocene, Victoria, Australia.

22 *G. pseudovenezuelana* Blow & Banner
22a–c Holotype. FCRM 1923, *C. danvillensis* Zone, Late Eocene, Lindi area, East Africa.

23 *G. yeguaensis* Weinzierl & Applin
23a–b Holotype. Subsurface, Rio Bravo Oil Company Deussen B-1, Upper Claiborne, Liberty County, Texas, USA.

Fig. 14. Oligocene *Globigerina* species (all figures × 60)

1–3 *G. ampliapertura* Bolli
1 Sample FCRM 1965, Zone P 19, Early Oligocene, Lindi area, East Africa (from Blow & Banner, 1962). **2** Paratype, **3** Holotype, Sample Bo 314A, *G. ampliapertura* Zone, Middle Oligocene. Cipero Fm., Cipero coast, Trinidad (2 from Bolli, 1957).

4 *G. pseudoampliapertura* Blow & Banner
Holotype. Sample FCRM 1923, *C. danvillensis* Zone, Late Eocene, Lindi area, East Africa.

5 *G. increbescens* Bandy
Holotype. Locality 58, Zone B, Late Eocene, Jackson Fm., Little Stave Creek, Alabama, USA.

6–10 *G. binaiensis* Koch
6–9 Topotypes, 9 with rudimentary last chamber. **10** Holotype. All from Sample M. Mühlberg Binai-Atingdunok 491, Middle Tertiary, Lower *Globigerina* Marls, Bulongan, East Borneo.

11 *G. sellii* (Borsetti)
Holotype. Oligocene, 'Marne variegate' di Vigoleno, Piacenza Province, Italy.

12 *G. tapuriensis* Blow & Banner
Holotype. Sample FCRM 1964, Zone P 19, Early Oligocene, Lindi area, East Africa.

13 *G. tripartita* Koch
Holotype. Sample M. Mühlberg Binai-Atingdunok 491, Middle Tertiary, Lower *Globigerina* Marls, Bulongan, East Borneo.

sis, whose species description was emended by Blow & Banner (1962).

G. pseudovenezuelana differs from *G. yeguaensis* in the chambers being more appressed, which results in a less lobate and elongate equatorial outline. Due to these features *G. pseudovenezuelana* occupies a position intermediate between *G. yeguaensis* and *G. venezuelana*.

No consistent characters exist to distinguish *G. euapertura* from *G. prasaepis* Blow, 1969. Blow voiced the opinion that *G. euapertura* 'seems to be limited to colder water environments and is unrelated to either *G. prasaepis* or *G. ampliapertura*'. However, he gives no evidence to justify his view. The two forms are therefore here considered as synonyms with *G. euapertura* Jenkins, 1960, having priority over *G. prasaepis* Blow, 1969.

G. euapertura differs from *G. venezuelana* in being usually slightly smaller, more loosely coiled, in having a larger, low arched aperture and in not having the circular to subquadrate equatorial outline typical for *G. venezuelana*. According to its author, *G. euapertura* is intermediate in morphology between *G. venezuelana* and *G. ampliapertura*.

G. euapertura is distinguished from *G. ampliapertura* by having the final chamber tending to embrace more of the earlier test resulting in the test becoming more globular. The aperture in the final chamber in *G. euapertura* is a lower arch compared with the almost circular opening of *G. ampliapertura*.

G. venezuelana differs from *G. pseudovenezuelana* in having a smooth test, still less inflated chambers and, as a consequence, less strongly depressed sutures. Furthermore, *G. venezuelana* is less hispid around the umbilical area and possesses a less well developed tooth-like lip partly covering the aperture.

(3) GLOBIGERINA PRAEBULLOIDES OCCLUSA Blow & Banner – G. PRAEBULLOIDES PRAEBULLOIDES Blow – G. PRAEBULLOIDES LEROYI Blow & Banner
Figures 13.12–14; 8

Type references: *Globigerina praebulloides occlusa* Blow & Banner, 1962, p. 93, pl. 9, figs. U–W.
Globigerina praebulloides Blow, 1959, p. 180, pl. 8, figs. 47a–c.
Globigerina praebulloides leroyi Blow & Banner, 1962, p. 93, pl. 9, figs. R–T.

Typical for the group is the small size compared with most other accompanying *Globigerina* species. The distinctly elongate test shape in equatorial view is caused by the comparatively large ultimate chamber. The subtle criteria by which the subspecies are distinguished and the presence of intermediate forms renders the identification of these long-ranging and stratigraphically insignificant forms difficult.

Blow & Banner (1962) describe the differences of their three subspecies as follows: *G. praebulloides leroyi* 'is distinguished from *praebulloides praebulloides* principally by its smaller umbilicus, lower and more symmetrical aperture, slower rate of chamber enlargement and greater degree of embrace between its chambers'. '*G. praebulloides occlusa* differs from *praebulloides* in possessing a smaller, shallower umbilicus, a smaller lower aperture which lacks a lip or rim and possessing a slightly thicker, rougher and more coarsely perforate wall. *G. praebulloides occlusa* is distinguished from *praebulloides leroyi* in possessing a shallower umbilicus and an asymmetrical aperture which lacks a lip. The chambers of *praebulloides occlusa* are more embracing than those of *praebulloides praebulloides*, but are less tightly embracing than those of *G. praebulloides leroyi*.'

(4) GLOBIGERINA TRIPARTITA Koch– G. TAPURIENSIS Blow & Banner–G. SELLII (Borsetti)–G. BINAIENSIS Koch
Figures 14.6–13; 8, 9, 12

Type references: *Globigerina bulloides* var. *tripartita* Koch, 1926, p. 746, p. 737, figs. 21a–b.
Globigerina tripartita tapuriensis Blow & Banner, 1962, p. 97, pl. 10, figs. H–K.
Globoquadrina sellii Borsetti, 1959, p. 209, pl. 1, figs. 3a–d.
Globigerina binaiensis Koch, 1935, *nom. nov.* for *G. aspera* Koch, 1926, p. 746, figs. 22a–c, 23a–c.

These large forms are characterized by the low trochospiral arrangement of the chambers increasing rapidly in size, which results in only three of them forming the last whorl. Except for *G. tapuriensis*, chambers are tangentially appressed, which results in a compact test shape, a somewhat closed umbilical area and low arched aperture. Due to the more globular shape of its chambers, *G. tapuriensis* is morphologically also close to *G. yeguaensis*. *G. sellii* (of which *G. oligocenica* Blow is a synonym) has a high, flattened, rather smooth apertural face.

The species description of *G. tripartita* was emended by Blow & Banner (1962). Compared with *G. sellii* the apertural face of *G. binaiensis* (*nomen novum* for *G. aspera* Koch) is still more distinctly flattened and smooth, set at about 90° and with an abrupt angle to the remaining surface of the final chamber.

The distinguishing characters of *G. tapuriensis* and *G. tripartita* are: *G. tapuriensis* has a more inflated last chamber making up about half of the total test size, its umbilicus is more open and the aperture a distinct low arch. In *G. tripartita* the last chamber is more appressed like a cap and makes up less than half of the total test, which results in a somewhat triangular aspect in equatorial view. The umbilicus is nearly closed and the aperture a very low slit.

(5) GLOBIGERINA OUACHITAENSIS OUACHITAENSIS Howe & Wallace– G. OUACHITAENSIS GNAUCKI Blow & Banner
Figures **13.15–16**; 8, 9, 12

Type references: *Globigerina ouachitaensis* Howe & Wallace, 1932, p. 74, pl. 10, figs. 7a–c.
Globigerina ouachitaensis gnaucki Blow & Banner, 1962, p. 91, pl. 9, figs. L–N.

The *G. ouachitaensis* subspecies are characterized by a last whorl composed of 4 globular chambers and a wide umbilicus of quadrate outline. The subspecies *gnaucki* differs from the subspecies *ouachitaensis* in a slightly more rapid growth rate of the chambers, resulting in a less distinct quadrate-shaped umbilicus.

The *G. ciperoensis* subspecies differ from the *G. ouachitaensis* subspecies in possessing 5 chambers in the last whorl instead of 4, in a lower spire (except for the subspecies *fariasi*), and in the umbilical area possessing a pentagonal outline instead of a quadrate one. Problems may arise in distinguishing *G. ouachitaensis* s.l. from *G. praebulloides* s.l., especially when intermediate forms are present. The *praebulloides* subspecies differ from those of *ouachitaensis* in the more rapidly increasing size in particular of the ultimate chamber, resulting in a distinctly elongated equatorial outline instead of a more quadrate one in *ouachitaensis*. Furthermore, the umbilicus in *praebulloides* does not possess the quadrate outline that is typical for *ouachitaensis*.

(6) GLOBOROTALIA INCREBESCENS Bandy– GLOBIGERINA PSEUDOAMPLIAPERTURA Blow & Banner– GLOBIGERINA AMPLIAPERTURA Bolli
Figures **14.1–5**; 8, 9, 12

Type references: *Globigerina increbescens* Bandy, 1949, p. 120, pl. 23, figs. 3a–c.
Globigerina pseudoampliapertura Blow & Banner, 1962, p. 95, pl. 12, figs. A–C.
Globigerina ampliapertura Bolli, 1957, p. 108, pl. 22, figs. 6a–c.

Globigerina ampliapertura has been established as a zonal marker in the Oligocene. Its correct identification and consequent application to biostratigraphy have been hampered by forms intermediate to *Globorotalia increbescens*, and by forms close in overall morphology to *Globigerina pseudoampliapertura*. Blow (1969) claims two evolutionary lineages based on different wall structures into which these morphologically similar taxa should fall:

(1) *Globorotalia increbescens–Globigerina ampliapertura– Globigerina prasaepis*
(2) *Globorotalia centralis–Globigerina pseudoampliapertura*
    The evolutionary trend for the aperture to move from

an umbilical–extraumbilical position in *increbescens* to an umbilical one in *G. ampliapertura* is an unusual one. The claim by Blow that *G. pseudoampliapertura* differs from *G. ampliapertura* in a smoother, non-granular, finely perforate wall is difficult to see in the often poorly preserved Oligocene specimens. An evolutionary lineage from *Globorotalia centralis* to *Globigerina pseudoampliapertura* has not been clearly proved. *G. centralis*, a synonym of *G. cerroazulensis*, develops into the carinate end form *G. cerroazulensis cunialensis* in the Late Eocene, so any development towards *G. pseudoampliapertura* or *G. ampliapertura* would have to branch off at an earlier stage.

G. pseudoampliapertura differs from *G. ampliapertura* in its laterally more appressed, reniform chambers, which in *G. ampliapertura* are more inflated. Furthermore, the test of *G. pseudoampliapertura* appears higher and more trochospiral in peripheral view, and the aperture of the last chamber is less regularly rounded.

(7) THE GLOBIGERINA CIPEROENSIS Bolli group
Figures **13.1–11**; 8, 9, 12

Type references: *Globigerina ciperoensis angustiumbilicata* Bolli, 1957, p. 109, pl. 22, figs. 12a–c.
Globigerina anguliofficinalis Blow, 1969, p. 379, pl. 11, figs. 1–2.
Globigerina ciperoensis ciperoensis Bolli, 1957, p. 109, pl. 22, figs. 10a–b.
Globigerina ciperoensis angulisuturalis Bolli, 1957, p. 109, pl. 22, figs. 11a–c.
Globigerina fariasi Bermudez, 1961, p. 1181, pl. 3, figs. 5a–c.

Bolli (1957) distinguished the three subspecies *ciperoensis*, *angustiumbilicata* and *angulisuturalis*. Considerable variability within each taxon exists. Some subsequent authors preferred to give them species rank, or placed them as subspecies into different species (*G. ouachitaensis ciperoensis*, Blow, 1969).

G. ciperoensis anguliofficinalis is an additional closely related taxon erected by Blow (1969), while *G. ciperoensis fariasi* was erected by Bermudez (1961).

All these taxa are characterized by a comparatively small size (about 0.3 mm), usually a low trochospiral test, and (with the exception of *G. ciperoensis angustiumbilicata*, which has 4–5) 5 chambers in the last whorl which increase only moderately in size as added. Of most significance for the distinction of the subspecies are the intercameral sutures between the chambers of the last whorl, the size of the umbilical opening and the height of the spire. All *ciperoensis* subspecies are characterized by a large umbilicus, except for the long-ranging *G. ciperoensis angustiumbilicata* where it is very narrow, almost closed.

G. ciperoensis angulisuturalis, often slightly smaller than *G. ciperoensis ciperoensis*, is characterized by intercameral sutures developed in the last whorl as narrow, U-shaped bands of less coarsely perforated wall (Figs. 13.5–7).

G. ciperoensis anguliofficinalis has almost identical features though the U-shaped bands are somewhat narrower and less distinct; furthermore the umbilicus is slightly narrower. This species is regarded by Blow as the ancestral form of *G. ciperoensis angulisuturalis*. The distinction of *G. ciperoensis fariasi* from *G. ciperoensis ciperoensis* lies in the higher trochospiral arrangement of the chambers in *fariasi*.

The ranges suggest that for stratigraphic work, the original three subspecies *angustiumbilicata*, *ciperoensis* and *angulisuturalis* remain the most useful taxa.

The genus *Globoquadrina*

The separation of *Globoquadrina* from *Globigerina* and *Neogloboquadrina* is somewhat subjective. *Globoquadrina* Finlay, 1947, was erected on *Globorotalia dehiscens* Chapman & Collins and has an open umbilicus, angular ventrally pointed chambers and a generally compact test. It was said to have a 'pronounced thin flap projecting inwards and downwards overhanging the aperture and largely concealing it'. Parker's emendation (1967) broadened the concept to include *Globorotaloides* and forms which were later separated as *Neogloboquadrina* Bandy, Frerichs & Vincent, 1967.

Among Oligocene species of *Globigerina* there are forms that could be included in *Globoquadrina* by reason of test shape. Specimens of some of them may even show apertural flaps in the umbilicus. Workers are divided in their use of the two generic names and certainly the distinction is not always clear.

We maintain only two species names in low latitudes: *G. dehiscens* and *G. altispira* with its subspecies. The main stratigraphic importance of this group is its rather sudden appearance near the base of the Miocene and its disappearance within the Middle Pliocene. Though the genus becomes important from near the base of the Miocene, it is also known from the Oligocene. Blow (1969) recorded *G. baroemoensis* (LeRoy), a form with wide umbilicus and rather compressed chambers, as low as P 21 Zone and possible even P 20 Zone. We have recently found quite high spired forms of *G. altispira* s.l. as low as the *G. opima opima* Zone in Barbados.

The status of the three species erected by Maiya, Saito & Sato (1976) from the Late Pliocene or Pleistocene is considered doubtful. These are *G. asanoensis*, *G. himiensis* and *G. kagaensis*. An examination of the types for most of the other species and subspecies that have been erected leads us to consider them as not within the genus or as probable synonyms of either the *dehiscens* group or the *altispira* group. Rather fuller synonymy lists may be found in Stainforth *et al.* (1975, pp. 245 and 266).

Taxonomic notes

GLOBOQUADRINA DEHISCENS (Chapman, Parr & Collins)
Figures 15.4–7; 6, 9, 12

Type reference: *Globorotalia dehiscens* Chapman, Parr & Collins, 1934, p. 569, pl. 11, fig. 6.

Synonyms: *Globorotalia quadraria* Cushman & Ellisor, 1939, p. 11, pl. 2, fig. 5.
Globoquadrina subdehiscens Finlay, 1947, p. 291.

The distinctive quadrate, compressed test with its rather angular chambers and flat spiral side distinguishes this species group from *G. altispira* s.l. Some less typical forms approach species of *Globigerina* found in the Oligocene but the latter are not so characteristically 4-chambered in the last whorl.

GLOBOQUADRINA ALTISPIRA (Cushman & Jarvis)
Figures 15.1–3; 6, 9, 12

GLOBOQUADRINA ALTISPIRA ALTISPIRA (Cushman & Jarvis)
Figures 15.1; 6

Type references: *Globigerina altispira* Cushman & Jarvis, 1936, p. 5, pl. 1, figs. 13, 14.
Globoquadrina altispira altispira Bolli, 1957, p. 111.

GLOBOQUADRINA ALTISPIRA CONICA Brönnimann & Resig
Figures 15.2; 8, 6

Type reference: *Globoquadrina altispira conica* Brönnimann & Resig, 1971, p. 1275, fig. 1.

GLOBOQUADRINA ALTISPIRA GLOBOSA Bolli
Figures 15.3; 6

Type reference: *Globoquadrina altispira globosa* Bolli, 1957, p. 111, pl. 24, figs. 9, 10.

The subspecies of *G. altispira* differ from *G. dehiscens* in having a less angular test without a flattened spiral side. The chambers are more globose with usually more than 4 in the last whorl. Well preserved specimens usually show prominent apertural flaps extending into the umbilicus.

The morphological range of the species has been used to erect the three subspecies. *G. altispira globosa* has a low spire with initially globular chambers becoming slightly laterally compressed. The central form, *G. altispira altispira*, has a higher spire with more laterally compressed chambers. High spired forms with drawn out, laterally compressed later chambers originally included by Bolli (1957) in *G. altispira altispira* (Fig. 15.2) were separated as *G. altispira conica* by Brönnimann & Resig (1971). We find all three subspecies to be long ranged whereas Brönnimann & Resig recorded their

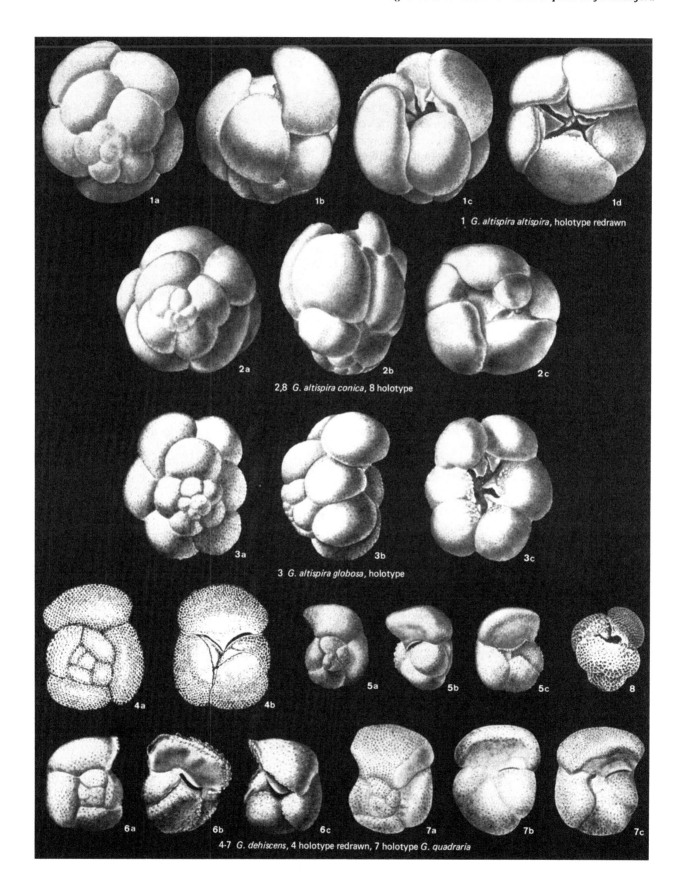

1 *G. altispira altispira*, holotype redrawn

2,8 *G. altispira conica*, 8 holotype

3 *G. altispira globosa*, holotype

4-7 *G. dehiscens*, 4 holotype redrawn, 7 holotype *G. quadraria*

new subspecies over only a short time interval from zones N 14 to base N 16.

The *G. altispira* group became extinct before the end of the Middle Pliocene and this is important stratigraphically.

The genus *Cassigerinella*

Two *Cassigerinella* species of stratigraphic significance are distinguished: an Early Oligocene to Middle Miocene *C. chipolensis* (Cushman & Ponton), of which *C. boudecensis* Pokorny and *C. globolocula* Ivanov are here regarded as junior synonyms, and a Late Eocene *C. winniana* (Howe) of which *C. eocaenica* Cordey is most likely a synonym. Because of their small size the two species are easily overlooked, the reason probably for their rare appearance in the literature.

The overlap in the Early Oligocene of *C. chipolensis* with *Hastigerina micra* was used by Blow & Banner (1962) as one of the criteria to define their *Globigerina oligocaenica* Zone. Bolli (1966a) termed this same stratigraphic interval *Cassigerinella chipolensis/Hastigerina micra* Zone, and Blow (1969) *Globigerina sellii/Pseudohastigerina barbadoensis* Zone (= P 19). The overlap of the ranges of *C. chipolensis* and *Pseudohastigerina micra* still appears to provide the best indicator for the Early Oligocene time span.

A basal Oligocene first appearance of *C. chipolensis* is generally agreed on, whereas data on its extinction level vary slightly. Bolli (1957) places this at the top of the *Globorotalia fohsi robusta* Zone in Trinidad, Blow (1969) very slightly higher, within his Zone N 13. The species was found in the Bodjonegoro-1 section of Java to extend nearly to the top of the *Globorotalia mayeri* Zone (Bolli, 1966b).

Taxonomic notes

CASSIGERINELLA CHIPOLENSIS (Cushman & Ponton)
Figures **16.1–2**; 7, 9, 12

Type reference: *Cassidulina chipolensis* Cushman & Ponton, 1932, p. 98, pl. 15, figs. 2a–c.
Synonyms: *Cassigerinella boudecensis* Pokorny, 1955, p. 136, text-figs. 1–3.
  Cassigerinella globolocula Ivanova, 1958, p. 57, pl. 11, figs. 1–3.

A characteristic difference between *C. chipolensis* and *C. winniana* lies in the chamber arrangement of the final whorl. In *chipolensis* it is alternating throughout, in *winniana* it changes from initially planispiral to alternating as clearly seen in the better drawn holotype figures of *C. eocaenica* (Figs. 16.5–6), considered a synonym of *C. winniana*. In addition, chambers in *chipolensis* are more inflated in the final whorl resulting in a more rounded periphery compared with *winniana*. The two species are also distinct in their sizes, *chipolensis* specimens measuring 0.13–0.19 mm against 0.10–0.12 mm in the smaller *winniana* (Cordey, 1968).

CASSIGERINELLA WINNIANA (Howe)
Figures **16.3–4**; 7, 9, 12

Type reference: *Cassidulina winniana* Howe, 1939, p. 82, pl. 11, figs. 7–8.
Synonym: *Cassigerinella eocaenica* Cordey, 1968, p. 368, text-figs. 1a–4.

For comparison of this species with *C. chipolensis* see above. Records of the species are too poor to provide reliable information on its total range. From data available it seems largely restricted to the Late Eocene, though the holotype was described from a Claiborne sample (i.e. probable late Middle Eocene), and Blow (1979) extends the range into the Early Oligocene (Zone P 19) and suggests that there *C. winniana* grades into *chipolensis*.

The genera *Catapsydrax*, *Globigerinita*, *Tinophodella* and *Turborotalita*

There is considerable variation in the way different workers use these designations, but it is felt that all four names can be used according to the following approach enabling *Catapsydrax*, particularly, to retain its stratigraphic value:

(1) A *Globigerina* that develops one bulla over the umbilicus. This is usually rather symmetrically rectangular but it may extend slightly along one or more sutures. Apertures round the bulla are found over the sutures or coalescing to form an open end to the bulla. They vary from 1 to 4 or 5 depending on the number of chambers in the last whorl..........................CATAPSYDRAX.

Fig. 15. *Globoquadrina* (*all figures* × 60)

1 *G. altispira altispira* Cushman & Jarvis
  1a–c Holotype redrawn (from Bolli, Loeblich & Tappan, 1957). Milestone No. 71. East of Port Antonio, published as Miocene, Jamaica.

2, 8 *G. altispira conica* Brönnimann & Resig
  2a–c From *G. fohsi robusta* Zone, Middle Miocene, Cipero Fm., Trinidad (from Bolli, 1957). **8** Holotype. DSDP Leg 7-62.1-30-1, 15–17 cm, Zone N 15, Middle Miocene, Eauripik Ridge, western equatorial Pacific.

  3 *G. altispira globosa* Bolli
  3a–c Holotype. Sample Bo 267, *C. dissimilis* Zone, Early Miocene, Cipero Fm., San Fernando area, Trinidad.

4–7 *G. dehiscens* (Chapman, Parr & Collins)
  4a–b Holotype of *G. subdehiscens* Finlay, 1947 (from Jenkins, 1971). Sample 5730, Altonian Stage (Early Miocene), Pakaurangi Point, New Zealand. **5a–c** Near topotype, Balcombian (Miocene) Balcombe Bay, Victoria, Australia (from Bolli *et al.*, 1957). **6a–c** *G. fohsi lobata* Zone, Cipero Fm., Middle Miocene, Trinidad (from Bolli, 1957). **7a–c** *G. quadraria* Cushman & Ellisor, 1939, Holotype, Core 10411-21 feet Humble Oil & Refining Co. well Ellender No. 1, Miocene, Terrebone Parish, Louisiana, USA.

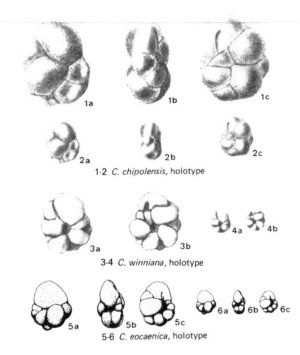

Fig. 16. *Cassigerinella* (figures 1, 3 and 5 × 150; 2, 4 and 6 × 60)

1–2 *C. chipolensis* (Cushman & Ponton)
    1a–c, 2a–c Holotype. Locality 19, Early Miocene, Chipola Fm., Calhoun County, Florida, USA.
3–4 *C. winniana* (Howe)
    3a–b, 4a–b Holotype. Eocene, Claiborne, Cook Mountain Fm., north of Dodson, Louisiana, USA.
5–6 *C. eocaenica* Cordey
    5a–c, 6a–c Holotype. Sample J.6B, Eocene, core hole on Blake Plateau, off Florida, USA.

(2) A *Globigerina* that develops one bulla that covers the umbilicus and extends in a sinuous manner along the sutures on the umbilical side and sometimes slightly onto the spiral side. Supplementary apertures are numerous around the edge of the bulla, of approximate equal size and not restricted to a position over the sutures between chambers of the last whorl . . . . TINOPHODELLA.

(3) A *Globigerina* where the last chamber extends back over the umbilicus covering the earlier *Globigerina* aperture and having a series of openings restricted to positions over the sutures . . . . GLOBIGERINITA.

(4) A *Globorotalia* where the last chamber extends across the umbilicus splaying out into short finger-like extensions along the sutures, each having an aperture at its end. The aperture for penultimate and earlier chambers is umbilical–extraumbilical . . . . . TURBOROTALITA.

The separation of the genera depends on the interpretation of what is a bulla and what is an extended chamber. The central problem is whether Brönnimann's holotype of *Globi-*

gerinita naparimaensis (genotype for *Globigerinita*) is congeneric with *Catapsydrax*, which would make the latter a junior synonym. Re-examination of the holotype (here given as Fig. 17.7) leads us to believe that Loeblich & Tappan (1957) were right in emending *Globigerinita* to include only forms where a modified last chamber extends across the umbilicus, as represented by Brönnimann's holotype. They erected a new genus *Tinophodella* removing Brönnimann's paratypes of *G. naparimaensis* to their species *T. ambitacrena*. Here, an irregular shaped, many apertured bulla is quite separate from the chambers of the last whorl.

The ranges of the various species of the bullate forms are given in Fig. 7. *Catapsydrax dissimilis* begins in the Eocene and both it and the much shorter-ranged *C. stainforthi* are most important for their last occurrences in the Early Miocene. The other genera and species are long-ranging and of small stratigraphic value.

Taxonomic notes

CATAPSYDRAX DISSIMILIS (Cushman & Bermudez)
Figures **17.1–4**; 7, 9, 12

Type reference: *Globigerina dissimilis* Cushman & Bermudez, 1937, p. 25, pl. 3, figs. 4–6.
Synonyms: *Catapsydrax unicavus* Bolli, Loeblich & Tappan, 1957, p. 37, pl. 7, figs. 9a–c.
    Globigerinita dissimilis ciperoensis Blow & Banner, 1962, p. 107, pl. 14, figs. A–C.

A valuable marker species with 4 chambers in the last whorl and a well developed bulla covering the umbilicus. The test is characteristically coarse-pored though the bulla is usually finer textured. The specimen chosen as holotype by Cushman & Bermudez has a less well developed bulla than is seen in the paratypes and in much additional material (for example, Fig. 17.2). Apertures are located at the ends of the bulla and may vary from one to four.

Various names have been given depending on the number of apertures. For example, *C. unicavus* Bolli, Loeblich & Tappan for one aperture (Fig. 17.4), *C. dissimilis dissimilis* for 2 apertures and *C. dissimilis ciperoensis* (Blow & Banner) for 3 or 4 apertures. However, the distribution of the various forms seems to be random throughout a similar total range and therefore we do not document them in the present work.

Caution is necessary, particularly in poorly preserved material, not to confuse *C. dissimilis* with some species of *Globigerina*, for example *G. venezuela*, which have a tendency to produce an aberrant small last chamber that can occupy a position partially covering the umbilicus. The more globular appearance of such a chamber is usually an indication of its origin. Also, a true bulla often exhibits a slightly different wall texture to that seen in the chambers of the test.

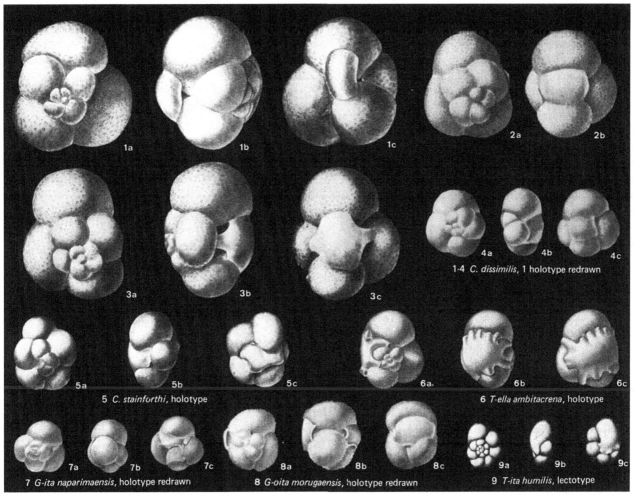

1-4 *C. dissimilis*, 1 holotype redrawn

5 *C. stainforthi*, holotype

6 *T-ella ambitacrena*, holotype

7 *G-ita naparimaensis*, holotype redrawn

8 *G-oita morugaensis*, holotype redrawn

9 *T-ita humilis*, lectotype

Fig. 17. *Catapsydrax, Globigerinita, Tinophodella, Turborotalita, Globigerinoita* (all figures × 60)

1-4 *Catapsydrax dissimilis* (Cushman & Bermudez)
    1a-c Holotype redrawn (from Bolli *et al.*, 1957). Eocene, 1 km north of Arroya Arenas, near Jaimanitas, Havana Province, Cuba. **2a-b** from Cipero Fm. Oligo–Miocene, Trinidad. **3a-c** from Cipero Fm., Oligo–Miocene, Trinidad. (2–3 from Bolli *et al.*, 1957.) **4a-c** Holotype of *C. unicavus* Bolli, Loeblich & Tappan, 1957, from *G. ciperoensis ciperoensis* Zone, Oligocene, Cipero Fm., San Fernando area, Trinidad.

    5 *Catapsydrax stainforthi* Bolli, Loeblich & Tappan
    5a-c Holotype. *Catapsydrax stainforthi* Zone, Miocene, Cipero Fm., Cipero coast, Trinidad.

6 *Tinophodella ambitacrena* Loeblich & Tappan
    6a-c Holotype. Albatross Station D2763, Recent, off east coast of Brazil.

7 *Globigerinita naparimaensis* Brönnimann
    7a-c Holotype redrawn (from Bolli *et al.*, 1957). *G. menardii* Zone, Middle Miocene, Lengua Fm., Trinidad.

8 *Globigerinoita morugaensis* Brönnimann
    8a-c Holotype redrawn (from Bolli *et al.*, 1957). *G. menardii* Zone, Middle Miocene, Lengua Fm., Trinidad.

9 *Turborotalita humilis* (Brady)
    9a-c Lectotype (from Banner & Blow, 1960a). *Challenger* Station 5, Recent, southwest of Canary Islands.

One small form with a semi-translucent, glassy wall and a small, flat bulla has been described as *C. parvulus* Bolli, Loeblich & Tappan (1957). It occurs as high as the *Globorotalia menardii* Zone in the Middle Miocene but its lower limit is uncertain as is its taxonomic status.

CATAPSYDRAX STAINFORTHI Bolli, Loeblich & Tappan
Figures **17.5**; 7, 9, 12

Type reference: *Catapsydrax stainforthi* Bolli, Loeblich & Tappan, 1957, p. 37, pl. 7, fig. 11.

A short-ranged species smaller than *C. dissimilis* and having 4

to 5 less inflated chambers in the last whorl giving a flatter appearance to the test. The bulla is flat and inclined to extend along the sutures. Up to 5 small apertures are confined to positions above the sutures, the maximum number being controlled by the number of chambers in the last whorl. The test does not have the coarse pores seen in *C. dissimilis*.

C. stainforthi praestainforthi (Blow) has a stated range beginning in the *Globigerinoides primordius* Zone which, if proved, would extend the range of the species downwards.

The relationship of *C. incrusta* (Akers) is not certain; both Blow (1969) and Stainforth *et al.* (1975) consider it as distinct from *C. stainforthi*.

GLOBIGERINITA NAPARIMAENSIS
Brönnimann
Figures **17.7**; 7, 9, 12

Type reference: *Globigerinita naparimaensis* Brönnimann, 1951a (holotype only), p. 18, figs. 1, 2.

The original illustration of the holotype and even Bolli, Loeblich & Tappan's redrawing are somewhat misleading because the last chamber, which wraps around from the left of the specimen (Fig. 17.7c) and crosses the umbilicus is so thin as to be almost invisible. Its 2 apertures over the earlier sutures are visible as is also an early *Globigerina* aperture opening into the umbilicus.

So far, this morphological type has not been found frequently and is of little stratigraphic value.

TINOPHODELLA AMBITACRENA Loeblich & Tappan
Figures **17.6**; 7, 9, 12

Type reference: *Tinophodella ambitacrena* Loeblich & Tappan, 1957, p. 114, fig. 3.
Synonym: *Globigerinita naparimaensis* Brönnimann, 1951a (figured paratypes only), p. 16, figs. 3–14.

Brönnimann's figured paratypes of *G. naparimaensis* are conspecific with *T. ambitacrena* and are now considered to be both specifically and generically different from the specimen he chose as the holotype.

    T. ambitacrena is a small species, usually with 3 chambers in the last whorl and with an irregular bulla covering a considerable percentage of the umbilical side and often extending onto the spiral side. Large numbers, up to 14, of small apertures rim the umbilicus.

    There can be some confusion with juvenile forms of *Globigerinatella insueta* but this is a more spherical form which may have more than one bulla and, even in young forms, usually shows some characteristic, circular areal apertures.

TURBOROTALITA HUMILIS (Brady)
Figures **17.9**; 7, 9, 12

Type reference: *Truncatulina humilis* Brady, 1884, p. 665, pl. 94, fig. 7.
Lectotype: *Globigerinita? humilis* (Brady), Banner & Blow, 1960a, p. 36, pl. 8, fig. 1.

The original illustration by Brady is not repeated here as the nature of the umbilical side was completely missed in this very small species. Many conspecific specimens were available in Brady's slides and Banner & Blow (1960a) chose a lectotype which they later (Blow & Banner, 1962, p. 122) placed in a new genus, *Turborotalita*.

    As in *Globigerinita naparimaensis*, the last chamber in this species is modified to form an extensive bulla that wraps over the umbilicus and ends in a series of finger-like extensions over the sutures. Apertures are usually separate at the end of each extension though they might combine in some instances. Earlier apertures are thought to be umbilical–extraumbilical. *Globigerinita parkerae* Loeblich & Tappan may be a junior synonym of *T. humilis*.

The genus *Globigerinoita*
Taxonomic notes

GLOBIGERINOITA MORUGAENSIS Brönnimann
Figures **17.8**; 7, 9, 12

Type reference: *Globigerinoita morugaensis* Brönnimann, 1952, p. 26, figs. 1a–c.

This small form has a *Globigerinoides* basic plan with 3 chambers in the last whorl and with one or two bullae developed over the apertures in the adult stage.

    The genus *Globigerinoita* remains monotypic and has not proved to be of much stratigraphic value.

The genus *Polyperibola*

    The monotypic genus *Polyperibola* is one of several in the Tertiary that appears suddenly as an already highly specialized form and becomes extinct again after only a short period. Like the well established *Globigerinatella insueta*, *P. christiani* could have considerable stratigraphic value. So far it is known only from within the *Globorotalia acostaensis* Zone. *Candeina nitida*, a form possibly related to *P. christiani*, first appears only at the base of the *Globorotalia margaritae* Zone, possibly in the uppermost part of the *Globorotalia humerosa* Zone. Presently known ranges thus indicate a gap between the two species. To date, the only other published possible occurrence of *P. christiani* is from Cyprus where Baroz & Bizon (1974) figured a specimen from within the *G. acostaensis* Zone termed 'indéterminé' and which Liska (1980) believes to be identical with *P. christiani*. The species is also now known to occur high in the Late Miocene of the Dominican Republic.

Taxonomic notes

POLYPERIBOLA CHRISTIANI Liska
Figures **18.1–2**; 7, 10, 12

Type reference: *Polyperibola christiani* Liska, 1980, p. 137, pl. 1, figs. 1–3.

Affinities in test morphology, chamber arrangement and presence of bullae exist between this species and the Early Miocene *Globigerinatella insueta*, which however differs in its larger size, coarser, more distinctly perforated walls and in having bullae that are either small and circular or narrow tube-like extensions along sutures, in contrast to the often broader and more inflated but very delicate bullae in *P. christiani*.

Because of a similar general chamber arrangement, closed umbilical area, similarly delicate type of wall structure and presence of tiny sutural apertures, *P. christiani* might be related to *Candeina nitida* and possibly be regarded as its ancestor. However, *P. christiani* differs from *C. nitida* in the presence of bullae and also in the generally smaller test size.

The genus *Globigerinatella*

The monotypic genus *Globigerinatella* was first described from the Cipero Formation of Trinidad. The authors, Cushman & Stainforth (1945), used its characteristic type species *G. insueta* as a marker for the middle of three zones based on planktic foraminifera by which they subdivided the Oligocene–Miocene Cipero Formation. This zone, at the time regarded as Oligocene in age, is now placed in the Early Miocene.

Brönnimann (1950) studied the ontogeny of the species in considerable detail and illustrated numerous growth stages and the variability of adult tests. Some authors (Brönnimann, 1950; Hofker, 1954b; Bolli *et al.*, 1957; Blow, 1969) debated possible origins of *G. insueta*. However, its abrupt appearance as a fully developed form makes any proposed evolutionary trends speculative. It has so far resisted subdivision by even the most avid 'splitters'.

The species has frequently been reported from lower latitudes, but only occasionally from more temperate areas like Spain (Soedino, 1970), Israel (Reiss, 1968), New Zealand (Jenkins, 1971), and the Mediterranean (Giannelli & Salvatorini, 1972). But even in optimum latitudes the distribution of the genus can be irregular, necessitating the use of other taxa to verify the stratigraphic level.

Taxonomic notes

GLOBIGERINATELLA INSUETA Cushman & Stainforth

Figures 18.3–6; 7, 9, 12

Type reference: *Globigerinatella insueta* Cushman & Stainforth, 1945, p. 69, pl. 13, figs. 6–9.

The species is readily recognizable by its more or less globular test with irregularly distributed bullae, which are either small, more or less circular in shape and covering individual sutural or areal apertures, or may extend, tube-like, along intercameral sutures. Superficially juveniles of the species are closest to *Globigerinita naparimaensis*, which however differs in a less globular test and in possessing only one bulla that extends along the sutures on the umbilical side. There are also no areal apertures with bullae in *G. naparimaensis*. Broken specimens of *Globigerinatella insueta* frequently show areal apertures on earlier chambers, often surrounded by small raised lips, a feature that is very characteristic for the species (Fig. 18.6).

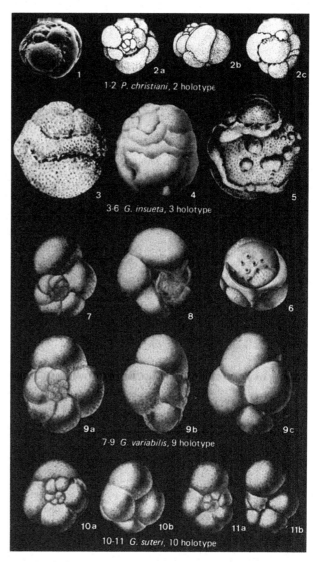

Fig. 18. *Polyperibola, Globigerinatella, Globorotaloides* (all figures × 60)

1–2 *Polyperibola christiani* Liska
1 Paratype, 2a–c Holotype. Both from sample TC 283/R.D.L.319, *G. acostaensis* Zone, Lengua Fm., Late Miocene, cut on Lizard road, Trinidad (1 from Liska, 1980).

3–6 *Globigerinatella insueta* Cushman & Stainforth
3 Holotype. Sample Rz 108, *G. insueta* Zone, Early Miocene, Cipero Fm., Cipero coast, Trinidad. 4, 6 Paratypes. *G. insueta* Zone, Early Miocene, Cipero Fm., Trinidad (from Bolli *et al.*, 1957). 6 Dissected specimen showing areal apertures of last chamber, covered by bulla which has been partly removed. Infralaminal accessory apertures visible at lower margin of remaining part of bulla. 5 Sample AGH 2695, *G. insueta* Zone, Early Miocene, Cipero Fm., Trinidad (from Postuma, 1971).

7–9 *Globorotaloides variabilis* Bolli
7–8 Paratypes, 9a–c Holotype. Sample Rz 502, *G. menardii* Zone, Middle Miocene, Lengua Fm., Pointe-à-Pierre, Trinidad (from Bolli, 1957).

10–11 *Globorotaloides suteri* Bolli
10a–b Holotype, 11a–b Paratypes. Sample Bo 314A, *G. ampliapertura* Zone, Middle Oligocene, Cipero Fm., Cipero coast, Trinidad (from Bolli, 1957).

The genus *Globorotaloides*

The genus originates in the Middle Eocene with *G. carcoselleensis* but is more typically developed in the Oligocene and particularly in the Miocene, with *G. suteri* and *G. variabilis* respectively. The generic character is quite distinct but because of the rather long-ranging species and morphologically intermediate forms the individual taxa are not of great stratigraphic significance.

Taxonomic notes

GLOBOROTALOIDES SUTERI Bolli
Figures **18.10, 11**; 7, 9, 12

Type reference: *Globorotaloides suteri* Bolli, 1957, p. 117, pl. 27, figs. 13a–c.

Compared with *G. variabilis*, *G. suteri* is in general smaller with early chambers more inflated and intercameral sutures on the spiral side more radial. As in *G. variabilis* the final bulla-like chamber is quite variable in size and shape; it may be low and deflated or quite distinctly inflated as in the holotype of the two species. Quite frequently this last bulla-like chamber may be absent.

GLOBOROTALOIDES VARIABILIS Bolli
Figures **18.7–9**; 7, 9, 12

Type reference: *Globorotaloides variabilis* Bolli, 1957, p. 117, pl. 27, figs. 20a–c.

G. variabilis is distinguished from *G. suteri* by its generally larger size and particularly by the more numerous and less inflated chambers of the early stage with their distinctly curved sutures and by the more rapidly increasing size of the last chambers.

The genus *Candeina*

In contrast to other planktic genera like *Globigerina* and *Globigerinoides*, *Candeina* has very thin, delicate, monolamellar, radial chamber walls with exceedingly fine pores and without spines, giving the test surface a smooth and often whitish, porcellaneous appearance.

In addition to the type species *nitida*, several other *Candeina* species and subspecies have been proposed. Some of them, like *C. milletti* Dollfuss, 1905, and probably *C. nitida triloba* Cushman, 1921, can be regarded as aberrant forms. Others, like *C. biloba* Jedlitschka, 1934, and *C. amicula* Takayanagi & Saito, 1962, apparently are related to *Orbulina* or *Globigerinoides*. Hornibrook & Jenkins (1965) described a *C. zeocenica* from the Oligocene of New Zealand. Compared with *C. nitida* this species is very small (maximum diameter 0.26 mm) or about half an adult *nitida*. Furthermore, the holotype figure of *zeocenica* shows a still tighter and lower trochospiral arrangement of chambers, and more distinct pores. Its range restricted to within the Oligocene makes it

unlikely that the species is directly related to *Candeina*, which ranges from Early Pliocene, or latest Miocene, to Recent.

Blow (1969) proposed an evolution of *Candeina nitida* through *C. nitida praenitida* from *Globigerinoides parkerae* Bermudez by 'a successive palingenetic development of the sutural apertures both spiral and intercameral, into the successively earlier parts of the test combined with a proliferation of the apertures. The loss of a primary aperture occurs within the continued morphogenetic development of *Candeina nitida praenitida*'.

The fact that the wall structure of *Candeina* is so different and distinct from *Globigerinoides* as pointed out above is not considered by Blow in his proposed evolution of *Candeina* from *Globigerinoides*. To check on the possible presence of intermediate forms, the Jamaica samples ER 307 (N 16, lower part Bowden Formation with *Globorotalia margaritae*) from which Blow selected the holotype of *Candeina nitida praenitida*, and ER 146/36 (basal N 16, without *Globorotalia margaritae*) from which he figured a hypotype of *Globigerinoides parkerae*, were examined. Neither of the two samples contained typical *G. parkerae* (described by Bermudez from the Holocene) or intermediate forms between *G. parkerae* and *C. nitida praenitida*.

According to Blow (1969) *C. nitida praenitida* differs from *C. nitida nitida* 'in lacking small supplementary apertures in the intercameral sutures prior to the suture between penultimate and antepenultimate chambers'. The presence of these apertures in earlier intercameral sutures is variable within the species. Occasionally they are restricted as in *praenitida* to only the suture between the last two chambers, but more often they are spread over more sutures. D'Orbigny's type figure and his original description of *C. nitida* ('Nous n'avons remarqué aucune ouverture centrale; mais, sur le retour de la spire, la dernière loge est percée, sur le bord, de six à sept petites ouvertures') show that it possesses apertures only along the suture between the last two chambers. As this is the characteristic claimed for *praenitida*, this subspecies has no foundation. We regard *Candeina* with the species *nitida* as still being monotypic.

Stratigraphic significance

There exists some disagreement among authors on the first appearance of *Candeina*. Numerous data, such as from the DSDP Legs 4 (Bolli, 1970) and 15 (Bolli & Premoli Silva, 1973) in the Caribbean, show no *Candeina* present prior to basal Pliocene (base of *Globorotalia margaritae* Zone). Some authors, like Stainforth *et al.* (1975) and Blow (1969), report the species as occurring already slightly earlier, from the Late Miocene *Globorotalia humerosa* Zone. For *C. nitida praenitida*, here included in *nitida*, Blow (1969) cites a range from late N 15 (*Globorotalia menardii* Zone). A re-examination of the holotype sample of *C. nitida praenitida*, ER 307, from the Bowden

Formation, Jamaica, has revealed the presence also of *C. nitida nitida* specimens with sutural apertures present not only between the last two but also between earlier chambers, and also of *G. margaritae*, thus giving the sample an Early Pliocene age. This fits the findings from numerous other sections that *C. nitida* first appears at about the base of the Pliocene.

Taxonomic notes

CANDEINA NITIDA d'Orbigny
Figures **19.1–2**; 7, 10, 11

Type reference: *Candeina nitida* d'Orbigny, 1839a, p. 108, pl. 2, figs. 27, 28.

This easily recognizable species with its shiny, porcellaneous-looking wall is tightly coiled with usually 3 chambers per whorl and a completely closed umbilicus, giving it a somewhat subglobular appearance. Numerous small circular to elongate apertures often surrounded by a rim are aligned along the sutures of the chambers of the last whorl in particular, but also occasionally along sutures of the penultimate one.

The genus *Globigerinoides*

The first appearance of *Globigerinoides* has widely been accepted in foraminiferal biostratigraphy as the datum for the Oligocene–Miocene boundary (Cita, 1968; Blow, 1969; Ikebe

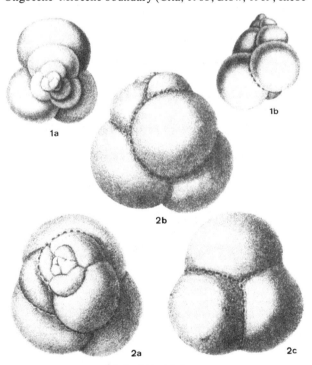

1-2 *C. nitida*, 1 holotype

Fig. 19. *Candeina* (all figures × 60)
1–2 *C. nitida* d'Orbigny
  1a–b Holotype. Probably from beach sands, Recent, Cuba. **2a–c** Probably from a *Challenger* sample, Recent, no locality given (from Brady, 1884).

et al., 1972). Other authors, in particular Lamb & Stainforth (1976) and Stainforth & Lamb (1981), report stratigraphic first occurrences of *Globigerinoides* as early as the base of the *Globigerina ciperoensis ciperoensis* Zone, and thus clearly within the Oligocene. The different levels at which *Globigerinoides* first appears may be related to the rise in ocean temperatures that occurred after the cool period in the Oligocene (Jenkins, 1971; Seiglie, 1973). Though this was a worldwide event, it took place at different times locally.

The unreliability of the first occurrence of the genus makes the application of the *Globigerinoides* 'datum' of doubtful value for the delineation of the Oligocene–Miocene boundary. As a substitute for this datum, Stainforth & Lamb (1981) tentatively suggest the first appearance of *Globoquadrina altispira* aff. *altispira* and *Globorotalia fohsi peripheroronda*. Unfortunately, *Globoquadrina altispira* aff. *altispira* is a difficult form to keep separate from the other subspecies and, for that reason, is far from ideal. As regards *Globorotalia fohsi peripheroronda*, this taxon does not appear in typical form in Trinidad, for instance, until the *Globigerinatella insueta* Zone. In the next older zone, the *Catapsydrax stainforthi* Zone, only forms considered as aff. *peripheroronda* are found.

We would still favour the use of *Globigerinoides* to delineate the Oligocene–Miocene boundary. Instead of using the appearance of rare and primitive *G. primordius* it would seem better to use the explosive appearance, apparently from a *G. primordius* ancestor of *G. trilobus* with its subspecies *trilobus*, *immaturus* and *sacculifer*. This seems to be an almost instantaneous event, at least in low latitudes.

Only very shortly after this event the characteristic *G. ruber* (*subquadratus* of some authors) and *G. altiaperturus* also appear. All these taxa, except for *G. altiaperturus* which is short lived, continue from Early Miocene to Recent. *G. diminutus*, a likely offspring of *G. ruber* and apparently restricted to low latitudes, is also short lived within the upper part of the Early Miocene. Appearing later in the Early Miocene and continuing into the Pliocene is *G. obliquus obliquus* from which *G. obliquus extremus* evolved in the Late Miocene.

Also in the Early Miocene, within the *Globigerinatella insueta* Zone in the low latitudes and the *Globigerinoides trilobus* Zone in the Mediterranean, *G. bisphericus* evolved from the *G. trilobus* stock to lead within the very short time of about 1.5 m.y. via a number of *Praeorbulina* subspecies to *Orbulina*.

G. bollii and *G. ruber seigliei* were erected in low latitudes but are preferably used for stratigraphic purposes in higher latitudes such as the Mediterranean. The first ranges there from within the Middle Miocene through the Pliocene, the second through the Late Miocene. *G. bulloideus*, a species first reported from the Mediterranean, ranges there through the Middle and Late Miocene. The only *Globigerinoides* species that appears in the Middle Miocene, and is restricted to this interval, is *G. mitra*, which has been found typically developed only in lower latitudes. It is a giant, apparently somewhat aberrant, form.

G. canimarensis, which leads to *G. conglobatus*, appeared within the Late Miocene. A very characteristic and short lived *Globigerinoides*, developing in low latitudes from *G. trilobus sacculifer*, is *trilobus fistulosus* s.s., ranging from within the Early Pliocene *Globorotalia margaritae evoluta* Subzone to the top of the Middle Pliocene *Globigerinoides trilobus fistulosus* Subzone. *G. tenellus*, which has been described from the Mediterranean, is documented in the area from Early Pleistocene to Recent.

In this chapter a number of other species of *Globigerinoides* are treated in the taxonomic notes because some of them, such as *bollii*, *bulloideus*, *seigliei*, *elongatus* and *tenellus*, are useful in Mediterranean biostratigraphy.

Summarizing, several *Globigerinoides* taxa are of considerable significance for Neogene stratigraphy, both in low as well as in mid-latitudes. Apart from the datum of the first appearance of the genus and the rapid, stratigraphically important evolution from *G. bisphericus* through *Praeorbulina* to *Orbulina*, there are several taxa which either by their short ranges, as *primordius*, *altiaperturus*, *diminutus*, *trilobus fistulosus*, *tenellus*, or because of their first or last occurrence, as *obliquus obliquus*, *obliquus extremus*, *conglobatus*, can be applied to biostratigraphy. Furthermore, *G. primordius*, *G. ruber* and *G. trilobus fistulosus* are used as zonal markers in low latitudes, *G. trilobus* and *G. altiaperturus* in mid-latitudes.

Taxonomic notes

GLOBIGERINOIDES ALTIAPERTURUS Bolli
Figures **20.10**; 7, 9, 12

Type reference: *Globigerinoides triloba altiapertura* Bolli, 1957, p. 113, pl. 25, figs. 7a–c.

The chamber arrangement and test shape of *G. altiaperturus* closely resembles that of *G. trilobus trilobus*, to which it appears to be related. It differs from it in having distinctly larger, higher arched semicircular apertures, which in *G. trilobus* are low arched, more slit-like. Its first appearance approximately coincides with that of the *G. trilobus* complex.

GLOBIGERINOIDES BISPHERICUS Todd
Figures **24.8**; 7, 9, 12

For taxonomic notes on this species see p. 199 (section on the Praeorbulina–Orbulina lineage).

GLOBIGERINOIDES BOLLII Blow
Figure **20.8**

Type reference: *Globigerinoides bollii* Blow, 1959, p. 189, pl. 10, figs. 65a–c.

This species is characterized by strongly embracing chambers and by an almost completely circular and rather small primary aperture. There is usually one small supplementary aperture; an additional second one may be present between the penultimate and antepenultimate chamber. Primary and supplementary apertures tend to become larger in stratigraphically younger specimens.

According to its author the species first appears within the *Globorotalia praefohsi* (N 11) and ranges up to the *Globorotalia tosaensis tenuitheca* (N 21) Zone. Therefore, the first

---

Fig. 20. *Globigerinoides* (all figures × 60)

1–2 *G. ruber* (d'Orbigny)
  1a–c Holotype, **2a–b** Lectotype. Both from Recent beach deposits, probably Cuba.

3 *G. diminutus* Bolli
  3a–c Holotype. UBOT well Penal 92, core 7419-39 feet, *G. insueta* Zone, Early Miocene, Cipero Fm., Trinidad.

4 *G. elongatus* (d'Orbigny)
  4a–c Lectotype. Recent marine deposits near Rimini, Adriatic Sea, Italy.

5 *G. seigliei* Bermudez & Bolli
  5a–c Holotype. Sample PJB 4/64, *G. altispira altispira* Zone (= *G. trilobus fistulosus* Subzone), Late Miocene (now placed into the Middle Pliocene), Cubagua Fm., Araya Peninsula, Venezuela.

6 *G. subquadratus* Brönnimann
  6a–c Holotype. Loc. C-85, *G. insueta* to *G. fohsi* s.l. Zone, Early to Middle Miocene, Saipan, Mariana Islands.

7 *G. tenellus* Parker
  7a Holotype, **7b–c** Paratypes. *Atlantis* Station 4711, Recent, midway between Greece and Libya, Mediterranean Sea.

8 *G. bollii* Blow
  8a–c Holotype. Sample RM 19697, *G. menardii* Zone, Middle Miocene, Pozon Fm., Eastern Falcon, Venezuela.

9 *G. bulloideus* Crescenti
  9a–c Holotype. Sample UC 10, *G. menardii* Zone, Middle Miocene, Ravenna Province, Italy.

10 *G. altiaperturus* Bolli
  10a–c Holotype. Sample Bo 267, *C. dissimilis* Zone, Early Miocene, Cipero Fm., San Fernando area, Trinidad.

11 *G. obliquus extremus* Bolli & Bermudez
  11a–c Holotype. Well Cubagua-1, core at 1029–34 feet, *G. margaritae* Zone, Early Pliocene, Cubagua Fm., Cubagua Island, Venezuela.

12 *G. obliquus obliquus* Bolli
  12a–c Holotype. Sample KR 23422, *G. mayeri* Zone, Middle Miocene, Lengua Fm., near Lengua Settlement, Trinidad.

13 *G. trilobus sacculifer* (Brady)
  13a–b Lectotype. Late Miocene or Pliocene?, from an exotic block, New Ireland, Bismark Archipelago, Indo-Pacific.

14 *G. trilobus immaturus* LeRoy
  14a–c Holotype. Sample Ho-862A, Miocene, Central Sumatra.

15 *G. trilobus trilobus* (Reuss)
  15a–c Holotype. Tertiary, Salzton, Wieliczka, Galicia.

16 *G. primordius* Blow & Banner
  16a–c Holotype. Sample Bo 274, *G. primordius* Zone, Early Miocene, Cipero Fm., San Fernando area, Trinidad.

17 *G. quadrilobatus* (d'Orbigny)
  17a–b Lectotype. Middle Miocene, near Nussdorf, Austria.

18 *G. quadrilobatus* (d'Orbigny)
  18a–c Holotype. Tertiary, Tegel marls, near Nussdorf, Austria.

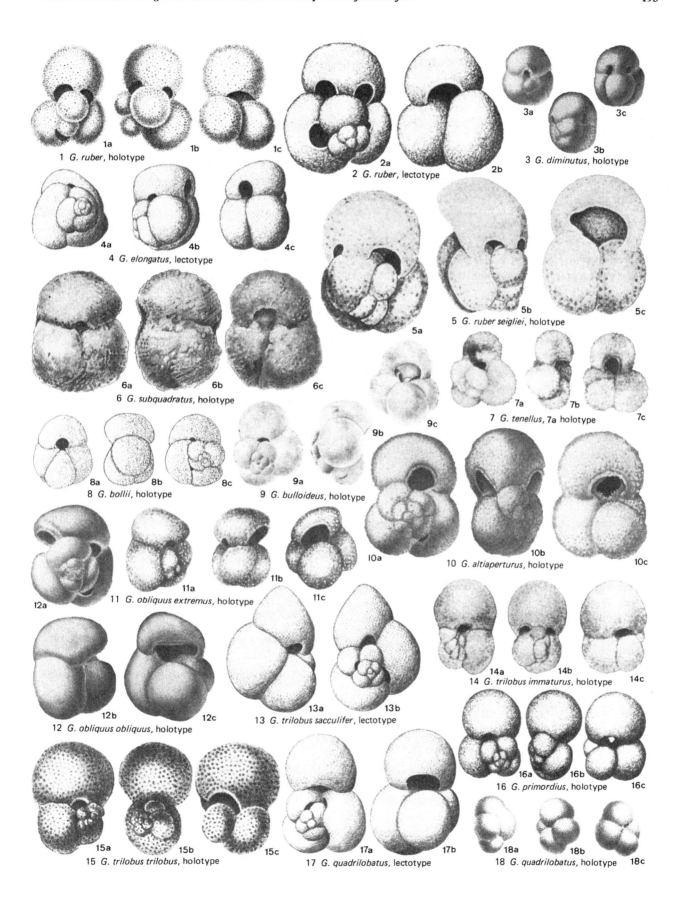

1 *G. ruber*, holotype

2 *G. ruber*, lectotype

3 *G. diminutus*, holotype

4 *G. elongatus*, lectotype

5 *G. ruber seigliei*, holotype

6 *G. subquadratus*, holotype

7 *G. tenellus*, 7a holotype

8 *G. bollii*, holotype

9 *G. bulloideus*, holotype

10 *G. altiaperturus*, holotype

11 *G. obliquus extremus*, holotype

12 *G. obliquus obliquus*, holotype

13 *G. trilobus sacculifer*, lectotype

14 *G. trilobus immaturus*, holotype

15 *G. trilobus trilobus*, holotype

16 *G. primordius*, holotype

17 *G. quadrilobatus*, lectotype

18 *G. quadrilobatus*, holotype

appearance of *G. bollii* could represent a useful datum in the Mediterranean region where the *G. fohsi* lineage is not developed. However, it is well documented in that area only from the *Sphaeroidinellopsis subdehiscens subdehiscens–Globigerina druryi* Zone (N 13). The species seems to have become extinct in the Mediterranean area in the Early Pleistocene (Borsetti *et al.*, 1979) but was already rare by the Late Pliocene.

GLOBIGERINOIDES BULLOIDEUS Crescenti
Figure **20.9**

Type reference: *Globigerinoides bulloideus* Crescenti, 1966, p. 43, pl. 9, figs. 9a–c.

The species is characterized by a wide primary aperture and only one small supplementary aperture. In its shape, arrangement of chambers and in size it is very close to *Globigerina bulloides*. It differs from *G. bollii* in possessing a wider primary aperture, a more lobate equatorial profile and less embracing chambers. *G. altiaperturus* differs from *G. bulloideus* in the higher and distinctly larger supplementary aperture. Compared with *G. primordius*, *G. bulloideus* possesses a larger primary aperture and a less lobate outline.

GLOBIGERINOIDES CONGLOBATUS (Brady)
Figures **21.1**; 7, 10, 11

Type reference: *Globigerina conglobata* Brady, 1879, p. 286, pl. 80, figs. 1–3 (not designated as holotype).
Lectotype: *Globigerinoides conglobatus* (Brady), 1879: Banner & Blow, 1960a, p. 6, pl. 4, figs. 4a–c.

This Pliocene to Holocene large sized subglobular species (lectotype 0.85 mm) with its later chambers distinctly depressed is easily recognizable.

    Bermudez (1961) erected a *G. canimarensis* (Fig. 21.3) as an ancestral form of *G. conglobatus*. It differs in smaller test size and higher arched primary and secondary apertures. Brady (1879) regarded *G. conglobatus* as a modification of *G. ruber*, in which the chambers of the final whorl become predominant in size. Another possibility is that *G. obliquus extremus* with its similarly depressed final chambers and similar position of the primary aperture, is the ancestral line.

GLOBIGERINOIDES DIMINUTUS Bolli
Figures **20.3**; 7, 9, 12

Type reference: *Globigerinoides diminutus* Bolli, 1957, p. 114, pl. 25, figs. 11a–c.

The species is morphologically similar to *G. ruber* but differs in its much smaller and more compact test. Chamber arrangement and position of the apertures symmetrically above the suture between earlier chambers are the same. Compared with *G. ruber* the aperture in *G. diminutus* is always small, higher and narrower, sometimes appearing almost as a vertical slit.

GLOBIGERINOIDES ELONGATUS (d'Orbigny)
Figure **20.4**

Type reference: *Globigerina elongata* d'Orbigny, 1826, p. 277, no figure.
Lectotype: *Globigerinoides elongatus* (d'Orbigny), 1826, Banner & Blow, 1960a, p. 12, pl. 3, figs. 10a–c.

The lectotype selected, described and figured by Banner & Blow (1960a) displays characters of both, *G. ruber* and *G. obliquus extremus*. Identical with *ruber* is the high arched primary aperture positioned symmetrically above the sutures between the two earlier chambers. As in *extremus* the final chamber is laterally distinctly depressed in an asymmetrical way. If attention is paid only to the shape of the last chamber but not to the position of the primary aperture, *elongatus* and *extremus* may readily be mistaken for each other.

    Whereas in low latitudes the index species *G. obliquus extremus* becomes extinct in the late Middle Pliocene, *G. elongatus* ranges throughout the Pliocene to Holocene. It is therefore particularly in the Late Pliocene and Holocene where misidentification of *G. elongatus* as *G. obliquus extremus* may lead to erroneous stratigraphic conclusions.

GLOBIGERINOIDES MITRA Todd
Figures **21.2**; 7, 9, 12

Type reference: *Globigerinoides mitra* Todd, 1957, p. 302, pl. 78, fig. 3.

This species occurs infrequently but is easily recognized because of its exceptionally large size (according to Todd up to 1.4 mm). Characteristic is the high spire formed by loosely arranged chambers, those of the last whorl often elongate, grape-shaped. As has been noted in Trinidad, *G. mitra* tests are often preserved in a pyritic state.

GLOBIGERINOIDES OBLIQUUS OBLIQUUS Bolli
Figures **20.12**; 7, 9, 12

Type reference: *Globigerinoides obliquus obliquus* Bolli, 1957, p. 113, pl. 25, figs. 10a–c.

The species is distinguished from *G. trilobus trilobus* and *G. altiaperturus* in having the ultimate chamber slightly compressed in a lateral, oblique way. Compared with *G. trilobus trilobus* the primary aperture of the last chamber is wider and somewhat higher, but not as high and semicircular as in *G. altiaperturus*.

GLOBIGERINOIDES OBLIQUUS EXTREMUS Bolli & Bermudez
Figures **20.11**; 7, 10, 11

Type reference: *Globigerinoides obliquus* Bolli subsp. *extremus* Bolli & Bermudez, 1965, p. 139, pl. 1, figs. 10–12.

G. *obliquus obliquus* gave rise to the subspecies *extremus* within the *Globorotalia humerosa* Zone, the last chamber becoming still more compressed laterally and, as a result, more asymmetrical. The earlier chambers, in particular those forming the last whorl, may also be compressed laterally and be of irregular shape, though less so than the final chamber.

The subspecies *extremus* ranges from the *Globorotalia humerosa* Zone into the lower part of the *Globorotalia exilis* Subzone. However, it has frequently been reported as occurring also higher stratigraphically. This is probably due to a confusion with *G. elongatus* (Fig. 20.4), present in the Pliocene to Holocene and, like *extremus*, possessing a similar asymmetrically depressed last chamber. The distinction between the two is best made by observing the position of the primary aperture which in *elongatus* (as in *G. ruber*) is symmetrical above the suture between the two earlier chambers, but in *extremus* is asymmetrical.

GLOBIGERINOIDES PRIMORDIUS Blow & Banner

Figures **20.6**; 7, 9, 12

Type reference: *Globigerinoides quadrilobatus* (d'Orbigny) subsp. *primordius* Blow & Banner, 1962, p. 115, pl. 9, figs. Dd–Ff.

This species is the immediate ancestor of the *Globigerinoides trilobus* stock, with overall morphological features already very close, in particular to *G. trilobus immaturus*. Typical *primordius* specimens differ in having more delicately built chamber walls, in the primary aperture being higher arched and in possessing only a single secondary opening, situated on the spiral side of the last chamber. It is generally agreed that the immediate ancestor of this first *Globigerinoides* species is a *Globigerina*. According to Blow & Banner it is very close to *G. praebulloides occlusa* (Fig. 13.12) from which it differs primarily only in the presence of a spiral secondary aperture.

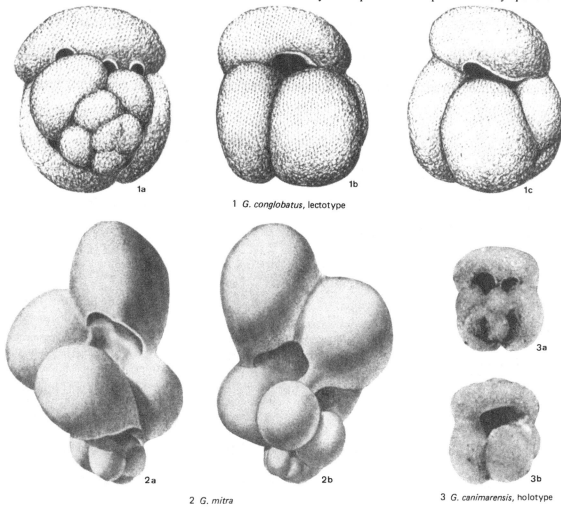

Fig. 21. *Globigerinoides* (all figures × 60)

1 *G. conglobatus* (Brady)
   1a–c Lectotype. *Challenger* Station 338, Recent, South Atlantic.
2 *G. mitra* Todd

2a–b *G. menardii* Zone, Middle Miocene, Lengua Fm., Trinidad (from Bolli, 1957).
3 *G. canimarensis* Bermudez
   3a–b Holotype. Pliocene, Matanzas Fm., Matanzas Province, Cuba.

The confusion caused by the lectotype of *Globigerinoides 'quadrilobatus'* erected by Banner & Blow is discussed in the taxonomic notes of the *Globigerinoides trilobus* group.

The difficulty of using the first occurrence of *G. primordius* to delineate the Oligocene–Miocene boundary is discussed on p. 191.

GLOBIGERINOIDES RUBER (d'Orbigny)
Figures **20.1, 2, 6**; 7, 9, 11

Type reference: *Globigerina rubra* d'Orbigny, 1839a, p. 82, pl. 4, figs. 12–14.

Lectotype: *Globigerinoides ruber* (d'Orbigny), 1839a; Banner & Blow, 1960a, p. 19, pl. 3, figs. 8a–b.

Synonym: *Globigerinoides subquadrata* Brönnimann, *in* Todd *et al.* 1954, p. 680, pl. 1, figs. 8a–c.

This species with 3 chambers forming the last whorl is morphologically similar to *G. trilobus trilobus*, except for the position of the apertures, which are higher arched as in *G. altiapertura*, and positioned symmetrically above the sutures between earlier chambers. The species appeared first at about the same time as the *G. trilobus* complex, to continue virtually unchanged to the Holocene, except for an interruption from the Middle Miocene to within the Early Pliocene. It is possible that the later forms represent a new development of a morphologically very similar test shape. In fact, some authors place the Miocene forms of *G. ruber* in a separate subspecies, *G. subquadratus*. In a comparison of the morphological characters of the two taxa, Cordey (1967) came to the conclusion that separation is only possible when the tests are dissected and the ontogenetic development is taken into account. Following Stainforth *et al.* (1975), it seems preferable to use the name *ruber* also for the Miocene forms, even though a preservation of the two names would aid in the distinction of the stratigraphically older and younger forms.

GLOBIGERINOIDES SEIGLIEI Bermudez & Bolli
Figure **20.5**

Type reference: *Globigerinoides rubra* (d'Orbigny) subsp. *seigliei* Bermudez & Bolli, 1969, p. 164, pl. 8, figs. 10–12.

As in *G. ruber* the primary aperture is symmetrically placed above the suture between the penultimate and antepenultimate chambers, but is larger and wider; the test wall is described as more finely spinose.

GLOBIGERINOIDES TENELLUS Parker
Figure **20.7**

Type reference: *Globigerinoides tenellus* Parker, 1958, p. 280, pl. 6, fig. 7 (holotype), figs. 8–11 (paratypes).

The major features of the species are its small size, strongly lobate equatorial outline, large primary aperture almost circular in outline, small and narrow supplementary apertures on the spiral side of the last chamber, more rarely in earlier chambers, and a wall surface roughened by dense, short spines.

The species was first described from the Holocene but is also documented throughout the Pleistocene. In the Mediterranean area it is considered a good marker for this interval (Blow, 1969; Colalongo & Sartoni, 1977).

Subspecies of GLOBIGERINOIDES TRILOBUS

GLOBIGERINOIDES TRILOBUS TRILOBUS (Reuss)
Figures **20.15**; 7, 9, 11

Type species: *Globigerina triloba* Reuss, 1850, p. 374, pl. 47, figs. 11a–c.

GLOBIGERINOIDES TRILOBUS IMMATURUS LeRoy
Figures **20.14, 17**; 7, 9, 11

Type species: *Globigerinoides sacculifer* (Brady) var. *immaturus* LeRoy, 1939, p. 263, pl. 3, figs. 19–21.

Synonym: *Globigerinoides quadrilobatus* (d'Orbigny) 1846, lectotype: Banner & Blow, 1970a, p. 17, pl. 4, fig. 3a–b; not *Globigerina quadrilobata* d'Orbigny, 1846.

GLOBIGERINOIDES TRILOBUS SACCULIFER (Brady)
Figures **20.13**; 7, 9, 11

Type species: *Globigerina sacculifera* Brady, 1877, p. 535, no figure.

Lectotype: *Globigerinoides quadrilobatus* (d'Orbigny) subsp. *sacculifer* (Brady): Banner & Blow, 1970a, p. 21, pl. 4, figs. 1a–b.

GLOBIGERINOIDES TRILOBUS FISTULOSUS (Schubert)
Figures **22.5–11**; 7, 10, 11

Type species: *Globigerina fistulosa* Schubert, 1910, p. 323, text-fig. 2, figs. 13a–c (no holotype designated).

The reason why the long-ranging and stratigraphically insignificant *G. trilobus* subspecies are included here is because they are often dominant in tropical/subtropical assemblages and because they form the stock from which some shorter-ranging index forms have branched off, such as the *G. bisphaericus–Praeorbulina–Orbulina* lineage and, in the Pliocene, *G. trilobus fistulosus*.

Banner & Blow (1960) designated a typical *Globigerinoides* specimen, which in the literature would be regarded as *G. trilobus*, as a lectotype for the apparently lost holotype of *Globigerina quadrilobata* d'Orbigny, 1846 (Fig. 20.18). According to d'Orbigny's original description and figures his taxon possessed no secondary aperture. For direct comparison we reproduce here the original illustrations (Fig. 20.18) and those of the lectotype (Fig. 20.17) at approximately the same magnification. The lectotype was selected by Banner & Blow from a heterogeneous topotypic collection deposited in Paris. Subsequently some authors followed Banner & Blow's new concept of *G. quadrilobata*. Bandy (1964) in discussing and illustrating this case concluded that the lectotype designated by Banner & Blow for *G. quadrilobata* is invalid in that it represents not only a different species but also a different genus compared with the original type figure. We agree with Bandy and consequently place Banner & Blow's lectotype of *G. quadrilobata* (Fig. 20.17) in synonymy with *Globigerinoides trilobus immaturus* (Fig. 20.14).

Bolli (1957) distinguished the subspecies *trilobus* from the subspecies *immaturus* by the possession of a final chamber larger than all earlier chambers combined. A comparison of

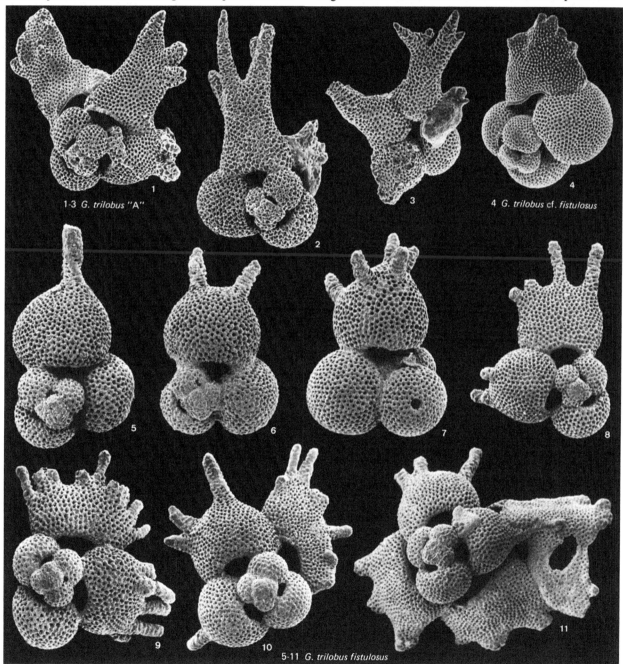

Fig. 22. *Globigerinoides* (all figures × 60)

1–3 *G. trilobus* 'A'
    From DSDP Leg 4-29-1-2, 4–6 cm, *G. truncatulinoides truncatulinoides* Zone, Pleistocene, Venezuela Basin, Caribbean Sea (from Bolli, 1970).
    4 *G. trilobus* cf. *fistulosus* (Schubert)

From DSDP Leg 4-29-2-3, 146–148 cm, *G. tosaensis tosaensis* Zone, Pliocene, Venezuela Basin, Caribbean Sea (from Bolli, 1970).

5–11 *G. trilobus fistulosus* (Schubert)
    From DSDP Leg 4-25-3-1, 10–20 and 150 cm, *G. margaritae evoluta* Subzone or *G. trilobus fistulosus* Subzone, Pliocene, off northeast coast of Brazil, South Atlantic.

Reuss' holotype figures of *triloba* and LeRoy's *immaturus* (Figs. 20.15 and 20.14) makes evident that, though in the holotype figures of *immaturus* the final chamber is slightly smaller in relation to the early test compared with *trilobus*, the difference is minimal. Blow's lectotype for *quadrilobatus* (Fig. 20.17) with its much larger sized early stage compared with the last chamber, reflects a typical *immaturus* subspecies in the sense of Bolli (1957). Forms with an ultimate elongate sack-like chamber are placed in *G. trilobus sacculifer*.

In the Pliocene and Pleistocene there evolve from *G. trilobus sacculifer* some bizarre forms where the final one to three chambers form at their peripheral ends fistule-like extensions varying in shape, number and length. Such forms are generally placed in *G. trilobus fistulosus* (synonym: *G. quadrilobatus hystricosus* Belford, 1962). On closer examination the following three types can be distinguished, two of which are of stratigraphic significance:

(1) *G. trilobus fistulosus* s.s. (Figs. 22.5–11): fistules distinct, finger-like, with rounded ends, one to several in last chamber, often present in penultimate, less frequently in earlier chambers. Chambers which possess more than two to three fistules are usually distinctly flattened and of a cockscomb shape. Where several fistules occur they usually are in line along the periphery of chambers, like fingers on a hand. Figs. 22.5–11 illustrate some of the considerable variability of the taxon as regards number of fistules and shape of chambers. Range: within *Globorotalia margaritae evoluta* Zone to top of *Globigerinoides trilobus fistulosus* Subzone.

(2) *Globigerinoides trilobus* cf. *fistulosus* (Fig. 22.4): Fistulose extensions generally confined to the last chamber, but much less extended and not as finger-like as in *G. trilobus fistulosus*. Observed range: *Globorotalia margaritae margaritae* Zone to Holocene.

(3) *Globigerinoides trilobus* 'A' (Figs. 22.1–3): Final 1–3 chambers strongly elongated, somewhat irregular in shape, fistules of varying length and diameter, often with a pointed end, not arranged in line, may branch off from chamber at different points and angles. Individual fistules may also branch. Such forms have been observed in the Caribbean restricted to the Late Pleistocene to ?Holocene.

The *Praeorbulina–Orbulina* lineage

More than 40 species and varieties of '*Orbulina*' have been recorded since d'Orbigny described the genus in 1839; they range in age from Cambrian to Recent. It is now recognized that pre-Miocene occurrences are unlikely to have any relation to the taxon, being broken chambers of other foraminifera, radiolaria, algal reproductive bodies and other spherical bodies (Loeblich & Tappan, 1961).

Orbulina universa d'Orbigny (1839a) was described from the Recent of Cuba. Unfortunately the author illustrated a specimen with a large, round aperture as well as the usual pores. The type being no longer available, Le Calvez (1974) picked a lectotype also with 'rounded aperture'. Examination of very large numbers of tests suggests that any large single aperture is most likely to be an artifact caused by damage to the test. Two sizes of areal pores are the usual condition of the adult test (see, for example, Bé, Harrison & Lott, 1973, pl. 9, 10). Bé *et al.* (1977) demonstrated that it is through the larger of these that the symbiotic dinoflagellates move diurnally in response to light intensity.

Jedlitschka (1934) proposed a new genus and species, *Candorbulina universa*, for forms that show a series of small apertures outlining the sutures formed where the chambers of the early, trochospiral test lie against the wall of the final, globular chamber; the use of the specific name *universa* was most unfortunate and has caused much confusion in synonymy lists. Brönnimann (1951b) considered the two genera too close for separation and placed *C. universa* Jedlitschka in his new species *Orbulina suturalis*. Present-day workers are split in their opinions as to the validity of *Candorbulina* and this fact alone shows that the two genera are too close for useful separation. Blow (1956) emended *Orbulina* d'Orbigny specifically to include the presence of areal apertures which had not been mentioned in the original description but had already been noted in the emendation by Brönnimann (1951b). Blow described and illustrated the evolution of *Orbulina* from *Globigerinoides bisphericus* by way of *G. glomerosa*.

Bolli (1957) removed *G. glomerosa* to *Porticulasphaera* Bolli, Loeblich & Tappan, 1957, noting the final, strongly embracing chamber and the lack of distinct primary umbilical aperture. He recognized that the similarity is morphological and probably has no genetic basis as *Porticulasphaera* is a short-ranged Eocene genus.

The stratigraphic significance of the development of *Orbulina* through *Globigerinoides bisphericus* and *G. glomerosa* led Olsson (1964) to propose a new genus, *Praeorbulina* with type species *P. glomerosa glomerosa* Blow, 1956. Though a near homeomorph of the Eocene genus, *Praeorbulina* shows slight differences listed by Olsson. This generic assignment has generally been accepted (e.g. Blow, 1969; Stainforth *et al.*, 1975).

A polyphyletic origin for *Orbulina* has been proposed by Hofker (1954a, 1969) who considered the number and size of pores to be important. However, it has been shown in living specimens that these characters depend on water temperature (Bé *et al.*, 1973; Hecht, Bé & Lott, 1976).

Two-chambered forms have been separated under the species *O. bilobata* (d'Orbigny), 1846, and even given generic rank (*Biorbulina* Blow, 1956) but their relationship to *O. universa* is close and they are best regarded as variants of the species.

The lineage from *Globigerinoides bisphericus* through *Globigerinoides* (= *Praeorbulina*) *transitoria* to *B. bilobata* proposed by Blow (1956) seems less certain than his lineage from *G. bisphericus* through *Globigerinoides* (= *Praeorbulina*) *glomerosa* to *O. suturalis* and *O. universa*. Jenkins, Saunders & Cifelli (1981) have shown that *Globigerinoides bisphericus* Todd and *Praeorbulina sicana* (de Stephani) can be kept separate, the latter being morphologically intermediate between *G. bisphericus* and *P. glomerosa*.

The evolutionary lineage culminating in *Orbulina* had been suggested as early as 1940 by Cushman & Dorsey who stated: 'it would seem therefore that *Candorbulina* originated from *Globigerinoides* in the Miocene, and gave rise to *Orbulina* but itself becoming extinct. If this be the case, *Candorbulina* should make a good index fossil for the Miocene.' *Candorbulina* was then being used in the way that we should use *Praeorbulina*. Since then, the *Orbulina* datum has been discussed by many workers (e.g. Finlay, 1947; LeRoy, 1948; Blow, 1956; Bandy, 1966; Jenkins, 1968; Bandy, Vincent & Wright, 1969). In tropical faunas the evolutionary development of *Orbulina* through *Praeorbulina* takes place within the upper part of the *P. glomerosa* Zone with the appearance of *O. universa* marking the base of the *Globorotalia fohsi peripheroronda* Zone. This is particularly useful in the Caribbean, Atlantic and Mediterranean regions and less so in the Pacific where the occurrence of *Orbulina* is more sporadic. When used in temperate faunas, the *Orbulina* datum has to be treated with caution as *O. universa* has been reported as appearing as late as the *Globorotalia mayeri* Zone (see, for example, Bandy, 1966).

Taxonomic notes

GLOBIGERINOIDES BISPHERICUS Todd
Figures **24.8**; 7, 9, 12

Type reference: *Globigerinoides bispherica* Todd, 1954, p. 681, pl. 1, figs. 1a–c.

A short-lived species differing from its probable ancestor, *G. trilobus* (Reuss), in its more enveloping last chamber which almost completely hides the umbilicus. Only 2 apertures are present along the suture between the last and earlier chambers though more apertures may remain open on the spire.

PRAEORBULINA SICANA (de Stefani)
Figures **24.7**; 7, 9, 12

Type references: *Globigerinoides conglobatus* in Cushman & Stainforth, 1945, pl. 13, fig. 6. Not *G. conglobatus* (Brady).
    Globigerinoides sicana de Stefani, 1952, p. 9. Not re-illustrated.

A re-study of the holotype of this species has shown that it has 4 apertures around the base of the last chamber (Jenkins *et al.*, 1981 and holotype refigured here as Fig. 24.7). For this reason

It is removed to the genus *Praeorbulina*. Also for this reason and because of the greater envelopment of the test by the last chamber with subsequent complete loss of the umbilicus, the species is not considered to be conspecific with *G. bisphericus* as has been suggested (Blow, 1969). It is believed to be an intermediate form in the lineage between *G. bisphericus* and *P. glomerosa curva*.

PRAEORBULINA GLOMEROSA CURVA Blow
Figures **23.5, 24.6**; 7, 9, 12

Type reference: *Globigerinoides glomerosa curva* Blow, 1956, p. 64, text-figs. 1.10–11.

P. glomerosa curva differs from *P. sicana* in having 7 instead of 4 apertures around the base of the last chamber in the holotype. It also shows greater envelopment of the test by the last chamber.

PRAEORBULINA GLOMEROSA GLOMEROSA Blow
Figures **23.4, 24.5**; 7, 9, 12

Type reference: *Globigerinoides glomerosa glomerosa* Blow, 1956, p. 65, text-figs. 1.18–19.

P. glomerosa glomerosa differs from *P. glomerosa curva* in having more apertures around the base of the last chamber and in possessing a test that is almost spherical in outline due to the high degree of envelopment by the last chamber.

PRAEORBULINA GLOMEROSA CIRCULARIS Blow
Figures **23.3, 24.4**; 7, 9, 12

Type reference: *Globigerinoides glomerosa circularis* Blow, 1956, p. 65, text-figs. 2.3–4.

P. glomerosa circularis differs from *P. glomerosa glomerosa* in having many circular, pore-like apertures between the last chamber and earlier ones rather than the slit-like apertures of the latter subspecies. It is almost completely spherical with sutures hardly incised. It differs from *Orbulina suturalis* in having all the apertures confined to the sutures.

PRAEORBULINA TRANSITORIA Blow
Figures **23.6**; 7, 9, 12

Type reference: *Globigerinoides transitoria* Blow, 1956, p. 65, text-figs. 2.12–13.

The presence of more than 2 apertures between the last chamber and earlier ones places this species in the genus *Praeorbulina*. It differs from *G. bisphericus* in being more symmetrically bilobate with the penultimate chamber occupying proportionally more of the test; morphologically intermediate forms can be found.

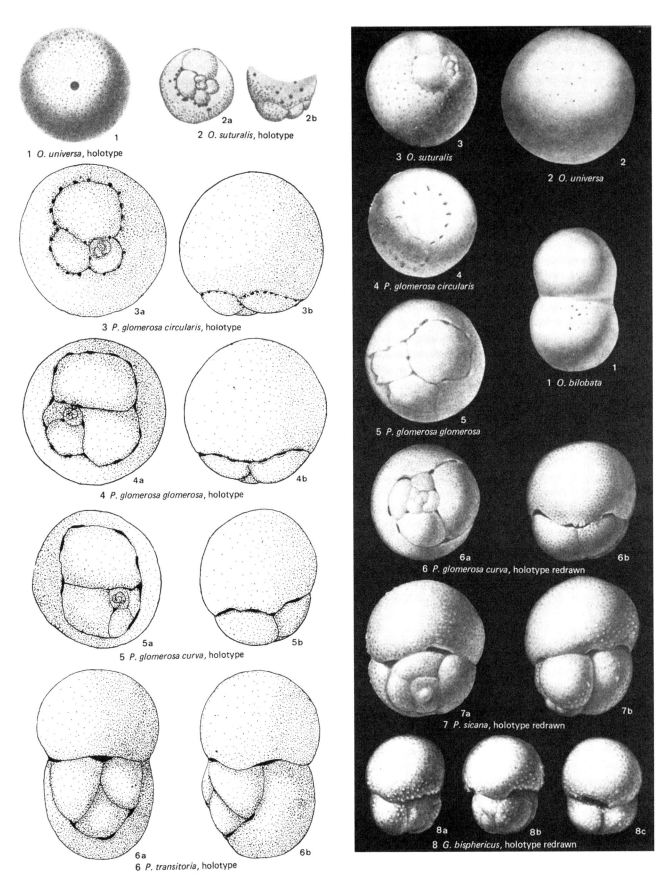

1 *O. universa*, holotype

2 *O. suturalis*, holotype

3 *P. glomerosa circularis*, holotype

4 *P. glomerosa glomerosa*, holotype

5 *P. glomerosa curva*, holotype

6 *P. transitoria*, holotype

Fig. 23.

3 *O. suturalis*

2 *O. universa*

4 *P. glomerosa circularis*

1 *O. bilobata*

5 *P. glomerosa glomerosa*

6 *P. glomerosa curva*, holotype redrawn

7 *P. sicana*, holotype redrawn

8 *G. bisphericus*, holotype redrawn

Fig. 24.

ORBULINA SUTURALIS Brönnimann
Figures **23.2, 24.3**; 7, 9, 11

Type reference: *Orbulina suturalis* Brönnimann, 1951b, p. 135, figs. 2–4.

Areal as well as sutural apertures distinguish this species from *Praeorbulina glomerosa circularis*. It differs from the closely related *O. universa* in having the earlier chambers of the test breaking the outline of the sphere.

ORBULINA UNIVERSA d'Orbigny
Figures **23.1, 24.2**; 7, 9, 11

Type reference: *Orbulina universa* d'Orbigny, 1839a, p. 2, pl. 1, fig. 1.

The species normally has a completely spherical test which only shows the early chambers internally when broken, though the trace of some sutures may be picked out by small, circular, pore-like apertures. Apertural pores are also widely scattered across the surface of the last chamber. The bilobate form *O. bilobata* (d'Orbigny) is considered by some workers as a distinct genus and species but we believe it to be a variant of *O. universa*, which does not appear to have stratigraphic value.

Globigerina nepenthes

G. nepenthes is used as a zonal marker, for example by Blow (1959 and 1969, definition of N 14), Cati *et al.* (1968) and Kennett (1973), but it is erratic in its appearance, and may be virtually absent from whole sections. However, it is a distinctive, easily recognizable form which, when present, is of value.

The coiling pattern of *G. nepenthes* suggests a distinction between it and other members of that genus but also does not fit elsewhere. There is some tendency towards *Sphaeroidinellopsis* but the species of that genus had formed distinct cortexes before the first appearance of *G. nepenthes*.

Taxonomic notes

GLOBIGERINA NEPENTHES Todd
Figures **25**; 7, 10, 11

Type reference: *Globigerina nepenthes* Todd, 1957, p. 301, pl. 78, fig. 7.

The coiling of *G. nepenthes* begins tightly but the last few chambers become more spiral producing the characteristic elongate outline. The high, cap-shaped last chamber has a large aperture usually bordered by a thickened rim. Some workers record transitions at the beginning of the range of *G. nepenthes* between it and such forms as *G. druryi* (Akers) and *G. woodi* Jenkins.

Within its range, *G. nepenthes* can show considerable morphological variation in closeness of coiling and thickness of test wall, which has led to a number of subspecies being proposed; for example, *G. nepenthes delicatula* Brönnimann & Resig (1971) here illustrated as Figs. 25.2–4. As variations can often be ascribed to environmental conditions and as they

---

Fig. 23. *Praeorbulina, Orbulina* (all figures × 60)

1 *O. universa* d'Orbigny
Holotype. Recent, marine sands. Type locality not designated.

2 *O. suturalis* Brönnimann
2a–b Holotype. *G. menardii* Zone, Middle Miocene, Lengua Fm., Trinidad.

3 *P. glomerosa circularis* (Blow)
3a–b Holotype. Sample RM 19285, *Siphogenerina transversa* Zone (= *P. glomerosa* Zone), Early Miocene, Pozon, Eastern Falcon, Venezuela.

4 *P. glomerosa glomerosa* (Blow)
4a–b Holotype. Information as for Fig. 23.3.

5 *P. glomerosa curva* (Blow)
5a–b Holotype. Information as for Fig. 23.3.

6 *P. transitoria* Blow
6a–b Holotype. Sample RM 19280, *Siphogenerina transversa* Zone (= *P. glomerosa* Zone), Early Miocene, Pozon, Eastern Falcon, Venezuela.

Fig. 24. *Globigerinoides, Praeorbulina, Orbulina* (all figures × 60)

1 *O. bilobata* (d'Orbigny) considered to be a variant of *O. universa*. From the *G. fohsi peripheroronda* Zone, Middle Miocene, Cipero Fm., Trinidad (from Bolli, 1957).

2 *O. universa* d'Orbigny
From the *G. mayeri* Zone, Middle Miocene, Lengua Fm., Trinidad (from Bolli, 1957c).

3 *O. suturalis* Brönnimann
From the *G. menardii* Zone, Middle Miocene, Lengua Fm., Trinidad (from Bolli, 1957c).

4 *P. glomerosa circularis* (Blow)
From the *P. glomerosa* Zone, Early Miocene, Cipero Fm., Trinidad (from Bolli, 1957c).

5 *P. glomerosa glomerosa* (Blow)
Information as for Fig. 24.4.

6 *P. glomerosa curva* (Blow)
6a–b Holotype redrawn (from Jenkins *et al.*, 1981). Information as for Fig. 24.4.

7 *P. sicana* (Blow)
7a–b Holotype redrawn (from Jenkins *et al.*, 1981). Zone II (*G. insueta* Zone), Early Miocene, Cipero Fm., Trinidad.

8 *G. bisphericus* Todd
8a–c Holotype redrawn (from Jenkins *et al.*, 1981). Locality C-85, *P. glomerosa* Zone, Early Miocene, 1.6 miles northeast of seaward tip of the southwestern point of Saipan, Mariana Islands.

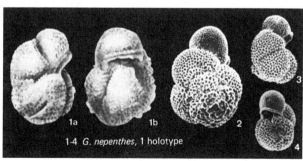

Fig. 25. *Globigerina nepenthes* (all figures × 60)

1–4 *G. nepenthes* Todd
1a–b Holotype. Locality S 621, Miocene, Tagpochau Limestone, northeast central Saipan, Mariana Islands. **2–3** DSDP Leg 7-62.1-14-1, 60–62 cm. **4** DSDP Leg 7-62.1-12-2, 36–38 cm. All Zone N 18, Early Pliocene, West-Central Pacific (from Brönnimann & Resig, 1971).

seem of little stratigraphic value, we retain these forms under the name *nepenthes* as do Stainforth *et al.*, 1975.

Oligocene to Miocene non-carinate *Globorotalia* species

Despite their rather inconspicuous features, primarily due to their lack of a peripheral keel, four out of the following seven taxa are sufficiently distinct both morphologically and in their ranges, to have been selected as zonal markers. The seven are: *Globorotalia opima nana*, *opima opima*, *kugleri*, *mayeri*, *continuosa*, *obesa*, and *fohsi peripheroronda*. Among themselves they differ in overall test size, number and shape of chambers forming the last whorl, direction of intercameral sutures, shape and size of apertures and peripheral outline in both axial and peripheral view.

Because of the stratigraphic significance of these taxa and the importance therefore of correct identification, their distinctive characters are pointed out below in some detail. Reference is made to the section 'The *Globorotalia fohsi* lineage' for the discussion of *Globorotalia fohsi peripheroronda* and its comparison with other Oligocene and Miocene non-carinate *Globorotalia* species.

Taxonomic notes

GLOBOROTALIA OPIMA OPIMA Bolli and
GLOBOROTALIA OPIMA NANA Bolli
Figures **26.15–30**; 6, 9, 12

Type references: *Globorotalia opima opima*, Bolli, 1957, p. 117, pl. 28, figs. 1a–c.
Globorotalia opima nana, Bolli, 1957, p. 118, pl. 28, figs. 3a–c.

The larger subspecies *opima* and the smaller sized *nana* are distinguished by size only. Number and arrangement of chambers in the last whorl, robust wall structure with coarse pores and low aperture are common to both. With its more restricted range, *G. opima opima* has turned out to be a valuable and widely used index fossil within the Oligocene, and a zone based on its total range has been established.

As only the large sized *opima* subspecies has a restricted, stratigraphically valuable range, its clear separation from *nana* is essential. Originally, only 'small' as against 'large' size was cited as a distinction for the two subspecies with measurements only given for the holotypes. Investigators therefore had to decide arbitrarily whether to place intermediate specimens into one or the other subspecies, with the result that broadening of the concept of *G. opima opima* sometimes resulted in an over-extension of the length of the zone.

To determine whether the two subspecies can be divided more accurately by size we have measured the greatest diameter of a sequence of different sized specimens from the type locality of the two subspecies, with the result that the following

four size intervals can be quite clearly distinguished: 0.25–0.32, 0.32–0.38, 0.39–0.43 and 0.45–0.5 mm or slightly larger. The smallest group belongs unquestionably to the subspecies *nana*, the largest to *opima*. Of the two intermediate size groups the larger, 0.39–0.43 mm can also be assigned to *opima* as it represents specimens equivalent to the largest group but apparently not fully grown and therefore lacking the final chamber.

The three groups 0.25–0.32, 0.39–0.43 and 0.45–0.5 mm consist of specimens having the last whorl formed by 4 chambers. Only rarely does one find here specimens with 5 chambers. Such an unusual specimen was unfortunately chosen as the holotype for *G. opima opima*, while the paratype figured along with it is the far more common 4-chambered form.

Specimens of the size range 0.32–0.38 mm often possess 5 chambers in the last whorl instead of 4, unlike the smaller and larger groups. This is also the size interval that divides the smaller, typical subspecies *nana* from the larger *opima*. Because of its intermediate size, this group with its 5-chambered specimens is difficult to assign to either of the two subspecies. Population studies in more samples from different levels within the zone and different areas should eventually aid in deciding whether these intermediate sized 5-chambered forms belong to the *opima* complex at all or whether they might possibly be ancestral to *Globorotalia mayeri*.

Based on the measurements made on the populations at the type locality, the following subdivision into the subspecies *nana* and *opima* is indicated:

0.25–0.32 mm: *G. opima nana*
0.32–0.38 mm: intermediate sized specimens often with 5 chambers, not typical for *nana* or *opima*, probably restricted to the *Globorotalia opima opima* Zone
0.39–0.5 mm or slightly larger: *G. opima opima*

Typical *G. opima opima* differ from *G. mayeri* in having the last whorl usually formed by 4 instead of 5 or 6 chambers. In side view *opima* tests are higher than those of *mayeri*. The equatorial peripheral outline appears more circular in *opima* compared with an often slightly elongate one in *mayeri*. While the last chamber in *G. mayeri* specimens may be sack-like and drawn back ('sacculiferid'), resulting in a curved suture between the last two chambers, this feature is unknown in *G. opima opima*. The aperture in *G. opima opima* is a low slit compared with a distinctly higher arched one in *G. mayeri*. The differences between *G. opima nana* and *G. continuosa* are virtually the same as those between *G. opima opima* and *G. mayeri*.

It is possible that the 5-chambered, medium to small sized atypical *G. opima* may represent the ancestral stock for *G. mayeri*, which develops more typically above this stratigraphic level. Likewise it is difficult to draw a distinct limit between the 4-chambered *G. opima nana* and *G. continuosa*.

Specimens that cannot clearly be assigned to one or the other taxon are present from the *Globigerina ciperoensis ciperoensis* to the *Catapsydrax stainforthi* Zone (Bolli & Saunders, 1982a, pl. 4).

The two subspecies *opima* and *nana* have been recorded from low and mid-latitudes including the Mediterranean area. To what extent ecological factors controlled the presence of the two subspecies is not yet fully understood. There are indications for a temperature control as pointed out by Stainforth *et al.* (1975) for Ecuador, where the subspecies *nana* is said to be abundant in a cool-water fauna but *opima* is absent.

Bolli (1957), Blow (1969) and Stainforth *et al.* (1975) quote *G. opima nana* as occurring from the Middle Eocene *Truncorotaloides rohri* Zone to the top of the Middle Oligocene *Globigerina ciperoensis ciperoensis* Zone. Re-examination of samples through the Middle and Late Eocene and Early Oligocene confirm the presence as early as the *Globorotalia lehneri* Zone of small, 4-chambered specimens morphologically very close to the type *nana*. Some of the specimens in the type locality sample of the *Truncorotaloides rohri* Zone in Trinidad closely resemble *G.* cf. *bolivariana*, a near homeomorph of *G. opima opima*, though they are smaller in size. It would appear that at least some of the Eocene and Early Oligocene small 4-chambered specimens that are morphologically close to *G. opima nana* may in fact be homeomorphs and not be related genetically.

GLOBOROTALIA KUGLERI Bolli
Figures **26.1–6**; 6, 9, 12

Type reference: *Globorotalia kugleri* Bolli, 1957, p. 118, pl. 28, figs. 5a–c.
Synonyms: *Globorotalia (Turborotalia) pseudokugleri* Blow, 1969, p. 391, pl. 10, figs. 4–6 (from Bolli, 1957, pl. 28, figs. 7a–c).
Globorotalia (Turborotalia) mendacis Blow, 1969, p. 390, pl. 30, figs. 5–6.

Typical *G. kugleri* specimens are comparatively small in size (around 0.3 mm) with 7–8 chambers forming the last whorl, equatorial periphery almost circular, spiral intercameral sutures variable, distinctly curved to straight but swept back as in Figs. 26.1a, 2, 4, or near radial as in Fig. 26.3a. Depending on the degree of curvature of the intercameral sutures, individual chambers may be of subglobular to distinctly elongate-curved shape.

For specimens with more radial sutures and more globular chambers Blow (1969) proposed the separate species *G. pseudokugleri*, citing *G.* cf. *kugleri* of Bolli (1957, pl. 28, figs. 7a–c, here shown on Fig. 26.3) as holotype. As an intermediate between *kugleri* forms with distinctly curved sutures and *G. pseudokugleri* the same author erected a *G. mendacis*. Apart from extreme forms these three taxa, which

virtually have the same range, are difficult to distinguish. As the subdivision appears to have no stratigraphic significance, we agree with Stainforth *et al.* (1975) that a subdivision of the *G. kugleri* plexus into these three taxa is not necessary.

G. kugleri has found wide application as an index form in the uppermost Oligocene and lowermost Miocene (range: *Globorotalia kugleri* to *Globigerinoides primordius* Zone). Certain affinities to *Globorotalia fohsi peripheroronda* and *G. mayeri* may readily lead to misidentification of *G. kugleri*, resulting in incorrect stratigraphic interpretations. It is essential therefore that the taxon is used in a very restricted sense. Here, the number of chambers in the last whorl (7–8) is of prime importance in separating the species from *G. fohsi peripheroronda* and *G. mayeri* which both display 5–6 chambers. Other distinguishing characters of *G. kugleri* are its small size and its more circular equatorial circumference.

GLOBOROTALIA MAYERI Cushman & Ellisor
Figures **26.31–43**; 6, 9, 12

Type reference: *Globorotalia mayeri* Cushman & Ellisor, 1939, p. 11, pl. 2, figs. 4a–c.
Synonym: *Globorotalia siakensis* LeRoy, 1939, p. 39, pl. 3, figs. 30–31.

This species of wide geographic distribution in both low and mid-latitudes and of extended stratigraphic range is widely used as an index form because of the value of its extinction level in the late Middle Miocene. The question as to whether or not *G. siakensis* LeRoy, originally described from Sumatra, and *G. mayeri* from Louisiana, are synonyms, has come under discussion in recent years particularly since Blow (1969) described them as being different, based on refigured holotypes.

To obtain a better understanding of the *mayeri/siakensis* problem for this book, Bolli & Saunders (1982a) investigated and compared a large number of specimens from the zonal type localities of the Cipero and Lengua formations of Trinidad, West Indies, ranging from the *Globigerina ciperoensis ciperoensis* to the *Globorotalia mayeri* Zone, and from well Bodjonegoro-1 in Java (Bolli, 1966b), from the *Globigerinatella insueta* (bottom of hole) to the *Globorotalia mayeri* Zone. Furthermore, the holotype of *G. mayeri*, previously illustrated by the original authors and re-illustrated in Blow (1969) was again carefully re-drawn and is here reproduced (Fig. 26.31).

The principal criteria for a distinction of *G. mayeri* from *G. siakensis* as given by Blow are the distinctly curved spiral intercameral sutures and less lobate equatorial periphery in *G. mayeri* compared with the more radial sutures and more strongly lobate periphery in *G. siakensis*. Examination and re-illustration of the *G. mayeri* holotype shows that the sutures are much less curved than in Blow's (1969) illustration and closer to Cushman & Ellisor's original figure.

The variability studies furthermore revealed that when

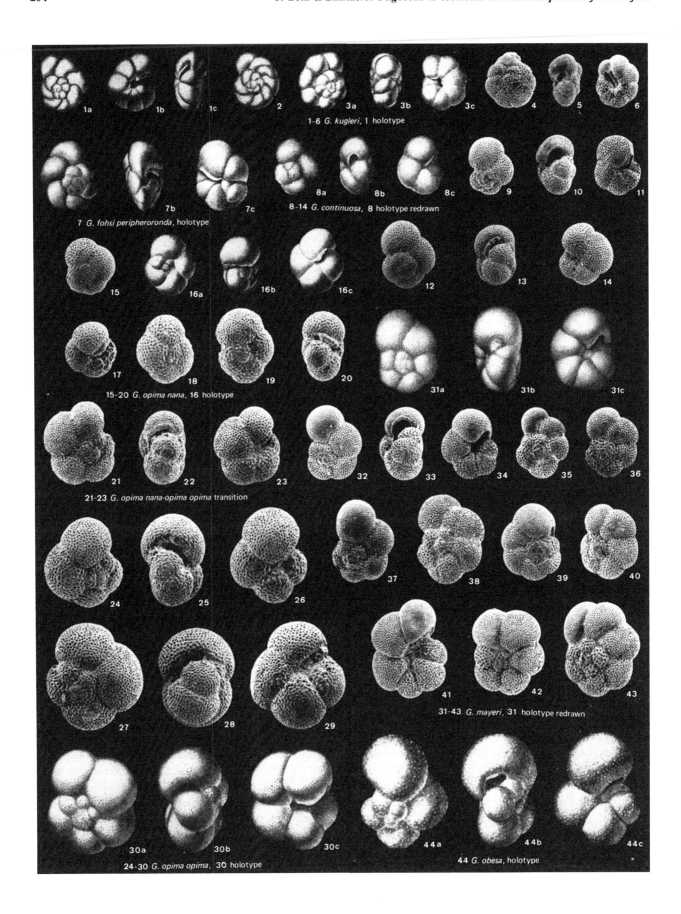

1-6 *G. kugleri*, 1 holotype

7 *G. fohsi peripheroronda*, holotype

8-14 *G. continuosa*, 8 holotype redrawn

15-20 *G. opima nana*, 16 holotype

21-23 *G. opima nana-opima opima* transition

31-43 *G. mayeri*, 31 holotype redrawn

24-30 *G. opima opima*, 30 holotype

44 *G. obesa*, holotype

curvature of intercameral sutures occurs it is usually only between the last two chambers with the last chamber having a drawn backward, sack-like (sacculiferid) shape. Specimens with this feature may occur throughout most of the range of the taxon.

Presentation of these *mayeri/siakensis* investigations including the figuring of variabilities from all zones throughout the range became too voluminous for inclusion in full here. Results, conclusions and extensive illustrations therefore had to be published separately (Bolli & Saunders, 1982a). Their conclusion was 'that neither on morphologic nor on stratigraphic distribution criteria can a clear distinction between *G. mayeri* and *G. siakensis* be maintained' and that they fall within the variability of one species. Both taxa were published in 1939, *G. mayeri* has priority over *G. siakensis* because its publication month is March 1939, whereas *G. siakensis*, originally thought to have been undated within 1939, has now been found to carry a November date (Bolli & Saunders, 1982b).

Contrary to the findings in low latitudes as discussed above and in more detail by Bolli & Saunders (1982a), some workers in higher latitudes, such as Cita *et al.* (1978) and Borsetti *et al.* (1979) believe that they are in a position to

---

Fig. 26. Oligocene to Miocene non-carinate *Globorotalia* species (all figures × 60)

1–6 *Globorotalia kugleri* Bolli
1a–c Holotype, **2**, **3a–c** Paratypes. Sample B 8672, *G. primordius* Zone, Early Miocene, Cipero Fm., San Fernando area, Trinidad. **4–6** Sample JS 899, *G. primordius* Zone, Early Miocene, Cipero Fm., north of Mosquito Creek, Trinidad.

7 *Globorotalia fohsi peripheroronda* Blow
7a–c Holotype. Sample RM 19304, *G. fohsi peripheroronda* Zone, Middle Miocene, Pozon Fm., Eastern Falcon, Venezuela.

8–14 *Globorotalia continuosa* Blow
8a–c Holotype redrawn. Sample RM 19542, *G. mayeri* Zone, Middle Miocene, Pozon Fm., Eastern Falcon, Venezuela. **9–11** Sample Bo 79/32, *G. mayeri* Zone, Middle Miocene, Lengua Fm., Golconda area, Trinidad. **12–14** Sample Bo 267, *C. dissimilis* Zone, Early Miocene, Cipero Fm., San Fernando area, Trinidad.

15–20 *Globorotalia opima nana* Bolli
16a–c Holotype; **15**, **17–20** Paratypes. Sample JS 20, *G. opima opima* Zone, Late Oligocene, Cipero Fm., Cipero coast, Trinidad.

21–23 *Globorotalia opima nana–opima opima* transition
5-chambered specimens of intermediate size. Information as for Figs. 26.15–20.

24–30 *Globorotalia opima opima* Bolli
24–26 Paratypes, medium sized specimens; **27–29** Paratypes, large specimens; **30a–c** Holotype. Information as for Figs. 26.15–20.

31–43 *Globorotalia mayeri* Cushman & Ellisor
31a–c Holotype redrawn (from Bolli & Saunders, 1982a). Ellender well 1, at 9612 feet, *G. fohsi fohsi* Zone, Louisiana, USA. **32–36** Sample Bo 79/31, *G. fohsi robusta* Zone, Middle Miocene, Cipero Fm., Trinidad. **37–38** Sample JS 32, *G. fohsi lobata* Zone, Middle Miocene, Cipero Fm., Cipero coast, Trinidad. **39–40** Sample Bo 267, *C. dissimilis* Zone, Early Miocene, Cipero Fm., San Fernando area, Trinidad. **41–43** Sample Bo 184A, *G. fohsi fohsi* Zone, Middle Miocene, Cipero Fm., Golconda Estate, Trinidad.

44 *Globorotalia obesa* Bolli
44a–c Holotype. Sample JS 16, *G. fohsi robusta* Zone, Middle Miocene, Cipero Fm., Cipero coast, Trinidad.

---

distinguish clearly a *Globorotalia siakensis* type, i.e. forms with more radially arranged spiral sutures, more open umbilicus and more distinctly lobate equatorial outline, and a *G. mayeri* type which, compared with *G. siakensis*, possesses more oblique to slightly curved sutures on the spiral side, a very narrow umbilical area and a less lobate equatorial outline.

These workers also believe they have evidence that in their areas these two morphologically distinguishable forms have also distinctly different stratigraphic significance in that *G. siakensis* ranges from the *Globigerina ciperoensis ciperoensis* Zone to the *Globorotalia menardii* Zone, *G. mayeri* from the upper part of the *Globorotalia fohsi lobata* Zone to the top of the *Globorotalia mayeri* Zone. Extensive faunal populations from, for example, the Mediterranean sections should be directly compared with those from low latitude sections to arrive at a reliable evolution of the variabilities in different latitudes.

Differing views on origin and evolutionary trends of *G. mayeri* have been expressed. From observations in low latitudes *G. mayeri* appears to have evolved from the *G. opima s.l.* stock within the *Catapsydrax stainforthi* Zone. *G. mayeri* continued unchanged parallel to the evolving *fohsi* lineage and even survived it for a short time after the extinction of *G. fohsi robusta*, the last subspecies in the *fohsi* lineage.

Some workers studying higher latitude faunas, where *G. fohsi peripheroronda* is also present but where the later *fohsi* subspecies did not develop, believe they have evidence that instead of *G. mayeri* giving rise to *G. fohsi peripheroronda*, this subspecies is instead the ancestor of *G. mayeri* (Jenkins, 1960, 1966; Blow, 1969). These questions of origin and evolution are also discussed in Bolli & Saunders (1982a). They concluded that still more complete documentation is needed from higher latitudes, where *G. mayeri* is thought to have evolved from *peripheroronda*, before such an interpretation can be considered proved.

For the differences between *G. mayeri* and *G. opima opima* reference is made to the taxonomic notes on *G. opima opima*. *G. mayeri* differs from *G. kugleri* s.s. in its larger size, fewer chambers (5–6 instead of 7–8) in the last whorl, in the less curved intercameral sutures on the spiral side, which results in more globular chambers, and in a higher arched aperture. In axial view the chambers are more symmetrical compared with *G. kugleri*.

G. mayeri differs from *G. fohsi peripheroronda* in possessing more distinctly radial intercameral sutures on the spiral side, in the chambers being more regularly inflated, which again clearly shows in axial view, and in the aperture being distinctly higher arched compared with the low slit in *G. fohsi peripheroronda*. In the *Catapsydrax stainforthi* Zone intermediate forms may occasionally make a clear distinction between *G. mayeri* and *G. fohsi peripheroronda* difficult.

Compared with the 4–4½-chambered *G. obesa, G. mayeri*

possesses 5–6 less globular chambers increasing less rapidly in size. Sutures in *G. obesa* are more deeply incised, resulting in a more lobate periphery in equatorial view.

In Trinidad *G. mayeri* coils at random from its appearance in the *Globigerina ciperoensis ciperoensis* to the *Globigerinatella insueta* Zone, to become strongly predominantly sinistral from the *Globorotalia fohsi peripheroronda* Zone onwards until its extinction at the top of the *Globorotalia mayeri* Zone. In Bodjonegoro-1, Java (Bolli, 1966b), the change from random to sinistral coiling also sets in within the *Globorotalia fohsi peripheroronda* Zone but, in contrast to Trinidad, continues to fluctuate on several occasions from random to sinistral through the lower part of the *Globorotalia fohsi fohsi* Zone, from where coiling remains exclusively sinistral until the extinction of the taxon.

GLOBOROTALIA CONTINUOSA Blow
Figures **26.8–14**; 6, 9, 12

Type reference: *Globorotalia opima* Bolli subsp. *continuosa* Blow, 1959, p. 218, pl. 19, figs. 125a–c.

This small, 4-chambered species was regarded by its author as closely related to the ancestral *G. opima nana* and to be itself the ancestor of *G. acostaensis*, which appears at the base of the Late Miocene. Origin, relations and range of *G. continuosa* were re-investigated by Bolli & Saunders (1982a) in conjunction with their study on *G. mayeri*. Here also their observations are based on a sequence of specimens throughout the range of the species. *G. continuosa* appears to have developed gradually from the *G. opima nana* stock through the *Globigerina ciperoensis ciperoensis* Zone to the *Globigerinatella insueta* Zone, becoming more lobate and gradually acquiring a higher arched aperture (Bolli & Saunders, 1982a, pl. 4).

From *G. mayeri* the species is distinguished by its smaller size and by possessing only 4 instead of 5–6 chambers in the last whorl. Identical with *G. mayeri* is the overall morphology, the range and the coiling pattern. We do not agree with Blow that *G. continuosa* is the ancestor of *G. acostaensis* for the following reasons: In Trinidad and Java *G. continuosa* became extinct together with *G. mayeri* at the top of the *G. mayeri* Zone. *G. acostaensis* appeared on the other hand only at the base of the zone of the same name. Both species are therefore absent in the intermediate *Globorotalia menardii* Zone. This interval lacking both *G. continuosa* (and also *G. mayeri*) and *G. acostaensis* is particularly well documented in the continuously cored section of well Bodjonegoro-1 in Java (Bolli, 1966b), where the *Globorotalia menardii* Zone is 214 metres thick.

In addition, the aperture of *G. continuosa* is a high, comma-shaped arch while that of *G. acostaensis* is, in contrast, a low slit.

From the above cited observations it is therefore concluded that *G. continuosa* is in its overall morphology and range closely related to *G. mayeri* and that the species is not the ancestor of the *G. acostaensis–G. humerosa–G. dutertrei* plexus, which first appears with *G. acostaensis* only sometime after the extinction of *G. continuosa*. From the stratigraphic point of view it is considered that there is little value in separating *G. continuosa* from *G. mayeri*.

GLOBOROTALIA OBESA Bolli
Figures **26.44**; 6, 9, 12

Type reference: *Globorotalia obesa* Bolli, 1957, p. 119, pl. 29, figs. 2a–c.

The last whorl of *G. obesa* is formed by 4 globular chambers, rapidly increasing in size and separated by deeply incised sutures resulting in a distinctly lobate equatorial periphery. These are the criteria by which the species is distinguished from other Oligocene–Miocene non-carinate *Globorotalia* species, including *G. continuosa*; this species also possesses 4 chambers but these are more ovate and less inflated.

The *Globorotalia acostaensis–Globorotalia humerosa–Neogloboquadrina dutertrei* lineage

Wherever possible we follow the principle that established lineages should be treated in such a way that related taxa are readily identifiable as such, usually as subspecies of a species. An example of this is the *Globorotalia fohsi* lineage. One would expect that the *acostaensis–humerosa–dutertrei* lineage, including the closely related additional taxa, could be treated in a similar way. However, the principal forms of stratigraphic significance (*acostaensis, humerosa, dutertrei*) are today generally placed in at least two different genera (*Globorotalia, Neogloboquadrina*). The principle of using subspecies of a single species under one genus for all taxa of the lineage is therefore not applicable here. As a consequence, we distinguish in the lineage the three *Globorotalia* species *acostaensis, humerosa, pseudopima* and the *Neogloboquadrina* species *dutertrei*.

The taxa *trochoidea* and *prehumerosa* subsequently erected as subspecies of *acostaensis* and *humerosa* respectively are retained as such. The taxon *pseudopima* originally described as a subspecies of *acostaensis* differs so much that it is here given species rank. In *N. dutertrei* are included the three subspecies *dutertrei, blowi* and *tegillata*.

Taxonomic problems

G. acostaensis Blow, *G. humerosa* Takayanagi & Saito and *N. dutertrei* (d'Orbigny) have in recent years repeatedly been discussed as regards their generic assignment and for their use as index fossils within the Late Miocene to Holocene. In addition to the original species descriptions, reference is made to Banner & Blow (1960a), Parker (1962, 1967), Bandy *et al.*

(1967), Blow (1969), Bolli (1970), Lamb & Beard (1972), Rögl & Bolli (1973), Stainforth *et al.* (1975) and Srinivasan & Kennett (1976). Where evolutionary trends have been discussed by these authors there is good agreement among them that these species represent an evolutionary lineage leading by gradual morphological changes from *G. acostaensis* through *G. humerosa* to *N. dutertrei*. Blow (1969) followed by Stainforth *et al.* (1975), Srinivasan & Kennett (1976) and others proposed a development of *G. acostaensis* from an ancestral *Globorotalia continuosa* Blow. Convincing documentation for such an assumption is still wanting.

Subdividing lineages with gradual, flowing, morphological changes into a number of clearly defined taxa is often subjective because of the difficulty of placing transitional forms. The interval occupied by such forms between typical specimens of two successive taxa is also one within which a zonal boundary can fluctuate to some degree pending on the observers' interpretation.

In addition to defining individual taxa as exactly as possible in the *G. acostaensis–G. humerosa–N. dutertrei* lineage, one is also faced with some additional problems, though these are less important for their biostratigraphic application. They concern the generic position of the various species. The great variability in characters of generic significance within this group of species has resulted in successive assignments to *Globigerina*, *Globorotalia*, *Globoquadrina* and *Neogloboquadrina*. The type species of *Neogloboquadrina* is *Globigerina dutertrei* d'Orbigny. As regards the arrangement of chambers, position of apertures and the tooth-like umbilical flaps, *Neogloboquadrina* combines features of the three genera *Globigerina*, *Globorotalia* and *Globoquadrina*.

The early forms of the lineage as represented by *acostaensis* and its subspecies *acostaensis* and *trochoidea*, and *humerosa* with the subspecies *humerosa* and *praehumerosa* possess a narrow umbilicus and an aperture which is distinctly umbilical–extraumbilical. Together with the 4-chambered species *pseudopima* with its nearly closed umbilicus, these taxa are retained in *Globorotalia*. In *Neogloboquadrina dutertrei* with its subspecies *dutertrei*, *blowi* and *tegillata* are here included Late Miocene to Holocene low to high trochospiral forms which display an open umbilicus, with or without tooth-like flaps extending from the last chambers into the umbilical area, and with an umbilical position of the aperture.

A second problem, also taxonomic in nature, concerns the original type species of *Globigerina dutertrei*, described and figured by d'Orbigny (1839a). The original type can no longer be found and therefore Banner & Blow (1960a) erected a lectotype which they chose from a tube found in the d'Orbigny collection which according to them is labelled in d'Orbigny's own handwriting as containing '*Globigerina dutertrei*'. According to Banner & Blow, of the eight specimens originally in the tube only a single undamaged specimen remained and this they selected as a lectotype.

A comparison of d'Orbigny's figured type (Figs. 27.3a–c) with the lectotype (Figs. 27.4a–c) reveals such distinctive morphological differences that by today's standards the two specimens would be assigned to two different species. In fact, Banner & Blow's lectotype of *G. dutertrei* is much closer and probably conspecific with *Globigerina eggeri* Rhumbler, 1901 (Figs. 27.1–2), rather than with d'Orbigny's figured type of *G. dutertrei*. However, the lectotype clearly falls within the variability of *Neogloboquadrina dutertrei* as generally accepted today and figured in Parker (1962), Rögl & Bolli (1973), Stainforth *et al.* (1975), and others.

In contrast the type figured by d'Orbigny does not fit into the variability of today's *dutertrei* concept for the following reasons: Its chambers in the last whorl increase more rapidly in size, in particular the final one. This results in a more lobate periphery and elongated test shape in spiral and umbilical view. The last chamber in d'Orbigny's *dutertrei* is distinctly larger than in the lectotype, or in today's concept of *dutertrei*, where it is often of the same size or even smaller than the penultimate chamber, often laterally also more appressed. D'Orbigny's figures also indicate that his specimen was spinose, in contrast to the *Neogloboquadrina dutertrei* as understood today which displays a pitted surface without spines. Finally, Recent *dutertrei* are almost exclusively dextrally coiling, while d'Orbigny's specimen coils sinistrally.

From this it is apparent that today's concept of *dutertrei* differs from the type figured by d'Orbigny, and that it is much closer or identical with that of *Globigerina eggeri*. However, in view of the present-day widespread usage and understanding of *Neogloboquadrina dutertrei*, this species name is maintained here.

Stratigraphic significance

The principal biostratigraphic value of the discussed taxa lies in the distinction of the two Late Miocene zones *G. acostaensis* and *G. humerosa*. From its first occurrence as small 4-chambered forms, *G. acostaensis* evolves within the *Globorotalia acostaensis* Zone to larger, 5-chambered forms, allowing a possible recognition of a lower and upper part. Because of the gradual evolution from *G. acostaensis* through *G. praehumerosa* to *G. humerosa*, placing of the boundary between the two zones (defined by the first appearance of *G. humerosa*) remains somewhat arbitrary. Similar transitions occur between *G. humerosa* and *Neogloboquadrina dutertrei*, but here the top of the *Globorotalia humerosa* Zone is defined by the first occurrence of *Globorotalia margaritae*. First occurrences of taxa within the lineage discussed are difficult to place unanimously while last occurrences within this group are, on our present knowledge, still more vague because juvenile

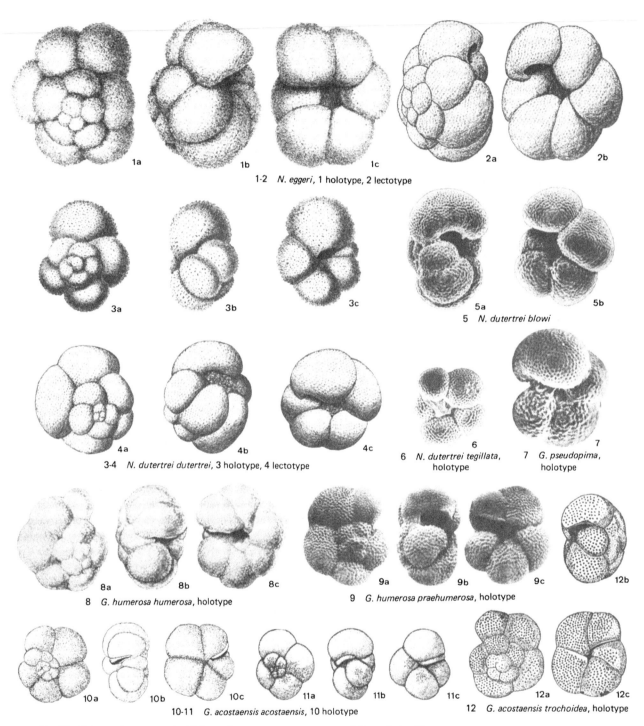

1-2   N. eggeri, 1 holotype, 2 lectotype

5   N. dutertrei blowi

3-4   N. dutertrei dutertrei, 3 holotype, 4 lectotype

6   N. dutertrei tegillata,   7   G. pseudopima,
holotype   holotype

8   G. humerosa humerosa, holotype

9   G. humerosa praehumerosa, holotype

10-11   G. acostaensis acostaensis, 10 holotype

12   G. acostaensis trochoidea, holotype

Fig. 27. *Globorotalia acostaensis–Globorotalia humerosa–Neogloboquadrina dutertrei* and related taxa (all figures × 60)

1–2 *N. eggeri* (Rhumbler)
   1a–c Holotype, **2a–b** Lectotype (re-illustrations of side and umbilical views of 1a). *Challenger* Station 300, Recent, north of Juan Fernandez, Atlantic.

3–4 *N. dutertrei dutertrei* (d'Orbigny)
   3a–c Holotype, **4a–c** Lectotype. Both from recent beach sands, Cuba.

   5 *N. dutertrei blowi* Rögl & Bolli
   5a–b. DSDP Leg 15-147-4-cc. *G. bermudezi* Subzone, Late Pleistocene, Cariaco Trough, Caribbean Sea (from Rögl & Bolli, 1973).

   6 *N. dutertrei tegillata* Brönnimann & Resig
   Holotype. DSDP Leg 7-62.1-15-1, 15–17 cm, Zone N 18, Early Pliocene, Pacific.

   7 *G. pseudopima* Blow
   Holotype. Sample Ba 14, Zone N 21, Late Pliocene, Otim River section, Sarmi Fm., West Irian.

   8 *G. humerosa humerosa* Takayanagi & Saito
   8a–c Holotype. Sample A-16, *G. cultrata cultrata/G. nepenthes* Zone (equivalent to *G. humerosa* Zone), Late Miocene, Nobori, Japan.

   9 *G. humerosa praehumerosa* Natori
   9a–c Holotype. Sample S-30, Zone N 30, Middle Pliocene, Shimajiri section, Yonabaru Fm., Okinawa, Japan.

10–11 *G. acostaensis acostaensis* Blow
   10a–c Holotype. Sample RM 19791, *S. semiinvoluta* Zone (equivalent to *G. acostaensis* Zone), Late Miocene, Pozon Fm., Eastern Falcon, Venezuela. **11a–c** CAP core 38BP, 860–862 cm, Zone N 17, Late Miocene, Pacific Ocean (from Parker, 1967, illustrated as *Globoquadrina continuosa*).

12 *G. acostaensis trochoidea* Bizon & Bizon
   12a–c Holotype. Middle Miocene (probably Late Miocene in our terminology), Western Epirus, Greece.

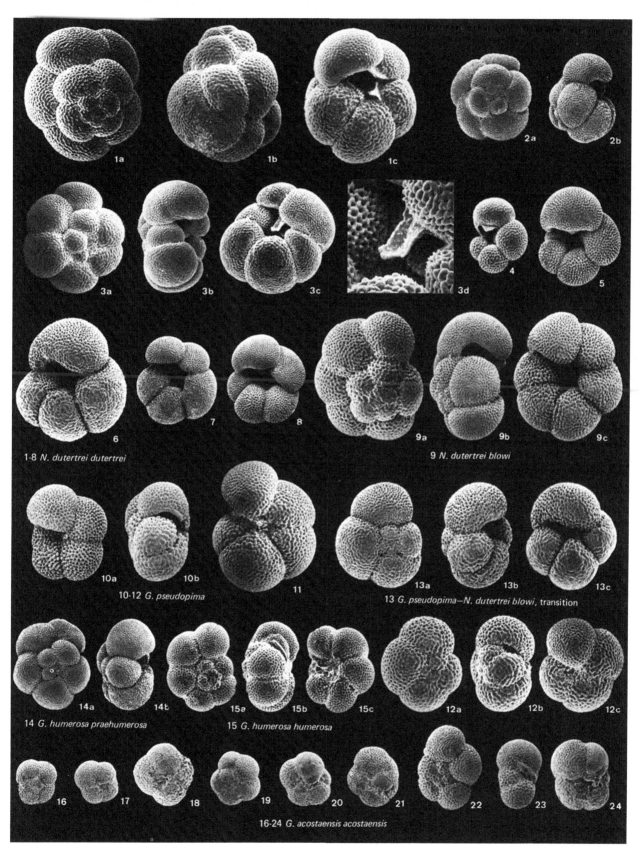

1-8 *N. dutertrei dutertrei*   9 *N. dutertrei blowi*

10-12 *G. pseudopima*   13 *G. pseudopima—N. dutertrei blowi*, transition

14 *G. humerosa praehumerosa*   15 *G. humerosa humerosa*

16-24 *G. acostaensis acostaensis*

Fig. 28. For caption see p. 210.

stages of later forms such as *N. dutertrei* may be difficult to distinguish from adult earlier ones.

Of the taxa discussed under the taxonomic notes only *pseudopima* and the subspecies *acostaensis*, *humerosa* and *dutertrei* are considered to be of practical biostratigraphic significance and their ranges only are included on the range chart.

Coiling trends

Coiling patterns of *acostaensis–humerosa–dutertrei* were followed in some detail in subsurface sections such as Bodjonegoro-1 in Java (Bolli, 1966b), Cubagua-1 in the Caribbean (Bermudez & Bolli, 1969) and in a number of DSDP sites from Legs 4 and 15 (Sites 29, 30, 31, 149, 154; Bolli, 1970; Bolli & Premoli Silva, 1973). *Globorotalia acostaensis* shows in all sections a strong preference for sinistral coiling within the *Globorotalia acostaensis* Zone. The coiling of *G. acostaensis* and *G. humerosa* is irregular in the *Globorotalia humerosa* Zone and lower part of the *Globorotalia margaritae* Zone. It is dextral throughout in Bodjonegoro-1, whereas in the Caribbean sections the pattern remains mainly sinistral with occasional excursions to random and dextral. The *N. dutertrei* group and *G. pseudopima* coil dextrally throughout, from the higher part of the *Globorotalia margaritae* Zone to the Holocene.

In the Mediterranean area *G. acostaensis* is dominantly sinistrally coiling, except for a short interval in its early range within the *Globorotalia acostaensis* Zone where it coils at random. It changes to dominantly dextral within the *Globorotalia conomiozea* Zone. The coiling change from sinistral to dextral occurs above the first appearance of *G. conomiozea* and just before the first appearance of *G. multiloba* and then remains constant throughout the Pliocene (Zachariasse, 1975; Stainforth *et al.*, 1975; Bossio *et al.*, 1976; Civis *et al.*, 1977; Manuputty, 1977; Cita & Ryan, 1978).

Taxonomic notes

An important difference between *G. acostaensis*, *G. humerosa* and *N. dutertrei* lies in the increase of test size, increase in number of chambers forming the last whorl and a trend towards a more open umbilicus. *G. acostaensis* and *G. humerosa* are usually very low trochispiral, *N. dutertrei* low to medium to high trochispiral. Apertures are umbilical–extra-umbilical in *G. acostaensis* and *G. humerosa* and mainly umbilical in *N. dutertrei*, where the last chambers may form an elongated lip or tooth-like flap. Such flaps, or a tendency to cover the aperture by a rudimentary last chamber, may occur already in *G. acostaensis*, which prompted Brönnimann & Resig to erect a *G. acostaensis tegillata*.

GLOBOROTALIA ACOSTAENSIS
ACOSTAENSIS Blow
Figures **27.10–11, 28.16–24**; 6, 10, 11

Type reference: *Globorotalia acostaensis* Blow, 1959, p. 208, pl. 17, figs. 106a–c.

The variability of *G. acostaensis* within the *Globorotalia acostaensis* Zone, for example in the well section Bodjonegoro-1 of Java (Bolli, 1966b), is such that early specimens in the lowermost part of the zone are very small, mostly 4-chambered and with only shallow intercameral sutural depressions. A gradual development within the zone into larger forms with a dominance of 5 chambers takes place. The size range of the early 4-chambered specimens varies from 0.2 to 0.3 mm. The later, 5-chambered specimens may reach 0.4 mm or very slightly more.

GLOBOROTALIA ACOSTAENSIS
TROCHOIDEA Bizon & Bizon
Figure **27.12**

Type reference: *Globorotalia acostaensis* Blow subsp. *trochoidea* Bizon & Bizon, 1965, p. 246, pl. 4, figs. 12a–c.

The subspecies differs from *G. acostaensis acostaensis* in a slightly more convex spiral side. For stratigraphic purposes it is included in *G. acostaensis acostaensis*.

Fig. 28. *Globorotalia acostaensis–Globorotalia humerosa–Neogloboquadrina dutertrei* and related taxa (all figures × 60, except 3d × 180). Figs. 28.1, 3, 11, 12 and 15 from Bolli (1970)

1–8 *N. dutertrei dutertrei* (d'Orbigny)
 1a–c DSDP Leg 4-30-1-5, 1–3 cm, *G. truncatulinoides* Zone, Pleistocene, Aves Ridge, Caribbean Sea. **2a–b** Small specimen from DSDP Leg 15-147-2-cc, *G. fimbriata* Subzone, Holocene, Cariaco Trough, Caribbean Sea. **3a–d** Specimen with umbilical tooth in penultimate chamber. Vema V26, core 119 (same locality as DSDP Site 29), Venezuela Basin, Caribbean Sea. **4–8** DSDP Leg 15-147-2-1-top, *G. fimbriata* Subzone, Holocene, Cariaco Trough, Caribbean Sea. 4–5 small specimens with umbilical teeth.
 9 *N. dutertrei blowi* Rögl & Bolli
 9a–c DSDP Leg 4-31-4-cc, *G. miocenica* Zone, Middle Pliocene, Beata Ridge, Caribbean Sea.
 10–12 *G. pseudopima* Blow
 10a–b, 12a–c Small specimens, 11 Large specimen. Information as for Fig. 28.9.
 13 *G. pseudopima–N. dutertrei blowi* transition
 13a–c DSDP Leg 4-31-4-cc, *G. miocenica* Zone, Middle Pliocene, Beata Ridge, Caribbean Sea.
 14 *G. humerosa praehumerosa* Natori
 14a–b Information as for Figs. 28.4–8.
 15 *G. humerosa humerosa* Takayanagy & Saito
 15a–c DSDP Leg 4-25-3-5, at 75 cm, *G. humerosa* Zone, Late Miocene, Atlantic Ocean.
 16–24 *G. acostaensis acostaensis* Blow
 16–17 Small 4-chambered specimens, **18** Larger 4-chambered specimen, **19–21** Small 5-chambered specimens, 21 with final chamber extending over umbilical area. **22–24** Large 5-chambered specimens. 16–24 from cores of well Bodjonegoro-1, Java (Bolli, 1966b), 16–17 at 641 m, 18 at 562.5 m, 19–20 at 578 m, 21 at 565.5 m, 22–24 at 407–408 m.

GLOBOROTALIA HUMEROSA
PRAEHUMEROSA Natori
Figures **27.9, 28.14**

Type reference: *Globorotalia (Turborotalia) humerosa prae-humerosa* Natori, 1976, p. 227, pl. 2, figs. 1a–c.

This subspecies was proposed as intermediate between *G. acostaensis* and *G. humerosa*. Comparing holotypes, the 5-chambered *praehumerosa* differs from *acostaensis* merely in its slightly larger size and from *humerosa* in possessing one chamber less in the last whorl. For biostratigraphic purposes Natori's *praehumerosa* is here included in *humerosa*.

GLOBOROTALIA HUMEROSA HUMEROSA
Takayanagi & Saito
Figures **27.8, 28.15**; 6, 10, 11

Type reference: *Globorotalia humerosa* Takayanagi & Saito, 1962, p. 78, pl. 28, figs. 1a–c.

The morphological features of *G. humerosa* are largely those of *G. acostaensis*, except for its larger test size and higher number of chambers forming the last whorl (6–7 instead of 4–5), resulting in a less rapid increase of the individual chambers. The division between *G. acostaensis* and *G. humerosa*, using number of chambers and test size, is most conveniently placed between 4–5 chambered forms of a maximum diameter up to about 0.4 mm, and 6–7 chambered forms of a size range from about 0.4 to 0.5 mm.

GLOBOROTALIA PSEUDOPIMA Blow
Figures **27.7, 28.10–12**; 6, 10, 11

Type reference: *Globorotalia (Turborotalia) acostaensis pseudopima* Blow, 1969, p. 387, pl. 35, figs. 1–3.

Compared with 4-chambered *G. acostaensis* the test of *G. pseudopima* is distinctly larger and the wall more robust and with coarser pores. The umbilicus is nearly closed in typical specimens. Forms transitional towards *N. dutertrei* display a more open umbilicus and a tendency for the apertures of the last chamber to become more umbilical. Except for their distinctly different stratigraphic ranges and different accompanying fauna, *G. pseudopima* and the Oligocene *G. opima opima* are difficult to distinguish from each other. Both possess usually 4, more rarely 4½–5 robust and inflated chambers in the last whorl, with about equally sized and spaced pores. For comparison with *G. opima opima* see Figs. 26.24–30.

NEOGLOBOQUADRINA DUTERTREI
DUTERTREI (d'Orbigny)
Figures **27.1–4, 28.1–8**; 6, 10, 11

Type reference: *Globigerina dutertrei* d'Orbigny, 1839a, p. 84, pl. 4, figs. 19–21.

Lectotype: *Globigerina dutertrei* d'Orbigny: Banner & Blow, 1960a, p. 11, pl. 2, figs. 1a–c.

Synonyms: *Globigerina eggeri* Rhumbler, 1901, p. 19, pl. 20, figs. 20a–c (after Brady, 1884, pl. 79, figs. 17a–c, there designated).

Globigerina dubia Egger. According to Banner & Blow (1961a) Brady's figures represent three different specimens.

Lectotype of *Globigerina eggeri* Rhumbler, 1901: Banner & Blow, 1960a, p. 11, pl. 2, figs. 4a–c (4a after Brady, 1884, pl. 79, fig. 17a; 4b and 4c are other views of the same specimen).

With the base of the Pliocene begin to appear low, medium and high spired forms of considerable size range, with 4–7 chambers in the last whorl, with apertures restricted mainly to an umbilical position. Such forms may also possess umbilical tooth-like flaps.

Bandy *et al.* (1967) subdivide *N. dutertrei* into the subspecies *dutertrei* with tooth-like apertural flaps and *subcretacea* (Lomnicki) (=subspecies *blowi*, see Rögl & Bolli, 1973, p. 570) without these features. On fig. 1, p. 154, they demonstrate that the subspecies *dutertrei* developed from the subspecies '*subcretacea*' in the Late Pliocene and remained restricted to tropical latitudes through to the present day thus constituting a good environmental indicator. Contrary to other authors, Bandy *et al.* believe that the subspecies '*subcretacea*' developed in Late Miocene from *Globorotalia globorotaloidea* (Colom).

In the Mediterranean area medium spired forms, commonly with 5 chambers in the last whorl and an umbilical–extraumbilical aperture bordered by a thick lip, occur in the latest Miocene. They are included in *N. dutertrei* even though the distinguishing features (tooth-like flaps and umbilical position of aperture) of the genus are not present.

NEOGLOBOQUADRINA DUTERTREI
TEGILLATA Brönnimann & Resig
Figure **27.6**

Type reference: *Globorotalia (Turborotalia) acostaensis* Blow *tegillata* Brönnimann & Resig, 1971, p. 1277, pl. 33, fig. 3.

Number of chambers in the last whorl and overall test size of the holotype of *Neogloboquadrina tegillata* closely compares with larger sized specimens of *Globorotalia acostaensis*. It differs from that species in possessing an umbilical flap extending from the last chamber to cover part of the umbilical area. Such flaps are frequently present in *N. dutertrei* and there form a criterion for its generic assignment. Brönnimann & Resig's taxon is therefore also placed in this genus. It shows that such umbilical flaps may occur in the lineage as early as the Late Miocene.

NEOGLOBOQUADRINA DUTERTREI BLOWI
Rögl & Bolli
Figures **27.5, 28.9**

Type reference: Neogloboquadrina dutertrei blowi Rögl &
Bolli, 1973, p. 570, pl. 9, figs. 19–22. *Nomen novum* for
Blow's (1969) invalid lectotype of *Globigerina sub-
cretacea* Lomnicki.

The subspecies *blowi* falls within the variability of *N. dutertrei*.
Typical specimens are often fairly large with 5–6 chambers
forming the last whorl and possessing a fairly wide umbilical
area. The aperture of the last chamber has the tendency to
become umbilical–extraumbilical in position but the apertural
flaps typical in many *dutertrei* specimens are absent. Coiling
usually low trochospiral. Widespread in tropical and temperate
waters.

The *Globorotalia fohsi* lineage

The Early–Middle Miocene *G. fohsi* represents one of
the best-documented evolutionary sequences of gradual morphological change known in the planktic foraminifera. To
stress the close relationship between the taxa into which the
lineage is split it is natural to give them subspecific rank within
the species *fohsi*. Such treatment is also advantageous for their
use in stratigraphy in that it offers more flexibility in their
application. Such taxonomic treatment on the subspecific level
was proposed by Bolli (1950, and later publications) and
followed e.g. by Stainforth *et al.* (1975). In contrast Blow &
Banner (1966) gave species rank to the taxa distinguished
within the lineage and, furthermore, placed them in the two
subgenera *Turbororotalia* (forms without peripheral keel) and
Globorotalia (forms with a peripheral keel).

The principal morphological changes that characterize
the subspecies of the *fohsi* lineage from the earliest *peripheroronda* to the end-form *robusta* are:

(1) Increase in test size from about 0.3 mm in *peripheroronda*
    to about 0.7 mm in *lobata* and *robusta*.
(2) Development of a peripheral keel from the non-carinate
    peripheroronda via *peripheroacuta* to the keeled *fohsi,
    praefohsi, lobata* and *robusta*.
(3) Increase in the number of chambers forming the last
    whorl from 5–6 in *peripheroronda, peripheroacuta* and
    fohsi to $6\frac{1}{2}$–8 in *praefohsi, lobata* and *robusta*.
(4) Distinctly lobate periphery in *lobata*.

Following their recognition as morphological stages of
an evolutionary sequence, and their application to biostratigraphy in Trinidad (Bolli, 1950, 1951, 1957), the *Globorotalia
fohsi* subspecies have found wide application for the subdivision
of the tropical–subtropical Middle Miocene. After having
been successfully used in this sense for some 15 years, the *G.
fohsi* group became the subject of some controversy as to the
best method of taxonomic subdivision (Banner & Blow,

1965b; Blow & Banner, 1966; Jenkins, 1966; Bolli, 1967;
Beckmann *et al.*, 1969; Olsson, 1972). These differing views
and interpretations were clearly reviewed in Stainforth *et al.*
(1975).

Bolli (1950) distinguished four *fohsi* subspecies. Blow &
Banner (1966) proposed for *G. fohsi barisanensis*, as interpreted
by Bolli, the new species *G. (T.) peripheroronda*, and a *G. (T.)
peripheroacuta* for specimens with an acute periphery that
previously were regarded as transitional between *G. fohsi
barisanensis* and *G. fohsi fohsi. G. fohsi fohsi* was subdivided
by these authors into *G. (G.) fohsi* and *G. (G.) praefohsi*. The
subspecies *lobata* and *robusta* were by them relegated to
'*formae*' of *G. fohsi*. Bolli (1967) agreed to the additional
splitting of the early part of the *fohsi* lineage into four instead
of the original two taxa, but maintained as subspecies the
morphologically characteristic, readily recognizable and stratigraphically significant *lobata* and *robusta*.

Stainforth *et al.* (1975) described and figured the six taxa
on a subspecies level, as did Bolli (1967), but preferred to
recognize only the four subspecies *peripheroronda, fohsi, lobata*
and *robusta*, and treat *peripheroacuta* and *praefohsi* as
infrasubspecific.

The morphological changes referred to above under (1)
to (4) are gradual through time. Therefore, limits between
recognized taxa remain to some extent subjective, progressively
more so the closer the lineage is subdivided into separate taxa.
Such tolerances must also be taken into account when delimit-
ing the zonal boundaries that are determined by these taxa.

For example, the presence or absence of an imperforate
peripheral keel in the chambers of the last whorl would appear
to be a distinct and easily recognizable feature. However,
transitional stages do occur between forms with an acute
periphery and those with an imperforate keel as they do
between forms with rounded and acute peripheries. Particularly
in poorly preserved specimens the presence or absence of an
imperforate partial or even full keel may be difficult to determine. This is also the opinion of Stainforth *et al.* (1975).

Stratigraphic significance

The value of the *Globorotalia fohsi* series for stratigraphic
subdivisions in the lower and middle part of the Middle
Miocene is generally agreed, especially by those working on
low latitude faunas. As reviewed above, divergent opinions
concern only the nomenclature of the taxonomic units to be
used.

In the present work we retain the four-fold zonation
established by Bolli, as this has proved of value not only in the
Caribbean but also in the Gulf of Mexico, Java and in Atlantic
and Indo-Pacific DSDP sites. As regards the two subspecies
lobata and *robusta*, these have proved to have virtually
worldwide distribution with *lobata* appearing first and *robusta*
marking the final phase of the plexus when this is ideally

developed. Comparison with nannofossil zonal schemes has shown the two forms to have real time significance and not to be 'highly variable in the time sense from one area to another' as Blow (1969) stated. Thus we retain them, as do Stainforth *et al.* (1975) and also continue to advocate the use of the two zones named after them.

G. fohsi peripheroronda, the earliest of the subspecies, has a distinctly wider latitudinal distribution than the later subspecies. It is present not only in today's tropical–subtropical belt but extends in the northern hemisphere as far as the Mediterranean area, where the later forms are no longer present. From this it has been deduced that the later, and particularly the keeled forms, were restricted to tropical–subtropical waters and that water temperatures of, for example, the Middle Miocene Mediterranean Sea, became too low for their survival. Whether the non-keeled *peripheroronda* persisted longer under cooler water conditions during at least part of the time when in warmer waters the acute and keeled forms began to develop has still to be verified. A re-study of zonal type locality material from Trinidad, made by us revealed that the first distinct representatives of the *G. fohsi* lineage appear no earlier than the *Globigerinatella insueta* Zone and not, as had previously been thought as early as the *Globorotalia kugleri* Zone (Bolli, 1957) or the N 4 *Globigerinoides primordius* Zone (Blow, 1969). Seemingly transitional specimens that possibly could have branched off from the *G. mayeri* stock do occur in the underlying *Catapsydrax stainforthi* Zone.

Coiling trends

The worldwide coiling pattern of the *G. fohsi* subspecies is random in *peripheroronda* but changes with the appearance of *peripheroacuta* to a strong preference for sinistral coiling, a direction that persists until the extinction of the last subspecies, *robusta* (Bolli, 1950, 1971).

Taxonomic notes

GLOBOROTALIA FOHSI PERIPHERORONDA
Blow & Banner
Figures **29.6, 14**; 6, 9, 12

Type reference: *Globorotalia (Turborotalia) peripheroronda* Blow & Banner, 1966, p. 294, pl. 1, figs. 1a–c.

This is the earliest and most primitive subspecies of the *fohsi* lineage. It possesses 5–6 chambers in the last whorl, is slightly elongate in equatorial view, as is typical for all *fohsi* subspecies except for the end-form *robusta*, which tends to be more circular in outline. Characteristic for the subspecies is the bluntly rounded periphery.

Prior to the erection of *peripheroronda*, specimens of this kind were widely recorded under the name *barisanensis*, based on the description and holotype figure of its author LeRoy. In rechecking the holotype, Blow & Banner (1966) determined its

periphery, described by LeRoy as subrounded and lobulate, to be in fact fully carinate. This prompted Blow & Banner to erect the new taxon *peripheroronda* for specimens with a bluntly rounded periphery.

The species with which the subspecies *peripheroronda* can most easily be mistaken are its direct descendant *peripheroacuta*, and also *G. mayeri* and *G. kugleri*. From *peripheroacuta* it differs in the rounded instead of acutely compressed periphery, in particular in the later chambers of the last whorl. *G. mayeri*, which is often larger in size, has a slower rate of growth of the chambers of the last whorl and therefore has a more circular equatorial outline. the intercameral sutures on the spiral side of *G. mayeri* are usually close to radial, not swept back as in *peripheroronda*. In peripheral view the spiral and umbilical sides are practically equally inflated in *G. mayeri*, whereas in *peripheroronda* the spiral side appears slightly more flattened compared with the umbilical side. *G. fohsi peripheroronda* differs from *G. kugleri* in possessing only 5–6 chambers in the last whorl against 7–8 in *kugleri*, and in the equatorial outline tending to be slightly more elongate, less circular. In size *peripheroronda* is often slightly larger than *kugleri*.

GLOBOROTALIA FOHSI PERIPHEROACUTA
Blow & Banner
Figures **29.5, 13**; 6, 9, 12

Type reference: *Globorotalia (Turborotalia) peripheroacuta* Blow & Banner, 1966, p. 294, pl. 1, figs. 2a–c.

The morphological features of this subspecies differ from its ancestor *peripheroronda* in its average test size being slightly larger and in the later chambers of the last whorl possessing an acutely compressed periphery. As all intermediate stages from a rounded to an acute periphery occur during evolution, assignment of transitional specimens to one or the other of these two subspecies is often subjective.

GLOBOROTALIA FOHSI FOHSI Cushman &
Ellisor
Figures **29.4, 12**; 6, 9, 12

Type reference: *Globorotalia fohsi* Cushman & Ellisor, 1939, p. 12, pl. 2, figs. 6a–c.

This subspecies differs from its ancestor *peripheroacuta* in a slightly larger average test size and in the presence of a delicate peripheral keel which is often not readily visible in the early chambers, particularly in more poorly preserved specimens.

GLOBOROTALIA FOHSI PRAEFOHSI Blow &
Banner
Figures **29.3, 11**; 6, 9, 12

Type reference: *Globorotalia (Globorotalia) praefohsi* Blow & Banner, 1966, p. 295, pl. 1, figs. 4a–c.

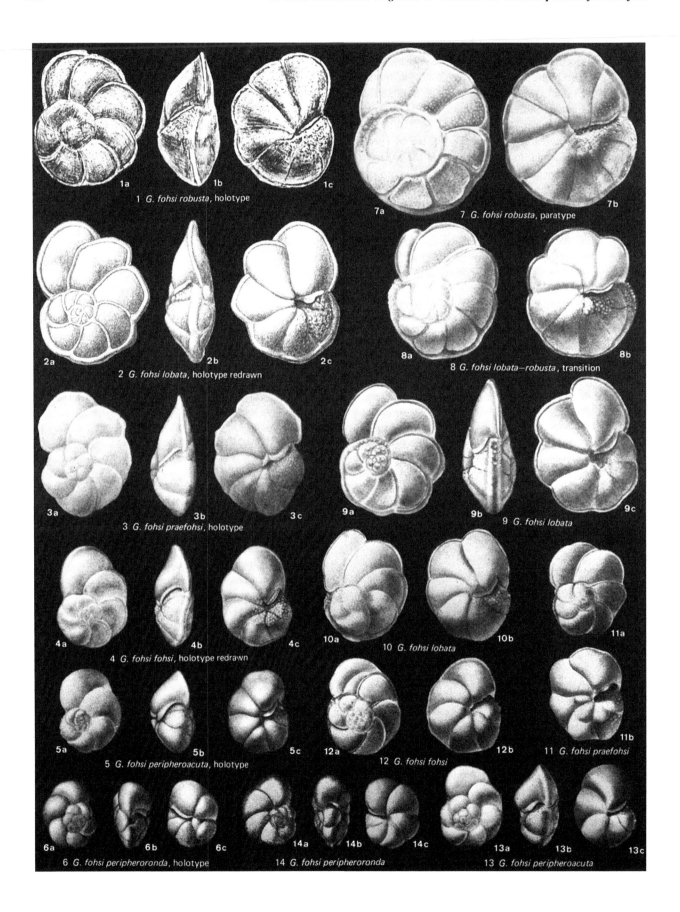

1 *G. fohsi robusta*, holotype

7 *G. fohsi robusta*, paratype

2 *G. fohsi lobata*, holotype redrawn

8 *G. fohsi lobata–robusta*, transition

3 *G. fohsi praefohsi*, holotype

9 *G. fohsi lobata*

4 *G. fohsi fohsi*, holotype redrawn

10 *G. fohsi lobata*

5 *G. fohsi peripheroacuta*, holotype

11 *G. fohsi praefohsi*

12 *G. fohsi fohsi*

6 *G. fohsi peripheroronda*, holotype

14 *G. fohsi peripheroronda*

13 *G. fohsi peripheroacuta*

Blow & Banner erected *praefohsi* as a morphologically intermediate form between *peripheroacuta* and *fohsi*. The last two chambers of the fairly large holotype are already of the elongate cockscomb shape, typical for *lobata*. Based on these features one would rather regard the holotype as a typically transitional form between the subspecies *fohsi* and *lobata* and not as intermediate between *peripheroacuta* and *fohsi*. Such a stratigraphically younger position is confirmed by the *Globorotalia fohsi lobata* Zone age of the Venezuelan sample from which the *praefohsi* holotype originates.

In addition to the holotype, Blow & Banner also figure a paratype which they selected from a Sumatra sample and whose morphological features fit well between those of the subspecies *peripheroacuta* and *fohsi*. Had Blow & Banner made the Sumatra paratype the holotype of their subspecies, its name *praefohsi* would have been better justified as an intermediate between the subspecies *peripheroacuta* and *fohsi*. Instead, it now represents a transition between the subspecies *fohsi* and *lobata*.

GLOBOROTALIA FOHSI LOBATA Bermudez
Figures **29.2, 9**; 6, 9, 12

Type reference: *Globorotalia lobata* Bermudez, 1949, p. 286, pl. 22, figs. 15–17.

This subspecies is a further development of *praefohsi*, manifested in a larger average test size and strongly elongate last

---

Fig. 29. *Globorotalia fohsi* lineage (all figures × 60)

1, 7 *G. fohsi robusta* Bolli
   1a–c Holotype, **7** Sample JS 46, *G. fohsi robusta* Zone, Middle Miocene, Cipero Fm., Trinidad (7 from Bolli, 1957).

   8 *G. fohsi lobata–G. fohsi robusta*, transition
   Sample JS 46, *G. fohsi robusta* Zone, Middle Miocene, Cipero Fm., Lengua area, Trinidad (from Bolli, 1957).

2; 9–10 *G. fohsi lobata* Bermudez
   2a–c Holotype. Bravo well No. 5, core 5, at 74–84 feet, *G. fohsi lobata* Zone, Middle Miocene, Trinchera Fm., Trujillo Province, Dominican Republic. **9a–b, 10a–b** Sample JS 32, *G. fohsi lobata* Zone, Middle Miocene, Cipero Fm., Cipero coast, Trinidad (from Bolli, 1957).

3, 11 *G. fohsi praefohsi* Blow & Banner
   3a–c Holotype. Sample RM 19410, *G. fohsi fohsi* Zone, Middle Miocene, Pozon Fm., Eastern Falcon, Venezuela. **11a–b** *G. fohsi fohsi* Zone, Middle Miocene, Cipero Fm., Trinidad (from Bolli, 1957).

4, 12 *G. fohsi fohsi* Cushman & Ellisor
   4a–c Holotype (redrawn). Ellender well No. 1, at 9612 feet, *G. fohsi fohsi* Zone, Middle Miocene, Terrebone Parish, Louisiana, USA. **12a–b** Sample Bo 185A, *G. fohsi fohsi* Zone, Middle Miocene, Cipero Fm., Trinidad (from Bolli, 1957).

5, 13 *G. fohsi peripheroacuta* Blow & Banner
   5a–c Holotype. Sample RM 19367, *G. fohsi fohsi* Zone, Middle Miocene, Pozon Fm., Eastern Falcon, Venezuela. **13a–c** Sample Bo 185A, *G. fohsi fohsi* Zone, Middle Miocene, Cipero Fm., Golconda Estate, Trinidad (from Bolli, 1957).

6, 14 *G. fohsi peripheroronda* Blow & Banner
   6a–c Holotype. Sample RM 19304, *G. fohsi peripheroronda* Zone, Middle Miocene, Pozon Fm., Eastern Falcon, Venezuela. **14a–c** Sample Bo 202, *G. fohsi peripheroronda* Zone, Middle Miocene, Cipero Fm., Hermitage Village, Trinidad (from Bolli, 1957).

2–3 chambers forming a still more strongly lobate cockscomb-like equatorial periphery. As in *praefohsi* the peripheral keel is often only delicately developed and may not readily be visible in the early chambers of the last whorl. The wall of the cockscomb-like peripheral portion, particularly in the last chamber, is thin and is therefore often broken away.

GLOBOROTALIA FOHSI ROBUSTA Bolli
Figures **29.1, 7**; 6, 9, 12

Type reference: *Globorotalia fohsi robusta* Bolli, 1950, p. 89, pl. 15, figs. 3a–c.

This is the end-form of the *fohsi* lineage. It differs from *lobata* in having a more compact, circular outline in equatorial view, instead of being more elongate like the earlier *fohsi* subspecies. The last chambers are generally less elongate compared with *lobata* with the periphery not lobate or, if it is, only very slightly so and only in the last few chambers. A distinct peripheral keel is present throughout the last whorl which tends to have more chambers (7–8) than does *lobata*. The spiral side of the test is usually lower, the umbilical side higher.

Globorotalia lenguaensis

This small, compact species first described from the Middle Miocene Lengua Formation (*Globorotalia mayeri* and *Globorotalia menardii* zones) of Trinidad has subsequently been reported from Venezuela (Blow, 1959), where it was used as a subzonal marker, from Japan (Saito, 1963), Java (Bolli, 1966b), Italy (Cita & Premoli Silva, 1967), Gulf of Mexico (Stainforth *et al.*, 1975), Sakhalin (Serova, 1977) and from DSDP sites.

The general consensus is that the range of *G. lenguaensis* in low latitudes is limited mainly to the Middle Miocene *Globorotalia mayeri* and *menardii* zones, but continues into the *Globorotalia acostaensis* Zone (Bolli, 1966b) and, according to Blow (1969), even in his N 17 Zone (= Globorotalia humerosa Zone). According to the latter author, the species appears as early as within N 12 (*Globorotalia fohsi robusta* Zone). This cosmopolitan species may be regarded as a useful additional index for the late Middle and early Late Miocene; this has been confirmed particularly for the Mediterranean area.

Taxonomic notes
GLOBOROTALIA LENGUAENSIS Bolli
Figures **30.25**; 6, 10, 12

Type reference: *Globorotalia lenguaensis* Bolli, 1957, p. 120, pl. 29, figs. 5a–c.

G. lenguaensis is a small, rather flat species with a circular, barely lobate outline and with a subangular to angular periphery with only a faint keel, 6–7 chambers in the last whorl. The test wall is smooth and glassy with distinct, well separated pores. It is randomly coiled in its early range and becomes strongly sinistral later.

Stainforth *et al.* regard *G. lenguaensis* as intermediate between *G. minima* Akers and *G. merotumida* Blow & Banner. We doubt its origin from *G. canariensis* var. *minima* of Akers as the main development of the latter form is in the Gulf Coast equivalent of the *G. fohsi peripheroronda* Zone after which it becomes very rare according to its author. In contrast, *G. lenguaensis* begins in the *G. mayeri* Zone or, at the earliest, in the *G. fohsi robusta* Zone.

In our opinion *G. lenguaensis* does not belong to the *menardiform* plexus and therefore did not give rise to *G. merotumida*, from which it differs in its smaller size, nearly circular against elongated equatorial circumference, in not being as high on the umbilical side and in possessing a much weaker peripheral keel.

The range of Blow's *G. paralenguaensis*, which he considered as a link with *G. merotumida*, falls within that of *G. lenguaensis*. We consider it a variant within *G. lenguaensis*.

Globorotalia margaritae, G. scitula, G. juanai and G. hirsuta

Species that show certain morphological similarities and therefore may be mistaken for the Early Pliocene index species *Globorotalia margaritae* are *G. juanai*, *G. scitula* with its subspecies and *G. hirsuta*. Based on some superficial similarities, Blow (1969) proposed an evolutionary lineage beginning with the Early Miocene *G. scitula praescitula*, leading through *G. scitula scitula* to *G. margaritae*, which in turn was supposed to have evolved through *G. praehirsuta* into *G. hirsuta*. Stainforth *et al.* (1975) followed Blow's concept and, in addition, regarded *G. juanai* as an intermediate between *G. scitula* and *G. margaritae*.

On morphological grounds the bioseries as proposed by Blow and enlarged by Stainforth *et al.* seems at first sight quite feasible. However, in the absence of clearly recognizable intermediate forms and taking into account the established stratigraphic ranges of the individual taxa, we have strong reservations against accepting such an evolutionary lineage.

The assumption that *Globorotalia juanai* is an intermediate between *G. scitula* and *G. margaritae* seems unlikely. In the type area for *G. juanai* (Cubagua well 1, Bermudez & Bolli, 1969), this species became extinct within the *Globorotalia acostaensis* Zone, well below the first appearance of *G. margaritae*. The proposal that *G. margaritae* branched off from the long-ranging *G. scitula scitula/gigantea* as proposed by Blow (1969) is likewise not supported by the presence of transitional, intermediate forms.

Blow (1969) erected a *Globorotalia hirsuta praehirsuta* with a range from mid N 18 (within the *Globorotalia margaritae margaritae* Subzone) to early N 22 (Early Pleistocene) where, according to him, it evolved into *Globorotalia hirsuta*. He went to some length to describe apparently minimal morphological

differences between *G. hirsuta praehirsuta* and *G. margaritae*. The figures of the holotype (Fig. 30.1) and its stratigraphic position in the earliest part of Zone N 19 (= *Globorotalia margaritae evoluta* Subzone) lead us to consider *G. hirsuta praehirsuta* as a junior synonym of *G. margaritae* and, as regards size, close to the holotype of the subspecies *evoluta*. Blow did not provide any documentation from which the full range of his *praehirsuta* could be ascertained. In fact, no such continuation of *praehirsuta* (= *margaritae*) forms above the established Early Pliocene *Globorotalia margaritae* Zone has to our knowledge been reported. Therefore, one has no reason to assume a connection through intermediate forms between *G. margaritae*, which is restricted to the Early Pliocene, and *G. hirsuta*, which is restricted to the Pleistocene and Holocene.

To aid in the correct identification of the Early Pliocene index species *G. margaritae* and its subspecies, it is important that the taxa under discussion can be reliably identified and distinguished from other, morphologically somewhat similar forms.

During its life span of about two million years, *G. margaritae* underwent morphological changes that prompted Cita (1973) to subdivide the species into an early *G. margaritae primitiva*, followed by *G. margaritae margaritae* and a final *G. margaritae evoluta*. She also used the evolutionary changes to divide the original *Globorotalia margaritae* Zone into a *Globorotalia margaritae margaritae* Subzone below and a *Globorotalia margaritae evoluta* Subzone above.

Characteristic for *G. margaritae*, especially for specimens from tropical and subtropical areas, is the thin, delicately built chamber wall; this makes the species particularly susceptible to solution in assemblages deposited in deep waters.

Stratigraphic significance of Globorotalia margaritae

When Bolli & Bermudez (1965) erected *Globorotalia margaritae* from the Los Hernandez beds on Margarita Island (Venezuela) they assigned to it a Late Miocene age, based on mollusc evidence from the age-equivalent Cubagua Formation of Eastern Venezuela. This assignment came under discussion during the sessions of the Mediterranean Neogene Congress held in Bologna in 1967. Based on evidence from a number of Italian sections where the species is also present but up to that time had been reported as *Globorotalia hirsuta*, it was agreed that the range of *G. margaritae* is, in fact, restricted to the Early Pliocene. Subsequently this view has widely been accepted with few exceptions, where the opinion is that the earliest *margaritae* appear already in the uppermost Miocene. For practical reasons, placing the Miocene–Pliocene boundary at the first appearance of the geographically widespread *G. margaritae* seems practical. Further information on the stratigraphic significance of *Globorotalia margaritae* and its subspecies can be found in Cita (1973), Stainforth *et al.* (1975) and Bolli & Bermudez (1978).

Coiling trends

G. margaritae shows a strong preference for sinistral coiling throughout its range. This is verified from many sections of the Caribbean, Gulf of Mexico, Atlantic, Mediterranean and Pacific. A very short swing to dextral coiling has been seen in a few Caribbean and Gulf Coast sections near the base of the Pliocene. Such a swing has also been reported from the Late Miocene but this needs careful checking as other morphologically similar forms such as *G. scitula* or *G. juanai* might be involved, both of which show strong dextral preference at this time. The sinistral coiling pattern of *G. margaritae* therefore is an additional criterion to distinguish this species from the predominantly dextrally coiling *G. juanai* and *G. scitula scitula*. Both the holotype and the neotype of *G. hirsuta* also coil dextrally.

Taxonomic notes

GLOBOROTALIA MARGARITAE MARGARITAE Bolli & Bermudez
Figures **30.9–14**; 6, 10, 11

Type reference: *Globorotalia margaritae* Bolli & Bermudez, 1965, p. 132, pl. 1, figs. 16–18.
Neotype: *Globorotalia margaritae* Bolli & Bermudez: Bolli & Bermudez, 1978, p. 139, pl. 1, figs. 1–3.

GLOBOROTALIA MARGARITAE PRIMITIVA Cita
Figures **30.15–19**; 6, 10, 11

Type reference: *Globorotalia margaritae primitiva* Cita, 1963, p. 1352, pl. 2, figs. 1–3.

GLOBOROTALIA MARGARITAE EVOLUTA Cita
Figures **30.1–5**; 6, 10, 11

Type reference: *Globorotalia margaritae evoluta* Cita, 1973, p. 135, pl. 1, fig. 1.

G. margaritae s.l. is a widely recognized index species for the Pliocene. On paleomagnetic evidence quoted by Cita (1975) its absolute range is ± 5.25 to 3.32 m.y. Compared with other Pliocene planktic index species, *G. margaritae* is remarkably cosmopolitan with a worldwide distribution through the tropical, subtropical and temperate Atlantic, Pacific and Indian Ocean provinces, and the Mediterranean area. However, both in deep water and in shallow water, more marginal assemblages it may be very infrequent.

Though morphological differences between the earliest and the latest *G. margaritae* are considerable, it is often difficult to place intermediate specimens with certainty into one of the three subspecies *primitiva*, *margaritae* and *evoluta* proposed by Cita (1973). The reason for this is the continuous variability of specimens from the small non-keeled and almost non lobate *primitiva* to the large, keeled and strongly lobate end-form *evoluta*. While the distinction between *G. margaritae primitiva* and *G. margaritae margaritae* (acquisition of a peripheral keel) is usually not difficult, that between *G. margaritae margaritae* and *G. margaritae evoluta* on the other hand is more difficult as keeled forms vary in amount of keel and lobateness of test; while the absolute size of the test can be affected by environmental factors.

Cita offers the following characters of distinction between the subspecies:

G. margaritae primitiva differs from *G. margaritae margaritae* in its smaller size, less elongate equatorial outline, slower increase in size of chambers, more symmetrical axial profile, and complete lack of a distinct keel.

Globorotalia margaritae evoluta differs from *G. margaritae margaritae* by its greater size, less elongated equatorial outline, slower increase in height of the chambers, and more symmetrical axial profile.

Figs. 30.17–19, 12–14, 6–8, 2–4 show a sequence of *G. margaritae* subspecies arranged in stratigraphic order, taken from Krasheninnikov & Pflaumann (1977).

Catalano & Sprovieri (1969) described a *G. praemargaritae* (Fig. 31.7) from the Late Miocene (N 17, *Globorotalia tumida plesiotumida* Zone) of Central Sicily. With its distinctly lobate, nearly circular equatorial periphery and fairly inflated chambers we consider the species to fall within the *G. scitula* plexus. Compared with *G. margaritae*, in particular with the early subspecies *primitiva*, chambers in the last whorl of *G. praemargaritae* increase less rapidly in size, resulting in a more circular peripheral outline of the test.

GLOBOROTALIA SCITULA SCITULA (Brady)
Figures **30.26–29, 31.3–4**; 6, 9, 11

Type reference: *Pulvinulina scitula* Brady, 1882, p. 716, Brady, 1884, pl. 103, figs. 7a–c (given as *Pulvinulina patagonica*).
Lectotype: *Globorotalia (Turborotalia) scitula* (Brady): Banner & Blow, 1960a, p. 27, pl. 5, figs. 5a–c.

GLOBOROTALIA SCITULA PRAESCITULA Blow
Figures **31.6**; 6, 9, 12

Type reference: *Globorotalia scitula praescitula* Blow, 1959, p. 221, pl. 19, figs. 128a–c.

GLOBOROTALIA SCITULA GIGANTEA Blow
Figures **31.5**; 6, 9, 12

Type reference: *Globorotalia scitula gigantea* Blow, 1959, p. 220, pl. 16, figs. 127a–c.

Globorotalia scitula s.l. ranges from Middle Miocene to Recent, an exceptionally long range for a Tertiary *Globobotalia* species. Blow (1959) separated a forerunner, *G. scitula praescitula* from the typical form.

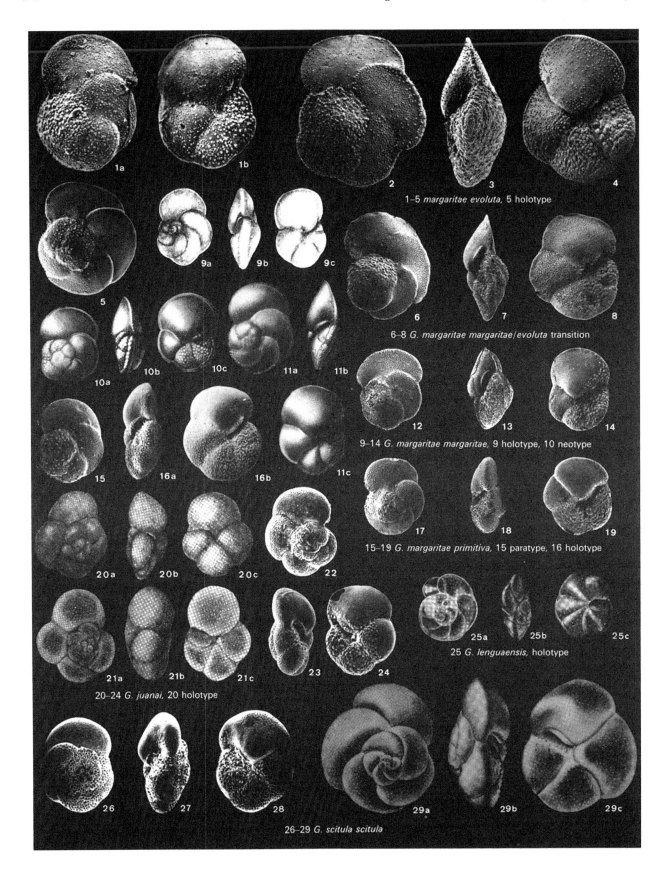

1–5 *margaritae evoluta,* 5 holotype

6–8 *G. margaritae margaritae/evoluta* transition

9–14 *G. margaritae margaritae,* 9 holotype, 10 neotype

15–19 *G. margaritae primitiva,* 15 paratype, 16 holotype

20–24 *G. juanai,* 20 holotype

25 *G. lenguaensis,* holotype

26–29 *G. scitula scitula*

Brady erected *Pulvinulina scitula* in 1882 as 'a variety of *Pulvinulina canariensis*' but did not illustrate it. In his *Challenger* report of 1884 he reviewed his earlier conclusions and placed this taxon in synonymy with *P. patagonica* (d'Orbigny). The specimen he illustrated (here reproduced as Fig. 31.3) is apparently 0.6 mm in diameter though his earlier description drew attention to its minute dimensions quoting a diameter of 0.25 mm. Banner & Blow's lectotype from Brady's syntypic

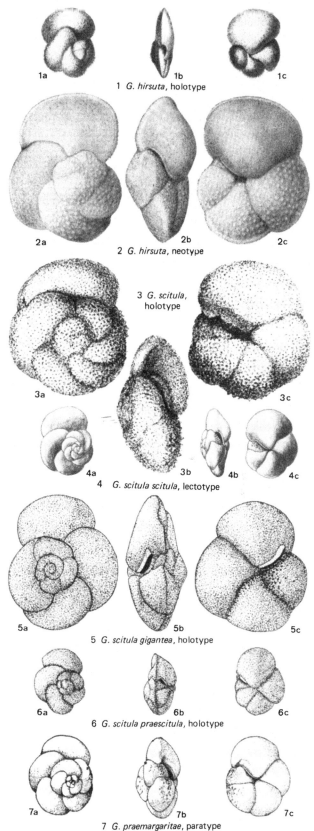

Fig. 30. *Globorotalia margaritae, Globorotalia juanai, Globorotalia lenguaensis, Globorotalia scitula* (all figures × 60)

1–5 *G. margaritae evoluta* Cita
1a–b Sample By 61, Zone N 19 (earliest part), Early Pliocene, Anzio, Italy (holotype of *G. praehirsuta* Blow, 1969). **2–4** DSDP Leg 41-366A-6-1, 83–85 cm, *G. margaritae evoluta* Zone, Early Pliocene, Eastern Atlantic (from Krasheninnikov & Pflaumann, 1977). **5** Holotype. DSDP Leg 13-132-14-cc, *G. margaritae evoluta* Zone, Early Pliocene, Mediterranean Sea.

6–8 *G. margaritae margaritae–G. margaritae evoluta* transition DSDP Leg 41-366A-8-6, 64–66 cm, *G. margaritae evoluta* Zone, Early Pliocene, Eastern Atlantic (from Krasheninnikov & Pflaumann, 1977).

9–14 *G. margaritae margaritae* Bolli & Bermudez
9a–c Holotype. *G. margaritae margaritae* Zone, Early Pliocene, Los Hernandez Beds, Cubagua Fm., Margarita Island, Venezuela. **10a–c** Neotype. Same origin as 9a–c. **11a–c** Sample PJB 7/64, *G. margaritae margaritae* Zone, Early Pliocene, Cubagua Fm., west end of Araya Peninsula, Venezuela (from Bolli & Bermudez, 1978). **12–14** Information as for Figs. 30.6–8.

15–19 *G. margaritae primitiva* Cita
15 Paratype, **16a–b** Holotype. Same origin as Fig. 30.5. **17–19** DSDP Leg 41-366A-11-1, 105–107 cm, *G. margaritae margaritae* Zone, Early Pliocene, Eastern Atlantic (from Krasheninnikov & Pflaumann, 1977).

20–24 *G. juanai* Bermudez & Bolli
20a–c Holotype. Late Miocene, Carenero Fm. near Carenero, State of Miranda, Venezuela. **21a–c** Cubagua well 1, core 2691–2703 feet, *G. acostaensis* Zone, Late Miocene, Cubagua Fm., Cubagua Island, Venezuela (from Bermudez & Bolli, 1969). **22–24** Core hole A, core 14, *G. acostaensis* Zone, Late Miocene, Gulf of Mexico (from Stainforth *et al.*, 1975).

25 *G. lenguaensis* Bolli
25a–c Holotype. Sample KR 23425, *G. menardii* Zone, Middle Miocene, Lengua Fm., Lengua Settlement, Trinidad.

26–29 *G. scitula* (Brady)
26–28 Sample FC 24, Late Pliocene, North Atlantic (from Cifelli & Glaçon, 1979). **29a–c** *G. mayeri* Zone, Middle Miocene, Lengua Fm., Trinidad (from Bolli, 1957).

Fig. 31. *Globorotalia hirsuta, Globorotalia scitula, Globorotalia praemargiritae* (all figures × 60)

1–2 *G. hirsuta* (d'Orbigny)
1a–c Holotype. Recent, marine sands, Teneriffe Island. **2a–c** Neotype. *Challenger* Station 8, Recent, off Gomera, Canary Islands.

3–4 *G. scitula* (Brady)
3a–c Holotype. *Challenger* Station 302, Recent, West Coast of Patagonia (specimen figured by Brady, 1884, as *Pulvinulina patagonica* d'Orbigny). **4a–c** Neotype. *Knight Errant* Station 7, Recent, Faröe Channel.

5 *G. scitula gigantea* Blow
5a–c Holotype. Sample RM 19480, *G. fohsi robusta* Zone, Middle Miocene, Pozon Fm., Eastern Falcon, Venezuela.

6 *G. scitula praescitula* Blow
6a–c Holotype. Sample RM 19152, *G. insueta* Zone, Middle Miocene, Pozon Fm., Eastern Falcon, Venezuela.

7 *G. praemargaritae* Catalano & Sprovieri
7a–c Paratype. Late Miocene, Saheliana series, Rossi River, Enna, Sicily, Italy.

Figure captions (right column, beside illustrations):
1 *G. hirsuta*, holotype
2 *G. hirsuta*, neotype
3 *G. scitula*, holotype
4 *G. scitula scitula*, lectotype
5 *G. scitula gigantea*, holotype
6 *G. scitula praescitula*, holotype
7 *G. praemargaritae*, paratype

series at 0.28 mm was presumably chosen to be close to the originally quoted size.

Blow (1959) also erected *G. scitula gigantea*. It differs from the *G. scitula scitula* lectotype in its distinctly larger size (0.59 mm), but is virtually identical in size with the original specimen figured by Brady.

G. scitula scitula differs from *G. margaritae* in having a more circular outline in equatorial view, in having a higher umbilical side, that is, in being more distinctly biconvex, in the absence of a peripheral keel, in its more robust, thicker chamber walls, and in showing a dextral coiling pattern.

GLOBOROTALIA JUANAI Bermudez & Bolli
Figures **30.20–24**; 6, 10, 12

Type reference: *Globorotalia juanai* Bermudez & Bolli, 1969, p. 171, pl. 14, figs. 4–6.

The species differs from *G. scitula scitula* in the more inflated chambers, less shiny surface and in the intercameral sutures on the spiral side being nearly radial or only very slightly curved backwards, giving the individual chambers a more globular aspect.

The Early Pliocene specimen figured in Stainforth *et al.* (1975, p. 364, figs. 3–5) as *G. juanai* has strongly curved oblique sutures and is very close to the holotype figure of *G. margaritae primitiva*, which is also figured by Stainforth *et al.* (1975, p. 364, figs. 7a–b). It is placed in *G. juanai* as these authors consider *G. margaritae primitiva* to be a junior synonym of that species. A comparison with the holotype of *G. juanai* clearly shows how it differs from the Early Pliocene *G. margaritae* specimens illustrated by Stainforth *et al.* on that page, and included by them in *G. juanai*.

G. juanai shows a very distinct preference for dextral coiling throughout its range (Bermudez & Bolli, 1969, fig. 2), whereas *G. margaritae* has a sinistral preference. *G. juanai* is known only from the Late Miocene *Globorotalia acostaensis* Zone, while *G. margaritae* is confined to the Early Pliocene *Globorotalia margaritae* Zone. The dextrally coiling specimen figured in Stainforth *et al.* (1975, fig. 174) from the *Globorotalia acostaensis* Zone is a typical *G. juanai*.

GLOBOROTALIA HIRSUTA (d'Orbigny)
Figures **31.1–2**; 6, 10, 11

Type reference: *Rotalina hirsuta* d'Orbigny, 1839b, p. 131, pl. 1, figs. 37–39.

Neotype: *Globorotalia (Globorotalia) hirsuta hirsuta* (d'Orbigny): Blow, 1969, p. 398, pl. 8, figs. 1–3.

The species which may be mistaken most easily for the Early Pliocene *G. margaritae* is *G. hirsuta*, known from the Pleistocene and Recent. The lost original type figured by d'Orbigny measured only about $\frac{1}{3}$ mm. It is unfortunate that Blow (1969) selected a neotype with three chambers more, measuring 0.59 mm or about double the original type. *G. margaritae* and

G. hirsuta possess morphological similarities, but an evolutionary connection cannot be recognized on present evidence. The reasons for this are pointed out above where it is shown that Blow's *G. praehirsuta* is in fact a *G. margaritae* with a range no higher than Early Pliocene and therefore cannot serve as an intermediate between *G. margaritae* and *G. hirsuta* in the Middle and Late Pliocene.

A characteristic difference between *G. hirsuta* and *G. margaritae* lies in the presence of a more convex umbilical surface in *hirsuta*, resulting in a deeper umbilical pit. In addition, chambers in *G. hirsuta* are more loosely coiled and are finely spinose as the name implies. The equatorial periphery in *G. hirsuta* is more lobate than in *G. margaritae*, except for the last chamber or two in highly evolved forms of the latter: the peripheral keel is less pronounced in *G. hirsuta*.

G. hirsuta is a temperate form, originally described by d'Orbigny from near the Canary Islands. In its typical development it is largely absent in tropical–subtropical areas. *G. margaritae* on the other hand is much more cosmopolitan, known to occur in the Early Pliocene of both tropical and temperate regions. The geographic concurrence of the two morphologically somewhat similar species in, for example, the Mediterranean area is the reason why in some publications *G. margaritae* has been erroneously identified as *G. hirsuta*.

The menardiform *Globorotalia* species

Stainforth *et al.* (1975) discuss under 'Menardiform Globorotalias' 'a group of Late Neogene Globorotalias having many features in common, yet differing widely in other aspects'. Species of this complex constitute an important element for the biostratigraphic subdivision of the Middle Miocene to Recent, particularly in the tropical/subtropical Atlantic province.

An extensive literature exists in particular on *Globorotalia menardii* and related forms. Reference is made to historical reviews on the complex taxonomic problems of this species and related forms by Banner & Blow (1960a) and by Stainforth *et al.* (1975). These authors also discuss characteristic morphological features and evolutionary relationships of the taxa concerned.

No general agreement exists on how individual species of the menardiform *Globorotalia* fit into a phylogenetic lineage. Because of the stratigraphic significance of several of the species, proposed relationships are briefly discussed here.

Evolutionary trends

When Stainforth (1948) established a Middle Miocene *Globorotalia menardii* Zone in Trinidad (emended by Bolli, 1957), the concept of this zonal marker was still a rather wide one. It was assumed that the Middle Miocene *G. menardii* evolved from the Early Miocene *G. archeomenardii* via *G. praemenardii* and continued to Recent without further significant morphological changes.

Subsequent studies of *G. menardii* and related forms in the Caribbean/Gulf of Mexico/tropical Atlantic have shown that the continuation of the Middle Miocene forms to the Holocene is not straightforward and in certain aspects differs from that of the Indo/Pacific realm.

After a period of about 10 million years during which little change in morphology occurred in the *G. menardii* group, a spectacular proliferation resulting in a number of different species commenced in the low latitude Late Miocene and continued into the Middle Pliocene, where most of these short-lived species became extinct again.

For some of the taxa included here in the group of menardiform species, an interrelation may at first sight not appear obvious. However, detailed studies through a number of Caribbean sections provide strong evidence for the following lineages that developed from the *menardii* stock: *G. pseudomiocenica–G. miocenica, G. multicamerata–G. pertenuis–G. exilis*, and *G. merotumida–G. plesiotumida–G. tumida tumida–G. tumida flexuosa*.

The *G. merotumida–plesiotumida–tumida tumida—tumida flexuosa* group has a worldwide distribution, but others, such as *G. multicamerata, G. pertenuis, G. exilis* and *G. miocenica* remained largely restricted to the Atlantic province. Several of these species are morphologically very distinct, have short ranges within the Pliocene and consequently provide, where present, excellent index species.

Stainforth *et al.* (1975) include in their 'menardiform' Globorotalias Late Neogene species which they regard as being closely related or having similar morphological features to the central species *Globorotalia menardii*. They place 30 taxa in this group of which 15 warm water forms are described and discussed individually, the other 15 are regarded as not typical warm water species and are therefore not utilized.

The following 17 menardiform taxa are here considered significant and are included in the discussion:

Globorotalia archeomenardii
        praemenardii
        menardii 'A'
        menardii 'B'
        menardii menardii
        menardii cultrata
        fimbriata
        pseudomiocenica
        miocenica
        multicamerata
        pertenuis
        exilis
        merotumida
        plesiotumida
        tumida
        tumida flexuosa
        ungulata

Relations between groups of above taxa and the re-sultant evolutionary lineages are indicated by gradual morphological changes between established species. Stainforth *et al.* (1975) include additional species in the group of menardiform Globorotalias such as *G. scitula, G. lenguaensis, G. margaritae* and *G. hirsuta*. We have not been able to observe intermediate forms suggesting that these species branched off from a menardiform stock and we think it more probable that they had different origins.

For a better understanding of relations and evolutionary trends within the many menardiform species more research needs to be directed in particular towards the complex in the Middle and Late Miocene that was provisionally subdivided into *G. menardii* 'A' and 'B' (Bolli, 1970). Furthermore, the branching off from these forms of the Pliocene lineages *G. pseudomiocenica–G. miocenica, G. multicamerata–G. pertenuis–G. exilis*, and the Late Miocene to Holocene *G. merotumida–G. plesiotumida–G. tumida tumida–G. tumida flexuosa* requires more investigation. Also, a close morphologic/biometric comparison of the Middle to Late Miocene *G. menardii* 'A' with the Pleistocene to Holocene *G. menardii menardii* and *G. menardii cultrata* is needed for a more complete understanding of the evolution of these forms from Miocene to Recent.

Stainforth *et al.* (1975) regard *Globorotalia scitula* as the rootstock from which the earliest menardiform line (*G. archeomenardii–G. praemenardii*) has developed, and from which in the later Miocene a second line (*G. juanai–G. margaritae–G. praehirsuta–G. hirsuta*) evolved. According to them the early *menardii* species branched off from *G. scitula* by the development of an acute, slightly carinate periphery. On evidence from Caribbean sections, *G. scitula scitula* (without peripheral keel and with only 4 chambers in the last whorl) appears only within the *Globorotalia fohsi peripheroronda* Subzone, while *G. archeomenardii*, with peripheral keel and 5 chambers in the last whorl, is known as early as the *Praeorbulina glomerosa* Zone. On such evidence a branching off in the Middle Miocene of the *menardii* complex from *G. scitula* is not likely. Blow (1969) and Stainforth *et al.* (1975) recognize a *G. praescitula*, a forerunner of *G. scitula*, which made its first appearance in the *Catapsydrax stainforthi* Zone.

We also do not consider that there is sufficient evidence from the presence of transitional or intermediate forms in the Late Miocene that *Globorotalia scitula* has given rise to a bioseries *G. juanai–G. margaritae–G. praehirsuta–G. hirsuta* as postulated by Stainforth *et al.* (1975). *G. scitula* has, in fact, undergone almost no morphological changes in its long range from the Middle Miocene to Recent.

Similarly, the interpretation of Banner & Blow (1965a) and Blow (1969) of an evolutionary sequence *Globorotalia lenguaensis–G. paralenguaensis–G. merotumida–G. plesiotumida–G. tumida* is not maintained here. From our investigations, *G. tumida* evolved via *G. merotumida–G. plesiotumida* directly from the *G. menardii* stock. This is confirmed by biometric studies on *G. menardii* and *G. tumida* by Schmid

(1934), who showed a continuous transition between the two species.

Stratigraphic significance of menardiform taxa

Because of their characteristic morphology and restricted ranges, numerous menardiform species are useful as index forms and several of them have become zonal markers. Their application to biostratigraphy is of particular value in the Pliocene. The slow and gradual evolution and general scarceness of the plexus from *G. archeomenardii* to *G. praemenardii* and *G. menardii* s.l. (*menardii* 'A' and 'B') has prevented their use as index species through the Middle and Late Miocene. Though there exists a Middle Miocene *Globorotalia menardii* Zone, it is defined as an interval zone and the zonal marker is not used in its definition.

Menardiform species become useful only with the rapid development through the Pliocene of a number of short-lived forms like *G. multicamerata*, *G. miocenica*, *G. exilis* and *G. pertenuis*. This plexus became typically developed however, only in the low latitude Atlantic province, where *G. miocenica* and *G. exilis* are used as zonal markers. Again, within the Atlantic province they occur only in ecologically favourable open basins. Within the Caribbean/Gulf of Mexico area, for example, they are well developed in numerous DSDP sites and also in Jamaica. In more marginal areas to which belong the island of Cubagua and the Araya Peninsula of Venezuela and the Dominican Republic, these forms may not be typically developed or be absent altogether. The same goes for the whole Indo-Pacific province, where these taxa are also not typically developed and are therefore difficult to apply for stratigraphic purposes.

The *G. merotumida–plesiotumida–tumida* lineage, originating in the Late Miocene, has been used by Banner & Blow (1965b) and Blow (1969) to define their zones N 16 and N 18. Morphological changes are difficult to define and recognize, particularly in the early part of the lineage, where the taxa are used in the zonal scheme. Placement of the zonal boundaries is therefore to some extent subjective. Furthermore, the lineage is continuously developed in the Indo/Pacific province only, whereas there exists a widespread interruption in the Atlantic province, from within the Early Pliocene *Globorotalia margaritae margaritae* Subzone to near the top of the Pliocene.

Of more local stratigraphic significance is *G. tumida flexuosa*, which apparently develops under certain favourable conditions from the *G. tumida* stock. Such forms are already common and typically developed in the Early Pliocene of the Indo/Pacific, from where the original type was described. In the low latitude Atlantic province, typical *flexuosa* are known from the Late Pleistocene, where they disappeared again some 80 000 years ago. Locally, or within certain more extended areas, *G. tumida flexuosa* may serve as a useful index.

Two characteristic species of the menardiform group are

G. fimbriata and *G. ungulata*. Both are known only from the Holocene, questionably also from the Late Pleistocene.

Coiling trends

The coiling pattern in menardiform species, in particular in *Globorotalia menardii* s.l., has been documented in numerous sections, primarily in the Caribbean region (e.g. Bolli, 1950, 1970; Bermudez & Bolli, 1969; Robinson, 1969; Lamb & Beard, 1972; Bolli & Premoli Silva, 1973), but also in the Indo-Pacific (Bandy, 1963; Bolli, 1964, 1966b). From these data the lineage *G. archeomenardii–praemenardii–menardii* s.l. coils at random from its origin in the *Catapsydrax stainforthi* Zone until within the *Globorotalia fohsi robusta* Zone, where a change to preferred sinistral coiling takes place which continues in the Caribbean area to near the top of the *Globorotalia menardii* Zone and in the Indo-Pacific, based on the section Bodjonegoro-1, into the *Globorotalia humerosa* Zone, where an abrupt change to dextral coiling takes place which continues into the Late Pliocene. In the Caribbean, based on the Cubagua-1 section (Bermudez & Bolli, 1969), the preferred direction of coiling of *G. menardii* s.l. fluctuates several times between sinistral and dextral through the *Globorotalia acostaensis* and *Globorotalia humerosa* zones. As in Bodjonegoro-1 it becomes dextral through most of the Early and Middle Pliocene to change to sinistral in the upper part of the Late Pliocene. This pattern of frequent changes was also reported by Robinson (1969) from Jamaica and has recently been found in the Dominican Republic. It is apparently typical at least for the Caribbean area of the Atlantic region. Recent *G. menardii* s.l. coil sinistrally worldwide.

G. multicamerata–pertenuis–exilis–pseudomiocenica and *miocenica*, all restricted to the Early and Middle Pliocene or part of it, coil dextrally throughout their ranges in the Atlantic region. These taxa are not typically developed in the Indo-Pacific. In contrast, the *G. tumida* lineage prefers sinistral coiling worldwide, from its first appearance in the Late Miocene to Recent.

Rare exceptions to these constant trends may, however, occur. One is reported for *G. exilis* in Jamaica, where immediately before its extinction the species reverts to sinistral coiling (Robinson, 1969). An exception to the sinistral coiling habit of *G. tumida* is found in the Java well Bodjonegoro-1, where the species briefly changes to a dextral mode in the boundary area of the *Globorotalia humerosa* and *Globorotalia margaritae* zones.

In summary, the taxa here placed into the menardiform *Globorotalia* species maintain, with the exception of the *G. tumida* group, the following general coiling trends: random from the Early Miocene *Catapsydrax stainforthi* to within the *Globorotalia fohsi robusta* Zone, change to sinistral coiling from within that zone to near the top of the *Globorotalia menardii* Zone in the Atlantic realm and to within the *Globo-*

rotalia humerosa Zone in the Indo-Pacific. Repeated fluctuations between sinistral and dextral in the Atlantic (Caribbean) during the Late Miocene *Globorotalia acostaensis* and *humerosa* zones. Worldwide preferred dextral coiling through the Early Pliocene *Globorotalia margaritae margaritae* Subzone to near the top of the Late Pliocene *Globorotalia tosaensis tosaensis* Zone. There, abrupt change to sinistral coiling that continues to Recent. The exception is the *G. tumida* group which, with few and only brief exceptions (see above), coils sinistrally throughout its range.

Globorotalia menardii and its subspecies

It is generally agreed that the *Globorotalia menardii* specimen originally described from Rimini beach sands (Adriatic coast) is not a Recent form but is reworked from nearby coastal Miocene exposures. Stainforth, Lamb & Jeffors (1978) requested the Commission of Zoological Nomenclature to suppress the lectotype of *Globorotalia menardii* (Parker, Jones & Brady, 1865) selected by Banner & Blow (1960a) from Recent Brady material from off the Isle of Man. They propose to replace this Recent lectotype from the Atlantic by a neotype from the Late Miocene of the Senigallia section, south of Rimini. Both for stratigraphic and geographic reasons, the request appears justified. However, as the Zoological Commission has not yet taken a decision on the matter, we continue here to subdivide the *menardii* complex into the Miocene to Early Pliocene *Globorotalia menardii* 'A' and 'B' and Pliocene to Holocene *G. menardii menardii* and *G. menardii cultrata*; this grouping is discussed in the following taxonomic notes.

Should the request by Stainforth *et al.* eventually be accepted by the Commission of Zoological Nomenclature, it would be necessary to consider whether the Miocene *G. menardii* 'A' becomes a synonym of the newly proposed neotype of *G. menardii*. Furthermore, if the younger *menardii* forms were considered to be sufficiently different, a new name would have to be found for them; Banner & Blow's lectotype from the Recent might belong here.

Taxonomic notes

GLOBOROTALIA ARCHEOMENARDII Bolli
Figures 32.6; 9, 12

Type reference: *Globorotalia archeomenardii* Bolli, 1957c, p. 119, pl. 28, figs. 11a–c.

This small species with a diameter of only about 0.3 mm possesses thin chamber walls and a delicate keel. Furthermore, compared with the considerably larger *G. praemenardii*, its spiral side appears more convex and the equatorial periphery less lobate.

GLOBOROTALIA PRAEMENARDII Cushman & Stainforth
Figures 32.7; 6, 9, 12

Type reference: *Globorotalia praemenardii* Cushman & Stainforth, 1945, p. 70, pl. 13, figs. 14a–c.

Representative specimens of this species measuring about 0.5 mm are distinctly larger than *G. archeomenardii* and have a more lobate equatorial periphery. In Trinidad, from where *G. praemenardii* was described, the taxon is usually rare, and in the *Globorotalia fohsi* zones immediately following the presence of its assumed ancestor, *G. praemenardii*, is often absent. A relationship based on transitional forms between the two taxa is therefore difficult to prove.

GLOBOROTALIA MENARDII 'A'
Figures 34.1; 6, 10, 12

Type reference: *Globorotalia menardii* 'A' Bolli, 1970, p. 582, pl. 5, figs. 1–4.

The Middle Miocene *Globorotalia menardii* (here termed *G. menardii* 'A', introduced by Bolli (1970)) differs from *G. praemenardii* in a more strongly developed peripheral keel, in possessing on the spiral side limbate intercameral sutures, which are non-limbate in *G. praemenardii*, and in displaying in the last whorl 6 against only 5 chambers. Furthermore, individual chambers of *G. menardii* 'A' are slightly less inflated on the spiral side and the test in equatorial view slightly less lobate, compared with *G. praemenardii*. The Middle Miocene *G. menardii* display considerable variability as is shown, for example, by Bolli (1957).

GLOBOROTALIA MENARDII 'B'
Figures 34.2–4; 6, 10, 11

Type reference: *Globorotalia menardii* 'B' Bolli, 1970, p. 582, pl. 5, figs. 5–7.

Globorotalia menardii 'B' was introduced by Bolli (1970) as a form intermediate between *G. menardii* 'A' and *G. multicamerata*. *G. menardii* 'B' first appears in the *G. acostaensis* Zone, probably as the ancestral form of *G. multicamerata*, which first occurs in the lower part of the *Globorotalia margaritae* Zone. It differs from *G. menardii* 'A' in the higher number of chambers forming the last whorl, 7–7½ against 5–6. As a result, specimens are usually also slightly larger in diameter and display somewhat thicker peripheral keels and intercameral sutures.

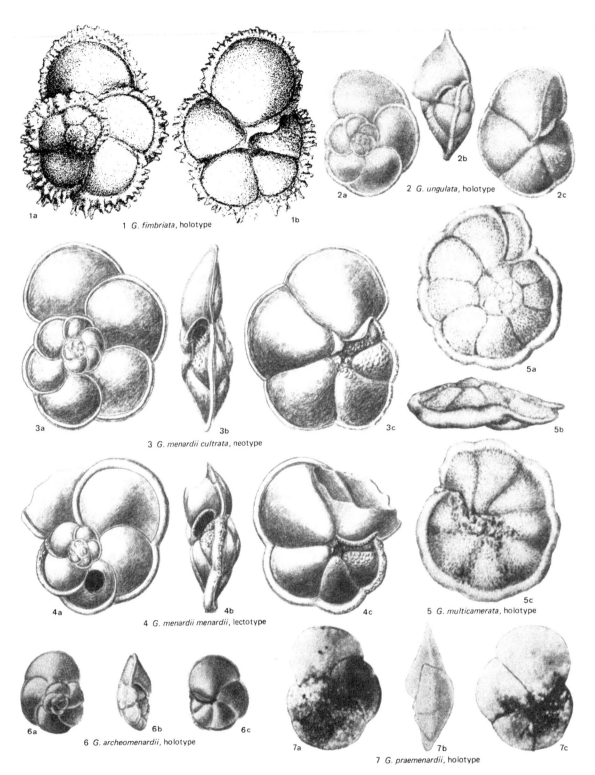

Fig. 32. Menardiform *Globorotalia* species (all figures × 60)

1 *G. fimbriata* (Brady)
  1a–b Holotype. *Challenger* Station 24, Recent, off Culebra Island, West Indies.

2 *G. ungulata* Bermudez
  2a–c Holotype. *Atlantis* Station 2953, Recent, 21°47′ N, 84°32′30″ W, Caribbean Sea.

3 *G. menardii cultrata* (d'Orbigny)
  3a–c Neotype. Recent, off St Helena, Atlantic.

4 *G. menardii menardii* (Parker, Jones & Brady)
  4a–c Lectotype. Recent, off Laxey, Isle of Man, Irish Sea.

5 *G. multicamerata* Cushman & Jarvis
  5a–c Holotype. Pliocene (originally described as Miocene), ½ mile east of Buff Bay, Jamaica.

6 *G. archeomenardii* Bolli
  6a–c Holotype. Sample Bo 202, *G. fohsi peripheroronda* Zone, Middle Miocene, Cipero Fm., Hermitage Quarry, Trinidad.

7 *G. praemenardii* Cushman & Stainforth
  7a–c Holotype. Sample Rz 425, *G. fohsi* Zone, Middle Miocene, Cipero Fm., Cipero coast, Trinidad.

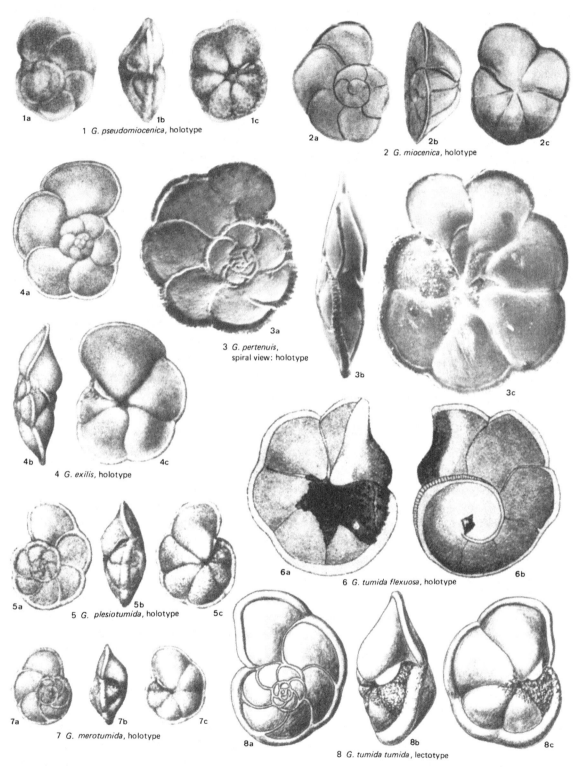

Fig. 33. Menardiform *Globōrotalia* species (all figures × 60)

1 *G. pseudomiocenica* Bolli & Bermudez
 1a–c Holotype. *G. menardii* Zone, Middle Miocene, Carenero Fm., road-cut between Higuerote and Chirimena, Miranda, Venezuela.

2 *G. miocenica* Palmer
 2a–c Holotype. K. V. Palmer Station 1, Pliocene (originally described as Miocene), Bowden Fm., Port Morant, Jamaica.

3 *G. pertenuis* Beard
 3a Holotype, **3b–c** Paratypes. Texas A & M Sigsbee Knolls core 64-A-9-5E at 3536 m, Pliocene (originally described as Pleistocene), Sigsbee Knolls, Gulf of Mexico.

4 *G. exilis* Blow
 4a–c Holotype. Sample ER 156, Zone N 19, Pliocene, Bowden Fm., Bowden Warf, Bowden, Jamaica.

5 *G. plesiotumida* Banner & Blow
 5a–c Holotype. Cubagua well-2, core at 1400 feet, *G. margaritae* Zone, Early Pliocene, Cubagua Island, Venezuela.

6 *G. tumida flexuosa* (Koch)
 6a–b Holotype, Late Tertiary, Surabaja, Java.

7 *G. merotumida* Banner & Blow
 7a–c Holotype. Cubagua well-2, core at 2700 feet, *G. humerosa* Zone, Late Miocene, Cubagua Island, Venezuela.

8 *G. tumida tumida* (Brady)
 8a–c Lectotype. Late Miocene or Pliocene, New Ireland.

GLOBOROTALIA MENARDII MENARDII
(Parker, Jones & Brady)
Figures 32.4, 34.5–7; 6, 10, 11

Type reference: *Rotalia menardii* Parker, Jones & Brady, 1865, p. 20, pl. 3, fig. 81.

Lectotype: *Globorotalia menardii* (Parker, Jones & Brady): Banner & Blow, 1960a, p. 31, pl. 6, figs. 2a–c.

Our investigations, mainly in the tropical Caribbean/Atlantic province, have led to the distinction on the Quaternary–Holocene of two *G. menardii* types, a robust, thick-walled form, fairly circular in equatorial outline and with a thick peripheral keel, and a more delicate, thinner-walled form, with a less marked peripheral keel and a more elongate outline in equatorial view. When comparing these two types with Banner & Blow (1960a) the robust form fits well with their lectotype of *G. menardii menardii* and the delicate, more elongate form with their neotype of *G. menardii cultrata*. Such a distinction was successfully followed by Bolli (1970) in the Caribbean DSDP sites and specimens figured by him. A comparison of the Pleistocene to Holocene *G. menardii* with forms from the Middle Miocene showed the latter on the average to be smaller in test size and with a somewhat slower increase in chamber size. As is the case for the Pleistocene to Holocene, two groups of specimens, a more robust and a more delicate one, can also be distinguished in the Middle Miocene. To distinguish the Middle Miocene forms from the Pleistocene–Holocene ones the older ones were designated as *G. menardii* 'A' by Bolli as discussed above.

GLOBOROTALIA MENARDII CULTRATA
(d'Orbigny)
Figures 32.3, 34.8–10; 6, 10, 11

Type reference: *Rotalina cultrata* d'Orbigny, 1839a, p. 76, pl. 5, figs. 7–9.

Neotype: *Globorotalia cultrata* (d'Orbigny): Banner & Blow, 1960a, p. 34, pl. 6, figs. 1a–c.

For distinction of this subspecies from *G. menardii menardii* see the discussion for that taxon.

GLOBOROTALIA FIMBRIATA (Brady)
Figures 32.1; 6, 10, 11

Type reference: *Pulvinulina menardii* (d'Orbigny) var. *fimbriata* Brady, 1884, p. 691, pl. 103, figs. 3a–b (same figures as used for lectotype designated by Banner & Blow, 1960a).

Lectotype: *Globorotalia cultrata* (d'Orbigny) subsp. *fimbriata* (Brady): Banner & Blow, 1960a, p. 25, pl. 5, figs. 2a–b.

G. fimbriata is easily distinguished from *G. menardii menardii*, of which it is apparently a variant, by the small, radially arranged spines along the peripheral keel. The species is known only from the latest Pleistocene and Holocene.

GLOBOROTALIA MULTICAMERATA Cushman & Jarvis
Figures 32.5, 35.16–19; 6, 10, 11

Type reference: *Globorotalia menardii* (d'Orbigny) var. *multicamerata* Cushman & Jarvis, 1930, p. 367, pl. 34, figs. 8a–c.

The species differs from its assumed ancestral form *G. menardii* 'B' in possessing more chambers in the last whorl, 8–10, and in a more robust test, including a thicker peripheral keel. Intermediate forms, difficult to assign to one or other of the two taxa, are frequent.

GLOBOROTALIA PERTENUIS Beard
Figures 33.3, 35.12; 6, 10, 11

Type reference: *Globorotalia pertenuis* Beard, 1969, p. 552, pl. 1, fig. 1.

Specimens of the species have a similar overall morphology to *G. multicamerata*; intermediate forms are frequent (Figs. 35.13–15). Typical specimens of *G. pertenuis* differ in having distinctly more delicate walls and less developed peripheral keels. With 6–8 the number of chambers forming the last whorl is often slightly less when compared with the 8–10 of *G. multicamerata*.

GLOBOROTALIA EXILIS Blow
Figures 33.4, 35.9–11; 6, 10, 11

Type reference: *Globorotalia* (*Globorotalia*) *cultrata exilis* Blow, 1969, p. 396, pl. 7, figs. 1–3.

The species possesses a wall and peripheral keel as delicate or more so than *G. pertenuis*. It differs in having only 5–6 against 6–8 chambers in the last whorl. These two species are very closely related and intermediate specimens are difficult to separate. Based on the distribution in Caribbean sections, *G. exilis* developed from *G. pertenuis* by a reduction in number of chambers towards the end of the *Globorotalia margaritae* Zone, to continue to the top of the *Globorotalia miocenica* Zone, in the upper part of its range without *G. pertenuis*. In how far the reduction of chambers, often coupled with irregular growth of chambers (Olsson, 1973), is a gerontic phenomenon or possibly caused locally by adverse conditions still remains to be clarified.

According to Blow (1969) *G. exilis* branched off from *G. cultrata* and, based on the thin, delicate type of wall structure, is the ancestor of the Pleistocene *G. fimbriata*. Such a view is not maintained here. As pointed out above, *G. exilis* is closely related to *G. pertenuis* by intermediate forms and from there to *G. menardii* B. On present evidence the dextral coiling *G. exilis* became extinct within the Pliocene at the top of the *Globorotalia miocenica* Zone, whereas *G. fimbriata*, which coils sinistrally, is known only from the Holocene. A relationship between these two species is therefore not established.

G. exilis is further characterized by slightly inflated chambers on the spiral side, slightly more so than in *G. pertenuis*. Consequently, intercameral sutures appear distinctly incised with early ones not limbate; later ones are limbate often only in their external half. This is explained by Stainforth *et al.* (1975), who showed that sutures became partially buried or overlapped by succeeding chambers.

Intermediate forms in the *Globorotalia margaritae* Zone between *G. menardii* 'B', *G. multicamerata* and *G. pertenuis*, and in the *Globorotalia miocenica* Zone between *G. pertenuis* and *G. exilis* point to a close relationship between these species. A detailed study of this group of species, taking into account all typical and transitional forms is still needed to improve on the present understanding of the relationship and stratigraphic ranges of these taxa.

GLOBOROTALIA MEROTUMIDA Blow & Banner
Figures **33.7**; 6, 10, 12

Type reference: *Globorotalia (Globorotalia) merotumida* Blow & Banner, *in* Banner & Blow, 1965a, p. 1352, figs. 1a–c.

Banner & Blow (1965a) proposed a lineage *G. merotumida–G. tumida plesiotumida–G. tumida tumida*, that developed according to Blow (1969) from *G. paralenguaensis* in the earliest part of Zone N 16 (= *Globorotalia acostaensis* Zone). Based on our observations this lineage is not related to the *G. lenguaensis* plexus. Instead, it developed only in the Late Miocene *Globorotalia humerosa* Zone (= N 17) from the *G. menardii* stock.

In areas where the lineage is continuously developed as is the case in the low latitude Indo-Pacific realm, all gradations from the *G. menardii* ancestor via *G. merotumida–G. plesiotumida* to *G. tumida tumida* occur in the highest Miocene within a comparatively short time interval. It is for these reasons and the fact that the evolutionary lineage is usually only poorly or incompletely developed in the low latitude Atlantic realm, from where it is usually absent within the Early Pliocene *Globorotalia margaritae margaritae* Zone to near the top of the Pliocene (Bolli & Krasheninnikov, 1977), that their use in biostratigraphy is limited. This applies in particular to the use of *G. merotumida*, *G. plesiotumida* and *G. tumida* to define the zones N 16, and N 17 and N 18. Stainforth *et al.* (1975) arrive at the same conclusion in their discussions of the plexus.

According to Banner & Blow (1965a) *G. merotumida* differs from *G. plesiotumida* 'in possessing (1) a test which is smaller in size at the same growth stage as measured by the number of chambers present, (2) a slower increase in whorl height as seen dorsally, (3) more uniformly enlarging chambers, (4) more consistently oblique dorsal intercameral sutures, (5) a thinner and more finely perforate test wall, (6) a thinner carina, (7) a relatively greater ventral convexity, and (8) a relatively broader apertural face'. According to the same authors 'G. merotumida has a much more convex and tumid test than any subspecies of *G. (G.) cultrata* (s.l.) (d'Orbigny)'.

GLOBOROTALIA PLESIOTUMIDA Blow & Banner
Figures **33.5**; 6, 10, 12

Type reference: *Globorotalia (Globorotalia) tumida plesiotumida* Blow & Banner, *in* Banner & Blow, 1965a, p. 1353, figs. 2a–c. For taxonomic remarks see discussion under *G. merotumida*.

GLOBOROTALIA TUMIDA TUMIDA (Brady)
Figures **33.8, 34.11–13**; 6, 10, 11

Type references: *Pulvinulina menardii* (d'Orbigny) var. *tumida* Brady, 1877, p. 535, no figures.
Pulvinulina tumida Brady: Brady, 1884, p. 692, pl. 103, figs. 4–5.
Lectotype: *Globorotalia tumida* (Brady): Banner & Blow, 1960a, p. 26, pl. 5, figs. 1a–c.

Compared with its ancestral forms *G. merotumida* and *G. plesiotumida*, *G. tumida tumida* is still more biconvexly tumid, and its wall and peripheral keel more robust. In equatorial view the taxon is also more elongate. All these features as well as test size are subject to considerable variability probably caused by changing environmental conditions. For further discussions of *G. tumida tumida*, including its use as an index form, see under *G. merotumida*.

GLOBOROTALIA TUMIDA FLEXUOSA (Koch)
Figures **33.6, 34.14–16**; 6, 10, 11

Type reference: *Pulvinulina tumida* var. *flexuosa* Koch, 1923–4, p. 357, text-figs. 9, 10.

This subspecies is a variant of *G. tumida tumida* with an additional chamber or two which become saddle-like, strongly bent down or flexed towards the umbilical side. Such forms are not necessarily restricted to certain stratigraphic levels but appear to be controlled rather by particular environmental conditions. *G. tumida flexuosa* was common in the younger Pleistocene of the Atlantic province, becoming extinct there some 80 000 years ago. However, virtually identical forms already occurred much earlier, for example in the Early Pliocene *Globorotalia margaritae* Zone in well Bodjonegoro-1 of Java (Bolli, 1966b). More locally, or within a given province, *G. tumida flexuosa* may serve as a useful index.

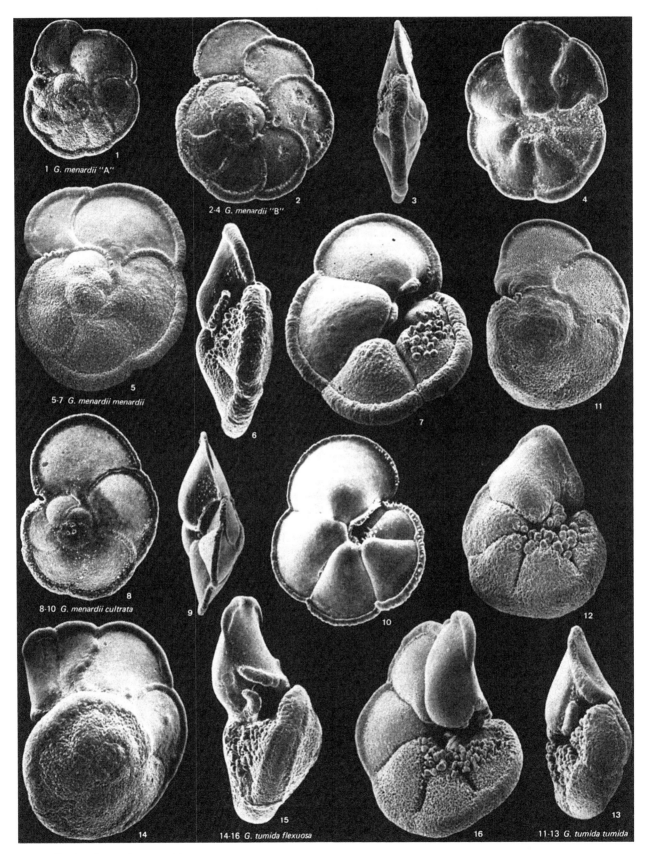

Fig. 34. For caption see p. 230.

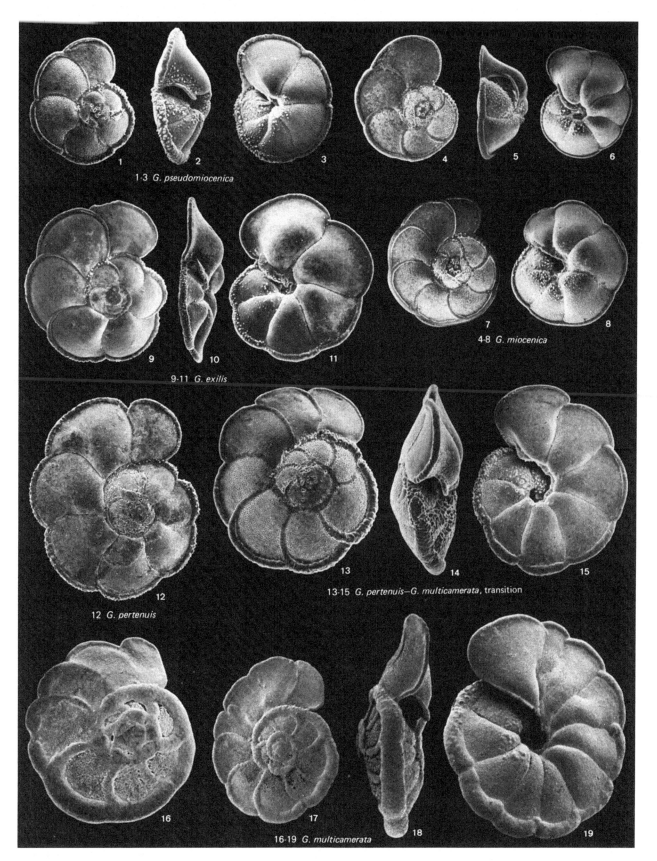

1-3 *G. pseudomiocenica*

4-8 *G. miocenica*

9-11 *G. exilis*

12 *G. pertenuis*

13-15 *G. pertenuis–G. multicamerata*, transition

16-19 *G. multicamerata*

Fig. 35. For caption see p. 230.

GLOBOROTALIA UNGULATA Bermudez
Figures **32.2**; 6, 10, 11

Type reference: *Globorotalia ungulata* Bermudez, 1961, p. 1304, pl. 15, figs. 13–15.

With few exceptions (Stainforth *et al.*, 1975) the species has been recorded only from the Holocene. Because of its tumid, somewhat elongate test it was considered to be related to *G. tumida*. However, the test, including the peripheral keel, is much more delicately built. specimens when present are as a rule considerably less frequent than *G. tumida*.

GLOBOROTALIA PSEUDOMIOCENICA Bolli & Bermudez
Figures **33.1, 35.1–3**; 6, 10, 11

Type reference: *Globorotalia pseudomiocenica* Bolli & Bermudez, 1965, p. 140, pl. 1, figs. 13–15.

The species appears to have evolved from *G. menardii* 'A' by the acquisition of a more convex umbilical side and by the

---

Fig. 34. Menardiform *Globorotalia* species (all figures × 60). All figures from Bolli, 1970

1 *G. menardii* 'A'
DSDP Leg 4-30-10-2, 56–58 cm, *G. mayeri* Zone, Middle Miocene, Aves Ridge, Caribbean Sea.

2–4 *G. menardii* 'B'
DSDP Leg 4-25-3-3, 75 cm, *G. margaritae* Zone, Early Pliocene, off northeast coast of Brazil, Atlantic.

5–7 *G. menardii menardii* (Parker, Jones & Brady)
DSDP Leg 4-29-1-2, 71–73 cm, *G. truncatulinoides truncatulinoides* Zone, Pleistocene, Venezuela Basin, Caribbean Sea.

8–10 *G. menardii cultrata* (d'Orbigny)
DSDP Leg 4-2-1-2, 71–73 cm, *G. truncatulinoides truncatulinoides* Zone, Pleistocene, Venezuela Basin, Caribbean Sea.

11–13 *G. tumida tumida* (Brady)
DSDP Leg 4-29-1-3, 10–12 cm, *G. truncatulinoides truncatulinoides* Zone, Pleistocene, Venezuela Basin, Caribbean Sea.

14–16 *G. tumida flexuosa* (Koch)
DSDP Leg 4-29-1-cc, *G. truncatulinoides truncatulinoides* Zone, Pleistocene, Venezuela Basin, Caribbean Sea.

Fig. 35. Menardiform *Globorotalia* species (all figures × 60). 1–16 from Bolli, 1970; 17–19 from Krasheninnikov & Pflaumann, 1977

1–3 *G. pseudomiocenica* Bolli & Bermudez
DSDP Leg 4-29-4-3, 71–73 cm, *G. margaritae* Zone, Early Pliocene, Venezuela Basin, Caribbean Sea.

4–8 *G. miocenica* Palmer
DSDP Leg 4-30-7-cc, *G. miocenica* Zone, Middle Pliocene, Aves Ridge, Caribbean Sea.

9–11 *G. exilis* Blow
DSDP Leg 4-31-3-2, 2–4 cm, *G. miocenica* Zone, Middle Pliocene, Beata Ridge, Caribbean Sea.

12 *G. pertenuis* Beard
Information as for Figs. 35.9–11.

13–15 *G. pertenuis–G. multicamerata*, transition
DSDP Leg 4-29-4-4, 110–112 cm, *G. margaritae* Zone, Early Pliocene, Venezuela Basin, Caribbean Sea.

16–19 *G. multicamerata* Cushman & Jarvis
16 Information as for Figs. 35.13–15. **17–19** DSDP Leg 41-366A-6-5. 33–35 cm, *G. margaritae* Zone, Early Pliocene, Sierra Leone Rise, Atlantic.

---

spiral side becoming more flattened or even nearly planar. *G. pseudomiocenica*, like *G. miocenica*, is more delicately built and has a more delicate keel than *Globorotalia menardii* 'A'. In equatorial view the periphery remains more lobate compared with that in typical *G. miocenica*. *G. pseudomiocenica* may be regarded as the link between *G. menardii* 'A' and *G. miocenica*. Consequently it displays considerable variability.

GLOBOROTALIA MIOCENICA Palmer
Figures **33.2, 35.4–8**; 6, 10, 11

Type reference: *Globorotalia menardii* (d'Orbigny) var. *miocenica* Palmer, 1945, p. 70, pl. 1, figs. 10a–c.

G. miocenica differs from *G. pseudomiocenica* in an absolutely flat spiral side. Occasionally the surface may even tend to become slightly concave. Compared with *G. pseudomiocenica*, the umbilical side is still higher and in equatorial view the periphery more circular and less lobate. Because the transitional forms are often erroneously identified as *G. miocenica* it must be stressed that correct species identification allows for little variation, one of the prime characters being the absolutely flat spiral side. Despite its name, *G. miocenica* is confined to the late Early and the Middle Pliocene.

Glorobotalia crassaformis and related taxa

Opinions are divided on the origin of *Globorotalia crassaformis* and related taxa and on their evolutionary trends and usefulness as stratigraphic indices within the Pliocene and Pleistocene. Comparison of published work actually shows a reversal of trends depending on the authority. Certain authors regard *crassaformis* as the end form of an evolutionary development, others as the initial species from which other morphologically distinguishable taxa have developed. Evolution of the non-carinate *crassaformis* from distinctly carinate ancestors has been suggested, but also the view that *crassaformis* gave rise to keeled forms has been stated. There is also no general agreement on ranges of individual taxa. The use and acceptance of taxa of the *crassaformis* plexus as index forms in biostratigraphy thus remains incompletely understood, controversial, and largely limited to restricted areas, such as the Mediterranean or the Caribbean.

Since the erection of *G. crassaformis* (Galloway & Wissler) in 1927 from the Pleistocene of California, several other taxa regarded as closely related have been proposed. *G. crassaformis* is known to be a cosmopolitan form, occurring in both warm and temperate waters and displaying considerable variability through time and in different environments; this has led to the erection of related taxa.

Morphological changes in the closely related taxa are fluid and their clear differentiation therefore difficult. Stainforth *et al.* (1975), are of the opinion that 'distinctions between adjacent forms are tenuous and arbitrary and not expressible in objective terms'. They therefore restrict themselves to describing and figuring only *G. crassaformis* with *G. aemiliana* and *G. ronda* as recognizably more primitive and more advanced forms, respectively.

Opinions on how valid the *crassaformis* plexus is for biostratigraphy are divided. Data from different regions indicate that in geographically and paleoenvironmentally restricted areas, certain taxa of the plexus may successfully be applied as index forms and zonal markers within the Pliocene–Pleistocene. It is for these reasons that the species and subspecies and their evolutionary trends as described in particular for New Zealand, Italy and the Caribbean are here briefly discussed.

Proposed lineages

The following bioseries or evolutionary trends have been proposed:

Kennett (1966) for New Zealand: *G. miozea–G. conomiozea–G. crassaformis.*

Colalongo & Sartoni (1967) for Italy: *G. hirsuta* (= *G. margaritae*)–*G. aemiliana–G. crassaformis.*

Conato & Follador (1967) for Italy: *G. crotonensis–G. crassacrotonensis–G. crassaformis.*

Blow (1969) general: *G. crassaformis crassaformis–G. crassaformis ronda–G. crassaformis oceanica.*

Blow (1969) general: *G. crassaformis crassaformis–G. crassula conomiozea–G. crassula crassula–G. crassula viola.*

Lamb & Beard (1972) Gulf of Mexico, Caribbean, Italy: *G. aemiliana–G. crassacrotonensis–G. crassaformis.*

The conflict in evolutionary trends postulated from different areas is apparent from the above list. It is in part because morphological changes within the plexus were controlled by different and regionally restricted environmental conditions. A study by Lidz (1972) of Late Pleistocene *G. crassaformis* populations from different latitudes has shown these to be largely controlled by temperature.

Some of the conflicting views are difficult to reconcile even when taking the above controls into account, and in our opinion require revision or further investigation. One of them is the view of Blow (1969), who regarded the non-carinate and pustulose *G. crassaformis* as the ancestor of the fully carinate and smooth surfaced *G. conomiozea*, which he considered to have branched off from it in the uppermost Miocene. In addition to his lineage *crassaformis–conomiozea–crassula–viola*, Blow proposed a second line of non-carinate forms, *crassaformis–ronda–oceanica* from which in turn developed *G. tosaensis tosaensis* and eventually *G. truncatulinoides pachytheca.*

Kennett (1966) believed that in this bioseries the ancestors of the non-carinate *crassaformis* are the fully keeled and smooth Late Miocene *G. miozea* and *G. conomiozea* with the latter restricted to the uppermost Miocene where it develops into *crassaformis*. This view remains problematical and is in contrast to Blow's opinion. Investigations on Pleistocene specimens (Lidz, 1972) have shown that a trend to form an imperforate rim or keel occurs in warm water low latitude specimens, with tests still retaining their pustulose surface character.

The lineage from *G. aemiliana* to *G. crassaformis* proposed by Colalongo & Sartoni (1967) from the Mediterranean area seems to be the most straightforward and therefore acceptable; it is based on a gradual increase in height of the umbilical side. Later in 1967 Conato & Follador proposed the same lineage using the additional names *G. crotonensis* and *G. crassacrotonensis* which are now considered as junior synonyms of *G. aemiliana* and *G. crassaformis* respectively.

The sequence of taxa given by Lamb & Beard (1972) for the Gulf of Mexico and Italy (*aemiliana–crassacrotonensis–crassaformis*) is the same as that used by the Italian workers. The difference lies in the earlier appearance of the series (within the Early Pliocene *Globorotalia margaritae* Zone) as compared with the Italian authors, whose sequences begin stratigraphically later; that is, in the Middle Pliocene.

Rögl (1974) closely followed the distribution of the *crassaformis* plexus in the extended Middle Pliocene to Holocene section of DSDP Site 262 in the Timor Trough (lat. 11° S). The fact that the ranges given by him differ in part from those for the Caribbean can probably be attributed to different environmental conditions that existed in the two areas.

Stratigraphic significance

Considerable discrepancies among authors also exist in their interpretation of ranges of individual taxa. Except for Blow, according to whom the *crassaformis* plexus originates in the Late Miocene, all other authors place this event either at the base or within the Early Pliocene *Globorotalia margaritae margaritae* Zone or, in the Mediterranean area, somewhat later; that is in the mid-Pliocene. Ranges of all the taxa distinguished by Blow run from the Late Miocene or basal Pliocene to Holocene and are thus of no stratigraphic significance within this interval. Data from a number of DSDP Caribbean sections (Bolli, 1970; Bolli & Premoli Silva, 1973) indicate for that region a much more restricted distribution of the subspecies *viola* and *hessi*. The taxa recognized as evolutionary stages by Italian workers, and adapted by Lamb & Beard (1972) are also stratigraphically restricted.

From the ranges reported it can be concluded that the *G. crassaformis* plexus is limited to the Pliocene–Holocene and that certain subspecies are significant for stratigraphically restricted intervals at least within certain areas.

G. crassaformis and subspecies have been proposed as zonal markers by a number of authors. Several Italian workers (see Lamb & Beard, 1972) subdivide the Middle Pliocene into

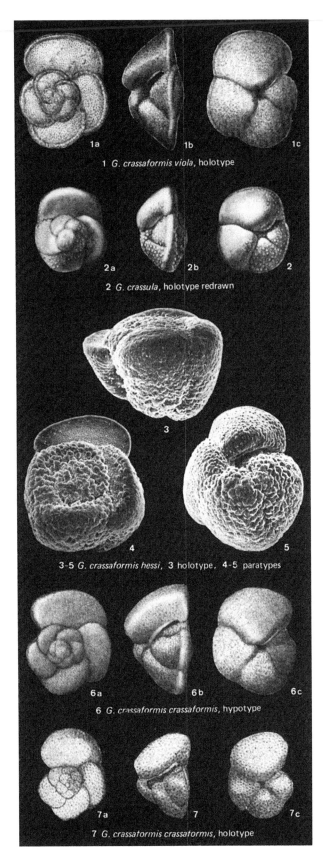

a *Globorotalia aemiliana* Zone (first occurrence of zonal marker to first occurrence of *G. crassaformis*) and a *Globorotalia crassaformis* Zone (first occurrence of zonal marker to first occurrence of *G. inflata*). In Chapter 8 of this volume Iaccarino combines these two zones into the *Globorotalia aemiliana* Zone (first occurrence of *G. aemiliana* to first occurrence of *G. inflata*). Based on evidence from several DSDP sites of Legs 4 and 15, Bolli & Premoli Silva (1973) introduced a *Globorotalia crassaformis viola* Subzone (first *G. truncatulinoides truncatulinoides* to first *G. crassaformis hessi*) and a *Globorotalia crassaformis hessi* Subzone (first zonal marker to first *G. calida calida*), subdividing the Early Pleistocene in the Caribbean.

Coiling trends

The preferred coiling direction in the *crassaformis* plexus is sinistral. Swings to dextral coiling are rare and mostly limited in time; locally they may, however, be of stratigraphic significance.

In the Pacific realm temporary dextral coiling occurs in the Late Pliocene of New Zealand (Jenkins, 1971), and in the Middle Pliocene of Java (Bolli, 1966b). Coiling directions have been followed in several Caribbean DSDP sites by Bolli (1970) and Bolli & Premoli Silva (1973). All taxa of the plexus recognized show a nearly exclusive preference for sinistral coiling throughout the Pliocene and Pleistocene. There are only a few instances of brief swings to dextral coiling: *G. crassaformis hessi* in the upper part of the Pleistocene in Site 30; *G. crassaformis crassaformis* in the upper part of the *Globorotalia margaritae evoluta* and lower part of the *Globigerinoides trilobus fistulosus* zones in the Early to Middle Pliocene and *G. crassaformis* cf. *viola* within the *Globorotalia tosaensis* Zone in the Late Pliocene, both in Site 148.

Lidz (1972) in her study on Late Pleistocene Atlantic/Caribbean *G. crassaformis* specimens from piston cores at three different latitudes (5, 15, 34° N) concluded that 'coiling direction is not influenced by temperature and appears to be neither consistent nor interregionally correlative among the

---

Fig. 36. *Globorotalia crassaformis, Globorotalia crassula* (all figures × 60)

1 *G. crassaformis viola* Blow
   1a–c Holotype. Sample WHB 181B, Zone N 19, Early Pliocene, near Buff Bay, Jamaica.

2 *G. crassula* Cushman & Stewart
   2a–c Holotype redrawn. Station 7, Pliocene, Humboldt, County, California, USA (from Blow, 1969).

3–5 *G. crassaformis hessi* Bolli & Premoli Silva
   3 Holotype, **4–5** Paratypes. DSDP Leg 4-29-1-3, 10–12 cm, *Globorotalia truncatulinoides truncatulinoides* Zone, Pleistocene, Venezuela Basin, Caribbean Sea.

6–7 *G. crassaformis crassaformis* (Galloway & Wissler)
   6a–c Hypotype. Core Nr. 258, 15–20 cm, Zone N 23, Pleistocene, offshore Niger Delta (from Blow, 1969). **7a–c** Holotype. Pleistocene, Lower San Pedro Limestone, middle level, Lomita Quarry, California, USA,

cores discussed'. Her text-figure 1 'Percentage of right-coiled specimens of *Globorotalia crassaformis*' shows a preference for sinistral coiling. It is interrupted in the 5° latitude core by only one brief swing to nearly 100% dextral. More frequent dextral forms recurring within shorter intervals are present in the 15° latitude core, and still more in that from 34°. Contrary to Lidz' conclusions these patterns therefore might rather be interpreted as reflecting fluctuations related to temperature, i.e. a distinct preference for sinistral coiling in warmer and a trend towards dextral coiling in cooler waters.

Taxonomic notes

All taxa discussed below are here treated as subspecies of *Globorotalia crassaformis*. They include: *crassaformis*, *oceanica*, *ronda*, *hessi* and *viola*.

In specimens of the *crassaformis* plexus, the last whorl consists basically of 4 chambers rather rapidly increasing in size and fairly loosely arranged. The spiral side of the test is more or less flat, the umbilical side convex but vaulted to a variable degree. In equatorial view the periphery is distinctly lobate, in edge view the peripheral area varies from rounded through acute to carinate. Prime characters in distinguishing taxa are the degree of ventral vaulting and presence of an imperforate keel. Shape of chambers, depth of intercameral sutures, shape, size and depth of umbilicus, peripheral outline and wall surface (density of pustules) are other characters used in distinguishing the taxa.

GLOBOROTALIA CRASSAFORMIS CRASSAFORMIS (Galloway & Wissler)
Figures **36.6–7**; 6, 10, 11

Type reference: *Globigerina crassaformis* Galloway & Wissler, 1927, p. 41, pl. 7, fig. 12.
Synonym: *Globorotalia crassacrotonensis* Conato & Follador, 1967, p. 557, pl. 4, figs. 3a–c.

The subspecies is characterized by a highly vaulted form with conically shaped umbilical side. The peripheral margin may vary from subacute to acute, or, in low latitude specimens, show a trace of an imperforate rim. The other non-carinate taxa of the plexus are either less highly vaulted umbilically or possess distinctly rounded peripheral margins with more inflated umbilical sides as in *G. ronda*, *G. oceanica* and *G. hessi*.

G. crassula Cushman & Stewart (Fig. 36.2) belongs to the *crassaformis* complex based on the original description. Blow's (1969) refigured holotype of *G. crassula* shows a peripheral keel, which makes its close relationship to *G. crassaformis* doubtful. According to the same author the holotype is strongly recrystallized and somewhat abraded. He regards it as an intermediate between *G. conomiozea* and *G. viola* but morphologically closer to *G. viola*. Its poor preservation and the fact that it appears more tightly coiled than is typical for the *crassaformis* plexus leaves its relationship in doubt.

GLOBOROTALIA CRASSAFORMIS OCEANICA
Cushman & Bermudez
Figures **37.11**; 6, 10, 11

Type reference: *Globorotalia (Turborotalia) oceanica* Cushman & Bermudez, 1949, p. 43, pl. 8, figs. 13–15

GLOBOROTALIA CRASSAFORMIS RONDA
Blow
Figures **37.10**; 6, 10, 11

Type reference: *Globorotalia (Turborotalia) crassaformis ronda* Blow, 1969, p. 388, pl. 4, figs. 4–6.

Both subspecies have a virtually identical test morphology with rounded peripheral margin and high, inflated umbilical side. According to Blow the most distinct character by which *ronda* differs from *oceanica* is the thickened test wall formed by densely packed sclerites. Both subspecies differ from the *crassaformis* subspecies in the distinctly rounded peripheral margin and the more inflated umbilical portion of the test and also by more closely appressed chambers as a result of tighter coiling. Because of its consistently smaller size, Lidz (1972) regarded *ronda* as an early ontogenetic stage found throughout the *crassaformis* lineage but maintained it as an ecological morphotype predominant at lower temperatures.

GLOBOROTALIA CRASSAFORMIS HESSI Bolli & Premoli Silva
Figures **36.3–5**; 6, 10, 11

Type reference: *Globorotalia hessi* Bolli & Premoli Silva, 1973 (= *Globorotalia crassaformis* 'B' Bolli, 1970, p. 580, pl. 4, fig. 13, holotype, figs. 15, 16, paratypes).

Compared with the subspecies *crassaformis*, *hessi* is more distinctly quadrangular and less lobate in equatorial view and the spiral surface often appears somewhat concave instead of planar. The last chamber is usually not larger than the penultimate, which is in contrast to the holotype of *G. crassaformis crassaformis*, where it is distinctly larger. The peripheral margin is rounded and the umbilical portion high and slightly inflated though less so than in *oceanica* and *ronda*. The last chamber is often slightly offset obliquely downwards and has an acute peripheral margin or a faint keel. The wall surface is rough as in *ronda* covered by pustules except for the last chamber which may be more delicate. *Hessi* is close to *ronda* but differs from it in the more quadrangular periphery in equatorial view, the slightly concave spiral side and in the above-mentioned characters of the last chamber. The spiral view of the Middle Pleistocene specimen figured in Stainforth *et al.* (1975, fig. 159/5) as *G. crassaformis* represents a typical *G. crassaformis hessi*.

GLOBOROTALIA CRASSAFORMIS VIOLA Blow
Figures **36.1**; 6, 10, 11

Type reference: *Globorotalia (Globorotalia) crassula viola*
Blow, 1969, p. 397, pl. 5, figs. 4–6.

Intermediate forms indicate that the subspecies *viola* is closely
related to the *G. crassaformis* plexus. In Caribbean DSDP
sections (Legs 4, 15) it branched off from *crassaformis* in the
upper part of the Early Pliocene through gradual acquisition
of a peripheral keel. Typical *viola* possess a complete peripheral
keel. In side view they are more angular compared with
crassaformis and not as highly conical. Intercameral sutures
on the spiral side may also be limbate in contrast to *crassaformis*. A particular feature is that early parts of the test remain
distinctly pustulose. Intermediate forms between non-carinate
crassaformis and wholly carinate *viola* forms, also still with the
pustulose surface, occur in several Pliocene–Pleistocene Caribbean DSDP sections; a close relation is therefore regarded as
established between the subspecies *crassaformis* and *viola*.

The *Globorotalia tosaensis–Globorotalia truncatulinoides* plexus

An evolutionary sequence originating from the *Globorotalia crassaformis* stock and leading through *G. tosaensis* to
the *G. truncatulinoides* end-form was postulated by Blow
(1969). He distinguished two separate but contemporaneous
lineages. One was characterized by forms whose tests are
thick-walled as a result of closely arranged fused pustules that,
at least in early chambers, obstruct the original perforation. It
commences with *G. crassaformis ronda* and evolves through *G.
tosaensis tosaensis* into the carinate *G. truncatulinoides pachytheca*. The second line develops from *G. crassaformis oceanica*
through *G. tosaensis tenuitheca* into the keeled *G. truncatulinoides truncatulinoides* and is characterized by forms with
thinner perforated walls even in the earlier chambers. Corresponding taxa of these two lineages, *G. crassaformis ronda/
oceanica*, *G. tosaensis tosaensis/teniutheca* and *G.
truncatulinoides pachytheca/truncatulinoides* have virtually
identical ranges except that, according to Blow (1969), the
thicker-walled subspecies *tosaensis* and *pachytheca* appear
very slightly later. Distinction of these taxa is primarily based
on the development of a peripheral keel, thickness of walls and
the tests becoming more compact and angular, and more
highly conical on the umbilical side.

It would seem logical to treat such lineages on a tri-
nominal basis using subspecies assigned to only one species,
as has been done for the *G. cerroazulensis* or *G. fohsi* lineages
for example.

However, morphological changes in investigated sections are often rather abrupt and gradual developments not as
clearly recognizable as one would wish to see. For these
reasons and until better evidence becomes available and

because most authors maintain *crassaformis*, *tosaensis* and
truncatulinoides as separate species we follow this practice
here.

Globorotalia truncatulinoides was erected in 1839 by
d'Orbigny from Recent sediments near the island of Teneriffe
(Canary Islands) under the genus *Rotalina*. Brady (1884)
described and figured the species again, this time from sediments from the South Atlantic, but erroneously called it
Pulvinulina micheliniana (d'Orbigny), a name previously applied to a Cretaceous species. Cushman (1931) re-instated the
name *truncatulinoides* and re-assigned it to the genus *Globorotalia*. As no holotype could be located from the d'Orbigny
collection, Blow (1969) described a neotype from off Gomera,
Canary Islands. Bermudez (1949) proposed a subgenus *Truncorotalia* for which he selected *truncatulinoides* as genotype.

Not until the value of Tertiary planktic foraminifera
began to be recognized did *G. truncatulinoides* attract attention
as a possible stratigraphic and ecologic index form. Data from
numerous marine and land based sections soon indicated that
typical representatives of this readily recognizable species are
present only from about the base Pleistocene upwards to the
Recent. Recognizing this, some authors began to investigate
morphological variation within the species, possible evolutionary trends and ancestral forms, to check whether the taxon
could be used for a further subdivision of the Pleistocene and
possibly latest Pliocene.

Evolutionary trends, ontogeny and variability

Takayanagi & Saito (1962) were the first to propose the
non-carinate *G. tosaensis* as a likely ancestral form to the
carinate *G. truncatulinoides*. They were followed in 1969 by
Blow, who introduced two more taxa, *G. tosaensis tenuitheca*
as an intermediate between the thin-walled *G. crassaformis
oceanica* and *G. truncatulinoides truncatulinoides*, and *G.
truncatulinoides pachytheca* as a thick-walled equivalent of *G.
truncatulinoides truncatulinoides*. Sprovieri, Ruggieri & Unti
(1980) proposed *G. truncatulinoides excelsa*, a very high conical
and thin-walled form whose first appearance according to its
authors marks the base of the Pleistocene in Southern Italy,
whereas *G. truncatulinoides truncatulinoides* is present there
already in the latest Pliocene.

To obtain a better knowledge of the distinguishing
characters, growth patterns and presence of intermediate
forms of the contemporaneous subspecies *truncatulinoides*,
pachytheca and *excelsa*, we investigated three samples. Two
are from the Pleistocene of the Caribbean DSDP Leg 4-29-1-1,
20–22 cm, and from Leg 15-154A-1-5, 80–82 cm. The third is
from a South Pacific piston core Downwind HG-74, 3–5 cm,
water depth 3030 m, lat. 28°29′ S, long. 106°30′ W.

In all three samples the general trend was found to be
the same. The smallest recognized specimens (0.175 mm)
consist exclusively of the thin-shelled *excelsa* type whereas the

smallest thick-shelled *pachytheca* specimens measure 0.35–0.375 mm. Both types, still clearly separable, are present as intermediate sized specimens (0.4–0.5 mm). Large specimens (0.5–0.8 mm) possess predominantly a thickened shell with distinctly thin-walled *excelsa* types being rare. Specimens transitional between the thick-shelled *pachytheca* and the thin-shelled *excelsa* type, which would be included in the subspecies *truncatulinoides*, become more frequent with increasing test size.

Based on the associations in these three samples it can be concluded that all three types lived at the same time in the same area. Different test thicknesses are thus probably related to the position of the specimen within the water column. Youngest and smallest specimens, all of the *excelsa* type, lived close to the surface. During continuing growth, and probably to facilitate sinking to deeper water levels, wall thickness increased by the addition of pustules resulting in the production of predominantly thicker-walled specimens.

In comparing the holotypes and the neotype of the three subspecies with the populations of the three samples examined it is presumed that the holotype of *excelsa* (0.56 mm) belongs to the thin-shelled group that lived close to the surface while the holotype of *pachytheca* (0.58 mm) with its thickened test lived some way below the surface. The 0.78 mm neotype of *truncatulinoides* is a large, adult specimen that, compared with *pachytheca*, has added 2 additional chambers which are less strongly calcified, thus giving the test a thinner overall appearance by comparison with the holotype of *pachytheca*. Variability, particularly in large, adult specimens exists, which may make it difficult to clearly divide thick *pachytheca* and thin *excelsa* type forms. The neotype of *truncatulinoides* apparently belongs to such an intermediate form.

Stratigraphic significance

While for many years the first occurrence of typical keeled *G. truncatulinoides* was widely accepted as indicative for the Pliocene/Pleistocene boundary, this has more recently come under some criticism as a consequence of a differentiation into finer morphological stages and a confrontation with other fossils and criteria used to define this boundary. Thus, according to Blow (1969), Lamb & Beard (1972) and Stainforth *et al.* (1975), *G. truncatulinoides truncatulinoides* first occurred only within the Early Pleistocene, whereas following Sprovieri *et al.* (1980) on evidence from Capo Rossella, Sicily, Italy, the taxon is present as early as basal Late Pliocene and *G. truncatulinoides excelsa* from base Pleistocene. As the presence of *G. truncatulinoides* s.l. is dependent on ecological factors, its first occurrence for example in the Mediterranean and the Gulf of Mexico/Caribbean is not necessarily synchronous.

G. truncatulinoides s.l. is a fairly cosmopolitan species whose present-day distribution has been studied in all oceans by numerous authors like Ericson, Wollin & Wollin (1954),

Bradshaw (1959), Parker (1971) and others. It occurs from low latitude equatorial waters to 50–60° N and S. Its maximum development however lies between 20 and 50°, becoming scarcer in lower and higher latitudes. In certain low latitudinal areas between 0 and 20°, such as the central Pacific, the species may even be virtually absent.

Within Pleistocene sections the presence of *G. truncatulinoides* may also be erratic. Examples for this are DSDP Site 262 in the Timor Trough, where throughout the 332 m of Pleistocene the species occurs irregularly and is, in particular, frequently absent in the Late Pleistocene (Rögl, 1974). In the classical South Italian sections where the Pliocene/Pleistocene boundary is defined, and particularly in the La Vrica section in Calabria (considered the most suitable by Selli *et al.*, 1977; Pelosio, Raffi & Rio, 1980), *G. truncatulinoides* occurs only very sporadically. This fact, in addition to its suspected relatively late appearance in the Mediterranean as a whole, means that its first occurrence in that general area can no longer be regarded as a reliable index for the Pliocene/Pleistocene boundary (Pelosio *et al.*, 1980).

These examples illustrate the problem of applying the first occurrence of *G. truncatulinoides* on a worldwide basis as an index for the Pliocene/Pleistocene boundary, particularly in areas of lower and higher latitudes where at that time ecological conditions for the species were not very favourable. Over considerable intermediate areas, on the other hand, conditions were more stable and the development of *G. truncatulinoides* more or less uninterrupted, as is recorded from a great number of DSDP sections. Here the initial occurrence of *G. truncatulinoides* serves as a valuable index for the Pliocene/Pleistocene boundary, as was originally proposed by Ericson, Ewing & Wollin (1963).

However, even in areas of 'normal' development of the species views regarding its first occurrence have not always agreed. In Blow (1969) *G. truncatulinoides truncatulinoides* first appears at the base of N 22 which is placed within the Early Pleistocene, and *G. truncatulinoides pachytheca* just above the base. Lamb & Beard (1972) and Stainforth *et al.* (1975), place the Pliocene/Pleistocene boundary considerably lower, at the extinction level of *G. altispira/G. venezuelana*. Here, a considerable gap exists between this event and the first *G. truncatulinoides*, consisting of the lower part of the *Globorotalia tosaensis* Zone of Lamb & Beard, or the *Globorotalia exilis* and the *Globorotalia* cf. *tosaensis* zones of Bolli & Premoli Silva (1973). On the other hand, many reports dealing with DSDP sections, including Bolli (1970) and Bolli & Premoli Silva (1973), place the Pliocene/Pleistocene boundary at the first appearance of *Globorotalia truncatulinoides* s.l.

Sprovieri *et al.* (1980) claim that in southern Italy the delicate *G. truncatulinoides excelsa* appears stratigraphically slightly after the thicker-walled *G. truncatulinoides truncatulinoides*. Also according to Blow (1969) and Rögl (1974)

comparatively thin-walled *G. truncatulinoides truncatulinoides* (in which they presumably also included specimens of the delicate *excelsa* type) appear slightly earlier than the still thicker-walled *G. truncatulinoides pachytheca*. These observations indicate that the stratigraphic significance of first appearance of thin as against thick-walled forms of *G. truncatulinoides* is at best of local significance.

G. tosaensis s.l. is widely accepted as the form from which the carinate *G. truncatulinoides* s.l. developed at about the Pliocene/Pleistocene boundary. This ancestral form thought by Blow (1969) to have branched off in mid-Pliocene time from the *G. crassaformis* stock overlaps with *G. truncatulinoides* in the Early Pleistocene where it becomes extinct. Situations where all intermediate stages of the proposed lineage *crassaformis* — *tosaensis* — *truncatulinoides* are clearly recognizable are dificult to find among the numerous DSDP sections.

Coiling trends in *Globorotalia truncatulinoides*

Coiling trends in *G. truncatulinoides* were first discussed by Ericson *et al.* (1954). Their investigations were based on deep sea cores from the Atlantic, including the Gulf of Mexico and the Caribbean, between approximately lats. 0° and 55° N. Their fig. 1 shows the dominant coiling directions in these areas based on counts from the tops of over 100 cores. Dextral coiling is dominant in low latitudes, continuing to about 40° N along the east coast of the United States, probably caused by the Gulf Stream. Sinistral coiling is strongly predominant in the mid-latitudes, 20–40° N, again swinging further north to about 50° N in the westernmost Atlantic. A second area where dextral coiling is predominant exists in the eastern Atlantic, from about 35–55° N. According to the authors the ratio of preferred coiling increases with the distance from the boundaries of water masses; at and near them dextral and sinistral specimens occur in about equal numbers. In addition to the coiling patterns deduced from the topmost samples from cores from the north Atlantic, Ericson, *et al.* also investigated coiling changes through time as documented in their piston cores. They found that patterns of apparently correlatable changes could be followed over considerable distances. Four cores from a profile 475 km long, crossing the Mid-Atlantic Ridge at the equator are used as an example.

Parker (1962, 1973) offered some preliminary data on the coiling pattern of *G. truncatulinoides* for the Pacific and Indian oceans. There exists a left-coiling province south of a line between Australia and South America that varies from lats. 25 to 55° S. North of it and including the North Pacific there is a preference for dextral coiling. A similar trend with a dividing line at approximately 30° S is also indicated for the Indian Ocean.

Bolli (1970) and Bolli & Premoli Silva (1973) followed coiling trends in *G. truncatulinoides* through the Pleistocene of

Caribbean DSDP sections 29, 30, 31, 148 and 154A. There, the general pattern is random in the Early Pleistocene (*Globorotalia crassaformis viola* to lower part of the *Globorotalia crassaformis hessi* Zone, to become predominantly dextral from there on to the Holocene. Short term fluctuations mainly between random and dextral, very occasionally also with short swings to sinistral, do occur and are difficult to correlate between the sites.

The observations of Rögl (1974) on the Pleistocene of DSDP Site 262 in the Timor Trough indicate a more stable behaviour of *G. truncatulinoides*. It is dextral throughout, except for a short period of sinistral coiling in the basal Pleistocene (lower part of the *Globorotalia crassaformis viola* Zone).

From these few cited observations alone it is evident that coiling patterns in present-day *G. truncatulinoides* are more complex than in any other known planktic species. In contrast, Recent *G. menardii*, for instance, coil exclusively sinistrally wherever present, whether in the Atlantic or Indo-Pacific province.

Temperature is regarded at least as one determining factor for coiling preference, though based on the findings of Ericson *et al.* (1954) in the North Atlantic, coiling directions in *G. truncatulinoides* are apparently controlled in addition by other still unknown factors. The effects of temperature are also seen in *Globigerina pachyderma*, where dextral coiling indicates warmer waters, sinistral coiling cooler conditions (Ericson, 1959; Bandy, 1972).

Available evidence indicates that coiling patterns in *G. truncatulinoides* are significant and reliable for correlation within limited areas where ecological stability persists. Over longer distances and particularly across water mass boundaries on the other hand the application of coiling patterns of *G. truncatulinoides* to correlation and stratigraphy becomes speculative and unreliable. This is confirmed by Ericson & Wollin (1956) who show that no close correlation in the coiling patterns of *G. truncatulinoides* exists between equatorial Atlantic and Caribbean cores.

Taxonomic notes

GLOBOROTALIA TRUNCATULINOIDES TRUNCATULINOIDES (d'Orbigny)
Figures 37.4–5; 6, 10, 11

Type reference: *Rotalina truncatulinoides* d'Orbigny, 1839b, p. 132, pl. 2, figs. 25–27.

Neotype: *Globorotalia (G.) truncatulinoides truncatulinoides* (d'Orbigny), Blow, 1969, p. 403, pl. 5, figs. 10–12.

The neotype shows, according to its author Blow, 'a normal wall texture and structure which is not thickened by a sheath-like structure (cf. *G. (G.) truncatulinoides pachytheca*). The pustules also densely packed over the earlier parts of the test

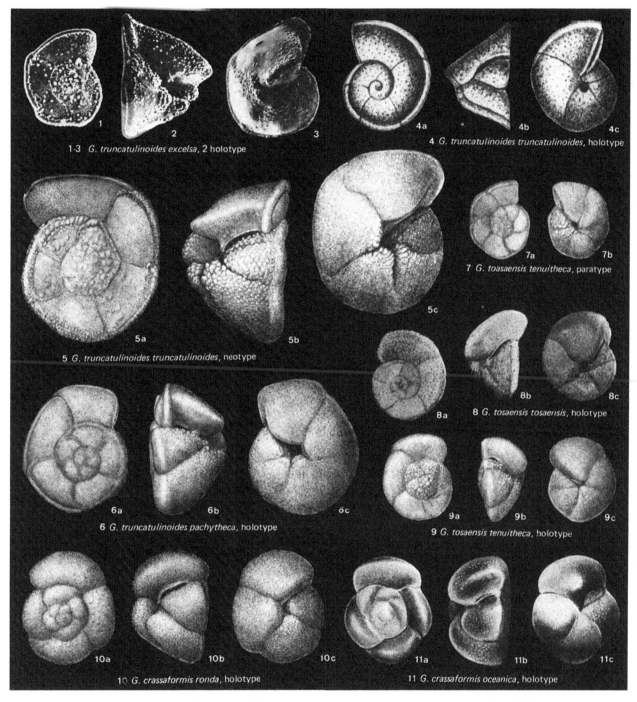

1-3 *G. truncatulinoides excelsa*, 2 holotype

4 *G. truncatulinoides truncatulinoides*, holotype

5 *G. truncatulinoides truncatulinoides*, neotype

7 *G. toasaensis tenuitheca*, paratype

8 *G. tosaensis tosaensis*, holotype

6 *G. truncatulinoides pachytheca*, holotype

9 *G. tosaensis tenuitheca*, holotype

10 *G. crassaformis ronda*, holotype

11 *G. crassaformis oceanica*, holotype

Fig. 37. *Globorotalia truncatulinoides, Globorotalia tosaensis, Globorotalia crassaformis* (all figures × 60)

1–3 *G. truncatulinoides excelsa* Sprovieri & Ruggieri
1, 3 Paratypes. **2** Holotype. Level 11, Early Pleistocene, Porto Palo, Agrigento, Sicily.

4–5 *G. truncatulinoides truncatulinoides* (d'Orbigny)
4a–c Holotype, Recent, coast of Teneriffe, Canary Islands, Atlantic.
5a–c Neotype. *Challenger* Station 8, Holocene, off Gomera, Canary Islands, Atlantic.

6 *G. truncatulinoides pachytheca* Blow
6a–c Holotype. Sample WHB 187A, Zone N 22, Pleistocene, Manchioneal Fm., San San Bay road cutting, Jamaica.

7, 9 *G. tosaensis tenuitheca* Blow
7a–b Paratype. Sample BA 40, Zone N 21, Late Pliocene, Sarmi Fm., West Irian, specimen transitional to *G. truncatulinoides truncatulinoides*. **9a–c** Holotype. Sample PEK/F.9, Zone N 21, Late Pliocene, Fiji Islands.

8 *G. tosaensis tosaensis* Takayanagi & Saito
8a–c Holotype. Sample A-17, Pliocene (originally given as Miocene), Nobori Fm., east of Nobori, Japan.

10 *G. crassaformis ronda* Blow
10a–c Holotype. Sample WHB 181A, lower part Zone N 19, Early Pliocene, near Buff Bay village, Jamaica.

11 *G. crassaformis oceanica* Cushman & Bermudez
11a–c Holotype. *Atlantis* Station 2980, Recent, off north coast of Cuba.

are clearly not fused together and they still show their individuality.' Furthermore, following Blow, the subspecies *truncatulinoides* when compared with *pachytheca* possesses 'more laterally compressed chambers in the final whorl, a more widely open umbilicus, a somewhat less tightly coiled test, less embracing chambers and more completely radial dorsal intercameral sutures. Further, the ventral side as a whole and the individual chambers on the ventral side are both more sharply and acutely conical in *truncatulinoides* (s.s.) than in *pachytheca*.' These distinguishing criteria, except perhaps for the last-mentioned, are difficult to recognize clearly on the type figures, and in practice it would not be easy to distinguish the two taxa on such criteria alone.

GLOBOROTALIA TRUNCATULINOIDES PACHYTHECA Blow
Figures **37.6**; 6, 10, 11

Type reference: *Globorotalia (G.) truncatulinoides pachytheca* Blow, 1969, p. 405, pl. 5, figs. 13–15.

In addition to the distinctions from the neotype of the subspecies *truncatulinoides* as pointed out above, the holotype of *pachytheca* differs in its smaller size (0.58 against 0.78 mm). Not counting the last two chambers, the neotype of *truncatulinoides* is exactly the same size as the holotype of *pachytheca*. It is possible therefore that the latter is not a fully grown specimen, and with two thin-walled chambers added would be difficult to distinguish from a *G. truncatulinoides truncatulinoides*.

To differentiate the two taxa merely on the presence or absence of a sheath-like structure covering particularly the earlier chambers is also not satisfactory because such growth must progress gradually, from almost non-existent in *excelsa* via distinct but still isolated pustules in *truncatulinoides* to fused pustules in *pachytheca*.

GLOBOROTALIA TRUNCATULINOIDES EXCELSA Sprovieri, Ruggieri & Unti
Figures **37.1**; 6, 10, 11

Type reference: *Globorotalia truncatulinoides excelsa* Sprovieri, Ruggieri & Unti, 1980, p. 3, pl. 1, figs. 3, 5, 7.

The holotype differs from the neotype of *G. truncatulinoides truncatulinoides* mainly in its smaller size (0.5 mm against 0.78 mm) and by a more strongly elevated umbilical cone. Furthermore, in side view the acute terminal point of each chamber is clearly higher compared with the preceding one and is not bent towards the umbilical funnel. In adult *G. truncatulinoides truncatulinoides* on the other hand the last 2 to 3 chambers are usually of about equal height around the umbilicus, or the last chamber may even be lower compared with the penultimate. Furthermore, in the subspecies *excelsa* the peripheral keel is more acute, the umbilicus more open and

the granules covering the surface rarer compared with *G. truncatulinoides truncatulinoides*, particularly in the earlier chamber of the test.

Judging from the figure of d'Orbigny, *Rotalina truncatulinoides* is a typical, delicate, small (0.5 mm), high conical *excelsa* with the characteristic increase in height of the succeeding last chamber on the umbilical side. The neotype selected by Blow is a considerably larger specimen (0.78 mm) with the last 3 chambers having about the same height on the umbilical side; in lateral view the umbilical cone is not as high and is more distinctly rounded. This is typical for an adult specimen with an already rather thickened wall at least in the early portion of the test.

GLOBOROTALIA TOSAENSIS TOSAENSIS Takayanagi & Saito
Figures **37.8**; 6, 10, 11

Type reference: *Globorotalia tosaensis* Takayanagi & Saito, 1962, p. 81, pl. 28, figs. 11a–c.

The subspecies is regarded by Blow (1969) as intermediate between the thick-walled *G. crassaformis ronda* and *G. truncatulinoides pachytheca*. It differs from *ronda* in its more compact and less lobate test and from *pachytheca* primarily in the absence of an imperforate peripheral keel. At 0.39 mm the holotype is also considerably smaller than the holotype of *G. pachytheca* which measures 0.58 mm. The subspecies *tosaensis* differs from the subspecies *tenuitheca* primarily in its thick, sheathed walls as a result of densely fused pustules. Blow (1969) quotes some smaller differences between the two taxa that in our opinion may fall largely within the variability of each taxon and be difficult to use as distinguishing criteria. They include a sharper curvature and less deep incision of the spiral intercameral sutures in *tosaensis* compared with *tenuitheca*, and a spiral peripheral margin in *tenuitheca* which is usually subacute against more broadly rounded in *tosaensis*.

GLOBOROTALIA TOSAENSIS TENUITHECA Blow
Figures **37.7, 9**; 6, 10, 11

Type reference: *Globorotalia (Turborotalia) tosaensis tenuitheca*, Blow, 1969, p. 394, pl. 4, figs. 13–15.

According to its author the subspecies is intermediate between the thin-walled *G. crassaformis oceanica* and *G. truncatulinoides truncatulinoides*. The holotype which is a phylogenetically primitive form of the taxon differs from *oceanica* in its much less inflated and slightly more embracing chambers and in a greater degree of planoconvexity of the test. The paratype figured by Blow (1969, pl. 4, figs. 16–17) is according to him a more advanced form with 'a clear area over the dorsal peripheral margin' which, however, contains normal wall pores and is therefore not yet an imperforate keel as is typical in *G. truncatulinoides truncatulinoides*. Both, the holo- and paratype of *tenuitheca* are only half the size of the succeeding form; that is 0.39 mm against 0.78 mm for the neotype of *G. truncatulinoides truncatulinoides*.

GLOBOROTALIA cf. TOSAENSIS

Type reference: *Globorotalia* cf. *tosaensis*, in Bolli, 1970, p. 583, pl. 3, figs. 16–18.

Several sites of DSDP Legs 4 and 15 (29, 30, 31, 148, 154A) drilled in the Caribbean recorded stratigraphic intervals of varying magnitude between the last occurrence of the index forms *G. miocenica/G. exilis* and the first *G. truncatulinoides*. Forms occur within this interval that at the time were regarded as being identical or very close to the *G. tosaensis* as published by Berggren (1968) from a North Atlantic piston core (Chain 61, Station 171). At the time Bolli (1970) regarded these forms as probably related to Berggren's specimens but atypical because of their exceedingly thick walls.

To name the interval, Bolli (1970) used this form as a zonal marker and called it *Globorotalia truncatulinoides* cf. *tosaensis*. The zone itself was defined as 'interval with zonal marker from the extinction of *Globorotalia exilis/Globorotalia miocenica* to the first occurrence of *Globorotalia truncatulinoides truncatulinoides*'. Bolli & Premoli Silva (1973) slightly altered it to 'interval from the extinction of *Globorotalia miocenica* to first occurrence of *Globorotalia truncatulinoides truncatulinoides*'. The zone was regarded as representing the topmost Pliocene.

Recent work has shown typical *G. tosaensis tosaensis* to be rare and often absent in the Atlantic province. Re-examination of Caribbean specimens of forms originally called *G. truncatulinoides* cf. *tosaensis* by Bolli (1970) are better referred to *G. truncatulinoides pachytheca*. They are, however, somewhat atypical in that compared with the holotype they are exceedingly thick-walled, more inflated on the umbilical side and with the peripheral keel often overgrown by pustules making it difficult to recognize. The specimen figured by Bolli (1970, pl. 3, figs. 16–18) is of Pleistocene age and occurs with typical *G. truncatulinoides truncatulinoides* and *G. truncatulinoides excelsa*.

The Caribbean specimens from DSDP Site 31-3-1, 100–102 cm and Site 148-15-6, 88–90, 106–108 and 127–129 cm, have again been compared with paratypes of *G. tosaensis tosaensis*, and with *G. crassaformis ronda* specimens received from W. H. Blow from Jamaican material, from where he described the subspecies. The thick-walled Caribbean specimens are now regarded as being intermediate between the two taxa. They possess the strongly rounded periphery of *ronda* but tests are more compact, less lobate and somewhat higher on the umbilical side as in *tosaensis*.

We here propose that the *G. truncatulinoides* cf. *tosaensis* Zone by Bolli in 1970 be renamed the *G. tosaensis tosaensis*

Zone. The interval certainly occurs in the Atlantic and Caribbean provinces but, due to the scarcity of the nominate species, usually has to be identified on indirect evidence in those areas (absence of *G. miocenica*, *G. exilis*, *G. truncatulinoides truncatulinoides*).

The genera *Sphaeroidinellopsis* and *Sphaeroidinella*

The genus *Sphaeroidinellopsis* was introduced for forms that differ from *Sphaeroidinella* by the absence of supplementary sutural apertures. The species of *Sphaeroidinellopsis* are characterized by great variability in test form, number of chambers forming the last whorl, shape of umbilical area, kind of lips formed and degree of cortex formation. The variability in *Sphaeroidinella* lies mainly in the test size and the number and shape of additional apertures.

The critical review and comparison of the existing *Sphaeroidinellopsis* species necessitated prolonged discussions and illustrations that eventually became too voluminous to fit into the concept of this volume. They therefore had to be published in full in a separate paper (Bolli & Saunders, 1981) to which reference is made. We thus can restrict ourselves here to a brief review of the problems discussed, to taxonomic notes and illustrations of these species that are considered valid, and to the stratigraphic significance of individual species.

We regard all *Sphaeroidinellopsis* species as belonging to a single evolutionary lineage, leading through the formation of additional apertures within the Early Pliocene to the genus *Sphaeroidinella*. At certain levels, however, as in the uppermost Miocene/basal Pliocene, the amount of variability may be such that the distinction of more than one partially contemporaneous taxon is justified. This is particularly so when certain forms are stratigraphically restricted as is the case of *S. hancocki* in the Early Pliocene.

Sphaeroidinellopsis

The principal problem in the taxonomic treatment of *Sphaeroidinellopsis* and its species definitions lies in the great variability of a number of test characters which in part not only change in time but may also be influenced by changing environmental conditions. This great variability has led to the erection of a number of species that in practical work are difficult to distinguish from each other and therefore to use stratigraphically. The following characters determine this great variability:

(1) Number, rate of growth and shape (globular, radially elongate, sacculiferid) of chambers forming the last whorl.

(2) Degree of cortex formation. It is absent or only slightly developed in the early stages of the lineage, increasing gradually to reach its full development in the latest Miocene–Early Pliocene. However, even at this

advanced evolutionary stage the degree of cortex formed may be very variable within the same population and can therefore not be used for distinction of taxa.

Due to the formation of the cortex, pores become progressively narrower and eventually become virtually closed. However, a process persists to keep the peripheral ends of the pores open, in particular in the ultimate and often also penultimate chambers (Figs. 38.8, 9, 15–17). At certain advanced stages this results in the formation around the pores of cone-shaped mounds that on superficial examination may resemble short spines. Subbotina (1958) erected *S. spinulosa*, based on such specimens, here placed in synonymy with *S. seminulina*.

The apertures change during evolution in that in the early stage they are bordered by a rounded lip as in *S. disjuncta* (Figs. 38.20, 22a). Gradually these lips become more acute and eventually in specimens with a well developed cortex take the shape of irregular, sharp edged flanges (Figs. 38.5, 7, 8). In specimens of similar ages but without a cortex, a few blunt isolated spines may take the place of flanges (Fig. 38.1).

The number and shape of chambers in the ultimate whorl are of some significance during the evolution of *Sphaeroidinellopsis*. The genus commences in the *Globorotalia fohsi peripheroronda* Zone with the earliest species *S. disjuncta*, comprising small, thick-walled 3–4-chambered specimens without smooth cortex (Figs. 38.18–22). In the following younger zones (*Globorotalia fohsi fohsi* to *Globorotalia acostaensis*) some specimens display a trend to grow larger with up to 5 chambers in the last whorl, divided by distinctly incised sutures. The ultimate chambers may become radially extended. Such forms are placed in *S. multiloba*. This development also coincides with increased cortex formation. Higher in this interval also appear the thin flanges around the umbilical area.

A rather sudden and distinct further change takes place in the *Globorotalia humerosa* Zone where a trend sets in towards a more spherical test form of specimens with only 3 chambers in the last whorl, characterized by a still thicker cortex and by sharp edged flanges surrounding the umbilical area. They are placed in *S. sphaeroides*. This species already represents a transitional form towards *Sphaeroidinella*, which appears in the later part of the Early Pliocene by the addition of extra apertures along the sutures.

Taking into account these general evolutionary trends within the genus it is rather unexpected to find a form again with a higher number of chambers (up to 6) in the last whorl. Such specimens placed in *S. hancocki* are restricted to the Early Pliocene. Here the last chambers are not only radially somewhat extended as in many late Middle Miocene specimens but also have a tendency to be drawn slightly backward (Fig. 38.2). As in *S. seminulina* the degree of cortex formation is also very variable (Figs. 38.1–3).

Sphaeroidinella

The genus *Sphaeroidinella* developed within the Early Pliocene from *Sphaeroidinellopsis* ancestors (*S. sphaeroides, S. seminulina*) by the acquisition of secondary apertures in the later stage of growth. Juvenile or immature specimens of *Sphaeroidinella* often remain without such additional apertures throughout the range of the genus and can then not be distinguished from *Sphaeroidinellopsis*.

Great variability exists in *Sphaeroidinella* as regards overall test size and shape, and number, size and shape of primary and secondary apertures. Based on these features the following taxa have been proposed in addition to *S. dehiscens*: *S. immaturus, S. excavata, S. ionica ionica* and *S. ionica evoluta*. Like the first appearance of the genus, their presence is determined to a considerable extent by locally prevailing ecological conditions and is therefore of limited stratigraphic significance.

Several workers have proposed that *Sphaeroidinella dehiscens* is a variant of one or more species of *Globigerinoides* that have developed a cortex usually in response to life in deeper parts of the water column (e.g. Bé, 1965; Bé & Hemleben, 1970). It is difficult to follow this line of reasoning when one considers the evolution through time of *Sphaeroidinellopsis* and *Sphaeroidinella*. It is even more difficult to follow Bandy, Ingle & Frerichs (1967), who maintain two lineages with *Globigerinoides sacculifer* and *G. conglobatus* both producing *Sphaeroidinellopsis seminulina* until the end of the Miocene and then producing *Sphaeroidinella dehiscens* from base Pliocene to Recent.

The stratigraphic significance of the *Sphaeroidinellopsis* and *Sphaeroidinella* species

Species of *Sphaeroidinellopsis* have been used repeatedly as zonal markers. Blow (1959) introduced a *Sphaeroidinella seminulina* Zone, approximately representing the *Globorotalia acostaensis* Zone as used here. He abandoned it again in his 1969 paper where the zone became part of his N 16 and N 17 zones. Lamb & Beard (1972) subdivided their *Globorotalia acostaensis* Zone (equivalent to the *Globorotalia acostaensis* and *humerosa* zones as used here) into a lower *Sphaeroidinellopsis seminulina* Subzone and an upper *Sphaeroidinellopsis sphaeroides* Subzone. Cita (1973) introduced two zones based on *Sphaeroidinellopsis*: a basal Pliocene *Sphaeroidinellopsis* Acme-Zone characterized by a peak of abundance of *Sphaeroidinellopsis*, and a Middle Pliocene *Sphaeroidinellopsis subdehiscens* Zone, ranging from the extinction of *Globorotalia margaritae* to the extinction of *Sphaeroidinellopsis*. Salvatorini & Cita (1979) proposed a Late Miocene *Sphaeroidinellopsis seminulina paenedehiscens* Zone defined by the first occurrence of the zonal marker to the first occurrence of *Globorotalia margaritae*. This is the same definition as previously given by Lamb & Beard (1972) for their *Sphaeroidinellopsis sphaeroides* Subzone.

The great variability and flowing evolutionary changes in *Sphaeroidinellopsis* render an accurate biostratigraphic application of individual species difficult. Nevertheless, and as evident from the distribution chart, the taxa discussed here can be of value particularly when established zonal markers or other index forms are missing. The most characteristic datums are (a) the first occurrence of 5–6-chambered *S. disjuncta*, (b) the first occurrence of *S. multiloba* (with smooth cortex) and (c) the first occurrence of *S. seminulina* and *S. sphaeroides* (with flanges and thick glassy cortex).

The species with the shortest range, easily recognizable and diagnostic for the Early Pliocene *Globorotalia margaritae margaritae* Subzone is *S. hancocki*.

Because juvenile stages of *Sphaeroidinella* may be without secondary apertures, the extinction level of *Sphaeroidinellopsis* is a less reliable stratigraphic datum. For example, Srinivasan & Srivastava (1974) state that the datum is a marker for the Miocene/Pliocene boundary in the tropics. However, their range chart for the Andaman–Nicobar Islands shows *Sphaeroidinella* appearing before *G. margaritae*. This is in contrast to the position in the Caribbean and other tropical areas where the base of *Sphaeroidinella* is distinctly above the base of *G. margaritae* s.l., the latter marking the base of the Pliocene while the former does not appear until within the *Globorotalia margaritae evoluta* Subzone or near the top of the Early Pliocene, as is for example the case in the Caribbean DSDP Sites 29, 31, 148 and 154. It is for these reasons that we are also reluctant to follow Banner & Blow (1965b) and Blow (1969, emended definition) in the use of the first appearance of *Sphaeroidinella dehiscens* to define the base of their Zone N 19.

Taxonomic notes

SPHAEROIDINELLOPSIS DISJUNCTA (Finlay)
Figures **38.18–22**; 7, 9, 12

Type reference: *Sphaeroidinella disjuncta* Finlay, 1940, p. 469, fig. 226.
Synonym: *Globigerina grimsdalei* Keijzer, 1945, p. 205, figs. 33a–c.

S. disjuncta is the stratigraphically earliest species of the lineage. In the early part of its range specimens are smaller compared with later ones, mostly with 3 chambers in the last whorl, occasionally with 4. The average test size increases gradually during the *Globorotalia fohsi* zones; 4-chambered specimens become more frequent and occasionally even 5-chambered forms appear. The shape of the chambers is subglobular, the last chambers have a tendency to become radially somewhat elongated. Sutural incisions in 3-chambered forms are slight; they are deeper in 4-chambered specimens. The aperture of the last chamber leading into the umbilicus is often protected by a distinct rounded lip, particularly in the larger specimens. Higher in the lineage, beginning with *S. multiloba*, the rounded lips gradually take on the flange-like shape typical for the

forms in the later part of the lineage (*S. seminulina, S. sphaeroides*). The walls are massive and distinctly pitted. Increased cortex formation that eventually leads to the smooth glassy surface characterizing *S. multiloba* gradually develops through transitional forms from *S. disjuncta*.

SPHAEROIDINELLOPSIS MULTILOBA (LeRoy)
Figures **38.14–17**; 7, 9, 12

Type reference: *Sphaeroidinella multiloba* LeRoy, 1944, p. 91, pl. 4, figs. 7–9.
Synonym: *Sphaeroidinella dehiscens subdehiscens* Blow, 1959, p. 195, pl. 12, figs. 71a–c.

This species, described from the Late Miocene of Java, was originally characterized as possessing 3 to 4 greatly inflated chambers in the last whorl, distinct sutures, strongly depressed, coarsely perforate glassy wall, large irregular aperture with crenulated edge. This description suggests evolution from *S. disjuncta* by the formation of a cortex. In the Caribbean/Gulf of Mexico/Atlantic province it begins at about the level of the *Globorotalia fohsi lobata* Zone. As in *S. disjuncta*, forms with a smaller, radially extended, occasionally sacculiferid end-chamber are often present, particularly in the 4-chambered specimens. Due to the formation of a glassy cortex with smooth surface, the walls of the last formed chambers become translucent in well preserved specimens with the pore channels penetrating the walls visible as thin white lines. Examples for this are Figs. 38.12–13 (*S. seminulina*) and Fig. 38.14 (*S. multiloba*).

Specimens with 5 and, more seldom, 6 chambers in the last whorl begin to occur from about the *Globorotalia fohsi fohsi* Zone upwards, usually the early forms without a cortex, the later forms with one. Because of their later appearance compared with the 3- and 4-chambered forms they are of some stratigraphic significance.

SPHAEROIDINELLOPSIS SEMINULINA
(Schwager)
Figures **38.6–13**; 7, 10, 11

Type reference: *Globigerina seminulina* Schwager, 1866, p. 256, pl. 7, fig. 112.
Neotype: *Sphaeroidinellopsis seminulina* (Schwager), Banner & Blow, 1970a, p. 24, pl. 7, figs. 2a–b.
Synonyms: Sphaeroidinella spinulosa Subbotina, 1958, p. 61, pl. 11, figs. 6a–c.
Sphaeroidinellopsis subdehiscens paenedehiscens Blow, 1969, p. 386, pl. 30, figs. 4, 9.

In this species a further thickening of the glassy cortex is seen compared with *S. disjuncta* and *S. multiloba*. This results in the intercameral sutures becoming less incised and sometimes hardly distinguishable. The aperture of the last chamber becomes more irregular-shaped and slit-like and is bordered by a thin, irregularly edged flange.

In specimens with a poorly developed cortex the penultimate and particularly the antepenultimate chambers form a small number of blunt spine-like extensions pointing inwards around the umbilical rim.

With progressive cortex building these spines coalesce and become flanges. The last whorl usually consists of 3 to 4 chambers, occasionally a small abortive sacculiferid chamber

may be added. In the 3-chambered specimens a trend develops to increase the test size and to form a comparatively large and rounded end-chamber with virtually no sutural incisions left between the individual chambers. These forms eventually lead to *S. sphaeroides*, which in turn gives rise to *Sphaeroidinella* within the Early Pliocene.

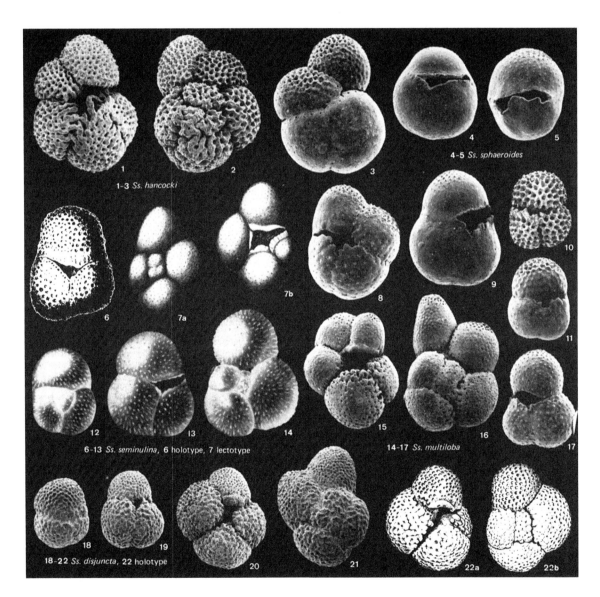

Fig. 38. *Sphaeroidinellopsis* (all figures × 60)

1–3 *S. hancocki* Bandy
   1–2 DSDP Leg 4-31-6-cc. **3** DSDP Leg 4-31-6-2, 50–52 cm. All from the *G. margaritae* Zone, Early Pliocene, Beata Ridge, Caribbean Sea (from Bolli & Saunders, 1981).

4–5 *S. sphaeroides* Lamb
   DSDP Leg 4-29-6-1, 8–10 cm, *G. humerosa* Zone, Late Miocene, Venezuela Basin, Caribbean Sea (from Bolli & Saunders, 1981).

6–13 *S. seminulina* (Schwager)
   6 Holotype. Late Tertiary, Kar Nikobar, Indian Ocean. **7a–b** Neotype. Pliocene, Kar Nikobar, Indian Ocean. **8, 10** DSDP Leg 4-31-6-2, 50–52 cm. **9** DSDP Leg 4-29-6-1, 8–10 cm. **11** DSDP Leg 4-29-4-4, 110–112 cm. **8–11** *G. humerosa* Zone, Late Miocene, Beata Ridge (Site 31) and Venezuela Basin (Site 29), Caribbean Sea (from

Bolli & Saunders, 1981). **12–13** *G. menardii* Zone, Lengua Fm., Middle Miocene, Trinidad (from Bolli, 1957).

14–17 *S. multiloba* (LeRoy)
   14 *G. menardii* Zone, Lengua Fm., Middle Miocene, Trinidad (from Bolli, 1957). **15–16** Sample HMB 79/31, *G. fohsi robusta* Zone, Middle Miocene, Cipero Fm., Golconda area, Trinidad (from Bolli & Saunders, 1981). **17** Well Bodjonegoro-1, core at 390 m, *G. acostaensis* Zone, Late Miocene, Java (from Bolli & Saunders, 1981).

18–22 *S. disjuncta* (Finlay)
   18–21 Sample Bo 202, *G. fohsi peripheroronda* Zone, Middle Miocene, Cipero Fm., Hermitage Quarry, Trinidad (from Bolli & Saunders, 1981). **22a–b** Holotype, redrawn. Locality 4270, Early Miocene, Ihungia Marl, Tangihanga, New Zealand.

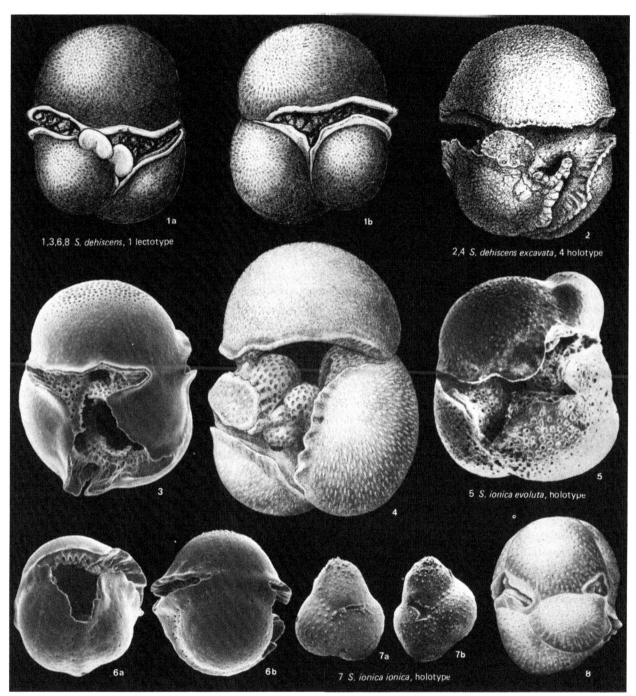

Fig. 39. *Sphaeroidinella* (all figures × 60)

1,3,6,8 *S. dehiscens* (Parker & Jones)
  1a–b Lectotype, 1a spiral view, 1b umbilical view. Recent, Atlantic.
  3, 6a–b Specimens with interior portion largely destroyed by calcium carbonate dissolution, on **3** pores towards apex of last chamber gradually less closed by glassy cortex. DSDP Leg 4-2-1, 100–102 cm, *G. truncatulinoides truncatulinoides* Zone, Pleistocene, Venezuela Basin, Caribbean Sea (from Bolli, 1970). **8** Specimen with small abortive chamber, Recent, Pacific (from Bolli, 1957).

2, 4 *S. dehiscens excavata* Blow
  2 Paratype. Recent (from Brady, 1884). **4** Holotype. *Challenger* Station 224, Zone N 23, Recent, west Pacific.

5 *S. ionica evoluta* Cita & Ciaranfi
  Holotype, umbilical view. DSDP Leg 13-125-4-4, 6–8 cm, *G. truncatulinoides truncatulinoides* Zone, Pleistocene, Ionian Basin, Mediterranean Sea.

7 *S. ionica ionica* Cita & Ciaranfi
  7a–b Holotype. 7a spiral view, 7b umbilical view. DSDP Leg 13-125-4-6, 80–82 cm, *G. inflata* Zone, Late Pliocene, Ionian Basin, Mediterranean Sea.

the spiral side where irregularly shaped secondary apertures form.

References

Bandy, O. L. 1949. Eocene and Oligocene foraminifera from Little Stave Creek, Clarke County, Alabama. *Bull. Am. Paleontol.*, **32** (131), 1–210.

Bandy, O. L. 1963. Miocene–Pliocene boundary in the Philippines as related to late Tertiary stratigraphy of deep-sea sediments. *Science*, **142** (3597), 1290–2.

Bandy, O. L. 1964. The type of *Globigerina quadrilobata* d'Orbigny. *Contrib. Cushman Found. foramin. Res.*, **15**, 36–7.

Bandy, O. L. 1966. Restrictions of the '*Orbulina*' datum. *Micropaleontology*, **12**, 79–86.

Bandy, O. L. 1972. Origin and development of *Globorotalia (Turborotalia) pachyderma* (Ehrenberg). *Micropaleontology*, **18**, 294–318.

Bandy, O. L. 1975. Messinian evaporite deposition and the Miocene/Pliocene boundary, Pasquasia–Capodarso sections, Sicily. In: Late Neogene Epoch Boundaries, ed. T. Saito & L. H. Burkle, *Micropaleontology*, Special Publication 1, 49–63.

Bandy, O. L., Frerichs, W. E. & Vincent, E. 1967. Origin, development, and geologic significance of *Neogloboquadrina* Bandy, Frerichs, and Vincent, gen. nov. *Contrib. Cushman Found. foramin. Res.*, **18**, 152–7.

Bandy, O. L., Ingle, J. C. & Frerichs, W. E. 1967. Isomorphism in '*Sphaeroidinella*' and '*Sphaeroidinellopsis*'. *Micropaleontology*, **13**, 483–8.

Bandy, O. L., Vincent, E. & Wright, R. C. 1969. Chronologic relationships of Orbulines to the *Globorotalia fohsi* lineage. *Rev. Esp. Micropaleontol.*, **1**, 131–45.

Banner, F. T. 1982. A classification and introduction to the Globigerinacea. In: *Aspects of Micropaleontology*, ed. F. T. Banner & H. H. Lord, pp. 142–239. George Allen & Unwin, London.

Banner, F. T. & Blow, W. H. 1959. The classification and stratigraphical distribution of the Globigerinacea. *Paleontology*, **2** (2), 1–27.

Banner, F. T. & Blow, W. H. 1960a. Some primary types of the species belonging to the superfamily Globigerinaceae. *Contrib. Cushman Found. foramin. Res.*, **11**, 1–41.

Banner, F. T. & Blow, W. H. 1960b. The taxonomy, morphology and affinities of the genera included in the subfamily Hastigerininae. *Micropaleontology*, **6**, 19–31.

Banner, F. T. & Blow, W. H. 1965a. Two new taxa of the Globorotaliinae (Globigerinacea, Foraminifera) assisting determination of the late Miocene/middle Miocene boundary. *Nature*, **207**, 1351–4.

Banner, F. T. & Blow, W. H. 1965b. Progress in the planktonic foraminiferal biostratigraphy of the Neogene. *Nature*, **208**, 1164–6.

Banner, F. T. & Blow, W. H. 1965c. *Globigerinoides quadrilobatus* (d'Orbigny) and related forms; their taxonomy, nomenclature and stratigraphy. *Contrib. Cushman Found. foramin. Res.*, **16**, 105–15.

Banner, F. T. & Blow, W. H. 1967. The origin, evolution and taxonomy of the foraminiferal genus *Pulleniatina* Cushman, 1927. *Micropaleontology*, **13**, 133–62.

Baroz, F. & Bizon, G. 1974. Le Néogène de la chaîne du Pentadaktylos et de la partie nord de la Messoria (Chypre);

étude stratigraphique et micropaleontologique. *Rev. Inst. Fr. Pet.*, **29**, 327–59.

Bé, A. W. H. 1965. The influence of depth on shell growth in *Globigerinoides sacculifer* (Brady). *Micropaleontology*, **11**, 81–97.

Bé, A. W. H., Harrison, S. M. & Lott, L. 1973. *Orbulina universa* d'Orbigny in the Indian Ocean. *Micropaleontology*, **19**, 150–92.

Bé, A. W. H. & Hemleben, C. 1970. Calcification in a living planktonic foraminifer *Globigerinoides sacculifer* (Brady). *Neues Jahrb. Geol. Palaeontol. Abhandlungen*, **134** (3), 221–34.

Bé, A. W. H., Hemleben, C., Anderson, O. R., Spindler, M., Hacunda, J. & Tuntivate-Choy, S. 1977. Laboratory and field observations of living planktonic foraminifera. *Micropaleontology*, **23**, 155–79.

Beard, J. H. 1969. Pleistocene paleotemperature record based on planktonic foraminifers, Gulf of Mexico. *Trans. Gulf Coast Assoc. geol. Soc.*, **19**, 535–53.

Beckmann, J. P. 1971. The foraminifera of Sites 68–75. *Initial Rep. Deep Sea drill. Proj.*, **8**, 713–25.

Beckmann, J. P. 1972. The foraminifera and some associated microfossils of Sites 135–144. *Initial Rep. Deep Sea drill. Proj.*, **14**, 389–420.

Beckmann, J. P. *et al.* 1969. Standard planktonic zones in Egypt. *Proceedings First International Conference on Planktonic Microfossils, Geneva, 1967*, **1**, 92–103.

Belford, D. J. 1962. Miocene and Pliocene planktonic foraminifera, Papua–New Guinea. *Bull. Bur. Miner. Resour. Geol. Geophys. Canberra*, **62**, 1–51.

Berggren, W. A. 1968. Micropaleontology and the Pliocene/Pleistocene boundary in a deep-sea core from the south-central North Atlantic. *G. Geol.*, **35** (2), 291–311.

Berggren, W. A. 1973. The Pliocene time-scale: calibration of planktonic foraminiferal and calcareous nannoplankton zones. *Nature*, **243**, 391–7.

Berggren, W. A. 1978. Recent advances in Cenozoic planktonic foraminiferal biostratigraphy, biochronology, and biogeography: Atlantic Ocean. *Micropaleontology*, **24**, 337–70.

Berggren, W. A. & Van Couvering, J. A. 1974. The late Neogene biostratigraphy, geochronology and paleoclimatology of the last 15 million years in marine and continental sequences. *Palaeogeogr. Palaeoclimatol. Palaeoecol.*, **16** (1–2), 1–216.

Bermudez, P. J. 1949. Tertiary smaller foraminifera of the Dominican Republic. *Contrib. Cushman Lab. foramin. Res.*, Special Publication **25**, 1–322.

Bermudez, P. J. 1961. Contribucion al estudio de las Globigerinidea de la region Caribe-Antillana (Paleoceno-Reciente). *Memoria Tercer Congreso Geologico Venezolano*, **3**, Special Publication 3, 1119–1393.

Bermudez, P. J. & Bolli, H. M. 1969. Consideraciones sobre los sedimentos del Mioceno medio al Reciente de las costas central y oriental de Venezuela. *Boletin de Geologia, Direccion de Geologia, Ministerio de Minas e Hidrocarbures*, **10** (20), 137–223.

Bizon, J. J. & Bizon, G. 1965. L'Hélvetien et le Tortonien de la région de Parga (Epire occidentale, Grèce). *Rev. Micropaleontol. Paris*, **7**, 242–56.

Blow, W. H. 1956. Origin and evolution of the foraminiferal genus *Orbulina* d'Orbigny. *Micropaleontology*, **2**, 57–70.

Blow, W. H. 1959. Age, correlation, and biostratigraphy of the

Upper Tocuyo (San Lorenzo) and Pozon formations, Eastern Falcon, Venezuela. *Bull. Am. Paleontol.*, **39** (178), 67–251.

Blow, W. H. 1965. *Clavatorella*, a new genus of the Globorotaliidae. *Micropaleontology*, **11**, 365–8.

Blow, W. H. 1969. Late Middle Eocene to Recent planktonic foraminiferal biostratigraphy. *Proceedings First International Conference on Planktonic Microfossils Geneva*, 1967, **1**, 199–422.

Blow, W. H. 1970. Validity of biostratigraphic correlations based on the Globigerinacea. *Micropaleontoogy*, **16**, 257–68.

Blow, W. H. 1979. *The Cainozoic Globigerinida*. E. J. Brill, Leiden (3 vols.), 1413 pp.

Blow, W. H. & Banner, F. T. 1962. Part 2: The Tertiary (Upper Eocene to Aquitanian) Globigerinaceae. In: *Fundamentals of Mid-Tertiary Stratigraphical Correlation*, ed. F. E. Eames *et al.*, pp. 61–151. Cambridge University Press, Cambridge.

Blow, W. H. & Banner, F. T. 1966. The morphology, taxonomy and biostratigraphy of *Globorotalia barisanensis* LeRoy, *Globorotalia fohsi* Cushman and Ellisor, and related taxa. *Micropaleontology*, **12**, 286–303.

Bolli, H. M. 1950. The direction of coiling in the evolution of some Globorotaliidae. *Contrib. Cushman Found. foramin. Res.*, **1**, 82–9.

Bolli, H. M. 1951. Notes on the direction of coiling of ratalid foraminifera. *Contrib. Cushman Found. foramin. Res.*, **2**, 139–43.

Bolli, H. M. 1957. Planktonic foraminifera from the Oligocene–Miocene Cipero and Lengua formations of Trinidad, B. W. I. *Bull. U.S. natl Mus.*, **215**, 97–123.

Bolli, H. M. 1959. Planktonic foraminifera as index fossils in Trinidad, West Indies and their value for worldwide stratigraphic correlation. *Eclog. geol. Helv.*, **52** (2), 627–37.

Bolli, H. M. 1964. Observations on the stratigraphic distribution of some warm water planktonic foraminifera in the young Miocene to Recent. *Eclog. geol. Helv.*, **57** (2), 541–52.

Bolli, H. M. 1966a. Zonation of Cretaceous to Pliocene marine sediments based on planktonic foraminifera. *Boletin Informativo, Asociacion Venezolana de Geologia, Mineria y Petroleo*, **9** (1), 3–32.

Bolli, H. M. 1966b. The planktonic foraminifera in well Bodjonegoro-1 of Java. *Eclog. geol. Helv.*, **59** (1), 449–65.

Bolli, H. M. 1967. The subspecies of *Globorotalia fohsi* Cushman and Ellisor and the zones based on them. *Micropaleontology*, **13**, 502–12.

Bolli, H. M. 1970. The foraminifera of Sites 23–31, Leg 4. *Initial Rep. Deep Sea drill. Proj.*, **4**, 577–643.

Bolli, H. M. 1971. The direction of coiling in planktonic foraminifera. In: *The Micropaleontology of Oceans*, ed. B. M. Funnell & W. R. Riedel, pp. 639–49. Cambridge University Press, Cambridge.

Bolli, H. M. & Bermudez, P. J. 1965. Zonation based on planktonic foraminifera of Middle Miocene to Pliocene warm-water sediments. *Bolletin Informativo, Asociacion Venezolana de Geologia, Mineria y Petroleo*, **8** (5), 119–49.

Bolli, H. M. & Bermudez, P. J. 1978. A neotype for *Globorotalia margaritae* Bolli and Bermudez. *J. foramin. Res.*, **8**, 138–42.

Bolli, H. M. & Krasheninnikov, V. A. 1977. Problems in Paleogene and Neogene correlations based on planktonic foraminifera. *Micropaleontology*, **23**, 436–52.

Bolli, H. M., Loeblich, A. R. & Tappan, H. 1957. Planktonic foraminiferal families Hantkeninidae, Orbulinidae,

Globorotaliidae and Globotruncanidae. *Bull. U.S. natl Mus.*, **215**, 3–50.

Bolli, H. M. & Premoli Silva, I. 1973. Oligocene to Recent planktonic foraminifera and stratigraphy of the Leg 15 Sites in the Caribbean Sea. *Initial Rep. Deep Sea drill. Proj.*, **15**, 475–97.

Bolli, H. M. & Saunders, J. B. 1981. The species of *Sphaeroidinellopsis* Banner and Blow, 1959. *Cah. Micropaleontol.* Fasc. **4**, 13–25.

Bolli, H. M. & Saunders, J. B. 1982a. *Globorotalia mayeri* and its relationship to *Globorotalia siakensis* and *Globorotalia continuosa*. *J. foramin. Res.*, **12**, 39–50.

Bolli, H. M. & Saunders, J. B. 1982b. *Globorotalia mayeri* and *Globorotalia siakensis*: Priorities. *J. foramin. Res.*, **12**, 369.

Borsetti, A. M. 1959. Tre nuovi foraminiferi planctonici dell' Oligocene piacentino. *G. Geol.*, **27**, 205–12.

Borsetti, A. M., Cati, F., Colalongo, M. L. & Sartoni, S. 1979. Biostratigraphy and absolute ages of the Italian Neogene. *Ann. Geol. Hellen.*, 7th. *Internat. Congr. Medit. Neogene*, Athens, pp. 183–97.

Bossio, A. *et al.* 1976. Corrélation de quelques sections stratigraphiques de Mio-Pliocène de la zone atlantique du Maroc avec les stratotypes de bassin Mediterranéen sur la base des foraminifères planctoniques, nannoplancton calcaire et ostracodes. *Atti della Societa Toscana di Scienze Naturali*, **83**, 121–37.

Bradshaw, J. S. 1959. Ecology of living planktonic foraminifera in the North and Equatorial Pacific Ocean. *Contrib. Cushman Found. foramin. Res.*, **10**, 25–64.

Brady, H. B. 1877. Supplementary notes on the foraminifera of the Chalk (?) of the New Britain group. *Geol. Mag.*, new series, decade 2, **4**, 534–6.

Brady, H. B. 1879. Notes on some reticularian Rhizopoda of the Challenger Expedition. *Q. J. Microscopical Soc. London*, **19**, 20–6, 261–99.

Brady, H. B. 1882. Report on the Foraminifera. *Proc. R. Soc. Edinburgh*, **11**, 708–17.

Brady, H. B. 1884. Report on the Foraminifera dredged by H.M.S. *Challenger*, during the years 1873–1876. *Rep. Sci. Results H.M.S. Challenger* 1873–6, **9** (Zoology), 1–814.

Brönnimann, P. 1950. Occurrence and ontogeny of *Globigerinatella insueta* Cushman and Stainforth from the Oligocene of Trinidad, B. W. I. *Contrib. Cushman Found. foramin. Res.*, **1**, 80–2.

Brönnimann, P. 1951a. *Globigerinita naparimaensis* n. gen., n. sp., from the Miocene of Trinidad, B. W. I. *Contrib. Cushman Found. foramin. Res.*, **2**, 16–18.

Brönnimann, P. 1951b. The genus *Orbulina* d'Orbigny in the Oligo-Miocene of Trinidad, B. W. I. *Contrib. Cushman Found. foramin. Res.*, **2**, 132–8.

Brönnimann, P. 1952. *Globigerinoita* and *Globigerinatheka*, new genera from the Tertiary of Trinidad, B. W. I. *Contrib. Cushman Found. foramin. Res.*, **3**, 25–8.

Brönnimann, P. & Resig, J. 1971. A Neogene globigerinacean biochronologic time-scale of the southwestern Pacific. *Initial Rep. Deep Sea drill. Proj.*, **7**, 1235–1469.

Carpenter, W. B. 1862. *Introduction to the Study of the Foraminifera*. Ray Society, London, 319 pp.

Catalano, R. & Sprovieri, R. 1969. Stratigrafia e micropaleontologia dell'intervallo tripolaceo di torrente Rossi (Enna). *Atti Accad. Gioenia Sci. Nat. Catania*, ser. 7, **1**, 513–27.

Cati, F. *et al.* 1968. Biostratigraphia del Neogene mediterraneo basata sui foraminiferi planctonici. *Boll. Soc. geol. Ital.*, **87**, 491–503.

Caudri, C. M. B. 1934. *Tertiary Deposits of Soemba.* Dissertation Leiden, published by H. J. Paris, Amsterdam, 224 pp.

Chapman, F., Parr, W. J. & Collins, A. C. 1934. Tertiary foraminifera of Victoria, Australia. The Balcombian deposits of Port Phillip; Part III. *J. Linn. Soc. London, Zool.*, **38**, 553–77.

Cifelli, R. & Glaçon, G. 1979. New late Miocene and Pliocene occurrences of *Globorotalia* species from the North Atlantic; and a paleogeographic review. *J. foramin. Res.*, **9**, 210–27.

Cita, M. B. 1968. Report of the Working Group Micropaleontology. *G. Geol.*, ser. 2, **35** (2), 1–22.

Cita, M. B. 1973. Pliocene biostratigraphy and chronostratigraphy. *Initial Rep. Deep Sea drill. Proj.*, **13**, 1343–79.

Cita, M. B. 1975. The Miocene/Pliocene boundary: History and definition. *Micropaleontology*, Special Publication **1**, 1–30.

Cita, M. B. & Blow, W. H. 1969. The biostratigraphy of the Langhian, Serravallian and Tortonian stages in the type-sections in Italy. *Riv. Ital. Paleontol.*, **75** (3), 549–603.

Cita, M. B. & Ciaranfi, N. 1972. A new species of *Sphaeroidinella* from Late Neogene deep-sea Mediterranean sediments (DSDP Leg XIII). *Riv. Ital. Paleontol.*, **78** (4), 693–710.

Cita, M. B., Colalongo, M. L., d'Onofrio, S., Iaccarino, S. & Salvatorini, G. 1978. Biostratigraphy of Miocene deep sea sediments (Sites 372 and 375), with special reference to the Messinian/Pre-Messinian interval. *Initial Rep. Deep Sea drill. Proj.*, **13**, 671–85.

Cita, M. B. & Premoli Silva, I. 1967. Evoluzione delle fauna planctoniche nell'intervallo stratigrafico compreso fra il Langhiano-tipo ed il Tortoniano-tipo e zonazione del Miocene Piemontese. *Istituto de Paleontologia dell'Universita di Milano*, 1–25.

Cita, M. B. & Ryan, W. B. F. 1978. Studi sul Pliocene e sugli strati di passagio dal Miocene al Pliocene. XI. The Bou Regreg section of the Atlantic coast of Marocco. Evidence, timing and significance of a Late Miocene regressive phase. *Riv. Ital. Paleontol.*, **84** (4), 1051–82.

Civis, J. *et al.* 1977. *Bioestratigrafia del Messiniense de la Rambla de Arejos (Almeria) – Analisis faunistica.* Messinian Seminar no. 3, Abstracts, Malaga.

Colalongo, M. L. *et al.* 1979. A proposal for the Tortonian/Messinian boundary. *Ann. Geol. Hellen.*, *7th Internat. Congr. Medit. Neogene*, pp. 285–94.

Colalongo, M. L. & Sartoni, S. 1967. *Globorotalia hirsuta aemiliana* nuova sottospecie cronologica del Pliocene in Italia. *G. Geol.*, **34** (1), 1–15.

Colalongo, M. L. & Sartoni, S. 1977. *Globigerina calabra* nuova specie presso il limite Plio-Pleistoceno della sezione della Vrica (Calabria). *G. Geol.*, **42**, 205–20.

Conato, V. & Follador, U. 1967. *Globorotalia crotonensis* e *Globorotalia crassacrotonensis* nuove specie del Pliocene italiano. *Boll. Soc. geol. Ital.*, **86**, 555–63.

Cordey, W. G. 1967. The development of *Globigerinoides ruber* (d'Orbigny, 1839) from the Miocene to Recent. *Palaeontology*, **10**, 647–59.

Cordey, W. G. 1968. A new Eocene *Cassigerinella* from Florida. *Paleontology*, **11** (3), 368–70.

Crescenti, U. 1966. Sulla biostratigrafia del Miocene affiorante al confine marchigiano-abruzzese. *Geologia Romana*, **5**, 1–54.

Cushman, J. A. 1917. New species and varieties of foraminifera from the Philippines and adjacent waters. *Proc. U.S. natl Mus.*, **51**, 651–62.

Cushman, J. A. 1919. Fossil foraminifera from the West Indies. *Carnegie Institute, Washington Publ.*, **291**, 21–71.

Cushman, J. A. 1921. Foraminifera of the Philippine and adjacent seas. *Bull. U.S. natl Mus.*, **100** (4), 1–608.

Cushman, J. A. 1927. An outline of a re-classification of the foraminifera. *Contrib. Cushman Lab. foramin. Res.*, **3**, 1–107.

Cushman, J. A. 1931. The foraminifera of the Atlantic Ocean, Part 8 – Rotaliidae, Amphisteginidae, Calcarinidae, Cymbaloporetidae, Globorotaliidae, Anomalinidae, Planorbulinidae, Rupertiidae and Homotremidae. *Bull. U.S. natl Mus.*, **104**, 1–179.

Cushman, J. A. & Bermudez, P. J. 1937. Further new species of foraminifera from the Eocene of Cuba. *Contrib. Cushman Lab. foramin. Res.*, **13**, 1–29.

Cushman, J. A. & Bermudez, P. J. 1949. Some Cuban species of *Globorotalia*. *Contrib. Cushman Lab. foramin. Res.*, **25**, 26–48.

Cushman, J. A. & Dorsey, A. L. 1940. Some notes on the genus *Candorbulina*. *Contrib. Cushman Lab. foramin. Res.*, **16**, 40–2.

Cushman, J. A. & Ellisor, A. O. 1939. New species of foraminifera from the Oligocene and Miocene. *Contrib. Cushman Lab. foramin. Res.*, **15**, 1–14.

Cushman, J. A. & Jarvis, P. W. 1930. Miocene foraminifera from Buff Bay, Jamaica. *J. Paleontol.*, **4** (4), 353–68.

Cushman, J. A. & Jarvis, P. W. 1936. Three new Foraminifera from the Miocene Bowden Marl, of Jamaica. *Contrib. Cushman Lab. foramin. Res.*, **12**, 3–5.

Cushman, J. A. & Ponton, G. M. 1932. The foraminifera of the Upper, Middle and part of the Lower Miocene of Florida. *Bull. Florida State geol. Surv.*, **9**, 7–147.

Cushman, J. A. & Renz, H. H. 1941. New Oligocene–Miocene Foraminifera from Venezuela. *Contrib. Cushman Lab. foramin. Res.*, **17**, 1–27.

Cushman, J. A. & Renz, H. H. 1947. The foraminiferal fauna of the Oligocene, Ste. Croix Formation of Trinidad, B. W. I. *Spec. Publ. Cushman Lab.*, **22**, 1–46.

Cushman, J. A. & Stainforth, R. M. 1945. The foraminifera of the Cipero Marl Formation of Trinidad, British West Indies. *Spec. Publ. Cushman Lab.*, **14**, 1–75.

de Stefani, T. 1952. Su alcune manifestazioni di idrocarburi in provincia di Palermo e descrizione di foraminiferi nuovi. *Plinia, Palermo, Italy*, **3** (1950–51), Nota 4, 1–12.

Dollfus, G. F. 1905. Review of: Millett, F. W., Report on the Recent foraminifera of the Malay Archipelago. *Rev. crit. Palaeozool. Paris*, **9**, 222–3.

d'Orbigny, A. 1826. Tableau méthodique de la classe des céphalopodes. *Annales des Sciences Naturelles, Paris*, ser. 1, **7**, 96–314.

d'Orbigny, A. 1839a. Foraminifères. In: *Histoire physique, politique et naturelle de l'île de Cuba*, ed. R. de la Sagra. Bertrand, Paris, 224 pp.

d'Orbigny, A. 1839b. Foraminifères des îles Canaries. In: *Histoire Naturelle des îles Canaries*, vol. 2, part 2, ed. P. Barker-Webb & S. Berthelot, pp. 119–46. Bethune, Paris.

d'Orbigny, A. 1846. *Foraminifères fossiles du bassin tertiaire de Vienne (Austriche)*. Gide et Comp, Paris, 312 pp.

Ericson, D. B. 1959. Coiling direction of *Globigerina pachyderma* as a climatic index. *Science*, **130**, 219–20.

Ericson, D. B., Ewing, W. M. & Wollin, G. 1963.

Pliocene–Pleistocene boundary in deep-sea sediments. *Science*, **139**, 727–37.

Ericson, D. B. & Wollin, G. 1956. Correlation of six cores from the equatorial Atlantic and the Caribbean. *Deep Sea Res. oceanogr. Abstr.*, **3**, 104–25.

Ericson, D. B., Wollin, G. & Wollin, J. 1954. Coiling direction of *Globorotalia truncatulinoides* in deep-sea cores. *Deep Sea Res. oceanogr. Abstr.*, **2**, 152–8.

Finlay, H. J. 1940. New Zealand Foraminifera: Key species in stratigraphy – No. 4. *Trans. R. Soc. N.Z.*, **69** (4), 448–72.

Finlay, H. J. 1947. New Zealand foraminifera: Key species in Stratigraphy – No. 5. *N.Z. J. Sci. Technol.*, sect. B., **28**, 259–92.

Galloway, J. J. & Wissler, S. G. 1927. Pleistocene foraminifera from the Lomita Quarry, Palos Verdes Hills, California. *J. Paleontol.*, **1** (1), 35–87.

Giannelli, L. & Salvatorini, G. 1972. I foraminiferi planctonici dei sedimenti terziari dell'Archipelago Maltese. I. Biostratigrafia del 'Globigerina limestone'. *Atti della Societa Toscana di Scienze Naturali*, ser. A, **79**, 49–74.

Giannelli, L. & Salvatorini, G. 1975. I foraminiferi planctonici dei sedimenti terziari dell'Archipelago Maltese. II. Biostratigrafia di: 'Blue Clay', 'Greensand' e 'Upper Coralline Limestone'. *Atti della Societa Toscana di Scienze Naturali*, **82**, 1–24.

Giannelli, L. & Salvatorini, G. 1976. Due nuove specie di foraminiferi planctonici del Miocene. *Bollettino della Societa Paleontologica Italiana*, **15**, 167–73.

Giannelli, L., Salvatorini, G. & Sampò, M. 1976. Segnalazione di *Hastigerinella digitata* (Rhumbler) in sedimenti del Miocene superiore del Bacino Piemontese. *Bollettino della Societa Paleontologica Italiana*, **15**, 159–66.

Grimsdale, T. F. 1951. Correlation, age determination and the Tertiary pelagic foraminifera. *Proc. Third World Petroleum Congress, The Hague*, sec. 1, 463–75.

Hecht, A. D., Bé, A. W. & Lott, L. 1976. Ecologic and paleoclimatic implications of morphologic variation of *Orbulina universa* in the Indian Ocean. *Science*, **194**, 422–4.

Hedberg, H. D. 1937. Foraminifera of the Middle Tertiary Carapita Formation of Northeastern Venezuela. *J. Paleontol.*, **11** (8), 661–97.

Hofker, J. 1954a. *Candorbulina universa* Jedlitschka and *Orbulina universa* d'Orbigny. *The Micropaleontologist*, **8** (2), 38–9.

Hofker, J. 1954b. Morphology of *Globigerinatella insueta* Cushman & Stainforth. *Contrib. Cushman Found. foramin. Res.*, **5**, 151–2.

Hofker, J. 1969. Have the genera *Porticulasphaera, Orbulina (Candorbulina)* and *Biorbulina* a biologic meaning? In: *Proceedings First International Conference on Planktonic Microfossils*, Geneva, 1967, **2**, 279–86.

Hofker, J. 1972. The *Sphaeroidinella* genus from Miocene to Recent. *Rev. Esp. Micropaleontol.*, **4** (2), 119–40.

Hornibrook, N. de B. & Jenkins, D. G. 1965. *Candeina zeocenica* Hornibrook and Jenkins, a new species of foraminifera from the New Zealand Eocene and Oligocene. *N.Z. J. Geol. Geophys.*, **8** (5), 839–42.

Howe, H. V. 1939. Louisiana Cook Mountain Eocene foraminifera. *Geol. Bull. Dep. Conserv. Louisiana*, **14**, 1–122.

Howe, H. V. & Wallace, W. E. 1932. Foraminifera of the Jackson Eocene at Danville Landing on the Ouachita Catahoula Parish, Louisiana. *Geol. Bull. Dep. Conserv. Louisiana*, **2**, 1–118.

Ikebe *et al.* 1972. Neogene biostratigraphy and radiometric time scale of Japan – An attempt at intercontinental correlation. *Pacific Geology*, **4**, 39–78.

Ivanova, L. V. in Bykova, N. K. *et al.* 1958. New genera and species of foraminifera. *Trudy, VNIGRI*, new series 115. Microfauna of the USSR, **9**, 5–106 (in Russian).

Jedlitschka, H. 1934. Ueber *Candorbulina*, eine neue *Candeina*-Foraminiferen-Gattung, und zwei neue *Candeina*-Arten. *Verh. naturforsch. Ver. Brunn*, **65**, 17–26.

Jenkins, D. G. 1960. Planktonic foraminifera from the Lakes Entrance oil shaft, Victoria, Australia. *Micropaleontology*, **6**, 345–71.

Jenkins, D. G. 1966. Standard Cenozoic stratigraphical zonal scheme. *Nature*, **211**, 178.

Jenkins, D. G. 1968. Acceleration of the evolutionary rate in the *Orbulina* lineage. *Contrib. Cushman Found. foramin. Res.*, **19**, 133–9.

Jenkins, D. G. 1971. New Zealand Cenozoic planktonic foraminifera. *Palaeont. Bull. geol. Surv. N.Z.*, **42**, 1–278.

Jenkins, D. G., Saunders, J. B. & Cifelli, R. 1981. The relationship of *Globigerinoides bisphericus* Todd 1954 to *Praeorbulina sicana* (de Stefani) 1952. *J. foramin. Res.*, **11**, 262–7.

Keijzer, F. G. 1945. Outline of the geology of the eastern part of the Province of Oriente, Cuba (E of 76° WL), with notes on the geology of other parts of the island. *Geographische en Geologische Mededeelingen*. Publicaties uit het Geographisch en uit het Mineralogisch-Geologisch Instituut der Rijksuniversiteit te Utrecht. Physiographisch-Geologische Reeks. Serv. II, no. 6, 1–239.

Keller, G. 1981. The genus *Globorotalia* in the early Miocene of the equatorial and northwestern Pacific. *J. foramin. Res.*, **11**, 118–32.

Kennett, J. P. 1966. The *Globorotalia crassaformis* bioseries in north Westland and Marlborough, New Zealand. *Micropaleontology*, **12**, 235–45.

Kennett, J. P. 1973. Middle and Late Cenozoic foraminiferal biostratigraphy of the southwest Pacific – DSDP Leg 21. *Initial Rep. Deep Sea drill. Proj.*, **21**, 575–639.

Koch, R. 1923. Die jungtertiäre Foraminiferenfauna von Kabu (Res. Surabaja, Java). *Eclog. geol. Helv.*, **18** (2), 342–61.

Koch, R. 1926. Mitteltertiäre Foraminiferen aus Bulongan, Ost-Borneo. *Eclog. geol. Helv.*, **19**, 722–51.

Koch, R. E. 1935. Namensänderung einiger Tertiär-Foraminiferen aus Niederländisch Ost-Indien. *Eclog. geol. Helv.*, **28** (2), 557–8.

Krasheninnikov, V. A. & Pflaumann, U. 1977. Zonal stratigraphy of Neogene deposits of the eastern part of the Atlantic Ocean by means of planktonic foraminifers, Leg 41, Deep Sea Drilling Project. *Initial Rep. Deep Sea drill. Proj.*, **41**, 613–57.

Lamb, J. L. 1969. Planktonic foraminiferal datums and Late Neogene epoch boundaries in the Mediterranean, Caribbean and Gulf of Mexico. *Trans. Gulf Coast Assoc. geol. Soc.*, **19**, 559–78.

Lamb, J. L. & Beard, J. H. 1972. Late Neogene planktonic foraminifers in the Caribbean, Gulf of Mexico, and Italian stratotypes. *Univ. Kansas Paleontol. Contrib.*, article **57** (Protozoa 8), 1–67.

Lamb, J. L. & Stainforth, R. M. 1976. Unreliability of *Globigerinoides* datum. *Bull. Am. Assoc. Petrol. Geol.*, **60** (9), 1564–9.

Le Calvez, Y. 1974. Révision des Foraminifères de la Collection d'Orbigny. I – Foraminifères des Iles Canaries. *Cah. Micropaleontol.*, **2**, 1–108.

LeRoy, L. W. 1939. Some small foraminifera, ostracoda and otoliths from the Neogene ('Miocene') of the Rokan-Tapanoeli area, Central Sumatra. *Natuurkd. Tijdschr. Ned. Indie*, **99**, 215–96.

LeRoy, L. W. 1944. Miocene foraminifera from Sumatra and Java, Netherlands East Indies. *Q. Colorado Sch. Mines*, **39** (3), 1–113.

LeRoy, L. W. 1948. The Foraminifer *Orbulina universa* d'Orbigny, a suggested Middle Tertiary time indicator. *J. Paleontol.*, **22**, 500–8.

Lidz, B. 1972. *Globorotalia crassaformis* morphotype variations in Atlantic and Caribbean deep-sea cores. *Micropaleontology*, **18**, 194–211.

Lipps, J. H. 1964. Miocene planktonic foraminifera from Newport Bay, California. *Tulana Stud. Geol.*, **2** (4), 109–33.

Liska, R. D. 1980. *Polyperibola*, a new planktonic foraminiferal genus from the late Miocene of Trinidad and Tobago. *J. foramin. Res.*, **10**, 136–42.

Loeblich, A. R. Jr & Tappan, H. 1957. The new planktonic foraminiferal genus *Tinophodella*, and an emendation of *Globigerinita* Brönnimann. *J. Washington Acad. Sci.*, **47**, 112–16.

Loeblich, A. R. Jr & Tappan, H. 1961. A vindication of the *Orbulina* time surface in California. *Contrib. Cushman Found. foramin. Res.*, **12**, 1–4.

Loeblich, A. R., Jr & Tappan, H. 1964. Foraminiferida, In: *Treatise on Invertebrate Paleontology, C. Protista*, ed. R. C. Moore. The Geological Society of America and the University of Kansas Press, 2 vols., 868 pp.

Lomnicki, J. R. 1901. Einige Bemerkungen zum Aufsatze: Die miocänen Foraminiferen in der Umgebung von Kolomea. *Verh. naturforsch Ver. Brunn*, **38**, 15–18.

Maiya, S., Saito, T. & Sato, T. 1976. Late Cenozoic planktonic foraminiferal biostratigraphy of northwest Pacific sedimentary sequences. *Progress in Micropaleontology*, Special Publication of Micropaleontology Press, pp. 395–422.

Manuputty, J. A. 1977. *Notes on the Late Miocene planktonic foraminiferal associations of the Lorca basin (Province of Murcia, SE Spain).* Messinian Seminar no. 3, Abstracts, Malaga.

Natland, M. L. 1938. New species of foraminifera from off the west coast of North America and from the later Tertiary of the Los Angeles Basin. *University of California, Scripps Institution of Oceanography, Tech. Ser. Bull.*, **4**, 137–52.

Natori, H. 1976. Planktonic foraminiferal biostratigraphy and datum planes in the Late Cenozoic sedimentary sequence in Okinawa-jima, Japan. *Progress in Micropaleontology*, Special Publication of Micropaleontology Press, pp. 214–43.

Ness, G., Levi, S. & Couch, R. 1980. Marine magnetic anomaly timescales for the Cenozoic and Late Cretaceous: A précis, critique, and synthesis. *Rev. Geophys. Space Phys.*, **18** (4), 753–70.

Olsson, R. K. 1964. *Praeorbulina* Olsson, a new foraminiferal genus. *J. Paleontol.*, **38**, 770–1.

Olsson, R. K. 1972. Growth changes in the *Globorotalia fohsi* lineage. *Eclog. geol. Helv.*, **65** (1), 165–84.

Olsson, R. K. 1973. Growth studies on *Globorotalia exilis* Blow and *Globorotalia pertenuis* Beard in the Hole 154A section,

Leg 15, Deep Sea Drilling Project. *Initial Rep. Deep Sea drill. Proj.*, **15**, 617–24.

Palmer, D. K. 1945. Notes on the foraminifera from Bowden, Jamaica. *Bull. Am. Paleontol.*, **29**, 5–82.

Parker, F. L. 1958. Eastern Mediterranean foraminifera. *Rep. Swedish Deep-Sea Exp.*, **8**, Sediment cores from the Mediterranean Sea and the Red Sea, no. 4, 219–83.

Parker, F. L. 1962. Planktonic foraminiferal species in Pacific sediments. *Micropaleontology*, **8**, 219–54.

Parker, F. L. 1965. A new planktonic species (Foraminiferida) from the Pliocene of Pacific deep-sea cores. *Contrib. Cushman Found. foramin. Res.*, **16**, 151–2.

Parker, F. L. 1967. Late Tertiary biostratigraphy (planktonic foraminifera) of tropical Indo-Pacific deep sea cores. *Bull. Am. Paleontol.*, **52**, 115–208.

Parker, F. L. 1971. Distribution of planktonic foraminifera in Recent deep-sea sediments. In: *The Micropalaeontology of Oceans*, ed. B. M. Funnell & W. R. Riedel, pp. 289–307. Cambridge University Press, Cambridge.

Parker, F. L. 1973. Late Cenozoic biostratigraphy (planktonic foraminifera) of tropical Atlantic deep-sea sections. *Rev. Esp. Micropaleontol.*, **5**, 253–89.

Parker, W. K. & Jones, T. R. 1865. On some foraminifera from the North Atlantic and Arctic Oceans, including Davis Straits and Baffin's Bay. *Philos. Trans. R. Soc. London*, **155**, 325–441.

Parker, W. K., Jones, T. R. & Brady, H. B. 1865. On the nomenclature of the foraminifera, Part XII (misprinted as Part X continued) – The species enumerated by d'Orbigny in the Annales des Sciences Naturelles, vol. 7, 1826: *Ann. Mag. Nat. hist. London*, ser. 3, **16**, 15–41.

Pelosio, G., Raffi, S. & Rio, D. 1980. *The Plio-Pleistocene boundary controversy. Status in 1979 at the light of the International Stratigraphic Guide.* Volume dedicated to Sergio Venzo, Grafiche STEP, Parma, pp. 131–40.

Poag, C. W. & Valentine, P. C. 1976. Biostratigraphy and ecostratigraphy of the Pleistocene basin Texas–Louisiana continental shelf. *Trans. Gulf Coast Assoc. geol. Soc.*, **26**, 185–256.

Pokorny, V. 1955. *Cassigerinella boudecensis* n. gen., n. sp. (Foraminifera, Protozoa) z oligocénu zdanickeho flyse. *Vesn. ustred. Ustavu geol.*, **30**, 136–40.

Postuma, J. A. 1971. *Manual of Planktonic Foraminifera.* Elsevier Publishing Company, 420 pp.

Reiss, Z. 1968. Planktonic foraminiferids, stratotypes, and a reappraisal of Neogene chronostratigraphy in Israel. *Israel J. earth Sci.*, **17**, 153–169.

Reuss, A. E. 1850. Neue Foraminiferen aus den Schichten des Oesterreichischen Tertiärbeckens. *Denkschriften der Akademie der Wissenschaften, Wien, Math.-Natw. Kl.*, **1**, 365–90.

Rhumbler, L. 1901. Foraminiferen. In: *Nordisches Plankton*, ed. K. Brandt. Lipsius und Tischer, Klel, **1** (14), 32 pp.

Rhumbler, L. 1911. Die Foraminiferen (Thalamophoren) der Plankton-Expedition; Part I – Die allgemeinen Organisationsverhältnisse der Foraminiferen. *Ergebnisse der Plankton-Expedition der Humboldt-Stiftung*, Kiel & Leipzig, vol. 3, 331 pp.

Robinson, E. 1969. Coiling directions in planktonic foraminifera from the coastal group of Jamaica. *Trans. Gulf Coast Assoc. geol. Soc.*, **19**, 555–8.

Rögl, F. 1974. The evolution of the *Globorotalia truncatulinoides* and *Globorotalia crassaformis* group in the Pliocene and Pleistocene of the Timor Trough, DSDP Leg 27, Site 262. *Initial Rep. Deep Sea drill. Proj.*, **27**, 743–67.

Rögl, F. & Bolli, H. M. 1973. Holocene to Pleistocene planktonic foraminifera of Leg 15, Site 147 (Cariaco Basin (Trench), Caribbean Sea) and their climatic interrelation. *Initial Rep. Deep Sea drill. Proj.*, **15**, 553–615.

Saito, T. 1963. Miocene planktonic foraminifera from Honshu, Japan. *Sci. Rep. Tohoku imp. Univ.*, ser. 2 (Geology), **35** (2), 123–209.

Saito, T. 1976. Geologic significance of coiling direction in the planktonic foraminifera *Pulleniatina*. *Geology*, **4**, 305–9.

Salvatorini, G. & Cita, M. B. 1979. Miocene foraminiferal stratigraphy, DSDP Site 397 (Cape Bojador, North Atlantic). *Initial Rep. Deep Sea drill. Proj.*, **47**, 317–73.

Saunders, J. B. *et al.* (1973). Paleocene to Recent planktonic microfossil distribution in the marine and land areas of the Caribbean. *Initial Rep. Deep Sea drill. Proj.*, **15**, 769–71.

Saunders, J. B. & Bolli, H. M. 1983. Trinidad's contribution to world biostratigraphy. *Transactions of the Fourth Latin American Congress, Trinidad*, 1979 (in press).

Schmid, K. 1934. Biometrische Untersuchungen von Foraminiferen. *Eclog. geol. Helv.*, **27**, 45–134.

Schubert, R. J. 1910. Ueber Foraminiferen und einen Fischotolithen aus dem fossilen Globigerinenschlamm von Neu-Guinea. *Verhandlungen der Geologischen Reichsanstalt, Wien*, 318–28.

Schwager, C. 1866. Fossile Foraminiferen von Kar Nikobar. In: *Novara Expedition 1857–1859*, vol. 2, part 2, pp. 187–268. Wien, Oesterreich, 1866, Geol. Theil.

Seiglie, G. A. 1963. Una nueva especie del género *Globigerina* del Reciente de Venezuela. *Boletin del Instituto Oceanografico, Cumana, Venezuela*, **2** (2), 90–1.

Seiglie, G. A. 1973. Revision of Mid-Tertiary stratigraphy of southwestern Puerto Rico. *Bull. Am. Assoc. Petrol. Geol.*, **57**, 405–6.

Selli, R. *et al.* 1977. The Vrica section (Calabria, Italy). A potential Neogene/Quaternary boundary stratotype. *G. Geol.*, **42** (1), 181–204.

Serova, M. Ya. 1977. Stratigraphic significance of the *Globorotalia lenguaensis* Bolli species for Miocene deposits of Sakhalin. *Voprosy Micropaleontologii*, **19**, 99–103 (in Russian).

Soedino, H. 1970. Planktonic foraminifera from the Velez Rubio region, S.E. Spain. *Rev. Esp. Micropaleontol.*, **2** (3), 215–34.

Sprovieri, R., Ruggieri, G. & Unti, M. 1980. *Globorotalia truncatulinoides excelsa* n. subsp., foraminifero planctonico guida per il Pleistocene inferiore. *Boll. Soc. geol. Ital.*, **99**, 3–11.

Srinivasan, M. S. & Kennett, J. P. 1976. Evolution and phenotypic variation in the Late Cenozoic *Neogloboquadrina dutertrei* plexus. *Progress in Micropaleontology*, Special Publication of Micropaleontology Press, pp. 329–55.

Srinivasan, M. S. & Kennett, J. P. 1981. A review of Neogene planktonic foraminiferal biostratigraphy: Applications in the equatorial and South Pacific. In: *The Deep Sea Drilling Project. A Decade of Progress*, ed. J. E. Warner, R. G. Douglas & E. L. Winterer, pp. 395–432. Society of Economic Mineralogists and Paleontologists, Special Publication 32.

Srinivasan, M. S. & Srivastava, S. S. 1974. *Sphaeroidinella dehiscens* datum and Miocene–Pliocene boundary. *Bull. Am. Assoc. Petrol. Geol.*, **58**, 304–11.

Stainforth, R. M. 1948. Description, correlation, and paleoecology of Tertiary Cipero Marl Formation, Trinidad, B. W. I. *Bull. Am. Assoc. Petrol. Geol.*, **32**, 1292–1330.

Stainforth, R. M. *et al.* 1975. Cenozoic planktonic foraminiferal zonation and characteristics of index forms. *Univ. Kansas Paleontol. Contrib.*, article **62**, 1–425, in 2 parts.

Stainforth, R. M. & Lamb, J. L. 1981. An evaluation of planktonic foraminiferal zonation of the Oligocene. *Univ. Kansas Paleontol. Contrib.*, **104**, 1–34.

Stainforth, R. M., Lamb, J. L. & Jeffors, R. M. 1978. *Rotalia menardii* Parker, Jones & Brady, 1865 (Foraminiferida): Proposed suppression of lectotype and designation of neotype Z.N. (S.) 2145. *Bull. zool. Nomencl. London*, **34** (4), 252–60.

Subbotina, N. N. 1953. Fossil foraminifera of the USSR; Globigerinidae, Hantkeninidae and Globorotaliidae. *Trudy, VNIGRI*, new series, **76**, 1–296 (in Russian). Translated into English by E. Lees, Collet's Ltd., London and Wellingborough, 321 pp.

Subbotina, N. N. in Bykova, N. K. *et al.* 1958. New genera and species of foraminifera. *Trudy, VNIGRI*, new series, 115. Microfauna of the USSR, **9**, 5–106 (in Russian).

Takayanagi, Y. & Saito, T. 1962. Planktonic foraminifera from the Nobori Formation, Shikoku, Japan. *Sci. Rep. Tohoku imp. Univ.*, ser. 2 (Geology), special volume **5**, 67–105.

Thomson, W. 1876. Comments. In: Murray, J., Preliminary reports to Professor Wyville Thomson, F.R.S., director of the civilian scientific staff, on work done aboard the 'Challenger'. *Proc. R. Soc. London*, **24**, 534.

Todd, R. 1954. Appendix in: Todd, R., Cloud, P. E. Jr, Low, D. & Schmidt, R. G. Probable occurrence of Oligocene on Saipan. *Am. J. Sci.*, **252**, 673–82.

Todd, R. 1957. Smaller Foraminifera. In: Geology of Saipan Mariana Islands, Part 3. Paleontology. *Prof. Pap. U.S. geol. Surv.*, **280-H**, 265–320.

Todd, R. 1964. Planktonic foraminifera from deep-sea cores off Eniwetok Atoll. *Prof. Pap. U.S. geol. Surv.*, **260-CC**, 1067–1111.

Todd, R. *et al.* 1954. Probable occurrence of Oligocene in Saipan. *Am. J. Sci.*, **252**, 673–82.

Toumarkine, M. 1978. Planktonic foraminiferal biostratigraphy of the Paleogene of Sites 360 to 364 and the Neogene of Sites 362A, 363, and 364 Leg 40. *Initial Rep. Deep Sea drill. Proj.*, **40**, 679–721.

Ujiié, H. 1976. *Prosphaeroidinella*, n.gen.: Probable ancestral taxon of *Sphaeroidinellopsis* (foraminifera). *Bull. natl. Sci. Mus. Tokyo*, ser. C (Geology & Paleontology), **2** (1), 9–26.

Weinzierl, L. L. & Applin, E. R. 1929. The Claiborne Formation on the coastal domes. *J. Paleontol.*, **3**, 384–410.

Zachariasse, W. J. 1975. Planktonic foraminiferal biostratigraphy of the Late Neogene of Crete (Greece). *Utrecht Micropaleontological Bulletins*, **11**, 1–171.

7

Southern mid-latitude Paleocene to Holocene planktic foraminifera

D. GRAHAM JENKINS

CONTENTS

Introduction

This chapter discusses Cenozoic planktic foraminifera from the Southern Hemisphere in areas south of latitude 30° S. Prior to the Deep Sea Drilling Project (DSDP), which began in 1968, nearly all of the documentation of species had come from the marine rocks of Australia and New Zealand with some additional information from southern Chile and Argentina. A misconception by early workers, that some of the species were indigenous to the area, led to species names like *Hantkenina australis* Finlay, *Globorotalia miozea* Finlay and *G. zealandica* Hornibrook. Another parochial attitude has been the use of locally developed stages as stratigraphic frameworks in Australia and New Zealand, and avoidance of the use of planktic foraminiferal zones. At least, this was the case until the work of Jenkins (1960) on the Oligo–Miocene sediments of Southeastern Australia. Another change was heralded with the entry of the *Glomar Challenger* into the area on Legs 3, 21, 25, 26, 28, 29, 35, 36, 39 and 40: this has led to a mass of new data of varying quality on both geographic and stratigraphic ranges of species.

Previous work

The first record of fossil planktic foraminifera was by Jones (*in* Mantell, 1850) working on Oligocene sediments from New Zealand. There were only a few subsequent works until the important papers by Finlay (1939–1947b). He recorded and described new species in a series of five papers on 'key' species and Hornibrook (1958b) added the sixth and final paper to the series. Hornibrook (1958b–1981) continued the work of documenting species within the framework of New Zealand stages, while Scott (1966–1980) undertook biometric studies of selected evolutionary lineages. Jenkins (1963–1974) worked on the stratigraphic ranges and described new species as well as refining and enlarging a set of zones for the New Zealand Cenozoic, initially developed in Southeastern Australia. Others who have contributed important papers include Vella (1961, 1964), Geiger (1962), McInnes (1965), Walters (1965) and Kennett (1966). Work on the correlation of planktic foraminifera with paleomagnetic stratigraphy in the Late Cenozoic was done by Kennett & Watkins (1974).

In Australia important works include those by Carter (1958, 1964), Jenkins (1958, 1960), Lindsay (1967, 1969), Ludbrook & Lindsay (1969), McGowran (1964, 1965) and Taylor (1966). With regard to South America the following are important: in southernmost Chile Todd & Knicker (1952) described faunas from the Early–Middle Eocene; Bertels (1970) described faunas from the Tertiary of Patagonia (Argentina); and Malumian (1970) and with colleagues (1971, 1973) described faunas from the Tertiary of Argentina.

In between these land areas in the Southern Hemisphere, Cenozoic faunas have been described from south of latitude 30° S in the *Initial Reports of the Deep Sea Drilling Project*: Atlantic Leg 3, Tasman Sea Leg 21 (Kennett, 1973), Indian Ocean Leg 25 (Sigal, 1974), Indian Ocean Leg 26 (Boltovskoy, 1974), Southern Ocean Leg 28 (Kaneps, 1975), Southwest Pacific Leg 29 (Jenkins, 1975), Southeast Pacific Leg 35, South Atlantic Legs 36 (Tjalsma, 1977), 39 (Berggren, 1977; Boersma, 1977), and 40 (Jenkins, 1978c; Toumarkine, 1978). Neogene species have been recently described and illustrated by Kennett and Srinivasan (1983).

An important paper was published by Shackleton & Kennett (1975) on oxygen isotope analyses of both benthic and planktic foraminifera from DSDP Leg 29, which resulted in a paleotemperature plot for the Cenozoic. This tended to confirm results obtained by similar analyses by Devereux (1967) on planktic foraminifera from New Zealand.

Paleoceanographic setting

The changing configuration of the continents during the Cenozoic has had a big influence on water masses and on the position and intensity of oceanic currents; this is especially true in the Southern Hemisphere. The two maps (Figs. 1 and 2) show the positions of the continents in the Paleocene and at the present day. The oceanic regime shown for the Paleocene existed with little change through the Eocene until the Late Oligocene. The major change in the oceanic currents of the area was brought about by the development of the Circum-Antarctic Current. Evidence from the stratigraphic range of *Jenkinsina samwelli* (Jenkins) indicates that Australia finally parted from Antarctica in the Late Oligocene (Jenkins, 1974b, 1978a). *J. samwelli* lived in the gulf between Antarctica and Australia from Late Eocene to Late Oligocene. Once the two continents had finally parted at the South Tasman Rise, *J. samwelli* spread out into the Southwestern Pacific, where it lived briefly and became extinct. It has been documented from DSDP Leg 29 Sites 282, 277 and 276, the South Island of New Zealand and from Leg 40 Site 360 in the Southeastern Atlantic (Jenkins, 1978b). By correlating the extinction of *J. samwelli* against that of *C. cubensis* it is possible to estimate that the initiation of the Circum-Antarctic Current took place 28–30 m.y. ago. This means that the Drake Passage had opened and was deep enough for seawater bearing plankton to flow through the gap between South America and Antarctica at this time. Since then the passage has widened and with the onset of the Late Cenozoic glaciations on Antarctica, cold surface currents have transported Sub-Antarctic faunas northwards into

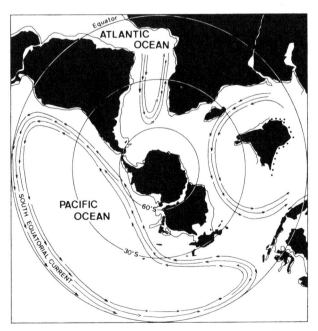

Fig. 1. The Southern Hemisphere in the Paleocene (after Smith & Briden, 1977); three major currents are shown

warmer water areas as shown in Fig. 2. Examples include the northward displacement of the sub-tropical convergence along the east Coast of New Zealand as compared with its much more southern position in the Tasman Sea, the influence of the cold Benguela Current along the west coast of South Africa (Jenkins, 1978c), and the Humboldt Current along the west coast of South America. From Miocene to Pleistocene the intensity of the cold surface currents has fluctuated, which is reflected in the faunal associations. The effects of the cold Benguela Current can be seen when comparing the Miocene to Pleistocene faunas at DSDP Leg 40, Site 360, with the more northern Site 362 (Jenkins, 1978c). It can be seen that *Globigerina pachyderma*, *G. bulloides* and *Globorotalia miozea* were transported northwards and were mixed with such warm water species as *Globigerinatella insueta*, *Clavatorella bermudezi*, *Globorotalia fohsi lobata* and *Globigerinoides sacculifer* at Site 362. Conversely, tropical and subtropical faunas have been brought southwards into the higher latitudes by currents like the Eastern Australian and Brazil Currents.

Therefore, in summary it can be said that the faunas of the Southern Hemisphere were fairly cosmopolitan in the Paleocene and Eocene when there was a southern extension of the warm waters of the Atlantic, Indian and Pacific oceans. With the development of the Circum-Antarctic Current in the Late Oligocene, faunas were spread in a clockwise direction eastwards and well-defined cold currents also carried the faunas northwards in certain areas. As a counterbalance,

warm currents brought subtropical faunas southwards into the area. For example, the following species were brought into New Zealand by warm currents: *Globigerinatella insueta* in the Early Miocene and *Globorotalia menardii* in the Middle Miocene. Unfortunately, most of these tropical migrations in the Cenozoic were brief affairs and the stratigraphic ranges of the tropical species are therefore only partial ranges; consequently they have only a limited value for correlation in the southern mid-latitudes.

Paleotemperature and diversity

The numbers of fossil species or the species diversity in a sample is dependent on many factors, including preservation and location. The importance of the latitudinal position of the fauna was established long ago by Wiseman & Ovey (1950) and confirmed by Bé (1967). Species diversity has been found to increase from one species in the near-freezing waters of the polar zones to about 22 in the tropics. There appears to be a direct correlation between temperature and diversity in living faunas and this also can be assumed for fossil species and paleotemperatures.

When the species diversity curve for the whole of the New Zealand Cenozoic is compared with the paleotemperature curve for New Zealand plotted by Devereux (1967), there seems to be a general similarity in so far as when the paleotemperature curve falls or rises the diversity also changes (Jenkins, 1968, 1973). The diversity curve for the Oligo–Miocene of the Lakes Entrance Oil shaft in Australia generally parallels that for the same period in New Zealand. Species counts are lower in the Australian locality but this probably reflects the shallow continental shelf environment that existed at that site.

Zones used in this chapter

Fig. 3 gives the zonal scheme used in this chapter together with the first and last occurrences of the marker species on which the zones are based. The zonal fossils are usually common throughout their zones, their ranges are based on those in southeastern Australia, New Zealand and on sites of DSDP Legs 29 and 40 (Jenkins, 1960, 1966a, 1967, 1975, 1978c). Fig. 4 provides a correlation of the zonal scheme used in this chapter with other schemes proposed within the Southern Hemisphere.

GLOBIGERINA PAUCILOCULATA ZONE

Category: Taxon range zone
Age: Early Paleocene
Author: Jenkins (1966a)
Definition: Range of zonal marker.
Remarks: The zonal marker, which in some ways resembles
Globigerina eugubina, has been found only in the Middle

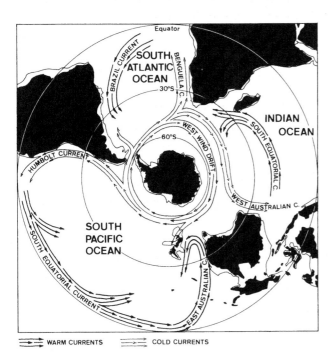

Fig. 2. Present-day disposition of continental areas in the Southern Hemisphere with major currents (after Smith & Briden, 1977)

Waipara River section of New Zealand. In addition to the taxa shown in Fig. 5 *Chiloguembelina waiparaensis* also occurs in the zone.

GLOBIGERINA TRILOCULINOIDES ZONE

Category: Interval zone
Age: Late Paleocene
Author: Jenkins (1966a)
Definition: Interval with zonal marker, from last occurrence of *Globigerina pauciloculata* to first occurrence of *Pseudohastigerina wilcoxensis*.
Remarks: In addition to the taxa shown on Fig. 5, *Globigerina spiralis* and *Globorotalia reissi* occur in the zone. *Chiloguembelina crinata* is restricted to it.

PSEUDOHASTIGERINA WILCOXENSIS ZONE

Category: Interval zone
Age: Early Eocene
Author: Jenkins (1966a)
Definition: Interval with zonal marker, from its first occurrence to first occurrence of *Morozovella crater*.
Remarks: In addition to the taxa shown on Fig. 5, *Globigerina aquiensis* and *Globorotalia esnaensis* are limited to the

zone. The extinction of *Globorotalia gracilis* coincides with the upper zonal boundary, while *G. reissi* and *Chiloguembelina wilcoxensis* become extinct within the zone.

MOROZOVELLA CRATER ZONE

Category: Taxon range zone
Age: Early Eocene
Author: Jenkins (1966a)
Definition: Range of zonal marker.
Remarks: In addition to the taxa shown on Fig. 5, *Morozovella dolabrata* becomes extinct in the upper part of the zone.

ACARININA PRIMITIVA ZONE

Category: Interval zone
Age: Middle Eocene
Author: Jenkins (1966a)
Definition: Interval with zonal marker, from last occurrence of *Morozovella crater* to first occurrence of *Globigerinatheka index*.
Remarks: In addition to the taxa shown on Fig. 5, *Planorotalites australiformis* becomes extinct in the zone.

GLOBIGERINATHEKA INDEX ZONE

Category: Interval zone
Age: Middle Eocene
Author: Jenkins (1966a)
Definition: Interval from first occurrence of zonal marker to first occurrence of *Chiloguembelina cubensis*.
Remarks: In addition to the taxa shown on Fig. 5, the significant species *Globigerina ouachitaensis* and *Candeina zeocenica* make their first appearances in the zone.

TESTACARINATA INCONSPICUA ZONE

Category: Concurrent range zone
Age: Late Eocene
Author: Jenkins (1966a)
Definition: Interval from first occurrence of *Chiloguembelina cubensis* to last occurrence of *Globorotalia aculeata*.
Remarks: *Zeauvigerina zealandica* becomes extinct within the zone.

GLOBIGERINA LINAPERTA ZONE

Category: Interval zone
Age: Late Eocene
Author: Jenkins (1966a)
Definition: Interval with zonal marker between last occurrence of *Globorotalia aculeata* and first occurrence of *Globigerina brevis*.
Remarks: In addition to the taxa shown on Fig. 5, *Hantkenina alabamensis* becomes extinct within the zone and *Globigerina praeturritilina* occurs in its upper part.

AGE		PLANKTIC FORAMINIFERAL ZONES	DATUM MARKERS
PLEIST.		*Globorotalia truncatulinoides*	F *G. truncatulinoides*
PLIO-CENE	L	*Globorotalia inflata*	F *G. inflata*
	E	*Globorotalia puncticulata*	F *G. puncticulata*
MIOCENE	L	*Globorotalia conomiozea*	F *G. conomiozea*
		Globorotalia miotumida	L *G. mayeri*
	M	*Globorotalia mayeri*	F *G. mayeri*
		Orbulina suturalis	F *O. suturalis*
	E	*Praeorbulina glomerosa curva*	F *P. glomerosa curva*
		Globigerinoides trilobus	F *G. trilobus*
OLIGOCENE	L	*Globigerina woodi connecta*	F *G. woodi connecta*
		Globigerina woodi woodi	F *G. woodi woodi*
		Globoquadrina dehiscens	F *G. dehiscens*
	E	*Globigerina euapertura*	L *G. angiporoides*
		Globigerina angiporoides	L *G. brevis*
		Globigerina brevis	F *G. brevis*
EOCENE	L	*Globigerina linaperta*	L *G. aculeata*
		Testacarinata inconspicua	F *C. cubensis*
	M	*Globigerinatheka index*	F *G. index*
		Acarinina primitiva	L *M. crater*
	E	*Morozovella crater*	F *M. crater*
		Pseudohastigerina wilcoxensis	F *P. wilcoxensis*
PALEO-CENE	L	*Globigerina triloculinoides*	L *G. pauciloculata*
	E	*Globigerina pauciloculata*	F *G. pauciloculata*

Fig. 3. Planktic foraminiferal zonal scheme and datum markers used in this chapter

GLOBIGERINA BREVIS ZONE

Category: Taxon range zone
Age: Late Eocene/Early Oligocene
Author: Jenkins (1966a)
Definition: Range of zonal marker.
Remarks: The following become extinct in the lowermost part of the *G. brevis* Zone: *Pseudohastigerina micra* (Cole), *Globigerinatheka index* (Finlay), *G. semiinvoluta* (Keijzer). *Globorotalia gemma* Jenkins makes its initial appearance within the zone.

GLOBIGERINA ANGIPOROIDES ZONE

Category: Interval zone
Age: Early Oligocene
Author: Jenkins (1966a)
Definition: Interval with zonal marker from last occurrence of *Globigerina brevis* to last occurrence of zonal marker.
Remarks: In addition to the taxa shown on Fig. 5, *Globigerina praeturritilina* becomes extinct within the zone and *G. ciperoensis angustiumbilicata* makes its first appearance.

GLOBIGERINA EUAPERTURA ZONE

Category: Interval zone
Age: Late Oligocene
Author: Jenkins (1966a)
Definition: Interval with zonal marker, from last occurrence of *Globigerina angiporoides* to first occurrence of *Globoquadrina dehiscens*.
Remarks: The upper part of the zone was called pre-*Globoquadrina dehiscens dehiscens* Zone by Jenkins (1960). In addition to the taxa shown on Fig. 5, *Globorotaloides testarugosa* makes its first appearance within the zone.

GLOBOQUADRINA DEHISCENS ZONE

Category: Interval zone
Age: Late Oligocene
Author: Jenkins (1960)
Definition: Interval from first occurrence of zonal marker to first occurrence of *Globigerina woodi woodi*.

GLOBIGERINA WOODI WOODI ZONE

Category: Interval zone
Age: Late Oligocene
Author: Jenkins (1960), redefined by Jenkins (1967)
Definition: Interval from first occurrence of zonal marker to first occurrence of *Globigerina woodi connecta*.

GLOBIGERINA WOODI CONNECTA ZONE

Category: Interval zone
Age: Late Oligocene/Early Miocene
Author: Jenkins (1967)
Definition: Interval from first occurrence of zonal marker to first occurrence of *Globigerinoides trilobus*.
Remarks: In addition to the taxa shown on Fig. 5, *Globigerina angulisuturalis* becomes extinct within the zone; *Globigerina falconensis* and *Globorotalia minutissima* appear for the first time within the zone.

GLOBIGERINOIDES TRILOBUS ZONE

Category: Interval zone
Age: Early Miocene
Author: Jenkins (1960), redefined by Jenkins (1967)
Definition: Interval from first occurrence of zonal marker to first occurrence of *Praeorbulina glomerosa curva*.
Remarks: In addition to the taxa shown on Fig. 5, *Globoquadrina tripartita* becomes extinct in the lower part of the zone; *Globorotalia zealandica incognita* is limited to its lower part and *Globigerina foliata* and *Globorotalia bella* appear first in the upper part.

PRAEORBULINA GLOMEROSA CURVA ZONE

Category: Interval zone
Age: Early Miocene
Author: Jenkins (1960), redefined by Jenkins (1967)
Definition: Interval from first occurrence of zonal marker to first occurrence of *Orbulina suturalis*.
Remarks: In addition to the taxa shown on Fig. 5, *Globorotaloides suteri* becomes extinct within the zone, *Beella digitata* is confined to it and *Globorotalia praemenardii* appears first in the zone.

ORBULINA SUTURALIS ZONE

Category: Interval zone
Age: Middle Miocene
Author: Jenkins (1966a)
Definition: Interval from first occurrence of zonal marker to first occurrence of *Globorotalia mayeri*.
Remarks: In addition to the taxa shown on Fig. 5, the following taxa become extinct within the zone: *Globorotalia bella*, *G. nana pseudocontinuosa* and *Globoquadrina larmeui*. The following appear first in the zone: *Globigerina woodi decoraperta* and *Globorotalia panda*.

GLOBOROTALIA MAYERI ZONE

Category: Taxon range zone
Age: Middle Miocene
Author: Jenkins (1960)
Definition: Range of zonal marker.
Remarks: In addition to the taxa shown on Fig. 5, *Globorotalia obesa* and *G. praemenardii* become extinct within the zone and *G. mayeri nympha* is limited to it.

AGE		NEW ZEALAND AND SOUTHWEST PACIFIC Jenkins 1971, 1975, this chapter	NEW ZEALAND PLANKTIC FORAMINIFERAL ZONES Jenkins 1966a, 1967	SOUTHWEST PACIFIC DSDP LEG 29 Jenkins 1975	EASTERN EQUATORIAL PACIFIC Jenkins & Orr 1972	SOUTHEAST ATLANTIC DSDP LEG 40 SITES 360, 362 Jenkins 1978c	TASMAN SEA AND SOUTHWEST PACIFIC Kennett 1973
PLEIST.		G. truncatulinoides		G. (G.) truncatulinoides	P. obliquiloculata	G. truncatulinoides	G. truncatulinoides-G. tosaensis
PLIOCENE	L	G. inflata	G. inflata	G. (T.) inflata	G. fistulosus	G. inflata	G. tosaensis / G. inflata
PLIOCENE	E	G. puncticulata		G. (T.) puncticulata	S. dehiscens	G. puncticulata	G. crassaformis / G. puncticulata / G. margaritae
MIOCENE	L	G. conomiozea	G. miozea sphericomiozea	G. (G.) conomiozea	G. tumida	G. conomiozea	G. conomiozea
MIOCENE	L	G. miotumida	G. miotumida miotumida	G. (G.) miotumida miotumida	G. plesiotumida	G. miotumida	G. nepenthes / G. continuosa
MIOCENE	M	G. mayeri	G. mayeri mayeri	G. (T.) mayeri mayeri	G. altispira	G. mayeri mayeri	G. mayeri
MIOCENE	M	O. suturalis	O. suturalis	O. suturalis	G. fohsi lobata-G. peripheroacuta / G. peripheroacuta	O. suturalis	O. suturalis
MIOCENE	E	P. glomerosa curva	P. glomerosa curva	P. glomerosa curva	P. glomerosa curva	P. glomerosa curva	G. trilobus
MIOCENE	E	G. trilobus	G. trilobus trilobus	G. trilobus trilobus	G. bisphericus / G. venezuelana / G. dissimilis	G. trilobus trilobus	
OLIGOCENE	L	G. woodi connecta	G. woodi connecta	G. (G.) woodi connecta	G. kugleri	G. woodi connecta	
OLIGOCENE	L	G. woodi woodi	G. woodi woodi	G. (G.) woodi woodi	G. angulisuturalis	G. woodi woodi	
OLIGOCENE	L	G. dehiscens	G. dehiscens	G. dehiscens	G. opima	G. dehiscens	
OLIGOCENE	L	G. euapertura	G. euapertura	G. (G.) euapertura	C. cubensis / G. ampliapertura	G. euapertura	
OLIGOCENE	E	G. angiporoides	G. angiporoides angiporoides	G. (S.) angiporoides angiporoides	P. barbadoensis	G. angiporoides	
EOCENE	L	G. brevis	G. brevis	G. (G.) brevis	G. insolita	G. brevis	
EOCENE	L	G. linaperta	G. linaperta	G. (S.) linaperta		G. linaperta	
EOCENE	M	T. inconspicua	G. inconspicua	G. (T.) aculeata		G. luterbacheri	
EOCENE	M	G. index	G. index index	G. (G.) index		G. index	
EOCENE	M	A. primitiva	P. primitiva	P. primitiva		P. primitiva	
EOCENE	E	M. crater	G. crater crater	G. (M.) crater crater			
EOCENE	E	P. wilcoxensis	G. wilcoxensis	G. wilcoxensis			
PALEOCENE	L	G. triloculinoides	G. triloculinoides	G. (S.) triloculinoides			
PALEOCENE	E	G. pauciloculata	G. pauciloculata				

Fig. 4. Correlation of zonal schemes within the Southern Hemisphere

SOUTH AUSTRALIA after Ludbrook & Lindsay 1969	SOUTHWESTERN AUSTRALIA OTWAY BASIN after Ludbrook in Wopfner & Douglas 1971	SOUTH AUSTRALIA ADELAIDE PLAINS SUBBASIN after Lindsay 1969	AUSTRALIA, VICTORIA after Taylor 1966 Singleton 1968 in Wopfner & Douglas 1971	AUSTRALIA, VICTORIA LAKES ENTRANCE OIL SHAFT Jenkins 1960
			G. inflata	
				— — — undefined — — —
				G. menardii miotumida
O. universa	*O. universa*		*G. lenguaensis* *G. mayeri*	
				G. mayeri mayeri
O. suturalis	*O. suturalis*	*O. suturalis*	*O. universa*	*O. universa*
				C. glomerosa circularis
P. glomerosa curva	*P. glomerosa curva*	*P. glomerosa curva*	*G. glomerosa* *O. suturalis*	*C. glomerosa curva*
				G. menardii praemenardii-
G. bisphericus *G. trilobus trilobus*	*G. sicanus* *G. trilobus* s.s.	*G. bisphericus* *G. trilobus trilobus*	*G. bisphericus* *G. trilobus* s.s.	*G. bispherica*
				G. triloba triloba
G. woodi	*G. woodi* s.s.	*G. woodi woodi*	*G. woodi* s.l. *G. kugleri* *G. dehiscens*	*G. woodi*
G. dehiscens	*G. dehiscens*	*G. euapertura*		*G. dehiscens dehiscens*
G. euapertura | *G. stavensis*	*G. euapertura* | *G. stavensis*	*G. stavensis*	*G. euapertura* *G. T. opima* s.l.	
G. labiacrassata	*G. labiacrassata*		*C. cubensis* *G. testarugosa*	
G. angiporoides angiporoides | *C. cubensis*	*S. angiporoides* | *C. cubensis*	*C. cubensis*	*G. (T.) gemma* *G. angiporoides* *G. brevis*	
G. linaperta	*S. linaperta*	*G. linaperta*		
			G. ampliapertura/ *G. linaperta*	
		T. aculeata	*G. index*	
G. aculeata	"T" *aculeata* | *H. primitiva*		*H. primitiva*	
			H. australis	
G. index index	*G. index index* s.s.			
P. primitiva	*T. primitiva*			
P. australiformis	*P. australiformis*			
			— — ?–?–?–?–?— —	
			G. pseudomenardii *G. chapmani*	
			T. aequa	
			T. aff. *acuta*	
			T. chapmani-ehrenbergeri	

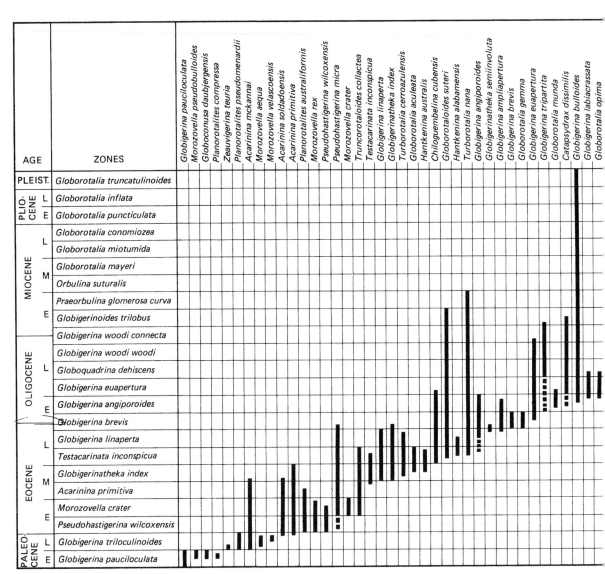

Fig. 5. Distribution of southern mid-latitude Paleocene to
Pleistocene planktic foraminifera

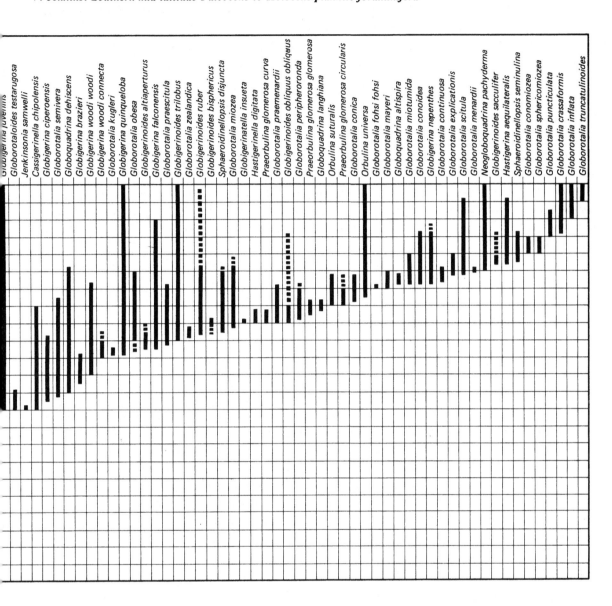

GLOBOROTALIA MIOTUMIDA ZONE

Category: Interval zone

Age: Late Miocene

Author: Jenkins (1960), redefined by Jenkins (1967)

Definition: Interval with zonal marker, from last occurrence of *Globorotalia mayeri* to first occurrence of *Globorotalia conomiozea* or *Globorotalia sphericomiozea*.

Remarks: In addition to the taxa shown on Fig. 5, *Globigerina ciperoensis angustiumbilicata, Globorotalia continuosa* and *Globigerinoides apertasuturalis* become extinct within the zone while *Hastigerina aequilateralis* and *Globigerinoides sacculifer* appear for the first time.

GLOBOROTALIA CONOMIOZEA ZONE

Category: Interval zone

Age: Late Miocene

Author: Kennett (1973), redefined by Jenkins (1975)

Definition: Interval from first occurrence of zonal marker to first occurrence of *Globorotalia puncticulata*.

Remarks: The zone seems to represent a relatively short period of time. The presence of *Globorotalia sphericomiozea* is here regarded as being an additional diagnostic marker for the zone.

GLOBOROTALIA PUNCTICULATA ZONE

Category: Interval zone

Age: Early Pliocene

Author: Kennett (1973), redefined by Jenkins (1975)

Definition: Interval from first occurrence of zonal marker to first evolutionary occurrence of *Globorotalia inflata*.

GLOBOROTALIA INFLATA ZONE

Category: Interval zone

Age: Late Pliocene

Author: Kennett (1973), redefined by Jenkins (1975)

Definition: Interval from first evolutionary occurrence of zonal marker to first occurrence of *Globorotalia truncatulinoides*.

Remarks: In addition to the taxa shown on Fig. 5, *Globigerina woodi decoraperta* becomes extinct within the zone.

GLOBOROTALIA TRUNCATULINOIDES ZONE

Category: Taxon range zone

Age: Pleistocene

Author: Kennett (1973), redefined by Jenkins (1975)

Definition: Range of zonal marker.

Discussion of taxa

This chapter discusses in the taxonomic notes and illustrates on Figs. 6 and 7 only the typical southern mid-latitude forms originally described from the New

Fig. 6. Southern mid-latitude planktic foraminifera (all figures × 60)

1 *Globigerina pauciloculata* Jenkins
   1a–c Holotype. Sample F 5667, S 68/197, *G. pauciloculata* Zone, Early Paleocene, Waipara River section, New Zealand.

2 *Globigerina linaperta* Finlay
   2a–c Holotype, redrawn (*in* Hornibrook, 1958a). Sample 5179A, Middle Eocene, Hampden section, N of Kakaho Creek, New Zealand.

3–5 *Globigerina angiporoides* Hornibrook
   3 Holotype, **4–5** Paratypes. Sample F 6502, S 136/777, Late Eocene or Early Oligocene, N end of Campbells Beach, All Day Bay, Kakanui, New Zealand.

6 *Globigerinatheka index* (Finlay)
   6a–c Holotype, redrawn (*in* Hornibrook, 1958a). Sample 5179A, Middle Eocene, Hampden section, N of Kakaho Creek, New Zealand.

7 *Acarinina primitiva* (Finlay)
   7a–c Holotype, redrawn (*in* Jenkins, 1966a). Sample 517B, Middle Eocene, Hampden section, N of Kakaho Creek, New Zealand.

8 *Planorotalites australiformis* (Jenkins)
   8a–c Holotype. Sample F 5671, S 68/201, Early Eocene, Middle Waipara River section, New Zealand.

9 *Morozovella crater* (Finlay)
   9a–c Holotype (figured in Hornibrook, 1958a). Sample 5570, Middle Eocene, Hurunui River section, New Zealand.

10 *Truncorotaloides collactea* (Finlay)
   10a–c Holotype, redrawn (*in* Jenkins, 1966a). Sample 5540, Middle Eocene, Hampden Beach section, N of Kakaho Creek, New Zealand

11 *Testacarinata inconspicua* (Howe)
   11a–c Holotype. Middle Eocene, Claiborne, Cook Mountain Fm., Couley Creek, Winn Parish, Louisiana, USA.

12 *Globorotalia aculeata* Jenkins
   12a–c Holotype. Sample F 5179A, S 146/662, *G. index* Zone, Middle Eocene, Hampden section, N of Kakaho Creek, New Zealand.

13 *Chiloguembelina cubensis* (Palmer)
   13a–b (Palmer did not designate a holotype from her specimens figured; the figures given here correspond with her figs. 1 and 4). Sample Palmer station 1163, Early Oligocene, water well SE of Cartagena, Santa Clara Province, Cuba.

14 *Jenkinsina samwelli* (Jenkins)
   Holotype. Sample NZMS S2, *G. euapertura* Zone, Early Oligocene, 34 miles SE of Cape Farewell, coast of South Island, New Zealand.

15 *Zeauvigerina teuria* Finlay
   15a–b Holotype, redrawn (*in* Jenkins, 1971). Sample Hu 12, Late Cretaceous, Senonian (determined by Finlay), *G. triloculinoides* Zone, Late Paleocene (this chapter), Te Uri Stream section, New Zealand.

16 *Hantkenina australis* Finlay
   16a–b Holotype, redrawn (*in* Jenkins, 1965). Sample 5179B, Late Middle Eocene, Hampden section, N of Kakaho Creek, New Zealand.

17 *Globigerina brevis* Jenkins
   17a–c Holotype. Sample F 9398, S 136/694, *G. brevis* Zone Early Oligocene, Kakanui River section, New Zealand.

18 *Globigerina euapertura* Jenkins
   18a–c Figured specimens from the type Whaingaroan of New Zealand (from Jenkins, 1971).

19 *Globigerina labiacrassata* Jenkins
   19a–c Holotype. Sample F 11447, S 127/425, *G. angiporoides* Zone, Early Oligocene, Earthquake section, Waitaki Valley, New Zealand.

20 *Globigerina brazieri* Jenkins
   20a–c Holotype. Sample F 15836, S 111/602, *G. woodi connecta* Zone, Early Miocene, Blue Cliffs section, New Zealand.

21 *Globigerina woodi woodi* Jenkins
   21a–c Figured specimen from Old Rifle Butts section, New Zealand (from Jenkins, 1971).

22 *Globigerina woodi connecta* Jenkins
   22a–c Holotype. Sample F 14851, N 2/554, *G. woodi connecta* Zone, Oligocene–Early Miocene, Parengarenga section, N Auckland, New Zealand.

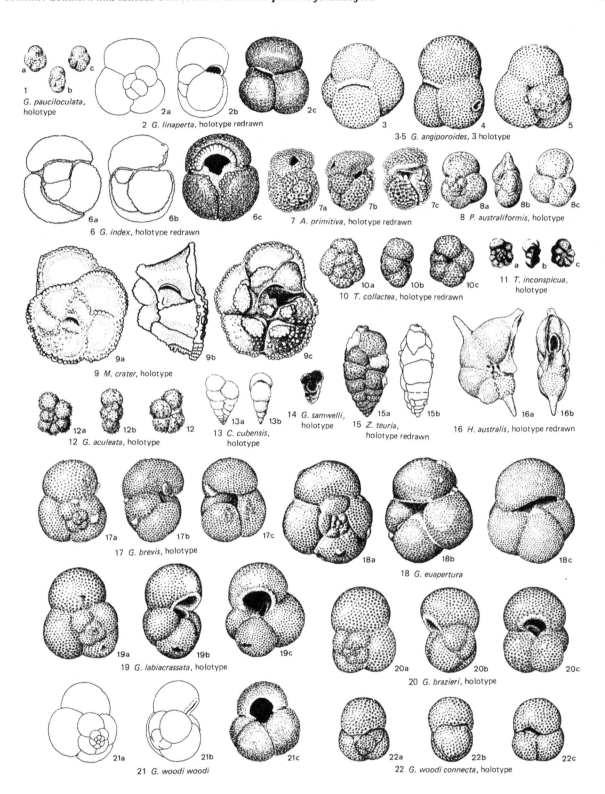

1 *G. pauciloculata,* holotype

2 *G. linaperta,* holotype redrawn

3-5 *G. angiporoides,* 3 holotype

6 *G. index,* holotype redrawn

7 *A. primitiva,* holotype redrawn

8 *P. australiformis,* holotype

9 *M. crater,* holotype

10 *T. collactea,* holotype redrawn

11 *T. inconspicua,* holotype

12 *G. aculeata,* holotype

13 *C. cubensis,* holotype

14 *G. samwelli,* holotype

15 *Z. teuria,* holotype redrawn

16 *H. australis,* holotype redrawn

17 *G. brevis,* holotype

18 *G. euapertura*

19 *G. labiacrassata,* holotype

20 *G. brazieri,* holotype

21 *G. woodi woodi*

22 *G. woodi connecta,* holotype

Zealand–Austral region. The numerous characteristic low latitude forms which are also present in the southern mid-latitudes and as such appear on the range chart Fig. 5, are discussed and illustrated in Chapters 5 and 6.

The only exception is *Globigerina juvenilis*, a species originally described from Trinidad. Because it is of stratigraphic significance in southern mid-latitudes but less used in low latitude biostratigraphy, it is discussed and figured in the present chapter.

ACARININA PRIMITIVA (Finlay)
Figures **6.7**; 5

Type reference: *Globoquadrina primitiva* Finlay, 1947a, p. 291, pl. 8, figs. 129–134.

The small subquadrate test of *A. primitiva* distinguishes it from other *Acarinina* species. Because of this test shape it was placed in *Pseudogloboquadrina* by Jenkins (1966a). In New Zealand *A. primitiva* has a long range from the Late Paleocene to the Middle Eocene. A similar range was recorded at DSDP Leg 29 Site 277 south of New Zealand (Jenkins, 1975), and ranges of Middle Eocene at DSDP Leg 40 Site 360 and Late Paleocene to Early Eocene at Site 362 (Toumarkine, 1978).

CHILOGUEMBELINA CUBENSIS (Palmer)
Figures **6.13**; 5

Type reference: *Guembelina cubensis* Palmer, 1934, p. 73, text-figs. 1–6 (no holotype designated).

C. cubensis ranges through the Middle Eocene to Late Oligocene; within its range in New Zealand it is the only *Chiloguembelina*. It has been recorded in South Australia and Victoria in the Late Eocene–Oligocene (Carter, 1958; Ludbrook & Lindsay, 1969), and is also present at DSDP Leg 29 Site 277 (Jenkins, 1975) and Leg 40 Site 362A (Toumarkine, 1978). The extinction level of *C. cubensis* appears to be a widespread marker in the Late Oligocene.

GLOBIGERINA ANGIPOROIDES Hornibrook
Figures **6.3–5**; 5

Type reference: *Globigerina angiporoides* Hornibrook, 1965b, p. 835, figs. 1–2.

G. angiporoides differs from *G. linaperta* in having spherical chambers, with the final one resembling a bulla with a very low arched aperture. *G. minima* Jenkins, which is regarded as the ancestral form of *G. angiporoides*, is smaller in size and has a smoother wall surface.

G. angiporoides ranges in New Zealand from the Late Eocene to Early Oligocene and is common; its ancestor *G. minima* Jenkins which is smaller and has a smoother wall, ranges from Middle to Late Eocene. *G. angiporoides* was recorded at DSDP Leg 29 Sites 277, 278, 281 and 282 (Jenkins,

1975) and by Toumarkine (1978) from Leg 40 Site 360, all from the Late Eocene to Early Oligocene, but only in the Early Oligocene at Site 362A of Leg 40. In South Australia, *G. angiporoides* appears to have a similar stratigraphic range (Lindsay, 1969).

GLOBIGERINA BRAZIERI Jenkins
Figures **6.20**; 5

Type reference: *Globigerina brazieri* Jenkins, 1966a, p. 1098, fig. 6, nos. 43–51.

G. brazieri is distinguished from its contemporary *G. woodi* in possessing a higher and more rounded aperture with a smoother rim. In New Zealand (Jenkins, 1971) it has a relatively short range in the Late Oligocene.

GLOBIGERINA BREVIS Jenkins
Figures **6.17**; 5

Type reference: *Globigerina brevis* Jenkins, 1966a, p. 1100, fig. 7, nos. 58–63.

G. brevis most closely resembles *G. tapuriensis* Blow & Banner from the Early Oligocene of East Africa (*in* Eames *et al.*, 1962), but differs from it in having a smaller test (0.44 mm), a more coarsely ornamented and non-hispid wall and more deeply incised umbilical sutures. Within its range *G. brevis* can be accompanied by *G. euapertura*, which has a low arched aperture and by *G. ampliapertura* with its arched oblong-shaped aperture. From both it is easily distinguished in test and apertural characteristics.

GLOBIGERINA EUAPERTURA Jenkins
Figures **6.18**; 5

Type reference: *Globigerina euapertura* Jenkins, 1960, p. 351, pl. 1, figs. 8a–c.

G. euapertura is limited to the Oligocene; its low arched aperture and tightly coiled test distinguish it from *G. ampliapertura* which has a higher arched and more oblong-shaped aperture. Within its Early Oligocene stratigraphic range *G. euapertura* differs from *G. brevis*, which has a more compact test with deep sutures. In the upper part of its range in the Late Oligocene it is distinguished from *G. woodi*, *G. brazieri* and *G. labiacrassata* in possessing a much lower aperture.

GLOBIGERINA JUVENILIS Bolli
Figures **7.1**; 5

Type reference: *Globigerina juvenilis* Bolli, 1957, p. 110, pl. 24, figs. 5a–c.

G. juvenilis is a small, low-spired, smooth-walled species with a lipped, umbilical aperture. Forms with an umbilical bulla have been referred to *Globigerina incrusta* and *G. glutinata* (Jenkins, 1971, 1978c). In the present author's opinion *G.*

juvenilis evolved from *Globorotalia munda* Jenkins in the late Early Oligocene of New Zealand (Jenkins, 1966a). This event was later observed also at DSDP Leg 29 Sites 277 and 282 as well as at Leg 40 Site 360 (Jenkins, 1975, 1978c). The first evolutionary appearance of *G. juvenilis* in these areas is therefore considered to be a good stratigraphic marker.

GLOBIGERINA LABIACRASSATA Jenkins
Figures **6.19**; 5

Type reference: *Globigerina labiacrassata* Jenkins, 1966a, p. 1102, fig. 8, nos. 64–71.

G. labiacrassata, with its distinctive very thick lipped aperture, is restricted in New Zealand to the late Early to early Late Oligocene and to the Late Oligocene at both DSDP Leg 29 Sites 277 and 282 (Jenkins, 1971, 1975) and Leg 40, Site 360 (Toumarkine, 1978).

GLOBIGERINA LINAPERTA Finlay
Figures **6.2**; 5

Type reference: *Globigerina linaperta* Finlay, 1939, p. 125, pl. 23, figs. 54–57.

G. linaperta has distinctive compressed chambers and a lipped aperture which distinguishes it from other species in the Middle and Late Eocene. In New Zealand *G. linaperta* ranges from the Middle Eocene *Globigerinatheka index* Zone to the *Globigerina linaperta* Zone. It has a similar range at DSDP Leg 29 Sites 277 and 282 (Jenkins, 1971, 1975), and in South Australia (Lindsay, 1969).

GLOBIGERINA PAUCILOCULATA Jenkins
Figures **6.1**; 5

Type reference: *Globigerina pauciloculata* Jenkins, 1966a, p. 1106, fig. 3, nos. 7–9.

Distinctive features of *G. pauciloculata* include its relatively small size (diameter 0.17 mm) and simple globigerine morphology reminiscent of *G. eugubina* to which it may be related. The wide range of variation in test morphology is illustrated in Jenkins (1966a). This small, distinctive species has only been recorded in the Early Paleocene of the Waipara section in New Zealand where it initially occurs below the first appearance of *Globigerina daubjergensis*.

GLOBIGERINA QUINQUELOBA Natland
Figures **7.2**; 5

Type reference: *Globigerina quinqueloba* Natland, 1938, p. 149, pl. 6, fig. 7.

G. quinqueloba has a distinctive hispid final chamber and a low aperture. It occurs today from the arctic and antarctic faunal provinces to the subtropical provinces (Bé, 1977) and first appeared in the Late Oligocene of the Southern Hemisphere.

This initial appearance has been recorded in New Zealand, DSDP Leg 29 Site 281 on the South Tasman Rise and Leg 40 Site 360 in the Southeastern Atlantic (Jenkins, 1966a, 1975, 1978c).

GLOBIGERINA WOODI CONNECTA Jenkins
Figures **6.22**; 5

Type reference: *Globigerina woodi connecta* Jenkins, 1964, p. 72, text-figs. 1a–c.

G. woodi connecta is distinguished from its immediate ancestor *G. woodi woodi* in possessing a lower aperture and a smaller test and from its descendant *Globigerinoides trilobus* in not having an aperture on the spiral side. The close relationship of the three taxa is confirmed by their similar wall structure (Jenkins, 1978c). This robust zonal species has a short range in the Late Oligocene–Early Miocene of southeastern Australia, New Zealand, DSDP Leg 29 Site 279, and Leg 40 Sites 360 and 362 (Jenkins, 1960, 1971, 1975, 1978c).

GLOBIGERINA WOODI WOODI Jenkins
Figures **6.21**; 5

Type reference: *Globigerina woodi* Jenkins, 1960, p. 352, pl. 2, figs. 2a–c.

G. woodi woodi is distinguished from *G. bulloides* by its coarser wall and higher arched aperture; from *G. woodi decoraperta* in not having a high spired test and from *G. apertura* in not having such a large aperture. In the Early Miocene, low arched apertural forms appear in the populations of *G. woodi* which increase in number with time and have been named *G. woodi connecta* (Jenkins, 1964). It seems that *G. woodi woodi*, which appeared cryptogenically in the Late Oligocene, was ancestral to the above-named taxa (except *G. bulloides*) as well as being ancestral to the Early Miocene *Globigerinoides trilobus* and *G. altiaperturus*, which are all linked in having a similar wall ornamentation. *G. woodi woodi* first appeared in the Late Oligocene and ranged to the Middle Miocene in both New Zealand and southeast Australia. On DSDP Leg 29 it was recorded at Sites 278 and 279 and on Leg 40 at Sites 360 and 362 (Jenkins, 1975, 1978c).

GLOBIGERINATHEKA INDEX (Finlay)
Figures **6.6**; 5

Type reference: *Globigerinoides index* Finlay, 1939, p. 125, pl. 14, figs. 85–88.

G. index has a subspherical test with fairly large, high arched apertures at the base of the final chamber; the holotype figured here has only one. Within its Middle to Late Eocene range, *G. index* is a fairly common species in New Zealand. It has also been recorded in the Middle–Late Eocene of South Australia (Ludbrook & Lindsay, 1969), south of New Zealand at DSDP Leg 29 Site 277 (Jenkins, 1975) and at the Leg 40 Sites 360 and 362A (Toumarkine, 1978).

GLOBOROTALIA ACULEATA Jenkins
Figures **6.12**; 5

Type reference: *Globorotalia inconspicua aculeata* Jenkins, 1966a, p. 1118, fig. 1, nos. 119–125.

G. aculeata is very small (0.2 mm) with a delicate test. It appears to be related to *T. inconspicua* Howe in New Zealand, from which it is distinguished in not possessing a keeled periphery. It ranges from the Middle to Late Eocene in New Zealand and has been recorded at similar levels south of New Zealand at DSDP Site 277 (Jenkins, 1971, 1975). In South Australia it appears to have become extinct later in the Late Eocene (Lindsay, 1969).

GLOBOROTALIA CONICA Jenkins
Figures **7.8–9**; 5

Type reference: *Globorotalia conica* Jenkins, 1960, p. 358, pl. 4, figs. 3a–c.

G. conica is distinguished from *G. mayeri* in possessing only 4–5 chambers in the final whorl, a distinctly higher umbilical side, strongly curved spiral sutures and a low arched aperture. It is distinguished from its Early Miocene homeomorph *G. zealandica* in having a smoother wall, more strongly recurved sutures and a less rounded periphery (Jenkins, 1971). Records of *G. conica* are confined to New Zealand, Australia and DSDP Leg 29 Sites 278, 279 and 281 in the South Tasman Sea (Jenkins, 1971, 1975). In the Leg 29 area it is much more common than elsewhere and could be a colder water species, with an extended stratigraphic range at Site 281, where it first occurs at the top of the *Globigerinoides trilobus* Zone.

GLOBOROTALIA CONOIDEA Walters
Figures **7.14–15**; 5

Type reference: *Globorotalia miozea conoidea* Walters, 1965, p. 124, fig. 8, 1–M.

G. conoidea is distinguished from *G. miozea* in having a more conical-shaped test with a sharper keel developed in the later chambers of the test. *G. conoidea* is thick walled like its ancestor *G. miozea* and ranges in New Zealand from Middle Miocene to Early Pliocene (Jenkins, 1971). It is present at DSDP Leg 29 Sites 279, 281 and 284 where it is restricted to the Middle–Late Miocene (Jenkins, 1960, 1975). At DSDP Leg 40 Site 360 it ranges into the Early Pliocene but is restricted to the Miocene at the more northern Site 362 (Jenkins, 1978c). *G. conoidea* is a thick-walled form of *Globorotalia miotumida* Jenkins (see Kennett and Srinivason, 1983).

GLOBOROTALIA CONOMIOZEA Kennett
Figures **7.18**; 5

Type reference: *Globorotalia conomiozea* Kennett, 1966, p. 235, text-figs. 10a–c.

G. conomiozea has an asymmetrical test with a flat spiral and a high umbilical side which gives it a conical shape. It is distinguished from *G. sphericomiozea*, which is smaller and has a thicker-walled test and from *G. conoidea*, which also has a much thicker-walled test (Jenkins, 1971). The evolutionary relationship of *G. conomiozea* to *G. puncticulata* at DSDP Leg 29 Site 284 was discussed by Malmgren & Kennett (1981). *G.*

Fig. 7. Southern mid-latitude planktic foraminifera (all figures × 60)

1 *Globigerina juvenilis* Bolli
1a–c Holotype. Sample JS 16, *G. fohsi robusta* Zone, Middle Miocene, Cipero Fm., Trinidad.

2 *Globigerina quinqueloba* Natland
2a–c Holotype. Recent, off Long Beach, California, USA.

3 *Globorotaloides testarugosa* (Jenkins)
3a–c Holotype. Sample at 1192 feet of Lakes Entrance shaft, *G. euapertura* Zone, Late Oligocene, NE of Lakes Entrance, Victoria, Australia.

4 *Globorotalia gemma* Jenkins
4a–c Holotype. Sample F 9398, S136/694, *G. brevis* Zone, Early Oligocene, Kakanui River section, New Zealand.

5 *Globorotalia munda* Jenkins
5a–c Holotype. Sample F 11444, S127/422, *G. angiporoides* Zone, Early Oligocene, Campbells Beach, All Day Bay, Kakanui, New Zealand.

6 *Globorotalia zealandica* Hornibrook
6a–c Holotype. Sample F 5125, S 137/499, *g. trilobus* Zone, Early Miocene, Pukeuri road cutting, Oamaru, New Zealand.

7 *Globorotalia semivera* (Hornibrook)
7a–c Holotype. Early Miocene, Campbells Beach, All Day Bay, Otepopo SD, New Zealand.

8–9 *Globorotalia conica* Jenkins
8a–c Holotype. Sample at 576 feet of Lakes Entrance shaft, Middle Miocene, NE of Lakes Entrance, Victoria, Australia. **9a–c** Sample F 5557, Middle Miocene, Cheviot Survey District, New Zealand (from Jenkins, 1971).

10–11 *Globorotalia sphericomiozea* Walters
10a–c Holotype. Sample GRW 100, Late Miocene, Waimata River section, New Zealand. **11a–c** Sample N 165/553, *G. conomiozea* Zone, Late Miocene, Palliser Bay section, New Zealand (from Jenkins, 1971).

12–13 *Globorotalia miotumida* Jenkins
12a–c Holotype. Sample at 380 feet of Lakes Entrance shaft, *G. miotumida* Zone, Late Miocene, NE of Lakes Entrance, Victoria, Australia (from Jenkins, 1971). **13a–c** Sample S 62/24, *G. miotumida* Zone, Late Miocene, Gore Bay section, New Zealand (from Jenkins, 1971).

14–15 *Globorotalia conoidea* Walters
14a–c Holotype. Middle–Late Miocene, Nine Fords Stream, N of Gisborne, New Zealand. **15a–c** Sample N80/430, Miocene, Muddy Creek, New Zealand (from Jenkins, 1971).

16 *Globorotalia explicationis* Jenkins
16a–c Holotype. Sample F 11070, N 100/504, *G. miotumida* Zone, Late Miocene, Tongaporutu River section, New Zealand.

17 *Globorotalia miozea* Finlay
17a–c Holotype, redrawn (in Hornibrook, 1958). Sample F 11070, N 100/504, Early–Middle Miocene, Tongaporutu River section, New Zealand.

18 *Globorotalia conomiozea* Kennett
18a–c Holotype. Sample S 41 F 648, *G. conomiozea* Zone, Late Miocene, Kapitea Creek section, New Zealand.

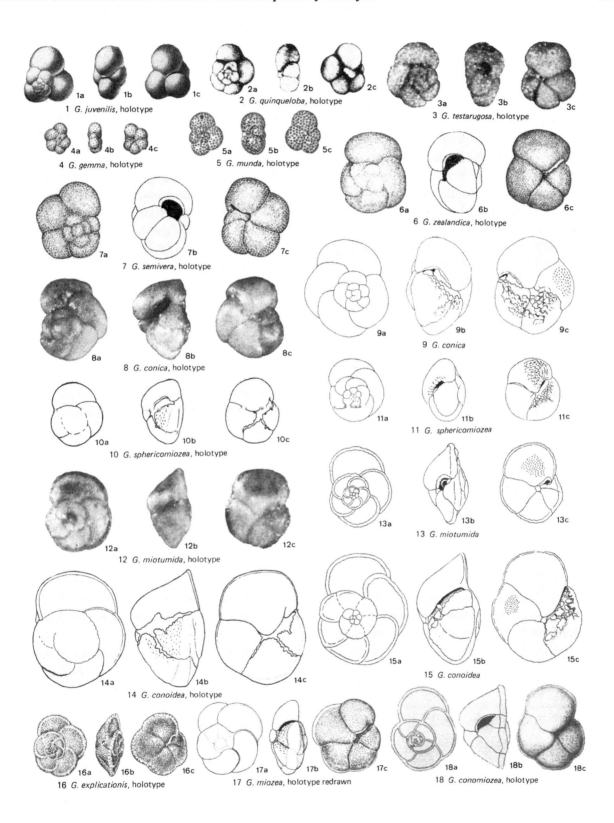

1 *G. juvenilis*, holotype

2 *G. quinqueloba*, holotype

3 *G. testarugosa*, holotype

4 *G. gemma*, holotype

5 *G. munda*, holotype

6 *G. zealandica*, holotype

7 *G. semivera*, holotype

8 *G. conica*, holotype

9 *G. conica*

10 *G. sphericomiozea*, holotype

11 *G. sphericomiozea*

12 *G. miotumida*, holotype

13 *G. miotumida*

14 *G. conoidea*, holotype

15 *G. conoidea*

16 *G. explicationis*, holotype

17 *G. miozea*, holotype redrawn

18 *G. conomiozea*, holotype

conomiozea has a short stratigraphic range in the Late Miocene in New Zealand but it appears to range into the Early Pliocene at DSDP Leg 29 Site 284, while at Leg 40 Sites 360 and 362 it is restricted to the Late Miocene (Jenkins, 1975, 1978c).

GLOBOROTALIA GEMMA Jenkins
Figures 7.4; 5

Type reference: *Globorotalia gemma* Jenkins, 1966a, p. 1115, fig. 11, nos. 97–103.

The small distinctive low trochospiral test (diameter 0.16 mm) of *G. gemma* can be easily overlooked if the fine fractions of Late Eocene–Early Oligocene *Globigerina brevis* Zone samples are not examined. With $4\frac{1}{2}$–5 chambers in the final whorl and a low arched aperture, it cannot be confused with any other species within its stratigraphic range except possibly for *Globanomalina barbadoensis* (Blow) which is larger, has 8 chambers in the final whorl and is planispirally coiled. In the Austral region there is no record of the latter species but they occur together in the Southeastern Atlantic at DSDP Leg 40 Site 362 (Toumarkine, 1978). *G. gemma* has a very short range in the Late Eocene–Early Oligocene of New Zealand and was used at DSDP Leg 29 Sites 277 and 282 to identify the *Globigerina brevis* Zone in absence of the zonal marker (Jenkins, 1975).

GLOBOROTALIA EXPLICATIONIS Jenkins
Figures 7.16; 5

Type reference: *Globorotalia miotumida explicationis* Jenkins, 1967, p. 1073, fig. 4, nos. 14–19.

G. explicationis is closely related to *G. miotumida* but is distinguished from it in possessing a final chamber which appears to be uncoiling as seen in the side-view of the holotype. *G. explicationis* evolved from *G. miotumida* in the Middle Miocene and ranges into the Late Miocene (Jenkins, 1971). In New Zealand it is fairly rare but at DSDP Leg 29 Site 284 it is quite common, and specimens referred to *G. explicationis* were found at DSDP Leg 40 Sites 360 and 362 (Jenkins, 1975, 1978c).

GLOBOROTALIA MIOTUMIDA Jenkins
Figures 7.12–13; 5

Type reference: *Globorotalia menardii miotumida* Jenkins, 1960, p. 362, pl. 4, figs. 9a–c.

G. miotumida has $4\frac{1}{2}$–5 chambers in the final whorl and is thin-walled with a keeled periphery. In New Zealand the species appears to be restricted to the Middle and Late Miocene as in southeastern Australia, but at DSDP Leg 29 Site 284 in the Tasman Sea it ranges into the Early Pliocene. On Leg 40 Site 360, *G. miotumida* has the same range as in New Zealand but at the more northern Site 362 ranges again into the Early Pliocene (Jenkins, 1975, 1978c).

GLOBOROTALIA MIOZEA Finlay
Figures 7.17; 5

Type reference: *Globorotalia miozea* Finlay, 1939d, p. 326, pl. 29, figs. 159–161.

The distinguishing features of *G. miozea* are a thick-walled and relatively narrow biconvex test, 5 chambers in the final whorl with strongly curved sutures on the spiral side and a low arched aperture. I believe *G. miozea* to have evolved from *G. praescitula* in the upper part of the Early Miocene by acquisition of a keel and a larger test; in New Zealand it ranges into the Middle Miocene (Jenkins, 1971). In a shallow water continental shelf deposit of the same age of southeastern Australia, *G. miozea* is not present and has been considered to be a deeper-water thick-walled ecophenotype of *G. miotumida* (Jenkins, 1975). *G. miozea* is common at DSDP Leg 29 Site 279 and also present at Site 281, while on Leg 40 Sites 360 and 362 it ranged into the Early Pliocene (Jenkins, 1975, 1978c).

GLOBOROTALIA MUNDA Jenkins
Figures 7.5; 5

Type reference: *Globorotalia munda* Jenkins, 1966a, p. 1121, fig. 14, nos. 126–133, fig. 15, nos. 152–166.

G. munda is a small species (holotype 0.22 mm) with 4 chambers in the final whorl and a low lipped, slit aperture. It is distinguished from *G. nana* which has a larger test (holotype 0.30 mm), a thicker lipped aperture and a non-hispid test surface. There is some evidence that *G. munda* may have evolved from *G. gemma*, and itself gave rise to the *Globigerina juvenilis* populations in the Late Oligocene *Globigerina euapertura* Zone, which eventually evolved into *G. glutinata* and possibly *G. bradyi*. *G. munda* has a short range and has been recorded in the *Globigerina angiporoides* to lower *Globigerina euapertura* zones of New Zealand and at DSDP Leg 29 Site 277 (Jenkins, 1971, 1975).

GLOBOROTALIA SEMIVERA (Hornibrook)
Figures 7.7; 5

Type reference: *Globigerina semivera* Hornibrook, 1961, p. 149, pl. 23, figs. 455–457.

G. semivera with 5 chambers in the final whorl and a fairly large aperture probably evolved from the 4-chambered *G. pseudocontinuosa* in the Late Oligocene *Globigerina euapertura* Zone of New Zealand, and became extinct in the Middle Miocene (Jenkins, 1971). On sites drilled on Leg 29 south of New Zealand, *G. semivera* has a range of Early Miocene in Site 279, Late Oligocene–Early Miocene in Site 282 (Jenkins, 1975) and in the Southeastern Atlantic Leg 40 Site 360 it ranges from Late Oligocene to Early Miocene (Jenkins, 1975). The relationship of *G. semivera* to the warmer water *G. mayeri* of Bolli & Saunders (1982) is not understood, but *G. semivera* has also been identified in the Early Miocene of France and the English Channel (Jenkins, 1966c, 1977).

GLOBOROTALIA SPHERICOMIOZEA Walters
Figures **7.10–11**; 5

Type reference: *Globorotalia miozea sphericomiozea* Walters, 1965, p. 104, fig. 3, 1A–6C.

G. sphericomiozea has a small thick-walled test with 4 chambers in the final whorl and a low arched aperture and is regarded as the immediate ancestor of *G. puncticulata* in the New Zealand region. This evolutionary lineage which led to *G. inflata* has recently been discussed by Scott (1980) and Malmgren & Kennett (1981). In New Zealand it ranges from the Late Miocene into basal Pliocene, and on DSDP Leg 29 Site 284 it is very rare and was not identified over the same interval at Leg 40 Sites 360 and 362 (Jenkins, 1975, 1978c).

GLOBOROTALIA ZEALANDICA Hornibrook
Figures **7.6**; 5

Type reference: *Globorotalia zealandica* Hornibrook, 1958b, p. 667, figs. 18, 19, 30.

The only species with which *G. zealandica* can be confused in the Early Miocene of New Zealand is *G. pseudocontinuosa*, whose holotype is smaller (0.23 mm as opposed to 0.43 mm). Both have 4 chambers in the final whorl but *G. zealandica* has a squarer peripheral outline with recurved sutures on the spiral side. *G. zealandica* is distinguished from its immediate ancestor *G. incognita* which is smaller (0.36 mm), has a more compact test and is thicker walled. *G. zealandica* has been recorded in the Early Miocene *Globigerinoides trilobus* Zone in New Zealand, southeastern Australia (Jenkins, 1960), DSDP Leg 29 (Jenkins, 1975), the English Channel (Jenkins, 1977) and by Poore (1979) in the North Atlantic DSDP Leg 49 Site 408 in zones N 7 and N 8.

GLOBOROTALOIDES TESTARUGOSA (Jenkins)
Figures **7.3**; 5

Type reference: *Globorotalia testarugosa* Jenkins, 1960, p. 368, pl. 5, figs. 8a–c.

During its ontogeny the aperture of *G. testarugosa* is umbilical–extraumbilical, then becomes umbilical and is finally hidden by a bulla in the larger specimens; the holotype illustrates the first stage of growth; all stages are illustrated in Jenkins (1971). *G. testarugosa* with its coarsely pitted wall is a distinctive species in the Late Oligocene of southeastern Australia and New Zealand (Jenkins, 1960, 1966a). It also has a similar range in DSDP Leg 29 Site 277 (Jenkins, 1975) and Leg 40 Site 360 (Toumarkine, 1978) and in South Australia (Lindsay, 1969).

HANTKENINA AUSTRALIS Finlay
Figures **6.16**; 5

Type reference: *Hantkenina australis* Finlay, 1939d, p. 538.

H. australis is distinguished from other species of the genus by recurved spines. It is found sporadically in the Middle–Late Eocene *Globigerinatheka index–Globorotalia inconspicua* zones in New Zealand. The only other good record of *H. australis* is in the early Middle Eocene of DSDP Leg 40 Site 362A in the South Atlantic (Toumarkine, 1978).

JENKINSINA SAMWELLI (Jenkins)
Figures **6.14**; 5

Type reference: *Guembelitria samwelli* Jenkins, 1978b, p. 132, figs. 1–3.

The small triserial test of *J. samwelli* easily distinguishes it from other *Jenkinsina* species because it possesses supplementary apertures (Jenkins, 1978b). In South Australia *J. samwelli* (as *G. stavensis*) ranges from the Early Oligocene to the early Late Oligocene (Lindsay, 1969). In New Zealand and at DSDP Leg 29 Sites 279 and 282 and the Southeastern Atlantic Leg 40 Site 360 it has a very short range at the base of the Late Oligocene (Jenkins, 1974, 1978c).

MOROZOVELLA CRATER (Finlay)
Figures **6.9**; 5

Type reference: *Globorotalia crater* Finlay, 1939d, p. 125.

M. crater differs from *M. caucasica* in having 5 chambers in the final whorl as opposed to $6\frac{1}{2}$ in the holotype. The latter form is rare in New Zealand. *M. crater* has a short range in the Early Eocene of New Zealand (Jenkins, 1971) and at DSDP Leg 29 Site 277 (Jenkins, 1975). Toumarkine (1978) recorded *Globorotalia aragonensis caucasica* from the Early Eocene of Leg 40 Sites 361 and 363 in the Southeast Atlantic, but did not record *M. crater*.

PLANOROTALITES AUSTRALIFORMIS (Jenkins)
Figures **6.8**; 5

Type reference: *Globorotalia australiformis* Jenkins, 1966a, p. 112, fig. 11, nos. 92–96.

The small test (holotype 0.26 mm) with 4 chambers in the final whorl distinguishes *P. australiformis* from other species of the genus. It has been recorded in New Zealand and on DSDP Leg 29 from the Paleocene *Globigerina triloculinoides* Zone into the Middle Eocene *Acarinina primitiva* Zone (Jenkins, 1966a, 1975). It also occurs in the equivalent of the *Acarinina primitiva* Zone in South Australia (Ludbrook & Lindsay, 1969).

TESTACARINATA INCONSPICUA (Howe)
Figures **6.11**; 5

Type reference: *Globorotalia inconspicua* Howe, 1939, p. 85, pl. 12, figs. 20–22.

This minute low trochospiral Eocene species is characterized by the large number of chambers (7–8) forming the last whorl. They increase only moderately in size and are of an angular-conical shape. The carinate periphery may occasionally be projected into short spine-like extensions.

TRUNCOROTALOIDES COLLACTEA (Finlay)
Figures **6.10**; 5

Type reference: *Globorotalia collactea* Finlay, 1939d, p. 327, pl. 29, figs. 164–165.

This small distinctive species occasionally has small sutural apertures on the spiral side at the base of the final chamber. *T. collactea* differs from *T. rohri* in being smaller, having a less spinose wall and a more compact test. It has been recorded from the Early to Late Eocene of New Zealand with a similar range at DSDP Leg 29 Site 277 (Jenkins, 1971, 1975). In South Australia *T. collactea* has only a short range in the Late Eocene (Lindsay, 1969).

ZEAUVIGERINA TEURIA Finlay
Figures **6.15**; 5

Type reference: *Zeauvigerina teuria* Finlay, 1947, pl. 4, figs. 49–54.

Z. teuria is a distinctive species, distinguished from Paleocene *Chiloguembelina* in possessing an aperture at the terminal end of a neck in the final chamber. It seems that the Paleocene *Z. aegyptica* recorded from the Esna Shale of Sinai (Said & Kenawy, 1956), Trinidad (Beckmann, 1957) and from the King's Park Shale, Perth, Australia, is a junior synonym of *Z. teuria*.

References

Bé, A. W. H. 1967. Foraminifera. Families Globigerinidae and Globorotaliidae; Fiches d'identification du zooplankton. *Conseil permenent international pour l'exploration de la mer. Zooplankton Sheet*, **108**, 1–8.

Bé, A. W. H. 1977. An ecological, zoogeographic and taxonomic review of Recent planktonic foraminifera. In: A. T. S. Ramsey (ed.) *Oceanic Micropaleontology*, vol. 1, pp. 1–100, Academic Press, London.

Beckmann, J. P. 1957. *Chiloguembelina* Loeblich and Tappan and related foraminifera from the Lower Tertiary of Trinidad, B.W.I. *Bull. U.S. natl. Mus.*, **215**, 83–95.

Berggren, W. A. 1977. Late Neogene planktonic foraminiferal biostratigraphy of Site 357 (Rio Grande Rise). *Initial Rep. Deep Sea drill. Proj.*, **39**, 591–614.

Bertels, A. 1970. Los foraminiferos planctonicos de la cuenca Cretacico–Tertiaria en Patagonia septentrional (Argentina),
con consideraciones sobre la estratigrafia de Fortin General Roca (Provincia de Rio Negro). *Ameghiniana*, **7**, 1–54.

Boersma, A. 1977. Cenozoic planktonic foraminifera – DSDP Leg 39 (South Atlantic). *Initial Rep. Deep Sea drill. Proj.*, **39**, 567–90.

Bolli, H. M. 1957. Planktonic foraminifera from the Oligocene–Miocene Cipero and Lengua formations of Trinidad, B.W.I. *Bull. U.S. natl. Mus.*, **215**, 97–123.

Bolli, H. W. & Saunders, J. B. 1982. *Globorotalia mayeri* and its relationship to *Globorotalia siakensis* and *Globorotalia continuosa*. *J. foramin. Res.*, **12**, 39–50.

Boltovskoy, E. 1974. Neogene planktonic foraminifera of the Indian Ocean (DSDP, Leg 26). *Initial Rep. Deep Sea drill. Proj.*, **26**, 675–741.

Carter, A. N. 1958. Tertiary foraminifera from the Aire District, Victoria. *Bull. geol. Surv. Victoria*, **55**, 1–76.

Carter, A. N. 1964. Tertiary foraminifera from Gippsland, Victoria, and their stratigraphical significance. *Mem. geol. Surv. Victoria*, **23**, 1–154.

Devereux, I. 1967. Oxygen isotope paleotemperature measurements of New Zealand Tertiary fossils. *N.Z. J. Sci. Technol.*, **10**, 988–1011.

Eames, F. E., Banner, F. T., Blow, W. H. & Clarke, W. J. 1962. *Fundamentals of Mid-Tertiary Stratigraphical Correlation*. Cambridge University Press, Cambridge, 163 pp.

Finlay, H. J. 1939a. New Zealand foraminifera: Key species in stratigraphy – No. 1. *Trans. R. Soc. N.Z.*, **68**, 504–33.

Finlay, H. J. 1939b. New Zealand foraminifera: The occurrence of *Rzehakina, Hantkenina, Rotaliatina*, and *Zeauvigerina*. *Trans. R. Soc. N.Z.*, **68**, 534–43.

Finlay, H. J. 1939c. New Zealand foraminifera: Key species in stratigraphy – No. 2. *Trans. R. Soc. N.Z.*, **69**, 89–128.

Finlay, H. J. 1939d. New Zealand foraminifera: Key species in stratigraphy – No. 3. *Trans. R. Soc. N.Z.*, **69**, 309–29.

Finlay, H. J. 1940. New Zealand foraminifera: Key species in stratigraphy – No. 4. *Trans. R. Soc. N.Z.*, **69**, 448–72.

Finlay, H. J. 1947a. New Zealand foraminifera: Key species in stratigraphy – No. 5. *N.Z. J. Sci. Technol.*, section B, **28**, 259–92.

Finlay, H. J. 1947b. The foraminiferal evidence for Tertiary trans-Tasman correlation. *Trans. R. Soc. N.Z.*, **76**, 327–52.

Geiger, M. E. 1962. Planktonic foraminiferal zones in the Upper Tertiary of Taranaki, New Zealand. *N.Z. J. Geol. Geophys.*, **5**, 304–8.

Hornibrook, N. de B. 1958a. New Zealand Upper Cretaceous and Tertiary foraminiferal zones and some overseas correlations. *Micropaleontology*, **4**, 25–38.

Hornibrook, N. de B. 1958b. New Zealand foraminifera: Key species in stratigraphy – No. 6. *N.Z. J. Geol. Geophys.*, **4**, 653–76.

Hornibrook, N. de B. 1961. Tertiary foraminifera from Oamaru District (N.Z.) Part 1 – Systematics and distribution. *Bull. N.Z. geol. Surv.*, **34**, 1–192.

Hornibrook, N. de B. 1962. The Cretaceous–Tertiary boundary in New Zealand. *N.Z. J. Geol. Geophys.*, **5**, 295–303.

Hornibrook, N. de B. 1964. A record of *Globigerinatella insueta* Cushman and Stainforth from New Zealand. *N.Z. J. Geol. Geophys.*, **7**, 891–2.

Hornibrook, N. de B. 1965a. A preliminary statement on the types of the New Zealand Tertiary foraminifera described in the reports of the Novara Expedition in 1865. *N.Z. J. Geol. Geophys.*, **8**, 530–6.

Hornibrook, N. de B. 1965b. *Globigerina angiporoides* n. sp. from the Upper Eocene and Lower Oligocene of New Zealand and the status of *Globigerina angipora* Stache 1865. *N.Z. J. Geol. Geophys.*, **8**, 834–83.

Hornibrook, N. de B. 1966. The *Orbulina* bioseries in the Clifden section, New Zealand. *Proceedings International Union of Geological Sciences. Commission on Mediterranean Stratigraphy*, 3rd Session, 21–2.

Hornibrook, N. de B. 1977. The Neogene (Miocene–Pliocene) of New Zealand. In: T. Saito & H. Ujiie (ed.), *Proceedings First International Congress on the Pacific Neogene Stratigraphy, Science Council of Japan and Geological Society of Japan*, pp. 145–50.

Hornibrook, N. de B. 1981. *Globorotalia* (planktonic Foraminiferida) in the Late Pliocene and Early Pleistocene of New Zealand. *N.Z. J. Geol. Geophys.*, **24**, 263–92.

Howe, H. V. 1939. Louisiana Cook Mountain Eocene foraminifera. *Louisiana Geological Survey, Geological Bulletin*, **14**, 1–122.

Jenkins, D. G. 1958. Pelagic foraminifera in the Tertiary of Victoria. *Geol. Mag.*, **95**, 438–9.

Jenkins, D. G. 1960. Planktonic foraminifera from the Lakes Entrance Oil shaft, Victoria, Australia. *Micropaleontology*, **6**, 345–71.

Jenkins, D. G. 1963. New Zealand mid-Tertiary stratigraphical correlation. *Nature*, **200** (4911), 1087.

Jenkins, D. G. 1964. A new planktonic foraminiferal subspecies from the Australasian Lower Miocene. *Micropaleontology*, **10**, 72.

Jenkins, D. G. 1965. The genus *Hantkenina* in New Zealand. *N.Z. J. Geol. Geophys.*, **8**, 518–26.

Jenkins, D. G. 1966a. Planktonic foraminiferal zones and new taxa from the Danian to Lower Miocene of New Zealand. *N.Z. J. Geol. Geophys.*, **8**, 1088–1126.

Jenkins, D. G. 1966b. Planktonic foraminiferal datum planes in the Pacific and Trinidad Tertiary. *N.Z. J. Geol. Geophys.*, **9**, 424–7.

Jenkins, D. G. 1966c. Planktonic foraminifera from the type Aquitanian–Burdigalian of France. *Contrib. Cushman Found. foramin. Res.*, **17**, 1–15.

Jenkins, D. G. 1967. Planktonic foraminiferal zones and new taxa from the Lower Miocene to the Pleistocene of New Zealand. *N.Z. J. Geol. Geophys.*, **10**, 1064–78.

Jenkins, D. G. 1968. Planktonic Foraminiferida as indicators of New Zealand Tertiary paleotemperatures. *Tuatara*, **16**(1), 32–7.

Jenkins, D. G. 1971. New Zealand Cenozoic planktonic foraminifera. *Palaeontol. Bull. N.Z. geol. Surv.*, **42**, 1–278.

Jenkins, D. G. 1973. Diversity changes in the New Zealand Cenozoic planktonic foraminifera. *J. foramin. Res.*, **3**(2), 78–88.

Jenkins, D. G. 1974a. Paleogene planktonic foraminifera of New Zealand and the Austral Region. *J. foramin. Res.*, **4**, 155–70.

Jenkins, D. G. 1974b. Initiation of the proto circum-Antarctic current. *Nature*, **252** (5482), 371–3.

Jenkins, D. G. 1975. Cenozoic planktonic foraminiferal biostratigraphy of the southwestern Pacific and Tasman Sea – DSDP Leg 29. *Initial Rep. Deep Sea drill. Proj.*, **29**, 449–67.

Jenkins, D. G. 1977. Lower Miocene planktonic foraminifera from a borehole in the English Channel. *Micropaleontology*, **23**, 297–318.

Jenkins, D. G. 1978a. *Guembelitria* aff. *stavensis* Bandy, a

paleooceanographic marker of the initiation of the circum-Antarctic current and the opening of Drake Passage. *Initial Rep. Deep Sea drill. Proj.*, **40**, 687–93.

Jenkins, D. G. 1978b. *Guembelitria samwelli* Jenkins, a new species from the Oligocene of the Southern Hemisphere. *J. foramin. Res.*, **8**, 132–7.

Jenkins, D. G. 1978c. Neogene planktonic foraminifers from DSDP Leg 40 sites 360 and 362 in the southeastern Atlantic. *Initial Rep. Deep Sea drill. Proj.*, **40**, 723–39.

Jenkins, D. G. & Orr, W. 1972. Planktonic foraminiferal biostratigraphy of the Eastern Equatorial Pacific – DSDP Leg 9. *Initial Rep. Deep Sea drill. Proj.*, **9**, 1059–1193.

Kaneps, G. G. 1975. Cenozoic planktonic foraminifera from Antarctic deep-sea sediments, Leg 28, DSDP. *Initial Rep. Deep Sea drill. Proj.*, **28**, 573–83.

Kennett, J. P. 1966. The *Globorotalia crassaformis* bioseries in North Westland and Marlborough, New Zealand. *Micropaleontology*, **12**, 235–45.

Kennett, J. P. 1973. Middle and Late Cenozoic planktonic foraminiferal biostratigraphy of the Southwest Pacific – DSDP Leg 21. *Initial Rep. Deep Sea drill. Proj.*, **21**, 575–640.

Kennett, J. P. & Srinivasan, M. S. 1983. *Neogene Planktonic Foraminifera*. Hutchinson Ross Pub. Comp. 265 pp.

Kennett, J. P. & Watkins, N. D. 1974. Late Miocene–Early Pliocene paleomagnetic stratigraphy, paleoclimatology and biostratigraphy in New Zealand. *Bull. geol. Soc. Am.*, **85**, 1385–98.

Lindsay, J. M. 1967. Foraminifera and stratigraphy of the type section of Port Willunga Beds, Aldinga Bay, South Australia. *Trans. R. Soc. South Aust.*, **91**, 93–110.

Lindsay, J. M. 1969. Cainozoic foraminifera and stratigraphy of the Adelaide Plains sub-basin, South Australia. *Bull. geol. Surv. South Aust.*, **42**, 1–60.

Ludbrook, N. H. & Lindsay, J. M. 1969. Tertiary foraminiferal zones in South Australia. *Proceedings First International Conference on Planktonic Microfossils, Geneva*, **2**, 366–75.

McGowran, B. 1964. Foraminiferal evidence for the Paleocene age of the Kings Park Shale (Perth Basin, Western Australia). *J. Roy. Soc. Western Australia*, **47**, 81–6.

McGowran, B. 1965. Two Paleocene foraminiferal faunas from the Wangerip Group, Pebble Point coastal section, Western Victoria. *Proc. R. Soc. Victoria*, **79**, 9–74.

McInnes, B. A. 1965. *Globorotalia miozea* Finlay as an ancestor of *Globorotalia inflata* (d'Orbigny). *N.Z. J. Geol. Geophys.*, **8**, 104–8.

Malmgren, B. A. & Kennett, J. P. 1981. Phyletic gradualism in a Late Cenozoic planktonic foraminiferal lineage; DSDP Site 284, Southwest Pacific. *Paleobiology*, **7**, 230–40.

Malumian, N. 1970. Foraminiferos Danianos de la Formacion Pedro Luro, Provincia de Buenos Aires, Argentina. *Ameghiniana*, **7**, 355–67.

Malumian, N. & Masiuk, V. 1973. Associaciones foraminiferologicas fossiles de la Republica Argentina. *Actas Quinto Congreso Geologico, Argentina*, **3**, 433–53.

Malumian, N., Masiuk, V. & Riggi, J. C. 1971. Micropaleontologia y sedimentologia de la perforacion SC-1 Provincia Santa Cruz, Republica Argentina. *Revista de la Asociacion Geologica Argentina*, **26**, 175–208.

Mantell, G. A. 1850. Notice of the remains of the *Dinornis* and other birds and of fossil and rock specimens, recently collected by Mr Walter Mantell in the middle island of New

Zealand; with additional notes on the northern island. *Q. J. geol. Soc. London*, **6**, 319–42.

Natland, M. L. 1938. New species of foraminifera from off the west coast of North America and from the later Tertiary of the Los Angeles Basin. *University of California, Scripps Institution of Oceanography, Bulletin, Tech. ser. Bull.*, **4**(5), p. 14.

Palmer, D. K. 1934. The foraminiferal genus *Guembelina* in the Tertiary of Cuba. *Memorias de la Sociedad Cubana de Historia Natural*, **8**, 73–6.

Poore, R. Z. 1979. Oligocene through Quaternary planktonic foraminiferal biostratigraphy of the North Atlantic: DSDP Leg 49. *Initial Rep. Deep Sea drill. Proj.*, **49**, 447–517.

Said, R. & Kenawy, A. 1956. Upper Cretaceous and Lower Tertiary foraminifera from northern Sinai, Egypt. *Micropaleontology*, **2**, 105–73.

Scott, G. H. 1966. Description of an experimental class within the Globigerinidae (Foraminifera). Parts I and II. *N.Z. J. Geol. Geophys.*, **9**, 513–40.

Scott, G. H. 1968a. Comparison of the primary apertures of *Globigerinoides* from the Lower Miocene of Trinidad and New Zealand. *N.Z. J. Geol. Geophys.*, **11**, 356–75.

Scott, G. H. 1968b. Comparison of Lower Miocene *Globigerinoides* from the Caribbean and New Zealand. *N.Z. J. Geol. Geophys.*, **11**, 376–90.

Scott, G. H. 1968c. Stratigraphic variation in *Globigerinoides trilobus trilobus* (Reuss) from the Lower Miocene of Europe, Trinidad and New Zealand. *N.Z. J. Geol. Geophys.*, **11**, 391–404.

Scott, G. H. 1975. Variation in *Globorotalia miozea* (Foraminiferida) from the New Zealand Neogene. *N.Z. J. Geol. Geophys.*, **18**, 865–80.

Scott, G. H. 1979. The Late Miocene to Early Pliocene history of the *Globorotalia miozea* plexus from Blind River, New Zealand. *Mar. Micropaleontol.*, **4**, 341–61.

Scott, G. H. 1980. *Globorotalia inflata* lineage and *G. crassaformis* from Blind River, New Zealand: recognition, relationship, and use in uppermost Miocene–Lower Pliocene stratigraphy. *N.Z. J. Geol. Geophys.*, **23**, 665–77.

Shackleton, N. J. & Kennett, J. P. 1975. Paleotemperature history of the Cenozoic and the initiation of the Antarctic glaciation: oxygen and carbon isotope analysis in DSDP Sites 277, 279 and 281. *Initial Rep. Deep Sea drill. Proj.*, **29**, 743–56.

Sigal, J. 1974. Comments on Leg 25 sites in relation to the Cretaceous and Paleocene stratigraphy in the eastern and southeastern Africa coast and Madagascar regional setting. *Initial Rep. Deep Sea drill. Proj.*, **25**, 687–723.

Singleton, O. P. 1968. Otway Region. In: J. McAndrew & M. A. H. Marsden (eds.), A regional guide to Victorian geology. *Publication of Melbourne University, Department of Geology*, 117–31.

Smith, A. G. & Briden, J. C. 1977. *Mesozoic and Cenozoic Paleoenvironmental Maps*. Cambridge University Press, 63 pp.

Taylor, D. J. 1966. Upper Cretaceous and Tertiary subsurface biostratigraphic scheme for Gippsland, Bass and Otway Basins. Mines Department, Victoria, Australia. Report 1966/30 (unpublished).

Tjalsma, R. C. 1977. Cenozoic foraminifera from the South Atlantic, DSDP Leg 36. *Initial Rep. Deep Sea drill. Proj.*, **36**, 493–517.

Todd, R. & Knicker, H. T. 1952. An Eocene foraminiferal fauna from the Agua Fresca Shale of Magallenes Province, southernmost Chile. *Spec. Publ. Cushman Found. foramin. Res.*, **1**, 1–28.

Toumarkine, M. 1978. Planktonic foraminiferal biostratigraphy of the Paleogene of sites 360 to 364 and the Neogene of sites 362A, 363 and 364, Leg 40. *Initial Rep. Deep Sea drill. Proj.*, **40**, 679–721.

Vella, P. 1961. Upper Oligocene and Miocene uvigerinid foraminifera from Raukumara Peninsula, New Zealand. *Micropaleontology*, **7**, 467–83.

Vella, P. 1964. Correlation of New Zealand and European Middle Tertiary. *Bull. Am. Assoc. Petrol. Geol.*, **48**, 1938–41.

Walters, R. 1965. The *Globorotalia zealandica* and *G. miozea* lineages. *N.Z. J. Geol. Geophys.*, **8**, 109–27.

Wiseman, J. D. H. & Ovey, C. D. 1950. Recent investigations on the deep-sea floor. *Proc. Geol. Assoc. London*, **61**, 28–84.

Wopfner, H. & Douglas, J. G. 1971. The Otway Basin of Southeastern Australia. *South Australia and Victoria Geol. Surveys Spec. Bull.*, 1–464.

8

Mediterranean Miocene and Pliocene planktic foraminifera

SILVIA IACCARINO

CONTENTS

Introduction

The geodynamic evolution of the Mediterranean during the Neogene and its effects on the planktic foraminiferal faunas was the main cause that necessitated the erection of a biostratigraphic framework distinct from that of the low latitudes. Earlier attempts to apply low latitude zonations to the Mediterranean sequences only partially solved the biostratigraphic problems of this area.

The planktic foraminiferal faunas of the Mediterranean Cenozoic can be divided into the following two stratigraphic groups, separated approximately at the time of the *Orbulina* datum:

(1) During the Paleogene and the early part of the Neogene (up to the *Orbulina* datum) open connections existed

between the Mediterranean and the Atlantic to the west, and the Mediterranean and Indian Ocean to the east. The planktic assemblages at that time still showed strong similarities with those of tropical regions. Until that time the biozones established for the low latitudes are therefore still largely recognizable in the Mediterranean region.

(2) At about the level of the evolutionary appearance of *Orbulina* at the Early/Middle Miocene boundary, a noticeable change took place in the composition of the Mediterranean planktic foraminiferal assemblages and their affinities with tropical areas from this level onwards become progressively more reduced. The closing of the Tethys to the east, the rotation of Spain towards the North-African continent, the formation of the Gibraltar sill and the subsequent modification of circulation patterns in the Mediterranean were the main causes for this significant change (Berggren & Philips, 1971; Berggren & Van Couvering, 1974).

In addition, the Mediterranean was strongly affected by the climatic deterioration which set in during the Middle Miocene and culminated in the glacial phases of the Pleistocene. Furthermore, eustatic changes of sea-level (Vail, Mitchum & Thompson, 1977) played an important role in the history of the Mediterranean. The interaction of the events resulted in a still stronger climatic control of the foraminiferal faunas and contributed to the isolation of the Mediterranean as a distinct bioprovince. This isolation reached its maximum in the latest Miocene when the Mediterranean was affected by the salinity crisis, an episode of extensive evaporite deposition.

Several biostratigraphic schemes referring solely to the Mediterranean have been proposed since 1967, either for the entire Neogene or for parts of it (Fig. 2). The zonal scheme adopted in this chapter is with few alterations that proposed by Iaccarino & Salvatorini (1982). On the distribution chart (Fig. 4) are shown the ranges of the common and significant Mediterranean Neogene planktic foraminifera. They are based on evidence from land sections in Italy, Spain, Greece and on deep-sea sites in the Mediterranean region.

Calibration of the biostratigraphic zonation with the stratotypes of the standard Neogene stages used in the Mediterranean area

The number of separate biostratigraphic schemes for the Mediterranean Neogene represents the best evidence of the need to utilize the stratigraphic philosophy of Hedberg (1976). The essence of this philosophy is the concept of chronostratigraphy, the element of stratigraphy which deals with the time relations of strata. The biostratigraphy as well as magnetostratigraphy, climatostratigraphy, etc. are the methods for time evaluation. These methods are extremely useful for corre-

lation and recognition of a time interval and/or horizon but none of them can constitute an objective standard definition.

Stages are stratigraphic units widely used by stratigraphers working on Mediterranean sequences. It is therefore important to determine the biostratigraphic position of stratotypes and boundary stratotypes of the classical stages in relation to the biostratigraphic scale adopted in this chapter (Fig. 1). The biostratigraphy of the stratotypes is mainly based on planktic foraminifera and calcareous nannofossils. In recent years most of the planktic foraminiferal biostratigraphic subdivisions first recognized in the type sections have been reviewed in the light of the advances in micropaleontology, particularly from tropical regions and from deep-sea sediments rich in planktic foraminifera. Moreover, a better knowledge of the total range of fossil taxa used in Mediterranean sequences and the stratigraphic significance of local events has led to further accuracy in the biostratigraphic record of the sequences.

Many of the original type sections yield scarce faunas, and often do not possess the most suitable requisites for good stratigraphic resolution. They were originally selected by using criteria very different from those required by modern concepts and recommended today by the *International Stratigraphic Guide*. As a consequence, some of the stratotypes do not accurately represent the time interval that is considered to be encompassed for stages, for example the Aquitanian, Burdigalian and Serravallian, which 'were originally units based on lithologic changes, regressions and transgressions, facies changes in faunas and floras or taxon range-zones' (Hedberg, 1971).

Furthermore, the original stratotypes of the stages were established in different localities and the boundary between two stages is therefore often affected by gaps or overlaps.

To avoid difficulties in time correlation, the principle that the top of a stage is defined by the base of the following stage (Hedberg, 1976) is followed here. The boundaries of chronostratigraphic units are better defined by mutual boundary stratotypes, which serve both as top of one stage and bottom of the following one. In the Mediterranean successions the following three boundary stratotypes have been proposed: Tortonian/Messinian (Colalongo *et al.*, 1979a); Miocene/ Pliocene (Cita, 1975a) and Pliocene/Pleistocene (Colalongo *et al.*, 1982). An international working group is currently also studying the Oligocene/Miocene boundary stratotype.

The Oligocene–Miocene boundary

The Oligocene–Miocene boundary is drawn at the base of the Aquitanian stage. Following Iaccarino & Salvatorini (1982) this boundary is placed informally at the first occurrence of *Globoquadrina dehiscens dehiscens*, since the first appearance of *Globigerinoides primordius* has been reported from the Mediterranean and other areas as occurring already within the

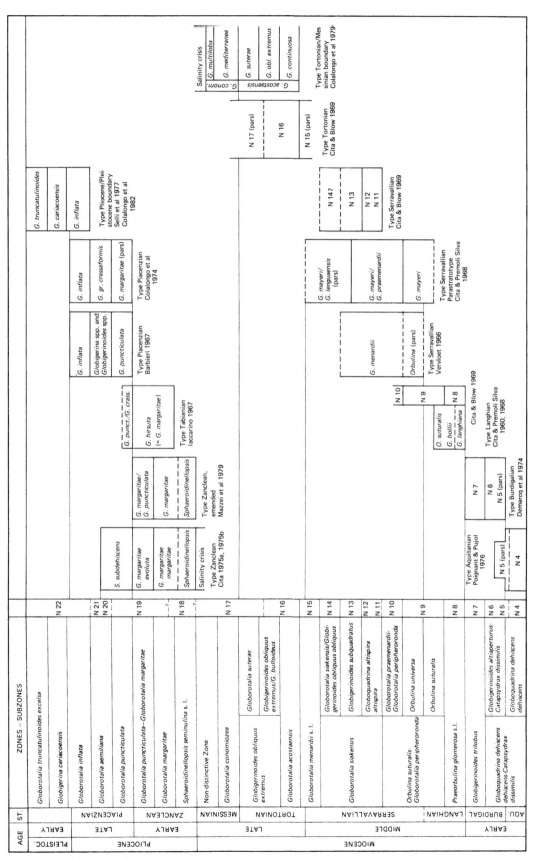

Fig. 1. Correlation of Mediterranean Neogene stages with planktic foraminiferal zones

Late Oligocene (Shafik & Chapronière, 1978; Biolzi *et al.*, 1981; Vismara Schilling, 1981).

Aquitanian

The stage was erected by Mayer (1858) but the stratotype was only later designated by Dollfus (1909) who, following the detailed information given by Mayer, located it 'dans le vallon de Saint-Jean-d'Etampes (Gironde) entre le Moulin de Bernachon et le Moulin de l'Eglise (Commune de Saucats)' (Vigneaux & Marks, 1971, for complete description). The type section contains a poor planktic foraminiferal assemblage unsuitable for reliable biostratigraphic resolution. The most recent biostratigraphic subdivision of the Aquitanian stratotype was carried out by Poignant & Pujol (1976) who recognized Zone N 4 and the lower part of Zone N 5.

The correlation of the lower boundary of Zone N 4 with the base of the *Globoquadrina dehiscens dehiscens–Catapsydrax dissimilis* Zone is based on the occurrence of *Globoquadrina dehiscens dehiscens* in the lowest sample; this also contains *Globigerinoides primordius*. Even if the lowest finding of *G. dehiscens dehiscens* in the type Aquitanian is not demonstrated to coincide with its first appearance, it is now well known that *G. primordius* appears stratigraphically earlier than the base of the Aquitanian and is found before the first occurrence of *G. dehiscens dehiscens* (Biolzi *et al.*, 1981; Vismara Schilling, 1981).

Burdigalian

As with the Aquitanian, the Burdigalian has never been defined properly in the Mediterranean area. It was first mentioned by Depéret in 1892 and only later Dollfus (1909) indicated as stratotype the 'gisement de Coquillat' in the Aquitaine Basin. This designation was recognized by Vigneaux (1971). Nevertheless, Demarcq *et al.* (1974) following the information given by Depéret proposed a second stratotype in the Rhône valley.

In the original stratotype section of Coquillat, Poignant & Pujol (1978) recognized Zone N 5 and perhaps the base of Zone N 6, while in the stratotype section of Carry-le-Rouet in the Rhône valley the Zones N 5 to N 7 were recognized. Whichever of the two stratotypes is adopted, the top has to be defined by the base of the Langhian.

Langhian

The Langhian stage was erected by Pareto (1865) who did not designate a definite section but only a type area. The Bricco della Croce was designated subsequently, as the type section for the stage by Cita & Premoli Silva (1960) who tentatively recognized three biozones: a lower *Globoquadrina dehiscens* Zone, which later was named *Globoquadrina langhiana* Zone (Cita & Gelati, 1960), a middle *Globigerina bollii* Zone, and an upper *Orbulina suturalis* Zone. The lower

boundary of the *Globoquadrina langhiana* Zone corresponds to the first occurrence of *Praeorbulina glomerosa* and this coincides with the lower boundary of the type Langhian (Cita & Premoli Silva, 1968). The first evolutionary occurrence of *Orbulina suturalis* occurs within the upper part of the type Langhian and the extinction of *Praeorbulina glomerosa*, which falls at the top of the section, was taken to coincide with the upper boundary of the Langhian. *Orbulina universa* is absent in the Langhian stratotype.

Subsequently, the type Langhian was identified with the Cessole Formation (Vervloet, 1966) corresponding to the 'main body of the stratotype of the Langhian'. However, the type section designated by Vervloet for the formation does not correspond to the Langhian as defined by Cita & Premoli Silva (1960) but is more reduced (Cita & Blow, 1969). However, according to Cita & Blow its base approximates the base of the Langhian and is taken to coincide with the first occurrence of *Praeorbulina glomerosa*. The top of the Cessole Formation corresponds to the top of the type Langhian as defined by Cita & Premoli Silva (1960) and therefore it would coincide with the top of their *Orbulina suturalis* Zone. Cita & Blow erroneously correlated the *Orbulina suturalis* Zone (without *Orbulina universa*) with N 9 and the earlier part of N 10, which would place the top of the Langhian appreciably higher, i.e. within N 10 and not within N 9. The top of the *Orbulina suturalis* Zone is here considered not to extend to the top of N 9. The last occurrence of *Praeorbulina glomerosa* is taken here to indicate the top of the Langhian, an event which approximates the first occurrence of *Orbulina universa* and thus would identify the Langhian–Serravallian boundary.

Serravallian

The term Serravallian was re-introduced by Cita & Elter (1960) and Cita (1964) as a substitute for the outdated and ill-chosen Helvetian. When defining the Serravallian, Pareto (1865) first described 'sandy marls, which contain sandstones in increasing quantity and thickness towards the top' outcropping near Serravalle Scrivia in Piedmont. At this locality, Vervloet (1966) formalized this lithologic succession as the Serravalle Formation and formally defined it as the type section of the Serravallian. He recognized in the type Serravallian part of his 'Orbulina Zone' and the 'menardii Zone'. The base of the stage appears to coincide with the first occurrence of *Orbulina universa*.

The lower boundary of the 'menardii Zone' is defined by the first occurrence of the nominate taxon which is apparently used in a broad sense to include *Globorotalia praemenardii*. In this sense the base of the 'menardii Zone' of Vervloet is correlated with the first occurrence of *G. praemenardii*. The upper boundary of the same zone, defined by the first occurrence of *Globigerina nepenthes* falls within the overlying formation which does not belong to the Serravallian.

The Serravallian as stratotypified by Vervloet is far from ideal for biostratigraphic dating; gaps in the evolutionary sequences and reworking almost certainly occur.

For a better definition of the Serravallian stage, Boni (1967) and Mosna & Micheletti (1968) proposed the Gavi section as stratotype in substitution for that designated by Vervloet. The section, which according to authors is more suitable for biostratigraphic resolution, ends below the first occurrence of *Globorotalia lenguaensis*. Cita & Premoli Silva (1968) proposed as reference section (parastratotype) the Arguello-Lequio section outcropping in the same area as the Langhian stratotype. This section is less sandy than the stratotype and has the advantage of offering a more complete biostratigraphic record, particularly for the upper part of the stage. The authors were able to distinguish three biozones: the *Globorotalia mayeri* Zone, at the very base of which they recognized the first occurrence of *Orbulina universa*, the *Globorotalia mayeri–Globorotalia praemenardii* Zone and the *Globorotalia mayeri–Globorotalia lenguaensis* Zone.

The marls with interbedded graded sandstones which occur in the lower part of the sequence were included in the original definition of the Serravalle Formation. Cita & Blow (1969) considered these alternations as a transitional unit occurring between the Cessole and Serravalle formations and recognized within them the biostratigraphic interval N 9–N 10. These alternations were referred to the Cessole Formation and were included within the Langhian.

As a consequence, the Serravallian and the Langhian stratotypes as represented by the planktic foraminifera occurring in the Cessole and Serravalle formations overlap partly within Zone N 10. On this basis Cita & Blow (1969) proposed to define the base of the Serravallian as coincident with the base of Zone N 11. Such a proposal appears to be inconsistent with the definition of the Serravallian given by Cita & Premoli Silva (1968), where the base was correlated with the disappearance of *Praeorbulina glomerosa*, which occurs within N 9.

Tortonian

The Tortonian stage was proposed by Mayer (1857) to comprise marls yielding *Ancillaria glandiformis* outcropping near Tortona (Piedmont). Gianotti (1953) described and designated the type section along the Rio Mazzapiedi–Rio Castellania valley, east of Tortona. It is characterized by marls in the lower and upper part, and by sands in the middle part.

In Gianotti's definition the type section includes the so-called 'marne tabacco a cerizi' underlying the evaporitic sequence and yielding a poor and oligotypic foraminiferal assemblage. After the definition of the Messinian neostratotype by Selli (1960), the 'marne tabacco a cerizi' was considered to be correlatable with that stage (Cita, Premoli Silva & Rossi, 1965; Cita, 1968; Blow, 1969; Cita & Blow, 1969; D'Onofrio

et al., 1975). In fact, Mayer himself (1868, 1889) had already referred it to the Messinian after he had proposed the Messinian stage (1857). Therefore, an overlap occurs between the type Tortonian (*sensu* Mayer, 1857, and Gianotti, 1953) and the type Messinian (*sensu* Selli). The biostratigraphic extent of the overlap varies according to different authorities; for example, Cita *et al.* (1965), Blow (1969) and D'Onofrio *et al.* (1975). Recently, Colalongo *et al.* (1979a) suggested an emendation of the Tortonian and Messinian stages and proposed that the boundary between the two stages be placed at the lithostratigraphic level of the first appearance of *Globorotalia conomiozea* in the Falconara boundary stratotype (Sicily). This event, in the Tortonian type section, occurs lower than the boundary fixed by Gianotti. Cita *et al.* (1965), who first investigated the planktic foraminifera of the Tortonian stratotype, recognized three zones: a *mayeri–lenguaensis* Zone (part), a *mayeri–nepenthes* Zone and a *menardii–nepenthes* Zone. This zonal subdivision was reviewed by Cita & Blow (1969), who recognized part of Zone N 15, the entire Zone N 16 and part of Zone N 17.

Messinian

The Messinian stage was introduced by Mayer (1867) to comprise the marine, brackish and fresh-water deposits occurring between the Tortonian and the Pliocene. The definition of the stage and the subsequent additional remarks given by Mayer himself (1867, 1868) were proved to be partially incorrect and imprecise. Selli (1960) re-examined and re-defined the Messinian stage. He designated as neostratotype the composite section of Capodarso-Pasquasia (Sicily).

However, the criteria adopted by him did not follow those suggested by the *International Stratigraphic Guide* (Hedberg, 1976). An overlap occurs between the Early Messinian and the Late Tortonian as stratotypified by Mayer (1857) and later by Gianotti (1953). D'Onofrio *et al.* (1975) re-investigated the Tortonian stratotype and the Messinian neostratotype to identify planktic events suitable for the recognition of the lower boundary of the Messinian. They recognized that the *Globorotalia conomiozea* datum is the best event which approximates the Tortonian/Messinian boundary in the respective stratotype sections. Colalongo *et al.* (1979a) proposed the Falconara section in Sicily as Tortonian/Messinian boundary stratotype and proposed that the boundary be placed at the lithostratigraphic level of the first appearance of *Globorotalia conomiozea*.

Pliocene

The subdivision into Early and Late Pliocene adopted here is widely used by stratigraphers working in the Mediterranean area, but an informal three-part subdivision is also used. Early and Late Pliocene are related to the two stages of the Pliocene, the Zanclean or Tabianian and the Piacenzian, while Early, Middle and Late are related to the planktic zones occurring in these two stages. In the first case the Early Pliocene coincides with the Zanclean and its upper boundary with the last occurrence of *Globorotalia margaritae* (Mazzei *et al.*, 1978). In the second case the upper boundary of the Early Pliocene coincides with the first occurrence of *Globorotalia aemiliana* (Borsetti *et al.*, 1979; Colalongo & Sartoni, 1979) and therefore the lower part of the Piacenzian is also included in the Early Pliocene. The Middle and Late Pliocene coincide with the ranges of the *Globorotalia aemiliana* and *G. inflata* zones respectively.

Tabianian

The Tabianian was introduced by Mayer (1867) to comprise the 'Couches de Tabiano à *Ficula undata*'. The stratotype section was designated and described by Iaccarino (1967) at Tabiano, northern Italy. The vertical extent of the Tabianian as stratotypified by Iaccarino is slightly more restricted compared with Mayer's definition. However, Barbieri (1967) maintained that the upper boundary of the stage is defined by the base of the Piacenzian which has priority, having been erected by Mayer in 1858.

The base of the Piacenzian and therefore of the Tabianian/Piacenzian boundary in its stratotype coincides with the disappearance of *Globorotalia margaritae* and *Uvigerina rutila*. In the Tabianian type section, however, *G. margaritae* disappears slightly below the last occurrence of *U. rutila*. The *Globorotalia puncticulata* and *Globorotalia crassaformis* Zone recognized by Iaccarino (1967) above the *Globorotalia margaritae* Zone correlates with the *Globorotalia puncticulata* Zone of the present zonal scheme because *Globorotalia crassaformis* has to be interpreted as *G. puncticulata*, whereas *G. puncticulata* has to be regarded at least in part as *G. puncticulata padana* (see taxonomic notes).

In the present chapter the Zanclean is used instead of the Tabianian because in the latter the very base of the Pliocene is not represented, as the *Sphaeroidinellopsis seminulina* Zone has not been identified. However, it is possible that the lower part of the Tabiano stratotype represented by 15 metres of conglomerate may correspond to that zone.

Zanclean

The Zanclean stage was introduced by Seguenza (1868) to comprise the 'Marnes blanches à foraminifères'. The Trubi marl section of Capo Rossello near Agrigento, Sicily, was designated and proposed by Cita (1975a) as the type section of the Miocene/Pliocene boundary as well as for the Zanclean stage.

According to Cita's definition, the Miocene/Pliocene boundary corresponds to 'the first appearance of permanent open marine conditions in the Mediterranean after the Late Miocene salinity crisis and desiccation'. According to the

original description of the stratotype and later references (Cita & Gartner, 1973; Cita, 1975b, c; Cita & Decima, 1975) the Zanclean extended upwards well after the extinction of *Globorotalia margaritae* and included the time interval represented by the stratotypes of the Tabianian (Iaccarino, 1967) and part of the Piacenzian (Barbieri, 1967). Subsequently Mazzei *et al.* (1978) re-defined the upper boundary of the Zanclean. They followed the rules of the *International Stratigraphic Guide* (Hedberg, 1976) that the top of a stage is defined by the base of the overlying one, and therefore located the top of the Zanclean at the level of the last occurrence of *Globorotalia margaritae* by correlation with the base of the Piacenzian stratotype.

Piacenzian

The Piacenzian was introduced by Mayer (1858) to comprise the blue marls outcropping in the Piacenzian Apennines. Following the suggestions of Pareto (1865), the stratotype was designated by Barbieri (1967) between Lugagnano and Castell'Arquato near Piacenza in northern Italy. According to Barbieri, the base of the Piacenzian coincides with the last occurrence of *Globorotalia hirsuta* (= *margaritae*), while in Colalongo, Elmi & Sartoni (1974) the *Globorotalia margaritae* Zone is extended up to the Early Piacenzian. Actually the upper part of the *Globorotalia margaritae* Zone of Colalongo *et al.* (defined as the interval from the first occurrence of the zonal marker to the first occurrence of *Globorotalia* gr. *crassaformis*) lacks the nominate taxon. Therefore, the part of the zone which postdates the extinction of *G. margaritae* correlates with the *Globorotalia puncticulata* Zone of Barbieri and the *G. puncticulata* Zone of the present zonal scheme which is Piacenzian in age.

The *Globigerina* spp. and *Globigerinoides* spp. Zone distinguished by Barbieri in the Piacenzian stratotype above the *Globorotalia puncticulata* Zone would correspond to the *Globorotalia aemiliana* Zone.

The Pliocene–Pleistocene boundary

The position of the Plio-Pleistocene boundary is a problem which has been debated for a long time. The controversy still existing arises essentially as a consequence of different stratigraphic philosophies.

A critical review of the historical concepts of the Plio-Pleistocene boundary and the chronostratigraphic subdivisions of the Quaternary, in the light of the *International Stratigraphic Guide* (Hedberg, 1976), are documented and discussed in Pelosio, Raffi & Rio (1980) and more recently in Colalongo *et al.* (1982). Reference is made to these authors for the detailed literature. The arrival of colder water forms in the Mediterranean has been the major criterion adopted by Italian stratigraphers to define the Plio-Pleistocene boundary since De Stefani (1876, 1891) first emphasized the stratigraphic importance of the appearance of the boreal elements in the Pleistocene Mediterranean faunas as the consequence of the first glacial expansion.

This criterion ratified by numerous congresses held on this topic (London, 1948; Algiers, 1952; Denver, 1965) is still reliable and in agreement with the modern concepts of stratigraphy. Obviously it must be fixed in a well-defined stratotype.

The section of Santa Maria di Catanzaro, stratotype of the Calabrian (the lowest stage of the Pleistocene), and the section of Le Castella, stratotype for the Plio-Pleistocene boundary, do not represent the very base of the Quaternary (Sprovieri, D'Agostino & Di Stefano, 1973; Selli *et al.*, 1977; Pelosio *et al.*, 1980) and have to be abandoned. Colalongo *et al.* (1982) considered that the Vrica section, also located in Calabria, southern Italy, is much more suitable for defining the Plio-Pleistocene boundary and formally proposed that it be defined by the lithologic level of the first appearance of the ostracod *Cyteropteron testudo* in the Vrica section. According to Ruggieri (1961) and Ruggieri & Sprovieri (1975) *C. testudo* first entered the Mediterranean together with *Arctica islandica*.

The first appearance of *Globorotalia truncatulinoides* widely considered as the best paleontological criterion for recognizing the Pleistocene no longer appears to be the most suitable event. As documented by Sprovieri, Ruggieri & Unti (1980), the entrance of *G. truncatulinoides* occurs in levels close to the first appearance of *Globorotalia inflata*, which are Late Piacenzian in age. In terms of absolute age an error of 0.2–0.3 m.y. is introduced if the appearance of *G. truncatulinoides* is used to define the base of the Quaternary. As suggested by Rio *et al.* (1984) the use of the first occurrence of *G. truncatulinoides* for the recognition of the base of the Pleistocene must therefore be abandoned in the Mediterranean area whenever possible.

The chronostratigraphic units of the Early Pleistocene which were used until some years ago were: Calabrian, Emilian and Sicilian. According to the proposal of Ruggieri *et al.* (1975), the beginning of the three stages was equated to the appearance of *Arctica islandica*, *Hyalinea balthica* and *Globorotalia truncatulinoides* respectively. The use of all three taxa as stage boundary markers has since been criticized on the grounds that the first two are environmentally controlled while *G. truncatulinoides* is too uncommon in the Mediterranean region.

In 1975 Ruggieri & Sprovieri proposed a new stage, the Santernian, to replace the Calabrian stage, which had been found to be a junior synonym of the Sicilian. On the basis of the evolution of *Gephyrocapsa*, Rio (1982) assigned to the Santernian an age of 1.6–1.7 to about 1.3 m.y. and to the Emilian an age of 1.3–1.2 m.y. The base of the Sicilian is dated at 1.1–1.2 m.y. whereas the age of the top has not yet been determined.

Rio (1982) suggested that the three stages mentioned above appear to be too short to be maintained in any international geochronologic scale. It would be more convenient to consider them as chronostratigraphic units of lower rank and to adopt only one stage for the whole Lower Pleistocene. Because of this uncertainty, no Pleistocene stage names are used in this paper.

Zones used in this chapter

GLOBOQUADRINA DEHISCENS DEHISCENS–CATAPSYDRAX DISSIMILIS ZONE

Category: Concurrent range zone

Age: Aquitanian–Burdigalian, Early Miocene

Author: Iaccarino & Salvatorini, 1982

Definition: Interval from first occurrence of *Globoquadrina dehiscens dehiscens* to last occurrence of *Catapsydrax dissimilis*.

Remarks: When defining the zone Iaccarino & Salvatorini followed the concept of Blow who considered *Catapsydrax ciperoensis* the only species of the *Catapsydrax dissimilis* stock ranging into the early Miocene, and therefore they named the zone *Globoquadrina dehiscens Catapsydrax ciperoensis* Zone. Such a distinction has been abandoned in this volume and *C. ciperoensis* has been placed back in the *C. dissimilis* group.

This zone was established as a substitute for the *Globigerinoides primordius* Zone still used for defining the very base of the Miocene. In fact, the first appearance of *G. primordius* is now known to be a late Oligocene event and therefore unreliable as a marker for the Oligocene–Miocene boundary. The choice of *G. dehiscens dehiscens* is mainly for the following reasons: (1) it occurs in the Aquitanian stratotype from the lowest fossiliferous sample; (2) the diagnostic morphologic features make it a taxon easily recognizable; (3) it is strongly resistant to dissolution and not particularly controlled by ecological factors.

GLOBOQUADRINA DEHISCENS DEHISCENS SUBZONE

Category: Interval subzone

Age: Aquitanian, Early Miocene

Author: Iaccarino & Salvatorini, 1982

Definition: Interval from first occurrence of *Globoquadrina dehiscens dehiscens* to first occurrence of *Globigerinoides altiaperturus*.

Remarks: *Globorotalia kugleri* and the last representatives of the *Globigerina tripartita* plexus disappear within the subzone. *Globigerina woodi* and *Globigerinoides altiaperturus* first appear in the upper part of the subzone (Biolzi *et al.*, 1981).

AGE		ST.	ZONES – SUBZONES		DATUM MARKERS
PLEISTOC.	EARLY		*Globorotalia truncatulinoides excelsa*		F *Globorotalia truncatulinoides excelsa*
PLEISTOC.	EARLY		*Globigerina cariacoensis*		F *Globigerina cariacoensis*
PLIOCENE	LATE	PIACENZIAN	*Globorotalia inflata*		F *Globorotalia inflata*
PLIOCENE	LATE	PIACENZIAN	*Globorotalia aemiliana*		F *Globorotalia aemiliana*
PLIOCENE	LATE	PIACENZIAN	*Globorotalia puncticulata*		L *Globorotalia margaritae*
PLIOCENE	EARLY	ZANCLEAN	*Globorotalia puncticulata–Globorotalia margaritae*		F *Globorotalia puncticulata*
PLIOCENE	EARLY	ZANCLEAN	*Globorotalia margaritae*		F *Globorotalia margaritae*
PLIOCENE	EARLY	ZANCLEAN	*Sphaeroidinellopsis seminulina* s. l.		First permanent open marine conditions after the Late Miocene salinity crisis
MIOCENE	LATE	MESSINIAN	Non-distinctive Zone		Coiling change of *G. acostaensis* from sinistral to dextral
MIOCENE	LATE	MESSINIAN	*Globorotalia conomiozea*		F *Globorotalia conomiozea*
MIOCENE	LATE	TORTONIAN	*Globigerinoides obliquus extremus*	*Globorotalia suterae*	F *Globorotalia suterae*
MIOCENE	LATE	TORTONIAN	*Globigerinoides obliquus extremus*	*Globigerinoides obliquus extremus/G. bulloideus*	F *Globigerinoides obliquus extremus*
MIOCENE	LATE	TORTONIAN	*Globorotalia acostaensis*		F *Globorotalia acostaensis*
MIOCENE	MIDDLE	SERRAVALLIAN	*Globorotalia menardii* s. l.		L *Globorotalia siakensis*
MIOCENE	MIDDLE	SERRAVALLIAN	*Globorotalia siakensis*	*Globorotalia siakensis/Globigerinoides obliquus obliquus*	L *Globigerinoides subquadratus*
MIOCENE	MIDDLE	SERRAVALLIAN	*Globorotalia siakensis*	*Globigerinoides subquadratus*	F *Globorotalia partimlabiata*
MIOCENE	MIDDLE	SERRAVALLIAN	*Globorotalia siakensis*	*Globoquadrina altispira altispira*	L *Globorotalia peripheroronda*
MIOCENE	MIDDLE	SERRAVALLIAN	*Orbulina suturalis-Globorotalia peripheroronda*	*Globorotalia praemenardii-Globorotalia peripheroronda*	F *Globorotalia praemenardii*
MIOCENE	MIDDLE	SERRAVALLIAN	*Orbulina suturalis-Globorotalia peripheroronda*	*Orbulina universa*	F *Orbulina universa*
MIOCENE	MIDDLE	LANGHIAN	*Orbulina suturalis-Globorotalia peripheroronda*	*Orbulina suturalis*	F *Orbulina suturalis*
MIOCENE	MIDDLE	LANGHIAN	*Praeorbulina glomerosa* s. l.		F *Praeorbulina glomerosa*
MIOCENE	EARLY	BURDIGAL.	*Globigerinoides trilobus*		L *Catapsydrax dissimilis*
MIOCENE	EARLY	BURDIGAL.	*Globoquadrina dehiscens dehiscens-Catapsydrax dissimilis*	*Globigerinoides altiaperturus-Catapsydrax dissimilis*	F *Globigerinoides altiaperturus*
MIOCENE	EARLY	AQU.	*Globoquadrina dehiscens dehiscens-Catapsydrax dissimilis*	*Globoquadrina dehiscens dehiscens*	F *Globoquadrina dehiscens dehiscens*

Fig. 2. Miocene to Pleistocene planktic foraminiferal zonal scheme and datum markers used in this chapter

GLOBIGERINOIDES ALTIAPERTURUS– CATAPSYDRAX DISSIMILIS SUBZONE

Category: Concurrent range subzone

Age: Burdigalian, Early Miocene

Author: Bizon & Bizon, 1972

Definition: Interval from first occurrence of *Globigerinoides altiaperturus* to last occurrence of *Catapsydrax dissimilis*.

Remarks: *Globigerinoides primordius* disappears within the lower part of the subzone but the genus *Globigerinoides* becomes more common, being represented by an increasing number of species (*G. trilobus, G. subquadratus, G. sacculifer*). *Globigerinatella insueta* a taxon always rare and sporadic in Mediterranean sequences, is documented first from the upper part of the subzone.

GLOBIGERINOIDES TRILOBUS ZONE

Category: Interval zone

Age: Burdigalian, Early Miocene

Author: Bizon & Bizon, 1972

Definition: Interval from last occurrence of *Catapsydrax dissimilis* to first occurrence of *Praeorbulina glomerosa* s.l.

Remarks: Bizon (1979) subdivided this interval into a lower *Globigerinoides trilobus* Zone and an upper *Globigerinoides bisphaericus* Zone with the boundary between the two zones marked by the appearance of *G. bisphaericus*. A *Globigerinoides bisphaericus* or *Globigerinoides sicanus* Zone appears in most of the Mediterranean zonal schemes (Fig. 3). In the light of recent investigations carried out on the holotypes of *sicanus, bisphaericus* and *glomerosa curva*, Jenkins, Saunders & Cifelli (1981) emphasized that *sicanus* and *bisphaericus* are two different species and more precisely *sicanus* belongs to the genus *Praeorbulina* and *bisphaericus* is a true *Globigerinoides*. Therefore, since *G. bisphaericus* and *G. sicanus* have never been distinguished as separate species in the Mediterranean region, this zone is not adopted here, even if the presence of true *bisphaericus* is recorded from the upper part of the zone.

The *Globoquadrina dehiscens dehiscens* Zone of Cati *et al.* (1968) corresponds to our *Globigerinoides trilobus* Zone and is defined by the same zonal criteria. It should not be confused with the *Globoquadrina dehiscens dehiscens* Subzone as used in the present work. The *G. trilobus* Zone of Borsetti *et al.* (1979), the lower boundary of which coincides with the first occurrence of the zonal marker, covers a shorter time interval because, according to these authors, *G. trilobus* appears later than *G. altiaperturus*.

PRAEORBULINA GLOMEROSA s.l. ZONE

Category: Lineage zone

Age: Langhian, Early Miocene

Author: Bizon & Bizon, 1972

Definition: Interval from first occurrence of *Praeorbulina glomerosa* s.l. to first occurrence of *Orbulina suturalis*.

Remarks: The evolutionary appearance of the subspecies of *Praeorbulina glomerosa* (*sicana, glomerosa, circularis*) takes place within this zone. Iaccarino & Salvatorini (1982) subdivided it into two subzones: a lower *Praeorbulina glomerosa sicana* Subzone and an upper *Praeorbulina glomerosa circularis* Subzone.

These subzones are easily applicable in suitable land and deep-sea sections where the populations of *Praeorbulina* are rich. After the revision of Jenkins *et al.* (1981) and following Blow's (1956) criteria in distinguishing the individual taxa within the genus *Praeorbulina*, *P. glomerosa curva* has to be considered as a junior synonym of *P. glomerosa sicana*. The appearance of *Orbulina suturalis* represents the final stage in the lineage.

ORBULINA SUTURALIS–GLOBOROTALIA PERIPHERORONDA ZONE

Category: Concurrent range zone

Age: Langhian–Serravallian, Early to Middle Miocene

Author: Bizon & Bizon, 1972

Definition: Interval from first occurrence of *Orbulina suturalis* to last occurrence of *Globorotalia peripheroronda*.

Remarks: *Globorotalia peripheroronda* is the only step within the *Globorotalia fohsi* lineage which is well developed in the Mediterranean region. However, sporadic specimens of *Globorotalia peripheroacuta* are recorded from the same area (Ruggieri & Sprovieri, 1970; Cita, 1976). The extinction of *G. peripheroronda*, an event easily detectable and widely documented in the Mediterranean, seems to occur sooner than at low latitudes probably due to environmental controls which prevented the entry or the evolution of the younger taxa of the lineage. In Malta (Giannelli & Salvatorini, 1975) and in the Balearic Basin, *G. peripheroronda* disappears within Zone N 10 instead of close to the N 11/N 12 zonal boundary (Blow, 1969). Iaccarino & Salvatorini (1982) distinguished within the zone three subzones on the basis of the first appearance of *Orbulina universa* and *Globorotalia praemenardii*.

ORBULINA SUTURALIS SUBZONE

Category: Lineage subzone

Age: Langhian, Middle Miocene

Author: Cita & Premoli Silva, 1971–1973 (in Cita, 1976)

Definition: Interval from first occurrence of *Orbulina suturalis* to first occurrence of *Orbulina universa*.

Remarks: Numerous species, among them *Globigerinoides diminutus*, *G. bisphaericus* and *Praeorbulina glomerosa* become extinct within the zone.

ORBULINA UNIVERSA SUBZONE

Category: Interval subzone

Age: Serravallian, Middle Miocene

Author: Iaccarino & Salvatorini, 1982

Definition: Interval from first occurrence of *Orbulina universa* to first occurrence of *Globorotalia praemenardii*.

Remarks: *Globigerina woodi woodi* becomes extinct; *Globorotalia miozea*, *Sphaeroidinellopsis disjuncta* and *Globigerina regina* appear within the zone.

GLOBOROTALIA PRAEMENARDII– GLOBOROTALIA PERIPHERORONDA SUBZONE

Category: Concurrent range subzone

Age: Serravallian, Middle Miocene

Author: Iaccarino & Salvatorini, 1982

Definition: Interval from first occurrence of *Globorotalia praemenardii* to last occurrence of *Globorotalia peripheroronda*.

Remarks: *Globigerinopsis aguasayensis* and *Globigerinoides bollii* appear within the zone.

GLOBOROTALIA SIAKENSIS ZONE

Category: Interval zone

Age: Serravallian, Middle Miocene

Author: Bizon & Bizon, 1972; renamed by Iaccarino & Salvatorini, 1982

Definition: Interval from last occurrence of *Globorotalia peripheroronda* to last occurrence of *Globorotalia siakensis*.

Remarks: Bizon & Bizon (1972) and Bizon (1979) named this interval *Globorotalia mayeri* Zone, following the concept of Bolli (1957, 1966) that *G. mayeri* and *G. siakensis* are the same species. Iaccarino & Salvatorini (1982) in agreement with Blow (1969), Stainforth *et al.* (1975) and others consider *G. mayeri* a taxon distinct from *G. siakensis* (see taxonomic notes in this chapter) and named the zone *Globorotalia siakensis* Zone.

The interval represented by this zone is defined by different authors in a number of ways as is evident from Fig. 3. In fact, significant and easily recognizable events characterizing this interval are absent. Iaccarino &

Salvatorini (1982) distinguished three subzones based on two events: the first event is the initial appearance of *Globorotalia partimlabiata*, a taxon disregarded up to now by most authors; this event is particularly reliable for the Mediterranean area, and can be used as an indicator for the N 12/N 13 zonal boundary (Salvatorini & Cita, 1979). The second event is the extinction of *Globigerinoides subquadratus*, a datum never adopted for the Mediterranean biozonations but used for low latitudes (also in this volume) as 'the last Miocene occurrence of *Globigerinoides ruber*' to characterize the lower boundary of the *G. mayeri* Zone.

GLOBOQUADRINA ALTISPIRA ALTISPIRA SUBZONE

Category: Interval subzone

Age: Serravallian, Middle Miocene

Author: Iaccarino & Salvatorini, 1982

Definition: Interval from last occurrence of *Globorotalia peripheroronda* to first occurrence of *Globorotalia partimlabiata*.

Remarks: The first occurrence of *Globigerinoides obliquus obliquus* is recorded in this subzone.

GLOBIGERINOIDES SUBQUADRATUS SUBZONE

Category: Interval subzone

Age: Serravallian, Middle Miocene

Author: Iaccarino & Salvatorini, 1982

Definition: Interval from first occurrence of *Globorotalia partimlabiata* to last occurrence of *Globigerinoides subquadratus*.

Remarks: The stratigraphic usefulness of the last frequent occurrence of *Globigerinoides subquadratus/ruber* in the Middle Miocene used as an index particularly in low latitude areas has more recently also been recognized in the Mediterranean area by Martinotti (1981). *Sphaeroidinellopsis seminulina* first develops close to the base of the subzone.

GLOBOROTALIA SIAKENSIS/ GLOBIGERINOIDES OBLIQUUS OBLIQUUS SUBZONE

Category: Interval subzone

Age: Serravallian, Middle Miocene

Author: Iaccarino & Salvatorini, 1982

Definition: Interval from last occurrence of *Globigerinoides subquadratus* to last occurrence of *Globorotalia siakensis*.

Remarks: *Globigerina nepenthes* first occurs close to the base of the subzone; the taxon has been used in the past as a useful event in biostratigraphic zonation. However, there is no general agreement on the timing of its appearance particularly as it seems to be strongly environment controlled.

Globigerinoides obliquus obliquus becomes consistently common from this subzone onwards. The lower boundary of the *Globigerinoides obliquus obliquus* Zone of Borsetti *et al.* (1979), defined by the first occurrence of the zonal marker, has been tentatively correlated with the base of this subzone.

GLOBOROTALIA MENARDII s.l. ZONE

Category: Interval zone
Age: Serravallian–Tortonian, Middle to Late Miocene
Author: Bolli, 1957
Definition: Interval from last occurrence of *Globorotalia siakensis* to first occurrence of *Globorotalia acostaensis*.
Remarks: *Globorotalia miozea* and *Sphaeroidinellopsis disjuncta* become extinct within, *Globigerinopsis aguasayensis* at the top of the zone.

GLOBOROTALIA ACOSTAENSIS ZONE

Category: Interval zone
Age: Tortonian, Late Miocene
Author: Iaccarino & Salvatorini, 1982
Definition: Interval from first occurrence of *Globorotalia acostaensis* to first occurrence of *Globigerinoides obliquus extremus*.
Remarks: D'Onofrio *et al.* (1975) recognized within the long-ranging *Globorotalia acostaensis* Zone three subzones: a lower *Globorotalia continuosa* Subzone, a middle *Globigerinoides obliquus extremus* Subzone and an upper *Globorotalia suterae* Subzone. Iaccarino & Salvatorini (1982) considered the first occurrence of *Globigerinoides obliquus extremus* as a significant and easily recognizable event, useful for worldwide correlation and to be emphasized as zonal marker. As a consequence, the *Globorotalia acostaensis* Zone is shortened and here corresponds to the *Globorotalia continuosa* Subzone of D'Onofrio *et al.* (1975). The *G. acostaensis* Zone of Borsetti *et al.* (1979), whose upper boundary is defined by the first appearance of *Globorotalia merotumida*, encompasses an interval too short to be considered a separate zone. In fact, *G. merotumida* is recorded as occurring first together with, or very close to *G. acostaensis* (Cita, 1976; Colalongo *et al.*, 1979b).

GLOBIGERINOIDES OBLIQUUS EXTREMUS ZONE

Category: Interval zone
Age: Tortonian, Late Miocene
Author: Iaccarino & Salvatorini, 1982
Definition: Interval from first occurrence of *Globigerinoides obliquus extremus* to first occurrence of *Globorotalia conomiozea*.
Remarks: As noted by Iaccarino & Salvatorini (1982), the first appearance of *Globorotalia mediterranea* at the very top of the zone can be a useful event for the recognition of the upper boundary in the absence of *Globorotalia conomiozea*. Bizon (1979) and Borsetti *et al.* (1979) used this event for the recognition of the Tortonian/Messinian boundary. Actually, in the stratotype of Falconara, where the Tortonian/Messinian boundary has been defined, the first appearance of *G. mediterranea* precedes that of *G. conomiozea*.

GLOBIGERINOIDES OBLIQUUS EXTREMUS/ GLOBIGERINOIDES BULLOIDEUS SUBZONE

Category: Interval subzone
Age: Tortonian, Late Miocene
Author: D'Onofrio *et al.*, 1975, renamed by Iaccarino & Salvatorini, 1982
Definition: Interval from first occurrence of *Globigerinoides obliquus extremus* to first occurrence of *Globorotalia suterae*.
Remarks: In the zonal scheme of D'Onofrio *et al.* (1975) this subzone was named *Globigerinoides obliquus extremus* Subzone. As this taxon is now considered as a zonal marker, *Globigerinoides bulloideus* has been associated as an additional taxon to define this subzone (Iaccarino & Salvatorini, 1982).

GLOBOROTALIA SUTERAE SUBZONE

Category: Interval subzone
Age: Tortonian, Late Miocene
Author: D'Onofrio *et al.*, 1975
Definition: Interval from first occurrence of *Globorotalia suterae* to first occurrence of *Globorotalia conomiozea*.
Remarks: *Globorotalia suterae* is a taxon well documented in the Mediterranean; therefore this subzone is useful for the recognition of the Late Tortonian in the area.

GLOBOROTALIA CONOMIOZEA ZONE

Category: Interval zone
Age: Messinian, Late Miocene
Author: Iaccarino & Salvatorini, 1982
Definition: Interval from first occurrence of *Globorotalia conomiozea* to first coiling change of *Globorotalia acostaensis* from sinistral to dextral (post *Globorotalia conomiozea* datum).
Remarks: The definition of the zone given by Iaccarino & Salvatorini (1982) has the advantage that the upper boundary represents an event also well documented in extra-Mediterranean successions and therefore useful

			Iaccarino this volume		Cati et al 1968		Bizon & Bizon 1972	Cita 1972-1975
PLEISTOC.	EARLY		*Globorotalia truncatulinoides excelsa*		*G. truncatulinoides*		*G. truncatulinoides*	
			Globigerina cariacoensis					
PLIOCENE	LATE	PIACENZIAN	*Globorotalia inflata*		*G. inflata*	*G. inflata/ G. tosaensis*	*G. inflata/ G. tosaensis*	MPL-6 *G. inflata*
			Globorotalia aemiliana		*G. crassaformis*	*G. crassaformis s. l.*	*G. crassaformis*	MPL-5 *G. elongatus*
					G. aemiliana			
			Globorotalia puncticulata		*G. bononiensis*	*G. puncticulata*	*G. puncticulata*	MPL-4 *S. subdehiscens*
	EARLY	ZANCLEAN	*Globorotalia puncticulata–Globorotalia margaritae*		*G. margaritae* — *G. puncticulata*			MPL-3 *G. margaritae/ G. puncticulata*
			Globorotalia margaritae		*Sphaeroidinellopsis* spp.	*G. margaritae*	*G. margaritae*	MPL-2 *G. margaritae ma*
			Sphaeroidinellopsis seminulina s. l.			*Sphaeroidinellopsis*		MPL-1 *Sphaeroidinellop*
MIOCENE	LATE	MESSINIAN	Non-distinctive Zone				*G. dutertrei/ G. humerosa*	D'Onofrio et al — *G. multiloba*
			Globorotalia conomiozea		*G. miocenica* s. l.			*G. mediterran*
		TORTONIAN	*Globigerinoides obliquus extremus*	*Globorotalia suterae*	*G. ventriosa/ G. nepenthes*		*G. acostaensis*	*G. suterae*
				Globigerinoides obliquus extremus/G. bulloideus				*G. obliquus extremus*
			Globorotalia acostaensis					*G. continuos*
	MIDDLE	SERRAVALLIAN	*Globorotalia menardii* s. l.				*G. menardii*	
			Globorotalia siakensis	*Globorotalia siakensis/Globigerinoides obliquus obliquus*	*G. obliquus/G. lenguaensis*		*G. mayeri*	
				Globigerinoides subquadratus				
				Globoquadrina altispira altispira	*G. altispira/ G. miozea*			
			Orbulina suturalis– Globorotalia peripheroronda	*Globorotalia praemenardii– Globorotalia peripheroronda*			*G. peripheroronda/ O. suturalis*	
				Orbulina universa				
		LANGHIAN		*Orbulina suturalis*	*O. suturalis*			
			Praeorbulina glomerosa s.l.		*P. glomerosa* s. l.		*P. glomerosa*	
	EARLY	BURDIGAL.	*Globigerinoides trilobus*		*G. bisphericus*		*G. trilobus*	
					G. dehiscens			
		AQU.	*Globoquadrina dehiscens dehiscens-Catapsydrax dissimilis*	*Globigerinoides altiaperturus- Catapsydrax dissimilis*	*G. altiaperturus/ G. trilobus*		*G. dissimilis/ G. altiaperturus*	
				Globoquadrina dehiscens dehiscens	*G. primordius*		*G. primordius*	
OLIG.	LATE		*Globorotalia kugleri*				*G. kugleri*	

Fig. 3. Correlation of Mediterranean zonal schemes

Zachariasse 1975	Cita & Premoli Silva 1971-1973 (in Cita 1976)	Bizon 1979	Borsetti et al 1979	Iaccarino & Salvatorini 1982	
	G. truncatulinoides	G. truncatulinoides	G. truncatulinoides / G. cariacoensis		
inflata	G. inflata	MPL 6 — G. inflata	G. inflata	Globorotalia inflata	
bononiensis	G. obliquus extremus	5 — G. crassaformis/ G. aemiliana	G. ex gr. crassaformis	Globorotalia aemiliana	G. crassaformis crassaformis / G. aemiliana-S. seminulina s. l.
	S. subdehiscens	4 — G. puncticulata	G. puncticulata	Globorotalia puncticulata	G. bononiensis / G. apertura
puncticulata	G. margaritae evoluta	3 — G. margaritae/ G. puncticulata		Globorotalia puncticulata—Globorotalia margaritae	
margaritae	G. margaritae margaritae	2 — G. margaritae	G. margaritae	Globorotalia margaritae	
Sphaeroidinellopsis	Sphaeroidinellopsis	1 — Sphaeroidinellopsis	Sphaeroidinellopsis	Sphaeroidinellopsis seminulina s. l.	
conomiozea	G. plesiotumida	G. mediterranea	atypical zone	Non-distinctive Zone	
			G. conomiocea/ G. mediterranea	Globorotalia conomiozea	
		G. humerosa	G. merotumida	Globigerinoides obliquus extremus	Globorotalia suterae / Globigerinoides obliquus extremus/G. bulloideus
acostaensis	G. acostaensis/ G. merotumida	G. acostaensis	G. acostaensis	Globorotalia acostaensis	
continuosa	G. lenguaensis/ G. obliquus	G. menardii	G. obliquus obliquus	Globorotalia menardii s. l.	
	G. nepenthes/G. druryi				Globorotalia siakensis/Globigerinoides obliquus obliquus
	O. universa	G. mayeri		Globorotalia siakensis	Globigerinoides subquadratus
			O. universa		Globoquadrina altispira altispira
		G. peripheroronda			Globorotalia praemenardii-Globorotalia peripheroronda
				Orbulina suturalis-Globorotalia peripheroronda	Orbulina universa
	O. suturalis		O. suturalis		Orbulina suturalis
	P. glomerosa	P. glomerosa	Praeorbulina spp. / Praeorbulina glomerosa s.l.	Praeorbulina glomerosa s.l.	P. glomerosa circularis / P. glomerosa sicana
	G. bisphericus (= G. sicanus)	G. bisphericus	G. sicanus	Globigerinoides trilobus	
	G. trilobus/G. praescitula	G. trilobus	G. trilobus		
	G. altiapertura	G. dissimilis/ G. altiaperturus	G.woodi/G.altiaperturus	Globoquadrina dehiscens dehiscens-Catapsydrax ciperoensis	Globigerinoides altiaperturus-Catapsydrax ciperoensis
	G. primordius	G. kugleri	G. primordius		Globoquadrina dehiscens dehiscens
	G. kugleri		G. gr. tripartita		

for correlations with areas which were not affected by the salinity crisis.

Close to the change from sinistral to dextral coiling of *Globorotalia acostaensis*, *Globigerina multiloba* first occurs. This event, which is ecologically controlled, is commonly recorded from Messinian sequences of the Mediterranean and can be useful in the recognition of the upper boundary of the zone. D'Onofrio *et al.* (1975) and Colalongo *et al.* (1979a, b) used this event to distinguish the *Globigerina multiloba* Subzone.

NON-DISTINCTIVE ZONE

Category: Interval zone

Age: Messinian, Late Miocene

Author: Iaccarino & Salvatorini, 1982

Definition: Interval from first coiling change of *Globorotalia acostaensis* from sinistral to dextral after *Globorotalia conomiozea* datum to first appearance of permanent open marine conditions in the Mediterranean after the Late Miocene salinity crisis.

Remarks: Iaccarino & Salvatorini (1982) called as 'non-distinctive Zone' the interval poorly or not characterized by planktic foraminifera and/or other fossils. During this time interval the Mediterranean was affected by a salinity crisis characterized by a pre-evaporite sequence (Tripoli Formation) in which foraminifera are still common (*Globigerina multiloba* Subzone of D'Onofrio *et al.*, 1975) followed by an evaporitic sequence (Gessoso-solfifera Formation) and finally by a brackish water facies with the return of faunas (Lago-mare facies).

SPHAEROIDINELLOPSIS SEMINULINA s.l. ZONE

Category: Interval zone

Age: Zanclean, Early Pliocene

Author: Iaccarino & Salvatorini, 1982

Definition: Interval from first appearance of permanent open marine conditions in the Mediterranean after the Late Miocene salinity crisis to first occurrence of *Globorotalia margaritae*.

Remarks: The zone represents the interval after the end of the salinity crisis that precedes the first arrival of *Globorotalia margaritae* in the Mediterranean. *Sphaeroidinellopsis* is not always common and generally is absent at the very base of the zone where only *Orbulina*, *Globigerina* and *Globigerinoides* characterize the foraminiferal assemblage.

The *Sphaeroidinellopsis* Acme Zone of Cita (1973, 1975a) is by definition an ecozone which, according to its author, falls within the range of *Globorotalia mar-*

garitae, even if this taxon is only occasionally recorded within it. The *Sphaeroidinellopsis* spp. Zone of Borsetti *et al.* (1979) corresponds to the same interval but its lower boundary is defined by the first occurrence of *Globigerinoides ruber elongatus* and *Globigerinoides ruber parkeri* or by *Sphaeroidinellopsis*.

GLOBOROTALIA MARGARiTAE ZONE

Category: Interval zone

Age: Zanclean, Early Pliocene

Author: Cita, 1975a

Definition: Interval from first occurrence of *Globorotalia margaritae* to first occurrence of *Globorotalia puncticulata*.

Remarks: No other criteria than those used for its definition aid in the recognition of this zone.

GLOBOROTALIA PUNCTICULATA–GLOBOROTALIA MARGARITAE ZONE

Category: Concurrent range zone

Age: Zanclean, Early Pliocene

Author: Cita, 1975a

Definition: Interval from first occurrence of *Globorotalia puncticulata* to last occurrence of *Globorotalia margaritae*.

Remarks: *Globigerina nepenthes* disappears within, *Globoquadrina altispira altispira* at the top of the zone.

GLOBOROTALIA PUNCTICULATA ZONE

Category: Interval zone

Age: Piacenzian, Late Pliocene

Author: Iaccarino & Salvatorini, 1982

Definition: Interval from last occurrence of *Globorotalia margaritae* to first occurrence of *Globorotalia aemiliana*.

Remarks: The *Globorotalia puncticulata* Zone of Colalongo & Sartoni (1979) and Borsetti *et al.* (1979) encompasses a longer time interval and includes both the *Globorotalia puncticulata–Globorotalia margaritae* Zone and the *Globorotalia puncticulata* Zone as here defined.

Iaccarino & Salvatorini (1982) subdivided the zone into two subzones. The boundary between them is marked by the first occurrence of *Globorotalia bononiensis*. This event, according to Colalongo & Sartoni (1979) falls within the range of *G. margaritae*. On the basis of my experience and from the literature the two species have never been found together.

GLOBOROTALIA AEMILIANA ZONE

Category: Interval zone

Age: Piacenzian, Late Pliocene

Author: Colalongo *et al.*, 1974, renamed by Iaccarino & Salvatorini, 1982

Definition: Interval from first occurrence of *Globorotalia aemiliana* to first occurrence of *Globorotalia inflata*.

Remarks: Colalongo *et al.* (1974) named this zone *Globorotalia* gr. *crassaformis* Cenozone (= assemblage zone). Borsetti *et al.* (1979) distinguished within their *Globorotalia* gr. *crassaformis* Zone a lower *Globorotalia aemiliana* Subzone and an upper *Globorotalia crassaformis crassaformis* Subzone, basing their subdivision on the appearance of *G. crassaformis*. Iaccarino & Salvatorini (1982) renamed the zone and recognized within it a lower *Globorotalia aemiliana–Sphaeroidinellopsis seminulina* Subzone and an upper *Globorotalia crassaformis crassaformis* Subzone. The boundary between them is defined by the extinction horizon of *Sphaeroidinellopsis*. Therefore, the *Sphaeroidinellopsis subdehiscens* Zone of Cita within which *Globorotalia aemiliana* is recorded, correlates with the lower subzone of Iaccarino & Salvatorini.

The subzones have not been retained in the present zonation since there is no general agreement on the range of *Globorotalia crassaformis crassaformis*, which according to some authors appears together with *Globorotalia aemiliana*. On the other hand *Sphaeroidinellopsis* spp. extinction which has been seen to be a more reliable bioevent is not always recognizable, particularly in shallower water sediments. Therefore the *Globorotalia aemiliana* Zone is more easily detectable in most sediments outcropping in the Mediterranean area as documented by the numerous works on the Pliocene sequences.

Globigerinoides obliquus extremus becomes rare within the upper part of the zone. According to Cita (1973) the extinction level of this taxon is close to the first occurrence of *Globorotalia inflata*, whereas according to Selli *et al.* (1977), Colalongo & Sartoni (1977) and Colalongo *et al.* (1981) the last *Globigerinoides obliquus extremus* occur close to the top of the *Globorotalia inflata* Zone in the Vrica section.

GLOBOROTALIA INFLATA ZONE

Category: Interval zone

Age: Piacenzian, Late Pliocene

Author: Colalongo & Sartoni, 1979

Definition: Interval from first occurrence of *Globorotalia inflata* to first occurrence of *Globigerina cariacoensis*.

Remarks: The *Globorotalia inflata* Zone represents the latest Pliocene in the Mediterranean. In the original definition

of Cati *et al.* (1968) the upper boundary of the zone was marked by the first appearance of *Globorotalia truncatulinoides*. This event was considered a reliable datum plane for the Plio-Pleistocene boundary.

Sprovieri (1976) proved that *G. truncatulinoides* appears well after the very early Pleistocene and suggested that this event could no longer be used to recognize the Plio-Pleistocene boundary. Later, the definition of the upper boundary of the zone was changed by Colalongo & Sartoni (1979), and the first occurrence of *Globigerina cariocoensis* which is close to the Plio-Pleistocene boundary in the Vrica type section, was taken as indicating the zonal boundary.

Globorotalia truncatulinoides truncatulinoides as distinct from *G. truncatulinoides excelsa* (Sprovieri *et al.*, 1980), first occurs together with *G. inflata* or within the *Globorotalia aemiliana* Zone, immediately before the advent of *G. inflata*.

GLOBIGERINA CARIACOENSIS ZONE

Category: Interval zone

Age: Early Pleistocene

Author: Colalongo & Sartoni, 1979, redefined by Iaccarino, this volume

Definition: Interval from first occurrence of *Globigerina cariacoensis* to first occurrence of *Globorotalia truncatulinoides excelsa*.

Remarks: The change in the zonal definition concerns the taxon which characterizes the upper boundary. *Globorotalia truncatulinoides truncatulinoides* first occurs, according to Sprovieri *et al.* (1980), within the latest part of *Globorotalia aemiliana* Zone or together with *G. inflata*.

GLOBOROTALIA TRUNCATULINOIDES EXCELSA ZONE

Category: Interval zone

Age: Early Pleistocene

Author: Ruggieri & Sprovieri, 1975; renamed by Iaccarino, this volume

Definition: Interval from first occurrence of *Globorotalia truncatulinoides excelsa* to Recent.

Remarks: The *Globorotalia truncatulinoides* Zone of Ruggieri & Sprovieri (1975) was established to define the base of the Sicilian stage. The zone is here re-named *Globorotalia truncatulinoides excelsa* Zone following the findings of Sprovieri *et al.* (1980) in order to avoid confusion with the *Globorotalia truncatulinoides* Zone as used by those authors who define the top of the *Globorotalia inflata* Zone by the first appearance of *Globorotalia truncatulinoides*. The *Globorotalia truncatulinoides* Zone of Colalongo & Sartoni corresponds to this zone.

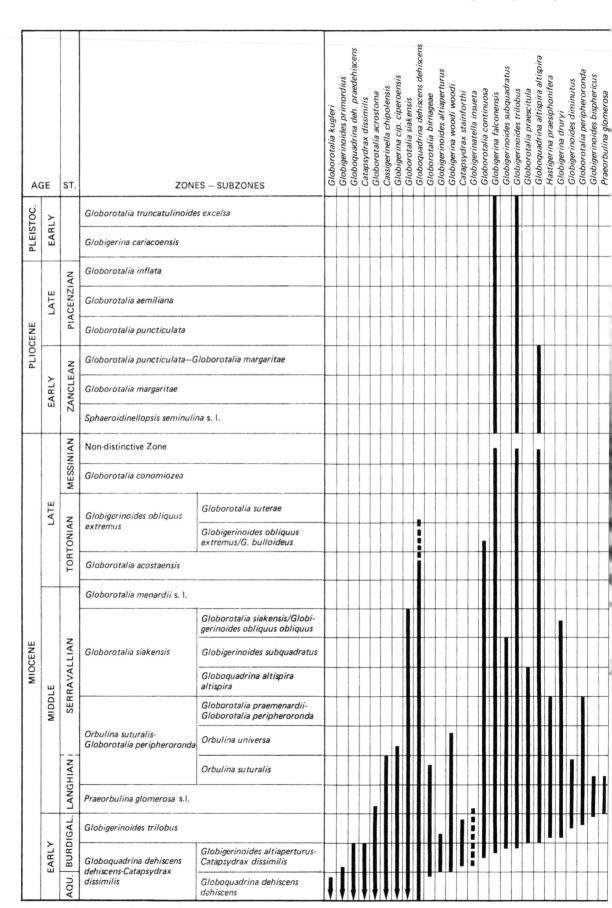

Fig. 4. Distribution of Miocene to Pleistocene
Mediterranean planktic foraminifera

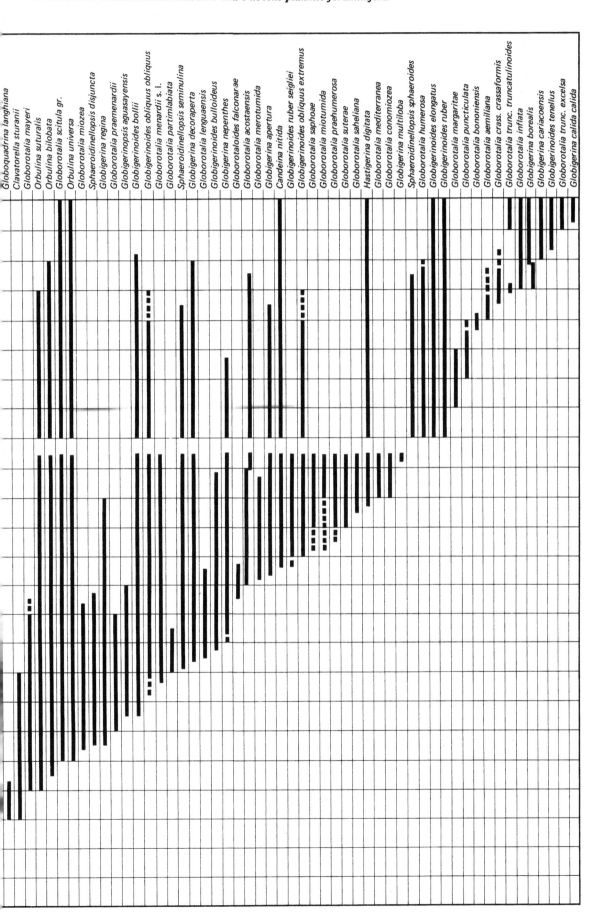

Discussion of taxa

This chapter deals only with typical Mediterranean taxa with few exceptions. For the stratigraphically significant low latitude and southern mid-latitude forms which are also present in the Mediterranean area and which appear as index forms on the range chart Fig. 4, reference is made to Chapter 6.

Exceptions from the above are the inclusion in this chapter of:

(a) *Globigerinopsis aguasayensis*, *Globoquadrina dehiscens praedehiscens*, *Globorotalia birnageae* and *G. siakensis*, forms used in Mediterranean biostratigraphy but originally described from low latitudes, where they are considered to be of lesser stratigraphic significance.

(b) *Globigerina calida calida* which is an index form in the Mediterranean and also a subzonal marker in low latitudes.

(c) The temperate to cold water species *Globigerina apertura*, *G. borealis*, *G. decoraperta*, *G. druryi* and *Globorotalia inflata*, originally described from outside the Mediterranean but characteristically developed and of stratigraphic significance in the Mediterranean.

On the other hand, all *Globigerinoides* and *Clavatorella* taxa, including those erected in the Mediterranean area (*G. tenellus*, *G. bollii*, *G. bulloideus*, *C. sturanii*) are dealt with in Chapter 6. Finally, *Globorotalia crassaformis* is figured here for comparison with other Mediterranean taxa, but taxonomic notes are also found in Chapter 6.

GLOBIGERINA APERTURA Cushman
Figures 5.9; 4

Type reference: *Globigerina apertura* Cushman, 1918, p. 57, pl. 12, figs. 8a–c.

The species is recognizable by its characteristically large, semicircular aperture and low trochospire. It was considered as a subspecies of *Globigerina bulloides* by Blow (1969), but its large aperture is a stable distinguishing feature which justifies its retention as a separate species. *G. riveroae* Bolli & Bermudez, which according to Blow (1969) is a junior synonym of *G. apertura* due to its similarly large aperture, is here considered as a separate species because it differs from *G. apertura* in having a more elevated trochospire. The shape and size of the aperture of *G. apertura* recalls that of *G. woodi* and the analogies are so close that Jenkins (1960) reported the existence of intermediate forms. Colalongo *et al.* (1979b) illustrated forms called by them *G. woodi-apertura*.

GLOBIGERINA BOREALIS Brady
Figures 5.6; 4

Type reference: *Globigerina bulloides* d'Orbigny var. *borealis* Brady, 1881, p. 412.

Type figure: *Globigerina bulloides*, arctic variety, Brady, 1878, p. 435, pl. 21, figs. 10a–c.

The species is better known as *Globigerina pachyderma*. Banner & Blow (1960) proposed the adoption of *G. borealis* instead of *G. pachyderma* (Ehrenberg), which has to be regarded as a 'nomen dubium' (Blow, 1969). In fact the taxon given by Brady (1878, pl. 21, fig. 10) as *Globigerina bulloides* arctic variety, corresponds well to *G. pachyderma*, while the figure given by Ehrenberg as *Aristerospira pachyderma* is not clearly recognizable.

Brady (1878) regarded *G. borealis* as 'a thick-walled, cold water variety of *Globigerina bulloides*' and in 1881 named it '*borealis*' but did not figure it. Subsequently, in 1884, he illustrated it (pl. 114, figs. 19–20) and considered it as a synonym of *G. pachyderma*. From Ehrenberg's original illustration (1873, pl. 1, fig. 4) it may be seen that *Aristerospira pachyderma* is a thick-walled form but not so clearly representative of the species as the specimens of Brady.

In my opinion the type figures given by Brady both in 1881 and in 1884 correspond well to the species concept while the lectotype illustrated by Banner & Blow (1960, pl. 3, figs.

Fig. 5. *Globigerina, Globorotaloides, Globoquadrina, Globigerinopsis* (all figures × 60)

1 *Globigerina druryi* Akers
  Holotype. Humble Oil & Refining Co. Ellender well-1, core 11, 257–67 feet, Middle Miocene, Terrebone Parish, Louisiana, USA.

2 *Globigerina falconensis* Blow
  Holotype. *Globigerinatella insueta* Zone, Miocene, Pozon Fm., near Pozon, Eastern Falcon, Venezuela.

3 *Globigerina multiloba* Romeo
  Holotype. Sample no. 2, *Orbulina suturalis* Zone, 'Tripoli' beds, Vallone di Casino section, Cosenza Prov., Calabria, Italy.

4 *Globigerina regina* Crescenti
  Holotype. Sample 32b, *Globorotalia menardii* Zone, Middle Miocene, Frosolone Fm., Aquevive section, central–southern Apennines, Italy.

5 *Globigerina decoraperta* Takayanagi & Saito
  Holotype. Sample A-16, *Globorotalia cultrata cultrata/Globigerina nepenthes* Zone, Miocene, Nobori Fm., cliff 100 m E of Nobori, Kochi Pref-, Shikoku, Japan.

6 *Globigerina borealis* Brady
  Holotype. Recent, North-polar area.

7 *Globigerina cariacoensis* Rögl & Bolli
  7a Holotype, 7b–c Paratypes. *Globorotalia fimbriata* Subzone, Holocene, DSDP Leg 15-147-2-1, top, Cariaco trough, Caribbean Sea.

8 *Globorotaloides falconarae* Giannelli & Salvatorini
  Holotype. Giannelli & Salvatorini, 1976, sample IPG7338, Tortonian, Falconara (Sicily).

9 *Globigerina apertura* Cushman
  Holotype. Miocene, Yorktown Fm., Suffolk, Virginia, USA.

10 *Globoquadrina langhiana* Cita & Gelati
  Holotype. Level 53, *Globoquadrina langhiana* Zone, Middle Miocene, Rio Marruco series, Gorgenzo, Cuneo Prov., Italy.

11 *Globoquadrina dehiscens praedehiscens* Blow & Banner
  Holotype. *Globorotalia kugleri* Zone, Oligocene, Cipero Fm., San Fernando Bypass Road, Trinidad.

12 *Globigerinopsis aguasayensis* Bolli
  Holotype. Sinclair Oil & Refining Co. Aguasay well 3, cutting at 13 390 feet, probably *Globorotalia fohsi robusta* Zone, Middle Miocene, Monagas, Eastern Venezuela.

13 *Globigerina calida calida* Parker
  Holotype. LSDH core 78P at 100–102 cm, Zone N 19, Pliocene, lat. 4°31′ S, long. 168°02′ E, Pacific.

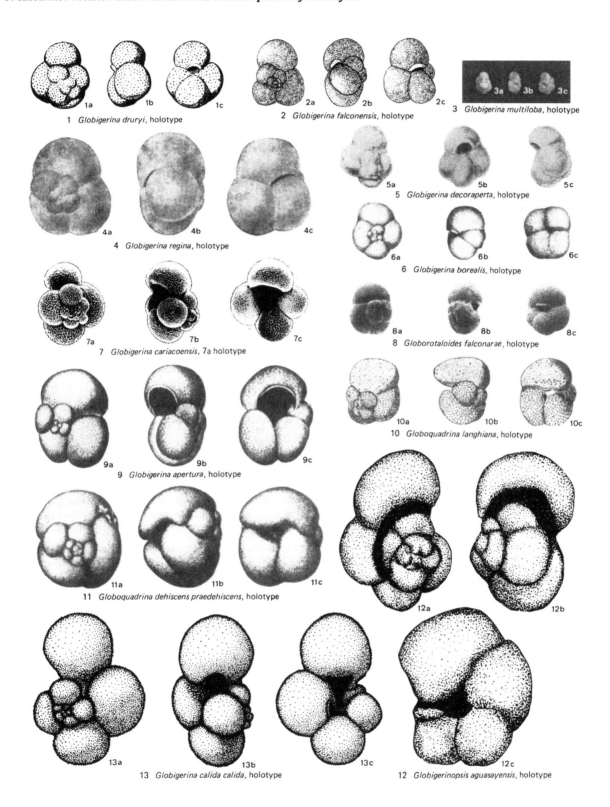

1 *Globigerina druryi*, holotype

2 *Globigerina falconensis*, holotype

3 *Globigerina multiloba*, holotype

4 *Globigerina regina*, holotype

5 *Globigerina decoraperta*, holotype

6 *Globigerina borealis*, holotype

7 *Globigerina cariacoensis*, 7a holotype

8 *Globorotaloides falconarae*, holotype

9 *Globigerina apertura*, holotype

10 *Globoquadrina langhiana*, holotype

11 *Globoquadrina dehiscens praedehiscens*, holotype

12 *Globigerinopsis aguasayensis*, holotype

13 *Globigerina calida calida*, holotype

4a–c) does not fully reflect Brady's illustrations. *G. borealis* shows great variability both in shape and in the calcification of the test. In particular, the last chamber is often variable in size and sometimes appears to be no more than a broad flap covering the aperture of the penultimate chamber. In other specimens it is abortive and may resemble a bulla. The aperture in such specimens may be umbilical–extraumbilical in position, giving a *Globorotalia*-like appearance to the test. Bandy & Theyer (1971) observed that there is a relationship between growth, development of the abortive last chamber and calcification. Normal development of the last chamber and thin wall are typical for the upper water column specimens. From bottom sediments come specimens with moderate calcification with an abortive last chamber becoming dominant. The advanced stage of calcification gives rise to specimens with a thickened wall, very restricted aperture and oval to subglobose test shape. Studies of this species suggest restriction of sinistral populations to Antarctic and Arctic waters, and dextral forms to warmer areas (Bandy, 1960, 1968). Bandy (1960, 1968) emphasized the stratigraphic significance of the coiling ratios in *G. borealis*; sinistral forms, limited to cold waters, can be used to separate Pleistocene from Pliocene sediments.

GLOBIGERINA CALIDA CALIDA Parker
Figures 5.13; 4

Type reference: *Globigerina calida* Parker, 1962, p. 221, figs. 9a–c.

Blow (1969), in an attempt 'to recognize the biostratigraphic value of the extreme morphological forms characterized by the holotype of *Globigerina calida*', restricted Parker's concept of her taxon, gave it subspecies rank and distinguished it from his *G. calida praecalida*, which stratigraphically precedes *G. calida calida*.

Compared with *G. calida praecalida*, the large-sized (up to 0.8 mm), low trochospiral and loosely coiled *G. calida calida* is characterized by a tendency towards radial elongation of the later chambers, in particular the final one, and thus approaches the condition seen in *Beella*. There is a trend for the chambers forming the last whorl to become more numerous, often numbering 5 or, more rarely, 6. *G. calida calida* differs from *G. bulloides* in the radially more elongated chambers resulting in a more lobate equatorial periphery.

Forms called *G.* aff. *calida calida* with intermediate morphological features between *G. calida calida* and *G. calida praecalida* have been observed within the Late Pliocene (Colalongo & Sartoni, 1977, pl. 12, fig. 2). These forms show more globular chambers, more open coiling and a more rapid increase in the size of the last chamber. The typical elongation of the last chamber is absent.

GLOBIGERINA CARIACOENSIS Rögl & Bolli
Figures 5.7; 4

Type reference: *Globigerina megastoma cariacoensis*, Rögl & Bolli, 1973, p. 564, pl. 2, text-figs. 4a–c.

The major morphological features of this species are the high to very high trochospire, fairly lobate equatorial periphery, loosely arranged globular chambers increasing regularly in size except for the last one which may be tilted over the umbilicus, a large umbilicus nearly quadrangular in outline, a wide and low-arched aperture without lip and a transparent and delicate wall densely and regularly covered by pores.

G. cariacoensis differs from *G. megastoma* (see lectotype description in Banner & Blow, 1960) only in the wider aperture without lip and in the greater number of chambers (13–14 against 11). The other features such as the height of the trochospire, the loose arrangement of the chambers, the shape of the last chamber and the structure of the wall are closely related.

The specimens of *G. cariacoensis* with more elongated chambers suggest a relationship with *G. bermudezi*, which differs in the much lower trochospire and in the final chamber being distinctly elongated in a spiro-umbilical direction which gives rise to a more extraumbilical position of the aperture. This species, described from the Caribbean, has also been recorded recently from the Mediterranean (Selli *et al.*, 1977; Colalongo & Sartoni, 1977; Colalongo & Sartoni, 1979). However, the specimens from this area do not show the typical wall structure of the Caribbean forms. The appearance of *G. cariacoensis* is inferred to be an important stratigraphic event in recognizing the Plio-Pleistocene boundary. Poore & Berggren (1975) have already tentatively utilized this event for this purpose. In fact, the first occurrence of *G. cariacoensis* is the event which most closely approximates the Plio-Pleistocene boundary in the Vrica stratotype section in Calabria.

GLOBIGERINA DECORAPERTA Takayanagi & Saito
Figures 5.5; 4

Type reference: *Globigerina druryi decoraperta*, Takayanagi & Saito, 1962, p. 85, pl. 28, figs. 10a–c.

This species, evolving from *G. druryi*, was originally described as a subspecies of that species, having a larger, higher arched aperture with a thickened rim instead of a thin lip and a low to very high spire. The morphologic features of *G. decoraperta* make it an easily recognizable taxon. Nevertheless it has been seen to show a great variability: high to moderately high arched aperture, low to high trochospire, thickened to thin rim. Also, in size this taxon varies remarkably.

In the Late Pliocene *G. decoraperta* seems to evolve into *G. rubescens*. In fact, *G. decoraperta* is well documented from Middle Miocene to Late Pliocene but is never recorded from the Pleistocene.

GLOBIGERINA DRURYI Akers
Figures **5.1**; 4

Type reference: *Globigerina druryi* Akers, 1955, p. 654, pl. 65, figs. 1a–c.

According to Aker's description the small size, the rounded and lobate periphery, the coarsely perforate wall and the small, low-arched and thin-lipped aperture are the major distinctive features of the species.

According to Blow (1969), *G. druryi* is the ancestor of *G. nepenthes*, while Brönnimann & Resig (1971) consider *G. druryi* closer to *G. falconensis*. *G. falconensis* as described and figured by Blow (1959) differs from *G. druryi* in its larger size, in the much more developed apertural lip and in the flatter trochospire.

G. pseudodruryi and *G. nepenthoides* both described by Brönnimann & Resig (1971) are here included in *G. druryi*. The creation of the two taxa hardly appears justified, taking into account the variability of the species.

Early forms of *G. druryi* range stratigraphically lower than recorded by Blow (1969). Their occurrence is consistent with the range of *G. pseudodruryi* and *G. nepenthoides*.

GLOBIGERINA FALCONENSIS Blow
Figures **5.2**; 4

Type reference: *Globigerina falconensis* Blow, 1959, p. 177, pl. 9, figs. 40a–c.

This species is readily recognizable by its well developed apertural lip which may entirely cover the umbilical area. Through time the lip becomes less thickened and less well developed and, as a result, the umbilicus more open. The result is that specimens from the Pliocene and Late Miocene possess thinner tests and thinner lips and the umbilicus appears wider. On the basis of this trend Blow (1969) distinguished a *Globigerina falconensis* forma *typica* for the oldest forms and a forma *atypica* for the youngest.

GLOBIGERINA MULTILOBA Romeo
Figures **5.3**; 4

Type reference: *Globigerina multiloba* Romeo, 1965, p. 1266, pl. 118, figs. 1a–c.

Typical specimens of this species have been recorded only from Messinian sediments in the Mediterranean Basin. Small sizes (maximum diameter 0.20 mm), 6–7 (exceptionally only 5) chambers in the last whorl, weakly convex spiral side, lobate periphery and coarsely perforate test are the diagnostic features. The latter is the most distinctive one because it allows *G. multiloba* to be separated from other similar forms like *G. concinna, G. quinqueloba* (which also has the last chambers covering the umbilicus), and from young specimens of *Globorotalia humerosa* (Catalano & Sprovieri, 1971, pl. 4, fig. 11). *Globigerina multiloba* recorded by Brönnimann & Resig (1971)

from the Oligocene to the Early Miocene in the Pacific is not referable to this species. When correctly interpreted, it is inferred to be restricted to the Mediterranean area.

The distribution of *G. multiloba* is thought to be dependent on the environmental conditions during the latest Miocene, which could explain its absence outside the Mediterranean, where the sea-water mass remained practically unchanged (D'Onofrio *et al.,* 1975).

GLOBIGERINA REGINA Crescenti
Figures **5.4**; 4

Type reference: *Globigerina regina* Crescenti, 1966, p. 34, text-fig. 4.

Globigerina regina is not well known in the literature but is a taxon commonly recorded in the Mediterranean area. The major morphologic features are a very low trochospire, a finely perforate wall surface, rounded axial periphery and flat or slightly convex spiral side. The chambers have a subspherical shape: the three chambers forming the last whorl have approximately the same size. The most peculiar feature is the aperture, which is a very elongated slit with a thin lip at the base of the last chamber. The aperture of *G. regina* is very similar to that of *Globigerina juvenilis* which, however, shows a distinctly lobate periphery.

GLOBIGERINOPSIS AGUASAYENSIS Bolli
Figures **5.12**; 4

Type reference: *Globigerinopsis aguasyensis* Bolli, 1962, p. 282, pl. 1, figs. 2a–c.

The main diagnostic feature of this species is the aperture which is arched-interiomarginal, umbilical in the early stage, in the last chambers of adult specimens extending further towards the periphery and the spiral side. The different development of the aperture gives rise to a large variability of morphotypes so that some adult individuals approach the shape of *Hastigerina* from which *G. aguasayensis* differs in remaining distinctly trochospiral throughout the various stages. Specimens with the *Globigerinopsis*-type aperture involving only the ultimate chamber occur commonly in the Mediterranean area. *G. aguasayensis* is not consistently recorded in the Mediterranean. When present, it is mostly represented by specimens referable to the hypotypes figured by Bolli (1962, pl. 1, figs. 5–7). It is here considered a good index species because its range is limited to the Middle Miocene. Its extinction precedes the first appearance of *Globigerina nepenthes*.

GLOBOQUADRINA DEHISCENS PRAEDEHISCENS Blow & Banner
Figures **5.11**; 4

Type reference: *Globoquadrina dehiscens praedehiscens* Blow & Banner, 1962 (*in* Eames *et al.*, 1962), p. 116, pl. 15, figs. Q–S.

Blow & Banner considered this taxon as the immediate ancestor of *Globoquadrina dehiscens dehiscens*. The major distinguishing features according to them are the number of chambers in the last whorl (only three are visible in the umbilical side of adult specimens), the globosely subconical axial profile, the smoothly and broadly rounded axial periphery, and the small but deep and open umbilicus. Steeply sloping umbilical margins characterize the chambers of the last whorl. The aperture is a very low arch bordered by a flap-like triangular lip which projects into the umbilicus as a tooth.

The change in the position of the aperture from umbilical-extraumbilical in *G. dehiscens praedehiscens* to umbilical in *G. dehiscens dehiscens* has been considered by Blow & Banner as the evolutionary character which distinguishes the two species. Such an evolutionary lineage is not accepted by all stratigraphers and has given rise to controversial opinions. According to Stainforth *et al.* (1975) *G. dehiscens praedehiscens* is a 'transitional variant' of *Globigerina tripartita*. This species does show great variability but *G. dehiscens praedehiscens* can be included in it only if interpreted in a broad sense (three chambers in the last whorl, elongated aperture, globose and appressed chambers in the last whorl).

GLOBOQUADRINA LANGHIANA Cita & Gelati
Figures **5.10**; 4

Type reference: *Globoquadrina langhiana* Cita & Gelati, 1960, p. 242, text-figs. 1a–c.

G. langhiana, first described from the lower part of the Langhian in the Piedmont Basin, has been documented by many authors. The major features which distinguish it from other species of *Globoquadrina* are the globose and rapidly increasing chambers: the last one being much larger than the previous ones. The equatorial periphery is moderately lobate and the general shape is subquadrate; the aperture is narrow, elongated and bordered by a distinct lip. According to Cita & Gelati, *G. langhiana* shows close analogies with *G. larmeui* Akers in its globose shape and in the reduced number of chambers. *G. larmeui* is much higher and the last chamber is more globose on the umbilical side.

GLOBOROTALIA ACROSTOMA Wezel
Figures **6.13**; 4

Type reference: '*Globorotalia*' *acrostoma* Wezel, 1966, p. 1298, text-figs. 1a–c.

This species is poorly known in the literature and since it shows a restricted distribution within the Mediterranean Early Miocene a more detailed description of its features appears here justified.

It is characterized by a subcircular equatorial profile, distinctly lobate periphery, a rounded axial profile, a convex ventral side and a generally flattened spiral side. The test is a narrow trochospire formed by subglobular to ovoid moderately appressed chambers separated on the spiral side by curved sutures. The aperture extending from the narrow umbilicus to the peripheral margin is a high vaulted arch developed perpendicular to the coiling direction. The apertural face is generally flattened, large and slightly concave. The range of the maximum diameter for the taxon is 0.23–0.36 mm.

G. acrostoma shows relationships with *G. mayeri* as illustrated by Cushman & Ellisor (1939). It differs from the holotype of this species in possessing fewer ($4\frac{1}{2}$–5 against 6) less elongated chambers in the last whorl, a higher aperture, a flattened and large apertural face and a smaller test size.

With *G. siakensis*, *G. acrostoma* has in common the rounded axial profile and the number of chambers in the last

---

Fig. 6. *Globorotalia* (all figures × 60)

1 *G. saheliana* Catalano & Sprovieri
Holotype. Late Miocene, Early Messinian, Sahelian, Section near Castello di Falconara, Southern Sicily.

2 *G. mediterranea* Catalano & Sprovieri
Holotype. Late Miocene, Sahelian, T. Rossi section (Enna), Sicily.

3 *G. saphoae* Bizon & Bizon
Holotype. Late Miocene, Tortonian, Section at Strelia, western Epirus, northwestern Greece.

4 *G. crassaformis* (Galloway & Wissler)
Holotype. Pleistocene, Lomito Quarry, near Lomito, California, USA.

5 *G. suterae* Catalano & Sprovieri
Holotype. Late Miocene, Early Messinian, Sahelian, Section near Sutera, southwestern Sicily.

6 *G. aemiliana* Colalongo & Sartoni
Holotype. Middle Pliocene, Piacenzian, Santerno section, near Borgo Tossignano, Emilia, northern Italy.

7–8 *G. partimlabiata* Ruggieri & Sprovieri
7 Paratype, 8 Holotype. Middle Miocene, Serravallian, Marne di San Cipirello, Sicily.

9 *G. inflata* (d'Orbigny)
Neotype. Recent, off Gomera, Canary Islands.

10 *G. bononiensis* Dondi
Holotype. Late Early to basal Middle Pliocene, T. Savena, E of Bologna, Italy.

11 *G. siakensis* (LeRoy)
Holotype. Auger hole sample, Miocene (?), Locality Ho-528, Rokam-Tapanoeli Region, Central Sumatra.

12 *G. puncticulata* (Deshayes)
Lectotype. Recent, near Rimini, Adriatic Sea.

13 *G. acrostoma* Wezel
Holotype. *Globigerinoides trilobus* Cenozone, upper part, Miocene, 'Zona a scaglie tettoniche', right bank of the Tempio River, Mirabella area, Sicily.

14 *G. birnageae* Blow
Holotype. *Globigerinatella insueta* Zone, Miocene, Pozon Fm., near Pozon, Eastern Falcon, Venezuela.

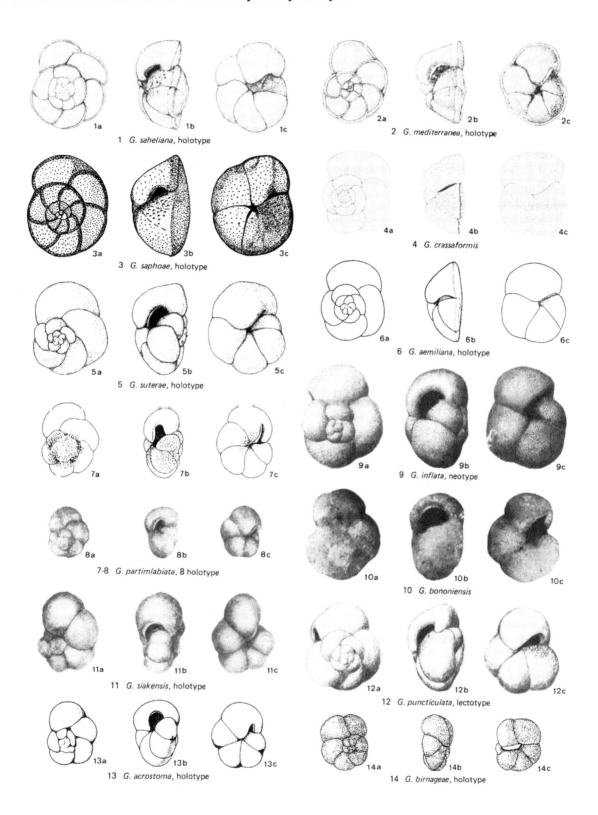

1a 1b 1c
1 *G. saheliana*, holotype

2a 2b 2c
2 *G. mediterranea*, holotype

3a 3b 3c
3 *G. saphoae*, holotype

4a 4b 4c
4 *G. crassaformis*

5a 5b 5c
5 *G. suterae*, holotype

6a 6b 6c
6 *G. aemiliana*, holotype

7a 7b 7c

8a 8b 8c
7-8 *G. partimlabiata*, 8 holotype

9a 9b 9c
9 *G. inflata*, neotype

10a 10b 10c
10 *G. bononiensis*

11a 11b 11c
11 *G. siakensis*, holotype

12a 12b 12c
12 *G. puncticulata*, lectotype

13a 13b 13c
13 *G. acrostoma*, holotype

14a 14b 14c
14 *G. birnageae*, holotype

whorl, but differs in possessing curved instead of radial inter-cameral sutures on the spiral side. Moreover, *G. siakensis* possesses a distinct open, deep umbilicus, a fairly low elongated arched aperture always with an apertural lip. It is not un-common however, to find specimens of *G. siakensis* with an apertural face similar to that of *G. acrostoma*.

Minor analogies exist between *G. acrostoma* and *G. continuosa* in that *G. acrostoma* differs in the higher and more transverse aperture, in the absence of the apertural lip, in having 4½–5 against 4 chambers in the last whorl and in the more inflated final chamber and greater test size.

G. acrostoma is mainly recorded from the Mediterra-nean area and has only recently been found in the Atlantic (Iaccarino & Salvatorini, 1979; Salvatorini & Cita, 1979).

GLOBOROTALIA AEMILIANA Colalongo & Sartoni
Figures **6.6**; 4

Type reference: *Globorotalia hirsuta aemiliana* Colalongo & Sartoni, 1967, p. 3, pl. 30, figs. 1a–c.
Synonym: *Globorotalia crotonensis* Conato & Follador, 1967, p. 556, fig. 4, nos. 1a–c.

In the original description, *G. aemiliana* was considered as a subspecies intermediate between *G. hirsuta* (the specimens identified by Colalongo & Sartoni as *G. hirsuta* were subse-quently found to belong to *G. margaritae*) and *G. crassaformis* (Fig. 6.4). When comparing the two forms it is difficult to relate *G. aemiliana* to *G. margaritae* and to accept that they belong to the same lineage as suggested by Colalongo & Sartoni. The morphologic features characterizing the two taxa make it necessary to consider them as two different species.

On the other hand, *G. aemiliana*, with its low umbilical vaulting may represent an early stage which leads through transitional forms to the high vaulted *G. crassaformis crassa-formis*. Such an evolutionary trend is postulated by Conato & Follador for the lineage *G. crotonensis* (a junior synonym of *G. aemiliana)–G. crassacrotonensis–G. crassaformis*. According to Barbieri (1971) *G. aemiliana* is a synonym of *G. crassula*. Similarities between the two species are evident but *G. crassula* differs in having a keeled peripheral margin.

GLOBOROTALIA BIRNAGEAE Blow
Figures **6.14**; 4

Type reference: *Globorotalia birnageae* Blow, 1959, p. 210, pl. 17, figs. 108a–c.

This species is characterized by its small size, closed umbilicus, tight coiling and well developed apertural lip. The last chamber is often smaller than the previous one. It is very useful in the Mediterranean area for recognizing the Early to early Middle Miocene.

GLOBOROTALIA BONONIENSIS Dondi
Figures **6.10**; 4

Type reference: *Globorotalia bononiensis* Dondi, 1962, p. 162, fig. 41–1.

The morphologic features of this species have been documented in Stainforth *et al.* (1975) and are here summarized as follows: globular shape, 4–5 chambers in the last whorl, slightly inflated spiral side, radial intercameral sutures, lobate and broadly rounded periphery, large and high-arched aperture often with a thin but distinct rim.

Colalongo & Sartoni (1967) suggested that *G. bononiensis* is to be regarded as a chronologic subspecies of *G. puncticulata* without explaining which characters have evolved. In their opinion *G. bononiensis* and *G. puncticulata* belong to the same evolutionary trend, *G. bononiensis* representing the immediate descendant of *G. puncticulata*. In favour of this opinion, some advanced specimens of *G. puncticulata* show a higher aperture, and a more inflated spiral side. On the other hand the radial intercameral sutures and the rim which borders the high aperture occurring in *G. bononiensis* appear to be typical of this species only.

Dondi (personal communication) is of the opinion that the two species do not belong to the same lineage and that *G. bononiensis* has evolved from the *G. opima–G. continuosa*-group. Gradstein (1974) and Zachariasse (1975) do not accept Dondi's view of the origin of *G. bononiensis* and suggest that the species is indigenous to the Mediterranean Basin. Actually, the species also occurs in the North Atlantic as has recently been recorded by Iaccarino & Salvatorini (1979). Furthermore, some specimens identified as *G. puncticulata* or *G. inflata* have probably to be regarded as *G. bononiensis* (Kennett, 1970; Berggren, 1977a, b).

GLOBOROTALIA INFLATA (d'Orbigny)
Figures **6.9**; 4

Type reference: *Globigerina inflata* d'Orbigny, 1839, p. 134, pl. 2, figs. 7–9.
Neotype: *Globorotalia (Turborotalia) inflata* (d'Orbigny). Banner & Blow, 1967, p. 144, pl. 4, figs. 1a–c.

The morphologic features of the species are fully described and discussed in Banner & Blow (1967) and in Stainforth *et al.* (1975).

Controversial opinions concern the evolution of *G. inflata*. According to Colalongo & Sartoni (1967) it evolved from *G. bononiensis*, which in turn derived from *G. puncticulata*. Also Gradstein (1974) and Zachariasse (1975) recorded a transition from *G. bononiensis* to *G. inflata*, the main morpho-logic change concerning the height of the aperture, the shape of the chambers and their numbers. This transition according to the latter authors could reflect a more widespread evolu-tionary change. Contrary to Colalongo & Sartoni, however, they did not record an evolutionary trend from *G. puncticulata* to *G. bononiensis* or from *G. puncticulata* to *G. inflata* as suggested by Kennett (1973) and by Cita (1973). In fact, Kennett believes that *G. inflata* evolved directly from *G. puncticulata*. If this were so, an overlap of the two taxa would be expected, which however does not occur in the Mediterranean Basin, where *G. puncticulata* disappears before the first occurrence of *G. inflata*.

In order to clear up the controversy, the most significant morphologic features of the three species and the major differences must be compared. *G. puncticulata* and *G. inflata* by definition possess curved intercameral sutures differing from those of *G. bononiensis*, which are radial. On the other hand the illustration of the neotype of *G. inflata* by Banner & Blow does not distinctly show this feature. In *G. inflata* 'the early chambers, in dorsal view, are reniform in outline but become increasingly larger than broad during ontogeny' (Banner & Blow, 1967); in *G. puncticulata* they are 'lunate in shape being much larger circumferentially than broad radially, a feature which becomes accentuated during ontogeny' (Banner & Blow, 1960). In *G. bononiensis* the shape of the chambers is ovate.

Following the concept that in an evolutionary trend a morphologic character acquired by a species cannot be reversed, it results that:

(1) If *G. inflata* has derived from *G. bononiensis* the latter cannot evolve from *G. puncticulata* because the intercameral sutures are curved in *puncticulata*, straight in *bononiensis* and curved again in *inflata*.

(2) If *G. inflata* has derived directly from *G. puncticulata*, *G. bononiensis* has to be regarded as belonging to a different lineage or to another branch of *G. puncticulata*.

The first appearance of *G. inflata* in the Mediterranean area is considered an important biostratigraphic event marking the base of the last zone of the Pliocene. It is recorded up to the Holocene. *G. inflata* is very common in the western Mediterranean and is scarce or absent in the eastern Mediterranean (Bizon & Müller, 1978). On the basis of its scarcity in this part of the Mediterranean, Bizon & Müller infer that *G. inflata* is not a useful index fossil for the uppermost Pliocene.

GLOBOROTALIA MEDITERRANEA Catalano & Sprovieri
Figures **6.2**; 4

Type reference: *Globorotalia miocenica mediterranea* Catalano & Sprovieri, 1969, p. 522, pl. 2, figs. 6a–c.

This species was regarded by its authors as a subspecies of *G. miocenica* Palmer. Both forms have a flat spiral side. They differ in the aperture which is bordered by a distinct lip in *G. mediterranea*, in the ventral side which is more conical in *G. miocenica* and in the angle formed by the peripheral margin

with the ventral side which is more acute in *G. miocenica*. *G. mediterranea* shows close similarities to *G. conomiozea*, which may make a separation difficult. Zachariasse (1975), in fact, included both forms under *G. conomiozea*. Berggren (1977b), instead, separated the two taxa primarily on the basis that the predominantly 5-chambered *G. mediterranea* disappears near the Miocene–Pliocene boundary on the Rio Grande Rise, whereas the 5-chambered *conomiozea* continues well into the Pliocene.

Cita & Ryan (1978) and Colalongo *et al.* (1979b) retained the two taxa as distinct entities and, in agreement with Chapronière (1973), they support the evolutionary lineage of *G. conomiozea* from *G. miotumida*. This trend, even though not well marked, is also recorded in the Mediterranean area.

Berggren (1977b) suggested that the low, moderately conical 5–6-chambered *G. miozea* is replaced by a relatively high conical 5–6-chambered form assigned to *G. mediterranea*. Nevertheless the specimens figured by him as *G. miozea* are not typical because the ventral side is too high, especially the last chamber. At least some of the specimens ascribed to *G. miozea* have probably to be included in *G. mediterranea*. In the Mediterranean area transitional forms from *G. miozea* to *G. mediterranea* have not been observed. The suggestion that *G. mediterranea* is a variant or morphotype of *G. conomiozea* is not supported by the stratigraphic record because the appearance of *G. mediterranea* precedes that of *G. conomiozea*. For a comparison of *G. mediterranea* with *G. miozea* and *G. conomiozea* reference is made to Chapter 7 of this volume, where the two species are figured.

The stratigraphic range of *G. mediterranea* in the Mediterranean Basin is restricted to the Late Miocene (latest Tortonian–Early Messinian). Its appearance is used by several authors (Bizon, Bizon & Montenat, 1975; Iaccarino *et al.*, 1975; Zachariasse, 1975; Berggren, 1977b) for recognition of the Tortonian–Messinian boundary. Actually *G. mediterranea* appears immediately prior to *G. conomiozea* in the upper part of the *G. suterae* Subzone (Colalongo *et al.*, 1979b).

GLOBOROTALIA PARTIMLABIATA Ruggieri & Sprovieri
Figures **6.7–8**; 4

Type reference: *Globorotalia acrostoma partimlabiata* Ruggieri & Sprovieri, 1970, p. 23, text-fig. 3.

G. partimlabiata was considered by its authors as a subspecies of *G. acrostoma*. However no transitional forms have been seen in the long interval from the last occurrence of *G. acrostoma* to the first occurrence of *G. partimlabiata*. For this reason the two taxa are here considered as separate species.

G. partimlabiata differs from *G. acrostoma* in having a more narrowly rounded periphery. Its aperture is higher and arranged perpendicular to the coiling direction and bordered by a distinct but incomplete lip. Giannelli & Salvatorini (1976) also recorded specimens with lips completely bordering the aperture. The latter authors consider *G. partimlabiata* as a junior synonym of their *G. melitensis* which they illustrated but did not describe in 1975. *G. partimlabiata* shows similarities with *G. mayeri*, which, however, has a more rounded profile and an aperture that is not perpendicular to the coiling direction.

According to its authors it is possible that *G. partimlabiata* could be considered as a morphotype of *G. peripheroronda*. However, the two taxa have morphologic features sufficiently different not to be confused.

GLOBOROTALIA PUNCTICULATA (Deshayes)
Figures **6.12**; 4

Type reference: *Globigerina puncticulata* Deshayes, 1832, p. 170; figures in Fornasini, 1899, p. 210, text-fig. 5.

Lectotype: *Globigerina puncticulata* Deshayes 1832 (= *Globorotalia (Turborotalia) puncticulata* (Deshayes)): Banner & Blow, 1960, p. 15, pl. 5, figs. 7a–c.

An extensive description and discussion of this species was made by Banner & Blow (1960) who carried out an accurate revision of the taxonomy of *Globigerina puncticulata* and, following the description of Deshayes, designated a lectotype selected from the material preserved in the d'Orbigny collection. Barbieri (1971), in agreement with Banner & Blow, considered *G. puncticulata* Deshayes as valid but stated that the lectotype chosen by Banner & Blow did not correspond to the original description given by Deshayes and did not solve the taxonomic problem of the species.

The controversy raised by Barbieri mainly concerns the form of the aperture which in the lectotype of Banner & Blow is a fairly high arch set into a slightly concave part of the apertural face, while according to Deshayes' description it is a small 'apertura rotunda'.

Within any population of *G. puncticulata* there exists a great variability in the shape of the aperture, from rounded to arched. Stainforth *et al.* (1975), who consider the lectotype of Banner & Blow a form morphologically distinct enough for biostratigraphic use, give a wide documentation of the species and point out the variability of the aperture.

The *G. puncticulata* concept adopted by Barbieri restricts it to forms having rounded peripheries and rounded apertures, while forms having rounded peripheries and elongated apertures extending from the umbilicus towards the peripheral margin are included by Barbieri in *G. crassaformis*. It follows that *G. crassaformis sensu* Barbieri (1967, 1971), Barbieri & Petrucci (1967) and Iaccarino (1967) has to be referred to *G. puncticulata* because it shows features consistent with the description of the lectotype and with the variability of the species.

Many divergencies of opinion concern the evolutionary ancestors of *G. puncticulata*. According to Kennett (1973), *G. puncticulata* is a form which has derived from *G. conomiozea* by reduction in the axial angularity and loss of the peripheral keel. Such a trend is extremely rare within the planktic foraminifera and was doubted by Blow.

Blow (1969), in turn, interpreted *G. puncticulata* as developing from *G. subscitula* Conato (1964) within the interval of the upper part of N 19 and/or the early part of N 20.

Berggren's (1977b) opinion was that the earliest forms of *G. puncticulata* are morphologically similar to forms referred to as *G. praeoscitans* Akers (1972) but are markedly smaller and generally lack the arched semicircular aperture of *G. praeoscitans*. In the Early Pliocene of the Rio Grande Rise the earliest forms of *G. puncticulata* are associated with *G. cibaoensis* into which it appears to grade imperceptibly.

Berggren suggested that *G. puncticulata* evolved in the Early Pliocene from *G. cibaoensis*. This lineage is, however, not sufficiently supported by the arguments put forward by him, such as the narrow sulcus near the peripheral margin of the last chamber above the aperture present in *G. cibaoensis* and in some early individuals of *G. puncticulata*. *G. cibaoensis* in fact has a sharp angular peripheral margin, limbate intercameral sutures on the spiral side and a less convex ventral side.

In the Mediterranean area *G. puncticulata* appears within the Early Pliocene, well after the re-establishment of open marine conditions which mark the Mio-Pliocene boundary.

The possible ancestors *G. conomiozea* and *G. cibaoensis* suggested by Kennett and Berggren respectively did not occur in the early Pliocene of the Mediterranean.

The only one of the above-mentioned species which occurs in the Mediterranean Early Pliocene is *G. subscitula*. A gradual evolution from this species into *G. puncticulata* has so far not been documented. According to Conato & Follador (1967), *G. subscitula* appears later than *G. puncticulata* and therefore cannot be its evolutionary ancestor.

The first appearance of *G. puncticulata* in the Mediterranean Basin could reflect a later migration into warmer waters after its initial evolutionary appearance in the cooler waters of the nearby Atlantic Ocean. If this suggestion, already emphasized by Kennett (1973) for New Zealand, is right, the possibility that *G. puncticulata* evolved from *G. sphaericomiozea* cannot be excluded. Kennett (1973) on the other hand included *G. puncticulata* in *G. sphaericomiozea* and Walters (1965) stated that *G. sphaericomiozea* appears to grade into *G. inflata*; therefore, *G. puncticulata* could represent an intermediate form between the two extreme taxa.

G. puncticulata padana Dondi & Papetti (1968), which is not considered here, differs from *G. puncticulata* in having less depressed sutures, less globular and more appressed chambers and only 3½ chambers in the final whorl. Its distribution extends higher than that of the type species.

GLOBOROTALIA SAHELIANA Catalano & Sprovieri
Figures **6.1**; 4

Type reference: *Globorotalia saheliana* Catalano & Sprovieri, 1971, p. 240, pl. 1, figs. 3a–c.

G. saheliana is a typical species of the Late Miocene in the Mediterranean area. It first occurs within the *Globorotalia suterae* Subzone and disappears before the salinity crisis when the environmental conditions drastically changed.

The species shows morphologic variabilities especially with regard to the number of chambers in the last whorl (4½ to 6) and whether the peripheral margin is acute and weakly keeled or rounded and not keeled (Colalongo *et al.*, 1979b). *G. saheliana* shows similarities with *G. saphoae* from which it differs in possessing a less globular test, a less lobate periphery, a thicker wall, less deeply incised sutures on the ventral side and a more vaulted aperture.

According to D'Onofrio *et al.* (1975) this species shows more consistent similarities with *G. suterae*, and it is here suggested that *G. saheliana* may have branched off from *G. suterae*. However, transitional forms have not been documented.

GLOBOROTALIA SAPHOAE Bizon & Bizon
Figures **6.3**; 4

Type reference: *Globorotalia saphoae* Bizon & Bizon, 1965, p. 248, pl. 4, figs. 9a–c.

The main features characterizing the species are the plano-convex shape of the test with the ventral side strongly convex, a thickened peripheral margin which looks like a keel and the fine pustules that cover the test. The ultimate chamber may be reduced in size and its position is elevated in respect to the plane of the spire.

Jenkins (*in* Bizon & Bizon, 1966) related *G. saphoae* to *G. miozea conoidea* and *G. miozea sphaericomiozea*, which belong to the lineage of *G. miozea*. According to Walters (*in* Bizon & Bizon, 1966) *G. saphoae* differs from *G. miozea conoidea* in being smaller in size and more hemispherical than conical on the umbilical side, and from *G. miozea sphaericomiozea* in being more keeled and more circular in equatorial periphery. After examining a population of *G. miozea* s.l. from an equivalent interval of New Zealand, Bizon & Bizon (1966) affirmed that it is quite impossible to distinguish young specimens of *G. miozea* from *G. saphoae* and proposed that the latter can be considered as a subspecies of *G. miozea*; they named it *G. miozea saphoae*. However, it is here preferred to maintain *G. saphoae* as a separate species.

GLOBOROTALIA SIAKENSIS (LeRoy)
Figures **6.11**; 4

Type reference: *Globigerina siakensis* LeRoy, 1939, p. 262, pl. 4, figs. 20–22.

Globorotalia siakensis, a species widely used in the biostratigraphic record was regarded for a long time as a taxon not discernible from *G. mayeri*. Blow (1969) re-examined the holotype of Cushman & Ellisor (1939) and reached the conclusion that the two species are not co-specific. Of the same opinion were Stainforth *et al.* (1975) and many other authors. More recently Bolli & Saunders (1982a) in a revision of *G. mayeri* concluded that *G. mayeri* and *G. siakensis* cannot be maintained as separate taxa. The conclusions reached by these authors, discussed in the taxonomic notes of *G. mayeri* in this volume (Chapter 6), are based on the direct examination of the holotype of *G. mayeri* and on illustrations and documentations of *G. siakensis*. To demonstrate that their opinion is correct, they had the holotype redrawn. Bolli & Saunders (1982, pl. 1) provide three illustrations of the holotype of *G. mayeri*: the original one of Cushman & Ellisor (1939, figs. 7–9), that redrawn for Blow (1969, figs. 10–12) and, that redrawn for Bolli & Saunders (1982a, figs. 13–15). When comparing the three illustrations of the holotype it appears that they are closely comparable but are distinctly different from the holotype of *G. siakensis*, both as figured originally by LeRoy (1939) and later by Blow (1969). My direct examination of the holotype of *G. siakensis* deposited in Bandung (Java) confirms that the illustrations correspond well to the holotype.

It is characterized by globose chambers not appressed against each other, a distinctly lobate equatorial periphery and an equally convex axial periphery. In *G. mayeri*, conversely, the chambers of the last whorl are radially elongated, appressed against each other, the equatorial periphery is distinctly weakly lobate and the axial periphery distinctly plano-convex. The only deviation is that the intercameral sutures are not distinctly curved as stated by Blow (1969). It is therefore considered here that the two species have different features which justify their separation. This view is independent of the intercameral sutures. All the considerations reported by Bolli & Saunders for *G. mayeri* are applicable to *G. siakensis*. In fact, the variability observed in *G. mayeri sensu* Bolli & Saunders is also typical of *G. siakensis*, such as the shape of the aperture which can be a low or a high arch.

G. siakensis also shows great variability in the number of chambers (from 4½ to 6–7) with generally few in the lower part of its range. A trend could, however, not be established. According to Bolli & Saunders, *G. continuosa* is considered a variant of *G. mayeri* (*G. siakensis* for us) (see pl. 4 of Bolli & Saunders, 1982), but as a variant it should have the same range. On the contrary *G. continuosa* continues well above the extinction of *G. siakensis*.

The features of *G. mayeri* in my concept remain stable during the total range of the species. *G. mayeri* is restricted to the Middle Miocene (Fig. 4) while *G. siakensis* ranges from Oligocene to Middle Miocene.

GLOBOROTALIA SUTERAE Catalano & Sprovieri
Figures **6.5**; 4

Type reference: *Globorotalia suterae* Catalano & Sprovieri, 1971, p. 241, pl. 1, figs. 1a–c.

The authors compared this species with the *G. crassaformis* group as illustrated by Blow (1969), from which it differs in having more chambers in the last whorl ($4\frac{1}{2}$–6 against 4), in the shape of the chambers and in the broadly rounded axial profile.

No comparison with the *G. scitula* plexus to which it appears closer, has been made so far. The two taxa seem close in the arrangement of the chambers. *G. suterae* may in fact have derived from *G. scitula* by a gradual increase of its size, the inflation of the chambers and the roundness of the axial periphery.

This species, considered here to be related to the *G. scitula* plexus, is common from within the Late Tortonian to the Early Messinian, predating the salinity crisis.

GLOBOROTALOIDES FALCONARAE Giannelli & Salvatorini
Figures **5.8**; 4

Type reference: *Globorotaloides falconarae* Giannelli & Salvatorini, 1976, p. 170, pl. 2.

This species is not well known in the literature and therefore a description of the type is given here. Small size (maximum diameter 0.25 mm), low trochospire with 12 chambers arranged in $2\frac{1}{2}$ whorls and with 4 chambers in the last whorl; weakly lobate equatorial periphery, ovate chambers ventrally slightly compressed, gradually increasing; slightly depressed sutures both ventrally and dorsally; low arched aperture extending from the umbilicus towards the peripheral margin and bordered by a protruding imperforate lip; cancellate surface and finely perforate wall; fairly wide umbilicus. The variations described by Bolli (1957) for the genus *Globorotaloides* are recognizable in the different growth stages of this species. In the early stage the aperture is an umbilical–extraumbilical slit, the umbilicus is small, the test is more compressed and the chambers are tangentially elongated.

In any growth stage specimens with an anomalous last chamber are common. This chamber is characterized by a more delicate structure, a smaller size than the previous one and an extension towards the umbilical area, completely or partially covering it. In small specimens it is bulla-like and, very unusually, has two infra-laminar apertures. *G. falconarae* shows strong similarities with *G. suteri*, which differs by possessing a more lobate periphery, more spherical chambers, and greater size (maximum diameter 0.35 mm). Close relationships with *G. suteri* are seen also in the different ontogenic stages. Therefore it is suggested that *G. falconarae* may be the direct descendant of *G. suteri*. Specimens closely related to *G. falconarae* are recorded below its range as given by Giannelli & Salvatorini. The species is known from the Late Miocene of the Mediterranean area and more recently has also been recognized from the same stratigraphic interval in the Atlantic (Iaccarino & Salvatorini, 1979; Salvatorini & Cita, 1979).

Acknowledgements

I wish to thank M. B. Cita for reading and discussing this chapter.

References

Akers, W. H. 1955. Some planktonic foraminifera of the American Gulf Coast and suggested correlations with the Caribbean Tertiary. *J. Paleontol.*, **29**, 647–64.

Akers, W. H. 1972. Planktonic foraminifera and biostratigraphy of some Neogene formations, northern Florida and Atlantic Coastal Plain. *Tulane Stud. Geol. Paleontol.*, **9**, 1–139.

Bandy, O. L. 1960. The geologic significance of coiling ratios in the foraminifer *Globigerina pachyderma* (Ehrenberg). *J. Paleontol.*, **34**, 671–81.

Bandy, O. L. 1968. Paleoclimatology and Neogene planktonic foraminiferal zonation. *G. Geol.*, **35**, 277–90.

Bandy, O. L. & Theyer, F. 1971. Growth variation in *Globorotalia pachyderma* (Ehrenberg). *Antarctic Journal United States*, **6**, 172–4.

Banner, F. T. & Blow, W. H. 1960. Some primary types of species belonging to the superfamily *Globigerinaceae*. *Contrib. Cushman Found. foramin. Res.*, **11**, 1–41.

Banner, F. T. & Blow, W. H. 1967. The origin, evolution and taxonomy of the foraminiferal genus *Pulleniatina* Cushman, 1927. *Micropaleontology*, **13**, 133–62.

Barbieri, F. 1967. The Foraminifera in the Pliocene section Vernasca-Castell'Arquato including the Piacenzian stratotype (Piacenza Province). *Mem. Soc. Ital. Sci. nat. Milano*, **15**, 145–63.

Barbieri, F. 1969. Planktonic foraminifera in Western Emily Pliocene (North Italy). *Proceedings First International Conference on Planktonic Microfossils*, Geneva, 1967, **1**, 66–80.

Barbieri, F. 1971. Comments on some Pliocene stages and on the taxonomy of a few species of *Globorotalia*. *L'Ateneo Parmense-Acta Naturalia*, **7**, 1–24.

Barbieri, F. & Petrucci, F. 1967. La série stratigraphique du Messinien au Calabrien dans la vallée du T. Crostolo (Reggio Emilia-Italie sept.). *Mem. Soc. Ital. Sci. nat. Milano*, **15**, 181–8.

Berggren, W. A. 1977a. Late Neogene planktonic foraminiferal biostratigraphy of DSDP Site 357 (Rio Grande Rise). *Initial Rep. Deep Sea drill. Proj.*, **39**, 591–614.

Berggren, W. A. 1977b. Late Neogene planktonic foraminiferal biostratigraphy of the Rio Grande Rise (South Atlantic). *Mar. Micropaleontol.*, **2**, 265–313.

Berggren, W. A. & Phillips, J. D. 1971. Influence of continental drift on the distribution of the Tertiary benthonic

Foraminifera in the Caribbean and Mediterranean regions. *Symposium on the Geology of Libya*, 263–99.

Berggren, W. A. & Van Couvering, J. A. 1974. Neogene biostratigraphy, geochronology and paleoclimatology of the last 15 million years in marine and continental sequences. *Palaeogeogr. Palaeoclimatol. Palaeoecol.*, **16**, 1–216.

Bertolino, V. *et al.* 1968. Proposal for a biostratigraphy of the Neogene in Italy based on planktonic foraminifera. *G. Geol.*, **35**, 23–30.

Biolzi, M., Bizon, G., Borsetti, A. M., Cati, F., Radovisc, A., Rögl, F. & Zachariasse, J. W. 1981. Planktonic foraminifera. In: In search of the Paleogene/Neogene boundary stratotype. Part 1. Potential boundary stratotype sections in Italy and Greece and a comparison with results from the deep-sea. *G. Geol.*, **44**, 165–72.

Bizon, G. 1979. Planktonic foraminifera. In: G. Bizon *et al.* Report of the Working Group on Micropaleontology. *Ann. Geol. Hellen.*, *7th Internat. Congr. Medit. Neogene, Athens*, pp. 1340–3.

Bizon, G. & Bizon, J. J. 1972. *Atlas des principaux foraminifères planctoniques du bassin méditerranéen. Oligocène à Quaternaire*. Editions Technip, Paris, 316 pp.

Bizon, G., Bizon, J. J. & Montenat, C. 1975. Definition biostratigraphique du Messinien. *C.r. Seances Acad. Sci. Paris*, **281**, 359–62.

Bizon, G. & Müller, C. 1978. Remarks on the determination of the Pliocene/Pleistocene boundary in the Mediterranean. *Initial Rep. Deep Sea drill. Proj.*, **42**, 847–53.

Bizon, G. & Müller, C. 1979. Report of the working group on micropaleontology. *Ann. Geol. Hellen. 7th Internat. Congr. Medit. Neogene, Athens*, 1335–64.

Bizon, J. J. & Bizon, G. 1965. L'Helvetien et le Tortonien de la region de Parga (Epire occidentale, Grece). *Rev. Micropaleontol.*, **1**, 242–56.

Bizon, J. J. & Bizon, G. 1966. *Globorotalia miozea saphoae*, nouveau nom pour *Globorotalia saphoae*. *Rev. Micropaleontol.*, **9**, 55.

Blow, W. H. 1956. Origin and evolution of the foraminiferal genus *Orbulina* d'Orbigny. *Micropaleontology*, **2**, 57–70.

Blow, W. H. 1959. Age, correlation and biostratigraphy of the upper Tokuyo (San Lorenzo) and Pozon Formations, Eastern Falcon, Venezuela. *Bull. Am. Paleontol.*, **39**, 67–252.

Blow, W. H. 1969. Late Middle Eocene to Recent planktonic foraminiferal biostratigraphy. *Proceedings First International Conference on Planktonic Microfossils, Geneva, 1967*, **1**, 199–442.

Bolli, H. M. 1957. Planktonic foraminifera from the Oligocene–Miocene Cipero and Lengua Formations of Trinidad, B.W.I. *Bull. U.S. natl Mus.*, **215**, 97–123.

Bolli, H. M. 1962. *Globigerinopsis*, a new genus of the foraminiferal family Globigerinidae. *Eclog. geol. Helv.*, **55**, 281–4.

Bolli, H. M. 1966. Zonation of Cretaceous to Pliocene marine sediments based on planktonic foraminifera. *Boletin Informativo, Asociacion Venezolana de Geologia, Mineria y Petroleo*, **9**, 3–32.

Bolli, H. M. & Bermudez, P. J. 1965. Zonation based on planktonic foraminifera of Middle Miocene to Pliocene warm-water sediments. *Boletin Informativo, Asociacion Venezolana de Geologia, Mineria y Petroleo*, **8**, 119–49.

Bolli, H. M., Loeblich, A. R. & Tappan, H. 1957. Planktonic foraminiferal families Hantkeninidae, Orbulinidae, Globorotaliidae and Globotruncanidae. *Bull. U.S. natl Mus.*, **215**, 3–50.

Bolli, H. M. & Saunders, J. B. 1982. *Globorotalia mayeri* and its relationship to *Globorotalia siakensis* and *Globorotalia continuosa*. *J. foramin. Res.*, **12**, 39–50.

Boni, A. 1967. Notizie sul Serravalliano tipo. In: R. Selli, *Guida alle escursioni del IV Congr. C.M.N.S.*, 47–63.

Borsetti, A. M., Cati, F., Colalongo, M. L. & Sartoni, S. 1979. Biostratigraphy and absolute ages of the Italian Neogene. *Ann. Geol. Hellen., 7th Internat. Congr. Medit. Neogene, Athens*, pp. 183–97.

Brady, H. B. 1878. On the reticularian and radiolarian Rhizopoda (Foraminifera and Polycystina) of the North-Polar Expedition of 1875–1876. *Ann. Mag. Nat. Hist. London*, ser. 5, **1**, 425–40.

Brady, H. B. 1881. On some Arctic Foraminifera from soundings obtained on the Austro-Hungarian North-Polar Expedition of 1872–1874. *Ann. Mag. Nat. Hist. London*, ser. 5, **8**, 393–418.

Brady, H. B. 1884. Report on the foraminifera dredged by H.M.S. *Challenger* during the years 1873–1876. *Rep. Sci. Results H.M.S. Challenger 1873–6*, **9** (zoology), 1–814.

Brönnimann, P. & Resig, J. 1971. A Neogene globigerinacean biochronologic time-scale of the Southwestern Pacific. *Initial Rep. Deep Sea drill. Proj.*, **7**, 1235–1470.

Catalano, R. & Sprovieri, R. 1969. Stratigrafia e micropaleontologia dell'intervallo tripolaceo di torrente Rossi (Enna). *Atti Accad. gioenia Sci. nat. Catania*, ser. 7, **1**, 513–27.

Catalano, R. & Sprovieri, R. 1971. Biostratigrafia di alcune serie saheliane (Messiniano inferiore) in Sicilia. *Proceedings II Planktonic Conference, Roma, 1970*, 211–49.

Cati, F. *et al.* 1968. Biostratigrafia del Neogene Mediterraneo basata sui foraminiferi planctonici. *Boll. Soc. geol. Ital.*, **87**, 491–503.

Chapronière, G. C. H. 1973. On the origin of *Globorotalia miotumida conomiozea* Kennett 1966. *Micropaleontology*, **18**, 461–8.

Cita, M. B. 1964. Considérations sur le Langhien des Langhe et sur la stratigraphie miocène du bassin Tertiaire du Piémont. *Proc. II Sess., Sabadell-Madrid, Com. Medit. Neog. Strat., Inst. 'Lucas Mallada'*, **9**, 203–10.

Cita, M. B. 1968. Report of the working group micropaleontology. *G. Geol.*, **35**, 1–22.

Cita, M. B. 1973. Pliocene biostratigraphy and chronostratigraphy. *Initial Rep. Deep Sea drill. Proj.*, **13**, 1343–79.

Cita, M. B. 1975a. Studi sul Pliocene e sugli strati di passaggio dal Miocene al Pliocene. VIII. Planktonic foraminiferal biozonation of the Mediterranean Pliocene deep sea record. A revision. *Riv. Ital. Paleontol.*, **81**, 527–44.

Cita, M. B. 1975b. The Miocene/Pliocene boundary: history and definition. In: T. Saito & L. H. Burkle (eds.) Late Neogene Epoch Boundaries, *Micropaleontology*, Special Publication, **1**, 1–30.

Cita, M. B. 1975c. Zanclean (additional remarks). In: F. Steininger (ed.) *Stratotypes of Mediterranean Neogene Stages*, vol. 2, Bratislava.

Cita, M. B. 1976. Planktonic foraminiferal biostratigraphy of the Mediterranean Neogene. *Progress in Micropaleontology*, pp. 47–68, Special Publication, Micropaleontology Press, The American Museum of Natural History, New York.

Cita, M. B. & Blow, W. H. 1969. The biostratigraphy of the Langhian, Serravallian and Tortonian stages in the type-sections in Italy. *Riv. Ital. Paleontol.*, **75**, 549–603.

Cita, M. B., Colalongo, M. L., D'Onofrio, S., Iaccarino, S. & Salvatorini, G. 1978. Biostratigraphy of Miocene deep-sea sediments (Sites 372 and 375), with special reference to the Messinian/Pre-Messinian interval. *Initial Rep. Deep Sea drill. Proj.*, **42**, 671–85.

Cita, M. B. & Decima, A. 1975. Rossellian: proposal of a superstage for the marine Pliocene. *VIth Congr. Reg. Com. Med. Neog. Strat., Bratislava*, 217–27.

Cita, M. B. & Elter, P. 1960. La posizione stratigrafica delle Marne a Pteropodi delle Langhe e della Collina di Torino ed il significato cronologico del Langhiano. *Acc. Naz. Lincei, Rend. Cl. Sc. Fis. Mat. Nat.*, **29**, 360–9.

Cita, M. B. & Gartner, S. 1973. Studi sul Pliocene e sugli strati di passaggio dal Miocene al Pliocene. IV. The stratotype Zanclean, foraminiferal and nannofossil biostratigraphy. *Riv. Ital. Paleontol.*, **79**, 503–58.

Cita, M. B. & Gelati, R. 1960. *Globoquadrina langhiana* n.sp. del Langhiano tipo. *Riv. Ital. Paleontol.* **66**, 241–6.

Cita, M. B. & Premoli Silva, I. 1960. Pelagic Foraminifera from the type Langhian. *Proc. Int. Paleontol. Union, Norden*, 1960, part XXII, 39–50.

Cita, M. B. & Premoli Silva, I. 1968. Evolution of the planktonic foraminiferal assemblages in the stratigraphical interval between the type-Langhian and the type-Tortonian and biozonation of the Miocene of Piedmont. *G. Geol.*, **35**, 1–27.

Cita, M. B., Premoli Silva, I. & Rossi, R. 1965. Foraminiferi planctonici del Tortoniano-tipo. *Riv. Ital. Paleontol.* **71**, 217–308.

Cita, M. B. & Ryan, B. F. 1978. Studi sul Pliocene e sugli strati di passaggio dal Miocene al Pliocene. XI. The Bou Regreg section of the Atlantic coast of Morocco. Evidence of a Late Miocene regressive phase. *Riv. Ital. Paleontol.*, **84**, 1051–82.

Colalongo, M. L. *et al.* 1979a. A proposal for the Tortonian/Messinian boundary. *Ann. Geol. Hellen., 7th Internat. Congr. Medit. Neogene, Athens*, 285–94.

Colalongo, M. L., Di Grande, A., D'Onofrio, S., Giannelli, L., Iaccarino, S., Mazzei, R., Romeo, M. & Salvatorini, G. 1979b. Stratigraphy of Late Miocene Italian sections straddling the Tortonian/Messinian boundary. *Boll. Soc. Paleontol. Ital.*, **18**, 258–302.

Colalongo, M. L., Elmi, C. & Sartoni, S. 1974. Stratotypes of the Pliocene and Santerno River Section. *Mem. Bur. Rech. geol. minieres*, **78**, 603–24.

Colalongo, M. L., Pasini, G., Pelosio, G., Raffi, S., Rio, D., Ruggieri, G., Sartoni, S., Selli, R. & Sprovieri, R. 1982. The Neogene/Quaternary boundary definition: a review and proposal. *Geogr. Fis. Dinam. Quat.*, **5**, 59–68.

Colalongo, M. L., Pasini, G. & Sartoni, S. 1981. Remarks on the Neogene/Quaternary boundary and the Vrica section (Calabria, Italy). *Boll. Soc. Paleontol. Ital.*, **20**, 99–120.

Colalongo, M. L. & Sartoni, S. 1967. *Globorotalia hirsuta aemiliana* nuova sottospecie cronologica del Pliocene in Italia. *G. Geol.*, **34**, 265–84.

Colalongo, M. L. & Sartoni, S. 1977. *Globigerina calabra* nuova specie presso il limite Plio-Pleistocene della sezione della Vrica (Calabria). *G. Geol.*, **42**, 205–20.

Colalongo, M. L. & Sartoni, S. 1979. Schema biostratigrafico per il Pliocene e il basso Pleistocene in Italia. *Contrib. Carta Neotettonica Italia*, **251**, Progetto Finalizzato Geodinamica, pp. 645–54.

Conato, V. 1964. Alcuni Foraminiferi nuovi nel Pliocene nordappenninico. *Geologica Romana*, **3**, 279–308.

Conato, V. & Follador, U. 1967. *Globorotalia crotonensis* e *Globorotalia crassacrotonensis* nuove specie del Pliocene italiano. *Boll. Soc. geol. Ital.*, **86**, 555–63.

Crescenti, U. 1966. Sulla biostratigrafia del Miocene affiorante al confine marchigiano-abruzzese. *Geologica Romana*, **5**, 1–54.

Cushman, J. A. 1918. Some Miocene foraminifera of the Coastal Plain of the United States. *Bull. U.S. geol. Surv.*, **676**, 1–100.

Cushman, J. A. & Ellisor, A. C. 1939. New species of foraminifera from the Oligocene and Miocene. *Contrib. Cushman Lab. foramin. Res.*, **15**, 1–14.

De Stefani, C. 1876. Sedimenti sottomarini dell'epoca post-pliocenica in Italia. *Boll. R. Com. geol. Ital.*, **7**, 272–89.

De Stefani, C. 1891. Les terrains tertiaires supérieurs du bassin de la Mediterranée. *Ann. Soc. geol. Belg.*, **18**, 201–419.

Demarcq, G., Magné, J., Anglada, R. & Carbonnel, G. 1974. Le Burdigalien stratotypique de la Vallée du Rhône: sa position biostratigraphique. *Bull. Soc. geol. Fr.*, **16**, 509–15.

Depéret, M. 1892. Note sur la classification et le parallélisme du Système Miocène. *C.R. Soc. geol. Fr.*, **13**, 145–56.

Deshayes, G. P. 1832. *Encyclopédie méthodique; Histoire naturelle des vers*, Mme. v. Agasse, vol. 2, pp. 1–594; vol. 3, pp. 595–1152.

Dollfus, G. F. 1909. Essai sur l'étage aquitanien. *Bull. Serv. Carte geol. Fr.*, **19**, 379–508.

Dondi, L. 1962. Nota paleontologica–stratigrafica sul pedeappennino padano. *Boll. Soc. geol. Ital.*, **81**, 113–229.

Dondi, L. & Papetti, I. 1968. Biostratigraphical zones of Po valley Pliocene. *G. Geol.*, **35**, 63–98.

D'Onofrio, S., Giannelli, L., Iaccarino, S., Morlotti, E., Romeo, M., Salvatorini, G., Sampò, M. & Sprovieri, R. 1975. Planktonic foraminifera from some Italian sections and the problem of the lower boundary of the Messinian. *Boll. Soc. Paleontol. Ital.*, **14**, 177–96.

D'Orbigny, A. D. 1839. Foraminifères des Iles Canaries. In: P. Barker-Webb & S. Berthelot, *Histoire naturelle des Iles Canaries*, vol. 2, pp. 120–46, Béthune, Paris.

Eames, F. E., Banner, F. T., Blow, W. H. & Clarke, W. J. 1962. *Fundamentals of Mid-Tertiary Stratigraphical Correlation*. Cambridge University Press, Cambridge, 151 pp.

Ehrenberg, C. G. 1873. Mikrogeologische Studien über Kleinste Leben der Meeres-Tiefgründe aller Zonen und dessen geologischen Einfluss. *Abh. K. Akad. Wiss. Berlin*, 1872, 131–397.

Fornasini, C. 1898. Globigerina adriatiche. *R. Acc. Sci. Inst. Bologna, Mem. Sci. Nat.*, **7**, 575–86.

Fornasini, C. 1899. Le Globigerine fossili d'Italia. *Paleontogr. Ital.*, **4**, 203–16.

Giannelli, L. & Salvatorini, G. 1975. I foraminiferi planctonici dei sedimenti terziari dell'arcipelago maltese. II. Biostratigrafia di: 'Blue clay', 'Greensand' e 'Upper Coralline Limestone'. *Atti della Societa Toscana Scienze Naturali Memorie*, **82**, 1–24.

Giannelli, L. & Salvatorini, G. 1976. Due nuove specie di foraminiferi planctonici del Miocene. *Boll. Soc. Paleontol. Ital.*, **15**, 167–73.

Gianotti, A. 1953. Microfaune della serie tortoniana del Rio Mazz-apiedi-Castellania (Tortona-Alessandria). *Riv. Ital. Paleontol. Mem.*, **6**, 167–308.

Glaçon, G., Vergnaud Grazzini, C., Leclaire, L. & Sigal, J. 1973. Presence des foraminifères: *Globorotalia crassula* Cushman et Stewart et *Globorotalia hirsuta* (D'Orbigny) en Mer Mediterranéé. *Rev. Esp. Micropaleontol.*, **5**, 373–401.

Gradstein, F. M. 1974. Mediterranean Pliocene *Globorotalia* – a biometrical approach. *Utrecht Micropaleontol. Bull.*, **7**, 1–128.

Hedberg, H. D. 1971. Preliminary report on chronostratigraphic units. *Internat. Subcommis. Stratigr. Classification*, Report no. 6, XXIV Internat. Geol. Congress Montreal, 39 pp.

Hedberg, H. D. 1976. *International Stratigraphic Guide*, Wiley, New York, 200 pp.

Iaccarino, S. 1967. Les Foraminifères du stratotype du Tabianien (Pliocène inférieur) du Tabiano Bagni (Parme). *Mem. Soc. Ital. Sci. nat. Milano*, **15**, 164–80.

Iaccarino, S., Morlotti, E., Papani, G., Pelosio, G. & Raffi, S. 1975. Litostratigrafia e biostratigrafia di alcune serie neogeniche della provincia di Almeria (Andalusia orientale-Spagna). *L'Ateneo Parmense, Acta Nat.*, **11**, 237–313.

Iaccarino, S. & Salvatorini, G. 1979. Planktonic foraminiferal biostratigraphy of Neogene and Quaternary of Site 398 of DSDP Leg 47B. *Initial Rep. Deep Sea drill. Proj.*, **47**, 255–85.

Iaccarino, S. & Salvatorini, G. 1982. A framework of planktonic foraminiferal biostratigraphy for Early Miocene to Late Pliocene Mediterranean area. *Paleontol. Stratigr. Evol.*, **2**, 115–25.

Jenkins, D. G. 1960. Planktonic foraminifera from the Lakes Entrance oil shaft, Victoria, Australia. *Micropaleontology*, **6**, 345–71.

Jenkins, D. G., Saunders, J. B. & Cifelli, R. 1981. The relationship of *Globigerinoides bisphericus* Todd 1954 to *Praeorbulina sicana* (De Stefani) 1952. *J. foramin. Res.*, **11**, 262–7.

Kennett, J. P. 1970. Pleistocene paleoclimates and foraminiferal biostratigraphy in subantarctic deep-sea cores. *Deep Sea Res.*, **17**, 125–40.

Kennett, J. P. 1973. Middle and Late Cenozoic planktonic foraminiferal biostratigraphy of the Southwest Pacific – DSDP Leg 21. *Initial Rep. Deep Sea drill. Proj.*, **21**, 575–639.

LeRoy, L. W. 1939. Some small foraminifera, ostracoda and otoliths from the Neogene ('Miocene') of the Rokan-Tapanoeli area, Central Sumatra. *Natuurkd. Tijdschr. Ned. Indie*, **99**, 215–96.

Martinotti, G. M. 1981. Biostratigraphy and planktonic Foraminifera of the Late Eocene to ?Pleistocene in the Ashqelon 2 well (southern coastal plain, Israel). *Rev. Esp. Micropaleontol.*, **13**, 343–81.

Mayer, Ch. 1867–1870. *Catalogue systematique et descriptif des fossiles des terrains tertiaires qui se trouvent au Musée Fédéral de Zürich.* 4 parts, 1–37, 1–65, 1–124, 1–54. Schabelitz Zurich.

Mayer, Ch. 1868. *Tableau Synchronistique des terrains Tertiaires supérieurs*, 4th edn, Zurich (table only).

Mayer, K. 1858. Versuch einer neuen Klassification der Tertiär-Gebilde Europas. *Verhandlungen der allgemeinen schweizerischen Gesellschaft fur die gesammten Naturwissenschaften*, **42**. Versammlung 1857, 195–9.

Mayer-Eymar, K. 1889. *Tableau des terrains de sédiment.* Zagreb, 35 pp.

Mazzei, R., Raffi, I., Rio, D., Hamilton, N. & Cita, M. B. 1978. Calibration of Late Neogene calcareous nannoplankton datum planes with the paleomagnetic record of Site 397 and correlation with Moroccan and Mediterranean sections. *Initial Rep. Deep Sea drill. Proj.*, **42**, 375–89.

Mosna, S. & Micheletti, A. 1968. Microfaune del 'Serravalliano'. C.M.N.S. Proc. IV Sess., *G. Geol.*, **35**, 183–9.

Pareto, L. 1865. Note sur les subdivisions que l'on pourrait établir dans les terrains tertiaires de l'Apennin septentrional. *Bull. Soc. geol. Fr.*, **22**, 210–77.

Parker, F. L. 1958. Eastern Mediterranean foraminifera. *Rep. Swedish Deep-Sea Exp.*, **8**, 394–594.

Parker, F. L. 1962. Planktonic foraminiferal species in Pacific sediments. *Micropaleontology*, **8**, 219–54.

Pelosio, G., Raffi, S. & Rio, D. 1980. The Plio-Pleistocene boundary controversy. Status in 1979 at the light of International Stratigraphic Guide *Volume dedicato a Sergio Venzo*, Grafiche Step, Parma, pp. 131–40.

Poignant, A. & Pujol, C. 1976. Nouvelles données micropaléontologiques (foraminifères planctoniques et petits foraminifères benthiques) sur le stratotype de l'Aquitanien. *Geobios*, **9**, 577–607.

Poignant, A. & Pujol, C. 1978. Nouvelles données micropaléontologiques (foraminifères planctoniques et petits foraminifères benthiques) sur le stratotype Bordelais du Burdigalien. *Geobios*, **11**, 655–713.

Poore, R. Z. & Berggren, W. A. 1975. Late Cenozoic planktonic foraminiferal biostratigraphy and paleoecology of Hatton-Rockall Basin. DSDP Site 116. *J. foramin. Res.*, **5**, 270–93.

Rio, D. 1982. The fossil distribution of Coccolithophore genus *Gephyrocapsa* Kamptner and related Plio-Pleistocene chronostratigraphic problems. *Initial Rep. Deep Sea drill. Proj.*, **68**, 325–43.

Rio, D., Sprovieri, R., Di Stefano, E. & Raffi, I. 1984. *Globorotalia truncatulinoides* (D'Orbigny) in the Mediterranean Upper Pliocene record, *Micropaleontology*, **30**, 121–37.

Rögl, F. & Bolli, H. M. 1973. Holocene to Pleistocene planktonic foraminifera of Leg 15, Site 147 (Cariaco Basin (Trench), Caribbean sea) and their climatic interpretation. *Initial Rep. Deep Sea drill. Proj.*, **15**, 553–616.

Romeo, M. 1965. '*Globigerina multiloba*' nuova specie del Messiniano della Calabria e Sicilia. *Riv. Ital. Paleontol.*, **71**, 1265–8.

Ruggieri, G. 1961. Alcune zone biostratigrafiche del Pliocene e Pleistocene italiano. *Riv. Ital. Paleontol.*, **67**, 405–17.

Ruggieri, G., Buccheri, G., Greco, A. & Sprovieri, R. 1975. Un affioramento di Siciliano nel quadro della revisione della stratigrafia del Pleistocene inferiore. *Boll. Soc. geol. Ital.*, **94**, 889–914.

Ruggieri, G. & Sprovieri, R. 1970. I microforaminiferi delle 'Marne di S. Cipirello'. *Lavori dell'Istituto di Geologia della Universita di Palermo*, **10**, 1–26.

Ruggieri, G. & Sprovieri, R. 1975. La definizione dello stratotipo del Piano Siciliano e le sue conseguenze. *Riv. Miner. Siciliana*, **151–153**, 7 pp.

Salvatorini, G. & Cita, M. B. 1979. Miocene foraminiferal stratigraphy, DSDP Site 397 (Cape Bojador, North Atlantic). *Initial Rep. Deep Sea drill. Proj.*, **47**, 317–73.

Seguenza, G. 1868. La formation zancléenne, ou recherches sur une nouvelle formation tertiaire. *Bull. Soc. geol. Fr.*, **25**, 465–86.

Selli, R. 1960. The Mayer-Eymar Messinian, 1867, proposal for a neostratotype. *Internat. Geol. Congr. Rep. 21st Sess., Norden*, 1960, 311–33.

Selli, R. *et al.* 1977. The Vrica Section (Calabria, Italy). A potential Neogene/Quaternary boundary stratotype. *G. Geol.*, **42**, 181–204.

Shafik, S. & Chapronière, G. H. 1978. Nannofossil and planktic foraminiferal biostratigraphy around the Oligocene–Miocene boundary in parts of the Indo-Pacific region. *BMR J. Australia Geol. Geophys.*, **3**, 135–51.

Spaak, P. 1981. The distribution of the *Globorotalia inflata* group in the Mediterranean Pliocene. *Proc. Kon. Ned. Akad. Wetensch.*, **84**, 201–15.

Sprovieri, R. 1976. Il datum plane di '*Globorotalia truncatulinoides*' (D'Orbigny) e il limite plio-pleistocenico. *Boll. Soc. geol. Ital.*, **95**, 1101–14.

Sprovieri, R., D'Agostino, S. & Di Stefano, E. 1973. Giacitura del Calabriano nei dintorni di Catanzaro. *Riv. Ital. Paleontol.*, **79**, 127–40.

Sprovieri, R., Ruggieri, G. & Unti, M. 1980. *Globorotalia truncatulinoides excelsa* n.subsp., foraminifero planctonico guida per il Pleistocene inferiore. *Boll. Soc. geol. Ital.*, **99**, 3–11.

Stainforth, R. M., Lamb, J. L., Luterbacher, H., Beard, J. H. & Jeffords, R. M. 1975. Cenozoic planktonic foraminiferal zonation and characteristics of index forms. *Univ. Kansas Paleontol. Contrib.*, **62**, 1–425.

Takayanagi, Y. & Saito, T. 1962. Planktonic foraminifera from the Nobori Formation, Shikoku, Japan. *Sci. Rep. Tohoku imp. Univ.*, **5**, 67–105.

Vail, P. R., Mitchum, T. H. & Thompson, S. 1977. Seismic stratigraphy and global changes of sea level, Part 3: Relative changes of sea level from coastal onlap. *Mem. Am. Assoc. Petrol. Geol.*, **26**, 63–98.

Vervloet, C. C. 1966. *Stratigraphical and micropaleontological data on the Tertiary of Southern Piemont (Northern Italy)*, Schotanus & Jens, Utrecht, 88 pp.

Vigneaux, M. 1971. Burdigalian. In: Stratotypes of Mediterranean Neogene Stages. *G. Geol.*, **37**, 49–54.

Vigneaux, M. & Marks, P. 1971. Aquitanian. In: Stratotypes of Mediterranean Neogene Stages. *G. Geol.*, **37**, 23–31.

Vismara Schilling, A. 1981. Biostratigraphic investigations on the Casa di Tosi section (Marche Region, Italy) with special reference to the Oligocene/Miocene boundary. *Riv. Ital. Paleontol.*, **87**, 227–44.

Walters, R. 1965. The *Globorotalia zealandica* and *G. miozea* lineages. *N.Z. J. Geol. Geophys.*, **8**, 109–27.

Wezel, C. F. 1966. '*Globorotalia*' *acrostoma*, nuova specie dell'Oligomiocene italiano. *Riv. Ital. Paleontol.*, **72**, 1298–1306.

Zachariasse, W. J. 1975. Planktonic foraminiferal biostratigraphy of the Late Neogene of Crete (Greece). *Utrecht Micropaleontol. Bull.*, **11**, 1–171.

9

Late Oligocene and Miocene planktic foraminifera of the Central Paratethys

FRED RÖGL

CONTENTS

Introduction

The collision between the Eurasian continent and the African and Indian plates caused the break-up of the circum-equatorial ocean of the Tethys into separate sedimentary basins interspersed between micro-plates. The development of the Cenozoic Mediterranean and the Paratethyan Seas is strongly connected with the formation of the Alpine–Himalayan mountain chain (Hsü & Bernoulli, 1978).

The Paratethys comprises the Alpine and Carpathian foredeep, the Pannonian and Dacian basins, and the Euxinian basin, which is the area of the Black Sea, Caspian Sea and Lake Aral (Fig. 1). The concept of the Paratethys as the Trans-European bioprovince was introduced by Laskarev (1924), and its subdivision into a western, central and eastern part by Seneš (1959).

The Paratethys came into existence in the Late Oligocene as an epicontinental marginal sea with E–W-stretching deep, narrow troughs and intermontane basins, connected to the Mediterranean and Indopacific. A limitation of the open marine connections resulted in far-reaching regressions, salt deposition, brackish-water facies, and extensive limnofluviatile sedimentation. The peculiar aquatic conditions caused rapid evolution of endemic taxa and communities.

The changing biological history and sedimentation in the Paratethys, and also in other European epicontinental sedimentation areas were the reason that, in the last century and in the beginning of this century, the Cenozoic system was subdivided into stages. The uncoordinated usage of these stages, based mainly on marine molluscan assemblages influenced by ecology, has caused deposits of varying ages to be wrongly correlated.

For this reason a new chronostratigraphic stage concept was developed for the Central Paratethys (Cicha & Seneš, 1968; Papp *et al.*, 1968; Baldi, 1969). It is based on the biostratigraphic concept of the 'integrated assemblage zone' (Steininger, 1977). According to its definition an integrated assemblage zone is a biostratigraphically characterized interval defined by the first appearance, the first concurrent appearances or the total or partial ranges of taxa belonging to various evolutionary lineages, derived *in situ* or appearing by migration. All these taxa cannot be expected to occur within a single rock unit as they are part of different environments.

In the changing environment of a marginal basin all available biostratigraphic aids have to be used. Most important are molluscs, foraminifera, ostracods and siliceous nannofossils. By means of palynomorphs correlation may be made between the marine time scale and the mammal ages of the continental facies. A few datum levels allow a general correlation with the regional subdivisions of the standard plankton

zonation. The local biozonations, such as for the Vienna Basin, are mainly dependent on changing ecological factors. Later workers have tried to correlate one with another by the use of evolutionary lineages (Grill, 1941, 1943; Papp, 1959).

Chronostratigraphic and biostratigraphic concept of the Central Paratethys

The newly introduced regional stages are based on reference sections with a stratotype and supplementary hypostratotypes, to allow for different types of facies. These stages have been extensively studied by the Paratethys Working Group of the Regional Committee on Mediterranean Neogene Stratigraphy and by the co-workers of the International Geological Correlation Programme 'Tethys–Paratethys Neogene', taking into account existing results and zonations. Correlations of these stages with the standard plankton zonations, and correlations with the Mediterranean regional stages are summarized by Cicha (1970), Cicha, Hagn & Martini (1971), Steininger, Rögl & Martini (1976), Seneš (1979) and Steininger & Papp (1979).

An attempt was made by Cicha, Marinescu & Seneš (1975) and Jiriček (1975) to achieve a biozonation using several different fossil groups combined.

A correlation of the regional stages of the Central Paratethys with the Eastern Paratethys and the Mediterranean is given in Fig. 2. In the Western Paratethys the stage systems of either the Mediterranean or of the Central Paratethys have

Fig. 1. European Neogene sedimentary basins and bioprovinces. Not included are overthrusted and faulted Neogene sedimentation areas in young mountain ranges

to be used depending on which influences are predominant at the time in question.

The Central Parathethyan system of stages will be discussed together with a short comment on the paleogeographic and geodynamic evolution of the area (Seneš & Marinescu, 1974; Rögl, Steininger & Müller, 1978; Steininger, Rögl & Müller, 1978; Steininger & Rögl, 1979; Rögl & Steinninger, 1983).

Egerian

Throughout the Oligocene the Eurasian and African continents were separated by relics of the Tethyan Ocean. The Alps began to appear in the Early to Middle Oligocene, but had not become high mountain ranges at that time. They functioned as a barrier between the Mediterranean and a central European sea. In the earlier Oligocene this sea was connected to the North Sea along the Danish–Polish trough, to the Northern Oceans by the Turgai Strait, and to the Eastern Mediterranean–Indopacific realm. Micro-plates, now included in the Carpathian, Balkan and Dinarides mountain ranges, split the sea into a series of troughs and intermontane basins. Beginning with the Egerian/Late Oligocene the tectonic isolation of this sea from the old Tethys created the 'Parathethys'.

The Egerian stage defined and described by Baldi &

Seneš (1975) was formerly considered to be an equivalent of the Chattian and Aquitanian. The Chattian, a stage of the North Sea realm, terminated by regression without a defined upper boundary. The Aquitanian, a west European Atlantic stage, began with a transgression; there is no continuity in space and time between these two stages. The use of these stages in the Central Parathethys has not been time-equivalent to the type areas.

The Egerian comprises the upper NP 24 to NN 1 nannoplankton zones, corresponding to part of the *Globorotalia opima opima* Zone and the *Globigerina ciperoensis* and *Globorotalia kugleri* zones. Important planktic foraminiferal species are: *Globigerinoides primordius, Globorotalia continuosa, G. semivera, G. mayeri, Globigerina woodi woodi* and *Cassigerinella boudecensis. Globorotalia kugleri* has not been reported anywhere in the Parathethyian region (Pishvanova, 1968; Cicha *et al.*, 1971; Rögl, 1975; Sztrakos, 1979).

Additional important biostratigraphic markers for the Egerian stage are benthic larger foraminifera with *Lepidocyclina* and the lineage of *Miogypsina* (*M. complanata* Schlumberger, *M. septentrionalis* Drooger, *M. formosensis* Yabe & Hanzawa, *M. gunteri* Cole). Also mollusc faunas with pectinids, and vertebrates characterize the stage.

The Egerian ended with far-reaching regressions and discordances.

Eggenburgian

From the Indian Ocean a transgression spread through the entire Parathethys, bringing warm-water molluscan faunas from the Iranian Qum Basin as far as the Bavarian Molasse Basin in the west. The transgression prograded to the south-west from Switzerland to the Rhône Valley, connecting for some time the western Mediterranean and the Parathethys by a new seaway. The time of this connection was the 'Helvetian', part of the Eggenburgian, and ranging up into the Ottnangian.

The occurrences of *Sphenolithos belemnos* and *Helicosphaera carteri* are indicators for the nannoplankton zones NN 2 to early NN 3. A zonation by means of planktic foraminifera is only partially successful. The planktic foraminiferal fauna diminishes progressively from east to west, with the best assemblages in the Chechis clay of Roumania (Popescu, 1975), in the Stebnik beds of Ukraine (Pishvanova, 1968), and also in some layers of the Zdanice Unit in Moravia (Cicha, 1970) and the Waschberg Unit of eastern Austria. The occurrences of *Globigerinoides trilobus, Globoquadrina dehiscens* and *Globoquadrina langhiana* make a comparison possible with the *Globigerinoides trilobus* Zone of the Mediterranean. A good marker for the late Eggenburgian is *Miogypsina intermedia* (Papp, 1960). A comprehensive description of this stage is given by Steininger & Seneš (1971).

Nearly identical planktic foraminiferal faunas continue throughout the Early Miocene Eggenburgian–Ottnangian

M.Y.	EPOCHS		MEDITERRANEAN	CENTRAL PARATETHYS	EASTERN PARATETHYS
			PLIOCENE	Dacian	Kimmerian
-5-			Messinian	Pontian	Bosphorian
					Portaferrian
		L			Novorossian
-10-			Tortonian	Pannonian	Maeotian
					Chersonian
					Bessarabian
				Sarmatian	Volhynian
	MIOCENE	M	Serravallian	Badenian	Konkian
-15-					Karaganian
					Tshokrakian
			Langhian		Tarkhanian
				Karpatian	
			Burdigalian	Ottnangian	Kozachurian
-20-		E		Eggenburgian	Sakaraulian
			Aquitanian		
				Egerian	Caucasian
-25-	OLIGO-CENE		Chattian		

Fig. 2. Neogene stage systems of middle and southern Europe. The correlation of Mediterranean, Central and Eastern Parathethys regional stages is according to Papp (1969), Rögl *et al.* (1978), Steininger & Papp (1979) and Paramonova *et al.* (1979).

stages. In addition to the above-listed markers the fauna consists mainly of small globigerinids like *Globigerina praebulloides*, *G. ciperoensis ottnangiensis*, *G. bollii lentiana*, *G. angustiumbilicata*, *G. dubia* and *G. obesa*. *Globorotalia acrostoma*, *G. opima nana* and sometimes floods of *Cassigerinella* also occur.

Ottnangian

The Ottnangian continues from the Eggenburgian without discontinuity. Foraminiferal assemblages are still dominated by small, planktic species, becoming less diverse in the younger parts. Very common is *Globigerina bollii lentiana*, giving the name to the *Globigerina bollii* Zone in the Carpathian foredeep (Pishvanova, 1968). Molluscs are of special biostratigraphic significance. The stage definition and an extensive description of the fossil content are given by Papp, Rögl & Seneš (1973).

Beginning in the late Eggenburgian, a restriction of the connection to the Indopacific caused a decrease of salinity in the Eastern Paratethys. The Kozachurian stage developed with its brackish-water facies an endemic molluscan fauna characterized by the bivalve *Rzehakia* ('Oncophora'). A landbridge came into existence between Africa and Eurasia, enabling mammal migrations, e.g. of Proboscidea. With the closure of the seaway to the western Mediterranean, the only marine connection remaining was from the Pannonian Basin to the southwest. The environments were changing by regression to the brackish *Rzehakia* facies also in the Central Paratethys, extending from Bavaria to southern Russia.

Karpatian

A new marine transgression extended from the Mediterranean in the south, with marine conditions continuing in a restricted part of the Pannonian Basin (Kokay, 1973). The sea formed a large bay-like extension northwards from the Pannonian Basin, extending into the Molasse Basin north of the Danube, up to the Carpathian foredeep in Poland. The Carpathian foredeep from eastern Poland to Roumania, and the Transsylvanian Basin became desiccated, whereas the Kozachurian facies persisted in the Eastern Paratethys (Rögl et al., 1978; Seneš, 1978).

The Karpatian was the first stage to be defined in the Central Paratethys, being used for beds formerly called Helvetian and Tortonian (Cicha & Tejkal, 1959; Cicha, Seneš & Tejkal, 1967).

The marine molluscan faunas show distinct Mediterranean influences. Nannofossil determinations correlate the Karpatian to the upper part of NN 4 and lower NN 5 (Martini & Müller, 1975). The foraminiferal fauna is characterized by uvigerinids, especially by *Uvigerina graciliformis* (Papp & Turnovsky, 1953). Planktic assemblages contain large *Globigerina* in the lower part (Vašiček, 1952; Rögl, 1969a) with

Globigerina praebulloides, *G.* cf. *concinna*, *Globorotalia obesa* and *Hastigerinella clavacella*. *Globigerinoides bisphericus*, *G. trilobus* and *Globorotalia mayeri* are common only in the late Karpatian.

Badenian

An Indopacific transgression flooded the Mediterranean and most parts of the Central and Eastern Paratethys at the beginning of the Middle Miocene (Langhian, Badenian and Tarkhanian stages). A Tethyan-like circum-equatorial warmwater circulation at the *Praeorbulina* level provided southern and middle Europe with a short subtropical climate. For the first time since the Eocene conditions were again favourable in central European seas for a reasonable planktic foraminiferal development.

The planktic assemblages include *Praeorbulina glomerosa*, *Orbulina suturalis*, *Globigerinoides trilobus*, *Globorotalia mayeri*, *G. acrostoma*, *Globoquadrina altispira*, *G. advena* and many species of *Globigerina*. The peculiarity of the Paratethys is expressed by the appearance of many species in this area: *Globigerina concinna*, *G. diplostoma*, *G. opinata*, *G. regularis*, *G. subcretacea*, *G. tarchanensis*, *Globigerinoides quadrilobatus*, *G. grilli*, *Globorotalia bykovae*, *G. transsylvanica* and the endemic genus *Velapertina*. The abundant foraminiferal fauna was described by numerous authors, many dating back to the last century: e.g. d'Orbigny (1846), Neugeboren (1847, 1856), Cžjžek (1848), Reuss (1850, 1867), Karrer (1861, 1868). A comprehensive description is given in the series *Chronostratigraphie und Neostratotypen* (Papp et al., 1978).

During the Badenian, the conditions of the Paratethys changed rapidly as a result of geodynamic processes. The eastern parts of the realm, including the Carpathian foredeep and parts of the Transsylvanian Basin, were isolated. A salinity crisis with the deposition of thick evaporites in the Carpathian foredeep and parts of the intermontane basins anticipated the Mediterranean Messinian event. In contrast a strong decrease in salinity occurred in southern Russia. Marine connections existed from the Pannonian Basin southwest to the Langhe Basin in Italy (Rögl et al., 1978; Steininger et al., 1978).

The closure of the Mediterranean connection by the northward movement of the Dinarides coincided with a last opening of a seaway to the Persian Gulf in the Late Badenian. A far-reaching transgression brought about a surprisingly similar facies all over the Eastern and Central Paratethys. On top of the diminished Kozachurian of southern Russia, and covering the salt deposits of the Ukraine and Poland, fine-laminated, brown radiolarian and pteropod marls were deposited. Nannofossils and Radiolaria species present here but not found in the Mediterranean demonstrate Indopacific affinities (Dumitrica, Gheta & Popescu, 1975; Rögl & Müller, 1976). The planktic foraminiferal fauna is dominated by

Globigerina and characterized by the occurrence of the genus *Velapertina*.

This succession of changes within the Badenian has resulted in a stratigraphic subdivision by substages (Cicha & Seneš, 1975), which is similar to the eco-zonation of the Vienna Basin (Grill, 1941, 1943).

After this last impressive cycle, the total isolation of the Paratethys took place. Only very short marine intervals in the Eastern Paratethys interrupted the gradual change from a marine to a fluviatile–terrestrial environment.

Sarmatian

The brackish facies, characteristic for the Ponto-Caspian region, begins with discordances and transgressions in the marginal parts of the basins. Planktic foraminifera are absent in the Central Paratethys while nannofossils, particularly *Braarudosphaera*, are abundant in distinct layers (Stradner & Fuchs, 1980). The benthic foraminiferal assemblages indicate a gradual reduction in salinity throughout the Sarmatian. The marine molluscan faunas were also reduced and developed endemic populations. Stronger marine influxes in the Eastern Paratethys indicate a marine connection to the Mediterranean in the region of the Dardanelles.

A detailed description of this stage of the Central Paratethys is given by Papp, Marinescu & Seneš (1974).

Pannonian

A progressive change in environment produced the Pannonian lake system. Vast areas of oligohaline lakes with up to 7‰ salinity existed from the Pannonian Basin to the Caspian Sea. Foraminifera ceased to exist in the western basins, but rich, endemic faunas of molluscs and ostracods evolved. Well-known endemic mollusc genera are *Congeria*, *Limnocardium* and *Melanopsis*.

Only in the Eastern Paratethys, in the Bagerov horizon of early Maeotian age were marine conditions re-established for a very short time, with foraminifera and nannofossils of Zone NN 10 (Semenenko & Ljulieva, 1978).

The invasion of the horse *Hipparion* reached Asia, Europe and Africa at about the same time (Berggren & Van Couvering, 1974). It is one of the most useful time levels correlated to the marine scale within NN 10, Tortonian, marking the beginning of the Pannonian.

Pontian

The final regression destroyed the lakes of the Central Paratethys, which were filled up by sands and fluviatile gravels. In the late Pontian a sudden change of the climate to dry conditions is connected with the Messinian event (Steininger & Papp, 1979).

Discussion of taxa

The taxa reviewed in the present chapter are almost entirely those that were originally described from the Paratethys region. Numerous other taxa first described from low latitudes, the Mediterranean region and the southern mid-latitudes occur in the Paratethys as characteristic forms and are therefore included on the range chart (Fig. 3). The taxonomic notes and illustrations for these are found in Chapters 6, 7 and 8.

CASSIGERINELLA BOUDECENSIS Pokorny
Figures **5.21–24**; 3

Type reference: *Cassigerinella boudecensis* Pokorny 1955, p. 138, text-figs. 1a–b.

The wall surface of this species is covered with coarse pustules, which are more prominent in the earlier part of the test. The opinion of Blow & Banner (1962) that these pustules are an artifact of preservation is contradicted by their ubiquitous presence, even in well preserved material. For this reason, we separate *C. boudecensis* from *C. chipolensis* (Cushman). In the latter species the chambers are also more globular and increase more rapidly in size.

There is a tendency in the Early Miocene (Ottnangian) of the Paratethys for the pustules on *C. boudecensis* to become spine-like in the distal part of the test. This perhaps warrants the erection of a new subspecies.

In low latitudes the genus *Cassigerinella* first appeared in the Late Eocene but is not found in the Mediterranean and the Paratethys until the Late Oligocene.

CASSIGERINELLA GLOBULOSA (Egger)
Figures **5.18–20**; 3

Type reference: *Cassidulina globulosa* Egger, 1857, p. 296, pl. 11, figs. 4–7.

This small, Early Miocene *Cassigerinella* from the Central Paratethys has a smooth chamber surface without spines and without coarse pustules. In contrast to *C. chipolensis*, the chambers are compressed and ovate.

GLOBIGERINA BREVISPIRA Subbotina
Figures **5.6**; 3

Type reference: *Globigerina brevispira* Subbotina, *in* Subbotina, Pishvanova & Ivanova, 1960, p. 56, pl. 11, figs. 4a–c.

The aperture of the very small and low trochospiral species measuring only 0.1–0.2 mm, has an umbilical–extraumbilical position. Krasheninnikov & Pflaumann (1977) for this reason placed it into the genus *Globorotalia*. However, the wall texture shows more resemblance to that of *Globigerina*, for which reason it is here left in this genus.

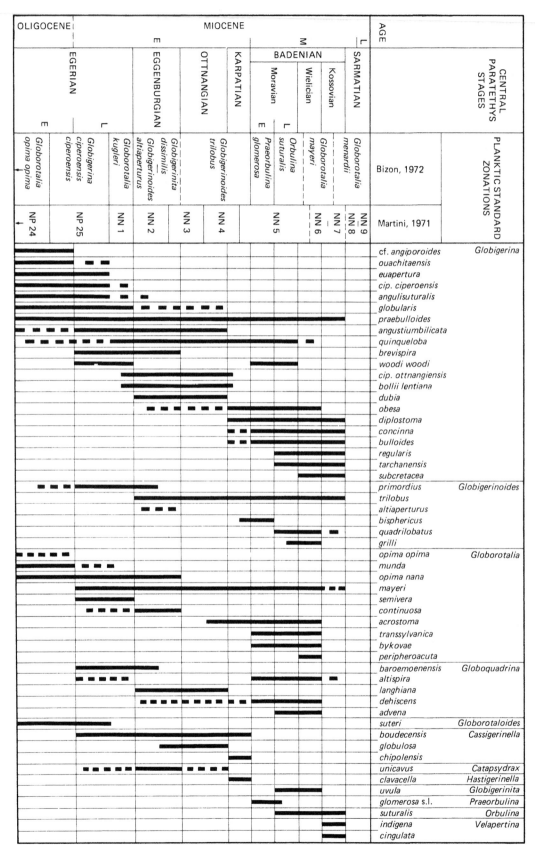

Fig. 3. Range chart showing significant Late Oligocene to Middle Miocene planktic foraminifera from the Central Paratethys

GLOBIGERINA BOLLII LENTIANA Rögl
Figures **4.11**; 3

Type reference: *Globigerina bollii lentiana* Rögl, 1969b, p. 220, pl. 2, figs. 1a–c, pl. 3, fig. 1.

The last chamber of the small 4-chambered subspecies is usually of the same size or smaller than the penultimate one. Compared with the generally larger *G. bollii* s.s., its descendant in the Middle Miocene, the chambers of *lentiana* are separated by more deeply incised sutures, which results in a more lobate periphery. Both taxa are abundant in certain horizons of the Ottnangian in the Paratethys and in the Langhian of the Mediterranean Neogene.

GLOBIGERINA BULLOIDES d'Orbigny
Figures **4.1–2**; 3

Type reference: *Globigerina bulloides* d'Orbigny, 1826, p. 277, Mod. no. 17 (juv.), Mod. no. 76 (adult).

Lectotype: *Globigerina bulloides* d'Orbigny, 1826: Banner & Blow, 1960, p. 3, pl. 1, figs. 1a–c.

To show the original species concept of *G. bulloides* the model no. 76 of d'Orbigny is figured here along with Banner & Blow's lectotype, which displays a somewhat stouter test but falls within the variability of the species. *G. bulloides* developed in the Middle Miocene, where it overlaps with its ancestor *G. praebulloides*. In that species chambers increase more rapidly in size, which results in a more elongate test shape with a higher and more asymmetrical arched aperture.

GLOBIGERINA CIPEROENSIS OTTNANGIENSIS Rögl
Figures **5.5**; 3

Type reference: *Globigerina ciperoensis ottnangiensis* Rögl, 1969b, p. 221, pl. 2, figs. 7a–c, pl. 3, fig. 3.

The last of the 5 chambers forming the final whorl is highly variable in size and position; it may be smaller or larger than the penultimate and may also be tilted towards the umbilicus. The subspecies *ottnangiensis* differs from *G. ciperoensis ciperoensis* mainly in the low arched and asymmetrically-shaped aperture, which extends somewhat towards the periphery. *G. praebulloides pseudociperoensis* Blow differs in having only 4 instead of 5 chambers in the penultimate whorl and also in the more regular, symmetrical aperture.

GLOBIGERINA CONCINNA Reuss
Figures **4.17–20**; 3

Type reference: *Globigerina concinna* Reuss, 1850, p. 373, pl. 47, figs. 8a–b.

G. concinna is similar to *G. ciperoensis ciperoensis* but differs in the chambers increasing in size slightly more rapidly, in a larger test size, 0.5–0.6 against 0.3–0.4 mm, and in a distinctly different stratigraphic range, Middle Miocene compared with Middle to Late Oligocene. Furthermore, *G. concinna* has a more open umbilical area, and its aperture is often slightly asymmetrical, similar to that in *G. diplostoma*. The names *concinna* or cf. *concinna* were originally used in Central America and the Caribbean area to name the similar Oligocene forms but were later separated by Bolli (1954) as *G. ciperoensis ciperoensis*.

GLOBIGERINA DIPLOSTOMA Reuss
Figures **4.3–8**; 3

Type reference: *Globigerina diplostoma* Reuss, 1850, p. 373, pl. 47, figs. 9a–b, 10; pl. 48, fig. 1.

The 4-chambered *G. diplostoma* differs from the similar *G. bulloides* in a slower increase of chamber size; the last chamber is often of the same size or even smaller compared with the penultimate. The ranges of the two apparently closely related species are virtually the same. Some authors therefore regard them as synonymous.

However, the populations of this species in the Middle Miocene of the Vienna Basin give the impression that it is a distinct valid species, distinguishable from *G. bulloides* co-occurring in the same samples, and not synonymous as maintained by Blow (1959).

GLOBIGERINA DUBIA Egger
Figures **5.1**; 3

Type reference: *Globigerina dubia* Egger, 1857, p. 281, pl. 9, figs. 7–9.

This small 5-chambered Early Miocene species measuring 0.25–0.3 mm is characterized by its high trochospire. The comparatively large final chamber partly overlaps the umbilical cavity. The Oligocene *G. ciperoensis fariasi* is of about the same size and shape but has a more open umbilicus. *G. dubia* differs from *G. ciperoensis ciperoensis* and *G. ciperoensis ottnangiensis* in the high trochospire with the last chamber partly covering the umbilicus.

GLOBIGERINA GLOBULARIS Roemer
Figures **4.9–10**; 3

Type reference: *Globigerina globularis* d'Orb.: Roemer, 1838, p. 390, pl. 1, fig. 57.
non *Globigerina globularis* d'Orbigny, 1826, p. 277, no. 3 (nomen nudum).

Description and illustrations by Roemer are not satisfactory. In figuring a comparative specimen (Fig. 4.10) Drooger (1956) attempted to clarify the concept of this species. According to the original description it is a stout, 3–4-chambered form with globular, somewhat embracing chambers. This comparatively large species originally described from the Late Oligocene resembles *G. euapertura*, but is not so distinctly 4-chambered.

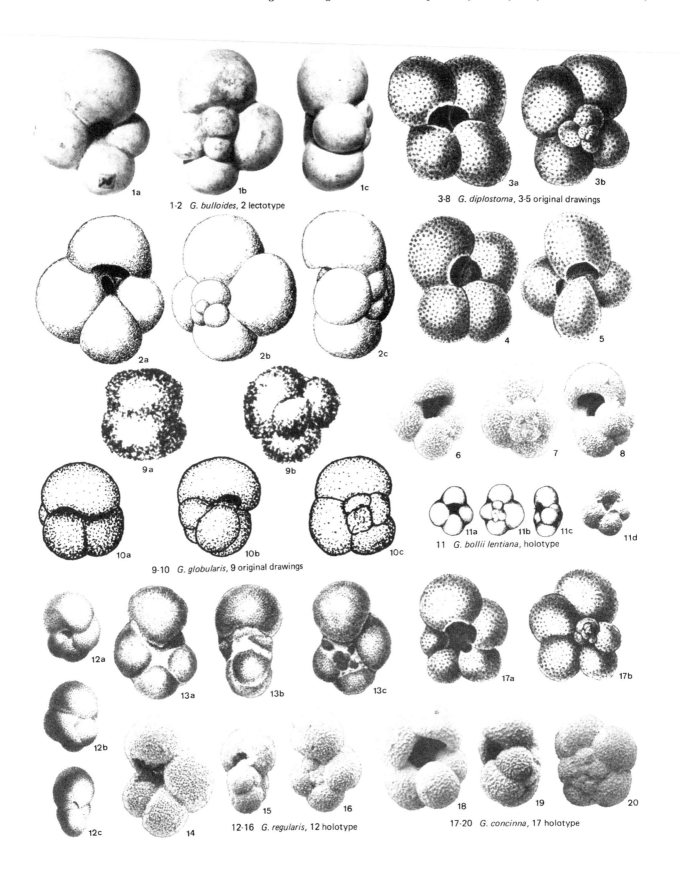

1-2   *G. bulloides*, 2 lectotype

3-8   *G. diplostoma*, 3-5 original drawings

9-10   *G. globularis*, 9 original drawings

11   *G. bollii lentiana*, holotype

12-16   *G. regularis*, 12 holotype

17-20   *G. concinna*, 17 holotype

GLOBIGERINA REGULARIS d'Orbigny
Figures **4.12–16**; 3

Type reference: *Globigerina regularis* d'Orbigny, 1846, p. 162, pl. 9, figs. 1–3.

non *Baggina regularis* (d'Orbigny, 1846): Popescu, 1982, p. 25, pl. 1, fig. 4 (not valid as neotype).

Synonym: *Globigerina opinata* Pishvanova, 1959, p. 18, pl. 4, figs. 2a–c.

This very low trochospiral species with a depressed spire has a bi-umbilicate, *Hastigerina*-like appearance. The spinose, *Globigerina*-like 4 chambers of the last whorl increase rapidly in size. The aperture is a low umbilical–extraumbilical arch extending towards the periphery. The relationship of *G. regularis* to *Globorotalia obesa* is not clear but it is possible that the two may be synonyms. Similarities exist also with *Hastigerina praesiphonifera*, which has a similar range.

GLOBIGERINA SUBCRETACEA Lomnicki
Figures **5.7**; 3

Type reference; *Globigerina* (*cretacea*, d'Orgigny): Lomnicki, 1900, p. 223, pl. 1, figs. 2a–c.

Globigerina subcretacea Lomnicki, 1901, p. 17; non *Globigerina subcretacea* Chapman, 1902, p. 410, pl. 36, fig. 16 (younger homonym); non *Globorotalia* (*Turborotalia*) *subcretacea* (Lomnicki): Blow, 1969, p. 392, pl. 4, figs. 18–20 (erroneously designated as lectotype).

This small, very low trochospiral species has 5 chambers moderately increasing in size, which form the last whorl. The aperture is umbilical–extraumbilical in position. *G. subcretacea*

Fig. 4. Late Oligocene to Middle Miocene *Globigerina* species of the Central Paratethys (all figures × 60)

1–2 *Globigerina bulloides* d'Orbigny
1a–c model no. 76 of d'Orbigny (1826). Recent, Adriatic near Rimini. **2a–c** Lectotype. Recent, Adriatic near Rimini.

3–8 *Globigerina diplostoma* Reuss
3a–b, 4–5 Original drawings of Reuss. Middle Miocene, Badenian, salt mine at Wieliczka, Poland. **6–8** Specimens from a paratype locality. Middle Miocene, Badenian, brick-yard Sooss, Baden, Austria.

9–10 *Globigerina globularis* Roemer
9a–b Original drawings of Roemer. Late Oligocene, Chattian, 'Meeressand', Kassel and Osnabrück, northern Germany. **10a–c** Locality BZ 34. Oligocene, Phare Saint-Martin, Biarritz, France (from Drooger, 1956).

11 *Globigerina bollii lentiana* Rögl
11a–d Holotype. 11a–c Original drawings, 11d SEM micrograph of umbilical side. Early Miocene, Ottnangian, Phosphoritsande, Plesching near Linz, Austria.

12–16 *Globigerina regularis* d'Orbigny
12a–c Holotype. Middle Miocene, Badenian, Nussdorf, Vienna, Austria. **13a–c** Synonym: *Globigerina opinata* Pishvanova. Middle Miocene, Badenian, Carpathian foredeep, Ukraina, USSR. **14–16** Specimens from Middle Miocene, Badenian, Walbersdorf, Austria.

17–20 *Globigerina concinna* Reuss
17a–b Holotype. Middle Miocene, Badenian, Grinzing, Vienna, Austria. **18–20** Near-topotypes, coll. A. E. Reuss, Middle Miocene, Badenian, Nussdorf, Vienna, Austria.

---

of Lomnicki is not to be mistaken for *G. subcretacea* Chapman 1902, which is a much larger Recent form. The lectotype of *G. subcretacea* erected by Blow is invalid because of his erroneous assumption that Lomnicki placed his species in direct synonymy with *G. cretacea* d'Orbigny? of Brady (1884, p. 596, pl. 82, fig. 10). Both forms are distinctly different and do not even belong to the same genus. Rögl & Bolli (1973) therefore introduced the new name *Neogloboquadrina dutertrei blowi* for Chapman's species.

GLOBIGERINA TARCHANENSIS Subbotina & Chutzieva
Figures **5.2**; 3

Type reference: *Globigerina tarchanensis*, in Bogdanowicz, 1950, p. 173, pl. 10, figs. 5a–c.

Globigerina tarchanensis Subbotina & Chutzieva, 1950: Subbotina, 1953, p. 61, pl. 3, figs. 13a–c (refigured).

G. tarchanensis differs from *G. concinna* and the *G. ciperoensis* group in its smaller size and in having the aperture bordered by a lip. From *G. c. ottnangiensis* it differs in the smaller chambers increasing more gradually in size. In comparison, *Globigerina subcretacea* is smaller, has a very flat trochospire and an aperture extending towards the periphery.

GLOBIGERINITA UVULA (Ehrenberg)
Figures **5.25–26**; 3

Type reference: *Pylodexia uvula* Ehrenberg, 1862, p. 308.

Pylodexia uvula Ehrenberg, 1873, tab. p. 241 (no. 423), pl. 2, figs. 24–25.

Synonyms: *Globigerina bradyi* Wiesner, 1931, p. 133.

Globigerina bradyi Wiesner: Banner & Blow, 1960, p. 5, pl. 3, fig. 1 (lectotype).

This minute species is readily recognizable by its closely coiled, very high trochospire, which usually has 3 to 3½ chambers per whorl. In low latitudes it ranges from Early Miocene to Holocene, but immigrated into the Central Paratethys only in the Middle Miocene.

GLOBIGERINOIDES GRILLI (Schmid)
Figures **5.3–4**; 3

Type reference: *Globigerinopsis grilli* Schmid, 1967, p. 349, figs. 2a–c.

Synonym: *Globigerinoides kuehni* Schmid, 1967, p. 347, figs. 1a–c.

The test has an interiomarginal–spiroumbilical aperture extending along the spiral suture for one or two, seldom more chambers and is without a lip. In addition, individual sutural apertures may be present on the spiral side on earlier chambers of the last whorl. The spiroumbilical aperture is typical for the genus *Globigerinopsis*, the additional secondary ones for *Globi-*

gerinoides, into which genus the taxon is placed here. It occurs sparsely in the Middle Miocene, Upper Lagenidae Zone of the Vienna Basin and in greater numbers in the zone with arenaceous foraminifera, where it represents approximately the *Globorotalia fohsi fohsi* Zone.

GLOBOROTALIA BYKOVAE (Aisenstat)
Figure **5.13**; 3

Type reference: *Turborotalia bykovae* Aisenstat *in* Subbotina *et al.*, 1960, p. 69, pl. 13, figs. 7a–c.

Synonym: *Globorotalia bykovae minoritesta* Papp: *in* Papp *et al.*, 1978, p. 278, pl. 7, figs. 1–3.

G. bykovae is similar to *G. peripheroronda* but differs in usually being smaller (0.2–0.36 mm) and in having chambers arranged in a slightly imbricated mode resulting in an undulating test surface. The umbilical side is fairly high with a small cavity which is still less developed in *G. peripheroronda*, where in comparison the peripheral angle is higher and more prominent.

The species is small in the Early Badenian, becoming larger in the Late Badenian (Early Kosovian) of the Central Paratethys. A subspecific distinction on size does not seem to be justified.

GLOBOROTALIA TRANSSYLVANICA Popescu
Figures **5.14–17**; 3

Type reference: *Globorotalia (Turborotalia) transsylvanica* Popescu, 1970, p. 200, pl. 30a–c.

Number, shape and arrangement of the chambers and form and position of the aperture are all closely comparable with *G. mayeri* from which *G. transsylvanica* is said to differ only in its very thin, smooth and translucent wall.

HASTIGERINELLA CLAVACELLA Rögl
Figures **5.27–28**; 3

Type reference: *Hastigerinella clavacella* Rögl, 1969a, p. 95, pl. 9, figs. 4a–c.

H. clavacella differs from *Clavatorella bermudezi* in its larger test size, up to 0.9 mm, with chambers arranged in a medium to high trochospire, becoming streptospiral in the final stage. In comparison, *C. bermudezi* has a very flat trochospire of nearly planispiral aspect. Chamber forms in the two species are similar with a trend for later ones to become more elongate and club-shaped. Chambers in *H. digitata* are more finger-like and clavate and are arranged in a lower trochospire compared with *H. clavacella*.

VELAPERTINA CINGULATA Popescu
Figures **5.11–12**; 3

Type reference: *Velapertina cingulata* Popescu, 1969, p. 106, pl. 1, figs. 1 and 3.

Synonyms: *Velapertina iorgulescui* Popescu, 1969, p. 105, pl. 2, figs. 5–6.
Catapsydrax prahovensis Popescu, 1969, p. 104, pl. 2, figs. 7–9.

We consider all three of Popescu's species (*V. cingulata*, *V. iorgulescui*, *C. prahovensis*) to be variants of a single species

---

Fig. 5. Late Oligocene to Middle Miocene species of *Cassigerinella*, *Globigerina*, *Globigerinita*, *Globigerinoides*, *Globorotalia*, *Hastigerinella* and *Velapertina* (all figures × 60)

1 *Globigerina dubia* Egger
1a–c Holotype. Early Miocene, Eggenburgian, Mairhof near Ortenburg, Bavaria.

2 *Globigerina tarchanensis* Subbotina & Chutzieva
2a–c Holotype. Middle Miocene, Tarkhanian, Tamansk Peninsula, USSR.

3–4 *Globigerinoides grilli* (Schmid)
3a–c Holotype. **4a–c** Holotype of synonym *Globigerinoides kuehni* Schmid. Both from Middle Miocene, Badenian, brick-yard Sooss, Baden, Austria.

5 *Globigerina ciperoensis ottnangiensis* Rögl
5a–d Holotype. 5a–c Original drawing, 5d SEM micrograph of umbilical side. Early Miocene, Ottnangian, Phosphoritsand, Plesching near Linz, Austria.

6 *Globigerina brevispira* Subbotina
6a–c Holotype. Early Miocene, Eggenburgian, Vorotichenian Formation, Carpathian foredeep, River Velikij Lukavec, USSR.

7 *Globigerina subcretacea* Lomnicki
7a–c Holotype. Middle Miocene, Badenian, Wielicka, Poland.

8–10 *Velapertina indigena* (Luczkowska)
8a–c Holotype. Middle Miocene, Badenian, Grabowiec Beds, near Bochnia, Poland. **9–10** Specimens from the Middle Miocene, Badenian, Valea Morilor, Roumania.

11–12 *Velapertina cingulata* Popescu
11a–c Holotype. **12a–c** Holotype of synonym *Velapertina iorgulescui* Popescu. Both from the Middle Miocene, Badenian, Spirialis marl horizon, Piatra, Roumania.

13 *Globorotalia bykovae* (Aisenstat)
13a–c Holotype. Middle Miocene, Badenian, Bogorodchanian Formation, Carpathian foredeep, Bogorodchany, Ukraina, USSR.

14–17 *Globorotalia transsylvanica* Popescu
14a–c Holotype. Middle Miocene, Badenian, Dej Beds, Ciceu-Giurgesti, Roumania. **15–17** Specimens from the Middle Miocene, Badenian, Bresnita, Roumania.

18–20 *Cassigerinella globulosa* (Egger)
18a–c Original drawings of Egger. Early Miocene, Ottnangian, Habühl near Ortenburg, Bavaria. **19–20** Specimens from the Early Miocene, Ottnangian. 19 from top marls, Mairhof near Ortenburg. 20 from Blättermergel, Neustift near Vilshofen, Bavaria.

21–24 *Cassigerinella boudecensis* Pokorny
21a–b Holotype. Late Oligocene, upper Pouzdrany marls, Boudky, CSSR. **22, 24** Topotypes, coll. V. Pokorny. **23** Near topotype, coll. P. Ctyroky. Pouzdrany, CSSR.

25–26 *Globigerinita uvula* (Ehrenberg)
25a–b Holotype. Recent, North Atlantic, Davis Street, 6000 ft. **26** Synonym: *Globigerina bradyi* Wiesner, lectotype. Recent, southern Indian Ocean, *Challenger* station no. 144, 1570 fathoms, figured by Brady, 1884, pl. 82, fig. 8.

27–28 *Hastigerinella clavacella* Rögl
27a–c Holotype. **28** Topotype. Late Early Miocene, Karpatian, Laa an der Thaya, Austria.

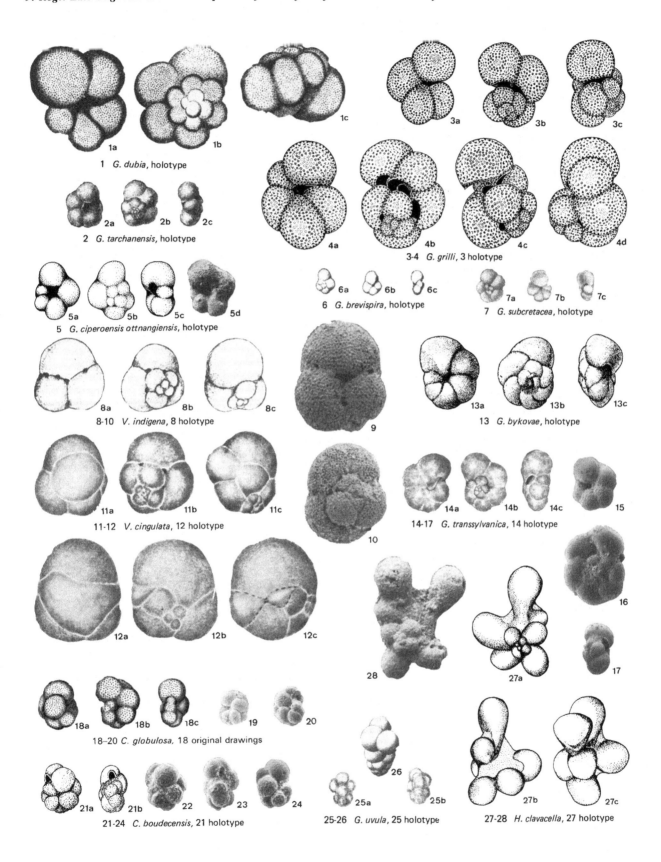

1   *G. dubia*, holotype

2   *G. tarchanensis*, holotype

3-4   *G. grilli*, 3 holotype

5   *G. ciperoensis ottnangiensis*, holotype

6   *G. brevispira*, holotype

7   *G. subcretacea*, holotype

8-10   *V. indigena*, 8 holotype

13   *G. bykovae*, holotype

11-12   *V. cingulata*, 12 holotype

14-17   *G. transsylvanica*, 14 holotype

18–20 *C. globulosa*, 18 original drawings

21-24 *C. boudecensis*, 21 holotype

25-26 *G. uvula*, 25 holotype

27-28 *H. clavacella*, 27 holotype

and place them here in *V. cingulata*. The species differs from *V. indigena* in its larger umbilical primary aperture and in the presence of small bullae covering the secondary sutural apertures. The two taxa have the same stratigraphic range.

Note: The genus *Velapertina* was described as possessing a trochospiral to globular test with the interiomarginal umbilical aperture covered by a bulla. Secondary sutural apertures are also present and in *V. cingulata* these may be covered by small bullae. Luczkowska (1971) stated that the occurrence of a bulla covering the aperture is not a constant feature and that only few specimens possess such structures. She therefore did not recognize the genus and referred *Globigerinoides indigena* Luczkowska to the genus *Praeorbulina*.

The only distinct feature we regard as typical for *Velapertina* is the occurrence of sutural openings similar to those in *Candeina*. Because of the occasional presence of bullae in *Velapertina* and its coarser pored wall surface, we retain the genus as separate from *Candeina*, while its lack of an embracing final chamber is the distinction from the stratigraphically older *Praeorbulina*. *Velapertina* has so far only been reported from the Late Badenian/Kossovian of the Parathethys, corresponding to the nannofossil Zone NN 7.

VELAPERTINA INDIGENA (Luczkowska)
Figures 5.8–10; 3

Type reference: *Globigerinoides indigena* Luczkowska, 1955, p. 152, pl. 10, figs. 5–7.

Synonym: *Velapertina luczkowskae* Popescu, *in* Popescu & Cioflica, 1973, p. 197, pl. 8, figs. 75–77.

V. indigena differs from *V. cingulata* in possessing a smaller primary aperture, varying from pore-like to a low arch, and in the secondary sutural apertures not being covered by bullae.

References

Baldi, T. 1969. On the Oligocene and Miocene stages of the Central Parathethys and on the formations of the Egerian in Hungary. *Ann. Univ. Sci. Budapest., Sec. Geol.*, **12** (1968), 19–28.

Baldi, T. & Seneš, J. 1975. OM–Egerien. Die Egerer, Pouzdřaner, Puchkirchener Schichtengruppe und die Bretkaer Formation. *Ser. Chronostratigraphie Neostratotypen*, vol. 5, pp. 1–577, Slov. Acad. Sci., Bratislava.

Banner, F. T. & Blow, W. H. 1960. Some primary types of species belonging to the superfamily Globigerinaceae. *Contrib. Cushman Found. foramin. Res.*, **11**, 1–41.

Berggren, W. A. & Van Couvering, J. A. 1974. Biostratigraphy, geochronology and paleoclimatology of the last 15 million years in marine and continental sequences. *Palaeogeogr. Palaeoclimatol. Palaeoecol.*, **16**, 1–216.

Blow, W. H. 1959. Age, correlation and biostratigraphy of the Upper Tocuyo (San Lorenzo) and Pozon formation; Eastern Falcon, Venezuela. *Bull. Am. Paleontol.*, **39**, 1–251.

Blow, W. H. 1969. Late Middle Eocene to Recent planktonic foraminiferal biostratigraphy. *Proceedings First International Conference on Planktonic Microfossils, Geneva*, 1967, **1**, 199–422.

Blow, W. H. & Banner, F. T. 1962. The Mid-Tertiary (Upper Eocene to Aquitanian) Globigerinaceae. In: F. E. Eames, F. T. Banner, W. H. Blow & W. J. Clarke, *Fundamentals of Mid-Tertiary Stratigraphic Correlation*, pp. 61–151, Cambridge University Press, Cambridge.

Bogdanowicz, A. K. 1950. Chokrakian foraminifera of Western Precaucasia (Russ.). *Trudy VNIGRI*, new series 51, Microfauna of the USSR, **4**, 129–76.

Bolli, H. M. 1954. Note on *Globigerina concinna* Reuss 1850. *Contrib. Cushman Found. foramin. Res.*, **5**, 1–3.

Brady, H. B. 1884. Report on the Foraminifera collected by H.M.S. *Challenger* during the years 1873–1876. *Rep. Sci. Res. H.M.S. Challenger*, 1873–6, **9** (Zoology), 1–814.

Chapman, F. J. 1902. On the foraminifera collected round the Funafuti Atoll from shallow and moderately deep water. *J. Linn. Soc. London, Zool.*, **28** (184), 379–417.

Cicha, I. 1970. Stratigraphical problems of the Miocene in Europe. *Rozpr. ustred. Ustavu Geol.*, **35**, 1–134.

Cicha, I., Hagn, H. & Martini, E. 1971. Das Oligozän und Miozän der Alpen und der Karpaten. Ein Vergleich mit Hilfe planktonischer Organismen. *Mitt. Bayer. Staatssamml. Palaeontol. hist. Geol.*, **11**, 279–93.

Cicha, I., Marinescu, F. & Seneš, J. 1975. *Correlation du Néogène de la Parathethys Centrale*. Geol. Surv., Praha, 33 pp.

Cicha, I. & Seneš, J. 1968. Sur la position du Miocène de la Parathethys Centrale dans le cadre du Tertiaire de l'Europe. *Geologicky Zbornik-Geologica Carpathica*, **19**, 95–116. Slov. Akad. Vied, Bratislava.

Cicha, I. & Seneš, J. 1975. Vorschlag zur Gliederung des Badenien der Zentralen Parathethys. *Proc. 6th Congr. RCMNS*, part 1, pp. 241–5, Slov. Acad. Sci., Bratislava.

Cicha, I., Seneš, J. & Tejkal, J. 1967. M 3 (Karpatien). Die Karpatische. *Ser. Chronostratigraphie Neostratotypen*, vol. 1, pp. 1–312, Slov. Acad. Sci., Bratislava.

Cicha, I. & Tejkal, J. 1959. Zum Problem des sog. Oberhelvets in dem Karpatischen Becken. *Vesn. ustred. Ustavu geol.*, **34**, 141–4.

Cita, M. B. & Premoli Silva, I. 1960. *Globigerina bollii*, nuova specie del Langhiano delle Langhe. *Riv. Ital. Paleontol.*, **46**, 119–26.

Cžjžek, J. 1848. Beitrag zur Kenntiss der fossilen Foraminiferen des Wiener Beckens. *Haidingers Naturwissenschaftliche Abhandlungen*, vol. 2, pp. 137–50. Braumüller & Seidel, Wien.

d'Orbigny, A. 1826. Tableau méthodique de la classe des Céphalopodes. *Annales des sciences Naturelles Paris*, ser. 1, 7, 245–314.

d'Orbigny, A. 1846. *Foraminifères fossiles du Bassin Tertiaire de Vienne (Autriche)*. Gide et Comp, Paris, 312 pp.

Drooger, C. W. 1956. Transatlantic correlation of the Oligo-Miocene by means of foraminifera. *Micropaleontology*, **2**, 183–92.

Dumitrica, P., Gheta, N. & Popescu, Gh. 1975. New data on the biostratigraphy and correlation of the Middle Miocene in the Carpathian area. *Dari Seama Sedint.*, **61** (1973–1974), 65–84.

Egger, J. G. 1857. Die Foraminiferen der Miocän-Schichten bei Ortenburg in Nieder-Bayern. *Neues Jahrb. Mineral. Geognos. Geol. Petrefacten-Kunde*, 1857, 266–311.

Ehrenberg, Ch. G, 1862. Über die Tiefgrund-Verhältnisse des Oceans am Eingang der Davisstraße und bei Island. *Monatsber. Preuss. Akad. Wiss. Berlin*, 1861, 275–315.

Ehrenberg, Ch. G. 1873. Mikrogeologische Studien über das kleinste Leben der Meeres-Tiefgründe aller Zonen und dessen geologischen Einfluss. *Abh. k. Akad. Wiss. Berlin*, 1872, 131–398.

Grill, R. 1941. Stratigraphische Untersuchungen mit Hilfe von Mikrofaunen im Wiener Becken und den benachbarten Molasse-Anteilen. *Oel und Kohle*, **31**, 595–602.

Grill, R. 1943. Über mikropaläontologische Gliederungsmöglichkeiten im Miozän des Wiener Beckens. *Mitt. Reichsamt Bodenf. Wien*, **6**, 33–44.

Hsü, K. J. & Bernoulli, D. 1978. Genesis of the Tethys and the Mediterranean. *Initial Rep. Deep Sea drill. Proj.*, **42** (1), 943–9.

Jiriček, R. 1975. *Biozonen der Zentralen Paratethys*. NAFTA, Gbely, ČSSR, 20 pp.

Karrer, F. 1861. Über das Auftreten der Foraminiferen in dem marinen Tegel des Wiener Beckens. *Sitz. Ber. Akad. Wiss. Wien, math.-naturwiss. Cl.*, **44**, 427–58.

Karrer, F. 1868. Die miocene Foraminiferenfauna von Kostej im Banat. *Sitz. Ber. Akad. Wiss. Wien, math.-naturwiss. Cl.*, **58**, 121–93.

Kokay, J. 1973. Faziostratotypen der Bantapusztaer Schichtengruppe. In: A. Papp, F. Rögl & J. Seneš, M2 – Ottnangien. *Ser. Chronostratigraphie Neostratotypen*, vol. 3, pp. 227–43. Slov. Acad. Sci., Bratislava.

Krasheninnikov, V. A. & Pflaumann, U. 1977. Zonal stratigraphy and planktonic foraminifers of Paleogene deposits of the Atlantic Ocean to the west of Africa. *Initial Rep. Deep Sea drill. Proj.*, **41**, 581–611.

Laskarev, V. 1924. Sur les équivalents du Sarmatien supérieur en Serbie. In: P. Vujevic (ed.) *Recueil de travaux offert à M. Jovan Cvijic par ses amis et collaborateurs*, pp. 73–85. Belgrade.

Lomnicki, J. L. M. 1900. Przyczynek do znajomosci fauny otwornic Miocenu Wieliczki. *Kosmos* (Lemberg), **24**, 220–8.

Lomnicki, J. L. M. 1901. Einige Bemerkungen zum Aufsatze: Die miozänen Foraminiferen in der Umgebung von Kolomea. *Verh. naturforsch. Ver. Brunn*, **39** (1900), 15–18.

Luczkowska, E. 1955. Tortonian foraminifera from the Chodenice and Grabowiec beds in the vicinity of Bochnia. *Rocz. Pol. Tow. geol.*, **23** (1953), 77–156.

Luczkowska, E. 1971. A new zone with *Praeorbulina indigena* (Foraminiferida, Globigerinidae) in the Upper Badenian (Tortonian s.s.) of Central Paratethys. *Rocz. Pol. Tow. geol.*, **40** (1970), 445–8.

Martini, E. & Müller, C. 1975. Calcareous nannoplankton and silicoflagellates from the type Ottnangian and equivalent strata in Austria (Lower Miocene). *Proc. 6th Congr. RCMNS*, part 1, pp. 121–37, Slov. Acad. Sci., Bratislava.

Neugeboren, J. L. 1847. Über die Foraminiferen des Tegels von Felsö-Lapugy unweit Dobra in Siebenbürgen. *Ber. Mitt. Freund. naturwiss. Wien*, **2**, 163–4.

Neugeboren, J. L. 1856. Die Foraminiferen aus der Ordnung der Stichostegier von Ober-Lapugy in Siebenbürgen. *Denkschr. Akad. Wiss. Wien, math.-naturwiss. Cl.*, **12** (2), 65–108.

Papp, A. 1959. Tertiär. Erster Teil. Grundzüge regionaler Stratigraphie. In: F. Lotze (ed.) *Handbuch Stratigr. Geologie*, vol. 3, part 1, 411 pp. F. Enke, Stuttgart.

Papp, A. 1960. Das Vorkommen von *Miogypsina* in Mitteleuropa

und dessen Bedeutung für die Tertiärstratigraphie. *Mitt. geol. Ges. Wien*, **51** (1958), 219–28.

Papp, A. 1969. Die Koordinierung des Miozäns der Paratethys. *Verh. geol. Bundesanst. Wien*, 1969, 2–6.

Papp, A., Grill, R., Janoschek, R., Kapounek, J., Kollmann, K. & Turnovsky, K. 1968. Zur Nomenklatur des Neogens in Österreich. *Verh. geol. Bundesanst. Wien*, 1968, 9–27.

Papp, A., Marinescu, F. & Seneš, J. 1974. M5 – Sarmatien (sensu E. Suess, 1866). *Ser. Chronostratigraphie Neostratotypen*, vol. 4, 707 pp. Slov. Acad. Sci., Bratislava.

Papp, A., Rögl, F., Cicha, I., Čtyroka, J. & Pishvanov, L. S. 1978. Planktonische Foraminiferen im Badenien. In: A. Papp *et al.* M 4 – Badenien. *Ser. Chronostratigraphie Neostratotypen*, vol. 6, pp. 268–78. Slov. Acad. Sci., Bratislava.

Papp, A., Rögl, F. & Seneš, J. 1973. M 2 – Ottnangien. Die Innviertler, Salgotarjaner, Bantapusztaer Schichtengruppe und die Rzehakia Formation. *Ser. Chronostratigraphie Neostratotypen*, vol. 3, 841 pp. Slov. Acad. Sci., Bratislava.

Papp, A. & Turnovsky, K. 1953. Die Entwicklung der Uvigerinen im Vindobon (Helvet und Torton) des Wiener Beckens. *Jahrb. geol. Bundesanst. Wien*, **46**, 117–42.

Paramonova, N. P., Ananova, E. N., Andreeva-Grigorivic, A. S., Belokrys, L. S., Gabunia, L. K. *et al.* 1979. Paleontological characteristics of the Sarmatian s.l. and Maeotian of the Ponto-Caspian area and possibilities of correlation to the Sarmatian s.str. and Pannonian of the Central Paratethys. *Ann. Geol. Pays Hellen.*, hors ser., fasc. 3, 961–71.

Pishvanova, L. S. 1959. Markiruyushie gorizonty planktonnykh foraminifer v miocenovyck otloshenijakh Predkarpatskogo progiba. *Trudy, Ukr. VNIGRI, Voprosy strat. lit. paleont. neft. raj. Ukrajiny*, **1**.

Pishvanova, L. S. 1968. On the zonation of the Miocene by means of planktonic foraminifera. *G. Geol.*, ser. 2, **35**, 233–54.

Pokorny, V. 1955. *Cassigerinella boudecensis* n.gen. n.sp. (Foraminifera, Protozoa) z oligocenu ždanickeho flyše. *Vesn. ustred. Ustravu geol.*, **30**, 136–40.

Popescu, Gh. 1969. Some new *Globigerina* (Foraminifera) from the Upper Tortonian of the Transylvanian Basin and the Subcarpathians. *Rev. Roum. Geol. Geophys. Geogr.*, ser. Geol., **13**, 103–6.

Popescu, Gh. 1970. Planktonic foraminiferal zonation in the Dej tuff complex. *Rev. Roum. Geol. Geophys. Geogr.*, ser. Geol., **14**, 189–203.

Popescu, Gh. 1975. Etudes des foraminiferes du Miocène inférieur et moyen du nord-ouest de la Transylvanie. *Mem. Inst. Geol. Geophys. Bucharest*, **23**, 1–121.

Popescu, Gh. 1976. Phylogenetic remarks on the genera *Candorbulina*, *Velapertina* and *Orbulina*. *Dari Seama Sedint.*, **62** (1974–75), 3. Paleont., 161–7.

Popescu, Gh. 1982. Note on *Globigerina regularis* d'Orbigny. *Dari Seama Sedint.*, **66** (1979), 23–6.

Popescu, Gh. & Cioflica, G. 1973. Contributii la microbiostratigrafia Miocenului mediu din nordul Transilvaniei. *Stud. cerc. geol. geof. geogr.*, ser. geol., **18**, 187–218.

Reuss, A. E. 1850. Neue Foraminiferen aus den Schichten des österreichischen Tertiärbeckens. *Denkschr. Akad. Wiss. Wien, math.-naturwiss. Cl.*, **1**, 365–90.

Reuss, A. E. 1867. Die fossile Fauna der Steinsalzablagerungen von Wieliczka in Galizien. *Sitz. Ber. Akad. Wiss. Wien, math.-naturwiss. Cl.*, **55**, 17–182.

Roemer, F. A. 1838. Die Cephalopoden des Nord-Deutschen

tertiären Meeressandes. *Neues Jahrb. Mineral. Geognos. Geol. Petrefacten-Kunde*, 1838, 381–94.

Rögl, F. 1969a. Die miozäne Foraminiferenfauna von Laa an der Thaya in der Molassezone von Niederöstereich. *Mitt. geol. Ges. Wien*, **61** (1968), 63–123.

Rögl, F. 1969b. Die Foraminiferenfauna aus den Phosphoritsanden von Plesching bei Linz (Oberösterreich) – Ottnangien (Untermiozän). *Naturkd. Jahrb. Stadt Linz*, 1969, 213–34.

Rögl, F. 1975. Die planktonischen Foraminiferen der Zentralen Paratethys. *Proc. 6th Congr. RCMNS*, part 1, pp. 113–20, Slov. Acad. Sci., Bratislava.

Rögl, F. & Bolli, H. M. 1973. Holocene to Pleistocene planktonic foraminifera of Leg 15, Site 147 (Cariaco Basin/Trench, Caribbean Sea) and their climatic interpretation. *Initial Rep. Deep Sea drill. Proj.*, **15**, 553–615.

Rögl, F. & Müller, C. 1976. Das Mittelmiozän und die Baden-Sarmat Grenze in Walbersdorf (Burgenland). *Ann. Naturhist. Mus. Wien*, **80**, 221–32.

Rögl, F. & Steininger, F. 1983. Vom Zerfall der Tethys zu Mediterran und Paratethys. *Ann. Naturhist. Mus. Wien*, **85/A**, 135–63

Rögl, F., Steininger, F. & Müller, C. 1978. Middle Miocene salinity crisis and paleogeography of the Paratethys (Middle and Eastern Europe). *Initial Rep. Deep Sea drill. Proj.*, **42**, 985–90.

Schmid, M. E. 1967. Zwei neue planktonische Foraminiferen aus dem Badener Tegel von Sooß, N.Ö. *Ann. Naturhist. Mus. Wien*, **71**, 347–52.

Semenenko, V. N. & Ljulieva, S. A. 1978. Opyt prjamoj korreljacii Mio-Pliocena vostočnogo Paratetisa i Tetisa (Attempt of a direct correlation of the Mio-Pliocene of the Eastern Paratethys and Tethys). *Sborn. 'Stratigr. Kainozoa Sever. Pricern. Kryma'*, Vyp., **2**, 95–105 (in Russian).

Seneš, J. 1959. Unsere Kenntnisse über die Paläogeographie der Zentral Paratethys. *Geol. Prace*, **55**, 83–108.

Seneš, J. 1978. Theoretische Erwägungen der zeitlichen Äquivalenz des Tarchanien mit den chronostratigraphischen Einheiten der Zentralen Paratethys. *Geol. Zborn., Geol. Carpat.*, **29**, 177–8.

Seneš, J. 1979. Correlation du Néogéne de la Tethys et de la Paratethys – Base de la reconstitution de la géodynamique récente de la region de la Méditerranèe. *Geol. Zborn., Geol. Carpat.*, **30**, 309–19.

Seneš, J. & Marinescu, F. 1974. Cartes paléogéographiques du Néogène de la Paratethys centrale. *Mem. Bur. Rech. geol. minieres*, **78**, 785–92.

Steininger, F. 1977. Integrated assemblage-zone biostratigraphy at marine–nonmarine boundaries: examples from the Neogene of Central Europe. In: E. G. Kauffman & J. H. Hazel (eds.) *Concepts and Methods of Biostratigraphy*, pp. 235–56, Dowden, Hutchinson & Ross Inc., Strondsburg.

Steininger, F. & Papp, A. 1979. Current biostratigraphic and radiometric correlations of Late Miocene Central Paratethys stages (Sarmatian s.str., Pannonian s.str., and Pontian) and Mediterranean stages (Tortonian and Messinian) and the Messinian event in the Paratethys. *Newsl. Stratigr.*, **8**, 100–10.

Steininger, F. & Rögl, F. 1979. The Paratethys history – a contribution towards the Neogene geodynamics of the Alpine orogene (an abstract). *Ann. Geol. Pays Hellen.*, hors ser., fasc. 3, 1153–65.

Steininger, F., Rögl, F. & Martini, E. 1976. Current Oligocene/Miocene biostratigraphic concept of the Central Paratethys (Middle Europe). *Newsl. Stratigr.*, **4**, 174–202.

Steininger, F., Rögl, F. & Müller, C. 1978. Geodynamik und paläogeographische Entwicklung des Badenien. In: A. Papp *et al.*, M 4 – Badenien (Moravien, Wielicien, Kosovien), *Ser. Chronostratigraphie Neostratotypen*, vol. 6, pp. 110–16. Slov. Acad. Sci., Bratislava.

Steininger, F. & Seneš, J. 1971. M 1 – Eggenburgien. Die Eggenburger Schichtengruppe und ihr Stratotypus. *Ser. Chronostratigraphie Neostratotypen*, vol. 2, 827 pp. Slov. Acad. Sci., Bratislava.

Stradner, H. & Fuchs, R. 1980. Über Nannoplanktonvorkommen im Sarmatien (Ober-Miozän) der Zentralen Paratethys in Niederösterreich und im Burgenland. *Beitr. Palaont. Osterr.*, **7**, 251–79.

Subbotina, N. N. 1953. Fossil foraminifera of the USSR. Globigerinidae, Hantkeninidae and Globorotaliidae. *Trudy VNIGRI*, **76**, 296 pp. (English translation by E. Lees, 1971, 321 pp., Collet's Ltd., London, Wellingborough.)

Subbotina, N. N., Pishvanova, L. S. & Ivanova, L. V. 1960. Stratigraphia oligocenovih i miocenovih otlozhenii Predkarpatja po forminiferam. *Trudy VNIGRI*, **153**, Microfauna of the USSR 9, pp. 5–127.

Sztrakos, K. 1979. La stratigraphie, paléoécologie, paléogéographie et les foraminifères de l'Oligocène du nord-est de la Hongrie. *Cah Micropaleontol.*, 1979, no. 3, 95 pp.

Vašiček, M. 1952. The contemporary state of the microbiostratigraphic research of the Miocene sedimentary deposits in the Out-Carpathian Neogene basin in Moravia. *Sb. ustred. Ustava geol.*, **18** (1951), ser. Paleontol., 145–95.

Wiesner, H. 1931. Die Foraminiferen. In: E. Drygalski, *Deutsche Südpolar Expedition* 1901–1903, **20** (Zool. **12**), 53–165, de Gruyter, Berlin-Leipzig.

10

Mesozoic calcareous nannofossils

KATHARINA PERCH-NIELSEN

CONTENTS

Introduction

Coccoliths are the minute calcite plates produced by unicellular marine algae, the coccolithophorids. Fossil coccoliths, together with small calcite bodies called nannoliths by some writers, constitute the calcareous nannofossils. Nannoliths are certainly organic but are of uncertain origin. Calcareous nannofossils proved to be extremely useful for the biostratigraphy of Jurassic through Pleistocene marine sediments. Their small size (1–25 μm) allows for age determinations of even very small samples such as from ditch cuttings, sidewall cores, etc. Biostratigraphic investigations of the Mesozoic were pioneered by Stradner (1961), who suggested specific assemblages for Jurassic and Cretaceous stages. While pre-Jurassic occurrences of calcareous nannofossils are scarce and often doubtful, zonal schemes have been developed for the Jurassic and the Cretaceous. Of these, the Jurassic scheme proposed by Barnard & Hay (1974) and slightly adapted by Moshkovitz & Ehrlich (1976a), Hamilton (1977, 1979) and Medd (1979) is still not satisfactory. In the Cretaceous, however, the sequence of FOs and LOs (first and last occurrences) of many species is quite well known and various authors have used different combinations of these FOs and LOs to establish zonal schemes for different geographic or climatic areas.

Introductions to calcareous nannoplankton have been given by Reinhardt (1972), Haq (1978) and Tappan (1980). The present chapter concentrates on those aspects of calcareous nannofossils that are useful for biostratigraphy. While it is not intended as a complete systematic treatise, it nevertheless gives an overview over a large number of Mesozoic species and genera which have not so far been used as stratigraphic markers. For a more complete overview of the calcareous nannofossils the *Catalogue of Calcareous Nannofossils* by Farinacci (1969–83) and the 'Synopsis der Gattungen und Arten der mesozoischen Coccolithen' by Reinhardt (1970a, b, 1971) can be consulted. Coccolithophorids are classified according to the International Code of Botanical Nomenclature (ICBN) and that convention is largely followed here. Bibliographies and indexes were published by Loeblich & Tappan (1966, 1968, 1969, 1970a, b, 1971, 1973) and by van Heck (1979–82) and Steinmetz (1983, in the biannual newsletter of the International Nannoplankton Association, INA).

Sampling and sample treatment

Whenever possible, marls should be preferred to limestones when sampling for calcareous nannofossils. No coccoliths will be found in clays which do not contain at least a small percentage of calcium carbonate or in more than very slightly metamorphosed sediments. Coccoliths are better preserved in slightly marly chalks than in pure ones.

The samples can be very small – the amount of a few rice grains is sufficient to prepare several smear-slides and/or to prepare slides from centrifuged samples. Because of the extremely small size of coccoliths every care has to be taken to avoid contamination at the outcrop or in the laboratory. Even the dust escaping from textile bags can be a cause of contamination making storage in plastic and paper bags advisable.

Preparation techniques

For routine biostratigraphic work, smear-slides are prepared in the manner outlined below. A skilled technician should be able to prepare some 10 to 20 per hour, depending on the sediment involved. To prepare a smear-slide, the sample is cleaned, if possible, by removing the outermost layer of material with a clean knife or razor blade before the sample is broken up and a small amount scratched off onto a glass slide. Hard samples can be gently crushed in a mortar. Cuttings are picked as clean as feasible and one to a few are spread on the glass slide. A few drops of distilled water and a flat toothpick or match serve to smear the sediment over the slide. If the sample is sandy or silty, the coarse fraction should be isolated and removed lest the slide be too thick for observation with the × 100 oil immersion objective. Coarse fraction removal is best accomplished by retaining all of the smeared sediment and water at one end of the slide by using the flat toothpick as a dam. The slide is then inclined and the toothpick lifted slightly so that the water containing the finest fraction (coccoliths and clay) flows under the toothpick. In this way, the sand and silt particles are retained at the raised end of the slide. It is an advantage to have the material smeared out unevenly, so, depending on the abundance of calcareous nannofossils, one can study them in an area of the slide where their distribution is ideal. Alternatively the sample can be smeared onto the cover-glass. In such a preparation the coccoliths lie near the objective and the few large particles do not hinder observation with a high-power objective. When the sediment is dry, the mounting medium (artificial, fluid Canada balsam or Canada balsam) is best applied to the surface that does not have the dried smear after which the two surfaces are pressed together. The slide may then be cooked on a hot-plate for approximately half a minute to one minute to make the slide more durable and easier to store. Cooking is not essential for the fast observation and age determination that might be desirable or necessary at a well site. Coccoliths can be observed with a light microscope (LM) with × 10 or × 12.5 occulars and a × 40 or × 100 objective, the latter requiring oil immersion. Polarization equipment and/or phase contrast equipment is necessary for the study of most of the forms and especially for the small ones measuring less than 10 μm; that is, for most of the stratigraphic markers. The use of a gypsum plate is recommended and described in detail by Reinhardt (1972) and Romein (1979).

For more detailed observations, for samples with rare but well preserved coccoliths and for observation with the electron microscope (EM), samples are treated ultrasonically for a few seconds and centrifuged for 30 seconds at 2000 r.p.m. to remove clay particles which are less than 2 μm in diameter. Ultrasonic treatment and centrifuging are repeated several times until the distilled water above the sediment in the centrifuge glass remains clear after centrifuging, which may take up to 5 or 6 times in very clay-rich samples. The treatment just described is known to break some of the more delicate forms, and very small forms may be lost. On the other hand, robust marker species that may be rare get concentrated. The use of chemicals is limited to those sediments where the above method gives unsatisfactory results, such as clays still sticking to the coccoliths and obstructing the view of details of the structure in the scanning or transmission electron microscope (SEM, TEM). In these cases, small amounts of Calgon are added to the first and second round of ultrasonic/centrifuge treatment and then washed out by the subsequent treatments. Slides for viewing with the LM are then prepared in the way outlined above. For observation with the SEM, a round cover-glass or broken piece of a cover-glass is mounted with glue or double tape on a SEM stub and a drop of the water with the cleaned coccoliths left to dry on the cover-glass. The preparation is ready for viewing with a SEM after coating with gold in a vacuum. Magnification of about × 2000 is used for scanning through the preparation and × 5000 to × 30 000 for photography of large and small forms, respectively. The preparation technique for TEM observations was described in detail by Noël (1965). Perch-Nielsen (1967) described a method for viewing the same specimen first by LM and then by TEM, and Thierstein *et al.* (1972) and Smith (1975a) described methods for LM and subsequent SEM observation of the same specimen.

Data presentation

The presentation of the data varies from one investigator to another. Range charts should include an indication of the abundance of calcareous nannofossils, their state of preservation and the abundance of the species found. The abundance is usually only estimated. Actual counts are only done when a special study is made for the recognition of changes in paleoenvironment. In this case, care has to be taken that the material studied is well preserved or at least of uniform, yet reasonable, preservation.

Illustrations are usually light micrographs. Magnifications of × 1500 to × 3000 are recommended; × 2000 is commonly used. Illustrations are not necessary where biostratigraphic interpretations are based on marker species illustrated by Barnard & Hay (1974), by Thierstein (1976) or in this chapter. Instead of new illustrations, a list of published figures used for identification can be given. In the case of calcareous nannofossils, these may not necessarily be the holotype, since some holotypes have been very poorly illustrated (see several

examples in this chapter, e.g. *Arkhangelskiella cymbioformis* and *Micula decussata*). Other holotypes are based on EM pictures and it may be safer to refer to a LM picture as an indication to understand the author's concept of the species. In the ideal case new species are shown as LM and as SEM pictures, but this has not been the case for all species described.

Preservation, reworking and caving

Problems of preservation are the same for Mesozoic and Cenozoic calcareous nannofossils and are discussed in Chapter 11. Thierstein (1980) studied the dissolution susceptibility of Late Cretaceous and Early Cenozoic calcareous nannofossils experimentally. According to his findings, the relative proportions of only a few taxa change significantly with increasing dissolution. *Micula decussata* was found to be the most resistant species, followed by *Quadrum trifidum, Lucianorhabdus cayeuxii, Kamptnerius magnificus, Braarudosphaera bigelowii* and *Markalius inversus*. Below the Turonian, *Watznaueria barnesae* is usually the most solution resistant species in the Cretaceous.

Reworking of calcareous nannofossils is known to have taken place extensively in areas where older sediments with coccoliths were available for submarine or subareal erosion. Reworking is relatively easy to detect in those cases where, for example, Cretaceous forms were reworked into Neogene sediments, since we know the ranges of most Cretaceous and Neogene forms reasonably well and can thus recognize reworking with some degree of confidence. But within the Jurassic, or even from the Jurassic into the Lower Cretaceous, our knowledge about the real ranges of the many species is as yet far too limited to be certain whether a given species was reworked from the Middle Jurassic to the Upper Jurassic or whether the species lived during the Middle as well as the Late Jurassic. Unfortunately, coccoliths usually do not show that they have been transported once or even repeatedly, since they are probably often transported mainly in silt- or sand-sized pieces of sediment and thus are not mechanically or chemically attacked. In the case of transport of single coccoliths, it is likely that smaller, delicate forms were dissolved while larger forms survived but may have broken to pieces, particularly in the case of forms with a wide, partly open central area. As the same can happen *in situ* during compaction and lithification it cannot be used as an argument for reworking. This implies that FOs should be used whenever possible for stratigraphy, though this is unsatisfactory when working with cuttings from boreholes.

Caving of younger coccoliths into older sediments in boreholes is not uncommon with the same problems of recognition as for reworking. The present uncertainty of true ranges makes the use of LOs difficult at any time.

Paleoenvironments

Due to their predominantly planktic mode of life, coccolithophorids do not give a clear indication of depth of deposition of the sediments in which they are found. Due to their small size, they have usually been removed from deposits of nearshore, high energy environments.

Shallow depth indicators

Generally it can be said that holococcoliths (coccoliths consisting of minute calcite rhombs of the same size, family Calyptrosphaeraceae) are only found in sediments deposited at shallow depths up to a few hundred metres. They are very rare in the Jurassic sediments, not necessarily because they were deposited in greater depths, but possibly due to their destruction through diagenesis. Thus, whereas the absence of holococcoliths is inconclusive, their presence is an indication of relatively shallow deposition. However, care must be exercised because holococcoliths are known to survive turbidity transport to great depths.

Deposition in calm waters

Deposition in very calm waters is indicated by the presence of many coccospheres in the sediment especially if coccospheres of coccoliths that do not overlap each other on the test are also found. Thus, Noël (1973) illustrated coccospheres of *Stradnerlithus* from laminated Kimmeridgian sediments.

Restricted environment

Some solution resistant, common coccoliths are useful facies indicators in the Upper Jurassic according to Keupp (1976). He found that a dominance of *Cyclagelosphaera margerelii* is characteristic of a restricted environment, while the dominance of *Ellipsagelosphaera* species suggests a direct connection to the open sea.

Marginal seas – open ocean

Thierstein (1976) found that forms common in the Cretaceous Tethys were absent in Deep Sea Drilling Project (DSDP) sites in the Pacific. He related this to the general paleogeographic setting of the two oceans at that time: marginal for the Tethys and oceanic for the Pacific. He found the following forms to be favoured by epicontinental seas and large shelf areas: *Nannoconus* spp., *Conusphaera mexicana*, *Calcicalathina oblongata*, *Lithraphidites bollii*, *Micrantholithus obtusus* and *Lithastrinus floralis*. He observed a similar preference for marginal conditions for *Lucianorhabdus cayeuxii*, *Tetralithus obscurus* and *Braarudosphaera bigelowii*. *Arkhangelskiella cymbiformis* and *Kamptnerius magnificus* should be added to this list as both species are usually more common in marginal than in oceanic environments.

Biogeography through time

Roth & Bowdler (1981) studied the calcareous nannoplankton biogeography of the Middle Cretaceous of the Atlantic Ocean. They found that paleogeographic patterns of coccolith distribution show weak latitudinal gradients and more pronounced neritic–oceanic gradients. Boreal and Austral assemblages were first observed in the later half of the Mid Cretaceous and are restricted to latitudes greater than 40°. Rapid spatial and temporal fluctuations in coccolith abundance, including monospecific blooms, were found to be typical of higher latitudes and restricted basins. Neritic assemblages were found along the continental margin of the eastern Atlantic and over the Walvis–Rio Grande Ridge and are possibly indicative of increased advection of nutrient-rich waters. Oceanic assemblages characterize areas far removed from continents. They include abundant *Watznaueria barnesae*, *Rhagodiscus splendens* and *R. asper*. *Seribiscutum primitivum* was found to be five times as abundant in the Falkland Plateau as in the contemporaneous Albian of England.

Thierstein (1981) studied the coccolith distribution during the Late Campanian and the Maastrichtian and found provincialism for many species including the markers *Micula decussata* (*M. staurophora* in Thierstein, 1981), *Quadrum trifidum* (*Tetralithus trifidus*), *Nephrolithus frequens*, *Lithraphidites quadratus* and *Micula murus* (Fig. 1).

Biostratigraphic zonation

Both for the Jurassic and the Cretaceous, FOs are mainly used for subdivision and zonation. No zonations are available for pre-Jurassic times and the Jurassic zonations are not yet satisfactory. The sequence of stratigraphic events in the Cretaceous is relatively well established in the Late Cretaceous whereas it is less well known in the Early Cretaceous, where the time intervals between the events are much longer than in the Late Cretaceous. Direct and indirect correlations with paleomagnetic data were presented by Thierstein (1976), Alvarez *et al.* (1977) and Channel *et al.* (1978) and are shown in Fig. 1 in Chapter 2.

Pre-Jurassic calcareous nannofossils and the Triassic/Jurassic boundary
(Figure 2)

Scientists have debated the appearance of the first coccolithophorids for some time. The forms illustrated by Noël (1961, 1965) from the Pennsylvanian near Tulsa and from other localities of Paleozoic age are believed to be contaminants from Mesozoic samples, since they look very much like them. Deflandre (1970) reported Silurian–Devonian forms from North America which do not seem to represent coccolithophorids, Pirini-Radrizzani (1971) illustrated probable coccoliths from the Permian of Turkey (Figs. 2.23, 24) and Di Nocera & Scandone (1977) and Wiedmann *et al.* (1979) gave

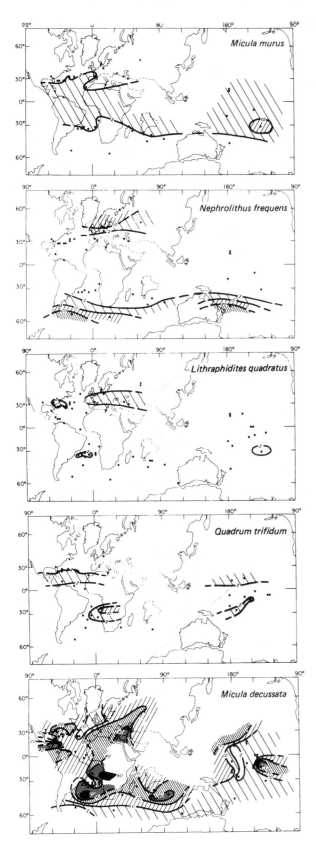

Fig. 1. Average per cent abundance of some Late Cretaceous marker species. After Thierstein (1981)

illustrations from the Triassic of Greece, the Southern Alps and Southern Germany. All these forms cannot be related directly to the Jurassic forms. Also the forms illustrated and named by Gartner & Gentile (1973) from the Pennsylvanian of Missouri and those described from Paleozoic limestones by Minoura & Chitoku (1979) cannot be related to the younger forms.

Jafar (1983) and Janofske (personal communication, 1982) describe and illustrate very small coccoliths and other calcareous nannofossils from the Late Triassic Norian and Rhaetian, most of which are shown in Figs. 2.1–22, 25–28. Jafar (1983) found very small (1.6–2.4 µm) specimens of *Crucirhabdus* and *C. minutus* (Figs. 2.5, 6) in the Norian. *C. minutus* is characterized because 'the two knobs visible along the minor axis of the coccolith are somewhat rounded and distinct extinction lines demarcate equally large knobs at both ends of the major axis'. It is found in most of the Norian *Rhabdoceras suessi* ammonite Zone. *C. curvatus* (Figs. 2.7–9)

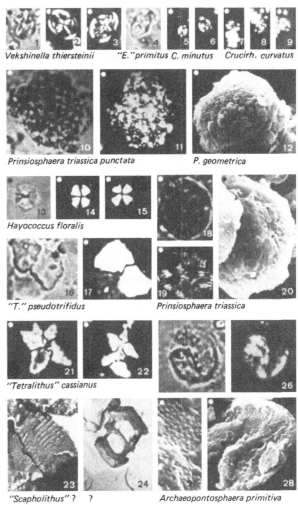

Fig. 2. LM and EM of pre-Jurassic calcareous nannofossils. LM: × 1800. Figs. 1–22, 25–28 from the Late Triassic, 23 and 24 from the Permian. * = holotype.

is slightly larger than *C. minutus* (2.6–2.8 μm) and yields 'under crossed nicols distinctly curved extinction lines at both ends of the coccolith aligned to the major axis'. It is the first genuine coccolith in the Norian. *C. primulus* (Figs. 5.34, 35) measures about 3–8(?) μm, the specimens increasing in size from the early Late Norian, when they appear, to the Hettangian, when they reach about 5 μm and when their 'FO' was used by Barnard & Hay (1974) for the definition of a zone. *Vekshinella thiersteinii* (Figs. 2.1–3) and *Chiastozygus?* (*Ellipsochiastozygus*) *primitus* (Fig. 2.4) are the first representatives of the Ahmullerellaceae and the Chiastozygazeae respectively. They were found in a sample of questionable Rhaetian or Hettangian age (Jafar, 1983, pages 222 and 257).

Jafar (1983) also described several species and subspecies of *Prinsiosphaera*, a genus he described as comprising 'spherical to hemispherical solid nannofossils often containing a depression at one end and consisting of parallely stacked groups of calcite plates oriented in a random fashion'. Such forms were found from the Karnian to the Rhaetian, as were rare to very rare *Archaeopontosphaera primitiva* (Figs. 2.25–28). The two species assigned to *Tetralithus* (Figs. 2.16, 17, 21, 22) may or may not be of organic nature. *Hayococcus floralis* (Figs. 2.13–15) is based on a single specimen from the Upper Norian.

Pieces of *Schizosphaerella* or whole specimens of this genus are found together with the species mentioned above and forms similar to *Annulithus arkellii*. Moshkovitz (1982) described *Conusphaera zlambachensis* (syn. *Eoconusphaera tollmanniae*) from the Rhaetian of Austria.

There is no sudden appearance of a large number of taxa at the Triassic/Jurassic boundary, whether it is placed below or above the Rhaetian. New coccoliths probably appeared throughout the Early Jurassic, but they usually were not preserved and thus we find sediments with a highly diverse assemblage some species of which have no known ancestors.

Similarly, there is no sudden disappearance of species and genera at the Triassic/Jurassic boundary. This is unlike the situation at the Cretaceous/Tertiary boundary, where some 30 of nearly 50 extant genera disappear. Only few genera of calcareous nannofossils are known from the Triassic perhaps because only a few existed or perhaps because they were not preserved in the Triassic environment?

Jurassic

Barnard & Hay (1974) suggested a 'tentative zonation of the Jurassic of Southern England and North France'. Their marker species and zones are shown in Figs. 3 and 4 together with the markers used by Hamilton (1977, 1979), Moshkovitz & Ehrlich (1976a), Manivit *et al.* (1979), Medd (1979), Roth, Medd & Watkins (1983) and Thierstein (1976), the last having been submitted for publication before that of Barnard & Hay

(1974) appeared. Hamilton (1982) reviewed studies on Jurassic calcareous nannofossils from the United Kingdom.

Most of the zones are interval zones defined from the first occurrence (FO) of one species to the FO of another species. Extensive descriptions of the zones by Barnard & Hay (1974) can be found in their publication, from which several of the illustrations of marker species were taken. The definition of the zones can be seen in Figs. 3 and 4, and is discussed below. The ranges of the markers and some other species are given in Fig. 4, whereas the species are illustrated in Fig. 5. The genera and families to which they belong are discussed and further illustrated in the taxonomic notes.

Diversity and zonation

Theoretically we should be in a very good position to offer a detailed zonation of the Jurassic: more than 70 genera and more than 150 species have been described from the Jurassic. Since only about a third of the genera pass the Jurassic/Cretaceous boundary, a sufficient number of datums occur to provide as detailed a zonation as that for the Tertiary or the Late Cretaceous. For example, considerable differences in species diversity exist between marls, where it is often high, and limestones and sandstones, where it is usually low. This is due to the better preservation of coccoliths in marly sediments than in other lithologies. Thus the known ranges of most coccoliths are probably too short on one or both ends of their distribution and ranges appear different from one area to another, not necessarily because the coccolithophorids disappeared earlier in one area than in the other, but because of selective preservation.

Ten years ago, before the description of many new species from the Toarcian by Goy, Noël & Busson (1979) and from mainly the Oxfordian and Kimmeridgian by Medd (1979) and Wise & Wind (1977), the Toarcian and the Oxfordian were the stages with the highest diversity. Grün & Zweili (1980) recently showed that there were also many previously unknown forms in the Middle Jurassic. I am convinced that further studies will show that the differences in diversity are not as considerable as they appear now.

For stratigraphic work this means that it is worth taking closely spaced samples in marls, studying them first in smear-slides with the LM and then using those samples with reasonable coccolith assemblages for detailed EM studies.

Tentative Jurassic zones

The zonations by Medd (1982) and Roth *et al.* (1983) are included in Fig. 3 but not discussed further in the following text.

ANNULITHUS ARKELLII ZONE

Definition: FO of *Annulithus arkellii* to FO of *Crucirhabdus primulus*.

Authors: Barnard & Hay (1974)

Age: Late Rhaetian to Early Hettangian

Remarks: The assemblage is sparse and consists mainly of *Schizosphaerella punctulata* with rare *A. arkellii*. The *A. arkellii* Zone corresponds to the lower part of the *Crucirhabdus* Zone of Prins (1969). Barnard & Hay (1974) also noted the presence of rare *Crepidolithus cavus* and *Vekshinella quadriarculla*. Small specimens of *C. primulus* were found by Jafar (1983) in the Upper Triassic. The FO of *C. primulus* in the Hettangian takes into account only the specimens larger than about 5 μm.

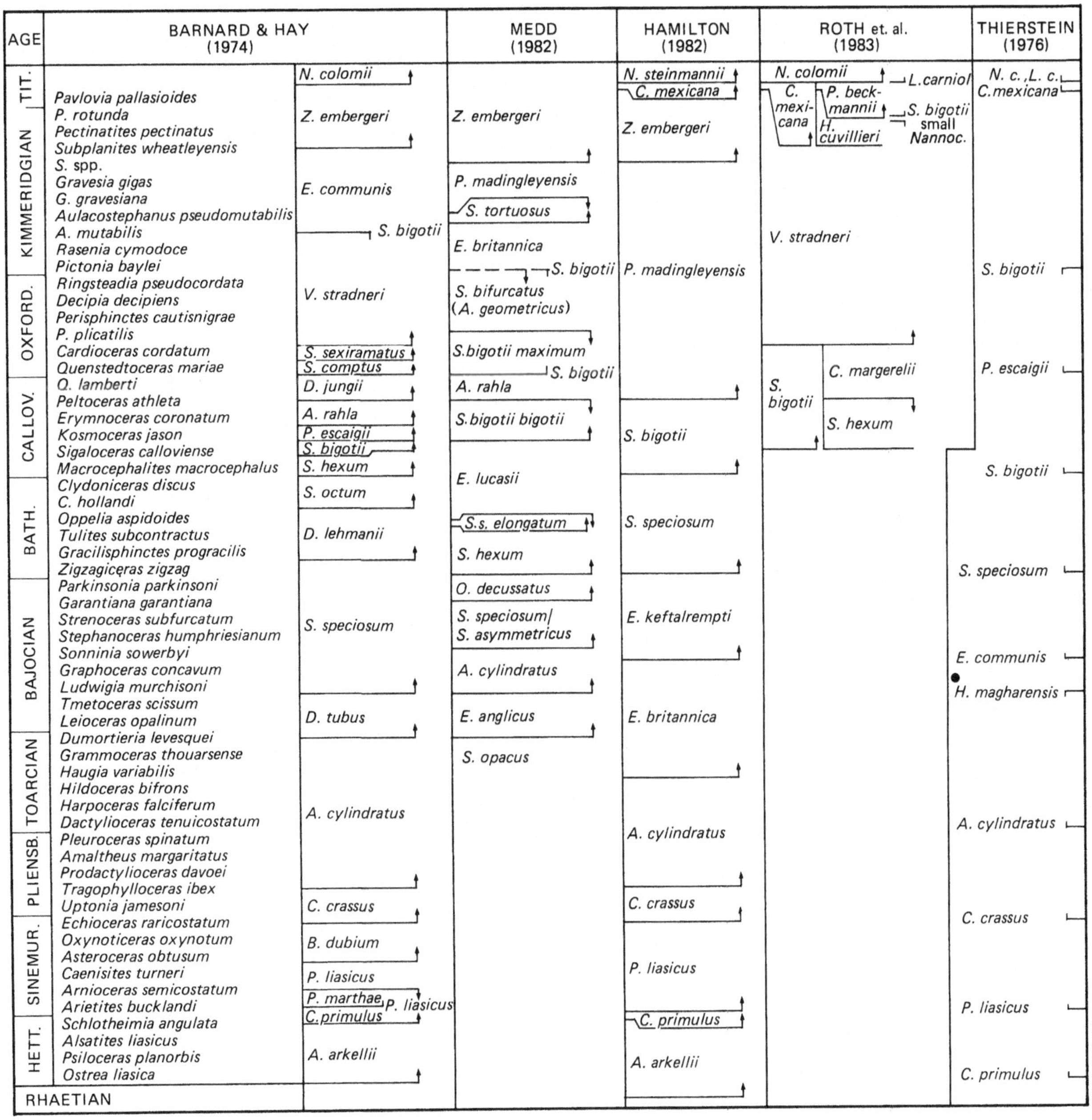

Fig. 3. Jurassic zonal schemes and their correlation to the classical British ammonite zonation of the Jurassic and a marker (●) of Moshkovitz & Ehrlich (1976a)

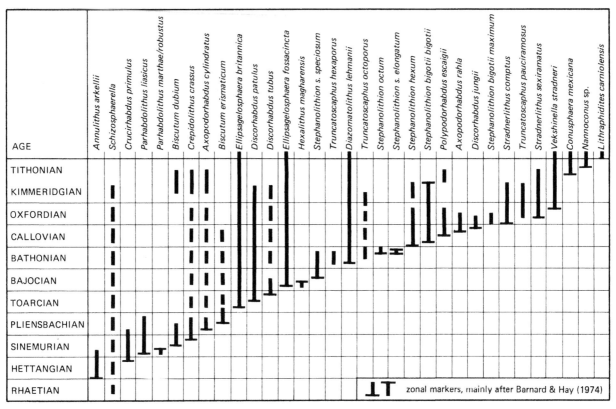

Fig. 4. Ranges of Jurassic markers and a few other
calcareous nannofossils

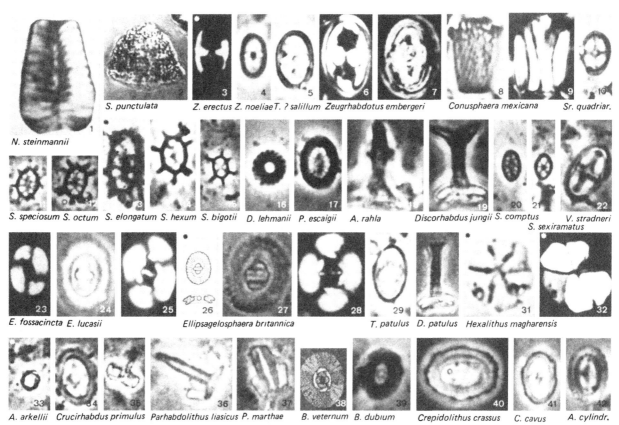

Fig. 5. LM of Jurassic markers and a few other species. LM:
c. × 2600.

CRUCIRHABDUS PRIMULUS ZONE

Definition: FO of *Crucirhabdus primulus* to FO of *Parhabdolithus liasicus.*

Authors: Barnard & Hay (1974)

Age: Late Hettangian to Early Sinemurian

Remarks: *S. punctulata* typically dominates the assemblage. Other species include *C. primulus, V. quadriarculla, Zeugrhabdotus erectus, T. patulus, C. cavus, C. crucifer* and *A. arkellii.* The *C. primulus* Zone corresponds to the upper part of the *Crucirhabdus* Zone of Prins (1969). As stated above, this usage of the FO of *C. primulus* takes into account only those specimens larger than about 5 μm.

PARHABDOLITHUS MARTHAE ZONE

Definition: FO of *Parhabdolithus liasicus* to LO of *Parhabdolithus marthae.*

Authors: Prins (1969), emended by Barnard & Hay (1974)

Age: Sinemurian

Remarks: *C. primulus, P. marthae* and *P. liasicus* are the most common species in the *P. marthae* Zone, which includes also the species already known in the *C. primulus* Zone.

PARHABDOLITHUS LIASICUS ZONE

Definition: LO of *Parhabdolithus marthae* to the FO of *Biscutum dubium* (*Paleopontosphaera dubia* of Barnard & Hay, 1974)

Authors: Prins (1969), emended by Barnard & Hay (1974)

Age: Late Sinemurian

Remarks: *P. liasicus* often dominates the assemblage of the *P. liasicus* Zone; *C. primulus* and *T. patulus* are other sometimes relatively common species. The species of the *P. marthae* Zone continue into the *P. liasicus* Zone with the exception of the marker species.

Hamilton (1982) defined her *P. liasicus* Zone from the FO of *P. liasicus* to the FO of *Crepidolithus crassus,* thus including the *P. marthae* Zone, the present *P. liasicus* Zone and also the overlying *B. dubium* Zone.

BISCUTUM DUBIUM ZONE

Definition: FO of *Biscutum dubium* to FO of *Crepidolithus crassus.*

Authors: Barnard & Hay (1974)

Age: Late Sinemurian

Remarks: The assemblage is similar to that found in the *P. liasicus* Zone, but includes relatively common *B. dubium. A. arkellii* is not found. The *B. dubium* Zone is the upper part of the *P. liasicus* Zone of Hamilton (1982).

CREPIDOLITHUS CRASSUS ZONE

Definition: FO of *Crepidolithus crassus* to FO of *Axopodorhabdus cylindratus.*

Authors: Prins (1969), emended by Barnard & Hay (1974)

Age: late Late Sinemurian to Early Pliensbachian

Remarks: The assemblage is similar to the one of the *B. dubium* Zone, but includes relatively common *C. crassus. P. liasicus* becomes rare towards the top of the zone; *C. primulus* may commonly occur in the lower part and disappears in the upper part of the zone.

AXOPODORHABDUS CYLINDRATUS ZONE

Definition: FO of *Axopodorhabdus cylindratus* to FO of *Discorhabdus tubus.*

Authors: Barnard & Hay (1974)

Age: Late Pliensbachian and Toarcian (part)

Remarks: *A. cylindratus* is rare in the lower part of the zone. Other species include *C. crassus, C. cavus, B. dubium, V. quadriarculla, Z. erectus, T. patulus* and *S. punctulata. P. liasicus* disappears towards the top of the *A. cylindratus* Zone, which corresponds to part or all of Prins' (1969) *C. crassus* Subzone and the *Striatococcus opacus* Subzone. The *A. cylindratus* Zone of Hamilton (1982) is defined from the FO of *A. cylindratus* to the FO of *Ellipsagelosphaera britannica,* a species which is thought to appear shortly after *D. tubus.*

Hamilton (1977, 1979) suggested a subdivision of the Pliensbachian–Toarcian interval as follows: the FO of *B. veternum* followed by the LO of *P. liasicus* and the LO of *C. primulus,* the FO of *E. britannica* and the FO of *Discorhabdus patulus.* This sequence is included in the *A. cylindatus* Zone as defined here.

DISCORHABDUS TUBUS ZONE

Definition: FO of *Discorhabdus tubus* to FO of *Stephanolithion speciosum.*

Authors: Barnard & Hay (1974)

Age: Late Toarcian to Early or all the Bajocian(?)

Remarks: Whereas many species from the subjacent zone are present in the *D. tubus* Zone, the general aspect of the assemblage changes with the appearance of the marker species and *Podorhabdus macrogranulatus, Striatomarginis primitivus, Mitrolithus elegans* and *Biscutum erismatum.*

The *D. tubus* Zone includes part of Hamilton's (1979) *A. cylindratus* and the entire *E. britannica* zones and the *Ellipsagelosphaera keftalrempti* Zone of Hamilton (1982). The latter was defined from the FO of *E. keftalrempti* (here *E. fossacincta*) to the FO of *S. speciosum.* The LO of *Hexalithus magharensis* was used as a marker event by Moshkovitz & Ehrlich (1976a) in the Bajocian of Israel.

STEPHANOLITHION SPECIOSUM ZONE

Definition: FO of *Stephanolithion speciosum* to FO of *Diazomatolithus lehmanii*.

Authors: Barnard & Hay (1974)

Age: Late Bajocian (?) to Early Bathonian

Remarks: *S. speciosum* is the first of a series of small but useful species of *Stephanolithion* to appear. Hamilton (1982) suggested a *S. speciosum* Zone defined from the FO of the namegiving species to the FO of *S. bigotii*, thus including the *S. speciosum* Zone of Barnard & Hay (1974) and their *D. lehmanii*, *S. octum* and *S. hexum* Zones. The Early Jurassic forms become rare in this Middle Jurassic zone, which is characterized by rare specimens of *Ellipsagelosphaera*, which later become more common. Barnard & Hay (1974) reported the LO of *M. elegans* from this zone.

DIAZOMATOLITHUS LEHMANII ZONE

Definition: FO of *Diazomatolithus lehmanii* to FO of *Stephanolithion octum*.

Authors: Barnard & Hay (1974)

Age: Late Bathonian (& Early Callovian?)

Remarks: The *D. lehmanii* Zone falls within the *S. speciosum* Zone of Hamilton (1982). *Stradnerlithus asymmetricus* and *Cyclagelosphaera margerelii* appear in this zone.

STEPHANOLITHION OCTUM ZONE

Definition: FO of *Stephanolithion octum* to FO of *Stephanolithion hexum*.

Authors: Barnard & Hay (1974)

Age: Late Bathonian and ? Early Callovian

Remarks: The *S. octum* Zone falls within the *S. speciosum* Zone of Hamilton (1982). Specimens of *Ellipsagelosphaera* are still rare. The zone can only be distinguished in well preserved material, where the delicate forms of *Stephanolithion* are well enough preserved to be identified. Medd (1979) noted that the range of *Stephanolithion elongatum* coincided with the Middle Bathonian *Prohecticoceras retrocostatum* Zone. Medd (1979, *in* Penn, Dingwall & Knox) stated, that 'neither first nor last occurrence of the various species is consistent from borehole to borehole in this closely correlated sequence and at present no zonal subdivision of the (type) Bathonian can be attempted on the basis of coccolith distribution'.

STEPHANOLITHION HEXUM ZONE

Definition: FO of *Stephanolithion hexum* to FO of *Stephanolithion bigotii*.

Authors: Barnard & Hay (1974)

Age: Early Callovian

Remarks: The *S. hexum* Zone represents the uppermost part of the *S. speciosum* Zone of Hamilton (1982). The recognition of this zone depends on the preservation of the small marker species in a state that makes it possible to distinguish it from the other species of the genus. *Ellipsagelosphaera* species dominate the assemblage. *S. speciosum* disappears in this zone.

STEPHANOLITHION BIGOTII ZONE

Definition: FO of *Stephanolithion bigotii* to FO of *Polypodorhabdus escaigii*.

Authors: Barnard & Hay (1974)

Age: Callovian

Remarks: *Ellipsagelosphaera* species also dominate the assemblages in this zone whereas *S. punctulata* becomes rare.

POLYPODORHABDUS ESCAIGII ZONE

Definition: FO of *Polypodorhabdus escaigii* to FO of *Axopodorhabdus rahla*.

Authors: Barnard & Hay (1974), nom. corr.

Age: Callovian

Remarks: *Ellipsagelosphaera* species dominate the assemblage of this zone, which represents the lowermost part of the *S. bigotii* Zone of Hamilton (1982).

AXOPODORHABDUS RAHLA ZONE

Definition: FO of *Axopodorhabdus rahla* to FO of *Discorhabdus jungii*.

Authors: Barnard & Hay (1974), nom. corr.

Age: Callovian

Remarks: Species of *Ellipsagelosphaera* dominate the assemblage, in which also *S. hexum* and *S. bigotii* are common. The *A. rahla* Zone represents the lower part of the *S. bigotii* Zone of Hamilton (1982).

DISCORHABDUS JUNGII ZONE

Definition: FO of *Discorhabdus jungii* to FO of *Stradnerlithus comptus* (syn. *Diadozygus dorsetense*)

Authors: Barnard & Hay (1974), nom. corr.

Age: Late Callovian

Remarks: Species of *Ellipsagelosphaera* dominate the assemblage of this zone. *S. bigotii*, *Z. erectus*, *A. cylindratus*, *D. tubus*, *D. jungii* and *P. escaigii* are common. *Truncatoscaphus pauciramosus* appears near the top of this zone. The *D. jungii* Zone correlates approximately with the uppermost part of the *S. bigotii* Zone of Hamilton (1982).

STRADNERLITHUS COMPTUS ZONE

Definition: FO of *Stradnerlithus comptus* to FO of *Stradner-lithus sexiramatus* (syn. *Actinozygus geometricus*).

Authors: Barnard & Hay (1974), nom. corr.

Age: Early Oxfordian

Remarks: The *S. comptus* Zone falls within the lower part of the *Polypodorhabdus madingleyensis* Zone of Hamilton (1982), which is defined from the FO of *P. madingleyensis* to the FO of *Zeugrhabdotus embergeri*. Species of *Ellipsagelosphaera* dominate the assemblage, in which *S. bigotii* and *Z. erectus* are also common. Goy (1981) found *S. comptus* in the Toarcian and so did Crux (personal communication); the zone as defined by Barnard & Hay (1974) is thus not useful.

STRADNERLITHUS SEXIRAMATUS ZONE

Definition: FO of *Stradnerlithus sexiramatus* to FO of *Vekshi-nella stradneri*.

Authors: Barnard & Hay (1974), nom. corr.

Age: Late Oxfordian

Remarks: Species of *Ellipsagelosphaera* dominate the assemblage. Other relatively common species are *C. margerelii*, *D. tubus*, *B. dubium*, *S. bigotii* and *Z. erectus*. The *S. sexiramatus* Zone corresponds to the lower part of the *P. madingleyensis* Zone of Hamilton (1982).

VEKSHINELLA STRADNERI ZONE

Definition: FO of *Vekshinella stradneri* to LO of *Stephano-lithion bigotii*.

Authors: Barnard & Hay (1974)

Age: Late Oxfordian to Early Kimmeridgian

Remarks: The assemblage of the *V. stradneri* Zone is essentially the same as in the *S. sexiramatus* Zone. *A. rahla* disappears towards the top of the zone. The *V. stradneri* Zone corresponds to the upper part of the *P. madingleyensis* Zone of Hamilton (1982). *S. bigotii* and *Conusphaera mexicana* overlap in the uppermost part of this zone according to Wind (1978).

ELLIPSAGELOSPHAERA COMMUNIS ZONE

Definition: LO of *Stephanolithion bigotii* to FO of *Zeugrhab-dotus embergeri*.

Authors: Barnard & Hay (1974), nom. corr.

Age: Kimmeridgian

Remarks: Wind (1978) observed the following regarding this zone: The evolution of *Z. embergeri* from *Zeugrhabdotus salillum* (here *Tranolithus? salillum*) during the Kimmeridgian and the LO of *S. bigotii* in the Late Kimmeridgian or Portlandian suggest that the datum planes defining this zone may be reversed. Unfortunately, the LO of *S. bigotii* is often governed by dissolution

effects, while the *Z. embergeri* datum is probably regulated, in part, by overgrowth phenomena, as extensive re-precipitation will lower the level at which *T.? salillum* looks like *Z. embergeri*. The *E. communis* Zone is dominated by species of *Ellipsagelosphaera*, as are the other Late Jurassic zones, and correlates with the uppermost part of the *P. madingleyensis* Zone of Hamilton (1982). *B. dubium* and *C. margarelii* are common whereas *Z. erectus* is rare.

ZEUGRHABDOTUS EMBERGERI ZONE

Definition: FO of *Zeugrhabdotus embergeri* to FO of *Nanno-conus steinmannii*.

Author: Worsley (1971) nom. corr.

Age: Late Kimmeridgian to Early Tithonian

Remarks: The *Z. embergeri* Zone is widely used with the definition given above. Some authors consider *Nanno-conus steinmannii* as a synonym of *N. colomii*. Hamilton (1982) defined her *Z. embergeri* Zone with the same base but with the FO of *Conusphaera mexicana* defining the top. According to Wind (1978) *C. mexicana* appears before *Z. embergeri* in DSDP Sites 105 and 391 in the North Atlantic. Wind (1978) thus suggested a *C. mexicana* Zone defined from the FO of *C. mexicana* and/or the FO of *Polycostella beckmannii* to the FO of *Lithraphidites carniolensis*. In Thierstein's (1975) *C. mexicana* Zone with the same definition, the FO of *N. colomii* could be used as an additional event to recognize the top of the zone.

The assemblage is dominated by species of *Ellipsagelo-sphaera*. *C. crassus* and *A. cylindratus* have their LO in this zone.

Jurassic/Cretaceous boundary

The first occurrence of the genus *Nannoconus* is almost synchronous with the Tithonian/Berriasian boundary and thus the Jurassic/Cretaceous boundary. Where *Nannoconus* is found consistently in the Lower Cretaceous, several species will also be found in the uppermost Jurassic: *N. quadratus*, *N. colomii*, *N. steinmannii*, *N. dolomiticus*, *N. globulus* and *N. broennimannii* according to Deres & Acheriteguy (1980). The same authors do not show any species of *Nannoconus* as appearing directly at the Jurassic/Cretaceous boundary.

Black (1971a) noted 'a substantial break between Upper Jurassic and Lower Cretaceous assemblages with very few species crossing the boundary' and, in a range chart, shows 5 species in the 'Jurassic' and 17 species in the Berriasian (Ryazanian) from Speeton, United Kingdom. Of the Jurassic species, none has its last occurrence at the Jurassic/Cretaceous boundary and several of the species shown by Black (1971a) as occurring only in the Berriasian or younger sediments have since been shown to occur also in the Jurassic. While Black

(1971a) reported *Stephanolithion laffittei* (here *Rotelapillus*) also from the Jurassic, both Thierstein (1975, 1976) and Rood & Barnard (1972) stated its first occurrence at the base of the Cretaceous. Another species to appear near the base of the Cretaceous is *Lithraphidites carniolensis*, and *Polycostella beckmannii* ranges only into the lowermost part of the Berriasian.

For practical purposes, the first evolutionary appearance of *R. laffittei* and *L. carniolensis* or, where consistently found in the Lower Cretaceous, of *Nannoconus*, approximates the Jurassic/Cretaceous boundary reasonably well. Of last occurrences, the disappearance of *P. beckmannii* probably best approximates it.

Cretaceous

The calcareous nannofossil zonation of the Cretaceous (especially the Upper Cretaceous) is well advanced compared to the zonation of the Jurassic. Perch-Nielsen (1979a) and Doeven (1983) gave overviews of several previous zonations and the former suggested evolutionary lineages for some of the stratigraphically important genera and families. The sequence of nannofossil events is shown in Fig. 6 while evolutionary lineages are discussed in the taxonomic notes. The definitions of the zones of Sissingh (1977) are used as a scale in many illustrations in this chapter. They can be derived from Figs. 6 and 7 and are discussed below. Manivit *et al.* (1979) correlated the *Nannoconus* events with Sissingh's zonation. Slightly different zonal schemes were proposed by Perch-Nielsen (1977) and Wise (1983) for the South Atlantic, by Roth (1978) and Doeven (1983) for the North Atlantic area, by Roth (1973) and Martini (1976) for the Pacific Ocean and by Verbeek (1976a, b, 1977a, b) for Tunisia and Europe. Australian Cretaceous sediments were discussed and/or zoned by Rade (1979) and Shafik (1978), the latter also presenting LM photographs of most Santonian coccoliths from the Gingin Chalk of Western Australia. Pienaar (1968) and Siesser (1982) discussed Cretaceous coccoliths from South Africa. Other papers having LM illustrations of Cretaceous coccoliths include Stover (1966: most of the Cretaceous, France and The Netherlands), Bramlette & Martini (1964: Maastrichtian), Thierstein (1971, 1973: Lower Cretaceous), Roth & Thierstein (1972: Atlantic), Thierstein (1976: entire Mesozoic), Hill (1976: Albian–Cenomanian of Texas and Oklahoma), Manivit (1971: entire Cretaceous of France), Risatti (1973: Upper Cretaceous of Mississippi), Deres & Acheriteguy (1980: *Nannoconus*), Taylor (1982: Lower Cretaceous of the United Kingdom) and Crux (1982: Upper Cretaceous of the United Kingdom).

While the sequence of calcareous nannofossil events is not exactly the same all over the world and in all facies, the zonation of the Cretaceous is usually easier than the zonation of the Jurassic. In many cases, the zones are defined by the FO

or the LO of a calcareous nannofossil of unknown ancestor, and thus there are no problems of species distinction in these cases. We have only recently begun to observe more closely the evolution within some genera and families and to use this knowledge for a refinement of the zonal scheme. The LM can be used for some of these detailed investigations, for others an EM is necessary. The ranges of many species as given in Figs. 6 and 7 are subject to controversy not only because of different species concepts of different authors – partly due to real differences of opinion, partly due to the different state of preservation of their material – but also due to uncertainties in the age assignment of the samples studied. The correlation of coccolith zones and the classical stages has been studied repeatedly from material from the stratotype localities of the stages, but since their coccolith content is sometimes very poor or coccoliths are absent, correlations have had to be made via other fossils with the evident possibilities of shifting boundaries higher or lower depending on one's own preferences, tradition or wishful thinking. For an overview see Perch-Nielsen (1979a) and Fig. 6.

Cretaceous zones

NANNOCONUS STEINMANNII ZONE (CC 1)

Definition: FO of *Nannoconus steinmannii* to FO of *Stradneria crenulata*.
Authors: Worsley (1971), emended by Thierstein (1971) and Sissingh (1977)
Age: Latest Portlandian/Tithonian to Early Berriasian
Remarks: See discussion of the Jurassic/Cretaceous boundary above.

STRADNERIA CRENULATA ZONE (CC 2)

Definition: FO of *Stradneria crenulata* to FO of *Calcicalathina oblongata*.
Author: Thierstein (1971), nom. corr.
Age: Late Berriasian to Early Valanginian
Remarks: Taylor (1978) suggested the FO of *Speetonia colligata* and/or the FO of *Paleopontosphaera salebrosa* (here *Biscutum salebrosum*) as a useful calcareous nannofossil event for the zonation in the Lower Cretaceous of the United Kingdom, where *C. oblongata* was not found. *C. oblongata* is a useful marker species in the Tethyan realm, but was also found in the northern North Sea.

CALCICALATHINA OBLONGATA ZONE (CC 3)

Definition: FO of *Calcicalathina oblongata* to FO of *Cretarhabdus loriei*.
Authors: Thierstein (1971), emended by Sissingh (1977)
Age: Late Valanginian
Remarks: In the North Sea area, the FO of *C. oblongata* can be substituted by the FO of *S. colligata* and/or the FO

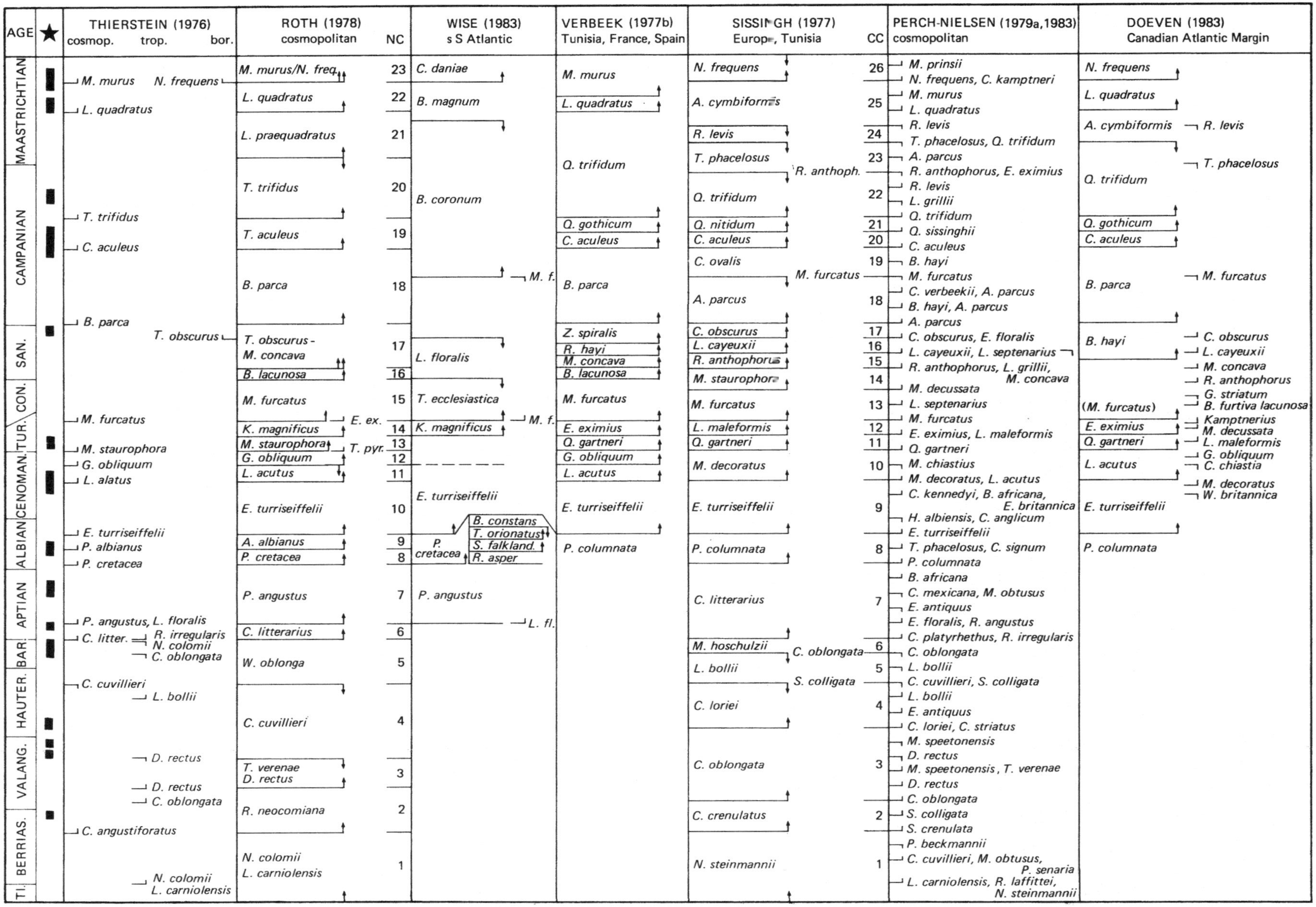

AGE	★	THIERSTEIN (1976) cosmop. trop. bor.	ROTH (1978) cosmopolitan	NC	WISE (1983) s S Atlantic	VERBEEK (1977b) Tunisia, France, Spain	SISSINGH (1977) Europe, Tunisia	CC	PERCH-NIELSEN (1979a,1983) cosmopolitan	DOEVEN (1983) Canadian Atlantic Margin
MAASTRICHTIAN	■	M. murus N. frequens	M. murus/N. freq.	23	C. daniae	M. murus	N. frequens	26	M. prinsii / N. frequens, C. kamptneri	N. frequens
		L. quadratus	L. quadratus	22	B. magnum	L. quadratus	A. cymbiformis	25	M. murus / L. quadratus	L. quadratus
			L. praequadratus	21		Q. trifidum	R. levis	24	R. levis / T. phacelosus, Q. trifidum	A. cymbiformis R. levis
CAMPANIAN	■	T. trifidus	T. trifidus	20	B. coronum		T. phacelosus	23	A. parcus / R. anthophorus, E. eximius	Q. trifidum T. phacelosus
							R. anthoph.		R. levis	
	■	C. aculeus	T. aculeus	19		Q. gothicum	Q. trifidum	22	L. grillii	Q. gothicum
	■					C. aculeus	Q. nitidum	21	Q. trifidum / Q. sissinghii	C. aculeus
							C. aculeus	20	C. aculeus	
			B. parca	18	M. f.	B. parca	C. ovalis	19	B. hayi / M. furcatus	B. parca M. furcatus
	■	B. parca					A. parcus	18	C. verbeekii, A. parcus / B. hayi, A. parcus	
SAN.	■	T. obscurus	T. obscurus –	17	L. floralis	Z. spiralis	C. obscurus	17	A. parcus / C. obscurus, E. floralis	B. hayi C. obscurus
			M. concava			R. hayi	L. cayeuxii	16	L. cayeuxii, L. septenarius	L. cayeuxii
CON.			B. lacunosa	16		M. concava	R. anthophorus	15	R. anthophorus, L. grillii, M. concava	M. concava
TUR.		M. furcatus	M. furcatus	15	T. ecclesiastica	B. lacunosa	M. staurophora	14	M. decussata	R. anthophorus
						M. furcatus	M. furcatus	13	L. septenarius / M. furcatus	G. striatum / B. furtiva lacunosa
	■	M. staurophora	K. magnificus E. ex.	14	K. magnificus M. f.	E. eximius	L. maleformis	12	E. eximius, L. maleformis	(M. furcatus) / Kamptnerius M. decussata
			M. staurophora T. pyr.	13		Q. gartneri	Q. gartneri	11	Q. gartneri	E. eximius / L. maleformis
		G. obliquum	G. obliquum	12		G. obliquum	M. decoratus	10	M. chiastius	Q. gartneri / G. obliquum
		L. alatus	L. acutus	11		L. acutus			M. decoratus, L. acutus	L. acutus C. chiastia
CENOMAN.					E. turriseiffelii	E. turriseiffelii	E. turriseiffelii	9	C. kennedyi, B. africana, E. britannica	E. turriseiffelii M. decoratus / W. britannica
		E. turriseiffelii	E. turriseiffelii	10					H. albiensis, C. anglicum / E. turriseiffelii	
ALBIAN		P. albianus	A. albianus	9	B. constans T. orionatus S. falkland. R. asper	P. columnata	P. columnata	8	T. phacelosus, C. signum / P. columnata	P. columnata
		P. cretacea	P. cretacea	8	P. cretacea				B. africana	
APTIAN			P. angustus	7	P. angustus		C. litterarius	7	C. mexicana, M. obtusus / E. antiquus	
		P. angustus, L. floralis	C. litterarius	6	L. fl.				E. floralis, R. angustus	
BAR.		C. litter. R. irregularis N. colomii C. oblongata					M. hoschulzii C. oblongata	6	C. platyrhethus, R. irregularis / C. oblongata	
			W. oblonga	5			L. bollii S. colligata	5	L. bollii / C. cuvillieri, S. colligata	
HAUTER.		C. cuvillieri / L. bollii	C. cuvillieri	4			C. loriei	4	L. bollii / E. antiquus / C. loriei, C. striatus	
VALANG.	■ ■	D. rectus	T. verenae D. rectus	3			C. oblongata	3	M. speetonensis / D. rectus / M. speetonensis, T. verenae / D. rectus / C. oblongata	
		D. rectus / C. oblongata								
BERRIAS.	■	C. angustiforatus	R. neocomiana	2			C. crenulatus	2	S. colligata / S. crenulata / P. beckmannii	
TI.		N. colomii L. carniolensis	N. colomii L. carniolensis	1			N. steinmannii	1	C. cuvillieri, M. obtusus, P. senaria / L. carniolensis, R. laffittei, N. steinmannii	

Fig. 6. Correlation of Cretaceous zonal schemes and their correlation with the stratotypes (*) of the stages

342

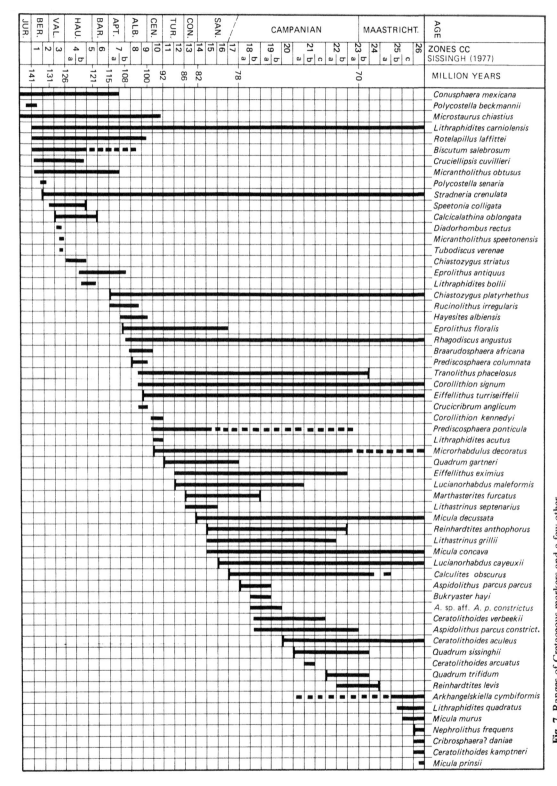

Fig. 7. Ranges of Cretaceous markers and a few other calcareous nannofossils

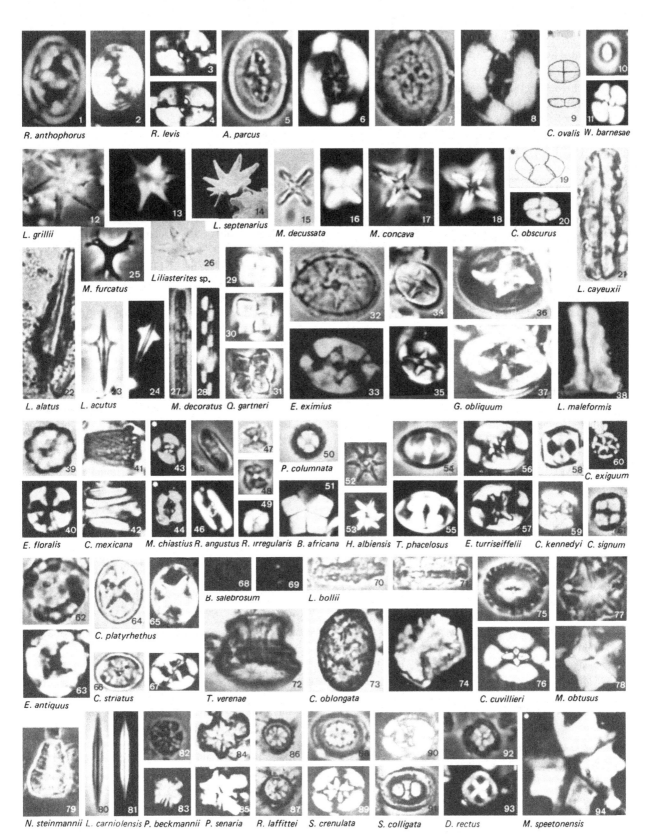

Fig. 8. LM of Cretaceous marker species. LM: *c*. × 2000

of *B. salebrosum*. In the Boreal region, CC 3 can be subdivided by the range of *Micrantholithus speetonensis* into CC 3a and CC 3b (with *M. speetonensis*) and CC 3c. Perch-Nielsen (1979a) suggested *Chiastozygus striatus* as a substitute marker for *C. loriei* in the Boreal region.

CRETARHABDUS LORIEI ZONE (CC 4)

Definition: FO of *Cretarhabdus loriei* to LO of *Speetonia colligata*.

Authors: Sissingh (1977)

Age: Early Hauterivian

Remarks: The FO of *Chiastozygus striatus* is used in the Boreal realm as a substitute marker for *C. loriei*. Sissingh (1977) suggested a subdivision of CC 4 by the LO of *Biscutum salebrosum*. This has been found to be an unreliable event, since *B. salebrosum* was found by several authors in the Barremian and the Aptian/Albian. Perch-Nielsen (1979a) suggested the FO of *Eprolithus antiquus* and the LO of *Cruciellipsis cuvillieri* as additional events to subdivide the Hauterivian in the Boreal realm. She also used the LO of *Chiastozygus striatus* as a substitute marker event for the top of CC 4. Thierstein (1976) had used the FO of *Lithraphidites bollii* and the LO of *C. cuvillieri* for the subdivision of the Hauterivian in the Tethyan realm. *L. bollii* was not found in the Boreal realm.

LITHRAPHIDITES BOLLII ZONE (CC 5)

Definition: LO of *Speetonia colligata* to LO of *Calcicalathina oblongata*.

Authors: Thierstein (1971), emended by Sissingh (1977)

Age: Late Hauterivian to Early Barremian

Remarks: The LO of *Chiastozygus striatus* was used by Perch-Nielsen (1979a) as a substitute event for the LO of *S. colligata*, especially in the Boreal realm, where *S. colligata* is not always present. The nominate species, *L. bollii*, is also not found in the Boreal realm. Taylor (1982) noted the FO of 'Dodekapodorhabdus noeliae' (her inverted commas) and the FO of *Nannoconus abundans* in the Late Hauterivian in the United Kingdom.

MICRANTHOLITHUS HOSCHULZII ZONE (CC 6)

Definition: LO of *Calcicalathina oblongata* to FO of *Chiastozygus litterarius* (*C. platyrhethus* of some authors)

Authors: Thierstein (1971), emended by Thierstein (1973) and Sissingh (1977)

Age: Late Barremian

Remarks: Thierstein (1976) noted the FO of *Rucinolithus irregularis* and the LO of *Nannoconus colomii* shortly

before the FO of *C. litterarius*. Perch-Nielsen (1979a) used the LO of *Nannoconus bermudezii* and the FO of *Chiastozygus* spp. as an approximation for the FO of *C. litterarius* (see taxonomic notes). Taylor (1982), on the other hand, used the LO of *Nannoconus abundans* as being approximately equivalent to the FO of *C. litterarius* in the Boreal realm.

CHIASTOZYGUS LITTERARIUS ZONE (CC 7)

Definition: FO of *Chiastozygus litterarius* to FO of *Prediscosphaera columnata*.

Authors: Thierstein (1971), emended by Manivit *et al.* (1977)

Age: Aptian and early Early Albian

Remarks: Sissingh (1977) suggested a subdivision of CC 7 into CC 7a and CC 7b by the LO of *Micrantholithus hoschulzii/ M. obtusus*. Perch-Nielsen (1979a) further subdivided CC 7a by the LO of *Nannoconus steinmanii* and the LO of *Conusphaera mexicana*. She used the LO of *Nannoconus kamptneri* and the FOs of *Eprolithus floralis* and *Braarudosphaera africana* to approximate the LO of *M. obtusus*. She also used the LOs of *Nannoconus wassallii* and *Eprolithus antiquus* as well as the FO of typical *Rhagodiscus angustus* in the Late Aptian to subdivide CC 7b. *Nannoconus quadriangulus* was found to appear about at the level of *P. columnata*. Thierstein (1976) suggested an earlier FO of *R. angustus* and *E. floralis*, in the Early Aptian, but still in CC 7. Doeven (personal communication, 1984) found intermediate forms between *B. africana* and *B. hockwoldensis* in samples of probable Valanginian age on the Scotian Shelf.

For the distinction between *Prediscosphaera columnata*, used as a zonal marker here, and *P. cretacea*, as used by many authors, see taxonomic notes.

PREDISCOSPHAERA COLUMNATA ZONE (CC 8)

Definition: FO of *Prediscosphaera columnata* to FO of *Eiffellithus turriseiffelii*.

Authors: Thierstein (1971), emended by Manivit *et al.* (1977)

Age: Albian

Remarks: Perch-Nielsen (1979a) suggested a subdivision of this zone by the FOs of *Tranolithus phacelosus*, *Corollithion signum* and the genus *Cribrosphaera*. For the use of *P. columnata/P. cretacea* see taxonomic notes. Thierstein (1976) used the FO of *Axopodorhabdus albianus* to subdivide the Albian. *A. albianus*, however, has been reported by several authors from as early as the Barremian and therefore is no longer a marker for the Albian.

EIFFELLITHUS TURRISEIFFELII ZONE (CC 9)

Definition: FO of *Eiffellithus turriseiffelii* to FO of *Microrhabdulus decoratus*.

Authors: Thierstein (1971), emended by Sissingh (1977)

Age: Late Albian and Early Cenomanian

Remarks: Whereas the FO of *E. turriseiffelii* has been used as a zonal marker event by many authors, the top of their *E. turriseiffelii* Zone varies. Manivit *et al.* (1977) used the FO of *Lithraphidites acutus* and this has been widely followed. The FO of *L. acutus* falls within CC 9 as defined here according to Manivit *et al.* (1977), but within CC 10 according to Doeven (1983).

The LO of *Hayesites albiensis* was used by Manivit *et al.* (1977) to subdivide their *E. turriseiffelii* Zone into a lower *H. albiensis* Subzone and an upper *Prediscosphaera spinosa* Subzone. Perch-Nielsen (1979a) proposed the LOs of *Ellipsagelosphaera britannica* and *Braarudosphaera africana* for a subdivision of CC 9. The FO of *Corollithion? completum* has to be replaced by the FO of *Corollithion kennedyi*, since *C.? completum*, which was described from the Maastrichtian, is not identical with the forms attributed to this species in the Cenomanian. According to Doeven (personal communication, 1984) *B. africana* persists into CC 10 on the Scotian Shelf.

MICRORHABDULUS DECORATUS ZONE (CC 10)

Definition: FO of *Microrhabdulus decoratus* to FO of *Quadrum gartneri*.

Authors: Sissingh (1977)

Age: Late Cenomanian

Remarks: Manivit *et al.* (1977) subdivided their *Lithraphidites acutus* Zone which spans from the FO of *L. acutus* to the FO of *Q. gartneri* and thus roughly coincides with CC 10, by the LO of *Microstaurus chiastius* into a *M. chiastius* Subzone below and a *Gartnerago obliquum* Subzone above. Perch-Nielsen (1979a) noted the subsequent FO of *Ahmuellerella octoradiata* and LO of *Gartnerago nanum* within CC 10. The FO of *Gartnerago obliquum* falls within CC 9 according to some authors, and within CC 10 according to others, depending on their taxonomic concept of the species.

QUADRUM GARTNERI ZONE (CC 11)

Definition: FO of *Quadrum gartneri* to FO of *Lucianorhabdus maleformis*.

Authors: Čepek & Hay (1969, 1970), emended by Sissingh (1977), nom. corr.

Age: Early and Middle Turonian (possibly also including late Late Cenomanian)

Remarks: Čepek & Hay (1969, 1970) and many later authors used the FO of *Eiffellithus eximius* as the top of their

Q. gartneri Zone. Doeven (1983) suggested the LO of *Axopodorhabdus albianus* as coinciding with the FOs of *Q. gartneri* and *L. maleformis* and predating the FO of *E. eximius*. The latter two species were used as interchangeable markers for the bottom of CC 12 by Perch-Nielsen (1979a).

LUCIANORHABDUS MALEFORMIS ZONE (CC 12)

Definition: FO of *Lucianorhabdus maleformis* to FO of *Marthasterites furcatus*

Authors: Sissingh (1977)

Age: Late Turonian to early Early Coniacian

Remarks: The zone below the FO of *M. furcatus* was named *E. eximius* by Verbeek (1976b), who used the FO of *E. eximius* as its base. Roth (1978) suggested a *K. magnificus* Zone with the FO of the nominate species as its lower boundary. According to Doeven (1983), the FO of *Kamptnerius* post-dates the FO of *E. eximius* and the FO of *L. maleformis*.

M. furcatus is a good marker species in low to mid latitudes, but occurs only sporadically in high latitudes. *L. maleformis* is very rare or absent in open ocean sections, where *E. eximius* is a better marker.

MARTHASTERITES FURCATUS ZONE (CC 13)

Definition: FO of *Marthasterites furcatus* to FO of *Micula decussata*.

Authors: Čepek & Hay (1969), emended by Sissingh (1977)

Age: Early Coniacian

Remarks: The name *M. furcatus* Zone has been used by several authors with different definitions for the top of the zone. *M. decussata* (*M. staurophora* of some authors) was considered synonymous with *Quadrum gartneri* by Roth & Bowdler (1979), a view which I do not hold (see taxonomic notes). *Q. gartneri* appears already in the Early Turonian according to Manivit *et al.* (1977). But also Doeven (1983) reported the FO of *M. decussata* before the FO of *M. furcatus* and thus the top for CC 13 as chosen by Sissingh (1977) is not reliable. Perch-Nielsen (1979a) noted the FO of *Lithastrinus septenarius* after the FO of *M. furcatus* and before the FO of *M. decussata*. Doeven (1983) correlated the FO of *Broinsonia* ex gr. *furtiva-lacunosa* and the subsequent LO of *Gartnerago striatum* (here *Allemannites striatus*) more or less with the FO of *M. decussata* as dated by Sissingh (1977) and Perch-Nielsen (1977). Roth (1978) and Verbeek (1977b) used the FO of *B. lacunosa* (here *B. enormis*) for the upper boundary of their *M. furcatus* Zone, where Hattner & Wise (1980) used the FO of *Broinsonia parca constricta* (here *Aspidolithus parcus constrictus*).

MICULA DECUSSATA ZONE (CC 14)

Definition: FO of *Micula decussata* to FO of *Reinhardtites anthophorus*.

Authors: Manivit (1971), emended Sissingh (1977)

Age: Late Coniacian and Early Santonian

Remarks: For lower boundary see above. The FO of *R. anthophorus* was also suggested by Doeven (1983) as an additional marker to those he used in his zonation, with the remark that forms like *R. anthophorus* seem to appear earlier, between the FO of *M. furcatus* and the LO of *A. striatus*. Perch-Nielsen (1979a) noted the FO of *Lithastrinus grillii* and the FO of *Micula concava* together with the FO of *R. anthophorus*, whereas Verbeek (1977b) and Roth (1978) used *M. concava* as a zonal marker in the Santonian.

REINHARDTITES ANTHOPHORUS ZONE (CC 15)

Definition: FO of *Reinhardtites anthophorus* to FO of *Lucianorhabdus cayeuxii*.

Author: Sissingh (1977)

Age: late Early Santonian

Remarks: Doeven (1983) followed Verbeek (1977) in using the FO of *Rucinolithus hayii* as a zonal marker in the Early Santonian and correlated this event with the interval between the markers for CC 15. Perch-Nielsen (1979a) suggested that the LO of *Lithastrinus septenarius* coincides with the FO of *L. cayeuxii*.

LUCIANORHABDUS CAYEUXII ZONE (CC 16)

Definition: FO of *Lucianorhabdus cayeuxii* to FO of *Calculites obscurus*.

Author: Sissingh (1977)

Age: Late Santonian

Remarks: Wind (1975a) and Wind & Wise (1978) discussed the possibility that *Calculites* includes forms of *Lucianorhabdus* that do not have or possess only a very short stem, and that *L. cayeuxii* and *C. obscurus* are the same species. If they are indeed the same species, Sissingh's zones CC 16 and CC 17 would have to be combined. CC 16 falls within Doeven's (1983) *Rucinolithus hayii* Zone. *Placozygus fibuliformis* (*Zygodiscus spiralis* in Verbeek, 1977b) has its FO in CC 16 according to the correlation of zonations by Doeven (1983). Perch-Nielsen (1979a) noted the LO of *Eprolithus floralis* as a substitute marker event for the FO of *C. obscurus*. Thierstein (1976) included *C. ovalis* in *C. obscurus* and correlated its FO with the base of the Santonian. He considered it the best marker for the Coniacian/Santonian boundary.

CALCULITES OBSCURUS ZONE (CC 17)

Definition: FO of *Calculites obscurus* to FO of *Aspidolithus* ex. gr. *parcus*.

Author: Sissingh (1977)

Age: Late Santonian/Early Campanian

Remarks: The FO of *A. parcus* (*Broinsonia parca* of many authors) is an event that has been used for zonation and coincides reasonably well with the Santonian/Campanian boundary, if an early form of *A. parcus*, *Aspidolithus* sp. 1 in Perch-Nielsen (1979a, fig. 9) or *A. parcus parcus* is considered (see taxonomic notes on Arkhangelskiellaceae). For the lower boundary see above. Valentine (1980) found the LO of *Chiastozygus cuneatus* to coincide with the FO of *A. parcus*.

ASPIDOLITHUS PARCUS ZONE (CC 18)

Definition: FO of *Aspidolithus* ex gr. *parcus* to LO of *Marthasterites furcatus*.

Author: Sissingh (1977)

Age: Early Campanian

Remarks: Perch-Nielsen (1977) used the same definition for her Early Campanian *Eiffellithus eximius* Zone. Verbeek (1977b), Roth (1978) and Doeven (1983) defined a *B. parca* Zone from the FO of *B. parca* to the FO of *Ceratolithoides aculeus*, the event used by Sissingh to define the top of his Zone CC 19. The interval where *A. parcus* and *M. furcatus* occur together is usually short and is often missed in condensed sequences which were not sampled close enough. In extended sections, however, this interval can be subdivided further by the FO of *Bukryaster hayii* into CC 18a and CC 18b. *Ceratolithoides verbeekii* appears in CC 18b according to Perch-Nielsen (1979a).

CALCULITES OVALIS ZONE (CC 19)

Definition: LO of *Marthasterites furcatus* to FO of *Ceratolithoides aculeus*.

Author: Sissingh (1977)

Age: late Early Campanian

Remarks: Sissingh (1977) suggested a subdivision of CC 19 by the LO of *Bukryaster hayii*. Doeven (1983) noted the successive disappearance of *Corollithion signum* and *Gephyrorhabdus coronadventis* within CC 19. *M. furcatus* is only occasionally present and thus not a reliable marker in high northern latitudes, while *C. aculeus* is absent. Accordingly Crux (1982) used neither and defined his *Broinsonia parca* Zone from the FO of *B. parca* to the FO of *Prediscosphaera stoveri*, an interval which includes the LO of *M. furcatus*.

CERATOLITHOIDES ACULEUS ZONE (CC 20)

Definition: FO of *Ceratolithoides aculeus* to FO of *Quadrum sissinghii* (*Tetralithus nitidus* and *Quadrum nitidum* and *Tetralithus/Quadrum gothicus/m.* partim of previous authors; see taxonomic notes)

Authors: Čepek & Hay (1969), emended by Martini (1976), nom. corr.

Age: late Early Campanian.

Remarks: Neither marker of this zone is to be found in high northern latitudes (North Sea area). On the Scotian Shelf the LO of relatively common *Eiffellithus eximius*, which is usually found within this zone can give an indication of the age according to Doeven (1983). Roth (1978) defined his *T. aculeus* Zone from the FO of *T. aculeus* to the FO of *T. trifidus*, thus including CC 20 and CC 21.

QUADRUM SISSINGHII ZONE (CC 21)

Definition: FO of *Quadrum sissinghii* (see above and taxonomic notes) to FO of *Quadrum trifidum* (*Tetralithus trifidus* of other authors).

Author: Sissingh (1977), nom. corr.

Age: early Late Campanian

Remarks: Sissingh (1977) suggested a subdivision into three subzones using the range of *Ceratolithoides arcuatus*. This subdivision has only been possible in the El Kef section in Tunisia. Neither marker is to be found in high northern latitudes (North Sea area), where the FO of *Arkhangelskiella cymbiformis* can best be used as an approximation of the base of CC 21. For the species concept of *A. cymbiformis* as used here see taxonomic notes.

QUADRUM TRIFIDUM ZONE (CC 22)

Definition: FO of *Quadrum trifidum* to LO of *Reinhardtites anthophorus*.

Authors: Bukry & Bramlette (1970), emended by Sissingh (1977), nom. corr.

Age: late Late Campanian

Remarks: Whereas all authors agree on *Q. trifidum* at the base of their *Q. trifidum* Zone, they use other markers for its top: the LO of *Q. trifidum* (Bukry & Bramlette, 1970; Martini, 1976; Roth, 1978; Doeven, 1983), the FO of *Lithraphidites praequadratus* (Roth, 1978); the FO of *Lithraphidites quadratus* (Verbeek, 1976b). Whereas *Q. trifidum* is not found in high northern latitudes, *R. anthophorus* is a reliable marker species in the North Sea area and can be found also in rather poorly preserved assemblages. The LO of *E. eximius* coincides with the LO of *R. anthophorus* well enough to be used as a substitute marker species according to Perch-Nielsen (1979a). Sissingh (1977) suggested the subdivision of CC 22 by the FO of *Reinhardtites levis*, a subdivision which is useful mainly in high northern latitudes, where *Reinhardtites* is more common than in lower latitudes.

TRANOLITHUS PHACELOSUS ZONE (CC 23)

Definition: LO of *Reinhardtites anthophorus* to LO of *Tranolithus phacelosus*.

Author: Sissingh (1977)

Age: latest Campanian to Early Maastrichtian

Remarks: The *T. phacelosus* (*T. orionatus* of some authors) Zone corresponds to the upper part of the *Q. trifidum* Zone of many authors (see above). According to Doeven (1983) *T. phacelosus* disappears slightly before *Q. trifidum*. *T. phacelosus* is useful in low as well as in high latitudes. The LO of *Aspidolithus parcus* subdivides CC 23 according to Sissingh (1977). This event can also be observed in low and high latitudes.

Ceratolithoides verbeekii   C. aculeus   C. kamptneri   C. arcuatus   B. hayi   M. murus   M. prinsii

Q. sissinghii   Q. trifidum   Arkhangelskiella cymbiformis   L. quadratus   C. daniae   N. frequens

Fig. 9. LM of Late Cretaceous marker species. LM: *c.* × 2000

REINHARDTITES LEVIS ZONE (CC 24)

Definition: LO of *Tranolithus phacelosus* to LO of *Reinhardtites levis*.

Author: Sissingh (1977)

Age: Early Maastrichtian

Remarks: Sissingh (1977) remarked, that 'the LO of *R. levis* virtually coincides with a distinct and interregional increase in number of large *Arkhangelskiella* representatives' (see taxonomic notes on *Arkhangelskiella*). This zone is recognizable in low and in high latitudes.

ARKHANGELSKIELLA CYMBIFORMIS ZONE (CC 25)

Definition: LO of *Reinhardtites levis* to FO of *Nephrolithus frequens*.

Authors: Perch-Nielsen (1972), emended by Sissingh (1977)

Age: Late Maastrichtian

Remarks: There are several definitions attached to the name *A. cymbiformis* Zone. Perch-Nielsen (1972) defined it from the LO of *R. anthophorus* (meaning the form now described as *R. levis*) to the FO of *M. murus* or *N. frequens*. This upper boundary provides a marker for low latitudes (*M. murus*) and one for high latitudes (*N. frequens*). Martini (1976) defined it from the LO of *Q. trifidum* to the FO of *Lithraphidites quadratus*, the way it is also used by Doeven (1983). Sissingh (1977) suggested a subdivision of CC 25 by the FO of *Arkhangelskiella cymbiformis* and the FO of *Lithraphidites quadratus*. Sissingh's concept of *A. cymbiformis* is more restricted than the species concept of most other authors, who plot the FO of *A. cymbiformis* at about the base of CC 22 in the Late Campanian. See taxonomic notes on *Arkhangelskiella*.

The FO of *L. quadratus* is a good marker event in low latitudes, but usually cannot be recognized in high northern latitudes. It was used by several authors as the base of a *L. quadratus* Zone.

NEPHROLITHUS FREQUENS ZONE (CC 26)

Definition: FO to LO of *Nephrolithus frequens*.

Authors: Čepek & Hay (1969)

Age: late Late Maastrichtian

Remarks: This zone definition works well in high latitudes, where *N. frequens* is relatively common. Problems arise with the top of the zone, since *N. frequens* is found reworked into the overlying Paleocene. Thus Perch-Nielsen, McKenzie & He (1982) suggested the FO of *Thoracosphaera* as the top of the Maastrichtian in high latitudes. In low latitudes, *N. frequens* is very rare and here the FO of *Micula murus* and the subsequent FO of *Micula prinsii* can be used to subdivide the interval

between the FO of *L. quadratus* (base CC 25c) and the top of the Maastrichtian. For a discussion of the Cretaceous/Tertiary boundary see Chapter 11.

Classification, taxonomic and other notes
(Figure 10)

There is no general agreement on the classification of Mesozoic calcareous nannofossils. Many authors therefore prefer to arrange the genera in alphabetical order for their systematic descriptions. Also species lists are often presented in alphabetical order by species name, because many species are assigned to different genera by different authors and thus are easier to find by this means.

Comprehensive classifications were last presented by Hay (1977) and Tappan (1980), both of whom include assignments not followed here. Tappan (1980) is followed here, where her suggestions coincide with my own philosophy of family concepts. There are families like, for example, the Cretaceous Arkhangelskiellaceae, which are relatively well defined and agreed upon and which disappear at the end of the Mesozoic. Members of the equally well defined Braarudosphaeraceae appear in the Early Cretaceous and some species are still alive. In a family such as the Polycyclolithaceae (syn. Lithastrinaceae, Eprolithaceae), my ideas about the evolution from *Eprolithus* to *Lithastrinus* and from *Eprolithus* to *Quadrum* and *Micula* prevent me from also including *Lapideacassis*, which seems to be unrelated to the other forms in this family. Instead *Quadrum* and *Micula*, which were included by Tappan (1980) in the well defined, now monogeneric, Tertiary family Sphenolithaceae are added here.

In general an attempt is made to keep families in which evolutionary lineages between several genera seem possible free from other genera which only look like those genera in an 'accidental' way. On the other hand families like the Microrhabdulaceae are retained as including all rod-like nannoliths with no or only a minuscule disk, even though no relationships can be established between them. This leaves some genera as 'incertae sedis', when there is no obvious connection to genera in a well defined family or they do not fit into any of those existing families that are here treated as repositories for 'similar objects'. Another problem is the distinction of separate families, in the Mesozoic, for genera with a wall of inclined elements (zeugoid) and a central bridge, + or X as Ahmuellerellaceae, Chiastozygaceae and Zygodiscaceae. Due to my feeling, that the Cenozoic genus *Neochiastozygus* (Chapter 11, Fig. 79), which could be assigned to the Chiastozygaceae, evolved from *Chiastozygus ultimus*, a form with a central + which would be assigned to the Ahmuellerellaceae in the Mesozoic (and the genus *Vekshinella*) I left all Cenozoic forms with a zeugoid wall in the Zygodiscaceae. Both the forms with a central X and those with a bridge double their wall in the

Early Paleocene. It seems that inconsistencies of this sort will remain for some time to come.

Generally, I have relied more on the holotype of the generotype of a genus than on the definition of the genus. This is practical in those cases, where the generotype was described from a EM picture. In the case of generotypes which were based on schematical drawings like *Arkhangelskiella* and *Placozygus*, this becomes unsatisfactory and the genus remains a doubtful one.

No extensive synonymy lists are given, but a few generic transfers are listed in the Alphabetic list of taxa. Some genera, mainly those including stratigraphic markers, are treated in detail, others only briefly and only the generotype is illustrated. The ranges given must be regarded as tentative in many cases. I here generally followed the 'splitters' rather than the 'lumpers', mainly because by doing so and illustrating the different holotypes in the same figure, there is a greater chance that investigators trying to distinguish these species will find them useful.

Ahmuellerellaceae Reinhardt (1965)
(Figures 11, 12)

The Ahmuellerellaceae include elliptical coccoliths with a wall of inclined elements and a central area bridged by a cross aligned with the axes of the ellipse. The genera included are shown in Fig. 11. Tappan (1980) also included *Helicolithus*, here included in the Chiastozygaceae, since its central structure is an X and not a +, and *Placozygus*, here referred to the Zygodiscaceae, because its central structure is a simple bridge and not a +.

The genera of the Ahmuellerellaceae are assigned to two groups in Fig. 11, on the basis of the construction of the central area. It is relatively simple in the first group and composite in the second. The genera are discussed in alphabetical order within these groups.

Ahmuellerella

The generotype of *Ahmuellerella*, *A. octoradiata* (Figs. 12.15, 16, 24), is characterized by a central cross consisting of 4 pairs of slightly diverging elements supporting a central cone or stem. *A. octoradiata* ranges from the Late Cenomanian to the top of the Maastrichtian.

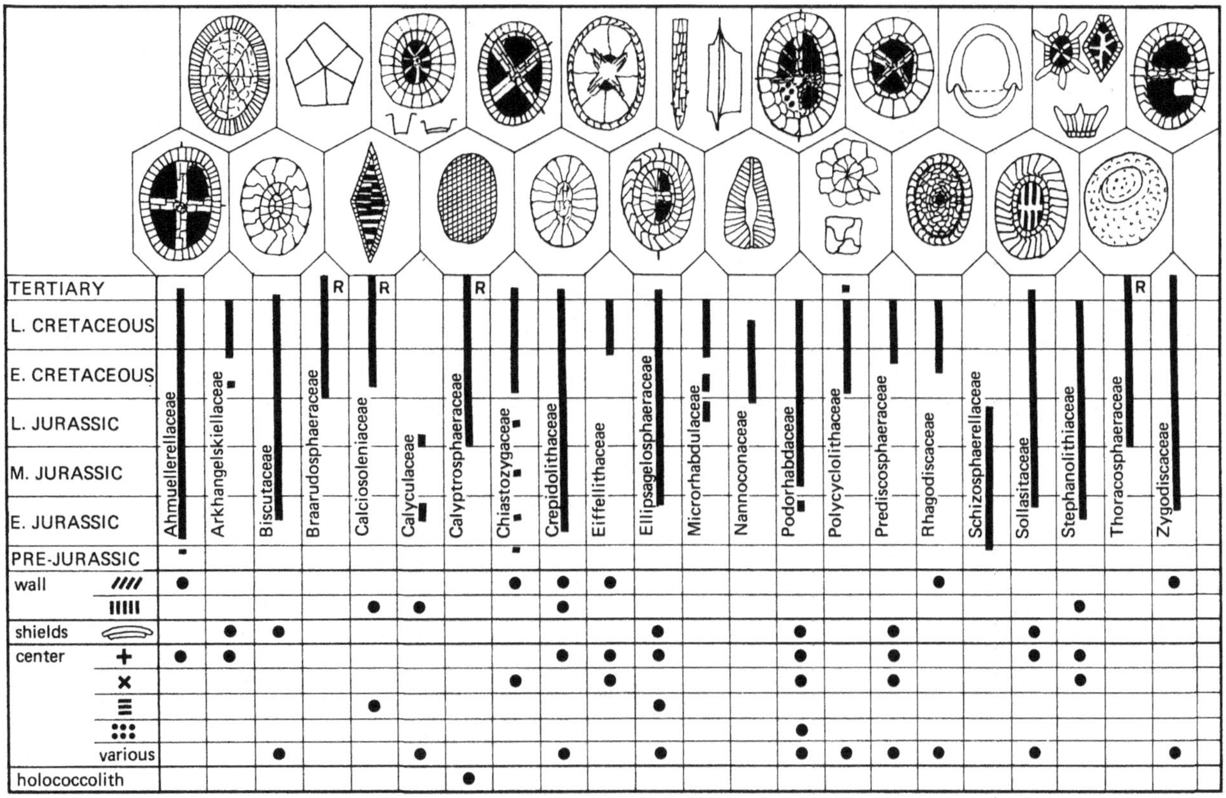

Fig. 10. Overview of the most important Mesozoic families.
R = Recent families

Ahmueller-ellaceae															
Tertiary															
L.Cretaceous															
E.Cretaceous															
L.Jurassic	*Ahmuellerella*	*Crucirhabdus*	*Saeptella*	*Staurolithites*	*Vekshinella*	*Staurorhabdus*	*Vagalapilla*	*Tubirhabdus*	*Vacherauvillius*	*Bukrylithus*	*Haslingfieldia*	*Heteromarginatus*	*Mitrolithus*	*Monomarginatus*	*Pontilithius*
M.Jurassic															
E.Jurassic															

Fig. 11. Overview of the genera of the Ahmuellerellaceae

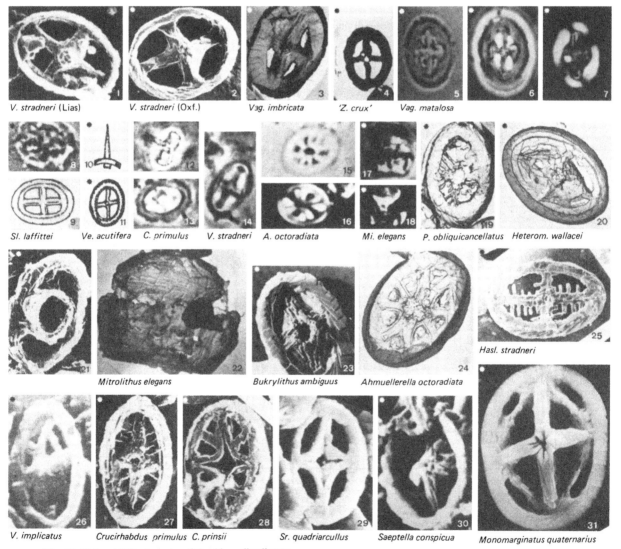

V. stradneri (Lias)   *V. stradneri* (Oxf.)   *Vag. imbricata*   *'Z. crux'*   *Vag. matalosa*

Sl. laffittei   *Ve. acutifera*   *C. primulus*   *V. stradneri*   *A. octoradiata*   *Mi. elegans*   *P. obliquicancellatus*   *Heterom. wallacei*

Mitrolithus elegans   *Bukrylithus ambiguus*   *Ahmuellerella octoradiata*   *Hasl. stradneri*

V. implicatus   *Crucirhabdus primulus*   *C. prinsii*   *Sr. quadriarcullus*   *Saeptella conspicua*   *Monomarginatus quaternarius*

Fig. 12. LM and EM of species of the Ahmuellerellaceae.
LM: *c.* ×2000

Crucirhabdus

In *Crucirhabdus*, the central cross is 'supported by one or more diagonal bars in each quadrant'. The generotype, *C. primulus* (Figs. 12.12, 13, 27), is used as a zonal marker in the Early Jurassic. It is characterized by 3 to 5 diagonal bars. Its central process is often broken off. It differs from *C. prinsii* (Fig. 12.28), which has only a diagonal bar.

Saeptella

Saeptella, as defined by its generotype *S. conspicua* (Fig. 12.30), includes coccoliths with a wall of inclined to vertical elements, a central cross and a grid covering the 4 quadrants. It was described from the Toarcian.

Staurolithites, Staurorhabdus, Vagalapilla and Vekshinella

The genera *Staurolithites*, *Staurorhabdus*, *Vagalapilla* and *Vekshinella* are regarded as synonyms by many authors. If this is so, *Staurolithites* is the oldest and thus valid form. They all include coccoliths with a simple central cross aligned with the axes. The species described in these genera are difficult to distinguish with the LM, and many more species have been described than can be distinguished according to Grün & Zweili (1980). Since they have little or no stratigraphic potential, they are not further discussed but the generotype of each genus is illustrated in Fig. 12.

The FO of *Vekshinella stradneri* (Figs. 12.1, 2, 14) has been used by Barnard & Hay (1974) as a zonal marker event in the Oxfordian. But already Rood, Hay & Barnard (1973) had noted that specimens of *V. stradneri* 'from the Lias and the Middle Jurassic closely resemble those previously figured from the Oxford Clay and the Lower Cretaceous'. Recently, Jafar (1983) described *V. thiersteinii* (Figs. 2.1–3) from the Rhaetian, and this species also seems impossible to distinguish from younger representatives of the genus. Similarly forms have been assigned by many authors to '*Zygolithus crux*', a species name serving as a waste basket for elliptical coccoliths with a wall of inclined elements and a central cross aligned with the axes.

Tubirhabdus

Tubirhabdus is characterized by a central, massive but hollow stem based on a central cross. Its generotype *T. patulus* (Figs. 12.21) ranges from the Lias (Prins, 1969) to the Middle Jurassic.

Vacherauvillius

The generotype of *Vacherauvillius*, *V. implicatus* (Fig. 12.26) is characterized by a low wall, a central cross and 2 curved bars and a short central process. Other species have more bars and/or additional radial bars. The genus is only known from the Toarcian.

Bukrylithus

B. ambiguus (Fig. 12.23) features a 'floor with a circular opening at the centre and a broad fibrous cross covering most of the floor'. The structure of the cross, which consists of parallel laths is more complex than that of *Staurolithites/Vekshinella* but similar to that of *Staurorhabdus* (compare with Fig. 12.29). *B. ambiguus* was described from the Hauterivian.

Haslingfieldia

In *Haslingfieldia*, the central cross is straight and slim and has laths attached at right angles to the cross. *H. stradneri* (Fig. 12.25), the generotype, was described from the Albian, other species were reported from the Late Cretaceous by Bukry (1969). The laths are normally not visible with the LM and the specimens would be assigned to '*Zygolithus crux*' by many authors (see above).

Heteromarginatus

The generotype of *Heteromarginatus*, *H. wallacei* (Fig. 12.20), features two (?) wall cycles according to Bukry (1969), an outer one consisting of radially arranged elements and an inner one consisting of inclined elements. The central cross bears a stem, and 4 auxiliary bars connect the bars of the cross. The arrangement of the central structure resembles the one in *Sollasites*, which, however, has two shields and no central stem. Similar arrangements are also known from *Prediscosphaera rhombica* (Fig. 61.25) and *Nodosella perch-nielseniae* (Fig. 73.16). *H. wallacei* is known from the Campanian.

Mitrolithus

Mitrolithus includes coccoliths with a high wall of inclined elements and a mushroom-shaped central process. *M. elegans* (Figs. 12.17, 18, 22) was described from the Oxfordian and was also found in the Cretaceous (reworked?), where it was described as *Alvearium dorsetense*, *Hexangulolithus primus* also may represent the central process of *M. elegans*. The central process is too wide to allow the recognition of the structure of the centre supporting it in distal view. No certain proximal views of *M. elegans* are available but it is assumed that it shows a cross aligned with the axes. This assumption, and consequently the assignment of *Mitrolithus* to the Ahmuellerellaceae, is based on proximal views of elliptical coccoliths with a high wall of inclined to vertical elements and a diamond-shaped central structure aligned with the axes. These were found in samples with relatively common *M. elegans*.

Monomarginatus

Monomarginatus is characterized by an elevated central cross and an additional diagonal bar in each quadrant. It differs from *Heteromarginatus* by having only one wall cycle. The generotype *M. quaternarius* (Fig. 12.31) and *M. pectinatus* were described from the Maastrichtian of the Falkland Plateau.

Pontilithus

The central area of *Pontilithus obliquicancellatus* (Fig. 12.19) is spanned by a central cross and more or less radial additional bars. The proximal views so far illustrated resemble the proximal views of *Tranolithus phacelosus* (Fig. 83.19). No distal views are available of the generotype of this Late Cretaceous genus.

Arkhangelskiellaceae Bukry (1969)
(Figures 13–17)

The Arkhangelskiellaceae include elliptical coccoliths with a rim consisting of 3 to 5 tiers of rim elements. Two to four tiers are visible in proximal view, usually 2 in distal view. Most genera and species lack a central process. Fig. 13 shows the terminology of the Arkhangelskiellaceae, Fig. 14 gives an overview over the genera assigned to the family and the structural elements that characterize them.

The first representatives of the Arkhangelskiellaceae appear in the Aptian and the family disappears at the top of the Cretaceous. No detailed studies are yet available on the development of the genera *Arkhangelskiella* (Campanian to Maastrichtian), *Gartnerago* and *Broinsonia* (Albian to Maastrichtian) or on the origin of the family in the Aptian with the appearance of the genus *Acaenolithus*. Special studies include that of Lauer (1975), who reported on evolutionary trends in the Santonian/Campanian, Verbeek (1977b), who showed measurements of *Broinsonia lacunosa* and *B. parca* and Hattner, Wind & Wise (1980) who described the lineage from *Broinsonia furtiva* through *B. parca parca* to *B. parca constricta*. Verbeek

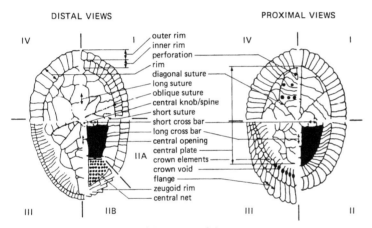

Fig. 13. Terminology of the genera of the Arkhangelskiellaceae

Genera \ structural elements	central cross	central plate	perforations	central net	flange	knob or spine	outer rim	inner rim	zeugoid rim	outer rim	crown	inner rim	1	2	3 or more
		general					distal			proximal			side tiers		
Acaenolithus	●		●			●	●	●		●		●	?		
Arkhangelskiella		●	●				●			●					●
Aspidolithus		●	●				●	●		●				●	
Broinsonia	●						●	●		●				●	
Cribricatillus	●			●					●						
Crucicribrum	●	●	●				●		●						●
Diloma	●			●						?	?	?	?		
Gartnerago		●	●				●	●		●	●	●		●	
Kamptnerius		●	●	●			●	●		●	●	●	●		
Misceomarginatus	●	●	●				●	●	●						
Thiersteinia	●	●	●			●	●			●					●

Fig. 14. Overview of the genera of the Arkhangelskiellaceae

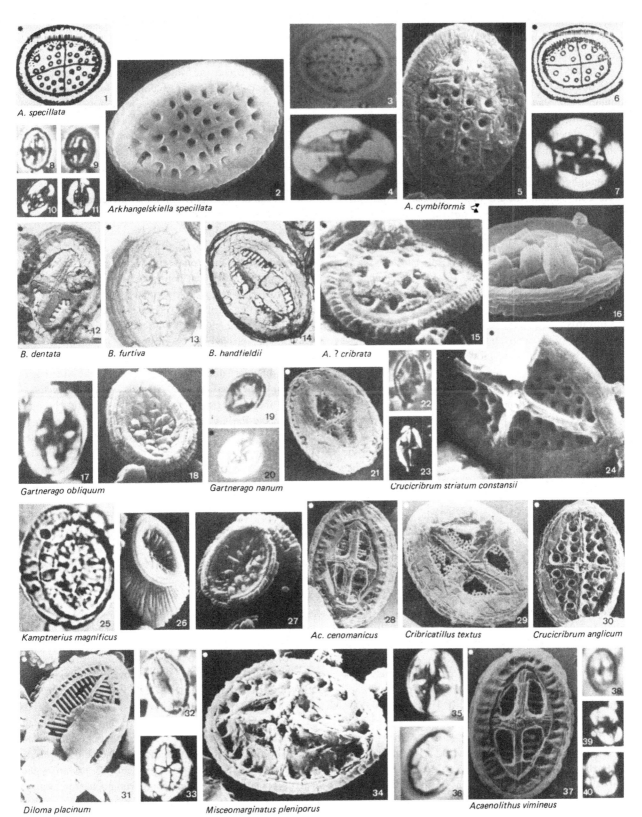

A. specillata

Arkhangelskiella specillata

A. cymbiformis

B. dentata   B. furtiva   B. handfieldii   A. ? cribrata

Gartnerago obliquum

Gartnerago nanum

Crucicribrum striatum constansii

Kamptnerius magnificus

Ac. cenomanicus   Cribricatillus textus   Crucicribrum anglicum

Diloma placinum   Misceomarginatus pleniporus   Acaenolithus vimineus

Fig. 15. LM and EM of Arkhangelskiellaceae species. LM:
c. ×2000

(1977b) found the width of the margin to increase relative to the width of the coccolith, and the total width of the coccolith to increase from 5–7 μm in the uppermost Santonian to 9–11 μm in the Lower Campanian; the width then remained the same through the rest of the range of *B. parca*, reaching a maximum in the *Quadrum gothicum* Zone of Verbeek. The same ratios were used by Lauer (1975), who, in addition, used the number of perforations in the central field to subdivide the Santonian/Campanian of Oman.

Only a few of the many species are generally used for biostratigraphy; these include the LO of *Gartnerago nanum* in the Late Cenomanian, the FO of *G. obliquum* in the Early Turonian, the FOs of *Aspidolithus furtivus*, *A. parcus parcus* and *A. parcus constrictus* around the Santonian/Campanian boundary, the LO of *A. parcus* near the Campanian/Maastrichtian boundary and the FO of *Arkhangelskiella cymbiformis* in the Late Campanian. The FO of *Kamptnerius magnificus* has been shown to be time transgressive by Thierstein (1976). It appears in the Turonian in high latitudes and moves towards the equator only during the Campanian. Like other members of the family, it is more common in shelf areas than in the open ocean sediments.

Acaenolithus

Acaenolithus was defined by Black (1973) as 'coccoliths in most respects resembling *Broinsonia* [see below], but differing in the presence of a knob' or stem at the centre of the distal side. The generotype, *A. cenomanicus* (Fig. 15.28), from the Albian has a very fine central net in each quadrant. For other species see Black (1973).

Arkhangelskiella

The genus *Arkhangelskiella* is characterized by a single distal rim and a single proximal rim and a relatively large central plate pierced by numerous perforations. Early forms have an inner and an outer rim visible from both sides (Figs. 15.2; 17). In well preserved specimens the perforations can be counted and their number and arrangement used for species definition. Such species cannot, however, be recognized in poorly preserved material, especially when heavy overgrowth has filled all perforations.

A. cymbiformis (Figs. 15.5–7, 16) has a relatively large central plate pierced by relatively few perforations when compared to *A. specillata* (Figs. 15.1, 2). The perforations are not visible in specimens from the chalk of northern Europe, due to slight overgrowth. Late Campanian specimens have a mean diameter of about 9 μm, the same as Early Maastrichtian ones. In the *L. quadratus* Subzone (CC 25c) the diameter is 10 μm in the early part and 11 μm in the later part. It decreases slightly again in the latest Maastrichtian to about 10 μm. The size varies from 7 μm to 14 μm, the end-members being very rare (unpublished data). Sissingh (1977) used the FO of *A. cymbiformis* for the subdivision of CC 25, showing a *A.*

cymbiformis with a relatively small central plate and defining it as 'characterised by a relatively broad rim, clearly subdivided in 2 or more layers. In plan view these layers are visible as 3 or more concentric rings. The plate area bears large perforations per quadrant i.e. 3 along the long suture, and 2 along the short suture'. Such forms are, however, also found together with *Reinhardtites levis* and *R. anthophorus* in CC 24 to CC 23 or CC 22. The FO of *A. cymbiformis* is therefore placed in the Late Campanian. A consistent distinction between *A. cymbiformis* and *A. specillata* is only possible in well preserved material, where the number of perforations per quadrant can be determined. In poorly preserved material, *A. cymbiformis* might be recognized by its relatively wider rim and narrower central plate.

Aspidolithus and Broinsonia

Aspidolithus is often considered a junior synonym of *Broinsonia*. This is not followed here, but the genera are distinguished by attributing species with a distinct central cross to *Broinsonia* and forms with a closed centre consisting of 8 segments pierced by perforations to *Aspidolithus*. Both genera have an outer and an inner rim in distal view and only an outer rim in proximal view. They appear in the Albian and disappear in the Maastrichtian. Albian species of *Aspidolithus* are small and such forms dominate the genus until they become extinct in the Early Campanian. The evolution of the large forms of the *A. parcus* group has been discussed in Lauer (1975), Perch-Nielsen (1979a), Hattner *et al.* (1980) and Wise (1983). It is summarized in Fig. 17, which follows Wise (1983).

Cribricatullus and Crucicribrum

Cribricatullus features a zeugoid rim, a central cross and a central net. It differs from *Crucicribrum* by the lack of a central knob or stem. *Cri. textus* (Fig. 15.29), the generotype of *Cribricatullus*, has a fine net; the type of *Crucicribrum*, *Cru. anglicum* (Fig. 15.30), has a perforated central plate and a central process. Both genera were described from the Albian. For other species see Black (1973).

Diloma

Diloma includes coccoliths with a 'complex rim of short radial elements' surrounding 'an area paved with thin parallel laths which extend from the rim to a thin crossbar on the long axis of the ellipse. A curved bar crosses all radial elements'. The authors Wind & Čepek (1979) noted that *Diloma* seemed unrelated to the podorhabdids or cretarhabdids. It is here tentatively assigned to the Arkhangelskiellaceae based on its aspect in the LM between crossed nicols. *D. primitiva*, the generotype has a central cross supporting a narrow central stem. *D. placinum* (Figs. 15.31–33) is long-elliptical and has only one additional bar along the longer axis. Both species were only reported from the Hauterivian according to Wind & Čepek (1979).

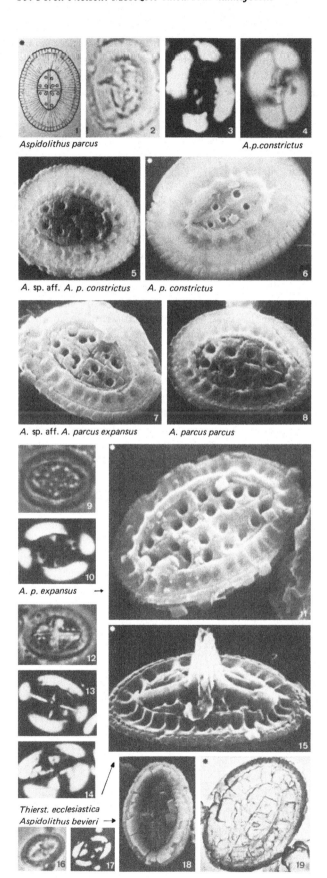

Fig. 16. LM and SEM of *Aspidolithus* and *Thiersteinia*. LM: *c.* ×2000

Gartnerago (syn. *Laffittius*) and *Kamptnerius*

Gartnerago differs from the previously discussed genera of the family by having a beaded crown (see Fig. 14) when seen from the distal side. It shares this characteristic structural element with *Kamptnerius*, which in addition features a more or less prominent flange. *G. nanum* (Figs. 15.19–21) is characterized by a narrow outer rim, possibly of zeugoid elements, a relatively small central plate with a cross and quadrants pierced by numerous small perforations. Between crossed nicols the central plate and the inner rim show the grey colour typical for species of *Gartnerago*. *G. obliquum* (Figs. 15.17, 18) probably includes many of the different large *Gartnerago* species described by Bukry (1969) as it is generally used now. It has a relatively large central plate that appears greyish and subdivided into 8 parts between crossed nicols. It has not such a distinct, small central cross as *G. nanum*, and only appears in the Early Turonian, after the LO of *G. nanum* in the Late Cenomanian. Similar forms have, however, been reported from the Middle Albian as *G. stenostaurion* by Hill (1976).

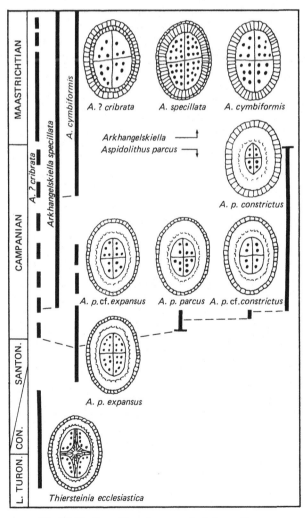

Fig. 17. Stratigraphic and presumed phylogenetic relationships among *Aspidolithus parcus* and related taxa. Distal views. Partly after Wise (1983)

The FO of *G. obliquum* in the Early Turonian, therefore, needs further evaluation in the Middle Cretaceous.

K. magnificus (Figs. 15.25–27) is the most common species of *Kamptnerius*. Like *Gartnerago*, the proximal rim is decorated by a crown. The outermost rim elements differ in length to form a flange. In well preserved specimens, perforations and a central cross are recognizable. They have been used to define various species besides the commonly used *K. magnificus*. The FO of *K. magnificus* used by Roth (1978) as a marker event in the Turonian, occurs considerably later in low latitudes and the species is often absent in the open ocean sections.

Misceomarginatus

M. pleniporus (Figs. 15.34–36), the only species of this high latitude Maastrichtian genus, is characterized by a zeugoid rim, a central plate and a central cross. Perforations are well developed along the rim.

Thiersteinia

Thiersteinia ecclesiastica (Figs. 16.12–15), the generotype and only species of this high latitude genus, is characterized by 'a three-tiered *Broinsonia*-like margin and a perforated central area surmounted on the distal surface by a solid spine. The spine is supported by buttresses which form a cross along the major and minor axes of the central area.' *T. ecclesiastica* ranges from the Turonian to the Coniacian/Santonian of the Falkland Plateau (Wise, 1983).

Biscutaceae Black (1971a)
(Figures 18–20)

The Biscutaceae include round and elliptical coccoliths consisting of two closely appressed shields, each shield constructed of radial, non-imbricate petaloid elements. Both shields usually remain dark between crossed nicols, whereas the central structure sometimes is birefringent.

Of the genera assigned to the Biscutaceae (syn. Discorhabdaceae Noël, 1973) by Tappan (1980), only three or four

are useful. The others have been regarded as synonyms by various authors. The genera *Biscutum*, *Discorhabdus*, *Noellithina* and *Seribiscutum* are discussed below. *Paleopontosphaera* is considered a synonym of *Biscutum* and *Bennocyclus* and *Bidiscus* synonyms of *Discorhabdus*. *Similicoronillithus* is regarded as a synonym of *Cyclagelosphaera*; the construction of the shields is the one typical for the Ellipsagelosphaeraceae and the genus should in any case be removed from the Biscutaceae.

Biscutum

In Chapter 11, *Biscutum* is tentatively assigned to the Tertiary family Prinsiaceae, since it is suggested, that this family evolved from the small forms of *Biscutum* that survived the Cretaceous/Tertiary boundary event(s). *Biscutum* is stratigraphically important in the Jurassic and Grün & Zweili (1980) have given extensive synonymy lists for the four species of *Biscutum* they consider useful: *B. castrorum*, *B. dubium*, *B. ellipticum* and *B. erismatum*. Grün & Zweili (1980) also transferred the generotype of *Paleopontosphaera*, *P. dubia*, which was used as a zonal marker by Barnard & Hay (1974), into *Biscutum*, commenting that *Paleopontosphaera* only differed from *Biscutum* by the former's central perforation (depression) around the central process, while the form and interlocking pattern typical of *Biscutum* (Fig. 19.23) were the same. They also noted that Hamilton's (1977) *B. ellipticum* should be included in their *B. dubium*. Grün, Prins & Zweili (1974) suggested that *Paleopontosphaera veterna* is a synonym of *Lotharingius sigillatus*. Grün & Zweili (1980) give the range of *L. sigillatus* as Toarcian to Oxfordian and the range of *B. erismatum*, in which the invalid *P. veterna* of Prins (1969), but not the valid *P. veterna* of Rood *et al.* (1973), is incorporated, as Pliensbachian to Late Jurassic. It can thus be assumed that *P. veterna*, the FO of which was used by Hamilton (1977) as a zonal marker in the Pliensbachian, should read *B. erismatum* and not *L. sigillatus* in Hamilton's zonal scheme. Also *P. salebrosa*, the LO of which was used by Sissingh (1977) as a zonal marker in the Lower Cretaceous, should be transferred

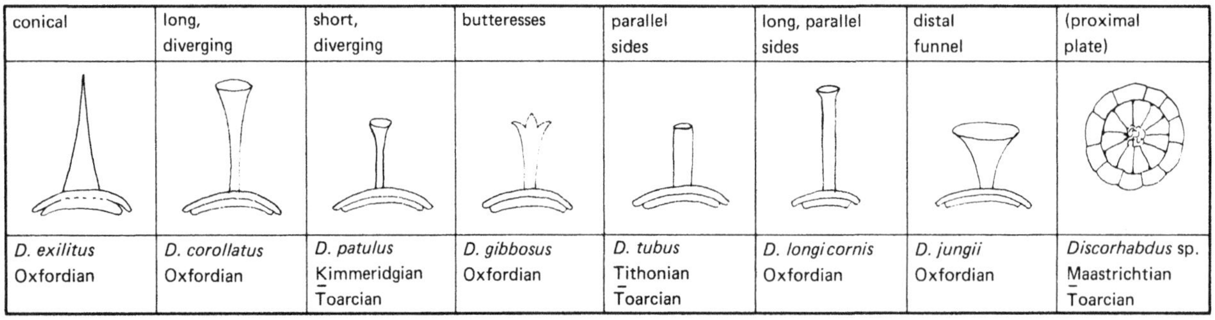

conical	long, diverging	short, diverging	butteresses	parallel sides	long, parallel sides	distal funnel	(proximal plate)
D. exilitus Oxfordian	*D. corollatus* Oxfordian	*D. patulus* Kimmeridgian — Toarcian	*D. gibbosus* Oxfordian	*D. tubus* Tithonian — Toarcian	*D. longicornis* Oxfordian	*D. jungii* Oxfordian	*Discorhabdus* sp. Maastrichtian — Toarcian

Fig. 18. Species of Jurassic *Discorhabdus* and their characteristic features

to *Biscutum*. It has since been found, however, that the LO of *B. salebrosum* (Fig. 19.24) occurred considerably later than assumed by Sissingh (1977).

In the LM, *Biscutum* is characterized by dark shields between crossed nicols, with only the central area showing

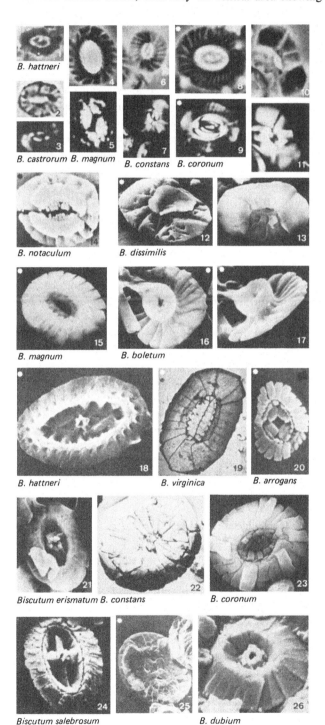

B. hattneri

B. castrorum   B. magnum   B. constans   B. coronum

B. notaculum   B. dissimilis

B. magnum   B. boletum

B. hattneri   B. virginica   B. arrogans

Biscutum erismatum   B. constans   B. coronum

Biscutum salebrosum   B. dubium

Fig. 19. LM and EM of species of *Biscutum*. LM: *c.* × 2000

birefringence. The central cross structures in *B. erismatum* and *B. salebrosum* can best be seen with phase contrast. *Biscutum* is known from the Jurassic through the Early Danian.

The range of *B. coronum* (Figs. 19.8, 9, 23) was used to define a zone in the high southern latitude Campanian/Maastrichtian (Fig. 6) by Wise (1983), who also used the names of *B. constans* (Figs. 19.6, 7, 22) and *B. magnum* (Figs. 19.4, 5, 15) for a subzone and zone respectively. The names of further species of *Biscutum* are given in the alphabetic list of species.

Discorhabdus

Whereas *Biscutum* includes the more or less elliptical coccoliths with radially arranged, interlocking elements in the distal shield, *Discorhabdus* includes the round forms. The FO of *D. patulus*, *D. tubus* and *D. jungii* have been used for the subdivision of the Jurassic, while the genus is stratigraphically unimportant in the Cretaceous, where mainly *D. ignotus*, a small form usually without a central process, is found. In Fig. 18 the side views of the Jurassic species are shown to facilitate determination and comparison. Basal disks of Jurassic forms of *Discorhabdus* that have lost their central process are quite common in many samples and cannot be determined to species level, since the species are distinguished by the outline and the length of the central process when seen in side view. *Discorhabdus biperforatus* of Rood *et al.* (1973) is considered to belong to *Podorhabdus*, since its basal disk is elliptical and its central process supported by a granulate central area with two perforations.

In order to distinguish the side views of *Discorhabdus* species from the side views of specimens of other genera, the shape of the basal disk can be taken into account. Two shields should be visible in *Discorhabdus* and Podorhabdaceae, whereas a single rim is visible in *Parhabdolithus*. While there are no perforations in the basal part of the stem in *Discorhabdus*, they are present and sometimes visible also with the LM in the Podorhabdaceae.

Noellithina

Noellithina is characterized by a distal shield very similar to the one of *Biscutum* and a central area spanned by a bridge, central cross and bars connecting the central structure with the wall or distal cycle of elements. The central area thus is more complex than in *Biscutum*, and seems comparable to the one in *Lotharingius* (Ellipsagelosphaeraceae). The two species assigned to *Noellithina*, *N. arcta* and *N. prinsii*, have been found in the Lias and are illustrated in Figs. 20.12, 13.

Seribiscutum

In *Seribiscutum* the shields are the same as in *Biscutum*, but the central area is partially covered by blocky elements and granules. In the generotype *S. bijugum* (Fig. 20.14), the central area is partially filled by two bars parallel to, but offset on either side of the minor axis. The only other species, *S. primitivum* (*Cribrosphaerella primitiva* of many authors; Fig. 20.15) shows 4 large elements in the central area. It is known from the Late Aptian to Santonian and seems to prefer high latitudes.

Braarudosphaeraceae Deflandre (1947)
(Figures 21, 22)

Genera and species of this family are found in the Cretaceous and in the Cenozoic. The family is also discussed and some species illustrated in Chapter 11. In the Cretaceous it is represented by the genera *Braarudosphaera*, *Bukryaster*, *Micrantholithus* and *Trapezopentus*. Of the other Cretaceous genera assigned to the Braarudosphaeraceae by Tappan (1980), *Hexalithus* is included in the Polycyclolithaceae rather

than as a member of this family, since it is not pentagonal. *Hexangulolithus* is thought to represent the distal part of a (reworked?) *Mitrolithus elegans*. *Octolithites* has not been found again since being described and might represent one of the rare occurrences of double *Braarudosphaera* specimens as found recently in Lower Danian samples from Tunisia (Figs. 22.1, 2).

The distribution and the terminology of the Cretaceous species are shown in Fig. 21. For the definitions of *Braarudosphaera* and *Micrantholithus* see Chapter 11. *Bukryaster* was previously considered to be related to *Lithastrinus*, but was found to represent a member of the Braarudosphaeraceae (B. Prins, personal communication, 1983). It is characterized by 5 segments with a small depression in every segment and a prominent ridge of each segment forming a long thorn. *Trapezopentus* consists of 'pentaliths constructed of overlapping trapezoidal elements surrounding a large central opening'. Its only species, *T. sarmatus* (Figs. 22.11–13), was described from the Hauterivian of the southern NE Atlantic.

Calciosoleniaceae Kamptner (1927)
(Figure 23)

The modern family Calciosoleniaceae is discussed and illustrated further in Chapter 11. *Scapholithus fossilis* has its FO in the Hauterivian and is found through to Recent times. Bukry (1969) reported a form not distinguishable from the Recent *Anoplosolenia brasiliensis* from the Late Cretaceous (Fig. 23.1) and described two new species which, however, have been considered synonymous with *S. fossilis*. Gorka (1957) described a nearly quadratic form as *Dictyolithus quadratus* from the Maastrichtian of Poland. Similar forms have been reported by Čepek (1970) from the Campanian and Maastrichtian of Germany (Figs. 23.2, 3). While of no stratigraphical interest, the genera *Anoplosolenia* and *Scapholithus* are remarkable because they survived the Cretaceous/Tertiary boundary without any change.

Calyculaceae Noël (1973)
(Figures 24, 25)

The Calyculaceae include the elliptical to subcircular coccoliths with a chalice-like shape which have a central area covered by a grid and subvertical wall elements turning horizontal distally. Tappan (1980) considered this Jurassic family to be synonymous with the Goniolithaceae, an assignment which seems unrealistic, since the latter family includes only pentagonal coccoliths forming a pentagondodecahedron.

Since the Calyculaceae are not stratigraphically important, only an overview of the genera assigned to it is given in Fig. 24 and a few species are illustrated in Fig. 25. Further illustrations can be found in Noël (1973), Goy *et al.* (1979) and Medd (1979).

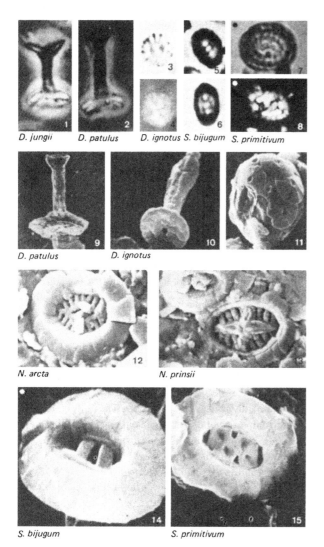

D. jungii   D. patulus   D. ignotus S. bijugum   S. primitivum

D. patulus   D. ignotus

N. arcta   N. prinsii

S. bijugum   S. primitivum

Fig. 20. LM and SEM of species of *Discorhabdus*, *Noellithina* and *Seribiscutum*. LM: *c.* × 2000

Calyptrosphaeraceae Boudreaux & Hay (1969)
(Figures 26–28)

The modern family Calyptrosphaeraceae includes coccolithophorids bearing holococcoliths (coccoliths consisting of uniform, small, calcite crystals) instead of elements of differing size and shape (heterococcoliths). Its Cenozoic representatives are discussed and illustrated in Chapter 11.

Perch-Nielsen (1979a) gave an overview of the Cretaceous holococcolith genera and suggested evolutionary lineages for the species of *Lucianorhabdus*. The characteristics of the genera are tabulated in Fig. 26. *Tetralithus* serves as a wastebasket for poorly defined forms. Its generotype, *T. pyramidus*, was described from the Miocene of Algeria and it seems questionable that it is a calcareous nannofossil at all.

Jurassic

The only holococcoliths known from the Jurassic are those illustrated by Medd (1979) from Upper Jurassic sediments of England (Figs. 28.25–27) as *Anfractus harrisonii* and *A. variabilis* and as '*Neococcolithes dubius*' (the name of an Eocene heterococcolith).

Cretaceous

Several holococcoliths are used for the subdivision of the Late Cretaceous: the FOs of *Lucianorhabdus maleformis* and *L. cayeuxii* in the Late Turonian and Santonian respectively, and *Calculites obscurus* in the Santonian. *L. cayeuxii* is distinguished from the older *L. maleformis* and the other older forms of the genus by the minute size or lack of a plug and by the minute remnant of the rim, which is often missing. *L. maleformis* has distinct remnants of the plug and of the rim.

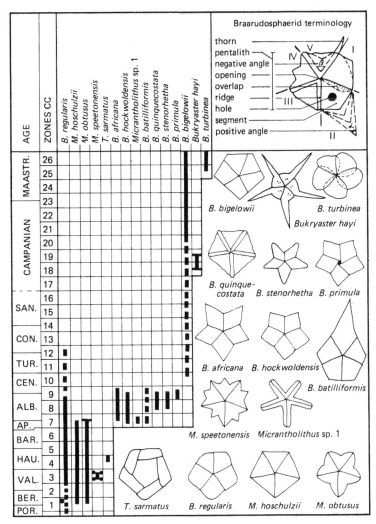

Fig. 21. The Cretaceous species of the Braarudosphaeraceae, their ranges and terminology

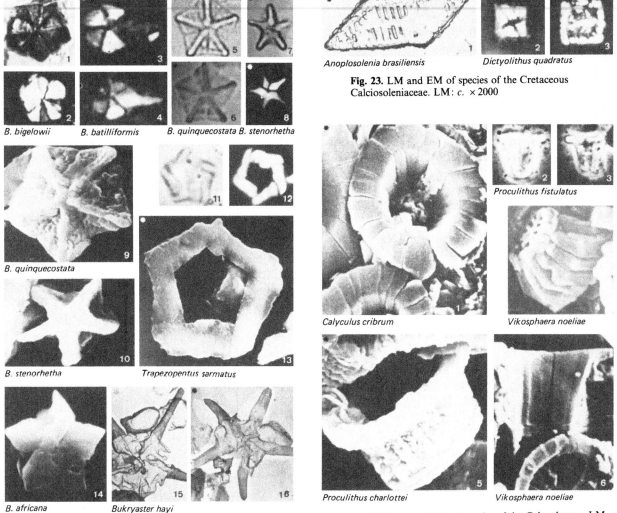

B. bigelowii   B. batilliformis   B. quinquecostata B. stenorhetha

B. quinquecostata

B. stenorhetha   Trapezopentus sarmatus

B. africana   Bukryaster hayi

Fig. 22. LM and EM of species of the Braarudosphaeraceae. LM: *c.* × 2000

Anoplosolenia brasiliensis   Dictyolithus quadratus

Fig. 23. LM and EM of species of the Cretaceous Calciosoleniaceae. LM: *c.* × 2000

Proculithus fistulatus

Calyculus cribrum   Vikosphaera noeliae

Proculithus charlottei   Vikosphaera noeliae

Fig. 25. LM and SEM of species of the Calyculaceae. LM: *c.* × 2000

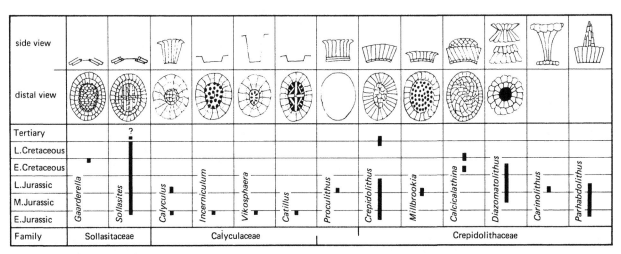

Fig. 24. The genera of the Calyculaceae, the Crepidolithaceae and the Sollasitaceae and their ranges. *Gaarderella*: proximal view

The latter form an angle of about 120° with the stem in side view. In *L. quadrifidus*, the plug is wider than the stem and the proximal disk well developed. *Isocrystallithus compactus* is characterized by the wide proximal disk and a stem which remains dark at 0° to the crossed nicols. Hattner & Wise (1980) suggested the inclusion of *L. arcuatus*, *L. maleformis*, *L. quadrifidus* and *I. compactus* in *L. cayeuxii*, since they found it difficult to distinguish them in their material. This is not followed here.

Calculites (syn. *Phanulithus*, see van Heck, 1980) was erected for 'heterococcoliths composed of a narrow rim and a broad wall consisting of a limited number of calcite blocks, but lacking a plate structure'. Whereas poorly preserved specimens of *C. obscurus* and other species of the genus show a few blocks as their main building material, well preserved specimens reveal the holococcolithic nature of the genus. *C. obscurus* differs from *C. ovalis* by the different direction of the sutures (see Figs. 27 and 28.3, 4, 9, 10). *Orastrum campanensis* (Figs. 28.1, 2), when observed between crossed nicols, shows extinction lines similar to those of *Calculites ovalis*, but has additional elements around the edges of the elliptical body composed of 4 blocky elements. They cause the characteristic bends at the outer end of the extinction lines. *O. campanensis* is more or less restricted to the Campanian. *O. asarotum* (Fig. 27), the genero-

type of *Orastrum*, features 'one or two pairs of segments' filling the central area and several smaller, outer segments. It was described from the Maastrichtian of the Falkland Plateau.

Many of the Late Cretaceous holococcoliths seem to be related to *I. compactus*, which appeared in the Cenomanian, and its descendants. The Early Cretaceous forms *Metadoga mercurius* (Figs. 28.31–33) and '*Lucianorhabdus*' *phlaskus* (definitely not a *Lucianorhabdus* species, Figs. 28.28–30) seem unrelated to the Late Jurassic or the Late Cretaceous forms. The only species persisting into the Cenozoic is *Octolithus multiplus* (see Chapter 11 and Figs. 28.16–18).

Holococcoliths are generally more common in nearshore and epicontinental seas than in open (deep) ocean sediments, where they are usually missing.

Chiastozygaceae Rood, Hay & Barnard (1973)
(Figure 29)

The Chiastozygaceae include elliptical coccoliths with a rim of inclined elements (=zeugoid) and a central structure forming an X or an H. The family is regarded as a synonym of the Zygodiscaceae by Tappan (1980). One could also include the genera assigned to the Chiastozygaceae in the Ahmuellerellaceae, since forms with an X and forms with a + in the centre develop one from the other.

Three genera are currently included in the Chiastozygaceae in the Mesozoic: *Chiastozygus*, *Helicolithus* and *Tegumentum*. *Chiastozygus* is characterized by a simple rim of inclined elements and a central X. *Helicolithus* also has a simple rim of inclined elements and a central X, but in addition has blocky plates derived from rim elements filling the area between the central structure and the rim (Fig. 29.15). *Tegumentum* has a wall consisting of a first wall of non-imbricate elements surrounding a second wall of strongly imbricate elements (for terminology see Chapter 11, Fig. 77).

C. striatus (Figs. 29.16–18) and *C. litterarius* (Figs. 29.5–9) are used as markers in the Cretaceous. *C. litterarius* was described from the Maastrichtian of Poland and may not be exactly the same form as the one used as a marker at the base of the Aptian, but so far it has not been possible to define the older form so as to be distinguishable from the poorly known younger one. *C. litterarius* has been replaced by some authors by *C. platyrhethum*, a species described from the Middle Cretaceous, or by *C. tenuis* (Figs. 29.22, 23), described from the Barremian. Noël (1980) illustrated a *C. litterarius* from the Late Hauterivian, making the use of its FO in the Late Barremian or Early Aptian doubtful. Generally specimens with a simple wall of inclined elements and a composite central X are assigned to *C. litterarius* but no biometric studies have been made to distinguish properly between the various slightly differing types of coccoliths that fit this broad definition.

C. striatus has a thin first wall and a blocky second wall, thus somewhat resembling species of *Eiffellithus*. The central

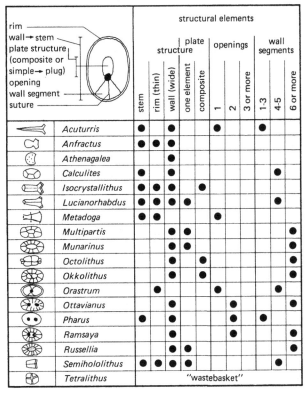

		stem	rim (thin)	wall (wide)	one element	composite	1	2	3 or more	1-3	4-5	6 or more
					plate structure		openings			wall segments		
	Acuturris	●		●			●			●		
	Anfractus	●	●	●								
	Athenagalea			●								
	Calculites	●		●							●	
	Isocrystallithus	●	●	●		●						
	Lucianorhabdus	●	●	●	●						●	
	Metadoga	●	●			●						
	Multipartis			●	●							●
	Munarinus			●	●							●
	Octolithus			●		●						●
	Okkolithus			●		●						●
	Orastrum		●			●				●		●
	Ottavianus			●				●				●
	Pharus	●		●				●	●			
	Ramsaya			●	●			●				●
	Russellia			●	●							●
	Semihololithus	●	●	●	●					●		
	Tetralithus				"wastebasket"							

Fig. 26. Holococcolithic genera, their terminology and characteristic elements

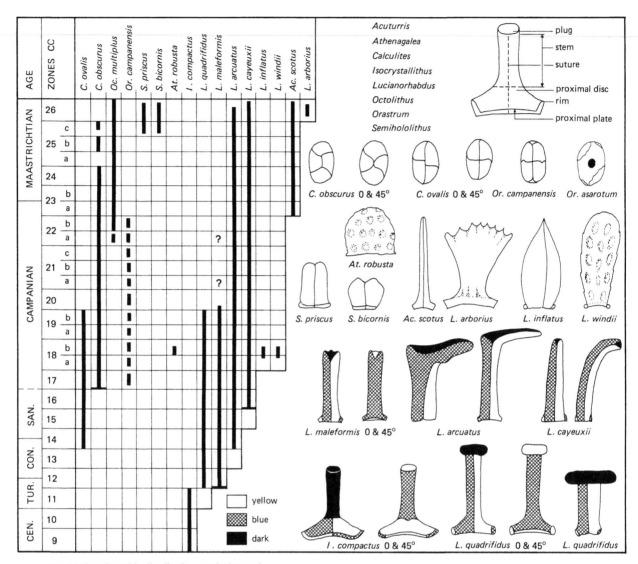

Fig. 27. Stratigraphic distribution, evolution and terminology of *Lucianorhabdus* and other genera of the Cretaceous Calyptrosphaeraceae

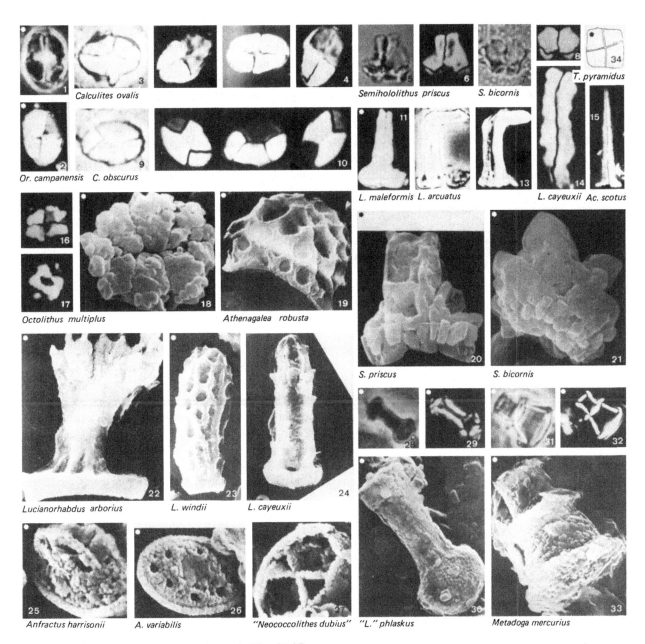

Calculites ovalis

Or. campanensis   C. obscurus

Semihololithus priscus   S. bicornis

T. pyramidus

L. maleformis   L. arcuatus   L. cayeuxii   Ac. scotus

Octolithus multiplus

Athenagalea robusta

S. priscus   S. bicornis

Lucianorhabdus arborius   L. windii   L. cayeuxii   "L." phlaskus   Metadoga mercurius

Anfractus harrisonii   A. variabilis   "Neococcolithes dubius"

Fig. 28. LM and SEM of species of Jurassic (Figs. 25–27) and Cretaceous holococcoliths. LM: *c.* ×2000

cross bears a central process which often is broken off. *C. tenuis* has a relatively narrow wall, a slender central X and a wide open central area. *C. fessus* (Figs. 29.24–26) is characterized by a central X formed by arms consisting of two parallel elements. It ranges from Cenomanian through Maastrichtian

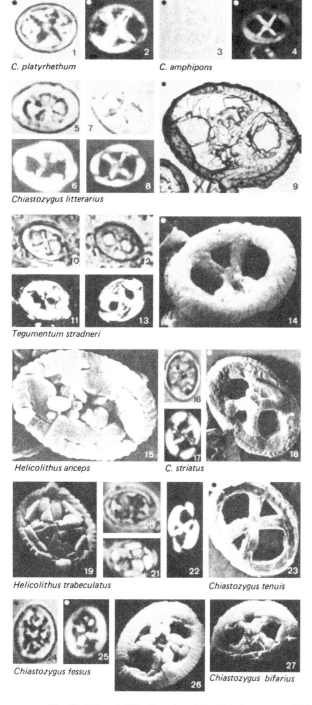

Fig. 29. LM and EM of species of the Chiastozygaceae. LM: *c.* × 2000

and differs from *Helicolithus anceps* (Fig. 29.15) by the absence of the blocky elements characteristic for the genus *Helicolithus*. Whereas the central X is small in *H. anceps*, it is prominent in *H. trabeculatus* (Figs. 29.19–21; syn. *Discolithus disgregatus*) which ranges from Albian to Maastrichtian according to Crux (1982). The latter also suggested the possibility of 'an evolutionary trend from *C. litterarius* through *C. bifarius* [Fig. 29.27] to *H. anceps*. This trend takes the form of reduction in number of laths forming the central cross and a change in nature of the rim from a simple single-cycle imbricating rim to a more complex two-cycle rim characteristic of the genus *Helicolithus*.' *C. bifarius* ranges from the Cenomanian to the Turonian, *H. anceps* from the Turonian to the Maastrichtian.

Other forms with a central X include *C. amphipons* (Figs. 29.34, 4) described from the Maastrichtian, and *Chiastozygus? primitus* (Figs. 2.4; 74.1) from the Late Triassic and Early Jurassic (see pre-Jurassic coccoliths). *C. ultimus* (Chapter 11, figs. 78.54, 55; 79) has a central cross at a slight angle with the axes, and was described from the Early Paleocene, where it gave rise to the genus *Neochiastozygus*.

Crepidolithaceae Black (1971a)
(Figures 30–33)

The Crepidolithaceae include 'elliptical coccoliths of substantial thickness consisting of a ring of calcite sectors without appreciable imbrication'. A proximal floor and distal massive process may be present. While Black (1971a) only included *Crepidolithus* and *Parhabdolithus* in this family, Medd (1979) is followed here in also assigning the genera shown in Fig. 24 as well as *Conusphaera* and *Eoconusphaera*.

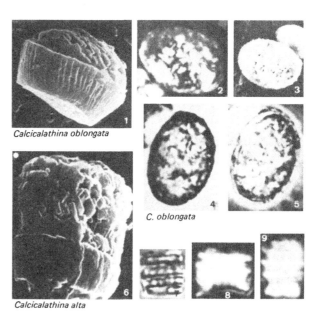

Fig. 30. LM and SEM of *Calcicalathina alta* and *C. oblongata*. LM: *c.* × 2000

Neocrepidolithus, which has a ring of inclined elements, is discussed briefly here, too. It is discussed further in Chapter 11. The genera are discussed in alphatical order below.

Calcicalathina

Calcicalathina is characterized by a high wall of vertical to slightly inclined elements. The central area is filled with calcite crystals of different orientation that rise above the distal margin of the wall. In *C. oblongata* (Figs. 30.1–5), the FO and LO of which define zonal boundaries in the Early Cretaceous, the central area rises only slightly above the wall. In *C. alta* (Figs. 30.6–9) from the Cenomanian, the central area rises higher than the height of the wall above it, and the elements are arranged in tiers. No intermediate forms have been reported from the Aptian or Albian.

Carinolithus

Carinolithus includes coccoliths with a relatively small, basket-like proximal part of more or less vertical elements and a central process topped by a distal, plate-like structure. The central process, which consists of long elements, is parallel-sided in *C. clavatus* (Figs. 33.3, 4, 13). In *C. sceptrum* (Figs. 33.1, 2, 12) it is short and funnel-shaped, whereas in *C. superbus* (Figs. 33.5, 6, 14) it is long and funnel-shaped. The species of *Carinolithus* are restricted to the Jurassic.

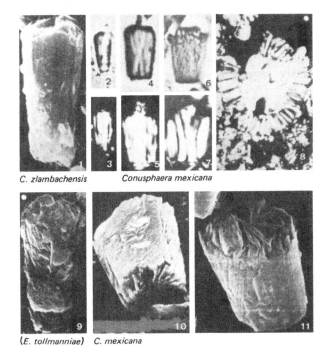

C. zlambachensis   Conusphaera mexicana

(E. tollmanniae)   C. mexicana

Fig. 31. LM and SEM of species of *Conusphaera* and *Eoconusphaera*. LM: *c.* × 2000

Conusphaera and Eoconusphaera

Conusphaera and *Eoconusphaera* include nannoliths described as elongated, truncated cones consisting of an outer layer of vertically arranged thin elements, and one or more inner cycles of inclined elements building the body. According to Jafar (1983), the author of *Eoconusphaera*, this genus differs from *Conusphaera* 'in having a dome-shaped broader end and lacking an axial canal'. The inclination of its central elements is inverse to that in the younger *C. mexicana*. *E. tollmanniae* (Fig. 31.9) is relatively higher than *C. mexicana* (syn. *Cretaturbella rothii*; Figs. 31.4–8, 10, 11). *C. mexicana* was found to form spheres (Fig. 31.8) in very thin thin-sections. It appears in the Tithonian and its LO is in the Early Aptian (CC 7a). Whereas it occurs as a rock-forming element in low latitude sediments, its distribution in high latitude sediments of appropriate age is patchy. *E. tollmanniae*'s FO lies in the Norian (Late Triassic) and it was also found in the Rhaetian. *E. tollmanniae* must be regarded as a junior synonym of *C. zlambachensis* (Figs. 31.1–3), described half a year earlier from the same region and age. Specimens of *C. zlambachensis* were reported from the eastern North Atlantic Sinemurian (?) and Pliensbachian (Crux, personal communication, 1983).

Crepidolithus

Crepidolithus is characterized by its thick wall of vertically to steeply arranged elements. The holotype of *C. cavus* (Fig. 33.27) is considered by Grün *et al.* (1974) to be a distal view of *Parhabdolithus marthae*, since the bridge bears a remnant of a spine. *C. crassus* (Figs. 33.15, 16, 23), the FO of which defines a zonal boundary in the Early Jurassic, is commonly used for specimens of *Crepidolithus* with a very wide margin and a narrow, featureless central area. *C. crucifer* is defined as a 'species of *Crepidolithus* with a cruciform structure in the central area', but no such structure is visible in the valid holotype (Fig. 33.28), only in the drawing of the invalid *C. crucifer* in Prins (1969). *C. impontus* (Figs. 32; 33.29) has a relatively wide open central area spanned by a bridge. Specimens of *Crepidolithus* are most common in the Liassic.

Neocrepidolithus has been described to include Maastrichtian and Paleocene coccoliths, previously assigned to *Crepidolithus*, that feature a wall of inclined elements. *N. cohenii* and *N. neocrassus* are the Maastrichtian representatives of *Neocrepidolithus* and are illustrated in Chapter 11 (Figs. 80.27, 28, 31, 32). *N. cohenii* is thought to be the ancestor of the Paleocene members of the genus. Whereas representatives of *Neocrepidolithus* are very rare in uppermost Maastrichtian sediments, they become more common in the lowermost Paleocene, especially in high latitudes.

Diazomatolithus

D. lehmanii (Figs. 33.19–22), the generotype and only species of the genus, is circular to subcircular and consists of a proximal, smaller cycle of elements arranged as a flat, truncated cone, and a distal cycle of elements arranged as a low funnel with a wide central opening. Its FO in the Bathonian is used as a zonal mark by Barnard & Hay (1974). It ranges up into the Late Albian. *Diazomatolithus* was provisionally placed in the Crepidolithaceae by Medd (1979), who suggested that it had evolved from *Proculithus* which, in turn, had evolved from *Carinolithus*.

central cross	central bridge	+− radial elements	perforated central plate
C. crucifer top Hettangian − basal Sinemurian (acc. to Barnard & Hay, 1974) but basal Pliensbachian (acc. to Prins, 1969)	*C. impontus* top Hettangian ? − top Jurassic ? +− radial elements	*C. crassus* * top Sinemurian − top Jurassic	*M. perforata* Upper Callovian − Upper Oxfordian

Fig. 32. Jurassic species of the Crepidolithaceae

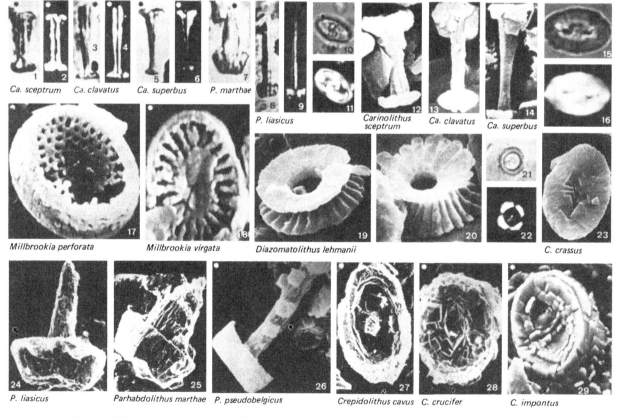

Ca. sceptrum   *Ca. clavatus*   *Ca. superbus*   *P. marthae*   *P. liasicus*   *Carinolithus sceptrum*   *Ca. clavatus*   *Ca. superbus*

Millbrookia perforata   *Millbrookia virgata*   *Diazomatolithus lehmanii*   *C. crassus*

P. liasicus   *Parhabdolithus marthae*   *P. pseudobelgicus*   *Crepidolithus cavus*   *C. crucifer*   *C. impontus*

Fig. 33. LM and EM of species of the Crepidolithaceae.
LM: *c.* ×2000

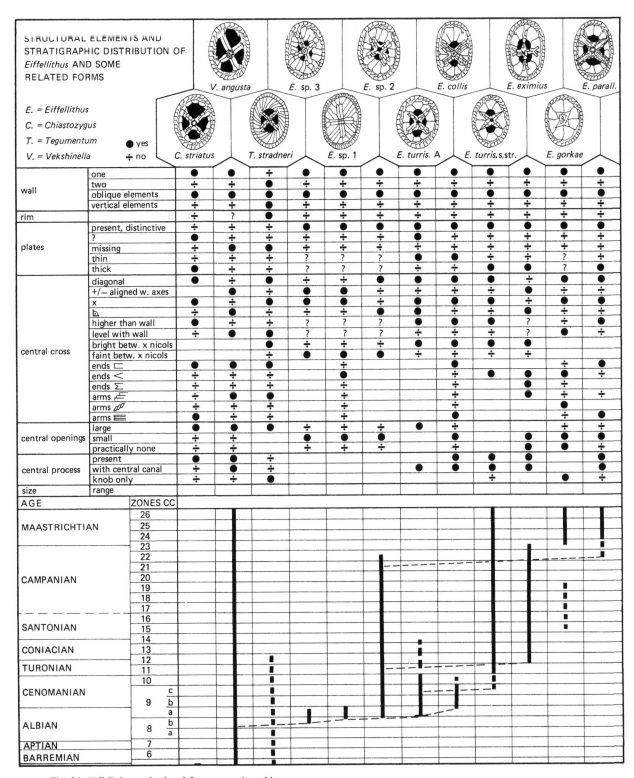

Fig. 34. *Eiffellithus* and related forms: stratigraphic distribution and structural elements. After Perch-Nielsen (1979a)

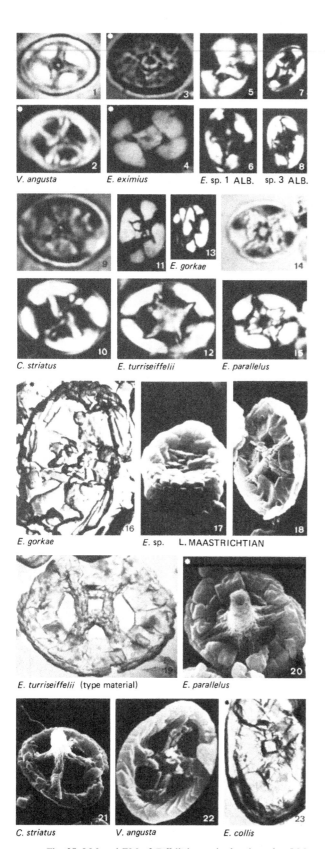

Fig. 35. LM and EM of *Eiffellithus* and related species. LM: *c.* × 2000

Millbrookia

Millbrookia was erected to include elliptical coccoliths with a low rim of vertical elements and a wide central area covered by a net-like floor. *M. perforata* (Fig. 33.17) is characterized by perforations of about equal size and *M. virgata* (Fig. 33.18) by radially arranged, elongate perforations. Both species have been reported from the Oxfordian only (Medd, 1979).

Parhabdolithus

Parhabdolithus is characterized by a massive wall of vertical elements and a prominent central process composed of rows of small, elongate elements. The shape of the central process (Figs. 33.7–9, 24, 25) defines the species, two of which serve as zonal markers in the Early Jurassic. *P. liasicus* (Figs. 33.8, 9, 24), the generotype, has a wide base and a relatively thin, long, conical central process. *P. marthae* (Figs. 33.7, 25) has a wide, short central process. *P. pseudobelgicus* has a long, slender stem with additional elements. *Zeugrhabdotus embergeri*, which has been assigned to *Parhabdolithus* by many authors, has a wall of distinctly inclined elements and a bridge with or without a central process. In side view, the species of *Parhabdolithus* are easily distinguished from, for example, those of *Discorhabdus* by the basket-like base (Figs. 33.24, 25) as compared to the two shields of the latter (Figs. 20.1, 2, 9).

Proculithus

This genus, with its high sides, may also be placed in the Calyculaceae (Fig. 24). It is defined to include coccoliths with 'one shield that comprises 2 or 3 cycles of plates with a wide central opening and a short bundle of axial rod-like plates that extend from the shield and diverge into a wide distal selvage margin'. *P. fistulatus* (Figs. 25.2, 3) is bucket-shaped, whereas *P. charlottei* (Fig. 25.5) has parallel sides and about equal height and width. The height is less than half the width in *P. expansus*. *P. fistulatus* is restricted to the Liassic, *P. charlottei* to the Oxfordian and *P. expansus* has been reported from both (Medd, 1979).

Eiffellithaceae Reinhardt (1965)
(Figures **34, 35**)

There is no agreement on the genera to be included in the Eiffellithaceae. While Medd (1979) used it as a wide family, which he subdivided into four subfamilies, I here only include *Eiffellithus* (syn. *Clinorhabdus*). The elliptical coccoliths have a thin rim of inclined elements, an inner cycle of usually plate-formed elements covering most of the central area and a central X or +, which may support a central process. The structural elements defining the species are tabulated in Fig. 34, where the ranges are also given, together with a few other, possibly related species. The generotype of *Eiffellithus*, *E. turriseiffelii*, is of Senonian age, but very similar forms are known as *Chiastozygus striatus* (Figs. 35.9, 10, 21) from the Hauterivian. The lineage proposed by Verbeek (1977b) from *Vekshinella angusta* (Figs. 35.1, 2, 22) to *E. turriseiffelii* in the Albian needs further verification.

E. turriseiffelii (Figs. 35.11, 12, 19), the FO of which defines the base of Zone CC 9, is characterized by a simple, oblique central cross. Its central process is long and tapers distally (Fig. 43.15). *E. eximius* (Figs. 35.3, 4), the FO of which occurs near the base of CC 12, has a central cross aligned with the axes. Both species are still distinguishable by LM, when only a part of the coccolith is preserved. The central cross in *E. eximius* leaves small incisions only in the flattest and most bent part of the elliptical coccolith, whereas the incisions are in between in *E. turriseiffelii*. The latter also has a wider central opening than *E. gorkae* (Figs. 35.13, 16), a relatively small species. In *E. parallelus* (Figs. 35.14, 15, 20), which only appears in the Late Campanian, the central cross is composed of rows of small parallel elements.

Ellipsagelosphaeraceae Noël (1965)
(Figures 36–40)

The Ellipsagelosphaeraceae include round and elliptical coccoliths with a distal shield of slightly overlapping elements and a single proximal shield of radially arranged elements. The central area is open, bridged or filled. Watznaueriaceae and Lotharingiaceae are considered synonyms of Ellipsagelosphaeraceae. Distal views of the generotypes of the genera included in this family are shown in Fig. 36. While only three species, *Ellipsagelosphaera britannica*, *E. communis* and *E. fossacincta* (syn. *E. keftalrempti*) are used as zonal markers, several species are often dominant in Late Jurassic and Early Cretaceous assemblages. This dominance has been shown by Keupp (1976) to have environmental significance, but can also

be caused by the high preservation potential of most species of *Ellipsagelosphaera*, *Cyclagelosphaera* and *Watznaueria*.

There is no general agreement among authors who have recently published on the Jurassic members of the Ellipsagelosphaeraceae as to which genera and which species are useful and really distinguishable. Most of the species recently erected were described and illustrated as seen with an SEM, and thus their use when working with the LM is limited, this despite the fact that some of the characteristics are also visible with the LM. Some are summarized here in alphabetical order of genus to encourage their separate recording.

Actinosphaera

Actinosphaera is considered a synonym of *Watznaueria*.

Ansulasphaera

Ansulasphaera includes circular to subcircular coccoliths with a distal shield and crown and a tube-like proximal shield. The only species, *A. helvetica* (Figs. 38; 39.3–8), was described from the Middle Callovian of Switzerland and has also been found in the Callovian of the western North Atlantic (Roth, 1983).

Calolithus

The genus *Calolithus* has often been disregarded as a product of dissolution of part of the coccolith. Black (1973, p. 72) claimed, as did the author of the genus, that the structural peculiarities that distinguish the elliptical *Calolithus* from *Watznaueria* and *Ellipsagelosphaera*, namely the perforations around the central area, are not the result of damage, as has been supposed by several later authors. They can be seen in perfectly preserved individuals and are merely emphasized and rendered more obvious by corrosion and other types of

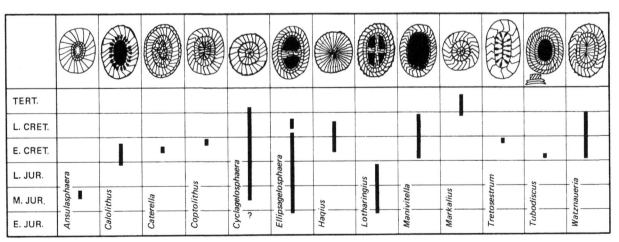

Fig. 36. The genera of the Ellipsagelosphaeraceae. Distal views of the generotypes. Proximal view of *Caterella*

damage. These structural peculiarities are, first, the contraction in width of the rays of the proximal shield with the formation of a slit-ring, and, second, the method of attachment between the two shields. The generotype of the genus, *C. martelae*, is illustrated in Figs. 40.36, 37. It is mainly found in the Late Jurassic.

Caterella

Caterella was introduced for elliptical coccoliths which only differ from *Ellipsageolosphaera* in 'having a large proximal central area, structurally independent of the rays of the proximal shield'. The only species, *C. perstrata* (Fig. 40.38) was found in the Hauterivian.

Coptolithus

Coptolithus was defined as an elliptical genus of Ellipsagelosphaeraceae with a large central opening covered by a bilaterally symmetrical grid. Here, too, one could argue that the space between the more or less radially oriented elements was due to dissolution rather than a primary feature. And even if it is a primary feature, the only species of the genus, *C. virgatus* (Fig. 40.39) could easily be included in *Watznaueria*. As with *Calolithus*, *Coptolithus* was described based on EM observations only. *C. virgatus* was described from the Albian.

Cyclagelosphaera

Cyclagelosphaera includes circular to subcircular coccoliths of the Ellipsagelosphaeraceae with a birefringent distal shield. Between crossed nicols, species of *Cyclagelosphaera* appear bright, whereas species of *Markalius* only show a partly bright central area. Many species have been described by Bukry (1969) and Black (1973), but none have been used stratigraphically. The species are distinguished by small differences in the fine structure of the central area (Fig. 38).

C. deflandrei (Figs. 40.10, 11) is a relatively thick and large representative of the genus. It often shows yellow colours between crossed nicols. In this it differs from *Haqius circumradiatus* which only shows weak birefringence. *C. bergeri* is very similar to *C. margerelii*, but has a wider central area and is larger. *C. margerelii* (Figs. 40.3, 18, 19), the generotype of *Cyclagelosphaera*, was described from the Upper Jurassic and persists into the Lower Tertiary, where two additional species evolve before the genus disappears. (See also Chapter 11, Figs. 23.37, 38, 51, 52.) The ranges of the other species are not well known, but *C. deflandrei* is mainly found in the lower Lower Cretaceous sediments.

Ellipsagelosphaera

Ellipsagelosphaera should include elliptical coccoliths with a distinct central tube (Fig. 40.41) between the two shields. In *Watznaueria*, there is no such structure. Between crossed nicols, the whole coccolith is bright. When viewed with a gypsum plate, the bridges of *E. britannica* and *E. frequens*, the generotype, have a colour different to that of the surrounding wall. In *W. biporta*, the bridge has the same colour as the wall. Fig. 37 gives an overview of the species of *Ellipsagelosphaera* and the structural elements which distinguish them. The stratigraphically important and some additional forms are illustrated in Figs. 5 and 40. *E. britannica* is characterized by a relatively massive bridge, but the distinction between *E.*

central area								overlapping elements	long-elliptical			thick distal shield
others								●				
open							●	■				
bridged	●	●	bifurcated bridge	many parallel br.	by randomly oriented elements				●	●		●
closed												
small	●						●	●				
large		●	●	●						●	very large	●
Ellipsagelosphaera	E. britannica	E. lucasii	E. reinhardtii	E. fasciata	E. plena	E. fossacincta	E. gresslyi	E. ovata	E. strigosa	E. ? tubulata
Synonyms	E. lucasii / E. frequens / E. communis		E. lucasii / E. britannica			E. keftalrempti / E. frequens / E. coronata / E. arata / W. barnesae			"Loxolithus armilla"	
Range	TOARCIAN – CAMPANIAN	BATHONIAN – TITHONIAN	OXFORDIAN – TITHONIAN	E.CRETACEOUS	CALLOVIAN – E. OXFORDIAN	M. BAJOCIAN – BARREMIAN	E. OXFORDIAN	CALLOVIAN – SANTONIAN	CALLOVIAN – E. OXFORDIAN	E.CALLOVIAN – E.OXFORDIAN

Fig. 37. Distal views of species of *Ellipsagelosphaera*, their characteristic features and range. Synonyms as suggested by Grün & Zweili (1980) and not necessarily including the holotypes of the species mentioned

britannica and *E. communis* and *E. frequens*, which also feature a bridge, is not clear and the species are not distinguished by many authors.

E. keftalrempti has no bridge and is here considered a junior synonym of *E. fossacincta*. Medd (1979) suggested the use of *E. britannica* for specimens with a small central opening, *E. lucasii* for forms with an opening width/total width ratio of 0.32–0.44, and *E. reinhardtii* for specimens with such a ratio greater than 0.44. In addition the bridge bifurcates laterally in *E. reinhardtii*. Medd (1971, 1979) also noted three size ranges for *E. britannica* s. ampl., but found no taxonomic or stratigraphic value in this observation. For further discussion see Black (1973) and Grün et al. (1974).

Esgia

The Late Jurassic *Esgia* was defined as 'circular to elliptical placoliths lacking a distinct central area and possessing many long radial to slightly curved elements'. The LM illustrations of the only species, *E. junior* suggest the genus to be synonymous with *Watznaueria*.

Haqius

Haqius includes 'circular placoliths composed of two shields with 40 or more slightly dextrally imbricated elements' in the distal shield and radially oriented elements in the proximal shield. The birefringence between crossed nicols is weak, with a diffuse extinction cross as seen in the only species, the generotype *H. circumradiatus* (Figs. 40.25–28). The imbrication of the distal shield elements is slighter than in *Markalius* and *Cyclagelosphaera*. *H. circumradiatus* ranges from the Hauterivian to the Campanian according to Crux (1982).

Lotharingius

A central cross supported by lateral bars and aligned with the axes of the elliptical coccoliths characterizes the Jurassic genus *Lotharingius* (Fig. 38) for which Noël (1973) also erected a new family. In her diagnoses she stressed that the distal shield was considerably smaller than the proximal one. This observation seems to have been based on poorly preserved specimens, where the overgrown distal crown was mistaken for the distal shield (Figs. 40.1, 2, 30).

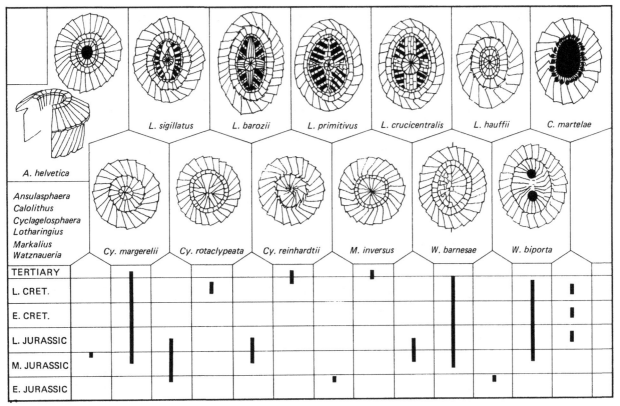

Fig. 38. Distal views of species of Ellipsagelosphaeraceae genera and their ranges

Manivitella

Manivitella unites elliptical coccoliths consisting of a two-layered ring surrounding a large and empty opening. Black (1973) distinguished *M. pemmatoidea* (Figs. 40.14, 15, 35) the generotype, as featuring a short spine or knob on each of the petaloid elements of the distal shield and *M. gronosa* with an unornamented distal shield. His *M. pecten* has a ring of centrally-directed projections lining the central opening and may or may not be a better preserved version of *M. gronosa*. *M. pemmatoidea* ranges from the Hauterivian through the Campanian. Some specimens assigned to *M. pemmatoidea* might represent empty *Gaarderella granulifera* (Fig. 65.1).

Margolatus

Margolatus, which is defined by the lack of a corona, is here considered a synonym of *Watznaueria*. Black (1973) regarded it as a genus intermediate between *Ellipsagelosphaera* and *Actinosphaera*, another synonym of *Watznaueria*.

Markalius

Markalius was described from the Tertiary, but a few species are also found in the Late Cretaceous. This round genus is further characterized by inversely imbricated elements in the distal shield. The latter remains dark between crossed nicols, whereas it is bright in species of *Cyclagelosphaera*. *M. inversus* (Figs. 40.20–22), the Tertiary generotype, was illustrated from the Maastrichtian by Perch-Nielsen (1968), who also described *M. perforatus* from the Middle Maastrichtian (1973). The species are distinguished by the fine structure of the central area. (See also Chapter 11, Figs. 23.25, 26, 49, 50.)

Tergestiella and Tremalithus

Like *Coccolithus*, the genus names *Tergestiella* and *Tremalithus* were used widely before the advent of the EM and thus many species were assigned to them that now are distributed to more narrowly defined genera. Whereas *Coccolithus* is a useful modern genus, the use of *Tergestiella* and *Tremalithus* has been more or less abandoned.

Tretosestrum

Tretosestrum includes long-elliptical coccoliths with two shields and a 'central area filled by elements which meet in the centre, and an axial line which ends at a large opening at both ends'. The only species, *T. perforatus* (Fig. 40.29) was described from the Albian.

Tubodiscus

Tubodiscus is characterized by its 'wide central tube rising distally, producing a distinct collar'. The generotype, *T. verenae* (Figs. 39.1, 2), has a short range in the Valanginian.

Watznaueria

Actinosphaera, *Colvillae*, *Esgia* and *Maslovella* are considered synonyms of *Watznaueria*, which is distinguished from *Ellipsagelosphaera* by the lack of a central tube. Many authors also regard *Calolithus* and *Ellipsageolsphaera* as synonyms of *Watznaueria*. Between crossed nicols this genus, like *Ellipsagelosphaera*, appears bright. The bridge in *W. biporta* (Figs. 40.16, 17) shows the same colour as the surrounding parts of the coccolith when viewed with the gypsum plate. *W. barnesae* (Figs. 40.23, 24) is the most common Cretaceous coccolith in poorly preserved assemblages.

Goniolithaceae Deflandre (1957)

Goniolithus fluckigeri, the only species of the only genus of the Goniolithaceae, a family with composite pentagonal coccoliths, was found in the type Maastrichtian by van Heck (1979). It is illustrated in Chapter 11, Fig. 41, since it was described and is better known from the Cenozoic. See also Black (1968, pl. 151, fig. 6) for a Maastrichtian form.

Microrhabdulaceae Deflandre (1963)
(Figures 41–43)

The Microrhabdulaceae include rod-like, cylindrical to spindle-shaped calcareous nannofossils of unknown origin. Some of the forms assigned to genera of this family may be stems of other coccoliths. Of the three genera proposed by Deflandre (1963) to belong to this family, *Microrhabdulinus*, *Microrhabduloidus* and *Microrhabdulus*, only the last one has been used repeatedly over the past decade, and only *M. decoratus* is used as a zonal marker. A few species of *Microrhabdulus* are shown on Figs. 41 and 43 together with the species of *Lithraphidites*, several of which are zonal markers,

Tubodiscus verenae

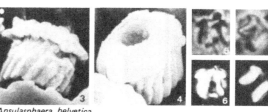

Ansulasphaera helvetica

Fig. 39. LM and SEM of *Ansulasphaera helvetica* and *Tubodiscus verenae*. LM: c. × 2000

and the species of *Pseudolithraphidites* and *Pseudomicula*. Side views of other coccoliths with a central process are shown for comparison in Fig. 43.

Lithraphidites and Pseudomicula

Lithraphidites includes calcareous rods with a more or less cruciform cross-section built of long blades of identical optical orientation. The species are distinguished by the shape of the blades. *L. acutus* (Figs. 42.7, 19) is characterized by a prominent spike near the centre of each blade and *L. alatus* (Figs. 42.6, 18) is closed umbrella-shaped. *L. bollii* (Figs. 42.1, 2, 14, 15) 'has irregular crystals arranged along an irregular wavy median axis' and its affiliation to *Lithraphidites* is questionable. The type species *L. carniolensis carniolensis*

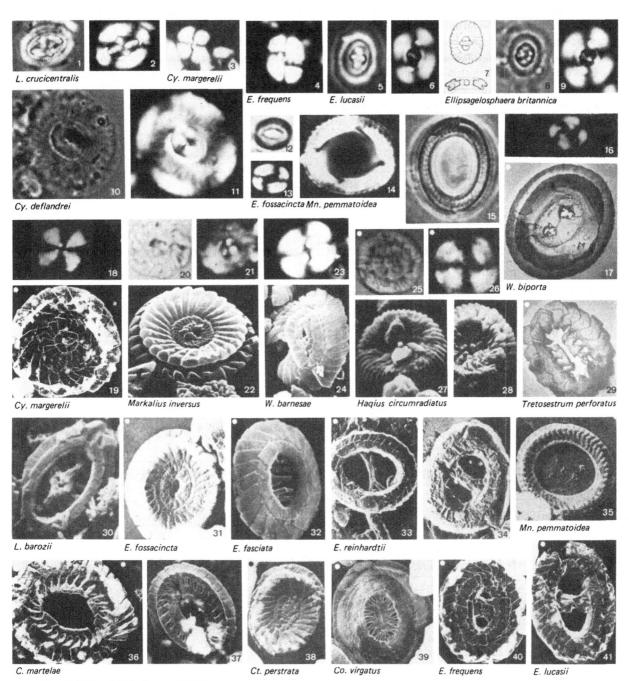

Fig. 40. LM and EM of species of the Ellipsagelosphaeraceae. LM: *c.* ×2000

(Figs. 42.3, 4, 16) is long and has slender, smooth blades. In *L. carniolensis serratus* (Fig. 42.20, syn. *Rhabdolithus aquitanicus* and *Rhabdophidites moeslensis*), the blades are slightly broader than in *L. carniolensis* and are finely serrate. *L. grossopectinatus* (Fig. 42.10) has coarsely serrate blades which are broader than in *L. carniolensis serratus*. *L. kennethii* (Fig. 42.11) has short broad blades with two spikes at the extremities. In *L. praequadratus* (Fig. 42.21) the wide part of the blades is longer than in *L. quadratus* (Figs. 42.8, 9, 22) but shorter than in *L. carniolensis*. The blades are wider in *L. quadratus*. The limit between the Maastrichtian *L. praequadratus* and the marker species *L. quadratus* can be drawn where the length of the wide part of the blades becomes less than two-thirds of the total length and where the blades are wider than the width of the rod itself. The Middle Cretaceous *L. pseudoquadratus* (Figs. 42.5, 17) has relatively wide but short blades reaching about one-third of the length of the rod.

The origin of *Lithraphidites* is unknown, the first *L. carniolensis* appearing quite suddenly at the base of the Cretaceous. It seems to have no relationship to the genus *Pseudolithraphidites* from the Tithonian or the similar Tertiary forms (see Chapter 11). The simple form *L. carniolensis* gave birth to more elaborate forms with wider blades in the Late Albian and again in the Campanian/Maastrichtian. The splitting of the short, wide blades of *L. quadratus* leads to the massive *Pseudomicula quadrata* (Figs. 42.23–25). Several specimens of species of *Lithraphidites* with small crowns of elements around one end of the rod suggest that these rods could represent central processes of extremely small coccoliths.

Microrhabdulus

Microrhabdulus is now used for all unattached Mesozoic rods with a more or less circular cross-section. The few Late Cretaceous species illustrated in Fig. 43.1–26 are shown for comparison with *M. decoratus* (Figs. 43.17–19), the FO of which marks the base of CC 10 in the Cenomanian. *M. decoratus* has parallel sides whereas *M. belgicus* (Figs. 43.20–23), *M. attenuatus* (Figs. 43.10–14; syn. *M. stradneri*) and *M. undosus* (Figs. 43.24–26) taper at both ends. Most central processes of coccoliths (Figs. 43.1–9, 15, 16) taper

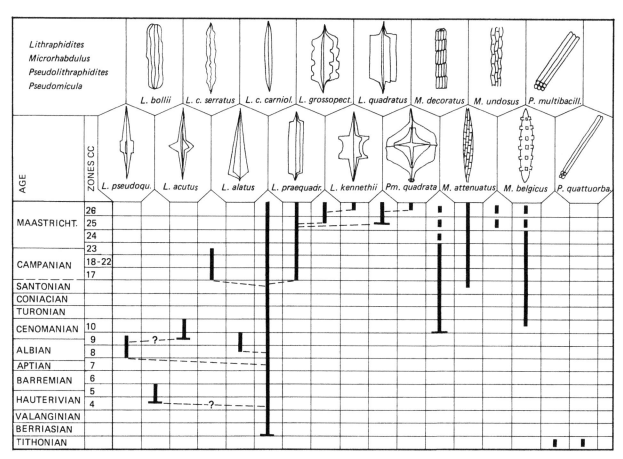

Fig. 41. Species of *Lithraphidites*, *Microrhabdulus*, *Pseudolithraphidites* and *Pseudomicula* and their tentative ranges and relationships

distally or have a completely different fine structure than *M. decoratus*, which is composed of tiers of laths of equal length.

Pseudolithraphidites

The two species of this genus, *P. multibacillatus* (Fig. 42.13) and *P. quattuorbacillatus* (Fig. 42.12) consist, respectively, of 6 and 4 circular rods with parallel sides; they were described from the Tithonian.

Nannoconaceae Deflandre (1959)
(Figures **44, 45**)

The Nannoconaceae include nannoliths with a thick wall of wedges or plates of calcite oriented perpendicular and spirally surrounding an axial cavity or canal (Fig. 44). Tappan (1980) included the genera *Nannoconus* (syn. *Brachiolithus*), *Conusphaera* (syn. *Cretaturbella*) and *Polycostella* in this family. The latter two are here included in the Crepidolithaceae and Polycyclolithaceae, respectively, since they are morphologically dissimilar to *Nannoconus*.

Nannoconus has its FO in the latest Jurassic and is particularly abundant in Lower Cretaceous sediments of the Tethyan realm. It has been found in Spitzbergen and on the Falkland Plateau and thus was a cosmopolitan genus at least during parts of the Early Cretaceous. *Nannoconus* was one of

the first calcareous nannofossils to be used in stratigraphy. Brönnimann (1955) suggested three distinct *Nannoconus* associations in the Upper Jurassic and Lower Cretaceous of Cuba: the *N. steinmannii*, the *N. kamptneri* and the *N. truittii* assemblages. Manivit *et al.* (1979) attempted a correlation between the *Nannoconus* zonation of Deres & Acheriteguy (1972) and the Coccolith zonation of Sissingh (1977). The correlation is difficult, because sediments rich in nannoconids often lack a diverse coccolith assemblage and vice versa. Perch-Nielsen (1979a) gave an overview of the distribution of nannoconids in the stratotypes of the Cretaceous stages. Deres & Archeriteguy (1980) gave an overview of all species of *Nannoconus* and their stratigraphic distribution (Fig. 44). Since these authors worked mainly with assemblages from exploration wells, they defined the age of their assemblages by using FOs and LOs of nannoconids as stage limits.

The species distinction in *Nannoconus* is based on the shape of the nannolith, its size, the shape and length of the central canal and the central cavity. For the study of the central canal and central cavity, the LM is preferable to the TEM or SEM.

Whereas smear-slides or strewn-preparations of cleaned samples are usually used for the determination of the nannoconids, the study of very thin thin-sections has allowed the

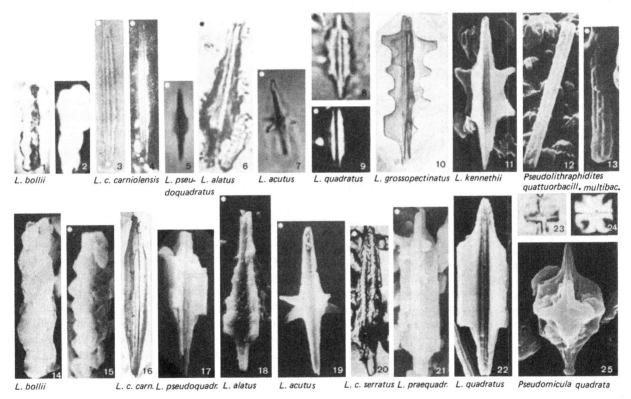

Fig. 42. LM and EM of species of *Lithraphidites*, *Pseudolithraphidites* and *Pseudomicula*. LM: *c.* × 2000

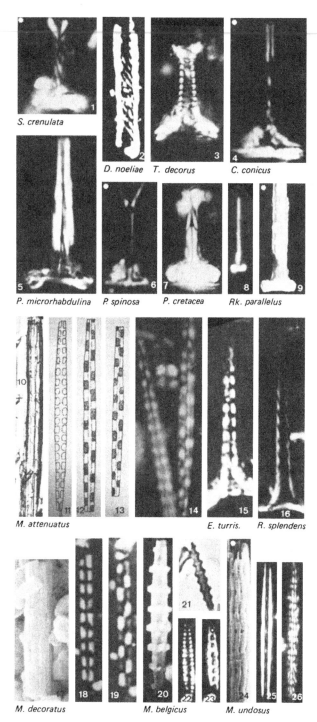

Fig. 43. LM and EM of species of *Microhabdulus* (Figs. 10–14, 17–26) and other genera with central processes, seen in side view. *S.* = *Stradneria*, *D.* = *Dodekapodorhabdus*, *T.* = *Tetrapodorhabdus*, *C.* = *Cretarhabdus*, *P.* = *Prediscosphaera*, *Rk.* = *Rhabdolekiskus*, *R.* = *Rhagodiscus*, *E.* = *Eiffellithus*. LM: ×2100

observation of rings of nannoconids (Fig. 45.45). These suggest the possibility of a similar origin for the nannoconids as for the coccoliths. Deres & Acheriteguy (1972, 1980), working with the LM, distinguished over 50 Early Cretaceous species of *Nannoconus*, some of them not yet formally described. Aubry (1974), working with the SEM, discussed and rejected the species concepts of the few species found in the French chalk (Turonian–Campian). The species are discussed below in order of their FO and are illustrated in Figs. 44 and 45.

Late Jurassic

The Late Jurassic assemblage includes seven species of *Nannoconus* according to Deres & Acheriteguy (1980; Fig. 44). Of these, *N. dolomiticus* (Fig. 45.49) is the first to appear. It is characterized by its long, parallel sides and narrow axial canal and has no central cavity as is found in *N. colomii*, which appears slightly later. *N. colomii* (Fig. 45.48) is a conical form with a central cavity in the basal part of the nannolith and a relatively narrow axial canal. The similar *N. steinmannii* (Fig. 45.41) lacks the central cavity and is less conical than *N. colomii* but more conical and wider than *N. dolomiticus*. *N. steinmannii minor* (Fig. 45.40) is shorter than *N. steinmannii* and has a relatively wider axial canal. The axial canal is even wider in the similar form *N. broennimannii* (Fig. 45.19), where it has a tendency to widen near the basal aperture. *N. globulus* (Fig. 44), which possibly appears already in the Late Jurassic, is characterized by its subspherical shape. The axial canal is short on both ends of the nannolith and the central cavity is subspherical. Basal and apical aperture are relatively narrow, whereas the apical one is wide in *N. circularis*, another subspherical form. *N. quadratus* (Fig. 45.5) is a very small species with a spindle-like shape. It is wider than high and has a subspherical central cavity. It was originally described as *Brachiolithus quadratus*, the generotype of *Brachiolithus*.

Berriasian–Hauterivian

According to Deres & Acheriteguy (1980) no new species appear in the Berriasian, while other authors have some or all Late Jurassic forms appearing in the Berriasian. In the Valanginian, *N. cornuta* (Fig. 45.6), a form with an irregular test and a large cylindrical central cavity and only short axial canals appears. *N. bermudezii* (Figs. 45.36, 37) is a large, long, slightly curved form with a long, slim axial canal and lacking a central cavity. *N. kamptneri* (Figs. 45.38, 39) is large, long and straight. Its central cavity is long and includes about one-third of the width of the only slightly conical test. *N. bonetii* resembles an elongate pear and its irregular central cavity is usually wider than the width of the wall. The only new species in the Hauterivian are *N. minutus* and *N. bucheri* (Fig. 45.21). *N. bucheri* is a medium-sized, conical form with truncated base and apex. The width of the central canal is about equal to the width of the wall. *N. minutus* (Fig. 45.2) is

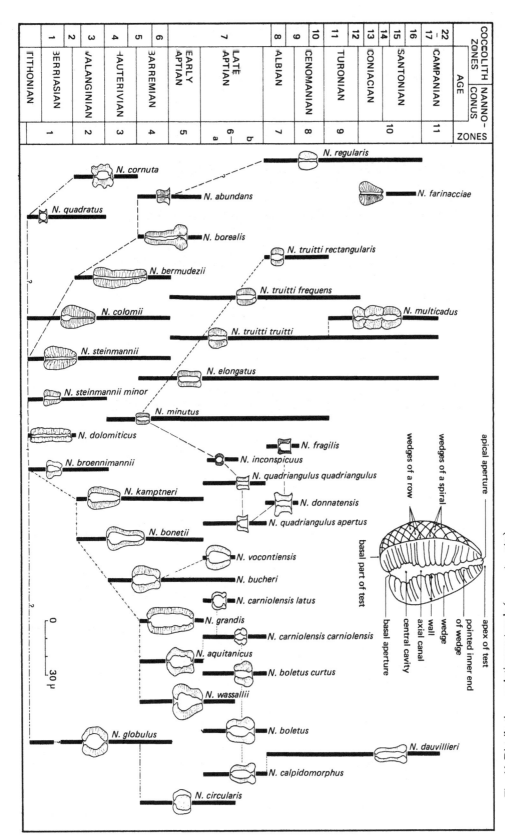

Fig. 44. Distribution, evolutionary trends (———, — · +, etc.) and terminology of species of *Nannoconus* mainly after Deres & Acheriteguy (1980)

very small (5–7 μm), has nearly parallel sides and is slightly higher than wide. The central cavity is as wide as the wall is thick.

Barremian

The nannoconids reached their highest diversity in Barremian–Aptian times, before most of the early Early Cretaceous species disappeared. *N. grandis* (Fig. 45.44) is a slightly conical, very large species with an often irregular, very wide central cavity and relatively wide apertures. *N. abundans* (Figs. 45.7–9, 51) is cylindrical with an abruptly expanding but thin apex and a thin axial canal. The sides can be slightly concave. The wedges forming the wall are thinner than in all other species, resulting in very fine or no striae being visible with the LM in side view. *N. aquitanicus* (Fig. 45.35) is about of equal height and width. One end is truncated, the other subspherical and the sides conical. The central cavity is wider than the width of the wall and is irregular in shape. *N. borealis* (Figs. 45.33, 50) is characterized by one massively enlarged side and one slightly enlarged side with a slightly curved, cylindrical part between. The axial canal varies in width. The enlargement is more pronounced than in *N. boletus* and *N. calpidomorphus*, both of which have wide central cavities. *N. wassallii* (Fig. 45.20) is pear-shaped with a truncated smaller end and a pear-shaped central cavity. *N. circularis* (Fig. 45.10) is a cylindrical form with slightly convex sides. The thin walls thin at both sides. The central cavity is nearly spherical. One aperture is wide, the other narrow. *N. elongatus* (Fig. 45.22, 23) is an elongate, cylindrical form of medium size (10–16 μm high) and has a central cavity about the width of the walls or slightly wider.

Aptian–Albian

Most of the nannoconids appearing in the Aptian and later are smaller than the bulk of early Early Cretaceous tests which, however, persisted into the Early Aptian. *N. truitti* (Figs. 45.28, 29) is quadrangular to subspherical, slightly conical and about as wide as high. The axial canal is thinner than the wall. Whereas *N. truitti rectangularis* (Fig. 45.26) is wider than high, *N. truitti frequens* (Fig. 45.27) is higher than wide. The central canal is wider than in the similar *N. regularis* (Fig. 45.25). *N. inconspicuus* (Fig. 45.3) is very small (3.5–4 μm high and 4–7 μm wide), has slightly convex sides and a central cavity of the same width as the wall thickness. *N. vocontiensis* (Fig. 45.24) is ovoid with truncated ends. It is slightly higher than wide and has a wide central cavity; the apertures are distinct. *N. boletus* (Fig. 45.32) is fungiform, higher than wide, and has a roughly cylindrical, wide central cavity which is wider than in the higher *N. borealis*. *N. boletus curtus* (Fig. 44) is similar to, but shorter than, *N. boletus*. *N. carniolensis carniolensis* (Figs. 45.16, 17) has the shape of a pot with a wide rim and has a wide central cavity tapering where the rim becomes wider. The central cavity is wider in *N. carniolensis latus* (Fig. 45.18), where the walls are unusually thin. The latter species has a very short range in the early Late Aptian. *N. quadriangulus quadriangulus* (Figs. 45.11–13) is a very small (5–8 μm) quadrangular form with a cylindrical central cavity of the same width as the wall. In *N. quadriangulus apertus* (Fig. 45.14) the sides diverge distally as does the very wide and open-ended central cavity. *N. calpidomorphus* (Fig. 45.34) is amphora-shaped and has a relatively wide, sausage-shaped central cavity, widening at both ends after an incision.

Besides *N. regularis* and *N. truitti rectangularis*, *N. fragilis*, *N. donnatensis* and *N. dauvillieri* also have their FOs in the Albian. *N. fragilis* (Fig. 45.15) is small and cylinder-shaped with slightly concave sides and has a very wide, distally open central cavity. *N. donnatensis* (Fig. 45.4) is a small form with strongly concave sides. It is wider than high and the axial canal is thinner than the walls. *N. donnatensis* is restricted to the Albian according to Deres & Acheriteguy (1980) and was previously referred to as *Nannoconus* sp. 92 by the same authors (1972). *N. fragilis* (Fig. 45.15; previously *Nannoconus* sp. 93) is smaller than *N. donnatensis* and has only slightly concave sides. Its central cavity is wider than the walls and distally open. It is also restricted to the Albian. *N. dauvilleri* (Fig. 45.30, 31) is the only Albian species that is considerably higher than wide. It has a wide central cavity, slightly concave sides and is somewhat thicker at one end than at the other. It is longer than *N. boletus* and has a wider central cavity than *N. borealis*.

Late Cretaceous

No new species appear in the Cenomanian. In the Turonian *N. multicadus* (Figs. 45.42, 43, 46) appears, a unique long form with 1 or 2 constrictions and a central cavity varying in width. Wise & Wind (1977) illustrated *N. multicadus* from the Albian of the south Atlantic. *N. farinacciae* (Fig. 45.47) was described from the Santonian and is pear-shaped. The few species persisting from the Early Cretaceous and *N. multicadus* disappear during the Late Cretaceous, none reaching the Maastrichtian (Fig. 44).

Podorhabdaceae Noël (1965)
(Figures 46–53)

The Podorhabdaceae include elliptical coccoliths with a 'narrow rim constructed of 2 to 3 layers of jointive petaloid or subquadrate elements; in polarized light this part shows a black cross with straight arms; petaloid rim surrounds a wide central area, spanned by various types of structures composed of many fine crystallites, with a solid or hollow' central process commonly arising from the centre (Tappan, 1980). Despite the many genera (Figs. 47, 50, 52) and species that have been described in this family, the Podorhabdaceae have furnished relatively few stratigraphically important species: *Axopodo-*

rhabdus cylindratus in the Pliensbachian, *Polypodorhabdus escaigii* and *A. rahla* in the Callovian, *Stradneria crenulata*, *Cretarhabdus loriei*, *Microstaurus chiastius*, *Cruciellipsis cuvillieri*, *Speetonia colligata* and *A. albianus* in the Early Cretaceous, and *Nephrolithus frequens* in the Late Cretaceous.

Tappan (1980), who is generally followed here, used a three-fold subdivision of the Podorhabdaceae (syn. Cretarhabdaceae, Retecapsaceae) consisting of the subfamilies Podorhabdoideae, Retecapsoideae and Cribrosphaerelloideae, which are discussed briefly below.

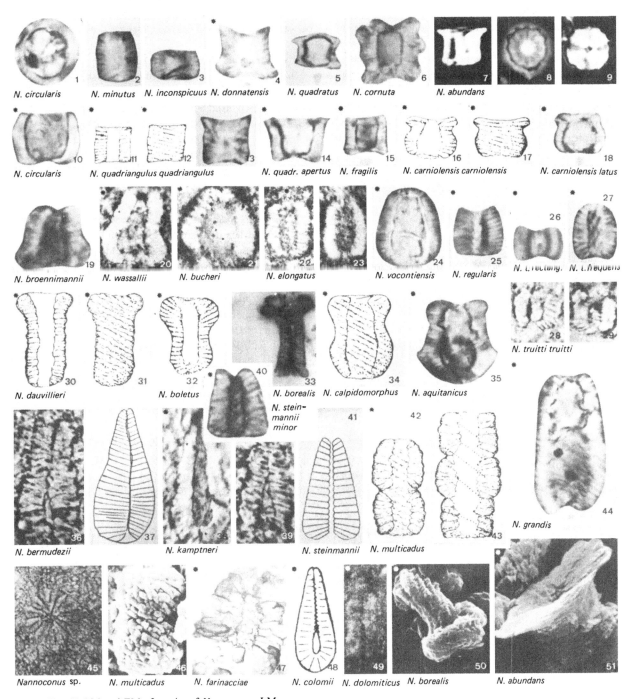

Fig. 45. LM and EM of species of *Nannoconus*. LM: *c.* × 1700

Podorhabdoideae

The genera grouped in this subfamily have a tubular central process. The genera are distinguished by the number and arrangement of openings in the central area, the species by details in the arrangement of the elements of the central area and the stem. An overview is given in Fig. 47.

Axopodorhabdus

Axopodorhabdus was erected for coccoliths with 4 openings and 4 stem-supporting bars arranged parallel to the axes of the ellipse. *A. cylindratus* (Figs. 49.5,6,16), the Pliensbachian marker, is the first representative of the genus. It differs from the Cretaceous species *A. albianus* (Figs. 49.8, 18) and similar forms only by the relatively wider elements building the shields and its range which is restricted to the Jurassic. *A. rahla* (Figs. 49.21, 22) is characterized by the four-fold buttress arrange-

ment on the stem and can only be determined in side view. According to Black (1972), *A. albianus*, the FO of which was used by Thierstein (1976) as a zonal marker in the Albian, but which has since been reported from the Barremian, can be distinguished from similar Middle Cretaceous forms by the different spacing of the openings in the central area (Fig. 48).

Since only *A. albianus* usually appears on range charts, it must be assumed that most authors do not distinguish between these species. Their exact ranges thus are unknown.

Boletuvelum

Boletuvelum is characterized by a two-shield base and a 'broad-based flaring hollow stem terminating in a massive, totally enclosed distal bulb'. The only species, *B. candens* (Fig. 49.19) was found in the Maastrichtian of the Falkland Plateau. *Boletuvelum* is not a typical representative of the Podorhabdaceae, to which it was assigned by Tappan (1980), but might belong to the Biscutaceae.

Cleistorhabdus

Cleistorhabdus has no openings except the central hole which is left when the stem has broken off. *C. williamsii* (Fig. 49.15) was described from the Albian and was also reported from the Cenomanian.

Dekapodorhabdus

The single illustration of *D. typicus* (Fig. 49.32), the generotype of *Dekapodorhabdus* from the Oxfordian, shows a coccolith with 10 openings in the central area.

Dodekapodorhabdus

The species of *Dodekapodorhabdus* have 12 openings in a cycle adjacent to the crown. The central process is usually

TERMINOLOGY PODORHABDACEAE

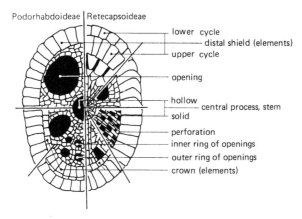

Podorhabdoideae | Retecapsoideae

- lower cycle
- distal shield (elements)
- upper cycle
- opening
- hollow
- central process, stem
- solid
- perforation
- inner ring of openings
- outer ring of openings
- crown (elements)

Fig. 46. Terminology of Podorhabdaceae

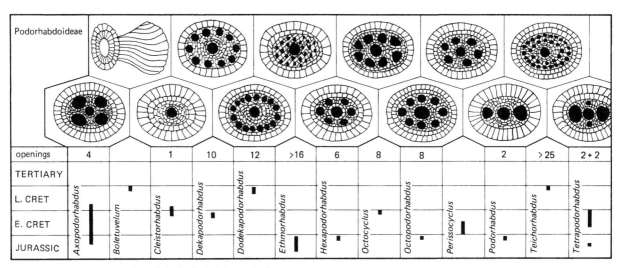

Podorhabdoideae												
openings	4	1	10	12	>16	6	8	8	2	> 25	2 + 2
TERTIARY												
L. CRET												
E. CRET												
JURASSIC	*Axopodorhabdus*	*Boletuvelum*	*Cleistorhabdus*	*Dekapodorhabdus*	*Dodekapodorhabdus*	*Ethmorhabdus*	*Hexapodorhabdus*	*Octocyclus*	*Octopodorhabdus*	*Perissocyclus*	*Podorhabdus*	*Teichorhabdus*	*Tetrapodorhabdus*

Fig. 47. Genus overview of the Podorhabdaceae I: the Podorhabdoideae

broken off in *D. noeliae* (Figs. 43.2, 49.13, 33) and leaves a large hole in the elevated central area. *D. noeliae* is restricted to the Late Campanian and Early Maastrichtian.

Ethmorhabdus

Ethmorhabdus is characterized by a larger number ($\geqslant 20$) of perforations around a central hole or process. *E. gallicus* (Figs. 49.1–4, 31), the generotype, ranges from the Bajocian to the top of the Kimmeridgian.

Hemipodorhabdus

Hemipodorhabdus is considered a junior synonym of *Podorhabdus*.

Hexapodorhabdus

Hexapodorhabdus is characterized by 6 openings surrounding the central hole or process. The generotype, *H. cuvillieri* (Fig. 49.27) was described from the Late Jurassic, but representatives of this genus are also known from the Cretaceous.

Octocyclus and Octopodorhabdus

The two genera are very similar, both featuring 8 openings surrounding the central hole or process. Whereas the Oxfordian generotype of *Octopodorhabdus*, *O. praevisus* (Fig. 49.28) has openings along the principal axes of the ellipse, the Albian *Octocyclus magnus* (Fig. 49.23) has the 4 principal buttresses in this position. Medd (1979) considered this to be unimportant and united the two genera into *Octopodorhabdus*. *O. decussatus* (Fig. 49.29) is considerably smaller than *O. magnus* and *O. plethotretus* has more than 8 openings and an inner cycle of openings. It only belongs to *Octopodorhabdus* when its wider definition as suggested by Wind & Čepek (1979) is accepted. For these authors forms with more than 8 openings and with an inner cycle of openings are also to be included in *Octopodorhabdus*. Such forms are included by other authors in *Perissocyclus* (see below). The Oxfordian *O. oculisminutis* has considerably smaller openings than the other species.

Perissocyclus

The definition of *Perissocyclus* includes coccoliths of the Podorhabdoideae with a wide central area which is 'pierced either by a single ring of windows which may be uneven in number, or by 2 concentric rings with fewer windows in the inner ring'. The generotype, the Hauterivian *P. noeliae* (Fig. 49.25) is reported to feature 5–9 openings, *P. fenestratus* (Fig. 49.24) 8–16 in the outer ring and 6–10 in the inner ring. *P. fletcheri* (Fig. 49.26) has one or two rings of small openings, the diameter of the outer ring being equal to half or less of the diameter of the coccolith. Black (1971a) thus included forms previously assigned to *Dodekapodorhabdus* in *Perissocyclus*, which is a mainly Early Cretaceous genus.

Podorhabdus

Wise & Wind (1977) have shown that *P. grassei* (Fig. 49.20), the generotype of *Podorhabdus*, only has 2 openings besides the central hole or process. *Hemipodorhabdus*, with the Albian type species *H. biforatus*, thus must be considered a junior synonym of *Podorhabdus*. The distinction between this genus and *Tetrapodorhabdus* becomes arbitrary where 2 minuscule openings appear along the minor axis of the ellipse.

Teichorhabdus

Teichorhabdus features 2 or more incomplete rings with 12 or more small openings and a massive central process. Its only species, *T. ethmos* (Fig. 49.30), was described from the Maastrichtian of the Falkland Plateau.

Tetrapodorhabdus

Tetrapodorhabdus is characterized by 2 large openings along the major axis and 2 small openings along the minor axis of the ellipse. In *Axopodorhabdus*, the other genus with 4 openings, these are arranged along the diagonals. *T. coptensis* (Fig. 49.17) the type species, was described from the Albian. *T. decorus* (Figs. 43.3; 49.11, 12) is best recognized in side view, where the complex helicoid character of the central process is conspicuous. The Late Cretaceous forms are up to 16 μm high, while similar Late Jurassic and Early Cretaceous forms are considerably smaller.

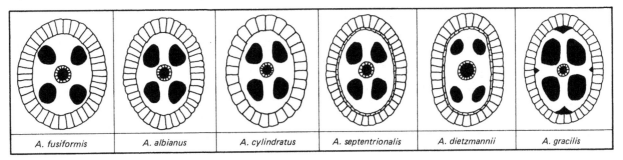

| A. fusiformis | A. albianus | A. cylindratus | A. septentrionalis | A. dietzmannii | A. gracilis |

Fig. 48. Jurassic (*A. cylindratus*) and Middle Cretaceous species of *Axopodorhabdus*

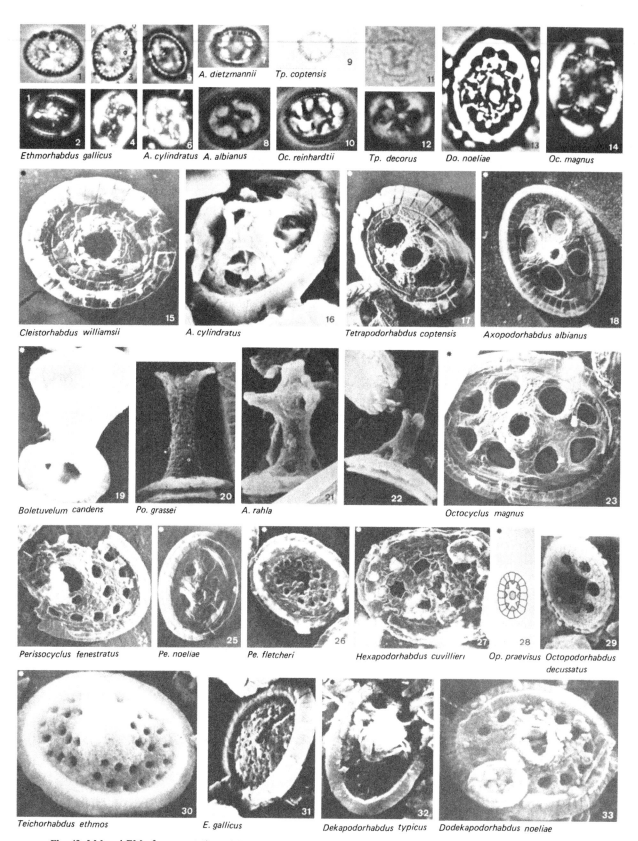

Fig. 49. LM and EM of representatives ot the Podorhabdoideae. LM: *c.* ×2000

Retecapsoideae

Black (1972) grouped in this subfamily genera having a distal shield with two cycles of non- or only slightly-imbricated elements and a proximal shield with one cycle of such elements. The central process (stem, knob, boss), if present, is solid, as opposed to the hollow one of the Podorhabdoideae. Whereas in the latter, a large central hole is left, when the central stem has broken off, a closed centre remains in the Retecapsoideae. An overview of the genera, many of them monospecific, and the structural elements that distinguish them is given in Fig. 50. This subfamily, which has furnished four markers, is discussed in detail by Grün & Allemann (1975).

Allemannites

The central area features a striated central cross aligned with the axes of the ellipse and topped by the base of a solid stem. Between the lateral bars are small irregularly-arranged openings and numerous small elements. The species are distinguished by the details of the arrangements of the openings. Grün & Allemann (1975) assumed *Cretarhabdus loriei* to be synonymous with *A. striatus* (Figs. 51.4, 21), the generotype of *Allemannites*, which differs from *Cretarhabdus* by the arrangement in cycles of the openings. This distinction is not made by many authors, who thus only use *Cretarhabdus* and regard *Allemannites* as synonymous. Wise (1983, p. 507) considered *A. striatus* synonymous with *Crucicribrum anglicum* and *Allemanites* thus a junior synonym of *Crucicribrum*.

Bipodorhabdus

Bipodorhabdus has 2 large openings and a central bridge featuring the remains of a solid central stem. The distal shield has a larger upper than lower cycle. This characteristic separates *Bipodorhabdus* from *Speetonia* with its larger lower cycle, a more typical genus of the subfamily. *Podorhabdus* has

a hollow stem. *B. tesselatus* (Fig. 51.24), the generotype, was described from the Campanian and might be a junior synonym of *Placozygus fibuliformis* (Fig. 82.15); *Zygodiscus spiralis* of many authors), which has its FO in the Late Santonian according to Verbeek (1977b), who used it as a zonal marker, and in the Middle Albian according to Hill (1976).

Cretadiscus

Cretadiscus, which belongs to the Retecapsoideae according to Tappan (1980), is a junior synonym of *Cribrosphaerella*, which belongs to the Cribrosphaerelloideae (see below).

Cretarhabdus (syn. *Cretarhabdella*)

Cretarhabdella was defined as 'coccoliths resembling *Cretarhabdus*' but differing 'in having an asymmetrical grid and the spine displaced towards the margin'. *C. lateralis* (Fig. 51.17), the generotype, should be regarded as an aberrant specimen of *Cretarhabdus conicus* (Figs. 51.1, 2, 22), the generotype of *Cretarhabdus*. In the latter, the upper cycle of the distal shield is smaller than the lower one, and concentric cycles of perforations appear in the conical central area. The central cross is aligned with the axes of the ellipse and supports a solid stem. The species are distinguished by the structural details of the central area.

Cruciellipsis

Cruciellipsis has an upper cycle of very long distal shield elements, a massive central cross along the axes of the ellipse and no central process. Similar forms with a central process are assigned to *Microstaurus* by Grün & Allemann (1975). *Helenea* has a less prominent upper cycle of distal shield elements. The generotype, *C. cuvillieri* (Figs. 51.14, 15, 33, 40), a marker species in the Early Cretaceous (Fig. 6), is

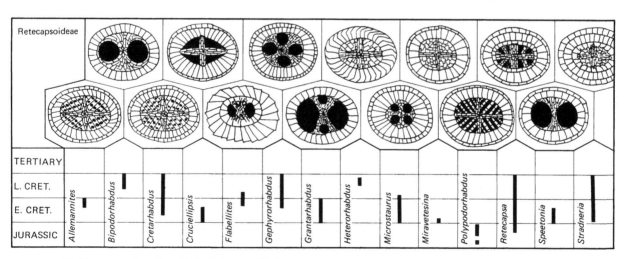

Fig. 50. Genus overview of the Podorhabdaceae II: the Retecapsoideae

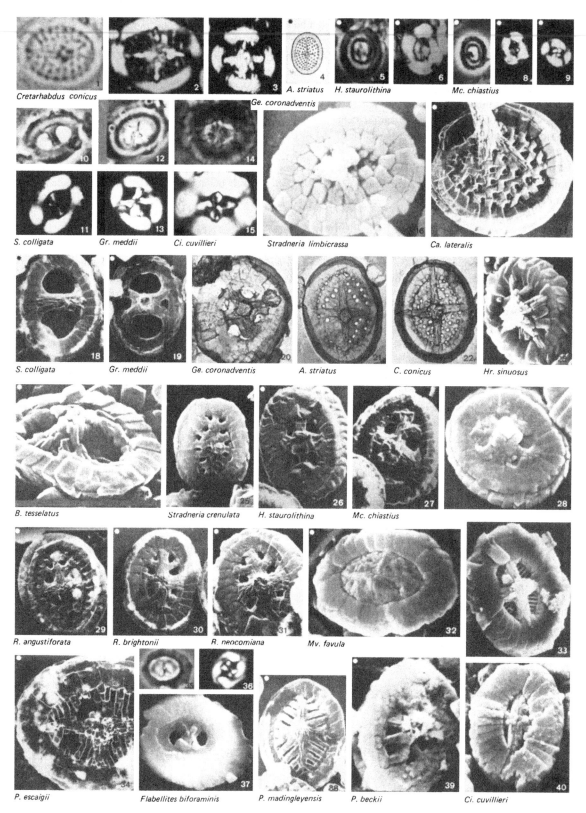

Fig. 51. LM and EM of representatives of the
Retecapsoideae. LM: *c.* ×2000

characterized by an elongate central area and a central cross of distally parallel laths.

Flabellites

Flabellites biforaminis (Fig. 51.35–37), the generotype, is characterized by a slightly asymmetrical distal shield of elongate elements. The central structure is a distal cross and is formed by two touching cycles of elements on the proximal side. *F. oblonga* is a long-elliptical form that ranges from the Cenomanian to the Santonian, *F. biforaminis* from the Aptian to the Cenomanian.

Gephyrorhabdus and Grantarhabdus

Four openings in the central area and a distal shield with a smaller upper and larger lower cycle of elements characterize *Gephyrorhabdus*. It differs from the older *Grantarhabdus* by the latter's larger upper cycle of distal shield elements. In both genera, the species are distinguished by the size and distribution of the openings in the central area. The solid stem distinguishes the two genera from *Axopodorhabdus* and *Tetrapodorhabdus*, which also have 4 openings, but a hollow stem.

Ge. coronadventis (Figs. 51.3, 20; syn. *Cretarhabdus unicornis*), the generotype, is a large, broad-elliptical form reported from the Middle Albian through the Campanian. *Gr. meddii* (Figs. 51.12, 13, 19), the generotype of *Grantarhabdus*, originates in the Berriasian.

Helenea

Helenea is regarded as a junior synonym of *Cretarhabdus* by Tappan (1980), but as a valid genus, characterized by a smaller upper cycle of radially arranged elements and a larger lower cycle of sickle-shaped, clockwise-curving elements by Grün & Allemann (1975). The central cross of the generotype, *H. staurolithina* (Figs. 51.5, 6, 26; syn. *Microstaurus lindensis* according to Grün & Allemann, 1975), is aligned with the axes of the ellipse. The complex structure of the distal shield distinguishes *Helenea* from *Cruciellipsis*, according to Grün & Allemann (1975).

Heterorhabdus

Heterorhabdus is a junior synonym of *Cretarhabdus* according to Tappan (1980), but differs from the latter by a much smaller central area with only one cycle with 8 perforations. The upper cycle of distal shield elements is smaller than the lower one and the two cycles can be seen connected to one cycle of oblique elements by overgrowth in the holotype of the generotype *H. sinuosus* (Fig. 51.23, from the Campanian). *Heterorhabdus* is very similar to *Stradneria* and might represent smaller representatives of the latter and thus be a junior synonym (see below). *H. primitivus* from the Maastrichtian has 8 perforations but no central cross.

Microstaurus

The central cross is aligned with the axes of the ellipse in the slightly broad-elliptical generotype of *Microstaurus*, *M. chiastius* (syn. *M. quadratus*; Figs. 51.7–9, 27, 28). The LO of *M. chiastius* is a well defined event in the Cenomanian Zone CC 10. The bars of the central cross have 'feet'; the short central process consists of few (4?) elements only and is easily visible with the LM. *Helenea* has curving distal shield elements in the upper cycle, whereas those in *Microstaurus* appear radially arranged. *Cruciellipsis* has no central process. According to Grün & Allemann (1975), *Helenea* was erected before *Microstaurus* and the latter before *Cruciellipsis*. Those authors who disagree with the separation of the above genera should use *Helenea* or go even further back to *Polypodorhabdus* or *Stradneria* (see below).

Miravetesina

Miravetesina from the early Early Cretaceous may be a predecessor of *Cretarhabdus* or a junior synonym of it. The central area of *Miravetesina* and the elements of the upper, smaller distal shield cycle are smaller than in *Cretarhabdus*, if the two type species are compared. The only species, *M. favula* (Fig. 51.32) ranges from the Early Berriasian through the Hauterivian or Barremian. Its central cross, which is aligned with the axes of the ellipse, is less prominent than in *C. conicus*, and no central process was observed.

Nephrolithus

Nephrolithus includes kidney-shaped coccoliths with a granular, perforated central area. The number of shield elements and perforations increases with the size of the coccolith (Perch-Nielsen, 1968). The upper cycle distal shield elements are longer than those of the lower cycle. The small forms are elliptical rather than kidney-shaped. *N. frequens* (Fig. 53.1, 2, 13; syn. see Perch-Nielsen, 1968), the generotype, appears in the Late Maastrichtian, where it defines the base of CC 26. It is common in high latitudes but very rare or absent in low latitudes. It can be distinguished from species of *Cribrosphaerella*, which also have a granular central area, by the thinner shield elements. *N. corystus* (Figs. 53.3, 4, 15) with a solid central stem, was also found in the Maastrichtian.

Polypodorhabdus

The genus *Polypodorhabdus* is characterized by a distal shield with a larger upper than lower cycle, a central cross topped by a solid stem and aligned with the axes of the ellipse and lateral bars leaving 3 to 10 elongate perforations in each quadrant. The FO of *P. escaigii* (Fig. 51.34), the generotype, is used as a stratigraphic event in the Callovian.

According to Barnard & Hay (1974), *P. escaigii* occurs before *A. rahla*, while Grün & Zweili (1980) found the opposite in Switzerland, and Thierstein (1976) reported *P. escaigii* only at the top of the Callovian. On the other hand, Penn *et al.* (1980) reported *P. escaigii* from the Bajocian of England, and *A. rahla* from the Callovian, at about the same level as Barnard & Hay (1974). The species probably most likely to be mistaken for *P. escaigii* is *Noellithina arcta* (Fig. 20.12), a form with a deeper and smaller central area than in *P. escaigii* but also with a central cross, short process and more or less radial bars in each quadrant. *N. arcta* was described from the Early Toarcian of France, where it is common. The assignment of *N. arcta* to the genus *Polypodorhabdus* seems questionable, since its central area is clearly depressed and a cycle of elements surrounds the central area distally, while this is not the case in the younger *P. escaigii*, which has a relatively flat central area. Black (1968) and Medd (1979) erected the species *P. madingleyensis* (Fig. 51.38, 20 bars), *P. paucisectus* (8–12 perforations) and *P. beckii* (Fig. 51.39, 4–8 bars) based on different numbers of bars or perforations. Grün & Zweili (1980) found specimens with 8–24 bars in their Jurassic material without being able to group them in the above-listed species.

Retecapsa

Retecapsa has a massive central cross of parallel elements along the axes of the ellipse and 4 lateral bars, thus leaving 8 perforations of slightly varying form and size. The central process is solid; this distinguishes *Retecapsa* from *Octocyclus* and *Octopodorhabdus*. The generotype, *R. brightonii* (Fig. 51.30), from the Hauterivian features a distinct ring of elements surrounding the central area and very long elements in the upper cycle of the distal shield. It is not clear from the illustration whether the elements of the upper or those of the lower cycle are longer. Black (1971a) distinguished several species using the size and form of the perforations and the lack of an additional ring of elements as criteria for the splitting.

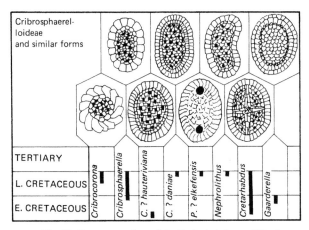

Fig. 52. Genus overview of the Podorhabdaceae III: the Cribrosphaerelloideae and other genera with a central area filled with small elements (cribroid)

Speetonia

Speetonia is characterized by its central bridge which is more or less aligned with the short axis of the ellipse and by having longer elements in the lower than in the upper cycle of the distal shield.

Black (1971a) erected a new family for this genus and was followed in this by Tappan (1980), but not by Grün & Allemann (1975), who included it in the Retecapsoideae.

Speetonia is very similar to *Grantarhabdus*, where the bridge bifurcates slightly to leave 2 large and 2 small openings. Whereas *Grantarhabdus* specimens have at least remains of a central process visible, none have been observed in *Speetonia*. *S. colligata* (syn. *Bipodorhabdus roegelii*, *S. nitida*; Figs. 51.10, 11, 18) appears in the Berriasian and has its LO in the Hauterivian, where it defines the base of Zone CC 5.

Stradneria

Stradneria has a conical central area with a central cross supporting a solid stem. A crown of relatively large elements surrounds the comparatively small distal part of the central area. The upper cycle of distal shield elements is smaller than the lower. *Stradneria* differs from *Cretarhabdus* in not having 2 or more concentric cycles of perforations in the central area. *S. limbicrassa* (Fig. 51.16), the generotype, described from the Maastrichtian, is larger than *S. crenulata* (Figs. 8.88, 89), a species described from the Late Maastrichtian. *Retecapsa angustiforata* (Fig. 51.29) the species used as a marker in the Berriasian, differs from the species in *Stradneria* and *Cretarhabdus* by having distinctly one cycle of 8 perforations in the central area. No distinct central cross is visible in *S. crenulata*.

Cribrosphaerelloideae

This subfamily includes coccoliths with a central area with a grid, but no process. The Cribrosphaerelloideae might be abandoned and the two genera it includes assigned to the Retecapsoideae, which already contains genera without a process.

Cribrocorona

The coccoliths of *Cribrocorona* are nearly as high as wide and have a central area spanned by a grid. *C. gallica* (Figs. 53.9–11), the generotype, ranges from the Late Campanian (?) through the Maastrichtian. *Cribrocorona* may have evolved from *Cribrosphaerella* during the Late Campanian.

Cribrosphaerella

Cribrosphaerella (syn. *Cretadiscus*, *Cribrosphaera*, *Favocentrum*) includes the round to polygonal to elliptical coccoliths with a central grid. The distal shield has an upper, larger cycle and a lower, smaller cycle. The proximal shield pierces the distal central area as a cycle of elements surrounding it. Many species have been distinguished by using the size of the coccoliths or their shape. The generotype, *C. ehrenbergii* s. ampl. (Figs. 53.7, 8, 16; syn. see Perch-Nielsen, 1968), shows an increasing number of shield elements and central area size with increasing size of the coccolith. *C. ehrenbergii* appears in the Albian and disappears at the end of the Cretaceous. It is quite common in the Campanian and Maastrichtian in high latitudes. *C.? daniae* (Figs. 53.5, 6, 14), a large species possibly belonging to this genus, is restricted to the high latitude Late Maastrichtian. *C.? hauteriviana* (Fig. 53.12) should be removed

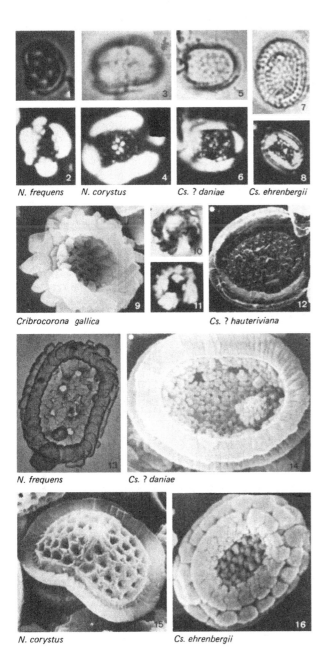

N. frequens   N. corystus   Cs. ? daniae   Cs. ehrenbergii

Cribrocorona gallica   Cs. ? hauteriviana

N. frequens   Cs. ? daniae

N. corystus   Cs. ehrenbergii

Fig. 53. LM and EM of representatives of the Cribrosphaerelloideae. LM: *c.* × 2000

from *Cribrosphaerella*, since its granulate central area is not surrounded by a podorhabdid type of shield.

Polycyclolithaceae Forchheimer (1972)
(Figures 54–58)

The Polycyclolithaceae contain mainly cylindrical and block, star or rosette-shaped, extinct nannoliths with or without obvious evolutionary links. Some may have covered a cell as did coccoliths of, for example, the Ellipsagelosphaeraceae, but others may have been included in a cell as, for example, the ceratoliths (Chapter 11). The Polycyclolithaceae (syn. Lithastrinaceae, Eprolithaceae) include genera such as *Eprolithus*, *Lithastrinus*, *Micula*, *Polycyclolithus*, *Quadrum* and *Radiolithus* for which relationships have been suggested (Prins *in* Perch-Nielsen, 1979a). For others like *Assipetra*, *Hayesites*, *Hexalithus*, *Pervilithus*, *Polycostella* and *Rucinolithus* relationships with the first group or among themselves seem questionable or even unlikely in some cases.

Many of the genera and species here attributed to the Polycyclolithaceae are shown in Fig. 54, where evolutionary links between the genera and species are indicated. Of the genera attributed to this family by Tappan (1980) *Cylindralithus* is here discussed with the Stephanolithiaceae, *Bukryaster* with the Braarudosphaeraceae and *Lapideacassis* is treated with the 'incertae sedis'. On the other hand, *Micula* and *Quadrum* are included (from the Sphenolithaceae) and so is *Hexalithus* (from the Braarudosphaeraceae). Many species of this family have been used as zonal markers or have short ranges and so are useful stratigraphically. The genera are discussed in approximate alphabetic order below.

Assipetra

Assipetra infracretacea (Figs. 55.1–6) is a nannolith 'having two sets of flat crystal plates, one set piercing the other at an obtuse angle'. In the LM, *A. infracretacea* appears as a subrectangular to suboval form with 4–6 radial sutures. It ranges from the Valanginian to the Aptian.

Eprolithus, *Lithastrinus*, *Polycyclolithus*, *Radiolithus* and *Rhombogyrus*

The genus *Eprolithus*, as characterized by its generotype, *E. floralis* (Figs. 56.19–24), includes more or less circular coccoliths with a wall and a wide central area located at approximately mid-height. The generotype of *Lithastrinus*, *L. grillii* (Figs. 56.1–4), features a narrow central area and the ray-shaped wall elements are arranged in two cycles joined at the bodice. In *Polycyclolithus* the central area is wide and the high wall consists of 9 pointed elements. The type species of *Radiolithus* (syn. *Rhombogyrus*), *R. planus* (Figs. 56.11–15), has a wide central area and a low wall. Crux (1982 and in Crux *et al.*, 1982) considered the three other genera to be synonymous with *Lithastrinus* while Perch-Nielsen (1979a) used all but

Polycyclolithus, since Forchheimer (1972) included *E. floralis*, the generotype of *Eprolithus*, in it.

E. antiquus (Figs. 56.25–27), *E. floralis*, *L. grillii* and *L. septenarius* (Figs. 56.5, 6) are used by various authors as markers or secondary markers in the Cretaceous. *E. antiquus* is characterized by a high number of wall elements (> 10). *E. floralis* and *E. apertior* both have 9 wall elements, but the wall is considerably higher in *E. apertior* than in *E. floralis*.

E. orbiculatus (Fig. 56.10) has a 'rim cycle composed of 9 zigzag elements' and a 'central area composed of 9 elements overlapping each other'. The difference from *E. floralis* is not clear. *E. septentrionalis* (Figs. 56.28, 29) from the Hauterivian has a narrow central area and 15–20 elements composing the wall. Its central area is considerably narrower than the one of the more or less contemporaneous *E. antiquus*.

Eprolithus sp. 1 (Figs. 56.17, 18) is considered to be a form between *E. floralis* and the first representative of the genus *Quadrum*, *Quadrum* sp. (Figs. 58.3, 32). *Eprolithus* sp. 2 has only 8 or 9 wall elements and a medium-sized central area and could represent the ancestor of the Late Cretaceous species of *Lithastrinus* (Fig. 56.9). *L. septenarius* (Figs. 56.5, 6), the 7-rayed species of *Lithastrinus*, appears in the Coniacian; *L. grillii*, the 6-rayed form, in the Santonian. Whereas the rays are long and slightly curved in the above two species, they are shorter in *L. moratus* (Figs. 56.7, 8), which appears as early as in the Cenomanian and features 7–9 rays.

Hayesites

Hayesites includes star-shaped nannoliths with 6–8 rays. *H. albiensis* (Figs. 55.10–13), the type species and a

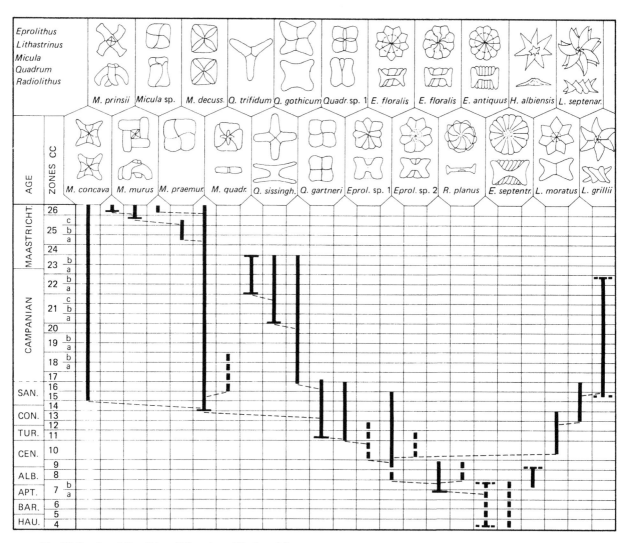

Fig. 54. Species of *Eprolithus, Lithastrinus, Micula* and ? related or similar genera and their stratigraphic distribution. Mainly after Perch-Nielsen (1979a)

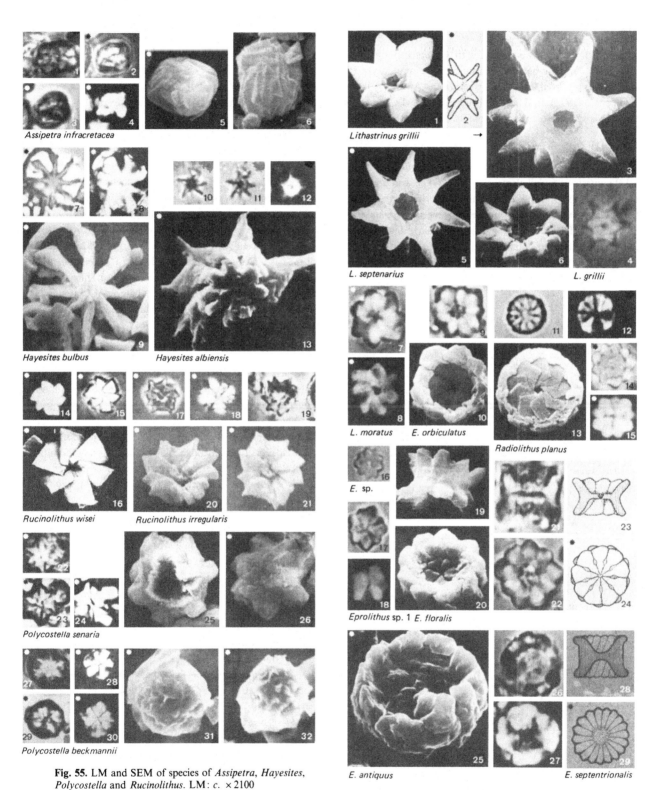

Fig. 55. LM and SEM of species of *Assipetra*, *Hayesites*, *Polycostella* and *Rucinolithus*. LM: *c.* ×2100

Fig. 56. LM and SEM of species of *Eprolithus*, *Lithastrinus* and *Radiolithus*. LM: *c.* ×2000

marker in the Albian, is conical. *H. bulbus* (Figs. 55.7–9), which is considered synonymous with *H. albiensis* by some authors, is flat and has long, distally expanding rays. *H. bulbosus* was reported from the Barremian and Aptian by Roth & Thierstein (1972). Both appear bright in the LM when lying flat and viewed between crossed nicols.

Hexalithus

The very small (1.7–3 μm) *Hexalithus gardetae* (Fig. 57.13) consists of 6 wall elements leaving no central area and has a relatively short range in the Campanian. The generotype, *H. lecaliae*, was described from the Miocene. *H. magharensis* (Figs. 5.31, 32) from the Bajocian features 6 triangular elements of different optical orientation and measures about 8–9 μm. *H. hexalithus* (Fig. 57.10) measures about 4 μm and was described from the Upper Jurassic.

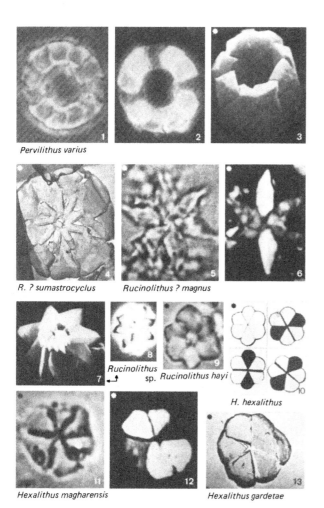

Pervilithus varius

R. ? sumastrocyclus   *Rucinolithus ? magnus*

Rucinolithus sp. *Rucinolithus hayi*

H. hexalithus

Hexalithus magharensis   *Hexalithus gardetae*

Fig. 57. LM and EM of species of *Hexalithus*, *Pervilithus* and *Rucinolithus*? LM: *c.* × 2000

Micula, Quadrum and Uniplanarius

Micula includes nannoliths consisting of interlocking calcite units. Early forms such as *M. decussata* (Figs. 58.6–12, 28), have the shape of a cube, later forms are conical or flat. *Quadrum* is characterized by one or two layers of usually 3 or 4 calcite blocks. The blocks do not interlock as in *Micula*. Roth & Bowdler (1979) considered *Quadrum* to represent overgrown specimens of *Micula*. Crux (1982) has illustrated a preservational series of *M. decussata* from etched to overgrown forms and shown that overgrown specimens of *M. decussata* do not resemble *Q. gartneri*. Hattner & Wise (1980) erected *Uniplanarius* for two forms previously assigned to *Quadrum*, arguing that its type, *Q. gartneri*, was an overgrown *Micula decussata*. Since it has been verified by Crux (1982) that this is not the case, *Uniplanarius* is superfluous. *M. decussata* is here used instead of the often cited *M. staurophora*, since the latter was so poorly described (from the Miocene) that it seems unlikely that it corresponds to what we find in the Late Cretaceous (see holotype in Fig. 58.5).

The distinction between *Q. gartneri* (*T. pyramidus* of Sissingh, 1977) and *M. decussata* is important, because the FO of *Q. gartneri* defines the base of CC 11 in the Early Turonian, while the FO of *M. decussata* defines the base of CC 14 in the Coniacian. *Q. gartneri* is assumed to be the ancestor of *M. decussata* (Fig. 54; Prins *in* Perch-Nielsen, 1979a) and itself has evolved from *Eprolithus floralis* by the loss of wall elements and by the closing of the central area. The distinction of the cubical forms can be made by the sutures, which are X-shaped in all *Micula* species but +-shaped in *Q. gartneri* and *Q. gothicum* (Figs. 58.17, 26). *Q. sissinghii* is characterized by two cycles of 4 long rays. *Q. sissinghii* (Fig. 58.19) is the *Q. nitidum* of previous authors; since the holotype of *Nannotetrina nitida* has been shown by Aubry (1983) to be a form restricted to the Middle Eocene, the Late Cretaceous forms have to be renamed. The species is often broken and appears then to have one cycle only. Specimens with short rays belong to *Q. gothicum*, which is more common and has an earlier FO. The FOs of *Q. sissinghii* and its younger 3-rayed successor, *Q. trifidum*, are used as zonal markers in the Late Campanian. However, they are not found in high latitudes.

M. decussata is cube-shaped and the centre of each side is crossed by sutures and is more or less funnel-shaped. *M. praemurus* (Figs. 58.13, 14, 21), *M. murus* (Figs. 58.15, 16, 22) and *M. prinsii* (Figs. 58.18, 23) evolved by the successive decrease of one cycle of elements. The latter is strongly bent and has bifurcations at the ends of the slender rays. The rays of *M. murus* are thicker than in *M. prinsii*. Both species are very rare or absent in high latitudes, but are useful markers for the Late Maastrichtian in low and mid latitudes. *M. concava* (Figs. 8.17, 18; 58.20) features strongly concave sides and the corners of the cube are extended to form horns. In some specimens, a minute central area can be seen. *M. cubiformis* (Fig. 58.27) may be a preservation stage of *M. decussata* or a predecessor of *M. concava*. It is cube-shaped with an appearance somewhat similar to a holococcolith, a situation often observed in *M. concava* also. Forchheimer (1972) gave its type level as Aptian, but the assemblage that she described must be regarded as Coniacian/Santonian due to the presence of *Marthasterites furcatus*.

Pervilithus

P. varius (Figs. 57.1–3), the only species of *Pervilithus*, is a cylindrical 'nannofossil with straight sides composed of 8–12 wall elements which are twisted in a spiral fashion'. Its range was given as Cenomanian–Campanian by Crux (1982).

Polycostella

The genus *Polycostella* was described as a 'flat conical pile of radially organised elements slightly sloping towards the proximal niche and forming 6–8 radial ridges on the distal side'. *P. senaria* (Figs. 55.22–26), the type species, has 6 ridges and is star-like. *P. beckmannii* (Figs. 55.27–32) is ball-shaped and has 6–8 ridges on the distal side. The two species have short ranges in the latest Jurassic/earliest Cretaceous, *P. beckmannii* appearing first (Fig. 7).

Rucinolithus

Rucinolithus includes 'calcareous nannofossils that have the appearance of a rosette of 5 or more inclined segments in

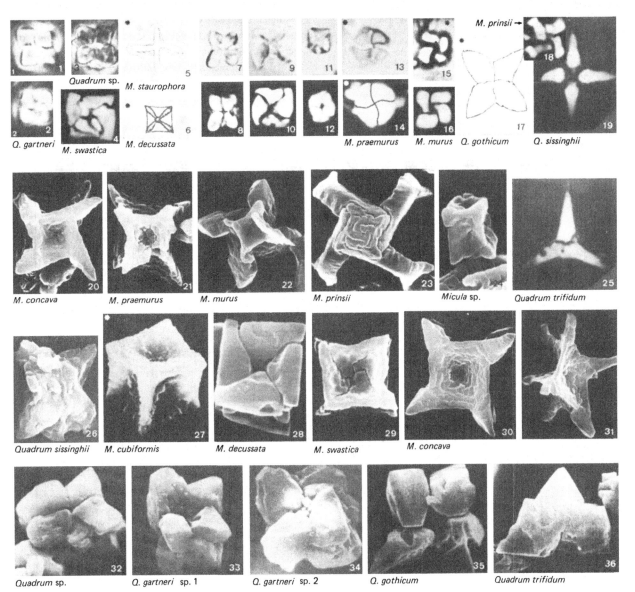

Fig. 58. LM and EM of species of *Micula* and *Quadrum*.
LM: *c.* × 2000

TERMINOLOGY PREDISCOSPHAERA

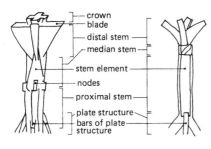

cretacea-type stem   spinosa-type stem

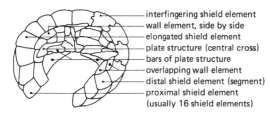

- interfingering shield element
- wall element, side by side
- elongated shield element
- plate structure (central cross)
- bars of plate structure
- overlapping wall element
- distal shield element (segment)
- proximal shield element
  (usually 16 shield elements)

Fig. 59. Terminology of *Prediscosphaera*

plan view'. Between crossed nicols each segment extinguishes separately like those of *Braarudosphaera*. The type species, *R. hayi* (Fig. 57.9), has 6 overlapping, slightly inclined elements. *R. wisei* (Figs. 55.14–16) has 6 strongly inclined elements forming a cone and *R. irregularis* (Figs. 55.17–21) has 9–11 imbricate elements forming a flat cone. *R.? magnus* (Figs. 57.5, 6) are 'large radial rosettes of 6 equant crystallites that are rhombic in outline and taper to points'; there is some doubt as to the biological origin of *R.? magnus* (Bukry, 1975). *R.? sumastrocyclus* (Fig. 57.4) is rosette-shaped and has 9–12 elements and an equal number of superimposed laths forming a small central star. Without those laths, such specimens resemble *Biantolithus sparsus* from the earliest Tertiary (Chapter 11, Fig. 11.11). *Rucinolithus* sp. (Figs. 57.7, 8) has '8 rays which radiate with only slight inclination and imbrication, a possible spine base is visible in distal view'. *Rucinolithus* sp. ranges from Santonian to Campanian, *R.? magnus* was described from the Campanian and *R.? sumastrocyclus* from the Santonian.

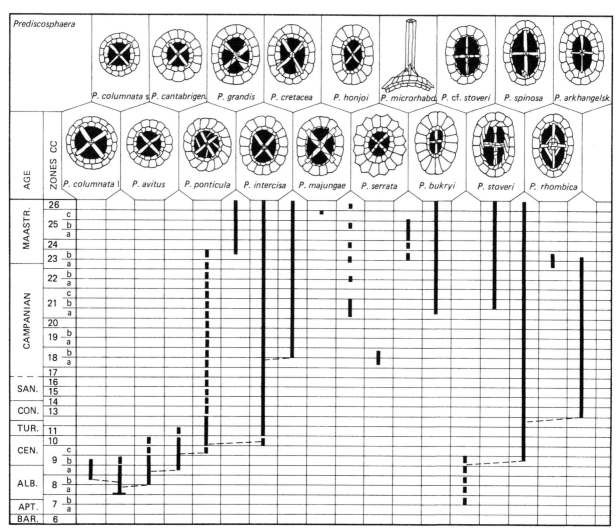

Fig. 60. Species of *Prediscosphaera* and their stratigraphic distribution. After Perch-Nielsen (1979a)

Prediscosphaeraceae Rood, Hay & Barnard (1971)

(Figures 59–61)

Prediscosphaera (syn. *Deflandrius*) is the only genus of this family which includes round and elliptical coccoliths with usually 16 elements to each of the two shields, a central + or

X and sometimes a central process. The species are distinguished by the orientation of the central structure, the relative widths of the wall and the distal shield, the shape of the coccolith and details of the construction of the wall and the central process. Unlike other genera, where the number of elements in the wall increases with the size of the specimen,

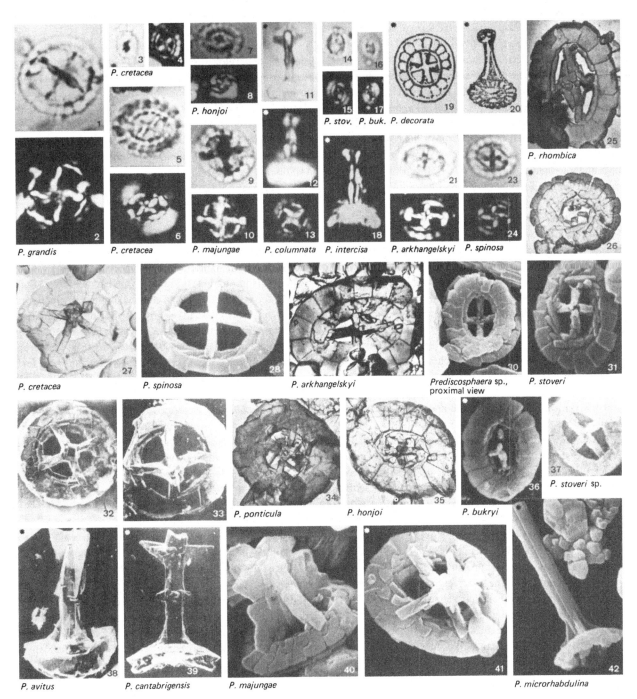

Fig. 61. LM and EM of species of *Prediscosphaera*. LM: *c.*
× 2000

both small and large *Prediscosphaera* have 16 elements. *P. cretacea* is generally used for coccoliths of this family with a central X, *P. spinosa* for those with a central +. However, in well preserved material, about a dozen additional species can be distinguished (Fig. 60) and some may be used for stratigraphy or for paleoenvironmental interpretations.

Early Cretaceous

The FO of *Prediscosphaerea* in the form of small, round specimens of *P. columnata* (Figs. 61.11–13) is used to define the base of CC 8 in the Early Albian. Later in CC 8 appear *P. avitus* (Figs. 61.32, 38), where the distal shield elements are wider than the wall elements, and larger specimens of *P. columnata*, *P. cantabrigensis* (Figs. 61.33, 39) appears in CC 9 in the Late Albian and is characterized by even more massive distal shield elements than in *P. avitus* and a slightly larger size than the latter. The first elliptical form appears in the Late Aptian and is assigned to *P.* cf. *P. stoveri* (Fig. 61.37). There is no ancestor known for either lineage of the genus (Fig. 60).

Late Cretaceous

In the Cenomanian, three new species appear: *P. ponticula* (Fig. 61.34) characterized by additional bars in the central structure and a subcircular shape, *P. intercisa* (Fig. 61.18), broad-elliptical with a narrow wall and a central X, and *P. spinosa* (Figs. 61.23, 24, 28), elliptical with a central + and a narrow wall. From the latter two species evolved two species with wider walls, *P. cretacea* (Figs. 61.3–6, 27) with a central X and *P. arkhangelskyi* (Figs. 61.21, 22, 29) with a central +. In the Late Campanian and Maastrichtian appear the very large *P. grandis* (Figs. 61.1, 2) and the very small *P. honjoi* (Figs. 61.7, 8, 35) with a central X, *P. bukryi* (Figs. 61.16, 17, 36) with a central + and a low wall and *P. stoveri* (syn. *P. germanica*; Figs. 61.14, 15, 26, 31) with a central + and a high, well developed wall. The FO of *P. stoveri* (? including *P. bukryi*) was used by Lambert (1980) to define a zonal boundary in the Late Campanian. *P. rhombica* (Fig. 61.25), a species with a central + and diagonal additional bars, was described from the Early Maastrichtian. *P. majungae* (Figs. 61.9, 10, 40, 41) is common in the Late Maastrichtian and is characterized by a central X, the bars of which overlap over the wall and the inner part of the distal shield. Several species of *Prediscosphaera* have a characteristic stem; the proximal stem is shorter in the *cretacea*-type stem than in the *spinosa*-type stem (Fig. 59) and the other way round for the median stem. *P. cantabrigensis* is characterized by prominent nodes. The coccoliths with heavy, short central processes (Fig. 61.40) that are typical for Late Maastrichtian assemblages can be assigned to *P. majungae*. Broken off distal stems of *cretacea*-type (Figs. 61.40, 41) can be and have sometimes been mistaken for species of *Micula* (see Polycyclolithaceae).

A complete list of the species assigned to *Prediscosphaera*, most of which have not been used since their description, is given in the alphabetic list of taxa.

Rhagodiscaceae Hay (1977)
(Figure 62)

The Rhagodiscaceae (syn. Actinozygaceae, Alfordiaceae) include elliptical coccoliths with a wall of inclined elements and a central area covered by small, approximately equal-sized granules.

Only one genus is here included in this family. The genera *Actinozygus*, *Alfordia*, *Rhabdolithina* and *Viminites* are regarded as superfluous, as suggested by Tappan (1980). *Reinhardtites* is removed from this family, since its generotype clearly has a central bridge and only evolves into a form with a completely covered central area during the Campanian. *Calcicalathina* could be added to the nominate genus *Rhagodiscus*, since it fits the requirements of this family about as well as it does those of the Crepidolithaceae, where it is discussed.

The genus *Rhagodiscus* is characterized by a relatively low wall of more or less inclined elements and a central area filled with granules.

Rhagodiscus angustus (Figs. 62.10–16) is used as a zonal marker in the Aptian. It is characterized by its long-elliptical outline and narrow bridge supporting a central process. *R. reniformis* (Figs. 62.1–4) is kidney-shaped and *R. asper* (Figs. 62.6, 7), the generotype, has an elliptical outline. The holotype of *R. elongatus* (Fig. 62.13) clearly has the same width/length proportions as the holotype of *R. angustus* (Fig. 62.10) and is considered a junior synonym. *R. eboracensis* (Fig. 62.5) from the Hauterivian has a wide central area without any base for a central process. Other Late Cretaceous species include *R. bispiralis*, which may represent a proximal view of *R. plebeius*, and *R. granulatus*, all described from the Maastrichtian. *R. plebeius* is characterized by 1 to 3 cycles of elements surrounding the central opening whereas the central area of *R. granulatus* is covered by unorderly arranged granules.

Schizosphaerellaceae Deflandre (1959)
(Figure 63)

The Schizosphaerellaceae include calcareous nannofossils consisting of subspherical tests constructed of two overlapping or interlocking hemispheres. The bipartite nature is unlike anything known in coccolithophores, but might represent an encysted or non-motile phase, similar to the braarudospheres according to Tappan (1980).

Schizosphaerella (syn. *Nannopatina*) is the only genus of the family and includes two species. *S. punctulata* (Figs. 63.1–4, 7, 8), is characterized by two subhemispherical valves, usually bell- or cup-like in shape. The valves are attached to each other by a simple hinge with a subperipheral groove. The ultrastructure of the wall was studied in detail by Aubry & Depeche (1974), Kälin (1980) and Kälin & Bernoulli (1984).

Elongate elements are arranged around a more or less quadrangular central void in pairs (Fig. 63.8). *S. punctulata* ranges from the latest Triassic to the end of the Jurassic and was most common in the Early Jurassic. *S. astraea* differs from *S. punctulata* by the simpler hinge which bears no subperipheral groove and by the arrangement of the radiating elongate elements building the wall (Figs. 63.5, 6, 9). *S. astraea* was reported in sediments of Sinemurian to Oxfordian age. Both species can be found in rock-forming quantities in various states of preservation.

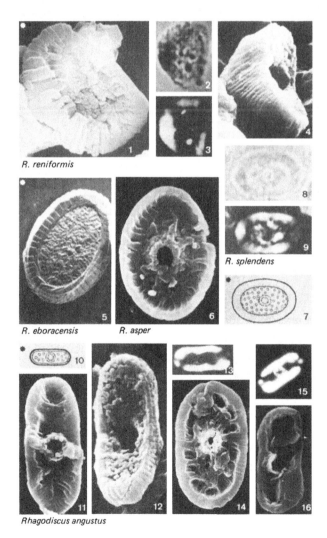

Fig. 62. LM and EM of species of *Rhagodiscus*. LM: *c.* × 2000

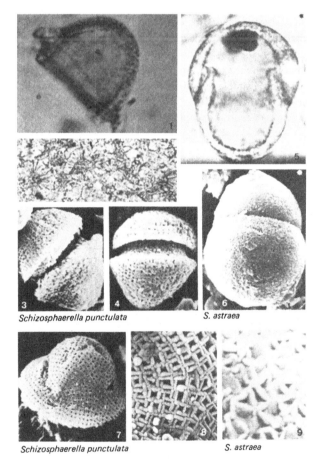

Fig. 63. LM and SEM of *Schizosphaerella astraea* and *S. punctulata*. LM: × 2000. Fig. 2: *c.* × 660

Sollasitaceae Black (1971a)
(Figures 64, 65)

The Sollasitaceae include elliptical coccoliths with two shields and a 'large central opening occupied by a grid or meshwork without a spine or central knob'. The family originally included *Sollasites* and *Gaarderella*. *Sollasites* has been assigned to different families by different authors: Biscutaceae by Grün *et al.* (1974) and Grün & Zweili (1980), Podorhabdaceae by Noël (1973), Rood *et al.* (1973) and Medd (1979) and to the Syracosphaeraceae by Grün & Allemann (1975). Whereas some Jurassic specimens seem nearer to the Podorhabdaceae, the Cretaceous ones show features of the Ellipsagelosphaeraceae. It has been suggested by Perch-Nielsen (1981) that *Sollasites* is the ancestor of the Tertiary genus *Cruciplacolithus*. Cretaceous forms very similar to *Cruciplacolithus* were assigned to that genus by Black (1971a) and others and are here assigned to *Sollasites*.

Most species of *Sollasites* (syn. *Costacentrum*) are difficult to distinguish (Fig. 64) with the LM, and thus some have probably been given too long a range. *S. falklandensis* has been used as a marker in the high southern latitude zonal scheme by Wise (1983; Fig. 6). Fig. 64 shows the species of *Gaarderella* and *Sollasites*, their structural differences and their tentative ranges. The species are distinguished by the different form and number of the bars additional to the central cross, which is aligned with the axes. *S. hayi* (Fig. 65.10) and *S. pinnatus* (Fig. 65.9) have a simple central cross and no additional bars; thus they do not fit exactly into the original definition of *Sollasites* as having a central opening spanned by one cross-bar along the shorter diameter and several bars along the length of the opening. In *S. pinnatus*, of which only a proximal view has been illustrated by Black (1971a), one could regard the two closely joined rows of elements of the bars as 2 bars but no such compromise can be used for *S. hayi*, where the bars are built of only one row of elements. The reason for assigning the two species to *Sollasites* rather than retaining them in *Cruciplacolithus* is because *Cruciplacolithus* has a proximal shield composed of two cycles, a feature not observed in the Cretaceous coccoliths. *S. pinnatus* was assigned to the genus *Seribiscutum* by Wise (1983). *Sollasites* should be used for forms with one or several longitudinal bars and a simple proximal shield (Jurassic and Cretaceous forms) and *Cruciplacolithus* for forms with one or several longitudinal bars and a double proximal shield (Tertiary forms). The presence of a species with additional more or less longitudinal bars in *Cruciplacolithus*, as for example *C. inseadus*, further supports the suggestion that *Sollasites* was the ancestor of *Cruciplacolithus*.

central cross	−	−	+	+	+	double	simple	+	+	wide	not distinct	−
additional bars	2, converging	4, undulating	4, converging	2, converging	2, diverging	−	−	2, parallel	2, parallel	"plate"	"plate"	− granules
size (μ)	∼3	∼3	2-3	5-6	5-7			∼3	5-6	3.5-5	10	8-13
Sollasites	S. pristinus	S. bipolaris	S. concentr.	S. lowei	S. arcuatus	S. pinnatus	S. hayi	S. horticus	S. barringt.	S. thierstein	S. falkland.	G. granulif.
L. CRET.												
E. CRET.												
L. JURASSIC												
M. JURASSIC												
E. JURASSIC												

Fig. 64. The species of *Sollasites* and *Gaarderella*, their ranges and characteristic features

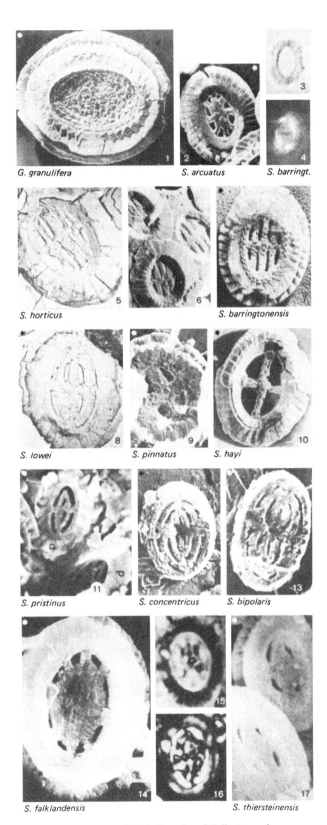

G. granulifera   *S. arcuatus*   *S. barringt.*

S. horticus   *S. barringtonensis*

S. lowei   *S. pinnatus*   *S. hayi*

S. pristinus   *S. concentricus*   *S. bipolaris*

S. falklandensis   *S. thiersteinensis*

Fig. 65. LM and EM of species of *Sollasites* and *Gaarderella*. LM: *c.* × 2000

Stephanolithiaceae Black (1968) and related genera
(Figures 66–80)

The Stephanolithiaceae are characterized by round, elliptical or polygonal coccoliths with an outer wall of vertically arranged elements. The central area is spanned by radially or otherwise arranged elements, which sometimes support a central process. Radial spines may project from the marginal ring or wall. The genera are distinguished by the outline, the composition of the wall, and the presence of spines; the species are mainly distinguished by the different arrangements of the bars in the central area.

The genera here included in the Stephanolithiaceae are more numerous than suggested by several authors. When an attempt was made to find the relationships between the various genera and to find their origin, it became evident that some genera usually not included in this family may be related to it. This is the case of *Scapholithus*, a genus known from the Hauterivian through the Pleistocene and with a Recent descendant, *Anoplosolenia* (see Calciosoleniaceae).

Fig. 66 is an overview of the morphological characteristics of the genera of, and related to, the Stephanolithiaceae. Evolutionary trends are also suggested. These are highly speculative, since they are based on the sequence of first occurrences as known to date, and not on the investigation of continuous sequences where the evolution could be followed step by step. The fossil record of the Jurassic and the Cretaceous calcareous nannoplankton is not readily available in continuous sections. Instead, we have to work with relatively few samples from areas relatively far apart both vertically and laterally. The attempt was made despite the shortcomings of this method, because in doing so and presenting the results, discussion may start and encourage more research to find the real relationships or prove the ones suggested here. Tappan (1980) also included the genera *Angulofenestrellithus* and *Pontilithus* in this family. This is not followed here, because the rim in both genera consists of strongly inclined elements.

The various genera included in the Stephanolithiaceae are treated individually below. Some of them have provided zonal markers in the Jurassic, some in the Cretaceous, but their biostratigraphic potential seems far from exhausted.

Corollithion (Figs. 68, 69)

Corollithion was defined by Stradner (1962) as a genus with hexagonal coccoliths and six radial bars as in its type species, *C. exiguum* (Figs. 69.5–7). Subsequently, Stradner (1963) also included a hexagonal form with only 4 radial bars in *Corollithion, C. signum* (Figs. 69.12, 13). Other authors have since included elliptical forms with 4 or 6 bars in *Corollithion*, when they had a wall of more or less vertical elements. Since there are other genera to accommodate these non-hexagonal forms, I prefer to limit *Corollithion* to hexagonal forms and

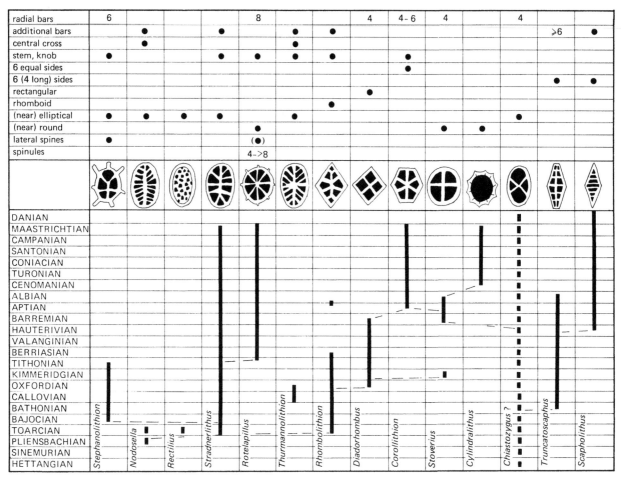

Fig. 66. Overview of the genera of the Stephanolithiaceae and similar genera, their ranges and characteristic features

TERMINOLOGY STEPHANOLITHIACEAE

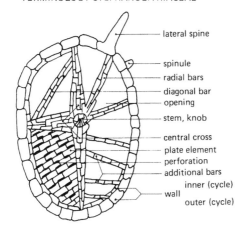

Fig. 67. Terminology of the Stephanolithiaceae

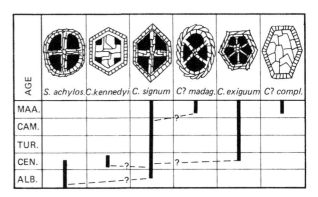

Fig. 68. The species of *Corollithion* and *Stoverius achylosus* and their ranges

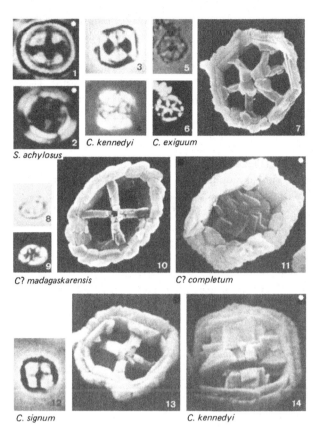

Fig. 69. LM and SEM of species of *Corollithion* and *Stoverius achylosus*. LM: *c.* ×2100

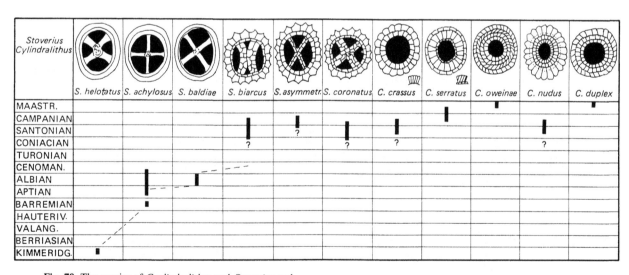

Fig. 70. The species of *Cyclindralithus* and *Stoverius* and their ranges

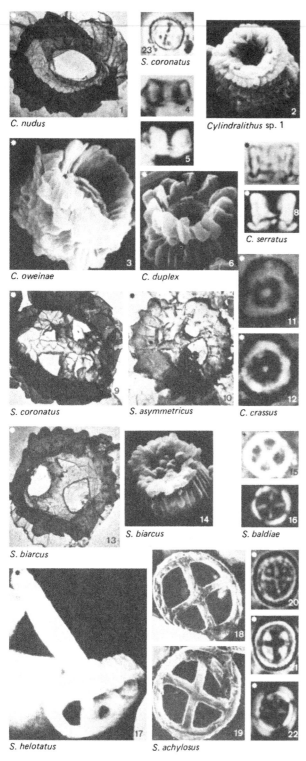

Fig. 71. LM and EM of species of *Cylindralithus* and *Stoverius*. LM: *c.* ×2000. + = *C. baldiae* from type material. * = holotype

thereby to the Cretaceous. In this case, there are only two more species which may belong to *Corollithion*, *C.? completum* (Fig. 69.11) and *C. kennedyi* (Figs. 69.3, 4, 14). The round to subcircular *Stoverius achylosus* (Figs. 69.1, 2), which has a double wall similar to *C. signum*, may be its ancestor. There seems to be no obvious relationship between *Corollithion* and *Truncatoscaphus*, where we see an increase in the number of bars from the first forms with 6 bars in the Bathonian to the last forms with more than 18 bars in the Albian. *C.? madagaskarensis* (Figs. 69.8–10) is elliptical like the forms in *Stradnerlithus*, but has a double wall like those of *Corollithion*.

Cylindralithus (Figs. 70, 71)

Cylindralithus includes the more or less circular forms with a relatively high wall of more or less vertical elements. The generotype, *C. serratus* (Figs. 71.7, 8), has an open central area, but several species assigned to the genus by Bukry (1969) have a central structure consisting of 4 bars. These forms have recently been assembled in *Stoverius*, and thus the genus is restricted again to forms with an open central area.

Diadorhombus (Figs. 72, 73)

Diadorhombus includes coccoliths with a rhomboid to square rim and an open central area spanned by bars aligned with the sides of the rim. This rather narrow definition of the genus results in only very few species having been included in it. These are shown in Figs. 72 and 73. *D. scutulatus* (Fig. 73.15) is smaller than *D. rectus* (Fig. 73.13) (1.5–2.1 μm against 6 μm). *D. minutus* may be a junior synonym of *S. scutulatus*, but two of its bars are not parallel to the rim.

Diadozygus

Diadozygus includes the generotype of *Stradnerlithus* and is thus its junior synonym.

Nodosella (Figs. 72, 73)

The genus *Nodosella* differs from *Stradnerlithus* by the increased number of bars in the central area. Goy *et al.* (1979) have re-defined the generotype *N. clatriata* (Fig. 73.18) and included it in *Stradnerlithus* since they felt that there was no need for this genus. I would like to use it to accommodate some species that were previously assigned to *Corollithion*, but which do not correspond to the narrower definition of that genus used here: these are *Nodosella silvaradion* (Fig. 73.17) and *Nodosella perch-nielseniae* (Fig. 73.16). In *N. silvaradion* the bars of the central area join in the centre of the coccolith to form a broad central platform sometimes supporting a central process. In *N. perch-nielseniae* a large diamond-shaped central area platform is attached to the rim by a central cross.

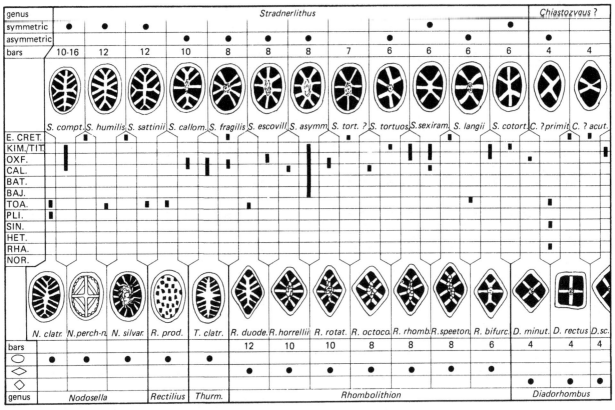

Fig. 72. The species of *Chiastozygus?*, *Diadorhombus*, *Nodosella*, *Rectilius*, *Rhombolithion*, *Stradnerlithus* and *Thurmannolithion*, the number of their bars and their tentative ranges

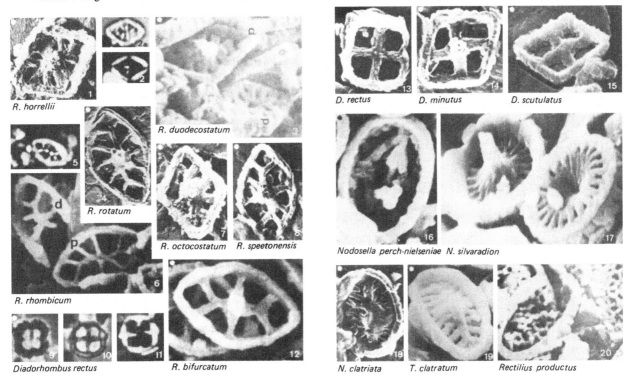

Fig. 73. LM and EM of species of *Diadorhombus*, *Nodosella*, *Rectilius*, *Rhombolithion* and *Thurmannolithion*. LM: *c.* × 2000

Rectilius (Figs. 72. 73)

Rectilius includes those forms with vertical wall elements, which have a central area spanned by a net-like structure, with perforations showing a polygonal outline. The generotype *R. productus* (Fig. 73.20) was described from the Toarcian.

wall	diverging	diverging	vertical	diverging	± vertical	± vertical
spinules	—	~8	~8	~ 8	>8	4-5
Rotela-pillus						
MAA.						
CAM.						
SAN.						
CON.						
TUR.						
CEN.						
ALB.						
APT.						
BAR.						
HAU.						
VAL.						
BER.						
TIT.						

R. radians · *R. caravacaensis* · *R. laffittei* · *R. crenulatus* · *R. octoradiatus* · *R. munitus*

Fig. 74. The species of *Rotelapillus*, their ranges and characteristics

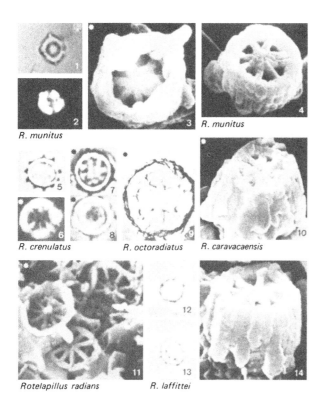

R. munitus

R. munitus

R. crenulatus   R. octoradiatus   R. caravacaensis

Rotelapillus radians   R. laffittei

Fig. 75. LM and EM of species of *Rotelapillus*. LM: *c*. × 2000

Rhombolithion (Figs. 72, 73)

The species of *Rhombolithion* include more or less rhomboid forms with a varying number of bars and a short central spine or knob. They are shown together with the species of *Stradnerlithus* in Fig. 72, since many authors consider the outline of a coccolith – rhomboid, hexagonal or elliptical – not to be a generic character. The number of bars define the species. Details of the arrangement of the bars are used to distinguish between species with an equal number of bars (e.g. symmetry or lack of it).

The oldest form belonging to this genus is *R. duodecostatum* (Fig. 73.3) described from the Early Toarcian. It has 12 bars. Forms with 10 bars are reported from the Callovian and Early Oxfordian, such with 8 bars from the Callovian through the Kimmeridgian and the Early Cretaceous and (reworked?) from the Late Cretaceous. Forms with 6 bars seem confined to the Oxfordian and Kimmeridgian. There thus seems to be a trend in the evolution of the species of *Rhombolithion* from forms with many bars in the Toarcian to forms with fewer bars in the Kimmeridgian, with only the forms with 8 arms passing the Jurassic/Cretaceous boundary. If the forms of *Diadorhombus*, with only 4 bars are included in the proposed evolutionary trend, we see those forms coming in at the same time as the 6-barred forms in the Early Oxfordian.

If the elliptical forms here assigned to *Stradnerlithus* are taken into account as well, no such trend as suggested above can be seen. In the Toarcian there are elliptical forms with more than 16, with 14, 12, 6 and even 4 bars, the oldest being a form with 4 bars from the Sinemurian, the next a form with more than 16 bars from the Pliensbachian.

Rotelapillus (Figs. 74, 75)

Noël (1973) erected *Rotelapillus* for broad-elliptical to round coccoliths with a narrow wall consisting of subvertical elements, a ring of proximal elements and a large central area with radially oriented bars and a central process. This definition also can include those forms usually assigned to *Stephanolithion*, which have a high rim and are here included in *Rotelapillus*: *R. laffittei*, *R. caravacaensis*, *R. crenulatus* and *R. munitus* (Figs. 74 and 75).

Whereas the Late Jurassic generotype of *Rotelapillus* has a wall of equally thin but not equally high elements, the walls of the other species, which are of Cretaceous age, have a wall of some thinner and some thicker elements, some of which develop into spines. Fig. 74 shows the tentative ranges of the species of *Rotelapillus* and the criteria used for their distinction. Note that all forms have 8 bars, but that the wall is more or less vertical in *R. laffittei*, while it diverges in *R. caravacaensis*, that there are 8 extended proximal elements (spinules) in *R. crenulatus* and only 4 to 5 extended wall-elements in *R. munitus*. The Late Cretaceous form *R. octoradiatus* seems to be

spines	? >10	>10	>10	>10	~8	6-10	6(-7)	6(-7)
radial bars	10-16	8	7-8	8	8	6	4	4
length in microns	~3	~4.5	3-4	>5	~6	7-8	3-5	>6
Stephanolithion	*Stradnerlithus comptus*	*S. elongatum*	*S. carinatum*	*S. speciosum speciosum*	*S. octum*	*S. hexum*	*S. bigotii bigotii*	*S. bigotii maximum*
TITHONIAN								
KIMMERIDGIAN								
OXFORDIAN								
CALLOVIAN								
BATHONIAN								
BAJOCIAN								
TOARCIAN								

Fig. 76. *Stradnerlithus comptus* and the species of *Stephanolithion*, their ranges and characteristics

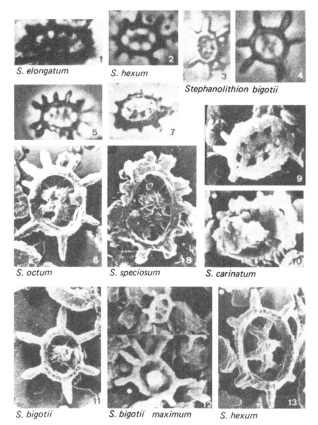

S. elongatum — 1 S. hexum — 2 — 3 — 4
Stephanolithion bigotii
— 5 — 7 — 9
S. octum — 6 S. speciosum — 8 S. carinatum — 10
S. bigotii — 11 S. bigotii maximum — 12 S. hexum — 13

Fig. 77. LM and EM of species of *Stephanolothion*. LM: *c.* ×2000

an intermediate form between *R. laffittei* and *R. munitus*, with fewer spinules than *R. laffittei* but with more than *R. munitus*, where they also are more prominent.

The species of *Rotelapillus* are best studied with phase contrast, but can also be distinguished with parallel/crossed nicols. The FOs of *R. laffittei* and *R. caravacaensis* occur very late in the youngest Jurassic and have been used to define the Jurassic/Cretaceous boundary.

Stephanolithion (Figs. 76, 77)

Stephanolithion is characterized by an elliptical to polygonal outline, more or less radially oriented lateral spines, and a central structure with a central knob or short process and more or less radially oriented additional bars (Figs. 67, 76). The number of bars and/or spines defines the species. Fig. 77 shows the four species of *Stephanolithion* used as zonal markers in the Middle and Late Jurassic, *S. elongatum*, *S. octum*, *S. hexum* and *S. bigotii* together with the other species of the genus. Not shown are the forms previously assigned to *Stephanolithion* but now regarded as belonging to the emended genus *Rotelapillus*, *R. laffittei*, *R. munitus*, *R. caravacaensis* and *R. crenulatus*.

With the LM, the usually rather small (3–6 μm) species of *Stephanolithion* are best observed with phase contrast.

The FO of *S. speciosum*, the oldest species of *Stephanolithion*, occurred in the Early Bajocian. No evident ancestor is yet known, but I would suggest a relationship to *Stradnerlithus comptus* as illustrated from the Toarcian by Goy (1981, pl. 12,

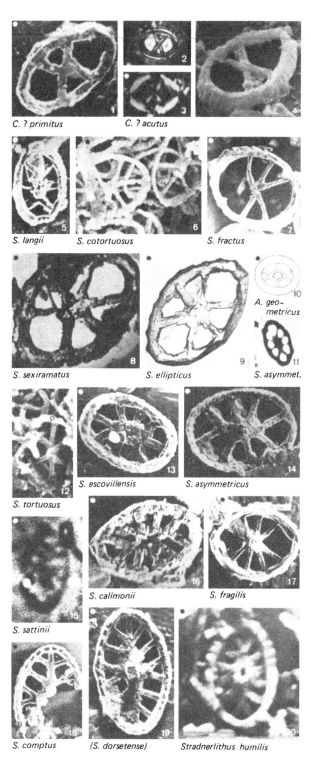

Fig. 78. LM and EM of species of *Actinozygus geometricus*, *Chiastozygus*? and *Stradnerlithus*. LM: c. × 2000

fig. 7), where some of the rim elements seem to have developed short extensions which could have evolved into spines by Bajocian times. At the same time, a reduction of bars in the central area should have taken place, from the 10 to 16 in *S. comptus* to 8 in *S. speciosum*. In the lineage leading to *S. bigotii*, the number of bars and spines decreases to 4 bars and 6 spines. There is no agreement yet whether *S. octum* or *S. hexum* appears first. Whereas Barnard & Hay (1974) suggested *S. octum* as the ancestor of *S. hexum*, Medd (1979) observed the FO of *S. hexum* below the level in the Bathonian where Barnard & Hay (1974) had found it, and thus indirectly below the FO of *S. octum*. If the trend from many spines and bars to fewer spines and bars worked also in the Bathonian, it would be more likely that *S. hexum* evolved from *S. octum* than the other way round. On the other hand Moshkovitz & Ehrlich (1976a) also reported *S. hexum* below *S. octum* and thus their use as zonal markers seems premature except maybe for local correlations in one basin.

Stoverius (Figs. 70, 71)

Stoverius includes species with a wall of more or less vertical elements, having a central structure of 4 bars and a round to broad-elliptical outline. *Stoverius* differs from *Cylindralithus* by featuring a central structure. In the Late Jurassic and Early Cretaceous forms such as *S. helotatus*, *S. achylosus* and *S. baldiae*, the wall is still relatively low and smooth; in the Late Cretaceous species *S. biarcus*, *S. asymmetricus* and *S. coronatus* the wall becomes higher and forms a serrate outline. The species are distinguished by the outline (smooth or serrate) and by the arrangement of the bars in the central cross.

Stradnerlithus (Figs. 72, 78)

Stradnerlithus (syn. *Diadozygus*) is characterized by an elliptical outline, and the varying number of bars in differing arrangements in the central area is used to define the species. A central knob or stem is present in most species. Its elliptical form is the characteristic used to differentiate it from *Rhombolithion*. This is not supported by all authors, since it is sometimes difficult to observe whether a form is rhombohedral or elliptical with somewhat accentuated 'corners'. Both genera are therefore shown together in Fig. 72. The oldest and most primitive species so far could be the Sinemurian *Chiastozygus*? *primitus* (*Ellipsochiastus primitus* in Jafar, 1983) with 4 bars, evolving into *S. langii* with its 6 bars in the Toarcian. But that cannot account for the up to 16 bars found in *S. comptus* from the Toarcian or the 14 and more bars in *Nodesella clatriata* (see above). *S. comptus*, which has 14 bars was originally described from the Late Kimmeridgian. It has not yet been illustrated from samples older than Early Oxfordian with the exception of the Toarcian occurrences reported by Goy *et al.* (1979). These authors found forms with 8 (*Stradnerlithus* sp., pl. 11, fig. 3) to more than 22 bars (*Stradnerlithus clatriatus*, pl. 11, figs. 10, 11) in the Toarcian. With these observations, any attempt to use the number of bars for species determination in *Stradnerlithus* seems questionable. On the other hand, we cannot make further progress if all forms are lumped in *S. comptus* without any mention of the number of bars.

The distribution of the species of *Stradnerlithus* in the Cretaceous is not yet well known. Since there are usually many much larger and more easily determinable forms at hand in the Cretaceous, *Stradnerlithus* was probably not always reported, even when present. We know that some species, especially those with relatively few bars, survived into the Cretaceous. Thus Wind & Wise *in* Wise & Wind (1977) reported specimens with 6, 7 and 8 bars from the Aptian of DSDP Site 327 in the South Atlantic.

Two species now assigned to *Stradnerlithus* have been used for the zonal scheme proposed by Barnard & Hay (1974): *S. comptus* (their *D. dorsetense*) and *S. sexiramatus* (their *A. geometricus*). Since *S. comptus* has since been reported from as low as the Toarcian by Goy *et al.* (1979), it seems a bad choice for a marker at the base of the Callovian, as suggested by Barnard & Hay (1974). The same applies for *S. sexiramatus*, where other 6-bar forms, labelled as *S. langii*, have been reported from the Toarcian by Rood & Barnard (1972). *S. langii* is, however, only about half the size of *S. sexiramatus* and the larger forms have not yet been reported from below the Oxfordian. *S. sexiramatus* thus seems useful after all.

Thurmannolithion (Figs. 72, 73)

Thurmannolithion includes elliptical coccoliths with a central cross aligned with the axes and supported by additional bars. In *Stradnerlithus* and *Nodosella* either the short axis or

the long axis or both are not paralleled by the central structure. *T. clatratum* (Figs. 72; 73.19), the generotype and only species, ranges from the Callovian to the Oxfordian.

Truncatoscaphus (Figs. 79, 80)

Truncatoscaphus includes forms with an 'eiffellithoid' rim and truncate ends, the longer sides being curved or broadly obtusely angled. The species are defined mainly by the number of bars in the central area. Several authors have incorporated *Truncatoscaphus* species in various other genera of the Stephanolithiaceae. The genus is used here, since it allows the grouping of several species into an evolutionary lineage: from *T. hexaporus* (Fig. 80.1) in the Bathonian through *T. octoporus* (Figs. 80.3, 4) in the Bathonian/Oxfordian to *T. pauciramosus* (Figs. 80.8, 9) in the Oxfordian/Kimmeridgian. In *T. pauciramosus* there are 10 openings, in *T. intermedius* 14 (Fig. 10.10) and in *T. delftensis* (Figs. 80.5–7, 11) over 14. The latter is the generotype of the genus which was described from the Albian. In it some of the bars seem to be arranged in pairs. In *T. senarius* (Fig. 80.2), the truncated ends do not form a right angle to the longer axis of the coccolith as in all other species. No ancestor of *Truncatoscaphus* is known. *Scapholithus* may have evolved from it.

Forms of *Truncatoscaphus* are best observed with phase contrast illumination if no EM is available, since the bars have to be counted to make a species determination. The species of *Truncatoscaphus* have not been used as zonal markers so far, but could prove useful in the future, since they are easy to determine.

number of bars	6	6	8	10	14	>14
Truncatoscaphus						
E. CRETACEOUS				■ HT		■HT
TITHONIAN KIMMERIDGIAN						
OXFORDIAN						
CALLOVIAN						
BATHONIAN						
BAJOCIAN	*T. hexaporus*	*T. senarius*	*T. octoporus*	*T. pauciramosus*	*T. intermedius*	*T. delftensis*

Fig. 79. The species of *Truncatoscaphus*, their ranges and characteristics. HT = Holotype

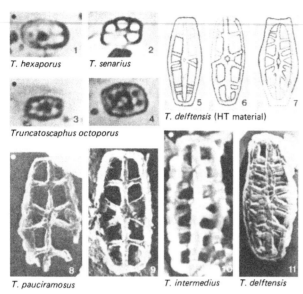

T. hexaporus   T. senarius

Truncatoscaphus octoporus   T. delftensis (HT material)

T. pauciramosus   T. intermedius   T. delftensis

Fig. 80. LM and EM of species of *Truncatoscaphus*. LM: *c.* ×2000

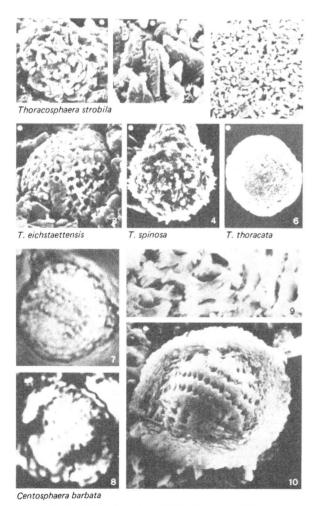

Thoracosphaera strobila

T. eichstaettensis   T. spinosa   T. thoracata

Centosphaera barbata

Fig. 81. SEM of species of Early Cretaceous *Thoracosphaera* and LM and SEM of *Centosphaera barbata*. LM: *c.* ×2000

Thoracosphaeraceae Schiller (1930)
(Figure **81**)

Three genera have been included in this family by Tappan (1980): *Thoracosphaera*, *Centosphaera* and *Brachiolithus*. Of these, *Brachiolithus* is a junior synonym of *Nannoconus*. *Centosphaera* was described by Wind & Wise *in* Wise & Wind (1977) from the Maastrichtian and includes large calcareous spheres constructed of hourglass-shaped blocks arranged in a patchwork pattern. The sphere is ringed by one or more keels composed of needle-like crystals.

Thoracosphaera, a calcareous dinoflagellate of nannofossil dimensions, is discussed further in Chapter 11. It is usually rare in Jurassic and Cretaceous sediments. *T. operculata* and *T. saxea* (see Chapter 11, Figs. 75.5, 6, 16, 17) are found mainly in Upper Cretaceous sediments. *T. eichstaettensis* (Fig. 81.3), *T. spinosa* (Fig. 81.4), *T. strobila* (Figs. 81.1, 2) and *T. thoracata* (Figs. 81.5, 6) were described from the Lower Cretaceous. *Thoracosphaera* becomes common and even dominant at the base of the Tertiary. Clearly, the absence of most coccolithophorids after the Cretaceous/Tertiary boundary events allowed for the bloom of this genus or rather one of its species, *T. operculata*.

Zygodiscaceae Hay & Mohler (1967)
(Figures **82–84**)

The Zygodiscaceae, which are based on the Paleocene/Eocene genus *Zygodiscus*, are here used only for the genera with a zeugoid wall of one or two cycles of inclined elements and a bridge aligned with the shorter axis. Genera with a zeugoid wall and a central + are grouped in the Ahmuellerellaceae and those with a central X in the Chiastozygaceae, whereas they are also included in the Zygodiscaceae in the discussion of the Cenozoic families (Chapter 11) (see also the introduction to these notes).

The seven genera briefly presented below have coccoliths with a wall of inclined elements and a bridge along the short axis of the ellipse. Only two are easily distinguishable. Coccoliths with massive blocks covering parts or all of the central area can be grouped in *Tranolithus*. Those with a simple bridge and none or some cover of the central area are grouped in *Zeugrhabdotus*. This is the oldest of the other genera (1965) if *Glaukolithus* (1964) is disregarded because of the difficulty in establishing the true nature of its generotype. Crux (1982) used only *Reinhardtites*, *Tranolithus* and *Zygodiscus*, the latter genus including mainly the species assigned by many other authors to *Zeugrhabdotus*, *Glaukolithus* and *Placozygus*. Since the generotype of *Zygodiscus*, *Z. adamas*, is a typical Paleocene/Eocene coccolith with a thin, high wall of two cycles and a thin floor and can easily be distinguished from the more sturdy Mesozoic forms, *Zygodiscus* should not be used for those. Finally, *Loxolithus* has no bridge, but consists only of a simple ring.

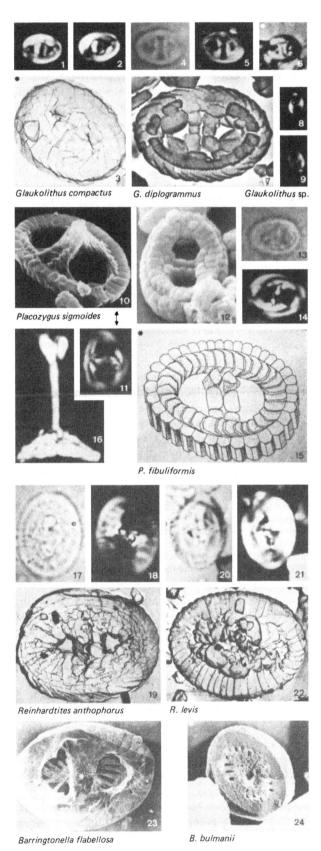

Glaukolithus compactus   G. diplogrammus   Glaukolithus sp.

Placozygus sigmoides ↕

P. fibuliformis

Reinhardtites anthophorus   R. levis

Barringtonella flabellosa   B. bulmanii

Fig. 82. LM and EM of species of *Barringtonella,*
Glaukolithus, Placozygus and *Reinhardtites.* LM: *c.* × 2000

Barringtonella

The proximal grid that characterizes *Barringtonella* consists of slender bars which converge from the wall towards the centre. A bridge along the short axis is visible on the distal side. Some well preserved specimens of *Tranolithus* have been found to have a similar proximal grid (Fig. 83.19). According to Black (1973), *Pontilithus,* another genus with a similar proximal grid, has a distinct central cross (see Ahmuellerellaceae). *B. bulmanii* (Fig. 82.24) and *B. flabellosa* (Fig. 82.23) were described from the Albian.

Glaukolithus

The generotype of *Glaukolithus, G. diplogrammus* (Figs. 82.4–7) features a bridge consisting of two parallel bars which meet the wall without additional lateral elements as, for example, seen in *Placozygus sigmoides* (Figs. 82.10, 11). It was described from the Neogene but is assumed to be reworked and has been reported from the Valanginian to the Maastrichtian. *G. compactus* (Figs. 82.1–3) has a more massive bridge consisting of interfingering elements. It is most common in the Late Cretaceous but is found from the Barremian onwards according to Crux (1982).

Loxolithus

The generotype and only species of *Loxolithus, L. armilla,* consists of a simple ring of inclined elements. The genus could thus be assigned to any of the families with coccoliths featuring a ring of inclined elements, but was assigned to the Zygodiscaceae in Tappan (1980). *L. armilla* has been reported from the Oxfordian to the Maastrichtian.

Placozygus

P. fibuliformis (Figs. 82.12–15; syn. *P. spiralis*?), the generotype of *Placozygus,* is characterized by a first wall of more or less vertical elements, a second, inner wall of inclined elements and a bridge of several elements. *P. sigmoides,* which was described from the Paleocene, has a simple wall of more or less inclined elements, but possible remains of an inner wall can be seen on the Maastrichtian specimen illustrated in Figs. 82.10, 11. With its wall of nearly vertical elements, *P. fibuliformis* shows features of *Bipodorhabdus.*

Reinhardtites

The species of *Reinhardtites* have a bridge which widens towards the wall. The bridge consists of many small, parallel laths. In the generotype, *R. anthophorus* (Figs. 82.16–19), a massive central process is usually present, and can be found still connected to the coccolith or broken off as a '*Microrhabdulus*' sp. The stem widens distally to form an inverted cone (Fig. 82.16). Whereas *R. anthophorus* features two large openings, the central area is filled with elements in *R. levis* (Figs. 82.20–22), where the bridge is less conspicuous. *R. levis* evolved from *R. anthophorus* in the Late Campanian through the successive closure of the central openings. The FO and LO of both species were used by Sissingh (1977) as zonal or subzonal boundaries. Whereas the LO of *R. levis* is still easy to recognize in poorly preserved material, the gradual change from *R. anthophorus* to *R. levis* makes the LO of *R. anthophorus* and the FO of *R. levis* difficult events to determine, especially in overgrown material or when the bridge of *R. anthophorus* has broken out. The FO of *R. anthophorus* in the Santonian also is not easy to determine in samples where specimens of *Placozygus* and *Zeugrhabdotus* species are present. *R. anthophorus* then can be recognized by the elements which surround the central area all the way round and by the somewhat elevated central area.

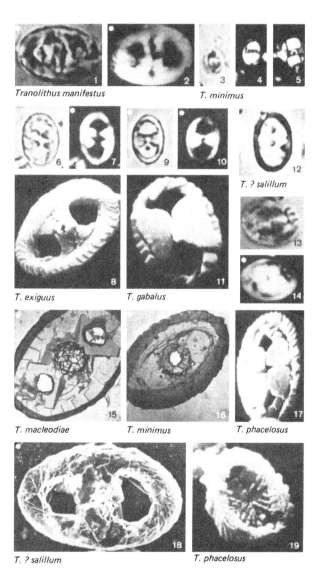

Tranolithus manifestus   *T. minimus*

T. ? salillum

T. exiguus   *T. gabalus*

T. macleodiae   *T. minimus*   *T. phacelosus*

T. ? salillum   *T. phacelosus*

Fig. 83. LM and EM of species of *Tranolithus*. LM: *c.* × 2000

Tranolithus

Tranolithus is characterized by usually 4, in some species only 2, massive blocks covering most of the central area. Only the FO and LO of the generotype *T. phacelosus* (syn. *T. orionatus*, *T. exiguus*?, Figs. 83.6–8; *T. manifestus*?, Figs. 83.1, 2) have been used stratigraphically. *T. orionatus* was described in the same year as *T. phacelosus*. *T. exiguus* and *T. manifestus* are considered by most authors as preservation synonyms of *T. phacelosus* but have slightly different ranges according to Stover (1966), their author. *T. phacelosus* (Figs. 83.13, 14, 17) with its 4 massive blocks is the first typical representative of the genus. It appears in the Albian and disappears in the Early Maastrichtian. The distal view of the holotype of *T.? salillum* (Figs. 83.12, 18), which was described from the Oxfordian, shows 4 slim blocks of differing size and an incomplete bridge forming the base for a central process. In proximal view it cannot be distinguished from a *Zeugrhabdotus* species. Crux (1982) illustrated proximal views of Cenomanian *Tranolithus* with a proximal grid of approximately radially arranged laths (Fig. 83.19). *T. gabalus* (Figs. 83.9–11), which ranges from the Hauterivian to the Campanian, has a simple wall and two large blocks. *T. minimus* (Figs. 83.3–5, 16; syn. *Z. tarboulensis*) and *T. macleodiae* (Figs. 83.15; syn. *Zygodiscus deflandrei*) feature a bridge and two large blocks covering the rest of the central area. The blocks are pierced by a large perforation in *Z. macleodiae*. The blocks are easy to recognize also with the LM and fill the ends of the ellipse in *T. minimus* and *T. macleodiae*. They are arranged along the minor axis in *T. gabalus*. *T. minimus* ranges from the Santonian to the Maastrichtian, one specimen having been observed by Crux (1982) in the Cenomanian. *T. macleodiae* seems restricted to the Campanian.

Zeugrhabdotus

The genus *Zeugrhabdotus* is based on the Late Jurassic species *Z. erectus* (Figs. 84.11–13), which is characterized by a simple bridge with or without a process. The width of the bridge, the two openings and the wall varies. While Medd (1979) regarded *Z. noeliae* (Fig. 84.1) as a junior synonym of *Z. erectus*, Wind (1978) saw it as a junior synonynm of *Tranolithus? sallillum* (Figs. 83.12, 18). They consider *T.? sallillum* to be an intermediate species between *Z. erectus* and the much larger and more solid *Z. embergeri* (Figs. 84.4, 6, 9, 10, 14, 15), the FO of which defines the base of the *Z. embergeri* Zone in the Kimmeridgian. *Z. embergeri* is slightly long-elliptical and has a high wall of very thin elements. This distinguishes *Z. embergeri* specimens from *P. sigmoides* and *Z. noeliae* specimens, where the elements of the wall are about twice as thick. *Z. embergeri* is more often found overgrown, even in samples where overgrowth otherwise is not common, than any other coccolith in the Late Jurassic or Early Cretaceous. Other species often assigned to this genus but which may also be included in *Placozygus* include *Z. acanthus* (Figs. 84.2, 3), characterized by a relatively wide open central area and typically two wall cycles, and *Z. theta* (Fig. 84.5) with similar features. The Late Cretaceous *Z. pseudanthophorus* (Figs. 84.7, 8; syn. *Z. lacunatus*) has a narrower bridge than *Z. embergeri* and is considered synonymous with *Z. embergeri* by many authors.

Zygolithites

Zygolithites is considered synonymous with *Zeugrhabdotus*.

Incertae sedis

Annulithus (Fig. 5.33)

A. arkellii, the only species of *Annulithus*, was defined as a 'coccolith-like object consisting of a simple ring of a few large, coarse, irregular elements'. The authors stated that *A.*

arkellii was rare in the lowermost Hettangian but common towards the upper part of the *A. arkellii* Zone. Hamilton (1982) reported it from the Rhaetian.

Ceratolithina (Figs. 85, 86)

C. hamata (Figs. 86.1, 2), the only species of the genus *Ceratolithina*, has two horns of different length and curvature. The apical spur of this Albian species is not well developed. The species of *Ceratolithoides*, which only evolved in the Campanian, differ from *C. hamata* in having more or less symmetrical horns.

Ceratolithoides (Figs. 85, 86)

The genus *Ceratolithoides* includes the horseshoe-shaped nannoliths and their cone-shaped ancestor of Late Cretaceous age. Cenozoic forms (Miocene/Pliocene) are assigned to *Amaurolithus* and *Ceratolithus*, two genera which are not related to the Cretaceous horseshoe-shaped forms according to Perch-Nielsen (1979a and Chapter 11), who suggested their evolution from the Miocene genus *Triquetrorhabdulus*. Crux & Lord (1982) were of the opinion that *Ceratolithina*, *Ceratolithoides* and the Cenozoic genera are related, based on the fact that 'the three genera all resemble one another'.

Four species are currently assigned to *Ceratolithoides*, two of which are used as markers in the low to mid latitude

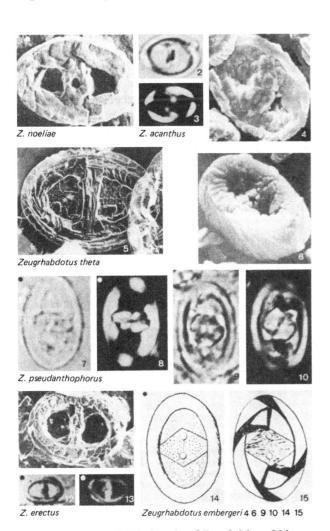

Z. noeliae   Z. acanthus

Zeugrhabdotus theta

Z. pseudanthophorus

Z. erectus   Zeugrhabdotus embergeri 4 6 9 10 14 15

Fig. 84. LM and EM of species of *Zeugrhabdotus*. LM: *c.* × 2000

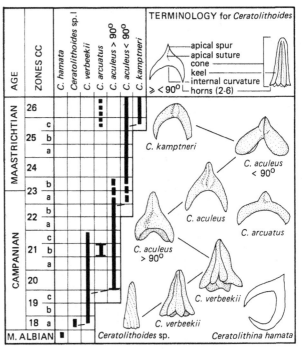

Fig. 85. Terminology and stratigraphic distribution of the species of *Ceratolithoides* and *Ceratolithina*. Maastrichtian range of *C. arcuatus* questionable. After Perch-Nielsen (1979a)

Campanian. *C. verbeekii*, the oldest distinguishable form, is cone-shaped, with 6 horns and keels (Figs. 86.16–20). *C. aculeus* (Figs. 86.5–14) is characterized by 2 horns and a well developed apical spur. Early forms have 4 short horns and a very long apical spur (Fig. 86.14), later forms have 2 longer horns and a shorter apical spur. In side view, the early forms of *C. aculeus* can be distinguished from the similar *C. verbeekii* by the fact that 3 horns are visible in the same focus in normal and polarized light in *C. verbeekii* (Figs. 86.16–18), and only 2 in *C. aculeus* (Figs. 86.5–8). The internal curvature is usually wider than 90° in the Campanian and around 90° or narrower in the Maastrichtian.

C. arcuatus (Fig. 86.15) was described from Tunisia and its range used for the definition of a subzone (CC 21b) in the Campanian. It has a short apical spur, two long horns and a very wide internal curvature. It appears at a time when the specimens of *C. aculeus* still have a very high apical spur.

C. kamptneri (Figs. 86.3, 4), the generotype, is characterized by two long horns and the lack of an apical spur. The horns meet at an angle of about 90°. *C. kamptneri* is only found in the Late Maastrichtian.

Laguncula (Figs. 87.1–3)

Laguncula was defined as 'hollow globular bodies with a thin shell composed of interpenetrating calcite rhombs and an aperture with an everted flange at the end of a short neck'. The only species was described from the Albian and forms assigned to *L. dorotheae* were found in the Hauterivian by Wind & Čepek (1979; Figs. 87.2, 3). These authors also noted 'the general appearance of the inflated portion of specimens of this species, and the possible presence of a basal plate suggests that *L. dorotheae* may be related to parhabdolithids with similarly inflated spines'.

Lapideacassis **and** *Scampanella* (Figs. 88.1–7)

Lapideacassis is characterized by more than one distal tier (see Chapter 11, Fig. 86 for terminology), whereas in *Scampanella*, the first and only tier constitutes more than half of the length of the body. The generotypes *L. mariae* (Fig. 88.7) and *S. cornuta* (Figs. 88.3, 6) are illustrated together with the other Cretaceous species, *L. glans* (Fig. 88.5), *L. tricornus* (Figs. 88.1, 2) and *S. magnifica* (Fig. 88.4). Paleogene forms are illustrated in Chapter 11 (Figs. 87.1–6) and in Perch-Nielsen & Franz (1977). In both genera, the species are distinguished by the number of spines, the presence of an apical process and, in *Lapideacassis*, also by the number of tiers. The stratigraphic distribution of the *Lapideacassis* and *Scampanella* species is not known in detail. They are very rare in Albian to Maastrichtian sediments.

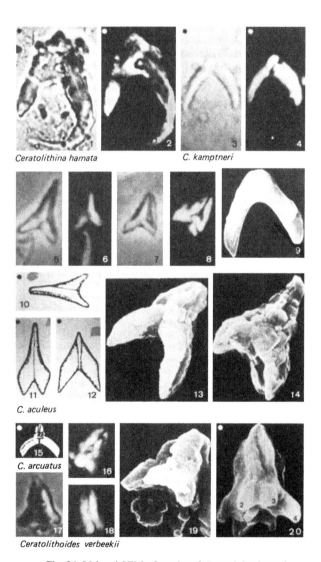

Ceratolithina hamata   *C. kamptneri*

C. aculeus

C. arcuatus

Ceratolithoides verbeekii

Fig. 86. LM and SEM of species of *Ceratolithoides* and *Ceratolithina*. LM: *c.* × 2000. Figs. 1, 2: × 1300

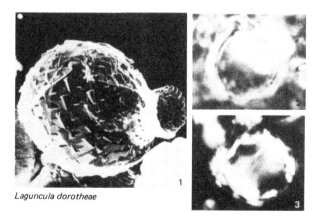

Laguncula dorotheae

Fig. 87. *Laguncula dorotheae*. LM and EM. LM: *c.* × 2000

Liliasterites and Marthasterites (Figs. 8, 89)

Marthasterites includes the Late Cretaceous, 3-armed nannoliths. Between crossed nicols, representatives of this genus remain dark, while one or two segments of the 3-armed *Quadrum trifidum* (Fig. 58.25) appear bright. Both the FO and the LO of *M. furcatus* (Figs. 89.2–4, 8–10), the generotype of *Marthasterites*, are used to define zonal boundaries (Fig. 6). The arms of *M. furcatus* typically bifurcate or show additional spines of varying length. Several subspecies have been described based on the details of the bifurcations and additional spines and in *M. jucundus* (Fig. 89.1) another species was created for more massive forms with 3 spines. *M. inconspicuus* (Figs. 89.5–7) is smaller than *M. furcatus*, has a triangular outline and ranges from the Turonian through the Maastrichtian. The subspecies of *M. furcatus* are not recognizable in poorly preserved material, especially when only overgrown specimens are available. They are not generally used. Whereas *M. furcatus* is a good marker species for Coniacian, Santonian and Early Campanian in low and mid latitudes, its occurrence in higher latitudes is patchy. Crux (1982) reported it from the Turonian *planus* Zone of England, suggesting an earlier appearance than generally assumed.

Marthasterites seems unrelated to the Early Eocene 3-armed nannoliths of *Tribrachiatus*, which were shown to have evolved from *Rhomboaster* (Chapter 11). Nothing is known about an ancestor of this very characteristic genus unless it evolved from *Liliasterites* (Figs. 8.26; 89), a genus where the bifurcations are longer (Stradner & Steinmetz,

1984). *Liliasterites* specimens reach greater sizes than *Marthasterites* specimens and were found in the Turonian, below the FO of *M. furcatus*. *L. angularis* (Figs. 89.11, 12) was found in samples from Bohemia and from the South Atlantic. Stradner & Steinmetz (1984) used its FO to define their *L. angularis* Zone (Late Turonian). *L. atlanticus* (Fig. 89.13), from the Upper Turonian of the South Atlantic has truncated arms, the ends of which are either notched or block-shaped.

Percivalia

The genus *Percivalia*, as characterized by its generotype *P. porosa*, has a proximal side composed of 6 to 11 narrow tiers of 30 to 42 elements surrounding the central area. The latter is pierced by two cycles of perforations in the generotype and is open but bridged by a broad bridge in *P. pontilitha*. Black (1973) erected a special family for this genus, which is left in 'incertae sedis' since it seems to have affinities both to the Arkhangelskiellaceae and the Cenozoic Pontosphaeraceae.

Petrarhabdus (Figs. 90.1–8)

The generotype of *Petrarhabdus*, *P. copulatus*, was described from the french chalk as *Tetralithus copulatus*. I did not find it again when I tried to re-illustrate many of the species described by Deflandre, by re-studying his material with the

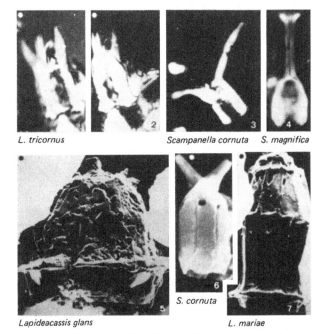

L. tricornus   Scampanella cornuta   S. magnifica

S. cornuta

Lapideacassis glans   L. mariae

Fig. 88. LM and EM of species of *Lapideacassis* and *Scampanella*. LM: *c.* × 2000

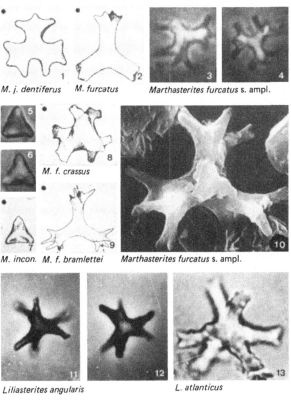

M. j. dentiferus   M. furcatus   Marthasterites furcatus s. ampl.

M. f. crassus

M. incon.   M. f. bramlettei   Marthasterites furcatus s. ampl.

Liliasterites angularis   L. atlanticus

Fig. 89. LM and SEM of species of *Liliasterites* and *Marthasterites*. LM: *c.* × 2000

SEM and TEM. Wind (1975b) found *P. copulatus* in the Indian Ocean Campanian and Maastrichtian, Wise (1983) on the Falkland Plateau. The elements of the distal shield are arranged inversely as in the Ellipsagelosphaeraceae and the proximal shield is only slightly smaller than the distal one. The central area is covered by a 'massive angular calcite cross' (Wind, 1975b).

Prolatipatella

Gartner (1968) erected the genus *Prolatipatella* for flat, thin, elliptical Late Cretaceous coccoliths with a wall of inclined elements and a central area built by concentrically arranged laths. With the LM, the generotype *P. multicarinata* (Chapter 11, Figs. 51.48, 49) looks like a Paleocene/Eocene specimen of *Pontosphaera*. It has been reported from the Maastrichtian and no similar forms have been found in the Early and early Late Paleocene (see discussion Chapter 11, Pontosphaeraceae).

Repagalum (Figs. 91.1–3)

Repagalum parvidentatum, the generotype and only species of the genus, has a distal and a proximal shield consisting of very narrow elements. These elements have about 1/2 to 1/4 of the width of the shield elements in the Ellipsagelosphaeraceae, the Podorhabdaceae or the Prediscosphaeraceae. The central area is occupied by a longitudinal bar from which

radial secondary bars connect to the shields. *R. parvidentatum* was described from the Campanian, but was also found in the Albian to Santonian, if *Tremalithus burwellensis* is considered a synonym as suggested by Noël (1970). Nothing is known about an ancestor of this genus. The only similar genus is the Paleocene *Ellipsolithus* (Chapter 11, Fig. 82), which may have evolved from *Repagalum*.

Tortolithus (Figs. 92.1–5)

Tortolithus includes coccoliths 'with an outer rim of large imbricating elements; the central area has a median suture, aligned parallel with the long axis of the elliptical coccolith, and around the suture lie approximately the same number of elements as in the rim in an imbricate relationship'.

The generotype, *T. caistorensis* (Figs. 92.3, 4), has 12 to 20 elements in its outer rim. *T. furlongii* (Fig. 92.1) is long-elliptical, *T. pagei* (Fig. 92.2) and *T. hallii* (Fig. 92.5) have a wider central area than the other two species and *T. hallii* seems to have an additional cycle of elements inside the rim. *T. caistorensis* was found in the Campanian, as were the other species.

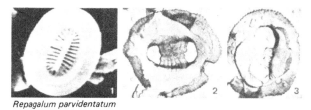

Repagalum parvidentatum

Fig. 91. EM of *Repagalum parvidentatum*

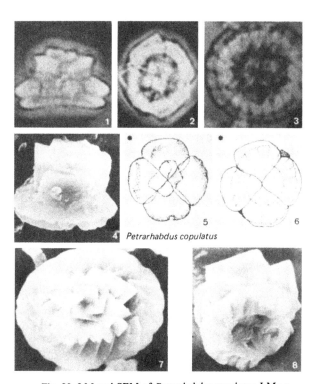

Petrarhabdus copulatus

Fig. 90. LM and SEM of *Petrarhabdus copulatus*. LM: *c.* × 2000

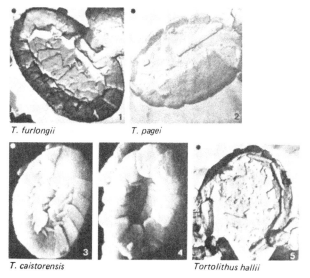

T. furlongii   *T. pagei*

T. caistorensis   *Tortolithus hallii*

Fig. 92. EM of species of *Tortolithus*

Acknowledgements

I gratefully acknowledge the cooperation and the support of the many colleagues with whom I discussed various aspects of Mesozoic calcareous nannofossils during the compilation of this chapter. D. Bukry, J. A. Crux, P. H. Doeven, M. E. Hill, A. W. Medd, H. Stradner and S. W. Wise read a semi-final version of the manuscript critically and made valuable suggestions for which I am very grateful. M. Biolzi helped with the arrangement of many of the figures. I thank her, my family and the co-editors for the patience and support they showed while I struggled to complete this last of my chapters.

Appendix: Alphabetic list of taxa

Acaenolithus Black (1973)
   A. cenomanicus Black (1973)
(*Actinosphaera* Noël, 1965) = *Watznaueria*
(*Actinozygus* Gartner, 1968) = *Stradnerlithus*
   (*A. geometricus* (Gorka, 1957) Rood *et al.*, 1971) = *Stradnerlithus sexiramatus*
Acuturris Wind & Wise *in* Wise & Wind (1977)
   A. scotus (Risatti, 1973) Wind & Wise *in* Wise & Wind (1977)
Ahmuellerella Reinhardt (1964)
   A. octoradiata (Gorka, 1957) Reinhardt (1964)
(*Alfordia* Black, 1975) = *Rhagodiscus*
Allemannites Grün *in* Grün & Allemann (1975)
   A. striatus (Stradner, 1963) Grün *in* Grün & Allemann (1975)
(*Alvearium* Black, 1967) = *Mitrolithus*
   (*A. dorsetense* Black, 1967) = *Mitrolithus elegans*
Amaurolithus Gartner & Bukry (1975)
Anfractus Medd (1979)
   A. harrisonii Medd (1979)
   A. variabilis Medd (1979)
Angulofenestrellithus Bukry (1969)
Annulithus Rood *et al.* (1973)
   A. arkellii Rood *et al.* (1973)
Anoplosolenia Deflandre *in* Grassé (1952)
   A. brasiliensis (Lohmann, 1919) Deflandre *in* Grassé (1952)
Ansulasphera Grün & Zweili (1980)
   A. helvetica Grün & Zweili (1980)
Archaeopontosphaera Jafar (1983)
   A. primitiva Jafar (1983)
Arkhangelskiella Vekshina (1959)
   A.? *cribrata* (Gazdzicka, 1978)
   A. cymbiformis Vekshina (1959)
   A. specillata Vekshina (1959)
Aspidolithus Noël (1969)
   A. bevieri (Bukry, 1969) Perch-Nielsen (1984a)
   A. furtivus (Bukry, 1969) Perch-Nielsen (1984a)
   A. parcus constrictus (Hattner *et al.*, 1980) Perch-Nielsen (1984a)
   A. parcus expansus (Wise & Watkins *in* Wise, 1983) Perch-Nielsen (1984a)
   A. parcus parcus (Stradner, 1963) Noël (1969)
   A. signatus Noël (1969)
Assipetra Roth (1973)
   A. infracretacea (Thierstein, 1973) Roth (1973)
Athenagalea Hattner & Wise (1980)
   A. robusta Hattner & Wise (1980)
Axopodorhabdus Wind & Wise *in* Wise & Wind (1977)

A. albianus (Black, 1967) Wind & Wise *in* Wise & Wind (1977)
A. cylindratus (Noël, 1965) Wind & Wise *in* Wise & Wind (1977)
A. dietzmannii (Reinhardt, 1965) Wind & Wise (1983)
A. rahla (Noël, 1965) Grün & Zweili (1980)

Barringtonella Black (1973)
   B. bulmanii Black (1973)
   B. flabellosa (Stradner *in* Stradner *et al.*, 1968) Black (1973)
(*Bennocyclus* Zweili & Grün *in* Grün *et al.*, 1974) = *Discorhabdus*
Biantholithus Bramlette & Martini (1964)
   B. sparsus Bramlette & Martini (1964)
(*Bidiscus* Bukry, 1969) = *Discorhabdus*
Bipodorhabdus Noël (1970)
   (*B. roeglii* Thierstein, 1971) = *Speetonia colligata*
   B. tesselatus Noël (1970)
Biscutum Black *in* Black & Barnes (1959)
   B. arrogans Perch-Nielsen (1973)
   B. blackii Gartner (1968)
   B. boletum Wind & Wise *in* Wise & Wind (1977)
   B. castrorum Black *in* Black & Barnes (1959)
   B. constans (Gorka, 1957) Black *in* Black & Barnes (1959)
   B. coronum Wind & Wise *in* Wise & Wind (1977)
   B. dissimilis Wind & Wise *in* Wise & Wind (1977)
   B. dubium (Noël, 1965) Grün *in* Grün *et al.* (1974)
   B. ellipticum (Gorka, 1957) Grün *in* Grün & Allemann (1975)
   B. erismatum (Wind & Wise *in* Wise & Wind, 1977) Grün & Zweili (1980)
   B. hattneri Wise (1983)
   B. magnum Wind & Wise *in* Wise & Wind (1977)
   B. notaculum Wind & Wise *in* Wise & Wind (1977)
   (*B. ornatum* Perch-Nielsen, 1973) = artefact?
   B. salebrosum (Black, 1971a) Perch-Nielsen (1984a)
   (*B. testudinarium* Black *in* Black & Barnes, 1959) = ?*B. constans*
   B. veternum (Prins, 1969 ex Rood *et al.*, 1973) Perch-Nielsen (1984a)
   B. virginica (Bukry, 1969) Wind & Wise *in* Wise & Wind (1977)
Boletuvelum Wind & Wise *in* Wise & Wind (1977)
   B. candens Wind & Wise *in* Wise & Wind (1977)
Braarudosphaera Deflandre (1947)
   B. africana Stradner (1961)
   B. batilliformis Troelsen & Quadros (1971)
   B. bigelowii (Gran & Braarud, 1935) Deflandre (1947)
   (*B. gartneri* Filewicz *et al. in* Wise & Wind, 1977) = *B. quinquecostata*
   B. hockwoldensis Black (1973)
   (*B. hoschulzii* Reinhardt, 1966a) = *Micrantholithus hoschulzii*
   (*B. imbricata* Manivit, 1966) = overgrown *B. bigelowii*
   (*B. minuta* Filewicz *et al. in* Wise & Wind, 1977) = *B. stenorhetha*
   B. primula Black (1973)
   B. quinquecostata Hill (1976)
   B. regularis Black (1973)
   B. stenorhetha Hill (1976)
   B. turbinea Stradner (1963)
(*Brachiolithus* Noël, 1959 ex Loeblich & Tappan, 1963) = *Nannoconus*
   (*B. quadratus* Noël, 1959 ex Loeblich & Tappan, 1963) = *N. quadratus*
Broinsonia Bukry (1969)
   (*B. bevieri* Bukry, 1969) = *Aspidolithus bevieri*
   (*B. cribrata* Gazdzicka, 1978) = *Arkhangelskiella*? *cribrata*
   B. dentata Bukry (1969)
   B. enormis (Shumenko, 1968) Manivit (1971)

B. furtiva Bukry (1969)
B. handfieldii Bukry (1969)
(*B. lacunosa* Forchheimer, 1972) = *B. enormis*
(*B. parca constricta* Hattner *et al.*, 1980) = *Aspidolithus parcus constrictus*
(*B. parca expansa* Wise & Watkins *in* Wise, 1983) = *Aspidolithus parcus expansus*
(*B. parca parca* (Stradner, 1963) Bukry, 1969) = *Aspidolithus parcus parcus*
Bukryaster Prins (1971)
   B. hayi (Bukry, 1969) Prins & Sissingh *in* Sissingh (1977)
Bukrylithus Black (1971a)
   B. ambiguus Black (1971a)

Calcicalathina Thierstein (1971)
   C. alta Perch-Nielsen (1979a)
   C. oblongata (Worsley, 1971) Thierstein (1971)
Calculites Prins & Sissingh *in* Sissingh (1977)
   C. additus (Wind & Wise *in* Wise & Wind, 1977) Perch-Nielsen (1984a)
   C. obscurus (Deflandre, 1959) Prins & Sissingh *in* Sissingh (1977)
   C. ovalis (Stradner, 1963) Prins & Sissingh *in* Sissingh (1977)
Calolithus Noël (1965)
   C. martelae Noël (1965)
Calyculus Noël (1973)
   C. cribrum Noël (1973)
Carinolithus Prins *in* Grün *et al.* (1974)
   C. clavatus (Deflandre *in* Deflandre & Fert, 1954) Medd (1979)
   C. sceptrum (Deflandre *in* Deflandre & Fert, 1954) Medd (1979)
   C. superbus (Deflandre *in* Deflandre & Fert, 1954) Prins *in* Grün *et al.* (1974)
Caterella Black (1971a)
   C. perstrata Black (1971a)
Catillus Goy *in* Goy *et al.* (1979)
Centosphaera Wind & Wise *in* Wise & Wind (1977)
   C. barbata Wind & Wise *in* Wise & Wind (1977)
Ceratolithina Martini (1967)
   C. hamata Martini (1967)
Ceratolithoides Bramlette & Martini (1964)
   C. aculeus (Stradner, 1961) Prins & Sissingh *in* Sissingh (1977)
   C. arcuatus Prins & Sissingh *in* Sissingh (1977)
   C. kamptneri Bramlette & Martini (1964)
   C. verbeekii Perch-Nielsen (1979a)
Ceratolithus Kamptner (1950)
Chiasmolithus Hay, Mohler & Wade (1966)
   (*C. parvus* Barrier, 1977a) = *Chiastoplacolithus parvus*
Chiastoplacolithus Barrier (1977a)
   C. parvus (Barrier, 1977a) Perch-Nielsen (1984a)
   C. quadratus (Worsley, 1971) Barrier (1977a)
Chiastozygus Gartner (1968)
   C.? acutus (Thierstein *in* Roth & Thierstein, 1972) Perch-Nielsen (1984a)
   C. amphipons (Bramlette & Martini, 1964) Gartner (1968)
   C. bifarius Bukry (1969)
   (*C. cuneatus* (Lyul'eva, 1967) Čepek & Hay, 1969) = *Helicolithus trabeculatus*
   C. fessus (Stover, 1966) Shafik (1979)
   C. litterarius (Gorka, 1957) Manivit (1971)
   C. platyrhethus Hill (1976)
   C.? primitus Prins (1969) ex Rood *et al.* (1973)
   C. striatus Black (1971a)
   C. tenuis Black (1971a)

(*C. trabeculatus* (Gorka, 1957) Risatti, 1973) = *Helicolithus trabeculatus*
   C. ultimus Perch-Nielsen (1981)
Cleistorhabdus Black (1972)
   C. williamsii Black (1972)
(*Clinorhabdus* Stover, 1966) = *Eiffellithus*
Coccolithus Schwarz (1894)
(*Colvillea* Black, 1964) = *Watznaueria*
Conusphaera Trejo (1969)
   C. mexicana Trejo (1969)
   C. zlambachensis Moshkovitz (1982)
Coptolithus Black (1973)
   C. virgatus Black (1973)
Corollithion Stradner (1961)
   C.? completum Perch-Nielsen (1973)
   C. exiguum Stradner (1961)
   C. kennedyi Crux (1981)
   C.? madagaskarensis Perch-Nielsen (1973)
   C. signum Stradner (1963)
(*Costacentrum* Bukry, 1969) = *Sollasites*
Crepidolithus Noël (1965)
   C. cavus Prins (1969) ex Rood *et al.* (1973)
   C. crassus (Deflandre *in* Deflandre & Fert, 1954) Noël (1965)
   C. crucifer Prins (1969) ex Rood *et al.* (1973)
   C. impontus Grün *et al.* (1974)
(*Cretadiscus* Gartner, 1968) = *Cribrosphaerella*
(*Cretarhabdella* Black, 1971a) = *Cretarhabdus*
Cretarhabdus Bramlette & Martini (1964)
   (*C. angustiforatus* (Black, 1971a) Bukry, 1973) = *Retecapsa angustiforata*
   C. conicus Bramlette & Martini (1964)
   (*C. crenulatus* Bramlette & Martini, 1964) = *Stradneria crenulata*
   C. lateralis (Black, 1971a) Perch-Nielsen (1984a)
   C. loriei Gartner (1968)
   (*C. unicornis* Stover, 1966) = *Gephyrorhabdus coronadventis*
(*Cretaturbella* Thierstein, 1971) = *Conusphaera*
   (*C. rothii* Thierstein, 1971) = *C. mexicana*
Cribricatullus Black (1973)
   C. textus Black (1973)
Cribrocorona Perch-Nielsen (1973)
   C. gallica (Stradner, 1963) Perch-Nielsen (1973)
(*Cribrosphaera* Arkhangelsky, 1912) = *Cribrosphaerella*
Cribrosphaerella Deflandre *in* Piveteau (1952)
   C.? daniae Perch-Nielsen (1973)
   C. ehrenbergii (Arkhangelsky, 1912) Deflandre *in* Piveteau (1952)
   C.? hauteriviana (Black, 1971a) Perch-Nielsen (1984a)
   (*C. primitiva* Thierstein, 1974) = *Seribiscutum primitivum*
Crucicribrum Black (1973)
   C. anglicum Black (1973)
   C. striatum constansii Wise & Parker *in* Wise (1983)
Cruciellipsis Thierstein (1971)
   C. cuvillieri (Manivit, 1966) Thierstein (1971)
   (*C. chiastia* (Worsley, 1971) Thierstein *in* Roth & Thierstein, 1972) = *Microstaurus chiastius*
Cruciplacolithus Hay & Mohler *in* Hay *et al.* (1967)
   C. inseadus Perch-Nielsen (1969)
Crucirhabdus Prins (1969) ex Rood *et al.* (1973)
   C. curvatus Jafar (1983)
   C. minutus Jafar (1983)
   C. primulus Prins (1969) ex Rood *et al.* (1973)
   C. prinsii Rood *et al.* (1973)

Cyclagelosphaera Noël (1965)
  C. bergeri Roth (1978)
  C. deflandrei (Manivit, 1966)
  C. margerelii Noël (1965)
  C. reinhardtii (Perch-Nielsen, 1968) Romein (1977)
  C. rotaclypeata Bukry (1969)
Cylindralithus Bramlette & Martini (1964)
  C. crassus Stover (1966)
  C. duplex Perch-Nielsen (1973)
  (*C. gallicus* (Stradner, 1963) Bramlette & Martini,
    1964) = *Cribrocorona gallica*
  C. nudus Bukry (1969)
  C.? oweinae Perch-Nielsen (1973)
  C. serratus Bramlette & Martini (1964)

(*Deflandrius* Bramlette & Martini, 1964) = *Prediscosphaera*
Dekapodorhabdus Medd (1979)
  D. typicus Medd (1979)
Diadorhombus Worsley (1971)
  D. minutus Rood *et al.* (1971)
  D. rectus Worsley (1971)
  D. scutulatus (Medd, 1971) Medd (1979)
(*Diadozygus* Rood *et al.*, 1971) = *Stradnerlithus*
  (*D. dorsetense* Rood *et al.*, 1971) = *Stradnerlithus comptus*
Diazomatolithus Noël (1965)
  D. lehmanii Noël (1965)
Dictyolithus Gorka (1957)
  D. quadratus Gorka (1957)
Diloma Wind & Čepek (1979)
  D. placinum Wind & Čepek (1979)
  D. primitiva (Worsley, 1971) Wind & Čepek (1979)
Discorhabdus Noël (1965)
  (*D. biperforatus* Rood *et al.*, 1973) = *Podorhabdus biperforatus*
  D. corollatus Noël (1965)
  D. exilitus Noël (1965)
  (*D. gibbosus* Noël, 1965) = nomen nudum
  D. ignotus (Gorka, 1957) Perch-Nielsen (1968)
  D. jungii Noël (1965)
  D. longicornis Medd (1979)
  D. patulus (Deflandre *in* Deflandre & Fert, 1954) Noël (1965)
  D. striatus Moshkovitz & Ehrlich (1976a)
  D. tubus Barnard & Hay (1974)
Dodekapodorhabdus Perch-Nielsen (1968)
  D. noeliae Perch-Nielsen (1968)
  (*D. reeschi* Medd, 1979) = nomen nudum

Eiffellithus Reinhardt (1965)
  E. collis Hoffmann (1970d)
  E. eximius (Stover, 1966) Perch-Nielsen (1968)
  E. gorkae Reinhardt (1965)
  (*E. multicostatus* Gazdzicka, 1978) = *E. parallelus*
  E. parallelus Perch-Nielsen (1973)
  E. turriseiffelii (Deflandre *in* Deflandre & Fert, 1954) Reinhardt
    (1965)
Ellipsagelosphaera Noël (1965)
  E. britannica (Stradner, 1963) Perch-Nielsen (1968)
  E. communis (Reinhardt, 1964) Perch-Nielsen (1968)
  (*E. crucicentralis* Medd, 1971) = *Lotharingius crucicentralis*
  E. fasciata (Wind & Čepek, 1979) Perch-Nielsen (1984a)
  E. fossacincta Black (1971a)
  E. frequens Noël (1965)
  E. gresslyi Grün & Zweili (1980)
  (*E. keftalrempti* Grün *in* Grün & Allemann, 1975) =
    E. fossacincta

E. lucasii Noël (1965)
E. ovata (Bukry, 1969) Black (1973)
E. plena Grün & Zweili (1980)
E. reinhardtii (Rood *et al.*, 1971) Noël (1973)
E. strigosa Grün & Zweili (1980)
E.? tubulata Grün & Zweili (1980)
(*Ellipsochiastozygus* Prins, 1969) = *Chiastozygus*
  (*E. primitus* Prins (1969) ex Rood *et al.*, 1973) = *Chiastozygus?*
    primitus
(*Eoconusphaera* Jafar, 1983) = *Conusphaera*
  (*E. tollmanniae* Jafar, 1983) = *Conusphaera zlambachensis*
Eprolithus Stover (1966)
  E. antiquus Perch-Nielsen (1979a)
  E. apertior Black (1973)
  E. floralis (Stradner, 1962) Stover (1966)
  E. orbiculatus (Forchheimer, 1972) Crux *in* Crux *et al.* (1982)
  E. septentrionalis (Stradner, 1963) Perch-Nielsen (1984a)
(*Esgia* Worsley, 1971) = *Watznaueria*
  (*E. junior* Worsley, 1971) = *Watznaueria barnesae*
Ethmorhabdus Noël (1965)
  E. gallicus Noël (1965)

(*Favocentrum* Black, 1964) = *Cribrosphaerella*
Flabellites Thierstein (1973)
  F. biforaminis Thierstein (1973)
  F. oblonga (Bukry, 1969) Crux *in* Crux *et al.* (1982)

Gaarderella Black (1973)
  G. granulifera Black (1973)
Gartnerago Bukry (1969)
  G. nanum Thierstein (1974)
  G. obliquum (Stradner, 1963) Noël (1970) or Reinhardt (1970a)
  G. stenostaurion (Hill, 1976) Perch-Nielsen (1984a)
  (*G. striatum* (Stradner, 1963) Forchheimer, 1972) = *Allemannites*
    striatus
Gephyrorhabdus Hill (1976)
  G. coronadventis (Reinhardt, 1966b) Hill (1976)
Glaukolithus Reinhardt (1964)
  G. compactus (Bukry, 1969) Perch-Nielsen (1984a)
  G. diplogrammus (Deflandre *in* Deflandre & Fert, 1954)
    Reinhardt (1964)
  (*G. fibuliformis* Reinhardt, 1964) = *Placozygus fibuliformis*
Goniolithus Deflandre (1957)
  G. fluckigeri Deflandre (1957)
Grantarhabdus Black (1971a)
  G. meddii Black (1971a)

Haqius Roth (1978)
  H. circumradiatus (Stover, 1966) Roth (1978)
Haslingfieldia Black (1973)
  H. stradneri Black (1973)
Hayesites Manivit (1971)
  H. albiensis Manivit (1971)
  H. bulbus Thierstein *in* Roth & Thierstein (1972)
Hayococcus Jafar (1983)
  H. floralis Jafar (1983)
Helenea Worsley (1971)
  H. staurolithina Worsley (1971)
Helicolithus Noël (1970)
  H. anceps (Gorka, 1957) Noël (1970)
  H. trabeculatus (Gorka, 1957) Verbeek (1977b)
(*Hemipodorhabdus* Black, 1971a) = *Podorhabdus*
Heteromarginatus Bukry (1969)
  H. wallacei Bukry (1969)

Heterorhabdus Noël (1970)
  H. primitivus Perch-Nielsen (1973)
  H. sinuosus Noël (1970)
Hexalithus Gardet (1955)
  H. gardetae Bukry (1969)
  H. lecaliae Gardet (1955)
  H. magharensis Moshkovitz & Ehrlich (1976a)
(*Hexangulolithus* Bukry, 1969) = distal part of *Mitrolithus elegans* (?)
Hexapodorhabdus Noël (1965)
  H. cuvillieri Noël (1965)

Incerniculum Goy *in* Goy *et al.* (1979)
Isocrystallithus Verbeek (1976b)
  I. compactus Verbeek (1976b)

Kamptnerius Deflandre (1959)
  K. magnificus Deflandre (1959)

(*Laffittius* Noël, 1969) = *Gartnerago*
Laguncula Black (1971b)
  L. dorotheae Black (1971b)
Lapideacassis Black (1971b)
  L. glans Black (1971b)
  L. mariae Black (1971b)
  L. tricornus Wind & Wise *in* Wise & Wind (1977)
Liliasterites Stradner & Steinmetz (1984)
  L. angularis Švabénická & Stradner *in* Stradner & Steinmetz (1984)
  L. atlanticus Stradner & Steinmetz (1984)
Lithastrinus Stradner (1962)
  (*L. floralis* Stradner, 1962) = *Eprolithus floralis*
  L. grillii Stradner (1962)
  L. moratus Stover (1966)
  L. septenarius Forchheimer (1972)
  (*L. septentrionalis* Stradner, 1963) = *Eprolithus septentrionalis*
  (*L. tesselatus* Stradner *in* Stradner, Adamiker & Papp, 1968) = *Eprolithus septentrionalis*
Lithraphidites Deflandre (1963)
  L. acutus Verbeek & Manivit *in* Manivit *et al.* (1977)
  L. alatus Thierstein *in* Roth & Thierstein (1972)
  L. bollii (Thierstein, 1971) Thierstein (1973)
  L. carniolensis carniolensis Deflandre (1963)
  L. carniolensis serratus Shumenko (1970)
  L. grossopectinatus Bukry (1969)
  L. kennethii Perch-Nielsen (1984b)
  L. praequadratus Roth (1978)
  L. pseudoquadratus Crux (1981)
  L. quadratus Bramlette & Martini (1964)
Lotharingius Noël (1973)
  L. barozii Noël (1973)
  L. crucicentralis (Medd, 1971) Grün & Zweili (1980)
  L. hauffii Grün & Zweili *in* Grün *et al.* (1974)
  L. primitivus (Prins, 1969 ex Rood *et al.*, 1973) Prins *in* Grün *et al.* (1974)
  L. sigillatus (Stradner, 1961) Prins *in* Grün *et al.* (1974)
Loxolithus Noël (1965)
  L. armilla (Black *in* Black & Barnes, 1959) Noël (1965)
Lucianorhabdus Deflandre (1959)
  L. arborius Wind & Wise *in* Wise & Wind (1977)
  L. arcuatus Forchheimer (1972)
  L. cayeuxii Deflandre (1959)
  L. inflatus Perch-Nielsen & Feinberg *in* Perch-Nielsen (1984b)
  L. maleformis Reinhardt (1966)

L. quadrifidus Forchheimer (1972)
L. windii Hattner & Wise (1980)
'*L.*' *phlaskus* Wind & Čepek (1979)

Manivitella Thierstein (1971)
  M. gronosa (Stover, 1966) Black (1973)
  M. pecten Black (1973)
  M. pemmatoidea (Deflandre *in* Manivit, 1965) Thierstein (1971)
(*Margolatus* Forchheimer, 1972) = *Watznaueria*
Markalius Bramlette & Martini (1964)
  M. inversus (Deflandre *in* Deflandre & Fert, 1954) Bramlette & Martini (1964)
  M. perforatus Perch-Nielsen (1973)
Marthasterites Deflandre (1959)
  M. furcatus (Deflandre *in* Deflandre & Fert, 1954) Deflandre (1959)
  M. furcatus var. *bramlettei* Deflandre (1959)
  M. furcatus var. *crassus* Deflandre (1959)
  M. inconspicuus Deflandre (1959)
  M. jucundus var. *dentiferus* Deflandre (1959)
(*Maslovella* Tappan & Loeblich, 1966) = *Watznaueria*
Metadoga Wind & Čepek (1979)
  M. mercurius Wind & Čepek (1979)
Micrantholithus Deflandre *in* Deflandre & Fert (1954)
  M. hoschulzii (Reinhardt, 1966a) Thierstein (1971)
  M. obtusus Stradner (1963)
  M. speetonensis Perch-Nielsen (1979a)
Microrhabdulinus Deflandre (1963) = *Microrhabdulus*
Microrhabduloides Deflandre (1963) = *Microrhabdulus*
Microrhabdulus Deflandre (1959)
  M. attenuatus (Deflandre, 1959) Deflandre (1963)
  M. belgicus Hay & Towe (1963)
  M. decoratus Deflandre (1959)
  (*M. stradneri* Bramlette & Martini, 1964) = *M. attenuatus*
  M. undosus Perch-Nielsen (1973)
Microstaurus Black (1971a)
  M. chiastius (Worsley, 1971) Grün *in* Grün & Allemann (1975)
  (*M. lindensis* Black, 1971a) = *Helenea staurolithina*
  (*M. quadratus* Black, 1971a) = *M. chiastius*
Micula Vekshina (1959)
  M. concava (Stradner *in* Martini & Stradner, 1960) Verbeek (1976b)
  M. cubiformis Forchheimer (1972)
  M. decussata Vekshina (1959)
  (*M. infracretacea* Thierstein, 1973) = *Assipetra infracretacea*
  M. murus (Martini, 1961) Bukry (1973)
  M. praemurus (Bukry, 1973) Stradner & Steinmetz (1984)
  M. prinsii Perch-Nielsen (1979a)
  M. quadrata (Stradner, 1961) Perch-Nielsen (1984a)
  (*M. staurophora* (Gardet, 1955) Stradner, 1963) = object of doubtful origin
  M. swastica Stradner & Steinmetz (1984)
Millbrookia Medd (1979)
  M. perforata Medd (1979)
  M. virgata Medd (1979)
Miravetesina Grün *in* Grün & Allemann (1975)
  M. favula Grün *in* Grün & Allemann (1975)
Misceomarginatus Wind & Wise *in* Wise & Wind (1977)
  M. pleniporus Wind & Wise *in* Wise & Wind (1977)
Mitrolithus Deflandre *in* Deflandre & Fert (1954)
  M. elegans Deflandre *in* Deflandre & Fert (1954)
Monomarginatus Wind & Wise *in* Wise & Wind (1977)

M. pectinatus Wind & Wise *in* Wise & Wind (1977)
M. quaternarius Wind & Wise *in* Wise & Wind (1977)
Multipartis Risatti (1973)
Munarinus Risatti (1973)

Nannoconus Kamptner (1931)
   N. abundans Stradner & Grün (1973)
   N. aquitanicus Deres & Acheriteguy (1980)
   N. bermudezii Brönnimann (1955)
   N. boletus Deflandre & Deflandre (1967)
   N. bonetii Trejo (1959)
   N. borealis Perch-Nielsen (1979a)
   N. broennimannii Trejo (1959)
   N. bucheri Brönnimann (1955)
   N. calpidomorphus Deflandre & Deflandre (1967)
   N. carniolensis carniolensis Deflandre & Deflandre (1967)
   N. carniolensis latus Deres & Achersiteguy (1980)
   N. circularis Deres & Acheriteguy (1980)
   N. colomii (de Lapparent, 1931) Kamptner (1938)
   N. cornuta Deres & Acheriteguy (1980)
   N. dauvillieri Deflandre & Deflandre (1959)
   N. dolomiticus Cita & Pasquaré (1959)
   N. donnatensis Deres & Acheriteguy (1980)
   N. elongatus Brönnimann (1955)
   N. farinacciae Bukry (1969)
   N. fragilis Deres & Acheriteguy (1980)
   N. globulus Brönnimann (1955)
   N. grandis Deres & Acheriteguy (1980)
   N. inconspicuus Deflandre & Deflandre (1967)
   N. kamptneri Brönnimann (1955)
   (*N. maslovii* Shumenko, 1969) = part of a rhabdolith
   N. minutus Brönnimann (1955)
   N. multicadus Deflandre & Deflandre (1959)
   N. planus Stradner (1963)
   N. quadratuis (Noël, 1959) Deres & Acheriteguy (1980)
   N. quadriangulus apertus Deflandre & Deflandre (1967)
   N. quadriangulus quadriangulus Deflandre & Deflandre (1967)
   N. regularis Deres & Acheriteguy (1980)
   (*N. spicatum* Shumenko, 1969) = part of a rhabdolith
   N. steinmannii Kamptner (1931)
   N. steinmannii minor Deres & Acheriteguy (1980)
   N. truitti frequens Deres & Acheriteguy (1980)
   N. truitti rectangularis Deres & Acheriteguy (1980)
   N. truitti truitti Brönnimann (1955)
   N. vocontiensis Deres & Acheriteguy (1980)
   N. wassallii Brönnimann (1955)
(*Nannopatina* Stradner, 1961) = *Schizosphaerella*
Nannotetrina nitida (Martini, 1961) Aubry (1983)
'*Neococcolithes dubius*' (Deflandre *in* Deflandre & Fert, 1954) Black (1967) – *N. dubius* is an Eocene species but was used by Medd (1979) for a Jurassic holococcolith
Neocrepidolithus Romein (1979)
   N. cohenii (Perch-Nielsen, 1968) Perch-Nielsen (1984a)
   N. neocrassus (Perch-Nielsen, 1968) Romein (1979)
Nephrolithus Gorka (1957)
   N. corystus Wind (1983)
   N. frequens Gorka (1957)
Nodosella Prins (1969) ex Rood *et al.* (1973)
   N. clatriata Prins (1969) ex Rood *et al.* (1973)
   N. perch-nielseniae (Filewicz *et al. in* Wise & Wind, 1977) Perch-Nielsen (1984a)
   N. silvaradion (Filewicz *et al. in* Wise & Wind, 1977) Perch-Nielsen (1984a)

Noellithina Grün & Zweili *in* Grün *et al.* (1974)
   N. arcta (Noël, 1973) Grün & Zweili *in* Grün *et al.* (1974)
   N. prinsii (Noël, 1973) Grün & Zweili *in* Grün *et al.* (1974)

Octocyclus Black (1972)
   O. magnus Black (1972)
(*Octolithites* Caratini, 1963) = *Braarudosphaera*?
Octolithus Romein (1979)
   O. multiplus (Perch-Nielsen, 1973) Romein (1979)
Octopodorhabdus Noël (1965)
   O. decussatus (Manivit, 1961) Rood *et al.* (1971)
   O. oculisminutus Grün & Zweili (1980)
   O. plethotretus Wind & Čepek (1979)
   O. praevisus Noël (1965)
Okkolithus Wind & Wise *in* Wise & Wind (1977)
Orastrum Wind & Wise *in* Wise & Wind (1977)
   O. asarotum Wind & Wise *in* Wise & Wind (1977)
   O. campanensis (Čepek, 1970) Wind & Wise *in* Wise & Wind (1977)
Ottavianus Risatti (1973)

(*Paleopontosphaera* Noël, 1965) = *Biscutum*
   (*P. dubia* Noël, 1965) = *Biscutum dubium*
   (*P. salebrosa* (Black, 1971a) Prins & Sissingh *in* Sissingh, 1977) = *Biscutum salebrosum*
   (*P. veterna* Prins (1969) ex Rood *et al.*, 1973) = *Biscutum veternum*
Parhabdolithus Deflandre (1952)
   (*P. angustus* (Stradner, 1963) Stradner *et al.*, 1968) = *Rhagodiscus angustus*
   (*P. asper* Stradner, 1963) = *Rhagodiscus asper*
   (*P. embergeri* (Noël, 1965) Stradner, 1963) = *Zeugrhabdotus embergeri*
   P. liasicus Deflandre (1952)
   P. marthae Deflandre *in* Deflandre & Fert (1954)
   (*P. splendens* Deflandre, 1953) = *Rhagodiscus splendens*
Percivalia Bukry (1969)
   P. pontilitha Bukry (1969)
   P. porosa Bukry (1969)
Perissocyclus Black (1971a)
   P. fenestratus (Stover, 1966) Black (1971a)
   P. fletcheri Black (1971a)
   P. noeliae Black (1971a)
Pervilithus Crux (1981)
   P. varius Crux (1981)
Petrarhabdus Wind & Wise *in* Wise (1983)
   P. copulatus (Deflandre, 1959) Wind & Wise *in* Wise (1983)
(*Phanulithus* Wind & Wise *in* Wise & Wind, 1977) = *Calculites*
Pharus Wind & Wise *in* Wise & Wind (1977)
Placozygus Hoffmann (1970b)
   P. fibuliformis (Reinhardt, 1964) Hoffmann (1970b)
   P. sigmoides (Bramlette & Sullivan, 1961) Romein (1979)
Podorhabdus Noël (1965)
   P. biperforatus (Rood *et al.*, 1973) Perch-Nielsen (1984a)
   (*P. cylindratus* Noël, 1965) = *Axopodorhabdus cylindratus*
   P.? elkefensis Perch-Nielsen (1981)
   P. grassei Noël (1965) emend. Wind & Wise *in* Wise & Wind (1977)
   P. macrogranulatus Prins (1969) ex Rood *et al.* (1973)
   (*P. rahla* Noël, 1965) = *Axopodorhabdus rahla*
Polycostella Thierstein (1971)
   P. beckmannii Thierstein (1971)
   P. senaria Thierstein (1971)

Polycyclolithus Forchheimer (1968) = *Eprolithus?*
  P. brotzenii Forchheimer (1968)
Polypodorhabdus Noël (1965)
  (*P. arctus* Noël, 1973) = *Noellithina arcta*
  P. beckii Medd (1979)
  P. escaigii Noël (1965)
  P. madingleyensis Black (1971a)
  P. paucisectus Black (1971a)
Pontilithus Gartner (1968)
  P. obliquicancellatus Gartner (1968)
Prediscosphaera Vekshina (1959)
  P. arkhangelskyi (Reinhardt, 1965) Perch-Nielsen (1984a)
  P. avitus (Black 1973) Perch-Nielsen (1984a)
  P. bukryi Perch-Nielsen (1973)
  P. cantabrigensis (Black, 1967) Reinhardt (1970b)
  P. columnata (Stover, 1966) Perch-Nielsen (1984a)
  P. cretacea (Arkhangelsky, 1912) Gartner (1968)
  (*P. decorata* Vekshina, 1959) = *P. cretacea*
  (*P. germanica* Bukry, 1969) = *P. stoveri*
  P. grandis Perch-Nielsen (1979a)
  P. honjoi Bukry (1969)
  P. implumis (Black, 1973) Perch-Nielsen (1984a)
  P. intercisa (Deflandre *in* Deflandre & Fert, 1954) Shumenko (1976)
  P. lata (Bukry, 1969) Bukry (1973)
  P. majungae Perch-Nielsen (1973)
  P. microrhabdulina Perch-Nielsen (1973)
  (*P.? orbiculofenestra* Gartner, 1968) = *Axopodorhabdus dietzmannii*
  P. ponticula (Bukry, 1969) Perch-Nielsen (1984a)
  P. propinquus (Gorka, 1957) Reinhardt (1970b)
  P. quadripunctata (Gorka, 1957) Reinhardt (1970b)
  P. rhombica (Perch-Nielsen, 1968) Reinhardt (1970b)
  P. serrata Noël (1970)
  P. solida (Shumenko, 1971) Shumenko (1976)
  P. spinosa (Bramlette & Martini, 1964) Gartner (1968)
  P. stoveri (Perch-Nielsen1968) Shafik & Stradner (1971)
  P. sp. cf. *P. stoveri in* Barrier (1977b)
Prinsiosphaera Jafar (1983)
  P. geometrica Jafar (1983)
  P. triassica Jafar (1983)
  P. triassica punctata Jafar (1983)
Proculithus Medd (1979)
  P. charlottei Medd (1979)
  P. expansus Medd (1979)
  P. fistulatus Prins (1969) ex Medd (1979)
Prolatipatella Gartner (1968)
  P. multicarinata Gartner (1968)
Pseudolithraphidites Keupp (1976)
  P. multibacillatus Keupp (1976)
  P. quattuorbacillatus Keupp (1976)
Pseudomicula Perch-Nielsen *in* Perch-Nielsen *et al.* (1978)
  P. quadrata Perch-Nielsen *in* Perch-Nielsen *et al.* (1978)
(*Pyrobolella* Black, 1971b) = *pyrites?*

Quadrum Prins & Perch-Nielsen *in* Manivit *et al.* (1977)
  Q. gartneri Prins & Perch-Nielsen *in* Manivit *et al.* (1977)
  Q. gothicum (Deflandre, 1959) Prins & Perch-Nielsen *in* Manivit *et al.* (1977)
  (*Q. nitidum* (Martini, 1961) Prins & Perch-Nielsen *in* Manivit *et al.*, 1977) = *Nannotetrina nitida*
  Q. sissinghii Perch-Nielsen (1984b, for Cretaceous forms previously assigned to *Q. nitidum*)

Q. trifidum (Stradner *in* Stradner & Papp, 1961) Prins & Perch-Nielsen *in* Manivit *et al.* (1977)

Radiolithus Stover (1966)
  R. planus Stover (1966)
Ramsaya Risatti (1973)
Rectilius Goy *in* Goy *et al.* (1979)
  R. productus Goy *in* Goy *et al.* (1979)
Reinhardtites Perch-Nielsen (1968)
  R. anthophorus (Deflandre, 1959) Perch-Nielsen (1968)
  R. levis Prins & Sissingh *in* Sissingh (1977)
Repagulum Forchheimer (1972)
  R. parvidentatum (Deflandre & Fert, 1954) Forchheimer (1972)
Retecapsa Black (1971a)
  R. angustiforata Black (1971a)
  R. brightonii Black (1971a)
  R. neocomiana Black (1971a)
(*Rhabdolekiskus* Hill, 1976) = *Lithraphidites*
  (*R. parallelus* Wind & Čepek, 1979) – should be assigned to another genus
(*Rhabdolithina* Reinhardt, 1967) = *Rhagodiscus*
  (*R. splendens* (Deflandre, 1953) Reinhardt, 1967) = *Rhagodiscus splendens*
(*Rhabdophidites* Manivit, 1971) = *Lithraphidites*
  (*R. moeslensis* Manivit, 1971) = *L. carniolensis*
Rhagodiscus Reinhardt (1967)
  R. angustus (Stradner, 1963) Reinhardt (1971)
  R. asper (Stradner, 1963) Reinhardt (1967)
  R. bispiralis Perch-Nielsen (1968)
  R. eboracensis Black (1971a)
  (*R. elongatus* Stover, 1966) = *R. angustus*
  R. granulatus Perch-Nielsen (1968)
  R. plebeius Perch-Nielsen (1968)
  R. reniformis Perch-Nielsen (1973)
  R. splendens (Deflandre, 1953) Verbeek (1977b)
Rhomboaster Bramlette & Sullivan (1961)
(*Rhombogyrus* Black, 1973) = *Radiolithus*
Rhombolithion Black (1973)
  R. bifurcatum (Noël, 1973) Grün & Zweili (1980)
  R. duodecostatum (Goy *in* Goy *et al.*, 1979) Perch-Nielsen (1984a)
  R. horrellii (Rood & Barnard, 1972) Perch-Nielsen (1984a)
  R. octocostatum (Rood & Barnard, 1972) Perch-Nielsen (1984a)
  R. rhombicum (Stradner & Adamiker, 1966) Black (1973)
  R. rotatum (Rood *et al.*, 1971) Black (1973)
  (*R. speetonensis* Rood & Barnard, 1972) = *R. rhombicum*
Rotelapillus Noël (1973)
  R. caravacaensis (Grün *in* Grün & Allemann, 1975) Perch-Nielsen (1984a)
  R. crenulatus (Stover, 1966) Perch-Nielsen (1984a)
  R. laffittei (Noël, 1957) Noël (1973)
  R. munitus (Perch-Nielsen, 1973) Perch-Nielsen (1984a)
  R. octoradiatus (Gartner, 1968) Perch-Nielsen (1984a)
  R. radians Noël (1973)
Rucinolithus Stover (1966)
  R. hayi Stover (1966)
  R. irregularis Thierstein *in* Roth & Thierstein (1972)
  R.? magnus Bukry (1975)
  R. sumastrocyclus Bukry (1969)
  R. wisei Thierstein (1971)
Russellia Risatti (1973)

Saeptella Goy *in* Goy *et al.* (1979) ex Goy (1981)
  S. conspicua Goy *in* Goy *et al.* (1979) ex Goy (1981)

Scampanella Forchheimer & Stradner (1973)
  S. cornuta Forchheimer & Stradner (1973)
  S. magnifica Perch-Nielsen *in* Perch-Nielsen & Franz (1977)
Scapholithus Deflandre *in* Deflandre & Fert (1954)
  S. fossilis Deflandre *in* Deflandre & Fert (1954)
Schizosphaerella Deflandre & Dangeard (1938)
  S. astraea Moshkovitz (1979)
  S. punctulata Deflandre & Dangeard (1938)
Semihololithus Perch-Nielsen (1971)
  S. bicornis Perch-Nielsen (1973)
  S. priscus Perch-Nielsen (1973)
Seribiscutum Filewicz *et al. in* Wise & Wind (1977)
  S. bijugum Filewicz *et al. in* Wise & Wind (1977)
  (*S. pinnatum* (Black, 1971a) Wise, 1983) = *Sollasites pinnatus*
  S. primitivum (Thierstein, 1974) Filewicz *et al. in* Wise & Wind (1977)
(*Similicoronilithus* Bukry, 1969) = *Cyclagelosphaera*
Sollasites Black (1967)
  S. arcuatus Black (1971a)
  S. barringtonensis Black (1967)
  S. bipolaris Rood *et al.* (1971)
  S. concentricus Rood *et al.* (1971)
  S. falklandensis Filewicz *et al. in* Wise & Wind (1977)
  S. hayi (Black, 1973) Perch-Nielsen (1984a)
  S. horticus (Stradner *et al. in* Stradner & Adamiker, 1966) Čepek & Hay (1969)
  S. lowei (Bukry, 1969) Roth (1970)
  S. pinnatus (Black, 1971a) Perch-Nielsen (1984a)
  S. pristinus Noël (1973)
  (*S. tardus* Roth, 1970) – holotype indistinct (Oligocene)
  S. thiersteinii Filewicz *et al. in* Wise & Wind (1977)
  (*S. wallacei* Bukry, 1969) Reinhardt, (1970b)
Speetonia Black (1971a)
  S. colligata Black (1971a)
  (*S. nitida* Black, 1971a) = *S. colligata*
Staurolithites Caratini (1963) = ?*Vekshinella*
  S. laffittei Caratini (1963)
Staurorhabdus Noël (1973) = ?*Vekshinella*
  (*S. quadriarcullus*) Noël, 1965) Noël, 1973) = *V. quadriarculla*
Stephanolithion Deflandre (1939)
  S. biogotii Deflandre (1939)
  S. bigotii maximum Medd (1979)
  S. carinatum Medd (1979)
  S. elongatum (Medd, 1979) Perch-Nielsen (1984a)
  (*S. laffittei* Noël, 1957) = *Rotelapillus laffittei*
  S. hexum Rood & Barnard (1972)
  S. octum (Rood & Barnard, 1972) Perch-Nielsen (1984a)
  S. speciosum Deflandre *in* Deflandre & Fert (1954)
Stoverius Perch-Nielsen (1984b)
  S. achylosus (Stover, 1966) Perch-Nielsen (1984b)
  S. asymmetricus (Bukry, 1969) Perch-Nielsen (1984b)
  S. baldiae (Stradner & Adamiker, 1966) Perch-Nielsen (1984b)
  S. biarcus (Bukry, 1969) Perch-Nielsen (1984b)
  S. coronatus (Bukry, 1969) Perch-Nielsen (1984b)
  S. helotatus (Wind & Wise *in* Wise & Wind, 1977) Perch-Nielsen (1984b)
Stradneria Reinhardt (1964)
  S. crenulata (Bramlette & Martini, 1964) Noël (1970)
  S. limbicrassa Reinhardt (1964)
Stradnerlithus Black (1971a)
  S. asymmetricus (Rood *et al.*, 1971) Medd (1979)
  S. callomonii (Rood *et al.*, 1971) Perch-Nielsen (1984a)

(*S. clatriatus* (Rood *et al.*, 1973) Goy *in* Goy *et al.* (1979) ex Goy, 1981) = *Nodosella clatriata*
S. comptus Black (1971a)
S. cotortuosus Medd *in* Gallois & Medd (1979)
S. ellipticus (Bukry, 1969) Perch-Nielsen (1984a)
S. escovillensis (Rood & Barnard, 1972) Medd (1979)
S. fractus (Black, 1973) Perch-Nielsen (1984a)
S. fragilis (Rood & Barnard, 1972) Perch-Nielsen (1984a)
S. humilis Goy *in* Goy *et al.* (1979)
S. langii (Rood & Barnard, 1972) Perch-Nielsen (1984a)
(*S. octoradiatus* Medd, 1979) = *Truncatoscaphus octoporus*
S. sattinii Medd (1979)
S. sexiramatus (Pienaar, 1969) Perch-Nielsen (1984a)
S. tortuosus Noël (1973)
(*Striatococcus* Prins, 1969; invalid) = (*Striatomarginis*)
  (*S. opacus* Prins, 1969, invalid)
(*Striatomarginis* Prins (1969) ex Rood *et al.*, 1973) = *Lotharingius*
  (*S. primitivus* Prins (1969) ex Rood *et al.*, 1973) = *L. primitivus*

Tegumentum Thierstein *in* Roth & Thierstein (1972)
  T. stradneri Thierstein *in* Roth & Thierstein (1972)
Teichorhabdus Wind & Wise *in* Wise & Wind (1977)
  T. ethmos Wind & Wise *in* Wise & Wind (1977)
Tergestiella Kamptner (1941) – based on modern species, should not be used for Cretaceous species
Tetralithus Gardet (1955) – wastebasket for poorly defined nannoliths of 4 pieces
  (*T. aculeus* Stradner, 1961) = *Ceratolithoides aculeus*
  T. cassianus Jafar (1983)
  (*T. gothicus* Deflandre, 1959) = *Quadrum gothicum*
  (*T. nitidus* Martini, 1961) = *Nannotetrina nitida* (Eocene species; Cretaceous forms = *Quadrum sissinghii*)
  (*T. obscurus* Deflandre, 1959) = *Calculites obscurus*
  T. pseudotrifidus Jafar (1983)
  T. pyramidus Gardet (1955)
  (*T. trifidus* Stradner *in* Stradner & Papp, 1961) = *Quadrum trifidum*
Tetrapodorhabdus Black (1971a)
  T. coptensis Black (1971a)
  T. decorus (Deflandre *in* Deflandre & Fert, 1954) Wind & Wise *in* Wise & Wind (1977)
Thiersteinia Wise & Watkins *in* Wise (1983)
  T. ecclesiastica Wise & Watkins *in* Wise (1983)
Thoracosphaera Kamptner (1927)
  T. eichstaettensis Keupp (1978)
  T. operculata Bramlette & Martini (1964)
  T. spinosa Keupp (1979)
  T. strobila Keupp (1979)
  T. thoracata Keupp (1979)
Thurmannolithion Grün & Zweili (1980)
  T. clatratum Grün & Zweili (1980)
Tortolithus Crux *in* Crux *et al.* (1982)
  T. caistorensis Crux *in* Crux *et al.* 1982)
  T. furlongii (Bukry, 1969) Crux *in* Crux *et al.* (1982)
  T. hallii (Bukry, 1969) Crux *in* Crux *et al.* (1982)
  T. pagei (Bukry, 1969) Crux *in* Crux *et al.* (1982)
Tranolithus Stover (1966)
  T. exiguus Stover (1977)
  T. gabalus Stover (1966)
  T. macleodiae (Bukry, 1969) Perch-Nielsen (1984a)
  T. manifestus Stover (1966)
  T. minimus (Bukry, 1969) Perch-Nielsen (1984a)
  (*T. orionatus* (Reinhardt, 1966a) Perch-Nielsen, 1968) = *T. phacelosus*

T. phacelosus Stover (1966)

T.? salillum (Noël, 1965) Crux (1981)

Trapezopentus Wind & Čepek (1979)

T. sarmatus Wind & Čepek (1979)

(*Tremalithus* Kamptner 1948 ex Deflandre *in* Piveteau, 1952) – wastebasket for nondescript, elliptical coccoliths

Tretosestrum Wilcoxon (1972)

T. perforatus Wilcoxon (1972)

Tribrachiatus Shamraï (1963)

Triquetrorhabdulus Martini (1965)

Truncatoscaphus Rood *et al.* (1971)

T. delftensis (Stradner & Adamiker, 1966) Rood *et al.* (1971)

T. hexaporus Moshkovitz & Ehrlich (1976a)

T. intermedius Perch-Nielsen (1984b)

T. octoporus Moshkovitz & Ehrlich (1976a)

T. pauciramosus (Black, 1973) Perch-Nielsen (1984a)

T. senarius (Wind & Wise *in* Wise & Wind, 1977) Perch-Nielsen (1984a)

Tubirhabdus Prins (1969) ex Rood *et al.* (1973)

T. patulus Prins (1969) ex Rood *et al.* (1973)

Tubodiscus Thierstein (1973)

T. verenae Thierstein (1973)

(*Uniplanarius* Hattner & Wise, 1980) = *Quadrum*

Vacherauvillius Goy *in* Goy *et al.* (1979) ex Goy (1981)

V. implicatus Goy *in* Goy *et al.* (1979) ex Goy (1981)

Vagalapilla Bukry (1969)

V. imbricata (Gartner, 1968) Bukry (1969)

V. matalosa (Stover, 1966) Thierstein (1973)

Vekshinella Loeblich & Tappan (1963)

V. acutifera (Vekshina, 1959) Loeblich & Tappan (1963)

V. angusta (Stover, 1966) Verbeek (1977b)

V. quadriarculla (Noël, 1965) Rood *et al.* (1971)

V. stradneri Rood *et al.* (1971)

V. thiersteinii Jafar (1983)

Vikosphaera Goy *in* Goy *et al.* (1979)

V. noeliae Goy *in* Goy *et al.* (1979)

(*Viminites* Black, 1975) = *Rhagodiscus*

Watznaueria Reinhardt (1964)

W. barnesae (Black *in* Black & Barnes, 1959) Perch-Nielsen (1968)

W. biporta Bukry (1969)

(*W. fasciata* Wind & Čepek, 1979) = *Ellipsagelosphaera fasciata*

(*W. ovata* Bukry, 1969) = *Ellipsagelosphaera ovata*

Zeugrhabdotus Reinhardt (1965)

Z. acanthus Reinhardt (1965)

Z. embergeri (Noël, 1959) Perch-Nielsen (1984a)

Z. erectus (Deflandre *in* Deflandre & Fert, 1954) Reinhardt (1965)

Z. noeliae Rood *et al.* (1971)

Z. pseudanthophorus (Bramlette & Martini, 1964) Perch-Nielsen (1984a)

(*Z. salillum* (Noël, 1965) Rood *et al.*, 1971) = *Tranolithus salillum*

Z. theta (Black *in* Black & Barnes, 1959) Black (1973)

Zygodiscus Bramlette & Sullivan (1961)

(*Z. spiralis* Bramlette & Martini, 1964) = *Placozygus fibuliformis*

(*Zygolithites* Black, 1972) = *Zeugrhabdotus*

'*Zygolithus crux*' (Deflandre & Fert, 1954) Bramlette & Sullivan, 1961) – used for any zeugoid coccolith with a +

References

Alvarez, W., Arthur, M. A., Fischer, A. G., Lowrie, W., Napoleone, G., Premoli Silva, I. & Roggenthen, W. M. 1977. Type section for the Late Cretaceous–Paleocene geomagnetic reversal time scale. *Bull. geol. Soc. Am.*, **88**, 383–9.

Arkhangelsky, A. D. 1912. Upper Cretaceous deposits of east European Russia. *Mater. Geol. Russ.*, **25**, 1–631.

Aubry, M. P. 1974. Remarques sur la systématique des *Nannoconus* de la craie. *Cah. Micropaleontol.*, **4**, 1–46.

Aubry, M. P. 1983. Corrélation biostratigraphiques entre les formations paléogènes épicontinentales de l'Europe du Nord-Ouest, basées sur le nannoplancton calcaire. Thèse Université Pierre et Marie Curie, Paris 6, 83-08, 1–208.

Aubry, M. P. & Depeche, F. 1974. Recherches sur les schizosphères: 1. Les schizosphères de Villers-sur-Mer. *Cah. Micropaleontol.*, **1**, 1–15.

Barnard, T. & Hay, W. W. 1974. On Jurassic coccoliths: A tentative zonation of the Jurassic of Southern England and North France. *Eclog. geol. Helv.*, **67** (3), 563–85.

Barrier, J. 1977a. Nannofossiles calcaires des marnes de l'Aptien inférieur type: Bédoulien de Cassis-La Bédoule (Bouches-du-Rhône). *Bull. Mus. natl. Hist. nat.*, 3/437, sc. de la Terre, **59**, 1–68.

Barrier, J. 1977b. Nannofossiles calcaires du Gargasien stratotypique. *Bull. Mus. natl. Hist. nat.*, 3/485, sc. de la Terre, **62**, 173–228.

Black, M. 1964. Cretaceous and Tertiary coccoliths from Atlantic seamounts. *Palaeontology*, **7**, 306–16.

Black, M. 1967. New names for some coccolith taxa. *Proc. geol. Soc. London*, **1640**, 139–45.

Black, M. 1968. Taxonomic problems in the study of coccoliths. *Palaeontology*, **11** (5), 793–813.

Black, M. 1971a. Coccoliths of the Speeton Clay and Sutterby Marl. *Proc. Yorkshire geol. Soc.*, **38** (3), 381–424.

Black, M. 1971b. Problematical microfossils from the Gault Clay. *Geol. Mag.*, **108**, 325–7.

Black, M. 1972, 1973, 1975. British Lower Cretaceous coccoliths. I. Gault Clay. 1, 2, 3. *Monogr. palaeontogr. Soc. London*, **126**, 1–48; **127**, 49–112; **129**, 113–42.

Black, M. & Barnes, B. 1959. The structure of coccoliths from the English Chalk. *Geol. Mag.*, **96**, 321–8.

Boudreaux, J. E. & Hay, W. W. 1969. Calcareous nannoplankton and biostratigraphy of the Late Pliocene–Pleistocene–Recent sediments of the Submarex cores. *Rev. Esp. Micropaleontol.*, **1**, 249–92.

Bramlette, M. N. & Martini, E. 1964. The great change in calcareous nannoplankton fossils between the Maestrichtian and Danian. *Micropaleontology*, **10**, 291–322.

Bramlette, M. N. & Sullivan, F. R. 1961. Coccolithophorids and related nannoplankton of the early Tertiary in California. *Micropaleontology*, **7**, 129–88.

Brönnimann, P. 1955. Microfossils incertae sedis from the Upper Jurassic and Lower Cretaceous of Cuba. *Micropaleontology*, **1**, 28–51.

Bukry, D. 1969. Upper Cretaceous coccoliths from Texas and Europe. *Univ. Kansas Paleontol. Contrib.*, **51** (Protista 2), 1–79.

Bukry, D. 1973. Phytoplankton stratigraphy, DSDP Leg 20, Western Pacific Ocean. *Initial Rep. Deep Sea drill. Proj.*, **20**, 307–17.

Bukry, D. 1974a. Cretaceous and Paleogene coccolith stratigraphy, DSDP, Leg 26. *Initial Rep. Deep Sea drill. Proj.*, **26**, 669–73.

Bukry, D. 1974b. Coccolith stratigraphy, offshore Western Australia, DSDP, Leg 27. *Initial Rep. Deep Sea drill. Proj.*, **27**, 623–30.

Bukry, D. 1975. Coccolith and silicoflagellate stratigraphy, Northwestern Pacific Ocean, DSDP, Leg 32. *Initial Rep. Deep Sea drill. Proj.*, **32**, 677–701.

Bukry, D. & Bramlette, M. N. 1970. Coccolith age determination Leg 3, Deep Sea Drilling Project. *Initial Rep. Deep Sea drill. Proj.*, **3**, 589–611.

Burns, D. A. 1976. Nannofossils from the Lower and Upper Cretaceous chalk deposits, Nettleton, Lincolnshire, England. *Rev. Esp. Micropaleontol.*, **8** (2), 279–300.

Caratini, C. 1963. Contribution à l'étude des coccolithes du Cénomanien supérieur et du Turonien de la région de Rouen. Thèse, Faculté des Sciences, Université d'Alger. (Publication du Laboratoire de Géologie Appliquée), 1–61.

Čepek, P. 1970. Zur Vertikalverbreitung von Coccolithen-Arten in der Oberkreide NW-Deutschlands. *Geol. Jahrbuch*, *Hannover*, **88**, 235–64.

Čepek, P. & Hay, W. W. 1969. Calcareous nannoplankton and biostratigraphic subdivision of the Upper Cretaceous. *Trans. Gulf Coast Assoc. geol. Soc.*, **19**, 323–36.

Čepek, P. & Hay, W. W. 1970. Zonation of the Upper Cretaceous using calcareous Nannoplankton. *Paläobotanik*, B, **3** (3–4), 333–40.

Channel, J. E. T., Lowrie, W., Medizza, F. & Alvarez, W. 1978. Paleomagnetism and tectonics in Umbria, Italy. *Earth planet. Sci. Lett.*, **39**, 199–210.

Cita, M. B. & Pasquaré, G. 1959. Studi stratigrafici sul sistema cretaceo in Italia. Nota IV. Osservazioni micropaleontologiche sul Cretaceo delle Dolomiti. *Riv. Ital. Paleontol. Stratigr.*, **65**, 385–442.

Crux, J. A. 1981. New calcareous nannofossil taxa from the Cretaceous of South East England. *Neues Jahrb. Geol. Palaeontol. Monatshefte*, **10**, 633–40.

Crux, J. A. 1982. Upper Cretaceous (Cenomanian to Campanian) calcareous nannofossils. In: A. R. Lord (ed.), *A Stratigraphical Index of Calcareous Nannofossils*, pp. 81–135. British Micropal. Soc.

Crux, J. A., Hamilton, G. B., Lord, A. R. & Taylor, R. J. 1982. *Tortolithus* gen. nov. Crux and new combinations of Mesozoic calcareous nannofossils from England. *INA Newsletter*, **4** (2), 98–101.

Crux, J. A. & Lord, A. R. 1982. Discussion. In: A. R. Lord (ed.), *A Stratigraphical Index of Calcareous Nannofossils*. pp. 168–73. British Micropal. Soc.

Deflandre, G. 1939. Les stéphanolithes, représentants d'un type nouveau de coccolithes du Jurassique supérieur. *C.r. Seances Acad. Sci. Paris*, **208**, 1331–3.

Deflandre, G. 1947. *Braarudosphaera* nov. gen., type d'une famille nouvelle de Coccolithophoridés actuels à éléments composites. *C.r. Seances Acad. Sci. Paris*, **225**, 439–41.

Deflandre, G. 1952. Classe des coccolithophoridés. In: P. P. Grassé (ed.), *Traité de Zoologie*, vol. 1, pp. 439–70.

Deflandre, G. 1953. Hétérogénéité intrinsèque et pluralité des éléments dans les coccolithes actuels et fossiles. *C.r. Seances Acad. Sci. Paris*, **237**, 1785–7.

Deflandre, G. 1957. *Goniolithus* nov. gen., type d'une famille nouvelle de Coccolithophoridés fossiles, à éléments pentagonaux non composites. *C.r. Seances Acad. Sci. Paris*, **244**, 2539–41.

Deflandre, G. 1959. Sur les nannofossiles calcaires et leur systématique. *Rev. Micropaleontol.*, **2**, 127–52.

Deflandre, G. 1963. Sur les Microrhabdulidés, famille nouvelle de nannofossiles calcaires. *C.r. Seances Acad. Sci. Paris*, **256**, 3484–6.

Deflandre, G. 1970. Présence de nannofossiles calcaires (coccolithes et incertae sedis) dans le Siluro-dévonien d'Afrique du Nord. *C.r. Hebd. Seances Acad. Sci. Paris*, **256**, 3484–6.

Deflandre, G. & Dangeard, L. 1938. *Schizosphaerella*, un nouveau microfossile méconnu du Jurassique moyen et supérieur. *C.r. Seances Acad. Sci. Paris*, **207**, 1115–17.

Deflandre, G. & Deflandre, M. 1959. Sur l'existence d'une association particulière de Nannoconidés dans le Crétacé supérieur du bassin de Paris. *C.r. Seances Acad. Sci. Paris*, **248**, 2372–4.

Deflandre, G. & Deflandre, M. 1962. Remarques sur l'évolution des Nannoconidés à propos de quelques nouveaux types du Crétacé inférieur de Haute-Provence. *C.r. Seances Acad. Sci. Paris*, **255**, 2638–40.

Deflandre, G. & Deflandre, M. 1967. *Fichier micropaléontologie général. Série 17, Nannofossiles calcaires I.* Centre Nat. Rech. Sci., pp. 3423–3830.

Deflandre, G. & Fert, C. 1954. Observations sur les Coccolithophoridés actuels et fossiles en microscopie ordinaire et électronique. *Ann. Paleontol.*, **40**, 115–76.

Deres, F. & Acheriteguy, J. 1972. Contribution à l'étude des Nannoconidés dans le Crétacé inférieur du Bassin d'Aquitaine. *Mem. Bur. Rech. geol. minieres*, **77**, 155–63.

Deres, F. & Acheriteguy, J. 1980. Biostratigraphie des Nannoconidés. *Bull. Cent. Rech. S.N.E.A.*, **4**, 1–53.

Di Nocera, S. & Scandone, P. 1977. Triassic nannoplankton limestones of deep basin origin in the Central Mediterranean region. *Palaeogeogr. Palaeoclimatol. Palaeoecol.*, **21**, 101–11.

Doeven, P. H. 1983. Cretaceous nannofossil stratigraphy and paleoecology of the Canadian Atlantic Margin. *Bull. geol. Surv. Can.*, **356**, 1–70.

Ehrenberg, C. C. 1836. Bemerkungen über feste mikroskopische, anorganische Formen in den erdigen und derben Mineralien. *Ber. Dtsch. Akad. Wiss.*, **1836**, 84–5.

Farinacci, A. 1969–83. *Catalogue of Calcareous Nannofossils*, 11 vols. Tecnoscienza, Roma.

Forchheimer, S. 1968. Die Coccolithen des Gault-Cenoman, Cenoman and Turon in der Bohrung Höllviken I, Südwest-Schweden. *Sver. geol. Unders.*, ser. C. 635, **62** (6), 1–84.

Forchheimer, S. 1972. Scanning electron microscope studies of Cretaceous coccoliths from the Köpingsberg Borehole No. 1, SE Sweden. *Sver. geol. Unders.*, ser. C. 668, **65** (14), 1–141.

Forchheimer, S. & Stradner, H. 1973. *Scampanella*, eine neue Gattung kretazischer Nannofossilien. *Verh. geol. Bundesanst.* *(Wien)*, 2, 285–9.

Gallois, R. W. & Medd, A. W. 1979. Coccolith-rich marker bands in the English Kimmeridge Clay. *Geol. Mag.*, **116**, 247–60.

Gardet, M. 1955. Contribution à l'étude des coccolithes des terrains Mesogènes de l'Algérie. *Publ. Serv. Carte Geol. Algerie*, ser. 2, Bull. 5, pp. 477–550.

Gartner, S. 1968. Coccoliths and related calcareous nannofossils from Upper Cretaceous deposits of Texas and Arkansas. *Univ. Kansas Paleontol. Contrib.*, **48**, 1–56.

Gartner, S. 1977. Nannofossils and biostratigraphy: An overview. *Earth Sci. Rev.*, **13**, 227–50.

Gartner, S. & Bukry, D. 1975. Morphology and phylogeny of the coccolithophycean family Ceratolithaceae. *J. Res. U.S. geol. Surv.*, **3**, 451–65.

Gartner, S. & Gentile, R. 1973. Problematic Pennsylvanian coccoliths from Missouri. *Micropaleontology*, **18**, 401–4.

Gazdzicka, E. 1978. Calcareous nannoplankton from the uppermost Cretaceous and Paleogene deposits of the Lublin Upland. *Acta geol. Pol.*, **28** (3), 335–75.

Geel, T. 1966. Biostratigraphy of the Upper Jurassic and Cretaceous sediments near Caravaca (SE Spain) with special emphasis on the Tintina and *Nannoconus*. *Geol. Mijnbouw*, **45**, 375–85.

Gorka, H. 1957. Les coccolithophoridés du Maestrichtien supérieur de Pologne. *Acta paleontol. Pol.*, **2**, 235–84.

Goy, G. 1981. *Nannofossiles calcaires des schistes carton (Toarcien Inférieur) du Bassin de Paris*. Doc. de la RCP 459, editions BRGM, 86 pp.

Goy, G., Noël, D. & Busson, G. 1979. Les conditions de sédimentation des schistes-carton (Toarcien Inf.) du Bassin de Paris déduites de l'étude des nannofossiles calcaires et des diagraphies. *Doc. Lab. Geol. Fac. Sci. Lyon*, **75**, 33–57.

Gran, H. H. & Braarud, T. 1935. A quantitative study of the phytoplankton in the Bay of Fundy and the Gulf of Maine (including observations on hydrography, chemistry and turbidity). *J. Biol. Board Canada*, **1**, 279–467.

Grassé, P. P. 1952. *Traité de Zoologie, Anatomie, Systématique, Biologie*. Vol. 1, fasc. 1: Phylogénie. Protozoaires: généralités. Flagellés. Masson, Paris, XII + 1071 pp., 830 figs. (Classe des Coccolithophoridés (Coccolithophoridae Lohmann, 1902), pp. 439–70, figs. 339–364 by G. Deflandre.)

Grün, W. & Allemann, F. 1975. The Lower Cretaceous of Caravaca (Spain): Berriasian calcareous nannoplankton of the Miravetes Section (Subbetic Zone, Prov. of Murcia). *Eclog. geol. Helv.*, **68** (1), 147–211.

Grün, W., Kittler, G., Lauer, G., Papp, A., *et al.* 1972. Studien in der Unterkreide des Wienerwaldes. *Jahrb. geol. Bundesanst. (Wien)*, **115**, 103–86.

Grün, W., Prins, B. & Zweili, F. 1974. Coccolithophoriden aus dem Lias epsilon von Holzmaden (Deutschland). *Neues Jahrb. Geol. Palaeontol. Abhandlungen*, **147** (3), 294–328.

Grün, W. & Zweili, F. 1980. Das kalkige Nannoplankton der Dogger-Malm-Grenze im Berner Jura bei Liesberg (Schweiz). *Jahrb. geol. Bundesanst. (Wien)*, **123** (1), 231–341.

Hamilton, G. 1977. Early Jurassic calcareous nannofossils from Portugal and their biostratigraphic use. *Eclog. geol. Helv.*, **70** (2), 575–97.

Hamilton, G. 1979. Lower and Middle Jurassic calcareous nannofossils from Portugal. *Eclog. geol. Helv.*, **72** (1), 1–17.

Hamilton, G. 1982. Triassic and Jurassic calcareous nannofossils. In: A. R. Lord (ed.), *A Stratigraphical Index of Calcareous Nannofossils*, pp. 136–67. British Micropal. Soc.

Haq, B. U. 1978. Calcareous nannoplankton. In: B. U. Haq & A. Boersma (eds.), *Introduction to Marine Micropaleontology*, pp. 79–107. Elsevier.

Hattner, J. G., Wind, F. H. & Wise, S. W., Jr. 1980. The Santonian–Campanian boundary: comparison of nearshore–offshore calcareous nannofossil assemblages. *Cah. Micropaleontol.*, **1980** (3), 9–26.

Hattner, J. G. & Wise, S. W., Jr. 1980. Upper Cretaceous calcareous nannofossil biostratigraphy of South Carolina. *South Carolina Geology*, **24** (2), 41–115.

Hay, W. W. 1977. Calcareous nannofossils. In: A. T. S. Ramsay (ed.), *Oceanic Micropalaeontology*, pp. 1055–1200. Academic Press, London.

Hay, W. W., Mohler, H. P. & Wade, M. E. 1966. Calcareous nannofossils from Nal'chik (Northwest Caucasus). *Eclog. geol. Helv.*, **59**, 379–99.

Hay, W. W. & Mohler, H. P. 1967. Calcareous nannoplankton from early Tertiary rocks at Pont Labau, France, and Paleocene–Eocene correlations. *J. Paleontol.*, **41**, 1505–41.

Hay, W. W., Mohler, H. P., Roth, P. H., Schmidt, R. R. & Boudreaux, J. E. 1967. Calcareous nannoplankton zonation of the Cenozoic of the Gulf Coast and Caribbean-Antillean area, and transoceanic correlation. *Trans. Gulf Coast Assoc. geol. Soc.*, **17**, 428–80.

Hay, W. W. & Towe, K. M. 1963. *Microrhabdulus belgicus*, a new species of nannofossil. *Micropaleontology*, **9**, 95–6.

Heck, S. van. 1979. Nannoplankton contents of the Type-Maastrichtian. *INA Newsletter*, **1** (1), N 5–N 6.

Heck, S. van. 1979–82. Bibliography and taxa of calcareous nannoplankton. *INA Newsletter*, **1** (1), AB 1–B 27; **1** (2), 13–42; **2** (1), 5–34; **2** (2), 43–80; **3** (1), 4–41; **3** (2), 51–86; **4** (1), 7–50; **4** (2), 65–96.

Herngreen, G., Randrianasolo, A. & Verbeek, J. W. 1982. Micropaleontology of Albian to Danian strata in Madagascar. *Micropaleontology*, **28** (1), 97–109.

Hill, M. E. 1976. Lower Cretaceous calcareous nannofossils from Texas and Oklahoma. *Palaeontographica B*, **156**, 103–79.

Hinte, J. E. van. 1976. A Cretaceous time scale. *Bull. Amer. Assoc. Petrol. Geol.*, **60**, 498–516.

Hoffmann, N. 1970a. Coccolithineen aus der weissen Schreibkreide (Unter Maastricht) von Jasmund auf Rügen. *Geologie*, **19**, 846–79.

Hoffmann, N. 1970b. *Placozygus* n.gen. (Coccolithineen) aus der Oberkreide des nördlichen Mitteleuropas. *Geologie*, **19**, 1004–9.

Hoffmann, N. 1970c. Elektronenmikroskopische Untersuchungen an stabförmigen Nannofossilien aus der Kreide und dem Paläogen Norddeutschlands. *Hercynia*, **7**, 131–62.

Hoffmann, N. 1970d. Taxonomische Untersuchungen an Coccolithen aus der Kreide Norddeutschlands anhand elektronenmikroskopischer Aufnahmen. *Hercynia*, **7**, 163–98.

Hoffmann, N. 1972a. Electronenoptische Untersuchungen an Coccolithineen aus der Kreide und dem Paläogen des nördlichen Mitteleuropas. *Hallesches Jahrb. mitteldtsch. Erdgesch.*, **11**, 41–60.

Hoffmann, N. 1972b. Coccolithen aus der Kreide und dem Paläogen des nördlichen Mitteleuropas. *Geologie*, **73**, 1–121.

Hojjatzadeh, M. 1977. Calcareous nannofossils from the Barremian to Coniacian of Western Morocco and their zonal use. *Newsl. Stratigr.*, **6** (3), 171–82.

International Code of Botanical Nomenclature. 1978. Bohn, Scheltema & Holkema, Utrecht, 457 pp.

Jafar, A. S. 1983. Significance of Late Triassic calcareous Nannoplankton from Austria and Southern Germany. *Neues Jahrb. Geol. Palaeontol. Abhandlungen*, **166** (2), 218–59.

Kälin, O. 1980. *Schizosphaerella punctulata* Deflandre & Dangeard: Wall ultrastructure and preservation in deeper-water carbonate sediments of the Tethyan Jurassic. *Eclog. geol. Helv.*, **73** (3), 983–1008.

Kälin, O. & Bernoulli, D. 1984. *Schizosphaerella* Deflandre & Dangeard in Jurassic deeper-water carbonate sediments, Site

547 (Mazagan continental margin) and Mesozoic Tethys. *Initial Rep. Deep Sea drill. Proj.* (in press).

Kamptner, E. 1927. Beitrag zur Kenntnis adriatischer Coccolithophoriden. *Arch. Protistenk.*, **58**, 173–84.

Kamptner, E. 1931. *Nannoconus steinmanni* nov. gen. nov. spec., ein merkwürdiges gesteinsbildendes Mikrofossil aus dem jüngeren Mesozoikum der Alpen. *Palaont. Z.*, **13**, 288–98.

Kamptner, E. 1938. Einige Bemerkungen über *Nannoconus*. *Palaont. Z.*, **20**, 249–57.

Kamptner, E. 1941. Die Coccolithineen der Südwestküste von Istrien. *Ann. Naturh. Mus. Wien*, **51**, 54–149.

Kamptner, E. 1948. Coccolithen aus dem Torton des Inneralpinen Wiener Beckens. *Sitzungsbericht Oesterr. Akad. Wiss., Math.-Naturw. Kl.*, Part I, **157**, 1–16.

Kamptner, E. 1950. Ueber den submikroskopischen Aufbau der Coccolithen. *Anzeiger Oesterr. Akad. Wiss., Math.-Naturw. Kl.*, **87**, 152–8.

Keupp, H. 1976. Kalikiges Nannoplankton aus den Solnhofener Schichten (Unter-Tithon, Südliche Frankenalb). *Neues Jahrb. Geol. Palaeontol. Monatshefte*, **6**, 361–81.

Keupp, H. 1978. Calcisphaeren des Untertithon der Südlichen Frankenalb und die systematische Stellung von *Pithonella* Lorenz 1901. *Neues Jahrb. Geol. Palaeontol. Monatshefte*, **2**, 87–98.

Keupp, H. 1979. Calciodinelloidea aus der Blätterton-Fazies des nordwest-deutschen Unter-Barremium. *Ber. naturhist. Ges. Hannover*, **122**, 7–69.

Lambert, B. 1980. Etude de la nannoflore calcaire du Campanien charentais. *Cah. Micropaleontol.* **1980** (3), 39–53.

Lapparent, J. de. 1931. Sur les prétendus 'embryons de Lagena'. *C.r. Soc. geol. Fr.*, **1931**, 222–3.

Lauer, G. 1975. Evolutionary trends in the Arkhangelskiellaceae (calcareous nannoplankton) of the Upper Cretaceous of Central Oman, SE Arabia. *Arch. Sci. Geneve*, **28**, 259–62.

Loeblich, A. R. & Tappan, H. 1963. Type fixation and validation of certain calcareous nannoplankton genera. *Proc. Biol. Soc. Wash.*, **76**, 191–6.

Loeblich, A. R. & Tappan, H. 1966, 1968, 1969, 1970a, b, 1971, 1973. Annotated index and bibliography of the calcareous nannoplankton. I: *Phycologia*, **5**, 81–216; II: *J. Paleontol.*, **42**, 584–98; III: *J. Paleontol.*, **43**, 568–88; IV: *J. Paleontol.*, **44**, 558–74; V: *Phycologia*, **9**, 157–74; VI: *Phycologia*, **10**, 315–39; VII: *J. Paleontol.*, **47**, 715–59.

Lohmann, H. 1919. Die Revölkerung des Ozeans mit Plankton nach den Ergebnissen der Zentrifugenfänge der 'Deutschland' 1911. *Arch. Biont.*, **4** (3), 1–617.

Lyul'eva, S. A. 1967. Coccolithophoridae in the Turonian strata of the Dnieper–Don Basins. *Geol. Zhurnal*, **27** (6), 91–8.

Manivit, H. 1961. Contribution à l'étude des coccolithes de l'Eocène. *Publ. Serv. Carte Geol. Algerie*, ser. 2, Bull. **25**, 331–82.

Manivit, H. 1965. Nannofossiles calcaires de L'Albo-Aptien. *Rev. Micropaleontol.*, **8**, 189–201.

Manivit, H. 1966. Sur quelques coccolithes nouveaux du Néocomien. *C.r. Somm. Soc. geol. Fr.*, **7**, 267–8.

Manivit, H. 1968. Nannofossiles calcaires du Turonien et du Sénonien. *Rev. Micropaleontol.*, **10**, 277–86.

Manivit, H. 1971. Les nannofossiles calcaires du Crétacé français (de l'Aptien au Danien). Essai de biozonation appuyée sur les stratotypes. Thèse, Université de Paris.

Manivit, H., Perch-Nielsen, K., Prins, B. & Verbeek, J. W. 1977. Mid Cretaceous calcareous nannofossil biostratigraphy. *Kon. Nederl. Akad. Wet.*, B **80**, (3), 169–81.

Manivit, H. *et al.* 1979. Calcareous nannofossil events in the Lower and Middle Cretaceous. *INA Newsletter*, **1**, N 7.

Martini, E. 1961. Nannoplankton aus dem Tertiär und der obersten Kreide von SW-Frankreich. *Senckenbergiana Lethaea*, **42**, 1–32.

Martini, E. 1965. Mid-Tertiary calcareous nannoplankton from Pacific deep-sea cores. In: W. F. Whittard & R. B. Bradshaw (eds.), *Submarine Geology and Geophysics. Proc. 17th Symp. Colston Res. Soc.*, London, pp. 393–411. Butterworths.

Martini, E. 1967. *Ceratolithina hamata* n.gen. n.sp., aus dem Alb von N-Deutschland (Nannoplankton, incertae sedis). *Neues Jahrb. Geol. Palaeontol. Abhandlungen*, **128** (3), 294–8.

Martini, E. 1976. Cretaceous to Recent calcareous nannoplankton from the Central Pacific Ocean (DSDP Leg 33). *Initial Rep. Deep Sea drill. Proj.*, **33**, 383–423.

Martini, E. & Stradner, H. 1960. *Nannotetraster*, eine stratigraphisch bedeutsame neue Discoasteridengattung. *Erdoel-Z.*, **76**, 266–70.

Medd, A. W. 1971. Some middle and upper Jurassic Coccolithophoridae from England and France. In: A. Farinacci (ed.), *Proceedings II Planktonic Conference Roma*, 1970, vol. 2, pp. 821–45. Edizioni Tecnoscienza.

Medd, A. W. 1979. The Upper Jurassic coccoliths from the Haddenham and Gamlingay boreholes (Cambridgeshire, England). *Eclog. geol. Helv.*, **72** (1), 19–109.

Medd, A. W. 1982. Nannofossil zonation of the English Middle and Upper Jurassic. *Marine Micropaleontol.*, **7** (1), 73–95.

Minoura, N. & Chitoku, T. 1979. Calcareous nannoplankton and problematic microorganisms found in the Late Paleozoic limestones. *J. Fac. Sci. Hokkaido Univ.*, ser. IV, **19** (1, 2), 199–212.

Moshkovitz, S. 1972. Biostratigraphy of the genus *Nannoconus* in the Lower Cretaceous sediments of the subsurface: Ashquelon-Helez area, central Israel. *Israel J. earth Sci.*, **21**, 1–28.

Moshkovitz, S. 1979. On the distribution of *Schizosphaerella punctulata* Deflandre & Dangeard and *Schizosphaerella astraea* n. sp. in the Liassic section of Stowell Park Borehole (Gloucestershire) and in some other Jurassic localities in England. *Eclog. geol. Helv.*, **72** (2), 455–65.

Moshkovitz, S. 1982. On the findings of a new calcareous nannofossil (*Conusphaera zlambachensis*) and other calcareous organisms in the Upper Triassic sediments of Austria. *Eclog. geol. Helv.*, **75** (3), 611–19.

Moshkovitz, S. & Ehrlich, A. 1976a. Distribution of Middle and Upper Jurassic calcareous nannofossils in the Northeastern Negev, Israel and in Gebel Maghara, Northern Sinai. *Bull. geol. Surv. Israel*, **69**, 1–47.

Moshkovitz, S. & Ehrlich, A. 1976b. *Schizosphaerella punctulata* Deflandre et Dangeard and *Crepidolithus crassus* (Deflandre) Noël, Upper Liassic Calcareous Nannofossils from Israel and Northern Sinai. *Israel J. Earth Sci.*, **25**, 51–7.

Neugebauer, J. 1975. Fossil-Diagenese in der Schreibkreide: Coccolithen. *Neues Jahrb. Geol. Palaeontol. Monatshefte*, **1975**, 489–502.

Noël, D. 1957. Coccolithes des terrains Jurassiques de l'Algérie. *Publ. Serv. Carte Geol. Algerie*, **2** (8), 303–85.

Noël, D. 1959. Etude de coccolithes du Jurassique et du Crétacé inférieur. *Publ. Serv. Carte Geol. Algerie*, **2** (20), 155–96.

Noël, D. 1961. Sur la présence de Coccolithophoridés dans des terrains primaires. *C.r. Seances Acad. Sci. Paris*, **252**, 3625–7.

Noël, D. 1965. *Sur les Coccolithes du Jurassique Européen et d'Afrique du Nord*. Edition du CNRS Paris, 209 pp.

Noël, D. 1969. *Arkhangelskiella* (coccolithes crétacés) et formes affines du Bassin de Paris. *Rev. Micropaleontol.*, **11** (4), 191–204.

Noël, D. 1970. *Coccolithes Crétacés. La Craie Campanienne du Bassin de Paris*. CNRS, Paris, 129 pp.

Noël, D. 1973. Nannofossiles calcaires de sédiments jurassiques finement laminés. *Bull. Mus. natl. Hist. nat.*, ser. 3, **75**, 95–156.

Noël, D. 1980. Niveaux d'apparition et de disparition relatifs de nannofossiles calcaires utilisables comme repères stratigraphiques dans le Crétacé inférieur et moyen. *Cah. Micropaleontol.*, **1980** (3), 57–83.

Penn, I. E., Dingwall, R. G. & Knox, R. W. O'B. 1980. The Inferior Oolite (Bajocian) sequence from a borehole in Lyme Bay, Dorset. *Rep. Inst. geol. Sci.*, **79** (3), 1–27.

Penn, I. E., Merriman, R. J. & Wyatt, R. J. 1979. The Bathonian strata of the Bath–Frome area. *Rep. Inst. geol. Sci.*, **78** (22), 1–88.

Perch-Nielsen, K. 1967. Eine Präparationstechnik zur Untersuchung von Nannoplankton im Lichtmikriskop und im Elektronenmikroskop. *Medd. Dansk. geol. Foren.*, **17**, 129–30.

Perch-Nielsen, K. 1968. Der Feinbau und die Klassifikation der Coccolithen aus dem Maastrichtien von Dänemark. *K. Dan. Vidensk. Selsk. Biol. Skr.*, **16** (1), 1–96.

Perch-Nielsen, K. 1969. Die Coccolithen einiger dänischer Maastrichtien- und Danienlokalitäten. *Bull. geol. Soc. Denmark*, **19**, 51–68.

Perch-Nielsen, K. 1971. Einige neue Coccolithen aus dem Paleozän der Bucht von Biskaya. *Bull. geol. Soc. Denmark*, **20**, 347–61.

Perch-Nielsen, K. 1972. Remarks on Late Cretaceous to Pleistocene coccoliths from the North Atlantic. *Initial Rep. Deep Sea drill. Proj.*, **12**, 1003–69.

Perch-Nielsen, K. 1973. Neue Coccolithen aus dem Maastrichtien von Dänemark, Madagaskar und Aegypten. *Bull. geol. Soc. Denmark*, **22**, 306–33.

Perch-Nielsen, K. 1977. Albian to Pleistocene calcareous nannofossils from the Western South Atlantic, DSDP Leg 39. *Initial Rep. Deep Sea drill. Proj.*, **39**, 699–825.

Perch-Nielsen, K. 1979a. Calcareous nannofossils from the Cretaceous between the North Sea and the Mediterranean. *IUGS Series A*, **6**, 223–72.

Perch-Nielsen, K. 1979b. Calcareous nannofossil zonation at the Cretaceous/Tertiary boundary in Denmark. *Proceedings Cretaceous–Tertiary Boundary Events symposium, Copenhagen*, **1**, 120–6.

Perch-Nielsen, K. 1981. New Maastrichtian and Paleocene calcareous nannofossils from Africa, Denmark, the USA and the Atlantic, and some Paleocene lineages. *Eclog. geol. Helv.*, **74** (3), 831–63.

Perch-Nielsen, K. 1983. Recognition of Cretaceous stage boundaries by means of calcareous nannofossils. In: T. Birkelund, *et al.* (eds.), *Symposium on Cretaceous Stage Boundaries, Copenhagen, Abstracts*, pp. 152–6.

Perch-Nielsen, K. 1984a. Validation of new combinations. *INA Newsletter*, **6** (1), 42–6.

Perch-Nielsen, K. 1984b. New Late Cretaceous and Paleogene calcareous nannofossils. *Eclog. geol. Helv.*, **77** (3) (in press).

Perch-Nielsen, K. & Franz, H. E. 1977. *Lapideacassis* and *Scampanella*, calcareous nannofossils from the Paleocene at Sites 354 and 356, DSDP Leg 39, South Atlantic. *Initial Rep. Deep Sea drill. Proj.*, **39**, 849–62.

Perch-Nielsen, K., McKenzie, J. A. & He, Q. 1982. Biostratigraphy and isotope stratigraphy and the 'catastrophic' extinction of calcareous nannoplankton at the Cretaceous/Tertiary boundary. *Spec. Pap. geol. Soc. Am.*, **190**, 353–71.

Perch-Nielsen, K., Sadek, A., Barakat, M. G. & Teleb, R. 1978. Late Cretaceous and Early Tertiary calcareous nannofossil and planktonic Foraminifera Zones from Egypt. *Actes du VIe Colloque Africain de Micropaleontologie, Tunis*, **2**, 337–403.

Pienaar, R. N. 1966. Microfossils from the Cretaceous System of Zululand studied with the aid of the electron microscope. *South Afr. J. Sci.*, **62**, 147–57.

Pienaar, R. N. 1969. Upper Cretaceous coccolithophorids from Zululand, South Africa. *Palaeontology*, **11**, 361–7.

Pirini-Radrizzani, C. 1971. Coccoliths from Permian deposits of eastern Turkey. In: A. Farinacci (ed.), *Proceedings II Planktonic Conference, Roma*, 1970, pp. 1017–37. Edizioni Tecnoscienza.

Piveteau, J. 1952. *Traité de Paléontologie*, vol. 1, pp. 107–15. Masson, Paris.

Prins, B. 1969. Evolution and stratigraphy of coccolithinids from the Lower and Middle Lias. In: P. Brönnimann & H. H. Renz, *Proceedings First International Conference on Planktonic Microfossils, Geneva*, **2**, 547–58.

Prins, B. 1971. Speculations on relations, evolution and stratigraphic distribution of discoasters. In: A. Farinacci (ed.), *Proceedings II Planktonic Conference, Roma*, 1970, pp. 1017–37. Edizioni Tecnoscienza.

Rade, J. 1979. Cretaceous biostratigraphic zonation based on calcareous nannoplankton in Middle East and offshore Australia. *Exogram and Oil & Gas*, **25** (4), 19–21.

Reinhardt, P. 1964. Einige Kalkflagellaten-Gattungen (Coccolithophoriden, Coccolithineen) aus dem Mesozoikum Deutschlands. *Monatsber. Dt. Akad. Wiss. Berlin*, **6**, 749–59.

Reinhardt, P. 1965. Neue Familien für fossile Kalkflagellaten (Coccolithophoriden, Coccolithineen). *Monatsber. Dt. Akad. Wiss. Berlin*, **7**, 30–40.

Reinhardt, P. 1966a. Zur Taxionomie und Biostratigraphie des fossilen Nannoplanktons aus dem Malm, der Kreide und dem Alttertiär Mitteleuropas. *Freiberger Forschungsh.*, C **196**, 1–109.

Reinhardt, P. 1966b. Fossile Vertreter coronoider und styloider Coccolithen (Familie Coccolithaceae Poche 1913). *Monatsber. Dt. Akad. Wiss. Berlin*, **8** (6), 513–24.

Reinhardt, P. 1967. Fossile Coccolithen mit rhagoidem Zentralfeld (Fam. Ahmuellerellaceae, Subord. Coccolithineae). *Neues Jahrb. Geol. Palaeontol. Monatshefte*, **1967**, 163–78.

Reinhardt, P. 1969. Neue Coccolithen-Arten aus der Kreide. *Monatsber. Dt. Akad. Wiss. Berlin*, **11**, 932–8.

Reinhardt, P. 1970a, b, 1971. Synopsis der Gattungen und Arten der mesozoischen Coccolithen und anderer kalkiger Nannofossilien. I, II, III. *Freiberger Forschungsh.*, C **260**, 5–32; C **265**, 41–111; C **267**, 19–41.

Reinhardt, P. 1972. Coccolithen. Kalkiges Nannoplankton seit Jahrmillionen. *Neue Brehm Bucherei*, **453**, 1–99.

Reinhardt, P. & Gorka, H. 1967. Revision of some Upper Cretaceous coccoliths from Poland and Germany. *Neues Jahrb. Geol. Palaeontol. Abhandlungen*, **129**, 240–56.

Risatti, J. B. 1973. Nannoplankton biostratigraphy of the Upper Bluffport Marl – Lower Prairie Bluff Chalk interval (Upper Cretaceous), in Mississippi. Proc. Symp. Calc. Nannofossils. *Gulf Coast Sect. Soc. Econ. Paleontol. Mineral.*, pp. 8–57.

Romein, A. J. T. 1977. Calcareous nannofossils from the Cretaceous/Tertiary boundary interval in the Barranco del Gredero I & II. *Proc. Kon. Ned. Akad. Wetensch*, ser. B, **80** (4), 256–68 and 269–79.

Romein, A. J. T. 1979. Lineages in early Paleogene calcareous nannoplankton. *Utrecht Micropaleontol. Bull.*, **22**, 1–231.

Rood, A. P. & Barnard, T. 1972. On Jurassic coccoliths: *Stephanolithion, Diadozygus* and related genera. *Eclog. geol. Helv.*, **65** (2), 327–42.

Rood, A. P., Hay, W. W. & Barnard, T. 1971. Electron microscope studies of Oxford clay coccoliths. *Eclog. geol. Helv.*, **64** (2), 245–72.

Rood, A. P., Hay, W. W. & Barnard, T. 1973. Electron microscope studies of Lower and Middle Jurassic coccoliths. *Eclog. geol. Helv.*, **66** (2), 365–82.

Roth, P. H. 1970. Oligocene calcareous nannoplankton biostratigraphy. *Eclog. geol. Helv.*, **63**, 799–881.

Roth, P. H. 1973. Calcareous nannofossils – Leg 17, DSDP. *Initial Rep. Deep Sea drill. Proj.*, **17**, 695–793.

Roth, P. H. 1978. Cretaceous nannoplankton biostratigraphy and oceanography of the Northwestern Atlantic Ocean. *Initial Rep. Deep Sea drill. Proj.*, **44**, 731–59.

Roth, P. H. 1983. Jurassic and Lower Cretaceous calcareous nannofossils in the western North Atlantic (Site 534): Biostratigraphy, preservation, and some observations on biogeography and paleoceanography. *Initial Rep. Deep Sea drill. Proj.*, **76**, 587–621.

Roth, P. H. & Bowdler, J. L. 1979. Evolution of the calcareous nannofossil genus *Micula* in the Late Cretaceous. *Micropaleontology*, **25** (3), 272–80.

Roth, P. H. & Bowdler, J. L. 1981. Middle Cretaceous calcareous nannoplankton biogeography and oceanography of the Atlantic Ocean. *Spec. Publ. Soc. Econ. Paleontol. Mineral.*, **32**, 517–46.

Roth, P. H., Medd, A. W. & Watkins, D. K. 1983. Jurassic calcareous nannofossil zonation, an overview with new evidence from Deep Sea Drilling Project Site 534. *Initial Rep. Deep Sea drill. Proj.*, **76**, 573–9.

Roth, P. H. & Thierstein, H. R. 1972. Calcareous nannoplankton: Leg 14 of the DSDP. *Initial Rep. Deep Sea drill. Proj.*, **14**, 421–85.

Schiller, J. 1930. Coccolithineae. In: Dr. L. Rabenhorst's Kryptogamen-Flora von Deutschland, Osterreich und der Schweiz, vol. 10, no. 2, pp. 89–267. Leipzig, Akad. Verlagsgesellschaft.

Schwarz, E. H. L. 1894. Coccoliths. *Ann. Mag. Nat. Hist.*, **6** (14), 341–6.

Shafik, S. 1978. A new nannofossil zone based on the Santonian Gingin Chalk, Perth Basin, Western Australia. *Bureau of Min. Res., J. Australian Geol. Geophys.*, **3**, 211–26.

Shafik, S. 1979. Validation of *Chiastozygus fessus* and *Reinhardtites biperforatus*. *INA Newsletter*, **1** (2), C 5.

Shafik, S. & Stradner, H. 1971. Nannofossils from the Eastern Desert, Egypt, with reference to Maastrichtian nannofossils from the USSR. *Jahrb. geol. Bundesanst. (Wien)*, special vol. 17, 69–104.

Shamraï, I. A. 1963. Certain forms of Upper Cretaceous and Paleogene coccoliths and discoasters from the southern Russian Platform. *Izv. Vyssh. Ucheb. Zaved. Geol. i Razv.*, **6** (4), 27–40.

Shumenko, S. I. 1968. Some aspects of the ontogenesis, variation and taxonomy of fossil coccolithophorids revealed by electron microscopic studies. *Paleont. Zhurnal*, **4**, 32–7.

Shumenko, S. I. 1969. The first electron microscope study of nannoconids from the Maastricht of the Ukraine. *Dopov. Akad. Nauk Ukrajins'k. RSR*, **7**, 606–8.

Shumenko, S. I. 1970. An electron microscopic study of Microrhabdulids and their taxonomic position. *Paleont. J.*, **2**, 161–8.

Shumenko, S. I. 1971. Lithology and rock-forming organisms (coccolithophorids) of Upper Cretaceous deposits of the eastern Ukraine and the region of the Kursk magnetic anomaly. *Kharkov, Kharkovskogo Univ.*, 1–163.

Shumenko, S. I. 1976. *Mesozoic Calcareous Nannoplankton of the European Part of the USSR. Nauka*, 140 pp. (in Russian).

Siesser, W. G. 1982. Cretaceous calcareous nannoplankton in South Africa. *J. Paleontol.*, **56** (2), 335–50.

Sissingh, W. 1977. Biostratigraphy of Cretaceous calcareous nannoplankton. *Geol. Mijnbouw.*, **56** (1), 37–65.

Sissingh, W. 1978. Microfossil biostratigraphy and stage-stratotypes of the Cretaceous. *Geol. Mijnbouw*, **57** (3), 433–40.

Smith, C. C. 1975a. Calcareous nannoplankton and stratigraphy of Late Turonian, Coniacian, and Early Santonian age of the Eagle Ford and Austin Groups of Texas. *Prof. Pap. U.S. Geol. Surv.*, **1075**, 1–98.

Smith, C. C. 1975b. Upper Cretaceous calcareous nannoplankton zonation and stage boundaries. *Trans. Gulf Coast Assoc. Geol. Soc.*, **25**, 263–78.

Steinmetz, J. 1983, 1984. Bibliography and taxa of calcareous nannoplankton. *INA Newsletter*, **5** (1), 4–13, **5** (2), 29–47 and **6** (1), 6–37.

Stover, L. E. 1966. Cretaceous coccoliths and associated nannofossils from France and the Netherlands. *Micropaleontology*, **12**, 133–67.

Stradner, H. 1961. Vorkommen von Nannofossilien im Mesozoikum und Alttertiär. *Erdoel-Z.*, **77**, 77–88.

Stradner, H. 1962. Uber neue und wenig bekannte Nannofossilien aus Kreide und Alttertiär. *Verh. geol. Bundesanst. (Wien)*, **1962**, 363–77.

Stradner, H. 1963. New contributions to Mesozoic stratigraphy by means of nannofossils. *Proceedings of the 6th World Petrol. congr. Sect.* 1, paper 4 (preprint), pp. 1–16.

Stradner, H. & Adamiker, D. 1966. Nannofossilien aus Bohrkernen und ihre elektronenmikroskopische Bearbeitung. *Erdoel-Erdgas Z.*, **82**, 330–41.

Stradner, H., Adamiker, D. & Papp, A. 1968. Electron microscope studies on Albian calcareous nannoplankton from the Delft 2 and Leidschendam 1 Deep wells, Holland. *Verh. Kon. Ned. Akad. Wetensch., Afd. Natuurk.*, I., **24** (4), 1–107.

Stradner, H. & Grün, W. 1973. On *Nannoconus abundans* nov. spec. and on laminated calcite growth in Lower Cretaceous nannofossils. *Verh. geol. Bundesanst. (Wien)*, **2**, 267–83.

Stradner, H. & Papp, A. 1961. Tertiäre Discoasteriden aus Osterreich und deren stratigraphische Bedeutung mit Hinweisen auf Mexico, Rumänien und Italien. *Jahrb. geol. Bundesanst. (Wien)*, special vol. 7, 1–159.

Stradner, H. & Steinmetz, J. 1984. Cretaceous calcareous nannofossils from the Angola Basin, Deep Sea Drilling Project Site 530, *Initial Rep. Deep Sea drill. Proj.*, **75**, 565–649.

Tappan, H. 1980. *The Paleobiology of Plant Protists*. Freeman & Co., San Francisco, 1028 pp.

Tappan, H. & Loeblich, A. R. Jr 1966. *Maslovella* nom. nud. pro *Colvillea* black, 1964 (Coccolithophorinae) non *Colvillea* Boj. ex Hook., 1834 (Leguminosae). *Taxon*, **15**, 43.

Taylor, R. 1978. The distribution of calcareous nannofossils in the Speeton Clay of Yorkshire and their biostratigraphical significance. *Proc. Yorkshire geol. Soc.*, **42**, 195–209.

Taylor, R. J. 1982. Lower Cretaceous (Ryazanian to Albian) calcareous nannofossils. In: A. R. Lord (ed.), *A Stratigraphical Index of Calcareous Nannofossils*, pp. 40–80. British Micropal. Soc.

Thierstein, H. R. 1971. Tentative Lower Cretaceous nannoplankton zonation. *Eclog. geol. Helv.*, **64**, 459–88.

Thierstein, H. R. 1973. Lower Cretaceous calcareous nannoplankton biostratigraphy. *Abh. Geol. Bundesanst. (Wien)*, **29**, 52 pp.

Thierstein, H. R. 1974. Calcareous nannoplankton: Leg 26, Deep Sea Drilling Project. *Initial Rep. Deep Sea drill. Proj.*, **26**, 619–67.

Thierstein, H. R. 1975. Calcareous nannoplankton biostratigraphy at the Jurassic–Cretaceous boundary. *Mem. Bur. Rech. geol. minieres*, **86**, 84–94.

Thierstein, H. R. 1976. Mesozoic calcareous nannoplankton biostratigraphy of marine sediments. *Marine Micropaleontol.*, **1**, 325–62.

Thierstein, H. R. 1980. Selective dissolution of Late Cretaceous and Earliest Tertiary calcareous nannofossils: experimental evidence. *Cret. Res.*, **2**, 165–76.

Thierstein, H. R. 1981. Late Cretaceous nannoplankton and the change at the Cretaceous–Tertiary boundary. In: J. E. Warme, *et al.* (eds.), The Deep Sea Drilling Project: a decade of progress. *Spec. Publ. Soc. Econ. Paleontol. Mineral.*, **32**, 355–94.

Thierstein, H. R., Franz, H. E. & Roth, P. H. 1972. Scanning electron and light microscopy of the same small object. *Micropaleontology*, **17**, 501–2.

Trejo, M. 1959. Dos nuevas especies del género *Nannoconus* (Protozoa, inc. sed.). *Ciencia (Mexico)*, **19**, 130–2.

Trejo, M. 1960. La familia Nannoconidae y su alcane estratigrafico en America (Protozoa, incertae sedis). *Bol. Asoc. Mex. Geol. Petrol.*, **12**, 259–314.

Trejo, M. 1969. *Conusphaera mexicana*, un nuevo coccolitoforido del Jurassico Superior de Mexico. *Revista Inst. Mexicano Petrol.*, **1** (4), 5–15.

Troelsen, J. C. & Quadros, L. P. 1971. Distribuicao bioestratigrafica dos nannofosseis em sedimentos marinhos (Aptiano-Mioceno) do Brasil. *Ann. Acad. Brasil. Cienc.*, **43**, 577–609.

Valentine, P. 1980. Calcareous nannofossil biostratigraphy, paleoenvironments, and post-Jurassic continental margin development. In: P. A. Scholle (ed.), Geological Studies of the COST No. B-3 Well, United States Mid-Atlantic Continental Slope Area. *Circ. U.S. geol. Surv.*, **833**, 67–83.

Vekshina, V. N. 1959. Coccolithophoridae of the Maastrichtian deposits of the west Siberian lowlands. *SNIIGGIMS*, **2**, 56–77.

Verbeek, J. W. 1976a. Upper Cretaceous calcareous nannoplankton from Ballon and Théligny, in the type area of the Cenomanian stage. *Proc. Kon. Ned. Akad. Wetensch.*, **B 79**, 69–82.

Verbeek, J. W. 1976b. Upper Cretaceous nannoplankton zonation in a composite section near El Kef, Tunisia. *Proc. Kon. Ned. Akad. Wetensch.*, **B 79**, 129–48.

Verbeek, J. W. 1977a. Late Cenomanian to Early Turonian calcareous nannofossils from a section SE of Javernant (Dépt. Aube, France). *Proc. Kon. Ned. Akad. Wetensch.*, **B 80**, 20–2.

Verbeek, J. W. 1977b. Calcareous nannoplankton biostratigraphy of Middle and Upper Cretaceous deposits in Tunisia, Southern Spain and France. *Utrecht Micropaleontol. Bull.*, **16**, 1–157.

Verbeek, J. W. & Wonders, A. A. H. 1977. The position of the Cenomanian and Turonian stratotypes in planktonic biostratigraphy. *Proceedings Kon. Ned. Akad. Wetensch.*, **B 80**, 16–19.

Wiedmann, J., Fabricius, G., Krystyn, L., *et al.* 1979. Ueber Umfang und Stellung des Rhaet. *Newsl. Stratigr.*, **8** (2), 133–52.

Wilcoxon, J. A. 1972. Upper Jurassic–Lower Cretaceous calcareous nannoplankton from the Western North Atlantic Basin. *Initial Rep. Deep Sea drill. Proj.*, **11**, 427–57.

Wind, F. H. 1975a. Affinity of *Lucianorhabdus* and species of *Tetralithus* in Late Cretaceous Gulf Coast samples. *Trans. Gulf Coast Assoc. geol. Soc.*, **25**, 350–61.

Wind, F. H. 1975b. *Tetralithus copulatus* Deflandre (Coccolithophyceae) from the Indian Ocean: a possible paleoecological indicator. *Antarctic Journal U.S.*, **10** (5), 265–8.

Wind, F. H. 1978. Western North Atlantic Upper Jurassic calcareous nannofossil biostratigraphy. *Initial Rep. Deep Sea drill. Proj.*, **44**, 761–73.

Wind, F. H. 1983. The genus *Nephrolithus* Gorka, 1957 (Coccolithophoridae). *J. Paleontol.*, **57** (1), 157–61.

Wind, F. H. & Čepek, P. 1979. Lower Cretaceous calcareous nannoplankton from DSDP Hole 397A (Northwest African Margin). *Initial Rep. Deep Sea drill. Proj.*, **47**, 221–55.

Wind, F. H. & Wise, S. W. 1978. Mesozoic holococcoliths. *Geology*, **6**, 140–2.

Wind, F. H. & Wise, S. W. Jr 1983. Correlation of Upper Campanian–Lower Maestrichtian calcareous nannofossil assemblages in drill and piston cores from the Falkland Plateau, Southwest Atlantic Ocean. *Initial Rep. Deep Sea drill. Proj.*, **71** (part 2), 551–64.

Wise, S. W. Jr 1983. Mesozoic and Cenozoic calcareous nannofossils recovered by Deep Sea Drilling Project Leg 71 in the Falkland Plateau Region, Southwest Atlantic Ocean. *Initial Rep. Deep Sea drill. Proj.*, **71**, 481–550.

Wise, S. W. Jr & Wind, F. H. 1977. Mesozoic and Cenozoic calcareous nannofossils recovered by DSDP Leg 36 drilling on the Falkland Plateau, SW Atlantic sector of the Southern Ocean. *Initial Rep. Deep Sea drill. Proj.*, **36**, 296–309.

Worsley, T. R. 1971. Calcareous nannofossil zonation of Jurassic and Lower Cretaceous sediments from the Western Atlantic. In: A. Farinacci (ed.), *Proceedings II Planktonic Conference Roma, 1970*, pp. 1301–22. Edizioni Tecnoscienza.

Worsley, T. & Martini, E. 1970. Late Maastrichtian nannoplankton provinces. *Nature*, **225** (5239), 1242–3.

11
Cenozoic calcareous nannofossils

KATHARINA PERCH-NIELSEN

CONTENTS

Introduction

Coccoliths are the minute calcite plates produced by unicellular marine algae, the coccolithophorids. The fossil coccoliths, together with small calcite bodies of organic, but otherwise unknown origin, called nannoliths by some, constitute the calcareous nannofossils. These have proven extremely useful for the biostratigraphy of marine sediments of Jurassic through Pleistocene age. Their small size allows for age determinations of even very small samples such as ditch cuttings, sidewall cores, etc. Biostratigraphic investigations were pioneered by Bramlette & Riedel (1954), Stradner (1959a, 1961), Bramlette & Sullivan (1961), Bramlette & Wilcoxon

(1967), Hay & Mohler (1967) and Hay et al. (1967). The Cenozoic has been subdivided into some 46 zones by Martini & Worsley (1970) and Martini (1971) and into 34 zones and 45 subzones by Okada & Bukry (1980, based on earlier papers by Bukry). Finer zonations are available for correlations within a basin or region.

An introduction to calcareous nannoplankton was given by Reinhardt (1972) and by Haq (1978). An even more detailed overview is available in Tappan (1980).

The present chapter concentrates on those aspects of the calcareous nannofossils that are useful for biostratigraphy. It is not intended as a complete systematic treatise. For a more complete overview of the calcareous nannofossils the Catalogue of Calcareous Nannofossils by Farinacci (1969–79) and the Handbook of Cenozoic Calcareous Nannoplankton by Aubry (1984) can be consulted. Coccolithophorids are classified according to the International Code of Botanical Nomenclature (ICBN) and that convention is largely followed here.

Sampling, sample treatment, preparation techniques, data presentation

The sampling, sample treatment, preparation techniques and the data presentation are the same for Cenozoic as for Jurassic or Cretaceous calcareous nannofossils and are discussed in Chapter 10. Abbreviations used are: LM = light microscope, TEM = transmission electron microscope, SEM = scanning electron microscope.

Preservation, reworking and caving

The reliability of any age assignment by means of calcareous nannofossils depends on the preservation of the coccoliths and nannoliths. It is therefore important to record this information together with an indication of their abundance and distribution. This can be done by an overall appraisal or by recording the degree of etching and/or overgrowth as proposed by Roth & Thierstein (1972).

Heavy overgrowth as well as strong dissolution can result in coccoliths and nannoliths that are still recognizable as such, but which cannot be placed into species. Most Cenozoic marker species are reasonably solution resistant. Those prone to dissolution will be discussed with the zone they define. Discoasters are usually less dissolution prone than the coccoliths associated with them. An assemblage very rich in discoasters thus may indicate heavy dissolution of the rest of the assemblage, particularly if coccoliths are represented mainly by damaged specimens such as single shields, broken shields or isolated central parts of Coccolithus or Prinsiaceae.

Reworking and caving are discussed in Chapter 10.

Paleoenvironments

Shallow depth indicators

Due to their planktic mode of life, coccolithophorids do not give a clear indication of depth of deposition of the sediments in which they are found. However, due to their small size they have usually been removed from deposits of nearshore, high energy environments. The Holococcoliths (coccoliths consisting of minute calcite rhombohedrons of uniform size) are usually not preserved in modern sediments deposited below more than a few hundred to about 1000 m waterdepth according to Edwards (1973b). They do, however, survive transport into greater depths by turbidity currents.

Several genera seem to be more common in shelf areas than in the open sea. These include *Pontosphaera* and related forms such as *Transversopontis* and *Scyphosphaera* as well as representatives of the Braarudosphaeraceae (*Braarudosphaera*, *Micrantholithus*, *Pemma*). Again, there are exceptions such as the Oligocene *Braarudosphaera* chalk reported from several open ocean sites of the South Atlantic. Generally, the shelf assemblages seem more diverse than the open ocean assemblages.

Restricted environment

Monospecific assemblages

Monospecific or monogeneric assemblages are rare in Cenozoic sediments. In the South Atlantic *Braarudosphaera* chalk, there are usually a few other calcareous nannofossils present, but the assemblage is dominated completely by *Braarudosphaera*. This must be interpreted as an unusually restricted environment in the open sea since no *Braarudosphaera* chalk has been found in contemporaneous sediments of other parts of the South Atlantic. For further discussion see Bukry (1981a).

Monogeneric or monospecific assemblages can also be derived by dissolution of all but the most solution-resistant coccoliths and nannoliths. In most of the Tertiary, discoasters are the most common solution-resistant taxa in low latitudes, whereas Prinsiaceae or *Coccolithus* species are the most common solution-resistant taxa in high latitudes. These assemblages usually contain many broken specimens.

Monospecific assemblages may be found in bays with a restricted environment.

Blooms

Blooms of certain species have been noted in many sections just above the Cretaceous/Tertiary boundary, where they include species of *Thoracosphaera* (a calcareous dinoflagellate) and *Braarudosphaera* plus *Biscutum*? *romeinii*, *B*? *parvulum* and *Prinsius petalosus* (Romein, 1982). They are considered to be due to unstable ecologic conditions after the Cretaceous/Tertiary boundary event(s) that caused the mass mortality of calcareous plankton at the end of the Cretaceous. *Braarudosphaera* blooms are reported from parts of the South Atlantic Oligocene, the Black Sea Quaternary (Percival, 1978) and the modern Panama Basin (Smayda, 1966).

Marginal seas – open ocean

Several species have been shown to be more common in marginal than in open seas. Whereas this contributes to paleo-environmental analysis, it hinders biostratigraphic correlation. Generally, the markers chosen by Bukry (1973a, 1975) for his zonal scheme are those usually encountered in low to mid latitude open ocean localities. Some of the markers for the zonation proposed by Martini (1971) are rare or absent in the open sea assemblages (see discussion of zones).

Okada & Honjo (1975) showed clearly the different community structure of the calcareous nannoplankton between the marginal seas along the western Pacific and the pelagic counterpart. Okada (1984) suggests *Florisphaera profunda* as a tool to recognize paleodepth. He found no *F. profunda* in shallow seas and an increase to over 50% of the assemblage in 2000 m. He found *G. oceanica* dominating the flora in shallow marginal seas and inland seas.

Hekel (1973) found less than 2% coccoliths in the fine fraction of the sediment in the Capricorn Basin in areas with a waterdepth above 40 m. More than 10% coccoliths were found in waterdepths above about 100 m. *Gephyrocapsa oceanica* and *Emiliania huxleyi* were found to dominate the assemblage to more than 95% in an area between the coast and 60 km offshore, in waterdepths of less than about 40 m. *Discoaster* was not found in the shallow Lower Pliocene of the Capricorn Basin.

Whereas the small Prinsiaceae *Gephyrocapsa* and *Emiliania* also dominate the calcareous nannoplankton assemblage in the Gulf of Elat, a total of 52 taxa were reported by Winter, Reiss & Luz (1979). They found high species diversity to correlate with high water temperature and low diversity with the relatively cool winter period. High salinities in the Gulf of Elat (41‰) seem to prevent the entrance of many common pelagic species such as *Calcidiscus leptoporus* and *Gephyrocapsa caribbeanica*. *Florisphaera profunda* was not found in the 1800 m deep Gulf. No blooms of *Braarudosphaera* were observed.

The distribution of the modern coccolithophorids of the Atlantic was investigated by McIntyre & Bé (1967a) and Okada & McIntyre (1977, 1979). The Pacific was treated by Okada & Honjo (1973, 1975), Honjo & Okada (1974), Okada & McIntyre (1977), Nishida (1979) and Reid (1980) among others. These oceanic assemblages are generally dominated by the small Prinsiaceae of the genera *Emiliania* and *Gephyrocapsa*, with the exception of tropical areas where various other species reach high abundances.

Biogeography through time

Generally, low latitude assemblages are more diverse than high latitude assemblages. From this it follows that fewer markers are available for the subdivision of high latitude sections. This has been demonstrated clearly where attempts have been made to apply the essentially low latitude zonations of Martini (1971) and Bukry (1973a, 1975) to high latitude Deep Sea Drilling Project (DSDP) sites. It is more evident in Oligocene and younger sections than in Paleocene and Eocene ones. Special zonations have been used for New Zealand and the Southwest Pacific (Edwards, 1971; Edwards & Perch-Nielsen, 1975; Waghorn, personal communication) and for the North Atlantic/Norwegian Sea (Müller, 1976). On the other hand, some high latitude markers, such as *Isthmolithus recurvus* and *Chiasmolithus oamaruensis* in the Late Eocene, are rare or absent in low latitudes. Details of these and other departures from the standard zonations of Martini (1971) and Okada & Bukry (1980) are discussed later in the present chapter.

McIntyre & Bé (1967a) and Ruddiman & McIntyre (1976) defined water masses of the North Atlantic using coccoliths as given below, with the additional remark, that all samples containing coccoliths are dominated by *E. huxleyi* and *G. caribbeanica*:

Polar: barren

Subpolar: *Coccolithus pelagicus* (species found today only in northern hemisphere)

Transitional: *Calcidiscus leptoporus*, type C

Subtropical: *Umbilicosphaera mirabilis, Rhabdosphaera claviger, Calciosolenia murrayi, Syracosphaera pulchra, Umbellosphaera irregularis, U. tenuis*

Tropical: *Umbellosphaera irregularis, Neosphaera coccolithomorpha, Oolithotus fragilis, U. tenuis, Discosphaera tubifer, Rhabdosphaera stylifer*

The mapping of the successions of these assemblages has allowed Ruddiman & McIntyre (1976) to recognize seven complete climatic cycles in the past 600000 years in the Northeast Atlantic.

Haq & Lohman (1976), Haq (1980) and Haq with various co-authors in several DSDP reports have investigated the calcareous nannoplankton biogeography of the Atlantic and adjacent seas. They constructed paleobiogeographic maps for Paleogene and Miocene time-slices and could show the latitudinal migrations of selected assemblages. So far their findings have not generally been taken into consideration by coccolith biostratigraphers.

Biostratigraphic zonation

The standard zonation of Martini (1971) and the zonation by Bukry (1973a, 1975; Okada & Bukry, 1980) have now been used for a decade and proved to be useful in many areas and for correlation between areas. Locally, finer subdivisions

have been proposed with, more often than not, the Martini or Bukry zonations as a framework. Figs. 1–9 show the zonations with their zonal markers and the ranges and illustrations of the zonal markers and a few other species. Martini (1971) used the abbreviations NP and NN (Nannoplankton Paleogene and Nannoplankton Neogene) and numbers to codify the zones. Okada & Bukry (1980) suggested CP and CN (Coccoliths Paleogene and Coccoliths Neogene) and numbers to codify the zones of Bukry (1973a, 1975). The zonation of Martini (1971) is here used as the framework for most ranges, since it has been and is widely used. Some of its marker species are not found in open ocean assemblages and here Okada & Bukry's zonation (1980) gives better results. Abbreviations used are: FO = first occurrence, LO = last occurrence.

Cretaceous/Tertiary boundary

The massive change in calcareous nannofossil assemblages at the Cretaceous/Tertiary boundary was first noted and illustrated by Bramlette & Martini (1964) and has since been described in greater detail (Perch-Nielsen, 1969, 1979a, b, 1981c; Percival & Fischer, 1977; Romein, 1977). The diverse Maastrichtian assemblage disappears suddenly – probably over a few thousands of years – after mass mortality at the boundary. It is then replaced by new species and genera evolving from some 15 to 18 genera that survived the Cretaceous/ Tertiary boundary events (Perch-Nielsen, 1982; Perch-Nielsen, McKenzie & He, 1982).

In the most complete section known across the Cretaceous/Tertiary boundary near El Kef, Tunisia, a clay layer exists in which the $CaCO_3$ content is very low and where coccoliths are absent. In low latitudes, *Thoracosphaera*, a calcareous dinoflagellate, bloomed shortly after the boundary after having occurred only very sporadically during the Cretaceous. In high latitudes, *Thoracosphaera* first occurs after the boundary and did not bloom until after NP 1. *Braarudosphaera* blooms have been observed at some localities and blooms of *Biscutum* (?) *romeinii, B.* (?) *parvulum* and *Prinsius petalosus* at low latitude sites; common *Cyclagelosphaera, Biscutum, Placozygus sigmoides, Markalius* and *Neocrepidolithus* occur in high latitude sites.

Paleocene

The Paleocene was a time of rapid evolution after the very heavy 'casualties' at the Cretaceous/Tertiary boundary. Based on the assumption that all Paleocene genera must have originated from Late Cretaceous forms, Perch-Nielsen (1981b) proposed the relationships shown in Fig. 5.

All zonal boundaries are based on first occurrences, sometimes of a genus, sometimes of the oldest species of a genus that was then described. It seems evident, that with some 25 new genera evolving during about 12 m.y., one should be able to do better, on a worldwide basis, than the nine-fold

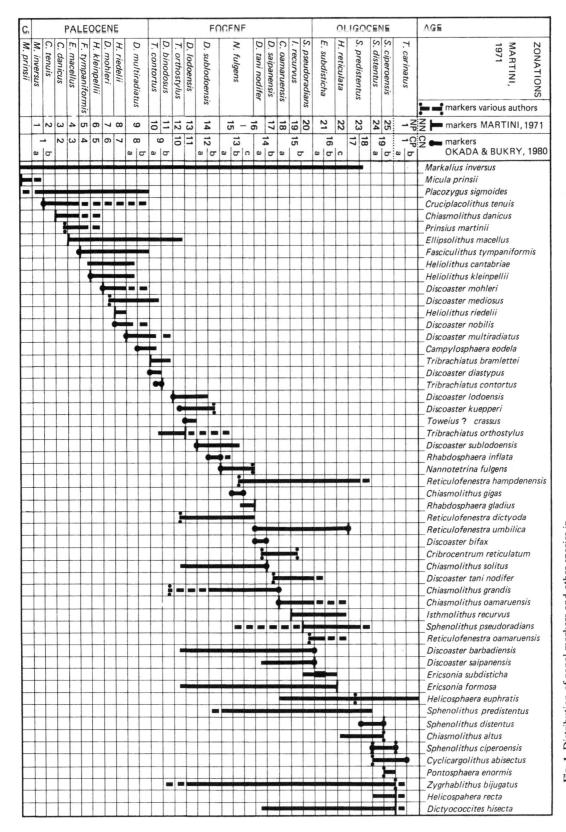

Fig. 1. Distribution of zonal markers and other species in the Paleogene

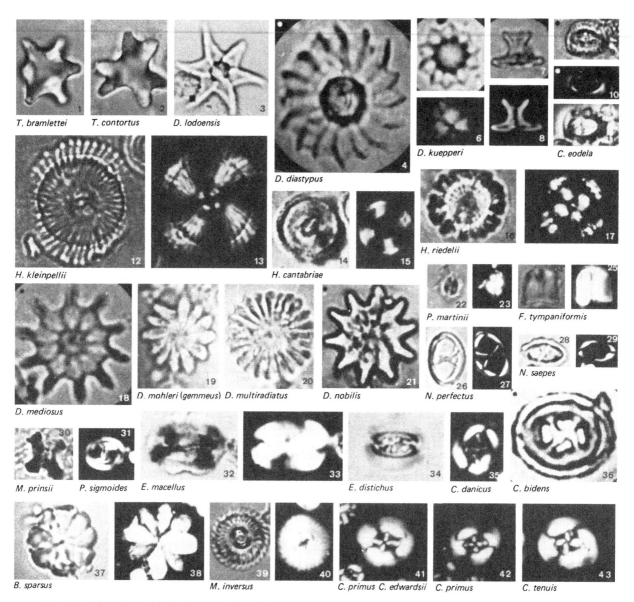

Fig. 2. Zonal markers of the Paleocene and Early Eocene.
Magnification of LM here and in following figures *c.* × 2000

Fig. 3. Zonal markers of the Early Eocene through Oligocene

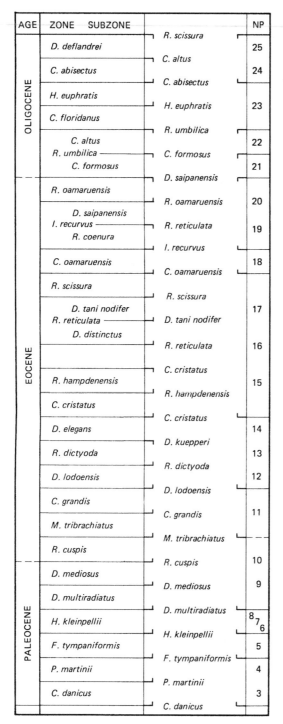

Fig. 4. New Zealand Paleogene zonation after Waghorn (personal communication) and approximate correlation with Martini's Standard Zonation (NP, 1971)

subdivision suggested by Martini (1971) or the ten-fold one by Okada & Bukry (1980). Preservation and biogeography may be limiting factors until more sections are available for study. Additional subdivisions have been proposed for parts of the Early Paleocene by Romein (1979) and Perch-Nielsen (1979a, 1981c) and are shown in Fig. 6. They are based on members of the genera *Biscutum?*, *Prinsius*, *Cruciplacolithus*, *Neochiastozygus* and *Chiasmolithus* that are discussed in detail in the taxonomic notes.

In the upper part of the Paleocene, the subdivision can be expected to be refined by using the same genera and the newly evolving *Heliolithus*, *Fasciculithus* and *Discoaster* s. ampl. that are found consistently in some areas. Other genera

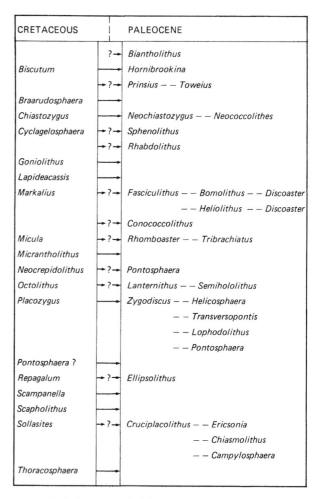

Fig. 5. Attempt to find the ancestors to the Paleocene genera. Cretaceous genera that are known (→) or thought (→?→) to have crossed the Cretaceous/Tertiary boundary and given rise to new genera after Perch-Nielsen (1981b)

such as *Hornibrookina, Conococcolithus, Lapideacassis, Scampanella, Zygodiscus, Lanternithus, Micrantholithus* and *Pontosphaera* occur only sporadically in time and space and thus show less promise for future worldwide correlations.

MARKALIUS INVERSUS ZONE (NP 1)

Definition: LO of Cretaceous coccoliths or FO of acme of *Thoracosphaera* to FO of *Cruciplacolithus tenuis*.

Authors: Mohler & Hay *in* Hay *et al.* (1967), emend. Martini (1970a)

Age: Early Paleocene (Early Danian)

Correlation: *Cruciplacolithus primus* Subzone, CP1a of Okada & Bukry (1980), where the base is defined by the LO of *Micula murus* and other Cretaceous species. *Biantholithus sparsus* Zone of Perch-Nielsen (1971c) emend. by Romein (1979) and defined as the interval from the massive occurrence or increased frequency of *Thoracosphaera operculata* to the FO of *C. primus*. For details see Fig. 6.

Remarks: The lower boundary of this zone poses no problems in most sections, since most sections are rather incomplete and the change from the rich Maastrichtian assemblage to a less diverse Tertiary assemblage is quite evident. In relatively complete sections, however, the presence of Maastrichtian coccoliths in varying amounts above the boundary can lead to problems in placing some samples and in exactly drawing the boundary. However, these problems are of an academic rather than a practical nature, since they are within a few centimetres of the boundary.

The upper boundary of NP 1 does pose problems, since *C. tenuis* has been shown by Romein (1979) to be a form of *Cruciplacolithus* with small 'feet' at the ends of the central bars. Such forms only evolved after the first forms with an oblique central cross, which are usually assigned to *Chiasmolithus danicus*, the FO of which defines the base of NP 3. Most authors put the base of NP 2 at the FO of any *Cruciplacolithus*, which is usually *C. primus*, a very small form of the genus.

NP 1 can be subdivided in high as well as in low latitudes (Fig. 6).

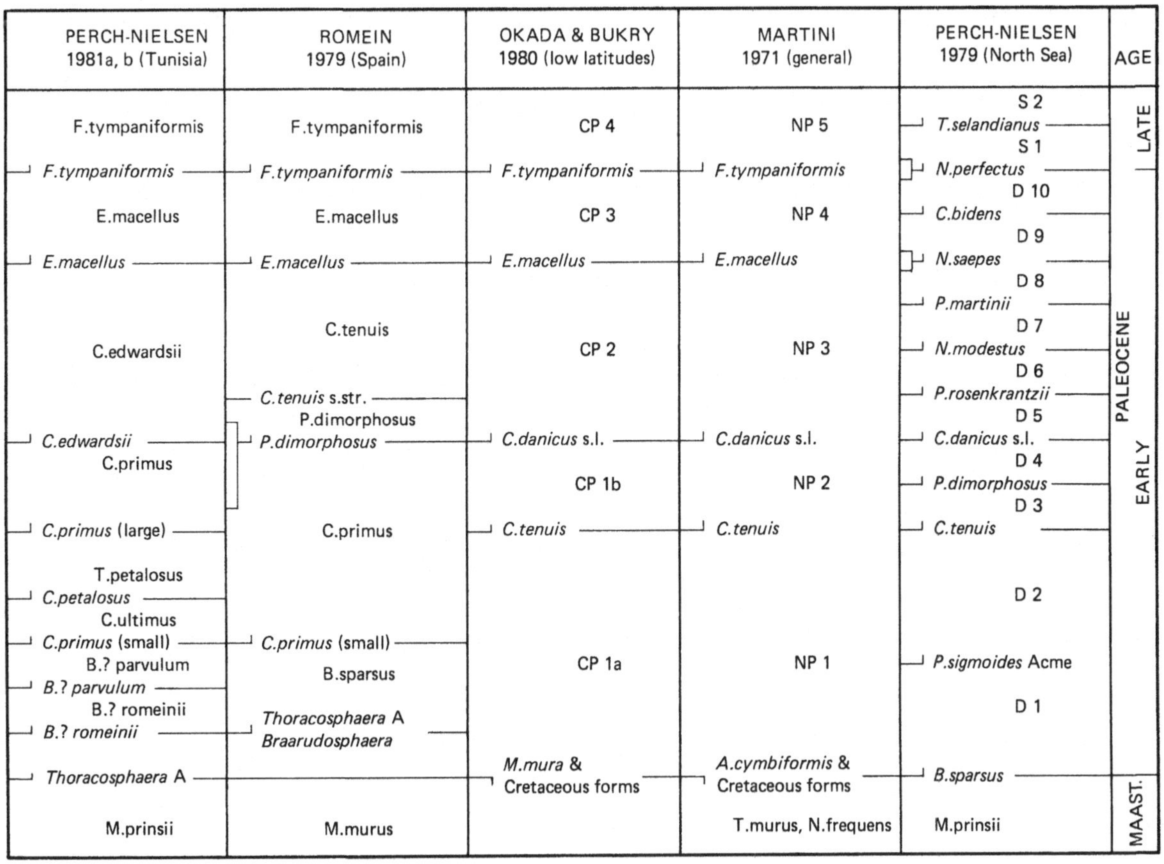

PERCH-NIELSEN 1981a, b (Tunisia)	ROMEIN 1979 (Spain)	OKADA & BUKRY 1980 (low latitudes)	MARTINI 1971 (general)	PERCH-NIELSEN 1979 (North Sea)		AGE
F.tympaniformis	F.tympaniformis	CP 4	NP 5	S 2 — T.selandianus — S 1		LATE
⌐ F.tympaniformis —	⌐ F.tympaniformis —	⌐ F.tympaniformis —	⌐ F.tympaniformis	⌐⌐ N.perfectus	D 10	
E.macellus	E.macellus	CP 3	NP 4	⌐ C.bidens	D 9	
⌐ E.macellus —	⌐ E.macellus —	⌐ E.macellus —	⌐ E.macellus	⌐⌐ N.saepes —	D 8	
				⌐ P.martinii	D 7	
C.edwardsii	C.tenuis	CP 2	NP 3	⌐ N.modestus	D 6	PALEOCENE
				⌐ P.rosenkrantzii —	D 5	
	— C.tenuis s.str. — P.dimorphosus			⌐ C.danicus s.l.	D 4	
⌐ C.edwardsii — C.primus	⌐ P.dimorphosus	⌐ C.danicus s.l. —	⌐ C.danicus s.l.			EARLY
		CP 1b	NP 2	⌐ P.dimorphosus —	D 3	
⌐ C.primus (large) —	C.primus	⌐ C.tenuis —	⌐ C.tenuis	⌐ C.tenuis —		
T.petalosus ⌐ C.petalosus — C.ultimus					D 2	
⌐ C.primus (small) — B.? parvulum	⌐ C.primus (small) —	CP 1a	NP 1	⌐ P.sigmoides Acme		
⌐ B.? parvulum — B.? romeinii	B.sparsus				D 1	
⌐ B.? romeinii —	Thoracosphaera A Braarudosphaera					
⌐ Thoracosphaera A —		⌐ M.mura & Cretaceous forms	A.cymbiformis & Cretaceous forms	⌐ B.sparsus		MAAST.
M.prinsii	M.murus		T.murus, N.frequens	M.prinsii		

Fig. 6. Definition and correlation of zones and subzones in the Early Paleocene

CRUCIPLACOLITHUS TENUIS ZONE (NP 2)

Definition: FO of *Cruciplacolithus tenuis* to FO of *Chiasmolithus danicus*.

Authors: Mohler & Hay *in* Hay *et al.* (1967), emend. Martini (1970a)

Age: Early Paleocene (Early Danian)

Correlation: *Cruciplacolithus tenuis* Subzone, CP1b of Okada & Bukry (1980). See also Fig. 6.

Remarks: The lower boundary is discussed above. Similar problems have occurred for the upper boundary since the definition of this zone. Generally, *Chiasmolithus danicus* has been used for coccoliths with an oblique cross in the centre. Romein (1979) has described a coccolith with a slightly turned central cross as *Cruciplacolithus edwardsii* and shown it to have evolved from *C. primus* considerably before 'typical' *Chiasmolithus danicus* with slightly curved bars evolved. *C. danicus* and related forms (see taxonomic notes) are more reliable markers in high than in low latitudes. Within NP 2, one can observe the FOs of *Ericsonia cava, Prinsius dimorphosus* and, depending on the usage of NP 2, *Cruciplacolithus edwardsii*.

CHIASMOLITHUS DANICUS ZONE (NP 3)

Definition: FO of *Chiasmolithus danicus* to FO of *Ellipsolithus macellus*.

Author: Martini (1970a)

Age: Early Paleocene (Late Danian)

Correlation: *Chiasmolithus danicus* Zone, CP2 of Okada & Bukry (1980). See also Figs. 4 and 6.

Remarks: The lower boundary is discussed above. The upper boundary is the FO of *E. macellus*, which is easy to recognize in well preserved material. In poorly preserved assemblages and/or in high latitudes, this species is missing but the zonal boundary can be identified by substitute evidence such as the absence of species of *Neochiastozygus* (*N. saepes, N. perfectus*), *Chiasmolithus bidens* and *Prinsius martinii*, all forms which are used to subdivide the high latitude Danian sections in the North Sea area (Fig. 6).

ELLIPSOLITHUS MACELLUS ZONE (NP 4)

Definition: FO of *Ellipsolithus macellus* to FO of *Fasciculithus tympaniformis*.

Author: Martini (1970a)

Age: Early Paleocene (Late Danian)

Correlation: *Ellipsolithus macellus* Zone, CP3 of Okada & Bukry (1980). See also Figs. 4 and 6.

Remarks: The lower boundary was discussed above. Instead of the FO of *E. macellus*, the FO of *E. distichus*, another species of *Ellipsolithus* appearing slightly after *E. macellus*

where the two are consistently found together, has been used as a marker by some authors. In high latitudes, the FO of *Neochiastozygus saepes* approximates the FO of *E. macellus* and thus can be used where *E. macellus* is missing. This allows us to correlate NP 4 to the Upper Danian of the type area (Perch-Nielsen, 1979a), despite the fact, that *E. macellus* is not found there normally.

The FO of *F. tympaniformis* is not the FO of the genus *Fasciculithus*, as was thought at the time when *F. tympaniformis* was chosen as a zonal marker. *F. magnus*, a very large species and the first representative of the genus can be found below the first *F. tympaniformis*, as can *F. ulii, F. magnicordis* and *F. janii* (see taxonomic notes, where also the lineages in the genus *Fasciculithus* are given). *Fasciculithus* is very rare or absent in the North Sea area. Here the base of NP 5 is correlatable to the FO of *Neochiastozygus perfectus*, which occurs above the FO of *Chiasmolithus bidens*. The latter species is usually observed within NP 4. Another substitute for *F. tympaniformis* is *Sphenolithus primus*, the first species of *Sphenolithus*, which usually appears just before the FO of *F. tympaniformis*.

FASCICULITHUS TYMPANIFORMIS ZONE (NP 5)

Definition: FO of *Fasciculithus tympaniformis* to FO of *Heliolithus kleinpellii*.

Authors: Mohler & Hay *in* Hay *et al.* (1967)

Age: Late Paleocene (Thanetian, Selandian)

Correlation: *Fasciculithus tympaniformis* Zone, CP4 of Okada & Bukry (1980). See also Fig. 6.

Remarks: The lower boundary of NP 5 was discussed above. Several new species of *Fasciculithus* and related genera such as *Heliolithus* and *Bomolithus* evolved before the FO of *H. kleinpellii* (Romein, 1979; Perch-Nielsen, 1981b; for details see taxonomic notes). *H. kleinpellii* was chosen as a zonal marker due to its wide distribution and easy recognition. In thick sections with well preserved material, this zone can be subdivided further by the FO of *B. elegans* and the subsequent FO of *H. cantabriae*.

HELIOLITHUS KLEINPELLII ZONE (NP 6)

Definition: FO of *Heliolithus kleinpellii* to FO of *Discoaster mohleri*.

Authors: Mohler & Hay *in* Hay *et al.* (1967)

Age: Late Paleocene (Thanetian, Selandian)

Correlation: *Heliolithus kleinpellii* Zone, CP5 of Okada & Bukry (1980). See also Fig. 4.

Remarks: The lower boundary of NP 6 was discussed above. After the evolution of *H. kleinpellii* several new, related forms appear and lead to the FO of *Discoaster bramlettei* in the upper part of the zone and to *Discoaster mohleri*, the FO of which defines the base of the following zone. Originally *Discoaster gemmeus* was used as a marker species for the upper boundary of this zone. But Bukry & Percival (1971) showed that *D. gemmeus* is, in fact, an Eocene species which is not found in the Paleocene and, therefore, erected *D. mohleri* based on a specimen illustrated by Hay & Mohler (1967) from their *D. gemmeus* Zone. *D. mohleri* is the oldest discoaster without any birefringent central area or stem.

DISCOASTER MOHLERI ZONE (NP 7)

Definition: FO of *Discoaster mohleri* (previously *D. gemmeus*, partim) to FO of *Heliolithus riedelii* (NP 7) or *Discoaster nobilis* (CP6).
Authors: Hay (1964) and Mohler & Hay *in* Hay *et al.* (1967)
Age: Late Paleocene (Thanetian, Selandian)
Correlation: *Discoaster mohleri* Zone, CP6 of Okada & Bukry (1980).
Remarks: The lower boundary of NP 7 was discussed above. The FO of *H. riedelii* has generally been found not to be a very reliable marker event, since *H. riedelii* is missing in many sections where it might be expected to be present. Thus the FO of *Discoaster nobilis* that occurs about at the same time, has been suggested as a zonal marker between the FO of *D. mohleri* and the FO of *D. multiradiatus* by Perch-Nielsen (1972). Both these species are widely distributed and easily recognizable. Other authors have used a combined NP 7/8 to get around the problem. *Neochiastozygus cearae* first occurs in NP 7.

HELIOLITHUS RIEDELII/DISCOASTER NOBILIS ZONE (NP 8)

Definition: FO of *Heliolithus riedelii* (NP 8) or *Discoaster nobilis* (CP7) to FO of *Discoaster multiradiatus*.
Authors: Bramlette & Sullivan (1961) and Perch-Nielsen (1972)
Age: Late Paleocene (Thanetian, Selandian)
Correlation: *Discoaster nobilis* Zone, CP7 of Okada & Bukry (1980). See also Fig. 4.
Remarks: The base of this zone was discussed above. NP 8 is often merged with NP 7 in sections where *H. riedelii* is not found. Several new species of *Discoaster* evolved in this interval (see taxonomic notes), before the FO of *D. multiradiatus*.

DISCOASTER MULTIRADIATUS ZONE (NP 9)

Definition: FO of *Discoaster multiradiatus* to FO of *Tribrachiatus bramlettei* (NP 9) or *Discoaster diastypus* (CP8).
Authors: Bramlette & Sullivan (1961) emend. Martini (1971) and Bukry & Bramlette (1970)

Age: Late Paleocene (Thanetian, Selandian)
Correlation: *Discoaster multiradiatus* Zone, CP8 with subzones CP8a, *Chiasmolithus bidens*, and CP8b, *Campylosphaera eodela*, of Okada & Bukry (1980). See also Fig. 4.
Remarks: The definition of the upper boundary of NP 9 is different from that of CP8, while the definition of the base is the same. The subdivision of CP8 into a and b is based on the FO of *C. eodela* and the FO of *Rhomboaster* spp. Within NP 9 several species of *Rhomboaster* evolve and can be used for a very fine subdivision of this interval (Gartner, 1971; Romein, 1979; see also taxonomic notes). Also several *Discoaster* species have their FO in NP 9.

With several new species also in *Fasciculithus*, *Zygodiscus* and *Neochiastozygus* and the new genera *Lophodolithus*, *Conococcolithus*, *Campylosphaera*, *Rhomboaster* and *Helicosphaera*, diversity reaches a first maximum in the Paleogene in NP 9. *Fasciculithus* becomes rare towards the top of the zone and disappears near the NP 9/NP 10 boundary. This can be used in sections, where *Tribrachiatus* is absent or very rare, to approximate the top of NP 9 and, at the same time, the Paleocene/Eocene boundary.

Paleocene/Eocene boundary

The Paleocene/Eocene boundary in terms of calcareous nannofossils is usually drawn at the top of NP 9/CP8, which corresponds to the *Pseudohastigerina* appearance among planktic foraminifera. Several genera disappear near the end of the Paleocene, namely *Fasciculithus*, *Hornibrookina*, *Placozygus*, *Rhomboaster*. *Fasciculithus* disappears 0.8 m.y. before the NP 9/NP 10 boundary according to Shackleton *et al.* (1984) and the last *Hornibrookina* was found in the lower part of NP 10 in the North Atlantic by Backman (1984). *Tribrachiatus* and *Rhabdolithus* appear close to the NP 9/NP 10 boundary.

Eocene

Diversity again increases during the Early Eocene but by Middle Eocene many species disappear and only few new ones appear in the remainder of the Eocene. Accordingly, FO and LO have been used for the definition of the zonal and subzonal boundaries (Figs. 1, 4). In the earliest part of the Eocene, the *Tribrachiatus* lineage furnishes excellent markers, followed by forms of *Discoaster*, *Nannotetrina*, *Chiasmolithus*, *Isthmolithus* and *Sphenolithus*.

TRIBRACHIATUS CONTORTUS ZONE (NP 10)

Definition: FO of *Tribrachiatus bramlettei* or *Discoaster diastypus* and *Tribrachiatus contortus* (CP9) to LO of *Tribrachiatus contortus*.
Authors: Hay (1964) and Bukry (1973a)

Age: Early Eocene (Ypresian)

Correlation: *Tribrachiatus contortus* Subzone, CP9a of Okada & Bukry (1980).

Remarks: NP 10 has not been found in several sequences where it might be expected to be present, probably due to the absence of the genus *Tribrachiatus* in certain areas for ecological reasons. When present, however, it is very easy to recognize NP 10 by the presence of *T. bramlettei* and/or *T. contortus*. Detailed studies of *T. bramlettei*, *T. contortus* and *T. orthostylus* were published by Hekel (1968) and Romein (1979). In overgrown material both species have a tendency to look like ordinary calcite rhombohedrons. In sections where *T. contortus* is very rare and thus its LO difficult to find, the FO of *Tribrachiatus orthostylus*, which usually occurs shortly before the LO of *T. contortus*, can be used as an approximation of the NP 10/NP 11 boundary. Many species continue from the Paleocene into the Eocene, including some of the species, the FO of which are used as zonal markers in the Paleocene, i.e. *Ellipsolithus macellus* and *Discoaster multiradiatus*. *Fasciculithus* is only found in the lowermost part of NP 10 according to Romein (1979), and its LO is used by other authors to define the Paleocene/Eocene boundary in sections where *Tribrachiatus* is absent or very rare.

DISCOASTER BINODOSUS ZONE (NP 11)

Definition: LO of *Tribrachiatus contortus* to FO of *Discoaster lodoensis*.

Authors: Mohler & Hay *in* Hay *et al.* (1967)

Age: Early Eocene (Ypresian)

Correlation: *Discoaster binodosus* Subzone, CP9b of Okada & Bukry (1980). See also Fig. 4.

Remarks: *Discoaster binodosus* may or may not be present in this zone which is defined by the absence of the marker species for the zones above and below. *Tribrachiatus orthostylus*, a species which evolved from *T. contortus* (see taxonomic notes), appears near the NP 10/NP 11 boundary and evolves from a form with a slight bifurcation at the end of the three arms to a form with three simple arms (Hekel, 1968; Romein, 1979) within NP 11. *Discoaster distinctus*, *Sphenolithus editus*, *S. radians* and *S. conspicuus* appear within NP 11. *Imperiaster obscurus* is a large species of unknown origin that typically appears in NP 11 and disappears in NP 12 or NP 13 of the North Sea and North Atlantic regions.

TRIBRACHIATUS ORTHOSTYLUS ZONE (NP 12)

Definition: FO of *Discoaster lodoensis* to LO of *Tribrachiatus orthostylus* (NP 12) or the FO of *Toweius*(?) *crassus* (CP10).

Authors: Brönnimann & Stradner (1960) & Bukry (1973a)

Age: Early Eocene (Ypresian)

Correlation: *Tribrachiatus orthostylus* Zone, CP10 of Okada & Bukry (1980), with a different definition (see above).

Remarks: The overlap of *D. lodoensis* and *T. orthostylus* is easily recognized in well preserved assemblages or in assemblages affected by dissolution. It is more difficult to recognize in assemblages affected by heavy overgrowth, where *T. orthostylus* and *D. lodoensis* are difficult to identify. In such cases, the presence of *Discoaster kuepperi* or *Rhabdosphaera truncata*, the FO of which approximates the FO of *D. lodoensis*, can be used, since *D. kuepperi* and *R. truncata* can still be recognized even in very poorly preserved material. The LO of *T. orthostylus* seems to occur too high in some sections; that is, it co-occurs with *Nannotetrina*. It is thus advisable to rely also on the FO of *T*? *crassus*, used to define the top of CP10 which approximates the top of NP 12 in normal sections. In the North Sea region, the LO of *Micrantholithus mirabilis* has been used by Perch-Nielsen (1971d) to subdivide NP 12. *Imperiaster obscurus* disappears in NP 12 or early NP 13.

DISCOASTER LODOENSIS ZONE (NP 13)

Definition: LO of *Tribrachiatus orthostylus* (NP 13) or FO of *Toweius*? *crassus* (CP11) to the FO of *Discoaster sublodoensis*.

Authors: Brönnimann & Stradner (1960) and Bukry (1973a)

Age: Early Eocene (Ypresian)

Correlation: *Discoaster lodoensis* Zone, CP11 of Okada & Bukry (1980), with a different definition (see above). See also Fig. 4.

Remarks: The lower boundary of NP 13 was discussed above. The number of arms in *Discoaster lodoensis* decreases from about 8 to usually 5–7 arms towards the FO of *D. sublodoensis*. The latter has usually 5, rarely 6, nearly straight arms as opposed to the strongly curved arms of *D. lodoensis*. *T*? *crassus* can reach high abundances in some samples and disappears in the lower part of CP12. *Discoaster diastypus* has its LO in NP 13 and *Discoaster nonaradiatus* appears in the upper part of this zone. *Sphenolithus conspicuus* and *S. editus* disappear near the base of NP 13, while *S. orphanknollensis* appears near it.

DISCOASTER SUBLODOENSIS ZONE (NP 14)

Definition: FO of *Discoaster sublodoensis* to FO of *Nannotetrina fulgens* (NP 14) and/or LO of *Rhabdosphaera inflata* (CP12).

Authors: Hay (1964) and Bukry (1973a)

Age: Middle Eocene

Correlation: *Discoaster sublodoensis* Zone, CP12 of Okada &

Bukry (1980), with a different definition for the top of the zone, when the LO of *R. inflata* is used as a substitute marker.

Remarks: *D. lodoensis* overlaps with *D. sublodoensis* in the lower part of the zone and the FO of *R. inflata* is used to subdivide the zone into subzones CP12a and CP12b. The top of the zone has sometimes been approximated by using the FO of any species of *Nannotetrina* instead of *N. fulgens* (or *N. alata* and *N. quadrata*, both probable synonyms of *N. fulgens*), which can be very rare in some sections where other species of *Nannotetrina* are found consistently to be present. In these cases, *R. inflata* has been found to overlap with *Nannotetrina*. *Sphenolithus furcatolithoides* and *S. spiniger* appear in the upper part of NP 14.

NANNOTETRINA FULGENS ZONE (NP 15)

Definition: FO of *Nannotetrina fulgens* to LO of *Rhabdolithus gladius* (FO of *N. quadrata* or LO of *Rhabdosphaera inflata* to FO of *Reticulofenestra umbilica* or FO of *Discoaster bifax* for CP13, the *N. quadrata* Zone).

Authors: Hay *in* Hay *et al.* (1967), emend. Martini (1970a) and Bukry (1973a)

Age: Middle Eocene

Correlation: The basal markers of NP 15 and CP13 are possibly synonymous or related and therefore equivalent, but the upper boundary markers are different. A three-fold subdivision of CP13 can be achieved by using the range of *Chiasmolithus gigas*.

Remarks: In many sections *N. fulgens* is not the first representative of *Nannotetrina* to appear. Also, it is usually less common than other species of the genus such as *N. cristata*, from which it probably evolved. In heavily overgrown assemblages the genus *Nannotetrina* can still be recognized and used as a marker, even when the species often can no longer be distinguished. Thus, the FO of *Nannotetrina* can be used to approximate the NP 14/NP 15 boundary in poorly preserved material (see also Fig. 4).

The range of *Chiasmolithus gigas* defines the limits of the subzone CP13b. The correlation of the top of NP 15, the LO of *Rhabdosphaera gladius*, and the base of CP14, the FO of *Reticulofenestra umbilica* is not clear. This is due partly to the fact that *R. gladius* is rare and not present in all sections where it might be expected to occur and partly due to the fact that *R. umbilica* is a species defined differently (if at all) by different authors (see taxonomic notes).

The FO of *D. bifax* was found to occur in NP 14 by Aubry (1983), who studied assemblages from the Paris Basin and the London Basin, and thus seems not to be a useful marker for a zone above the FO of *Nannotetrina* in hemipelagic or neritic environments.

The LO of *D. kuepperi*, *D. sublodoensis* and *D. nonaradiatus* are in this zone, while *R. gladius*, *D. tanii nodifer* and *Lanternithus minutus* (usually) first appear within NP 15. The genus *Nannotetrina* disappears near the LO of *R. gladius* so this LO can be used as an approximation for the NP 15/NP 16 boundary in the sections where *R. gladius* is very rare or missing and where *R. umbilica* seems to appear before *Nannotetrina*, as found in the North Atlantic by Perch-Nielsen (1972).

DISCOASTER TANII NODIFER ZONE (NP 16)

Definition: LO of *Rhabdolithus gladius* to LO of *Chiasmolithus solitus*.

Authors: Hay *et al.* (1967), emend. Martini (1970a)

Age: Middle Eocene

Correlation: *Discoaster bifax* Subzone, CP14a of Okada & Bukry (1980) corresponds approximately with NP 16, but has different boundary species: FO of *Reticulofenestra umbilica* or FO of *Discoaster bifax* to LO of *Chiasmolithus solitus* or LO of *Discoaster bifax*. See also Fig. 4.

Remarks: The lower boundary of NP 16 was discussed above. NP 16 is one of the zones of the middle Eocene which is well defined but difficult to use. *R. gladius*, the marker of the base of the zone, is often absent and thus of little use for zonation. The same is true of *C. solitus*, which is especially rare or absent in low latitude sections. Furthermore, *C. solitus* is often not determined correctly, in that any *Chiasmolithus* with slightly curved bars is assigned to *C. solitus*, while the type of *C. solitus* shows not only curved bars but also an overlapping distal cycle of elements (see taxonomic notes). Besides the LO of *D. bifax* used by Bukry (1973a) for the subdivision of his *Reticulofenestra umbilica* Zone (CP14), *Discoaster distinctus* also has its LO in the upper part of NP 16 and can be used as an approximation for the zonal boundary in low latitudes. In the South Atlantic, *Dictyococcites bisectus* has its FO in the upper part of NP 16, while in the North Sea area it does not appear sooner than in NP 19. There, Perch-Nielsen (1971d) has subdivided the NP 16/NP 17 interval by the subsequent FOs of *Clathrolithus spinosus*, *Cribrocentrum reticulatum* (NP 16) and *Helicosphaera compacta*.

DISCOASTER SAIPANENSIS ZONE (NP 17)

Definition: LO of *Chiasmolithus solitus* to FO of *Chiasmolithus oamaruensis*.

Author: Martini (1970a)

Age: Middle Eocene

Correlation: *Discoaster saipanensis* Subzone, CP14b, of Okada & Bukry (1980) corresponds approximately with NP 17, but has different boundary species: LO of *C.*

solitus or LO of *D. bifax* to LO of *C. grandis* or FO of *C. oamaruensis*. See also Fig. 4.

Remarks: The *D. saipanensis* Zone is difficult to recognize in many low latitude areas, where the genus *Chiasmolithus* is very rare or absent and thus the zone seems to be thicker than it should. In poorly preserved material, even fragments of the rim of *C. grandis* can still be recognized. In high latitudes, *D. saipanensis* and *D. barbadiensis* may be very rare or missing. *Sphenolithus furcatolithoides* has its LO near the base of this zone and *Helicosphaera compacta* its FO. *Cribrocentrum reticulatum* is cosmopolitan. Martini (1976) defined a combined NP 17/NP 18 as the interval from the LO of *Chiasmolithus solitus* to the LO of *C. grandis* in the Equatorial Pacific.

CHIASMOLITHUS OAMARUENSIS ZONE (NP 18)

Definition: FO of *Chiasmolithus oamaruensis* to FO of *Isthmolithus recurvus*.

Author: Martini (1970a)

Age: Late Eocene

Correlation: *Chiasmolithus oamaruensis* Subzone, CP15a, of Okada & Bukry (1980) with differing definition of the base of the subzone: LO of *C. grandis* or FO of *C. oamaruensis* to FO of *I. recurvus*. See also Fig. 4.

Remarks: The *C. oamaruensis* Zone is another zone which is difficult to recognize in low latitudes, since both *Chiasmolithus* and *Isthmolithus* are very rare or, especially the latter, often absent. *Cribrocentrum reticulatum* is found throughout NP 18 in high and in low latitudes. In high latitudes, *Chiasmolithus* and *Isthmolithus* are well represented and thus the recognition of this zone poses no problems.

ISTHMOLITHUS RECURVUS ZONE (NP 19)

Definition: FO of *Isthmolithus recurvus* to FO of *Sphenolithus pseudoradians*.

Authors: Hay, Mohler & Wade (1966), emend. Martini (1970a)

Age: Late Eocene

Correlation: Lower part of the *Isthmolithus recurvus* Subzone, CP15b of Okada & Bukry (1980). See also Fig. 4.

Remarks: As with the other Late Eocene zones, the *I. recurvus* Zone is difficult to recognize in low latitudes due to the scarcity of *I. recurvus*. *S. pseudoradians* seems to be restricted to certain areas and has been reported from older parts of the Eocene, all the way down to NP 15. This may be due to problems in distinguishing *S. pseudoradians* from large forms of the older *S. radians* (see taxonomic notes) or to different ranges of *S. pseudoradians* in different areas. The FO of *S. pseudoradians*

thus cannot be used as a marker event and Martini (1976) proposed a combined NP 19/NP 20 in the Equatorial Pacific. It is defined from the LO of *C. grandis* to the LO of *D. saipanensis*. The combined NP 19/NP 20 of Aubry (1983) is defined from the FO of *I. recurvus* to the LO of *D. saipanensis*. In high latitudes, *I. recurvus* overlaps only slightly with *Cribrocentrum reticulatum*. The LO of *C. reticulatum* can therefore be used in low latitudes to approximate the NP 18/NP 19 boundary.

SPHENOLITHUS PSEUDORADIANS ZONE (NP 20)

Definition: FO of *Sphenolithus pseudoradians* to LO of *Discoaster saipanensis* and/or *Discoaster barbadiensis* (CP15).

Author: Martini (1970a)

Age: Late Eocene

Correlation: Upper part of the *Isthmolithus recurvus* Subzone, CP15b, of Okada & Bukry (1980). See also Fig. 4.

Remarks: As discussed above, *S. pseudoradians* is an unreliable marker and NP 19 and NP 20 should not be distinguished (see discussion under NP 19). The LO of the disc-shaped discoasters *D. saipanensis* and *D. barbadiensis* is easily recognizable in low latitudes, but has been shown to occur earlier in high latitudes than in low latitudes (Cavelier, 1975, 1979). It has been used to define the Eocene/Oligocene boundary in terms of calcareous nannofossils.

Eocene/Oligocene boundary

The Eocene/Oligocene boundary in terms of calcareous nannofossils is usually drawn at the top of NP 20/CP15 defined by the LO of the disc-shaped discoasters represented by *D. barbadiensis* and *D. saipanensis*. This event has been found to slightly predate the disappearance of the planktic foraminifer genus *Hantkenina*. There are no other calcareous nannofossil events readily available to draw the Eocene/Oligocene boundary in terms of LO or FO. Reworking of Eocene material with disc-shaped discoasters may prevent us from recognizing the Eocene/Oligocene boundary. In some sections, *Ericsonia subdisticha* and/or *E. obruta* appear more common in the basal Oligocene than in the underlying Eocene and higher in the Oligocene.

Oligocene

Diversity is usually relatively low in most Oligocene samples. Some of the Eocene species disappear in the lower part of the Oligocene, while new sphenoliths appear in the upper part. No disc-shaped discoasters are found (except if reworked) and *Discoaster deflandrei*, *D. tanii nodifer* and *D. tanii* are usually the only discoasters present.

ERICSONIA SUBDISTICHA ZONE (NP 21)

Definition: LO of *Discoaster saipanensis* to LO of *Ericsonia formosa*.

Authors: Roth & Hay *in* Hay *et al.* (1967), emend. Martini (1970a)

Age: Early Oligocene

Correlation: *Coccolithus subdistichus* and *C. formosus* Sub-zones, CP16a and CP16b of Okada & Bukry (1980). See also Fig. 4.

Remarks: The LO of *D. saipanensis* and the LO of the other remaining disc-shaped discoaster, *D. barbadiensis*, are easily recognizable in low latitudes, where discoasters are usually quite common in the Upper Eocene. In high latitudes they are rare and disappear earlier if compared with the planktic foraminiferal zonation, that is, considerably before the Eocene/Oligocene boundary. Also, *E. formosa* sometimes occurs only sparsely before it becomes extinct. The end of the acme of *E. subdisticha* defines the base of CP16b. *E. subdisticha* is quite common in the basal Oligocene in some sections, while it can be very rare or absent in others. Its FO is in the Eocene, its LO higher up in the Oligocene.

Bybell (1982) observed the LO of the following species in NP 21 of Alabama and Mississippi: *Chiasmolithus titus* and *Pemma papillatum* followed by *Pedinocyclus larvalis*, *Calcidicus protoannulus*, *S. pseudoradians*, *S. tribulosus*, *I. recurvus* and *Bramletteius serraculoides* followed by *Coronocyclus serratus* and *Lithostromation operosum*. Some 40 species continued beyond the FO of *Sphenolithus distentus*, which here also was found to overlap with *R. umbilica*. In areas with reworked Eocene material, NP 21 might be impossible to recognize.

HELICOSPHAERA RETICULATA ZONE (NP 22)

Definition: LO of *Ericsonia formosa* to LO of *Reticulofenestra umbilica*.

Authors: Bramlette & Wilcoxon (1967), emend. Martini (1970a)

Age: Early Oligocene

Correlation: *Reticulofenestra hillae* Subzone, CP16c of Okada & Bukry (1980) with different definition of the top of the interval: LO of *E. formosa* to LO of *Reticulofenestra hillae* or LO of *R. umbilica*. See also Fig. 4.

Remarks: *R. umbilica* and *R. hillae* become rare towards their extinction at the end of NP 22. The LO of *Isthmolithus recurvus* occurs only slightly below the LO of *R. umbilica*. Occasional occurrences of *R. umbilica* in higher parts of the Oligocene as, for example, in the Bath Cliff section of Barbados can be due to reworking or to the survival of the species in this area. *Helicosphaera reticulata* may or may not be present in NP 22. *Lanternithus minutus* and *Rhabdolithus tenuis* are locally common.

SPHENOLITHUS PREDISTENTUS ZONE (NP 23)

Definition: LO of *Reticulofenestra umbilica* to FO of *Sphenolithus ciperoensis*.

Authors: Bramlette & Wilcoxon (1967), emend. Martini (1970a)

Age: Middle Oligocene (Rupelian)

Correlation: *Sphenolithus predistentus* and *Sphenolithus distentus* Zones, CP17 and CP18 of Okada & Bukry (1980) with the following definitions: LO of *R. hillae* or LO of *R. umbilica* to FO of *S. distentus* (CP17) and FO of *S. distentus* to FO of *S. ciperoensis* (CP18). The FO of *Cyclicargolithus abisectus* was used by Müller (1976) as a substitute marker for the FO of *S. ciperoensis* in mid to high latitude sections.

Remarks: *S. predistentus* is usually present in NP 23, but may be missing in very high latitude sections. The Middle and Late Oligocene sphenoliths are very small and therefore easily overlooked especially in poorly preserved material. Also, their determination may pose problems and therefore this is discussed in detail in the taxonomic notes. Generally, in high latitude Oligocene (and Miocene), one cannot use the sphenoliths for zoning and one is left with the absence of *Reticulofenestra umbilica* and *Cyclicargolithus abisectus* and the presence of *Dictyococcites bisectus* for the recognition of Middle Oligocene. The FO of *C. abisectus* and *Helicosphaera recta* and the LO of *Helicosphaera compacta* are usually found close to the FO of *S. ciperoensis* and thus can be used to roughly approximate the NP 23/NP 24 boundary.

SPHENOLITHUS DISTENTUS ZONE (NP 24)

Definition: FO of *Sphenolithus ciperoensis* to LO of *Sphenolithus distentus*.

Authors: Bramlette & Wilcoxon (1967)

Age: Late Oligocene

Correlation: *Cyclicargolithus floridanus* Subzone, CP19a of Okada & Bukry (1980). See also Fig. 4.

Remarks: The lower boundary of this zone was discussed above. *Sphenolithus predistentus* usually disappears before the LO of *S. distentus* and *Helicosphaera recta* and *H. euphratis* are usually also part of the assemblage. Rare *Triquetrorhabdulus carinatus* appear near the top of NP 24 in some sections. This Oligocene zone is also difficult or impossible to recognize in high latitudes or in areas with poor connections to the open ocean. In these circumstances, the FO of *Pontosphaera enormis* has been shown to be a useful event for the subdivision of the Upper Oligocene and is an approximation of the NP 24/NP 25 boundary (Martini, 1981a).

442

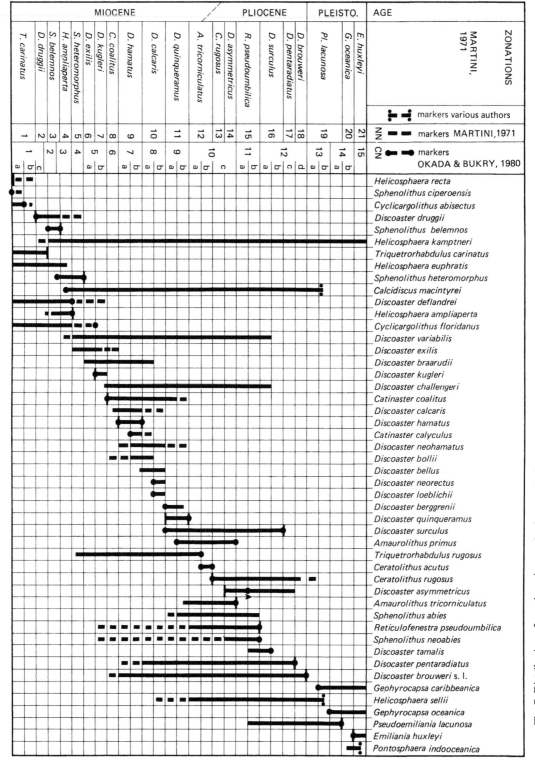

Fig. 7. Distribution of zonal markers and other species in the Neogene

SPHENOLITHUS CIPEROENSIS ZONE (NP 25)

Definition: LO of *Sphenolithus distentus* to the LO of *Helicosphaera recta* and/or the LO of *Sphenolithus ciperoensis*.

Authors: Bramlette & Wilcoxon (1967), emend. Martini (1976)

Age: Late Oligocene

Correlation: *Dictyococcites bisectus* Subzone, CP19b, of Okada & Bukry (1980). See also Fig. 4.

Remarks: While *H. recta* and *S. ciperoensis* are useful markers in many sections in low latitudes, the LO of the equally disappearing *Dictyococcites bisectus* and *Zygrhablithus bijugatus* (the latter at least in most sections where it is present consistently in the Late Oligocene) are the events used for an approximation of the NP 25/NN 1 boundary in higher latitudes. Here, too, the FO of *Pontosphaera enormis* has been used to define the base of NP 25 (Martini, 1981a). *H. recta* has been found in many sections to survive the extinction of *S. ciperoensis*, *D. bisectus* and *Z. bijugatus* and thus is not a very reliable marker species. *Triquetrorhabdulus carinatus* and *T. milowii* are sometimes found in NP 25.

Oligocene/Miocene boundary

The Oligocene/Miocene boundary in terms of calcareous nannofossils is placed at the top of NP 25 by some, and within NN 1 by other authors. The sequence of disappearance of *H. recta*, *S. ciperoensis*, *D. bisectus* and *Z. bijugatus* varies somewhat from section to section and the boundary is thus set differently by different authors, depending on their choice of marker species. In lower latitudes *S. ciperoensis* is usually used, in higher latitudes, *D. bisectus*.

Miocene

The subdivision of the Miocene is based mainly on species of *Discoaster* (Figs. 7–9) and thus is usually easily accomplished in low latitudes, where discoasters are common in open ocean assemblages. It is difficult in high latitudes, where discoasters are absent or very rare, and in assemblages from marginal seas, where discoasters and most other markers tend to be rare or missing. Also, the other marker species belong to genera that are more common or even restricted to low latitudes and thus, even more than in the Oligocene, the zonation is most reliable and correlatable over wide distances in low latitudes only. The overgrowth of discoasters in many assemblages can pose serious problems for their correct identification and thus affect the reliability of the resulting zonation.

TRIQUETRORHABDULUS CARINATUS ZONE (NN 1)

Definition: LO of *Helicosphaera recta* and/or *Sphenolithus ciperoensis* to FO of *Discoaster druggii*.

Authors: Bramlette & Wilcoxon (1967), emend. Martini & Worsley (1970)

Age: Early Miocene and/or latest Oligocene

Correlation: *Cyclicargolithus abisectus* and *Discoaster deflandrei* Subzones, CN1a and CN1b of Okada & Bukry (1980).

Remarks: NN 1 is usually regarded as the lowermost zone in the Miocene. Bukry (1973a) distinguishes a lower part of this zone with common *C. abisectus* (CN1a) and a middle part with low diversity and relatively common *D. deflandrei* (CN1b). The upper part (CN1c) corresponds to the following Zone NN 2. *Cyclicargolithus floridanus* and *T. carinatus* are usually present throughout, the former being especially common. In poorly preserved material and in samples with common detritus, it can be difficult to distinguish poorly preserved *T. carinatus* from other elongate calcite particles. *D. druggii* can be difficult to distinguish from other discoasters when the fossils are overgrown. But even then, the slightly tapering ends of the arms can be distinguished from the diverging ones of the more common and longer-ranging *D. deflandrei*.

DISCOASTER DRUGGII ZONE (NN 2)

Definition: FO of *Discoaster druggii* to LO of *Triquetrorhabdulus carinatus*.

Authors: Martini & Worsley (1970)

Age: Early Miocene

Correlation: The *Discoaster druggii* Subzone (CN1c) of Okada & Bukry (1980) is defined by the FO of *D. druggii* to the FO of *Sphenolithus belemnos*. The latter was found to overlap with *T. carinatus* and thus the correlation is not precise.

Remarks: *D. druggii* can be difficult to determine in overgrown material (see above). Early forms of *H. kamptneri* s.ampl. and *H. ampliaperta* are found in the upper part of the zone. The FO of *S. belemnos*, used by Bukry (1973a) to define the base of his *S. belemnos* Zone, usually is found slightly below the LO of *T. carinatus*. In high latitudes, *D. druggii*, *S. belemnos* and *T. carinatus* are very rare or absent and there are no convenient substitutes to subdivide the NN 1 to NN 3 interval. *T. carinatus* may occur sporadically also in mid and low latitudes.

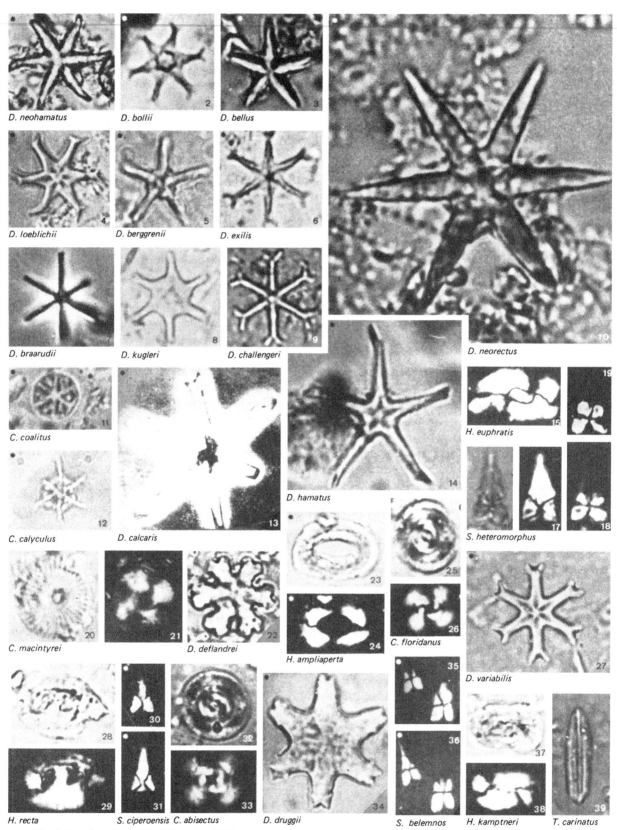

Fig. 8. Zonal markers of the Miocene and Pliocene

SPHENOLITHUS BELEMNOS ZONE (NN 3)

Definition: LO of *Triquetrorhabdulus carinatus* to LO of *Sphenolithus belemnos*.

Authors: Bramlette & Wilcoxon (1967), emend. Martini (1971)

Age: Early Miocene

Correlation: The *Sphenolithus belemnos* Zone of Bukry (1973a), emend. Bukry (1975) and CN2 of Okada & Bukry (1980) are defined from the FO of *S. belemnos* to the FO of *S. heteromorphus* or the LO of *S. belemnos*.

Remarks: Since *S. belemnos* and *S. heteromorphus* overlap in several sections where both species are found, a new subzone could be distinguished in such cases. In poorly preserved material, it can be difficult to distinguish *S. heteromorphus* from *S. conicus*, especially if one compares the forms to the holotypes of the two species (see taxonomic notes) and not to the common usage over the past decade of a *S. heteromorphus* with a low proximal column as illustrated by Bramlette & Wilcoxon (1967).

HELICOSPHAERA AMPLIAPERTA ZONE (NN 4)

Definition: LO of *Sphenolithus belemnos* to LO of *Helicosphaera ampliaperta*.

Authors: Bramlette & Wilcoxon (1967), emend. Martini (1971)

Age: Early Miocene; lowest part of the type Langhian

Correlation: The *H. ampliaperta* Zone of Bukry (1975) and CN3 of Okada & Bukry (1980) do not correlate exactly with NN 4. CN3 is defined from the FO of *S. heteromorphus* (or LO of *S. belemnos* in Bukry, 1973a) to the FO of *Calcidiscus macintyrei* or to the end of the acme of *Discoaster deflandrei* (or the LO of *H. ampliaperta* in Bukry, 1973a).

Remarks: The first rare specimens of *Calcidiscus macintyrei* are sometimes found in the upper part of the range of *H. ampliaperta*. In some areas of the South Atlantic and the Pacific, *H. ampliaperta* is very rare or absent and thus NN 4 and NN 5 cannot be distinguished (Martini, 1976). In some instances, the FO of *Discoaster exilis* can be used to approximate this boundary (Martini, 1980). *Helicosphaera euphratis* becomes rare and often disappears in the upper part of NN 4. *C. floridanus* becomes

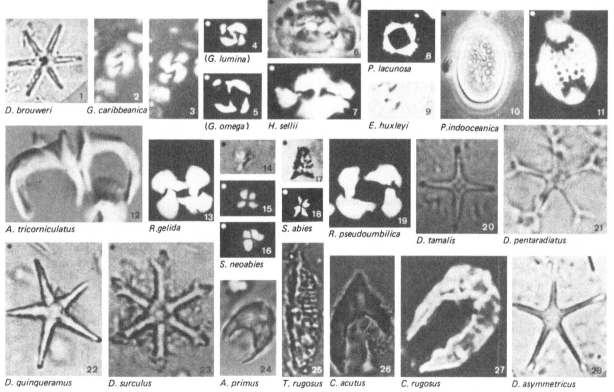

Fig. 9. Zonal markers of the Pliocene and Pleistocene

rare in many areas from the upper part of NN 4 upwards, though it may continue with high frequencies in, for example, the South Atlantic.

SPHENOLITHUS HETEROMORPHUS ZONE (NN 5)

Definition: LO of *Helicosphaera ampliaperta* to LO of *Sphenolithus heteromorphus*.

Authors: Bramlette & Wilcoxon (1967)

Age: Early/Middle Miocene

Correlation: *Sphenolithus heteromorphus* Zone (CN4) of Okada & Bukry (1980) does not correlate exactly with NN 5 (see correlations for NN 4).

Remarks: Within this zone we see the dominance of *D. deflandrei* with its short arms giving way to a dominance of slender-armed discoasters at the base of the zone. Elliptical forms of *C. macintyrei* (*C. leptoporus centrovalis* Stradner & Fuchs, 1980) become more common than below, and the FO of *Triquetrorhabdulus rugosus* occurs in the upper part of this zone. Bukry (1973a) also mentions the FO of *Sphenolithus abies* in the *S. heteromorphus* Zone, while many authors only report the FO of *S. abies* from the Upper Miocene. This may be due to different species concepts and/or regional variations of distribution. The first forms of *Reticulofenestra pseudoumbilica* s.ampl. are also reported from this zone (Bukry, 1973a). The species concepts in the Miocene and Pliocene *Reticulofenestra* and related forms are controversial and the whole family of the Prinsiaceae is discussed in the taxonomic notes.

DISCOASTER EXILIS ZONE (NN 6)

Definition: LO of *Sphenolithus heteromorphus* to FO of *Discoaster kugleri* and/or LO of *Cyclicargolithus floridanus* (CN5a).

Authors: Hay (1970), emend. Martini (1974)

Age: Middle Miocene

Correlation: Lower part of *Discoaster exilis* Zone, *Coccolithus miopelagicus* Subzone (CN5a) of Okada & Bukry (1980). Lowermost part of *Discoaster variabilis* Zone of temperate areas of Bukry (1973a).

Remarks: *Discoaster braarudii*, the slender, six-rayed discoaster with non-tapered arms and without a central knob appears near the base of this zone. *C. floridanus* decreases in abundance and is replaced by abundant *R. pseudoumbilica* near the top of this zone, which often is difficult to determine when the discoasters are overgrown and *D. kugleri* difficult to distinguish from similar other overgrown discoasters such as *D. variabilis* or *D. deflandrei*. The *D. variabilis* Zone is defined from the LO of *C. floridanus* to the FO of *D. mendomobensis*.

DISCOASTER KUGLERI ZONE (NN 7)

Definition: FO of *Discoaster kugleri* and/or LO of *Cyclicargolithus floridanus* (CN5b) to FO of *Catinaster coalitus*.

Authors: Bramlette & Wilcoxon (1967), emend. Martini (1971)

Age: Middle Miocene

Correlation: Upper part of the *Discoaster exilis* Zone, *Discoaster kugleri* Subzone (CN5b) of Okada & Bukry (1980). Lower part of the *D. variabilis* Zone (see NN 6).

Remarks: Besides *Discoaster kugleri*, also *D. challengeri, D.* aff. *brouweri* and *D. bollii*, all six-rayed discoasters, appear in this part of the Middle Miocene. *C. floridanus* definitely disappears and *D. deflandrei* becomes very rare and disappears near the top of this zone. The NN 7 Zone is difficult or impossible to distinguish in high latitudes, where discoasters and *C. coalitus* are very rare or missing. *D. kugleri* has its LO near the base of NN 8.

CATINASTER COALITUS ZONE (NN 8)

Definition: FO of *Catinaster coalitus* to FO of *Discoaster hamatus*.

Authors: Bramlette & Wilcoxon (1967), emend. Martini (1971)

Age: Middle Miocene

Correlation: *Catinaster coalitus* Zone, (CN6) of Okada & Bukry (1980). Middle part of the *D. variabilis* Zone (see NN 6).

Remarks: *C. coalitus* is easily recognizable and a good marker in low latitudes. The FO of *Discoaster pseudovariabilis* also approximates the FO of *C. coalitus*. *Discoaster exilis* and *Coccolithus miopelagicus* usually disappear before the top of this zone. The NN 8 Zone can be subdivided by the FO of *Catinaster calyculus*. The latter is very rare after its appearance but becomes more common than *C. coalitus* around the top of the zone. Again, this zone is difficult to recognize in high latitudes and no substitute markers can be recommended for high latitude use yet.

DISCOASTER HAMATUS ZONE (NN 9)

Definition: FO to LO of *Discoaster hamatus*.

Authors: Bramlette & Wilcoxon (1967), emend. Martini (1971)

Age: Middle/Late Miocene

Correlation: *Discoaster hamatus* Zone (CN7) of Okada & Bukry (1980) and middle part of the *D. variabilis* Zone (see NN 6).

Remarks: The most conspicuous change in this zone is the FO of five-rayed discoasters like the large *D. hamatus* (with tapering, asymmetrically bifurcated arms) and *D. bellus* (more or less straight, tapering arms, smaller than *D.*

hamatus), as well as *D. prepentaradiatus* with its bifurcated arms. Bukry (1973a) uses the FO of *C. calyculus* to subdivide the *D. hamatus* Zone into a lower *Helicosphaera kamptneri/carteri* Subzone (CN7a) and an upper *Catinaster calyculus* Subzone (CN7b). This subdivision is in disagreement with the observations by Thierstein (1974) and Salis (1984) who found that *C. calyculus* appears before *D. hamatus* in the Indian Ocean and the South Atlantic respectively. This disagreement may be due to differing species concepts of *C. calyculus* (see taxonomic notes).

DISCOASTER CALCARIS ZONE (NN 10)

Definition: LO of *Discoaster hamatus* to FO of *Discoaster quinqueramus* (NN 10) or FO of *Discoaster berggrenii* or FO of *Discoaster surculus* (CN8).

Author: Martini (1969)

Age: Late Miocene

Correlation: ± *Discoaster neohamatus* Zone (CN8) of Okada & Bukry (1980) and upper part of the *D. variabilis* Zone (see NN 6) and lowermost part of *D. mendomobensis* Zone of Bukry (1973a).

Remarks: *C. calyculus* disappears above the base of NN 10, usually after the LO of *C. coalitus*. This zone was subdivided by Bukry (1973a) into a lower *Discoaster bellus* Subzone (CN8a) and an upper *D. neorectus* Subzone (CN8b) by the FO of *D. neorectus* and/or the FO of *D. loeblichii*. *D. neorectus* is a very large, six-rayed dicoaster with tapering arms and a range restricted to CN8b. Early forms of *D. surculus* (*D. cf. D. pseudovariabilis*) appear in the upper part of NN 10. *D. berggrenii* appears before *D. quinqueramus* and thus the boundary is shifted slightly downwards, if the two species are considered as one species, and the top of CN8 is somewhat older than the top of NN 10. The FO of *D. mendomobensis* which defines the base of the zone of this name was correlated with the upper part of CN8b by Bukry (1981b).

DISCOASTER QUINQUERAMUS ZONE (NN 11)

Definition: FO to LO of *Discoaster quinqueramus* (NN 11) and FO of *Discoaster berggrenii* and/or FO of *Discoaster surculus* to LO of *Discoaster quinqueramus* (CN9).

Authors: Gartner (1969c), emend. Martini (1971)

Age: Late Miocene

Correlation: ± *Discoaster quinqueramus* Zone (CN9) of Okada & Bukry (1980) and most of the *D. mendomobensis* Zone of Bukry (1973a, emend. 1981b).

Remarks: The FO of *Amaurolithus primus*, the first horseshoe-shaped calcareous nannofossil in the Neogene, can be used to subdivide the *D. quinqueramus* Zone into a lower

D. berggrenii Subzone and an upper *A. primus* Subzone, that is, the CN9a and CN9b of Okada & Bukry (1980). *D. berggrenii* and *D. quinqueramus* are considered synonyms by some authors, others report *D. berggrenii* slightly below *D. quinqueramus* and distinguish the two species by the more prominent central knobs (on both sides) of *D. berggrenii*, which consequently seems to have relatively shorter arms. Several other species of *Amaurolithus*, the non-birefringent group of horseshoe-shaped nannofossils, appear shortly after *A. primus*, still in NN 11 (see taxonomic notes). *D. surculus* is present throughout the zone.

AMAUROLITHUS TRICORNICULATUS ZONE (NN 12)

Definition: LO of *Discoaster quinqueramus* to FO of *Ceratolithus rugosus* (NN 12) and/or LO of *Ceratolithus acutus* (CN10b).

Authors: Gartner (1969c), emend. Martini (1971)

Age: Late Miocene/Early Pliocene

Correlation: Lower and middle Subzones of *Amaurolithus tricorniculatus* Zone (CN10) of Okada & Bukry (1980).

Remarks: The FO of *Ceratolithus acutus* and/or the LO of the genus *Triquetrorhabdulus* define the boundary between the *T. rugosus* Subzone (CN10a) and the overlying *C. acutus* Subzone (CN10b) in Bukry (1973a). The upper part of the zone is characterized by the appearance of birefringent ceratoliths. Since ceratoliths are only common in low latitudes, the zonation of the uppermost Miocene and lowermost Pliocene is difficult in high latitudes, and no good alternative species seem to be available.

Miocene/Pliocene boundary

The Miocene/Pliocene boundary in terms of calcareous nannofossils has been placed in the *A. tricorniculatus* Zone e.g. by Okada & Bukry (1980) and in the overlying *C. rugosus* Zone (NN 13) by Martini (1971). The FO of *C. acutus* has been observed just above the Miocene/Pliocene boundary in Capo Rossello, the Zanclean stratotype in Sicily, and thus the Miocene/Pliocene boundary seems to fall within the *Triquetrorhabdulus rugosus* Subzone (CN10a) or within the *A. tricorniculatus* Zone (NN 12) (Mazzei *et al.*, 1979).

Pliocene

The zonation of the Pliocene is based mainly on discoasters, ceratoliths and sphenoliths, all forms which are rare or absent in high latitudes. There, often only the LO of *Reticulofenestra pseudoumbilica* and/or the FO of *Pseudoemiliania lacunosa* in the middle part of the Pliocene can be used for a subdivision. It has long been assumed that the

FO of the genus *Gephyrocapsa* coincided with the Pliocene/ Pleistocene boundary and thus this FO was used to exclude sediments with *Gephyrocapsa* from the Pliocene in high latitudes, where the Pliocene cannot always be recognized by the presence of discoasters. Samtleben (1980) has shown that the first *Gephyrocapsa* appear long before the LO of *R. pseudoumbilica* in the middle part of the Pliocene (see also taxonomic notes for the Prinsiaceae).

CERATOLITHUS RUGOSUS ZONE (NN 13)

Definition: FO of *Ceratolithus rugosus* to FO of *Discoaster asymmetricus* (NN 13) or FO of *C. rugosus* and/or LO of *C. acutus* to FO of *D. asymmetricus* (CN10c).

Author: Gartner (1969c)

Age: Early Pliocene (Zanclean)

Correlation: Lower part of *Ceratolithus rugosus* Subzone (CN10c) of Okada & Bukry (1980), or the entire CN10c of Bukry (1981b).

Remarks: The finding of this zone depends on the presence and the correct determination of *D. asymmetricus*. While this seems possible in some low latitude sections where typical *D. asymmetricus* appear in reasonable numbers, it is difficult or impossible in sections, where discoasters are rare and slightly irregular specimens of *D. pentaradiatus* with missing bifurcations can be mistaken for *D. asymmetricus*. Between crossed nicols the arms of *D. pentaradiatus* do not behave like a single crystal as in *D. asymmetricus* but show slight birefringence and extinction lines are visible between the arms. In cases where *D. asymmetricus* is very rare, it is advisable to use the combined NN 13/NN 14 as suggested by Martini (1980).

DISCOASTER ASYMMETRICUS ZONE (NN 14)

Definition: FO of *Discoaster asymmetricus* to LO of *Amaurolithus tricorniculatus*.

Author: Gartner (1969c)

Age: Early Pliocene (Zanclean)

Correlation: Upper part of *Ceratolithus rugosus* Subzone (CN10c) of Okada & Bukry (1980), where the top is defined by the LO of *A. tricorniculatus* and/or the LO of *A. primus*, or the *A. delicatus* Subzone (CN10d) of Bukry (1981b).

Remarks: For the lower boundary see remarks above. The upper boundary is defined by the sharp decrease in non-birefringent ceratoliths, the LO of *A. tricorniculatus* and/or the LO of *A. primus*, the last representatives of the non-birefringent ceratoliths with the exception of the very rare and conspicuous *A. bizarrus*. Birefringent ceratoliths continue through the Holocene. Again, this zone is difficult or impossible to recognize in high latitudes, where ceratoliths and discoasters are very rare or absent.

RETICULOFENESTRA PSEUDOUMBILICA ZONE (NN 15)

Definition: LO of *Amaurolithus tricorniculatus* to LO of *Reticulofenestra pseudoumbilica*.

Author: Gartner (1969c)

Age: Early Pliocene (Zanclean/Piacenzian thus Early/Late Pliocene?)

Correlation: *Reticulofenestra pseudoumbilica* Zone (CN11) of Okada & Bukry (1980).

Remarks: For the lower boundary see above. The upper boundary can be difficult to recognize in low latitude sections, where *R. pseudoumbilica* is very rare towards the middle part of the Pliocene. The remaining, long-ranging sphenoliths *Sphenolithus abies* and *S. neoabies* disappear near the LO of *R. pseudoumbilica* and can be used as markers where the latter is rare. *R. pseudoumbilica* and *S. abies* typically disappear together in open ocean low latitude sections, while *S. abies* persisted later than *R. pseudoumbilica* in marginal seas leading Boudreaux (1974) to recognize a local *S. abies* Zone. *S. neoabies* usually disappears shortly after typical *S. abies*. This LO could be used for a further subdivision of the NN 15. Bukry (1973a) used the acme of *D. asymmetricus* for a subdivision of this zone into a lower *S. neoabies* and an upper *D. asymmetricus* Subzone (CN11a, b). The distinction of *R. pseudoumbilica* from slightly smaller, younger, Quaternary forms, is discussed in the taxonomic notes on the Prinsiaceae. According to Samtleben (1980), the first, small representatives of *Geophyrocapsa* appear within this zone and also the FO of *Pseudoemiliania lacunosa* is found below the LO of *R. pseudoumbilica* in many sections. The top of NN 15 can usually be recognized easily in high latitudes and there it is often the only recognizable biohorizon in the Pliocene. Backman & Shackleton (1983) suggest an age of 3.56 ± 0.02 m.y. for the LO of *R. pseudoumbilica* and 3.45 ± 0.02 m.y. for the LO of *Sphenolithus* spp.

DISCOASTER SURCULUS ZONE (NN 16)

Definition: LO of *Reticulofenestra pseudoumbilica* to the LO of *Discoaster surculus*.

Authors: Hay & Schmidt *in* Hay *et al.* (1967), emend. Gartner (1969c)

Age: Late Pliocene (Piacenzian)

Correlation: Lower part of *Discoaster brouweri* Zone (CN12a, b) of Okada & Bukry (1980).

Remarks: For the lower boundary see above. The LO of *D. surculus* is easily recognizable in low and mid latitude sections. The bifurcation at the end of the arms becomes narrower towards the end of the range of *D. surculus*. The LO of *Discoaster tamalis*, a four-rayed discoaster with tapering arms, serves to subdivide this interval into a lower *D. tamalis* Subzone (CN12a) and an upper *D. surculus* Subzone according to Bukry (1973a). *P. lacunosa* becomes well established in this zone while *Discoaster challengeri* and *D. variabilis* disappear. In high latitudes, this zone cannot usually be recognized, since discoasters disappear before the LO of *R. pseudoumbilica*. Backman & Shackleton (1983) suggest an age of 2.41 ± 0.03 m.y. for the LO of *D. surculus*.

DISCOASTER PENTARADIATUS ZONE (NN 17)

Definition: LO of *Discoaster surculus* to LO of *Discoaster pentaradiatus*.

Authors: Martini & Worsley (1970)

Age: Late Pliocene (Piacenzian)

Correlation: *Discoaster pentaradiatus* Subzone (CN12c) of Okada & Bukry (1980).

Remarks: NN 17 usually is a very thin zone, since the discoasters disappear in close order towards the Pliocene/Pleistocene boundary. The LO of *D. pentaradiatus* is only slightly younger than that of *D. surculus*. *D. asymmetricus* disappears about at the same time as *D. pentaradiatus*. Backman & Shackleton (1983) studied the discoasters of the Late Pliocene quantitatively and suggest an age of 2.35 ± 0.04 m.y. for the LO of *D. pentaradiatus*.

DISCOASTER BROUWERI ZONE (NN 18)

Definition: LO of *Discoaster pentaradiatus* to the LO of *Discoaster brouweri*.

Authors: Boudreaux & Hay *in* Hay *et al.*, 1967), emend. Martini (1971)

Age: Late Pliocene

Correlation: *Calcidiscus macintyrei* Subzone (CN12d) of Okada & Bukry (1980).

Remarks: The NN 18 is usually rather thin. In sections with glacial debris as in the North Atlantic, the search for the last discoasters can be a time-consuming business, since discoasters were already rare in the original assemblage. Here, sometimes sideviews of single arms can indicate the presence of discoasters long before a whole specimen is found. The size of *Gephyrocapsa* specimens increases towards the top of this zone. Backman & Shackleton (1983) suggest an age of 1.88 ± 0.01 m.y. for the LO of *D. brouweri*.

Pliocene/Pleistocene boundary

The Pliocene/Pleistocene boundary in terms of calcareous nannofossils is usually set above the LO of *D. brouweri*. This approximates only poorly the findings of various authors who studied the sections in Calabria, Italy, that were considered to be designated as stratotypes for the Calabrian, the basal stage of the Pleistocene (Hay & Boudreaux, 1969;

Smith, 1969; Bandy & Wilcoxon, 1970; Takayama, 1970; Rio, 1974; Selli *et al.*, 1977), see discussions in Gartner (1973) and Berggren & Van Couvering (1974). The main problem is the reworking of discoasters, namely *D. brouweri*, in sections with a detrital component and even in pelagic sequences. In such cases, the appearance of larger species of *Gephyrocapsa* may give an indication of the younger age of the sediment. The LO of *Calcidiscus macintyrei* usually occurs definitely after the LO of *D. brouweri* and thus can give an indication, that one is above the Pliocene/Pleistocene boundary in sections where the distribution of the last discoasters suggests reworking.

Pleistocene

The Pleistocene is primarily subdivided by small representatives of the family Prinsiaceae and secondarily by *C. macintyrei*, *Helicosphaera sellii* and *Pontosphaera indooceanica*. Whereas the LM can be used for the subdivisions proposed by Martini (1971) and Bukry (1973a), more detailed and quantitative investigations almost always have to be done by TEM or SEM. The various species of *Gephyrocapsa* can hardly be distinguished without an electron microscope and also the recognition and definite determination of the marker of NN 21, *Emiliania huxleyi*, is best done with an EM. Quantitative investigations have been used to subdivide the Standard Zones or to define intervals that are correlatable with climatic changes (see taxonomic notes, Prinsiaceae).

PSEUDOEMILIANIA LACUNOSA ZONE (NN 19)

Definition: LO of *Discoaster brouweri* to LO of *Pseudoemiliania lacunosa*.

Author: Gartner (1969c)

Age: Early Pleistocene (Calabrian)

Correlation: *Crenalithus doronicoides* Zone (CN13 and lower part of the *Gephyrocapsa oceanica* Zone (CN14)), the *Gephyrocapsa caribbeanica* Subzone (CN14a) of Okada & Bukry (1980); *Calcidiscus macintyrei* Zone, *Helicosphaera sellii* Zone, interval of small *Gephyrocapsa* and *Pseudoemiliania lacunosa* Zone of Gartner (1977b); see also taxonomic notes on Prinsiaceae.

Remarks: The NN 19 as defined by Martini (1971) can be subdivided by different events which occurred in the following order: FO of *G. caribbeanica*, LO of *C. macintyrei* and LO of *Helicosphaera sellii*, FO of *G. oceanica*. Bukry (1973a) used the two *Gephyrocapsa* events to define the subzones '*Emiliania annula*' (=round *P. lacunosa*), CN13a, *G. caribbeanica*, CN13b and '*Emiliania ovata*' (=elliptical *P. lacunosa*), CN14a. The other events were used by Gartner (1977b) and others in different areas (see taxonomic notes, Prinsiaceae) and can be recognized over wide areas. *P. lacunosa* becomes very rare towards its LO.

Backman & Shackleton (1983) found the LO of *C. macintyrei* to have occurred some 1.45±0.01 m.y. ago. They also report the diachronous LO of *H. sellii* across latitudes, with its youngest age of 1.37 m.y. in the equatorial Pacific.

GEPHYROCAPSA OCEANICA ZONE (NN 20)

Definition: LO of *Pseudoemiliania lacunosa* to FO of *Emiliania huxleyi*.

Authors: Boudreaux & Hay *in* Hay *et al.* (1967), emend. Gartner (1969c) as the *Gephyrocapsa* Zone

Age: Late Pleistocene

Correlation: Upper part of *Gephyrocapsa oceanica* Zone (CN14), *Ceratolithus cristatus* Subzone (CN14b) of Okada & Bukry (1980).

Remarks: It is difficult to recognize *E. huxleyi* with the LM except in well preserved assemblages. Generally they look like 'empty, very small Prinsiaceae'. Thus it is recommended to use a TEM or a SEM for the verification of the presence of *E. huxleyi*. Thierstein *et al.* (1977), who studied the global synchroneity of late Quaternary coccolith datum levels, found the extinction of *P. lacunosa* to be a synchronous datum level. Also the FO of *E. huxleyi* was found to occur consistently in the same oxygen isotope stage (8). The FO of *Gephyrocapsa ericsonii* and *G. muellerae* approximates the NN 20/NN 21 boundary.

EMILIANIA HUXLEYI ZONE (NN 21)

Definition: Interval above the FO of *Emiliania huxleyi*.

Authors: Boudreaux & Hay *in* Hay *et al.* (1967)

Age: Late Pleistocene

Correlation: *Emiliania huxleyi* Zone (CN15) of Okada & Bukry (1980).

Remarks: Three events have been suggested to subdivide this interval further: the acme of *E. huxleyi* (Gartner, 1977b), the LO of *Pontosphaera indooceanica* by Čepek (1973), and the dominance reversal of *G. caribbeanica*/ *E. huxleyi* investigated and found to be time-transgressive by Thierstein *et al.* (1977). In tropical and subtropical waters the last-mentioned event occurred earlier than in transitional waters (oxygen isotope stages 5b and 5a against 4).

E. huxleyi is one of those coccoliths with the widest distribution at present. It is found in very high latitudes with low temperatures as well as in the tropics. It also is one of the last species to be dissolved when sinking to great oceanic depths.

Classification, taxonomic and other notes

There is no general agreement on the classification of Cenozoic calcareous nannofossils. Consequently, many authors arrange the genera in alphabetic order for their systematic descriptions. Species lists are often presented in alphabetic order of the species, because many species are assigned to different genera by different authors and thus are easier to find in this way.

Classifications have been attempted repeatedly. Hay (1977) and Tappan (1980) include assignments not acceptable to this author. In the following, the classification used by Perch-Nielsen (1971a) is followed with some minor changes. For the terminology, the suggestions by the Roundtable on Calcareous Nannoplankton in Rome (Farinacci, 1971) are followed with minor changes and additions. While the notes are not a complete taxonomic treatment of the more important and more commonly used calcareous nannofossils of the Cenozoic, they should facilitate the stratigraphic use of them by illustrating all the markers and those forms with which they occur and could be confused by LM and partly also by SEM. The families are arranged in alphabetic order.

The magnifications of the LM pictures are c. ×2000 where nothing else is indicated. SEM magnifications are not given, but can be estimated from the LM picture of the same species.

Quotations from the original species description are indicated by quotation marks without mention of the author. Author and year of publication of a given species can be found in the list of taxa at the end of this chapter.

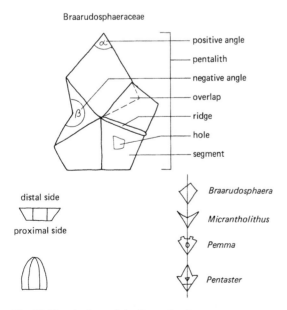

Fig. 10. Terminology of the Braarudosphaeraceae

Fig. 11. LM and SEM of representatives of the
Braarudosphaeraceae. * = Holotype, *B.* = *Braarudosphaera*,
M. = *Micrantholithus*, *P* = *Pemma*

Braarudosphaeraceae Deflandre (1947)
Figures 10, 11

The first four of the following genera can be assigned to the modern family of the Braarudosphaeraceae, which includes calcareous pentaliths:

Braarudosphaera
Micrantholithus
Pemma
Pentaster
(*Biantholithus* and *Eodiscoaster*)

The genera are distinguished by the form of the individual pentagonal nannoliths and by the presence or absence of a hole in each segment (see Fig. 10, for the terminology). Intermediate forms between *Braarudosphaera* and *Micrantholithus* exist and *Braarudosphaera* segments can be eroded so much that they look like *Micrantholithus* segments. The type species of *Braarudosphaera*, *B. bigelowii*, has trapezoidal segments composing the pentalith, while those of *Micrantholithus*, as in *M. flos*, are triangular. In *Pemma* the segments are either perforated by a hole, decorated by a button or show a central depression. *Pentaster* is defined as having pentaliths bearing a large radial ray on each segment, a smaller ray protruding radially between each pair of large rays (Fig. 11). According to Martini (1971), the author of the genus *Eodiscoaster*, this genus was based on an overgrown specimen of a *Braarudosphaera* and therefore should not be used.

The genus *Biantholithus* cannot be considered as belonging to the Braarudosphaeraceae but should be an 'incertae sedis', since it does not have pentaliths and also does not fit into other families such as the Discoasteraceae or Fasciculithaceae. The FO of *Biantholithus sparsus*, its type species, is often used as a marker for the Cretaceous/Tertiary boundary.

Since the numerous species of *Braarudosphaera*, *Micrantholithus* and *Pemma* are not generally used for biostratigraphy in the Cenozoic, they are not treated here in detail. Perch-Nielsen (1971a) and Aubry (1985) gave an overview of the species of these genera described before 1970 and 1982 respectively. Bybell & Gartner (1972) have described several species and the genus *Pentaster* and recognized provincialism among the Mid Eocene representatives of the family. The Braarudosphaeraceae of the Cenozoic are usually most common in some nearshore and hemipelagic deposits, where they bloomed repeatedly, especially shortly after the Cretaceous/Tertiary boundary event(s) and during the Middle Eocene. On the other hand, contemporaneous nearshore and hemipelagic sediments can be devoid of Braarudosphaeraceae and the presence of the Oligocene, pelagic *Braarudosphaera* chalk in the South Atlantic complicates the simple picture of the Braarudosphaeraceae as a nearshore nannoplanktic form.

Despite the fact that the Braarudosphaeraceae have not yet been used much stratigraphically in the Cenozoic, there seems no reason why they should not be used in the future. To do this, the maximum ranges of the various species have to be established where they occur in rich floras. This would allow us to date those meagre assemblages where representatives of the Braarudosphaeraceae dominate but where no marker species can be found. Thus, the Early Danian *Braarudosphaera alta* seems to have a very limited distribution in low to mid latitude sections in the middle part of NP 1.

Pemma and *Micrantholithus* seem most common in Mid Eocene sediments and the former may even be restricted to this time. Gartner (1971) used the FO of *Pemma papillatum* as the base of his *P. papillatum* Zone which would lie in NP 16 or CN14. Bybell & Gartner (1972) described provincialism among the Mid Eocene Braarudosphaeraceae.

Calciosoleniaceae Kamptner (1927)
Figure 12

The modern family of the Calciosoleniaceae includes several modern genera (see Tappan, 1980 p. 781) and the genus *Scapholithus*, which was erected by Deflandre (1954) to include the fossil rhomboid forms. *S. fossilis*, the type species of the genus (Fig. 12), has been found in Hauterivian (Early Cretaceous) right through to Recent and thus is one of the few species that survived the Cretaceous/Tertiary boundary event(s). *S. rhombiformis*, the other species often found in Tertiary sediments is known mainly from the Paleocene and Eocene. Other species (see Perch-Nielsen, 1971a and Aubry,

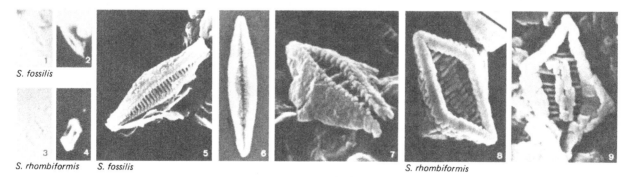

S. fossilis

S. rhombiformis   *S. fossilis*   *S. rhombiformis*

Fig. 12. LM and SEM of species of *Scapholithus*

1985) are either considered synonyms of the above species or are poorly defined and not generally used. The species of *Scapholithus* occur sporadically and usually in low numbers; they have not been used for biostratigraphy.

Calyptrosphaeraceae Boudreaux & Hay (1969)
Figures 13, 14

The modern family of Calyptrosphaeraceae includes coccolithophorids bearing holococcoliths (coccoliths consisting of uniform, small, calcite crystals) instead of elements of differing size and shape that compose most coccoliths (heterococcoliths). Most genera contain only one or few species (see overview by Tappan, 1980 p. 788; of the genera listed there, *Naninfula* and *Scampanella* are not holococcoliths and are considered to belong to the Rhabdosphaeraceae and 'incertae

sedis' respectively). The genera are distinguished by their shape or outline, the species by small differences in the outline or by the presence/absence of holes, depressions, knobs or processes. Fig. 13 gives an overview over the commonly reported holococcoliths of the Paleogene. Not included is the genus *Pseudozygrhablithus* with its species *P. altus*, *P. comprimus* and *P. latus*, since they are considered synonymous with *Zygrhablithus bijugatus*. Also, the genera *Quinquerhabdus* and *Polycladolithus* with their generotypes *Q. colossicus* and *P. operosus* are considered of limited stratigraphic value and are not included in Fig. 13. *Trochoaster* is assumed to be a junior synonym of *Lithostromation* and is included in the family Lithostromationaceae.

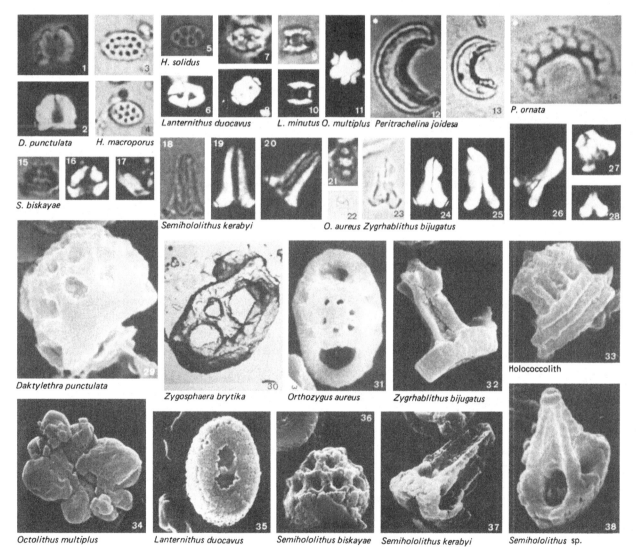

Fig. 13. LM and EM of representatives of the Calyptrosphaeraceae (holococcoliths). *H. = Holodiscolithus*

Paleocene

Octolithus multiplus, originally described from the Maastrichtian has also been found in the Lower Paleocene of many regions and thus is one of the few species that survived the Cretaceous/Tertiary boundary event(s). Besides *O. multiplus*, characterized by four large blocks and four small ones (Fig. 13.34), *Lanternithus duocavus* and *Lanternithus* sp. are found in the Early Paleocene (Perch-Nielsen, 1981c). *Semihololithus kerabyi* is distinguished by LM from *Zygrhablithus bijugatus* by its 'channel' visible both at 0° and at 45° to the crossed nicols (Figs. 13.19 and 13.20).

Eocene

Z. bijugatus is the most common holococcolith in Eocene sediments. It can be low or high, broad or slim (Figs. 13.23–28) and broken specimens have been described as *Sujkovskiella enigmatica*, which is a junior synonym of *Z. bijugatus*. Other synonyms are *Rhabdosphaera? semiformis*, *Isthmolithus claviformis* and *Rhabdolithus costatus*. *Lanternithus minutus* appears in the Middle Eocene and also can be common in the Late Eocene. The remaining holococcoliths shown in Fig. 13 are all also found in the Eocene but are usually rare. *Clathrolithus spinosus* and *Corannulus germanicus* (Fig. 14), two species and genera also belonging to the Calyptrosphaeraceae, are restricted to the Late Eocene.

Oligocene

L. minutus persists into the Early Oligocene and *Z. bijugatus* disappears only near the Oligocene/Miocene boundary and can be used to establish this boundary in sections where it is consistently present in the Oligocene. Other holococcoliths are usually very rare.

Neogene

Neogene sediments rich in holococcoliths seem to be relatively rare. The forms found are conspecific with some of the holococcolith species described from the modern seas by Heimdal & Gaarder (1980).

Conclusions

With the exception of the FO of *C. spinosus* and the FO of *C. germanicus* in the Late Eocene of the North Sea area (Perch-Nielsen, 1971d) and the LO of *Z. bijugatus* near the end of the Oligocene (Edwards, 1971), holococcoliths have not been used stratigraphically. The preservation of holococcoliths seems to be restricted to relatively shallow seas (c. less than 1000 m) or to sediments derived from such seas as turbidites, and thus they may have some potential in paleoenvironmental and paleoceanographic studies.

Ceratolithaceae Norris (1965)
Figures 15–17

Five genera have been assigned to the Ceratolithaceae, the modern family embracing the Neogene horseshoe-shaped nannoliths, namely *Amaurolithus*, *Angulolithina*, *Ceratolithus*, (*Ceratolithoides*) and (*Ceratolithina*).

Of these genera, *Ceratolithoides* and *Ceratolithina* are Cretaceous genera which have no apparent descendants in the Tertiary genera. Species of the true ceratoliths *Amaurolithus* and *Ceratolithus* are common in low latitude Upper Miocene and Lower Pliocene sediments and here are used for the definition of zonal boundaries. The terminology of the ceratoliths is given in Fig. 15 and an overview of the species and their ranges is shown in Fig. 16, where the presumed ancestors,

Clathrolithus spinosus

Corannulus germanicus

Fig. 14. LM and SEM of the holococcoliths *C. spinosus* and *C. germanicus*

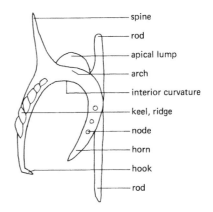

Ceratolithaceae

- spine
- rod
- apical lump
- arch
- interior curvature
- keel, ridge
- node
- horn
- hook
- rod

Fig. 15. Terminology of the Ceratolithaceae

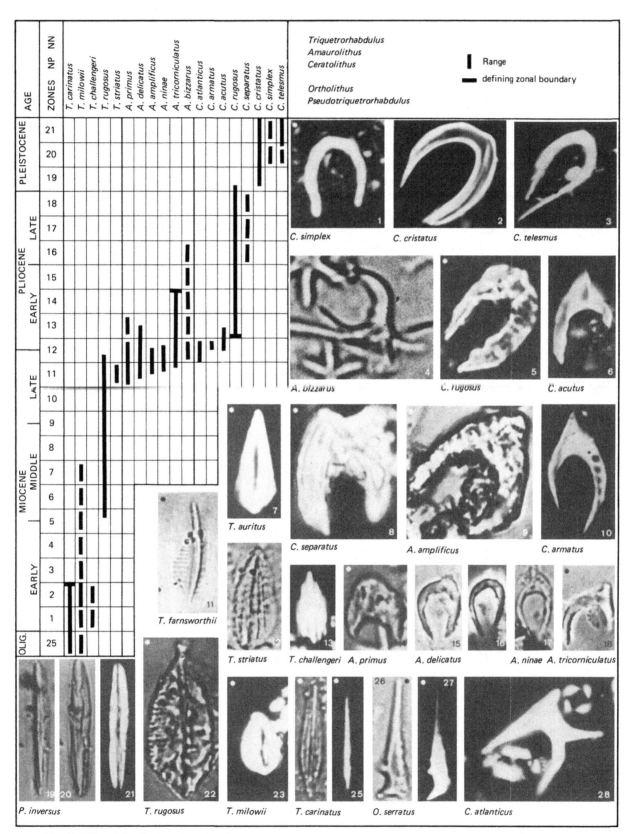

Fig. 16. LM and distribution of species of *Amaurolithus* (*A.*), *Ceratolithus* (*C.*), *Ortholithus* (*O.*), *Pseudotriquetrorhabdulus* (*P.*; *Wiseorhabdus* of some authors) and *Triquetrorhabdulus* (*T.*)

the forms of *Triquetrorhabdulus*, are also illustrated. Details of the fine structure are visible in the SEM pictures in Fig. 17, but ceratoliths are best studied with the LM to make the distinction between the non-birefringent forms of the genus *Amaurolithus* and the birefringent forms of the genera *Ceratolithus* and *Angulolithina*.

Amaurolithus contains the horseshoe-shaped coccoliths which show no birefringence between crossed nicols. The characteristics distinguishing the species are given below. *A. amplificus* is a robust, strongly asymmetrical form with a short but thick apical spine and an extended apical region. Rough nodes are visible in well preserved specimens on the large horn and on the apical region, but can be replaced by a ridge in overgrown specimens. *A. amplificus* is more asymmetrical than *C. acutus*, but is difficult to distinguish from the latter in an oblique position when it appears slightly birefringent. *A. bizarrus* features one or two horseshoe-shaped ceratolith structures attached to a long rod. *A. delicatus* is a delicate form with a rounded apex and no apical spine or lump and no rod. *A. ninae*, another relatively delicate form, has an apical lump and *A. primus*, the oldest ceratolith, is a relatively small, crescent-shaped form with short horns and an extended apical region. *A. tricorniculatus* has one long and one short horn and a spine more or less in line with the shorter horn. It is more

delicate and smaller than *A. amplificus* and clearly asymmetrical.

Angulolithina has been described as 'two thick limbs joined in the form of a V and making an angle of less than 90 degrees'. It shows birefringence in cross-polarized light, acting as a single optical unit and is distinguished from *Ceratolithus* by a continuous opening of the angle formed by two equal limbs. *Angulolithina* shows highest relief in the orientation where *Ceratolithus* shows lowest relief when viewed with a single polarizer (90° and 270°). Only one species has been described. This is *A. arca*, which is reported from the Late Miocene and Early Pliocene of the Pacific (Bukry, 1973c).

Ceratolithus contains the horseshoe-shaped coccoliths which show birefringence between crossed nicols. *C. acutus* is a relatively robust species with unequal or almost equal horns and a blunt apical spine terminating in an acute angle. This nearly symmetrical apical structure may bear a ridge-like thickening. *C. acutus* appears brightest at 45° to the polarizing directions of the crossed nicols. *C. rugosus* has no such apical spine. *C. acutus* appears slightly above the Miocene/Pliocene boundary and its FO approximates the boundary. *C. armatus* has a relatively narrow angle between the horns, which are distinctly curved, and usually unequal in length. There are nodes on both horns, those on the shorter arm continuing on

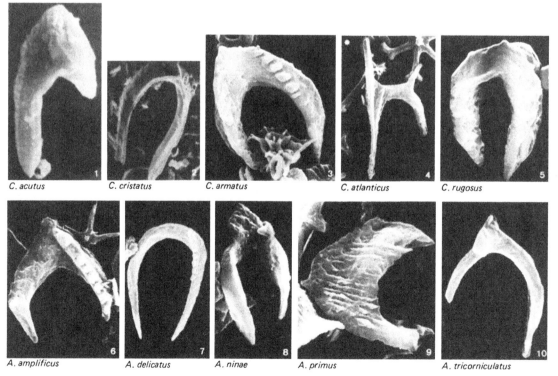

C. acutus   C. cristatus   C. armatus   C. atlanticus   C. rugosus

A. amplificus   A. delicatus   A. ninae   A. primus   A. tricorniculatus

Fig. 17. SEM of Ceratolithaceae. *A.* = *Amaurolithus*, *C.* = *Ceratolithus*

the well developed spine which lies slightly asymmetrically. The inner curvature in *C. atlanticus* is broad and the apical spine has the form of a long, straight rod which also forms one of the horns; another horn and a second apical spine do not necessarily lie in the same plane as the body of the ceratolith. *C. cristatus* is a relatively robust species with ornate horns of nearly equal length and only a very reduced apical region and no distinct apical spine. *C. cristatus* is more robust than the other Pleistocene species, *C. simplex* and *C. telesmus*. *C. dentatus* has been found to be a junior synonym of *A. amplificus* (Gartner & Bukry, 1975). *C. rugosus* is a robust form with horns that are parallel for most of their length. *C. rugosus* is usually heavily calcified, but the rows of nodes on the horns can be seen in well preserved specimens. *C. separatus* has a wide and flat, saddle-like arch between the two thick, tapering horns which are ornamented by nodes or keels. *C. separatus* has a wider arch than the contemporaneous *C. rugosus* or the slightly younger *C. cristatus*. It is brightest at 45° to the polarizing directions. *C. simplex* is a small, smooth, simple, hook-shaped form with an arch and horns of nearly equal diameter. In *C. simplex* the diameter of the arch and the horns is more uniform than in *C. cristatus* and the species also is smaller and has no keels or nodes. The horns are shorter and less convergent than in *C. telesmus*. *C. telesmus* has long, tapering and converging horns.

Ceratolithina? *vesca* is a small, asymmetric species with one straight and narrow arm and one slightly curved and thicker arm described from the late Middle Eocene of Trinidad.

Conclusions

Amaurolithus and *Ceratolithus* are very useful markers in the Late Miocene and Early Pliocene of low latitude open sea sections. Their absence in high latitudes and in nearshore sediments makes it necessary to adapt the low latitude zonations accordingly in such areas.

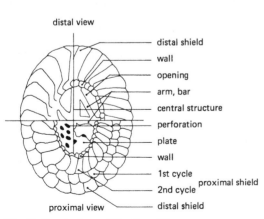

Fig. 18. Terminology of the Coccolithaceae

Coccolithaceae Poche (1913)
Figures 18–24

The family Coccolithaceae includes some very well known and well illustrated genera and species, but also a large number of dubious genera and species.

There is no general agreement as to which genera should be included in the family. A comparison of the classifications of Hay (1977) and Tappan (1980) reveals considerable differences and this is not the right place to express a third view. On the other hand, the genera *Chiasmolithus* and *Cruciplacolithus* play an important role in the subdivision of the Paleogene and several species of other genera of the Coccolithaceae have been used as zonal or subzonal markers. We shall concentrate on these and discuss only a few of the other species and genera.

The Coccolithaceae include elliptical and round coccoliths with a distal shield (for terminology see Fig. 18) of radiating petaloid elements. In most genera of the Coccolithaceae their optical orientation is such that the distal shield does not show birefringence when viewed between crossed nicols. Thus, most coccoliths of the Coccolithaceae appear smaller between crossed nicols than in normal light, since the proximal shield is usually smaller and shows strong birefringence in many species. In other species the proximal shield only shows slight or no birefringence. The proximal shield consists of two cycles of elements in some genera (*Chiasmolithus, Coccolithus, Cruciplacolithus, Ericsonia*) and of a single cycle of elements in others (*Birkelundia, Calcidicus, Cyclagelosphaera, Markalius*). The double proximal shield is typical for Tertiary coccoliths. Specimens of the Cretaceous species *Watznaueria barnesae* and *Markalius inversus* with double proximal shields have been illustrated by Perch-Nielsen (1979a) from the lowermost Paleocene.

The following three groups of Coccolithaceae are discussed separately:
  (1) elliptical or round Coccolithaceae with a distinct central structure,
  (2) elliptical Coccolithaceae with an open or closed central area,
  (3) round Coccolithaceae with an open or closed central area.

(1) Elliptical or round Coccolithaceae with a distinct central structure

The following genera are assigned to this group of Coccolithaceae: *Birkelundia, Bramletteius, Campylosphaera, Chiasmolithus* and *Cruciplacolithus*.

Birkelundia is characterized by a large central area spanned by central structure and with a single proximal shield. The genus name is not used widely and the species that were assigned to *Birkelundia* are usually assigned to *Cruciplacolithus* or to *Coccolithus/Ericsonia*: *B. arenosa, B. jugata, B. staurion* (Figs. 19.17, 18).

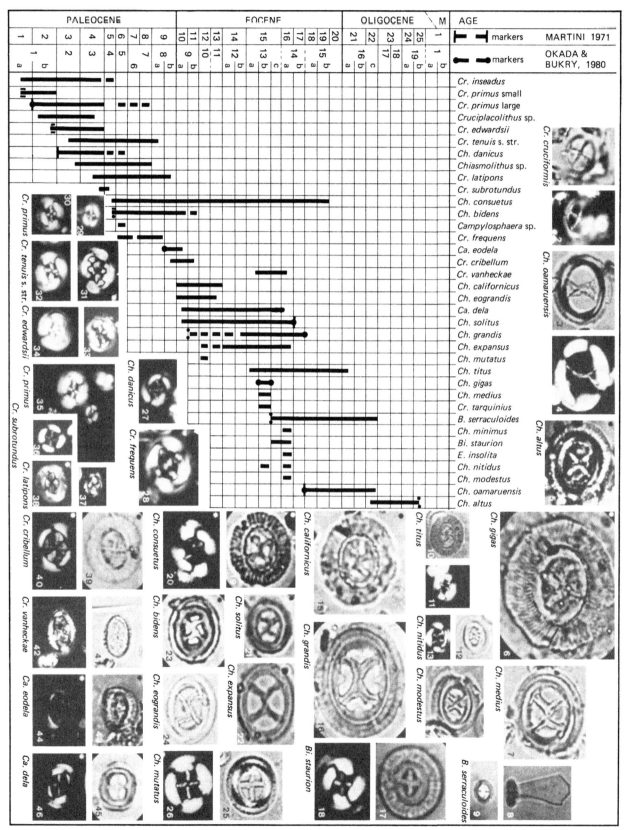

Fig. 19. LM and distribution of species of *Birkelundia* (*Bi.*), *Bramletteius* (*B.*), *Campylosphaera* (*Ca.*), *Chiasmolithus* (*Ch.*), *Cruciplacolithus* (*Cr.*) and *Ericsonia insolita*.

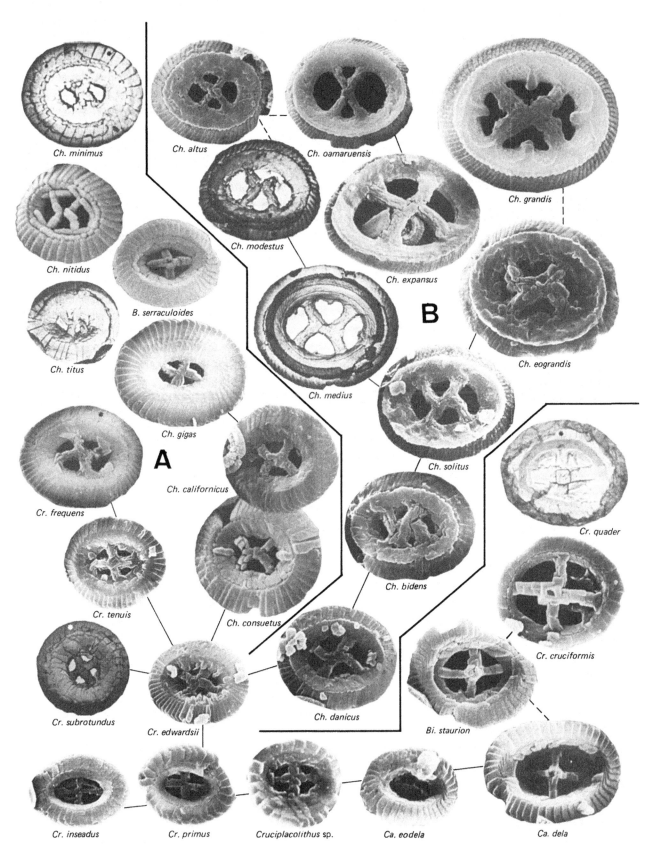

Fig. 20. EM and tentative evolutionary lineages in the genera *Campylosphaera* (Ca.), *Chiasmolithus* (Ch.) and *Cruciplacolithus* (Cr.)

Bramletteius features a proximal shield like *Cruciplaco-lithus* and a central process: *B. serraculoides* (Figs. 19.8, 9; rudder to paddle-shaped process; synonym *B. variabilis*) and *B.? duolatus* (two processes, Late Pliocene).

Campylosphaera is characterized by a long-elliptical outline and is convex. Its species are assigned by many authors to *Cruciplacolithus*, since the differences between the two genera are small and not accepted generally. Both genera have a central cross aligned with the axes, but while *Cruciplacolithus*

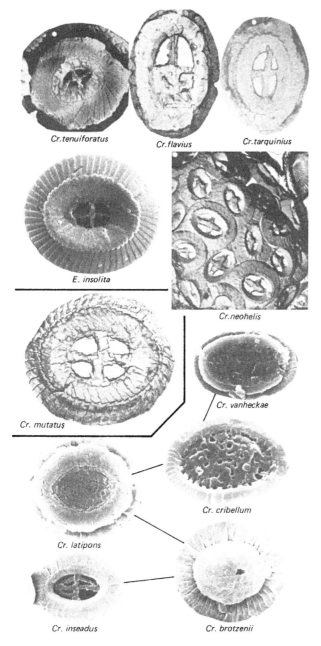

Cr.tenuiforatus

Cr.flavius

Cr.tarquinius

E. insolita

Cr.neohelis

Cr. vanheckae

Cr. mutatus

Cr. cribellum

Cr. latipons

Cr. inseadus

Cr. brotzenii

Fig. 21. EM of *Cruciplacolithus* (*Cr.*) species and *Ericsonia insolita*

has an elliptical to round outline, *Campylosphaera* is long-elliptical and its down-bent ends give it a nearly rectangular shape when seen with the LM. The elements surrounding the central area in *Cruciplacolithus* are in a shallower position than those in the deeper central area of *Campylosphaera*. *Ca. dela* (Figs. 19.45, 46; 20), the type species, is the largest form. *Ca. eodela* (Figs. 19.43, 44; 20) is somewhat narrower and smaller and the undescribed form illustrated in Fig. 20 as *Cruciplaco-lithus* sp. is small, but still shows the characteristic deep central area.

Species of *Chiasmolithus* and *Cruciplacolithus* are used as markers mainly in the Paleocene and Middle to Late Eocene, but many more potentially useful forms have been described than are presently used. Generally, *Cruciplacolithus* is used for forms with a central cross aligned with the axes and *Chiasmolithus* is used for forms with a central cross in X or H form. This is a practical genus concept which is easy to use, including when working with the LM. But when one tries to link the forms to ancestors and descendants one finds that one goes from *Cruciplacolithus* to *Chiasmolithus* at least twice and that another criterion, namely the construction of the row of elements surrounding the central area on the distal side of the coccolith, could be used as another character in the search for evolutionary trends. In most forms now assigned to *Cruci-placolithus*, including the generotype *Cr. tenuis*, this row of elements does not reach or only just reaches the crest of the distal shields (A forms in Figs. 20 and 22). In most *Chiasmolithus*, including the late forms of the generotype *Ch. danicus*, this row of elements builds the crest and thus partly covers the distal shield (B forms in Figs. 20 and 22). Both genera have a proximal shield consisting of two cycles (Fig. 21, e.g. *Cr. brotzenii*).

LM pictures of most species of *Chiasmolithus*, *Cruci-placolithus*, *Campylosphaera* and *Bramletteius* are assembled for easy comparison in Fig. 19 where their ranges are given though these are sometimes still tentative. Where possible and reasonable, the holotype or a specimen from the same material as the holotype is illustrated. Both Gartner (1970) and Romein (1979) have suggested evolutionary lineages for *Chiasmo-lithus*, the latter also for *Cruciplacolithus*. In Figs. 20 and 21 I have used some of their ideas but included also my own, supported by SEM pictures of the distal view of the species where available. To distinguish the various species which look rather similar, the distinguishing features are listed in Fig. 22. Note that distal and proximal views of the central structure are mirror images.

It has been suggested by Perch-Nielsen (1981b) that *Cr. primus*, the first typical species of *Cruciplacolithus*, evolved from *Cr. inseadus* (Fig. 20), which in turn may have evolved from the Cretaceous genus *Sollasites*. The earliest *Cr. primus* (Figs. 19.29 and 19.35 small form; 20) are very small, 3–5 μm. With time, larger forms evolve and when the largest forms reach 8–9 μm (Figs. 19.30 and 19.35 large form), the central cross turns slightly to form *Cr. edwardsii* (Figs. 19.33, 34; 20).

Cr. tenuis (Figs. 19.31, 32; 20), according to Romein (1979), who has recently studied the holotype, has small feet and evolved from *Cr. primus* later than *Cr. edwardsii*. Thus, if one takes the first form with an oblique cross as *Ch. danicus*, as has been commonly done, the base of NP 3 would lie below the base of NP 2 (defined by the FO of *Cr. tenuis*; see also Fig. 6 for detailed zonation of this interval). *Ch. danicus* (Figs. 19.27; 20) was described from the Upper Danian and we can assume that a more evolved form was described than that now known as *Cr. edwardsii*. The original illustrations of *Ch. danicus* do not allow a very accurate comparison with better preserved forms but it seems reasonable to use a form commonly found in the Upper Danian, i.e. one, where the central cross is already slightly curved or where two arms form a continuous structure and where the two other arms meet at slightly offset points. At this stage, the originally A-type central area develops into a B-type central area (Figs. 20 and 22). In *Ch. bidens* (Figs. 19.23; 20), the next species of *Chiasmo-lithus* to evolve, the central cross consists of two straight and two slightly bent arms which meet slightly offset. In *Ch. eograndis*, this offset takes the form of a rather long bridge parallel to the long axis. In *Ch. solitus* (Figs. 19.21; 20), where the central cross is much like the one of *Ch. bidens*, the distal shield has become narrower, and the two 'teeth', which gave *Ch. bidens* its name, are missing. Several other species can be distinguished in this B lineage in the Middle and Late Eocene, based on the symmetry or asymmetry of the central X, the relative size of the central area and of the entire coccolith itself. *Ch. grandis* (Figs. 19.16; 20) has four well developed 'teeth' and usually is very large. *Ch. oamaruensis* (Figs. 19.3, 4; 20; 22B) has a very narrow central X, which can pose problems in distinguishing it from *Ch. expansus* (Figs. 19.22; 20) with a wider, but equally symmetrical central X. *Ch. altus* (Figs. 19.5; 20) has a relatively small central area and a central cross with arms meeting at about right angles and of equal length.

The A-lineage includes forms with 'feet' like *Cr. tenuis* with a central cross (+), *Cr. frequens* (Figs. 19.28; 20) with a central X, and *Ch. nitidus* (Figs. 19.12, 13; 20) with feet that almost surround the whole central area. In another lineage, the slightly oblique cross of *Cr. edwardsii* evolves into the symmetrical X in *Ch. consuetus* (Figs. 19.19, 20; 20; 22A) and the larger form *Ch. californicus* (Figs. 19.15; 20). In the very large *Ch. gigas* (Figs. 19.6; 20), the central X is slightly asymmetrical, as it is also in the relatively small *Ch. titus* (Figs. 19.10, 11; 20).

The genus *Campylosphaera* is used for convex-rimmed forms with a central cross. A possible ancestor to *Ca. eodela* was found in Zone NP 5 in Denmark (*Cruciplacolithus* sp. in Fig. 20). *Ca. eodela* serves as a marker to subdivide the Late Paleocene NP 9 and no intermediate forms have been found. *Birkelundia*, a genus with only a single, simple proximal shield, has one form usually assigned to *Cruciplacolithus*, *B. staurion*, which is found mainly in the Middle Eocene and has no obvious ancestor and descendant.

Another group of forms has been assigned to *Cruciplaco-lithus*, since it has a central cross, but the rest of the central area is almost filled with elements protruding from the shields (Fig. 18). In the Danian *Cr. brotzenii* (Fig. 21), these elements protrude more or less radially from the shields. In *Cr. latipons* (Figs. 19.37, 38; 21), the central cross seems to fill almost the whole central area and in *Cr. cribellum* (Figs. 19.39, 40; 21), the central cross almost disappears in the perforated 'central plate'. *Cr. vanheckae* (Figs. 19.41, 42; 21) has evolved in the Middle Eocene from *Cr. cribellum* and is long-elliptical. In *Ericsonia insolita* (Fig. 21) the central area is extremely small, but features a minute central cross. It seems more likely that it evolved from an *Ericsonia* in the Eocene, *Ericsonia* itself having evolved from *Cr. primus* in the Early Paleocene.

There are additional forms where no obvious connections are evident and some others which have not yet been

GENUS	CENTRAL AREA	ELEMENTS SURROUNDING CENTRAL AREA DISTALLY: DO NOT REACH CREST OR JUST REACH CREST A	BUILD THE CREST B
mainly *Chiasmolithus*	X		*Ch. oamaruensis*
	X	*Cr. subrotundus*	*Ch. altus, Ch. modestus*
	X		*Ch. expansus*
	X		*Ch. grandis*
	X		*Ch. eograndis*
	X	*Ch. titus*	*Ch. solitus, Ch. bidens*
	X	*Ch. danicus* (early forms)	*Ch. danicus* (late forms)
	X	*Ch. consuetus, Ch. californicus*	*Ch. medius*
mainly *Cruciplacolithus*	X	*Ch. nitidus*	
	X	*Cr. frequens*	
	X	*Cr. edwardsii*	
	X	*Ch. gigas, Ch. minimus, Cr. edwardsii*	*Cr. tarquinius*
	+	*Cr. mutatus, Cr. cruciformis*	
	+	*Cr. tenuis (Cr. notus)*	
	⬦	*Cr. inseadus*	
	+	*Cr. primus, Cr. tenuis s. ampl., E. insolita, Cr. flavius Cr. quader, Ca. eodela, Ca. dela, Bi. staurion, B. serraculoides*	
		Cr. brotzenii, Cr. latipons, Cr. cribellum, Cr. vanheckae	

Fig. 22. Guide to the determination of *Chiasmolithus* and *Cruciplacolithus* species and some similar species of related genera in the Paleogene

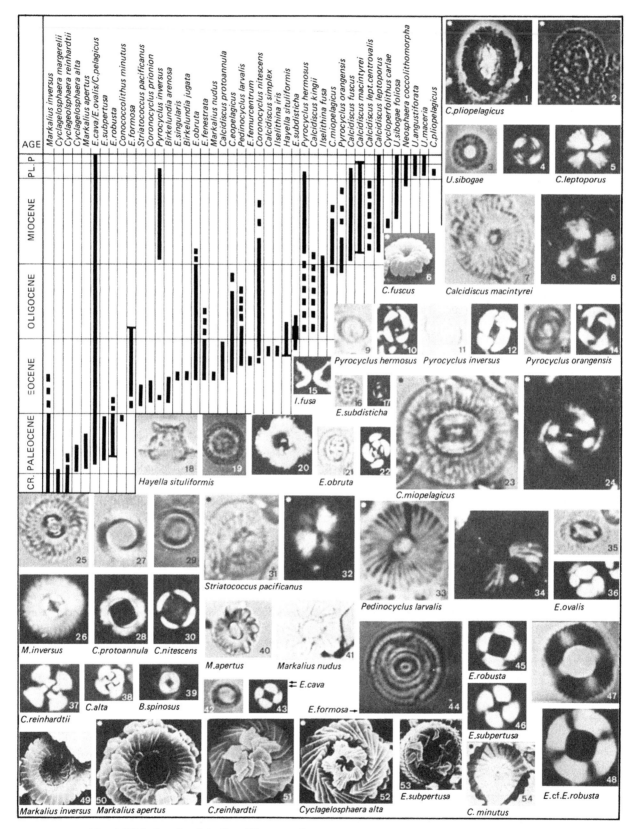

Fig. 23. LM and SEM of species of the Coccolithaceae and their distribution

described. *Chiasmolithus* and *Cruciplacolithus* are most common in high latitude assemblages and there they may serve as markers, when other forms like discoasters, which prefer warmer waters, are missing.

The last *Chiasmolithus*, *Ch. altus*, disappears towards the end of the Oligocene and this event has been used by Waghorn (1985) to define a zonal boundary corresponding roughly to the NP 24/NP 25 boundary in New Zealand and the surrounding oceans. Generally, the Oligocene is not as rich in *Chiasmolithus* and *Cruciplacolithus* as the Paleocene and especially the Eocene. But while *Chiasmolithus* disappears at the end of the Oligocene, *Cruciplacolithus* continues into the Miocene with *Cr. tenuiforatus* (Fig. 21), a form very similar to *E. insolita*, and finally the modern species *Cr. neohelis* (Fig. 21), a form rather similar in shape and size to *Cr. primus*.

(2) Elliptical Coccolithaceae with an open or closed central area

The following genera are assigned to this group: *Birkelundia*, *Clausicoccus*, *Coccolithus* and *Ericsonia*.

Birkelundia was discussed above, together with species with a central structure. Only the generotype, *B. arenosa*, has a granulate central area.

Clausicoccus was introduced for forms with a perforated central plate as in *C. fenestratus*, its type species. Other authors have included such forms in *Ericsonia/Coccolithus* or in *Cruciplacolithus*, from which *Clausicoccus* is thought to have evolved.

While *Coccolithus* and *Ericsonia* are regarded as two valid genera by some authors, others consider *Ericsonia* to be a junior synonym of *Coccolithus*. The generotypes of both genera certainly are very similar, if one accepts *C. pelagicus* (Figs. 24.1–3) as the type of *Coccolithus* and compares its proximal view with the proximal view which forms the basis of the type of *Ericsonia*, *E. occidentalis* (Fig. 24.17). The distal shield is larger than the proximal shield, and the proximal shield consists of two cycles. These are well developed and distinguishable due to a relatively large difference in size in *E. occidentalis*, while they are nearly of equal size and therefore hardly distinguishable in *C. pelagicus*. The central area of *E. occidentalis* is empty, while it is filled with a cycle of more or less radially oriented elements and sometimes a bridge in *C. pelagicus*. Since the erection of *Ericsonia* by Black (1964), many species have been assigned to it, mainly from the Paleogene. *Coccolithus* tends to be used more in the Neogene and, of course, was used for nearly all 'buttonshaped coccoliths' before the erection of such Tertiary genera as *Chiasmolithus*, *Cruciplacolithus*, etc. in the 1960s.

While representatives of elliptical Coccolithaceae with an open or closed central area are present in most Tertiary assemblages from high to mid latitudes, the presence/absence of *C. pelagicus*, considered to be a cold-water species, is used

for paleoecological interpretations in low latitudes. *C. pliopelagicus*, *C. miopelagicus*, *C. eopelagicus*, *E. ovalis*, *E. cava* (Fig. 23) and many more elliptical forms with an open or closed central area are distinguished by some authors and considered synonymous with *C. pelagicus* by others. Biometric studies by Backman (1980) showed that *C. pliopelagicus* could not be distinguished from *C. pelagicus* in the Miocene/Pliocene of the NE Atlantic, but that *C. miopelagicus* could be distinguished from *C. pelagicus* by its larger diameter ($>13 \mu m$).

Biostratigraphically important is the LO of *E. formosa* (Figs. 23.44; 24.20; *Coccolithus*, *Cyclococcolithus* or *Cyclococcolithina formosa* of some authors), a round form which disappears in the Early Oligocene.

(3) Round Coccolithaceae with an open or closed central area

The following genera are assigned to this group of genera: *Calcidiscus*, (*Cyclococcolithus/Cyclococcolithina/Cycloplacolithella/Cycloplacolithus/Tiarolithus*), *Conococcolithus*, *Coronocyclus*, *Cyclagelosphaera*, *Cycloperfolithus*, *Hayella*, *Ilselithina*, *Markalius*, *Neosphaera*, *Pedinocyclus*, *Striatococcolithus*, *Umbilicosphaera*.

There is no agreement on the value of the outline of a coccolith – round or elliptical in the case of the Coccolithaceae and many other families – as a factor determining a genus or species. All the above-listed genera were erected for more or less circular forms. Some genera look more or less similar to the elliptical Coccolithaceae, others are included in this family due to the lack of any more suitable family to assign them to (see Figs. 23 and 24).

Cyclagelosphaera and *Markalius* are survivors from the Cretaceous and are only found in the Paleocene and in the Paleocene through Oligocene respectively. *Cyclagelosphaera* is characterized by a high birefringence of the whole coccolith in crossed polarized light, whereas *Markalius* remains dark with the exception of the central area or parts of it. Some species of the two genera are shown in Fig. 23. NP 1, the lowermost Standard Zone of the Tertiary, was named after *M. inversus*, a species present in the Maastrichtian. Species of *Markalius* are quite common in high latitude Danian sediments, together with species of *Cyclagelosphaera*. Their occurrence has not yet been investigated in detail, but is likely to be useful in the subdivision of NP 1 and NP 2 in high latitudes. *M. inversus* (Figs. 23.25, 26, 49) (*M. astroporus* of some authors) has a central area covered by radial elements reaching the centre, while the central area is wide open in *M. apertus* (Fig. 23.50), where it is surrounded by a cycle of small elements. *C. reinhardtii* (Figs. 23.37, 51) has a central area topped by a cone of overlapping elements, while only a flat cycle of elements surrounds the radial central elements in *C. margerelii* and a high cone with depressions builds the central area of *C. alta*. In the LM, *C. alta* (Figs. 23.38, 52) appears very bright in the central area between crossed nicols. It is smaller than *C. reinhardtii*, which appears more uniformly bright. *C. margerelii* has a distal and proximal shield of about equal size, while they are different in size in the other two species. Between crossed nicols the central area seems less bright than the shields.

Cyclococcolithina, Cyclococcolithus, Cycloplacolithella and *Cycloplacolithus* have been designated synonyms of *Calcidiscus* by Loeblich & Tappan (1978) together with some other generic names that have not been used so widely. This was based on the assumption that *C. quadriforatus*, the type species of *Calcidiscus*, is a junior synonym of *C. leptoporus*. The illustration of the holotype of *C. quadriforatus* is such that the form described cannot with certainty be related to a modern species as we now know them. It is likely, but not certain, that *C. quadriforatus* represents a single shield of *C. leptoporus*. Nevertheless, *Calcidiscus* is here used for the mainly Neogene round forms with a distal shield and a simple proximal shield and only a small, open or closed central area. Illogically, we also include the elliptical forms of the Neogene, *C. leptoporus centrovalis* in the same genus, while the round Paleogene forms of *E. formosa* are assigned to the generally elliptical genera *Ericsonia* or *Coccolithus*, since they have a double proximal shield. This seems reasonable when one looks at the probable origin of these species or subspecies, but one cannot press them into a rigid definition of a genus.

Only the FO of *C. leptoporus* (Fig. 23.5) in the Early Miocene and the LO of *C. macintyrei* (Figs. 23.7, 8; 24.4) in the Early Pleistocene have been used for age determination, the former in poor assemblages and the latter for a zonal or subzonal boundary.

Janin (1981a, b) investigated *C. leptoporus* and *C. macintyrei* in manganese nodules and came to the conclusion that the subdivision of the two species as proposed by Bukry & Bramlette (1969b) was not applicable since intermediate forms existed between the *C. leptoporus* with 17–33 elements in the distal shield and a diameter of 4–8.5 µm (forms B and C of McIntyre, Bé & Preikstas, 1967) and the *C. macintyrei* with 38–42 elements and a diameter of about 7.4–11.8 µm (form A of McIntyre *et al.*, 1967). Instead she proposes three varieties:

(1) forms with 40 to 45 elements, distal shield 10–14 µm, *C. macintyrei*;

(2) forms with 29 to 38 elements, distal shield 5.5–10 µm, *C. leptoporus* J;

(3) forms with 18 to 26 or even 12 to 27 elements and 3.6–12 µm diameter of the distal shield, *C. leptoporus* K (the letters are my additions).

Thus, according to Janin we shoul use *C. macintyrei* only if we find more than 40 elements and not only by a size larger than 10 µm. In practice, and mainly with the LM, the size will probably be used for the determination.

C. leptoporus and *C. macintyrei* vary not only considerably in size, but also in form (round, elliptical) and in the relative size of the central area. According to unpublished data from the Gulf Coast, the following species and 'subspecies' can be distinguished:

(1) forms with 18 to 24 elements, distal shield 4–8 µm, and a O/S (opening to shield) ratio of 1/12–1/8, *C. leptoporus* M;

(2) forms with 18 to 24 elements, 4–8 µm and a O/S ratio of 1/20–1/12, *C. leptoporus* N;

(3) forms with 38 to 44 elements, 10–14 µm and a O/S ratio of 1/8–1/6, *C. macintyrei* O;

(4) forms with 38 to 44 elements, 7–9 µm, a O/S ratio of 1/8–1/6, *C. macintyrei* P;

(5) forms with 40 to 48 elements, 13 or 14 µm, a O/S ratio of 1/32–1/6, *C. macintyrei* Q;

(6) forms with 28 to 34 elements, 8–12 µm and a O/S ratio of 1/12–1/3, *C. leptoporus* R.

Elliptical forms can be assigned to *C. leptoporus centrovalis*, a subspecies which probably also could be further subdivided according to the criteria used for the distinction of the round *C. leptoporus* and *C. macintyrei*. With such a fine species concept it should be possible to use these forms for the subdivision of sequences in areas where other markers are missing due to latitudinal influences or due to dissolution.

C. kingii (Fig. 24.14) and *C. protoannula* (Figs. 23.27, 28; 24.19) are two of the species with a relatively wide central opening which have also been assigned to *Calcidiscus*.

Conococcolithus was described as 'circular placoliths having the form of a truncate cone' and its generotype, *C. minutus*, remains the only species included in it that fits this definition. *Conoccolithus panis* Edwards (1973a) is clearly elliptical and belongs to *Biscutum*.

Coronocyclus includes circular coccoliths 'with a wide central opening, constructed of a ring of imbricated wedge-shaped segments, bearing nodes or short spines directed proximally, peripherally and distally' (Hay *et al.*, 1966). Stradner (1968) emended this diagnosis requiring only a 'ring of bevelling wedge-shaped segments', since he noted that the nodes and short spines could be the result of overcalcification and therefore should not be included in the genus diagnosis. While such spines are present also in well preserved specimens of the generotype of the genus, *C. serratus*, they are absent or less prominent in other species assigned to this genus, like *C. nitescens* or *C. prionion* (Figs. 23.29, 30; 24.9, 15). *C. serratus* has been suggested to be a junior synonym of *C. nitescens*.

Cycloperfolithus is characterized by 'a central opening spanned by a sieve-plate with regularly arranged pores showing a honeycomb-pattern'. Its generotype and only species, *C. carlae* (Fig. 24.13), is best described as a *Calcidiscus leptoporus* with a central net. It was described from the Middle Miocene of the central Paratethys. It is the only round species of the Coccolithaceae with a central net and has not yet been reported from other areas.

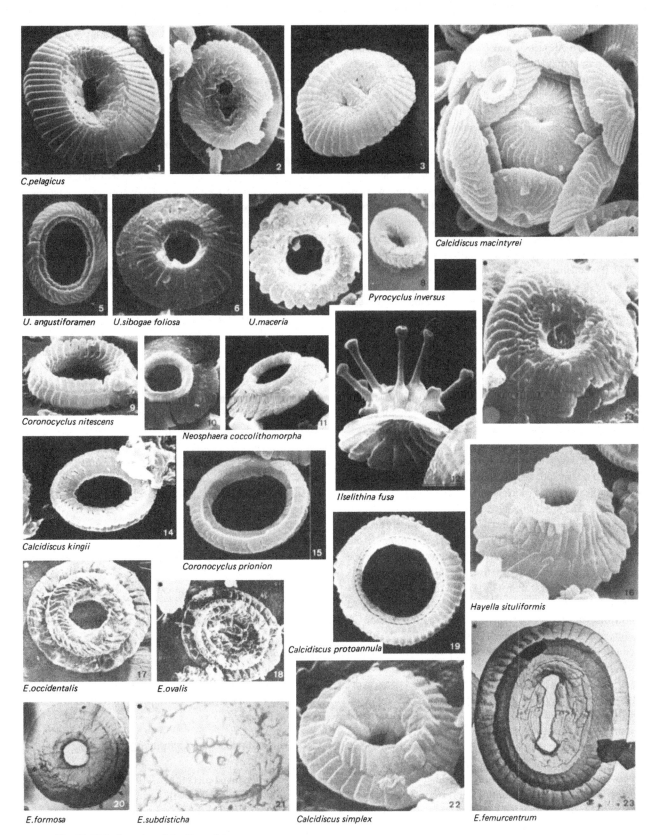

C.pelagicus

U. angustiforamen   U.sibogae foliosa   U.maceria

Pyrocyclus inversus

Calcidiscus macintyrei

Coronocyclus nitescens

Neosphaera coccolithomorpha

Ilselithina fusa

Calcidiscus kingii

Coronocyclus prionion

Calcidiscus protoannula

Hayella situliformis

E.occidentalis   E.ovalis

E.formosa   E.subdisticha   Calcidiscus simplex   E.femurcentrum

Fig. 24. EM of species of the Coccolithaceae

Hayella was proposed as a generic name almost simultaneously by Roth (1969) and by Gartner (1969a). Gartner (1969b) showed that *Hayella* Gartner had priority over *Hayella* Roth (which is a synonym of *Ilselithina* anyway) due to the fact that his publication was mailed (available) earlier in 1969 than the one of Roth dated 1968 but mailed in 1969. Two species have been assigned to the genus for which we know no ancestors or descendants: *H. situliformis* (Figs. 23.18–20; 24.16), the generotype, from the Upper Eocene and *H.? gauliformis*, a form from the Lower Eocene. *Hayella* was described as a 'calcareous body in the shape of a truncated cone with the smaller end closed, bearing a peripheral rimlike flange at both ends and a constricted lip at the wider, open end'.

Ilselithina has been classified as genus incertae sedis but is here included in the Coccolithaceae following Tappan (1980). It is constructed as a simple, round proximal shield and a distal or apical crown-like structure consisting of a varying number of arms branching out distally like a bunch of flowers. *Hayella elegans* of Roth (1968) has been considered a junior synonym of *I. iris*, the generotype of *Ilselithina* Stradner *in* Stradner & Adamiker (1966), by Gartner (1969b). It is known from the Late Eocene and Oligocene. The other species of this genus, *I. fusa* (Figs. 23.15; 24.12), was found in the Oligocene (Roth, 1970) and in the Early Miocene (Edwards & Perch-Nielsen, 1975; Perch-Nielsen, 1977). These very small coccoliths (2–5 μm) are easily overlooked with the LM and are best seen in side view.

Neosphaera with its type species *N. coccolithomorpha* (Fig. 24.10, 11) includes circular coccoliths with a large central opening, the margin of which is surrounded by a narrow circle of elements. The distal shield is composed of radially oriented elements. The size of the central opening varies widely according to Okada & McIntyre (1977). *Neosphaera* differs from other genera with round coccoliths by the presence of only one shield (?) and the narrow, relatively high circle of elements surrounding the central area. The genus *Craspedolithus* and its three species *C. declivus*, *C. ragulus* and *C. vidalii* are considered synonyms of *Neosphaera coccolithomorpha* by Okada & McIntyre (1977). The specimen illustrated in Fig. 24.10 was found in the uppermost Miocene of the Central Atlantic, thus providing a fossil record for this Recent genus and species.

Pedinocyclus replaces the name *Leptodiscus* because the latter is preoccupied by a dinoflagellate (Bukry & Bramlette, 1969b). *Pedinocyclus* includes 'circular coccoliths constructed as a single, thin, nearly flat shield, composed of slightly inclined to radial elements that are not clearly imbricate. Central opening relatively small' (Bukry & Bramlette, 1969b). *Pedinocyclus* can be distinguished from individual shields of *C. leptoporus* by the small amount of imbrication in the former. The curved sutures between the elements resemble those in the Pliocene *Deutschlandia gaarderae*, a form with a single shield and a central opening covered by a cone in well preserved speci-

mens. This brings up the question of assigning *Pedinocyclus* to the Coccolithaceae for convenience, when perhaps by assigning it to the Recent family of Deutschlandiaceae we would suggest ancestors to that family. *P. larvalis* (Figs. 23.33, 34) has been reported from the Late Eocene and the Early Oligocene, and no other species have been assigned to this genus.

Striatococcolithus has two shields composed of narrow, essentially radial elements. In cross-polarized light either both the shields and the small central area are dark to faintly visible, or the shields are dark except for the tiny central area which is bright. The proximal shield is distinctly smaller than the distal one. In *Calcidiscus* at least the proximal shield appears bright in cross-polarized light and *Markalius* has a larger and consistently bright central area. *S. pacificanus* (Figs. 23.31, 32), the generotype and so far the only species of the genus, was reported from the Early and Middle Eocene by Bukry (1971a) and Romein (1979) (Fig. 23).

Umbilicosphaera with its generotype *U. sibogae* (Figs. 23.3, 4) was described by Weber-van Bosse (1901). The original drawing shows a circular form with a relatively small central opening (about ⅓ of the diameter or less) but reveals nothing of the fine structure of the coccoliths. In many Neogene range-charts we find *U. sibogae* usually as the only fossil representative of the genus. It is here assumed that these determinations probably include a number of species, both described and yet undescribed. Therefore, only tentative ranges can be given for the species of this genus (Fig. 23). In the hope of facilitating future determinations, the EM photographs of some of the Recent and fossil species are given in Fig. 24 and the species are discussed below.

U. sibogae: round, relatively large central opening (Figs. 23.3, 4)

U. sibogae foliosa: round, relatively small central opening (Fig. 24.6)

U. angustiforamen: elliptical, very large central opening (Fig. 24.5)

U. maceria: elliptical, relatively large central opening (Fig. 24.7)

U. hulburtiana: elliptical, medium to very large central opening

According to Gaarder (1970), *U. mirabilis* is a junior synonym of *U. sibogae* and therefore the generotype of the genus should be *U. sibogae*.

Deutschlandiaceae Kamptner (1928)
Figure 25

The larger, distal coccoliths of *Deutschlandia* are circular and consist of a single shield with a small central area and with a small central cone. The fossil *D. gaarderae* (Figs. 25.1–3) has a more prominent central cone than the Recent *D. anthos*, and is about twice its size. The shield shows only faint birefringence between crossed nicols, but the ring of small elements surrounding the central cone is bright. The cone is often broken out. *D. gaarderae* was described from the Early Pliocene Zone NN 15 of the South Atlantic.

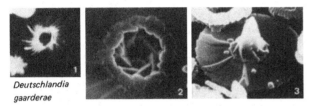

Fig. 25. LM and SEM of *Deutschlandia gaarderae* from the Pliocene

Discoasteraceae Tan (1927)
Figures 26–35; 38

The Discoasteraceae include the calcareous nannofossils of star or rosette shape (Fig. 26). Flat-lying specimens remain dark when viewed with crossed nicols since the C-axis of the calcite then is vertical. The family includes about a dozen genera, most of which are usually included in *Discoaster* except by Prins, who defined them in 1971. Theodoridis (1983) suggested *Eu-discoaster* and *Helio-discoaster* as replacements for the, according to him, illegitimate genus name *Discoaster*. This is not followed here. The genera generally used are: *Catinaster, Discoaster* (syn. *Agalmatoaster, Clavodiscoaster, Gyrodiscoaster, Heliodiscoaster, Hemidiscoaster, Radiodiscoaster, Truncodiscoaster, Turbodiscoaster*), *Discoasteroides*, (*Biantholithus, Ellipsodiscoaster, Imperiaster, Marthasterites, Tribrachiatus*).

Discoasteraceae

ridge
arm
interray area
central area
suture
knob
branch
distal knob
ray
proximal knob
node

side views

distal views   proximal views

Fig. 26. Terminology of the Discoasteraceae

Of these genera, *Catinaster* includes the basket-like forms with rays which may extend beyond the outer border of the central part, and *Discoaster* all forms with relatively flat, star- or rosette-shaped bodies. *Biantholithus* is here included with the Braarudosphaeraceae, *Ellipsodiscoaster* is based on micrococcoliths of *Umbellosphaera irregularis* and thus should not be used. *Imperiaster* seems to be unrelated to *Discoaster* or *Catinaster*, which derived from *Discoaster* in the Middle Miocene; it is here discussed with the 'incertae sedis group' of *Micula–Nannotetrina–Tribrachiatus* (syn. *Marthasterites*). As *Discoaster* is the most important genus of the family, it is discussed before *Catinaster* below.

Discoaster and *Discoasteroides*
Discoaster, the way it is generally used now, has over 100 species which can be distinguished without major problems when they are well preserved. Some of them have been used as zonal markers, but many more of them have relatively well known, and relatively short ranges and thus can be useful in biostratigraphy. Overviews are given in Prins (1971), Perch-Nielsen (1971a) and Aubry (1985) and also the *Catalogue* of Farinacci (1969–79) is easily consulted in the case of *Discoaster*, since only very few species that now are considered to belong to this genus were described under another generic name.

Discoasteroides was erected for discoasters with a 'large, terminally concave stem flaring out as radiate elements at the end' and is used by some and not by other authors.

The most important species are discussed below in the time-slices where they appear.

Paleocene
Discoasteroides bramlettei (Figs. 29.27; 38.20, 21) is the oldest discoaster and appears shortly before the marker species *D. mohleri*. Whereas *D. bramlettei* has a central area which shows birefringence when viewed with crossed nicols, *D. mohleri* remains dark.

D. mohleri (Figs. 29.28, 29; 38.22; *D. gemmeus* of many authors; but the holotype of *D. gemmeus* was described from Eocene material and is smaller than the Paleocene form, which was consequently named *D. mohleri* by Bukry & Percival, 1971) has no distinct central area or central knob. It has 9 to 16 rays of about equal length and often a somewhat irregular outline (Figs. 29.28, 29). Its FO defines the base of NP 7/CP6.

D. mediosus (Fig. 38.15), a large discoaster with parallel-sided rays with rounded, somewhat pointed ends also appears in NP 7.

D. megastypus (Figs. 38.5, 6) is another form with a bright central part consisting of a relatively high central knob, while the central part in *D. bramlettei* consists of radial elements just topping the cone built by the rays.

D. nobilis (Fig. 38.14) is characterized by the still relatively short, but slightly bent rays and its FO defines the

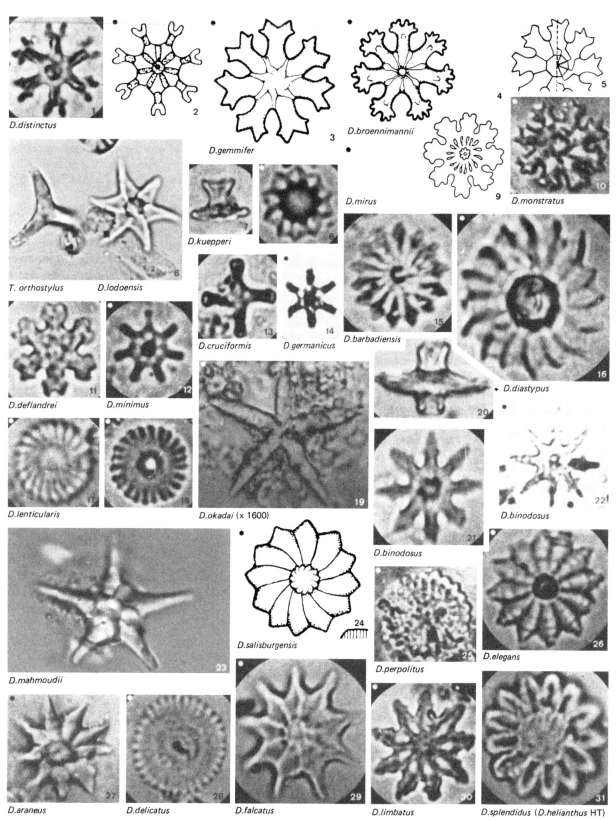

Fig. 27. LM of Paleocene through Middle Eocene
discoasters

base of CP7. The other species appearing in this zone are listed in alphabetic order below.

D. araneus (Fig. 27.27) is a species with a prominent central knob and usually 7–9 tapering rays that have a variable free length from $\frac{1}{3}$ to $\frac{2}{3}$ of the total ray length. The latter is not uniform, even on the same specimen and also the angle between the rays varies.

D. delicatus (Fig. 27.28) is a delicate, very thin form with 30 to 50 rays joined throughout their length and with a central area normally somewhat raised on one side, forming a ring around a depressed centre. *D. delicatus* usually has more rays than the somewhat younger *D. multiradiatus*.

D. falcatus (Figs. 27.29; 29.22) has 7–10, commonly 9 curving rays joined through half their length and featuring a ridge near the convexly curving side and a distinct, but not very high, central knob. *D. falcatus* usually has fewer rays than *D. mediosus*. Its curved rays are shorter and more regularly distributed than in *D. araneus*.

D. limbatus (Fig. 27.30) has '8–11, usually 9 rays joined through about half their length, curving slightly and commonly ornamented by small nodes towards the tips'. The central knob is relatively large and the rays are thickened around the margin and along the ridge or high part of each ray (preservation artefact?).

D. okadai (Fig. 27.19) 'is a very large 4 to 7 ray discoaster. The straight rays are long, thin, tapered, with simple pointed tips. There is no apparent central-area knob or ornamentation. Many specimens show asymmetry'. Five-rayed forms are most abundant, but 4-, 6- and 7-rayed forms also were found. *D. okadai* is the Paleocene discoaster with the longest free rays. It lacks the well developed central area of *D. mahmoudii*.

D. splendidus (syn. *D. helianthus*, Fig. 27.31) is 'sun-flower-shaped, with 13–14 straight-radiating rays joined through more than half their length and having rounded to somewhat pointed tips' and a relatively wide, but flat central knob. The rays have thickened borders.

D. multiradiatus (Figs. 29.30; 38.8), the FO of which defines the base of NP 9/CP8, is rosette-shaped and has about 16 to 35 marginally-pointed rays, which are joined along most of their length. Early specimens have a relatively large central area with remnants of the distal cycle and the column of *Heliolithus* causing a bright central area in cross-polarized light. In younger forms, a central stem evolves and the number of rays generally decreases from older to younger forms. Specimens without a central stem or knob are also found. The other species appearing in NP 9/CP8 are listed below in alphabetic order.

D. binodosus (Figs. 27.21, 22), a species characterized by usually 6 to 8 arms decorated with a pair of nodes and a relatively large central knob appears in NP 9 according to Prins (1971), but only later, in the Early Eocene according to

Aubry (1983) who cites many authors. Early specimens resemble *D. mediosus* but have longer free arms and the pairs of nodes are not present in *D. mediosus*. *D. limbatus* shows a tendency to develop nodes on its relatively short arms, but has a smaller central area than early *D. binodosus*. The number of rays is higher, up to 10 in early specimens.

D. elegans (Figs. 27.26; 29.31) appears in NP 9 according to Romein (1979), but only in the Early Eocene according to other authors. It has '11–15 pointed rays, joined through most of their length and concentrically lined' and a pronounced central knob. It differs from the Eocene *D. barbadiensis* by the concentric ornamentation and from the Eocene *D. boulangeri* (Figs. 28.13, 14) which also features concentric ornamentation by the radially symmetrical ray-tips.

D. lenticularis (Figs. 27.17, 18) is 'small, but centrally thick, with 20–26 rays joined throughout their length' and with bluntly rounded tips and a central knob. It differs from other, usually larger, forms by its curvature of the rays and the great increase in thickness towards the centre.

D. mahmoudii (Figs. 27.23; 29.20) is a star-shaped form with 'long arms, a prominent knob with a central depression on the proximal side and a flat, smaller star-shaped knob on the distal side. It usually has 5 rays, but specimens with 4 to 7 rays were observed'.

D. perpolitus (Fig. 27.25) has 22 to 30 rays joined over most of their length, with concentric ridges and a small central knob. It has more rays than *D. elegans*.

D. salisburgensis (Figs. 27.24; 29.21) is a rosette-shaped discoaster with a thick central part topped by a central stem. *D. barbadiensis* has a flatter central part whereas *D. diastypus* has a central stem on both sides of the disc.

Eocene

D. diastypus (Figs. 27.16, 20), the FO of which defines the base of CP9 at the beginning of the Eocene, is characterized by a stem on both sides of the disc and has 9 to 16 sharply pointed rays, joined through most of their length and curving sinistrally as viewed from the side with the larger stem.

D. binodosus (Fig. 27.22) is the nominate species for NP 11 and CP9b. Its presence or FO is not used to define this zone or subzone, however. It was discussed above. Species appearing in NP 11 are listed alphabetically below.

D. barbadiensis (Fig. 27.15) is a rosette-shaped form with 11 to 18 marginally rounded or pointed rays which are joined along most of their length, has a central stem and is distinguished from *D. multiradiatus* by the lower number of rays and by the higher stem.

D. deflandrei (Fig. 31.28) has its FO in NP 11 according to Romein (1979), while other authors find similar forms only in NP 12 or even higher. *D. deflandrei* was described from the Early Oligocene where it usually has 6, rarely 5 or 7 arms with a terminal bifurcation. The spaces between the arms are subcircular rather than angular and the central area is flat or only slightly elevated, without a distinct knob. While rare specimens may be found in the Eocene and Early Miocene, *D. deflandrei* is only common in the Oligocene, where it is often the only discoaster present.

D. minimus (Fig. 27.12) is a discoaster similar but smaller than *D. mediosus* and with only 7 to 9 parallel-sided rays. It has a small central knob. Since the holotype has no nodes, the forms with nodes, which were included by Sullivan (1964) in *D. minimus*, should be assigned to *D. binodosus*.

D. pacificus (Fig. 29.10) has 8 to 15 pointed, slightly curving rays joined together for about half their lengths. Both sides of the disk have a short stem built of the prolongation of the rays towards the centre. *D. pacificus* is a relatively large form and differs from *D. diastypus* by having less prominent stems and longer free rays.

D. robustus (Fig. 29.11) has short, robust rays and rises conically to the centre, which gives it a triangular outline in side view. It usually has 8 or 9 rays. *D. pacificus* and *D. robustus*, which were described from a Pacific sample of doubtful, mixed age, have their FO in NP 11 according to Romein (1979).

D. lodoensis (Figs. 27.6; 29.24) the FO of which defines the base of NP 12/CP10, is characterized by the curved arms. They usually number 6, rarely 5, 7 or more, and are joined for about $\frac{1}{2}$ to $\frac{1}{3}$ of their length. The arms taper gradually to a sharp point and many specimens have a prominent central stem or knob. Poorly preserved specimens show only short remnants of the arms, but are still recognizable by the curved ridge joining the central knob. In assemblages heavily affected by dissolution, single arms can still be recognized by their curved outline. The other discoasters appearing in NP 12/CP10 are discussed below in alphabetic order.

D. cruciformis (Figs. 27.13; 29.12) is a simple cross with 4 arms, often bent, and originally described from the early Late Eocene. The Late Pliocene *D. tamalis* has 4 slender, tapering arms which are not ornate as those of *D. cruciformis* often are.

D. distinctus (Figs. 27.1, 2) was also described from the early Late Eocene. It has usually 6 arms with bifurcated ends with a pair of nodes in the form of a screw-wrench and a distinct central knob. Its screw-wrench-formed arms distinguish it from all other discoasters.

D. gemmifer (Figs. 27.3; 29.3) is another discoaster with 6 to 8 arms with bifurcating ends. The tips of the bifurcating arms form an angle of about 90° and the terminal pair of nodes is separated from the tips of the arms by short incisions. *D. gemmifer* has no prominent central knob.

D. germanicus (Fig. 27.14) has usually 6, sometimes 5 or 7, arms with a thickening at about half their length and with a slightly bifurcated or incised tip. It has a central knob and is usually smaller than *D. tanii*.

D. kuepperi (Figs. 27.7, 8; 29.17, 18) usually has 9, sometimes 8 to 12 marginally-rounded or pointed rays which are joined along most of their length. The central stem is funnel-shaped and shows birefringence when viewed between crossed nicols. This is visible even in heavily calcified specimens which otherwise would not be determinable.

D. septemradiatus (Fig. 29.7) has 7 or 8 arms which bifurcate at the ends. According to its description, the holotype, however, only shows 6 rays with no bifurcations and subsequent illustrations (i.e. Prins, 1971) followed the paratype rather than the (mislabelled?) holotype. The species thus remains poorly defined, specially since the holotype resembles more a *Lithostromation* than a *Discoaster*.

According to Prins (1971), several discoasters have their FO in NP 13: *D. broennimannii*, *D. mirus*, *D. monstratus* and *D. munitus* (Figs. 27.4; 27.9; 27.5 & 10; 28.15). They are all characterized by having more than 6 arms and bifurcations and nodes in various combinations. Nine-armed forms have also been assigned to *D. nonaradiatus* (Figs. 28.19; 29.2).

D. sublodoensis (Figs. 28.7; 29.25), the FO of which defines the base of NP 14/CP12 is characterized by usually '5 sharply pointed rays joined through about half their length and straight-radiating in the separated outer part'. *D. sublodoensis* is smaller than *D. lodoensis* and its rays are straight, whereas they are distinctly curved in *D. lodoensis*. Other discoasters having their FO in this zone are *D. strictus* (Fig. 28.16), a form with usually 6 tapering arms and a doubly bent suture and *D. wemmelensis* (Figs. 28.5, 6; 29.4, 9), a small, rosette-shaped, double-layered form. Many Early Eocene discoasters disappear in this zone, among them *D. kuepperi*, the LO of which defines the top of CP12a.

In the Middle Eocene NP 15 to NP 16/CP13 to CP14a a number of new discoasters appear. They are discussed in alphabetic order below.

D. bifax (Figs. 28.9–12; 29.8) is a rosette-shaped form with typically 14 approximately radial rays terminating in broad points. It has a prominent stem on one side and a prominent, wider knob on the other side of the disk. Its presence defines Zone CP14a. However, Aubry (1983) found *D. bifax* ranging down into NP 14.

D. gemmeus (Fig. 28.20) is found here according to Bukry (1973a). It is characterized by 8 or 9 rounded rays which are united over most of their length. On one side, crooked sutures are visible, on the other side high ridges unite in the centre of the relatively thick form.

D. martinii (Figs. 28.17, 18; 29.6) is a species with usually 5 arms with a pair of nodes on each and with extreme bifurcation at the end. The tips of the bifurcated arms nearly touch each other. While Prins (1971) shows a distribution of *D. martinii* from NP 13 through NP 16, Aubry (1983) uses its presence to correlate with NP 15 and the basal part of NP 16 only.

D. nodifer (*D. tanii nodifer* of many authors; Fig. 28.1) is characterized by usually 6 straight arms with a pair of nodes and terminal notches. The similar *D. germanicus* has hardly any nodes and no notches and *D. distinctus* has a pronounced bifurcation at the end of the arm. The FO of *D. nodifer* has been used to define the base of NP 16 by some authors, since it seems to occur near the LO of *R. gladius*, the original marker, and the LO of *Nannotetrina*, another substitute event of the top of NP 15.

D. saipanensis (Figs. 28.8; 29.19) usually has 6 or 7, rarely 5 or 8, straight rays which are joined for approximately half their length and taper distally to sharp points. A central stem or knob is present on one side. Its rays are usually more numerous than those of *D. sublodoensis* and joined over more of their length.

The only FO of a discoaster in the late Middle and the Late Eocene seems to be the one of *D. tanii* (Fig. 28.2), which was described as having usually 5, sometimes 6, arms of almost uniform width which truncate abruptly and sometimes have terminal notches. Small nodes are sometimes present on the side of the arms, but they are seldom paired and not consistently developed as in *D. nodifer*. While the FO of *D. tanii* is reported

by Bukry (1973a) from CP14b/NP 17 and *D. tanii* was originally described from the Late Eocene, Prins (1971) shows a range starting from NP 9.

Oligocene

While not many Eocene discoasters continue into the Oligocene (Fig. 30), not many new ones develop either, and this makes the Oligocene a time with very low discoaster diversity.

D. adamanteus (Fig. 31.26) is a relatively small, 6-rayed form where the rays are joined over almost their complete length. Ridges are visible from the elevated centre to the point of the rays which makes the species look like the final overgrowth stage of any other discoaster with 6 rays or arms. Since *D. adamanteus* is smaller than all other discoasters in the assemblage where it is found, it can be assumed that it is a valid species.

D. calculosus (Figs. 29.1; 31.31) is a compact, 6-armed species with a short free length of the broad bifurcate arms, two lateral nodes near the distal end of each arm and no prominent central knob. The interarm area is rounded and shallow, the sutures straight and the surface pebbly.

D. tanii ornatus (Figs. 28.3, 4) has 5 arms and the notch at the tip of the arms is more conspicuous than in *D. tanii*.

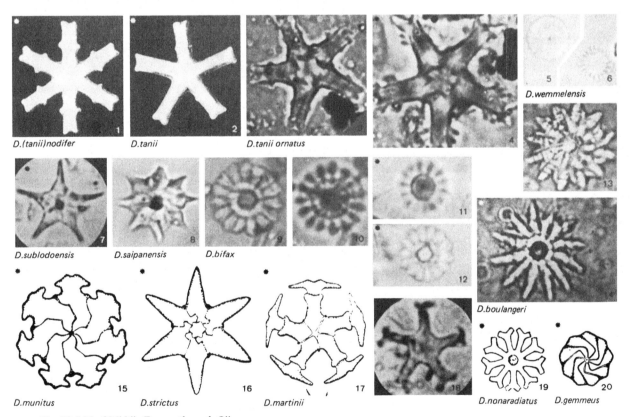

Fig. 28. LM of Middle Eocene through Oligocene discoasters

Fig. 29. EM of Paleogene discoasters

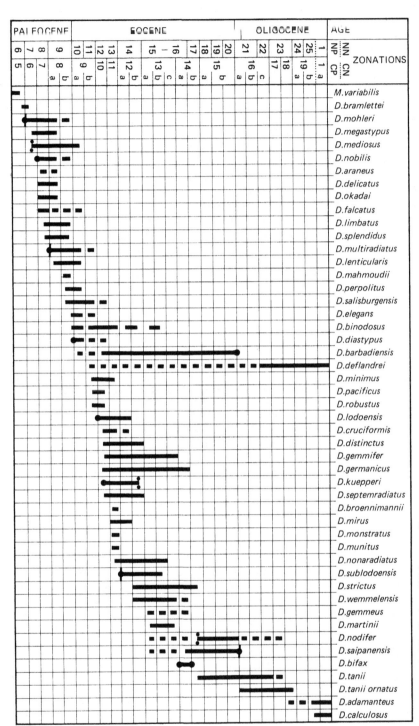

Fig. 30. Distribution of Paleogene discoasters

Also the nodes are conspicuous and commonly occur as a pair on each side of the arms. The central knob is prominent and more conspicuous than in *D. nodifer*, where it may or may not be present.

Early Miocene (CN1a to CN4/NN 1 to NN 5)

The Early Miocene sees the FO of a few new discoasters; one, *D. druggii* (Figs. 31.29, 30) is used to define the base of NN 2/CP1c. It usually has 6 arms 'that may be bluntly rounded, truncated or notched at the tips. The central area is broad and nearly flat' and the arms taper slightly, which distinguishes this species from *D. deflandrei* even when it is overgrown. The interarm area is wide in *D. druggii* and narrow in *D. deflandrei*, with which it occurs. Other species appearing in the Early Miocene are listed in alphabetic order below.

D. aulakos is a species often included in *D. deflandrei*, but differing from this species by its sharply pointed interarm area, while this is smoothly rounded in *D. deflandrei*.

D. exilis (Fig. 31.23) is characterized by usually 6, rarely 5, tapering arms. The tip of the arms are slightly notched or bifurcated. The central area features a central, stellate knob. *D. exilis* differs from *D. variabilis* by its more slender, less bifurcated arms.

D. formosus (Figs. 31.21, 22) has 6 or 7, rarely 4 or 5, arms and a relatively large central area with a prominent central knob. The arms are distributed somewhat irregularly in the forms which are not 6-armed and are slender, tapering towards the tip which is not bifurcated. It differs from *D. druggii* mainly through the presence of a prominent knob.

Hayaster perplexus (*Discoaster perplexus* of many authors; Fig. 31.27) is a disk with a 'circular outline modified only by the straight distal margins of the rays'. The usually 11 or 12 (occasionally 10 or 13) rays are joined throughout their length and separated by sutures radiating straight from the centre. See also *Hayaster* under 'Incertae sedis'.

D. petaliformis (Fig. 31.25) has 6 long, tapering arms with wide bifurcations at their tips and a very pronounced central stem in the form of a small star with petaloid, rounded tips. Its central area is distinctly smaller than in the younger *D. bollii*.

D. variabilis (Fig. 31.14) usually has 6, sometimes 5 and rarely 4 or 3, arms and a stellate central knob with the tips of the star extending to the margin between the arms on one side and along the median line of the arms on the other (concave) side. The arms terminate with a bifurcation forming an angle of about 90°, which is more or less filled with an additional web. The arms themselves are parallel-sided or slightly tapering and the interarm area is V-shaped. *D. variabilis* differs from *D. deflandrei* by the V-shaped interarm area, the longer arms and the web between the bifurcated ends of the arms. *D. exilis* has less prominent bifurcations without a web and more narrowly tapering arms than *D. variabilis*. The FO of *D. variabilis* and the FO of *D. exilis* or an early form of it fall in

the late Early Miocene Zone CN3/NN 4 according to Bukry (1973a), who also pointed out, that at the top of CN3, the dominance of *D. deflandrei* diminishes and longer-armed forms such as *D. exilis* and *D. variabilis* become more common. The long-armed forms are not determinable to species level in some overgrown assemblages, but the dominance change can still be observed.

Middle Miocene (CN4 to CN7/NN 5 to NN 9)

The FO of many discoasters and related forms (*Catinaster*) occur in the Middle Miocene. Four have been used as zonal markers: *D. kugleri* and *D. hamatus* and *C. coalitus* and *C. calyculus* (see *Catinaster*).

D. kugleri (Fig. 31.24) typically has 6 short, stubby arms. They are slightly notched and the central area does not carry a central knob. *D. kugleri* differs from *D. bollii*, which also has a relatively large central area, a knob on both sides and short arms, by the lack of a central knob and the lack of a bifurcation at the tip of the arms.

D. hamatus (Fig. 31.7) is a large, 5-armed form with a small knob in the centre. The arms are 'long, somewhat curved, and turn sharply clockwise and downward near the end, as viewed on the convex side. A much smaller knob is usual as a bifurcation, although it appears to be a continuation of the main part of the arm as it extends in the same direction'. *D. hamatus* differs from *D. neohamatus* and *D. calcaris* by having only 5 arms instead of 6. The other two species have tapering and terminally bent arms.

Other discoasters which have their FO in the Middle Miocene are discussed below in alphabetic order.

D. bellus (Fig. 31.12) is a relatively small, simple 5-armed form lacking a central knob. The arms taper slightly and terminate in points. *D. bellus* is consistently smaller than *D. hamatus* and lacks the tip turned to the side of that species. From *D. pentaradiatus*, it differs by the lack of slight birefringence and the lack of terminal bifurcations and the non-tapering arms in *D. pentaradiatus*. The younger 5-armed forms *D. berggrenii* and *D. quinqueramus* have prominent central knobs.

D. bollii (Figs. 31.20; 33.6) has usually 6, rarely 5, relatively short arms bifurcating into short terminations in different planes. Both sides of the large central area have a stellate stem, one small, one large, a structural element which distinguishes *D. bollii* 'from any other species of otherwise somewhat similar appearance'.

D. braarudii (Fig. 31.19) is a small to medium-sized discoaster with slender, untapering arms and a very small central area. The arms end rounded or blunt and lie in a single plane. *D. braarudii* differs from *D. brouweri* by the arms that bend down like umbrella ribs in the latter, while they are in the same plane in *D. braarudii*.

D. brouweri (Figs. 32.11; 33.3) as emended by Bramlette & Riedel (1954), includes forms with usually 6 long arms that

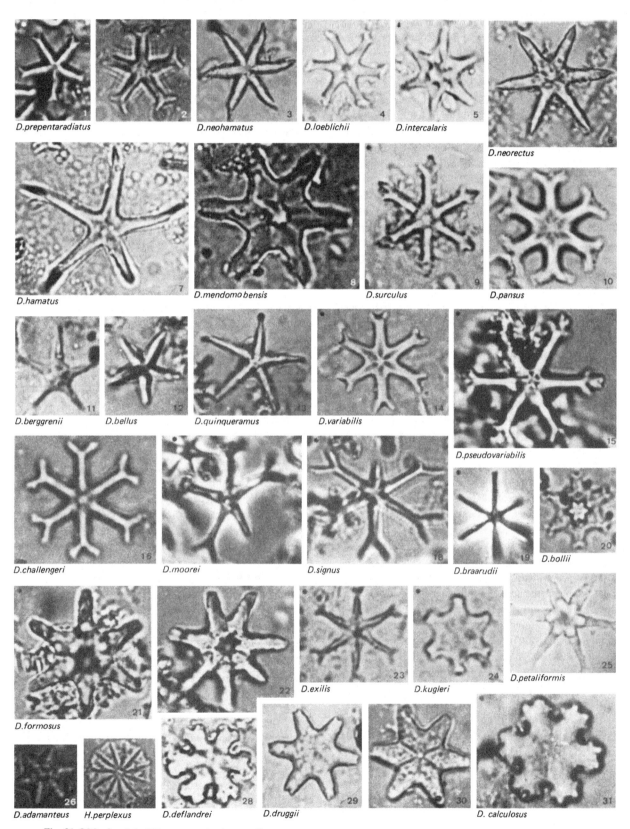

Fig. 31. LM of mainly Oligocene and Miocene discoasters

bend down like umbrella ribs with or without a central knob on the concave side. *D. brouweri* differs from other 6-armed forms by the absence of any bifurcation or sidewards bending of the tips of the arms.

D. calcaris (Fig. 32.10) has a central knob and usually 6 long, slightly tapering arms bifurcating asymmetrically at the tips. This unusual termination of the arms in different planes than the plane of the arms differentiates *D. calcaris* from other 6-armed species. In *D. neohamatus* the ends of the arms bend consistently in one direction and not as abruptly as in *D. calcaris*.

D. challengeri (Figs. 31.16; 33.1) usually has 6, occasionally 5 or rarely 7, arms which are subcylindrical and distally bifurcate into short, rounded terminations. The angle between these terminations is greater than 90°, usually about 120° and the central area is with or without a central knob. Other similar species either have tapering arms, bifurcations with a narrower angle or they have 5 arms.

D. moorei (Fig. 31.17) is a small species with 5 distinctly tapering arms that terminate in broad bifurcations enclosing an angle of about 90°. The arms are relatively short and asymmetrically arranged. The central area may or may not have a small knob. It differs from *D. asymmetricus*, the younger asymmetrical, 5-armed species by the presence of bifurcations. It differs from *D. pentaradiatus* by the asymmetrical arrangement of the arms and the lack of a slight birefringence and from *D. prepentaradiatus* by the asymmetrical arrangement of the arms and the longer bifurcations with wider angles.

D. neohamatus (Fig. 31.3) has 6 long, slender arms, the outer ends of which consistently are bent in one direction and terminate in points. The central area is very small and no knob is present. In overgrown specimens, the arms appear to be more bent than in better-preserved specimens. *D. calcaris* is similar but has asymmetrically bifurcating arms. *D. hamatus* has only 5 arms.

D. prepentaradiatus (Figs. 31.1, 2) has 5 arms in symmetrical arrangement and no central area developed. The arms bifurcate at the tips, arms and tips lying in the same plane. The angle between the bifurcating tips varies from less than 90° in the holotype to over 120° in a paratype. The arms are relatively longer than in *D. moorei*, where they are arranged asymmetrically. *D. prepentaradiatus* differs from the younger *D. pentaradiatus* by the lack of the typical downward bent arms and the slight birefringence of that species.

D. pseudovariabilis (Fig. 31.15) usually has 6 arms, rarely 5, which are bifurcated at the ends. Between the bifurcations a tongue-like node points slightly downwards. In *D. variabilis* the area between the bifurcation is filled with a 'web-like' rather than with a tongue-like projection. In *D. subsurculus* the bifurcation and projection sit on the end of a slightly shorter arm which extends from a relatively larger central area. All these forms have a more or less pronounced

central knob, as does *D. surculus*, from which *D. pseudovariabilis* hardly differs if the holotypes are compared, but whose later representatives have a distinctly less prominent bifurcation than *D. pseudovariabilis*.

D. signus (Fig. 31.18) has 6 slender, untapering to only slightly tapering arms with 'two even more slender and distinctly tapering bifurcating tips. The length of the bifurcation is equal to half or more of the unbifurcated arm length. Although no central area is developed, a prominent knob forms the hub for the equally spaced rays'. No other 6-armed species has such long bifurcations.

D. subsurculus has 6 arms with wide bifurcations and a small node within it. A small knob sits in the centre of the relatively large central area. For further differentiation see *D. pseudovariabilis* above.

Late Miocene (CN7 to CN10/NN 9 to NN 12)

Several discoasters are used as markers in the Late Miocene: the LO of *D. hamatus* (see above), the FO and LO of *D. quinqueramus* and the FO of *D. neorectus*.

D. quinqueramus (Figs. 31.13; 33.9) has 5 tapering, long non-bifurcating arms arranged symmetrically and a robust central area with a very prominent, 5-armed central knob. Early forms are difficult to distinguish from the older *D. berggrenii* (see below), which features an even more prominent central knob and shorter arms. Other 5-armed species either have bifurcating arms and/or no prominent central knob or bending arms.

D. neorectus (Fig. 31.6), the LO of which defines the base of CN8b, is a gigantic, 6-armed species with a small, twisted central knob but no separately marked central area. The arms are long and symmetrically arranged and tapering to a simple point. *D. neorectus* is 2 to 3 times as large as *D. intercalaris*, which also has a larger central area and less pointed ends of the arms.

Other species that have their FO in the Late Miocene are listed below in alphabetic order.

D. asymmetricus (Figs. 32.3; 33.2), the FO of which defines the base of NN 14 in the Pliocene, was claimed by Bukry (1973a) to appear already in the Late Miocene, with the added remark, that the specimens might be aberrant forms of *D. bellus*. In fact, it is difficult to distinguish between the typical 5-armed, asymmetric *D. asymmetricus* and aberrant forms of any 5-armed discoaster with non-bifurcated arms or where the bifurcation has been lost as is the case in many specimens of *D. pentaradiatus*. In this species, the bifurcation may be present on some and not on other arms due to breaking or dissolution. Only *D. pentaradiatus*, however, shows slight birefringence between crossed nicols.

D. berggrenii (Figs. 31.11; 33.8) is a symmetric, 5-armed form with tapering arms and a very prominent knob consisting of the inwardly directed prolongations of the ridges on the arms on one side and of ridges extending from the interarm area over a depressed central area to form a central knob (SEM in Perch-Nielsen, 1972) on the other side. The free length of the arms is about equal to the diameter of the central area, whereas it is longer in *D. quinqueramus*, which has thinner arms and less prominent knobs. The exact break between *D. berggrenii* and *D. quinqueramus* is difficult to determine and the species have therefore been put in synonymy by various authors, a practice which shifts the boundary of CN9/NN 11 downwards. The very well developed knobs of the central area differentiate *D. berggrenii* from all other Miocene 5-armed discoasters. The most similar, but unrelated, form is the Late Paleocene *D. mahmoudii* (see above).

D. icarus (Fig. 32.18) is a large species 'with rays terminating in a wide-angle of bifurcation of branches with "flaps" between the branches'. The 6-armed species has a central knob and straight suture lines.

D. intercalaris (Fig. 31.5) is a medium-sized, 6-rayed species with a large central area and a central stem. The symmetrically arranged arms show a distinct tapering and terminate in simple rounded points or have a small indentation. *D. intercalaris* is distinguished from the *D. variabilis* group by the single-pointed termination of the arms instead of the broadly flaring bifurcation. *D. intercalaris* is not umbrella-

shaped as *D. brouweri* and distinctly smaller than *D. neorectus*. Bukry (1971b) suggested, 'that *D. intercalaris* may be a cool-water relative of *D. variabilis* that failed to develop bifurcations. This possibility is indicated by the small size of the bifurcations of *D. variabilis* specimens associated with *D. intercalaris*'.

D. loeblichii (Fig. 31.4) is a small to medium-sized, 6-armed species with a central area occupying about a third of the total discoaster diameter and a small central knob. The arms are distinctly tapering and the tips of the arms 'have distinctive unequal bifurcations that are bent slightly out of the plane of the arms. Both limbs of the bifurcation taper to points, but all six sets show one limb that is consistently more than twice as long as the other'. *D. loeblichii* differs from other 6-armed forms by its unequal bifurcations, similar ones only being found in *D. calcaris*, which is larger and has a relatively smaller central area.

D. mendomobensis (Fig. 31.8) is a large form with 6 arms and a relatively large central area topped with a small central knob. The arms terminate in 'discernible wrench- or spade-shaped bifurcations with notched tips' and have more or less parallel sides. *D. mendomobensis* differs from other Miocene discoasters by its large central area and tapering, notched bifurcations.

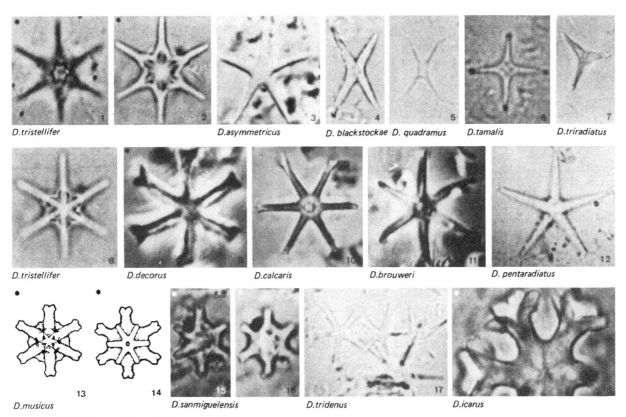

| *D.tristellifer* | | *D.asymmetricus* | *D. blackstockae* | *D. quadramus* | *D.tamalis* | *D.triradiatus* |

| *D.tristellifer* | *D.decorus* | *D.calcaris* | *D.brouweri* | *D. pentaradiatus* |

| *D.musicus* | | *D.sanmiguelensis* | *D.tridenus* | *D.icarus* |

Fig. 32. LM of mainly Late Miocene and Pliocene discoasters

D. pansus (*D. variabilis pansus* of some authors; Fig. 31.10) is a large, 6-armed discoaster that has slightly tapering arms terminating in large broadly bifurcate tips. The large bifurcate limbs are almost perpendicular to the rays, forming an angle of about 120° or more. There is a small knob on the medium-sized central area. *D. pansus* is distinguished from *D. decorus* by its large bifurcations nearly perpendicular to the moderately long arms, and from *D. variabilis* by the latter's 'web-like' node between the bifurcations. *D. challengeri* has parallel-sided arms and no central area development.

D. pentaradiatus (Figs. 32.12; 33.5) as emended by Bramlette & Riedel (1954) has 5 thin, tapering arms with a terminal bifurcation which characteristically is thin and delicate (and therefore often broken off). It shows slight birefringence when viewed between crossed nicols. *D. pentaradiatus* shows only little of the umbrella-like bending of *D. brouweri*, from which it differs by also having 5 instead of 6 arms. *D. prepentaradiatus* is flatter than *D. pentaradiatus* and has shorter and more robust bifurcations.

D. perclarus is a 6-armed form with a small central area, relatively short, very thin arms and a bifurcation with thin limbs forming an angle of about 120°. The similar *D. signus*

has bifurcations with an angle of about 60° and has relatively longer arms, but an equally small central area.

D. surculus (Figs. 31.9; 33.10) usually has 6 arms that are normally symmetrically arranged. The arms are more or less parallel-sided and bifurcate towards their ends. Within the bifurcation a tongue-like node protrudes further than the ends of the bifurcation. In early forms such as the holotype, the bifurcation is wide, in later forms it is narrower and only slightly wider than the tongue-like protrusion. Early forms are difficult to distinguish from *D. pseudovariabilis* (see above) and thus the FO of *D. surculus* depends more than usual on the taxonomic species concept of individual authors. *D. surculus* differs from the discoasters of the *D. variabilis* group by its tongue-like protrusion instead of the web-like one in *D. variabilis* and related forms.

Pliocene

The Pliocene is characterized by the gradual disappearance, one by one, of the Late Miocene and the few new Pliocene discoasters until, with the LO of *D. brouweri*, the discoasters die out near the Pliocene/Pleistocene boundary. Pliocene discoasters are all relatively slender and there is a relative increase in 3- and 4-armed forms. Backman &

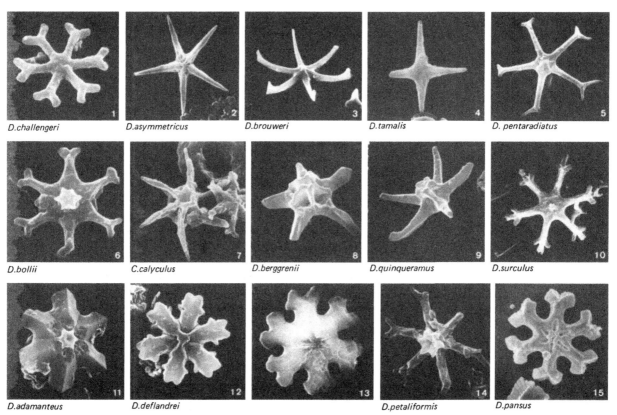

D.challengeri	*D.asymmetricus*	*D.brouweri*	*D.tamalis*	*D. pentaradiatus*
D.bollii	*C.calyculus*	*D.berggrenii*	*D.quinqueramus*	*D.surculus*
D.adamanteus	*D.deflandrei*		*D.petaliformis*	*D.pansus*

Fig. 33. SEM of Neogene discoasters and *Catinaster calyculus*

Shackleton (1983) have studied the extinction pattern of the discoasters in several oceans and found it to correlate well with paleomagnetic data. Besides the questionable FO of *D. asymmetricus* (see above), only LOs of discoasters have been used to subdivide the Pliocene. The species which have their FO in the Pliocene are discussed in alphabetic order below.

D. blackstockae (Fig. 32.4) is a 4-armed species with a small 'oblong central area. The 4 narrowly tapering, bladelike' arms end in simple points and form two wide and two narrow angles of about 60° and 120° respectively. *D. blackstockae* differs from the similar *D. quadramus* by its simply terminating arms (or are the bifurcations broken off?) and from *D. tamalis*, another 4-armed form by the different angles between the arms.

D. decorus (*D. variabilis decorus* of some authors; Fig. 32.9) is a large, symmetric, 6-armed (rarely 5-armed) species with a small central area. The arms are slightly tapering and have narrowly bifurcated tips. The bifurcation seems slightly narrower than in *D. variabilis*, which also has slightly shorter arms in relation to the diameter of the central area.

D. quadramus (Fig. 32.5) is a 4-armed species with slender arms bifurcating distally. The arms may embrace angles of 90° or 60° and 120°. Similar to *D. pentaradiatus*, *D. quadramus* can show faint birefringence in cross-polarized

light. The central area is small and has a faint 4 sided knob. *D. quadramus* is the only 4-armed discoaster with bifurcating arms.

D. tamalis (Figs. 32.6; 33.4) has 4 slender arms meeting at 90° and a small central area. The arms are tapering and have no bifurcation. *D. tamalis* is slightly bent, but not as strongly as the more robust Eocene species *D. cruciformis*, which also has additional nodes on arms and central area.

D. triradiatus (Fig. 32.7) has 3 slender, tapering arms and is by some authors considered an aberrant form of *D. brouweri*. It occurs with increased frequency shortly before its extinction according to Stradner (1973) and Backman & Shackleton (1983). The arms of *D. triradiatus* meet at about 120° and the form is not or only slightly bent. The Eocene triradiate species of *Tribrachiatus* are more robust and the early forms show bifurcations. Also the Late Cretaceous forms of *Marthasterites* are more robust and most of them show bifurcations at the ends of the arms.

D. tristellifer (Figs. 32.1, 2, 8) is typically 6-armed and characterized by star-shaped knobs of different diameter on opposite sides of the central area. The arms are long and tapering. The smaller knob has its points aligned between the arms of the discoaster, the points of the larger knob are aligned with the bays between the arms of the discoaster. *D. tristellifer* is the only form with both long arms and two central knobs of different size.

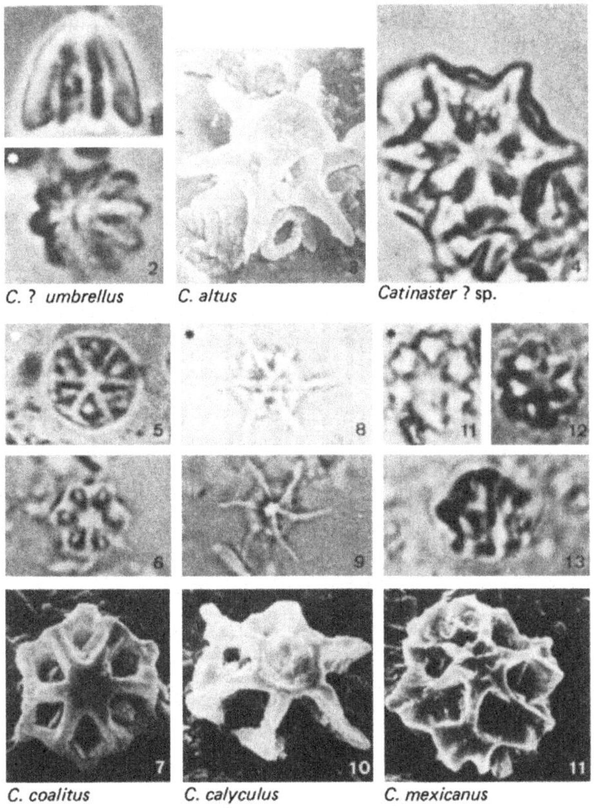

Catinaster

The species of *Catinaster* are easy to distinguish from discoasters by their relatively large, basket-shaped central part. In *C. coalitus* (Figs. 34.5–7), the type species, the 6 rays do not extend beyond the basket, while they do so and are bent in typical *C. calyculus*. Whereas the holotype of *C. coalitus* is nearly circular (Fig. 34.5), many specimens usually also assigned to *C. coalitus* are hexagonal in outline (Figs. 34.6, 7). These forms are somewhat smaller and are found in the older part of the short range of *C. coalitus* in the Middle Miocene. In early forms of *C. calyculus* (Figs. 34.8–10) the rays extend beyond the basket, but are straight. Such forms, when included in *C. calyculus*, extend the range of this species below the FO of *Discoaster hamatus*.

In *C. mexicanus* (Figs. 34.11–14) the rays bifurcate distally and in *C. altus* (Fig. 34.3) the rays extend into a basal (proximal?) 6-rayed body. *C.? umbrellus* (Figs. 34.1, 2) has more than 6 rays arranged in an umbrella shape and probably is not related to *Catinaster*. *Catinaster* ? sp. (Fig. 34.4) is a relatively large, 6-rayed form with a small central knob and relatively short ridges compared with the relatively long rays in the other *Catinaster* species. Species of *Catinaster* reach 'few' in low latitude to mid latitude sections, and are rare or absent in high latitudes.

C.? umbrellus   *C. altus*   *Catinaster ? sp.*

C. coalitus   *C. calyculus*   *C. mexicanus*

Fig. 34. LM and SEM of *Catinaster* species

Evolution and conclusions

Papers dealing with the evolutionary trends of discoasters were published in 1971 by Bukry and by Prins. Romein (1979) and Perch-Nielsen (1981b) contributed to the Paleogene part of the discussion about the early evolution of the Paleogene discoasters. Driever (1981) presented a quantitative study of Pliocene associations of Discoaster from the Mediterranean. Backman & Shackleton (1983) studied quantitative biochronology of Pliocene discoasters from the Atlantic, Indian and Pacific oceans.

No comprehensive evolutionary scheme based on LM and TEM/SEM observations and biometric analysis and fully illustrated with LM and SEM pictures has been presented for any part of the Cenozoic and nothing that could be attempted here would be an improvement on the views presented by Prins (1971) and Romein (1979).

The earliest true Cenozoic discoasters appear in the Late Paleocene and are 'compact, multirayed, heavily constructed forms with very short free length of the rays or very broad rays, producing a plan view in which the area encompassed by the greatest diameter of the fossil is largely a field of calcite. Mid-Tertiary discoasters show fewer rays which are relatively broad, so that a substantial part of the area encompassed by the greatest diameter is filled by calcite rays. In the final stage of development, just before extinction in Late Pliocene time,

discoasters are very narrow-rayed and include, in addition to the usual 6-rayed forms 3-, 4-, or 5-rayed' forms (Bukry, 1971b). Whereas there is a significant reduction in the amount of calcite involved to form a discoaster from the Paleocene through the Pliocene, the overall diameter of *Discoaster* species is only slightly reduced.

According to Haq (1971), the Early and Middle Eocene and the Middle Miocene were the times with the highest *Discoaster* diversity. Not enough new species have been described since to shift these diversity peaks although it would now seem, that the Neogene peak occurred in the late Middle to early Late Miocene (compare Figs. 30 and 35).

While discoasters are used widely in the subdivision of low and high latitude Paleogene sediments, a marked decrease in their worldwide diversity occurs in the Oligocene. In the Neogene they become sparse in high latitudes and are mainly useful in mid and low latitude sections. The zonation based exclusively on discoasters proposed by Prins (1971) has not gained acceptance, probably due to the fact that coccoliths are usually also present and add significantly to the age determination. However, it could be useful in assemblages where all coccoliths have been dissolved and only discoasters and arms of discoasters are left, as is the case in sediments deposited below the calcium carbonate compensation depth (CCD) for planktic foraminifera and coccolithophorids.

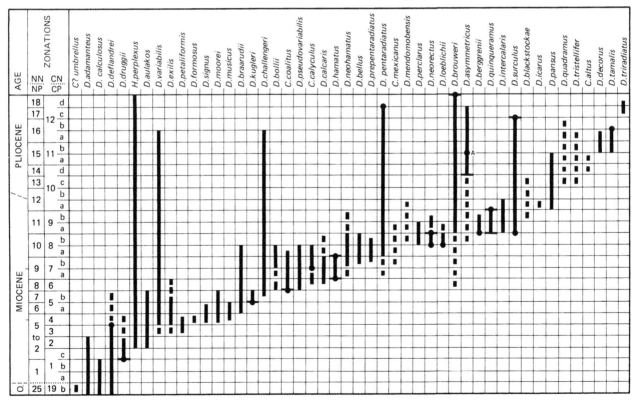

Fig. 35. Distribution of Neogene discoasters

Theoretically, the discoasters should be ideally suited to study evolution. Practically, attempts are often hampered by the poor preservation in the form of extensive overgrowth, partial dissolution and breakage and by downhole contamination that may not always be obvious. While even rather poorly preserved specimens can still be determined as markers, they cannot be used for the study of the evolution of the discoasters.

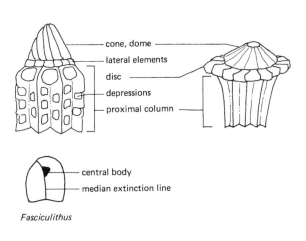

Fig. 36. Terminology of the Fasciculithaceae and Heliolithaceae

Fasciculithaceae Hay & Mohler (1967) and Heliolithaceae Hay & Mohler (1967)
Figures 36–40

The Fasciculithaceae and the Heliolithaceae are closely related Paleocene families. The Fasciculithaceae include *Fasciculithus*, the Heliolithaceae *Bomolithus* and *Heliolithus*. While *Fasciculithus* includes the cylinder-like nannoliths with a proximal column, a disk or lateral elements and a distal cone (Fig. 36), *Bomolithus* consists of three cycles of elements and *Heliolithus* of two in *H. riedelii*, the generotype, and of three in *H. kleinpellii* (Fig. 40). *Bomolithus* has been included by Romein (1979) in *Heliolithus*, but is here regarded as an early representative of the Heliolithaceae. Links to the Discoasteraceae have been suggested by Prins (1971), Romein (1979) and others and are shown in Fig. 37 but are not further discussed here. The three genera are fully illustrated by SEM in Perch-Nielsen (1971b, 1977) and Perch-Nielsen *et al.* (1978). Romein (1979) and Perch-Nielsen (1981b) have proposed evolutionary lineages for these two families.

The stratigraphically important forms are *F. tympaniformis*, *H. kleinpellii* and *H. riedelii*, each defining the base of a zone in the Late Paleocene. *F. tympaniformis* was assumed to be the first fasciculith when it was proposed as a marker by Hay & Mohler (1967) but is now known to have several predecessors. While many authors use *F. tympaniformis* as a kind of waste-paper basket for all fasciculiths, many of these can be distinguished also with the LM and thus can be used for the further subdivision of NP 4, NP 5 and NP 9 in extended sections. The presently described fasciculiths are illustrated in Fig. 38, which will facilitate comparison of many forms.

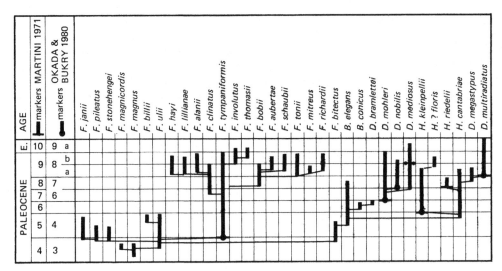

Fig. 37. Distribution and tentative evolutionary lineages in *Bomolithus* (B.), *Discoaster* (D.), *Fasciculithus* (F.) and *Heliolithus* (H.)

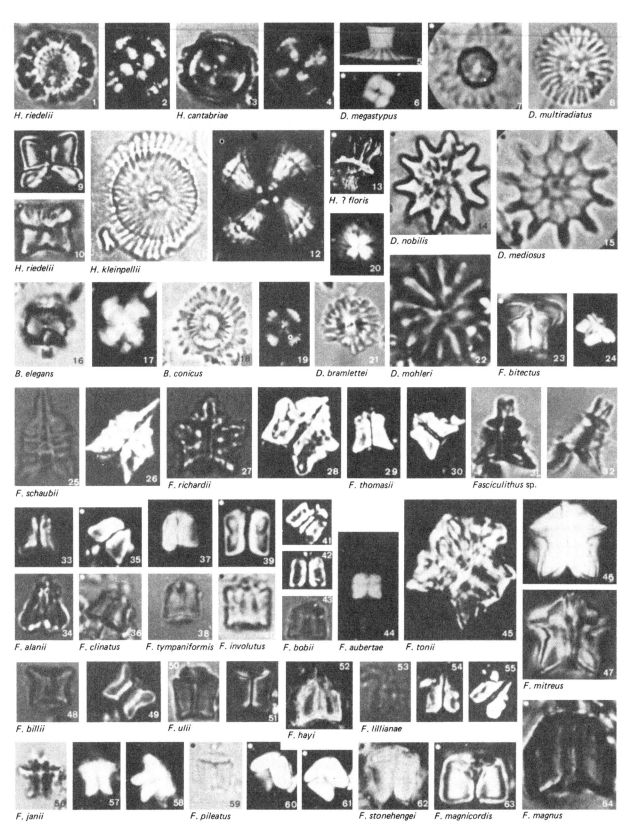

Fig. 38. LM of species of *Bomolithus* (*B.*), *Discoaster* (*D.*), *Fasciculithus* (*F.*) and *Heliolithus* (*H.*)

Fasciculithus

F. magnus, the oldest species known, is a large form which differs considerably from younger fasciculiths. Its body consists of a short, proximal part and a higher distal part (Figs. 38.64; 39.6, 7). The central structure of the distal side lies at the bottom of a depression in the distal part of the body and closely resembles the central area of *Markalius*, from which it might have evolved (Perch-Nielsen, 1977).

F. magnicordis (Fig. 38.63) is smaller than *F. magnus* and has a relatively larger central body according to Romein (1979).

In *F. ulii* (Figs. 38.50, 51; 39.8) and in all younger fasciculiths the proximal part of the body is higher than the distal part. *F. ulii* has a relatively low distal cone or dome and the cone is missing or very flat in *F. billii* (Figs. 38.48, 49) where the distal part of the column is slightly flaring.

F. stonehengei (Fig. 38.62) and *F. merloti* are probably junior synonyms of *F. pileatus* (Figs. 38.59–61), which together with *F. janii*, is characterized by a distal disk. In *F. pileatus* the distal disk does not overlap the proximal column or only slightly so, while it clearly does so in *F. janii* (Figs. 38.56–58; 39.9).

In *F. bitectus* (Figs. 38.23, 24) a flat dome and a cycle of lateral elements are present and well distinguishable as two superimposed, optically differently oriented structural elements when seen between crossed nicols in side view.

F. tympaniformis (Figs. 38.37, 38; 39.10) appears after the species discussed above and is the first form with a simple proximal column tapering distally. A central body (Fig. 36) is present in early forms but disappears in the upper part of NP 5 according to Romein (1979). The optical extinction line changes from vertical to oblique at the central body. No new species of *Fasciculithus* appear in NP 6, after the extinction of all early fasciculiths with the exception of *F. tympaniformis* in NP 5. The latter then gave rise to a second diversification within the genus in NP 8 and especially NP 9, where the genus included about a dozen species before it became extinct near the Paleocene/Eocene boundary.

F. clinatus (Figs. 38.35, 36) is thought to be the ancestor of the other strongly tapering forms like *F. alanii*, *F. lillianae* and *F. hayi*. It is distinguished from *F. tympaniformis* by its nearly triangular outline in side view. The median extinction line curves to the left or to the right distally, or bifurcates and forms a V-shaped pattern.

F. alanii (Figs. 38.33, 34; 39.1) is easily recognized by its high conical, but slightly concave outline. It is distinguished from *F. lillianae* (Figs. 38.53–55), which is usually smaller, by the nearly parallel-sided proximal part of the column in the latter. *F. hayi* (Fig. 38.52) and *F. lillianae* may be synonyms, though it does seem that they can be distinguished by the slightly conical proximal part of the column in *F. hayi* compared with the parallel sides in *F. lillianae*.

Another group of fasciculiths shows depressions on the proximal column: *F. schaubii*, *F. richardii*, *F. thomasii*, *F. involutus*, *F. bobii*, *F. tonii*, *F. mitreus* and *F. aubertae* (Fig. 38). *F. mitreus* and *F. tonii* may be considered synonyms. *F. tonii*, a very large species, is characterized by the diverging sides of

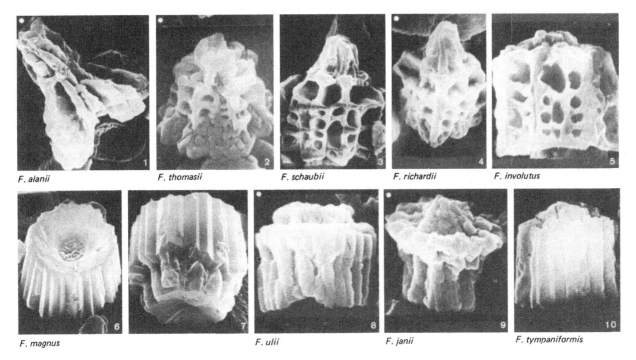

F. alanii　　　F. thomasii　　　F. schaubii　　　F. richardii　　　F. involutus

F. magnus　　　F. ulii　　　F. janii　　　F. tympaniformis

Fig. 39. SEM of species of *Fasciculithus*

the proximal column and a cone composed of spirally oriented ridges. *F. involutus* has parallel sides and a very low cone; the depressions are distributed in several cycles on the proximal column. *F. thomasii* is bell-shaped with slightly concave sides in the lower part of the column and convex sides in the upper part. *F. bobii* has parallel sides, a low cone and only one cycle of depressions in the distal part of the column. *F. aubertae* has a slightly conical outline, a very low cone and about two cycles of depressions. *F. schaubii* (Fig. 39.3) can be distinguished from other large species such as *F. tonii* and *F. richardii* (Fig. 39.4) by its relatively high cone above a high column with nearly parallel to slightly conical sides with at least three cycles of depressions. The holotype (SEM) of *F. richardii* has a cubical proximal column with a nearly square base. The sides of the column are subdivided vertically and horizontally by depressions. It is considered only a variant of the polygonal *F. schaubii* by Romein (1979).

Bomolithus and Heliolithus

Bomolithus includes forms with three cycles of elements, if the generotype *B. elegans* (Figs. 38.16, 17; 40.2) is regarded as typical. Only the central part of the nannolith, thus the column, appears bright when viewed in plane view between

H. cantabriae

B. elegans

H. riedelii

B. conicus

H. kleinpellii

Fig. 40. SEM of species of *Bomolithus* and *Heliolithus*

crossed nicols. In *Heliolithus*, which has three cycles of elements in *H. kleinpellii* (Figs. 38.11, 12; 40.5, 6) but only two cycles in its generotype *H. riedelii* (Figs. 38.1, 2, 9, 10; 40.3), all cycles appear bright between crossed nicols. This thus has to be used as a criterion for distinguishing *Bomolithus* and *Heliolithus*, if we do not want to transfer *H. kleinpellii* and *H. cantabriae* into a *Bomolithus* with three cycles of elements or put *Bomolithus* into *Heliolithus* as suggested by Romein (1979).

H. kleinpellii is a rather large, flat form which is very bright between crossed nicols. Column and median cycle appear still brighter than the much larger distal cycle and all three are of about equal thickness. The centre can be occupied by the radially to tangentially oriented elements of the column and the median cycle or it can be open. The extinction cross is straight and thin in the central part and curved and flaring over most of the heliolith, thus covering about half of each quadrant. *H. kleinpellii* is usually the largest nannofossil in the samples where it occurs.

H. riedelii is smaller than *H. kleinpellii* and much higher. In side view, the column is slightly higher than the distal cycle and no median cycle is present. Seen from above, both the column and the distal cycle show birefringence. The outline is serrate and the whole heliolith has a rugged appearance. The extinction cross is straight. *H. riedelii* can be distinguished from *H. cantabriae* and *H. kleinpellii* by its rugged appearance and straight extinction lines.

H. cantabriae has a column of tangentially oriented elements, a median cycle and a distal cycle. The column is considerably higher than in *H. kleinpellii*.

Bomolithus elegans and *B. conicus* (Figs. 38.18, 19; 40.4) as well as *Discoasteroides megastypus* show no birefringence in the distal cycle.

Conclusions

The genera *Bomolithus*, *Fasciculithus* and *Heliolithus* are rich in species, three of which are commonly used as zonal markers in the Late Paleocene. Whereas most species have been well illustrated by LM and by SEM, and Romein (1979) and Perch-Nielsen (1981b) have suggested evolutionary relations, many unpublished SEM pictures do not fit into the described species and cannot easily be assigned a place in the evolutionary schemes. Detailed work with the LM and the SEM in a stratigraphically expanded section is needed.

Goniolithaceae Deflandre (1957)
Figure 41

Tappan (1980) has assigned two other genera to the Goniolithaceae in addition to the original *Goniolithus*, which remains a monospecific genus. This is not followed here, since the two other (Jurassic) genera do not have pentagonal, composite coccoliths. The family includes pentagonal coccoliths with a wall of vertical elements and a granular centre. The coccoliths form a dodecahedron like *Braarudosphaera*, but have a much more complicated fine structure than that genus. *G. fluckigeri*, the generotype and only species, is found from the Maastrichtian through the Eocene and thus is one of the few species that survived the Cretaceous/Tertiary boundary event(s) and continued for a considerable time, but evidently did not give rise to other species during that time.

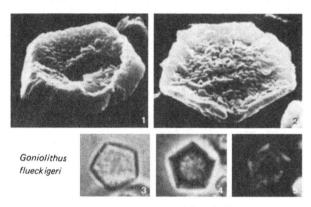

Goniolithus flueckigeri

Fig. 41. LM and SEM of *Goniolithus fluckigeri*

Helicosphaeraceae Black (1971a)
Figures 42–47

The Helicosphaeraceae include helical-rimmed coccoliths, usually with a flange. They may have a central bridge and an empty or a completely filled central area (Fig. 43). More than 40 species have been described in the genus *Helicosphaera/Helicopontosphaera* but only *H. recta*, *H. ampliaperta*, *H. compacta*, *H. reticulata* and *H. sellii* have been used as markers for the zonation of the Cenozoic. Many other species have relatively short ranges and thus are also biostratigraphically useful.

Helicosphaera / Helicopontosphaera

The original illustration of *H. carteri*, the generotype of *Helicosphaera*, is such that we shall never know exactly which Recent coccolith Wallich described (1877). The ovoid form of the coccolithophorid he illustrated (Fig. 45.1; figs. 3 and 4) suggests affinity with *Helicosphaera*. However, the coccoliths he illustrated (figs. 6, 7, 17) resemble *Coccolithus pelagicus* because they lack helical rims, have larger central area openings, and have a more prominent bridge than any modern *Helicosphaera* species. The apparent contradiction between coccolith and coccosphere morphology has lead to confusion over the validity of the name *Helicosphaera*. One school of thought holds that Wallich (1877) actually described a coccosphere bearing *Helicosphaera*-type coccoliths, as evidenced by the gross morphology of the cell, but was unable to discern their helical rims because of inadequate equipment and thus

favours the use of *Helicosphaera* over *Helicopontosphaera* and considers the latter a subjective junior synonym. The opposing school of thought holds that Wallich's *Coccosphaera carteri* may actually have been a coccosphere bearing *C. pelagicus*-type coccoliths, and when the generotype of *Helicosphaera* became *C. carteri* by monotypy (Kamptner, 1954) the concept of a helical-rimmed *Helicosphaera* genus was invalidated. This school thus rejects *Helicosphaera* in favour of the clearly described and illustrated *Helicopontosphaera* Hay & Mohler (1967). For further discussions see Jafar & Martini (1975) and Haq (1973). Species originally assigned to either genus are herein referred to *Helicosphaera*.

The oldest *Helicosphaera* has been reported by Haq (1971) from the Paleocene of Iran and by Perch-Nielsen (1977) from the Paleocene of the South Atlantic (Fig. 46.13). The first validly described forms are *H. seminulum* and *H. lophota* from the Early Eocene. Several species are still living today. An overview and suggestions for evolutionary lineages were given by Haq (1973), but since then, many new species have been described and a new overview is therefore given in Figs. 43–47.

The study of species of *Helicosphaera* with the LM is difficult. To facilitate the task, the forms have been distributed into four groups in Fig. 43 according to the characteristics of the central area. In group A, the central area is open or closed with only narrow slits or it is very small, with pores barely visible with the LM. In group B, the central area features an oblique, 'inversely' oriented bridge. Group C species have a vertically oriented bridge, composed of optically continuous proximal shield elements in subgroup Ca, and constructed of optically discontinuous elements in subgroup Cb. In group D, the bridge is oblique but 'normally' oriented. The bridge is optically continuous in subgroup Da and optically discontinuous in subgroup Db. Within these groups and subgroups, species can be distinguished by their outline (elliptical, ovoid, rhombohedral or rectangular), by the presence or absence, shape and size of the terminal flange, or by details of the fine structure and/or angle of the central bridge. Size also is useful for species identification, as can be seen in Fig. 42, where species are illustrated at the same magnification.

The ranges of many species as shown in Figs. 44 and 45 are tentative, as very few authors have used more than a few species and many species have only been described and not mentioned again. Increased usage of *Helicosphaera* species should improve biostratigraphic resolution in the Cenozoic.

In the remarks below, the species of *Helicosphaera* are compared to other species of similar age rather than of similar structure. Species are arranged according to age in Figs. 44–46.

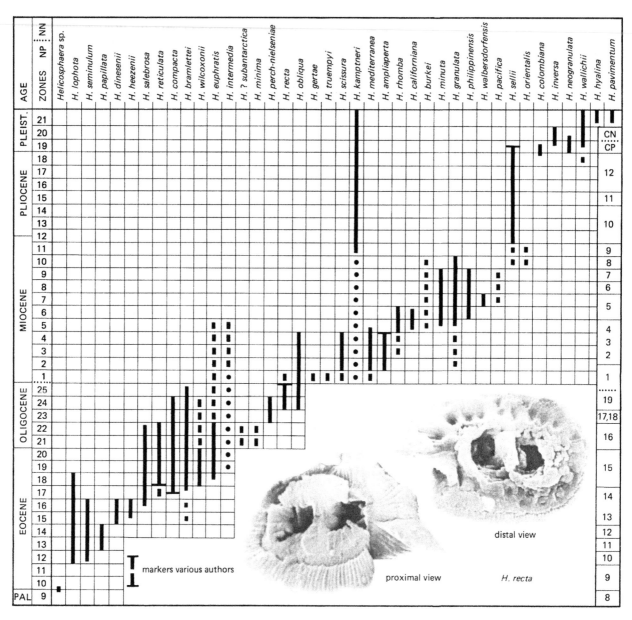

Fig. 42. Distribution of species of *Helicosphaera*

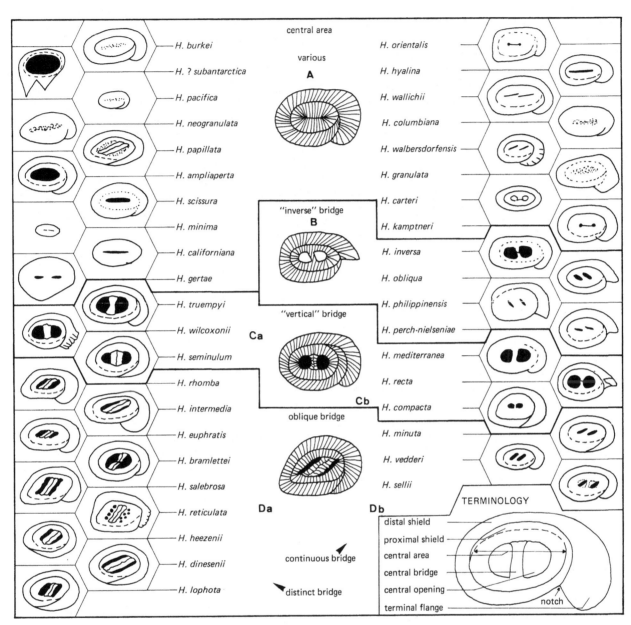

Fig. 43. Terminology and guide to the determination of species of *Helicosphaera* (proximal views)

Paleocene

No name has yet been attached to a Paleocene form of *Helicosphaera*. Haq (1973) illustrated a specimen under the name *Helicopontosphaera* cf. *H. lophota* without giving details of its age (NP 5 or NP 9 or Late Eocene). The SEM picture in Fig. 46.13 from the Paleocene (NP 9) of the South Atlantic, shows a form without a protruding flange, where the terminal flange joins the distal shield to form a long-elliptical to ovoid outline without a notch (see terminology, Fig. 43). The form is very rare and no better SEM could be made to determine the structure of the central area which features a bridge parallel to the minor axis and thus seems similar to *H. seminulum*. The ancestor of *Helicosphaera* is believed to be *Zygodiscus* or possibly *Neocrepidolithus* (Perch-Nielsen, 1981b) when viewed by a specialist in fossil coccolithophorids and to be *Coccolithus pelagicus* when viewed by a specialist of living forms (Gaarder, 1970).

Eocene

All Eocene species, with the exception of the Late Eocene *H. compacta*, have a bridge which appears optically independent from the adjacent parts of the coccolith (Fig. 47). The bridge is parallel to the minor axis in *H. seminulum* and in *H. wilcoxonii*. All Early and Middle Eocene species of *Helicosphaera* are relatively large and egg-shaped in outline with no notch or only a minor, shallow one. In the Late Eocene, several forms with a protruding flange appear and *H. wilcoxonii* evolves from *H. seminulum* through the increase in size of the flange to a sharply terminated, serrate flange with an acute notch. Forms with a spine-like flange appear in NP 15 (Fig. 46.14).

Another Early Eocene form, *H. lophota* has an oblique bridge at an angle of about 45° to the axes and also has an egg-shaped outline without a notch.

The first form with an angle to the major axis that is definitely narrower than 45° is *H. papillata*, which is characterized by a distal side completely covered with small depressions and a nearly horizontal bridge (parallel to the major axis); the outline is ovoid to elliptical without a notch. The species has its FO in NP 13/NP 14. *H. dinesenii* and *H. heezenii* appear in NP 15. *H. dinesenii* has a bridge forming a low angle with the major axis. *H. heezenii* has a bridge parallel or nearly parallel with the major axis. Both species have an egg-shaped to elliptical outline without a notch. They are both relatively large and are regarded by some authors as synonymous.

The protruding flange is evident in Late Eocene representatives of *H. lophota*. In *H. salebrosa* the bridge is oblique and the outline broadly elliptical to rhomboidal, with a nearly straight base-line, with which the side opposite the flange forms an angle of about 90°. The notch is very small and the bridge is connected to the shields not only at its ends but also by thin bars leaving a row of perforations at each side of the

bridge. *H. reticulata* is very similar to *H. salebrosa* and the latter is considered a junior synonym by many authors. Some observed differences between these species include: (1) notch more pronounced and central area larger in *H. reticulata* than in *H. salebrosa* (central area length is about one half the long axis in the former and only about one third the long axis in the latter); and (2) the notch is positioned above the base-line in *H. reticulata* and is an extension of the base-line in *H. salebrosa*. Since the illustration of the holotype is unclear (Bramlette & Wilcoxon, 1967), illustrations of *H. reticulata* by Gartner (1971) are considered typical here. *H. reticulata* seems to appear somewhat later than *H. salebrosa* but their ranges are shown to be about the same, since they have usually not been distinguished and more reliable data are lacking.

Also appearing in the Late Eocene are several broad elliptical species with a relatively small central area and a broad bridge running either parallel or slightly tilted to the minor axis. The flange is not very prominent and the notch is shallow. They can be assigned to *H. bramlettei*, which is characterized in the LM by a 'two-step' or s-shaped bridge oriented about 45° to the axes. Very small forms appear already in the Middle Eocene. Bramlette & Wilcoxon (1967) figured a specimen of *H. bramlettei* under the name of *H. intermedia*. Since their publication has been used by many authors to determine Late Eocene to Miocene *Helicosphaera*, the range of *H. intermedia*, as originally defined by Martini (1965) has probably been extended too far down.

H. euphratis (possible junior synonym: *H. parallela*) is a long-elliptical species with a prominent flange and a notch about at the level of the base-line. The holotypes of the two species are a TEM and a LM picture respectively. *H. parallela* appears to be relatively longer than *H. euphratis*, and its holotype comes from younger (Oligocene or Miocene) sediments than that of *H. euphratis* (Late Eocene, later re-correlated by Haq as Oligocene). The holotype of *H. intermedia* has a large, protruding flange extending below the base-line. The bridge is oblique and forms a steeper angle with the major axis than that in *H. euphratis/H. parallela*. As indicated above, the range of this species, which was described from the Miocene, probably does not go as far down as generally assumed due to misidentification of the species, when Bramlette & Wilcoxon (1967) are followed.

H. compacta is the oldest form without an optically independent bridge and appears in the Late Eocene. It has an egg-shaped form but without a protruding flange and thus without a notch. The base-line is slightly convex and the angle between the base-line and the side opposite the flange approximates 90°, as in *H. reticulata* and *H. salebrosa*, from which it might have evolved through the closing of the central area. Perhaps the bars between the bridge and the shields in *H. salebrosa* are a first step towards the closure of the central area, which is wide open in most older forms.

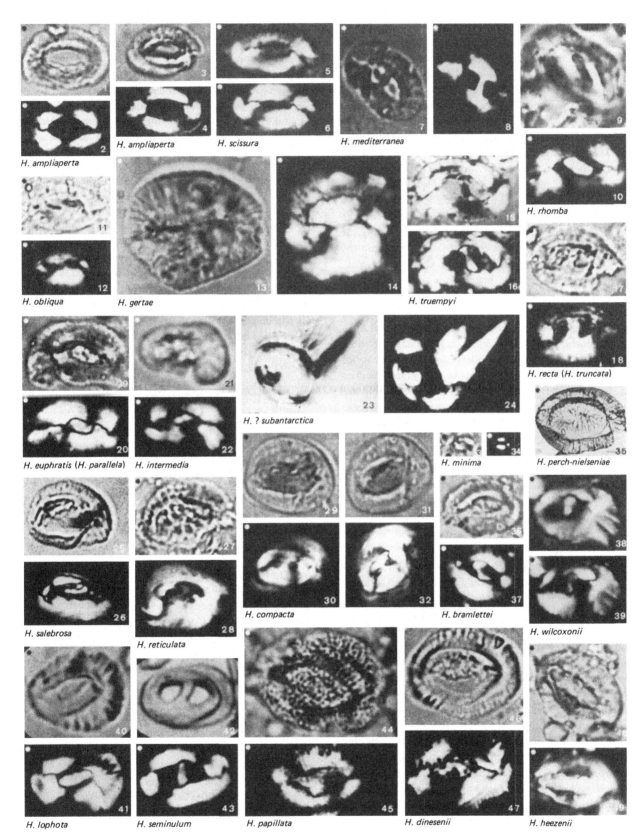

Fig. 44. LM and EM of Eocene through Middle Miocene species of *Helicosphaera*

Oligocene

Essentially two groups of species are present in the Oligocene: *H. compacta* and the species probably derived from it, *H. perch-nielseniae*, *H. obliqua* and *H. recta* (*H. truncata*), and the descendants of *H. lophota*. The coccoliths of the first group are characterized by a proximal central area built of more or less radial elements meeting along the major axis. In *H. compacta* there are only small perforations in the central area. The outline is smooth, without a notch. In *H. perch-nielseniae* there are two narrow slits more or less along the major axis. The flange forms a prominent spine and thus an acute notch above the base-line. The side opposite the flange is strongly convex forming an angle of almost 90° at about the level of the notch. In *H. obliqua* the two slits piercing the central area are oriented inversely to the usual orientation of bridges in *Helicosphaera*, and the flange is relatively small, forming a terminal spine and thus an acute notch in about the middle of the coccolith. *H. obliqua* is more elliptical and thinner than the other species of this group. *H. recta* has two large openings in the central area and a flange terminating in a very prominent, often heavy spine forming an angle of about 90° with the base-line. *H. recta* differs from the Eocene forms with two large openings by having a bridge optically continuous with the central area between the two openings. The bridge is discontinuous in the Eocene forms and in the Miocene species *H. truempyi*. Younger forms with two openings do not have such a prominent spine as termination of their flange. The position of the minute *H. minima* (3–4 μm) in the evolutionary scheme proposed in Fig. 47 is unknown since its fine structure is not known. *H. minima* can hardly be seen in normal light with the LM but shows the typical extinction lines of *Helicosphaera* when viewed between crossed nicols. It seems to have an elongate perforation in the central area and its flange protrudes only very little.

The other group of *Helicosphaera* species in the Oligocene includes forms with optically discontinuous bridges. These are *H. euphratis* (*H. parallela* ?), *H. intermedia* and *H. bramlettei*. Their ancestry is uncertain (see Fig. 47), however, they may have evolved from *H. lophota* either directly or indirectly via *H. bramlettei*. Their characteristics have been discussed with the Eocene forms above.

H.? subantarctica, a large coccolith described from the Lower Oligocene of the Southwest Pacific, has no bridge spanning the wide central opening, and therefore is assigned to *Helicosphaera* with some doubts. It has a large flange consisting of the distal shield only. It has not been reported from low latitudes or from high northern latitudes. Its position in the evolutionary schemes is highly speculative. *H. wilcoxonii* has both a relatively wide central opening and a larger than normal flange. It might be an ancestor of *H.? subantarctica*, though possibly not such a direct one as shown in Fig. 47.

Representatives of both major Oligocene groups cross the Oligocene/Miocene boundary. The LO of *H. recta* is used to define the top of NP 25/base NN 1. Forms assigned to *H. recta*, but possibly representing *H. mediterranea* (see below), have been found to range considerably higher than the LO of *S. ciperoensis* or *Dictyococcites bisectus*, two other species commonly used to approximate the Oligocene/Miocene boundary.

Miocene

The *Helicosphaera* species with an optically discontinuous bridge disappear during the Early or early Middle Miocene; *H. euphratis*/*H. parallela*/*H. intermedia* as well as *H. rhomba*, a species similar to *H. intermedia*. The flange in the holotype of *H. rhomba* is not as prominent as in the holotype of *H. intermedia* and the bridge and adjacent open parts of the central area are larger and straighter in *H. rhomba* than in *H. intermedia*. The correct determination of *H. euphratis*/*H. parallela*/*H. intermedia*/*H. rhomba* is difficult and probably only will be possible once detailed studies both with the LM and the SEM have mapped the changes in various characteristics such as the outline of the coccolith, the size, shape and angle of the bridge, the form and the size of the flange, the size, form and position of the notch and the overall size of the specimens. Until then it would be most suitable to use the oldest of these names, *H. intermedia*.

H. truempyi is the other Early Miocene form with an optically discontinuous bridge. It is a very large form with a wide central opening spanned by a bridge more or less parallel to the minor axis. The flange is well developed and a notch with an angle of about 120° is evident in proximal view between the base-line and the flange. Its origin is unknown, but it could have evolved from *H. bramlettei* via forms showing a widening of the central opening and an increase in the size of the flange. The bridge shows a step-like pattern between crossed nicols similar to *H. bramlettei*. It was described from NN 1 but is also present in NP 25 (KPN, unpublished observation, 1982).

The Early Miocene contains a relatively diverse *Helicosphaera* flora. Besides *H. truempyi*, the *H. intermedia* group and *H. obliqua*, several new forms appear: a group with an essentially elliptical to elongate elliptical outline and an 'open' central area includes: *H. scissura*, a medium-sized form with a small median furrow, *H. ampliaperta*, a large form with a relatively wide, long-elliptical central opening, and the small *H. californiana* with a narrow slit and a length to width ratio of 2 or more. They could have evolved from *H. obliqua* by the elimination of the bridge.

H. mediterranea (junior synonyms *H. crouchii* and *H. transylvania*) has two large openings in the central area and an optically continuous bridge like *H. recta* from which it probably evolved by the loss of the spine in the flange. The latter is quite prominent and forms a notch with the base-line. Reports of *H.*

recta from the Lower Miocene could include some misidentified *H. mediterranea*, in addition to genuine *H. recta*.

H. gertae is another large Early Miocene form. It has a base-line bent at an angle of about 120° and a long median slit. *H. compacta*, its presumed ancestor, has a simple convex base-line and two small perforations instead of a median slit. All these new Early Miocene species disappear during the Early or early Middle Miocene, together with the forms with an optically discontinuous bridge.

The only form surviving into the Late Miocene, and, in fact supposedly through the Holocene is *H. kamptneri/H. carteri*, which is reported to have its FO in NN 1 or NN 2. The early evolution of *H. kamptneri/H. carteri* has yet to be studied in detail. SEM pictures of *Helicosphaera* with a large flange and a long median slit in the proximally continuous but distally granulate central area have been illustrated from NN 1 and NN 2 by Perch-Nielsen (1977) who assigned them to *H. granulata*. Bramlette & Wilcoxon (1967) and Haq (1973) illustrated a similar form as *H.* sp. cf. *H. kamptneri*. Typically, living *H. kamptneri* have two small openings visible on the proximal side and two equivalent pierced depressions on the distal side. Clearly, the two holes are not artifacts of preservation but are integral structural features. Such distal pores are not present in illustrations of Early Miocene forms, which typically have a granular central structure. The proximal side of the central area is divided by a long median slit. Distal views of species belonging to the *H. obliqua* group also show a granulate central area. An independent small bridge is visible on the proximal side with the LM under crossed nicols. By accepting an Early Miocene FO for *H. kamptneri* we accept a wider species variability in this species than is useful.

H. burkei is another species in the size-range of *H. kamptneri*. It was described from the Middle Miocene and possibly is a synonym of *H. granulata*. Its holotype is the TEM picture of a proximal view (Fig. 45.33), while *H. granulata* was originally described by LM (Figs. 45.39, 40). Specimens of *H. granulata* from the type material were illustrated by SEM by Haq (1973) and show a granulate distal central area and a somewhat irregular arrangement of the radial elements of the proximal central area, not unlike the one in *H. burkei*. Whereas

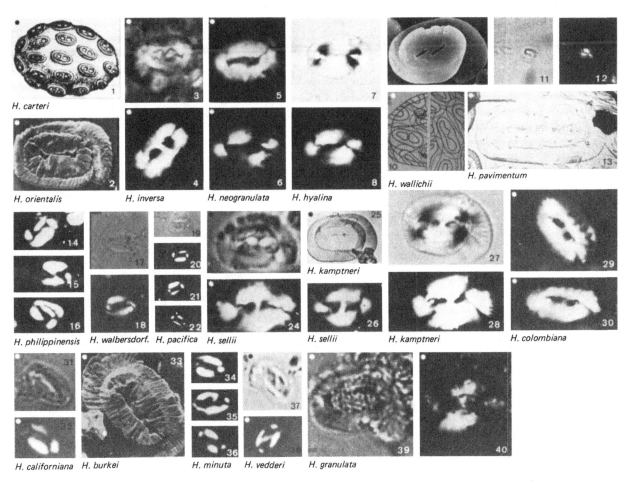

Fig. 45. LM and EM of Middle Miocene through Recent species of *Helicosphaera*

the flange is prominent in *H. granulata* and protrudes below the base-line, it extends only to about the base-line level in *H. burkei*. Thus we could consider a lineage from *H. obliqua* group to *H. granulata* with large flange and on to *H. burkei* with a smaller flange but still a median slit instead of the two openings of *H. kamptneri* and finally *H. kamptneri* with two small openings.

Three small, 3.5–7 μm long species have been described from the Middle Miocene. They are most easily visible in the LM with cross-polarized illumination but difficult to discern in bright field. *H. philippinensis* has a somewhat rectangular outline and a moderately protruding flange. The two central openings are difficult to see with the LM. *H. walbersdorfensis* has a flange extending distally to the side opposite the flange

and two slightly oblique slits which are not easily distinguishable by LM. *H. minuta* has an elliptical outline and two oblique slits like *H. walbersdorfensis*, but the flange is less prominent in *H. minuta*, in which there is no notch in the base-line. The bridge between the slits is continuous with the proximal part of the central area and is bright when the coccolith is in a position with the major axis at 45° to the crossed nicols. Also this form is difficult to view in normal light. *H. vedderi* has been considered a junior synonym of *H. minuta* by Bukry (1981b) but can be distinguished from *H. minuta* by its larger openings and better-developed bridge according to Prins (personal communication, 1983).

One small species, *H. orientalis* of about 4 μm in length and one middle-sized form, *H. sellii*, appear in the Late

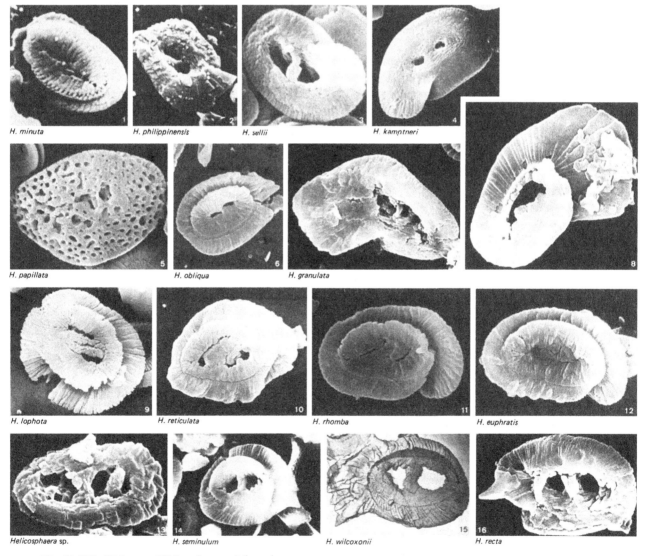

Fig. 46. EM of Paleocene (*Helicosphaera* sp.) through Recent species of *Helicosphaera*

Miocene. *H. orientalis* was illustrated by a single proximal view (TEM) and is a nearly rectangular form with a femur-like central opening and a moderately large flange forming a small notch on the level of the base-line. The only other such rectangular, small form is *H. philippinensis*, which seems to have a larger flange and inversely oriented slits on the proximal side. Like the latter species and many other forms of *Helicosphaera*, *H. sellii* also shows quite different structures on the distal and on the proximal side. Distally, two large openings are separated by a bridge which runs more or less parallel to the minor axis; proximally, an oblique bridge separates the two oblique slits. This can be demonstrated with the LM when the focus is carefully changed from the distal to the proximal side. The two bridges are illustrated in Fig. 46.3 in a distal view, where one can see the proximal bridge and the slits through the large holes of the distal side.

The relationship between these Middle and Late Miocene small species has not been worked out in detail. It seems possible that they derived from *H. granulata*, since it was the only *Helicosphaera* present in the early Middle Miocene besides the small *H. californiana*. In either case one would observe the re-occurrence of a bridge after the bridge had disappeared in both lineages earlier in the Miocene. *H. philippinensis*, which has an inversely oriented bridge, could have evolved from *H.*

obliqua although their ranges, as known today, are separated by a full zone (NN 5).

Pliocene

Usually only *H. kamptneri* and *H. sellii* are distinguished in Pliocene sections. Specimens of *H. kamptneri* with a reasonable resemblance to the (Recent) holotype only appear in the Late Miocene/Early Pliocene. They are characterized by two small perforations of equal position and orientation on both sides of the coccolith and a flange meeting the base-line with only a relatively small angle. *H. sellii* was discussed above. It is assumed here that some of the Pleistocene species, most likely *H. colombiana* and *H. wallichii*, already evolved from *H. sellii* during the Late Pliocene.

Pleistocene

The *Helicosphaera* species of the Pleistocene have been studied by Gartner (1977b), who introduced three new species: *H. colombiana*, *H. neogranulata* and *H. inversa*. *H. colombiana* 'is a species with a slightly asymmetrical ovoid outline. The centre of the species has two depressions aligned with the longitudinal axis' (Gartner, 1977b). The bridge between the depressions or perforations is optically continuous with the adjacent parts of the coccolith. The distinction from *H. sellii* is not clear (to me) and was not given by the author. The

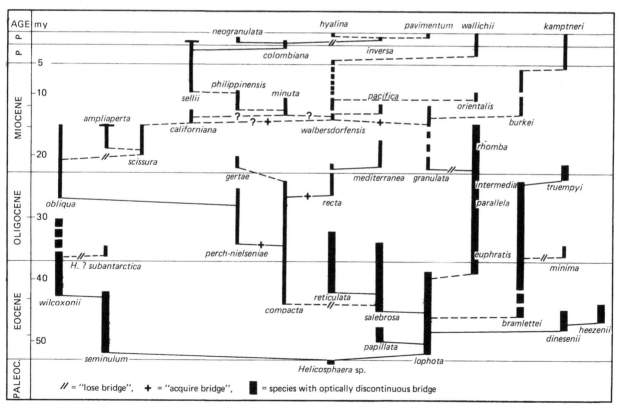

Fig. 47. Tentative evolutionary lineages in *Helicosphaera*

perforations seem smaller in *H. colombiana* than in *H. sellii*. *H. neogranulata* 'has an ovoid rather than elliptical outline with one end more broadly rounded than the other. A broad longitudinal slit is outlined at the centre of each specimen' (Gartner, 1977b). *H. neogranulata* is not closely related to the Early and Middle Miocene *H. granulata* but probably evolved from *H. colombiana* or *H. kamptneri*. It differs from *H. granulata* in having a much narrower granular area at the centre. *H. inversa* 'is an elliptical species with parallel sides and similarly rounded ends. The centre of the species has two large perforations which are separated by a diagonal bar' (Gartner, 1977b). The bar is optically continuous with the surrounding parts of the coccolith, but oriented inversely to the usual orientation of *Helicophaera* bridges.

While *H. sellii* disappears in the Early Pleistocene, *H. kamptneri* continues and is found in the modern plankton together with *H. wallichii*, a form with two oblique slits, *H. hyalina* with one, long central slit and the small *H. pavimentum* with two very small perforations or none at all.

Paleoecology

Many species of *Helicosphaera* seem to occur most commonly and consistently in hemipelagic sediments and are not found in pelagic sediments deposited at great distances from a land mass or from marine highs. It is possible that they are restricted to, or at least seem to prefer, upwelling areas. In several DSDP sites, *Helicosphaera* species are present, disappear and reappear without obvious reason. The best known case is the one of *H. ampliaperta*, the Early Miocene marker species, which is not found in large areas of the Pacific and is very rare in many other open ocean areas. Such patterns will have to be studied in order to utilize fully the stratigraphic potential of the many *Helicosphaera* species.

Conclusions

Relatively few of the 40 species of *Helicosphaera* have become officially used as markers in zonal schemes: *H. compacta* and *H. reticulata* were used by Gartner (1971), *H. recta* and *H. ampliaperta* by Martini (1971) and Bukry (1973a), *H. carteri* (*H. granulata?*) by Huang (1979) and *H. sellii* by Gartner (1977b). It seems likely that more species could be used, at least for local correlations.

Since Haq (1973) published an overview of the *Helicosphaera* species and made a first attempt at the recognition of evolutionary trends, the number of species has almost doubled. Their distinction can be difficult, especially when the assemblage is poorly preserved. Next to nothing is known about the variability of the species at any one level or through time. Since detailed studies of successions of assemblages with rich *Helicosphaera* floras are missing, it is important that the species concepts are based on the holotypes wherever possible, and not on subsequent illustrations and determinations.

The evolutionary trends shown in Fig. 47 are based partly on more or less obvious trends and similarities and partly on the possibilities available from the known ranges of species. Only in a very few cases have they been observed as transitions from one species to another. This important task still has to be done.

Lithostromationaceae Deflandre (1959)
Figure 48

The Lithostromationaceae include nannoliths that are similar to asteroliths; their outline is triangular, hexagonal or nearly circular and they are usually relatively large. They have symmetrically arranged small depressions of varying form and size. The genera *Lithostromation*, *Martiniaster* and *Trochoaster* have been included in this family by Tappan (1980). Bybell (1975) suggested that the species of *Trochoaster* be included in *Lithostromation*, because the species included all 'have numerous circular depressions with a surrounding hexagonal ridge, no discernible crystals, and three-way symmetry (or a multiple)' as is typical for *Lithostromation* and its type *L. perdurum* (Figs. 48.4, 12). *Martiniaster* (nom. substitutum pro *Coronaster*) with its type *M. fragilis* is characterized by 12 radially arranged ridges and symmetrically arranged depressions (Fig. 48.5).

The observation of the Lithostromationaceae species with the LM is difficult, since they are very high and thus their illustration with LM pictures is not satisfactory. Flat-lying specimens do not show birefringence between crossed nicols. Species of Lithostromationaceae are usually rare, with the exception of *L. simplex* in some Middle Eocene samples-according to Martini (1961b), who studied the stratigraphic distribution of the Lithostromationaceae. Aubry (1983) found relatively common *Lithostromation* in the Oligocene of the Paris Basin and fairly common occurrences are known from the Middle Miocene of some nearshore Gulf Coast sediments.

The oldest species of *Lithostromation*, *L. simplex*, is reported from the Early Eocene and the species illustrated in Fig. 48 all occur in the Middle Eocene. Late Eocene to Late Pliocene occurrences may or may not be due to reworking. An overview of the species of the Lithostromationaceae, many of which are doubtful, is given in Perch-Nielsen (1971a) and in Aubry (1985). The species of Lithostromationaceae are mainly found in hemipelagic sediments and are very rare (displaced?) in open ocean deposits.

Pontosphaeraceae Lemmermann (1908)
Figures 49–54

The Pontosphaeraceae are characterized by a usually large central area and a wall consisting of two cycles of elements (zygoid). The elements are arranged more or less radially on the proximal and more or less concentrically on the distal side (Fig. 49). The wall can be very low to very high and the floor pierced by perforations or holes or thinned by depressions. The following genera are commonly assigned to the Pontosphaeraceae:

Calciopilleus(?): calyptra-shaped with transverse ribs; Late Miocene (Fig. 50).

Crepidolithus/Neocrepidolithus: the Jurassic *Crepidolithus* has vertical wall elements, the Cretaceous/Paleocene *Neocrepidolithus* has oblique wall elements. The proximal side is covered by more or less radially oriented elements with or without additional elements in the central area, the distal side is completely filled with the wall elements in older forms and by a central area composed of tangentially or concentrically arranged elements in Late Paleocene forms. The species are distinguished by the arrangement of the elements of the central area and by the relative width of the two wall

cycles (for details see Perch-Nielsen, 1979a and 1981a); Jurassic (*Crepidolithus*) and Cretaceous to Late Paleocene (*Neocrepidolithus*).

Nannocorbis(?): tube-shaped, with rims turned outwards at both ends; Late Miocene to Early Pliocene (Fig. 50).

Pontosphaera. This includes *Crassapontosphaera*, a genus considered to be based on an overgrown form of *Pontosphaera*; *Discolithina* and/or *Discolithus*, two genera which long served for any flat coccolith not

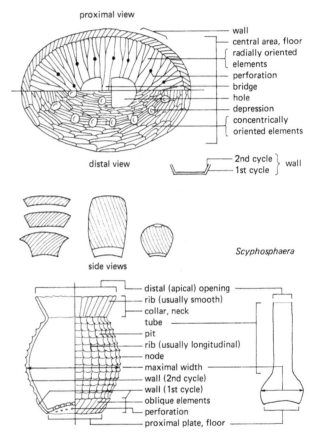

Pontosphaera, Transversopontis

Fig. 49. Terminology of the Pontosphaeraceae

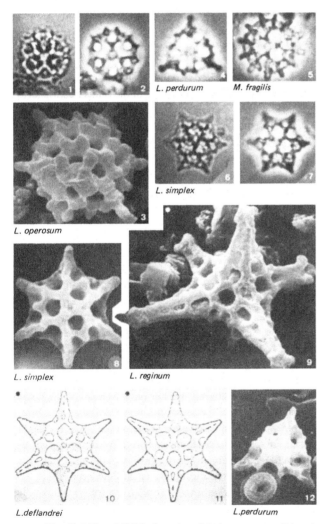

L. perdurum   *M. fragilis*

L. operosum

L. simplex

L. simplex   *L. reginum*

L.deflandrei   *L.perdurum*

Fig. 48. LM and SEM of species of *Lithostromation* (*L.*) and *Martiniaster* (*M.*)

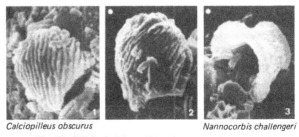

Calciopilleus obscurus   *Nannocorbis challengeri*

Fig. 50. SEM of *Calciopilleus obscurus* and *Nannocorbis challengeri*, two genera and species possibly belonging to the Pontosphaeraceae

one disregards modern forms with a similar wall as in the Zygodiscaceae, which are assigned to *Coronosphaera* Gaarder (*in* Gaarder & Heimdal, 1977), which was included in the Syracosphaeraceae by Tappan (1980). The genera and their species are discussed in alphabetic order below.

Chiastozygus

Chiastozygus is one of the few genera which survived the Cretaceous/Tertiary boundary event(s). Its wall is simple and strong and it bears a more or less X-shaped bridge. It is represented only by *C. ultimus* in the Early Paleocene (Figs. 78.54, 55; 79). *C. ultimus* has a central cross aligned with the axes or at a slight angle with them.

Chiphragmalithus

Chiphragmalithus is characterized by a high, indented wall, irregular outline and a central X or + reaching higher than the wall. The rim can be narrow but typically is flaring in *C. barbatus* (Figs. 78.7–10; 79) and to a lesser extent in *C. acanthodes* (Figs. 78.19; 79). *C. armatus*, the oldest form, is elliptical while the other species are circular to subcircular (Figs. 78.16–18). Species of *Chiphragmalithus* are mainly found in the Early Eocene NP 12/CP10 to NP 14, but their ranges are not very well known. *C. armatus* is thought to have evolved from *N. dubius*. The other species could have evolved from *C. armatus* (Fig. 78) by an increase in height of the flaring wall and a transformation of the central structure from the H of *C. armatus* to the cross in the other species. Alternatively, *C. calathus* could have evolved from *N. protenus* by acquiring a flaring, high wall as did *C. armatus* from *N. dubius*.

The genus *Nannotetrina*, which evolved late in NP 14, is thought to have evolved from *Chiphragmalithus* by the growth of the arms of the central cross beyond the wall and by a reduction of the wall (see *Nannotetrina*).

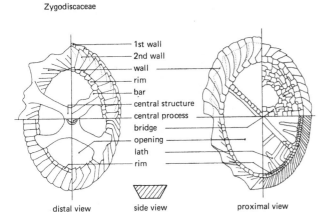

Zygodiscaceae

1st wall
2nd wall
wall
rim
bar
central structure
central process
bridge
opening
lath
rim

distal view   side view   proximal view

Fig. 77. Terminology of the Zygodiscaceae

Isthmolithus

Isthmolithus has a long elliptical to rhomboedrical outline and one or several bridges parallel to the minor axis. *I. recurvus*, the most common species of the genus and stratigraphically important in the Late Eocene/Early Oligocene has two parallel bridges each consisting of two parts joining along the major axis with a slight thickening. The identification of strongly overgrown specimens which tend to assume a calcite crystal-like shape with partly or wholly filled openings is still possible (Figs. 78.1, 2; 79). *I. rhenanus* has a central plate-like structure instead of parallel bridges and *I. unipons* has only a single bridge (Figs. 78.3 and 78.5, 6).

Lophodolithus

Lophodolithus is characterized by two thin walls and an asymmetrical outline. Most species have an I-shaped bridge as in the type species *L. mochloporus* (Figs. 80.5, 6). *L. mochloporus* has an ovoid outline with the wall elements of the wider end clearly visible with the LM. *L. nascens* (Figs. 80.12, 13; 81.9) is the only representative of *Lophodolithus* appearing in the Paleocene and has only a slightly asymmetrical outline with one end of the ellipse slightly narrower than the other. It is thus closely related to *Zygodiscus adamas* (see below), from which it evolved. *L. reniformis* (Figs. 80.14, 15) has a kidney-shaped outline and *L. acutus* (Figs. 80.3, 4; 81.7) has walls expanding into a flange on one side and one end, thus producing an asymmetric outline. The narrow side and end of the walls 'meet at a distinctive right angle or slightly acute angle that contrasts the broadly rounded rim at the diagonally opposite position of the rim' (Bukry & Percival, 1971). *L. rotundus* (Figs. 80.1, 2; 81.8) has an asymmetrical enlargement of the flange with well visible wall elements. The rim elements build a floor-like structure much like in a *Pontosphaera*. *L. rotundus* has a subcircular rim and a subcircular floor opening but no bridge as the other species of the genus.

Neochiastozygus

Neochiastozygus was erected for elliptical coccoliths with two walls of inclined and/or vertical elements and a central cross or X. Similar forms with only one wall are assigned to *Chiastozygus*, a Mesozoic genus only just reaching the Early Paleocene and *Neococcolithes*, a mainly Eocene genus. Some species of *Neochiastozygus* can be used for the detailed subdivision of the Early and early Late Paleocene: *N. modestus*, *N. saepes* and *N. perfectus* (Fig. 6). The distinction of the *Neochiastozygus* species by LM is usually best attempted with crossed or halfway crossed nicols. In *Neochiastozygus* the arms of the central structure consist of two parallel laths clearly visible with crossed nicols. The arms of species of *Neococcolithes* consist of a single lath and thus appear homogenous. LM pictures of all species and their ranges are assembled in Fig. 78 and SEM illustrations can be found in

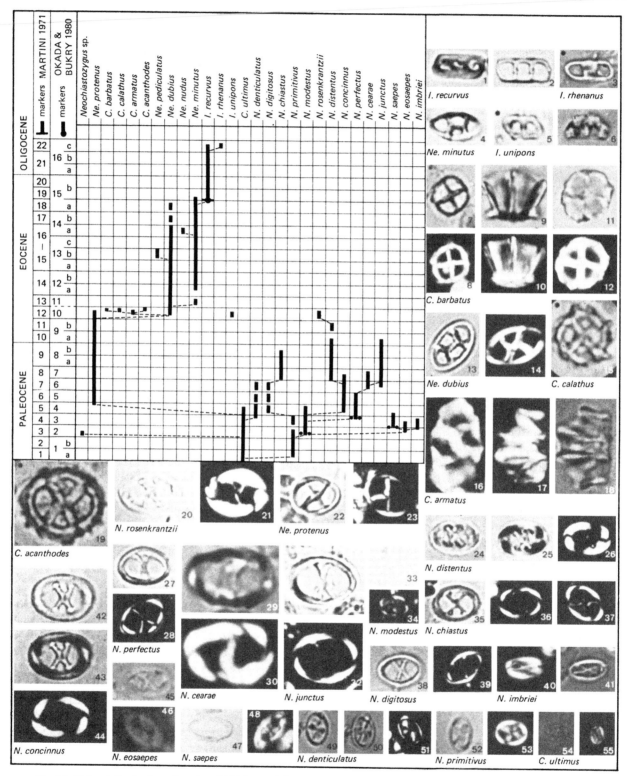

Fig. 78. LM and distribution of species of *Chiphragmalithus*
(*C.*), *Chiastozygus ultimus*, *Isthmolithus* (*I.*), *Neochiastozygus*
(*N.*) and *Neococcolithes* (*Ne.*) and their evolutionary
relationships (tentative; partly after Perch-Nielsen, 1981b)

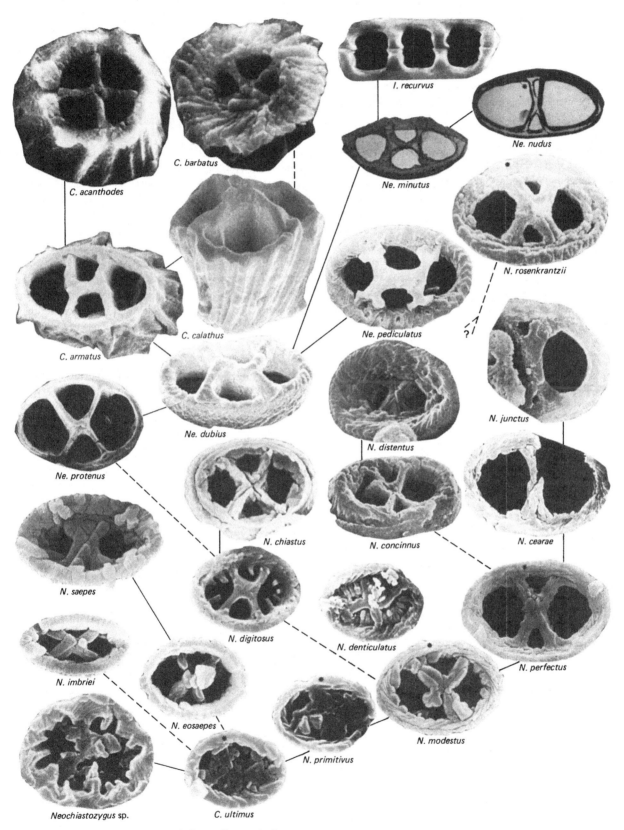

Fig. 79. EM and tentative evolutionary lineages in the genera *Chiphragmalithus* (*C.*), *Chiastozygus ultimus*, *Isthmolithus* (*I.*), *Neochiastozygus* (*N.*) and *Neococcolithes* (*Ne.*)

Perch-Nielsen (1969, 1971c, 1977, 1979a and 1981b, where also an evolutionary scheme was suggested). The species are discussed in the order in which they appear (Figs. 78 and 79).

N. primitivus (Figs. 78.52, 53; 79) includes small, elliptical coccoliths with a well developed first or outer wall of inclined elements and a less developed, lower, second or inner wall. The central cross is asymmetrical with bars of unequal length forming an angle of about 90° in the centre. The longer bars form an angle of a few degrees to about 40° with the long axis. The secondary laths extending from the bars are often overgrown giving the bars a bulky appearance. In *N. denticulatus*, a similar species, the laths extend from the rim of the coccolith and not from the bars as in *N. primitivus*. Together with *C. ultimus*, *N. primitivus* represents the cross-bearing coccoliths of the family in NP 1/CP1a.

N. modestus (Figs. 78.33, 34; 79), the FO of which defines a subzone of NP 3 is characterized by two equally well developed walls and a nearly, but not perfectly, symmetrical central X. Also, *N. modestus* is a small coccolith compared with the later forms of the genus (Fig. 78). It differs from *N. primitivus*, from which it evolved, by the nearly equal length of the arms of the central cross and the fully developed second, inner wall, which can be recognized also with the LM between crossed nicols. The inner wall appears brighter than the outer one. The general impression of this coccolith is coarser and less elegant, than the one of the younger *N. perfectus*, which is also larger.

N. imbriei, *N. eosaepes* and *N. saepes* (Figs. 78.40, 41, 45, 46, 47, 48; 79) all have a first wall consisting of vertical elements and a very low second wall consisting of slightly inclined elements or no second wall at all. In *N. imbriei* the outline is long-elliptical and the central cross consists of bars forming a very acute angle with the long axis. In *N. eosaepes* the central cross is asymmetrical, with bars of unequal length forming an angle of about 90° oriented more or less diagonally. The outline is elliptical. In *N. saepes*, the FO of which defines a subzone around the NP 3/NP 4 boundary, the outline is long-elliptical with pointed ends and the central cross symmetrical and narrow, the arms forming an angle of less than 90°. Whereas *N. eosaepes* measures about 4–5.6 μm, *N. saepes* reaches 5.4 to over 6 μm in the type material from Denmark. Between crossed nicols, the extinction pattern of the walls is the same for the three species, with straight instead of the curved extinction lines as in the other *Neochiastozygus* species; compare Figs. 78.48 and 78.32.

N. perfectus (Figs. 78.27, 28; 79), the FO of which defines a subzone near the NP 4/NP 5 boundary is characterized by a rim consisting of two fully developed, but relatively thin walls. The central cross is more or less symmetrical and narrow. The inner wall appears brighter between crossed nicols than the outer one. Most older *Neochiastozygus* species are smaller than *N. perfectus*, many younger ones of equal size

or larger. *N. perfectus* differs from the younger *N. chiastus* by its straight arms as opposed to the curved ones of *N. chiastus*. It differs from *N. distentus* by the much narrower second, inner wall (compare Fig. 78.28 with 78.26) in *N. perfectus* and the wider openings in that species. *N. junctus* is larger than *N. perfectus* and has a very thin first, outer wall, which is barely visible with the LM, while the outer wall can be seen in *N. perfectus*.

N. denticulatus (Figs. 78.49–51; 79) is another small form with a central cross forming a low angle with the axes. Laths extend from the rim towards the centre, which distinguishes this species from *N. primitivus*, where they extend from the bars of the central cross.

N. digitosus (Figs. 78.38, 39; 79) is a small, delicate form with two relatively thin, but fully developed walls and a central cross with slightly curved bars. Laths extend from the rim. The inner wall appears brighter between crossed nicols than the outer wall and is less prominent than in the younger and larger *N. chiastus*. The central cross is narrower in *N. digitosus* than in *N. chiastus*, where the bars meet at an angle of about 90°, while this angle is less in *N. digitosus*.

N. concinnus (Figs. 78.42–44; 79) is a relatively large form with a very thin outer wall and a prominent inner wall. The central cross is more or less symmetrical, the bars on each side of the minor axis forming an angle of usually less than 90°. Typically the bars meet the inner wall with a small bifurcation which is clearly visible with the LM (Figs. 78.42, 43), and unique to this species of *Neochiastozygus*.

N. junctus and *N. cearae* (Figs. 78.29–32; 79) are two large forms with narrow central crosses. Whereas the central cross is built by small elements in *N. junctus*, it consists of larger blocks in *N. cearae*. In the latter the cross is so narrow that it leaves only the two large openings instead of the four found in all other species of this genus. *N. cearae* shows features of *Zygodiscus* and early *Transversopontis*. In these genera, the connection between the bridge and the walls consists of bridge elements piercing the wall with an arrow-like structure, whereas in *N. cearae* they do not.

N. distentus (Figs. 78.24–26; 79) has a very prominent inner wall and only a very thin outer wall. The width of the inner wall is not a result of overgrowth but of the presence of many layers of more or less tangentially oriented elements forming it. Between crossed nicols the inner wall appears bright. The central cross accordingly is smaller than in other species of comparable size and consists of parallel blocks forming the bars. A slight bifurcation can be present but is not as well developed as in *N. concinnus*, where the openings are much larger than in *N. distentus*.

N. chiastus (Figs. 78.35–37; 79) has a narrow outer wall and a slightly thicker inner wall. At least two of the four bars of the central cross are curved, a structural feature which is clearly visible with the LM and gives the species its name. It is best seen when the specimen is oriented at 45° to the crossed nicols.

   N. rosenkrantzii (Figs. 78.20, 21; 79) is the only species of the genus *Neochiastozygus* to appear in the Eocene and the last one to disappear. It is characterized by a high outer and a relatively prominent, but low inner wall. The central cross is more or less symmetrical and consists of bars built of parallel blocks meeting at 90° or a lower angle. *N. rosenkrantzii* is very similar to *N. concinnus* in also showing a slight bifurcation at the end of the bars, but differs from the latter in having a lower

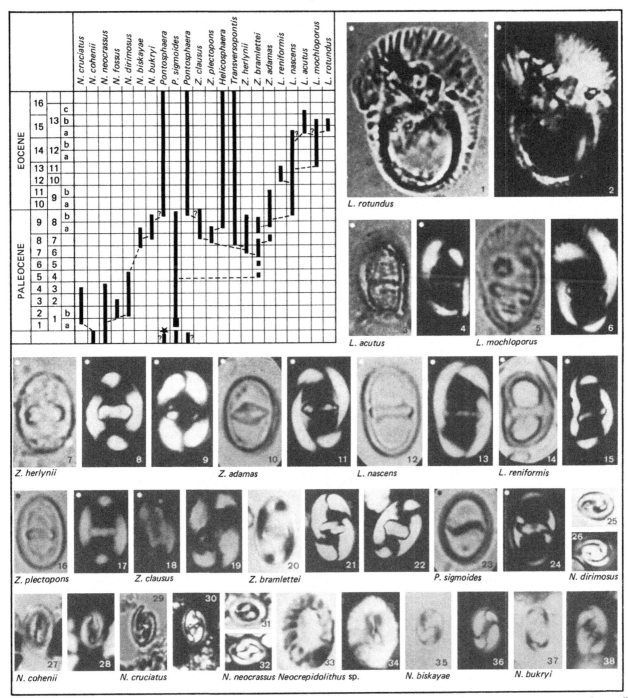

Fig. 80. LM and distribution of species of *Lophodolithus* (*L.*), *Neocrepidolithus* (*N.*), *Placozygus* (*P.*) and *Zygodiscus* (*Z.*) and their (tentative) evolutionary links with each other and with *Helicosphaera* and *Pontosphaera*. * = *Prolatipatella multicarinata*

inner wall of not so regularly arranged elements as in *N. concinnus.*

While the distinction of the *Neochiastozygus* species is not easy in poorly preserved assemblages, it should be attempted in extended Paleocene sections with well preserved calcareous nannofossils to refine the stratigraphic subdivision already established by means of other species.

Neococcolithes

Neococcolithes is a genus including elliptical coccoliths with a single wall and a central X- or H-shaped cross. The most common species is *N. dubius*, a coccolith with an H-shaped central structure (Figs. 78.13, 14; 79) which is known from the Early and Middle Eocene and disappears in the Late Eocene. Similar forms are *N. minutus* (Figs. 78.4; 79), a smaller form with a long-elliptical outline and pointed ends and an H-shaped central structure and *N. nudus* (Fig. 79) a form with a very narrow, X-shaped central cross. *N. pediculatus* (Fig. 79) has an H-shaped central structure with bifurcating bars near the wall and *N. pyramidus* represents an overgrown *N. dubius* and therefore should be suppressed. *N. protenus* (Figs. 78.22, 23; 79) is the only Paleocene species assigned to *Neococcolithes*. It has an X-shaped central cross which is more or less irregular with bars of equal or slightly unequal length. *N. protenus* is considered to be the ancestor of the Eocene representatives of the genus.

Placozygus

Placozygus is one of the few genera that survived the Cretaceous/Tertiary boundary event(s). It is represented by the species *P. sigmoides*, previously often listed as *Zygodiscus sigmoides*. *Placozygus* has a single wall constructed of inclined, block-shaped elements and an I-shaped bridge. *P. sigmoides* (Figs. 80.23, 24; 81.1, 2) which is found already in the Maastrichtian, is characterized by its I-shaped bridge. The bridge can bear a central process and joins the rim with an arrow-like extension on the proximal side (Fig. 81.2) and with a bifurcation in the form of 'feet' on the distal side (Fig. 81.1). Between crossed nicols, these 'feet' are bright when the long axis of the specimen is oriented at 30° to the crossed nicols (Fig. 80.24). *P. sigmoides*, while present in the Maastrichtian, seems to be extremely rare in the very basal Paleocene but reaches an acme in the middle of NP 1 in high latitudes. Its size increases from about 4 to 5 μm in NP 1 to over 10 μm in NP 9 where it disappears.

Zygodiscus

Zygodiscus includes those coccoliths with an I-shaped bridge that have a regular elliptical outline and two very thin walls. *Placozygus* has only one massive wall and *Lophodolithus* does not have a regular elliptical outline. *Zygodiscus* species are numerous in the Late Paleocene but only *Z. adamas*, the type species extends into the Early Eocene. Many Mesozoic species have also been assigned to *Zygodiscus* (see *Farinacci Catalogue*) but most of them only have one wall and should be assigned to other genera. *Z. bramlettei* (Figs. 80.20–22; 81.3, 4) has a very thin outer wall but a relatively thick inner wall and the openings are surrounded by a ring of elements. In *Z. herlynii* (Figs. 80.7–9) the openings are smaller than in *Z. bramlettei* and the bridge is diamond-shaped. *Z. plectopons*

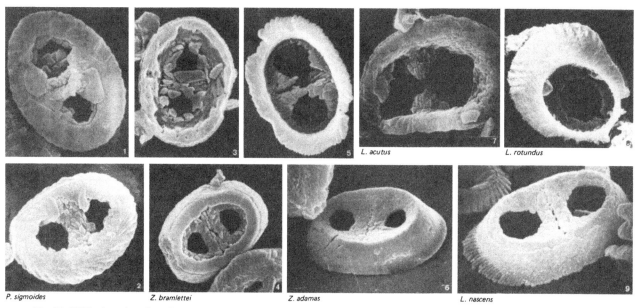

L. acutus   *L. rotundus*

P. sigmoides   *Z. bramlettei*   *Z. adamas*   *L. nascens*

Fig. 81. SEM of species of *Lophodolithus* (*L.*), *Placozygus* (*P.*) and *Zygodiscus* (*Z.*)

(Figs. 80.16, 17) has medium-sized openings but a relatively small diamond-shaped bridge and *Z. clausus* (Figs. 80.18, 19) has very small openings and a very wide bridge. *Z. adamas* (Figs. 80.10, 11; 81.6) has relatively high, thin walls, a well developed, diamond-shaped bridge and large openings. SEM illustrations of most of these species are given in Perch-Nielsen (1981a).

Evolution

Evolutionary lineages for *Neochiastozygus* and related forms were suggested by Perch-Nielsen (1981a) and are shown in Fig. 78 and illustrated in Fig. 79. Romein (1979) proposed some relationships in *Zygodiscus* and *Lophodolithus* as did Perch-Nielsen (1981a). A tentative scheme is given in Fig. 80, where also the questionable connection to the Late Paleocene forms of *Pontosphaera* is shown.

Incertae sedis

Besides the genera discussed above within the more or less well established Tertiary families, a few genera remain which seem unrelated to these families. They are discussed below in alphabetic order.

Ellipsolithus (Figs. 82.1–14)

The species of *Ellipsolithus* are characterized by their bright distal shield when viewed between crossed nicols, the elliptical to long-elliptical outline and the elongate central area. The proximal shield is usually broken off. The coccoliths are often difficult to see in normal transmitted light due to the thinness of the distal shield. The species are distinguished by the organization of the elements in the central area.

E. macellus (Figs. 82,7–9, 13, 14), the type species, is also the oldest form and its FO defines the base of NP 4 or CP3. Its central area is completely covered and the size and outline of the specimens vary considerably. *E. bollii* (Figs. 82.1–3, 10, 11), characterized by a central ridge and a high number of small depressions or holes in the central area, appears shortly after *E. macellus* and disappears again in NP 5. *E. distichus* (Figs. 82.4, 12) has a wider central area than the other species and relatively large elongate, radially oriented openings in it. It ranges from the lower part of NP 4 through NP 9. Since it seems to be more solution resistant than *E. macellus*, its FO has been used to approximate the lower boundary of NP 4. *E. lajollaensis* (Figs. 82.5, 6) has 'rounded ends and long parallel sides' and a relatively small, elongate central area with fewer perforations than in *E. distichus* or *E. bollii*. It is found in the Middle Eocene.

Florisphaera (Figs. 83.1–3)

Florisphaera includes plate-shaped coccoliths overlapping each other to about 2/3 or 1/2. The outline is generally rectangular, with a tendency to zig-zagging at the shorter ends. The plates are slightly bent. The single coccoliths of *F. profunda* look like thin, small plates and show only very faint birefringence when viewed between crossed nicols (Fig. 83.1). *F. profunda* was found in the Atlantic and the Pacific tropical to transitional watermasses by Okada & McIntyre (1977). Okada (1984) found *F. profunda* to be a depth-sensitive species, which became a major component of the flora in the Late Miocene.

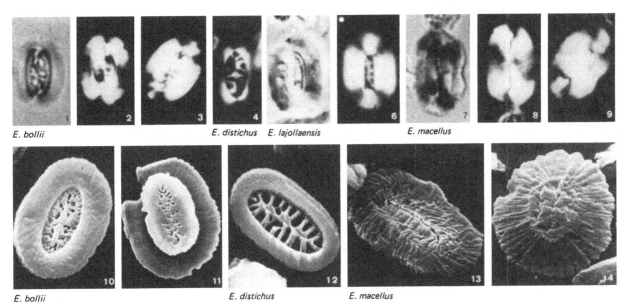

E. bollii   E. distichus   E. lajollaensis   E. macellus

E. bollii   E. distichus   E. macellus

Fig. 82. LM and SEM of species of *Ellipsolithus*

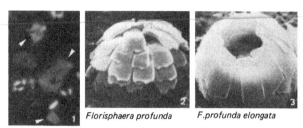

Florisphaera profunda   F.profunda elongata

Fig. 83. LM and SEM of *Florisphaera profunda*

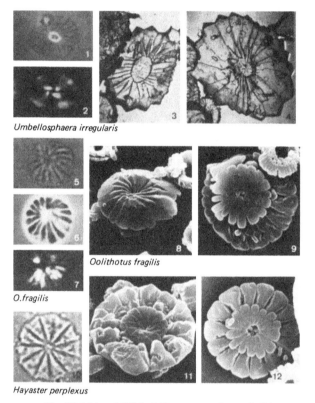

Umbellosphaera irregularis

Oolithotus fragilis

O.fragilis

Hayaster perplexus

Fig. 84. LM and SEM of *Hayaster perplexus, Oolithotus fragilis* and *Umbellosphaera irregularis*

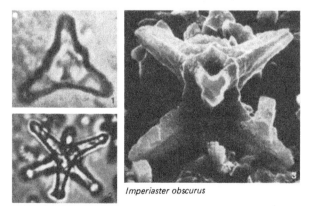

Imperiaster obscurus

Fig. 85. LM and SEM of *Imperiaster obscurus*

Hayaster (Figs. 84.10–12)

Hayaster includes flat coccoliths with an 'approximately circular outline modified by straight peripheral edges of several triangular-outlined crystallites composing' the distal shield. Its type species *H. perplexus* (Figs. 84.10–12) was originally described as a *Discoaster* species, but SEM investigations reveal a small proximal shield. Both shields remain dark between crossed nicols and no separate central area exists. *H. perplexus* appears in the Late Oligocene according to Ellis (1975) and was found in the North Pacific Central Gyre by Reid (1980) and Okada & McIntyre (1977).

Imperiaster (Figs. 85.1–3)

Imperiaster consists of calcareous nannofossils built of two superimposed groups of three arms. The generotype *I. obscurus* (Figs. 85.1–3; syn. *Marthasterites reginus*) was described from the Lower Eocene and has been found since in NP 11 and NP 12. *M. reginus*, which was also described from the Lower Eocene, was assigned to *Lithostromation* by Bybell (1975) based on the specimen illustrated (Fig. 48.9) which was found in the upper Middle Eocene NP 16. '*L. reginus*' thus is likely not to be related to *I. obscurus*, but rather to *L. deflandrei* (Figs. 48.10, 11).

Lapideacassis and Scampanella (Figs. 86; 87.1–6)

The species of *Lapideacassis* and *Scampanella* are rare in Paleocene sediments of the South Atlantic and very rare in Paleocene sediments of the North Sea area and the Eocene of the Atlantic. Originally described from the Cretaceous, the two genera cross the Cretaceous/Tertiary boundary without modifications.

In *Scampanella* only the first distal tier is present and constitutes more than half the length of the body of the fossil (see Fig. 86 for terminology). In *Lapideacassis* the number of distal tiers is greater than one. The species in both genera are distinguished by the number of spines, the presence of an apical process and in *Lapideacassis* also by the number of tiers. An overview was given by Perch-Nielsen & Franz (1977).

Minylitha (Figs. 88.2–4)

Minylitha includes calcareous nannofossils described as 'thick plate-like individual calcite elements with polygonal outline, depressed central area, and raised rim'. They appear only moderately bright in polarized light. *M. convallis*, the only species, is found mainly in NN 10 and NN 11.

Nannotetrina (Figs. 89.1–7; 90.1–6)

The genus *Nannotetrina* includes calcareous nannofossils with a central cross formed by four straight or bent arms. The inter-arm space can be empty or filled with basket-shaped elements. The distal side is dominated by the central cross, the proximal side by the basket and extensions of the basket.

The FO of *N. alata* (Fig. 89.1) was used by Martini (1971) to define the base of NP 15, the FO of *N. quadrata* (Fig. 89.3) by Bukry (1973a) to define the base of his *N. quadrata* Zone (CP13). These two species have been considered synonyms of *N. fulgens* (Figs. 89.2; 90.1, 2) by many authors, who accordingly used the FO of *N. fulgens* to define the base of CP13/NP 15. The illustrations of the holotypes of the three

Lapideacassis/Scampanella

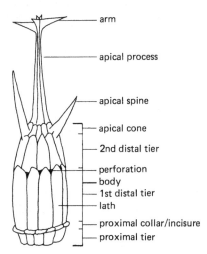

Fig. 86. Terminology of *Lapideacassis* and *Scampanella*

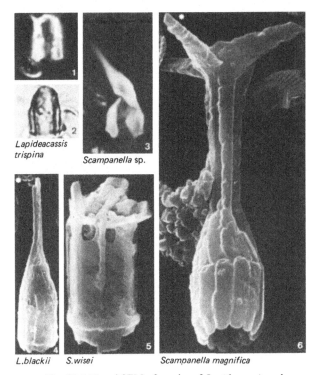

Lapideacassis trispina

Scampanella sp.

L.blackii *S.wisei* *Scampanella magnifica*

Fig. 87. LM and SEM of species of *Lapideacassis* and *Scampanella*

species are shown in Fig. 89. *N. fulgens* has a small central basket and long tapering arms. *N. alata* also has long arms, but they do not taper, instead they have wing-shaped sideways extensions to one side. *N. quadrata* has long arms and wing-shaped extensions to both sides of the arms and no central basket.

The first species of *Nannotetrina* to appear in NP 14 is *N. cristata* (Figs. 89.5, 6; 90.5), a relatively small species of the genus. It has a massive central cross and the basket is almost as wide as the arms of the cross span. There is a tendency to a slight bifurcation at the end of the arms. This bifurcation is extreme in *N. pappii* (Figs. 89.7; 90.3, 4), which appears later in NP 15. In proximal view, *N. pappii* resembles an eight-armed discoaster (Fig. 90.4). In distal view it can be observed, that the eight spines derive from a bifurcation of the four arms of the central cross (Fig. 90.3). *N. nitida* has been illustrated recently by Aubry (1983) and is an extremely slender form of the genus with, in the LM, a characteristically turned small central cross on top of the junction of the long, slender arms. The species, which was described from the Eocene of France and since 'adopted' by specialists of the Upper Cretaceous for a Late Campanian form, thus is restricted to the Middle Eocene NP 15.

The species of *Nannotetrina* are not always distinguishable in poorly preserved material, but the genus as such is still recognizable even in heavily overgrown material. This has led some authors to use the FO of the genus *Nannotetrina* as the marking event for the base of NP 15 or CP13. Also, whereas *N. cristata* may be present in many sections, *N. fulgens* (*N. quadrata*, *N. alata*) is often very rare or absent.

Romein (1979) suggested that *Nannotetrina* evolved from *Micula*. Therefore the species are illustrated together in

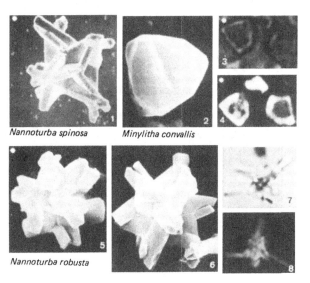

Nannoturba spinosa *Minylitha convallis*

Nannoturba robusta

Fig. 88. LM and SEM of *Minylitha convallis* and *Nannoturba* spp.

Fig. 89. Tappan (1980) included *Nannotetrina* in the Zygodiscaceae based on the central cross which it shares with some of the genera assembled in that family. In fact it seems likely, that *Nannotetrina* evolved from *Chiphragmalithus* (to which genus most species have been once assigned) in NP 14 by the loss of the wall.

Nannoturba (Figs. 88.1, 5–8)

Nannoturba is described as 'forms consisting of several sticks or plates of rectangular to quadratic shape'. In *N. robusta* (Figs. 88.5–8) the sticks 'are combined at an angle of 90°'. *N. robusta* has been found in the Lower Eocene (NP 12, NP 13) of Denmark, Northern Germany, Poland and the North Atlantic as well as the southwest Pacific and NP 15 of the South Atlantic. *N. spinosa* (Fig. 88.1) is characterized by its hexagonal shape and is 'constructed of rectangular plates which are combined at an angle of 120°'. It was reported from the Lower Eocene of the North Atlantic.

Oolithotus (Figs. 84.5–9)

Oolithotus is characterized by the asymmetrical position of the connection between the large distal and the small

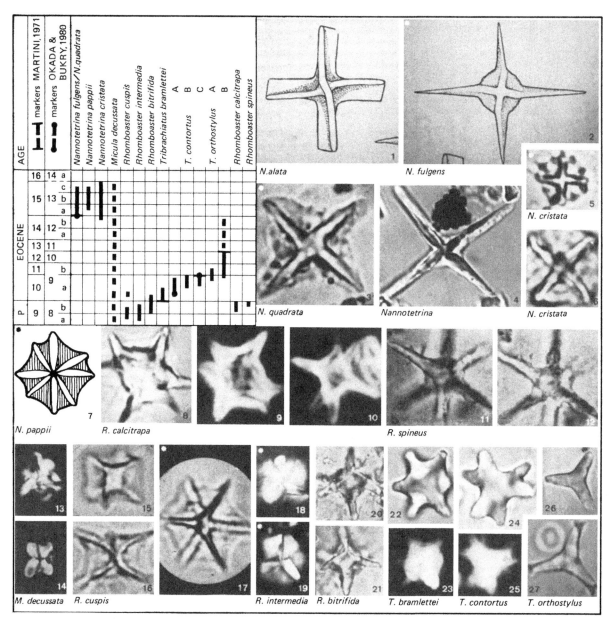

Fig. 89. Ranges and LM of species of *Nannotetrina* (*N.*), *Rhomboaster* (*R.*) and *Tribrachiatus* (*T.*)

proximal shield, which is unique. The type species *O. fragilis* (syn. *O. antillarum*; Figs. 84.5–9), appears in the Early Pliocene and is found in the present seas.

Rhomboaster (Figs. 89.8–12, 15–21; 90.9–11)

Rhomboaster includes calcareous nannofossils in the form of a rhomboedron with depressed faces. Similar forms are known from the Upper Cretaceous and assigned to *Micula* (Figs. 89.13, 14) and in the Middle Eocene, where they belong to *Nannotetrina* (Figs. 89; 90). *Rhomboaster* species are found in the uppermost Paleocene and lowermost Eocene.

R. cuspis (Figs. 89.15–17; 90.11), the type species of the genus, is a simple rhomboedron built of a single crystal (?). According to Romein (1979), *R. intermedia* (Figs. 89.18, 19) is characterized by its composition of different crystallographic units and *R. bitrifida* by two sets of three major arms and one minor arm. *R. calcitrapa* (Figs. 89.8–10) has relatively long, spine-like tapering arms and *R. spineus* (Figs. 89.11, 12; 90.9) has very long arms arranged in two sets of three tapering arms turned by 60°. The arms are decorated with a pair of short spines.

Romein (1979) suggested a lineage from *Micula*

decussata over *R. intermedia* to *R. bitrifida* and *R. calcitrapa*. The species of *Rhomboaster* are easily overlooked with the LM and taken as odd-shaped calcite rhomboedrons, from which they are difficult to distinguish when heavily overgrown. Their presence is a good indication for a latest Paleocene age.

Tribrachiatus (Figs. 89.22–27; 90.7, 8, 12, 13)

The genus *Tribrachiatus* is used for three-armed, Eocene calcareous nannofossils, whereas the similar Cretaceous forms are assigned to *Marthasterites*. Miocene and Pliocene triradiate forms are included in *Discoaster*. The triradiate forms of the Cretaceous, Eocene and Neogene are not related.

T. bramlettei (syn. *T. nunnii*; Figs. 89.22, 23; 90.13) consists of two sets of three arms with about equal angles between them. In *T. contortus* (Figs. 89.24, 25; 90.7, 8) Type A, the larger angle between two arms of different layers is 90° or smaller, in *T. contortus* Type B larger than 90° and in Type C the two layers are still recognizable, but they nearly overlap each other. *T. orthostylus* (Figs. 89.26, 27; 90.12) includes the forms of Type A with a slight bifurcation and Type B with no bifurcation. Quantitative studies have been made by Hekel (1968) and Romein (1979) and show the stratigraphically

Nannotetrina cristata

Nannotetrina pappii

Nannotetrina fulgens

Nannotetrina sp.

T. contortus

Rhomboaster spineus   *R. bitrifida*   *Rhomboaster cuspis*   *Tribrachiatus orthostylus*   *Tribrachiatus bramlettei*

Fig. 90. SEM of species of *Nannotetrina*, *Rhomboaster* and *Tribrachiatus*

useful distribution of the changing assemblages. The FO and LO of *T. contortus* are used to define CP9a, the FO of *T. bramlettei* defines the base and the LO of *T. contortus* the top of NP 10. Since the latter coincides almost with the FO of *T. orthostylus* (syn. *Marthasterites tribrachiatus*), this FO is often used for an approximation of the base of NP 11. The LO of *T. orthostylus* defines the top of NP 12.

Tribrachiatus is thought to have evolved from *Rhomboaster* (Romein, 1979) around the Paleocene/Eocene boundary and disappears already in the late Early Eocene.

Trochastrites (Fig. 91.1–4)

The genus *Trochastrites* includes forms with three bifurcating arms and the two species assigned to it could also be assigned to *Tribrachiatus*, with which, however, they do not seem to be related. *T. bramlettei* (Figs. 91.1, 2), and the generotype, *T. hohnensis* (Figs. 91.3, 4), were described from the 'Lower Upper Eocene' and occur in NP 14 to NP 16 in the Anglo-Parisian basin according to Aubry (1983).

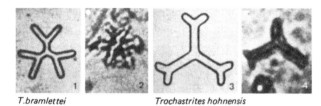

T.bramlettei   *Trochastrites hohnensis*

Fig. 91. LM of species of *Trochastrites*

Umbellosphaera (Figs. 84.1–4)

Umbellosphaera irregularis (syn. *Ellipsodiscoaster lidzii, U. tenuis*; Figs. 84.1–4) is characterized by the roughly elliptical outline of the distal shield and a stem-like, proximal central structure supported by radially to tangentially arranged joists. The shield is only faintly birefringent between crossed nicols and the central structure is strongly birefringent. Usually only the micrococcoliths of *U. irregularis* illustrated here are found in the sediments. *U. irregularis* appears in the Late Pliocene and is presently found in the Pacific and the Atlantic.

Acknowledgements

During the compilation of this chapter, which lasted – on and often off – over four years, many colleagues discussed various aspects of Cenozoic calcareous nannofossil biostratigraphy with me. M.-P. Aubry, J. Backman, D. Bukry, M. E. Hill, A. J. T. Romein, H. Stradner and S. W. Wise read a semi-final version of the manuscript and made valuable suggestions for which I am very grateful. I thank my family and the co-editors for the understanding and patience they showed while I struggled to complete the task.

Appendix: Alphabetic list of taxa

(*Agalmatoaster* Klumpp, 1953) = *Discoaster*
Amaurolithus Gartner & Bukry (1975)
  A. amplificus (Bukry & Percival, 1971) Gartner & Bukry (1975)
  A. bizarrus (Bukry, 1973c) Gartner & Bukry (1975)
  A. delicatus Gartner & Bukry (1975)
  A. ninae Perch-Nielsen (1977)
  A. primus (Bukry & Percival, 1971) Gartner & Bukry (1975)
  A. tricorniculatus (Gartner, 1967) Gartner & Bukry (1975)
Angulolithina Bukry (1973c)
  A. arca Bukry (1973c)
(*Apertapetra* Hay, Mohler & Wade, 1966) = *Reticulofenestra*

Biantholithus Bramlette & Martini (1964)
  B. sparsus Bramlette & Martini (1964)
Birkelundia Perch-Nielsen (1971d)
  B. arenosa Perch-Nielsen (1971d)
  B. jugata (Perch-Nielsen, 1967) Perch-Nielsen (1971d)
  B. staurion (Bramlette & Sullivan, 1961) Perch-Nielsen (1971d)
Biscutum Black *in* Black & Barnes (1959)
  B. constans (Gorka, 1957), Black *in* Black & Barnes (1959)
  B.? parvulum Romein (1979)
  B.? romeinii Perch-Nielsen (1981a)
  Biscutum sp. 1–4
Blackites Hay & Towe (1962)
  (*B. amplus* Roth & Hay *in* Hay *et al.*, 1967) = *B. spinosus*
  (*B. incompertus* Roth, 1970) = *R. tenuis*
  B. spinosus (Deflandre & Fert, 1954) Hay & Towe (1962)
  B. trochos Bybell (1982)
Bomolithus Roth (1973)
  B. conicus (Perch-Nielsen, 1971) Perch Nielsen (1984a)
  B. elegans Roth (1973)
Braarudosphaera Deflandre (1947)
  B. alta Romein (1979)
  B. bigelowii (Gran & Braarud, 1935) Deflandre (1947)
  B. stylifer Troelsen & Quadros (1971)
Brachiolithus Noël ex Loeblich & Tappan (1963)
Bramletteius Gartner (1969a)
  B.? duolatus Martini (1981b)
  B. serraculoides Gartner (1969a)
  (*B. variabilis* Roth, 1970) = *B. serraculoides*

Calcidiscus Kamptner (1950)
  C. kingii (Roth, 1970) Loeblich & Tappan (1978)
  C. leptoporus (Murray & Blackman, 1898) Loeblich & Tappan (1978)
  C. leptoporus centrovalis (Stradner & Fuchs, 1980) Perch-Nielsen (1984a)
  C. macintyrei (Bukry & Bramlette, 1969b) Loeblich & Tappan (1978)
  C. protoannulus (Gartner, 1971) Loeblich & Tappan (1978)
  C. quadriforatus Kamptner (1950)
Calciopilleus Müller (1974a)
  C. obscurus Müller (1974a)
Campylosphaera Kamptner (1963)
  C. dela (Bramlette & Sullivan, 1961) Hay & Mohler (1967)
  C. eodela Bukry & Percival (1971)
  Campylosphaera sp. 1
Caneosphaera Gaarder *in* Gaarder & Heimdal (1977)
Catinaster Martini & Bramlette (1963)
  C. altus (Müller, 1974a) Perch-Nielsen (1984a)
  C. calyculus Martini & Bramlette (1963)
  C. coalitus Martini & Bramlette (1963)

C. mexicanus Bukry (1971b)
C.? umbrellus Bukry (1971b)
Centosphaera Wind & Wise (1977)
Ceratolithina Martini (1967)
　C.? vesca Bukry & Percival (1971)
Ceratolithoides Bramlette & Martini (1964)
Ceratolithus Kamptner (1950)
　C. acutus Gartner & Bukry (1974)
　C. armatus Müller (1974a)
　C. atlanticus Perch-Nielsen (1977)
　C. cristatus Kamptner (1950)
　(*C. dentatus* Bukry, 1973c) = *A. amplificus*
　C. rugosus Bukry & Bramlette (1968)
　C. separatus Bukry (1979)
　C. simplex Bukry (1979)
　C. telesmus Norris (1965)
Chiasmolithus Hay, Mohler & Wade (1966)
　C. altus Bukry & Percival (1971)
　C. bidens (Bramlette & Sullivan, 1961) Hay & Mohler (1967)
　C. californicus (Sullivan, 1964) Hay & Mohler (1967)
　C. consuetus (Bramlette & Sullivan, 1961) Hay & Mohler (1967)
　C. danicus (Brotzen, 1959) Hay & Mohler (1967)
　C. eograndis Perch-Nielsen (1971d)
　C. expansus (Bramlette & Sullivan, 1961) Gartner (1970)
　C. gigas (Bramlette & Sullivan, 1961) Radomski (1968)
　C. grandis (Bramlette & Riedel, 1954) Radomski (1968)
　C. medius Perch-Nielsen (1971d)
　C. minimus Perch-Nielsen (1971d)
　C. modestus Perch-Nielsen (1971d)
　C. nitidus Perch-Nielsen (1971d)
　C. oamaruensis (Deflandre, 1954) Hay, Mohler & Wade (1966)
　C. solitus (Bramlette & Sullivan, 1961) Locker, 1968
　C. titus Gartner (1970)
Chiastozygus Gartner (1968)
　C. ultimus Perch-Nielsen (1981a)
Chiphragmalithus Bramlette & Sullivan (1961)
　C. acanthodes Bramlette & Sullivan (1961)
　C. armatus Perch-Nielsen (1971d)
　C. barbatus Perch-Nielsen (1967)
　C. calathus Bramlette & Sullivan (1961)
　(*C. cristatus* (Martini, 1958) Bramlette & Sullivan, 1961) = *Nannotetrina cristata*
　(*C. quadratus* Bramlette & Sullivan 1961) = *Nannotetrina quadrata*
Clathrolithus Deflandre (1954)
　C. spinosus Martini (1961a)
Clausicoccus Prins (1979)
　(*C. fenestratus* (Deflandre & Fert, 1954) Prins, 1979) = *Ericsonia fenestrata*
(*Clavodiscoaster* Prins, 1971) = *Discoaster*
Coccolithus Schwarz (1894)
　(*C. crassus* Bramlette & Sullivan, 1961) = *Toweius? crassus*
　C. eopelagicus (Bramlette & Riedel, 1954) Bramlette & Sullivan (1961)
　(*C. formosus* (Kamptner, 1963) Wise (1973)) = *Ericsonia formosa*
　C. miopelagicus Bukry (1971a)
　C. pelagicus (Wallich, 1877) Schiller (1930)
　C. pliopelagicus Wise (1973)
　(*C. subdistichus* (Roth & Hay *in* Hay *et al.*, 1967) Bukry *et al.* (1971)) = *Ericsonia subdisticha*
Coccosphaera Wallich (1877)
　(*C. carteri* Wallich, 1877) = *Helicosphaera carteri*

Conococcolithus Hay & Mohler (1967)
　C. minutus Hay & Mohler (1967)
　(*C. panis* Edwards, 1973a) = *Biscutum panis*
Corannulus Stradner (1962)
　C. germanicus Stradner (1962)
(*Coronaster* Martini, 1961b) = *Martiniaster*
Coronocyclus Hay, Mohler & Wade (1966)
　C. nitescens (Kamptner, 1963) Bramlette & Wilcoxon (1967)
　C. prionion (Deflandre & Fert, 1954) Stradner *in* Stradner & Edwards (1968)
　C. serratus Hay, Mohler & Wade (1966)
Coronosphaera Gaarder *in* Gaarder & Heimdal (1977)
　C. binodata (Kamptner, 1927) Gaarder *in* Gaarder & Heimdal (1977)
　C. mediterranea (Lohmann, 1902) Gaarder *in* Gaarder & Heimdal (1977)
(*Craspedolithus* Kamptner, 1963) = *Neosphaera*
　(*C. declivus* Kamptner, 1963) = *N. coccolithomorpha*
　(*C. ragulus* Kamptner, 1967) = *N. coccolithomorpha*
　(*C. vidalii* Bukry, 1975) = *N. coccolithomorpha*
(*Crassapontosphaera* Boudreaux & Hay, 1969) = *Pontosphaera*
Crenalithus Roth (1973)
　C. doronicoides (Black & Barnes, 1961) Roth (1973)
　(*C. productellus* Bukry, 1975) = *Dictyococcites productus*
Crepidolithus Noël (1965)
Cribrocentrum Perch-Nielsen (1971d)
　C. coenurum (Reinhardt, 1966) Perch-Nielsen (1971d)
　C. foveolatum (Reinhardt, 1966) Perch-Nielsen (1971d)
　C. martinii (Hay & Towe, 1962) Perch-Nielsen (1971d)
　C. reticulatum (Gartner & Smith, 1967) Perch-Nielsen (1971d)
Cruciplacolithus Hay & Mohler *in* Hay *et al.* (1967)
　C. brotzenii (Perch-Nielsen, 1969) Perch-Nielsen (1984a)
　C. cribellum (Bramlette & Sullivan, 1961) Romein (1979)
　C. cruciformis (Hay & Towe, 1962) Roth (1970)
　C. edwardsii Romein (1979)
　C. flavius Roth (1970)
　C. frequens (Perch-Nielsen, 1977) Romein (1979)
　C. inseadus Perch-Nielsen (1969)
　C. latipons Romein (1979)
　C. mutatus Perch-Nielsen (1971d)
　C. neohelis (McIntyre & Bé, 1967b) Reinhardt (1972)
　(*C. notus* Perch-Nielsen, 1977) = *C. tenuis*
　C. primus Perch-Nielsen (1977)
　C. quader Roth (1970)
　C. subrotundus Perch-Nielsen (1969)
　C. tarquinius Roth & Hay *in* Hay *et al.* (1967)
　C. tenuiforatus Clocchiatti & Jerkovic (1972)
　C. tenuis (Stradner, 1961) Hay & Mohler *in* Hay *et al.* (1967)
　C. vanheckae Perch-Nielsen (1984b)
　Cruciplacolithus sp. 1
Cyclagelosphaera Noël (1965)
　C. alta Perch-Nielsen (1979a)
　C. margerelii Noël (1965)
　C. reinhardtii (Perch-Nielsen, 1968) Romein (1977)
Cyclicargolithus Bukry (1971a)
　C. abisectus (Müller, 1970) Wise (1973)
　C. floridanus (Roth & Hay *in* Hay *et al.*, 1967) Bukry (1971a)
　C. marismontium (Black, 1964) n.comb.
(*Cyclococcolithina* Wilcoxon, 1970a) = *Calcidiscus*
(*Cyclococcolithus* Kamptner, 1954) = *Calcidiscus*
Cycloperfolithus Lehotayova & Priewalder (1978)
　C. carlae Lehotayova & Priewalder (1978)

(*Cycloplacolithella* Haq (1968) = *Calcidiscus*
(*Cycloplacolithus* Kamptner (1963) = *Calcidiscus*

Daktylethra Gartner *in* Gartner & Bukry (1969)
   D. punctulata Gartner *in* Gartner & Bukry (1969)
Deutschlandia Lohmann (1912)
   D. anthos Lohmann (1912)
   D. gaarderae Perch-Nielsen (1980)
(*Diademopetra* Hay, Mohler & Wade, 1966) = *Corannulus*
   (*D. luma* Hay, Mohler & Wade, 1966) = *C. spinosus*
Dictyococcites Black (1967)
   D. antarcticus Haq (1976)
   D. bisectus (Hay, Mohler & Wade, 1966) Bukry & Percival
      (1971)
   D. callidus Perch-Nielsen (1971d)
   D. danicus Black (1967)
   D. daviesii (Haq, 1968) Perch-Nielsen (1971d)
   D. hesslandii (Haq, 1966) Haq & Lohman (1976)
   D. onustus Perch-Nielsen (1971d)
   (*D. productellus* Bukry, 1975) = *D. productus*
   D. productus (Kamptner, 1963) Backman (1980)
   D. scrippsae Bukry & Percival (1971)
Discoaster Tan (1927)
   D. adamanteus Bramlette & Wilcoxon (1967)
   D. araneus Bukry (1971b)
   D. asymmetricus Gartner (1969c)
   D. aulakos Gartner (1967)
   D. barbadiensis Tan (1927)
   D. bellus Bukry & Percival (1971)
   D. berggrenii Bukry (1971b)
   D. bifax Bukry (1971a)
   D. binodosus Martini (1958)
   D. binodosus hirundinus Martini (1958)
   D. blackstockae Bukry (1973d)
   D. bollii Martini & Bramlette (1963)
   D. boulangeri Lezaud (1968)
   D. braarudii Bukry (1971b)
   (*D. bramlettei* Martini, 1958) = *Trochastrites bramlettei*
   D. bramlettei (Bukry & Percival, 1971) Romein (1979)
   D. broennimannii Stradner (1961)
   D. brouweri Tan (1927) emend. Bramlette & Riedel (1954)
   D. calcaris Gartner (1967)
   D. calculosus Bukry (1971b)
   D. challengeri Bramlette & Riedel (1954)
   D. cruciformis Martini (1958)
   D. decorus (Bukry, 1971b) Bukry (1973c)
   D. deflandrei Bramlette & Riedel (1954)
   D. delicatus Bramlette & Sullivan (1961)
   D. diastypus Bramlette & Sullivan (1961)
   D. distinctus Martini (1958)
   D. druggii Bramlette & Wilcoxon (1967)
   D. elegans Bramlette & Sullivan (1961)
   D. exilis Martini & Bramlette (1963)
   D. falcatus Bramlette & Sullivan (1961)
   D. formosus Martini & Worsley (1971)
   D. gemmeus Stradner (1959b)
   D. gemmifer Stradner (1961)
   D. germanicus Martini (1958)
   D. hamatus Martini & Bramlette (1963)
   (*D. helianthus* Bramlette & Sullivan, 1961) = *D. splendidus*
   D. icarus Stradner (1973)
   D. intercalaris Bukry (1971a)
   D. kuepperi Stradner (1959b)

D. kugleri Martini & Bramlette (1963)
D. lenticularis Bramlette & Sullivan (1961)
D. limbatus Bramlette & Sullivan (1961)
D. lodoensis Bramlette & Riedel (1954)
D. loeblichii Bukry (1971a)
D. mahmoudii Perch-Nielsen (1981a)
D. martinii Stradner (1959b)
D. mediosus Bramlette & Sullivan (1961)
D. megastypus (Bramlette & Sullivan, 1961) n.comb.
D. mendomobensis Wise (1973)
D. minimus Sullivan (1964)
D. mirus Deflandre *in* Deflandre & Fert (1954)
D. mohleri Bukry & Percival (1971)
D. monstratus Martini (1961a)
D. moorei Bukry (1971b)
D. multiradiatus Bramlette & Reidel (1954)
D. munitus Stradner (1961)
D. musicus Stradner (1959b)
D. neohamatus Bukry & Bramlette (1969b)
D. neorectus Bukry (1971a)
D. nobilis Martini (1961a)
D. nodifer (Bramlette & Riedel, 1954) Bukry (1973c)
D. nonaradiatus Klumpp (1953)
D. pacificus Haq (1969)
D. pansus (Bukry & Percival, 1971) Bukry (1973c)
D. pentaradiatus Tan (1927) emend. Bramlette & Riedel (1954)
D. perclarus Hay *in* Hay *et al.* (1967)
D. perplexus Bramlette & Riedel (1954)
D. perpolitus Martini (1961a)
D. petaliformis Moshkovitz & Ehrlich (1980)
D. prepentaradiatus Bukry & Percival (1971)
D. pseudovariabilis Martini & Worsley (1971)
D. quadramus Bukry (1973d)
D. quinqueramus Gartner (1969c)
D. robustus Haq (1969)
D. saipanensis Bramlette & Riedel (1954)
D. salisburgensis Stradner (1961)
(*D. sanmiguelensis* Bukry, 1981b)? = *D. musicus*
D. septemradiatus (Klumpp, 1953) Martini (1958)
D. signus Bukry (1971b)
D. splendidus Martini (1960)
D. strictus Stradner (1961)
D. sublodoensis Bramlette & Sullivan (1961)
D. subsurculus Gartner (1967)
D. surculus Martini & Bramlette (1963)
D. tamalis Kamptner (1967)
D. tanii Bramlette & Riedel (1954)
   (*D. tanii nodifer* Bramlette & Riedel, 1954) = *D. nodifer*
D. tanii ornatus Bramlette & Wilcoxon (1967)
D. triradiatus Tan (1927)
D. tristellifer Bukry (1976)
D. variabilis Martini & Bramlette (1963)
D. wemmelensis Achuthan & Stradner (1969)
(*Discoasteroides* Bramlette & Sullivan, 1961) = *Discoaster*
(*Discolithina* Loeblich & Tappan, 1963) = *Pontosphaera*
(*Discolithus* Huxley, 1868) = *Pontosphaera*
Discosphaera Haeckel (1894)
   D. tubifer (Murray & Blackman, 1898) Ostenfeld (1900)
(*Discoturbella* Roth, 1970) = *Naninfula*

(*Ellipsodiscoaster* Boudreaux & Hay, 1969) = *Umbellosphaera*
   (*E. lidzii* Boudreaux & Hay, 1969) = *Umbellosphaera irregularis*
Ellipsolithus Sullivan (1964)

E. bollii Perch-Nielsen (1977)
E. distichus (Bramlette & Sullivan, 1961) Sullivan (1964)
E. lajollaensis Bukry & Percival (1971)
E. macellus (Bramlette & Sullivan, 1961) Sullivan (1964)
Emiliania Hay & Mohler *in* Hay *et al.* (1967)
 (*E. annula* (Cohen, 1964) Bukry, 1971a) = *Pseudoemiliania
  lacunosa*
 E. huxleyi (Lohmann, 1902) Hay & Mohler *in* Hay *et al.* (1967)
 (*E. ovata* Bukry, 1973c) = *Pseudoemiliania lacunosa*
(*Eodiscoaster* Martini (1961a)) = *Braarudosphaera*
Ericsonia Black (1964)
 E. cava (Hay & Mohler, 1967) Perch-Nielsen (1969)
 E. fenestrata (Deflandre & Fert, 1954) Stradner *in* Stradner &
  Edwards (1968)
 E. formosa (Kamptner, 1963) Haq (1971)
 E. insolita Perch-Nielsen (1971d)
 E. obruta Perch-Nielsen (1971d)
 E. occidentalis Black (1964)
 E. ovalis Black (1964)
 E. subdisticha (Roth & Hay *in* Hay *et al.*, 1967) Roth *in*
  Baumann & Roth (1969)
 E. subpertusa Hay & Mohler (1967)

Fasciculithus Bramlette & Sullivan (1961)
 F. alanii Perch-Nielsen (1971b)
 F. aubertae Haq & Aubry (1981)
 F. billii Perch-Nielsen (1971b)
 F. bitectus Romein (1979)
 F. bobii Perch-Nielsen (1971b)
 F. clinatus Bukry (1971a)
 F. hayi Haq (1971)
 F. involutus Bramlette & Sullivan (1961)
 F. janii Perch-Nielsen (1971b)
 F. lillianae Perch-Nielsen (1971b)
 F. magnicordis Romein (1979)
 F. magnus Bukry & Percival (1971)
 (*F. merloti* Pavsič, 1977) = *F. pileatus*
 F. mitreus Gartner (1971)
 F. pileatus Bukry (1973d)
 F. richardii Perch-Nielsen (1971b)
 (*F. rotundus* Haq & Lohman, 1976) = *Bomolithus elegans*
 F. schaubii Hay & Mohler (1967)
 (*F. stonehengei* Haq & Aubry, 1981) = *F. pileatus*
 F. thomasii Perch-Nielsen (1971b)
 F. tonii Perch-Nielsen (1971b)
 F. tympaniformis Hay & Mohler *in* Hay *et al.* (1967)
 F. ulii Perch-Nielsen (1971b)
 Fasciculithus sp.
Florisphaera Okada & Honjo (1973)
 F. profunda Okada & Honjo (1973)
 F. profunda elongata Okada & McIntyre (1980)
(*Furcatolithus* Martini, 1965) = *Sphenolithus*

Gephyrocapsa Kamptner (1943)
 G. aperta Kamptner (1963)
 G. caribbeanica Boudreaux & Hay (1969)
 G. ericsonii McIntyre & Bé (1967)
 G. lumina Bukry (1973c)
 G. margarelii Bréhéret (1978a)
 G. mediterranea Pirini Radrizzani & Valleri (1977)
 G. muellerae Bréhéret (1978a)
 G. oceanica Kamptner (1943)
 G. omega Bukry (1973c)

G. ornata Heimdal (1973)
G. parallela Hay & Beaudry (1973)
G. pelta Samtleben (1980)
G. protohuxleyi McIntyre (1970)
G. rota Samtleben (1980)
G. sinuosa Hay & Beaudry (1973)
Goniolithus Deflandre (1957)
 G. fluckigeri Deflandre (1957)
(*Gyrodiscoaster* Stradner, 1961) = *Discoaster*

Hayaster Bukry (1973d)
 H. perplexus (Bramlette & Riedel, 1954) Bukry (1973d)
Hayella Gartner (1969a, b)
 H.? gauliformis Troelsen & Quadros (1971)
 H. situliformis Gartner (1969a, b)
(*Hayella* Roth, 1969) = *Ilselithina*
 (*H. elegans* roth, 1969) = *I. iris*
(*Helicopontosphaera* Hay & Mohler *in* Hay *et al.*,
  1967) = *Helicosphaera*
Helicosphaera Kamptner (1954) and *Helicopontosphaera*
 H. ampliaperta Bramlette & Wilcoxon (1967)
 H. bramlettei Müller (1970)
 H. burkei Black (1971a)
 H. californiana Bukry (1981b)
 H. carteri (Wallich, 1877) Kamptner (1954)
 H. colombiana Gartner (1977b)
 H. compacta Bramlette & Wilcoxon (1967)
 (*H. crouchii* Bukry, 1981b) = *H. mediterranea*
 H. dinesenii Perch-Nielsen (1971d)
 H. euphratis Haq (1966)
 H. gertae Bukry (1981b)
 H. granulata Bukry & Percival (1971)
 H. heezenii Bukry (1971a)
 H. hyalina Gaarder (1970)
 H. intermedia Martini (1965)
 H. inversa Gartner (1980)
 H. kamptneri Hay & Mohler *in* Hay *et al.* (1967)
 H. lophota Bramlette & Sullivan (1961)
 H. mediterranea Müller (1981)
 H. minima Martini (1974)
 H. minuta Müller (1981)
 H. neogranulata Gartner (1977b)
 H. obliqua Bramlette & Wilcoxon (1967)
 H. orientalis Black (1971a)
 H. pacifica Müller & Brönnimann (1974)
 H. papillata Bukry & Bramlette (1969b)
 (*H. parallela* Bramlette & Wilcoxon, 1967) = *H. euphratis*
 H. pavimentum Okada & McIntyre (1977)
 H. perch-nielseniae Haq (1971)
 H. philippinensis Müller (1981)
 H. recta Haq (1966)
 H. reticulata Bramlette & Wilcoxon (1967)
 H. rhomba Bukry (1971a)
 H. salebrosa Perch-Nielsen (1971d)
 H. scissura Miller (1981)
 H. sellii Bukry & Bramlette (1969b)
 H. seminulum Bramlette & Sullivan (1961)
 H.? subantarctica Edwards & Perch-Nielsen (1975)
 H. truempyi Biolzi & Perch-Nielsen (1982)
 (*H. truncata* Bramlette & Wilcoxon, 1967) = *H. recta*
 H. vedderi Bukry (1981b)
 H. walbersdorfensis Müller (1974b)
 H. wallichii (Lohmann, 1902) Boudreaux & Hay (1969)

H. wilcoxonii Gartner (1971)
Helicosphaera sp. (Paleocene, NP 9)
(*Heliodiscoaster* Tan, 1927) = *Discoaster*
Heliolithus Bramlette & Sullivan (1961)
  H. cantabriae Perch-Nielsen (1971c)
  H.? floris Haq & Aubry (1981)
  H. kleinpellii Sullivan (1964)
  H. riedelii Bramlette & Sullivan (1961)
(*Heliorthus* Brönnimann & Stradner, 1960) = *Neococcolithes*
(*Hemidiscoaster* Tan, 1927) = *Discoaster*
Holodiscolithus Roth (1970)
  H. macroporus (Deflandre *in* Deflandre & Fert, 1954) Roth (1970)
  H. solidus (Deflandre *in* Deflandre & Fert, 1954) Roth (1970)
Hornibrookina Edwards (1973a)
  H. australis Edwards & Perch-Nielsen (1975)
  H. edwardsii Perch-Nielsen (1977)
  H. teuriensis Edwards (1973a)

Ilselithina Stradner *in* Stradner & Adamiker (1966)
  I. fusa Roth (1970)
  I. iris Stradner *in* Stradner & Adamiker (1966)
Imperiaster Martini (1970b)
  I. obscurus (Martini 1958) Martini (1970b)
(*Indumentalithus* Vekshina, 1959) = *Neococcolithes*
Isthmolithus Deflandre (1954)
  (*I. claviformis* Brönnimann & Stradner, 1960) = *Zygrhablithus bijugatus*
  I. recurvus Deflandre (1954)
  I. rhenanus Martini (1973)
  I. unipons Bramlette & Sullivan (1961)

(*Koczyia* Boudreaux & Hay, 1969) = *Pontosphaera*

Lanternithus Stradner (1962)
  L. duocavus Locker (1967)
  L. minutus Stradner (1962)
Lapideacassis Black (1971b)
  L. blackii Perch-Nielsen *in* Perch-Nielsen & Franz (1977)
  L. trispina Perch-Nielsen *in* Perch-Nielsen & Franz (1977)
Lithostromation Deflandre (1942b)
  L. deflandrei (Stradner, 1959b) Perch-Nielsen (1984a)
  L. operosum (Deflandre, 1954) Bybell (1975)
  L. perdurum Deflandre (1942b)
  L. reginum (Stradner, 1962) Bybell (1975)
  L. simplex (Klumpp, 1953) Bybell (1975)
Lophodolithus Deflandre *in* Deflandre & Fert (1954)
  L. acutus Bukry & Percival (1971)
  L. mochloporus Deflandre *in* Deflandre & Fert (1954)
  L. nascens Bramlette & Sullivan (1961)
  L. reniformis Bramlette & Sullivan (1961)
  L. rotundus Bukry & Percival (1971)

Markalius Bramlette & Martini (1964)
  M. apertus Perch-Nielsen (1979a)
  (*M. astroporus* (Stradner, 1963) Mohler & Hay *in* Hay *et al.*, 1967) = *M. inversus*
  M. inversus (Deflandre *in* Deflandre & Fert, 1954) Bramlette & Martini (1964)
  M. nudus Perch-Nielsen (1971d)
Marthasterites Deflandre (1959)
  (*M. bramlettei* Brönnimann & Stradner, 1960) = *T. bramlettei*
  (*M. contortus* (Stradner, 1958) Deflandre (1959)) = *T. contortus*
  (*M. reginus* Stradner, 1962) = *I. obscurus*

(*M. tribrachiatus* (Bramlette & Riedel, 1954) Deflandre, 1959) = *T. orthostylus*
Martiniaster Loeblich & Tappan (1963)
  M. fragilis (Martini, 1961b) Loeblich & Tappan (1963)
Micrantholithus Deflandre *in* Deflandre & Fert (1954)
  M. altus Bybell & Gartner (1972)
  M. concinnus Bramlette & Sullivan (1961)
  M. entaster Bramlette & Sullivan (1961)
  M. flos Deflandre *in* Deflandre & Fert (1954)
  M. mirabilis Locker (1965)
  M. pinguis Bramlette & Sullivan (1961)
Micula Vekshina (1959)
  M. murus (Martini, 1961a) Bukry (1973c)
  M. prinsii Perch-Nielsen (1979c)
Minylitha Bukry (1973c)
  M. convallis Bukry (1973c)

Naninfula Perch-Nielsen (1968)
  N. deflandrei Perch-Nielsen (1968)
Nannocorbis Müller (1974a)
  N. challengeri Müller (1974a)
Nannotetrina Achuthan & Stradner (1969)
  N. alata (Martini, 1960) Haq & Lohmann (1976)
  N. cristata (Martini, 1958) Perch-Nielsen (1971d)
  N. fulgens (Stradner, 1960) Achuthan & Stradner (1969)
  N. nitida (Martini, 1961) Aubry (1983)
  N. pappii (Stradner, 1959) Perch-Nielsen (1971d)
  N. quadrata (Bramlette & Sullivan, 1961) Bukry (1973a)
  Nannotetrina sp.
Nannoturba Müller (1979)
  N. robusta Müller (1979)
  N. spinosa Müller (1979)
Neochiastozygus Perch-Nielsen (1971c)
  N. cearae Perch-Nielsen (1977)
  N. chiastus (Bramlette & Sullivan, 1961) Perch-Nielsen (1971c)
  N. concinnus (Martini, 1961a) Perch-Nielsen (1971c)
  N. denticulatus (Perch-Nielsen, 1969) Perch-Nielsen (1971c)
  N. digitosus Perch-Nielsen (1971c)
  N. distentus (Bramlette & Sullivan, 1961) Perch-Nielsen (1971c)
  N. eosaepes Perch-Nielsen (1981a)
  N. imbriei Haq & Lohmann (1976)
  N. junctus (Bramlette & Sullivan, 1961) Perch-Nielsen (1971c)
  N. modestus Perch-Nielsen (1971c)
  N. perfectus Perch-Nielsen (1971c)
  N. primitivus Perch-Nielsen (1981a)
  N. rosenkrantzii Perch-Nielsen (1971c)
  N. saepes Perch-Nielsen (1971c)
  Neochiastozygus sp.
Neococcolithes Sujkowski (1931)
  N. dubius (Deflandre, 1954) Black (1967)
  N. minutus (Perch-Nielsen, 1967) Perch-Nielsen (1971d)
  N. nudus Perch-Nielsen (1971c)
  N. pediculatus (Perch-Nielsen, 1967) Perch-Nielsen (1971d)
  N. protenus (Bramlette & Sullivan, 1961) Black (1967)
  (*N. pyramidus* (Perch-Nielsen, 1967) Perch-Nielsen, 1971d) = *N. dubius*
Neocrepidolithus Romein (1979)
  N. biskayae Perch-Nielsen (1981a)
  N. bukryi Perch-Nielsen (1981a)
  N. cohenii (Perch-Nielsen, 1968) Perch-Nielsen (1984a)
  N. cruciatus (Perch-Nielsen, 1979a) Perch-Nielsen (1981a)
  N. dirimosus (Perch-Nielsen, 1979a) Perch-Nielsen (1981a)
  N. fossus (Romein, 1977) Romein (1979)

N. neocrassus (Perch-Nielsen, 1968) Romein (1979)
Neocrepidolithus sp.
Neosphaera Lecal-Schlauder (1950)
   N. coccolithomorpha Lecal-Schlauder (1950)
Noelaerhabdus Jerkovič (1970)
   N. bozinovicae Jerkovič (1970)

Octolithus Romein (1979)
   O. multiplus (Perch-Nielsen, 1973) Romein (1979)
Oolithotus Reinhardt *in* Cohen & Reinhardt (1968)
   (*O. antillarum* (Cohen, 1964) Reinhardt *in* Cohen &
     Reinhardt, 1968) = *O. fragilis*
   O. fragilis (Lohmann, 1912) Martini & Müller (1972)
Orthorhabdus Bramlette & Wilcoxon (1967)
   O. serratus Bramlette & Wilcoxon (1967)
Orthozygus Bramlette & Wilcoxon (1967)
   O. aureus (Stradner, 1962) Bramlette & Wilcoxon (1967)

Pedinocyclus Bukry & Bramlette (1971)
   P. larvalis (Bukry & Bramlette, 1969b) Loeblich & Tappan
     (1973)
Pemma Klump (1953)
   P. angulatum Martini (1959c)
   P. balium Bybell & Gartner (1972)
   P. basquensis (Martini, 1959c) Báldi-Beke (1971)
   P. papillatum Martini (1959c)
   Pemma sp.
Pentaster Bybell & Gartner (1972)
   P. lisbonensis Bybell & Gartner (1972)
Peridinium Ehrenberg (1832)
   P. cf. *P. trochoideum* (Stein) Lemmermann
Peritrachelina Deflandre (1952)
   P. joidesa Bukry & Bramlette (1968)
   P. ornata Deflandre (1952)
Placozygus Hoffmann (1970)
   P. sigmoides (Bramlette & Sullivan, 1961) Romein (1979)
Pontosphaera Lohmann (1902)
   P. bicaveata (Perch-Nielsen, 1967) Romein (1979)
   P. callosa (Martini, 1969) Varol (1982)
   P. cribraria (Perch-Nielsen, 1967) Haq (1971)
   P. desueta (Müller, 1970) Perch-Nielsen (1984a)
   P. discopora Schiller (1925)
   P. distincta (Bramlette & Sullivan, 1961) Roth & Thierstein
     (1972)
   P. enormis Locker (1967) Perch-Nielsen (1984a)
   P. excelsa (Perch-Nielsen, 1971d) Perch-Nielsen (1977)
   P. formosa (Bukry & Bramlette, 1969b) Romein (1979)
   P. inconspicua (Sullivan, 1964) Perch-Nielsen (1984a)
   P. indooceanica Čepek (1973)
   P. japonica (Takayama, 1967) Nishida (1971)
   P. latelliptica Báldi-Beke & Baldi (1974) Perch-Nielsen (1984a)
   P. latoculata Bukry & Percival (1971) Perch-Nielsen (1984a)
   P. millepuncta Gartner (1967)
   P. multipora (Kamptner, 1948) Roth (1970)
   P. ocellata (Bramlette & Sullivan, 1961) Perch-Nielsen (1984a)
   P. pectinata (Bramlette & Sullivan, 1961) Sherwood (1974)
   P. plana (Bramlette & Sullivan, 1961) Haq (1971)
   P. punctosa (Bramlette & Sullivan, 1961) Perch-Nielsen (1984a)
   P. rimosa (Bramlette & Sullivan, 1961) Roth & Thierstein (1972)
   P. rothii Haq (1971)
   P. scissura (Perch-Nielsen, 1971d) Romein (1979)
   P. scutellum Kamptner (1952)
   P. syracusana Lohmann (1902)

P. versa (Bramlette & Sullivan, 1961) Sherwood (1974)
Pontosphaera sp.
Prinsius Hay & Mohler (1967)
   P. bisulcus (Stradner, 1963) Hay & Mohler (1967)
   P. dimorphosus (Perch-Nielsen, 1969) Perch-Nielsen (1977)
   P. martinii (Perch-Nielsen, 1969) Haq (1971)
   P. petalosus (Ellis & Lohmann, 1973) Romein (1979)
   (*P. rosenkrantzii* Perch-Nielsen, 1979a) = *P. tenuiculum*
   P. tenuiculum (Okada & Thierstein, 1979) Perch-Nielsen (1984a)
Prolatipatella Gartner (1968)
   P. multicarinata Gartner (1968)
Pseudoemiliania Gartner (1969c)
   P. lacunosa (Kamptner, 1963) Gartner (1969c)
Pseudotriquetrorhabdulus Wise *in* Wise & Constans (1976)
   P. inversus (Bukry & Bramlette, 1969b) Wise *in* Wise & Constans
     (1976)
(*Pseudozygrhablithus* Haq, 1971) = *Zygrhablithus*
   (*P. altus* Haq, 1971) = *Z. bijugatus*
   (*P. comprimus* Haq, 1971) = *Z. bijugatus*
   (*P. latus* Haq, 1971) = *Z. bijugatus*

(*Quinquerhabdus* Bukry & Bramlette, 1969b) invalid
   (*Q. colossicus* Bukry & Bramlette, 1969b) invalid

(*Radiodiscoaster* Prins, 1971) = *Discoaster*
Repagalum Forchheimer (1972)
Reticulofenestra Hay, Mohler & Wade (1966)
   (*R. coenura* (Reinhardt, 1966) Roth, 1970) = *Cribrocentrum*
     coenurum
   R. dictyoda (Deflandre *in* Deflandre & Fert, 1954) Stradner *in*
     Stradner & Edwards (1968)
   R. gelida (Geitzenauer, 1972) Backman (1978)
   R. hampdenensis Edwards (1973a)
   R. haqii Backman (1978)
   R. hillae Bukry & Percival (1971)
   R. minuta Roth (1970)
   R. minutula (Gartner, 1967) Haq & Berggren (1978)
   R. oamaruensis (Deflandre *in* Deflandre & Fert, 1954) Stradner *in*
     Haq (1968)
   (*R. placomorpha* (Kamptner, 1948) Stradner *in* Haq, 1968) = *R.*
     umbilicus
   R. pseudoumbilica (Gartner, 1967) Gartner (1969c)
   (*R. reticulata* (Gartner & Smith, 1967) Roth & Thierstein
     (1972)) = *Cribrocentrum reticulatum*
   R. samodurovii (Hay, Mohler & Wade, 1966) Roth (1970)
   (*R. scissura* Hay, Mohler & Wade, 1966) = *D. bisectus*
   R. umbilica (Levin, 1965) Martini & Ritzkowski (1968)
Rhabdolithus Kamptner ex Deflandre *in* Grassé (1952) and
Rhabdosphaera Haeckel (1894)
   R. claviger Murray & Blackman (1898)
   R. creber Deflandre *in* Deflandre & Fert (1954)
   R. gladius Locker (1967)
   R. inflata Bramlette & Sullivan (1961)
   R. morionum (Deflandre *in* Deflandre & Fert, 1954) Bramlette &
     Sullivan (1961)
   R. pannonicus Báldi-Beke (1960)
   R. perlongus Deflandre *in* Grassé (1952)
   R. pinguis Deflandre *in* Deflandre & Fert (1954)
   R. procera Martini (1969)
   R. pseudomorionum Locker (1967)
   R. rudis Bramlette & Sullivan (1961)
   R. scabrosa Deflandre *in* Deflandre & Fert (1954)
   R. siccus Stradner *in* Bachmann *et al.* (1963)

R. signatorius Bóna (1964)
R. solus Perch-Nielsen (1971d)
(*R. spinula* Levin, 1965) = *Blackites spinosus*
R. stylifer Lohmann (1902)
R. tenuis Bramlette & Sullivan (1961)
R. truncata Bramlette & Sullivan (1961)
R. vitreus Deflandre *in* Deflandre & Fert (1954)
Rhabdothorax Kamptner (1958) ex Gaarder & Heimdal (1973)
    R. serratus (Bramlette & Wilcoxon, 1967) Roth (1970)
Rhomboaster Bramlette & Sullivan (1961)
    R. bitrifida Romein (1979)
    R. calcitrapa Gartner (1971)
    R. cuspis Bramlette & Sullivan (1961)
    R. intermedia Romein (1979)
    R. spineus (Shafik & Stradner, 1971) Perch-Nielsen (1984a)

Scampanella Forchheimer & Stradner (1973) emend. Perch-Nielsen
    & Franz (1977)
    S. magnifica Perch-Nielsen *in* Perch-Nielsen & Franz (1977)
    S. wisei Perch-Nielsen *in* Perch-Nielsen & Franz (1977)
    Scampanella sp.
Scapholithus Deflandre *in* Deflandre & Fert (1954)
    S. fossilis Deflandre *in* Deflandre & Fert (1954)
    S. rhombiformis Hay & Mohler (1967)
Scyphosphaera Lohmann (1902)
    S. abelei Rade (1975)
    S. aequatorialis Kamptner (1963)
    S. amphora Deflandre (1942a)
    S. ampla Kamptner (1955)
    S. antilleana Boudreaux & Hay (1969)
    S. apsteinii Lohmann (1902)
    S. aranta Kamptner (1967)
    S. australensis Rade (1975)
    S. biarritzensis Lezaud (1968)
    S. campanula Deflandre (1942a)
    S. canescens Kamptner (1955)
    S. cantharellus Kamptner (1955)
    S. cohenii Boudreaux & Hay (1969)
    S. columella Stradner (1969)
    S. conica Kamptner (1955)
    S. cylindrica Kamptner (1955)
    S. darraghi Rade (1975)
    S. deflandrei Müller (1974a)
    S. elegans (Ostenfeld, 1910) Deflandre (1942a)
    S.? expansa Bukry & Percival (1971)
    S. galeana Kamptner (1967)
    S. gladstonensis Rade (1975)
    S. globulata Bukry & Percival (1971)
    S. globulosa Kamptner (1955)
    S. graphica Müller (1974a)
    S. halldalii Deflandre *in* Deflandre & Fert (1954)
    S. hemirana Kamptner (1967)
    S. intermedia Deflandre (1942a)
    S. kamptneri Müller (1974a)
    S. lagena Kamptner (1955)
    S. magna Kamptner (1967)
    S. martinii Jafar (1975b)
    S. oremesa Kamptner (1967)
    S. pacifica Rade (1975)
    S. penna Kamptner (1955)
    S. piriformis Kamptner (1955)
    S. porosa Kamptner (1967)
    S. procera Kamptner (1955)

S. pulcherrima Deflandre (1942a)
S. queenslandensis Rade (1975)
S. recta (Deflandre, 1942a) Kamptner (1955)
S. recurvata Deflandre (1942a)
S. rottiensis Jafar (1975b)
S. tercisensis Lezaud (1968)
S. tora Kamptner (1967)
S. tubicena Stradner (1969)
S. tubifera Kamptner (1955)
S. turris Kamptner (1955)
S. ventriosa Martini (1968)
Semihololithus Perch-Nielsen (1971b)
    S. biskayae Perch-Nielsen (1971b)
    S. kerabyi Perch-Nielsen (1971b)
    Semihololithus sp.
Sollasites Black (1967)
(*Sphenaster* Wilcoxon, 1970b) = *Sphenolithus*
    (*S. metula* Wilcoxon, 1970b)
Sphenolithus Deflandre *in* Grassé (1952)
    S. abies Deflandre *in* Deflandre & Fert (1954)
    S. anarrhopus Bukry & Bramlette (1969a)
    S. belemnos Bramlette & Wilcoxon (1967)
    S. capricornutus Bukry & Percival (1971)
    S. celsus Haq (1971)
    S. ciperoensis Bramlette & Wilcoxon (1967)
    S. compactus Backman (1980)
    S. conicus Bukry (1971a)
    S. conspicuus Martini (1976)
    S. delphix Bukry (1973c)
    S. dissimilis Bukry & Percival (1971)
    S. distentus (Martini, 1965) Bramlette & Wilcoxon (1967)
    S. editus Perch-Nielsen *in* Perch-Nielsen *et al.* (1978)
    S. elongatus Perch-Nielsen (1980)
    S. furcatolithoides Locker (1967)
    S. grandis Haq & Berggren (1978)
    S. heteromorphus Deflandre (1953)
    S. intercalaris Martini (1976)
    S. moriformis (Brönnimann & Stradner, 1960) Bramlette &
        Wilcoxon (1967)
    S. neoabies Bukry & Bramlette (1969a)
    S. obtusus Bukry (1971a)
    S. orphanknollensis Perch-Nielsen (1971c)
    S. pacificus Martini (1965)
    S. predistentus Bramlette & Wilcoxon (1967)
    S. primus Perch-Nielsen (1971b)
    S. pseudoradians Bramlette & Wilcoxon (1967)
    S. quadrispinatus Perch-Nielsen (1980)
    S. radians Deflandre *in* Grassé (1952)
    S. spiniger Bukry (1971a)
    S. stellatus Gartner (1971)
    S. tribulosus Roth (1970)
    S. verensis Backman (1978)
(*Stradnerius* Haq, 1968) = *Dictyococcites*
Striatococcolithus Bukry (1971a)
    S. pacificanus Bukry (1971a)
(*Sujkowskiella* Hay, Mohler & Wade, 1966) = *Zygrhablithus*
    (*S. enigmatica* Hay, Mohler & Wade, 1966) = *Z. bijugatus*
Syracosphaera Lohmann (1902)
    S. clathrata Roth & Hay *in* Hay *et al.* (1967)
    S. histrica Kamptner (1941)
    S. pulchra Lohmann (1902)

Thoracosphaera Kamptner (1921)
  T. albatrosiana Kamptner (1963)
  (*T. atlantica* Haq & Lohmann, 1976) = *T. operculata*
  (*T. candora* Kamptner, 1967) = *T. tuberosa*
  (*T. corsena* Kamptner, 1967) = *T. heimii*
  T. deflandrei Kamptner (1956)
  (*T. ellipsoidea* Kamptner, 1967) = *T. heimii*
  T. fossata Jafar (1975a)
  T. granifera Fütterer (1978)
  (*T. granulosa* Kamptner, 1963) = *T. heimii*
  T. heimii (Lohmann, 1919) Kamptner (1941)
  (*T. imperforata* Kamptner, 1952) = *T. heimii*
  (*T. narena* Kamptner, 1967) = *T. tuberosa*
  T. operculata Bramlette & Martini (1964)
  (*T. pelagica* Kamptner, 1927) = *T. heimii*
  T. prolata Bukry & Bramlette (1969a)
  (*T. reliana* Kamptner, 1967) = *T. heimii*
  (*T. ricoseta* Kamptner, 1967) = *T. albatrosiana*
  T. saxea Stradner (1961)
  (*T. sinensis* Zhong, 1982) = framboid pyrite
  (*T. subtilis* Kamptner, 1967) = *T. heimii*
  T. tesserula Fütterer (1978)
  T. tuberosa Kamptner (1963)
Tiarolithus Kamptner (1958)
Toweius Hay & Mohler (1967)
  T. africanus (Perch-Nielsen, 1980) Perch-Nielsen (1984a)
  T. callosus Perch-Nielsen (1971d)
  T.? crassus (Bramlette & Sullivan, 1961) Perch-Nielsen (1984a)
  (*T. craticulus* Hay & Mohler, 1967) = *T. pertusus*
  T. eminens (Bramlette & Sullivan, 1961) Perch-Nielsen (1971b)
  T.? gammation (Bramlette & Sullivan, 1961) Romein (1979)
  T.? magnicrassus (Bukry, 1971) Romein (1979)
  T. occultatus (Locker, 1967) Perch-Nielsen (1971d)
  T. pertusus (Sullivan, 1965) Romein (1979)
  (*T. petalosus* Ellis & Lohmann, 1973) = *Prinsius petalosus*
  T. selandianus Perch-Nielsen (1979a)
  T. tovae Perch-Nielsen (1971b)
Transversopontis Hay, Mohler & Wade (1966)
  T. duocavus (Bramlette & Sullivan, 1961) Locker (1973)
  T. exilis (Bramlette & Sullivan, 1961) Perch-Nielsen (1971d)
  T. fimbriatus (Bramlette & Sullivan, 1961) Locker (1968)
  T. latus Müller (1970)
  T. obliquipons (Deflandre *in* Deflandre & Fert, 1954) Hay, Mohler & Wade (1966)
  T. pax Stradner & Seifert (1980)
  T. pseudopulcher Perch-Nielsen (1967)
  T. pulcher (Deflandre *in* Deflandre & Fert, 1954) Perch-Nielsen (1967)
  T. pulcheroides (Sullivan, 1964) Báldi-Beke (1971)
  T. pygmaea (Locker, 1967) Perch-Nielsen (1984a)
  T. rectipons (Haq, 1968) Roth (1970)
  T. sigmoidalis Locker (1967)
Transversopontis sp.
Tribrachiatus Shamraï (1963)
  T. bramlettei (Brönnimann & Stradner, 1960) Proto Decima *et al.* (1975)
  T. contortus (Stradner, 1958) Bukry (1972)
  T. orthostylus Shamraï (1963)
Triquetrorhabdulus Martini (1965)
  T. auritusa Stradner & Allram (1982)
  T. carinatus Martini (1965)
  T. challengeri Perch-Nielsen (1977)

T. farnsworthii (Gartner, 1967) Perch-Nielsen (1984a)
  (*T. martinii* Gartner, 1967) = *Rhabdothorax serratus*
  T. milowii Bukry (1971a)
  T. rugosus Bramlette & Wilcoxon (1967)
  T. striatus Müller (1974a)
Trochastrites Stradner (1961)
  T. bramlettei (Martini, 1958) Stradner (1961)
  T. hohnensis (Martini, 1958) Bouché (1962)
(*Trochoaster* Klumpp, 1953) = *Lithostromation*
(*Truncodiscoaster* Prins, 1971) = *Discoaster*
(*Turbodiscoaster* Prins, 1971) = *Discoaster*

Umbellosphaera Paasche *in* Markali & Paasche (1955)
  U. irregularis Paasche *in* Markali & Paasche (1955)
  ((*U. tenuis* Kamptner 1937) Paasche *in* Markali & Paasche 1955) = *U. irregularis*
Umbilicosphaera Lohmann (1902)
  U. angustiforamen Okada & McIntyre (1977)
  U. hulburtiana Gaarder (1970)
  U. jafarii Müller (1974b)
  U. maceria Okada & McIntyre (1977)
  (*U. mirabilis* Lohmann, 1902) = *U. sibogae*
  U. sibogae (Weber-van Bosse, 1901) Gaarder (1970)
  U. sibogae foliosa (Kamptner, 1963) Okada & McIntyre (1977)

Watznaueria Reinhardt (1964)
  W. barnesae (Black *in* Black & Barnes, 1959) Perch-Nielsen (1968)

Zygodiscus Bramlette & Sullivan (1961)
  Z. adamas Bramlette & Sullivan (1961)
  Z. bramlettei Perch-Nielsen (1981a)
  Z. clausus Romein (1979)
  Z. herlynii Sullivan (1964)
  Z. plectopons Bramlette & Sullivan (1961)
  (*Z. sigmoides* Bramlette & Sullivan, 1961) = *Placozygus sigmoides*
(*Zygolithus* Kamptner ex Matthes, 1956) = *Neococcolithes*
Zygosphaera Kamptner (1936)
  Z. brytika Roth (1970)
Zygrhablithus Deflandre (1959)
  Z. bijugatus (Deflandre *in* Deflandre & Fert, 1954) Deflandre (1959)

References

Achuthan, M. V. & Stadner, H. 1969. Calcareous Nannoplankton from the Wemmelian stratotype. In: P. Brönnimann & H. H. Renz (eds.) *Proceedings First International Conference on Planktonic Microfossils, Geneva*, vol. 1, pp. 1–13. E. Brill, Leiden.

Aubry, M.-P. 1983. Corrélations biostratigraphiques entre les formations paléogènes épicontinentales de l'Europe du Nord-Ouest, basées sur la nannoplancton calcaire. Thèse Université Pierre et Marie Curie, Paris 6, 83–08, 208 pp.

Aubry, M.-P. 1985. *Handbook of Cenozoic Calcareous Nannoplankton*. Micropaleontology Press, American Museum of Natural History. (Preprint.)

Bachmann, A., Papp, A. & Stradner, H. 1963. Mikropaläontologische Studien im 'Badener Tegel' von Frättingsdorf N.O. *Mitt. Geol. Ges. Wien*, **56**, 117–210.

Backman, J. 1978. Late Miocene–Early Pliocene nannofossil

biochronology and biogeography in the Vera Basin, SE
Spain. *Acta Univ. Stockholm. Contrib. Geol.*, **32** (2), 93–114.

Backman, J. 1979. Pliocene biostratigraphy of DSDP Sites 111 and
116 from the north Atlantic Ocean and the age of northern
hemisphere glaciation. *Acta Univ. Stockholm. Contrib. Geol.*,
32 (3), 115–37.

Backman, J. 1980. Miocene–Pliocene nannofossils and
sedimentation rates in the Hatton-Rockall basin, NE
Atlantic Ocean. *Acta Univ. Stockholm. Contrib. Geol.*, **36**
(1), 1–91.

Backman, J. 1984. Cenozoic calcareous nannofossil biostratigraphy
from the northeastern Atlantic Ocean – Deep Sea Drilling
Project Leg 81. *Initial Rep. Deep Sea drill. Proj.*, **81** (in
press).

Backman, J. & Shackleton, N. J. 1983. Quantitative biochronology
of Pliocene and Early Pleistocene calcareous nannoplankton
from the Atlantic, Indian and Pacific oceans. *Mar.
Micropaleontol.*, **8** (2), 141–70.

Báldi-Beke, M. 1960. (Die stratigraphische Bedeutung miozäner
Coccolithophoren aus Ungarn.) *Földt. Közl.*, **90**, 213–23.

Báldi-Beke, M. 1971. The Eocene nannoplankton of the Bakony
Mountains, Hungary. *Ann. Inst. Geol. Publ. Hung.*, **54** (4),
13–38.

Báldi-Beke, M. & Baldi, T. 1974. Nannoplankton and molluscs of
the Novaj profile, a faciostratotype for Egerian. *Földt. Közl.,
Bull. Hung. Geol. Soc.*, **104**, 60–88.

Bandy, O. L. & Wilcoxon, J. A. 1970. The Pliocene–Pleistocene
boundary, Italy and California. *Bull. geol. Soc. Am.*, **81**,
2939–47.

Baumann, P. & Roth, P. H. 1969. Zonierung des Obereozäns und
Oligozäns des Monte Cagnero (Zentralappennin) mit
planktonischen Foraminiferen und Nannoplankton. *Eclog.
geol. Helv.*, **62**, 303–23.

Berggren, W. A. & Van Couvering, J. A. 1974. The Late Neogene –
biostratigraphy, geochronology and paleoclimatology of the
last 15 million years in marine and continental sequences.
Palaeogeogr. Palaeoclimatol. Palaeoecol., **16** (1/2) special
issue, 9–35.

Biolzi, M. & Perch-Nielsen, K. 1982. *Helicosphaera truempyi*, a
new Early Miocene calcareous nannofossil. *Eclog. geol.
Helv.*, **75** (1), 171–5.

Biolzi, M., Perch-Nielsen, K. & Ramos, I. 1981. *Triquetrorhabdulus*
– an Oligocene/Miocene calcareous nannofossil genus. *INA
Newsletter*, **3** (2), 89–92.

Black, M. 1964. Cretaceous and Tertiary coccoliths from Atlantic
seamounts. *Palaeontology*, **7**, 306–16.

Black, M. 1967. New names for some coccolith taxa. *Proc. geol.
Soc. London*, **1640**, 139–45.

Black, M. 1971a. The systematics of coccoliths in relation to the
palaeontological record. In: B. M. Funnell & W. R. Riedel
(eds.) *The Micropaleontology of Oceans*, pp. 611–24.
Cambridge University Press.

Black, M. 1971b. Problematical microfossils from the Gault Clay.
Geol. Mag., **108**, 325–7.

Black, M. & Barnes, B. 1959. The structure of coccoliths from the
English Chalk. *Geol. Mag.*, **96**, 321–8.

Black, M. & Barnes, B. 1961. Coccoliths and discoasters from the
floor of the South Atlantic Ocean. *J. R. Microsc. Soc.*, **80**,
137–47.

Bóna, J. 1964. (Coccolithophoriden-Untersuchungen in der
neogenen Schichtenfolge des Mecsekgebirges.) *Földt. Közl.*,
94, 121–31 (in Hungarian).

Borsetti, A. W. & Cati, F. 1972. Il nannoplancton calcareo vivente
nel Tirreno centromeridionale. *G. Geol.*, ser. 2a, **38**,
395–452.

Bouché, P. M. 1962. Nannofossiles calcaires du Lutétien du bassin
de Paris. *Rev. Micropaleontol.*, **5**, 75–103.

Boudreaux, J. E. 1974. Calcareous nannoplankton ranges, Deep
Sea Drilling Project Leg 23. *Initial Rep. Deep Sea drill.
Proj.*, **23**, 1073–90.

Boudreaux, J. E. & Hay, W. W. 1969. Calcareous nannoplankton
and biostratigraphy of the Late Pliocene–Pleistocene–Recent
sediments in the Submarex cores. *Rev. Esp. Micropaleontol.*,
1, 249–92.

Bramlette, M. N. & Martini, E. 1964. The great change in
calcareous nannoplankton fossils between the Maestrichtian
and Danian. *Micropaleontology*, **10**, 291–322.

Bramlette, M. N. & Riedel, W. R. 1954. Stratigraphic value of
discoasters and some other microfossils related to Recent
coccolithophores. *J. Paleontol.*, **28**, 385–403.

Bramlette, M. N. & Sullivan, F. R. 1961. Coccolithophorids and
related nannoplankton of the Early Tertiary in California.
Micropaleontology, **7**, 129–74.

Bramlette, M. N. & Wilcoxon, J. A. 1967. Middle Tertiary
calcareous nannoplankton of the Cipero Section, Trinidad,
W.I. *Tulane Stud. Geol.*, **5**, 93–131.

Bréhéret, J. G. 1978a. Formes nouvelles quaternaires et actuelles de
la famille des Gephyrocapsaceae (Coccolithophorides). *C.r.
Seances Acad. Sci. Paris*, **287D**, 447–9.

Bréhéret, J. G. 1978b. Biostratigraphie du Pleistocène supérieur et
de l'Holocène de deux carottes de l'Atlantique Nord à l'aide
des coccolithes. *C.r. Seances Acad. Sci. Paris*, **287D**,
599–601.

Brönnimann, P. & Stadner, H. 1960. Die Foraminiferen- und
Discoasteridenzonen von Kuba und ihre interkontinentale
Korrelation. *Erdoel-Z.*, **76**, 364–9.

Brotzen, F. 1959. On *Tylocidaris* species (Echinoidea) and the
stratigraphy of the Danian of Sweden. *Arsbok. Sver. geol.
Unders.*, **54** (2), 1–81.

Bukry, D. 1971a. Cenozoic calcareous nannofossils from the Pacific
Ocean. *Trans. San Diego Soc. nat. Hist.*, **16**, 303–27.

Bukry, D. 1971b. Discoaster evolutionary trends.
Micropaleontology, **17**, 43–52.

Bukry, D. 1972. Further comments on coccolith stratigraphy, Leg
12, Deep Sea Drillng Project. *Initial Rep. Deep Sea drill.
Proj.*, **12**, 1071–83.

Bukry, D. 1973a. Low-latitude coccolith biostratigraphic zonation.
Initial Rep. Deep Sea drill. Proj., **15**, 685–703.

Bukry, D. 1973b. Coccolith and silicoflagellate stratigraphy, Deep
Sea Drilling Project Leg 18, eastern North Pacific. *Initial
Rep. Deep Sea drill. Proj.*, **18**, 817–31.

Bukry, D. 1973c. Coccolith stratigraphy, eastern equatorial Pacific,
Leg 16, Deep Sea Drilling Project. *Initial Rep. Deep Sea
drill. Proj.*, **16**, 653–711.

Bukry, D. 1973d. Phytoplankton stratigraphy, Deep Sea Drilling
Project Leg 20, Western Pacific Ocean. *Initial Rep. Deep Sea
drill. Proj.*, **20**, 307–17.

Bukry, D. 1975. Coccolith and silicoflagellate stratigraphy,
northwestern Pacific Ocean, Deep Sea Drilling Project Leg
32. *Initial Rep. Deep Sea drill. Proj.*, **32**, 677–701.

Bukry, D. 1976. Coccolith stratigraphy of Manihiki Plateau,
central Pacific, Deep Sea Drilling Project Site 317. *Initial
Rep. Deep Sea drill. Proj.*, **33**, 493–501.

Bukry, D. 1979. Neogene coccolith stratigraphy, Mid Atlantic

Ridge, Deep Sea Drilling Project Leg 45. *Initial Rep. Deep Sea drill. Proj.*, **45**, 307–17.

Bukry, D. 1981a. Cenozoic coccoliths from the Deep Sea Drilling Project. *Spec. Publ. Soc. econ. Paleontol. Mineral. Tulsa*, **32**, 335–53.

Bukry, D. 1981b. Pacific coast coccolith stratigraphy between Conception and Cabo Corrientes, Deep Sea Drilling Project, Leg 63. *Initial Rep. Deep Sea drill. Proj.*, **63**, 445–71.

Bukry, D. & Bramlette, M. N. 1968. Stratigraphic significance of two genera of Tertiary calcareous nannofossils. *Tulane Stud. Geol.*, **6**, 149–55.

Bukry, D. & Bramlette, M. N. 1969a. Coccolith age determinations, Leg 1, Deep Sea Drilling Project. *Initial Rep. Deep Sea drill. Proj.*, **1**, 369–87.

Bukry, D. & Bramlette, M. N. 1969b. Some new and stratigraphically useful calcareous nannofossils of the Cenozoic. *Tulane Stud. Geol. Paleontol.*, **7**, 131–42.

Bukry, D. & Bramlette, M. N. 1970. Coccolith age determinations, Leg 3, Deep Sea Drilling Project. *Initial Rep. Deep Sea drill. Proj.*, **3**, 589–611.

Bukry, D. & Bramlette, M. N. 1971. Validation of *Pedinocyclus* and *Quinquerhabdus* new calcareous nannoplankton genera. *Tulane Stud. Geol. Paleontol.*, **8**, 122.

Bukry, D., Douglas, R. G., Kling, S. A. & Krasheninnikov, V. 1971. Planktonic microfossil biostratigraphy of the northwestern Pacific Ocean. *Initial Rep. Deep Sea drill. Proj.*, **6**, 1253–1300.

Bukry, D. & Percival, S. F. 1971. New Tertiary calcareous nannofossils. *Tulane Stud. Geol. Paleontol.*, **8**, 123–46.

Burns, D. A. 1973. Structural analysis of flanged coccoliths in sediments from the South West Pacific Ocean. *Rev. Esp. Micropaleontol.*, **5** (1), 147–60.

Bybell, L. M. 1975. Middle Eocene calcareous nannofossils at Little Stave Creek, Alabama. *Tulane Stud. Geol. Paleontol.*, **11** (4), 177–247.

Bybell, L. M. 1982. Validation of *Blackites trochos*. *INA Newsletter*, **4** (2), 101.

Bybell, L. M. & Gartner, S. 1972. Provincialism among mid-Eocene calcareous nannofossils. *Micropaleontology*, **18** (3), 319–36.

Cavelier, C. 1975. Le diachronisme de la zone à *Ericsonia subdisticha* (nannoplancton) et la position de la limite Eocène–Oligocène en Europe et en Amérique du Nord. *Bull. Bur. Rech. geol. min. Paris*, vol. 2, section 4, part 3, pp. 201–25.

Cavelier, C. 1979. La limite Eocène–Oligocène en Europe occidentale. *Mem. 54 Sci. Géol. Univ. L. Pasteur, Strassbourg, Inst. de Géol.*, 280 pp.

Čepek, P. 1973. Die Art *Pontosphaera indooceanica* n.sp. und ihre Bedeutung für die Stratigraphie der jüngsten Sedimente des Indischen Ozeans. '*Meteor' Forschungs Ergebnisse*, C/12, 1–8.

Clocchiatti, M. & Jerkovic, L. 1972. *Cruciplacolithus tenuiforatus*, nouvelle espèce de Coccolithophoridé du Miocène d'Algérie et de Yougoslavie. *Cah. Micropaleontol.*, **2** (2), (1970), 1–6.

Cohen, C. L. D. 1964. Coccolithophorids from two Caribbean deep-sea cores. *Micropaleontology*, **10**, 231–50.

Cohen, C. L. D. & Reinhardt, P. 1968. Coccolithophorids from the Pleistocene Caribbean deep-sea core CP-28. *Neues Jahrb. Geol. Palaeontol. Abhandlungen.*, **131** (3), 289–304.

Deflandre, G. 1942a. Coccolithophoridées fossiles d'Oranie. Genres *Scyphosphaera* Lohmann et *Thoracosphaera* Ostenfeld. *Bull. Soc. Hist. nat. Toulouse*, **77**, 125–37.

Deflandre, G. 1942b. Possibilités morphogénétiques comparées du calcaire et de la silice, à propos d'un nouveau type de microfossile calcaire de structure complex, *Lithostromation perdurum* n. gen. n. sp. *C.r. Seances Acad. Sci. Paris*, **214**, 917–19.

Deflandre, G. 1947. *Braarudosphaera* nov. gen., type d'une famille nouvelle de Coccolithophoridés actuels à éléments composites. *C.r. Seances Acad. Sci. Paris*, **225**, 439–41.

Deflandre, G. 1950. Observations sur les Coccolithophoridés, à propos d'un nouveau type de Braarudosphaeridé, *Micrantholithus*, à éléments clastiques. *C.r. Seances Acad. Sci. Paris*, **231**, 1156–8.

Deflandre, G. 1953. Hétérogénéité intrinsèque et pluralité des éléments dans les coccolithes actuels et fossiles. *C.r. Seances Acad. Sci. Paris*, **237**, 1785–7.

Deflandre, G. 1957. *Goniolithus* nov. gen., type d'une famille nouvelle de Coccolithophoridés fossiles, à éléments pentagonaux non composites. *C.r. Seances Acad. Sci. Paris*, **244**, 2539–41.

Deflandre, G. 1959. Sur les nannofossiles calcaires et leur systématique. *Rev. Micropaleontol.*, **2**, 127–52.

Deflandre, G. & Fert, C. 1954. Observations sur les Coccolithophoridés actuels et fossiles en microscopie ordinaire et électronique. *Ann. Paleontol.*, **40**, 115–76.

Driever, B. W. M. 1981. A quantitative study of Pliocene associations of *Discoaster* from the Mediterranean. *Proc. Kon. Ned. Akad. Wetensch.*, B. **84** (4), 437–55.

Duplessy, J. C., Moyes, J., Pujol, C., Pujos-Lamy, A. & Reyss, J. L. 1975. Stratigraphie des dépots quaternaires d'une carotte prélevée au N.E. des Açores. *C.r. Seances Acad. Sci. Paris*, **281**, ser. D, 1971–1974.

Edwards, A. R. 1971. A calcareous nannoplankton zonation of the New Zealand Paleogene. In: A. Farinacci (ed.), *Proceedings II Planktonic Conference Roma*, 1970, **1**, 381–419.

Edwards, A. R. 1973a. Key species of New Zealand calcareous nannofossils. *N. Z. J. Geol. Geophys.*, **16** (1), 68–89.

Edwards, A. R. 1973b. Calcareous nannofossils from the southwest Pacific, Deep Sea Drilling Project, Leg 21. *Initial Rep. Deep Sea drill. Proj.*, **21**, 641–91.

Edwards, A. R. & Perch-Nielsen, K. 1975. Calcareous nannofossils from the southern Southwest Pacific, Deep Sea Drilling Project, Leg 29. *Initial Rep. Deep Sea drill. Proj.*, **29**, 469–539.

Ehrenberg, C. G. 1832 (separate 1830). Beiträge zur Kenntnis der Organisation der Infusorien und ihrer geographischen Verbreitung, besonders in Sibirien. *Abh. Preuss. Akad. der Wissensch.*, 1830, 1–88.

Ellis, C. H. 1975. Calcareous nannofossil biostratigraphy Leg 31, DSDP. *Initial Rep. Deep Sea drill. Proj.*, **31**, 655–76.

Ellis, C. H. & Lohman, W. H. 1973. *Toweius petalosus* new species, a Paleocene calcareous nannofossil from Alabama. *Tulane Stud. Geol. Paleontol.*, **10**, 107–10.

Farinacci, A. 1969–1979. *Catalogue of Calcareous Nannofossils*. Ed. Tecnoscienza, Rome, 10 vols.

Farinacci, A. 1971. Roundtable on Calcareous Nannoplankton Roma. In: A. Farinacci (ed.) *Proceedings II Planktonic Conference, Roma*, 1970, **2**, 1343–69.

Forchheimer, S. 1972. Scanning electron microscope studies of Cretaceous coccoliths from the Köpingsberg borehole no. 1 SE Sweden. *Sver. Geol. Unders.*, C/668, 65/14, 1–141.

Forchheimer, S. & Stradner, H. 1973. *Scampanella*, eine neue Gattung kretazischer Nannofossilien. *Verh. geol. Bundesanst. (Wein)*, **2**, 285–9.

Fütterer, D. 1976. Kalkige Dinoflagellaten (Calciodinelloideae) und die systematische Stellung der Thoracosphaeridoideae. *Neues Jahrb. Geol. Palaeontol. Abhandlungen*, **151** (2), 119–41.

Fütterer, D. 1978. Distribution of calcareous dinoflagellates in Cenozoic sediments of Site 366, Eastern North Atlantic. *Initial Rep. Deep Sea drill. Proj.*, **41**, 709–37.

Gaarder, K. R. 1970. Three new taxa of Coccolithinae. *Nytt Mag. Bot.*, **17**, 113–26.

Gaarder, K. R. & Heimdal, B. R. 1973. Light and scanning electron microscope observations on Rhabdothorax regale (Gaarder) Gaarder nov. comb. *Norw. J. Bot.*, **20**, 89–97.

Gaarder, K. R. & Heimdal, B. R. 1977. A revision of the genus *Syracosphaera* Lohmann (Coccolithineae). '*Meteor' Forsch. Ergebnisse*, **D/24**, 54–71.

Gartner, S. Jr 1967. Calcareous nannofossils from Neogene of Trinidad, Jamaica, and Gulf of Mexico. *Univ. Kansas, Paleontol. Contrib.*, **29**, 1–7.

Gartner, S. Jr 1968. Coccoliths and related calcareous nannofossils from Upper Cretaceous deposits of Texas and Arkansas. *Univ. Kansas Paleontol. Contrib.*, **48**, Protista 1, 1–56.

Gartner, S. Jr 1969a. Two new calcareous nannofossils from the Gulf Coast Eocene. *Micropaleontology*, **15**, 31–4.

Gartner, S. Jr 1969b. Hayella Roth and Hayella Gartner. *Micropaleontology*, **15**, 490.

Gartner, S. Jr 1969c. Correlation of Neogene planktonic foraminifera and calcareous nannofossil zones. *Trans. Gulf Coast Assoc. geol. Soc.*, **19**, 585–99.

Gartner, S. Jr 1970. Phylogenetic lineages in the Lower Tertiary coccolith genus *Chiasmolithus*. *North Am. Paleontol. Convention Sept.* 1969, *Proc. G*, pp. 930–57.

Gartner, S. Jr 1971. Calcareous nannofossils from the JOIDES Blake Plateau cores and revision of Paleogene nannofossil zonation. *Tulane Stud. Geol.*, **8**, 101–21.

Gartner, S. Jr 1972. Late Pleistocene calcareous nannofossils in the Caribbean and their interoceanic correlation. *Palaeogeogr. Palaeoclimatol. Palaeoecol.*, **12**, 169–91.

Gartner, S. Jr 1973. Absolute chronology of the Late Neogene calcareous nannofossil succession in the Equatorial Pacific. *Bull. geol. Soc. Am.*, **84**, 2021–34.

Gartner, S. Jr 1977a. Nannofossils and biostratigraphy: an overview. *Earth Sci. Rev.*, **13**, 227–50.

Gartner, S. Jr 1977b. Calcareous nannofossil biostratigraphy and revised zonation of the Pleistocene. *Mar. Micropaleontol.*, **2**, 1–25.

Gartner, S. Jr 1980. Validation of *Helicopontosphaera inversa* (2). *INA Newsletter*, **2** (1), 35.

Gartner, S. Jr & Bukry, D. 1969. Tertiary holococcoliths. *J. Paleontol.*, **43**, 1213–21.

Gartner, S. Jr & Bukry, D. 1974. *Ceratolithus acutus* Gartner & Bukry n. sp. and *Ceratolithus amplificus* Bukry & Percival – nomenclatural clarification. *Tulane Stud. Geol. Paleontol.*, **11**, 115–18.

Gartner, S. Jr & Bukry, D. 1975. Morphology and phylogeny of the coccolithophycean family Ceratolithaceae. *J. Res. U. S. geol. Surv.*, **3**, 451–65.

Gartner, S. Jr & Smith, L. A. 1967. Coccoliths and related calcareous nannofossils from the Yazoo Formation (Jackson, late Eocene) of Louisiana. *Univ. Kansas Paleontol. Contrib.*, **20**, 1–7.

Geitzenauer, K. R. 1972. The Pleistocene calcareous nannoplankton of the subantarctic Pacific Ocean. *Deep Sea Res.*, **19**, 45–61.

Gran, H. H. & Braarud, T. 1935. A quantitative study of the phytoplankton in the Bay of Fundy and the Gulf of Maine (including observations on hydrography, chemistry and turbidity). *J. Biol. Board Canada*, **1**, 279–467.

Grassé, P. P. 1952. *Traité de zoologie*. Masson, Paris.

Haeckel, E. 1894. *Systematische Phylogenie der Protisten und Pflanzen*. Reimer, Berlin, 400 pp.

Haq, B. U. 1966. Electron microscope studies on some Upper Eocene calcareous nannoplankton from Syria. *Stockholm Contrib. Geol.*, **15**, 23–37.

Haq, B. U. 1968. Studies on Upper Eocene calcareous nannoplankton from NW Germany. *Stockholm Contrib. Geol.*, **18**, 13–74.

Haq, B. U. 1969. The structure of Eocene coccoliths and discoasters from a Tertiary deep-sea core in the Central Pacific. *Stockholm Contrib. Geol.*, **21**, 1–19.

Haq, B. U. 1971. Paleogene calcareous nannoflora Parts I–IV. *Stockholm Contrib. Geol.*, **25**, 1–158.

Haq, B. U. 1973. Evolutionary trends in the Cenozoic coccolithophore genus *Helicopontosphaera*. *Micropaleontology*, **19**, 32–52.

Haq, B. U. 1976. Coccoliths in cores from the Bellinghausen abyssal plain and Antarctic continental rise. *Initial Rep. Deep Sea drill. Proj.*, **35**, 557–67.

Haq, B. U. 1978. Calcareous nannoplankton. In: B. U. Haq & A. Boersma (eds.), *Introduction to Marine Micropaleontology*, pp. 79–107. Elsevier.

Haq, B. U. 1980. Biogeographic history of Miocene calcareous nannoplankton and paleoceanography of the Atlantic Ocean. *Micropaleontology*, **26** (4), 414–43.

Haq, B. U. & Aubry, M.-P. 1981. Early Cenozoic calcareous nannoplankton biostratigraphy and palaeobiogeography of North Africa and the Middle East and trans-Tethyan correlations. In: M. J. Salem & M. T. Busrewil (eds.), *Geology of Libya*, vol. 1, pp. 271–304. Academic Press, London.

Haq, B. U. & Berggren, W. A. 1978. Late Neogene calcareous plankton biochronology of the Rio Grande Rise (South Atlantic Ocean). *J. Paleontol.*, **52** (6), 1167–94.

Haq, B. U. & Lohman, G. P. 1976. Early Cenozoic calcareous nannoplankton biogeography of the Atlantic Ocean. *Mar. Micropaleontol.*, **1**, 119–94.

Hay, W. W. 1964. The use of the electron microscope in the study of fossils. *Annu. Rep. Smithsonian Inst.*, 1963, 409–15.

Hay, W. W. 1970. Calcareous nannofossils from cores recovered on Leg 4. *Initial Rep. Deep Sea drill. Proj.*, **4**, 455–501.

Hay, W. W. 1972. Probabilistic stratigraphy. *Eclog. geol. Helv.*, **65**, 255–66.

Hay, W. W. 1977. Calcareous nannofossils. In: A. T. S. Ramsay, *Oceanic Micropaleontology*, pp. 1055–1200, Academic Press.

Hay, W. W. & Beaudry, F. M. 1973. Calcareous nannofossils – Leg 15, Deep Sea Drilling Project. *Initial Rep. Deep Sea drill. Proj.*, **15**, 625–83.

Hay, W. W. & Boudreaux, J. E. 1969. Calcareous nannoplankton and biostratigraphy of the Late Pliocene–Pleistocene–Recent sediments in the Submarex cores. *Rev. Esp. Micropaleontol.*, **1** (3), 249–92.

Hay, W. W. & Mohler, H. P. 1967. Calcareous nannoplankton from early Tertiary rocks at Pont Labau, France, and Paleocene–Eocene correlations. *J. Paleontol.*, **41**, 1505–41.

Hay, W. W., Mohler, H. P., Roth, P. H., Schmidt, R. R. & Boudreaux, J. E. 1967. Calcareous nannoplankton zonation of the Cenozoic of the Gulf Coast and Caribbean–Antillean area, and transoceanic correlation. *Trans. Gulf Coast Assoc. geol. Soc.*, **17**, 428–80.

Hay, W. W., Mohler, H. P. & Wade, M. E. 1966. Calcareous nannofossils from Nal'chik (northwest Caucasus). *Eclog. geol. Helv.*, **59**, 379–99.

Hay, W. W. & Towe, K. M. 1962. Electron microscope examination of some coccoliths from Donzacq (France). *Eclog. geol. Helv.*, **55**, 497–517.

Heck, S. van 1979–82. Bibliography and taxa of calcareous nannoplankton. *INA Newsletter*, **1**(1), AB I–B27; **1** (2), 13–42; **2** (1), 5–34; **2** (2), 43–80; **3** (1), 4–41; **3** (2), 51–86; **4** (1), 7–50; **4** (2), 65–96.

Heimdal, B. R. 1973. Two new taxa of Recent coccolithophorids. *'Meteor' Forschungs Ergebnisse*, D/13, 70–5.

Heimdal, B. R. & Gaarder, K. R. 1980. Coccolithophorids from the northern part of the eastern central Atlantic. I. Holococcolithophorids. *'Meteor' Forschungs Ergebnisse*, D/32, 1–14.

Heimdal, B. R. & Gaarder, K. R. 1981. Coccolithophorids from the northern part of the eastern central Atlantic. II. Heterococcolithophorids. *'Meteor' Forschungs Ergebnisse*, D/33, 37–69.

Hekel, H. 1968. Nannoplanktonhorizonte und tektonische Strukturen in der Flyschzone nördlich von Wien (Bisambergzug). *Jb. Geol. Bundes Anstalt* III, 293–337.

Hekel, H. 1973. Late Oligocene to Recent nannoplankton from the Capricorn Basin (Great Barrier Reef Area). *Publ. Queensl. geol. Surv.* 359, paleont. papers **33**, 1–24.

Hoffman, N. 1970. Coccolithineen aus der weissen Schreibkreide (Unter-Maastricht) von Jasmund auf Rügen. *Geologie*, **19**, 846–79.

Honjo, S. & Okada, H. 1974. Community structure of coccolithophores in the photic layer of the mid-Pacific. *Micropaleontology*, **20** (2), 209–30.

Huang, T.-C. & Ting, J.-S. 1979. Calcareous nannofossil succession from the Oligo–Miocene Peikangchi section and revised stratigraphic correlation between Northern and Central Taiwan. *Proc. geol. Soc. China*, **22**, 105–20.

Huxley, T. H. 1868. On some organisms living at great depths in the North Atlantic Ocean. *Q. J. Microscop. Sci.* 2, **8**, 203–12.

International Code of Botanical Nomenclature. 1978. Bohn, Scheltema & Holkema, Utrecht, 457 pp.

Jafar, S. A. 1975a. Calcareous nannoplankton from the Miocene of Rotti, Indonesia. *Verh. Kon. Ned. Akad. Wetensch. afd. Nat.*, **28**, 1–99.

Jafar, S. A. 1975b. Some comments on the calcareous nannoplankton genus *Scyphosphaera* and the neotypes of *Scyphosphaera* species from Rotti, Indonesia. *Senckenbergiana Lethaea*, **56** (4/5), 365–79.

Jafar, S. A. 1979. Taxonomy, stratigraphy and affinities of calcareous nannoplankton genus *Thoracosphaera* Kamptner. *IV. Int. Palyn. Conf.*, Lucknow (1976/1977), **2**, 1–21.

Jafar, S. A. & Martini, E. 1975. On the validity of the calcareous nannoplankton genus *Helicosphaera. Senckenbergiana Lethaea*, **56** (4/5), 381–97.

Janin, M. C. 1981a. Essai de datation de concrétions polymétalliques et évolution quaternaire du coccolithe *Cyclococcolithus leptoporus–macintyrei. Bull. Soc. geol. Fr.*, ser. 7, **23** (3), 287–96.

Janin, M. C. 1981b. Etude micropaléontologique de quelques concrétions polymétalliques. *Mem. Sci. Terre. Paris*, 1–120.

Jerkovič, L. 1970. *Noelaerhabdus* nov. gen. type d'une nouvelle famille de coccolithophoridés fossiles: Noelaerhabdaceae du Miocène supérieur de Yougoslavie. *C.r. Hebd. Seances Acad. Sci.*, **270**, 468–70.

Kamptner, E. 1927. Beitrag zur Kenntnis adriatischer Coccolithophoriden. *Arch. Protistenk.*, **58**, 173–84.

Kamptner, E. 1928. Über das System und die Phylogenie der Kalkflagellaten. *Arch. Protistenk.*, **64**, 19–43.

Kamptner, E. 1936. Über die Coccolithineen der Südwestküste von Istrien. *Anz. Akad. Wiss. Wien, Math.-Naturw. Kl.*, **73**, 243–7.

Kamptner, E. 1937. Neue und bemerkenswerte Coccolithineen aus dem Mittelmeer. *Arch. Protistenk.*, **89**, 279–316.

Kamptner, E. 1941. Die Coccolithineen der Südwestküste von Istrien. *Ann. Naturh. Mus. Wien*, **51**, 54–149.

Kamptner, E. 1943. Zur Revision der Coccolithineen-Spezies *Pontosphaera huxleyi* Lohm. *Anz. Akad. Wiss. Wien, Math.-Naturw. Kl.*, **80**, 43–9.

Kamptner, E. 1948. Coccolithen aus dem Torton des inneralpinen Wiener Beckens. *Sitz. Ber. Österr. Akad. Wiss., Math.-Naturw. Kl.*, Part 1, **157**, 1–16.

Kamptner, E. 1950. Über den submikroskopischen Aufbau der Coccolithen. *Anz. Österr. Akad. Wiss., Math.-Naturw. Kl.*, **87**, 152–8.

Kamptner, E. 1952. Das mikroskopische Studium des Skelettes der Coccolithineen (Kalkflagellaten). *Mikroskopie*, **7**, 232–44 and 375–86.

Kamptner, E. 1954. Untersuchungen über den Feinbau der Coccolithen. *Arch. Protistenk.*, **100**, 1–90.

Kamptner, E. 1955. Fossile Coccolithineen-Skelettreste aus Insulinde. *Verh. Kon. Ned. Akad. Wetensch. afd. Nat.*, **50** (2), 1–105.

Kamptner, E. 1956. *Thoracosphaera deflandrei* nov. spec., ein bemerkenswertes Kalkflagellaten-Gehäuse aus dem Eocän von Donzacq (Dep. Landes, Frankreich). *Österr. Bot. Z.*, **103**, 142–63.

Kamptner, E. 1958. Betrachtungen zur Systematik der Kalkflagellaten, nebst Versuch einer neuen Gruppierung der Chrysomonadales. *Arch. Protistenk.*, **101**, 171–202.

Kamptner, E. 1963. Coccolithineen-Skelettreste aus Tiefseeablagerungen des Pazifischen Ozeans. *Ann. Naturh. Mus. Wien*, **66**, 139–204.

Kamptner, E. 1967. Kalkflagellaten-Skelettreste aus Tiefseeschlamm des Südatlantischen Ozeans. *Ann. Naturh. Mus. Wien*, **71**, 117–98.

Klumpp, B. 1953. Beitrag zur Kenntnis der Mikrofossilien des Mittleren und Oberen Eozän. *Palaeontographica*, **103A**, 377–406.

Lecal-Schlauder, J. 1950. Notes préliminaires sur les coccolithophoridés d'Afrique du Nord. *Bull. Soc. Hist. nat. Afr. Nord*, **40**, 160–7.

Lehotayova, R. & Priewalder, H. 1978. *Cycloperfolithus*, eine neue Nannofossil Gattung aus dem Badenien der Zentralen Paratethys. In: A. Papp, I. Cicha, J. Seneš & F. Steininger (eds.), *Chronostratigraphie und Neostratotypen*, 4, M$_4$ Badenien, pp. 486–9.

Lemmermann, E. 1908. Flagellatae, Chlorophyceae, Coccosphaerales und Silicoflagellatae. In: K. Brandt & C. Apstein (eds.), *Nordisches Plankton*, pp. 1–40. Lipsius & Tischer, Kiel and Leipzig.

Levin, H. L. 1965. Coccolithophoridae and related microfossils from the Yazoo formation (Eocene) of Mississippi. *J. Paleontol.*, **39**, 265–72.

Levin, H. L. & Joerger, A. P. 1967. Calcareous nannoplankton from the Tertiary of Alabama. *Micropaleontology*, **13**, 163–82.

Lezaud, L. 1968. Espèces nouvelles de nannofossiles calcaires (Coccolithophoridés) d'Aquitaine Sud-Ouest. *Rev. Micropaleontol.*, **11**, 22–8.

Lipps, J. H. 1969. *Triquetrorhabdulus* and similar calcareous nannoplankton. *J. Paleontol.*, **43**, 1029–32.

Locker, S. 1965. Coccolithophoriden aus Eozänschollen Mecklenburgs. *Geologie* (Berlin), **14**, 1252–65.

Locker, S. 1967. Neue Coccolithophoriden (Flagellata) aus dem Alttertiär Norddeutschlands. *Geologie* (Berlin), **16**, 361–4.

Locker, S. 1968. Biostratigraphie des Alttertiärs von Norddeutschland mit Coccolithophoriden. *Monatsber. Deutsch. Akad. Wiss. Berlin*, **10**, 220–9.

Locker, S. 1973. Coccolithineen aus dem Paläogen Mitteleuropas. *Paläobotanik*, **3**, 735–836.

Loeblich, A. R. Jr & Tappan, H. 1963. Type fixation and validation of certain calcareous nannoplankton genera. *Proc. Biol. Soc. Wash.*, **76**, 191–6.

Loeblich, A. R. Jr & Tappan, H. 1966, 1968, 1969, 1970a, 1970b, 1971, 1973. Annotated index and bibliography of the calcareous nannoplankton. I: *Phycologia*, **5**, 81–216; II: *J. Paleontol.*, **42**, 584–98; III: *J. Paleontol.*, **43**, 568–88; IV: *J. Paleontol.*, **44**, 558–74; V: *Phycologia*, **9**, 157–74; VI: *Phycologia*, **10**, 315–39; VII: *J. Paleontol.*, **47**, 715–59.

Loeblich, A. R. Jr & Tappan, H. 1978. The coccolithophorid genus *Calcidicus* Kamptner and its synonyms. *J. Paleontol.*, **52** (6), 1390–2.

Lohmann, H. 1902. Die Coccolithophoridae, eine Monographie der Coccolithen bildenden Flagellaten, zugleich ein Beitrag zur Kenntnis des Mittelmeerauftriebs. *Arch. Protistenk.*, **1**, 89–165.

Lohmann, H. 1912. Untersuchungen über das Pflanzen- und Tierleben der Hochsee. *Veroff. Inst. Meereskd. Univ. Berlin*, N.F., Geogr.-Naturw. Reihe, **1**, 1–92.

Lohmann, H. 1919. Die Bevölkerung des Ozeans mit Plankton nach den Ergebnissen der Zentrigugenfänge während der Ausreise der 'Deutschland' 1911. *Arch. Biont.* (Ges. Naturf. Freunde Berlin), **4** (3), 1–617.

McIntyre, A. 1970. *Gephyrocapsa protohuxleyi* sp. n. a possible phyletic link and index fossil for the Pleistocene. *Deep Sea Res.*, **17**, 187–90.

McIntyre, A. & Bé, A. W. H. 1967a. Modern Coccolithophoridae of the Atlantic Ocean. I. Placoliths and cyrtoliths. *Deep Sea Res.*, **14**, 561–97.

McIntyre, A. & Bé, A. W. H. 1967b. *Coccolithus neohelis* sp. n., a fossil coccolith type in contemporary seas. *Deep Sea Res.*, **14**, 369–71.

McIntyre, A., Bé, A. W. H. & Preikstas, R. 1967. Coccoliths and the Pliocene–Pleistocene boundary. *Prog. Oceanography*, **4**, 3–25.

McIntyre, A., Bé, A. W. H. & Roche, M. B. 1970. Modern Pacific Coccolithophorida: a paleontological thermometer. *Trans. N. Y. Acad. Sci.*, ser. II, **32**, 720–31.

Markali, J. & Paasche, E. 1955. On two species of *Umbellosphaera*, a new marine coccolithophorid genus. *Nytt Mag. Bot.*, **4**, 95–100.

Martini, E. 1958. Discoasteriden und verwandte Formen im NW-deutschen Eozän (Coccolithophorida). 1. Taxonomische Untersuchungen. *Senckenbergiana Lethaea*, **39**, 353–88.

Martini, E. 1959a. Discoasteriden und verwandte Formen im NW-deutschen Eozän (Coccolithophorida). 2. Stratigraphische Auswertung. *Senckenbergiana Lethaea*, **40**, 137–57.

Martini, E. 1959b. Der stratigraphische Wert von Nanno-Fossilien im nordwestdeutschen Tertiär. *Erdöl & Kohle*, **12**, 137–40.

Martini, E. 1959c. *Pemma angulatum* und *Micrantholithus basquensis*, zwei neue Coccolithophoriden-Arten aus dem Eozän. *Senckenbergiana Lethaea*, **40**, 415–21.

Martini, E. 1960. Braarudosphaeriden, Discoasteriden und Verwandte Formen aus dem Rupelton des Mainzer Beckens. *Notizbl. Hess. Landesamt. Bodenforsch.*, **88**, 65–87.

Martini, E. 1961a. Nannoplankton aus dem Tertiär und der obersten Kreide von SW-Frankreich. *Notizbl. Hess. Landesamt. Bodenforsch.*, **42**, 1–32.

Martini, E. 1961b. Der stratigraphische Wert der Lithostromationidae. *Erdöl-Z.*, **77**, 100–3.

Martini, E. 1965. Mid-Tertiary calcareous nannoplankton from Pacific deep-sea cores. In: W. F. Whittard & R. B. Bradshaw (eds.), Submarine Geology and Geophysics. *Proc. 17th Symp. Colston Res. Soc. London*, pp. 393–411. Butterworths.

Martini, E. 1967. *Ceratolithina hamata* n.g., n. sp., aus dem Alb von N-Deutschland (Nannoplankton, incertae sedis). *Neues Jahrb. Geol. Palaeontol. Abhandlungen*, **128**, 294–8.

Martini, E. 1968. Calcareous nannoplankton from the type Langhian. *G. Geol.*, **35**, 163–72.

Martini, E. 1969. Nannoplankton aus dem Miozän von Gabon (Westafrika). *Neues Jahrb. Geol. Palaeontol. Abhandlungen*, **132**, 285–300.

Martini, E. 1970a. Standard Palaeogene calcareous nannoplankton zonation. *Nature*, **226**, 560–1.

Martini, E. 1970b. *Imperiaster* n. g. aus dem europäischen Unter-Eozän (Nannoplankton, incertae sedis). *Senckenbergiana Lethaea*, **51**, 383–6.

Martini, E. 1971. Standard Tertiary and Quaternary calcareous nannoplankton zonation. In: A. Farinacci (ed.), *Proceedings II Planktonic Conference, Roma, 1970*, **2**, 739–85.

Martini, E. 1973. Nannoplankton-Massenvorkommen in den Mittleren Pechelbronner Schichten (Unter-Oligozän). *Oberrhein. Geol. Abh.*, **22**, 1–22.

Martini, E. 1974. In: E. Martini & V. Moisescu, Nannoplankton–Untersuchungen in oligozänen Ablagerungen zwischen Cluj und Huedin (NW Siebenbürgisches Becken, Rumänien). *Neues Jahrb. Geol. Palaeontol. Monatshefte*, **1**, 18–37.

Martini, E. 1976. Cretaceous to Recent calcareous nannoplankton from the Central Pacific Ocean (DSDP Leg 33). *Initial Rep. Deep Sea drill. Proj.*, **33**, 383–423.

Martini, E. 1979. Calcareous nannoplankton and silicoflagellate biostratigraphy at Reykjanes Ridge, North-eastern North Atlantic (DSDP Leg 49, Sites 407 and 409). *Initial Rep. Deep Sea drill. Proj.*, **49**, 533–50.

Martini, E. 1980. Oligocene to Recent calcareous nannoplankton from the Philippine Sea, Deep Sea Drilling Project Leg 59. *Initial Rep. Deep Sea drill. Proj.*, **59**, 547–65.

Martini, E. 1981a. Nannoplankton in der Ober-Kreide, im Alttertiär und im tieferen Jungtertiär von Süddeutschland und dem angrenzenden Österreich. *Geol. Bavar.*, **82**, 345–56.

Martini, E. 1981b. Oligocene to Recent calcareous nannoplankton from the Philippine sea, Deep Sea Drilling Project Leg 59. *Initial Rep. Deep Sea drill. Proj.*, **59**, 547–65.

Martini, E. & Bramlette, M. N. 1963. Calcareous nannoplankton from the experimental Mohole drilling. *J. Paleontol.*, **37**, 845–56.

Martini, E. & Müller, C. 1972. Nannoplankton aus dem nördlichen Arabischen Meer. '*Meteor*' *Forschungs Ergebnisse*, **C/10**, 63–74.

Martini, E. & Ritzkowski, S. 1968. Die Grenze Eozän/Oligozän in der Typus-Region des Unteroligozäns (Helmstedt-Egeln-Latdorf). *Mem. Bur. Rech. Geol. Mineral.*, **69** (Colloque sur l'Eocène, Paris, mai 1968), 233–7.

Martini, E. & Stradner, H. 1960. *Nannotetraster*, eine stratigraphisch bedeutsame neue Discoasteridengattung. *Erdöl-Z.*, **76**, 266–70.

Martini, E. & Worsley, T. 1970. Standard Neogene calcareous nannoplankton zonation. *Nature*, **225**, 289–90.

Martini, E. & Worsley, T. 1971. Tertiary calcareous nannoplankton from the western equatorial Pacific. *Initial Rep. Deep Sea drill. Proj.*, **7**, 1471–1507.

Matthes, H. I. 1956. *Einführung in die Mikropaläontologie*, Hirzel, Leipzig, 348 pp.

Mazzei, R., Raffi, I., Rio, D., Hamilton, N. & Cita, M. B. 1979. Calibration of Late Neogene calcareous plankton datum planes with the paleomagnetic record of Site 397 and correlation with Moroccan and Mediterranean sections. *Initial Rep. Deep Sea drill. Proj.*, **47**, 375–89.

Medd, A. W. 1971. Some middle and upper Jurassic Coccolithophoridae from England and France. In: A. Farinacci (ed.), *Proceedings II Planktonic Conference, Roma*, 1970, **2**, 821–45.

Medd, A. W. 1979. The Upper Jurassic coccoliths from the Haddenham and Gamlingay boreholes (Cambridgeshire, England). *Eclog. geol. Helv.*, **72** (1), 19–109.

Miller, P. 1981. Tertiary calcareous nannoplankton and benthic foraminifera biostratigraphy of the Point Arena area, California. *Micropaleontology*, **27** (4), 419–43.

Moshkovitz, S. & Ehrlich, A. 1980. Distribution of the calcareous nannofossils in the Neogene sequence of the Jaffa-1 Borehole, Central Coastal Plain, Israel. *Geol. Surv. Israel Report P.D./1/80*, 1–25.

Müller, C. 1970. Nannoplankton-Zonen der Unteren Meeresmolasse Bayerns. *Geol. Bavar.*, **63**, 107–18.

Müller, C. 1972. Kalkiges Nannoplankton aus Tiefseekernen des Ionischen Meeres. '*Meteor*' *Forschungs Ergebnisse*, **C/10**, 75–95.

Müller, C. 1974a. Calcareous nannoplankton, Leg 25 (Western Indian Ocean). *Initial Rep. Deep Sea drill. Proj.*, **25**, 579–633.

Müller, C. 1974b. Nannoplankton aus dem Mittel-Miozän von Walbersdorf (Burgenland). *Senckenbergiana Lethaea*, **55**, 389–405.

Müller, C. 1976. Tertiary and Quaternary calcareous nannoplankton in the Norwegian–Greenland Sea, DSDP Leg 38. *Initial Rep. Deep Sea drill. Proj.*, **38**, 823–41.

Müller, C. 1979. Calcareous nannofossils from the North Atlantic (Leg 48). *Initial Rep. Deep Sea drill. Proj.*, **48**, 589–639.

Müller, C. 1981. Beschreibung neuer *Helicosphaera*-Arten aus dem Miozän und Revision biostratigraphischer Reichweiten einiger neogener Nannoplankton-Arten. *Senckenbergiana Lethaea*, **61** (3/6), 427–35.

Müller, C. & Brönnimann, P. 1974. Eine neue Art der Gattung *Helicosphaera* Kamptner aus dem Pazifischen Ozean. *Eclog. geol. Helv.*, **67**, 661–2.

Murray, G. & Blackman, V. H. 1898. On the nature of the coccospheres and rhabdosphaeres. *Philos. Trans. R. Soc. London*, **190B**, 427–41.

Nishida, S. 1971. Nannofossils from Japan IV. *Trans. Proc. Palaeontol. Soc. Japan*, new ser., **83**, 143–61.

Nishida, S. 1979. Atlas of Pacific nannoplanktons. *News of Osaka Micropaleontologists, special paper*, **3**, 1–31.

Noël, D. 1965. *Sur les coccolithes du Jurassique Européen et d'Afrique du Nord*. Centre Nat. Rech. Sc., 209 pp.

Norris, R. E. 1965. Living cells of *Ceratolithus cristatus* (Coccolithophorineae). *Arch. Protistenk.*, **108**, 19–24.

Okada, H. 1984. Modern nannofossil assemblages in sediments of coastal and marginal seas along the Western Pacific Ocean. *Utrecht Micropaleontol. Bull.* **30**, 171–87.

Okada, H. & Bukry, D. 1980. Supplementary modification and introduction of code numbers to the low-latitude coccolith biostratigraphic zonation (Bukry, 1973; 1975). *Mar. Micropaleontol.*, **5** (3), 321–5.

Okada, H. & Honjo, S. 1973. The distribution of oceanic coccolithophorids in the Pacific. *Deep Sea Res.*, **20**, 355–74.

Okada, H. & Honjo, S. 1975. Distribution of coccolithophorids in marginal seas along the western Pacific Ocean and in the Red Sea. *Mar. Biol.*, **31**, 271–85.

Okada, H. & McIntyre, A. 1977. Modern coccolithophores of the Pacific and North Atlantic Oceans. *Micropaleontology*, **23**, 1–55.

Okada, H. & McIntyre, A. 1979. Seasonal distribution of modern coccolithophores in the Western North Atlantic Ocean. *Marine Biol.*, **54**, 319–28.

Okada, H. & McIntyre, A. 1980. Validation of *Florisphaera profunda* var. *elongata*. *INA Newsletter*, **2** (2), 81.

Okada, H. & Thierstein, H. R. 1979. Calcareous nannoplankton – Leg 43, DSDP. *Initial Rep. Deep Sea drill. Proj.*, **43**, 507–73.

Ostenfeld, C. H. 1900. Über *Coccosphaera*. *Zool. Anz.*, **23**, 198–200.

Ostenfeld, C. H. 1910. *Thorosphaera*, eine neue Gattung der Coccolithophoriden. *Ber. Deutsch. Bot. Ges.*, **28**, 397–400.

Pavšič, J. 1977. Nannoplankton from the Upper Cretaceous and Paleocene beds in the Gorica region. *Geol. Razprave in Poročila*, **20**, 33–83.

Perch-Nielsen, K. 1967. Nannofossilien aus dem Eozän von Dänemark. *Eclog. geol. Helv.*, **60** (1), 19–32.

Perch-Nielsen, K. 1968. *Naninfula*, genre nouveau des nannofossiles calcaires. *C.r. Seances Acad. Sci. Paris*, **267**, 2298–300.

Perch-Nielsen, K. 1969. Die Coccolithen einiger dänischer Maastrichtien- und Danienlokalitäten. *Bull. geol. Soc. Denmark*, **19**, 51–66.

Perch-Nielsen, K. 1971a. Durchsicht Tertiärer Coccolithen. *Proceedings II Planktonic Conference, Roma*, 1970, **2**, 939–80.

Perch-Nielsen, K. 1971b. Einige neue Coccolithen aus dem Paläozän der Bucht von Biskaya. *Bull. geol. Soc. Denmark*, **21**, 347–61.

Perch-Nielsen, K. 1971c. Neue Coccolithen aus dem Paläozän von Dänemark, der Bucht von Biskaya und dem Eozän der Labrador See. *Bull. geol. Soc. Denmark*, **21**, 51–66.

Perch-Nielsen, K. 1971d. Elektronenmikroskopische Untersuchungen an Coccolithen und verwandten Formen aus dem Eozän von Dänemark. *Det Kongelige Danske Videnskabernes Selskab Biol. Skrifter*, **18** (3), 1–76.

Perch-Nielsen, K. 1971e. Coccolith Terminology. In: A. Farinacci, Roundtable on Calcareous Nannoplankton. *Proceedings II Planktonic Conference, Roma*, 1970, **2**, 1348–59.

Perch-Nielsen, K. 1972. Remarks on Late Cretaceous to Pleistocene coccoliths from the North Atlantic. *Initial Rep. Deep Sea drill. Proj.*, **12**, 1003–69.

Perch-Nielsen, K. 1973. Neue Coccolithen aus dem Maastrichtian von Dänemark, Madagaskar und Aegypten. *Bull. geol. Soc. Denmark*, **22**, 306–33.

Perch-Nielsen, K. 1977. Albian to Pleistocene calcareous nannofossils from the western South Atlantic. *Initial Rep. Deep Sea drill. Proj.*, **39**, 699–823.

Perch-Nielsen, K. 1979a. Calcareous nannofossil zonation at the Cretaceous/Tertiary boundary in Denmark. *Proceedings Cretaceous–Tertiary Boundary Events Symposium, Copenhagen*, **1**, 115–35.

Perch-Nielsen, K. 1979b. Calcareous nannofossils in Cretaceous/Tertiary boundary sections in Denmark. *Proceedings Cretaceous–Tertiary Boundary Events Symposium, Copenhagen*, **1**, 120–6.

Perch-Nielsen, K. 1979c. Calcareous nannofossils from the Cretaceous between the North Sea and the Mediterranean. *IUGS Ser. A.*, **6**, 223–72.

Perch-Nielsen, K. 1980. New Tertiary calcareous nannofossils from the South Atlantic. *Eclog. geol. Helv.*, **73** (1), 1–7.

Perch-Nielsen, K. 1981a. New Maastrichtian and Paleocene calcareous nannofossils from Africa, Denmark, the USA and the Atlantic, and some Paleocene lineages. *Eclog. geol. Helv.*, **74** (3), 831–63.

Perch-Nielsen, K. 1981b. Les coccolithes du Paléocène près de El Kef, Tunisie et leurs ancêtres. *Cah. Micropaleontol.*, **1981** (3), 7–23.

Perch-Nielsen, K. 1981c. Les nannofossiles calcaires à la limite Crétacé–Tertiaire près de El Kef, Tunisie. *Cah. Micropaleontol.*, **1981** (3), 25–37.

Perch-Nielsen, K. 1982. Maastrichtian coccoliths in the Danian: survivors or reworked 'dead bodies'? *Abstract, IAS meeting, Copenhagen*, 122.

Perch-Nielsen, K. 1984a. Validation of new combinations. *INA Newsletter*, **6** (1), 42–6.

Perch-Nielsen, K. 1984b. New Late Cretaceous and Paleogene calcareous nannofossils. *Eclog. geol. Helv.* **77**.

Perch-Nielsen, K. & Franz, H. E. 1977. *Lapideacassis* and *Scampanella*, calcareous nannofossils from the Paleocene at Sites 354 and 356, DSDP Leg 39, South Atlantic. *Initial Rep. Deep Sea drill. Proj.*, **39**, 849–62.

Perch-Nielsen, K., McKenzie, J. A. & He, Q. 1982. Bio- and isotope-stratigraphy and the 'catastrophic' extinction of calcareous nannoplankton at the Cretaceous/Tertiary boundary. *Spec. Pap. geol. Soc. Am.*, **190**, 353–71.

Perch-Nielsen, K., Sadek, A., Barakat, M. G. & Teleb, F. 1978. Late Cretaceous and Early Tertiary calcareous nannofossil and planktonic foraminifera zones from Egypt. *Actes du VIe Colloque Africain de Micropaleontologie, Tunis, Ann. Mines et Géol.*, **28** (II), 337–403.

Percival, S. Jr 1978. Indigenous and reworked coccoliths from the Black Sea. *Initial Rep. Deep Sea drill. Proj.*, **42B**, 773–81.

Percival, S. F. & Fischer, A. G. 1977. Changes in calcareous nannoplankton in the Cretaceous–Tertiary crisis at Zumaya, Spain. *Evol. Theory*, **2**, 1–35.

Pirini Radrizzani, C. & Valleri, G. 1977. New data on calcareous nannofossils from the Pliocene of the Tyrrhenian Basin Site 132 DSDP, Leg 13. *Riv. Ital. Paleontol.*, **83** (4), 869–96.

Poche, F. 1913. Das System der Protozoa. *Arch. Protistenk.*, **30**, 125–321.

Prins, B. 1971. Speculations on relations, evolution, and stratigraphic distribution of discoasters. In: A. Farinacci (ed.), *Proceedings II Planktonic Conference, Roma, 1970*, **2**, 1017–37.

Prins, B. 1979. Notes on nannology – 1. *Clausicoccus*, a new genus of fossil coccolithophorids. *INA Newsletter*, **1** (1), N-2–N-4.

Proto Decima, F. & Masotti, C. 1981. The genus *Gephyrocapsa* (Coccolithophorales) in the Plio-Pleistocene of the Timor trough. *Mem. Sci. Geol. Univ. Padova*, **34**, 453–64.

Proto Decima, F., Roth, P. H. & Todesco, L. 1975. Nannoplancton calcareo del Paleocene e dell' Eocene della Sezione di Possagno. *Schweiz, Paläontol. Abh.*, **97**, 35–55.

Pujos-Lamy, A. 1977a. *Emiliania* et *Gephyrocapsa* (Nannoplancton calcaire): biometrie et intêret biostratigraphique dans le Pleistocène supérieur marin des Açores. *Rev. Esp. Micropaleontol.*, **9** (1), 69–84.

Pujos-Lamy, A. 1977b. Essai d'établissement d'une biostratigraphie du nannoplancton calcaire dans le Pleistocène de l'Atlantique Nord-oriental. *Boreas*, **6**, 323–31.

Rade, J. 1975. *Scyphosphaera* evolutionary trends with special reference to eastern Australia. *Micropaleontology*, **21**, 151–65.

Radomski, A. 1968. Calcareous nannoplankton zones in Palaeogene of the western Polish Carpatians. *Rocz. Pol. Tow. geol.*, **38**, 545–605.

Reid, F. M. H. 1980. Coccolithophorids of the North Pacific Central Gyre with notes on their vertical and seasonal distribution. *Micropaleontology*, **26**, 151–76.

Reinhardt, P. 1964. Einige Kalkflagellaten-Gattungen (Coccolithophoriden, Coccolithineen) aus dem Mesozoikum Deutschlands. *Monatsber. Deutsch. Akad. Wiss. Berlin*, **6**, 749–59.

Reinhardt, P. 1966. Fossile Vertreter coronoider und styloider Coccolithen (Familie Coccolithaceae Poche 1913). *Monatsber. Deutsch. Akad. Wiss. Berlin*, **8**, 513–24.

Reinhardt, P. 1972. Coccolithen. Kalkiges Plankton seit Jahrmillionen. *Die neue Brehm-Bücheri, A. Ziemsen Verlag*, 1–99.

Rio, D. 1974. Remarks on Late Pliocene–Early Pleistocene calcareous nannofossil stratigraphy in Italy (1). *L'Ateneo Parmense, Acta Nat.*, **10** (3), 409–49.

Rio, D. 1982. The fossil distribution of coccolithophore genus *Gephyrocapsa* Kamptner and related Plio-Pleistocene chronostratigraphic problems. *Initial Rep. Deep Sea drill. Proj.*, **68**, 325–43.

Romein, A. J. T. 1977. Calcareous nannofossils from the Cretaceous/Tertiary boundary interval in the Barranco del Gredero (Caravaca, prov. Murcia, SE Spain). *Proc. Kon. Ned. Akad. Wetensch.*, **80**, 256–79.

Romein, A. J. T. 1979. Lineages in early Paleogene calcareous nannoplankton. *Utrecht Micropalaeontol. Bull.*, **22**, 1–231.

Romein, A. J. T. 1982. The Cretaceous/Tertiary boundary: an astronomic or a sedimentary problem? *IAS 3rd Eur. Meeting, Copenhagen*, 123–7.

Roth, P. H. 1969. Calcareous nannoplankton zonation of Oligocene sections in Alabama (U.S.A.), on the islands of Trinidad and Barbados (W.I.), and the Blake Plateau (E coast of Florida, U.S.A.). *Eclog. geol. Helv.*, **61**, 459–65.

Roth, P. H. 1970. Oligocene calcareous nannoplankton biostratigraphy. *Eclog. geol. Helv.*, **63**, 799–881.

Roth, P. H. 1973. Calcareous nannofossils – Leg 17, Deep Sea Drilling Project. *Initial Rep. Deep Sea drill. Proj.*, **17**, 695–795.

Roth, P. H. & Thierstein, H. R. 1972. Calcareous nannoplankton: Leg 14 of the Deep Sea Drilling Project. *Initial Rep. Deep Sea drill. Proj.*, **14**, 421–85.

Ruddiman, W. F. & McIntyre, A. 1976. Northeast Atlantic paleoclimatic changes over the past 600000 years. *Mem. geol. Soc. Am.*, **145**, 111–46.

Salis, A. K. von 1984. Miocene calcareous nannofossil biostratigraphy of DSDP Hole 521A, SE Atlantic. *Initial Rep. Deep Sea drill. Proj.*, **73**, 425–7.

Samtleben, C. 1978. Pliocene–Pleistocene coccolith assemblages from the Sierra Leone Rise – Site 366, Leg 41. *Initial Rep. Deep Sea drill. Proj.*, **41**, supplement, 913–31.

Samtleben, C. 1980. Die Evolution der Coccolithophoriden-Gattung *Gephyrocapsa* nach Befunden im Atlantik. *Paläontol. Z.*, **54** (1, 2), 91–127.

Schiller, J. 1925. Die planktonischen Vegetationen des adriatischen Meeres. *Arch. Protistenk.*, **51**, 1–130.

Schiller, J. 1930. Coccolithineae. In: *Dr. L. Rabenhorst's Kryptogamen-Flora von Deutschland, Österreich und der Schweiz.*, vol. 10, part 2, pp. 89–267. Akad. Verlagsges., Leipzig.

Schwarz, E. H. L. 1894. Coccoliths. *Ann. Mag. Nat. Hist.*, ser. 6, **14**, 341–6.

Selli, R., Accordi, A. A., Bandini Mazzanti, M., *et al.* 1977. The Vrica section (Calabria, Italy). *G. Geol.*, **42** (2), 181–204.

Shackleton, N. J. *et al.* 1984. Accumulation rates in Leg 74 sediments. *Initial Rep. Deep Sea drill. Proj.*, **74**, 621–44.

Shafik, S. & Stradner, H. 1971. Nannofossils from the Eastern Desert, Egypt with reference to Maastrichtian nannofossils from the USSR. *Jahrb. geol. Bundesanst. (Wien)*, special volume **17**, 69–104.

Shamraï, I. A. 1963. Certain forms of Upper Cretaceous and Paleogene coccoliths and discoasters from the southern Russian Platform. *Izv. Vyssh. Ucheb. Zaved. Geol. i Razv.*, **6** (4), 27–40.

Sherwood, R. W. 1974. Calcareous nannofossil systematics, paleoecology, and biostratigraphy of the Middle Eocene Weches Formation of Texas. *Tulane Stud. Geol.*, **11**, 1–79.

Smayda, T. J. 1966. A quantitative analysis of the phytoplankton of the Gulf of Panama. *Inter-Amer. Trop. Tuna Comm.*, **11**, 355–612.

Smith, L. A. 1969. Pleistocene discoasters at the stratotype of the Calabrian Stage (Santa Maria di Catanzaro) and at Le Castella, Italy. *Trans. Gulf Coast Assoc. geol. Soc.*, **19**, 579–83.

Stradner, H. 1958. Die fossilen Discoasteriden Österreichs. I. *Erdöl-Z.*, **74**, 178–88.

Stradner, H. 1959a. First report on the discoasters of the Tertiary of Austria and their stratigraphic use. *Proc. 5th World Petrol. Congr. New York*, 1959, **1**, 1081–95.

Stradner, H. 1959b. Die fossilen Discoasteriden Österreichs. II. *Erdöl-Z.*, **75**, 472–88.

Stradner, H. 1961. Vorkommen von Nannofossilien im Mesozoikum und Alttertiär. *Erdöl-Z.*, **77**, 77–88.

Stradner, H. 1962. Über neue und wenig bekannte Nannofossilien aus Kreide und Alttertiär. *Verh. geol. Bundesanst. (Wien)*, 176–86.

Stradner, H. 1963. In: K. Gohrbrandt, Zur Gliederung des Paläogen im Helvetikum nördlich Salzburg nach planktonischen Foraminiferen. *Mitt. geol. Ges. Wien*, **56**, 1–116.

Stradner, H. 1969. The nannofossils of the Eocene flysch in the Hagenbach Valley (northern Vienna Woods), Austria. *Rocz. Pol. Tow. geol.*, **39**, 403–32.

Stradner, H. 1973. Catalogue of calcareous nannoplankton from sediments of Neogene age in the eastern North Atlantic and Mediterranean Sea. *Initial Rep. Deep Sea drill. Proj.*, **13** (2), 1137–99.

Stradner, H. & Adamiker, D. 1966. Nannofossilien aus Bohrkernen und ihre elektronenmikroskopische Bearbeitung. *Erdöl-Erdgas Z.*, **82**, 330–41.

Stradner, H. & Allram, F. 1982. The nannofossil assemblages of Deep Sea Drilling Project Leg 66, Middle America trench. *Initial Rep. Deep Sea drill. Proj.*, **66**, 589–639.

Stradner, H. & Edwards, A. R. 1968. Electron microscopic studies on Upper Eocene coccoliths from Oamaru Diatomite, New Zealand. *Jahrb. geol. Bundesanst. (Wien)*, special volume **13**, 1–66.

Stradner, H. & Fuchs, R. 1980. Über Nannoplanktonvorkommen im Sarmatien (Ober-Miozän) der Zentralen Paratethys in Niederösterreich und im Burgenland. *Beitr. Paläontol. Österr.*, **7**, 251–79.

Stradner, H. & Papp, A. 1961. Tertiäre Discoasteriden aus Österreich und deren stratigraphische Bedeutung mit Hinweisen auf Mexico, Rumänien und Italien. *Jahrb. geol. Bundesanst. (Wien)*, special volume **7**, 1–159.

Stradner, H. & Seifert, P. 1980. *Transversopontis pax* n. sp., ein neues Nannofossil aus dem basalen Oligozän des nördlichen Niederösterreich. *Beitr. Paläontol. Österr.*, **7**, 281–91.

Sujkowski, Z. 1931. (Etude pétrographique du Crétacé de Pologne.) *Spraw. Polsk. Inst. Geol.*, **6**, 485–628.

Sullivan, F. R. 1964. Lower Tertiary nannoplankton from the California Coast Ranges. I. Paleocene. *Univ. Calif. Publ. Geol. Sci.*, **44**, 163–227.

Sullivan, F. R. 1965. Lower Tertiary nannoplankton from the California Coast Ranges. II. Eocene. *Univ. Calif. Publ. Geol. Sci.*, **53**, 1–74.

Takayama, T. 1967. First report on nannoplankton of the upper Tertiary and Quaternary of the southern Kwanto Region, Japan. *Jahrb. geol. Bundesanst. (Wien)*, **110**, 169–98.

Takayama, T. 1970. The Pliocene–Pleistocene boundary in the Lamont core V21-98 and at Le Castella, southern Italy. *J. Mar. Geol.*, **6**, 70–7.

Tan, S. H. 1927. Discoasteridae incertae sedis. *Proc. Sect. Sc. K. Akad. Wet. Amsterdam*, **30**, 411–19.

Tangen, K., Brand, L. E., Blackwelder, P. L., *et al.* 1982. *Thoracosphaera heimii* (Lohmann) Kamptner is a Dinophyte: observations on its morphology and life cycle. *Mar. Micropaleontol.*, **7** (3), 193–212.

Tappan, H. 1980. *The Paleobiology of Plant Protists.* Freeman & Co., 1028 pp.

Theodoridis, S. A. 1983. On the legitimacy of the generic name *Discoaster* Tan, 1927 ex Tan, 1931. *INA Newsletter*, **5** (1), 15–21.

Thierstein, H. R. 1974. Calcareous nannoplankton – Leg 26, Deep Sea Drilling Project. *Initial Rep. Deep Sea drill. Proj.*, **26**, 619–67.

Thierstein, H. R., Geitzenauer, K. R., Molfino, B., *et al.* 1977. Global synchroneity of Late Quaternary coccolith datum levels: validation by oxygen isotopes. *Geology*, **5**, 400–4.

Towe, K. M. 1979. Variation and systematics in calcareous nannofossils of the genus *Sphenolithus*. *Am. Zool.*, **19**, 555–72.

Troelsen, J. C. & Quadros, L. P. 1971. Distribuição

Bioestratigráfica dos Nanofósseis em Sedimentos Marinhos (Aptian–Mioceno) do Brasil. *Ann. Acad. brasil. Ciênc.*, **43**, 577–609.

Varol, O. 1982. Calcareous nannofossils from the Antalya Basin, Turkey. *Neues Jahrb. Geol. Palaeontol. Monatshefte*, **4**, 244–56.

Vekshina, V. N. 1959. Coccolithophoridae of the Maastrichtian deposits of the west Siberian lowland. *Trudy Sib. nauchno-issled. Inst. Geol. Geofiz. mineral Syrya (SNIIGGIMS)*, **2**, 56–81.

Waghorn, D. B. 1984. A revision of New Zealand Late Eocene and Oligocene calcareous nannofossil zones (in press).

Wallich, G. C. 1877. Observations on the coccosphere. *Ann. Mag. nat. Hist.*, ser. 4, **16**, 322–39.

Weber-van Bosse, A. 1901. Etudes sur les Algues de l'Archipel Malaisien. *Ann. Jard. Bot. Buitenzorg*, **17** [ser. 2.2], 126–41.

Wilcoxon, J. A. 1970a. *Cyclococcolithina* Wilcoxon nom. nov. (nom. subst. pro *Cyclococcolithus* Kamptner, 1954). *Tulane Stud. Geol. Paleontol.*, **8**, 82–3.

Wilcoxon, J. A. 1970b. *Sphenaster*, new genus, a Pliocene calcareous nannofossil from the tropical Indo-Pacific. *Tulane Stud. Geol. Paleontol.*, **8**, 78–81.

Wind, F. H. & Wise, S. W. 1977. In: Wise & Wind, Mesozoic and Cenozoic calcareous nannofossils recovered by DSDP Leg 36 drilling on the Falkland Plateau, southwest Atlantic sector of the southern ocean. *Initial Rep. Deep Sea drill. Proj.*, **36**, 269–491.

Wise, S. W. 1973. Calcareous nannofossils from cores recovered during Leg 18, Deep Sea Drilling Project: biostratigraphy and observations on diagenesis. *Initial Rep. Deep Sea drill. Proj.*, **18**, 569–615.

Wise, S. W. & Constans, R. E. 1976. Mid Eocene plankton correlations Northern Italy–Jamaica, W.I. *Trans. Gulf Coast Assoc. geol. Soc.*, **26**, 144–55.

Winter, A., Reiss, Z. & Luz, B. 1978. Living *Gephyrocapsa protohuxleyi* McIntyre in the Gulf of Elat (Aqaba). *Mar. Micropaleontol.*, **3**, 295–8.

Winter, A., Reiss, Z. & Luz, B. 1979. Distribution of living coccolithophore assemblages in the Gulf of Elat (Aqaba). *Mar. Micropaleontol.*, **4**, 197–223.

Worsley, T. R. & Jorgens, M. L. 1974. Oligocene calcareous nannofossil provinces. *Spec. Publ. Soc. econ. Paleontol. Mineral.*, **21**, 85–108.

Zhong, S. L. 1982. Neogene calcareous nannofossils from the Huangliu Formation of the Yinggehai Basin, south China Sea. *Acta Palaeontol. Sin.*, **21** (2), 191–201.

12
Calpionellids

JÜRGEN REMANE

CONTENTS

Introduction

Calpionellids are a small group of Late Tithonian to late Early Valanginian planktic protozoans of unknown affinities. They may be very common in micritic limestones of pelagic facies (Tithonian facies, 'calpionella-limestones'). They have a very wide geographic distribution within the Tethyan province, reaching from eastern Mexico in the west at least to Iran in the east (Fig. 1). Heim & Gansser (1939) found *Calpionella alpina* in the Kiogar nappes of Tibet and, according to their drawings, this observation seems credible, whereas the presence of calpionellids in central New Guinea (Rickwood, 1955) is not yet substantiated by figured specimens. According to Brunnschweiler (personal communication) they are, however, true calpionellids. On the other hand, '*Calpionella*' *schneebergeri* Brunnschweiler from Western Australia is certainly not a calpionellid (Remane, 1971).

It should be emphasized that calpionellids have also been observed in DSDP Sites 367, 398, and 416 in the eastern North Atlantic and in sites 100, 105, 391, and 534A in the western North Atlantic. On the American side of the Atlantic, offshore wells have extended the occurrence of calpionellids as far north as the Scotian Shelf and the Grand Banks (Jansa, Remane & Ascoli, 1980). Other examples of the N–S extension of calpionellids are rather limited. Nothing has so far been discovered in the Boreal realm. My own repeated scrutiny of samples from Argentine Tithonian and Berriasian has also been unsuccessful.

Wide geographic range and rapid evolution make calpionellids a very strong tool in long-range chrono-correlation, with a greater power of resolution than any other group among Upper Tithonian – Berriasian micro or nannofossils. On a regional scale, independant correlations by ammonites and calpionellids (for example, Le Hégarat & Remane, 1968) have attained the same degree of precision. In one case, it was possible to place the Berriasian–Valanginian boundary within an interval of 1 m, where the Valanginian is 200 m thick (a result confirmed by ammonites).

The following publications provide more detailed information on calpionellids and groups thought to be related to them, that is on the whole complex of 'fossil tintinnids': Tappan & Loeblich (1968), Borza (1969), Remane (1969b, 1971), Trejo (1976) and Remane (1978). In addition, there is the classical monograph of Colom (1948).

Discussion of taxa

General remarks

The family Calpionellidae Bonet 1956 is here taken to have its original meaning, including all Late Tithonian to Valanginian genera with a hyaline wall. According to Borza (1979), the Aptian/Albian genus *Colomiella*, although very similar to Calpionellidae, evolved independently from the Aptian *Praecolomiella* with a microgranular wall structure. This means that *Colomiella* cannot be derived from Berriasian or Valanginian genera like *Calpionellopsis* or *Calpionellites*

and that hyaline forms belong to different families, Calpionellidae and Colomiellidae.

Contrary to a widespread opinion (Colom, 1948; Bonet, 1956; Tappan & Loeblich, 1968; Borza, 1969; Trejo, 1976), calpionellids are not fossil tintinnids (Remane, 1969b, 1971, 1978) because their test, the lorica, is primarily calcitic. Their taxonomic position remains unsettled. All calpionellids display a polar axial (i.e. radial) symmetry, with the oral opening at one extremity and the aboral pole at the other (fig. 2).

Routine stratigraphic work depends entirely on thin sections where calpionellids are encountered in random sections. Of course, only axial sections give an exact image of lorica size and proportions, but in most cases the exact orientation of a given section cannot be deduced from its outline.

The taxonomic problems resulting from this fact have been discussed at length by Remane (1963, 1964). The most important points to reiterate here are:

(1) In all species with a rounded aboral pole the variability of lorica size and proportions is exaggerated by oblique sections (apparent variability, Fig. 3).

(2) Oblique sections may add apparent transitional forms to the real ones, thus exaggerating their stratigraphic range (e.g. *Calpionella alpina–Calpionella elliptica*, Figs. 3, 6). They may even simulate a type of transition between two species which does not exist at all (e.g. *Crassicollaria intermedia–Crassicollaria brevis*, Figs. 4, 5).

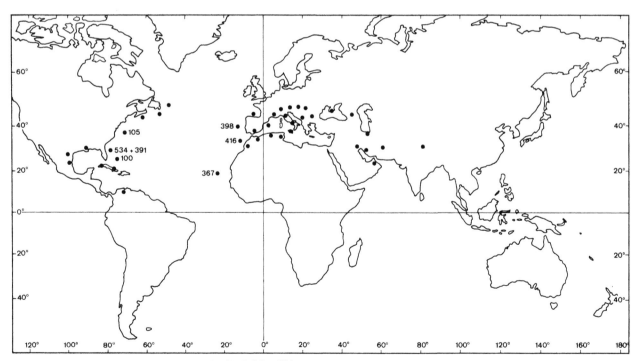

Fig. 1. Geographic distribution of the family Calpionellidae. Numbers indicate DSDP sites

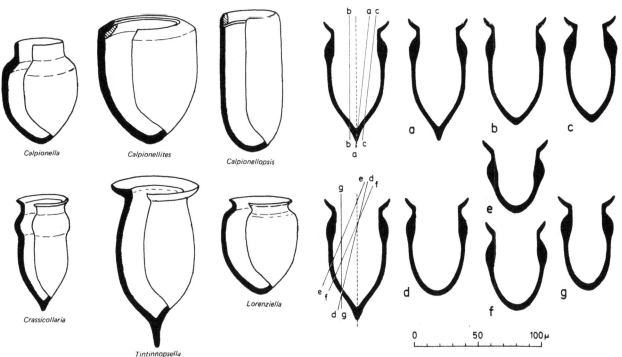

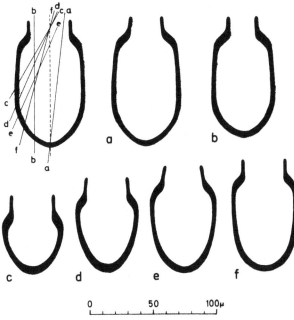

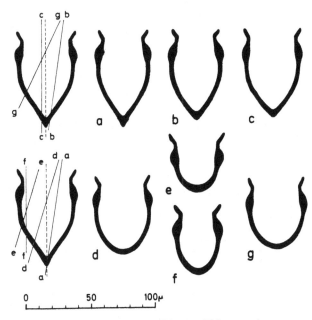

Fig. 2. Schematic drawings of the six stratigraphically most important calpionellid genera. Oblique hatching indicates the internal collars of *Calpionellites* and *Calpionellopsis*, which are distinguished from the lorica walls by an extinction at 45° between crossed Nicols. Modified after Remane (1969b)

Fig. 4. *Crassicollaria intermedia* (Durand Delga). Oblique sections obtained by geometric construction from the model to the left. After Remane (1963).

Fig. 3. *Calpionella elliptica* Cadisch. Oblique sections as they can be obtained by geometric construction from the model to the upper left. Note that marginal oblique sections (d and above all c) look exactly like axial sections of small *C. alpina*. After Remane (1962)

Fig. 5. *Crassicollaria brevis* Remane. Oblique sections obtained by geometric construction from the model to the left. Note that sections like d to g are indistinguishable from oblique sections of *Cr. intermedia* and that these sections establish a continuous transition from one species to the other which in reality does not exist. After Remane (1962)

(3) Collar morphology is still characteristic in oblique sections and therefore is most useful for generic definitions.

In the following descriptions, genera are listed in alphabetic order. For phylogenetic relations, see Remane (1971, 1978).

Unfortunately, the evolution of early representatives of *Remaniella* and related species, the origin of the genus and its relations to *Calpionellopsis* are still imperfectly known because these forms are rather rare. As soon as the actual taxonomic confusion is overcome they will certainly prove to be of great stratigraphic value.

Another problem is the gap in time between the Early to Mid-Tithonian microgranular *Chitinoidella* Doben 1963 and its Aptian homeomorphs *Deflandronella* Trejo 1972 and *Parachitinoidella* Trejo 1972. The question arises how far this gap is only apparent and if the separation of these forms on family level by Trejo (1976) is justified. In any case, the study of microgranular loricae in micritic limestones is particularly difficult and much has still to be learned about different species of *Chitinoidella* and their stratigraphic ranges.

CALPIONELLA Lorenz 1902
Figures 6, 18.1–4

The genus *Calpionella* is characterized by a short cylindrical collar, clearly narrower than the lorica. The base of the collar is well defined and a very typical 'shoulder' appears below the collar in all sections (Fig. 6).

The only possible confusion on generic level is between small, atypical specimens of *Calpionella alpina* with a less pronounced shoulder and oblique sections of *Crassicollaria parvula* during Early Berriasian (Fig. 7), but the separation of these forms is not important for stratigraphy.

There are two species: *Calpionella alpina* Lorenz 1902 and *Calpionella elliptica* Cadisch 1932. Both of them have a rounded aboral pole so that axial sections cannot be distinguished from moderately oblique sections. Certain oblique sections of the elongated *C. elliptica* look exactly like small *C. alpina* (Fig. 3) simulating the presence of transitional forms during the whole interval where both species occur together (Figs. 6, 7).

The range of *C. elliptica* is much more restricted than that of *C. alpina* (see Fig. 7) and therefore the quantitative separation of the two species is important. Real frequencies can, of course, not be determined under these conditions, but objective apparent frequencies can be obtained by a schematic distinction (Remane, 1963, 1964) based on the length (without collar) to width ratio (all sections with a ratio $> 1.35 = C.$ *elliptica*, $< 1.25 = C.$ *alpina*, with intermediate values classified as *Calpionella sp*).

Despite its long range, *Calpionella alpina* is important for stratigraphy due to significant morphologic changes in the course of evolution (Figs. 6, 18.1–3). Three 'forms' can be distinguished, a large, somewhat elongated variant in the subzones A2 and A3 (Fig. 18.1), an intermediary one, nearly spherical, mainly in the lower part of Zone B (Fig. 18.2), and finally a small variety, again slightly more elongated (Fig. 18.3). Unfortunately, the additional variation introduced by oblique sections makes a quantitative separation of these varieties impossible and it would be useless to give them a formal status, for example as subspecies.

CALPIONELLITES Colom 1948
Figures 18.5–6

The genus *Calpionellites* is characterized by an oral constriction formed by a 45° inward deflection of the lateral walls. The collar is internal, corresponding to a conical ring doubling the rim of the oral opening, with or without a funnel-shaped extension.

The phylogenetic derivation of this strange type of collar is documented by three successive species: *Remaniella murgeanui* (Pop, 1974a) = *R. 'dadayi'* auct. (Fig. 18.20), *Calpionellites darderi* (Colom, 1934), Figs. 18.5–6, and *Calpionellites coronata* Trejo 1975. *R. murgeanui* is transitional between typical *Remaniella* (Fig. 18.19) and *Ct. darderi*: the two branches of the collar become separated by the oral end of the lorica wall (Figs. 8, 18.20). In *Ct. darderi*, the outer branch has completely disappeared, whereas the inner one is overgrown by the lateral walls, which at the same time bend inwards, thus resulting in an oral constriction. In thin section, the whole appears as an inwardly directed bifurcation with parallel branches (Figs. 8, 18.5–6). Quite often the outer part, i.e. the rim of the oral opening breaks off so that the collar appears as an asymmetrical Y in thin section. In any case, the inner branch gives an extinction between crossed Nicols in a 45° position of the axis of the lorica (Fig. 8).

In *Ct. coronata* the inner branch of the collar develops a prolongation turning outwards around the edge of the oral opening (Fig. 8). Recognition of early *Ct. coronata* where this 'secondary collar' is still very thin is difficult. Moreover, all three species mentioned above have in common a large, bell-shaped lorica, broadly parabolic in more or less longitudinal sections. Even if the species cannot be distinguished, the stratigraphic error is not great: *R. murgeanui* appears only towards the end of Zone D (Fig. 7) and the range of *Ct. coronata* is largely parallel to that of *Ct. darderi*, whose appearance defines the base of Zone E.

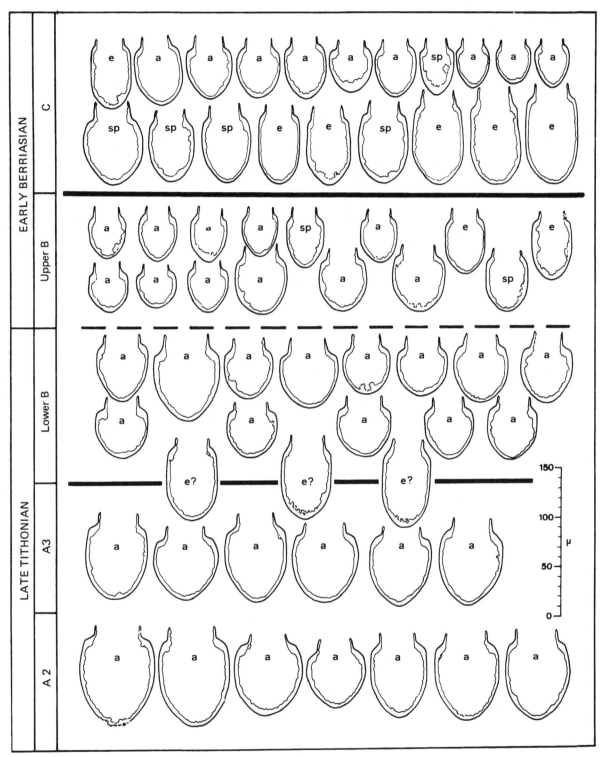

Fig. 6. Changes in the apparent variability (due to oblique sections) of the genus *Calpionella* Lorenz across the Jurassic–Cretaceous boundary. For a quantitative separation in frequency statistics the following schematic determinations have to be made: a: *C. alpina*, e: *C. elliptica*, sp: *Calpionella* sp. Letters at the left margin indicate Vocontian calpionellid zones (SE France). e? designates a form homeomorphous to *C. elliptica* which only occurs at the boundary between Zones A and B, there are definitely no *C. elliptica* during most of Zone B. Here, and in Figs. 7,

12 and 14–17, the old Tithonian–Berriasian boundary between the *jacobi* and the *grandis* zones has been retained. A majority of the participants in the 'Colloque International sur la limite Jurassique–Crétacé, Lyon–Neuchâtel 1973' was in favour of lowering the boundary to the base of the *jacobi* Zone (approximately coeval with the boundary between Zones A and B). So far, however, the new boundary has not been defined precisely in terms of Late Tithonian ammonite zones.

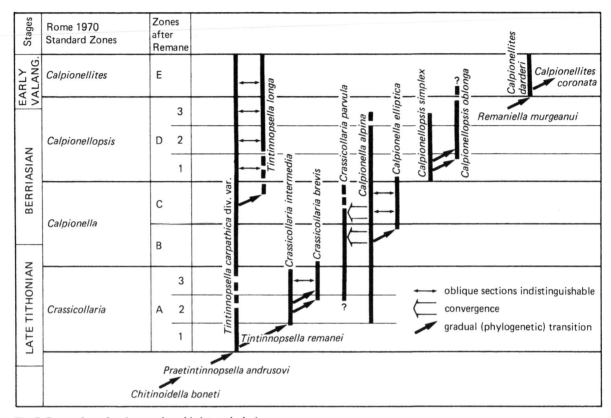

Fig. 7. Range chart showing stratigraphic intervals during which certain species may easily be confused

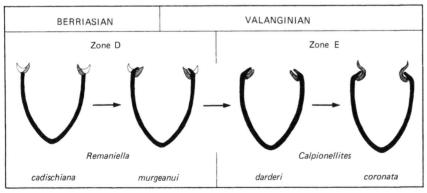

Fig. 8. Changes of collar anatomy in the course of the evolution from *Remaniella cadischiana* (Colom) to *Calpionellites coronata* Trejo. Schematic; the length of the loricae would be around 100 μm

CALPIONELLOPSIS Colom 1948

Figures **9, 10, 18.7–10**

As in *Calpionellites*, the inner rim of the opening is formed by a separate ring, showing extinction between crossed Nicols at 45°. Morphologically, this collar is not at all conspicuous and therefore the optical character is important in distinguishing *Calpionellopsis* from members of other genera in which the collar has been lost accidentally.

In some cases a fissure has formed between collar and lateral walls; in thin section, one has the impression of two points superposed to the walls (like the German ü, Fig. 18.10).

There are two species: *Calpionellopsis simplex* (Colom) 1939 (Fig. 18.9–10) and *Calpionellopsis oblonga* (Cadisch) 1932 (Fig. 18.7–8). Typical specimens are easily distinguished even in oblique sections: *Cs. simplex* (Fig. 9) is broader than *Cs. oblonga* (Fig. 10) and in *Cs. simplex* the lateral walls are parallel (curved in oblique sections), whereas in *Cs. oblonga* the straight lateral walls converge slightly towards the opening over two-thirds of the length of the lorica. More or less axial sections of *Cs. oblonga* show a conical aboral end (Figs. 10, 18.7).

A confusion between the two species is only possible where real transitional forms occur (passage D1–D2, see Fig. 7).

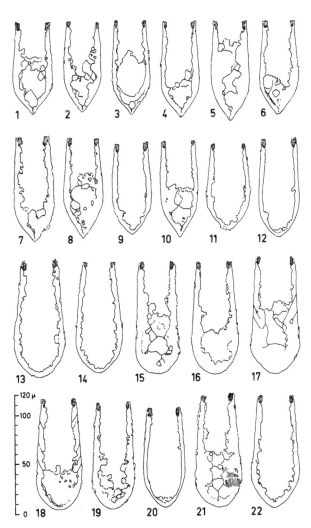

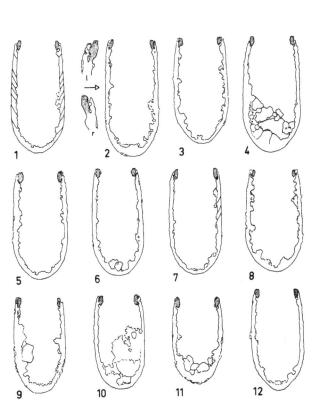

Fig. 9. Typical, more or less axial sections of *Calpionellopsis simplex* (Colom). After Remane (1965). Scale as in Fig. 10

Fig. 10. Typical representatives of *Calpionellopsis oblonga* (Cadisch); 1–10 are axial sections, 11–22 nearly so. After Remane (1965)

CRASSICOLLARIA Remane 1962

Figures **11, 18.11–15**

During early Late Tithonian (Zone A), there is a great proliferation of mostly elongate calpionellids of elliptical outline in more or less longitudinal sections ('*Calpionella elliptica*'

auct., see Remane, 1962, 1969a). This genus, *Crassicollaria*, differs from *Calpionella* by its wider opening and the smooth transition from the collar to the body of the lorica. Below the base of the collar, there is a more or less pronounced swelling, but never a well defined shoulder as in *Calpionella*.

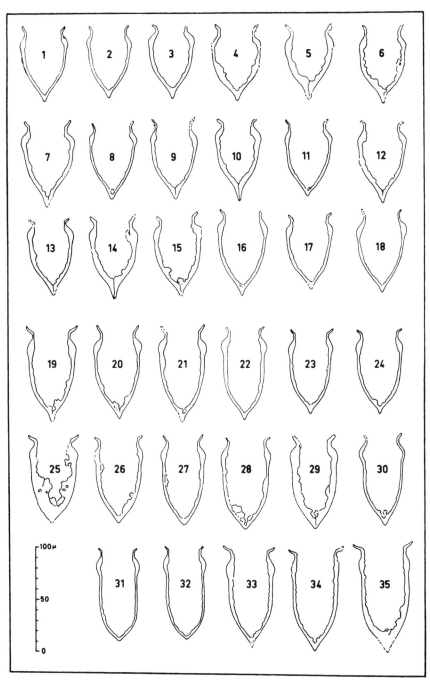

Fig. 11. Axial sections of *Crassicollaria brevis* Remane (1–18) and of *Crassicollaria intermedia* (Durand Delga) 19–35. The evolution from *C. intermedia* to *Cr. brevis* is very gradual, but note the difference between the initial and the final stage. Stratigraphic provenance of specimens: 1–2, 4–18, 20, 28 (Subzone A3); 3 (Zone B); 19, 21–27, 29–35 (Subzone A2). After Remane (1963)

The collar itself is variable in *Crassicollaria*; in some species (*Cr. intermedia*, *Cr. brevis*), the distal part shows a sharp outward deflection. This character is inherited from the genus *Tintinnopsella* and there is indeed a continuous transition from earliest *Tintinnopsella carpathica* through *Tintinnopsella remanei* Borza 1969 to *Crassicollaria intermedia*. The separation of the two genera is extremely difficult at this level (Fig. 7). The critical interval is, however, short and therefore stratigraphically significant.

During Zone A, a clear-cut separation of species within the genus *Crassicollaria* is impossible, except for the species pair *Cr. intermedia–Cr. brevis* which is distinguished from all others by the outward deflection of the distal part of its collar (Figs. 18.11–12, 14–15).

In both species, the conical aboral portion of the lorica even allows the identification of sections which are only approximately axial. Thus it is rather easy to follow the evolution of lorica proportions. An overall change from elongated (*Cr. intermedia*) to shorter, conical forms (*Cr. brevis*) is quite obvious (Fig. 11).

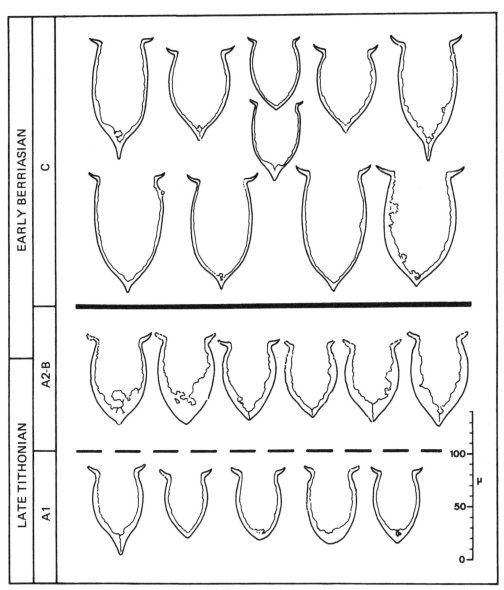

Fig. 12. Changes in the morphology of *Tintinnopsella carpathica* (Murgeanu & Filipescu) through Late Tithonian and Early Berriasian. These changes clearly reflect an evolutionary trend, as they are observable everywhere but, due to oblique sections, the different varieties cannot be separated as subspecies. From about mid-Berriasian to mid-Valanginian, the size of *T. carpathica* remains practically constant

LORENZIELLA Knauer & Nagy 1963

Figures **18.16–18**

The typical feature of the genus *Lorenziella* is a marked suboral constriction, forming a narrow sulcus below the collar. Otherwise *Lorenziella* is characterized by a progressive reduction of the collar so that in *Lorenziella hungarica* only the sulcus is left, corresponding in thin section to a crescentic rim around the opening, the convex side turned inwards (Figs. 18.16–18). The opening is rather narrow and the lower two-thirds of the lorica parabolic which gives a very characteristic outline to more or less longitudinal sections.

In badly-preserved material, the distinction from *Tintinnopsella carpathica* may become difficult. Records of *Lorenziella hungarica* from Early Berriasian strata are certainly due to confusion with a rather small spherical variety of *T. carpathica* which is relatively frequent during Zone C (Fig. 18.21).

TINTINNOPSELLA Colom 1948

Figures **12, 18.21–24**

The genus *Tintinnopsella* Colom 1948 has a very distinctive collar: it corresponds to a right angle outward deflection of the lateral wall, tapering distally and at the same time gently bent upwards. In thin sections this gives a slender, curved triangle with the distal end pointing obliquely upwards. The opening is always very large (Fig. 12).

There is no possibility of confusion on generic level except for real transitional forms: *Tintinnopsella remanei* leading to the genus *Crassicollaria* (Subzone A1) and forerunners of *Lorenziella hungarica* around the boundary of the subzones D2 and D3 (*L. plicata*, see Remane, 1968).

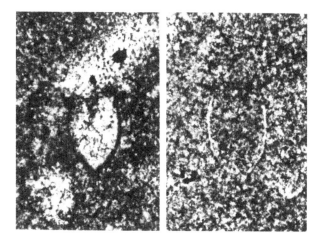

Fig. 13. *Chitinoidella boneti* Doben (left) and *Praetintinnopsella andrusovi* Borza (right) showing how the microgranular wall is progressively replaced by an inner, hyaline layer. The evolution thus leads to small *Tintinnopsella carpathica* (Murgeanu & Filipescu) at the beginning of Zone A (see Fig. 12). × 350

The main species is *Tintinnopsella carpathica* (Figs. 12, 18.21–24), characterized by an ovoid lorica with axial appendix (Fig. 18.24). Its stratigraphic range equals that of the whole family Calpionellidae (Late Tithonian–Early Valanginian, Fig. 14). *T. carpathica* is nevertheless of stratigraphic interest, because significant variations in size can be observed (Fig. 12), as in *Calpionella alpina*.

Evolution begins (Subzone A1) with a very small variety, homeomorphous to *Chitinoidella boneti* Doben 1963 of the Middle Tithonian. The latter has, however, a microgranular wall structure. There are intermediate forms (*Praetintinnopsella andrusovi* Borza 1969, see Fig. 13) with a double-layered wall: the outer layer plus collar microgranular, the inner layer hyaline calcitic as in *T. carpathica*.

There is little change during the rest of Zone A and Zone B, where *T. carpathica* is represented by a rare, somewhat larger form. Really large, typical forms appear only in the Early Berriasian. At the same time, variability increases considerably, but only the most elongated forms can be separated at a specific level (*Tintinnopsella longa*), Fig. 18.25. Even here, the distinction of oblique sections is impossible (Fig. 7).

Comments on zones

Specific problems in the determination of Calpionellid zones

If one wants to go beyond the simple recognition of zones, statistics of relative frequencies of genera and species are necessary, and the same is true if the lower and upper boundaries of Zone B are to be determined exactly.

After what has been said about the problem of oblique sections, it is evident that the delimitation of species has to be schematic if we want to obtain objectively quantifiable units (Remane, 1963, 1964, 1968).

The case of *Calpionella alpina* and *Calpionella elliptica* has already been mentioned (Fig. 6). All sections showing a ratio of length (without collar) to width of the lorica of more than 1.35 belong certainly to *C. elliptica*; for a ratio between 1.25 and 1.35 the designation *Calpionella* sp. should be employed no matter whether these might be real transitional forms between *C. alpina* and *C. elliptica* or only oblique sections of *C. elliptica*, and everything below 1.25 should be counted with *C. alpina* although some oblique sections of *C. elliptica* might be included.

The distinction of *Calpionellopsis simplex* and *Calpionellopsis oblonga* in statistics is less problematic (Remane, 1965, 1968) because only few marginal sections of *Cs. simplex* have an opening as narrow (28 μm or less) as have typical *Cs. oblonga* (Fig. 9). Uncertain sections which have to be labelled '*Calpionellopsis sp.*' are concentrated around the D1/D2 boundary where real transitional forms occur and are thus stratigraphically significant.

As to the genus *Crassicollaria*, only the distinction of *Crassicollaria intermedia* and *Crassicollaria brevis* is important and this only in determining the boundary between the subzones A2 and A3. Here it is necessary to limit specific attribution to sections which are so close to the axis of the lorica that the aboral portion still appears pointed. All the rest become 'Crassicollaria sp.'. Additional studies may separate typical *Cr. massutiniana* and *Cr. parvula*.

In *Tintinnopsella carpathica* the aboral portion is rounded and only very few sections touch the axial appendage. For statistical purposes it is therefore necessary to include *all* sections showing an elliptical outline, some of which may very well be oblique sections of *Tintinnopsella longa*. However, the quantitative separation of this latter species is not important for stratigraphy.

The calpionellid zonation established in SE France

Figures 14–17

Correlation with other fossil groups

One of the most detailed calpionellid zonations has been established in the Vocontian Trough of SE France (Remane, 1963, 1964, 1968, see Fig. 15). A detailed correlation with ammonite zones was accomplished in the same region by Le Hégarat & Remane (1968) for the Berriasian, and by Allemann & Remane (1979) for the Valanginian. Thierstein (1975) tied his nannoplankton zonation to the above-named Vocontian zones through the study of profiles dated by calpionellids and often also by ammonites. For the Valanginian, his results were confirmed by Manivit (1979). Finally, the Valanginian foraminiferal zones of Moullade in Busnardo *et al.* (1979) should be mentioned.

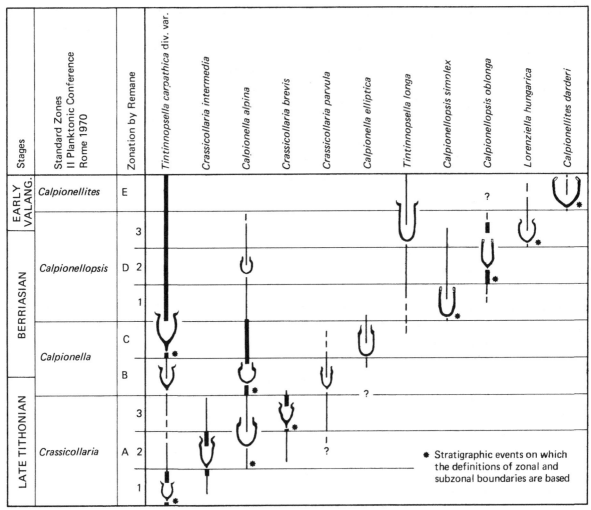

Fig. 14. Stratigraphic ranges of the most important calpionellid species. During Subzone A2, *Cr. parvula* cannot be distinguished from marginal sections of other species of *Crassicollaria* and hence its presence is uncertain. At the boundary between Zones A and B, homeomorphs of *C. elliptica* exist over a very short time interval (see Fig. 6), but during most of Zone B *C. elliptica* is definitely not present

Of all these correlations between different fossil groups, the one between calpionellid and nannoplankton zones is certainly the most reliable one. There is so far no general agreement on ammonite zones, not even for a limited area such as the Western Mediterranean realm (Wiedmann, 1975).

Comments on zones and zonal boundaries

Zone A can be rapidly determined on the quantitative predominance of the genus *Crassicollaria*. However, the genus is missing in the earliest part and therefore the best way to attain a precise definition of the lower boundary of Zone A is to use the appearance of the smallest variety of *Tintinnopsella carpathica*. This is also the time of appearance of Calpionellidae with a fully hyaline wall. The transition *Praetintinnopsella andrusovi – Tintinnopsella carpathica* provides an excellent phylogenetic control for this zonal boundary.

Zone B is characterized by a quasi constant predominance of *Calpionella alpina* whose relative frequency increases very rapidly at the base of the zone and may come to exceed 90%.

The 'explosion' of *C. alpina* is accompanied by a change from very large to somewhat smaller, spherical forms (Fig. 6).

Although this change is very difficult to express in terms of subspecies, it provides a phylogenetic control for the lower boundary of Zone B. Additionally, an elongated form, homeomorph of *Calpionella elliptica*, was observed to exist over a very short interval at the A/B boundary, from the E Alps down to the Subbetic Zone of S Spain.

Zone C is distinguished from Zone B by the predominance of *Tintinnopsella carpathica*. However, as this species remains frequent and sometimes even dominant during Zones D and E, the absence of *Calpionellopsis* becomes an important additional criterion. The range of *Calpionella elliptica* matches fairly well with Zone C, but in SE France this species is somewhat sporadic.

The best phylogenetic control of the lower boundary of Zone C is attained when it is made to coincide with the appearance of large, typical *T. carpathica*.

There is no fauna characteristic for the whole of *Zone D* so that we are left with the boundary definitions as given in Fig. 15. The ancestor of *Calpionellopsis simplex* is unknown. In SE France the genus *Calpionellopsis* is restricted to a part of Zone D (D1 to lower D3), while *Calpionellopsis oblonga* is frequent during D2 and lower D3.

The characteristic species of *Zone E, Calpionellites*

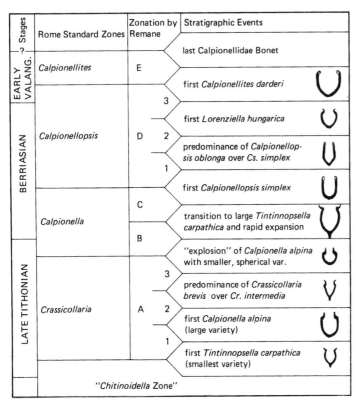

Fig. 15. Definition of Vocontian (SE France) calpionellid zones and subzones as established by Remane (1963, 1964, 1968, 1971).

darderi, is generally rare in SE France so that it is difficult to say if its range reaches the top of the zone. The upper boundary is anyway badly defined due to the gradual extinction of calpionellids. On the contrary, the transition *Remaniella murgeanui–Calpionellites darderi* provides an excellent phylogenetic control for the lower boundary of Zone E.

Comments on subzonal boundaries

The base of the Subzone A2 is defined by the appearance of *Calpionella alpina* (largest variety). The forerunners of *C. alpina* are unknown; the boundary is empirical.

The base of the Subzone A3 corresponds to the level where *Crassicollaria brevis* comes to predominate over *Crassicollaria intermedia*. There is thus good phylogenetic control, but the transition is, of course, gradual.

The base of the Subzone D2, as redefined by Allemann & Remane (1979) corresponds to the first occurrence of *Calpionellopsis oblonga*. *Calpionellopsis simplex* has a rounded aboral pole and axial sections are therefore impossible to identify. Apparent transitional forms between the two species are thus added to the real ones and it is often difficult to trace a precise boundary, although phylogenetic control is, of course, very good.

The base of the Subzone D3 is defined by the appearance of *Lorenziella hungarica* (for the problem of the most exact definition see Allemann & Remane, 1979).

The forerunner of *Lorenziella hungarica* is *Lorenziella plicata*, but unfortunately both species are rare which limits the practical value of this boundary.

Time stratigraphic units are normally defined by their *lower* boundaries only and preferably by the first appearance of a characteristic species or variant. The upper boundary is then automatically determined by the base of the following unit. We have followed the same principle in defining our calpionellid zones and therefore practically none of their boundaries coincides with a clear-cut extinction event. Moreover, extinctions of calpionellid species are mostly gradual and difficult to assign to a precise stratigraphic level.

In oil wells, where strata have to be dated from cuttings, extinctions of species are important stratigraphic events. Here they become 'appearances', and their depth can be determined with greater precision than real appearances which usually are blurred by caved material. Therefore it seems interesting to characterize zonal boundaries also in the opposite way, using those extinctions that allow the closest approximation of the top of a given zone.

The top of Zone E corresponds to the extinction of the family Calpionellidae. As calpionellids become very rare towards the end of their existence, at first only transverse sections will be encountered. They can be distinguished from the Aptian genus *Colomiella* by the orientation of the black cross visible when thin sections are observed between crossed

Nicols. In calpionellids this is always parallel to the Nicols, whereas in *Colomiella* some of the sections show the black cross at 45°. Of course, if *Calpionellites darderi* has not been specifically observed, the possibility remains that Zone E is absent and the association belongs to some older zone.

There is no extinction corresponding exactly to the top of Zone D, but the disappearance of *Calpionella alpina* and *Calpionellopsis simplex* during D3 (Fig. 14) provides a good approximation. The extinction of *Calpionellopsis simplex* is the more reliable of the two criteria as the last *Calpionella alpina* (*C.* aff. *alpina*) are atypic and difficult to determine. No subdivision of Zone D is possible on the basis of extinctions.

The top of Zone C does not correspond to an extinction event either, but it can be bracketed by the extinction of *Calpionella elliptica* just above (during D1) and of *Crassicollaria parvula* further down.

The top of Zone B cannot even be charactized approximately by an extinction. As a matter of fact, this boundary is defined by an important step in the re-diversification of calpionellid faunas after the crisis of the latest Tithonian – earliest Berriasian. On the other hand, the beginning of this crisis provides an excellent possibility to determine the top of Zone A, where *Crassicollaria intermedia*, *Cr. massutiniana*, and *Cr. brevis* vanish almost simultaneously. The base of Zone A is well defined by the extinction of the genus *Chitinoidella* and of *Praetintinnopsella andrusovi*, transitional between *Chitinoidella boneti* and the smallest variety of *Tintinnopsella carpathica*. These forms are, however, very difficult to observe and therefore the top of typical *Saccocoma* facies provides a good substitute which is very easy to recognize. Subzones of Zone A cannot be distinguished in cuttings.

As a whole, one can say that the Rome Standard Zones (Fig. 14) can be determined rather precisely in cuttings, but further distinctions remain more or less tentative and need great experience in calpionellid identification.

Stage boundaries and calpionellid zones

Neither of the two stage boundaries falling within the range of the family Calpionellidae corresponds to the zonal or subzonal boundaries described above, though very good approximations are possible.

The Tithonian-Berriasian boundary, which is also the Jurassic–Cretaceous boundary of the Tethys province, falls in the middle of Zone B. A lower and an upper B can be vaguely distinguished by the gradual passage of *Calpionella alpina* to the smallest variety (Fig. 6) and this provides a good estimate of the Tithonian/Berriasian boundary.

The Berriasian/Valanginian boundary as defined by ammonites in SE France lies in Subzone D3. The best approximation is ascertained by the appearance of *Remaniella murgeanui*. Also, in SE France, *Calpionellopsis oblonga* disappears abruptly before the end of D3, just above the Valanginian base.

In some cases, *R. murgeanui* (mostly called *R. 'dadayi'*) has been recorded from Early Berriasian strata, but this is almost certainly a confusion with *R. ferasini* (Catalano) which is similar in its general aspect but does not have the two branches of the collar separated by the lateral wall.

In oil wells, stage boundaries can be estimated in the following way. For the recognition of the Berriasian–Valanginian boundary the criteria indicated for the top of Zone D provide a very good estimate (as a matter of fact they correspond even better to the top of the Berriasian than to that of Zone D). As to the Tithonian–Berriasian boundary, species from Zone A provide the only proof that the Upper Tithonian has been reached (cf. Fig. 14).

Comparison of different calpionellid zonations

Figure 16

A detailed review of early, still contradictory zonations based on calpionellids was given by Remane (1969a). The situation changed dramatically when the four calpionellid specialists attending the Second Planktonic Conference in Rome during 1970 agreed on a standard zonation for the Western Mediterranean realm (Allemann, Catalano, Farès & Remane, 1971, see Figs. 14–17). This zonation is based on the

Vocontian zones proposed by Remane (1963, 1964, 1968), the only difference being that Zones B and C were 'lumped' together, because the expansion of *Tintinnopsella carpathica* is less pronounced in the southern parts of the Western Mediterranean realm, where *Calpionella elliptica* is more abundant.

The Rome Standard Zones were accepted immediately upon publication (Borza, 1974; Pop, 1974b, 1976; Dragastan, Mutiu & Vinogradov, 1975; Bakalova, 1977) so that they can be considered to be valid for the whole Mediterranean realm and Cuba, from the Black Sea to the Caribbean. Thus, at least for the time being, calpionellids offer much better possibilities for direct, long-distance correlations than ammonites. There remain, however, some problems concerning eastern Mexico (Trejo, 1975, 1980) and the northern Caucasus (Makarieva, 1979).

Some years ago, Trejo (1975) proposed a Mexican calpionellid zonation including Late Valanginian and Early Hauterivian, whereas, according to independent observations by Allemann and myself (Allemann & Remane, 1979), Calpionellidae disappear towards the end of the Early Valanginian. This fact is well documented by ammonites in the Subbetic of S Spain and SE France.

Quite recently, Trejo (1980, and personal communica-

Stages		Rome Standard Zones	Remane		Pop (1974b-1976)	Catalano & Liguori (1971)	Allemann, Grün & Wiedmann (1975)		Trejo (1980) revised zonation
VALANGINIAN	LATE								— ? —
					— ? —	— ? —			*Tintinnopsella carpathica*
	EARLY	*Calpionellites*	E			*Calpionellites darderi*			*Calpionellites darderi*
BERRIASIAN		*Calpionellopsis*	D	3	*Lorenziella*	*Calpionellopsis simplex-Calpionellopsis oblonga*	*Remaniella dadayi – Calpionellopsis oblonga*		*Remaniella dadayi*
									Calpionellopsis oblonga
				2	*Calpionellopsis oblonga*		*Calpionellopsis simplex*		*Calpionellopsis simplex*
				1	*Calpionellopsis simplex*				*Remaniella cadischiana*
		Calpionella	C		*Calpionella elliptica*	*Calpionella elliptica*	*Calpionella elliptica*		*Calpionella elliptica*
			B		*Remaniella*	*Calpionella alpina*	*Calpionella alpina*		*Calpionella*
					Calpionella alpina				
LATE TITHONIAN		*Crassicollaria*	A	3	*Crassicollaria brevis-parvula*	*Crassicollaria intermedia*	*Crassicollaria intermedia*		*Crassicollaria*
				2	*Crassicollaria intermedia*				
				1					

Fig. 16. Comparison of calpionellid zonations from recent publications by different authors. Except for Trejo (1980), the Rome Standard Zones constitute a common base

tion) adapted his zonation to the Late Tithonian–Early/Middle Valanginian interval where calpionellids occur in Europe. Ammonites can only be compared on generic level whereas calpionellids are represented by the same species on both sides of the Atlantic. Therefore they are certainly more reliable for intercontinental correlations. Also in his revised zonation, Trejo (1980) used a comprehensive Tintinnopsella Zone, corresponding closely to Zones C to E of Remane and subdivided into 7 subzones (Fig. 16).

Makarieva (1979) made a very interesting attempt to tie calpionellid Zones A–E to the ammonite zonation of the northern Caucasus. Unfortunately, a lot of species which were already abandoned some years ago (Remane, 1969a, 1971) were revived and were included in her stratigraphic table and zonal descriptions (e.g. *Tintinnopsella doliphormis*, *Amphorellina spp.*, *Salpingellina levantina*, *Favelloides balearica*). Line drawings are rather schematic and do not allow a re-determination of these forms so that it is impossible to say what is the exact significance of calpionellid zones as used in this publication.

Subdivisions of the Rome Standard Zones proposed by different authors are given on Fig. 16. The main difference between these subdivisions and the zonation of SE France is that *Calpionella elliptica* is preferred to *Tintinnopsella carpathica* (Zone C) as the index species of an Early Berriasian regional zone. This is quite logical where *C. elliptica* is frequent and constantly present throughout its whole range.

My study of material from Catalano & Liguori (1971), showed, however, that the transition to large, typical forms within *Tintinnopsella carpathica* is perfectly visible even though it is not accompanied by a rapid increase of relative frequency. The base of the *Calpionella elliptica* Zone is situated somewhat below that of Zone C.

As to subdivisions on subzonal level, those introduced by Pop were based on the comparison of sections from the S Carpathians (Pop, 1974b) and from Cuba (Pop, 1976). The results of Pop are important because they show that very detailed intercontinental correlations are possible if the bias due to different observers is eliminated. Among individual subzones, the *Remaniella* Subzone merits some discussion. According to ammonite finds (Pop, 1974b), its base seems to match very closely with the Tithonian–Berriasian boundary.

Stages		Ammonites after various authors		Calpionellids after Remane		Nannofossils after Thierstein (1975)	Foraminifera after Busnardo, Thieuloy & Moullade (1979)
VALANGINIAN	LATE	verrucosum				Calcicalathina oblongata	Lenticulina hauteriviana
	EARLY	campylotoxum		E			Lenticulina busnardoi
		pertransiensis					Lenticulina nodosa nodosa
		otopeta		3		Cretarhabdus crenulatus	
BERRIASIAN		bois-sieri	callisto		D		
			picteti	2			
			paramimounum	1			
		occi-tanica	dalmasi		C		
			privasensis			Nannoconus colomi	
			subalpina				
		grandis		B			
TITHONIAN	LATE	jacobi		3		?	
		?		2	A	Conusphaera mexicana	
				1			
	MIDDLE			Chitinoidella			

Fig. 17. Vocontian calpionellid zones and subzones correlated with zonations based on other guide fossils, established in the same sections of SE France. Ammonites: Berriasian according to Le Hégarat (1971), Valanginian according to Busnardo & Thieuloy in Busnardo *et al.* (1979); nannofossils according to Thierstein (1975); foraminifera according to Moullade in Busnardo *et al.* (1979).

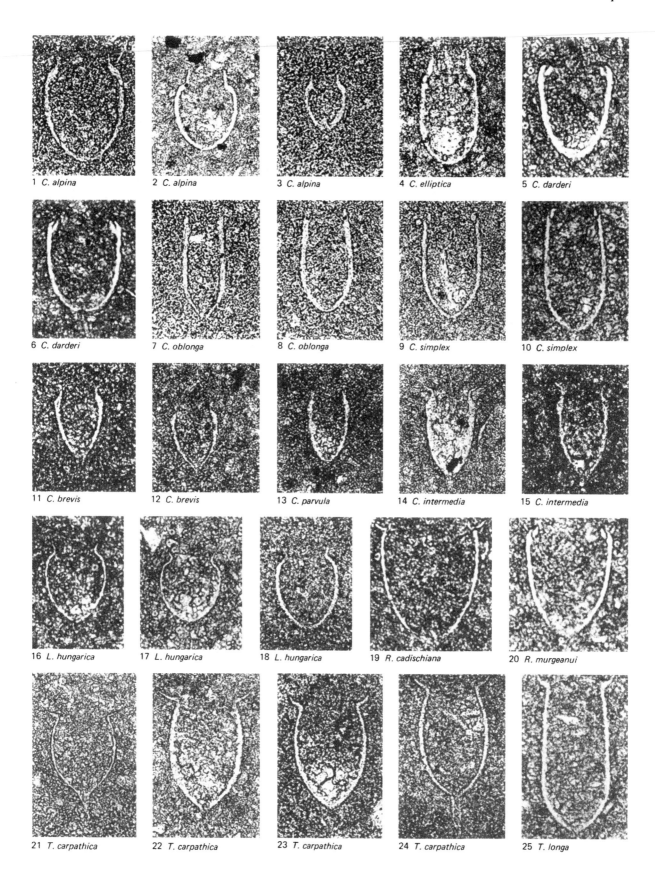

1 *C. alpina*   2 *C. alpina*   3 *C. alpina*   4 *C. elliptica*   5 *C. darderi*

6 *C. darderi*   7 *C. oblonga*   8 *C. oblonga*   9 *C. simplex*   10 *C. simplex*

11 *C. brevis*   12 *C. brevis*   13 *C. parvula*   14 *C. intermedia*   15 *C. intermedia*

16 *L. hungarica* 17 *L. hungarica* 18 *L. hungarica* 19 *R. cadischiana* 20 *R. murgeanui*

21 *T. carpathica* 22 *T. carpathica* 23 *T. carpathica* 24 *T. carpathica* 25 *T. longa*

Indeed, the first *Remaniella* also seem to appear at the very base of the Berriasian in SE France (Remane, 1963, 1964, 1968), but there, this genus is generally very rare.

The most detailed subzonal divisions so far attempted are the Early Cretaceous subzones of Trejo's *Tintinnopsella* Zone. There is good general correspondence to the succession of species in Europe, but it is of course difficult to establish a precise correlation with zonal or subzonal boundaries of other authors (Fig. 16).

Possible provincialism in calpionellid faunas

There is no real provincialism in calpionellid faunas and to the present day no endemic species are known. Regional differences between faunas are mostly only quantitative. Species of *Remaniella* and *Tintinnopsella longa*, although nowhere abundant, are more frequent in the southern Mediterranean realm (e.g. North Africa, Baleares) and the same is true for *Calpionellites darderi*, which is a zonal index.

Only two cases are of some importance for stratigraphy:
(1) *Calpionella elliptica* is less frequent and, above all, not constantly represented throughout its zone in SE France and probably there are minor regional differences in its stratigraphic range.
(2) *Calpionellopsis oblonga*, too, is less frequent in SE

France and never occurs together with *Calpionellites darderi* whereas very similar forms, so far not separable from *Cs. oblonga* on a specific level, occur together with *Ct. darderi* in the southern Mediterranean realm.

But, as a whole, we may say that biogeographic differentiation of calpionellid faunas was insignificant. The results of Pop (1976) prove that much more detailed interregional correlation than that obtained by the Rome Standard Zones is possible. However, this means that specialists have first to agree on some more taxonomic details.

References

Allemann, F. 1975. In: Allemann, Grün & Wiedmann.
Allemann, F., Catalano, R., Farès, F. & Remane, J. 1971. Standard calpionellid zonation (Upper Tithonian–Valanginian) of the western Mediterranean Province. *Proc. II Plankt. Conf.*, *Roma*, 1970, **2**, 1337–40.
Allemann, F., Grün, W. & Wiedmann, J. 1975. The Berriasian of Carvaca (Prov. of Murcia) in the subbetic zone of Spain and its importance for defining this stage and the Jurassic–Cretaceous boundary. Colloque sur la limite Jurassique–Crétacé, Lyon, Neuchâtel, sept. 1973. *Mem. Bur. Rech. geol. minieres*, **86**, 14–22.
Allemann, F. & Remane, J. 1979. In: Busnardo, Thieuloy, Moullade, *et al.*
Allemann, F. & Trejo, M. 1975. Two new species of *Calpionellites* from the Valanginian of Mexico and Spain. *Eclog. geol. Helv.*, **68**, 457–60.
Bakalova, D. G. 1977. La succession à calpionelles de la coupe près du village de Ginci, Bulgarie du Nord-Ouest. *C.r. Acad. bulgare Sci.*, **30**, 423–6.
Bonet, F. 1956. Zonificación microfaunística de las calizas cretácicas del Este de México. *Bol. Asoc. Mex. Geol. Petrol*, **8**, 389–488. Or: *Int. geol. Congr., Rep. 20th Sess. (Mexico) 1956.*
Borza, K. 1969. *Die Mikrofazies und Mikrofossilien des Oberjuras und der Unterkreide der Klippenzone der Westkarpaten.* Verl. Slow. Akad. Wiss., Bratislava, 302 pp.
Borza, K. 1974. Die stratigraphische Verwendung von Calpionelliden in den Westkarpaten. *Proc. Xth Congr. Carpatho-balkan. geol. Assoc.*, 1973, pp. 31–5.
Borza, K. 1979. Tintinnina aus dem oberen Apt und unteren Alb der Westkarpaten. *Geol. Zb., Geologica carpat.*, **30**, 341–61.
Busnardo, R., Thieuloy, J.-P., Moullade, M. *et al.* 1979. Hypostratotype mésogéen de l'étage valanginien (Sud-Est de la France). *Ed. Cent. natl. Rech. sci.*, **6**, 1–143.
Cadisch, J. 1932. Ein Beitrag zum Calpionellenproblem. *Geol. Rdsch.*, **23**, 241–57.
Catalano, R. & Liguori, V. 1971. Facies a calpionelle della Sicilia occidentale. *Proc. II Plankt. Conf.*, *Roma*, 1970, **1**, 167–210.
Colom, G. 1934. Estudios sobre las Calpionelas. *Bol. R. Soc. Esp. Hist. Nat.*, **34**, 379–89.
Colom, G. 1939. Tintínnidos fósiles (Infusorios Oligotricos). *Las Ciencias*, **4**, 815–25.
Colom, G. 1948. Fossil tintinnids: loricated Infusoria of the order of the Oligotricha. *J. Paleontol.*, **22**, 233–63.
Colom, G. 1965. Essais sur la biologie, la distribution géographique et stratigraphique des Tintinnoidiens fossiles. *Eclog. geol. Helv.*, **58**, 319–34.

Fig. 18 (all *c.* × 300); (5, 6, 10, 16, 17, 19, 20, 25, by F. Allemann, Bern)

1–3 1–3 Illustrate the evolution of *Calpionella alpina* Lorenz from Late Tithonian (Zone A, 1) to Late Berriasian (Zone D, 3); 2 shows the spherical variant characteristic of the Tithonian–Berriasian passage beds (Zone B).
4 *Calpionella elliptica* Cadisch Early Berriasian (Zone C).
5, 6 *Calpionellites darderi* (Colom) Early Valanginian (Zone E).
7, 8 *Calpionellopsis oblonga* (Cadisch) Late Berriasian (Zone D); 7 axial section showing the pointed aboral portion (collar badly preserved); 8 slightly oblique section with a well preserved collar.
9, 10 *Calpionellopsis simplex* (Colom) Late Berriasian (Zone D).
11, 12 *Crassicollaria brevis* Remane Late Tithonian (upper part of Zone A).
13 *Crassicollaria parvula* Remane Late Tithonian to Early Berriasian (Zones A to C).
14, 15 *Crassicollaria intermedia* (Durand Delga) Late Tithonian (lower part of Zone A).
16–18 *Lorenziella hungarica* Knauer Late Berriasian to Early Valanginian (Subzone D3 to Zone E).
19 *Remaniella cadischiana* (Colom) Berriasian to Early Valanginian (upper part of Zone B to Zone E).
20 *Remaniella murgeanui* (Pop) Uppermost Berriasian (Subzone D3).
21–24 *Tintinnopsella carpathica* (Murgeanu & Filipescu) Large forms as encountered in Berriasian to Early Valanginian strata (Zones C to E); note the long appendage visible in 24.
25 *Tintinnopsella longa* (Colom) Late Berriasian to Early Valanginian (Zones D and E).

Doben, K. 1963. Ueber Calpionelliden an der Jura/Kreide-Grenze. *Mitt. Bayer. Staatssamml. Palaeontol. hist. Geol.*, 3, 35–50.

Dragastan, O., Mutiu, R. & Vinogradov, C. 1975. Les zones micropaléontologiques et la limite Jurassique–Crétacé dans les Carpates orientales (massif du Haghimas) et sur la plate-forme moesienne. Colloque sur la limite Jurassique–Crétacé, Lyon, Neuchâtel, sept. 1973. *Mem. Bur. Rech. geol. minieres*, 86, 188–203.

Haq, B. U. & Boersma, A. (eds.) 1978. *Introduction to Marine Micropaleontology*, Elsevier, 376 pp.

Heim, A. & Gansser, A. 1939. Central Himalaya. Geological observations of the Swiss expedition 1939. *Denkschr. Schweiz. naturforsch. Ges.*, 73, 1–245.

Jansa, L. F., Remane, J. & Ascoli, P. 1980. Calpionellid and foraminiferal–ostracod biostratigraphy at the Jurassic–Cretaceous boundary, offshore eastern Canada. *Riv. Ital. Paleontol.*, 86, 67–126.

Knauer, J. & Nagy, I. 1963. *Lorenziella* nov. gen. uj Calpionellidea nemzetség. *Lorenziella* nov. gen. nouveau genre des Calpionellidés. *Foldt. Int. Evi Jel.*, 1961, 143–53.

Le Hégarat, G. 1971. Le Berriasien du Sud-Est de la France. *Doc. Lab. Geol. Fac. Sci. Lyon*, 43 (2 fasc.), 1–576.

Le Hégarat, G. & Remane, J. 1968. Tithonique supérieur et Berriasien de la bordure cévenole. Corrélation des ammonites et des calpionelles. *Geobios*, 1, 7–70.

Lorenz, Th. 1902. Geologische Studien im Grenzgebiet zwischen helvetischer und ostalpiner Fazies. II. Der südliche Rhätikon. *Ber. natf. Ges. Freiburg/Br.*, 12, 35–95.

Makarieva, S. F. 1979. Mesozoic Tintinids from the North Caucasus, and the Jurassic–Cretaceous boundary. In: *Upper Jurassic and the boundary to the Cretaceous system.* Academy Nauk, USSR, Siberian branch, Institute for Geology and Geophysics, Novosibirsk, 168–71.

Manivit. 1979. In: Busnardo, Thieuloy, Moullade, *et al.*

Murgeanu, G. & Filipescu, M. G. 1933. *Calpionella carpathica* n. sp. dans les Carpathes roumaines. *Notat. biol.*, 1, 63–4.

Pop, G. 1974a. Une nouvelle espèce néocomienne de Calpionellidés. *Rev. roumaine Geol. Geophys. Geogr.*, 18, 105–7.

Pop, G. 1974b. Les zones de Calpionellidés tithonique–valanginiennes du sillon de Resita (Carpates méridionales). *Rev. roumaine Geol. Geophys. Geogr.*, 18, 109–25.

Pop, G. 1976. Tithonian–Valanginian calpionellid zones from Cuba. *Dari Seama Sedint.*, 62 (1974–75), 237–66.

Remane, J. 1962. Zur Calpionellen-Systematik. *Neues Jahrb. Geol. Palaeontol. Monatshefte*, 1962, 8–24.

Remane, J. 1963. Les calpionelles dans les couches de passage Jurassique–Crétacé de la fosse vocontienne. *Trav. Lab. Geol. Fac. Sci. Univ. Grenoble*, 39, 25–82.

Remane, J. 1964. Untersuchungen zur Systematik und Stratigraphie der Calpionellen in den Jura-Kreide-Grenzschichten des Vocontischen Troges. *Palaeontographica A*, 123, 1–57.

Remane, J. 1965. Neubearbeitung der Gattung *Calpionellopsis* Col. 1948 (Protozoa, Tintinnina?). *Neues Jahrb. Geol. Palaeontol. Abhandlungen*, 122, 27–49.

Remane, J. 1968. In: Le Hégarat & Remane.

Remane, J. 1969a. Les possibilités actuelles pour une utilisation stratigraphique des calpionelles (Protozoa incertae sedis, Ciliata?). *Proc. 1st Internat. Conf. Plankt. Microfossils, Geneva, 1967*, 2, 559–73.

Remane, J. 1969b. Nouvelles données sur la position taxonomique des Calpionellidea Bonet (1956) et leurs rapports avec les Tintinnina actuels et les autres groupes de 'Tintinnoidiens' fossiles. *Proc. 1st Internat. Conf. Plankt. Microfossils, Geneva, 1967*, 2, 574–87.

Remane, J. 1971. Les calpionelles, protozoaires planctoniques des mers mésogéennes de l'époque secondaire. *Ann. Guebhard*, 47, 1–25.

Remane, J. 1978. In: Haq & Boersma (eds.).

Rickwood, F. K. 1955. The geology of the western highlands of New Guinea. *J. geol. Soc. Aust.*, 2, 63–82.

Tappan, H. & Loeblich, A. R. Jr 1968. Lorica composition of modern and fossil Tintinnida (ciliate Protozoa), systematics, geologic distribution, and some new Tertiary taxa. *J. Paleontol.*, 42, 1378–94.

Thierstein, H. R. 1971. Tentative Lower Cretaceous calcareous nannoplankton zonation. *Eclog. geol. Helv.*, 64, 459–88.

Thierstein, H. R. 1975. Calcareous nannoplankton biostratigraphy at the Jurassic–Cretaceous boundary. Colloque sur la limite Jurassique–Crétacé, Lyon, Neuchâtel, sept. 1973. *Mem. Bur. Rech. geol. minieres.*, 86, 84–94.

Trejo, M. 1972. Nuevos Tintinidos del Aptiano superior de México. *Rev. Inst. Mex. Pet.*, 4, 80–7.

Trejo, M. 1975. Los Tintinidos mesozoicos de México. Colloque sur la limite jurassique–Crétacé, Lyon, Neuchâtel, sept. 1973. *Mem. Bur. Rech. geol. minieres.*, 86, 95–104.

Trejo, M. 1976. Tintinidos mesozoicos de México (taxonomia y datos paleobiológicos). *Bol. Asoc. Mex. Geol. Petrol.*, 27, 329–449.

Trejo, M. 1980. Distribución estratigráfica de los Tintinidos mesozoicos mexicanos. *Rev. Inst. Mex. Pet.*, 12, 4–13.

Wiedmann, J. 1975. In: Allemann, Grün & Wiedmann.

Index

The index is divided into the following sections according to fossil groups:

The separation allows the groups to be treated differently and relevant notes will be found at the beginning of each section.

Contents lists at the beginning of each chapter give access to major subdivisions of the text.

Within the chapters, references to figures will be found immediately following taxonomic entries. They are arranged with the illustration of the taxon first, in bold type, followed by references to range charts, etc. in roman type.

Foraminifera (Chapters 4 to 9)

The page references are given as follows:

Bold – description of taxon. This reference will also give access to figure numbers for illustrations and range

Roman – other important references to taxa in the text

Italics – zonal definitions

Range charts will be found in the individual chapters as follows:

Chapter 4, Cretaceous taxa 36–41

Chapter 5, Paleocene and Eocene taxa 100–105

Chapter 6, Oligocene to Holocene low latitude taxa 168–175

Chapter 7, Southern mid-latitude Paleocene to Holocene taxa 270–271

Chapter 8, Mediterranean Miocene and Pliocene taxa 298–299

Chapter 9, Paratethyan Late Oligocene to Middle Miocene taxa 320

CALCAREOUS NANNOFOSSILS (Chapters 10 and 11)

The page references are given as follows:

Bold – remarks on taxa and references to illustrations and range
Roman – other references to taxa in the text
Italics – zonal definitions

Alphabetical lists of taxa with authors and years and indication by parentheses of those names that are known or thought to be invalid are located on pp. 413–420 for the Mesozoic and 538–545 for the Cenozoic.

An integrated range chart for marker species will be found for the Jurassic on p. 336, for the Cretaceous on 342, for the Paleogene on 431 and for the Neogene on 442.

CALPIONELLIDS (Chapter 12)

The page references are given as follows:
Bold – major references to taxa including descriptive notes
Roman – other references to taxa in the text
    A range chart of the most important species will be found on 565.